2018 UNIFORM PLUMBING CODE®

Illustrated Training Manual™

Copyright © 2018
International Association of Plumbing and Mechanical Officials
All Rights Reserved

No part of this work may be reproduced or recorded in any form or by any means, except as may be expressly permitted in writing by the publisher.

First Printing, January 2018

Published by the International Association of Plumbing and Mechanical Officials
4755 E. Philadelphia Street • Ontario, CA 91761-2816 – USA
Main Phone: (909) 472-4100 • Main Fax: (909) 472-4150

2018 UPC ITM Foreword

IAPMO is proud to present this 2018 edition of the Uniform Plumbing Code (UPC) Illustrated Training Manual (ITM). As in the previous edition, the ITM contains the entire 2018 UPC text as developed and published through the ANSI consensus process. The 2018 edition follows the double column text similar to the Code, and each commentary narrative is tagged with an icon (a wrench) underneath the applicable code section serving as a pointer to direct the reader toward important information.

A new feature brought to this edition is Learning Links. Learning Links are URLs to be typed into your internet browser that will link you with a video presentation to better illustrate the commentary. The Learning Links in the ITM eBook will be hyperlinks that your cursor can hover over and click on to automatically open your browser to the video presenation. A list of Learning Links follow the Table of Contents.

The commentary narrative is a living document that continues to grow and mature. This 2018 edition displays qualities that have progressed beyond its previous edition. Every narrative has been re-evaluated for its content both technical and grammatical. The rewriting has been extensive throughout the ITM and the reader will be pleasantly surprised to find fresh narrative in old places accompanied with new illustrations.

The development of the ITM is the responsibility of the Publication Development Committee and IAPMO staff. The committee has painstakingly reviewed the narrative to ensure technical accuracy.

This publication has many uses. Its value lies in its use as a reference document or a training document. However, care should be taken by users to differentiate between the code text and the narrative and remember that the narrative is not code, has not been legally adopted, and should not be applied as such.

The Publication Development Committee and IAPMO staff welcomes your comments and suggestions for this and future ITMs.

The 2016 and 2017 Publication Development Committees shared in the development of and final approval of the commentary narrative of this edition. Since that time, changes in the committee membership may have occurred.

2016 IAPMO
PUBLICATION DEVELOPMENT COMMITTEE

Martin Cooper, *Chairman*
City of Foster City

Keith Bonenfant, State of California
Kenneth Borski, City of Houston/Planning & Dev.
Dan Daniels, Pueblo Regional Bldg. Dept.
Ronald Rice, Sr. Mechanical Inspector - Retired
Jeremy Stettler, Davis School District, Utah
Daniel Cole, IAPMO Staff Liaison
Doug Kirk, IAPMO Staff

2017 IAPMO
PUBLICATION DEVELOPMENT COMMITTEE

Martin Cooper, *Chairman*
City of Foster City

Keith Bonenfant, State of California
Kenneth Borski, City of Houston
Shane Peters, City of Santa Monica
Bruce Pfeiffer, City of Topeka - Retired
Jeremy Stettler, Davis School District, Utah
Phillip White, Plumbers Local 68
Daniel Cole, IAPMO Staff Liaison
Doug Kirk, IAPMO Staff

TABLE OF CONTENTS

Chapter 1	Administration ...1	Appendix A	Recommended Rules for Sizing the Water Supply System ...521
Chapter 2	Definitions ...17		
Chapter 3	General Regulations ...41	Appendix B	Explanatory Notes on Combination Waste and Vent Systems ...541
Chapter 4	Plumbing Fixtures and Fixture Fittings ...65		
Chapter 5	Water Heaters ...89	Appendix C	Alternate Plumbing Systems ...543
Chapter 6	Water Supply and Distribution ...167	Appendix D	Sizing Storm Water Drainage Systems ...555
Chapter 7	Sanitary Drainage ...227	Appendix E	Manufacturered/Mobile Home Parks and Recreational Vehicle Parks ...563
Chapter 8	Indirect Wastes ...263		
Chapter 9	Vents ...277		
Chapter 10	Traps and Interceptors ...307	Appendix F	Firefighter Breathing Air Replenishment Systems ...575
Chapter 11	Storm Drainage ...327	Appendix G	Sizing of Venting Systems ...579
Chapter 12	Fuel Gas Piping ...351	Appendix H	Private Sewage Disposal Systems ...589
Chapter 13	Health Care Facilities and Medicial Gas and Vacuum Systems ...427		
		Appendix I	Installation Standard for PEX Tubing Systems for Hot- and Cold-Water Distribution ...615
Chapter 14	Firestop Protection ...459		
Chapter 15	Alternate Water Sources for Nonpotable Applications ...465	Appendix J	Combination of Indoor and Outdoor Combustion and Ventilation Opening Design ...623
Chapter 16	Nonpotable Rainwater Catchment Systems ...487	Appendix K	Potable Rainwater Catchment Systems ...625
Chapter 17	Referenced Standards ...497	Appendix L	Sustainable Practices ...629
		Appendix M	Peak Water Demand Calculator ...643
		Useful Tables	...649

LEARNING LINKS

These learning links to websites are not affiliated with IAPMO. IAPMO is not responsible for information housed on those sites and IAPMO does not endorse any linked websites or products. Use of information from third-party websites is at your risk. These links have been provided to supplement your knowledge on the subject matter.

Section 204.0
Bottle Water Filling Station: goo.gl/jfjqG6

Section 205.0
Clinical Sink: goo.gl/xSs74o

Section 208.0
Flushometer Tank: goo.gl/1Q2gPP

Section 223.0
Waterless Urinals: goo.gl/VeJhsk, goo.gl/QFxuQv

Section 225.0
Water Hammer Arrestor: goo.gl/jKiQMn

Section 411.3
Automatic Toilet Seat Cover: goo.gl/9gvmAb

Section 604.4
Hard-Drawn Copper or Copper Alloy Tubing: goo.gl/q3BR68

Section 605.1.3.2
Pressed Fittings: goo.gl/hBrJr4

Section 605.1.3.3
Push Fit Fittings: goo.gl/hBrJr4

Section 609.4
Plastic Plumbing Air Test: goo.gl/d1sywgJ

Section 723.1
Plastic DWV Air Test: goo.gl/d1sywgJ

CHAPTER 1
ADMINISTRATION

101.0 General.

101.1 Title. This document shall be known as the "Uniform Plumbing Code," may be cited as such, and will be referred to herein as "this code."

The Uniform Plumbing Code (UPC) is a document published and copy written by the International Association of Plumbing and Mechanical Officials (IAPMO). Within the UPC or this manual any reference to "this code" refers to the UPC. Some jurisdictions that adopt the code and then amend it also amend the name of the code correspondingly. For example, the state of California has renamed its amended version of the UPC, the California Plumbing Code (CPC). In this case the reference to "this code" in the CPC would refer only to the document published as the CPC.

101.2 Scope. The provisions of this code shall apply to the erection, installation, alteration, repair, relocation, replacement, addition to, use, or maintenance of plumbing systems within this jurisdiction.

The requirements of the code apply to all installations of plumbing within the jurisdiction having adopted the code, whether they are existing installations or under construction. They also apply to repairs, renovations or the maintenance of plumbing systems, whether completed by a homeowner or a licensed contractor.

101.3 Purpose. This code is an ordinance providing minimum requirements and standards for the protection of the public health, safety, and welfare.

It must be noted that the requirements contained in this code are minimum requirements. Any installation of plumbing that does not meet these minimum requirements could result in an unsanitary condition that can result in serious health problems leading to sickness and, in some cases, death. Improperly installed plumbing systems can also result in structural weakness or damage to the building with the accompanying risk of personal injury and needless costly repair. Therefore, the UPC clearly defines its purpose at the outset as "providing minimum requirements and standards for the protection of the public health, safety, and welfare." All of the provisions within the UPC reflect this purpose, either directly or indirectly.

101.4 Unconstitutional. Where a section, subsection, sentence, clause, or phrase of this code is, for a reason, held to be unconstitutional, such decision shall not affect the validity of the remaining portions of this code. The legislative body hereby declares that it would have passed this code, and each section, subsection, sentence, clause, or phrase thereof, irrespective of the fact that one or more sections, subsections, sentences, clauses, and phrases are declared unconstitutional.

It is possible that under the scrutiny of the courts, a portion of the code might be found to be unconstitutional. If that were to happen, the rest of the code would remain in force and valid.

101.5 Validity. Where a provision of this code, or the application thereof to a person or circumstance, is held invalid, the remainder of the code, or the application of such provision to other persons or circumstances, shall not be affected thereby.

If it can be legally proven that a provision in this code does not apply to a certain situation or individual, then that decision would have no effect on the validity of the code for other situations or individuals.

102.0 Applicability.

102.1 Conflicts Between Codes. Where the requirements within the jurisdiction of this plumbing code conflict with the requirements of the mechanical code, this code shall prevail. In instances where this code, applicable standards, or the manufacturer's installation instructions conflict, the more stringent provisions shall prevail. Where there is a conflict between a general requirement and a specific requirement, the specific requirement shall prevail.

Should the user feel there is a conflict between this code and the Uniform Mechanical Code (UMC) used in that jurisdiction, it is recommended that the user contact the Authority Having Jurisdiction (AHJ). The AHJ will then determine if there is a conflict and how best to resolve it. However, if there is a conflict between the UPC and the UMC as published by IAPMO, then the UPC should be followed. It is still advisable, however, to contact the AHJ for guidance or contact IAPMO for an interpretation of the sections in conflict.

102.2 Existing Installations. Plumbing systems lawfully in existence at the time of the adoption of this code shall be permitted to have their use, maintenance, or repair continued where the use, maintenance, or repair is in accordance with the original design and location and no hazard to life, health, or property has been created by such plumbing system.

This code is published and oftentimes adopted in three year cycles. There may be many changes in the code from code cycle to code cycle and it is not the intent of any jurisdiction to require existing buildings to continually meet the requirements of the newly adopted edition of the code. There are three conditions that will allow an existing plumbing system to survive changes in the code without being brought up to the requirements of the new code. These conditions are:

1. The existing system must have been properly installed according to the code in effect at the time of the installation;

2. The system must have been properly maintained according to the code in effect when periodic maintenance was performed; and

3. The system must not be dangerous, unsafe, insanitary, or a nuisance and menace to life, health, or property. An example of this would be a broken and leaking sewer pipe.

If it can be proved that any or all of the above provisions were not met, the AHJ is justified in requiring that system to be brought up to the requirements of the new edition of the code.

102.3 Maintenance. The plumbing and drainage system, both existing and new, of a premise under the Authority Having Jurisdiction shall be maintained in a sanitary and safe operating condition. Devices or safeguards required by this code shall be maintained in accordance with the code edition under which installed.

The owner or the owner's designated agent shall be responsible for maintenance of plumbing systems. To determine compliance with this subsection, the Authority Having Jurisdiction shall be permitted to cause a plumbing system to be reinspected.

All plumbing systems require some maintenance and, as systems age, they usually require more frequent and thorough attention. The best plumbing systems can become hazardous to public health and safety if left unmaintained; therefore, this code emphasizes the need of the building owner to properly maintain the plumbing system. The UPC requires building owners to be responsible for the safe and sanitary condition of their plumbing systems and that they adhere to the code while maintaining the plumbing in their buildings.

If the AHJ is given reason to suspect that a system is not being maintained properly, it may require a reinspection, and any violations found will need to be brought into compliance with the current adopted code.

102.4 Additions, Alterations, Renovations, or Repairs. Additions, alterations, renovations or repairs shall conform to that required for a new system without requiring the existing plumbing system to be in accordance with the requirements of this code. Additions, alterations, renovations, or repairs shall not cause an existing system to become unsafe, insanitary, or overloaded.

Additions, alterations, renovations, or repairs to existing plumbing installations shall comply with the provisions for new construction unless such deviations are found to be necessary and are first approved by the Authority Having Jurisdiction.

When plumbing is repaired, altered or added to, the new work must be according to the current edition of the code in force at the time the work is done. The remaining existing system does not have to be brought up to the requirements of the current code. However, if the repairs or alterations overload the existing system or cause it to become unsafe or insanitary, then the overloaded portion or the unsafe section would have to be made acceptable and brought into compliance with the current code.

When plumbing systems in existing buildings are repaired or altered, it is expected that every possible effort will be made to ensure that the affected plumbing complies with the current code. However, conditions, usually structural, sometimes make compliance highly impractical. In this event, the AHJ has discretionary power to allow a deviation from the code requirements.

102.4.1 Building Sewers and Drains. Existing building sewers and building drains shall be permitted to be used in connection with new buildings or new plumbing and drainage work where they are found on examination and test to be in accordance with the requirements governing new work, and the proper Authority Having Jurisdiction shall notify the owner to make changes necessary to be in accordance with this code. No building, or part thereof, shall be erected or placed over a part of a drainage system that is constructed of materials other than those approved elsewhere in this code for use under or within a building.

Existing building drains and building sewers are sometimes reused in new or renovated buildings in which all or most of the plumbing systems are being rebuilt. Buildings destroyed by fire or natural disaster and those changing occupancy would fall into this category. In order to have this reuse of the building sewer or building drain approved by the AHJ, the owner must prove that the material used for the sewer or drain complies with the current code. The reused piping must satisfactorily hold a pressure test as prescribed by the current code, and an examination of the existing piping must indicate that the material is sound and in satisfactory condition.

Existing building sewers and building drains may be used in connection with new buildings or new plumbing and drainage functions only when they are found, on examination and test, to conform in all respects to the requirements governing new work. For example, in **Figure 102.4.1** a new building or portion of a building is proposed on an area where

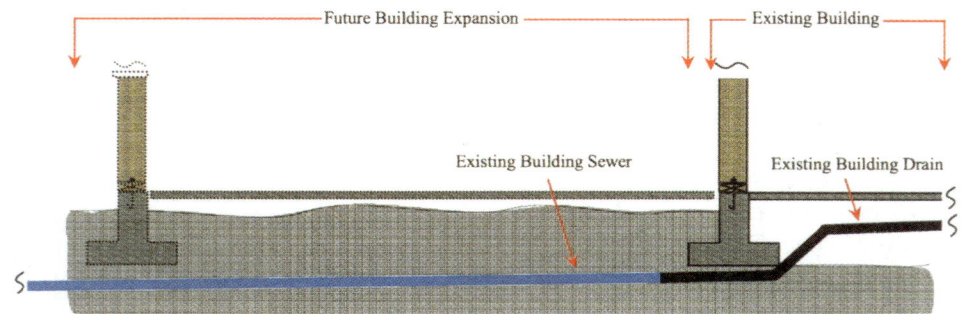

Whenever a building is built over an existing sewer, the existing sewer becomes a new building drain. This new building drain must be constructed of materials approved for use under or within a building in accordance with Section 701.2.

**FIGURE 102.4.1
BUILDING CONSTRUCTED OVER EXISTING SEWERS**

there is existing piping underground. The existing piping under the proposed work must meet the requirements for piping underneath a building. If the existing piping was plastic pipe SDR 35, it would not be allowed for use underneath a building and the pipe would have to be excavated and replaced with pipe that is for use underneath a building, such as cast iron pipe. It must also be of the proper size for the installation if fixtures are being added to the system.

102.4.2 Openings. Openings into a drainage or vent system, excepting those openings to which plumbing fixtures are properly connected or which constitute vent terminals, shall be permanently plugged or capped in an approved manner, using the appropriate materials in accordance with this code.

This section requires that when pipe stubs are left unused for any reason, such as when installed in anticipation of future needs, or left unused after a fixture has been removed and not replaced, the pipe openings must be securely plugged or capped. This not only prevents the undesirable and possibly dangerous leakage of liquids and sewer gases into the building, but it also discourages people from using these pipe stubs to make illegal connections to the plumbing systems (see **Figure 102.4.2**).

102.5 Health and Safety. Where compliance with the provisions of this code fails to eliminate or alleviate a nuisance, or other dangerous or insanitary condition that involves health or safety hazards, the owner or the owner's agent shall install such additional plumbing and drainage facilities or shall make such repairs or alterations as ordered by the Authority Having Jurisdiction.

Even if all the requirements of the UPC are followed, there may still be instances where a plumbing system or part of a system may remain unsafe or insanitary. If this were to happen, the AHJ may order the owner to correct the problem using methods prescribed by the AHJ.

102.6 Changes in Building Occupancy. Plumbing systems that are a part of a building or structure undergoing a change in use or occupancy, as defined in the building code, shall be in accordance with the requirements of this code that are applicable to the new use or occupancy.

A change in building occupancy refers to a building being used for a different purpose than originally intended or designed. A single-family residence converted into a restaurant or a store changed into a doctor's office is an example of a change in building occupancy or use. Whenever this happens, the plumbing systems must be suitable for the new use as defined in the building code adopted by the jurisdiction and the current plumbing code. This could involve separate or increased facilities for men and women, different types of fixtures, increase in pipe sizes, etc.

102.7 Moved Structures. Parts of the plumbing system of a building or part thereof that is moved from one foundation to another, or from one location to another, shall be in accordance with the provisions of this code for new installations and completely tested as prescribed elsewhere in this section for new work, except that walls or floors need not be removed during such test where other equivalent means of inspection acceptable to the Authority Having Jurisdiction are provided.

The practice of relocating structures has increased over the years. Sometimes buildings built where no code or a different code was followed will be moved into an area where the UPC is enforced. These systems may pose a health hazard if they do not comply with the code. Even though the structure may be several years old, it must now comply with the code as a new installation. Plumbing systems in these buildings are required to comply with the current code.

Plumbing systems in moved buildings must be tested and inspected as with new work. This is because during the process of moving, the integrity of the system may be compromised by the loosening of joints or by pipe separating or breaking. Pipes concealed within walls and floors do not need to be uncovered for these inspections, provided that the AHJ is satisfied the system can be adequately inspected by other means.

102.8 Appendices. The provisions in the appendices are intended to supplement the requirements of this code and shall not be considered part of this code unless formally adopted as such.

Valuable information is contained in these appendices such as the sizing of water piping and storm water drains, the installation of combination waste and vent systems, private sewage disposal systems and much more. This reference material, though highly valuable, is not legally a

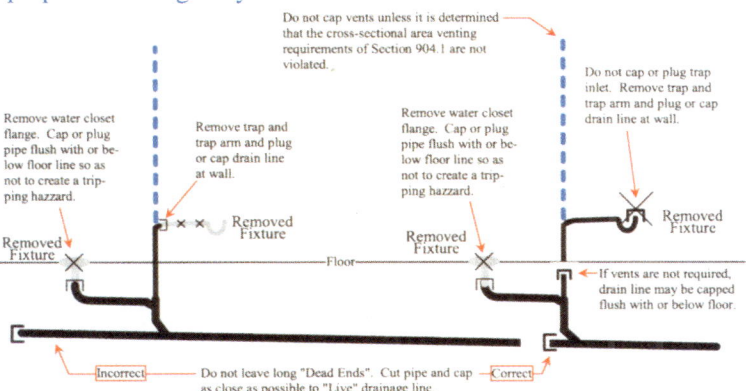

FIGURE 102.4.2
PLUGGING AND CAPPING DWV LINES

part of the code as it stands. If a local jurisdiction chooses to make all or part of these appendices code material, it may do so through its adoption process. It should also be noted that there may even be certain chapters of this code that are not adopted by some jurisdictions and that those chapters will also not be a part of the legal code document.

103.0 Duties and Powers of the Authority Having Jurisdiction.

103.1 General. The Authority Having Jurisdiction shall be the Authority duly appointed to enforce this code. For such purposes, the Authority Having Jurisdiction shall have the powers of a law enforcement officer. The Authority Having Jurisdiction shall have the power to render interpretations of this code and to adopt and enforce rules and regulations supplemental to this code as deemed necessary in order to clarify the application of the provisions of this code. Such interpretations, rules, and regulations shall comply with the intent and purpose of this code.

In accordance with the prescribed procedures and with the approval of the appointing authority, the Authority Having Jurisdiction shall be permitted to appoint a such number of technical officers, inspectors, and other employees as shall be authorized from time to time. The Authority Having Jurisdiction shall be permitted to deputize such inspectors or employees as necessary to carry out the functions of the code enforcement agency.

The Authority Having Jurisdiction shall be permitted to request the assistance and cooperation of other officials of this jurisdiction so far as required in the discharge of the duties in accordance with this code or other pertinent law or ordinance.

By definition, the AHJ is the individual or agency given the power and responsibility to ensure that this code is followed. The UPC is adopted by various and diverse communities. When adopted by a smaller community, the AHJ may be one individual or a small committee. When adopted by larger communities, such as cities, counties or states, the number of people involved in the enforcement of the code will increase proportionately. The organizations are usually building, safety or health departments and might even be state inspection agencies.

A governing body such as a city council or commission, or the state legislature will appoint a building official to administer the department. This individual will then hire or appoint assistants, inspectors, plans check personnel, and other staff as needed to enforce the various codes adopted by the governing body. The inspector is the individual that most plumbers deal with. He or she will be the one to visit the job site and inspect work for compliance with the code. They are also the officials who will interpret and enforce the code requirements.

103.2 Liability. The Authority Having Jurisdiction charged with the enforcement of this code, acting in good faith and without malice in the discharge of the Authority Having Jurisdiction's duties, shall not thereby be rendered personally liable for damage that accrues to persons or property as a result of an act or by reason of an act or omission in the discharge of duties. A suit brought against the Authority Having Jurisdiction or employee because of such act or omission performed in the enforcement of provisions of this code shall be defended by legal counsel provided by this jurisdiction until final termination of such proceedings.

This section declares that whenever damage occurs to a person or property as the result of something the AHJ or its representatives either did or failed to do while lawfully performing the duties of their position, the AHJ and its representatives are not to be held liable for the damages incurred. If a lawsuit is brought against the AHJ or its representatives, the jurisdiction will bear the cost of the defense. Without these safeguards it may be impossible to find someone who would take on the responsibility as an employee of the AHJ.

103.3 Applications and Permits. The Authority Having Jurisdiction shall be permitted to require the submission of plans, specifications, drawings, and such other information in accordance with the Authority Having Jurisdiction, prior to the commencement of, and at a time during the progress of, work regulated by this code.

The issuance of a permit upon construction documents shall not prevent the Authority Having Jurisdiction from thereafter requiring the correction of errors in said construction documents or from preventing construction operations being carried on thereunder where in violation of this code or of other pertinent ordinance or from revoking a certificate of approval where issued in error.

The clearest way to communicate the nature and scope of most plumbing projects is by the use of detailed plans, drawings and specifications. **Figure 103.3** is an example of a page of a detailed set of plans. Hence, the authority having jurisdiction (AHJ) must have the right to demand copies of all such documents for review prior to issuing a permit for work to commence. In addition, if needed at any time during the progress of the work, the AHJ has the right to require such documentation as needed. If a permit is granted and later it is found that there are mistakes in the documents provided, the AHJ will require that the errors be corrected and the plumbing installation meet the requirements of the code. The AHJ may also stop work on an installation in violation of the code until the errors are corrected.

103.3.1 Licensing. Provision for licensing shall be determined by the Authority Having Jurisdiction.

The licensing of plumbing journeyworkers and apprentices, as well as plumbing contractors, is the responsibility of the AHJ. Throughout the United States and Canada the requirements for licensing range from nonexistent to very rigorous. The United Association continues to support and promote the licensing of journeyworkers and apprentices in the plumbing industry, especially through its "Standard for Excellence" policy.

103.4 Right of Entry. Where it is necessary to make an inspection to enforce the provisions of this code, or where the Authority Having Jurisdiction has reasonable cause to believe that there exists in a building or upon premises a condition or violation of this code that makes the building or premises

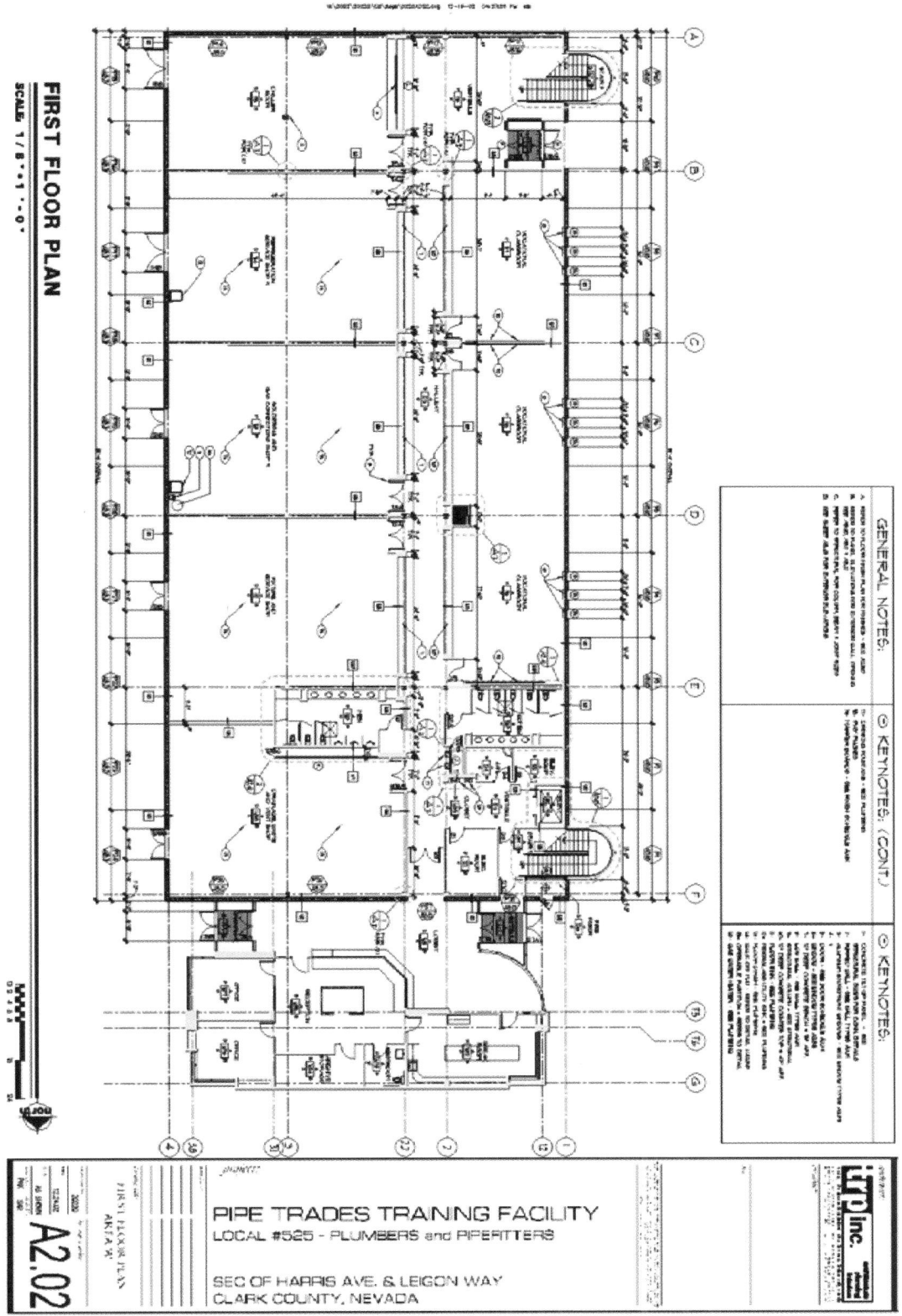

FIGURE 103.3
EXAMPLE OF SUBMITTED PLANS

unsafe, insanitary, dangerous, or hazardous, the Authority Having Jurisdiction shall be permitted to enter the building or premises at reasonable times to inspect or to perform the duties imposed by the Authority Having Jurisdiction by this code, provided that where such building or premises is occupied, the Authority Having Jurisdiction shall present credentials to the occupant and request entry. Where such building or premises is unoccupied, the Authority Having Jurisdiction shall first make a reasonable effort to locate the owner or other person having charge or control of the building or premises and request entry. Where entry is refused, the Authority Having Jurisdiction has recourse to every remedy provided by law to secure entry.

Where the Authority Having Jurisdiction shall have first obtained an inspection warrant or other remedy provided by law to secure entry, no owner, occupant, or person having charge, care or control of a building or premises shall fail or neglect, after a request is made as herein provided, to promptly permit entry herein by the Authority Having Jurisdiction for the purpose of inspection and examination pursuant to this code.

Protecting the health and safety of a community is the basis for enforcement. For that reason, the code has given the AHJ the right to enter a building to inspect the condition of the plumbing systems whenever there is a just cause to do so. If the plumbing inspector requests entry and the building owner refuses to allow the inspector to enter the building in question, then a suitable court-authorized warrant may be secured to which the owner is then legally obligated to allow the inspector to enter the building for inspection purposes.

104.0 Permits.

104.1 Permits Required. It shall be unlawful for a person, firm, or corporation to make an installation, alteration, repair, replacement, or remodel a plumbing system regulated by this code except as permitted in Section 104.2, or to cause the same to be done without first obtaining a separate plumbing permit for each separate building or structure.

Before the start of plumbing work, a plumbing permit must be obtained before anyone may install or alter a plumbing system. A permit is usually required for each separate building or structure, although on very large projects there may be more than one permit for the project. There may be different requirements on who may obtain a plumbing permit depending on the work to be done, the location where the work will be done, and local and state regulations. In some instances, there may be a requirement to have a journeyman or master plumber's license, or a plumbing contractor's license to obtain a plumbing permit. Plumbers should check with the local AHJ to determine the permitting requirements in their area. See **Figure 104.1** for a sample of a plumbing permit.

104.2 Exempt Work. A permit shall not be required for the following:

(1) The stopping of leaks in drains, soil, waste, or vent pipe, provided, however, that a trap, drainpipe, soil, waste, or vent pipe become defective, and it becomes necessary to remove and replace the same with new material, the same shall be considered as new work and a permit shall be procured and inspection made as provided in this code.

(2) The clearing of stoppages, including the removal and reinstallation of water closets, or the repairing of leaks in pipes, valves, or fixtures, provided such repairs do not involve or require the replacement or rearrangement of valves, pipes, or fixtures.

Exemption from the permit requirements of this code shall not be deemed to grant authorization for work to be done in violation of the provisions of the code or other laws or ordinances of this jurisdiction.

(1) There are limited minor repairs to plumbing systems that may be made without first obtaining a permit. For example, repair of leaking drains and vents may be made without a permit as long as pipe or fittings are not removed and replaced with new materials. However, if the leak is in a pipe and new piping material is needed for the repair, a permit and inspection are required. This may sound very restrictive but it must be remembered that even minor repairs to a plumbing system must be completed correctly or property damage, sickness or death may occur. Therefore, it becomes necessary to inspect these minor repairs to ensure compliance with the code and a safe plumbing system.

(2) Other plumbing repairs that don't require a permit include the clearing of stopped pipes; repairing leaks in pipes, valves and fixtures and the reinstallation of water closets (such as occurs after a wax seal is repaired or a new floor has been installed). However, if pipe, fittings or fixtures are replaced or rerouted, a permit must be obtained and the work then inspected for compliance with the code. The fact that a permit and the resulting inspection are not required does not mean that the work can be done in violation of the code. At all times all plumbing system work shall be in compliance to the code.

104.3 Application for Permit. To obtain a permit, the applicant shall first file an application therefore in writing on a form furnished by the Authority Having Jurisdiction for that purpose. Such application shall:

(1) Identify and describe the work to be covered by the permit for which application is made.

(2) Describe the land upon which the proposed work is to be done by legal description, street address, or similar description that will readily identify and locate the proposed building or work.

(3) Indicate the use or occupancy for which the proposed work is intended.

(4) Be accompanied by construction documents in accordance with Section 104.3.1.

(5) Be signed by the permittee or the permittee's authorized agent. The Authority Having Jurisdiction shall be permitted to require evidence to indicate such authority.

(6) Give such other data and information in accordance with the Authority Having Jurisdiction.

ADMINISTRATION

FIGURE 104.1
SAMPLE PERMIT - CLARK COUNTY, NEVADA

ADMINISTRATION

🔧 While particular permit applications may vary in form from one jurisdiction to another, the code requires that an application be filed and approved prior to the start of the work. The need for the above-cited information is so the inspector will know where the work is located, what type of work has been done, the occupancy requirements of the use of the system, who to contact, and ultimately who is responsible for the plumbing installation when he or she arrives for inspection. As work progresses or if there is a question concerning the compliance of a system to the code, the AHJ may require further information, such as design specifications or detailed drawings, so that a proper inspection of the system can be made.

104.3.1 Construction Documents. Construction documents, engineering calculations, diagrams, and other data shall be submitted in two or more sets with each application for a permit. The construction documents, computations, and specifications shall be prepared by, and the plumbing designed by, a registered design professional. Construction documents shall be drawn to scale with clarity to identify that the intended work to be performed is in accordance with the code.

Exception: The Authority Having Jurisdiction shall be permitted to waive the submission of construction documents, calculations, or other data where the Authority Having Jurisdiction finds that the nature of the work applied for is such that reviewing of construction documents is not necessary to obtain compliance with the code.

🔧 Unless the AHJ determines that plan review isn't necessary to ensure code compliance, all permit applications need to be accompanied by plans (usually two sets) and any other such data that is needed to determine that the proposed work will be in compliance with the code. The plans and specifications will go to plan review specialists who will examine them for compliance with the code. There may be several different specialists who will examine the plans such as plumbing, mechanical, and fire systems specialists, as well as various systems of the proposed project for code compliance. Small or minor installations may not be required to have plans submitted if the inspections and compliance to the code can be determined without them.

104.3.2 Plan Review Fees. Where a plan or other data is required to be submitted in accordance with Section 104.3.1, a plan review fee shall be paid at the time of submitting construction documents for review.

The plan review fees for plumbing work shall be determined and adopted by this jurisdiction.

The plan review fees specified in this subsection are separate fees from the permit fees specified in Section 104.5.

Where plans are incomplete or changed so as to require additional review, a fee shall be charged at the rate shown in Table 104.5.

🔧 The plan review fee may be a separate fee or a part of the plumbing permit fee. This fee pays for the time it takes for the review of a project's plans and specifications to ensure compliance with the code. A thorough plan review is an aid to both the AHJ and the plumber interested in complying with the code.

104.3.3 Time Limitation of Application. Applications for which no permit is issued within 180 days following the date of application shall expire by limitation, plans and other data submitted for review thereafter, shall be returned to the applicant or destroyed by the Authority Having Jurisdiction. The Authority Having Jurisdiction shall be permitted to exceed the time for action by the applicant for a period not to exceed 180 days upon request by the applicant showing that circumstances beyond the control of the applicant have prevented the action from being taken. No application shall be extended more than once. In order to renew action on an application after expiration, the applicant shall resubmit plans and pay a new plan review fee.

🔧 In the event that a permit application cannot be approved within 180 days because the application with pertinent plans and data are incomplete or contain uncorrected code violations, or for any other reason caused by the applicant, the application expires and is either returned or destroyed. The applicant may receive a one-time extension of 180 days if he or she can prove that the extension is needed for reasons beyond his or her control.

104.4 Permit Issuance. The application, construction documents, and other data filed by an applicant for a permit shall be reviewed by the Authority Having Jurisdiction. Such plans shall be permitted to be reviewed by other departments of this jurisdiction to verify compliance with applicable laws under their jurisdiction. Where the Authority Having Jurisdiction finds that the work described in an application for permit and the plans, specifications, and other data filed therewith are in accordance with the requirements of the code and other pertinent laws and ordinances and that the fees specified in Section 104.5 have been paid, the Authority Having Jurisdiction shall issue a permit therefore to the applicant.

104.4.1 Approved Plans or Construction Documents. Where the Authority Having Jurisdiction issues the permit where plans are required, the Authority Having Jurisdiction shall endorse in writing or stamp the construction documents "APPROVED." Such approved construction documents shall not be changed, modified, or altered without authorization from the Authority Having Jurisdiction, and the work shall be done in accordance with approved plans.

The Authority Having Jurisdiction shall be permitted to issue a permit for the construction of a part of a plumbing system before the entire construction documents for the whole system have been submitted or approved, provided adequate information and detailed statements have been filed in accordance with the pertinent requirements of this code. The holder of such permit shall be permitted to proceed at the holder's risk without assurance that the permit for the entire building, structure, or plumbing system will be granted.

🔧 Plumbing permits are granted after the AHJ has determined, by plan review and other research, that the proposed work conforms to the code and other pertinent laws and after the appropriate permit fees have been paid. If changes to the work are required after the plans and specifications have been reviewed and approved and a permit issued, such changes must be documented and new plans submitted and approved by the AHJ prior to the changed work being done.

Sometimes it is expedient to begin a part of a project before all plans and pertinent data have been developed. In this case, the AHJ may approve a permit application for a portion of a project as long as the review of the partial documentation reveals that the installation is in compliance with the code. Approval of the permit application for a partial system does not ensure that later applications for permits for subsequent work will be automatically approved; each application will be judged on its own merit.

104.4.2 Validity of Permit. The issuance of a permit or approval of construction documents shall not be construed to be a permit for, or an approval of, a violation of the provisions of this code or other ordinance of the jurisdiction. No permit presuming to give authority to violate or cancel the provisions of this code shall be valid.

The issuance of a permit based upon plans, specifications, or other data shall not prevent the Authority Having Jurisdiction from thereafter requiring the correction of errors in said plans, specifications, and other data or from preventing building operations being carried on thereunder where in violation of this code or of other ordinances of this jurisdiction.

It is possible that the plan review and permit application review may not reveal all possible code violations in the proposed work. Indeed, plans may be inadvertently stamped "APPROVED" even though they contain code violations. If this happens, the plumber must still install the system according to the provisions of the code. He or she is not excused from adhering to the code, even though the plans have been approved. In addition, the AHJ may require correction to the plans and construction data, even after they have been initially approved. The bottom line - no plumbing system may be installed that does not comply fully with the code even though the plans are approved.

104.4.3 Expiration. A permit issued by the Authority Having Jurisdiction under the provisions of this code shall expire by limitation and become null and void where the work authorized by such permit is not commenced within 180 days from the date of such permit, or where the work authorized by such permit is suspended or abandoned at a time after the work is commenced for a period of 180 days. Before such work is recommenced, a new permit shall first be obtained to do so, and the fee, therefore, shall be one-half the amount required for a new permit for such work, provided no changes have been made or will be made in the original construction documents for such work, and provided further that such suspensions or abandonment has not exceeded 1 year.

After a permit has been approved, the plumber has 180 days to begin work on the project. If he or she fails to do so, the permit becomes null and void, and a new permit must be approved before work may commence. Likewise, if the plumber suspends work on the project for 180 days, the permit is no longer valid, and a new permit is required for work to begin again.

If there are good reasons for a delay to the start of a project beyond the 180-day period allowed, the AHJ may, upon request, extend the time period by another 180 days. This extension may only be allowed once.

104.4.4 Extensions. A permittee holding an unexpired permit shall be permitted to apply for an extension of the time within which work shall be permitted to commence under that permit where the permittee is unable to commence work within the time required by this section. The Authority Having Jurisdiction shall be permitted to extend the time for action by the permittee for a period not exceeding 180 days upon written request by the permittee showing that circumstances beyond the control of the permittee have prevented the action from being taken. No permit shall be extended more than once. In order to renew action on a permit after expiration, the permittee shall pay a new full permit fee.

104.4.5 Suspension or Revocation. The Authority Having Jurisdiction shall be permitted to, in writing, suspend or revoke a permit issued under the provisions of this code where the permit is issued in error or on the basis of incorrect information supplied or in violation of other ordinance or regulation of the jurisdiction.

104.4.6 Retention of Plans. One set of approved construction documents and computations shall be retained by the Authority Having Jurisdiction until final approval of the work covered therein.

One set of approved construction documents, computations, and manufacturer's installation instructions shall be returned to the applicant and said set shall be kept on the site of the building or work at times during which the work authorized thereby is in progress.

Until an approved plumbing project is completed, the AHJ will keep a set of approved plans for reference. Another set of approved plans returned to the applicant must be kept at the job site at all times until final inspection and approval of the project. This set of plans will be used by the installer and the inspector. It is always good practice to have a set of "working plans" that the installer uses, as well as the signed and stamped set of plans set aside for the use of the code officials. This set will also go to the owner of the building for his or her use once the project is completed.

104.5 Fees. Fees shall be assessed in accordance with the provisions of this section and as set forth in the fee schedule, Table 104.5. The fees are to be determined and adopted by this jurisdiction.

Permit fees are the means by which the AHJ is funded to provide the services it renders. Without these fees there would be no plan reviews or inspections and compliance to the codes would be left up to the responsibility of the plumber or contractor. In areas where this has been attempted, invariably someone takes advantage of the fact that there is no "watchdog" and installs inferior systems to the detriment not only of the building owner or its residents but to the public at large. Permit fees are therefore necessary and may vary greatly among jurisdictions. Table 104.5 is a sampling of what may be included in a local jurisdiction's fee schedule.

TABLE 104.5
PLUMBING PERMIT FEES

Permit Issuance

1. For issuing each permit ....*_____
2. For issuing each supplemental permit ....*_____

Unit Fee Schedule (in addition to Items 1 and 2 above)

1. For each plumbing fixture on one trap or a set of fixtures on one trap (including water, drainage piping, and backflow protection therefore) ....*_____
2. For each building sewer and each trailer park sewer ....*_____
3. Rainwater systems – per drain (inside building)..*_____
4. For each cesspool (where permitted) ....*_____
5. For each private sewage disposal system...*_____
6. For each water heater, vent, or both ....*_____
7. For each gas piping system of one to five outlets ....*_____
8. For each additional gas piping system outlet, per outlet...*_____
9. For each industrial waste pretreatment interceptor, including its trap and vent, except kitchen-type grease interceptors functioning as fixture traps ....*_____
10. For each installation, alteration, or repair of water piping, water treating equipment, or both ....*_____
11. For each repair or alteration of drainage or vent piping, each fixture...*_____
12. For each lawn sprinkler system on one meter including backflow protection devices therefore............*_____
13. For atmospheric-type vacuum breakers not referenced in Item 12:
    One to 5 ....*_____
    over 5, each ....*_____
14. For each backflow protective device other than atmospheric-type vacuum breakers:
    Two inch (50 mm) diameter and smaller ....*_____
    over 2 inch (50 mm) diameter...*_____
15. For each gray water system...*_____
16. For initial installation and testing for a reclaimed water system..*_____
17. For each annual cross-connection testing of a reclaimed water system (excluding initial test) ....*_____
18. For each medical gas piping system serving one to five inlet(s)/outlet(s) for a specific gas................*_____
19. For each additional medical gas inlet(s)/outlet(s) ....*_____

Other Inspections and Fees

1. Inspections outside of normal business hours ....*_____
2. Reinspection fee ....*_____
3. Inspections for which no fee is specifically indicated..*_____
4. Additional plan review required by changes, additions, or revisions to approved plans (minimum charge – ½ hour)..*_____

For SI units: 1 inch = 25 mm

* Jurisdiction will indicate their fees here.

104.5.1 Work Commencing Before Permit Issuance. Where work for which a permit is required by this code has been commenced without first obtaining said permit, a special investigation shall be made before a permit is issued for such work.

104.5.2 Investigation Fees. An investigation fee, in addition to the permit fee, shall be collected whether a permit is then or subsequently issued. The investigation fee shall be equal to the amount of the permit fee that is required by this code if a permit were to be issued. The payment of such investigation fee shall not exempt a person from compliance with other provisions of this code, nor from a penalty prescribed by law.

There is always someone who will try to get away without paying a fee. The code takes this into consideration, and when someone has been found to have done plumbing work that requires a permit without first getting that permit, an investigation fee may be levied. This will pay for an investigation to take place to determine what work has been done and if it is code compliant. Application for a permit, with the resulting permit fee, will then be allowed to proceed. The fees will be required to be paid even if a permit is eventually not granted. The fees will also not preclude any other penalty that may be levied against the offender.

104.5.3 Fee Refunds. The Authority Having Jurisdiction shall be permitted to authorize the refunding of a fee as follows:

(1) The amount paid hereunder that was erroneously paid or collected.

(2) Refunding of not more than a percentage, as determined by this jurisdiction where no work has been done under a permit issued in accordance with this code.

The Authority Having Jurisdiction shall not authorize the refunding of a fee paid except upon written application filed by the original permittee not to exceed 180 days after the date of fee payment.

There may be instances where a mistake is made and an unnecessary fee is charged. The AHJ will usually refund that portion of the overcharged fee. When a permit fee is collected, but no work was done, the AHJ may refund all or a portion of the fee paid, depending on the policies in place. A refund will not be paid if a written application for refund has not been made or if the permit has expired.

105.0 Inspections and Testing.

105.1 General. Plumbing systems for which a permit is required by this code shall be inspected by the Authority Having Jurisdiction.

No plumbing system or portion thereof shall be covered, concealed, or put into use until inspected and approved as prescribed in this code. Neither the Authority Having Jurisdiction nor the jurisdiction shall be liable for expense entailed in the removal or replacement of material required to permit inspection. Plumbing systems regulated by this code shall not be connected to the water, the energy fuel supply, or the sewer system until authorized by the Authority Having Jurisdiction.

All plumbing installations requiring a permit must be inspected by the AHJ. The work must be exposed so that it can be inspected and tested. This includes work above ground and below ground. If work has been completed and subsequently covered over, the cost of the removal and replacement of any building material that needs to be removed to reveal the system for inspection will not be the responsibility of the jurisdiction. For example, if a water line is ready for inspection but it has been backfilled so that the inspector cannot see the individual joints in the system, the backfill will be required to be removed at the cost to the plumber or contractor.

Inspections are made before work is concealed - piping in walls covered or piping in trenches backfilled. This inspection ensures that the size, placement, and materials of the system are code compliant and that the system is water or airtight. Connections to water supplies, energy fuel supplies and sewers may not be made until this final inspection is completed or unless permitted by the AHJ.

105.2 Required Inspections. New plumbing work and such portions of existing systems as affected by new work, or changes, shall be inspected by the Authority Having Jurisdiction to ensure compliance with the requirements of this code and to ensure that the installation and construction of the plumbing system are in accordance with approved plans. The Authority Having Jurisdiction shall make the following inspections and other such inspections as necessary. The permittee or the permittee's authorized agent shall be responsible for the scheduling of such inspections as follows:

(1) The underground inspection shall be made after trenches or ditches are excavated and bedded, piping installed, and before backfill is put in place.

(2) Rough-in inspection shall be made prior to the installation of wall or ceiling membranes.

(3) Final inspection shall be made upon completion of the installation.

All plumbing work done under a permit must be inspected by the AHJ to make sure that it complies with this code and approved plans. There are typically three phases of plumbing installation that require inspection. There is the initial stage of underground plumbing systems consisting of drain, waste, vent, water, and gas supply. Secondly, there is the rough-in phase consisting of the same plumbing systems, but above ground. Thirdly, there is the final stage of installing plumbing fixtures and trim material. If there is an addition to an existing system, the portions of the existing system that are affected by the addition shall also be inspected and approved.

105.2.1 Uncovering. Where a drainage or plumbing system, building sewer, private sewage disposal system, or part thereof, which is installed, altered, or repaired, is covered or concealed before being inspected, tested, and approved as prescribed in this code, it shall be uncovered for inspection after notice to uncover the work has been issued to the responsible person by the Authority Having Jurisdiction.

ADMINISTRATION

The requirements of this section shall not be considered to prohibit the operation of plumbing installed to replace existing equipment or fixtures serving an occupied portion of the building in the event a request for inspection of such equipment or fixture has been filed with the Authority Having Jurisdiction not more than 72 hours after such replacement work is completed, and before a portion of such plumbing system is concealed by a permanent portion of the building.

The test, inspection and approval of all plumbing work under permit are required prior to covering or concealing the work. If any plumbing work under permit is covered or concealed prior to approval by the AHJ, it must be uncovered by the applicant to allow for an inspection.

When an existing plumbing fixture or device fails and needs to be replaced in an occupied portion of a building, it is often impractical to wait for an inspection and approval before putting the replacement into service. Therefore, the code allows that such replacement material may be used before the inspection, provided that a request for an inspection is made within 72 hours of the replacement and the replaced materials remain accessible for the inspection.

105.2.1.1 Water Supply System. No water supply system or portion thereof shall be covered or concealed until it first has been tested, inspected, and approved.

105.2.1.2 Covering or Using. No plumbing or drainage system, building sewer, private sewer disposal system, or part thereof, shall be covered, concealed, or put into use until it has been tested, inspected, and accepted as prescribed in this code.

105.2.2 Other Inspections. In addition to the inspections required by this code, the Authority Having Jurisdiction shall be permitted to require other inspections to ascertain compliance with the provisions of this code and other laws that are enforced by the Authority Having Jurisdiction.

105.2.3 Inspection Requests. It shall be the duty of the person doing the work authorized by a permit to notify the Authority Having Jurisdiction that such work is ready for inspection. The Authority Having Jurisdiction shall be permitted to require that a request for inspection be filed not less than 1 working day before such inspection is desired. Such request shall be permitted to be made in writing or by telephone, at the option of the Authority Having Jurisdiction.

It shall be the duty of the person requesting inspections in accordance with this code to provide access to and means for inspection of such work.

105.2.4 Advance Notice. It shall be the duty of the person doing the work authorized by the permit to notify the Authority Having Jurisdiction, orally or in writing that said work is ready for inspection. Such notification shall be given not less than 24 hours before the work is to be inspected.

105.2.5 Responsibility. It shall be the duty of the holder of a permit to make sure that the work will stand the test prescribed before giving the notification.

The equipment, material, and labor necessary for inspection or tests shall be furnished by the person to whom the permit is issued or by whom inspection is requested.

The person performing the work under a permit is responsible for requesting an inspection. Inspection requests can be made in several ways depending on the policies of the various AHJs. Phone, fax and now even Internet requests are made for inspections, but the plumber should inquire as to what method is used in his or her location.

Inspection requests are usually required to be placed at least 24 hours before the inspection is needed. The person requesting the inspection is responsible for providing access to the work to be inspected and a means whereby it may be inspected—for example, a lift or scaffolding may be required for the inspection of piping installed in a high location.

The permittee must test the system and make sure that it will hold the required pressure before calling for the inspection. If he or she fails to do this, leaks may prevent the test from passing and possibly resulting in another inspection and a reinspection fee. The permittee or the person requesting an inspection is responsible for providing all of the equipment needed for the test.

105.2.6 Reinspections. A reinspection fee shall be permitted to be assessed for each inspection or reinspection where such portion of work for which inspection is called is not complete or where required corrections have not been made.

This provision shall not be interpreted as requiring reinspection fees the first time a job is rejected for failure to be in accordance with the requirements of this code, but as controlling the practice of calling for inspections before the job is ready for inspection or reinspection.

Reinspection fees shall be permitted to be assessed where the approved plans are not readily available to the inspector, for failure to provide access on the date for which the inspection is requested, or for deviating from plans requiring the approval of the Authority Having Jurisdiction.

To obtain reinspection, the applicant shall file an application therefore in writing upon a form furnished for that purpose and pay the reinspection fee in accordance with Table 104.5.

In instances where reinspection fees have been assessed, no additional inspection of the work will be performed until the required fees have been paid.

Plumbers anticipating the need for an inspection might be tempted to call for an inspection before the work is completed. If the system is not completed and made ready for the test by the time specified for the inspection, the inspection will fail, and the system will have to be reinspected at a later date. To discourage this practice, the AHJ may charge a fee for the reinspection. The AHJ may also charge a fee when reinspection becomes necessary because access to the system is not made available at the time of the initial inspection, the approved plans are not available at the time of the inspection, or the plumber has deviated from the approved plans.

When the AHJ requires a reinspection, the applicant must apply in writing on a special reinspection form and pay the necessary fee. No other inspections of that work will be made until the reinspection fee has been paid.

105.3 Testing of Systems. Plumbing systems shall be tested and approved in accordance with this code or the Authority Having Jurisdiction. Tests shall be conducted in the presence of the Authority Having Jurisdiction or the Authority Having Jurisdiction's duly appointed representative.

No test or inspection shall be required where a plumbing system, or part thereof, is set up for exhibition purposes and has no connection with a water or drainage system. In cases where it would be impractical to provide the required water or air tests, or for minor installations and repairs, the Authority Having Jurisdiction shall be permitted to make such inspection as deemed advisable in order to be assured that the work has been performed in accordance with the intent of this code. Joints and connections in the plumbing system shall be gastight and watertight for the pressures required by the test.

If a test is required for a particular system, then the installation must be tested and withstand the test before that system is approved. When a system test is required, the inspector must be present during the test to observe it in progress and verify that the system has indeed withstood the test. If a plumbing system is set up for a display, such as in a convention or conference area, and will not be connected with water or sewer, it does not need to be tested. The test on a small plumbing job or a job where it is determined that it would be too difficult or costly to test the system, may be waived by the AHJ, if it can be validated that the installation meets the intent of the code. For example, a running water test on a drainage system or the pressurization of a water system without resulting in leaking may substitute for the formal test for that system, but only if first approved and witnessed by the AHJ.

All joints in the plumbing system whether, they are DWV, water or fuel gas joints, shall be water or airtight and free of leaks after the proper test is applied before the system is approved and placed into service.

105.3.1 Defective Systems. An air test shall be used in testing the sanitary condition of the drainage or plumbing system of building premises where there is reason to believe that it has become defective. In buildings or premises condemned by the Authority Having Jurisdiction because of an insanitary condition of the plumbing system, or part thereof, the alterations in such system shall be in accordance with the requirements of this code.

The AHJ has discretionary power to conduct other than ordinary inspections of plumbing systems if it suspects code or other regulation violations. If the AHJ suspects a system or part thereof is defective, it may require that system to undergo an air test to prove its integrity. When a condemned system is restored, the work must comply in all ways with the provisions of the code.

105.3.2 Retesting. Where the Authority Having Jurisdiction finds that the work will not pass the test, necessary corrections shall be made, and the work shall be resubmitted for test or inspection.

105.3.3 Approval. Where prescribed tests and inspections indicate that the work is in accordance with this code, a certificate of approval shall be issued by the Authority Having Jurisdiction to the permittee on demand.

105.4 Connection to Service Utilities. No person shall make connections from a source of energy or fuel to a plumbing system or equipment regulated by this code and for which a permit is required until approved by the Authority Having Jurisdiction. No person shall make connection from a water-supply line nor shall connect to a sewer system regulated by this code and for which a permit is required until approved by the Authority Having Jurisdiction. The Authority Having Jurisdiction shall be permitted to authorize temporary connection of the plumbing equipment to the source of energy or fuel for the purpose of testing the equipment.

Before connection of a plumbing system to an energy source, fuel source, water supply or sewer system can be legally made, it must first have passed inspection and been approved by the AHJ. This will preclude any contamination or leak that may happen when an incomplete system is connected to its source. Temporary connections during the progress of the work may be made to facilitate testing or to provide power to the job site but only after approval of the AHJ and after all precautions are made for safety.

106.0 Violations and Penalties.

106.1 General. It shall be unlawful for a person, firm, or corporation to erect, construct, enlarge, alter, repair, move, improve, remove, convert, demolish, equip, use, or maintain plumbing or permit the same to be done in violation of this code.

The installation, repair or maintenance of plumbing systems that do not comply with the various sections of the code or a violation of requirements found in other documents referenced by this code constitutes a violation of the code. Since the violation could cause property damage, sickness or even death, no one is allowed to violate the code or knowingly permit it to be violated by others. This code, once adopted, becomes law and just like any other law enacted by a governing body, enforcement and penalty provisions for violations are included in the code.

106.2 Notices of Correction or Violation. Notices of correction or violation shall be written by the Authority Having Jurisdiction and shall be permitted to be posted at the site of the work or mailed or delivered to the permittee or their authorized representative.

Refusal, failure, or neglect to comply with such notice or order within 10 days of receipt thereof, shall be considered a violation of this code and shall be subject to the penalties set forth by the governing laws of the jurisdiction.

Should the AHJ need to inform the plumber of a code violation or a need to make a correction, such notification may be made in person, by mail or by notice posted at the job site. If the permittee fails to comply with the notice within 10 days after receiving it, he or she will be in violation of the code and appropriately penalized. Plumbing systems that fail the pressure test will have to be repaired, and another request for inspection will be needed. Once the final test and inspection has been completed and approved, a certificate of approval will be issued for the work.

106.3 Penalties. A person, firm, or corporation violating a provision of this code shall be deemed guilty of a misdemeanor, and upon conviction thereof, shall be punishable by a fine, imprisonment, or both set forth by the governing laws of the jurisdiction. Each separate day or portion thereof, during which a violation of this code occurs or continues, shall be deemed to constitute a separate offense.

A violation of the code is a misdemeanor crime and those who violate the code may be fined and/or imprisoned accordingly. In addition, each day an offense continues to occur becomes a separate violation of the code and is likewise punishable by fine and/or imprisonment. The likelihood of someone going to jail for a violation of the code is quite small; however, it has happened. The violation usually entails a very serious and flagrant violation of the code and is accompanied by sickness or death. The violation is also usually found to be knowingly installed in violation of the code without concern for the consequences. Fines are much more common and are usually levied when the owner, contractor or plumber refuses to repair a violation.

106.4 Stop Orders. Where work is being done contrary to the provisions of this code, the Authority Having Jurisdiction shall be permitted to order the work stopped by notice in writing served on persons engaged in the doing or causing such work to be done, and such persons shall forthwith stop work until authorized by the Authority Having Jurisdiction to proceed with the work.

"Stop work orders," though rarely needed, are a tool the AHJ can use to shut down a plumbing job that is being performed contrary to the code. Such orders are usually employed only after other attempts have been made to force the contractor to comply with the code (see **Figure 106.4**).

106.5 Authority to Disconnect Utilities in Emergencies. The Authority Having Jurisdiction shall have the authority to disconnect a plumbing system to a building, structure, or equipment regulated by this code in case of emergency where necessary to eliminate an immediate hazard to life or property.

In the case where a plumbing system is causing a serious threat to life or property, such as a significant natural gas, sewer or water leak, the AHJ has the right to have the offending system disconnected to prevent injury or damage.

106.6 Authority to Condemn. Where the Authority Having Jurisdiction ascertains that a plumbing system or portion thereof, regulated by this code, has become hazardous to life, health, or property, or has become insanitary, the Authority Having Jurisdiction shall order in writing that such plumbing either be removed or placed in a safe or sanitary condition. The order shall fix a reasonable time limit for compliance. No person shall use or maintain defective plumbing after receiving such notice.

Where such plumbing system is to be disconnected, written notice shall be given. In cases of immediate danger to life or property, such disconnection shall be permitted to be made immediately without such notice.

The purpose of this authority is to provide measures to protect the "public health, safety, and welfare." Hence, when a plumbing system creates a hazard to public health, safety, and welfare, the code enforcement agency may condemn the system until it is made safe. During the time the system is condemned, no one may use it, and the building's occupants may be forced to leave the building. If the problem is not corrected in the time allowed by the AHJ, then it may have the system disconnected. The owner will be notified prior to the disconnection unless the situation presents an emergency. In that case, no notification is required before the disconnection.

107.0 Board of Appeals.

107.1 General. In order to hear and decide appeals of orders, decisions, or determinations made by the Authority Having Jurisdiction relative to the application and interpretations of this code, there shall be and is hereby created a Board of Appeals consisting of members who are qualified by experience and training to pass upon matters pertaining to plumbing design, construction, and maintenance and the public health aspects of plumbing systems and who are not employees of the jurisdiction. The Authority Having Jurisdiction shall be an ex-officio member and shall act as secretary to said board but shall have no vote upon a matter before the board. The Board of Appeals shall be appointed by the gov-

Any Jurisdiction

BUREAU OF BUILDINGS

STOP WORK

ON ALL ____________ WORK ON THIS STRUCTURE ____

THIS ORDER IS ISSUED BECAUSE OF VIOLATIONS OF THE ____________ CODE.

DATE POSTED ____________

THIS NOTICE MUST NOT BE REMOVED EXCEPT BY THE INSPECTOR POSTING SAME

BY ____________ INSPECTOR

FIGURE 106.4
SAMPLE STOP WORK ORDER

erning body and shall hold office at its pleasure. The board shall adopt rules of procedure for conducting its business and shall render decisions and findings in writing to the appellant with a duplicate copy to the Authority Having Jurisdiction.

107.2 Limitations of Authority. The Board of Appeals shall have no authority relative to interpretation of the administrative provisions of this code, nor shall the board be empowered to waive requirements of this code.

CHAPTER 2
DEFINITIONS

201.0 General.

201.1 Applicability. For the purpose of this code, the following terms have the meanings indicated in this chapter.

No attempt is made to define ordinary words, which are used in accordance with their established dictionary meanings, except where a word has been used loosely, and it is necessary to define its meaning as used in this code to avoid misunderstanding.

202.0 Definition of Terms.

202.1 General. The definitions of terms are arranged alphabetically according to the first word of the term.

203.0   – A –

ABS. Acrylonitrile-butadiene-styrene.

Accepted Engineering Practice. That which conforms to technical or scientific-based principles, tests, or standards that are accepted by the engineering profession.

Accessible. Where applied to a fixture, connection, appliance, or equipment, "accessible" means having access thereto, but which first may require the removal of an access panel, door, or similar obstruction.

Accessible, Readily. Having a direct access without the necessity of removing a panel, door, or similar obstruction.

Air Break. A physical separation which may be a low inlet into the indirect waste receptor from the fixture, appliance, or device indirectly connected.

See Figure 203.0a

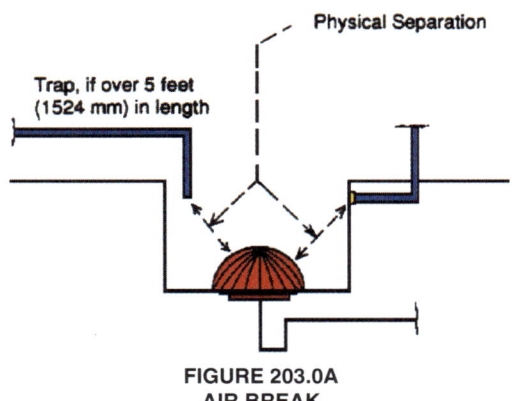

**FIGURE 203.0A
AIR BREAK**

Air Gap, Drainage. The unobstructed vertical distance through the free atmosphere between the lowest opening from a pipe, plumbing fixture, appliance, or appurtenance conveying waste to the flood-level rim of the receptor.

See Figure 203.0b

Air Gap, Water Distribution. The unobstructed vertical distance through the free atmosphere between the lowest opening from a pipe or faucet conveying potable water to the flood-level rim of a tank, vat, or fixture.

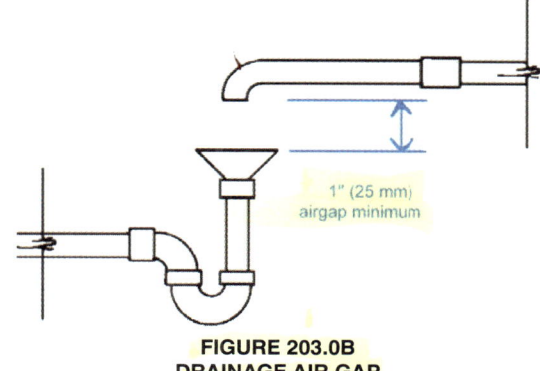

**FIGURE 203.0B
DRAINAGE AIR GAP**

Alternate Water Source. Nonpotable source of water that includes but not limited to gray water, on-site treated nonpotable water, rainwater, and reclaimed (recycled) water.

Anchors. See Supports.

Appliance. A device that utilizes an energy source to produce light, heat, power, refrigeration, air conditioning, or compressed fuel gas. This definition also shall include a vented decorative appliance.

Appliance, Low-Heat. A fuel-burning appliance that produces a continuous flue gas temperature, at the point of entrance to the flue, of not more than 1000°F (538°C).

Appliance, Medium-Heat. A fuel-burning appliance that produces a continuous flue gas temperature, at the point of entrance to the flue, of more than 1000°F (538°C) and less than 2000°F (1093°C).

Appliance Categorized Vent Diameter/Area. The minimum vent diameter/area permissible for Category I appliances to maintain a nonpositive vent static pressure when tested in accordance with nationally recognized standards. [NFPA 54:3.3.6]

NFPA 54 deals with the minimum vent diameter rather than the area of the vent. This minimum diameter will be found in the various sizing tables contained in this chapter.

Category I appliance vents are installed and sized per the requirements contained in NFPA 54. The four categories were created to differentiate appliances based on the pressure produced in the vent and the actual temperature and dew point of the flue gases. The category number is thus based on equipment design. All fuel-burning appliances have a rating plate that contains its category number. This category number, therefore, tells installers the venting requirements of each appliance. The appliance categories are as follows:

Category I is an appliance that operates with a nonpositive vent static pressure and with a vent gas temperature that avoids excessive condensate production in the vent. In these appliances the products of combustion rise up and out of the vent system by the heat in the flue gases. These are the appliances that will be discussed within this chapter. These drafthood equipped appliances as well as many mid-efficiency

DEFINITIONS

(fan-equipped but still with a nonpositive static pressure vent) models are the most traditional appliances designed for vertical venting.

The following category appliances are to be installed per the manufacturer's installation instructions for all parameters of the appliance itself and the venting system. NFPA 54 does not contain any additional venting information for these category appliances.

Category II is an appliance that operates with a nonpositive vent static pressure and with a vent gas temperature that may cause excessive condensate production in the vent.

Category III is an appliance that operates with a positive vent static pressure and with a vent gas temperature that avoids excessive condensate production in the vent.

Category IV is an appliance that operates with a positive vent static pressure and with a vent gas temperature that may cause excessive condensate production in the vent.

Appliance Fuel Connector. An assembly of listed semi-rigid or flexible tubing and fittings to carry fuel between a fuel-piping outlet and a fuel-burning appliance.

Approved. Acceptable to the Authority Having Jurisdiction.

Approved Testing Agency. An organization primarily established for purposes of testing to approved standards and approved by the Authority Having Jurisdiction.

Area Drain. A receptor designed to collect surface or storm water from an open area.

See **Figure 203.0c**

FIGURE 203.0C
AREA DRAIN - BELOW STREET LEVEL

Aspirator. A fitting or device supplied with water or other fluid under positive pressure that passes through an integral orifice or constriction, causing a vacuum.

See **Figure 203.0d**

Authority Having Jurisdiction. The organization, office, or individual responsible for enforcing the requirements of a code or standard, or for approving equipment, materials, installations, or procedures. The Authority Having Jurisdiction shall be a federal, state, local, or other regional department or an individual such as a plumbing official, mechanical official, labor department official, health department official, building official, or others having statutory authority. In the absence of statutory authority, the Authority Having Jurisdiction may be some other responsible party. This definition shall include the Authority Having Jurisdiction's duly authorized representative.

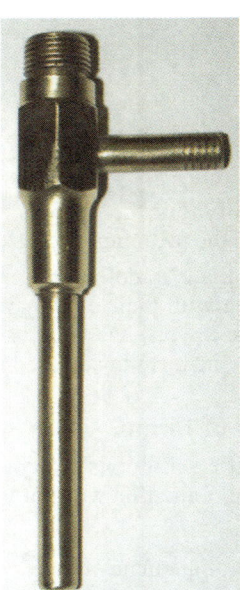

FIGURE 203.0D
ASPIRATOR

204.0 – B –

Backflow. The flow of water or other liquids, mixtures, or substances into the distributing pipes of a potable supply of water from sources other than its intended source. See Backpressure Backflow and Backsiphonage.

Backflow Connection. An arrangement whereby backflow can occur.

Backflow Preventer. A backflow prevention device, an assembly, or another method to prevent backflow into the potable water system.

Backpressure Backflow. Backflow due to an increased pressure above the supply pressure, which may be due to pumps, boilers, gravity, or other sources of pressure.

Backsiphonage. The flowing back of used, contaminated, or polluted water from a plumbing fixture or vessel into a water supply pipe due to a pressure less than atmospheric in such pipe. See Backflow.

Backwater Valve. A device installed in a drainage system to prevent reverse flow.

See **Figure 204.0a**

Bathroom. A room equipped with a shower, bathtub, or combination bath/shower.

Bathroom, Half. A room equipped with only a water closet and lavatory.

DEFINITIONS

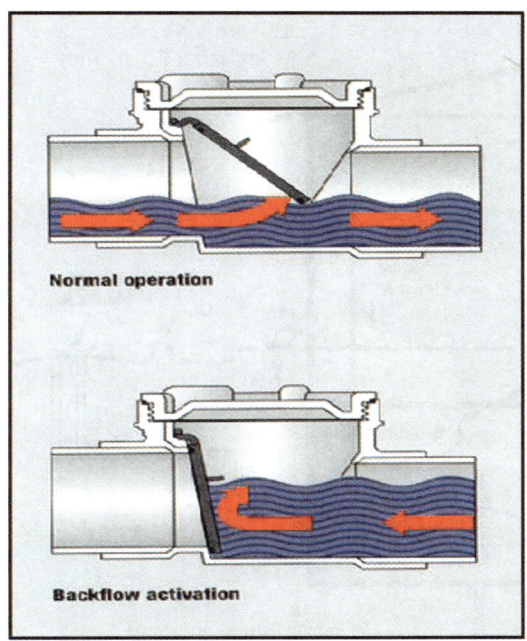

FIGURE 204.0A
BACKWATER VALVE OPERATION

FIGURE 204.0B
BOTTLE FILLING STATION

Bathroom Group. Any combination of fixtures, not to exceed one water closet, two lavatories, either one bathtub or one combination bath/shower, and one shower, and may include a bidet and an emergency floor drain.

Battery of Fixtures. A group of two or more similar, adjacent fixtures that discharge into a common horizontal waste or soil branch.

Bedpan Steamer. A fixture that is used to sterilize bedpans by way of steam.

Boiler Blowoff. An outlet on a boiler to permit emptying or discharge of sediment.

Bonding Conductor or Jumper. A reliable conductor to ensure the required electrical conductivity between metal parts required to be electrically connected. [NFPA 70:100(I)]

Bottle Filling Station. A plumbing fixture connected to the potable water distribution system and sanitary drainage system that is designed and intended for filling personal use drinking water bottles or containers not less than 10 inches (254 mm) in height. Such fixtures can be separate from or integral to a drinking fountain and can incorporate a water filter and a cooling system for chilling the drinking water.

See Figure 204.0b

See Learning Link goo.gl/jfjqG6

Branch. A part of the piping system other than a main, riser, or stack.

Branch, Fixture. See Fixture Branch.

Branch, Horizontal. See Horizontal Branch.

See Figure 204.0c

Branch Vent. A vent connecting one or more individual vents with a vent stack or stack vent.

See Figure 204.0c

Building. A structure built, erected, and framed of component structural parts designed for the housing, shelter, enclosure, or support of persons, animals, or property of any kind.

Building Drain. That part of the lowest piping of a drainage system that receives the discharge from soil, waste, and other drainage pipes inside the walls of the building and conveys it to the building sewer beginning 2 feet (610 mm) outside the building wall.

See Figure 204.0c

Building Drain (Sanitary). A building drain that conveys sewage only.

Building Drain (Storm). A building drain that conveys storm water or another drainage, but no sewage.

Building Sewer. That part of the horizontal piping of a drainage system that extends from the end of the building drain and that receives the discharge of the building drain and conveys it to a public sewer, private sewer, private sewage disposal system, or another point of disposal.

Building Sewer (Combined). A building sewer that conveys both sewage and storm water or other drainage.

Building Sewer (Sanitary). A building sewer that conveys sewage only.

Building Sewer (Storm). A building sewer that conveys storm water or another drainage, but no sewage.

Building Subdrain. That portion of a drainage system that does not drain by gravity into the building sewer.

See Figure 204.0c

Building Supply. The pipe is carrying potable water from the water meter or another source of water supply to a building or other point of use or distribution on the lot.

DEFINITIONS

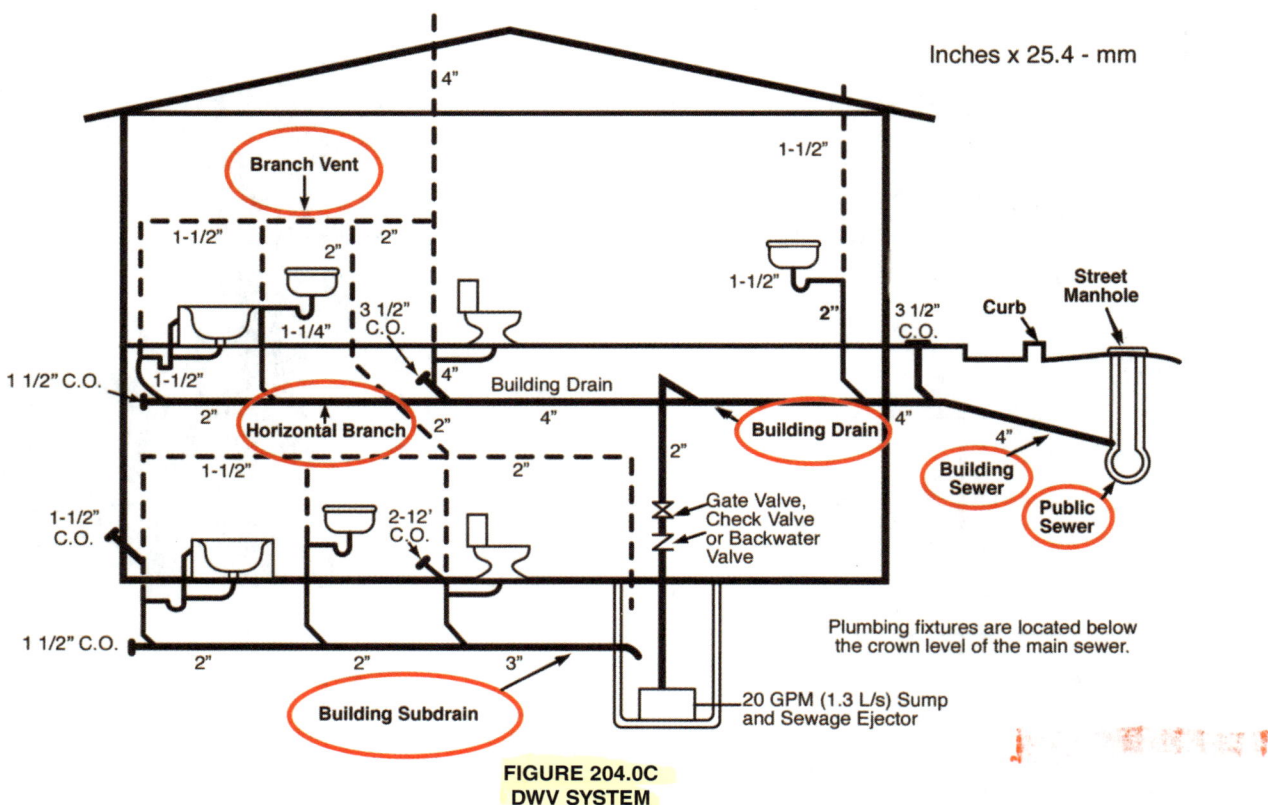

**FIGURE 204.0C
DWV SYSTEM**

205.0 – C –

Category 1. Activities, systems, or equipment whose failure is likely to cause major injury or death to patients, staff, or visitors. [NFPA 99:3.3.146.1]

Category 2. Activities, systems, or equipment whose failure is likely to cause minor injury to patients, staff, or visitors. [NFPA 99:3.3.146.2]

Category 3. Activities, systems, or equipment whose failure is not likely to cause injury to patients, staff, or visitors, but can cause discomfort. [NFPA 99:3.3.146.3]

Category 3 Vacuum System. A Category 3 vacuum distribution system that can be either a wet system designed to remove liquid, air-gas, or solids from the treated area; or a dry system designed to trap liquids and solids before the service inlet and to accommodate air-gas only through the service inlet. [NFPA 99:3.3.21]

Category 4. Activities, systems, or equipment whose failure would have no impact on patient care. [NFPA 99:3.3.146.4]

Certified Backflow Assembly Tester. A person who has shown competence to test and maintain backflow assemblies to the satisfaction of the Authority Having Jurisdiction.

Cesspool. A lined excavation in the ground that receives the discharge of a drainage system or part thereof, so designed as to retain the organic matter and solids discharging therein but permitting the liquids to seep through the bottom and sides.

See **Appendix H**

Chemical Waste. See Special Wastes.

Chimney. One or more passageways, vertical or nearly so, for conveying flue or vent gases to the outdoors. [NFPA 54:3.3.18]

Chimney, Factory-Built. A chimney composed of listed factory-built components assembled in accordance with the manufacturer's installation instructions to form the completed chimney. [NFPA 54:3.3.18.2]

Chimney, Masonry. A field-constructed chimney of solid masonry units, bricks, stones, listed masonry chimney units, or reinforced portland cement concrete, lined with suitable chimney flue liners. [NFPA 54:3.3.18.3]

See **Figure 205.0a**

Chimney, Metal. A chimney constructed of metal with a minimum thickness not less than 0.127 inches (3.23 mm) (No. 10 manufacturer's standard gauge) steel sheet.

Chimney Classifications:

Chimney, High-Heat Appliance-Type. A factory-built, masonry, or metal chimney suitable for removing the products of combustion from fuel-burning high-heat appliances producing combustion gases in excess of 2000°F (1093°C), measured at the appliance flue outlet.

Chimney, Low-Heat Appliance-Type. A factory-built, masonry, or metal chimney suitable for removing the products of combustion from fuel-burning low-heat appliances producing combustion gases not in excess of 1000°F (538°C) under normal operating conditions, but capable of producing combustion gases of 1400°F (760°C) during intermittent forced firing for periods up to one hour. Temperatures are measured at the appliance flue outlet.

Chimney, Medium-Heat Appliance-Type. A factory-built, masonry, or metal chimney suitable for removing the products of combustion from fuel-burning medium-heat appliances producing combustion gases, not in excess of 2000°F (1093°C), measured at the appliance flue outlet.

Chimney, Residential Appliance-Type. A factory-built or masonry chimney suitable for removing products of combustion from residential-type appliances producing combustion gases, not in excess of 1000°F (538°C), measured at the appliance flue outlet. Factory-built Type HT chimneys have high-temperature thermal shock resistance.

Clarifier. See Interceptor (Clarifier).

Clear Water Waste. Cooling water and condensate drainage from refrigeration and air-conditioning equipment; cooled condensate from steam heating systems, and cooled boiler blowdown water.

Clinical Sink. A fixture that has the same flushing and cleansing characteristics of a water closet that is used to receive the wastes from a bedpan. Also, known as a bedpan washer.

See **Figure 205.0b**

See Learning Link goo.gl/xSs74o

Coastal High Hazard Areas. An area within the flood hazard area that is subject to high-velocity wave action, and shown on a Flood Insurance Rate Map or other flood hazard map as Zone V, VO, VE or V1-30.

Code. A standard that is an extensive compilation of provisions covering broad subject matter or that is suitable for adoption into law independently of other codes and standards.

Combination Temperature and Pressure-Relief Valve. A relief valve that actuates when a set temperature, pressure, or both is reached. Also, known as a T&P Valve.

Combination Thermostatic/Pressure Balancing Valve. A mixing valve that senses outlet temperature and incoming hot and cold water pressure and compensates for fluctuations in incoming hot and cold water temperatures, pressures, or both to stabilize outlet temperatures.

Combination Waste and Vent System. A specially designed system of waste piping embodying the horizontal wet venting of one or more sinks or floor drains using a common waste and vent pipe adequately sized to provide free movement of air above the flow line of the drain.

See **Figure 205.0c**

Combined Building Sewer. See Building Sewer (Combined).

Combustible Material. A material that, in the form in which it is used and under the conditions anticipated, will ignite and burn; a material that does not meet the definition of noncombustible. [NFPA 54:3.3.64.1]

What is or is not combustible material will be very important for the installation of fuel-burning appliances and their venting systems. The objective is to keep the appliance and its vent or piping the proper distance away from combustible material in order to eliminate the possibility of fire being created by the close proximity of these materials.

Common. That part of a plumbing system that is so designed and installed as to serve more than one appliance, fixture, building, or system.

Condensate. The liquid phase produced by condensation of a gas or vapor.

Conductor. A pipe inside the building that conveys storm water from the roof to a storm drain, combined building sewer, or other approved point of disposal.

Confined Space. A room or space having a volume less than 50 cubic feet per 1000 British thermal units per hour (Btu/h) (4.83 m^3/kW) of the aggregate input rating of all fuel-burning appliances installed in that space.

Construction Documents. Plans, specifications, written, graphic, and pictorial documents prepared or assembled for

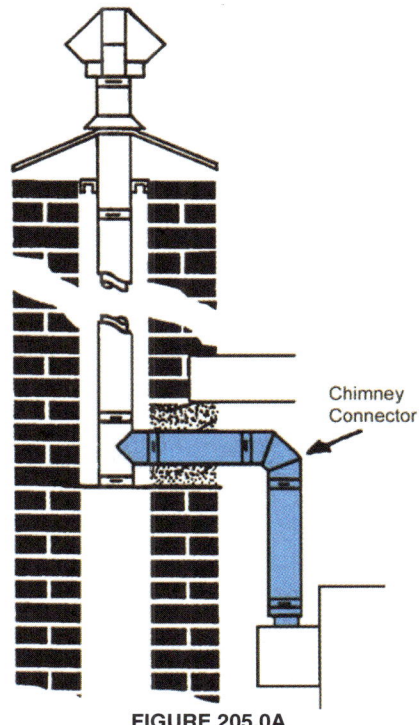

FIGURE 205.0A
MASONRY CHIMNEY WITH LINER

FIGURE 205.0B
CLINICAL SINK

DEFINITIONS

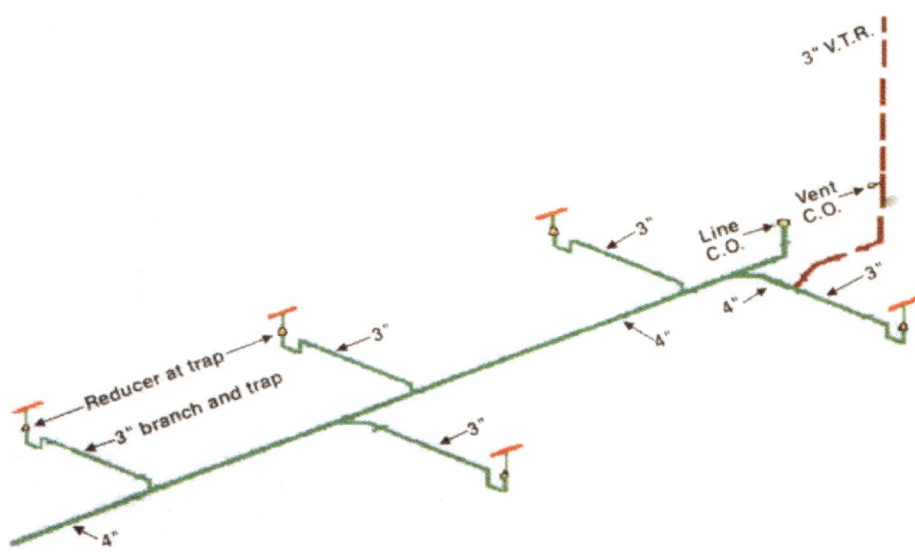

FIGURE 205.0C
COMBINATION WASTE AND VENT SYSTEM

describing the design, location, and physical characteristics of the elements of a project necessary for obtaining a permit.

Contamination. An impairment of the quality of the potable water that creates an actual hazard to the public health through poisoning or the spread of disease by sewage, industrial fluids, or waste. Also, defined as High Hazard.

Continuous Vent. A vertical vent that is a continuation of the drain to which it connects.

See Figure 205.0d

Continuous Waste. A drain is connecting the compartments of a set of fixtures to a trap or connecting other permitted fixtures to a common trap.

See Figure 205.0e

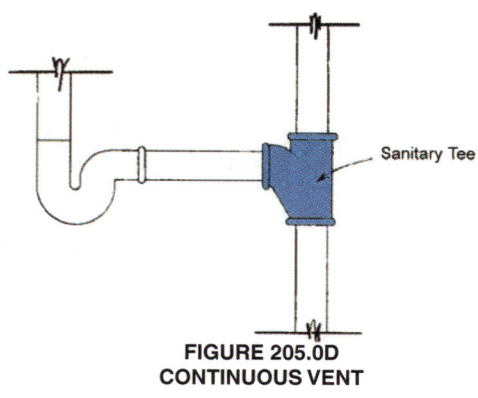

FIGURE 205.0D
CONTINUOUS VENT

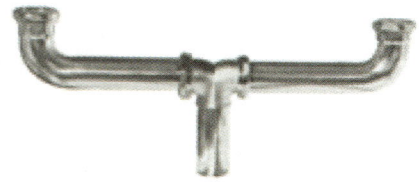

FIGURE 205.0E
CONTINUOUS WASTE

Copper Alloy. A homogenous mixture of two or more metals in which copper is the primary component, such as brass and bronze.

CPVC. Chlorinated Polyvinyl Chloride.

Critical Care Area. A room or space in which failure of equipment or a system is likely to cause major injury or death to patients or caregivers (Category 1). [NFPA 99:3.3.28]

Critical Level. The critical level (C-L or C/L) marking on a backflow prevention device or vacuum breaker is a point conforming to approved standards and established by the testing laboratory (usually stamped on the device by the manufacturer) that determines the minimum elevation above the flood-level rim of the fixture or receptor served at which the device may be installed. Where a backflow prevention device does not bear a critical level marking, the bottom of the vacuum breaker, combination valve, or the bottom of such approved device shall constitute the critical level.

See Figure 205.0f

Cross-Connection. A connection or arrangement, physical or otherwise, between a potable water supply system and a plumbing fixture or a tank, receptor, equipment, or device, through which it may be possible for nonpotable, used, unclean, polluted, and contaminated water, or other substances to enter into a part of such potable water system under any condition.

206.0 – D –

Debris Excluder. A device installed on the rainwater catchment conveyance system to prevent the accumulation of leaves, needles, or other debris in the system.

Department Having Jurisdiction. The Authority Having Jurisdiction, including any other law enforcement agency affected by a provision of this code, whether such agency is specifically named or not.

Design Flood Elevation. The elevation of the "design flood," including wave height, relative to the datum specified

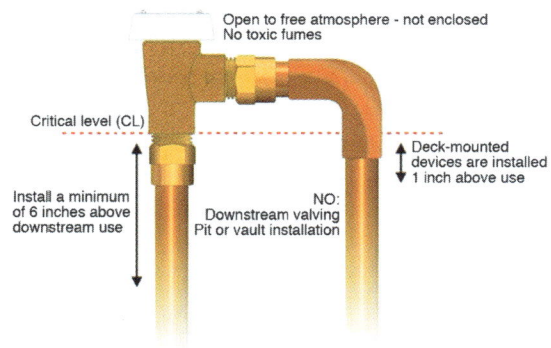

**FIGURE 205.0F
CRITICAL LEVEL OF AVB**

on the community's legally designated flood hazard map. In areas designated as Zone AO, the design flood elevation is the elevation of the highest existing grade of the building's perimeter plus the depth number (in feet) specified on the flood hazard map. In areas designated as Zone AO where a depth number is not specified on the map, the depth number is taken as being equal to 2 feet (610 mm).

Developed Length. The length along the centerline of a pipe and fittings.

See **Figure 206.0a**

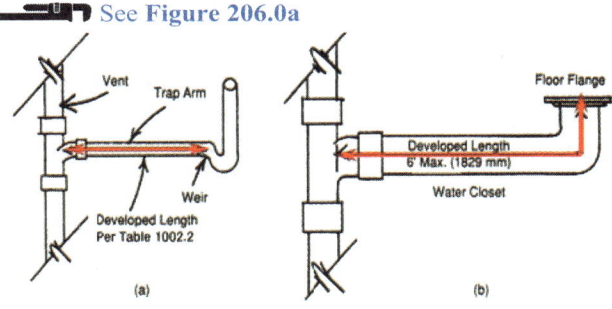

**FIGURE 206.0A
DEVELOPED LENGTH PER TABLE 1002.2**

Diameter. Unless specifically stated, "diameter" is the nominal diameter as designated commercially.

Direct-Vent Appliances. Appliances that are constructed and installed so that all air for combustion is derived directly from the outdoors and all flue gases are discharged to the outdoors. [NFPA 54:3.3.5.3]

Direct-vent gas appliances offer higher efficiency than natural vent gas appliances. Natural vent appliances use room air for combustion to provide air to fuel the fire. Direct-vent appliances utilize one of two styles of dual piping systems. The first method is the Co-Axial Pipe or Concentric Method, where a pipe within a pipe will bring combustion air from outside through the outer chamber and exhausts the products of combustion to the outside through the inner chamber (see **Figure 509.2.6**). The second method, the Co-Linear Pipe Method, uses two separate pipes. One pipe brings in combustion (supply) air while the other pipe carries away the products of combustion to the outside (see **Figure 206.0b**).

The objective of the direct-vent appliance is to have no openings within the building. No vent collars (draft hoods) or combustion air openings are allowable within the building as part of a direct-vent appliance installation; however, these appliances are not necessarily sealed combustion chamber appliances unless they are specifically listed as such.

**FIGURE 206.0B
DIRECT VENT WATER HEATER**

Domestic Sewage. The liquid and water-borne wastes derived from the ordinary living processes, free from industrial wastes, and of such character as to permit satisfactory disposal, without special treatment, into the public sewer or by means of a private sewage disposal system.

Downspout. The rain leader from the roof to the building storm drain, combined building sewer, or other means of disposal located outside of the building. See Conductor and Leader.

Drain. A pipe that carries waste or waterborne wastes in a building drainage system.

Drainage System. Includes all the piping within public or private premises that conveys sewage, storm water, or other liquid wastes to a legal point of disposal, but does not include the mains of a public sewer system or a public sewage treatment or disposal plant.

Drinking Fountain. A plumbing fixture connected to the potable water distribution system and sanitary drainage system that provides drinking water in a flowing stream so that the user can consume water directly from the fixture without the use of accessories. Drinking fountains should also incorporate a bottle filling station and can incorporate a water filter and a cooling system for chilling the drinking water.

DEFINITIONS

Dry Vent. A vent that does not receive the discharge of any sewage or waste.

See **Figure 204.0c**

Durham System. Soil or waste system in which all piping is threaded pipe, tubing, or other such rigid construction, using recessed drainage fittings to correspond to the types of piping.

See **Figure 206.0c**

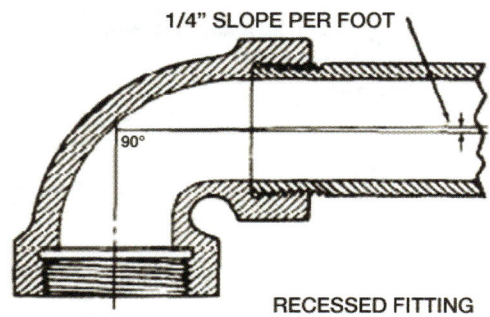

**FIGURE 206.0C
DURHAM DRAINAGE FITTING**

207.0 – E –

Effective Ground-Fault Current Path. An intentionally constructed, low-impedance electrically conductive path designed and intended to carry current under ground-fault conditions from the point of a ground fault on a wiring system to the electrical supply source and that facilitates the operation of the overcurrent protective device or ground-fault detectors on high-impedance grounded systems. [NFPA 54:3.3.34]

Effective Opening. The minimum cross-sectional area at the point of water supply discharge measured or expressed in terms of (1) diameter of a circle or (2) where the opening is not circular, the diameter of a circle of equivalent cross-sectional area. (This applies to an air gap).

See **Figure 207.0**

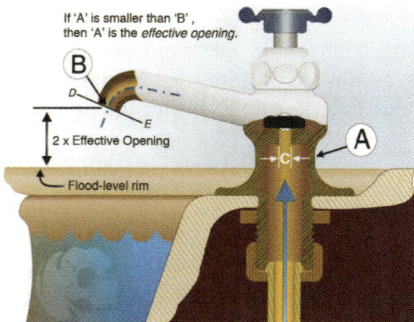

**FIGURE 207.0
EFFECTIVE OPENING**

Essentially Nontoxic Transfer Fluid. Essentially nontoxic at practically nontoxic, Toxicity Rating Class 1 (reference "Clinical Toxicology of Commercial Products" by Gosselin, Smith, Hodge, & Braddock).

Exam Room Sink. A sink used in the patient exam room of a medical or dental office with a primary purpose of the washing of hands.

Excess Flow Valve (EFV). A valve designed to activate when the fuel gas passing through it exceeds a prescribed flow rate. [NFPA 54:3.3.99.3]

Existing Work. A plumbing system or any part thereof that has been installed prior to the effective date of this code.

Expansion Joint. A fitting or arrangement of pipe and fittings that permit the contraction and expansion of a piping system.

208.0 – F –

F Rating. The time period that the penetration firestop system limits the spread of fire through the penetration, where tested in accordance with ASTM E814 or UL 1479.

A firestop assembly meeting the F-rating will remain in the opening during the fire test and the hose stream test within the following limitations:

- The firestop withstands the fire test for the rating period (1, 2, 3 or 4 hours) without allowing the passage of flame through the opening or flaming on any material on the unexposed side (see **Figure 208.0a**).

- The firestop does not develop any openings that would permit a projection of water on the unexposed side during the hose stream test.

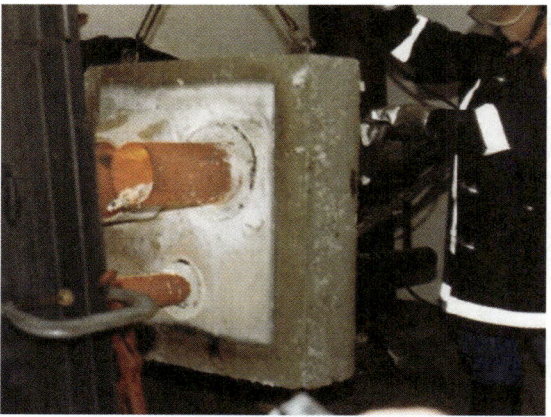

**FIGURE 208.0A
RESULTS OF TEMPERATURE TEST OF FIRESTOP**

Fixture Branch. A water supply pipe between the fixture supply pipe and the water distribution pipe.

Fixture Drain. The drain from the trap of a fixture to the junction of that drain with any other drain pipe.

Fixture Fitting. A device that controls and guides the flow of water.

Fixture Supply. A water supply pipe is connecting the fixture with the fixture branch.

Fixture Unit. A quantity in terms of which the load-producing effects on the plumbing system of different kinds of plumbing fixtures are expressed on some arbitrarily chosen scale.

Flammable Vapor or Fumes. The concentration of flammable constituents in the air that exceeds 25 percent of its lower flammability limit (LFL).

Flood Hazard Area. The greater of the following two areas:
(1) The area within a floodplain subject to a 1 percent or greater chance of flooding in any given year.
(2) The area designated as a flood hazard area on a community's flood hazard map, or otherwise legally designated.

Flood Level. See Flooded.

Flood-Level Rim. The top edge of a receptor from which water overflows.

See Figure 207.0

Flooded. A fixture is flooded where the liquid therein rises to the flood-level rim.

Flue Collar. That portion of an appliance designed for the attachment of a draft hood, vent connector, or venting system. [NFPA 54:3.3.44]

Flush Tank. A tank located above or integral with water closets, urinals, or similar fixtures for the purpose of flushing the usable portion of the fixture.

Flush Valve. A valve located at the bottom of a tank for flushing water closets and similar fixtures.

Flushometer Tank. A tank integrated within an air accumulator vessel that is designed to discharge a predetermined quantity of water to fixtures for flushing purposes.

See Figure 208.0b

See Learning Link goo.gl/1Q2gPP

**FIGURE 208.0B
FLUSHOMETER TANK**

Flushometer Valve. A valve that discharges a predetermined quantity of water to fixtures for flushing purposes and is actuated by direct water pressure.

FOG Disposal System. A grease interceptor that reduces nonpetroleum fats, oils, and grease (FOG) in the effluent by separation, mass, and volume reduction.

Fuel Gas. Natural, manufactured liquefied petroleum, or a mixture of these.

209.0   – G –

Gang or Group Shower. Two or more showers in a common area.

Gas Piping. An installation of pipe, valves, or fittings that are used to convey fuel gas, installed on a premise or in a building, but shall not include:
(1) A portion of the service piping.
(2) An approved piping connection 6 feet (1829 mm) or less in length between an existing gas outlet and a gas appliance in the same room with the outlet.

Gas Piping System. An arrangement of gas piping or regulators after the point of delivery and each arrangement of gas piping serving a building, structure, or premises, whether individually metered or not.

Governing Body. The person or persons who have the overall legal responsibility for the operation of a health care facility. [NFPA 99:3.3.62]

Grade. The slope or fall of a line of pipe in reference to a horizontal plane. In drainage, it is usually expressed as the fall in a fraction of an inch (mm) or percentage slope per foot (meter) length of pipe.

Gravity Grease Interceptor. A plumbing appurtenance or appliance that is installed in a sanitary drainage system to intercept nonpetroleum fats, oils, and greases (FOG) from a wastewater discharge and is identified by volume, 30 minute retention time, baffle(s), not less than two compartments, a total volume of not less than 300 gallons (1135 L), and gravity separation. [These interceptors comply with the requirements of Chapter 10 or are designed by a registered design professional.] Gravity grease interceptors are generally installed outside.

Gray Water. Untreated wastewater that has not come into contact with toilet waste, kitchen sink waste, dishwasher waste or similarly contaminated sources. Gray water includes wastewater from bathtubs, showers, lavatories, clothes washers, and laundry tubs. Also, known as grey water, graywater, and greywater.

Gray Water Diverter Valve. A valve that directs gray water to the sanitary drainage system or a subsurface irrigation system.

Grease Interceptor. A plumbing appurtenance or appliance that is installed in a sanitary drainage system to intercept nonpetroleum fats, oil, and greases (FOG) from a wastewater discharge.

Grease Removal Device (GRD). A hydromechanical grease interceptor that automatically, mechanically removes non-petroleum fats, oils and grease (FOG) from the interceptor, the control of which are either automatic or manually initiated.

Grounding Electrode. A conducting object through which a direct connection to earth is established. [NFPA 70:100(I)]

210.0   – H –

Hangers. See Supports.

Heat-Fusion Weld Joints. A joint used in some thermoplastic systems to connect the pipe to fittings or pipe lengths directly to one another (butt-fusion). This method of joining pipe to fittings includes socket-fusion, electro-fusion, and saddle-fusion. This method of welding involves the application of heat and pressure to the components, allowing them to fuse together forming a bond between the pipe and fitting.

DEFINITIONS

High Hazard. See Contamination.

Horizontal Branch. A drain pipe extending laterally from soil or waste stack or building drain with or without vertical sections or branches, which receives the discharge from one or more fixture drains and conducts it to the soil or waste stack or the building drain.

Horizontal Pipe. A pipe or fitting that is installed in a horizontal position or which makes an angle of less than 45 degrees (0.79 rad) with the horizontal.

See Figure 210.0

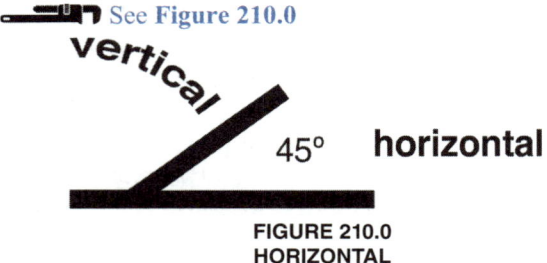

FIGURE 210.0
HORIZONTAL

Hot Water. Water at a temperature exceeding or equal to 120°F (49°C).

House Drain. See Building Drain.

House Sewer. See Building Sewer.

Hydromechanical Grease Interceptor. A plumbing appurtenance or appliance that is installed in a sanitary drainage system to intercept nonpetroleum fats, oil, and grease (FOG) from a wastewater discharge and is identified by flow rate, and separation and retention efficiency. The design incorporates air entrainment, hydromechanical separation, interior baffling, or barriers in combination or separately, and one of the following:

(1) External flow control, with an air intake (vent), directly connected.
(2) External flow control, without air intake (vent), directly connected.
(3) Without external flow control, directly connected.
(4) Without external flow control, indirectly connected.

These interceptors comply with the requirements of Table 1014.2.1. Hydromechanical grease interceptors are generally installed inside.

211.0 – I –

Indirect-Fired Water Heater. A water heater consisting of a storage tank equipped with an internal or external heat exchanger used to transfer heat from an external source to heat potable water. The storage tank either contains heated potable water or water supplied from an external source, such as a boiler.

Indirect Waste Pipe. A pipe that does not connect directly to the drainage system but conveys liquid wastes by discharging into a plumbing fixture, interceptor, or receptacle that is directly connected to the drainage system.

See Figure 211.0a

Individual Vent. A pipe installed to vent a fixture trap, and that connects with the vent system above the fixture served or terminates in the open air.

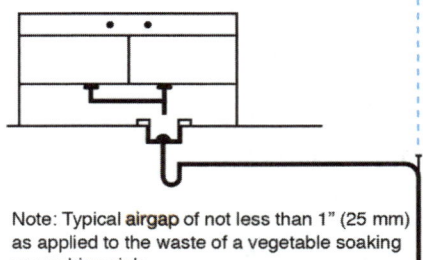

Note: Typical airgap of not less than 1" (25 mm) as applied to the waste of a vegetable soaking or washing sink.

FIGURE 211.0A
INDIRECT WASTE PIPE

Industrial Waste. Liquid or water-borne waste from industrial or commercial processes, except domestic sewage.

Insanitary. A condition that is contrary to sanitary principles or is injurious to health.

Conditions to which "insanitary" shall apply include the following:

(1) A trap that does not maintain a proper trap seal.
(2) An opening in a drainage system, except where lawful that is not provided with an approved liquid-sealed trap.
(3) A plumbing fixture or other waste discharging receptor or device that is not supplied with water sufficient to flush and maintain the fixture or receptor in a clean condition.
(4) A defective fixture, trap, pipe, or fitting.
(5) A trap, except where in this code exempted, directly connected to a drainage system, the seal of which is not protected against siphonage and backpressure by a vent pipe.
(6) A connection, cross-connection, construction, or condition, temporary or permanent that would permit or make possible by any means whatsoever for an unapproved foreign matter to enter a water distribution system used for domestic purposes.
(7) The preceding enumeration of conditions to which the term "insanitary" shall apply, shall not preclude the application of that term to conditions that are, in fact, insanitary.

Interceptor (Clarifier). A device designed and installed to separate and retain deleterious, hazardous, or undesirable matter from normal wastes and permit normal sewage or liquid wastes to discharge into the disposal terminal by gravity.

Invert. The lowest portion of the inside of a horizontal pipe.

212.0 – J –

Joint, Brazed. A joint obtained by joining of metal parts with alloys that melt at temperatures exceeding 840°F (449°C), but less than the melting temperature of the parts to be joined.

Joint, Compression. A multipiece joint with cup-shaped threaded nuts that, when tightened, compress tapered sleeves so that they form a tight joint on the periphery of the tubing they connect.

Joint, Flanged. One made by bolting together a pair of flanged ends.

🔧 See Figure 212.0a

Joint, Flared. A metal-to-metal compression joint in which a conical spread is made on the end of a tube that is compressed by a flare nut against a mating flare.

Joint, Mechanical. The general form for gas-tight or liquid-tight joints obtained by the joining of parts through a positive holding mechanical construction.

Joint, Press-Connect. A permanent mechanical joint incorporating an elastomeric seal or an elastomeric seal and corrosion resistant grip ring. The joint is made with a pressing tool and jaw or ring that complies with the manufacturer's installation instructions.

🔧 See Figure 212.0b

Joint, Soldered. A joint obtained by the joining of metal parts with metallic mixtures or alloys that melt at a temperature up to and including 840°F (449°C).

Joint, Welded. A gastight joint obtained by the joining of metal parts in the plastic molten state.

213.0 – K –
No definitions.

214.0 – L –

Labeled. Equipment or materials bearing a label of a listing agency (accredited conformity assessment body). See Listed (third-party certified).

🔧 See Figure 214.0a

Lavatories in Sets. Two or three lavatories that are served by one trap.

🔧 See Figure 214.0b

Leader. An exterior vertical drainage pipe for conveying storm water from roof or gutter drains. See Downspout.

Levels of Sedation.

Deep Sedation/Analgesia. A drug-induced depression of consciousness during which patients cannot be easily aroused but respond purposefully following repeated or painful stimulation. The ability to independently maintain ventilatory function may be impaired. Patients may require assistance in maintaining a patent airway, and spontaneous ventilation may be inadequate. Cardiovascular function is usually maintained. [NFPA 99:3.3.61.2]

General Anesthesia. A drug-induced loss of consciousness during which patients are not arousable, even by painful stimulation. The ability to independently maintain ventilatory function is often impaired. Patients often require assistance in maintaining a patent airway, and positive pressure ventilation may be required because of depressed spontaneous ventilation or drug-induced depression of neuromuscular function. Cardiovascular function may be impaired. [NFPA 99:3.3.61.1]

Minimal Sedation (Anxiolysis). A drug induced state during which patients respond normally to verbal commands. Although cognitive function and coordination may be impaired, ventilatory and cardiovascular functions are unaffected. [NFPA 99:3.3.61.4]

Moderate Sedation/Analgesia (Conscious Sedation). A drug-induced depression of consciousness during

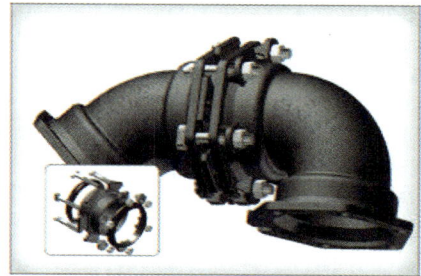

FIGURE 212.0A
JOINT, FLANGED

FIGURE 212.0B
PRESSED FITTING

FIGURE 214.0A
LABEL

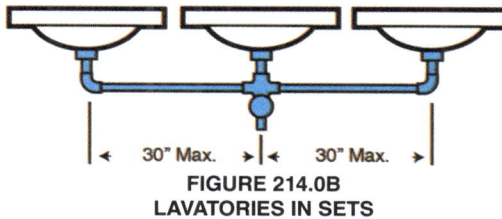

FIGURE 214.0B
LAVATORIES IN SETS

which patients respond purposefully to verbal commands, either alone or accompanied by light tactile stimulation. No interventions are required to maintain a patient airway, and spontaneous ventilation is adequate. Cardiovascular function is usually maintained. [NFPA 99:3.3.61.3]

Liquefied Petroleum Gas (LP-Gas) Facilities. Liquefied petroleum gas (LP-Gas) facilities include tanks, containers, container valves, regulating equipment, meters, appurtenances, or any combination thereof for the storage and supply of liquefied petroleum gas for a building, structure, or premises.

Liquid Waste. The discharge from a fixture, appliance, or appurtenance in connection with a plumbing system that does not receive fecal matter.

Listed (Third-party certified). Equipment or materials included in a list published by a listing agency (accredited conformity assessment body) that maintains periodic inspection of current production of listed equipment or materials and whose listing states either that the equipment or material complies with approved standards or has been tested and found suitable for use in a specified manner.

DEFINITIONS

Listing Agency. An agency accredited by an independent and authoritative conformity assessment body to operate a material and product listing and labeling (certification) system and that are accepted by the Authority Having Jurisdiction, which is in the business of listing or labeling. The system includes initial and ongoing product testing, a periodic inspection on current production of listed (certified) products, and that makes available a published report of such listing in which specific information is included that the material or product is in accordance with applicable standards and found safe for use in a specific manner.

Lot. A single or individual parcel or area of land legally recorded or validated by other means acceptable to the Authority Having Jurisdiction on which is situated a building or which is the site of any work regulated by this code, together with the yards, courts, and unoccupied spaces legally required for the building or works, and that is owned by or is in the lawful possession of the owner of the building or works.

Low Hazard. See Pollution.

215.0 – M –

Macerating Toilet System. A system comprised of a sump with macerating pump and with connections for a water closet and other plumbing fixtures, which is designed to accept, grind and pump wastes to an approved point of discharge.

Main. The principal artery of a system of continuous piping to which branches may be connected.

Main Sewer. See Public Sewer.

Main Vent. The principal artery of the venting system to which vent branches may be connected.

May. A permissive term.

Medical Air. For the purposes of this code, medical air is air supplied from cylinders, bulk containers, or medical air compressors, or reconstituted from oxygen USP and oil-free, dry nitrogen NF. [NFPA 99:3.3.96]

Medical Gas. A patient medical gas or medical support gas. [NFPA 99:3.3.99]

Manifold. A device for connecting the outlets of one or more gas cylinders to the central piping system for that specific gas. [NFPA 99:3.3.93]

Medical Gas System. An assembly of equipment and piping for the distribution of nonflammable medical gases such as oxygen, nitrous oxide, compressed air, carbon dioxide, and helium. [NFPA 99:3.3.100]

Medical Support Gas. Nitrogen or instrument air used for any medical support purpose (e.g., to remove excess moisture from instruments before further processing, or to operate medical-surgical tools, air-driven booms, pendants, or similar applications) and, if appropriate to the procedures, used in laboratories and are not respired as part of any treatment. Medical support gas falls under the general requirements for medical gases. [NFPA 99:3.3.101]

Medical-Surgical Vacuum. A method used to provide a source of drainage, aspiration, and suction in order to remove body fluids from patients. [NFPA 99:3.3.102]

Medical-Surgical Vacuum System. An assembly of central vacuum-producing equipment and a network of piping for patient suction in medical, medical-surgical, and waste anesthetic gas disposal (WAGD) applications. [NFPA 99:3.3.103]

Mobile Home Park Sewer. That part of the horizontal piping of a drainage system that begins 2 feet (610 mm) downstream from the last mobile home site and conveys it to a public sewer, private sewer, private sewage disposal system, or other point of disposal.

Mulch. Organic materials, such as wood chips and fines, tree bark chips, and pine needles that are used in a mulch basin to conceal gray water outlets and permit the infiltration of gray water.

Mulch Basin. A subsurface catchment area for gray water that is filled with mulch and of sufficient depth and volume to prevent ponding, surfacing, or runoff.

216.0 – N –

Nitrogen, NF. Nitrogen complying as a minimum, with nitrogen NF. [NFPA 99:3.3.109.1]

Nuisance. Includes, but is not limited to:

(1) A public nuisance known at common law or in equity jurisprudence.

(2) Where work regulated by this code is dangerous to human life or is detrimental to health and property.

(3) Inadequate or unsafe water supply or sewage disposal system.

217.0 – O –

Offset. A combination of elbows or bends in a line of piping that brings one section of the pipe out of line but into a line parallel with the other section.

Oil Interceptor. See Interceptor (Clarifier).

On-Site Treated Nonpotable Water. Nonpotable water, including gray water that has been collected, treated, and intended to be used on-site and is suitable for direct beneficial use.

218.0 – P –

Patient Care Space. Any space of a health care facility wherein patients are intended to be examined or treated. [NFPA 99:3.3.127]

    Category 1 Space. Space in which failure of equipment or a system is likely to cause major injury or death of patients, staff, or visitors. [NFPA 99:3.3.127.1]

    Category 2 Space. Space in which failure of equipment or a system is likely to cause minor injury to patients, staff, or visitors. [NFPA 99:3.3.127.2]

    Category 3 Space. Space in which the failure of equipment or a system is not likely to cause injury to patients, staff, or visitors but can cause discomfort. [NFPA 99:3.3.127.3]

    Category 4 Space. Space in which failure of equipment or a system is not likely to have a physical impact on patient care. [NFPA 99:3.3.127.4]

Patient Medical Gas. Piped gases such as oxygen, nitrous oxide, helium, carbon dioxide, and medical air that are used in the application of human respiration and the calibration of medical devices used for human respiration. [NFPA 99:3.3.131]

PB. Polybutylene.

PE. Polyethylene.

PE-AL-PE. Polyethylene-aluminum-polyethylene.

PE-RT. Polyethylene of raised temperature.

Penetration Firestop System. A specific assemblage of field-assembled materials, or a factory-made device, which has been tested to a standard test method and, where installed properly on penetrating piping materials, is capable of maintaining the fire- resistance rating of assemblies penetrated.

See **Figure 218.0a**.

Person. A natural person, his heirs, executor, administrators, or assigns and shall also include a firm, corporation, municipal or quasi-municipal corporation, or governmental agency. The singular includes the plural, male includes female.

PEX. Cross-linked polyethylene.

PEX-AL-PEX. Cross-linked polyethylene–aluminum-cross-linked polyethylene.

See **Figure 218.0b**

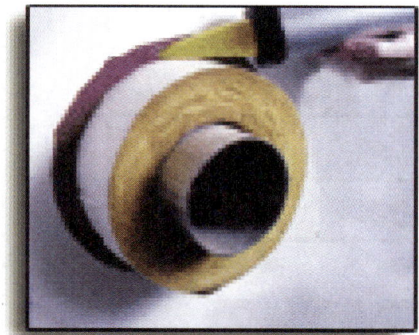

FIGURE 218.0A
PENETRATION FIRESTOP SYSTEM - INSULATED PIPE

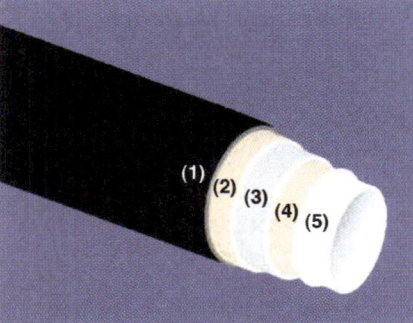

1. Cross-linked polyethylene (PEX)
2. Adhesive/bonding layer
3. Oxygen barrier (welded aluminum)
4. Adhesive
5. (PEX)

FIGURE 218.0B
PE & PEX-AL-PEX PIPE CONSTRUCTION

Pipe. A cylindrical conduit or conductor is conforming to the dimensions commonly known as "pipe size."

Plumbing. The business, trade, or work having to do with the installation, removal, alteration, or repair of plumbing systems or parts thereof.

Plumbing Appliance. A special class of device or equipment that is intended to perform a special plumbing function. Its operation, control, or both may be dependent upon one or more energized components, such as motors, controls, heating elements, or pressure- or temperature-sensing elements. Such device or equipment may operate automatically through one or more of the following actions: a time cycle, a temperature range, a pressure range, a measured volume or weight; or the device or equipment may be manually adjusted or controlled by the user or operator.

Plumbing Appurtenance. A manufactured device, a prefabricated assembly, or an on-the-job assembly of component parts that is an adjunct to the basic piping system and plumbing fixtures. An appurtenance demands no additional water supply, nor does it add a discharge load to a fixture or the drainage system. It performs some useful function in the operation, maintenance, servicing, economy, or safety of the plumbing system.

Plumbing Fixture. An approved type installed receptacle, device or appliance that is supplied with water or that receives liquid or liquid-borne wastes and discharges such wastes into the drainage system to which it may be directly or indirectly connected. Industrial or commercial tanks, vats, and similar processing equipment are not plumbing fixtures, but may be connected to or discharged into approved traps or plumbing fixtures where and as otherwise provided for elsewhere in this code.

Plumbing Official. See Authority Having Jurisdiction.

Plumbing System. Includes all potable water, alternate water sources, building supply, and distribution pipes; all plumbing fixtures and traps; all drainage and vent pipes; and all building drains and building sewers, including their respective joints and connections, devices, receptors, and appurtenances within the property lines of the premises and shall include potable water piping, potable water treating or using equipment, medical gas and medical vacuum systems, liquid and fuel gas piping, and water heaters and vents for same.

Plumbing Vent. A pipe provided to ventilate a plumbing system, to prevent trap siphonage and backpressure, or to equalize the air pressure within the drainage system.

See **Figure 218.0c**

Plumbing Vent System. A pipe or pipes installed to provide a flow of air to or from a drainage system or to provide a circulation of air within such system to protect trap seals from siphonage and backpressure.

See **Figure 218.0c**

Pollution. An impairment of the quality of the potable water to the degree that does not create a hazard to the public health but which does adversely and unreasonably affect the aesthetic qualities of such potable water for domestic use. Also, defined as "Low Hazard."

Potable Water. Water that is satisfactory for drinking, culinary, and domestic purposes and that meets the requirements of the Health Authority Having Jurisdiction.

PP. Polypropylene.

Pressure. The normal force exerted by a homogeneous liquid or gas, per unit of area, on the wall of the container.

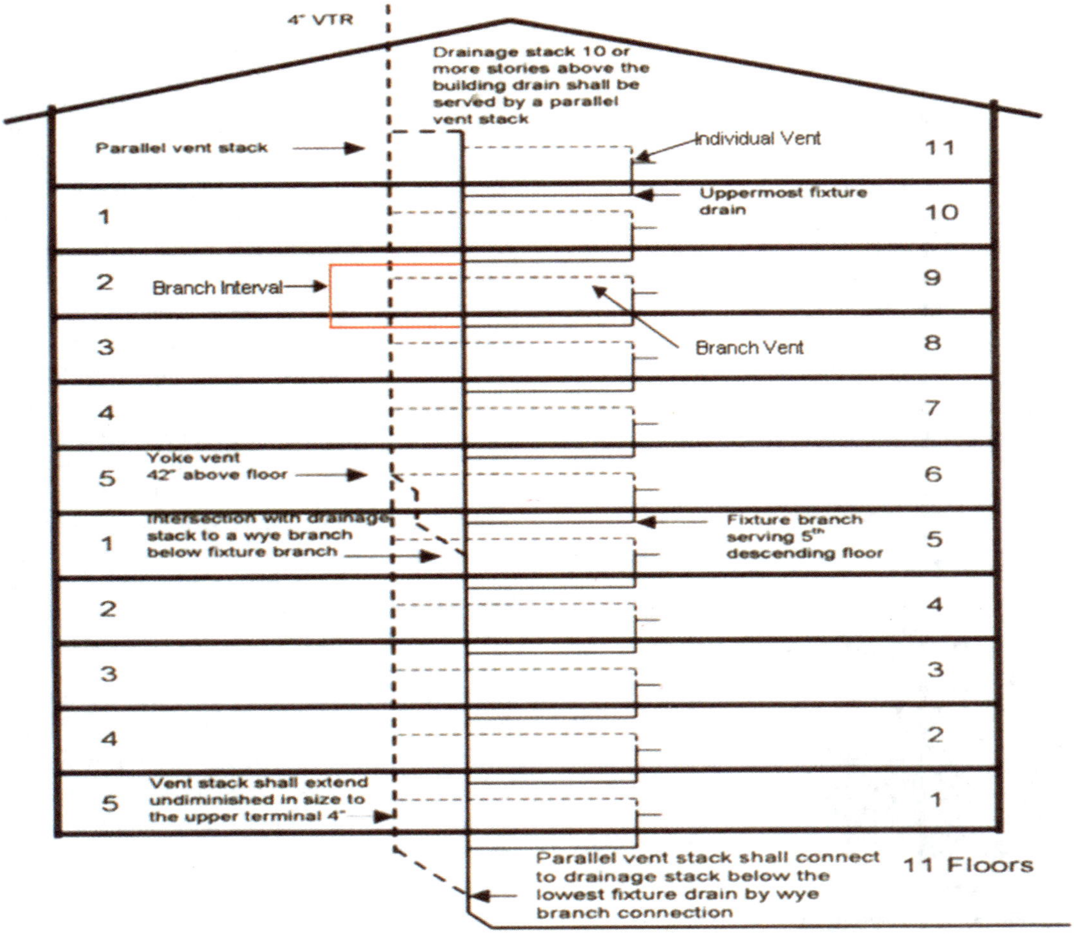

DEFINITIONS

FIGURE 218.0C
VENT SYSTEM

Residual Pressure. The pressure available at the fixture or water outlet after allowance is made for pressure drop due to friction loss, head, meter, and other losses in the system during maximum demand periods.

Static Pressure. The pressure is existing without any flow.

Pressure-Balancing Valve. A mixing valve that senses incoming hot and cold water pressures and compensates for fluctuations in either to stabilize outlet temperature.

See **Figure 218.0d**

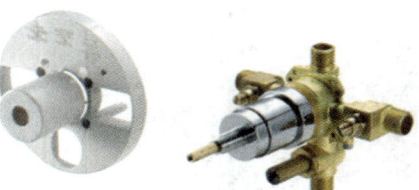

FIGURE 218.0D
PRESSURE BALANCING VALVE

Pressure-Lock-Type Connection. A mechanical connection that depends on an internal retention device to prevent pipe or tubing separation. The connection is made by inserting the pipe or tubing into the fitting to a prescribed depth.

Private or Private Use. Applies to plumbing fixtures in residences and apartments, to private bathrooms in hotels and hospitals, and to restrooms in commercial establishments where the fixtures are intended for the use of a family or an individual.

Private Sewage Disposal System. A septic tank with the effluent discharging into a subsurface disposal field, into one or more seepage pits, or into a combination of subsurface disposal field and seepage pit or of such other facilities as may be permitted under the procedures set forth elsewhere in this code.

See **Appendix H**

Private Sewer. A building sewer that receives the discharge from more than one building drain and conveys it to a public sewer, private sewage disposal system, or another point of disposal.

Proportioning System for Medical Air USP. A central supply that produces medical air (USP) reconstituted from oxygen USP and nitrogen NF by means of a mixer or blender. [NFPA 99:3.3.96.1]

Public or Public Use. Applies to plumbing fixtures that are not defined as private or private use.

DEFINITIONS

Public Sewer. A common sewer directly controlled by public authority.

See Figure 204.0c

Push Fit Fitting. A mechanical fitting where the connection is assembled by pushing the tube or pipe into the fitting and is sealed with an o-ring.

PVC. Polyvinyl Chloride.

PVDF. Polyvinylidene Fluoride.

219.0 – Q –

Quick-Disconnect Device. A hand-operated device that provides a means for connecting and disconnecting a hose to a water supply, and that is equipped with a means to shut off the water supply when the device is disconnected.

Quick-Disconnect Device (Fuel Gas). A hand-operated device that provides a means for connecting and disconnecting an appliance or an appliance connector to a gas supply, and that is equipped with an automatic means to shut off the gas supply when the device is disconnected. [NFPA 54:3.3.28.3]

220.0 – R –

Rainwater. Natural precipitation that has not been contaminated by use.

Rainwater Catchment System. A system that utilizes the principal of collecting, storing, and using rainwater from a rooftop or other manmade, aboveground collection surface. Also, known as a rainwater harvesting system.

Rainwater Storage Tank. The central component of the rainwater catchment system. Also, known as a cistern or rain barrel.

Receptor. An approved plumbing fixture or device of such material, shape, and capacity as to adequately receive the discharge from indirect waste pipes, so constructed and located as to be readily cleaned.

Reclaimed Water. Nonpotable water provided by a water/wastewater utility that, as a result of tertiary treatment of domestic wastewater, meets requirements of the public health Authority Having Jurisdiction for its intended uses.

Registered Design Professional. An individual who is registered or licensed by the laws of the state to perform such design work in the jurisdiction.

Regulating Equipment. Includes valves and controls used in a plumbing system that is required to be accessible or readily accessible.

Relief Vent. A vent, the primary function of which is to provide circulation of air between drainage and vent systems or to act as an auxiliary vent on a specially designed system.

Remote Outlet. Where used for sizing water piping, it is the furthest outlet dimension, measuring from the meter, either the developed length of the cold-water piping or through the water heater to the furthest outlet on the hot-water piping.

Rim. See Flood-Level Rim.

Riser. A water supply pipe that extends vertically one full story or more to convey water to branches or fixtures.

Roof Drain. A drain installed to receive water collecting on the surface of a roof and to discharge it into a leader, downspout, or conductor.

Roof Washer. A device or method for removal of sediment and debris from a collection surface by diverting initial rainfall from entry into the cistern(s). Also, known as a first flush device.

Roughing-In. The installation of all parts of the plumbing system that can be completed prior to the installation of fixtures. This includes drainage, water supply, gas piping, vent piping, and the necessary fixture supports.

221.0 – S –

Sand Interceptor. See Interceptor (Clarifier).

Scavenging. Evacuation of exhaled mixtures of oxygen and nitrous oxide. [NFPA 99:3.3.147]

Standard Cubic Feet per Minute (SCFM). Volumetric flow rate of gas in units of standard cubic feet per minute. [NFPA 99:3.3.156]

SDR. An abbreviation for "standard dimensional ratio," which is the specific ratio of the average specified outside diameter to the minimum wall thickness for outside controlled diameter plastic pipe.

Seam, Welded. See Joint, Welded.

Seepage Pit. A lined excavation in the ground which receives the discharge of a septic tank so designed as to permit the effluent from the septic tank to seep through its bottom and sides.

See Appendix H

Septic Tank. A watertight receptacle that receives the discharge of a drainage system or part thereof, designed and constructed so as to retain solids, digest organic matter through a period of detention, and allow the liquids to discharge into the soil outside of the tank through a system of open joint piping or a seepage pit meeting the requirements of this code.

See Appendix H

Service Piping. The piping and equipment between the street gas main and the gas piping system inlet that is installed by, and is under the control and maintenance of, the serving gas supplier.

Sewage. Liquid waste containing animal or vegetable matter in suspension or solution and that may include liquids containing chemicals in solution.

Sewage Ejector. A device for lifting sewage by entraining it on a high-velocity jet stream, air, or water.

Sewage Pump. A permanently installed mechanical device, other than an ejector, for removing sewage or liquid waste from a sump.

Shall. Indicates a mandatory requirement.

Shielded Coupling. An approved elastomeric sealing gasket with an approved outer shield and a tightening mechanism.

See Figure 221.0a

Shock Arrester. See Water Hammer Arrester.

Should. Indicates a recommendation or that which is advised but not required.

DEFINITIONS

**FIGURE 221.0A
SHIELDED COUPLING**

Single-Family Dwelling. A building designed to be used as a home by the owner of such building, which shall be the only dwelling located on a parcel of ground with the usual accessory buildings.

Size and Type of Tubing. See Diameter.

Slip Joint. An adjustable tubing connection, consisting of a compression nut, a friction ring, and a compression washer, designed to fit a threaded adapter fitting or a standard taper pipe thread.

See **Figure 221.0b**

**FIGURE 221.0B
SLIP JOINT FITTING**

Slope. See Grade.

Soil Pipe. A pipe that conveys the discharge of water closets, urinals, clinical sinks, or fixtures having similar functions of collection and removal of domestic sewage, with or without the discharge from other fixtures to the building drain or building sewer.

Special Wastes. Wastes that require some special method of handling, such as the use of indirect waste piping and receptors, corrosion-resistant piping, sand, oil or grease interceptors, condensers, or other pretreatment facilities.

Stack. The vertical main of a system of soil, waste, or vent piping extending through one or more stories.

Stack Vent. The extension of soil or waste stacks above the highest horizontal drain connected to the stack.

Standard. A document, the main text of which contains only mandatory provisions using the word "shall" to indicate requirements and which is in a form generally suitable for mandatory reference by another standard or code or for adoption into law. Nonmandatory provisions shall be located in an appendix, footnote, or fine print note and are not to be considered a part of the requirements of a standard.

Station Inlet. An inlet point in a piped medical/surgical vacuum distribution system at which the user makes connections and disconnections. [NFPA 99:3.3.157]

Station Outlet. An outlet point in a piped medical gas distribution system at which the user makes connections and disconnections. [NFPA 99:3.3.158]

Sterilizer. A piece of equipment that disinfects instruments and equipment by way of heat.

Storm Drain. See Building Drain (Storm).

Storm Sewer. A sewer used for conveying rainwater, surface water, condensate, cooling water, or similar liquid wastes.

Subsoil Drain. A drain that collects subsurface or seepage water and conveys it to a place of disposal.

See **Figure 221.0c**

Subsoil Irrigation Field. Gray water irrigation field installed in a trench within the layer of soil below the topsoil. This system is typically used for irrigation of deep rooted plants.

Subsurface Irrigation Field. Gray water irrigation field installed below finished grade within the topsoil.

Sump. An approved tank or pit that receives sewage or liquid waste and which is located below the normal grade of the gravity system and which must be emptied by mechanical means.

See **Figure 221.0c**

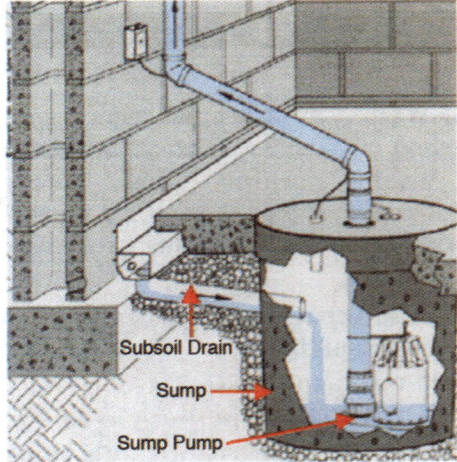

**FIGURE 221.0C
SUMP AND SUMP PUMP**

Supports. Supports, hangers, and anchors are devices for properly supporting and securing pipe, fixtures, and equipment.

Surge Tank. A reservoir to modify the fluctuation in flow rates to allow for uniform distribution of gray water to the points of irrigation.

222.0 – T –

T Rating. The time period that the penetration firestop system, including the penetrating item, limits the maximum temperature rise of 325°F (181°C) above its initial temperature through the penetration on the nonfire side, where tested in accordance with ASTM E814 or UL 1479.

🔧 ASTM E 814 was first published in 1981. This standard is used to evaluate firestops that are intended for use in openings in fire-resistive walls and floors that have been evaluated in accordance with test method ASTM E 119 (first published in 1917). The ASTM E 814 test method uses a smaller wall or floor unit with the penetrating material and firestop in place. The ASTM E 814 test is similar to the ASTM E 119 test in that the firestop assembly is exposed to a fire test along with a hose stream test. The difference between the two is that the ASTM E 814 test method is less costly due to the use of a smaller furnace and smaller wall or floor units (see **Figure 208.0a**).

Tailpiece. The pipe or tubing that connects the outlet of a plumbing fixture to a trap.

Thermostatic (Temperature Control) Valve. A mixing valve that senses outlet temperature and compensates for fluctuations in incoming hot or cold water temperatures.

Toilet Facility. A room or space containing not less than one lavatory and one water closet.

Transition Gas Riser. A listed or approved section or sections of pipe and fittings used to convey fuel gas and installed in a gas piping system to provide a transition from below-ground to aboveground.

Trap. A fitting or device so designed and constructed as to provide, where properly vented, a liquid seal that will prevent the back passage of air without materially affecting the flow of sewage or wastewater through it.

🔧 See **Figure 222.0a**

Trap Arm. Those portions of a fixture drain between a trap and the vent.

Trap Primer. A device and system of piping that maintains a water seal in a remote trap.

🔧 See **Figure 222.0b**

Trap Seal. The vertical distance between the crown weir and the top dip of the trap.

   Crown Weir (Trap Weir). The lowest point in the cross-section of the horizontal waterway at the exit of the trap.

🔧 See **Figure 222.0a**

   Top Dip (of the trap). The highest point in the internal cross-section of the trap at the lowest part of the bend (inverted siphon). By contrast, the bottom dip is the lowest point in the internal cross-section.

🔧 See **Figure 222.0a**

223.0 – U –

Unsanitary. See Insanitary.

Urinal, Hybrid. A urinal that conveys waste into the drainage system without the use of water for flushing and automatically performs a drain-cleansing action after a pre-determined amount of time.

🔧 When the cleansing cycle is activated, hybrid urinals emit one gallon of water in a drain cleansing action. This sends one gallon of water down the pipes (in about 20 seconds) to automatically flush the system. See Learning Links goo.gl/VeJhsk and goo.gl/QFxuQv. See **Figure 223.0**.

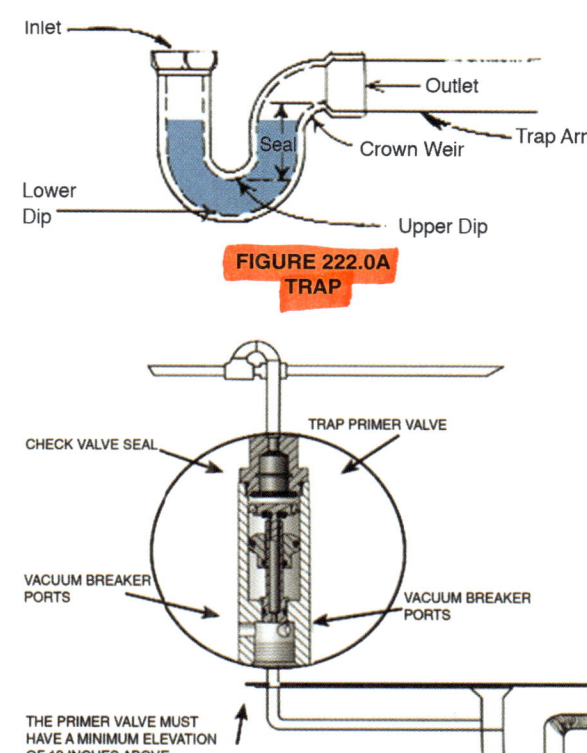

FIGURE 222.0A TRAP

FIGURE 222.0B TRAP PRIMER

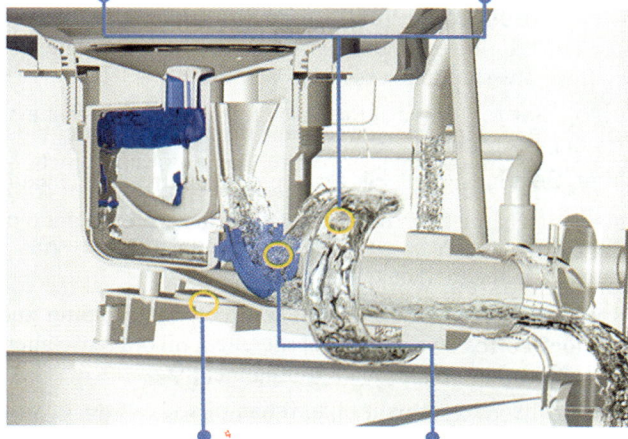

FIGURE 223.0 HYBRID URINAL (courtesy of Falcon Waterfree Technology)

DEFINITIONS

User Outlet. See Station Outlet.

224.0 – V –

Vacuum. A pressure less than that exerted by the atmosphere.

Vacuum Breaker. See Backflow Preventer.

Vacuum Relief Valve. A device that prevents excessive vacuum in a pressure vessel.

 See **Figure 224.0a**

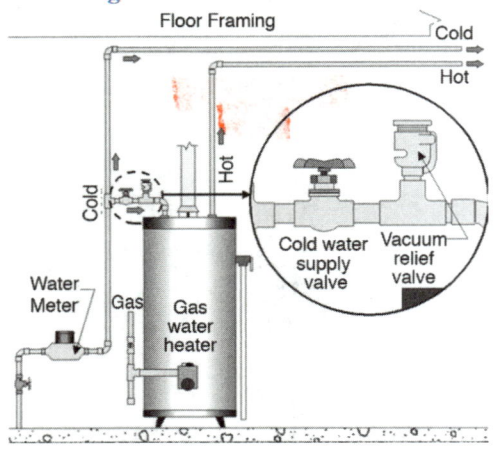

FIGURE 224.0A
VACUUM RELIEF VALVE

Vacuum System-Level 1. A system consisting of central vacuum-producing equipment with pressure and operating controls, shutoff valves, alarm warning systems, gauges, and a network of piping extending to and terminating with suitable station inlets at locations where patient suction could be required.

Valve, Isolation. A valve that isolates one piece of equipment from another.

Valve, Pressure-Relief. A pressure-actuated valve held closed by a spring or other means and designed automatically to relieve pressure in excess of its setting.

Valve, Riser. A valve at the base of a vertical riser that isolates that riser.

Valve, Service. A valve is serving horizontal piping extending from a riser to a station outlet or inlet.

Valve, Source. A single valve at the source that controls a number of units that makes up the source.

Valve, Zone. A valve that controls the gas or vacuum to a particular area.

Vent. See Plumbing Vent; Dry Vent; Wet Vent.

Vent Connector, Gas. That portion of a gas venting system that connects a listed gas appliance to a gas vent and is installed within the space or area in which the appliance is located.

See **Figure 224.0b**

Vent Offset. An arrangement of two or more fittings and pipe installed for the purpose of locating a vertical section of the vent pipe in a different but parallel plane with respect to an adjacent section of a vertical vent pipe. [NFPA 54:3.3.102]

Vent Pipe. See Plumbing Vent.

Vent Stack. The vertical vent pipe installed primarily for the purpose of providing circulation of air to and from any part of the drainage system.

See **Figure 218.0c**

Vent System. See Plumbing Vent System.

Vented Flow Control Device. A device installed upstream from the hydromechanical grease interceptor having an orifice that controls the rate of flow through the interceptor, and an air intake (vent) downstream from the orifice, which allows air to be drawn into the flow stream.

Venting System. A continuous open passageway from the flue collar or draft hood of an appliance to the outdoors for the purpose of removing flue or vent gases. [NFPA 54:3.3.95.7]

See **Figure 224.0b** and **224.0c**

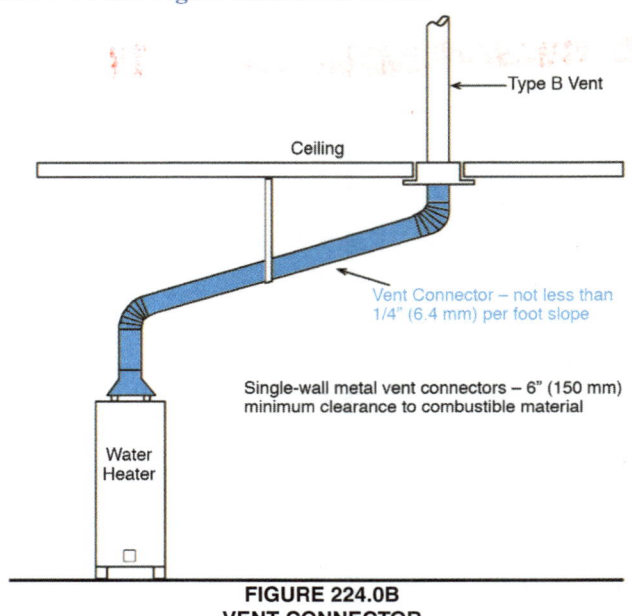

FIGURE 224.0B
VENT CONNECTOR

Vent, Gases. A listed factory-made vent pipe and vent fittings for conveying flue gases to the outdoors.

See **Figure 224.0c**

Type B Gas Vent. A factory-made gas vent listed by a nationally recognized testing agency for venting listed or approved appliances equipped to burn only gas.

Type B gas vents are double-wall-type vents with very low permissible temperature ratings. They incorporate an aluminum inner tube with an outer galvanized steel tube. The air between the tubes has insulating properties, which will minimize heat loss. Type B gas vents are the most popular vents for water heaters (see **Figure 224.0c**).

Type BW Gas Vent. A factory-made gas vent listed by a nationally recognized testing agency for venting listed or approved gas-fired vented wall furnaces.

Type L Gas Vent. A venting system consisting of listed vent piping and fittings for use with oil-burning appliances listed for use with Type L or with listed gas appliances.

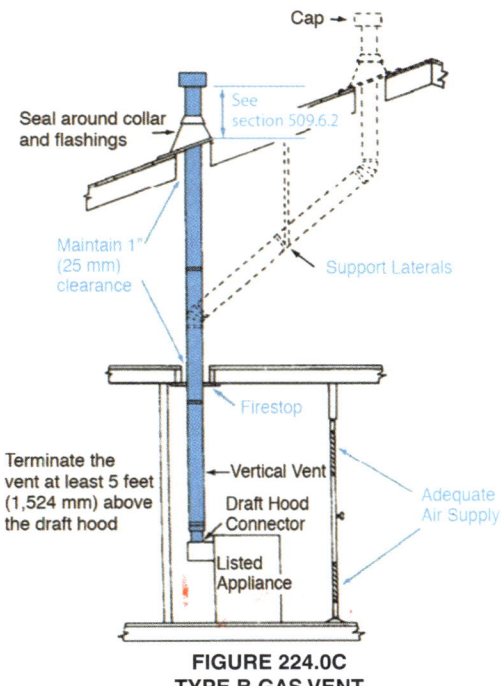

**FIGURE 224.0C
TYPE B GAS VENT**

🔧 Where higher temperatures are produced, the Type L vent must be used. The inner tube in these vents is stainless steel rather than aluminum. Type L vents can be used for Type B installations, but Type B cannot be used for Type L installations. Both are factory-made vent piping.

Vertical Pipe. A pipe or fitting that is installed in a vertical position or that makes an angle of not more than 45 degrees (0.79 rad) with the vertical.

🔧 See **Figure 224.0d**

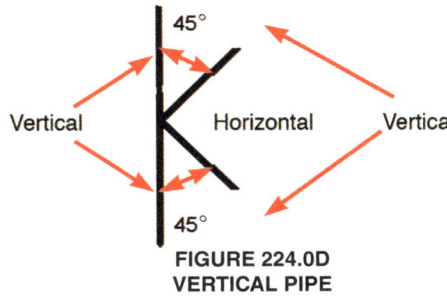

**FIGURE 224.0D
VERTICAL PIPE**

225.0 – W –

Wall-Hung Water Closet. A water closet installed in such a way that no part of the water closet touches the floor.

Waste. See Liquid Waste and Industrial Waste.

Waste Pipe. A pipe that conveys only liquid waste, free of fecal matter.

Water-Conditioning or Treating Device. A device that conditions or treats a water supply to change its chemical content or remove suspended solids by filtration.

Water Distribution Pipe. In a building or premises, a pipe that conveys potable water from the building supply pipe to the plumbing fixtures and other water outlets.

Water Hammer Arrester. A device designed to provide protection against hydraulic shock in the building water supply system.

🔧 See **Figure 225.0a**

See Learning Link goo.gl/jKiQMn

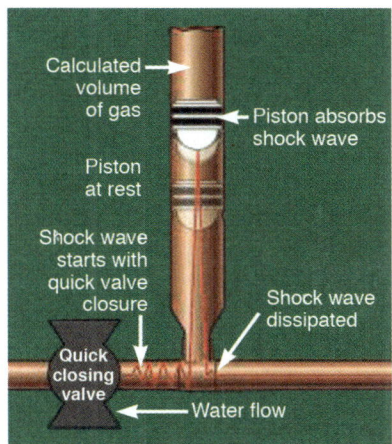

**FIGURE 225.0A
WATER HAMMER ARRESTOR**

Water Heater or Hot Water Heating Boiler. An appliance designed primarily to supply hot water for domestic or commercial purposes and equipped with automatic controls limiting water temperature to a maximum of 210°F (99°C).

🔧 See **Figure 225.0b**

Water Main (Street Main). A water supply pipe for public or community use.

Water Supply System. The building supply pipe, the water distribution pipes, and the necessary connecting pipes, fittings, control valves, backflow prevention devices, and all appurtenances carrying or supplying potable water in or adjacent to the building or premises.

Water/Wastewater Utility. A public or private entity which may treat, deliver or do both functions to reclaimed (recycled) water, potable water, or both to wholesale or retail customers.

Welder, Pipe. A person who specializes in the welding of pipes and holds a valid certificate of competency from a recognized testing laboratory, based on the requirements of the ASME Boiler and Pressure Vessels code, Section IX.

Wet Procedure Locations. The area in a patient care space where a procedure is performed that is normally subject to wet conditions while patients are present, including standing fluids on the floor or drenching of the work area, either of which condition is intimate to the patient or staff. [NFPA 99:3.3.171]

Wet Vent. A vent that also serves as a drain.

🔧 See **Figure 225.0c**

Whirlpool Bathtub. A bathtub fixture equipped and fitted with a circulating piping system designed to accept, circulate, and discharge bathtub water upon each use.

DEFINITIONS

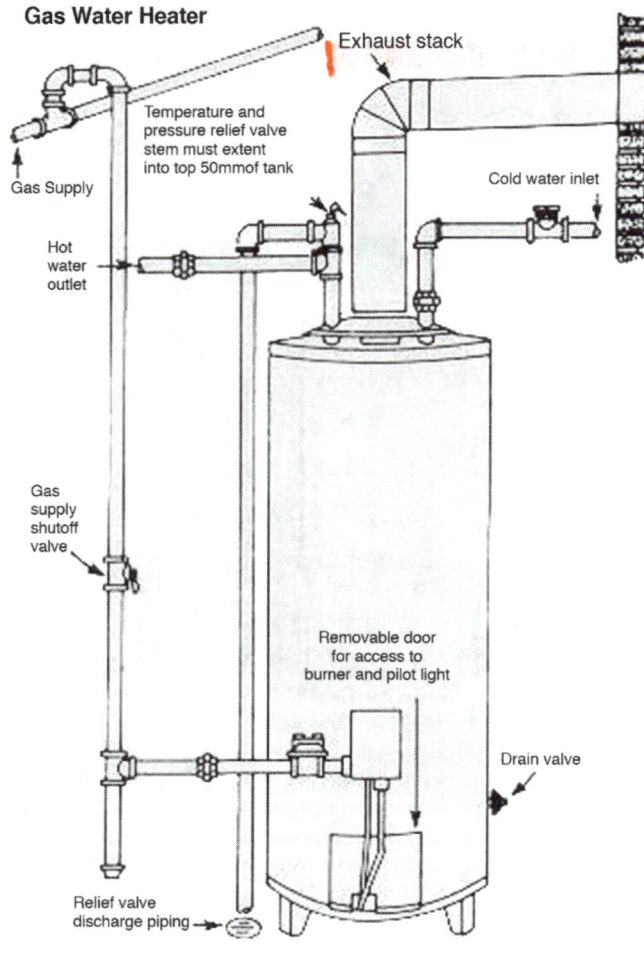

**FIGURE 225.0B
WATER HEATER INSTALLATION**

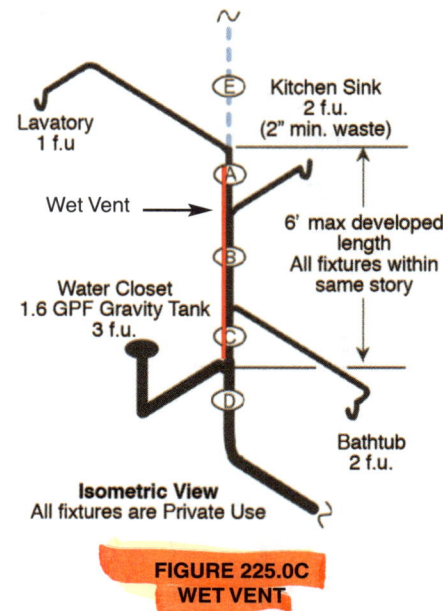

**FIGURE 225.0C
WET VENT**

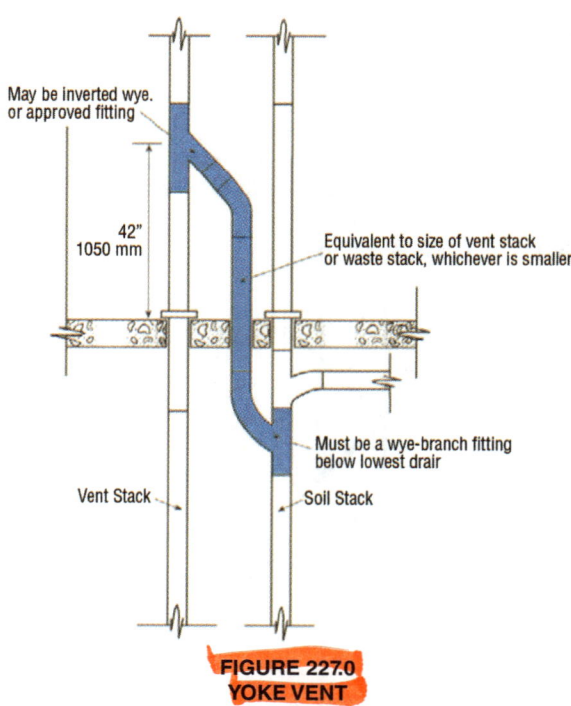

**FIGURE 227.0
YOKE VENT**

226.0   – X –

No definitions.

227.0   – Y –

Yoke Vent. A pipe connecting upward from soil or waste stack to a vent stack to prevent pressure changes in the stacks.

See Figure 227.0

228.0   – Z –

No definitions.

DEFINITIONS

ABBREVIATIONS

A

A	Area
Abs	Absolute
ABS	Acrilonitrile-Butadiene-Styrene
Acet.	Acetylene
ACP	Asbestos Cement Pipe
AD	Area Drain
AGA	American Gas Association
AHAM	Association of Home Appliance Manufacturers
AISI	American Iron and Steel Institute
Al	Aluminum
ANSI	American National Standards Institute
AP	Access Panel
APHA	American Public Health Association
API	American Petroleum Institute
ASA	American Standards Association
Asb	Asbestos
ASCE	American Society of Civil Engineering
ASHRAE	American Society of Heating, Refrigeration and Air Conditioning Engineers
ASME	American Society of Mechanical Engineers
ASSE	American Society of Sanitary Engineers
ASTM	American Society for Testing and Materials
Avg	Average
AWWA	American Water Works Association

B

Bbl	Barrel
B&S	Bell and Spigot
B&S	Brown and Sharpe Gauge
BBE	Ball Both Ends
BOE	Ball One End
BFP	Backflow Preventer
Bldg	Building
Blk	Black
BOCA	Building Officials Conference of America
BOD	Biochemical Oxygen Demand
BrT	Brine Tank
BS	Bar Sink
BT	Bathtub
BTB	Back to Back
BTU	British Thermal Unit

C

C	Centrigrade
CA	Combustion Air
Cap	Capacity
CB	Catch Basin
CD	Condensate Drain
CDA	Copper Development Association
CFM	Cubic Feet per Minute
CFS	Cubic Feet per Second
CI	Cast Iron
Cir.	Circumference
Circ	Circular
CISPI	Cast Iron Soil Pipe Institute
CKV	Check Valve
CL	Center Line
CL	Critical Level
Cl	Chlorine
cm	Centimeter
CO	Cleanout
Conc.	Concrete
CPE	Chlorinated Polyethylene
CPVC	Chlorinated Poly Vinyl Chloride
C.S.	Cast Steel
CS	Commercial Standard
CS&PS	Commercial Standards and Product Standards
C to B	Center to Back
C to C	Center to Center
C to T	Center to Throat
C x C	Copper to Copper
C x FIPS	Copper to Female Iron Pipe Size
C x MIPS	Copper to Male Iron Pipe Size
C to F	Center to Face
Cu	Copper
Cu Ft	Cubic Foot
Cu In	Cubic Inch
Cu Yd	Cubic Yard
CV	Check Valve
CW	Cold Water

D

D	Diameter
Dec. Lgth	Decreased Length
Deg or °	Degree
DF	Drinking Fountain
DH	Double Hub
Diam	Diameter
Dif	Difference
Di. W	Deionized Water
Dr	Drain or Drainage
DS	Downspout
DT	Drum Trap
DU	Disposal Unit
DW	Dishwasher

2018 UNIFORM PLUMBING CODE ILLUSTRATED TRAINING MANUAL

DEFINITIONS

Dwg	Drawing	Gl	Galvanized Iron
DWV	Drainage Waste and Vent	GP	Gauge Pressure
	E	GPD	Gallons Per Day
ECO	Energy Cut-off	GPM	Gallons Per Minute
EHCI	Extra-Heavy Cast Iron	GPS	Gallons Per Second
El	Elevation	GV	Gate Valve
Elec	Electric		**H**
E to B	End to Base	H or Ht	Height
E to C	End to Center	H	Hydrogen
E to E	End to End	HB	Hose Bibb
E to F	End to Face	Hd	Head
E to T	End to Throat	HDR	Header
EWC	Electric Water Cooler	Hor	Horizontal
	F	HP	Head Pressure
F	Fahrenheit	Hp	Horsepower
FAI	Fresh Air Inlet	Hr	Hour
FB	Foot Bath	HR	Hose Rack
FD	Floor Drain	HW	Hot Water
Fe	Iron	HWR	Hot Water Return
FF	Finish Floor		**I**
FG	Finish Grade	in	Inch
FHC	Fire Hose Cabinet	IAPMO	International Association of Plumbing and Mechanical Officials
Fig	Figure	ID	Inside Diameter
Flg	Flange	Inc	Increaser or Increased
FP	Fire Plug	Inv El	Invert Elevation
FPS	Feet Per Second	IPS	Iron Pipe Size
FS	Federal Specification	IS	Installation Standard
FS	Floor Sink	IWF	Iron Water Filter
FSPS	Female Standard Pipe Size		**L**
Ft	Foot or Feet	L or Lgth	Length
F to B	Face to Back	L	Liter
F to C	Face to Center	Lav	Lavatory
F to E	Face to End	Lb	Pound
F to F	Face to Face	LF	Lawn Faucet
F to T	Face to Throat	LH	Left Hand
Ftg	Fitting	LHGC	Lever Handled Gas Cock
Ftg	Footing	LHT	Left Hand Thread
FU	Fixture Unit	LIA	Lead Industries Association
FVT	Float Valve Tank	Liq	Liquid
FV	Foot Valve	LT	Laundry Tray
	G	LW	Light Weight
G	Gas		**M**
G	Gravity	M	Thousand
Gal	Gallons	M	Meter
Galv	Galvanized	Max	Maximum
GFD	Garage Floor Drain	MC	Medicine Chest
GFWH	Gas-Fired Water Heater		

DEFINITIONS

MCAA	Mechanical Contractors' Association of America	PB	Polybutylene
MFCC	Membrane Filter Coliform Count	PDI	Plumbing and Drainage Institute
Mfg	Manufacturing	PE	Polyethylene
Mfr	Manufacturer	pH	(See definition.)
MG	Mechanism Pressure Gap	PIV	Post Indicator Valve
MGD	Million Gallons Per Day	POC	Point of Connection
MH	Manhole	PP	Polypropylene
Ml	Malleable Iron	PPI	Plastics Pipe Institute
mm	Millimeter	PPM	Parts Per Million
MS	Material Standard	PS	Property Standard
MS	Mild Steel	PSI	Pounds Per Square Inch
Min	Minimum	PSIG	Pounds Per Square Inch Gage
Min	Minutes	PT	Plugged Tee
MP	Melting Point	PTRV	Pressure Temperature Relief Valve
MPN	Most Probable Number	PVC	Poly Vinyl Chloride
MS	Mop Sink		**Q**
MSPS	Male Standard Pipe Size	Qt	Quart
MSS	Manufacturers Standardization Society		**R**
	N	R	Hydraulic Radius
NAPHCC	National Association of Plumbing, Heating and Cooling Contractors	Rad	Radius
		RD	Rate of Demand
NBFU	National Board of Fire Underwriters	RD	Roof Drain
NBS	National Bureau of Standards	Red	Reducer
NC	National Coarse	Rl	Rough-in
NDTS	Not Drawn to Scale	RL	Roof Leader
NFPA	National Fire Protection Association	RP	Reduced Pressure Zone Backflow Preventer
NIC	Not in Contract	RS	Rate of Supply
Nip	Nipple		**S**
No.	Number	S	Hydraulic Slope (in inches per foot)
NPS	Nominal Pipe Size	San	Sanitary
NPS	National Pipe Thread Straight	Sb	Antimony
NPT	National Pipe Thread Tapered	SBCCI	Southern Building Code Conference International
NSF	National Sanitation Foundation		
NTS	Not to Scale	Sec	Second
	O	SH	Single Hub
O	Offset	SHGC	Square Head Gas Cock
O	Oxygen	SN	Tin
OD	Outside Diameter	Spec	Specification
OFD	Overflow Drain	Sq	Square
OS&Y	Outside Screw and Yoke	Sq Ft	Square Foot
OSY	Open Sight Yoke Valve (either gate or compression)	Sq In	Square Inch
		SR	Styrene Rubber
Oz	Ounce	SS	Service Sink or Slop Sink
	P	SS	Stainless Steel
P	Pitch	SS	Swingspout
P	Pressure	SSU	Seconds Saybolt Universal
Pb	Lead	Std	Standard S&W Soil and Waste

DEFINITIONS

	T
T	Temperature
T	Time
TBE	Thread Both Ends
TOE	Thread One End
	U
U or Urn	Urinal
UL	Underwriters Laboratory
USASI	United States of America Standards Institute (Obsolete – See ANSI)
	V
V	Valve
V	Velocity
V	Vent
V	Volume
VB	Vacuum Breaker
VCP	Vitrified Clay Pipe
VTR	Vent Through Roof
	W
WH	Water Heater
WQA	Water Quality Association
	Y
YB	Yard Box

CHAPTER 3
GENERAL REGULATIONS

301.0 General.

301.1 Applicability. This chapter shall govern the general requirements, not specific to other chapters, for the installation of plumbing systems.

301.2 Minimum Standards. Pipe, pipe fittings, traps, fixtures, material, and devices used in a plumbing system shall be listed (third-party certified) by a listing agency (accredited conformity assessment body) as complying with the approved applicable recognized standards referenced in this code, and shall be free from defects. Unless otherwise provided for in this code, materials, fixtures, or devices used or entering into the construction of plumbing systems, or parts thereof shall be submitted to the Authority Having Jurisdiction for approval.

All materials used in the plumbing system must have the approval of the Authority Having Jurisdiction (AHJ). This is accomplished by two methods. First, the architect or engineer will specify materials that meet or conform to an applicable standard for the use intended, which should be referenced in the code. For example, the plumbing engineer may specify IPEX DR 35 sewer pipe for the building sewer. The pipe is manufactured to ASTM D 3035, which is referenced in Chapter 17, and is marked (labeled) as such. The plumbing contractor, installer, and inspector only has to ensure that the pipe is indeed IPEX DR 35, with the proper markings on the pipe, to verify compliance with the minimum standards. The second method is for materials that are not made to standards referenced in the code. The architect or engineer must then submit these materials to the AHJ for approval. Sections 301.3 and 301.5 explain how these materials achieve approval from the AHJ.

Every pipe, fitting, fixture, etc., should be a good, quality product; however, this is not always the case. The material in question could be made to the proper standard and used for its intended use and still violate the requirement in Section 301.2 "... and shall be free from defects." This violation most often occurs because the item is damaged. For example, a pipe may be installed that is cracked. This is not only bad plumbing but it violates this section and several others in the code. To comply with this section of the code the materials must be in good order and installed per recognized standards and other applicable sections in the code.

301.2.1 Marking. Each length of pipe and each pipe fitting, trap, fixture, material, and device used in a plumbing system shall have cast, stamped, or indelibly marked on it any markings required by the applicable referenced standards and listing agency, and the manufacturer's mark or name, which shall readily identify the manufacturer to the end user of the product. Where required by the approved standard that applies, the product shall be marked with the weight and the quality of the product. Materials and devices used or entering into the construction of plumbing and drainage systems, or parts thereof shall be marked and identified in a manner satisfactory to the Authority Having Jurisdiction. Such marking shall be done by the manufacturer. Field markings shall not be acceptable.

Exception: Markings shall not be required on nipples created from cutting and threading of approved pipe.

In order for the contractor, installer or inspector to verify if a plumbing product meets the requirements of the code, the item must be marked or labeled (see **Figure 301.2.1a**). The code does not specifically state the marking requirements, except for the manufacturer's identification. The referenced standard states the minimum marking requirements. The identification requirements typically include any of the following: the name of the manufacturer or trademark; type or model number; maximum rated pressure and temperature; serial number; nominal size and standard designation.

Another type of marking on some plumbing products is the IAPMO product markings. The Research and Testing (R&T) Division of IAPMO tests and evaluates products to determine whether they comply with select standards and applicable codes. Each tested product is then assessed as meeting or not meeting the standards. Compliant products may qualify for the application of one of seven possible IAPMO marking labels.

By its shape or markings, an IAPMO label may denote that a product is either "listed" or "classified." Each of these labels carries considerably different implications. A product that bears the "listed" label has demonstrated compliance with the specific standard or standards to which it was tested. Listed products require no special approval or acceptance by local authorities when installed within the limits of their listing. Additionally, a listed product is defined as compliant with code stipulations, engineering concepts, or other fundamental principles found in the Uniform Plumbing Code (UPC) or Uniform Mechanical Code (UMC). Products bearing the labels shown below may be safely used in the plumbing or mechanical system.

Conversely, a "classified" label represents compliance with all standards to which the product was tested but does not ensure compliance with UPC or UMC code stipulations or requirements. Classified products are published in a separate Product Directory, the objective being to avoid user confusion regarding the propriety of their application. UPC Section 301.3 allows the use of these products, but only with specific approval from the local AHJ. Products bearing the classified label (see **Figure 301.2.1b**) must have approval from the AHJ before being used in the plumbing or mechanical system.

The use of specialized labeling is intended to provide the consumer with guidance regarding the limits of product application. Compliance with an evaluation standard does not suggest that a product is suitable for any specific application; nor

GENERAL REGULATIONS

**FIGURE 301.2.1A
PIPE MARKING**

**FIGURE 301.2.1B
IAPMO PRODUCT LABELS**

does it suggest that a product is code compliant or approved for use in a particular application. IAPMO R&T tests to standards specified by the person who submits the product for evaluation. Because all test standards are not necessarily related to code requirements or standard engineering practices, test compliance is essentially unrelated to the appropriate utilization of a particular product.

301.2.2 Standards. Standards listed or referred to in this chapter or other chapters cover materials that will conform to the requirements of this code, where used in accordance with the limitations imposed in this or other chapters thereof and their listing. Where a standard covers materials of various grades, weights, quality, or configurations, the portion of the listed standard that is applicable shall be used. Design and materials for special conditions or materials not provided for herein shall be permitted to be used by special permission of the Authority Having Jurisdiction after the Authority Having Jurisdiction has been satisfied as to their adequacy. A list of plumbing standards that appear in specific sections of this code is referenced in Table 1701.1. Standards referenced in Table 1701.1 shall be applied as indicated in the applicable referenced section. A list of additional standards, publications, practices, and guides that are not referenced in specific sections of this code appear in Table 1701.2. The documents indicated in Table 1701.2 shall be permitted in accordance with Section 301.3. An IAPMO Installation Standard is referenced in Appendix I for the convenience of the users of this code. It is not considered as a part of this code unless formally adopted as such by the Authority Having Jurisdiction.

A referenced standard is a standard that is referenced in the body of the code. Standards referenced in Table 1701.1 shall be applied as indicated in the applicable reference section. A list of additional standards, publications, and guides that are not referenced in specific sections of this code appear in Table 1701.2. The standards from Table 1701.2 shall be permitted only after they have been approved by the Authority Having Jurisdiction.

A standard is a set of technical definitions, requirements and guidelines for the manufacture of devices and products that establishes test methods, specifications, classifications, practices and scoping requirements. Their purpose is to ensure that products are made to be safe and efficient for the consumer.

Codes and standards work together to protect public health and safety. A standard is considered a basis of comparison or an approved model. Simply stating, codes tell the user what to do, when to do it and under what circumstances to do it. Codes are often legal requirements that are enforced by local jurisdictions that enforce their provisions. Standards provide the user with approved materials and tell the user that such products have been tested to performance requirements and are applicable only to the extent the code references the standard. For example, a standard may have performance specifications for materials and use of application or installation requirements. The standard is only applicable to the extent suggested in the text of the code. The code text takes precedence when the requirements of the standard conflict with the requirements of the code.

For example, Section 604.1 references that all building water piping material must comply with Table 604.1 and to the Standards listed in Table 1701.1. One of them, ASTM F 1281, is the standard for PEX-AL-PEX pressure pipe for water distribution systems. This standard regulates the performance requirements for materials that cover PEX-AL-PEX composite pressure pipe. In addition, the performance requirements include dimensions, burst and sustained pressure performance. Section 604.13, Water Heater Connectors, requires that PEX- AL-PEX must not be installed within the first 18 inches (46 cm) of piping connected to the water heater. The mandatory information for connectors in Section 604.13 suggests that if water heating equipment malfunctions, the assemblies (connectors) shall have satisfactory strength to allow short-term conditions. However, the standard ASTM F 1281 does not restrict installing PEX-AL-PEX within the first 18 inches of piping connected to the water heater; therefore, Section 604.13 would take precedence over the standard.

Another example, Section 604.1 requires that pipe, tube and fittings carrying water used in potable water systems intended to supply drinking water must meet the requirements of NSF 61. However, ASTM 1281 does not contain mandatory language that this material must conform to NSF 61 and, therefore, Section 604.1 takes precedence over the standard and requires conformance to NSF 61 for potable water piping used to supply drinking water. Remember, the referenced standard is a guide to aid the user in deciding whether a product, device, joining method or installation complies with the code.

301.2.3 Plastic Pipe, Plastic Pipe Fittings, and Components.
Plastic pipe, plastic pipe fittings, and components other than those for gas shall comply with NSF 14.

301.2.4 Cast-Iron Soil Pipe, Fittings, and Hubless Couplings.
Cast-iron soil pipe, fittings, and hubless couplings shall be third party certified in accordance with ASTM C1277 and CISPI 310 for couplings and ASTM A888, ASTM A74, and CISPI 301 for pipes and fittings.

301.2.5 Existing Buildings.
In existing buildings or premises in which plumbing installations are to be altered, repaired, or renovated, the Authority Having Jurisdiction has discretionary powers to permit deviation from the provisions of this code, provided that such proposal to deviate is first submitted for proper determination in order that health and safety requirements, as they pertain to plumbing, shall be observed.

Existing systems may remain in service when maintained in accordance with the code in effect when it was first installed. A repair to an existing system would require compliance to the code in effect at the time of repair. However, the AHJ per this section of the code has the ability to allow a deviation from the new code as long as the deviation is submitted for approval before hand and the installation meets the health and safety requirements of the code.

301.3 Alternate Materials and Methods of Construction Equivalency.
Nothing in this code is intended to prevent the use of systems, methods, or devices of equivalent or superior quality, strength, fire resistance, effectiveness, durability, and safety over those prescribed by this code. Technical documentation shall be submitted to the Authority Having Jurisdiction to demonstrate equivalency prior to installation. The Authority Having Jurisdiction shall have the authority to approve or disapprove the system, method, or device for the intended purpose.

However, the exercise of this discretionary approval by the Authority Having Jurisdiction shall have no effect beyond the jurisdictional boundaries of said Authority Having Jurisdiction. An alternate material or method of construction so approved shall not be considered as in accordance with the requirements, intent, or both of this code for a purpose other than that granted by the Authority Having Jurisdiction where the submitted data does not prove equivalency.

New materials and methods of construction are constantly being developed in this industry. It is not the intent of the UPC to prohibit the use of these new materials and methods before they are specified in the code as long as the AHJ approves their use. It should also be noted that the approval of the AHJ in one jurisdiction does not give approval for the use of these alternatives in other jurisdictions; nor does it give approval for other uses than the specific use that it was initially approved for. Each jurisdiction will have to evaluate this alternative material or method for its own use.

The AHJ may approve the alternative material or method of construction, provided the AHJ finds that the proposed design is satisfactory and in compliance with the intent of the UPC.

When using any new material, the installer should make sure that it is acceptable to the local AHJ and that it is installed in accordance with its conditions of approval.

An example would be if a new pipe material is introduced for natural gas systems but is not included in the materials listed in Chapter 12. The person who wishes to use this new material would have to submit the material plus information regarding its use to the AHJ for approval to use it in the new installation. The following sections of the code concerning testing of the material would then apply if the AHJ is not satisfied with the documentation provided by the proponent for use.

301.3.1 Testing.
The Authority Having Jurisdiction shall have the authority to require tests, as proof of equivalency.

301.3.1.1 Tests.
Tests shall be made in accordance with approved or applicable standards, by an approved testing agency at the expense of the applicant. In the absence of such standards, the Authority Having Jurisdiction shall have the authority to specify the test procedure.

301.3.1.2 Request by Authority Having Jurisdiction.
The Authority Having Jurisdiction shall have the authority to require tests to be made or repeated where there is reason to believe that a material or device no longer is in accordance with the requirements on which its approval was based.

301.4 Flood Hazard Areas.
Plumbing systems shall be located above the elevation in accordance with the building code for utilities and attendant equipment or the elevation of the lowest floor, whichever is higher.

Exception: Plumbing systems shall be permitted to be located below the elevation in accordance with the building code for utilities and attendant equipment or the elevation of the lowest floor, whichever is higher, provided that the systems are designed and installed to prevent water from entering or accumulating within their components, and the systems are constructed to resist hydrostatic and hydrodynamic loads and stresses, including the effects of buoyancy, during the occurrence of flooding to such elevation.

GENERAL REGULATIONS

🔧 Code changes resulting from the devastating flood caused by Hurricane Katrina in New Orleans in 2004 have been widespread. One example is the provision in the UPC requiring plumbing systems to be placed above flood elevations in accordance with the building code for utilities, or the elevation of the lowest floor, whichever is higher. This will ensure that the plumbing system will remain functional and not become contaminated. If the plumbing system is placed below flood elevations, then the system shall be required to be installed and supported to withstand all the loads, stresses, and buoyancy effects on the system that may occur as well as to prevent water from entering any part of the system. In this manner the system can be protected, remain intact, and not become contaminated or cause contamination.

301.4.1 Coastal High Hazard Areas. Plumbing systems in buildings located in coastal high hazard areas shall be in accordance with the requirements of Section 301.4, and plumbing systems, pipes, and fixtures shall not be mounted on or penetrate through walls that are intended to breakaway under flood loads in accordance with the building code.

301.5 Alternative Engineered Design. An alternative engineered design shall comply with the intent of the provisions of this code and shall provide an equivalent level of quality, strength, effectiveness, fire resistance, durability, and safety. Material, equipment, or components shall be designed and installed in accordance with the manufacturer's installation instructions.

301.5.1 Permit Application. The registered design professional shall indicate on the design documents that the plumbing system, or parts thereof, is an alternative engineered design so that it is noted on the construction permit application. The permit and permanent permit records shall indicate that an alternative engineered design was part of the approved installation.

301.5.2 Technical Data. The registered design professional shall submit sufficient technical data to substantiate the proposed alternative engineered design and to prove that the performance meets the intent of this code.

301.5.3 Design Documents. The registered design professional shall provide two complete sets of signed and sealed design documents for the alternative engineered design for submittal to the Authority Having Jurisdiction. The design documents shall include floor plans and a riser diagram of the work. Where appropriate, the design documents shall indicate the direction of flow, pipe sizes, grade of horizontal piping, loading, and location of fixtures and appliances.

301.5.4 Design Approval. An approval of an alternative engineered design shall be at the discretion of the Authority Having Jurisdiction. The exercise of this discretionary approval by the Authority Having Jurisdiction shall have no effect beyond the jurisdictional boundaries of said Authority Having Jurisdiction. An alternative engineered design so approved shall not be considered as in accordance with the requirements, intent, or both of this code for a purpose other than that granted by the Authority Having Jurisdiction.

301.5.5 Design Review. The Authority Having Jurisdiction shall have the authority to require testing of the alternative engineered design in accordance with Section 301.3.1, including the authority to require an independent review of the design documents by a registered design professional selected by the Authority Having Jurisdiction and at the expense of the applicant.

301.5.6 Inspection and Testing. The alternative engineered design shall be tested and inspected in accordance with the submitted testing and inspection plan and the requirements of this code.

🔧 The UPC allows the professional engineering community (plumbing or mechanical engineers) to design plumbing systems that may not strictly conform to the code. This may be due to new innovative methods or computer engineering that has not as yet been recognized by the code. These new systems must provide the equivalent protection of health and safety that is required from a standard plumbing system.

The professional engineer shall provide all the documentation and data to prove to the AHJ that the system meets the intent of the code. If the system is approved by the AHJ, then it must be accompanied by an inspection and testing plan designed by the engineer. The installation must meet the inspection and testing requirements of the engineer's plan and the pertinent requirements of the code.

302.0 Iron Pipe Size (IPS) Pipe.

302.1 General. Iron, steel, copper, and copper alloy pipe shall be standard-weight iron pipe size (IPS) pipe.

🔧 All tubing designated as pipe shall meet the sizing requirements of standard weight iron pipe size. Iron Pipe Size (IPS) represents the approximate inside diameter of the pipe in inches. To establish the inside diameter, the outside diameter (OD) of the pipe was standardized with a standard wall thickness, called standard weight. The ODs of IPS are also known as Nominal Pipe Size (NPS) and have the same OD standard. However, there are varying weights (designated as Schedule) of wall thicknesses for each pipe size (see **Figure 302.1**). Since the OD is standardized, as the wall thickens, the inside diameter decreases. Therefore the Code requires the standard weight (STD) for each pipe size so the inside diameter is the closest approximation to the pipe size. For example, one-inch pipe has an OD of 1.315 inches. The standard weight has a wall thickness of 0.133 inches. To calculate the inside diameter subtract two times the wall thickness from the OD, which yields an inside diameter of 1.049 inches.

303.0 Disposal of Liquid Waste.

303.1 General. It shall be unlawful for a person to cause, suffer, or permit the disposal of sewage, human excrement, or other liquid wastes, in a place or manner, except through and by means of an approved drainage system, installed and maintained in accordance with the provisions of this code.

🔧 This provision of the code requires the creation of drainage systems within the premises for the disposal of all wastes to a legal point of disposal such as a public sewer system or a private sewage disposal system. It also requires that

Pipe dimensions, imperial / Metric pipe chart

Nom. Pipe Sizes Inches	mm DN	OD inches	OD mm	Schedule Designations ANSI/ASME	Wall Thickn. inches	Wall Thickn. mm	Lbs/Ft	Kg/m
1/8"	6	0.405	10.29	10/10S	0.049	1.24	0.1863	0.28
1/8"	6	0.405	10.29	STD/40/40S	0.068	1.73	0.2447	0.36
1/8"	6	0.405	10.29	XS/80/80S	0.095	2.41	0.3145	0.47
1/4"	6	0.540	13.72	10/10S	0.065	1.65	0.3297	0.49
1/4"	8	0.540	13.72	STD/40/40S	0.088	2.24	0.4248	0.63
1/4"	8	0.540	13.72	XS/80/80S	0.119	3.02	0.5351	0.80
3/8"	8	0.675	17.15	10/10S	0.065	1.65	0.4235	0.63
3/8"	10	0.675	17.15	STD 40/40S	0.091	2.31	0.5676	0.84
3/8"	10	0.675	17.15	XS 80/80S	0.126	3.20	0.7388	1.10
1/2"	10	0.840	21.34	5/5S	0.065	1.65	0.5383	0.80
1/2"	15	0.840	21.34	10/10S	0.083	2.11	0.671	1.00
1/2"	15	0.840	21.34	STD 40/40S	0.109	2.77	0.851	1.27
1/2"	15	0.840	21.34	XS 80/80S	0.147	3.73	1.088	1.62
1/2"	15	0.840	21.34	160	0.188	4.78	1.309	1.95
1/2"	15	0.840	21.34	XX	0.294	7.47	1.714	2.55
3/4"	20	1.050	26.67	5/5S	0.065	1.65	0.684	1.02
3/4"	20	1.050	26.67	10/10S	0.083	2.11	0.857	1.28
3/4"	20	1.050	26.67	STD 40/40S	0.113	2.87	1.131	1.68
3/4"	20	1.050	26.67	XS 80/80S	0.154	3.91	1.474	2.19
3/4"	20	1.050	26.67	160	0.219	5.56	1.944	2.89
3/4"	20	1.050	26.67	XX	0.308	7.82	2.441	3.63
1"	25	1.315	33.40	5/5S	0.065	1.65	0.868	1.29
1"	25	1.315	33.40	10/10S	0.011	2.77	1.404	2.09
1"	25	1.315	33.40	STD 40/40S	0.133	3.38	1.679	2.50
1"	25	1.315	33.40	XS 80/80S	0.179	4.55	2.172	3.23
1"	25	1.315	33.40	160	0.250	6.35	2.844	4.23
1"	25	1.315	33.40	XX	0.358	9.09	3.659	5.45
1-1/4"	32	1.660	42.16	5/5S	0.065	1.65	1.107	1.65
1-1/4"	32	1.660	42.16	10/10S	0.109	2.77	1.806	2.69
1-1/4"	32	1.660	42.16	STD 40/40S	0.140	3.56	2.273	3.38
1-1/4"	32	1.660	42.16	XS 80/80S	0.191	4.85	2.997	4.46
1-1/4"	32	1.660	42.16	160	0.250	6.35	3.765	5.60
1-1/4"	32	1.660	42.16	XX	0.382	9.70	5.214	7.76
1-1/2"	40	1.900	48.26	5/5S	0.065	1.65	1.274	1.90
1-1/2"	40	1.900	48.26	10/10S	0.109	2.77	2.085	3.10
1-1/2"	40	1.900	48.26	STD 40/40S	0.145	3.68	2.718	4.05
1-1/2"	40	1.900	48.26	XS 80/80S	0.200	5.08	3.631	5.40
1-1/2"	40	1.900	48.26	160	0.281	7.14	4.859	7.23
1-1/2"	40	1.900	48.26	XX	0.400	10.16	6.408	9.54

**FIGURE 302.1
STANDARD WEIGHT IRON PIPE SIZES**

all systems be maintained in proper working condition. As a side note, the term suffer is an archaic term still used in legal expressions meaning "allow."

304.0 Connections to Plumbing System Required.

304.1 General. Plumbing fixtures, drains, appurtenances, and appliances, used to receive or discharge liquid wastes or sewage, shall be connected properly to the drainage system of the building or premises, in accordance with the requirements of this code.

Anything draining into the building drainage system must be connected in compliance with the other sections of the code. For example, Chapter 4 details how fixtures are connected to the drainage system; Chapter 7 explains how other drainage piping and systems are connected together; and Chapter 8 provides methods for indirect waste piping to drain to the drainage system. **Figure 304.1** shows a typical installation in which all plumbing fixtures are installed according to code requirements.

305.0 Damage to Drainage System or Public Sewer.

305.1 Unlawful Practices. It shall be unlawful for a person to deposit, by any means whatsoever, into a plumbing fixture, floor drain, interceptor, sump, receptor, or device, which is connected to a drainage system, public sewer, private sewer, septic tank, or cesspool, any ashes; cinders; solids; rags; inflammable, poisonous, or explosive liquids or gases; oils; grease; or any other thing whatsoever that is capable of causing damage to the drainage system or public sewer.

It is simply common sense that any of the items listed in Section 305.1 should not be placed into the drainage system; however, common sense is not always used and some unscrupulous individuals may try to do the convenient thing and deposit these materials into the drainage system. Therefore, Section 306.1 requires pretreatment of waste containing these items.

There are many ways to prevent the entry of materials or

GENERAL REGULATIONS

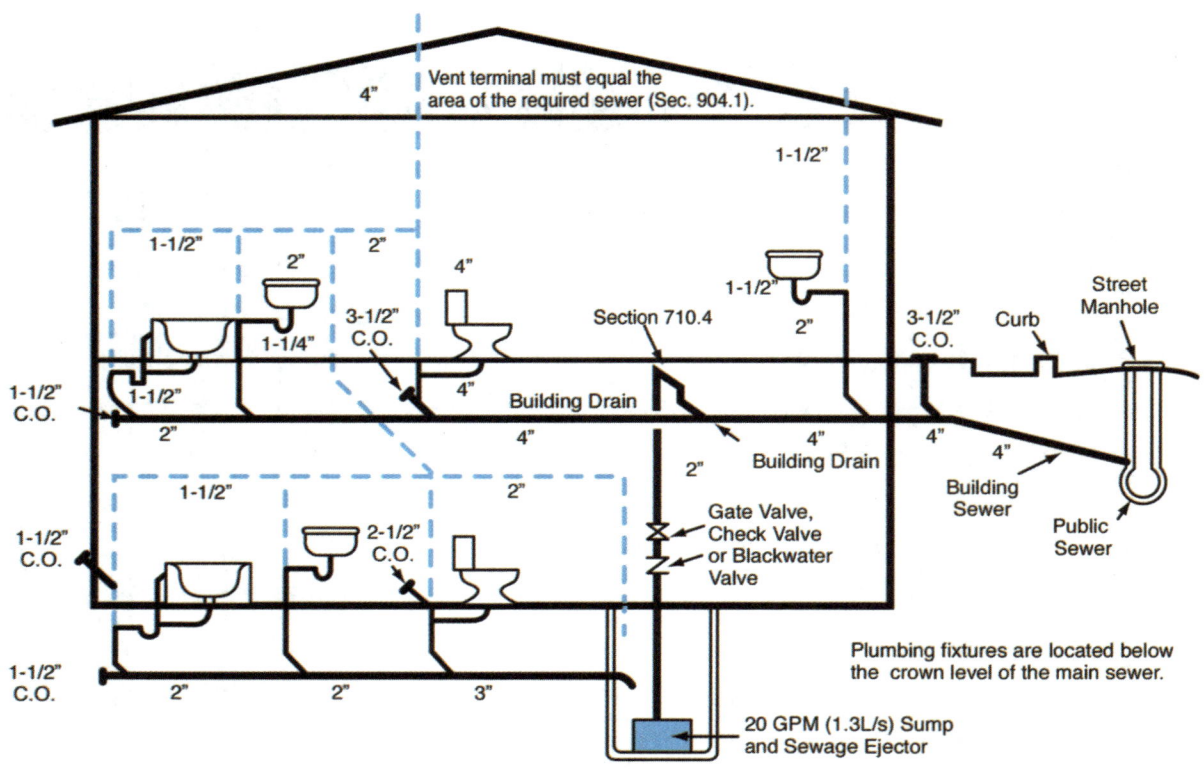

FIGURE 304.1
ALL FIXTURES CONNECTED TO DRAINAGE SYSTEM

products into the drainage system that would damage the system or public sewer. It is important to know what is going to be discharged into the drainage system. For example, when there are chemical, industrial or food processing within the premises, it must be determined if the products in the liquid waste may:

1. Cause blockage of sewer piping;
2. Cause corrosion or destruction of piping;
3. Contain flammable materials that may cause fires or explosions;
4. Contain products that may affect any form of life; or
5. Contain radioactive materials.

Methods to intercept the above materials must be used. Suspended solids can be removed from liquids by using catch basins, sand traps, or interceptors prior to entry into the drainage system. Floating solids, such as wood chips and leaves, can be prevented from entering the drainage system by installing a quarter bend that faces downward inside a catch basin as seen in **Figure 305.1**.

Chapter 8 contains requirements for chemical waste interception, and Chapter 10 contains requirements for grease, sand and oil interception. Some jurisdictions require that an accessible inspection sampling box be installed on the drainage system downstream from the last fixture inlet on a process, chemical or industrial drainage system (see **Figure 306.1**). This sampling box is located on private property and connected in a manner that prevents regular sanitary sewage from passing through it.

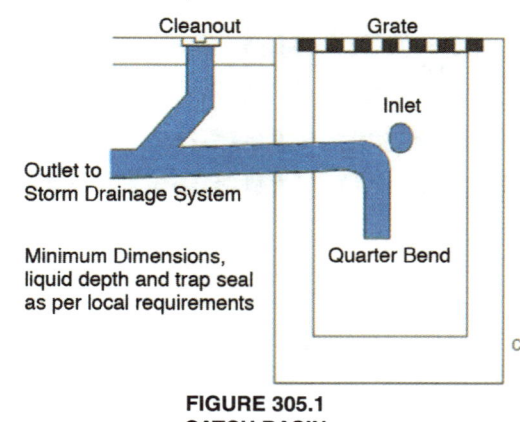

FIGURE 305.1
CATCH BASIN

306.0 Industrial Wastes.

306.1 Detrimental Wastes. Wastes detrimental to the public sewer system or detrimental to the functioning of the sewage treatment plant shall be treated and disposed of as found necessary and directed by the Authority Having Jurisdiction.

🔧 These conditions are usually regulated by both the plumbing official concerned about the drainage piping system and the public works official concerned about the operation of the public sewage treatment plant. The common practice is to intercept, neutralize or provide adequate treatment to liquid wastes on private property before such wastes enter any part of the drainage system or drain to an outside waterway, such as a stream, river, or lake (see **Figure 306.1**).

GENERAL REGULATIONS

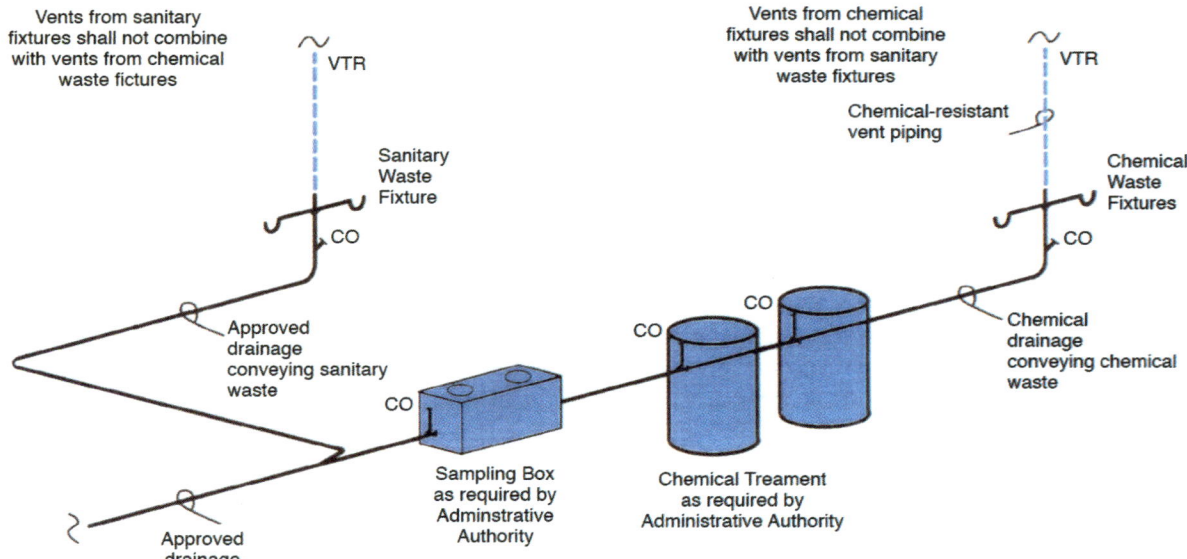

FIGURE 306.1
PRETREATMENT WITH SAMPLING BOX

Any drainage piping used to convey industrial wastes prior to treatment must be carefully selected and approved. Approved materials include lead, glass, extra-strength vitrified clay, stainless steel and certain types of plastic. Most pipe manufacturers will provide technical data that show acceptable concentrations of chemicals and temperature limits for the various products. Vent piping for a chemical waste system must be of the same or other approved material required for chemical wastes.

Petroleum products must not enter the drainage system. The main concern is the danger of combustion within the drainage piping. Also, a mix of sand and oil can form a solid mass that could cause blockages within the drainage piping system.

Some wastes may contain chemical content that may be detrimental to the functioning of the sewage treatment plant. If organic bacteria are killed by chemicals in the wastewater, the sewage treatment plant will be impaired in its ability to function.

Another concern is that chemicals may cause severe corrosion to metal piping or machinery associated with the sewage treatment system. Acid solutions can cause damage to concrete pipe or concrete tanks. In such cases, the AHJ will require an appropriate method of disposal.

306.2 Safe Discharge. Sewage or other waste from a plumbing system that is capable of being deleterious to surface or subsurface waters shall not be discharged into the ground or a waterway unless it has first been rendered safe by some acceptable form of treatment in accordance with the Authority Having Jurisdiction.

307.0 Location.

307.1 System. Except as otherwise provided in this code, no plumbing system, drainage system, building sewer, private sewage disposal system, or parts thereof shall be located in a lot other than the lot that is the site of the building, structure, or premises served by such facilities.

A "plumbing system" (as defined in UPC Section 218.0) includes water piping, fuel-gas piping and other plumbing components, none of which may cross a property line except to enter a public or private utility easement. Ownership of adjacent properties is not a consideration; crossing a property line requires a legally recorded easement in all cases. Also a townhouse (duplex) must have separate utilities (sewer, water, gas) for each unit, such that it is totally independent of any other adjacent dwelling.

307.2 Ownership. No subdivision, sale, or transfer of ownership of existing property shall be made in such manner that the area, clearance, and access requirements of this code are decreased.

Whenever a sewer or private disposal field is constructed on an adjacent lot under common ownership, the properties must be consolidated and recorded at the office of the county recorder (see **Figure 307.2a**). When a property is legally subdivided, all clearances of buildings, septic tanks, and disposal fields must be maintained from property boundaries as required by UPC Table H 101.8, Location of Sewage Disposal Systems (see **Figure 307.2b**). Some jurisdictions maintain the maximum ratio of buildings related to the area of the property in the zoning bylaws. Sometimes there is a minimum land area requirement for the installation of a private disposal system.

308.0 Improper Location.

308.1 General. Piping, fixtures, or equipment shall not be so located as to interfere with the normal use thereof or with the normal operation and use of windows, doors, or other required facilities.

One would think that a regulation such as this would not be necessary; however, there are installers who try to take

GENERAL REGULATIONS

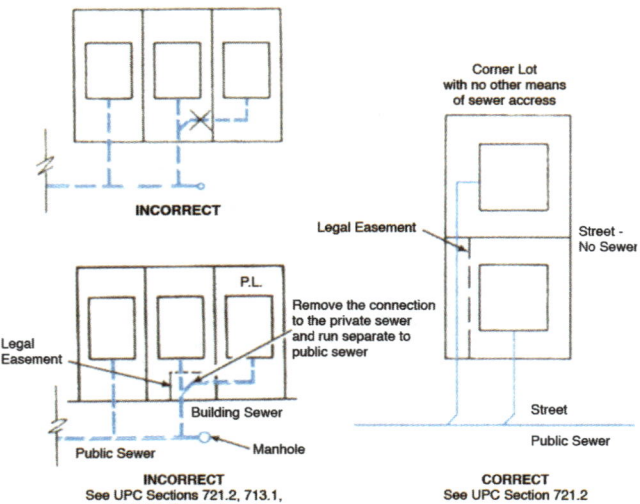

**FIGURE 307.2A
CORRECT AND INCORRECT LOCATIONS OF SEWER SYSTEMS**

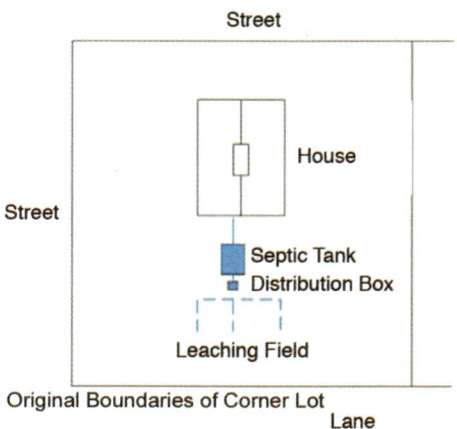

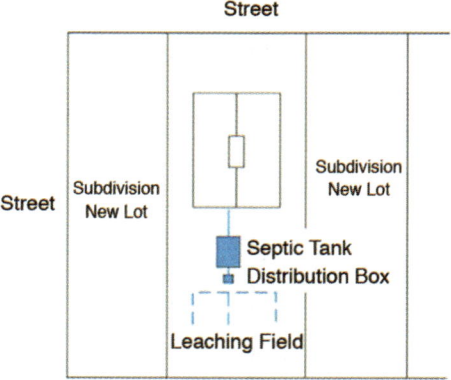

**FIGURE 307.2B
ORIGINAL AND SUBDIVIDED PROPERTY WITH SEPTIC SYSTEM**

shortcuts, save money or who just do not know any better that install fixtures, appliances and piping improperly so that the use of doors and windows are restricted. This commonly occurs during renovations or conversions of older buildings. See **Figure 308.1** for a drawing of a prohibited installation in which the piping blocks the door, hatch, and stairs.

309.0 Workmanship.

Workmanship is an important issue that concerns quality trade practices. The goal here is not only neat and professional-appearing plumbing, minimum amount of labor, and sustainability, but also code compliance. Evaluation of workmanship must never be allowed to become subjective. As it turns out, work that is fully compliant with all of the Code's provisions will exhibit good "workmanship" in both looks and function. Good workmanship results in the safe and satisfactory operation of a plumbing system for an expected duration of time, with maximum economy, and minimum maintenance.

The Code expects workmanship to be in accordance with accepted engineering practices as well as the manufacturer's installation instructions. Hiding imperfections or taking short cuts such as not taking the time to remove the rough edges from pipe ends are unacceptable practices and reflect poor workmanship.

309.1 Engineering Practices.
Design, construction, and workmanship shall be in accordance with accepted engineering practices and shall be of such character as to secure the results sought to be obtained by this code.

309.2 Concealing Imperfections.
It is unlawful to conceal cracks, holes, or other imperfections in materials by welding, brazing, or soldering or by using therein or thereon paint, wax, tar, solvent cement, or other leak-sealing or repair agent.

Temporary and unorthodox fixes are not acceptable, especially when trying to hide an imperfection. Dam-

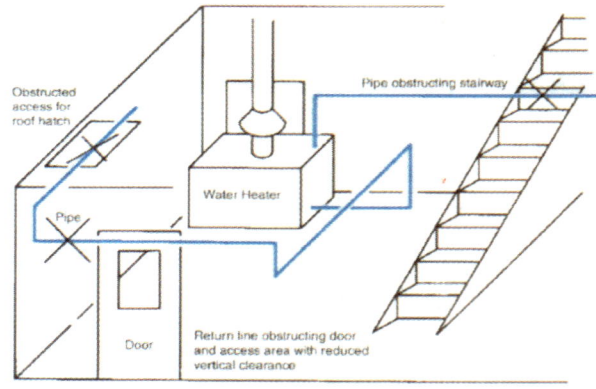

**FIGURE 308.1
IMPROPER LOCATION OF PIPING**

aged or defective pipe, fixtures and fittings must be replaced. These imperfections can be the result of damage during transit, defects in manufacturing, sand holes, damage due to freezing, corrosion of the system, or damage during installation.

309.3 Burred Ends.
Burred ends of pipe and tubing shall be reamed to the full bore of the pipe or tube, and chips shall be removed.

🔧 The presence of burrs in the pipe or tubing of a plumbing system is serious. Chips and burrs must be removed from piping systems to prevent blockage, flow restrictions, pressure drop, excessive turbulence and obstructions to control valves. Also remember that all pipe ends, no matter what the material, must be reamed or deburred. If the cutting wheels of tube cutters are dull, the pipe or tube will be excessively burred. The same condition will result if excessive pressure is placed on the cutting wheel. See **Figure 309.3** for an example of a reaming tool that removes burrs.

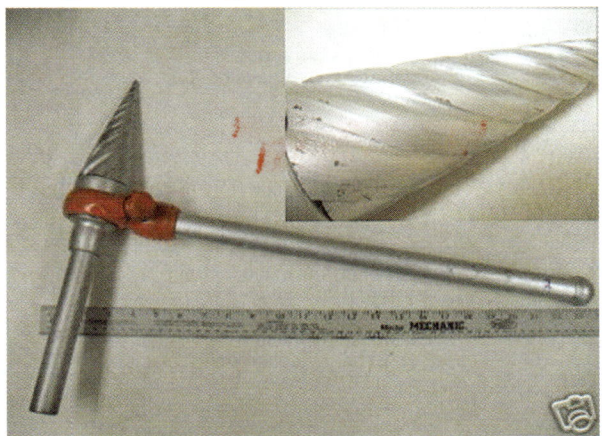

**FIGURE 309.3
PIPE SPIRAL REAMING TOOL**

309.4 Installation Practices. Plumbing systems shall be installed in a workmanlike manner which is in accordance with this code, applicable standards, and the manufacturer's installation instructions. All materials shall be installed so as not to adversely affect the systems and equipment or the structure of the building, and in compliance with all laws and other provisions of this code. All plumbing systems shall be in accordance with construction documents approved by the Authority Having Jurisdiction.

🔧 In this section, there are three guidelines to use for proper installation of the plumbing system. If a conflict occurs between the standards, instructions, or code the most stringent of the three provisions must be used. The example used in the commentary for Section 301.2.2 concerning PEX-AL PEX pipe is a perfect one. The code requires a minimum of 18 inches from the water heater for installation of this piping but the standard does not. One must use the more stringent provision. In this case, the more stringent provision is contained in the code, which supersedes the standard. This section also ensures that all work installed shall not adversely affect the existing building system or structure even if it was installed to the code, standard, and manufacture instruction. All changes to plans and specifications must be submitted to the Authority Having Jurisdiction. In addition, the term "workmanlike manner" is used to describe the quality of work that would be performed by a worker of average skill and ability or in a manner that is in line with other contractors in the same community.

309.5 Sound Transmission. Plumbing piping systems shall be designed and installed in conformance with sound limitations as required in the building code.

310.0 Prohibited Fittings and Practices.

310.1 Fittings. No double hub fitting, single or double tee branch, single or double tapped tee branch, side inlet quarter bend, running thread, band, or saddle shall be used as a drainage fitting, except that a double hub sanitary tapped tee shall be permitted to be used on a vertical line as a fixture connection.

🔧 **Figure 310.1** shows examples of the prohibited fittings and practices addressed in this section. These fittings are prohibited because either they are not true drainage fittings with contoured edges to allow proper flow into and down the pipe or they contain obstructions that may cause blockages.

The "tee branch" fittings referenced here, such as the cast iron twin tapped tee, are not "sanitary tees." Essentially, a flat surface was cast onto the walls of these branch or fixture fittings, then this flat surface was drilled and tapped to accommodate threaded pipe or trap adapters. Because the branch opening on these fittings did not extend beyond the walls of the fitting (as is the case with a sanitary branch), any pipe or fitting that was screwed into these sidewall taps extended into the barrel of the fittings, thus obstructing both air and waste flow. These fittings were prohibited many years ago.

310.2 Drainage and Vent Piping. No drainage or vent piping shall be drilled and tapped for the purpose of making connections thereto, and no cast-iron soil pipe shall be threaded.

🔧 The thickness or material in the wall of any type of drainage or vent piping is not sufficient enough to form a tapped thread. Approved adapter fittings are required whenever drainage pipe is joined with dissimilar materials.

310.3 Waste Connection. No waste connection shall be made to a closet bend or stub of a water closet or similar fixture.

🔧 A closet bend is an approved fitting that may include an integral or attached stub receiving vertically discharged waste from a water closet and changing its direction of flow, usually to horizontal. The closet bend is a trap arm up to the point of fixture vent connection. A waste connection shall not be made to a trap arm before its vent. The waste flow that would occur in the trap arm from this installation could cause venting problems. **Figure 310.3** shows an illustration of a prohibited closet bend with side inlet.

310.4 Use of Vent and Waste Pipes. Except as hereinafter provided in Section 908.0 through Section 911.0, no vent pipe shall be used as a soil or waste pipe, nor shall a soil or waste pipe be used as a vent. Also, single-stack drainage and venting systems with unvented branch lines are prohibited.

🔧 The UPC's basic principle for waste and vent is that the two shall never comingle. In other words, the vent pipe shall never be used for draining a fixture, and the drain pipe shall never be used for venting a fixture. There are four basic exceptions to this principle. Sections 908.0 Wet Vent-

GENERAL REGULATIONS

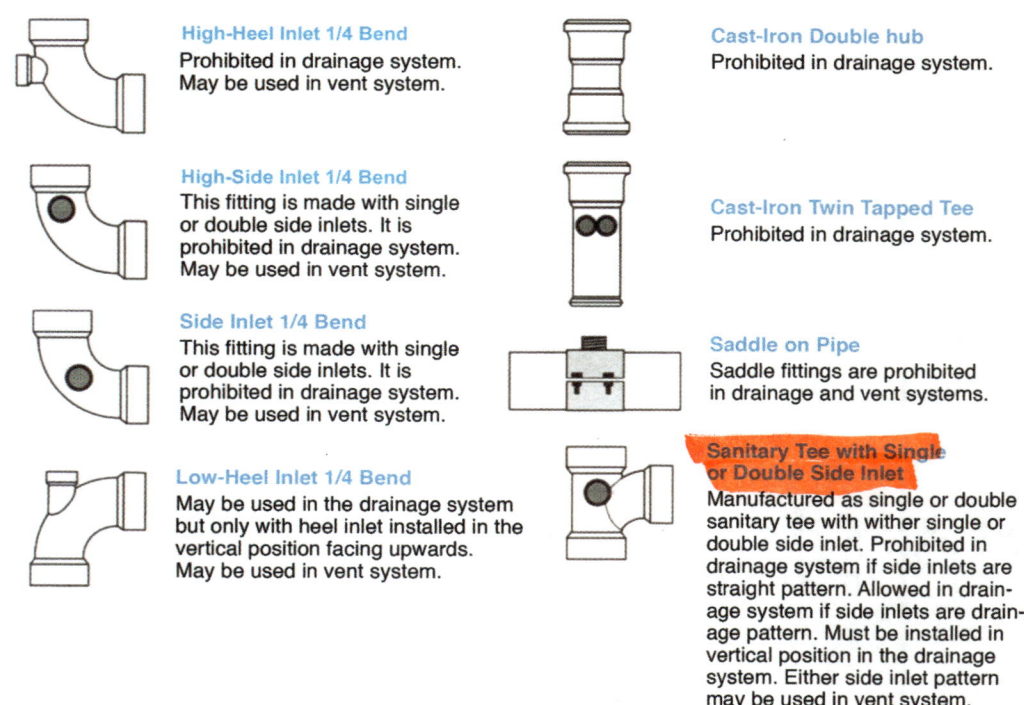

**FIGURE 310.1
PROHIBITED OR RESTRICTED FITTINGS**

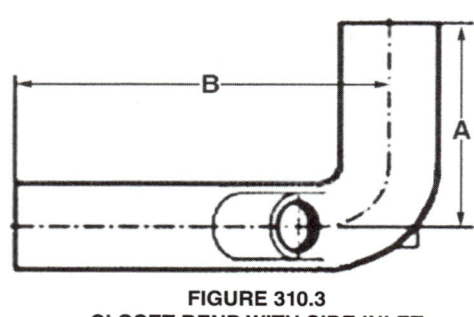

**FIGURE 310.3
CLOSET BEND WITH SIDE INLET**

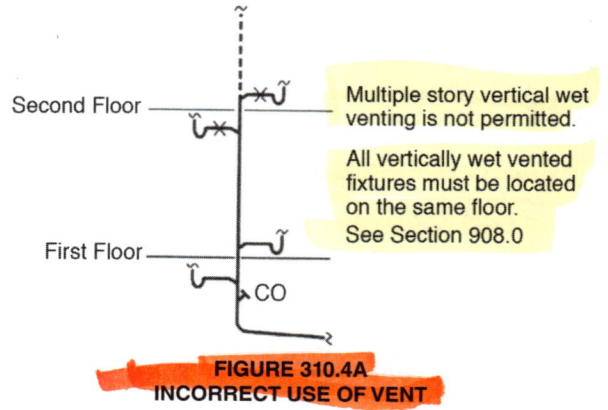

**FIGURE 310.4A
INCORRECT USE OF VENT**

ing, 909.0 Special Venting for Island Fixtures, 910.0 Combination Waste and Vent Systems, and 911.0 Circuit Venting in Chapter 9 discuss "special" venting systems that utilize different configurations and requirements, such as increased sizes and placement of fittings to accommodate venting. See Chapter 9 for explanations and illustrations. See **Figure 310.4a** for an example of incorrect use of a second floor vertical waste pipe for a first floor vent pipe.

Single-stack systems, such as the Philadelphia Stack System and the Sovent System, utilize one vertical stack for both sanitary waste and venting. Branches connecting to these systems carry waste from fixtures that are not individually vented. The design of these systems (fixtures without individual trap vents) does not meet the requirements of the code and therefore they are prohibited.

Single-stack systems such as those mentioned above have been used in the United States and around the world. The code recognizes this fact by placing a method of single-stack venting in Appendix C. The system must be designed by an engineering professional and approved by the AHJ. Other methods may be used but must adhere to the requirements of Sections 301.3 and 301.5. See **Figure 310.4b** for an example of one type of single-stack venting system.

310.5 Obstruction of Flow. No fitting, fixture and piping connection, appliance, device, or method of installation that obstructs or retards the flow of water, wastes, sewage, or air in the drainage or venting systems, in an amount exceeding the normal frictional resistance to flow, shall be used unless it is indicated as acceptable in this code or is approved in accordance with Section 301.2 of this code. The enlargement of a 3 inch (80 mm) closet bend or stub to 4 inches (100 mm) shall not be considered an obstruction.

Any installation that causes an obstruction to occur in any part of the plumbing system is prohibited. The obstruc-

GENERAL REGULATIONS

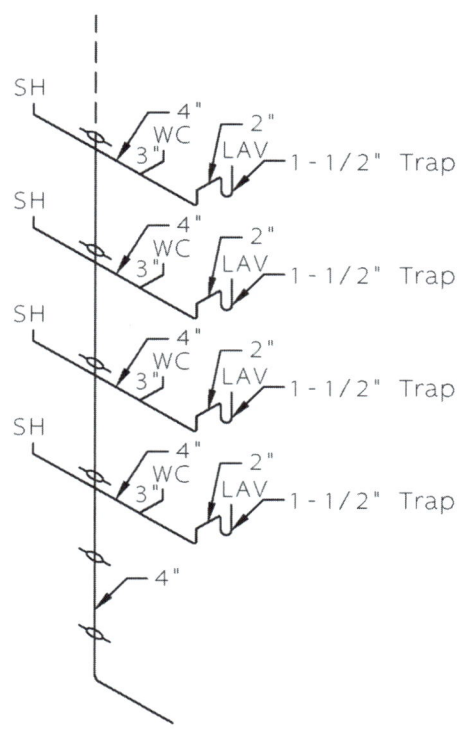

FIGURE 310.4B
AN ALTERNATE ENGINEERED SINGLE-STACK DRAINAGE SYSTEM (FROM ILLUSTRATED PLUMBING CODES DESIGN HANDBOOK, ASPE)

tion will cause problems such as reduction in flow, increased turbulence or pressures and blockages in the system, to name just a few. See **Figure 310.5a** for approved and prohibited installations that can cause a reduction in flow from fixtures and appliances.

An example of an approved fitting that intentionally creates an obstruction is a flow control device used on hydro-mechanical grease interceptors. The flow control is used to decrease flow at a designated amount for grease extraction. There shall be no solids in this effluent to cause a blockage when the flow control is used.

A four-inch by three-inch (100 mm by 80 mm) closet bend is not considered an obstruction in the direction of flow because the diameter of the outlet of the closet bowl normally varies between 2 ⅜ inches and 2 ¾ inches. A 4-inch closet ring is normally used to secure the bowl to the closet bend. The purpose for using a four-inch diameter closet ring is for stability when securing both the water closet bowl and the closet ring to the floor. It is imperative that there is no contact between the water closet outlet (horn) and the closet ring or the closet bend inlet. This would create an obstruction and would continuously cause a blockage in the system.

To ensure no contact between the water closet horn and its closet ring or closet bend, the inlet opening on a closet bend is increased two pipe sizes from 3 inches to 4 inches. The reduction in the diameter is in the vertical section of the pipe and does not cause an obstruction. It is also not necessary to maintain a four-inch diameter pipe to drain one water closet. The trap arm and drain may be 3 inches in size.

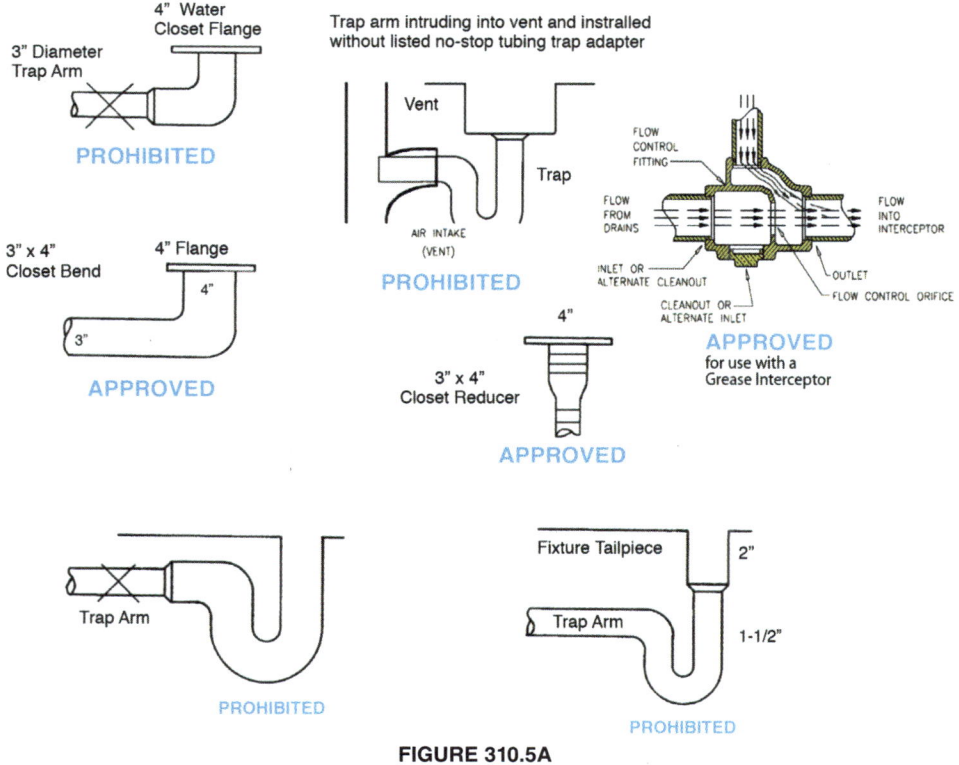

FIGURE 310.5A
APPROVED AND PROHIBITED INSTALLATIONS

GENERAL REGULATIONS

Offset closet flanges may obstruct flow if it does not comply with the applicable standard required by the UPC. For example, Table 701.2 requires PVC fittings to comply with ASTM D2665. In that standard, it references ASTM D3311, which is the standard for DWV plastic fittings patterns. The pattern for an offset closet flange must be free of ledges and corners that would obstruct flow (see **Figure 310.5b**).

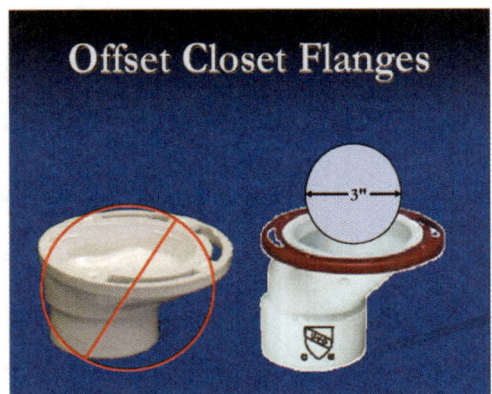

FIGURE 310.5B
OFFSET CLOSET FLANGES

310.6 Dissimilar Metals. Except for necessary valves, where intermembering or mixing of dissimilar metals occurs, the point of connection shall be confined to exposed or accessible locations.

Historically, mixing of metals in drainage systems has not been a problem, as evidenced by the traditional commingling of lead, tin, copper, brass and ferrous metals in the manufacturing and installation of traps, ferrules, cleanouts, drains, caulked and wiped joints, and so forth. There is, however, a concern when mixing dissimilar metals, such as copper and steel piping, in a water system. The connection creates the possibility of galvanic action occurring, which could cause corrosion of the system and leaks.

Galvanic action occurs because the difference in electrical potential between the two metals will always cause flow from the anode to the cathode when there is an electrolyte to facilitate this flow. In the case of connections between steel and copper water pipe, the steel becomes the anode, the copper becomes the cathode and the water serves as an electrolyte in this process of ionic transfer. The steel pipe sacrifices itself to the electrolytic process, eventually resulting in pipe failure and potential property damage. For this reason the transition shall be in an exposed and accessible area so that if a problem develops it will be seen and easily repaired (See **Figure 310.6**).

310.7 Direction of Flow. Valves, pipes, and fittings shall be installed in correct relationship to the direction of flow.

This requirement is necessary to avoid damage to the seats and discs in valves and to reduce noise or vibration in the piping system. Some appurtenances, such as strainers or pressure-reducing valves, would not function if the flow was reversed. The manufacturer usually places an arrow on the device to indicate the direction of flow. A check valve would neither operate in a reversed position, nor work in a vertical position, considering it was not spring-loaded and designed for that position. Globe valves are directional flow valves, designated by an arrow located on the body of the valve. Gate valves are not directional flow valves. The manufacturer's instructions, catalogue, or product literature will usually indicate if the device should be mounted either horizontally, or vertically and the proper direction of flow (See **Figure 310.7**).

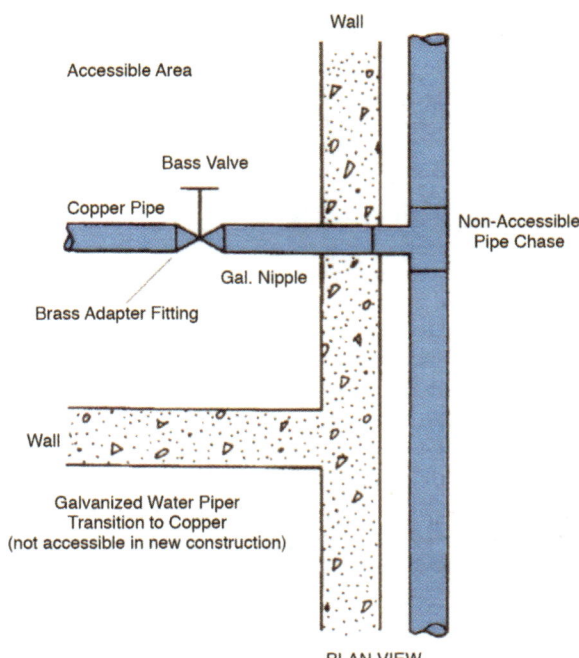

FIGURE 310.6
DISSIMILAR METALS EXPOSED

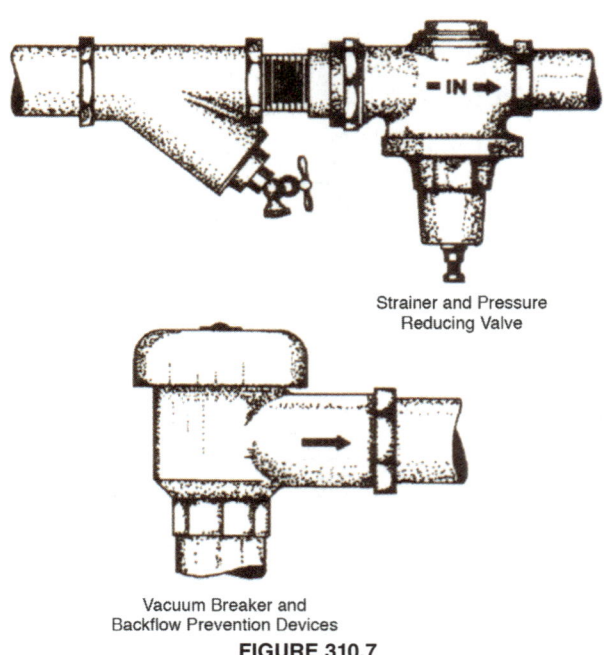

FIGURE 310.7
DIRECTION OF FLOW

310.8 Screwed Fittings. Screwed fittings shall be ABS, cast-iron, copper, copper alloy, malleable iron, PVC, steel, or other approved materials. Threads shall be tapped out of solid metal or molded in solid ABS or PVC.

🔧 Threaded pipe for plumbing systems is always Schedule 40 IPS size or above. This thickness allows enough material so that the threads may be cut or molded into the pipe and to remain strong enough to complete the joint and hold under pressure.

311.0 Independent Systems.

311.1 General. The drainage system of each new building and new work installed in an existing building shall be separate and independent from that of any other building, and, where available, every building shall have an independent connection with a public or private sewer.

Exception: Where one building stands in the rear of another building on an interior lot, and no private sewer is available or can be constructed to the rear building through an adjoining court, yard, or driveway, the building drain from the front building shall be permitted to be extended to the rear building.

🔧 The reason for this requirement is that if there is a blockage in the drainage system of one building, it will not affect the other buildings on the system. If each building has its own independent connection to the sewer, then only a blockage in the sewer itself will affect other buildings connected to that sewer. **Figure 311.1** shows independent sewer systems and their connections.

The exception allows a building drain to be extended to a new building on an interior lot when there is no alternative. The preferred method would be an independent building sewer to go around the front building. However, if this is impossible because of structural conditions, then an extension of the building drain from the front building to the rear building or lot will be allowed.

312.0 Protection of Piping, Materials, and Structures.

312.1 General. Piping passing under or through walls shall be protected from breakage. Piping passing through or under cinders or other corrosive materials shall be protected from external corrosion in an approved manner. Approved provisions shall be made for expansion of hot water piping. Voids around piping passing through concrete floors on the ground shall be sealed.

🔧 Care should always be taken to ensure that plumbing passing through or under a wall will be protected from the weight or the expansion and contraction of the wall, which may cause damage to the pipe (see **Figure 312.1**).

Corrosion of sewer or drainage piping may occur when the pipe is placed in soil containing cinders or other corrosive materials. Cinders from coal fires often contain sulfur. The sulfur dissolves in groundwater to form sulfuric acid, which is corrosive to many piping materials.

There are several methods to provide for the expansion of hot water piping systems such as expansion loops, expansion joints, swing joints, and expansion coils. Consult water supply and hydronic manuals for methods of piping for expansion and contraction.

Voids around piping passing through concrete floors on the ground must be sealed in a manner that is durable and waterproof to protect the interior of the building. This will also prevent the entry of vermin and insects into the room or building.

312.2 Installation. Piping in connection with a plumbing system shall be so installed that piping or connections will not be subject to undue strains or stresses, and provisions shall be made for expansion, contraction, and structural settlement. No plumbing piping shall be directly embedded in concrete or masonry. No structural member shall be seriously weakened or impaired by cutting, notching, or otherwise, as defined in the building code.

🔧 The installation of the plumbing system will require piping through walls and floors, in the ground and under footings, and in many other conditions that could exert pressure or strain on the piping. Care must be taken so that the piping is installed to move freely, preventing structural set-

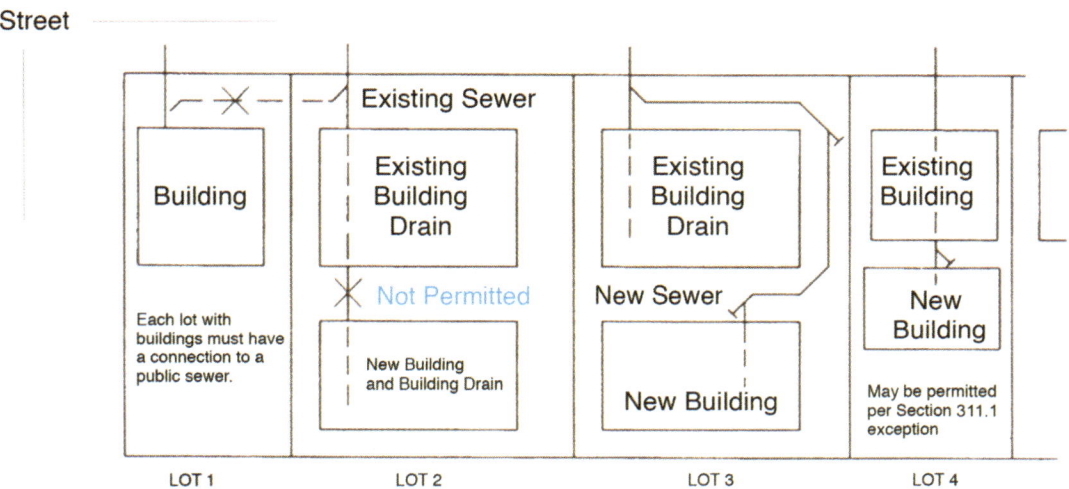

FIGURE 311.1
INDEPENDENT SEWER CONNECTIONS AND EXCEPTION

GENERAL REGULATIONS

tlement or expansion and contraction of piping from detrimentally affecting the system. Often, failure of the piping system is caused by the pipe being tightly secured in hangers, straps, or the structure itself. Expansion and contraction of either the pipe or the structure causes the pipe to crack or to slip out of its fittings.

Pipe passing perpendicularly through or parallel within a concrete or masonry wall or floor should always pass through a sleeve or box per Section 312.10. The intent is that piping shall not be located within or pass through concrete walls or slabs where a slight shifting or settlement can result in a major plumbing pipe failure. A pipe chase must be provided in the concrete wall when piping is to be recessed. Regardless of the method used to pass pipe through these conditions there must be provisions to allow movement of the pipe. This includes, but is not limited to, slabs, walls or footings. In no case shall the building impose a load on the pipe that could result in breakage or leaks in the piping system. **Figure 312.2** shows some recommended and prohibited piping installations.

When pipe passes or is placed within walls, the wall or structure should not be weakened by notching or cutting. The building code has provisions for how much material can be removed from the wall or stud before it is weakened to the point of failure.

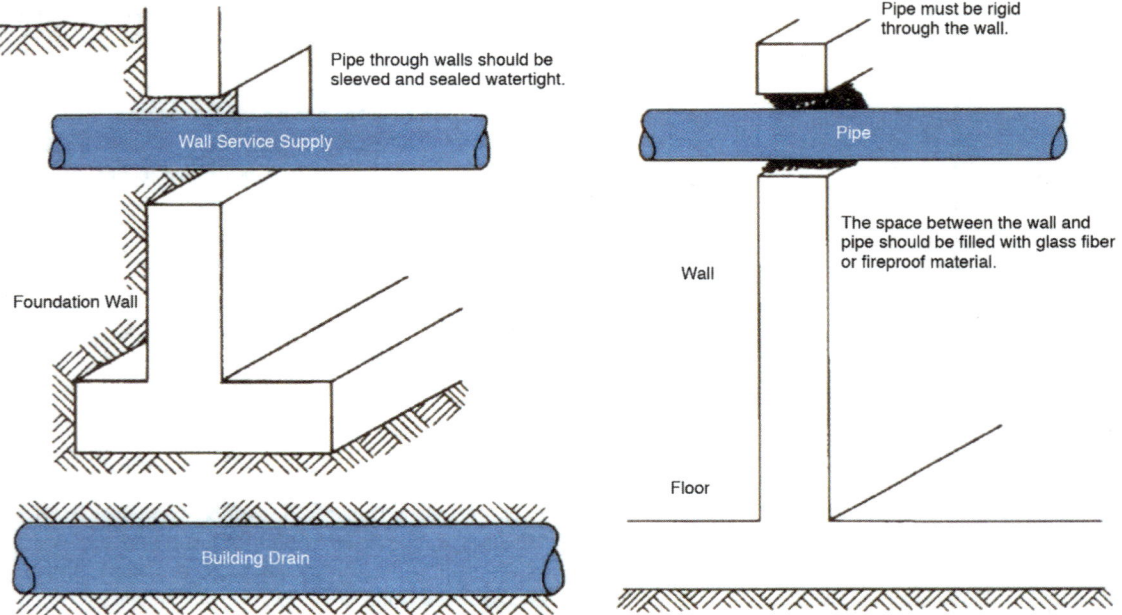

FIGURE 312.1
PROTECTION OF PIPING THROUGH WALLS

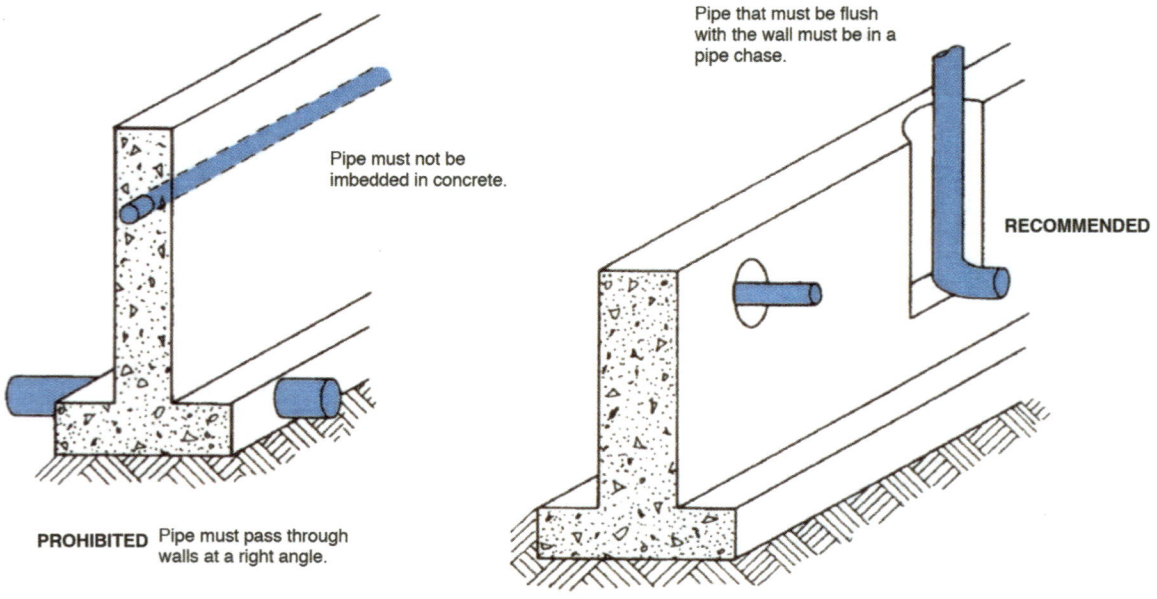

FIGURE 312.2
PROPER PLACEMENT OF PIPING WITHIN WALLS

312.3 Building Sewer and Drainage Piping. No building sewer or other drainage piping or part thereof, constructed of materials other than those approved for use under or within a building, shall be installed under or within 2 feet (610 mm) of a building or structure, or less than 1 foot (305 mm) below the surface of the ground.

🔧 The standard for a piping material will define its use for either within the building (the building drain) or outside the building or structural slab (the sewer or building sewer). Materials that may be used under the building will be strong enough to withstand the settlement or weight of the building. Piping materials such as cast iron and Schedule 40 PVC or ABS may be used for the building drain. **Figure 312.3** illustrates this requirement.

Sewers consisting of materials for use only outside of the building or structure, such as PVC SDR 35 or clay pipe, must be installed a minimum of two feet from the building or structure and one foot below grade. In this manner the piping will be protected from damage due to the building weight or settlement. If damage is likely to occur, vertical piping must also be protected.

312.4 Corrosion, Erosion, and Mechanical Damage. Piping subject to corrosion, erosion, or mechanical damage shall be protected in an approved manner.

🔧 Common methods of protecting piping from corrosion above and below ground include PVC sleeving or wrapping, painting, asphalt coating, or enclosing the pipe within approved material. Approved machine (factory) wrapped piping may be required. Coating ferrous pipe by galvanizing is an approved method for protecting pipe above ground only [see Sections 609.3(1) and 701.2(1)].

Protection from mechanical damage usually requires the installation of sleeves, steel barricades and guards, or placing the pipe in recesses, pipe chases or gutters. Suspended ceilings or raceways may be required to protect overhead piping.

312.5 Protectively Coated Pipe. Protectively coated pipe or tubing shall be inspected and tested, and a visible void, damage, or imperfection to the pipe coating shall be repaired in an approved manner.

312.6 Freezing Protection. No water, soil, or waste pipe shall be installed or permitted outside of a building, in attics or crawl spaces, or in an exterior wall unless, where necessary, adequate provision is made to protect such pipe from freezing.

🔧 All plumbing piping must be protected from freezing. The required depth of cover over underground piping will be determined by the AHJ based on the minimum anticipated temperature, the type of soil and porosity and the average moisture content of the soil. These factors determine the average depth of frost penetration. It is necessary to protect piping installed in locations such as attics and crawl spaces that may be susceptible to freezing temperatures. Adequate protection may include piping insulation, building insulation or other means necessary to prevent freezing.

Figure 312.6 shows methods of providing such protection.

312.7 Fire-Resistant Construction. Piping penetrations of fire-resistance-rated walls, partitions, floors, floor/ceiling assemblies, roof/ceiling assemblies, or shaft enclosures shall be protected in accordance with the requirements of the building code and Chapter 14, "Firestop Protection."

🔧 Piping penetrations of fire-resistance-rated walls must return the wall to its fire-rated integrity prior to the penetration. Not doing so will weaken the building's ability to resist the spread of fire within the building. The requirements for this firestopping are contained in the UPC Chapter 14.

312.8 Waterproofing of Openings. Joints at the roof around pipes, ducts, or other appurtenances shall be made watertight by the use of lead, copper, galvanized iron, or other approved flashings or flashing material. Exterior wall openings shall be made watertight. Counterflashing shall not restrict the required internal cross-sectional area of the vent.

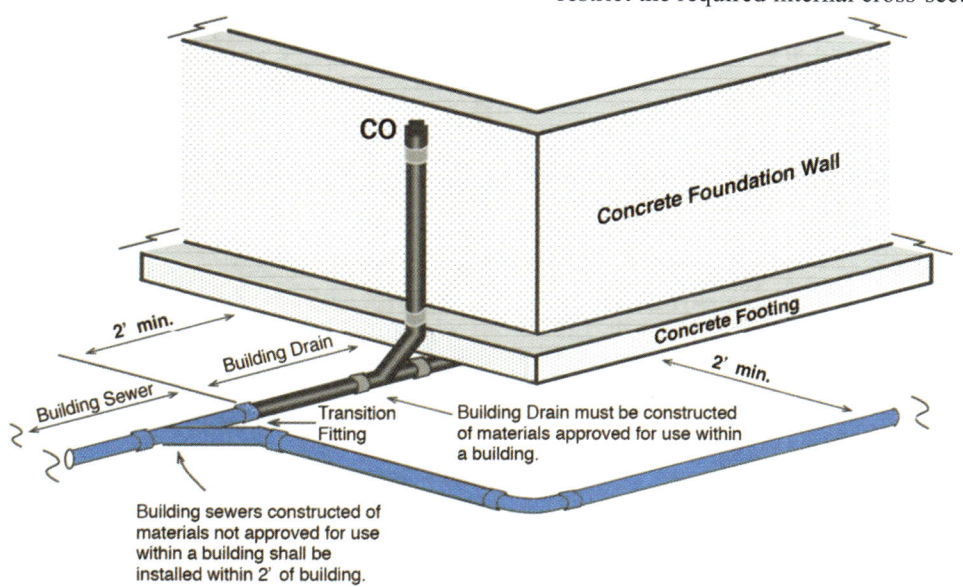

FIGURE 312.3
CLEARANCES FOR PIPING MATERIALS USED FOR BUILDING DRAIN AND BUILDING SEWERS

GENERAL REGULATIONS

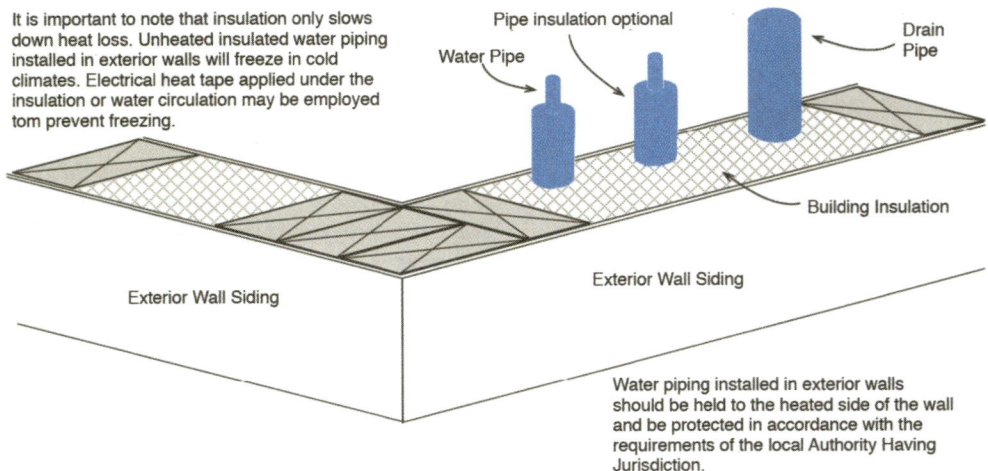

FIGURE 312.6
PROTECTION OF PIPING IN EXTERIOR WALLS

🔧 One of the biggest drawbacks for multiple vents and for the placement of air-conditioning units, exhaust fans or other items on a roof is the possibility of water leakage around the openings through the roof. The requirements for flashing materials to be used and the opening to be watertight are specified in this section of the UPC. The Authority Having Jurisdiction may also require that the penetration be tested to prove its soundness.

Counterflashing is a style of vent flashing that can be used to seal the opening around a vent pipe. This flashing fits over the vent pipe and covers the opening in the roof, but it also sheaths the pipe up to and over the top of the pipe. Part of the flashing riser is then in the topmost interior of the vent. As will be seen in Chapter 9, there is a requirement for certain cross-sectional area dimensions for venting. The counterflashing in the interior of the vent pipe shall not restrict this dimension. For example, if a cross-sectional area of 3.14 square inches is required for aggregate venting and the interior dimension of a two-inch vent pipe would be reduced when utilizing a counterflashing, then the pipe size would have to be increased (see **Figure 312.8**).

312.9 Steel Nail Plates. Plastic and copper or copper alloy piping penetrating framing members to within 1 inch (25.4 mm) of the exposed framing shall be protected by steel nail plates not less than No. 18 gauge (0.0478 inches) (1.2 mm) in thickness. The steel nail plate shall extend along the framing member not less than 1½ inches (38 mm) beyond the outside diameter of the pipe or tubing.

Exception: See Section 1210.3.3.

🔧 The protection of piping in the interior of walls is of utmost importance. The penetration of plastic and copper pipe by nails, screws or staples is very common. The methods provided here will ensure the pipe is not punctured either during construction or later by the owner of the building (see **Figure 312.9**).

Section 1210.3.3, Tubing in Partitions, discusses the requirements for fuel gas piping protection. These requirements will be more stringent because of the volatility of the contents of the pipe. The pipe manufacturer's installation requirements should also be taken into consideration. These requirements may be even more stringent and, if so, should therefore be followed instead.

FIGURE 312.9
USE OF NAIL PLATES

312.10 Sleeves. Sleeves shall be provided to protect piping through concrete and masonry walls, and concrete floors.

Exception: Sleeves shall not be required where openings are drilled or bored.

🔧 See **Figure 312.10**

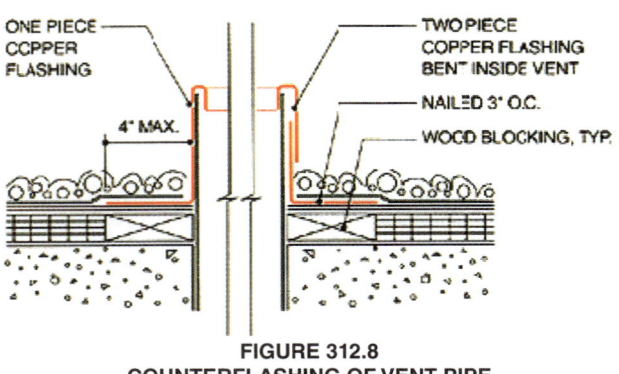

FIGURE 312.8
COUNTERFLASHING OF VENT PIPE

GENERAL REGULATIONS

**FIGURE 312.10
PIPING SLEEVE IN CONCRETE**

312.10.1 Building Loads. Piping through concrete or masonry walls shall not be subject to a load from building construction.

312.10.2 Exterior Walls. In exterior walls, annular space between sleeves and pipes shall be sealed and made watertight, as approved by the Authority Having Jurisdiction. A penetration through fire-resistive construction shall be in accordance with Section 312.7.

312.10.3 Firewalls. A pipe sleeve through a firewall shall have space around the pipe completely sealed with an approved fire-resistive material in accordance with other codes.

These provisions for piping sleeves are almost identical to the requirements in the previous sections concerning piping protection. The only difference is where walls are rated for fire resistance, the annular space between the pipe and the sleeve shall be filled with a fire resistant material having an F-rating as required by the building or fire code.

312.11 Structural Members. A structural member weakened or impaired by cutting, notching, or otherwise shall be reinforced, repaired, or replaced so as to be left in a safe structural condition in accordance with the requirements of the building code.

In Section 312.2 the code prohibits the weakening of a structural member. By the drilling, boring or notching, the weakening of plumbing wall supports is unavoidable. Because of this fact, this section adds the requirement to return the member to its original integrity. The building code will state the methods used to accomplish this.

312.12 Rodentproofing. Strainer plates on drain inlets shall be designed and installed so that no opening exceeds ½ of an inch (12.7 mm) in the least dimension.

312.12.1 Meter Boxes. Meter boxes shall be constructed in such a manner as to restrict rodents or vermin from entering a building by following the service pipes from the box into the building.

312.12.2 Metal Collars. In or on buildings where openings have been made in walls, floors, or ceilings for the passage of pipes, such openings shall be closed and protected by the installation of approved metal collars securely fastened to the adjoining structure.

312.12.3 Tub Waste Openings. Tub waste openings in framed construction to crawl spaces at or below the first floor shall be protected by the installation of approved metal collars or metal screen securely fastened to the adjoining structure with no opening exceeding ½ of an inch (12.7 mm) in the least dimension.

313.0 Hangers and Supports.

313.1 General. Piping, fixtures, appliances, and appurtenances shall be supported in accordance with this code, the manufacturer's installation instructions, and in accordance with the Authority Having Jurisdiction.

All parts of the plumbing system must be supported with reference to this code, the manufacturer's installation instructions, and any other requirements of the AHJ. Hangers must be located so that the pipe is supported independently of fixtures, appliances, and appurtenances; and fixtures, appliances, and appurtenances are supported independently of the pipe. Examples of appurtenances are meters, pressure-reducing valves, strainers, and backflow prevention devices.

313.2 Material. Hangers and anchors shall be of sufficient strength to support the weight of the pipe and its contents. Piping shall be isolated from incompatible materials.

Data regarding the weight of pipe is available in engineering manuals or manufacturers' literature. Plans and specifications must be checked to determine wall thickness of the pipe as per pipe schedule. The weight must include the contents of the pipe. This is calculated as the weight of water contained within a given size of pipe. The weight of flanges, bolts, valves and fittings (plus insulation) must also be calculated. The proper size of anchors, rods and hangers must be selected for the weight and size of the piping. See Table 313.6 for one method of selecting hanger rod size.

Where anchors and supports are attached to the structure, the structure must be of sufficient strength to support the additional load of the piping system, fixtures or appliances. This is most critical when renovating commercial and industrial buildings. **Figure 313.2** shows examples of anchors.

Pipe and its supporting hangers should be one compatible system. If the supports are of a different material than the pipe, care should be taken to isolate the pipe from incompatible materials; otherwise, corrosion may occur and cause pipe failure. The isolation may be something as simple as taping the pipe with 10- or 20-mil black tape or using an isolating hanger.

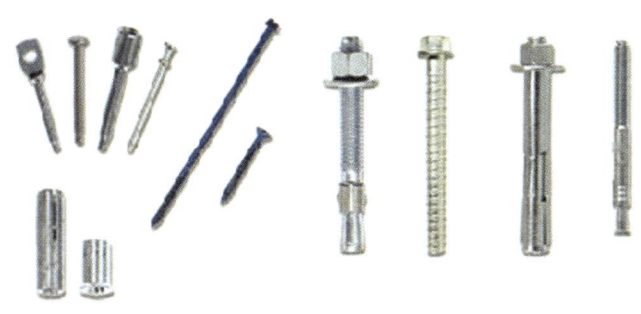

**FIGURE 313.2
ANCHOR SYSTEMS**

GENERAL REGULATIONS

313.3 Suspended Piping. Suspended piping shall be supported at intervals not to exceed those shown in Table 313.3.

🔧 Table 313.3 is a very extensive table containing the requirements for vertical and horizontal support for most of the piping materials and applications used in the plumbing system. Distances between supports will vary depending on the size, type of material, type of joint, and use of the pipe. For example, Schedule 40 PVC DWV pipe will have supports placed closer together than threaded steel water pipe.

The parameters of the table – distances between supports both vertical and horizontal – were developed based on the ability of the piping material and type of joint, when full, to remain level or plumb without sagging, and properly aligned. Therefore, the table should be followed at all times. The improper placement of the supports will cause the installation to fail inspection and worse, may cause the failure of the piping system itself. See **Figures 313.3a** and **313.3b** for examples of pipe hangers and supports.

One of the most important elements of Table 313.3 will be the notes at the bottom of the table. These five notes are very important to the proper support of piping materials in the plumbing system. Notes 1, 2, 3 and 4 pertain mostly to cast iron pipe. It is important to understand when installing either compression gasket or hubless cast iron pipe with shielded couplings (or no hub couplings) that if the pipe is over 4 feet (1,210 mm) in length, there should be one support at every joint (see **Figure 313.3c**).

An explanation for each note contained in Table 313.3 follows:

1. The hanger or support must be placed next to the pipe joint not more than 18 inches (457 mm) from the joint (see **Figure 313.3c**).
2. A brace to prevent horizontal movement of horizontal piping shall be placed at 40-foot intervals of piping. This is especially critical in areas of earthquake activity.

Buildings are now built to withstand extensive ground movement so the piping system must also be able to withstand this movement (see **Figure 313.3c**).

FIGURE 313.3A
PIPE HANGERS

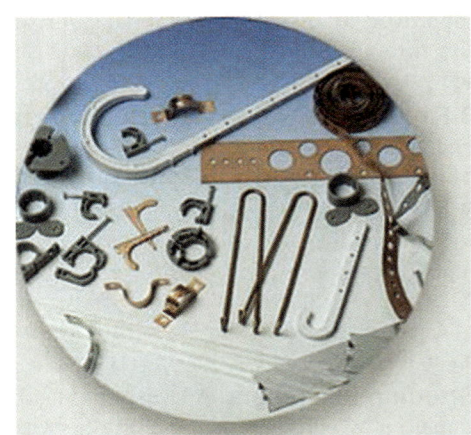

FIGURE 313.3B
PIPE SUPPORTS

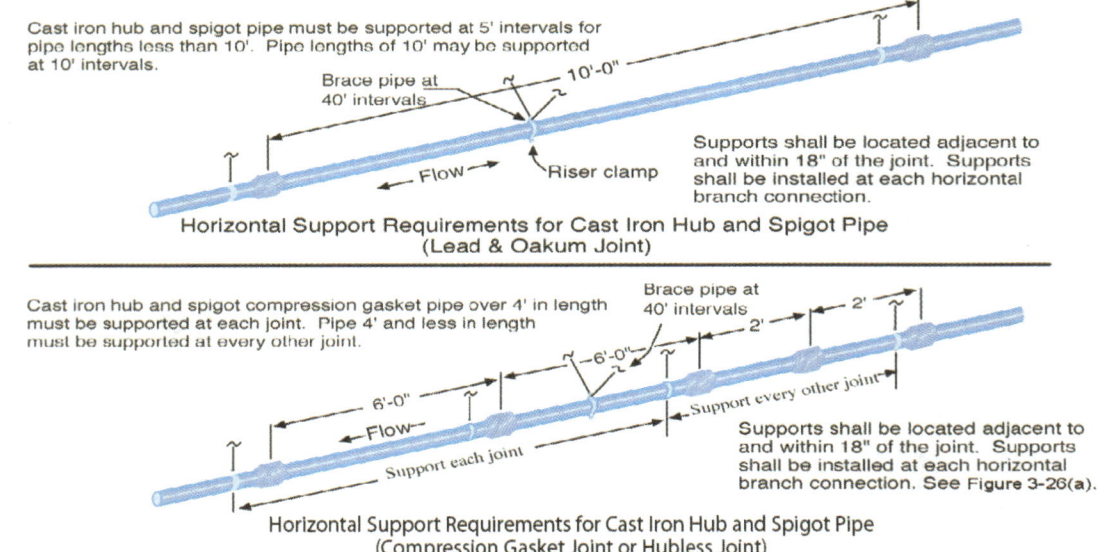

FIGURE 313.3C
CAST IRON SUPPORTS

GENERAL REGULATIONS

TABLE 313.3
HANGERS AND SUPPORTS

MATERIALS	TYPES OF JOINTS	HORIZONTAL	VERTICAL
Cast	Lead and Oakum	5 feet, except 10 feet where 10 foot lengths are installed[1, 2, 3]	Base and each floor, not to exceed 15 feet
Cast	Compression Gasket	Every other joint, unless over 4 feet then support each joint[1, 2, 3]	Base and each floor, not to exceed 15 feet
Cast-Iron Hubless	Shielded Coupling	Every other joint, unless over 4 feet then support each joint[1,2,3,4]	Base and each floor, not to exceed 15 feet
Copper & Copper Alloys	Soldered, Brazed, Threaded, or Mechanical	1½ inches and smaller, 6 feet; 2 inches and larger, 10 feet	Each floor, not to exceed 10 feet[5]
Steel Pipe for Water or DWV	Threaded or Welded	¾ inch and smaller, 10 feet; 1 inch and larger, 12 feet	Every other floor, not to exceed 25 feet[5]
Steel Pipe for Gas	Threaded or Welded	½ inch, 6 feet; ¾ inch and 1 inch, 8 feet; 1¼ inches and larger, 10 feet	½ inch, 6 feet; ¾ inch and 1 inch, 8 feet; 1¼ inches every floor level
Schedule 40 PVC and ABS DWV	Solvent Cemented	All sizes, 4 feet; allow for expansion every 30 feet[3]	Base and each floor; provide mid-story guides; provide for expansion every 30 feet
CPVC	Solvent Cemented	1 inch and smaller, 3 feet; 1¼ inches and larger, 4 feet	Base and each floor; provide mid-story guides
CPVC-AL-CPVC	Solvent Cemented	½ inch, 5 feet; ¾ inch, 65 inches; 1 inch, 6 feet	Base and each floor; provide mid-story guide
Lead	Wiped or Burned	Continuous Support	Not to exceed 4 feet
Steel	Mechanical	In accordance with standards acceptable to the Authority Having Jurisdiction	
PEX	Cold Expansion, Insert and Compression	1 inch and smaller, 32 inches; 1¼ inches and larger, 4 feet	Base and each floor; provide mid-story guides
PEX-AL-PEX	Metal Insert and Metal Compression	½ inch, ¾ inch, 1 inch — All sizes 98 inches	Base and each floor; provide mid-story guides
PE-AL-PE	Metal Insert and Metal Compression	½ inch, ¾ inch, 1 inch — All sizes 98 inches	Base and each floor; provide mid-story guides
PE-RT	Insert and Compression	1 inch and smaller, 32 inches; 1¼ inches and larger, 4 feet	Base and each floor; provide mid-story guides
Polypropylene (PP)	Fusion weld (socket, butt, saddle, electrofusion), threaded (metal threads only), or mechanical	1 inch and smaller, 32 inches; 1¼ inches and larger, 4 feet	Base and each floor; provide mid-story guides

For SI units: 1 inch = 25.4 mm, 1 foot = 304.8 mm

Notes:
[1] Support adjacent to joint, not to exceed 18 inches (457 mm).
[2] Brace not to exceed 40 foot (12 192 mm) intervals to prevent horizontal movement.
[3] Support at each horizontal branch connection.
[4] Hangers shall not be placed on the coupling.
[5] Vertical water lines shall be permitted to be supported in accordance with recognized engineering principles with regard to expansion and contraction, where first approved by the Authority Having Jurisdiction.

3. The intersection at tees or wyes in horizontal pipe must be supported by a hanger on the branch also placed within 18 inches of the joint.

4. It does little good to support piping on the joint itself. The hanger will not be taking the weight of pipe if placed on the joint. The joint itself will have all the weight, which will lead to the failure of that joint.

5. Consideration for expansion and contraction of piping is contained in other areas of the code; however, vertical water piping is sometimes forgotten. This note creates a requirement for the expansion and contraction of vertical water piping to be addressed by the design professional (see **Figure 313.3d**).

Another consideration for designing a system of pipe hangers, supports and riser clamps is to determine whether the pipe is scheduled to be insulated. Whenever insulation is

GENERAL REGULATIONS

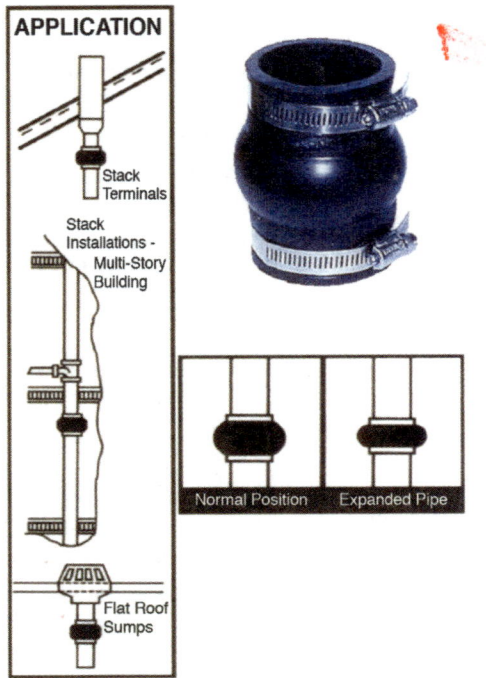

FIGURE 313.3D
APPLICATION OF DWV EXPANSION JOINT

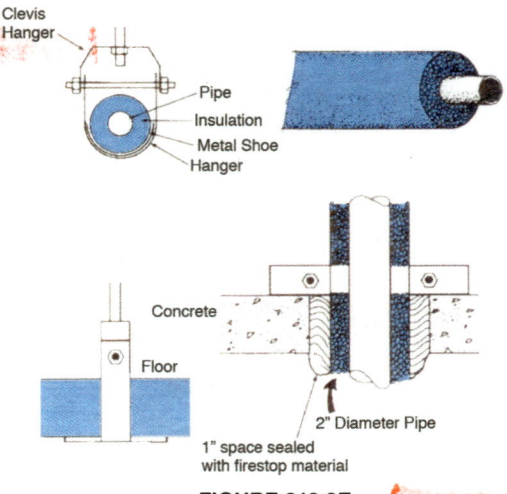

FIGURE 313.3E
PIPE SUPPORTS WITH INSULATION

applied, the type and thickness of the insulating material must be considered. The size of the hangers must be increased if the pipe is insulated, and the hangers are to be placed externally over the insulation. Also, there must be sufficient space for installers to apply and repair the insulation (see **Figure 313.3e**).

There is one last thing to consider when installing piping supports. There are other sections of the code that pertain to supporting piping or equipment. These depend on the type of installation – water, gas, DWV and medical gases. All of the code must be taken into account when installing the plumbing system.

313.4 Alignment. Piping shall be supported in such a manner as to maintain its alignment and prevent sagging.

313.5 Underground Installation. Piping in the ground shall be laid on a firm bed for its entire length; where other support is otherwise provided, it shall be approved in accordance with Section 301.2.

Care should be taken with underground piping to ensure that it is supported along its entire length. This is especially critical with underground drainage piping. There should be no sagging of the pipe. This will cause the drainage pipe to lose its grade and increase the possibility of blockage occurring in the installation. This is especially important with flexible piping such as plastic pipe materials (see **Figure 313.5**).

313.6 Hanger Rod Sizes. Hanger rod sizes shall be not smaller than those shown in Table 313.6.

TABLE 313.6
HANGER ROD SIZES

PIPE AND TUBE SIZE (inches)	ROD SIZE (inches)
½ – 4	⅜
5 – 8	½
10 – 12	⅝

For SI units: 1 inch = 25.4 mm

313.7 Gas Piping. Gas piping shall be supported by metal straps or hooks at intervals not to exceed those shown in Table 1210.2.4.1.

Gas piping shall be supported horizontally and vertically and not exceed the distances noted in Table 1210.2.4.1. **Figure 313.7** illustrates the support of piping, including copper natural gas piping, per the requirements of Table 1210.2.4.1. Even when piping is placed on supporting members, such as ceiling joists, the pipe must be anchored with straps or hooks. Further requirements are given in Chapter 12.

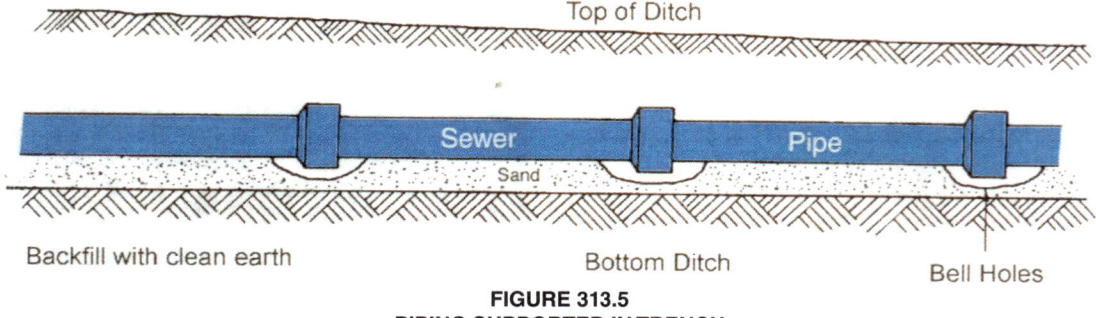

FIGURE 313.5
PIPING SUPPORTED IN TRENCH

GENERAL REGULATIONS

FIGURE 313.7
PIPING SUPPORT INSTALLATION

314.0 Trenching, Excavation, and Backfill.

314.1 Trenches. Trenches deeper than the footing of a building or structure, and paralleling the same, shall be located not less than 45 degrees (0.79 rad) from the bottom exterior edge of the footing, or as approved in accordance with Section 301.2.

🔧 The ability of a soil to support a load from a structural foundation is referred to as the bearing capacity of the soil. Analytic methods by geotechnical and structural engineers have determined that the failure zone (called radial shear zones), is the area under the footing having the ultimate soil bearing capacity. This area under the footing is roughly calculated by a 45-degree angle from the bottom exterior edge of the footing (see Figure 314.1). Any disturbance of the soil within this critical area reduces the bearing capacity of the soil for the foundation. Therefore, trenches are prohibited within the radial shear zone.

The structural engineer may request to have the ditch within the radial shear zone, provided the request complies with Section 301.5 Alternate Engineered Design.

314.2 Tunneling and Driving. Tunneling and driving shall be permitted to be done in yards, courts, or driveways of a building site. Where sufficient depth is available to permit, tunnels shall be permitted to be used between open-cut trenches.

Tunnels shall have a clear height of 2 feet (610 mm) above the pipe and shall be limited in length to one-half the depth of the trench, with a maximum length of 8 feet (2438 mm). Where pipes are driven, the drive pipe shall be not less than one size larger than the pipe to be laid.

🔧 It may be necessary at times to place piping under existing sidewalks and driveways. This is especially true in residential and repair installations. Tunneling or pipe driving may be required to complete the installation. In that case, the above requirements should be followed.

314.3 Open Trenches. Excavations required to be made for the installation of a building drainage system or part thereof, within the walls of a building, shall be open trench work and shall be kept open until the piping has been inspected, tested, and accepted.

🔧 Tunneling and driving are not permitted within the building. Instead trenches shall be used to install piping within the walls of a building. The trenches must be kept open and the pipe joints visible for inspection. Be sure to follow Occupational Safety and Health Administration (OSHA) safety requirements for trench work.

314.4 Excavations. Excavations shall be completely backfilled as soon after inspection as practicable. Precaution shall be taken to ensure compactness of backfill around piping without damage to such piping. Trenches shall be backfilled in thin layers to 12 inches (305 mm) above the top of the piping with clean earth, which shall not contain stones, boulders, cinder fill, frozen earth, construction debris, or other materials that will damage or break the piping or cause corrosive

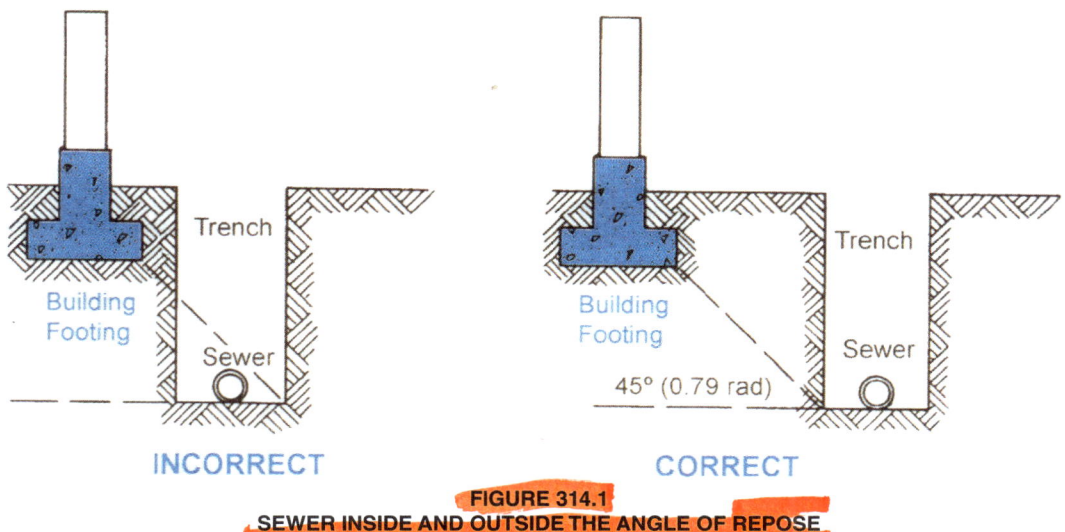

FIGURE 314.1
SEWER INSIDE AND OUTSIDE THE ANGLE OF REPOSE

action. Mechanical devices such as bulldozers, graders, etc., shall be permitted to be then used to complete backfill to grade. Fill shall be properly compacted. Precautions shall be taken to ensure permanent stability for pipe laid in filled or made ground.

Underground thermoplastic pipe and fittings for sewers and other gravity flow applications shall be installed in accordance with this code and Section 314.4.1.

This section prohibits the use of fill materials that, by their shape, size or chemical makeup, might cause damage or corrosion to the pipe. Backfill methods that could result in pipe failure over a period of time are also prohibited. Any material that will pass through a three-sixteenth of an inch sieve is typically acceptable as backfill material. Stones that are larger could damage buried piping.

It is also required to backfill the trench to 12 inches above the buried pipe before the use of any mechanical device to compact or to continue the backfill of the trench. The trench should always be compacted to the requirements of the building code or site specifications.

314.4.1 Installation of Thermoplastic Pipe and Fittings. Trench width for thermoplastic sewer pipe shall be not less than 1.25 times the outside diameter of the piping plus 12 inches (305 mm) or the outside diameter of the piping plus not less than 16 inches (406 mm). Thermoplastic piping shall be bedded in not less than 4 inches (102 mm) of granular fill supporting the piping. The backfill for thermoplastic piping shall be compacted along the sides of the piping in 6 inch (152 mm) layers and continue to not less than 12 inches (305 mm) above the piping. Compaction shall be not less than an 85 percent standard proctor density.

The standard for thermoplastic piping underground installation is ASTM D2321 which is found in Table 1701.1. This standard contains detailed requirements for the installation of thermoplastic pipe used for sewer and other gravity flow applications. Controlling the trench width enables the thermoplastic piping to gain side fill support from the compaction. If the trench width is too wide, the side fill support is more difficult to maintain and if it is too narrow then there is not enough room to compact the side fill support in 6 inch layers.

The granular fill support allows the thermoplastic piping to embed itself in the fill and increase the bedding angle which takes advantage of the side fill support to prevent the piping from flattening due to the load of the backfill on top of the pipe. A cubic foot of dirt weighs about 45 pounds and if wet the weight increases. Because plastic piping is flexible the weight exerted on the top of the pipe forces the sides of the pipe to deflect outward.

The compaction is necessary to prevent the deflection of the pipe from the weight of the backfill. Compaction of the side fill prevents the weight on top of the pipe from causing the pipe to become oval and leak or fail because of the fracture of pipe walls. Undisturbed native soil is considered to be 100% standard proctor density so the 85% compaction attempts to restore the soil back to its original state.

There is not a prescriptive methodology within this code section for multiple piping installations. Consult with your local authority having jurisdiction for trench widths on multiple piping installations.

315.0 Joints and Connections.

315.1 Unions. Approved unions shall be permitted to be used in drainage piping where accessibly located in the trap seal or between a fixture and its trap; in the vent system, except underground or in wet vents; at any point in the water supply system; and in gas piping as permitted by Section 1212.5.1.

Unions cannot be used indiscriminately anywhere in a plumbing system, but they are allowed to be used in drainage, vent, water supply and gas systems under the stated conditions. A union can be used on the drainage system only where the trap is located (in the trap seal or between the fixture and its trap) and must be accessible. A union can be used anywhere on the vent system except underground or on the wet vent. A union can be used at any point in the water supply system, either above ground or underground. A union can be used in the gas pipe system according to the referenced section for fuel gas piping.

315.2 Prohibited Joints and Connections. A fitting or connection that has an enlargement, chamber, or recess with a ledge, shoulder, or reduction of pipe area that offers an obstruction to flow through the drain shall be prohibited.

316.0 Increasers and Reducers.

316.1 General. Where different sizes of pipes and fittings are to be connected, the proper size increasers or reducers or reducing fittings shall be used between the two sizes. Copper alloy or cast-iron body cleanouts shall not be used as a reducer or adapter from cast-iron drainage pipe to iron pipe size (IPS) pipe.

Pipe sizes typically change as demand loads are increased or decreased. Usually these changes are made by the installation of a reducing tee at the point where the load on the trunk line is increased or decreased by connection of a branch line. However, if an appropriate reducing tee is unavailable, it is not uncommon to utilize a reducer downstream of the tee branch.

A reducer is designed to provide a smooth, even flow and has little adverse effect on the flow of liquids or gases as possible. The proper reducer for the type of use and installation, either eccentric or concentric, should be used (see **Figure 316.1**).

**FIGURE 316.1
ECCENTRIC STEEL AND CONCENTRIC COPPER REDUCERS**

317.0 Food-Handling Establishments.

317.1 General. Food or drink shall not be stored, prepared, or displayed beneath soil or drain pipes unless those areas are protected against leakage or condensation from such pipes reaching the food or drink as described below. Where building design requires that soil or drain pipes be located over such areas, the installation shall be made with the least possible number of joints and shall be installed to connect to the nearest adequately sized vertical stack with the provisions as follows:

(1) Openings through floors over such areas shall be sealed watertight to the floor construction.

(2) Floor and shower drains installed above such areas shall be equipped with integral seepage pans.

(3) Soil or drain pipes shall be of an approved material as listed in Table 1701.1 and Section 701.2. Materials shall comply with established standards. Cleanouts shall be extended through the floor construction above.

(4) Piping subject to operation at temperatures that will form condensation on the exterior of the pipe shall be thermally insulated.

(5) Where pipes are installed in ceilings above such areas, the ceiling shall be of the removable type or shall be provided with access panels to form a ready access for inspection of piping.

Special care should be taken when installing drainage piping over food handling areas (kitchens or food storage rooms). Contamination of the food being prepared or stored below the piping could lead to sickness or death. The methods of pipe installation and protection against leakage enumerated here are only some of the methods that may be required. The local health department may also have further requirements for drainage piping installed over food areas.

318.0 Test Gauges.

318.1 General. Tests in accordance with this code, which are performed utilizing dial gauges, shall be limited to gauges having the following pressure graduations or incrementations.

318.2 Pressure Tests (10 psi or less). Required pressure tests of 10 pounds-force per square inch (psi) (69 kPa) or less shall be performed with gauges of 0.10 psi (0.69 kPa) incrementation or less.

318.3 Pressure Tests (greater than 10 psi to 100 psi). Required pressure tests exceeding 10 psi (69 kPa) but less than or equal to 100 psi (689 kPa) shall be performed with gauges of 1 psi (7 kPa) incrementation or less.

318.4 Pressure Tests (exceeding 100 psi). Required pressure tests exceeding 100 psi (689 kPa) shall be performed with gauges incremented for 2 percent or less of the required test pressure.

318.5 Pressure Range. Test gauges shall have a pressure range not exceeding twice the test pressure applied.

The purpose of the gauge is not only to measure the pressure but to visually witness if the piping held its test or leaked out. Using too large a gauge will make this impossible (see Figure 318.5).

**FIGURE 318.5
TEST GUAGE**

319.0 Medical Gas and Vacuum Systems.

319.1 General. Such piping shall be in accordance with the requirements of Chapter 13. The Authority Having Jurisdiction shall require evidence of the competency of the installers and verifiers.

In jurisdictions where medical gas and vacuum systems are regulated under the UPC, this section and Chapter 13 would apply. This section requires evidence of competency of installers and verifiers. Such proof of evidence that the AHJ may require is meeting the requirements of ASSE 6010 Medical Gas Systems Installers Certification (see Section 1306.0 Qualifications of Installers).

320.0 Rehabilitation of Piping Systems.

320.1 General. Where pressure piping systems are rehabilitated using an epoxy lining system, it shall be in accordance with ASTM F2831.

Until the development of ASTM F2831 there has been no uniform enforcement of requirements for installing, testing, and quality control of epoxy lining systems used in the rehabilitation of pressure piping systems. Installers and inspectors will need to become familiar with the requirements contained in the standard. ASTM F2831 gives the inspector the authority to require third party testing to ensure the provisions of the standard are followed properly. Different types of epoxy coating include: Blow through, spin cast, or hand sprayed. Blow through application is limited to a maximum 6" diameter piping and must be performed from small pipe size to the larger pipe sizes. All epoxy coating applications must be performed by a trained and certified installer. Coating thickness shall be a minimum .01 inches and determined by a wet film thickness gauge. When required by the AHJ, the installer to provide spool pieces in random locations as requested.

Prior to beginning the epoxy coating process the installer must clean the pipe to be free of all visible oil, grease, dirt, mill, scale, rust and previously applied coatings. Each manufacturer's installation instructions for epoxy require visual inspection and documentation that this has been performed satisfactorily and only 33% of any coatings remain. NSF 61 cure times shall be used for epoxy lining in potable water systems or as specified by the manufacturer's installation instructions.

Third party testing includes visual inspection at the open- and exit of piping. Hand held probes may be used on piping up to 2" and any piping over 2" a CCTV shall be used to verify a 4' maximum length of piping. The pipe shall be tested to 150 psig or 1-1/2 times the working pressure. Flow rates must also be verified to ensure the piping meets the minimum flow rate requirements of local model codes. When used for potable water systems an annual chemical extraction test must be performed to verify continued compliance with Section 5 of NSF/ANSI 61.

To aid the end user and any future necessary system repairs all internally coated pipe and tubing shall be permanently and legibly marked. These markings shall be placed at each outlet and on the outside of exposed pipe at 20-ft intervals as follows: Manufacturer's name or trademark and coating designation and material with prohibition on the use of flame and heat to repair any part of system.

CHAPTER 4
PLUMBING FIXTURES AND FIXTURE FITTINGS

401.0 General.

401.1 Applicability. This chapter shall govern the materials and installation of plumbing fixtures, including faucets and fixture fittings, and the minimum number of plumbing fixtures required based on occupancy.

401.2 Quality of Fixtures. Plumbing fixtures shall be constructed of dense, durable, non-absorbent materials and shall have smooth, impervious surfaces, free from unnecessary concealed fouling surfaces.

🔧 Plumbing fixtures must collect waste water or sanitary wastes and convey it to the fixture outlet and ultimately to the drainage system. Flush-type fixtures empty the waste into the drainage system while leaving the fixture clean and ready for the next use. Therefore, it must be smooth, resistant to stains and, of course, able to last for a reasonable amount of time.

Much research has been conducted by engineers and product manufacturers over the last hundred years to accommodate this requirement. Flush-type fixtures are designed with just the right flow to empty and scour itself to leave the fixture clean and ready for the next use. Materials and the slope of fixtures are tested to ensure this occurs.

In Chapter 6, Section 601.2 requires the fixture to be supplied with water when necessary. To enhance this effect, the fixture must be designed so that there will be no part of the receiving surface of the fixture unwashed, thus creating a "fouling surface." No part of the fixture should be concealed so that it cannot be cleaned. Fixtures shall be manufactured to one of the referenced standards in Table 1701.1.

402.0 Installation.

402.1 Cleaning. Plumbing fixtures shall be installed in a manner to afford easy access for repairs and cleaning. Pipes from fixtures shall be run to the nearest wall.

🔧 This section contains the requirement for "easy access for repairs." One of the more common violations of this provision is the extension of countertops over the lids on toilet tanks. In all cases, the mechanism in a toilet tank shall be accessible for repair or replacement.

This section also requires that piping "shall be run to the nearest wall." Piping installed on the surface of a wall creates difficult cleaning challenges, thus increasing the difficulty in providing a sanitary environment. Piping is to be concealed within the walls when practicable to avoid this condition.

402.2 Joints. Where a fixture comes in contact with the wall or floor, the joint between the fixture and the wall or floor shall be made watertight.

🔧 When a fixture is mounted to a wall, the fixture/wall joint is to be sealed watertight, usually with a waterproof caulking. If not, water may run down between the fixture and the wall, causing an unhygienic condition leading to scum or mold buildup. See **Figure 402.2** for an installation without sealant.

**FIGURE 402.2
WALL JOINT NOT SEALED**

402.3 Securing Fixtures. Floor-outlet or floor-mounted fixtures shall be rigidly secured to the drainage connection and to the floor, where so designed, by screws or bolts of copper, copper alloy, or other equally corrosion-resistant material.

🔧 Fixtures that mount on the floor must be securely connected to the drainage system and floor. For example, a water closet must attach to the closet ring. The closet ring will attach by screws to the floor and also to the water closet with closet bolts. Both of which must be corrosion resistant. The joint between the water closet and the floor must also be sealed with a waterproof caulking.

402.4 Wall-Hung Fixtures. Wall-hung fixtures shall be rigidly supported by metal supporting members so that no strain is transmitted to the connections. Floor-affixed supports for off-the-floor plumbing fixtures for public use shall comply with ASME A112.6.1M. Framing-affixed supports for off-the-floor water closets with concealed tanks shall comply with ASME A112.6.2. Flush tanks and similar appurtenances shall be secured by approved non-corrosive screws or bolts.

🔧 Wall-hung water closet installation requirements are contained in Section 402.6 stating that they are to be connected to an approved carrier. This section refers to other fixtures that are wall mounted, such as lavatories, sinks or urinals. All must be supported by the proper mounting device, whether it is by mounting plates, brackets, or arms and not the drainage connection as in **Figures 402.4a**. For framing-affixed supports with concealed tank see **Figure 402.4b**.

402.5 Setting. Fixtures shall be set level and in proper alignment with reference to adjacent walls. No water closet or

PLUMBING FIXTURES AND FIXTURE FITTINGS

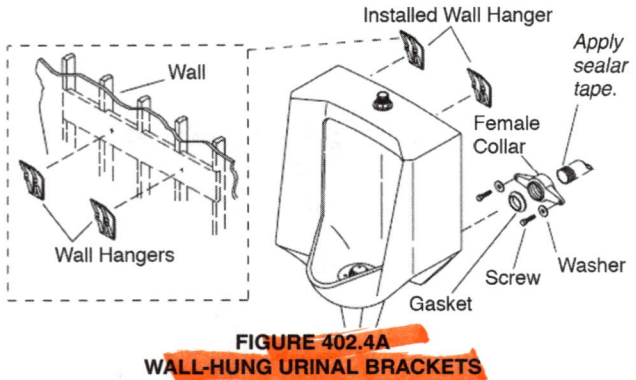

**FIGURE 402.4A
WALL-HUNG URINAL BRACKETS**

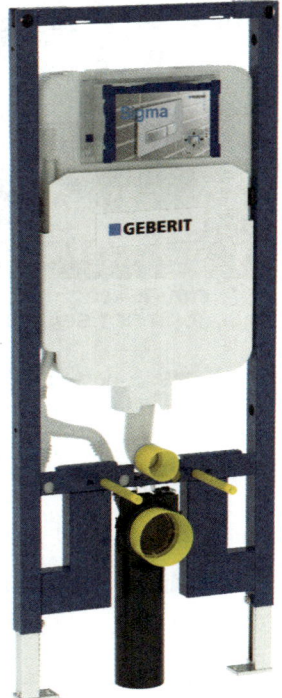

**FIGURE 402.4B
FRAMING AFFIXED SUPPORT WITH CONCEALED TANK**
(Courtesy of The Chicago Faucet Company)

bidet shall be set closer than 15 inches (381 mm) from its center to a side wall or obstruction or closer than 30 inches (762 mm) center to center to a similar fixture. The clear space in front of a water closet, lavatory, or bidet shall be not less than 24 inches (610 mm). No urinal shall be set closer than 12 inches (305 mm) from its center to a side wall or partition or closer than 24 inches (610 mm) center to center.

Exception: The installation of paper dispensers or accessibility grab bars shall not be considered obstructions.

The side-wall, center-to-center and front-clearance measurements for the water closet and urinal are some of the most important measurements in plumbing. These literally decide where piping will be placed in the building. One thing to remember, however, is that the measurements are not quite the same if stalls are being used or if handicap access is required. Always check the architectural print and specifications before laying out piping. Remember, these are minimum measurements (see **Figure 402.5**).

402.6 Flanged Fixture Connections. Fixture connections between drainage pipes and water closets, floor outlet service sinks and urinals shall be made using an approved copper alloy, hard lead, ABS, PVC, or iron flanges caulked, soldered, solvent cemented; rubber compression gaskets; or screwed to the drainage pipe. The connection shall be bolted with an approved gasket, washer, or setting compound between the fixture and the connection. The bottom of the flange shall be set on an approved firm base.

Wall-mounted water closet fixtures shall be securely bolted to an approved carrier fitting. The connecting pipe between the carrier fitting and the fixture shall be an approved material and designed to accommodate an adequately sized gasket. Gasket material shall be neoprene, felt, or similar approved types.

402.6.1 Closet Rings (Closet Flanges). Closet rings (closet flanges) for water closets or similar fixtures shall be of an approved type and shall be copper alloy, copper, hard lead, cast-iron, galvanized malleable iron, ABS, PVC, or other approved materials. Each such closet ring (closet flange) shall be approximately 7 inches (178 mm) in diameter and, where installed, shall, together with the soil pipe, present a 1½ inch (38 mm) wide flange or face to receive the fixture gasket or closet seal.

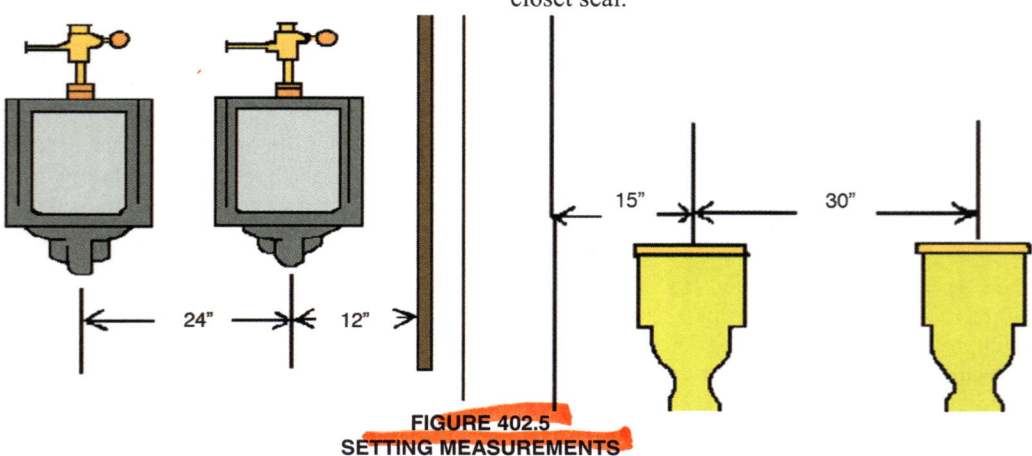

**FIGURE 402.5
SETTING MEASUREMENTS**

Caulked-on closet rings (closet flanges) shall be not less than ¼ of an inch (6.4 mm) thick and not less than 2 inches (51 mm) in overall depth.

Closet rings (closet flanges) shall be burned or soldered to lead bends or stubs, shall be caulked to cast-iron soil pipe, shall be solvent cemented to ABS and PVC, and shall be screwed or fastened in an approved manner to other materials.

Closet bends or stubs shall be cut-off to present a smooth surface even with the top of the closet ring before the rough inspection is called.

Closet rings (closet flanges) shall be adequately designed and secured to support fixtures connected thereto.

🔧 Before the rough-in inspection the closet ring or flange must be connected to the closet bend or stub. The bend or pipe must be cut flush with the top of the ring or flange. The flange and the pipe itself act as the sealing surface for the water closet bowl and the wax ring. The inspector will want to ensure that there is enough surface area to make a proper seal (see **Figure 402.6.1**).

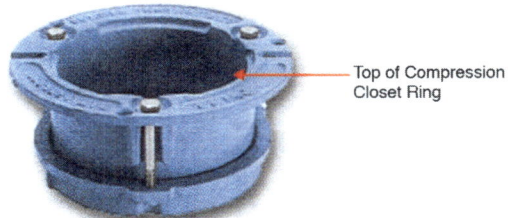

FIGURE 402.6.1
CUT PIPE TO TOP OF COMPRESSION CLOSET RING

402.6.2 Securing Closet Flanges. Closet screws, bolts, washers, and similar fasteners shall be of copper alloy, copper, or other listed equally corrosion-resistant materials. Screws and bolts shall be of a size and number to properly support the fixture installed.

402.6.3 Securing Floor-Mounted, Back-Outlet Water Closet Bowls. Floor-mounted, back-outlet water closet bowls shall be set level with an angle of 90 degrees (1.57 rad) between the floor and wall at the centerline of the fixture outlet. The floor and wall shall have a flat mounting surface not less than 5 inches (127 mm) to the right and left of the fixture outlet centerline. The fixture shall be secured to the wall outlet flange or drainage connection and the floor by corrosion-resistant screws or bolts. The closet flange shall be secured to a firm base.

Where floor-mounted, back-outlet water closets are used, the soil pipe shall be not less than 3 inches (80 mm) in diameter. Offset, eccentric, or reducing floor flanges shall not be used.

🔧 The code takes special care here to give instructions on installing floor-mounted back outlet water closets. These fixtures can be difficult to install properly because there are two mounting surfaces, the floor and the wall. Both have to be flat and create a 90-degree angle to be properly set. The closet has to mount rigidly to the back outlet and to the floor (see **Figure 402.6.3**).

FIGURE 402.6.3
BACK OUTLET, FLOOR-MOUNTED WATER CLOSET

402.7 Supply Fittings. The supply lines and fittings for every plumbing fixture shall be so installed as to prevent backflow in accordance with Chapter 6.

🔧 As will be seen in Chapter 6, all connections from fixtures to the water supply shall be protected from backflow by either an airgap or an approved backflow protection device.

402.8 Installation. Fixtures shall be installed in accordance with the manufacturer's installation instructions.

🔧 To ensure that the fixture and fixture fitting operate at their optimal performance, the journeyworker or apprentice must install the device per the manufacturer's instructions. It should also be tested to make certain the flow rates are correct.

402.9 Design and Installation of Plumbing Fixtures. Plumbing fixtures shall be installed in accordance with the manufacturer's installation instructions. The means of backflow prevention shall not be compromised by the designated fixture fitting mounting surface.

🔧 Where plumbing fixture fittings are installed with integral backflow prevention, care must be exercised that the mounting surface of the fixture does not interfere with the operation or requirements of the fitting or device. The manufacturer must identify or verify the critical level for installation and the type and location of the backflow protection system in the product literature or in the installation instructions.

402.10 Slip Joint Connections. Fixtures having concealed slip joint connections shall be provided with an access panel or utility space not less than 12 inches (305 mm) in its least dimension and so arranged without obstructions as to make such connections accessible for inspection and repair.

🔧 "Slip joints" are defined in Section 221.0 as three-piece assemblies involving the use of a friction ring and a compression washer, which is prone to be the weak link in this assembly (see **Figure 402.10a**). Potential failure of this washer may cause the joint to lose its gas-tight or water-tight status and leak. Thus, this joint, if concealed, shall be accompanied by a 12-inch by 12-inch access panel to allow for repair of the joint as in **Figure 402.10b**.

Two-piece assemblies are not slip joints by definition. Normally, these joints are comprised of a slip joint nut and a brass or nylon (plastic) ferrule. These materials have an anticipated lifespan that is equivalent to brass tubing. Therefore, no access panel is required. The same applies when tubing joints are assembled by soldering.

**FIGURE 402.10A
SLIP JOINT FITTING**

**FIGURE 402.10B
ACCESS PANEL**

402.11 Future Fixtures. Where provisions are made for the future installation of fixtures, those provided for shall be considered in determining the required sizes of the drain and water supply piping. Construction for future installations shall be terminated with a plugged fitting or fittings. Where the plugged fitting is at the point where the trap of a fixture is installed, the plumbing system for such fixture shall be complete and be in accordance with the plumbing requirements of this code.

Provisions for expansion or future additions should be part of the original installation. The future fixtures should be added to the DFUs and WFUs for correct sizing. However, since this may not always be possible, additions or remodels that occur after the owner occupies the building must comply with Section 102.4, Additions, Alterations, Renovations, or Repairs.

403.0 Accessible Plumbing Facilities.

403.1 General. Where accessible facilities are required in applicable building regulations, the facilities shall be installed in accordance with those regulations.

403.2 Fixtures and Fixture Fittings for Persons with Disabilities. Plumbing fixtures and fixture fittings for persons with disabilities shall be in accordance with ICC A117.1 and the applicable standards referenced in Chapter 4.

Requirements for fixtures for persons with disabilities are located in ANSI A117.1, Accessible and Usable Buildings and Facilities Standard and ADA American Disability Act and can be found at www.ada.gov.

403.3 Exposed Pipes and Surfaces. Water supply and drain pipes under accessible lavatories and sinks shall be insulated or otherwise be configured to protect against contact. Protectors, insulators, or both shall comply with ASME A112.18.9 or ASTM C1822.

404.0 Waste Fittings and Overflows.

404.1 Waste Fittings. Waste fittings shall comply with ASME A112.18.2/CSA B125.2, ASTM F409 or Table 701.2 for aboveground drainage piping and fittings.

404.2 Overflows. Where a fixture is provided with an overflow, the waste shall be so arranged that the standing water in the fixture shall not rise in the overflow where the stopper is closed or remain in the overflow where the fixture is empty. The overflow pipe from a fixture shall be connected to the house or inlet side of the fixture trap, except that overflow on flush tanks shall be permitted to discharge into the water closets or urinals served by them, but it shall be unlawful to connect such overflows with any other part of the drainage system.

Not all plumbing fixtures are required to have an overflow. For example, the applicable standards for plastic, enameled cast iron, enameled steel, and ceramic plumbing fixtures do not require overflows for lavatories, sinks and bidets. However, the four standards referenced in Section 409.1 require bathtubs and whirlpools constructed of ceramic, cast iron, enameled steel, stainless steel or plastic meet the minimum dimensions illustrated in the standard, which includes an opening for an overflow. When an overflow is provided, whether mandatory or optional, the fixture waste shall be so arranged that the standing water in the fixture cannot rise in the overflow when the stopper is closed or remain in the overflow when the fixture is empty. The only time water is permitted in the fixture overflow is when the fixture is filled to the point of overflow.

The overflow pipe from a fixture must connect only on the house or inlet side of a fixture trap except for the overflow on a flush tank. The flush tank may discharge into the water closet or urinal it serves. An overflow is not permitted to bypass the trap of the fixture it serves. It may either be an integral part of the fixture or a separate fitting (see **Figure 404.1**).

When waste and overflow fittings are used and are concealed, as in most bathtub installations, they must be constructed from materials approved for drainage pipe. When they are exposed or accessible, they may be of seamless drawn

PLUMBING FIXTURES AND FIXTURE FITTINGS

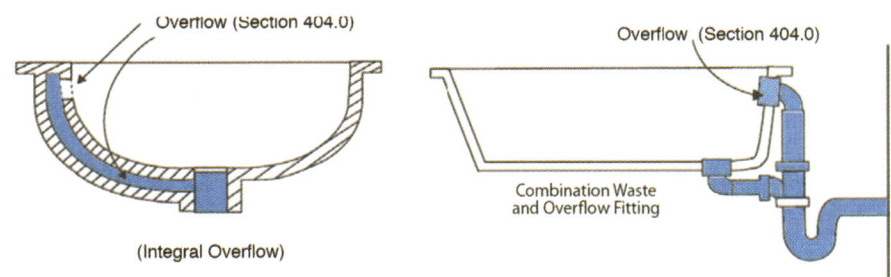

**FIGURE 404.1
OVERFLOWS**

brass, not less than No. 20 B & S gauge (0.032 in.) (0.8 mm) in thickness, or other approved material as per Section 701.4.

405.0 Prohibited Fixtures.

405.1 Prohibited Water Closets. Water closets having an invisible seal or an unventilated space or having walls which are not thoroughly washed at each discharge shall be prohibited. A water closet that might permit siphonage of the contents of the bowl back into the tank shall be prohibited.

🔧 Water closets, because they receive the most hazardous of waste, fecal matter, must have an interior that is cleaned with each flush. There can be no walls of the fixture that are not washed down with each flush. The water seal must be visible so that one can verify if the water closet seal is functioning properly.

405.2 Prohibited Urinals. Trough urinals and urinals with an invisible seal shall be prohibited.

🔧 Trough urinals are prohibited (see **Figure 405.2**). Trough urinals are typically provided with a spray bar that, at best, will provide no more than partial flushing of the fixture walls and waste water. Consequently, these fixtures are truly insanitary and, therefore, represent a potential health hazard. When they are installed, they are mostly in high use areas such as stadiums. When permitted by the Authority Having Jurisdiction (AHJ), they are installed per Section 301.3, Alternate Materials and Methods of Construction Equivalency.

405.3 Miscellaneous Fixtures. Fixed wooden, or tile wash trays or sinks for domestic use shall not be installed in a building designed or used for human habitation. No sheet metal-lined wooden bathtub shall be installed or reconnected. No dry or chemical closet (toilet) shall be installed in a building used for human habitation unless first approved by the Health Officer.

🔧 The wooden fixture is prohibited because of its tendency to become insanitary. It is not impervious or smooth and cannot be fully cleaned with each use.

The dry or chemical toilet should only be used if no other means exists to provide facilities for the occupants of the facility. They should only be used if the health officer, usually a local health department inspector, approves its use (see **Figure 405.3**).

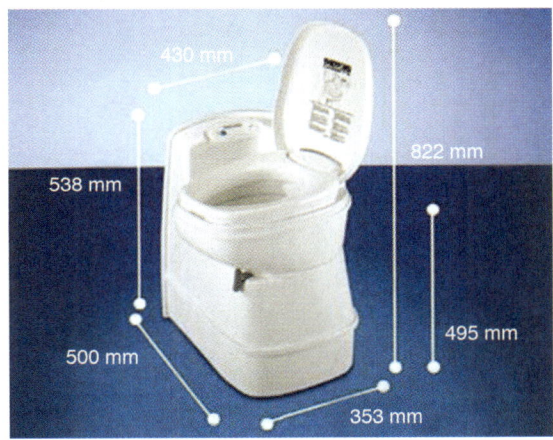

**FIGURE 405.3
CHEMICAL TOILET**

406.0 Special Fixtures and Specialties.

406.1 Water and Waste Connections. Baptisteries, ornamental and lily ponds, aquaria, ornamental fountain basins, and similar fixtures and specialties requiring water, waste connections, or both shall be submitted for approval to the Authority Having Jurisdiction prior to installation.

🔧 Fountains and other ornamental features are often built "on the job." These are plumbing fixtures and usually require water and waste connections to function properly. They will require the same backflow and trap seal protection as other plumbing fixtures. Their construction shall be approved by the AHJ (see **Figure 406.1**).

406.2 Special Use Sinks. Restaurant kitchen and other special use sinks shall be permitted to be made of approved-type bonderize and galvanized sheet steel of not less than No. 16 U.S. gauge (0.0635 inches) (1.6 mm). Sheet-metal plumbing fixtures shall be adequately designed, constructed, and braced in an approved manner to accomplish their intended purpose.

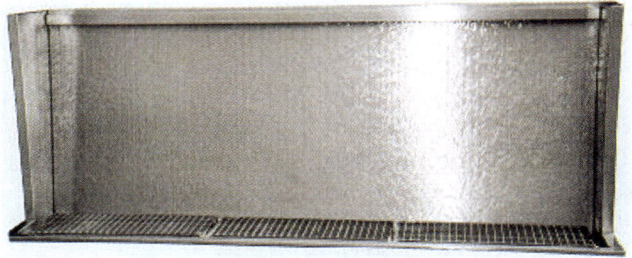

**FIGURE 405.2
FLOOR-MOUNTED TROUGH URINAL**

PLUMBING FIXTURES AND FIXTURE FITTINGS

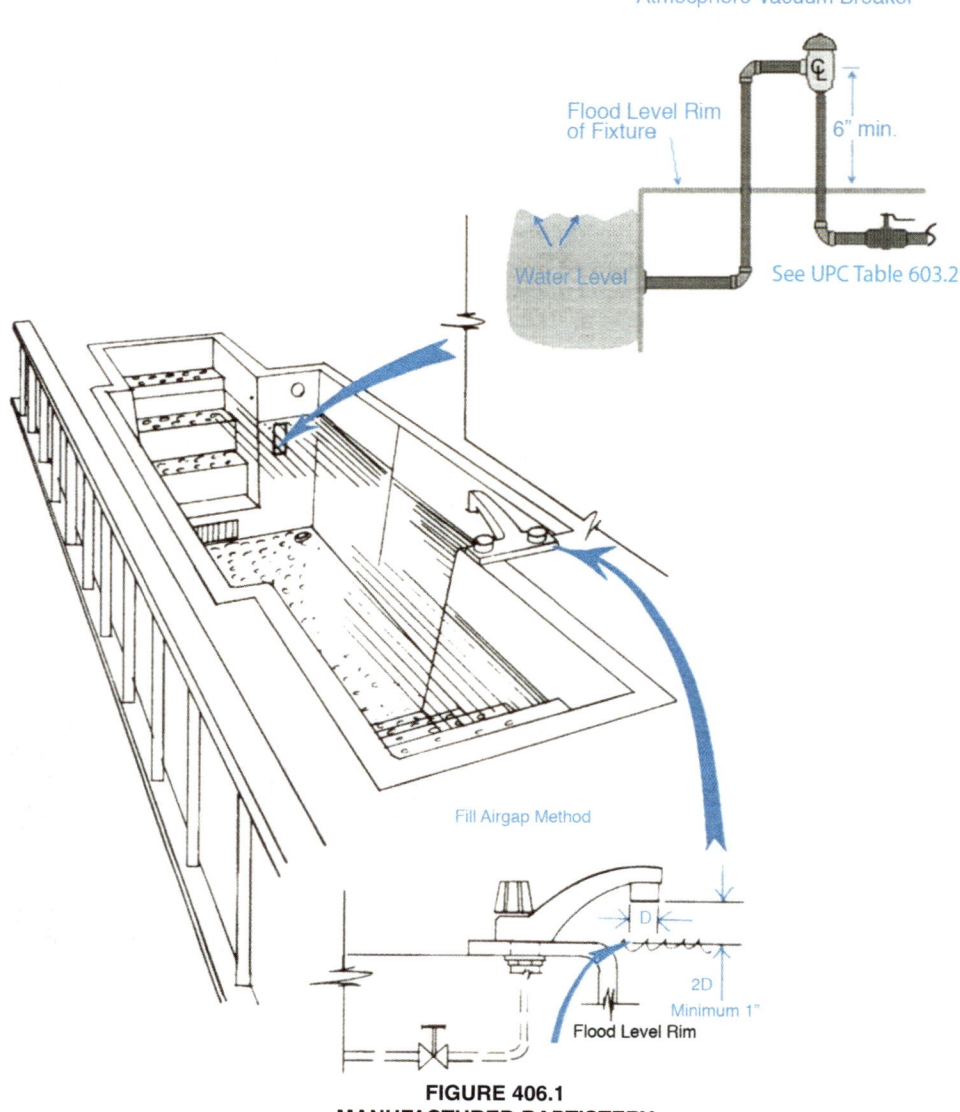

**FIGURE 406.1
MANUFACTURED BAPTISTERY**

🔧 Sinks, even if they are part of stainless steel kitchen counters, are still plumbing fixtures and shall be installed as such. They will still be part of the plumber's responsibility.

406.3 Special Use Fixtures. Special use fixtures shall be made of one of the following:

(1) Soapstone
(2) Chemical stoneware
(3) Copper-based alloy
(4) Nickel-based alloy
(5) Corrosion-resistant steel
(6) Other materials suited for the intended use of the fixture

🔧 These "special use" fixtures are usually installed because the waste that drains into them is caustic and needs special handling (see **Figure 406.3**). The materials are chosen depending on the resistance to the waste. They will almost always require some kind of interceptor to neutralize the waste.

406.4 Zinc Alloy Components. Zinc alloy components shall comply with applicable nationally recognized standards and shall be used in accordance with their listing.

🔧 The term "zinc alloy components" usually refer to brass, an alloy of zinc and copper.

407.0 Lavatories.

407.1 Application. Lavatories shall comply with ASME A112.19.1/CSA B45.2, ASME A112.19.2/CSA B45.1, ASME A112.19.3/CSA B45.4, ASME A112.19.12, CSA B45.5/IAPMO Z124, CSA B45.8/IAPMO Z403, CSA B45.11/IAPMO Z401 or CSA B45.12/IAPMO Z402.

407.2 Water Consumption. The maximum water flow rate of faucets shall comply with Section 407.2.1 and Section 407.2.2.

407.2.1 Maximum Flow Rate. The maximum flow rate for public lavatory faucets shall not exceed 0.5 gpm at 60 psi (1.9

**FIGURE 406.3
RESTAURANT AND SOAPSTONE SINKS**

L/m at 414 kPa) and 2.2 gpm at 60 psi (8.3 L/m at 414 kPa) for private lavatory faucets.

407.2.2 Metering Faucets. Metered faucets shall deliver a maximum of 0.25 gallons (1.0 L) per metering cycle.

🔧 Metered faucets (see **Figure 407.2.2**) are required in public areas so that an end-user will not be able to leave a faucet running, walk away and waste water. These faucets deliver a pre-set amount of water (0.25 gal per use) and then shut off. They can be either spring loaded or electronically controlled.

**FIGURE 407.2.2
METERED FAUCET**

407.3 Limitation of Hot Water Temperature for Public Lavatories. Hot water delivered from public-use lavatories shall be limited to a maximum temperature of 120°F (49°C) by a device that complies with ASSE 1070/ASME A112.1070/CSA B125.70. The water heater thermostat shall not be considered a control for meeting this provision.

🔧 In the case of lavatories for public places, the temperature of the hot water is limited to 120°F to avoid scalding. The code prohibits the use of a water heater thermostat for temperature control. There are many temperature and temperature/pressure limiting devices for faucets available for limiting temperatures for lavatories.

These harmonized standards control and limits the water temperature to fittings for fixtures such as sinks, lavatories or bathtubs and is intended to reduce the risk of scalding. These devices are intended to supply tempered water to plumbing fixture fittings, or be integral with plumbing fixture fittings supplying tempered water. The device is equipped with an adjustable and lockable means to limit the setting of the device towards the hot position.

407.4 Transient Public Lavatories. Self-closing or metering faucets shall be installed on lavatories intended to serve the transient public, such as those in, but not limited to service stations, train stations, airports, restaurants, and convention halls.

407.5 Waste Outlet. Lavatories shall have a waste outlet and fixture tailpiece not less than 1¼ inches (32 mm) in diameter. Continuous wastes and fixture tailpieces shall be constructed from the materials specified in Section 701.4. Waste outlets shall be provided with an approved stopper or strainer.

407.6 Overflow. Where overflows are provided, they shall be installed in accordance with Section 404.2.

408.0 Showers.

408.1 Application. Manufactured shower receptors and shower bases shall comply with ASME A112.19.1/CSA B45.2, ASME A112.19.2/CSA B45.1, ASME A112.19.3/CSAB45.4, CSA B45.12/IAPMO Z402, or CSA B45.5/IAPMO Z124.

408.2 Water Consumption. Showerheads shall have a maximum flow rate of not more than 2.5 gpm at 80 psi (9.5 L/m at 552 kPa).

408.3 Individual Shower and Tub-Shower Combination Control Valves. Showers and tub-shower combinations shall be provided with individual control valves of the pressure balance, thermostatic, or combination pressure balance/thermostatic mixing valve type that provide scald and thermal shock protection for the rated flow rate of the installed showerhead. These valves shall be installed at the point of use and comply with ASSE 1016/ASME A112.1016/CSA B125.16 or ASME A112.18.1/CSA B125.1.

Gang showers, where supplied with a single temperature-controlled water supply pipe, shall be controlled by a mixing valve that complies with ASSE 1069. Handle position stops shall be provided on such valves and shall be adjusted per the manufacturer's instructions to deliver maximum mixed water setting of 120°F (49°C). Water heater thermostats shall not be considered a suitable control for meeting this provision.

🔧 Many reported incidents of serious burns and

PLUMBING FIXTURES AND FIXTURE FITTINGS

shocks received while showering have prompted the requirement for anti-scald shower valves, thermostatic mixing or pressure-balancing shower control valves (see **Figure 408.3a**). Both types of valves regulate the water temperature by tempering hot water with cold water. The code requires that shower and tub-shower combinations in all occupancies shall have individual shower control valves of the pressure balance or the thermostatic mixing valve type.

Where multiple or gang showers are installed, such as in athletic facilities or work places, in lieu of individually controlled pressure balance or thermostatic mixing valves, master thermostatic valves controlling all the showerheads in the system may be used (see **Figure 408.3b**). In the master thermostatic valve, a moving piston responds to temperature fluctuations from an expanding or moving bellows, a bimetallic unit or heat-conducting fluid effecting hot and cold water input. Set screws or limit stops are pre-adjusted to maintain a constant outlet water temperature not to exceed 120°F. Upon cold water failure, the hot water is instantly shut down.

FIGURE 408.3A
TUB-SHOWER COMBINATION CONTROL VALVE

FIGURE 408.3B
MASTER MIXING VALVE

In addition to the pressure balance or thermostatic mixing valves required by the code, there are numerous temperature-sensitive devices available for installation at the water outlet. Temperature-sensitive aerators, showerheads, etc., are all "aftermarket" products that offer an excellent safety function for those systems that are not otherwise protected. However, these devices are not alternatives to the code-required pressure balance or thermostatic mixing valves.

It must be noted that an abrupt change in temperature (to either hot or cold) may cause an involuntary reflex that can result in a fall when a person is standing on a slippery surface. This hazard is not present when a person is sitting in a bathtub. Consequently, this code requirement is not applicable to bathtubs or other fixtures that are not also equipped with a shower riser. Handheld flexible showers are associated with bathtubs and other vessels where the bather is seated and are also not required to comply with this section.

408.4 Waste Outlet. Showers shall have a waste outlet and fixture tailpiece not less than 2 inches (50 mm) in diameter. Fixture tailpieces shall be constructed from the materials specified in Section 701.2 for drainage piping. Strainers serving shower drains shall have a waterway at least equivalent to the area of the tailpiece.

Strainers provide protection for the drainage system from solids that should not be placed in the system. The free area of the strainer, especially for a shower, should equal the area of the tailpiece. For example, the free area of the shower strainer should equal 3.14 (80 mm) square inches as in **Figure 408.4**.

FIGURE 408.4
STRAINER

408.5 Finished Curb or Threshold. Where a shower receptor has a finished dam, curb, or threshold, it shall be not less than 1 inch (25.4 mm) lower than the sides and back of such receptor. In no case, shall a dam or threshold be less than 2 inches (51 mm) or exceeding 9 inches (229 mm) in depth where measured from the top of the dam or threshold to the top of the drain. Each such receptor shall be provided with an integral nailing flange to be located where the receptor meets the vertical surface of the finished interior of the shower compartment. The flange shall be watertight and extend vertically not less than 1 inch (25.4 mm) above the top of the sides of the receptor. The finished floor of the receptor shall slope uniformly from the sides towards the drain not less than ⅛ inch per foot (10.4 mm/m), nor more than ½ inch per foot (41.6 mm/m).

PLUMBING FIXTURES AND FIXTURE FITTINGS

Thresholds shall be of sufficient width to accommodate a minimum 22 inch (559 mm) door. Shower doors shall open so as to maintain not less than a 22 inch (559 mm) unobstructed opening for egress. The immediate adjoining space to showers without thresholds shall be considered a wet location and shall comply with the requirements of the building, residential, and electrical codes.

Exceptions:

(1) Showers in accordance with Section 403.2.
(2) A cast-iron shower receptor flange shall be not less than 0.3 of an inch (7.62 mm) in height.
(3) For flanges not used as a means of securing, the sealing flange shall be not less than 0.3 of an inch (7.62 mm) in height.

The shower receptor is the floor or bottom of the shower. This section refers mainly to manufactured receptors (see **Figure 408.5**). The requirements of the receptor provide protection from leakage. If installed correctly, it will provide a solid ground for the shower and protect the floor below from leakage. As cited in this code section, exception (1) is for accessibility standards in regards to shower receptors only, which allows a shower with no threshold so a wheelchair can roll into the shower.

408.6 Shower Compartments. Shower compartments, regardless of shape, shall have a minimum finished interior of 1024 square inches (0.6606 m^2) and shall also be capable of encompassing a 30 inch (762 mm) circle. The minimum required area and dimensions shall be measured at a height equal to the top of the threshold and a point tangent to its centerline. The area and dimensions shall be maintained to a point of not less than 70 inches (1778 mm) above the shower drain outlet with no protrusions other than the fixture valve or valves, showerheads, soap dishes, shelves, and safety grab bars, or rails. Fold-down seats in accessible shower stalls shall be permitted to protrude into the 30 inch (762 mm) circle.

FIGURE 408.5
MANUFACTURED SHOWER RECEPTOR

Exceptions:

(1) Showers that are designed to be in accordance with ICC A117.1.
(2) The minimum required area and dimension shall not apply for a shower receptor having overall dimensions of not less than 30 inches (762 mm) in width and 60 inches (1524 mm) in length.

The shower, whether a one-piece stall or one built up on the site, is a plumbing fixture. It will be the responsibility of the plumber to ensure the shower is tested and is watertight. Install the manufactured shower per the manufacturer's instructions as well as all the requirements set forth in Chapter 4.

Figure 408.6 illustrates how to size the area of the shower. The dimensions must be maintained to a height of 70 inches in the shower.

Exception (1) to the requirements is for showers that meet ICC/ANSI A117.1, Standard on Accessible and Usable Buildings and Facilities which include accessible shower compartments such as roll-in or curbless types.

Exception (2) is for the retrofit of a bathtub to a shower. This will allow the tub to be replaced with a shower in the same dimensions.

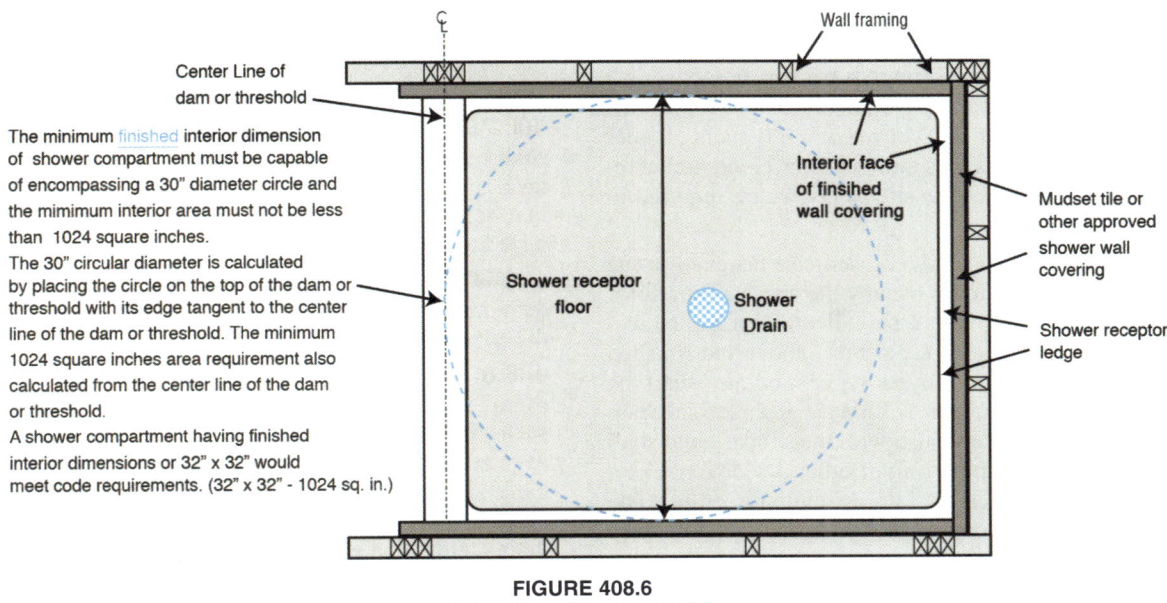

FIGURE 408.6
SHOWER RECEPTOR AREA

PLUMBING FIXTURES AND FIXTURE FITTINGS

408.7 Lining for Showers and Receptors. Shower receptors built on-site shall be watertight and shall be constructed from approved-type dense, nonabsorbent, and non-corrosive materials. Each such receptor shall be adequately reinforced, shall be provided with an approved flanged floor drain designed to make a watertight joint on the floor, and shall have smooth, impervious, and durable surfaces.

Shower receptors shall have the subfloor and rough side of walls to a height of not less than 3 inches (76 mm) above the top of the finished dam or threshold shall be first lined with sheet plastic, lead, or copper, or shall be lined with other durable and watertight materials. Showers that are provided with a built in place, permanent seat or seating area that is located within the shower enclosure, shall be first lined with sheet plastic, lead, copper, or shall be lined with other durable and watertight materials that extend not less than 3 inches (76 mm) above horizontal surfaces of the seat or the seating area.

Lining materials shall be pitched ¼ inch per foot (20.8 mm/m) to weep holes in the subdrain of a smooth and solidly formed subbase. Such lining materials shall extend upward on the rough jambs of the shower opening to a point not less than 3 inches (76 mm) above the horizontal surfaces of the seat or the seating area, the top of the finished dam or threshold and shall extend outward over the top of the permanent seat, permanent seating area, or rough threshold and be turned over and fastened on the outside face of both the permanent seat, permanent seating area, or rough threshold and the jambs.

Nonmetallic shower subpans or linings shall be permitted to be built up on the job site of not less than three layers of standard grade 15 pound (6.8 kg) asphalt impregnated roofing felt. The bottom layer shall be fitted to the formed subbase and each succeeding layer thoroughly hot-mopped to that below. Corners shall be carefully fitted and shall be made strong and watertight by folding or lapping, and each corner shall be reinforced with suitable webbing hot-mopped in place.

Folds, laps, and reinforcing webbing shall extend not less than 4 inches (102 mm) in all directions from the corner, and webbing shall be of approved type and mesh, producing a tensile strength of not less than 50 pounds per square foot (lb/ft^2) (244 kg/m^2) in either direction. Nonmetallic shower subpans or linings shall be permitted to consist of multilayers of other approved equivalent materials suitably reinforced and carefully fitted in place on the job site as elsewhere required in this section.

Linings shall be properly recessed and fastened to the approved backing so as not to occupy the space required for the wall covering, and shall not be nailed or perforated at a point that is less than 1 inch (25.4 mm) above the finished dam or threshold. An approved type subdrain shall be installed with a shower subpan or lining. Each such subdrain shall be of the type that sets flush with the subbase and shall be equipped with a clamping ring or other device to make a tight connection between the lining and the drain. The subdrain shall have weep holes into the waste line. The weep holes located in the subdrain clamping ring shall be protected from clogging.

Built on-site shower receptors may be a monolithic poured-in-place or equivalent receptor complete with integral threshold, sides and back directly supported by the underlying ground (see **Figure 408.7a**). The sides and back must extend at least three inches (80 mm) above the finished threshold before any wood superstructure or wall covering may be added. Not many of these receptors are built anymore. The receptor shown in **Figure 408.7b** is far more common.

The above instructions and referenced standards for constructing a built on-site shower receptor are quite specific with importance focusing on the liner. The liner used must comply with the appropriate standard listed, which will be verified by the AHJ. Even though the plumber often subcontracts out the installation of this type of shower to a tile company, the inspection is still the responsibility of the plumber because the shower is a plumbing fixture and is contained in the plumbing permit. The liability for the integrity of the shower will remain with the plumber even if he or she does not install the receptor.

408.7.1 PVC Sheets. Plasticized polyvinyl chloride (PVC) sheets shall conform to ASTM D4551. Sheets shall be joined by solvent cementing in accordance with the manufacturer's installation instructions.

408.7.2 Chlorinated Polyethylene (CPE) Sheets. Nonplasticized chlorinated polyethylene sheets shall conform to ASTM D4068. The liner shall be joined in accordance with the manufacturer's installation instructions.

408.7.3 Sheet Lead. Sheet lead shall weigh not less than 4 lb/ft^2 (19.5 kg/m^2) and shall be coated with an asphalt paint or other approved coating. The lead sheet shall be insulated from conducting substances, other than the connecting drain, by 15 pound (6.8 kg) asphalt felt or an equivalent. Sheet lead shall be joined by burning.

408.7.4 Sheet Copper. Sheet copper shall comply with ASTM B152 and shall weigh not less than 12 ounces per square foot (oz/ft^2) (3.7 kg/m^2) or No. 24 B & S Gauge (0.02 inches) (0.51 mm). The copper sheet shall be insulated from conducting substances, other than the connecting drain, by 15 pound (6.8 kg) asphalt felt or an equivalent. Sheet copper shall be joined by brazing or soldering.

408.7.5 Tests for Shower Receptors. Shower receptors shall be tested for watertightness by filling with water to the level of the rough threshold. The test plug shall be so placed that both upper and under sides of the subpan shall be subjected to the test at the point where it is clamped to the drain.

The test for the shower lining and receptor is to demonstrate no water loss. The lining should be tested before the concrete base is poured. See **Figure 408.7.5**.

408.8 Public Shower Floors. Floors of public shower rooms shall have a nonskid surface and shall be drained in such a manner that wastewater from one bather shall not pass over areas occupied by other bathers. Gutters in public or gang shower rooms shall have rounded corners for easy cleaning and shall be sloped not less than 2 percent toward drains. Drains in gutters shall be spaced at a maximum of 8 feet (2438 mm) from sidewalls nor more than 16 feet (4877 mm) apart.

PLUMBING FIXTURES AND FIXTURE FITTINGS

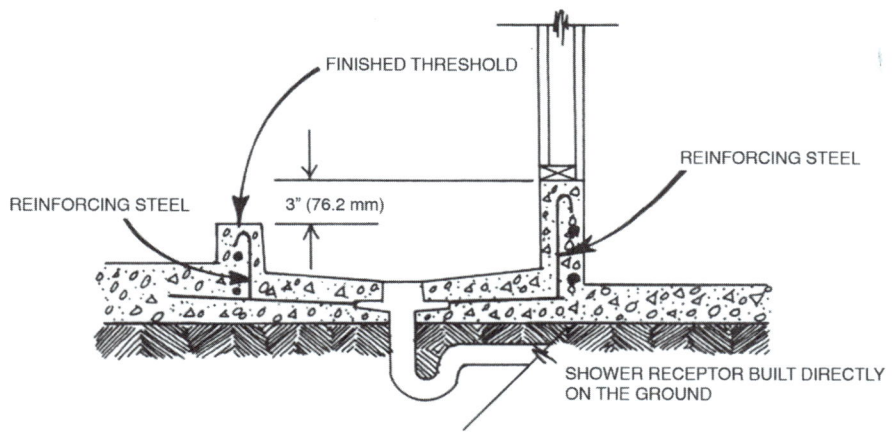

FIGURE 408.7A
SHOWER RECEPTOR BUILT UPON THE GROUND

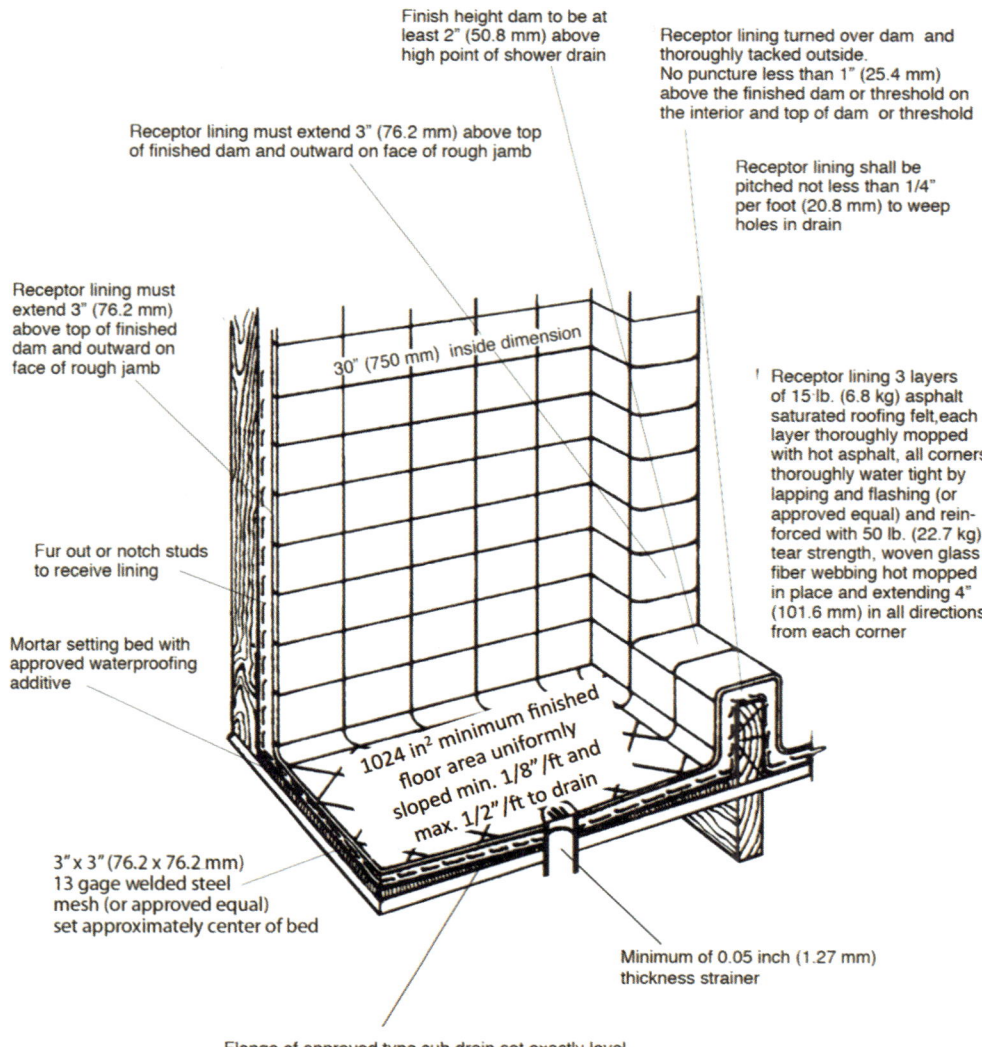

FIGURE 408.7B
SHOWER RECEPTOR BUILT ABOVE GROUND

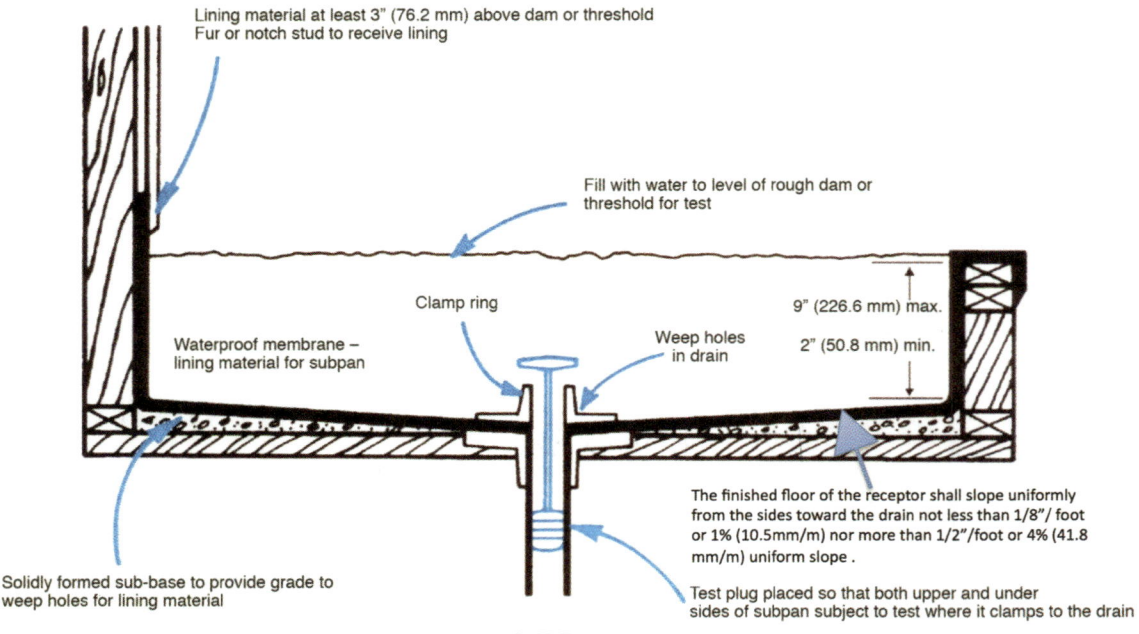

**FIGURE 408.7.5
SHOWER RECEPTOR WATER TEST**

🔧 Public shower rooms typically have several showerheads that must be spaced with enough separation to prevent unintended splashing from one bather to another. Showerheads may be placed on one or more walls. They may also be arranged around a pedestal or column to provide for more shower stations centered on a centrally located water source. In all cases, water must flow away from the feet of each bather in a way that does not flow over the other bathers. Sloping the shower floor to one or more troughs is a common means of directing water away from each bather (see **Figure 408.8**).

408.9 Location of Valves and Heads. Control valves and showerheads shall be located on the sidewall of shower compartments or otherwise arranged so that the showerhead does not discharge directly at the entrance to the compartment so that the bather can adjust the valves before stepping into the shower spray.

🔧 The location of the control valve on the side wall of the shower not only keeps the water inside the shower as it should but allows the bather to turn on the water and adjust the temperature of the water without getting wet. This protects the bather from scalding or from a sudden reaction to cold water, which could cause a slip or fall.

408.10 Water Supply Riser. A water supply riser from the shower valve to the showerhead outlet, whether exposed or not, shall be securely attached to the structure.

🔧 The riser must be secured to the structure. An unsecured riser will cause vibration and possible water hammer in the line. It must also be secured to allow the attachment of the shower head to the riser without damaging the riser, fitting, or stubout.

409.0 Bathtubs and Whirlpool Bathtubs.

409.1 Application. Bathtubs shall comply with ASME A112.19.1/CSA B45.2, ASME A112.19.2/CSA B45.1, ASME A112.19.3/CSA B45.4, CSA B45.5/IAPMO Z124, or CSA B45.12/IAPMO Z402. Whirlpool bathtubs shall comply with ASME A112.19.7/CSA B45.10. Pressure sealed doors within a bathtub or whirlpool bathtub enclosure shall comply with ASME A112.19.15.

409.2 Waste Outlet. Bathtubs and whirlpool bathtubs shall have a waste outlet and fixture tailpiece not less than 1½ inches (40 mm) in diameter. Fixture tailpieces shall be constructed from the materials specified in Section 701.2 for drainage piping. Waste outlets shall be provided with an approved stopper or strainer.

409.3 Overflow. Where overflows are provided, they shall be installed in accordance with Section 404.2.

409.4 Limitation of Hot Water in Bathtubs and Whirlpool Bathtubs. The maximum hot water temperature discharging from the bathtub and whirlpool bathtub filler shall be limited to 120°F (49°C) by a device that complies with ASSE 1070/ASME A112.1070/CSA B125.70. The water heater thermostat shall not be considered a control for meeting this provision.

🔧 Scalding injuries have occurred in bathtubs and whirlpool bathtubs over the last several years. ASSE 1070/ASME A112.1070/CSA B125.70 is standard for the temperature limiting devices to prevent scalding for these fixtures. The intent of this section is to provide a safe water temperature limiting device rather than the water heater thermostat as a device to prevent scalding.

409.5 Backflow Protection. The water supply to a bathtub and whirlpool bathtub filler valve shall be protected by an air gap or in accordance with Section 417.0.

409.6 Installation and Access. Bathtubs and whirlpool

PLUMBING FIXTURES AND FIXTURE FITTINGS

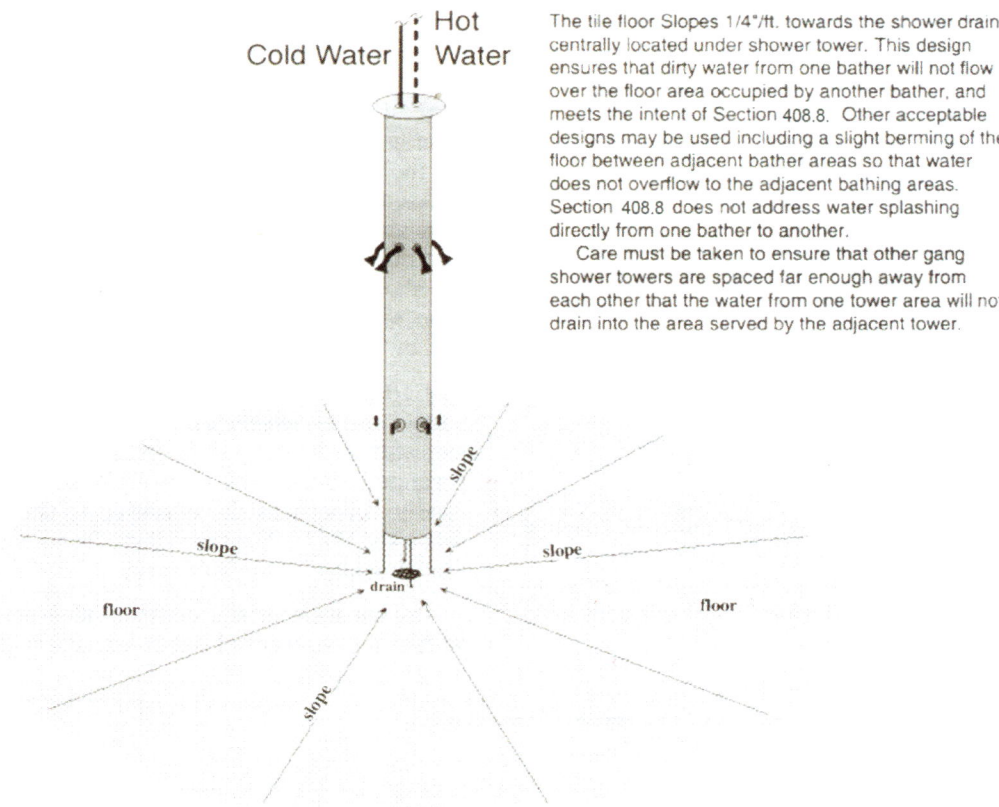

**FIGURE 408.8
GANG SHOWERS**

bathtubs shall be installed in accordance with the manufacturer's installation instructions. Access openings shall be of a size and opening to permit the removal and replacement of the circulation pump.

Whirlpool pump access located in the crawl space shall be located not more than 20 feet (6096 mm) from an access door, trap door, or crawl hole.

The circulation pump shall be located above the crown weir of the trap.

The pump and the circulation piping shall be self-draining to minimize water retention. Suction fittings on whirlpool bathtubs shall comply with ASME A112.19.7/CSA B45.10.

The criteria listed above for bathtubs and whirlpool bathtubs provide access for repair and maintenance. The location of the pump above the crown weir of the trap ensures that the water in the pump will drain with each use so that water from one bather will not be used for the next bather (see **Figure 409.6**).

Severe injuries and even deaths have occurred when suction fittings not listed for use in whirlpool bathtubs have been used. Injuries have included torn scalps and drowning deaths have occurred from hair being caught in the suction fitting not allowing the individual to rise out of the water. Take care that only listed suction fittings for use with whirlpool bathtubs are used.

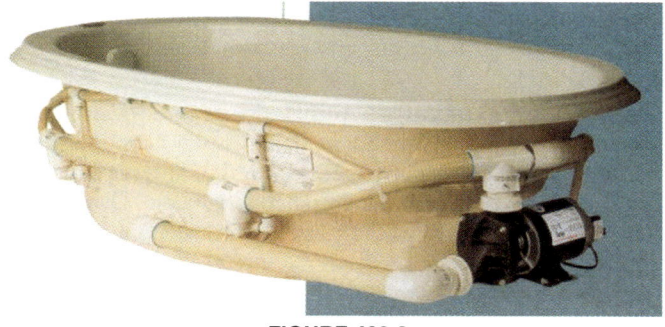

**FIGURE 409.6
WHIRLPOOL TUB**

409.6.1 Flexible PVC Hoses and Tubing. Flexible PVC hoses and tubing intended to be used on whirlpool bathtub water circulation systems or pneumatic systems shall comply with IAPMO Z1033.

410.0 Bidets.

410.1 Application. Bidets shall comply with ASME A112.19.2/CSA B45.1 or ASME A112.19.3/CSA B45.4.

410.2 Backflow Protection. The water supply to the bidet shall be protected by an air gap or in accordance with Section 603.3.2, Section 603.3.5, or Section 603.3.6.

410.3 Limitation of Water Temperature in Bidets. The maximum hot water temperature discharging from a bidet shall be limited to 110°F (43°C) by a device that complies with ASSE 1070/ASME A112.1070/CSA B125.70. The water heater thermostat shall not be considered a control for meeting this provision.

🔧 Water pressure fluctuations resulting in loss of cold water flow has and could cause significant risk of scalding to the bidet user. A temperature limiting device is required limiting the temperature to 110°F.

411.0 Water Closets.

411.1 Application. Water closets shall comply with ASME A112.19.2/CSA B45.1, ASME A112.19.3/CSA B45.4, or CSA B45.5/IAPMO Z124. Water closet bowls for public use shall be of the elongated type. In nurseries, schools, and other similar places where plumbing fixtures are provided for the use of children less than 6 years of age, water closets shall be of a size and height suitable for children's use.

411.2 Water Consumption. Water closets shall have a maximum consumption not to exceed 1.6 gallons (6.0 Lpf) of water per flush.

411.2.1 Dual Flush Water Closets. Dual flush water closets shall comply with ASME A112.19.14. The effective flush volume for dual flush water closets shall be defined as the composite, average flush volume of two reduced flushes and one full flush.

411.2.2 Flushometer Valve Activated Water Closets. Flushometer valve activated water closets shall have a maximum flush volume of 1.6 gallons (6.0 Lpf) of water per flush.

411.3 Water Closet Seats. Water closet seats shall be properly sized for the water closet bowl type, and shall be of smooth, non-absorbent material. Seats, for public use, shall be of the elongated type and either of the open front type or have an automatic seat cover dispenser. Plastic seats shall comply with IAPMO Z124.5.

🔧 Health department concerns regarding unintentional contact with either a toilet seat or toilet bowl has resulted in this code section. Transmittal of various diseases from one person to another may potentially occur when a person comes in contact with a fixture that was contaminated by the previous user. Open front seats and elongated bowls reduce the likelihood of user contact, as well as an automatic seat cover dispenser (see **Figure 411.3** and Learning Link **goo.gl/9gvmAb**).

411.4 Personal Hygiene Devices. Water closets with integral personal hygiene devices shall comply with ASME A112.4.2/CSA B45.16.

🔧 Water closets that that utilize sprays, dryers, heated seats or other personnel hygiene devices may have installation requirements such as dedicated circuits or ground fault circuit interrupters per local electrical codes. In addition, the potable water supply shall be protected in compliance with Section 603.2 for the degree of hazard.

412.0 Urinals.

412.1 Application. Urinals shall comply with ASME A112.19.2/CSA B45.1, ASME A112.19.19, or CSA B45.5/IAPMO Z124. Urinals shall have an average water consumption not to exceed 1 gallon (3.8 Lpf) of water per flush.

**FIGURE 411.3
AUTOMATIC SEAT COVER DISPENSER**

412.1.1 Nonwater Urinals. Nonwater urinals shall have a liquid barrier sealant to maintain a trap seal. Nonwater urinals shall permit the uninhibited flow of waste through the urinal to the sanitary drainage system. Nonwater urinals shall be cleaned and maintained in accordance with the manufacturer's instructions after installation. Where nonwater urinals are installed, not less than one water supplied fixture rated at not less than 1 water supply fixture unit (WSFU) shall be installed upstream on the same drain line to facilitate drain line flow and rinsing. Where nonwater urinals are installed, they shall have a water distribution line rough-in to each individual urinal location to allow for the installation of an approved backflow prevention device in the event of a retrofit.

🔧 Nonwater urinals are designed to allow urine to pass through a trap or trap like device without the use of water. This was designed to allow significant water savings, especially in the areas with significant water shortages. Nonwater urinals are popular in meeting requirements in "green" building projects. The UPC requires a water distribution line to be roughed in for each nonwater urinal in the event a water supplied urinal replaces the nonwater urinal. This rough-in water distribution line may requre a downstream shutoff valve to prevent dead ends.

The upstream fixture required in this code section must be 1 WSFU or greater. Water is needed to clean the waste line from the nonwater urinals. Technically, a WSFU assumes a fixture that has a probability distribution for frequency of use. With an upstream fixture, the amount of water used to facilitate drain rinsing will correspond to the frequency of use. There will be greater frequency of use during times of congestion. When there is congestion, there is more use of the upstream fixture and therefore greater volume is provided to rinse the drain. When there is no congestion, there will be less use of the upstream fixture, reducing the volume amount needed to rinse the drain.

412.2 Backflow Protection. A water supply to a urinal shall be protected by an approved-type vacuum breaker or other approved backflow prevention device in accordance with Section 603.5.

PLUMBING FIXTURES AND FIXTURE FITTINGS

🔧 Experiments have proven that siphonage can occur in a urinal even up the side wall of the urinal; therefore, each urinal water supply shall be protected by an approved vacuum breaker. Usually the urinal is supplied water by a flushometer valve (see **Figure 412.2**). Chapter 6 states the requirement for the flushometer valve – installation to the critical level of the valve at six inches (153 mm) above the top of the fixture. Also take care in observing the critical level of the valve; most are marked with a "CL" on the valve riser just below the integral vacuum breaker. This is the critical level and not the center line of the valve.

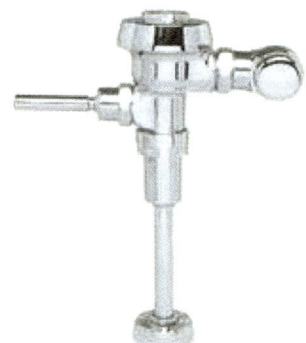

FIGURE 412.2
FLUSHOMETER VALVE

413.0 Flushing Devices.

413.1 Where Required. Each water closet, urinal, clinical sink, or other plumbing fixture that depends on trap siphonage to discharge its waste contents shall be provided with a flushometer valve, flushometer tank, or flush tank designed and installed so as to supply water in sufficient quantity and rate of flow to flush the contents of the fixture to which it is connected, to cleanse the fixture, and to refill the fixture trap, without excessive water use. Flushing devices shall comply with the antisiphon requirements in accordance with Section 603.5.

🔧 The water closet, urinal, and clinical sink are all fixtures that may have integral traps. The trap is siphoned rather than drained, and the trap is refilled by the flushing valve refilling the bowl area of the trap seal. Whichever method of valve is used – flushometer, flushometer tank, or flush tank (see **Figure 413.1**) – all must be protected from backflow.

413.2 Flushometer Valves. Flushometer valves and flushometer tanks shall comply with ASSE 1037/ASME A112.1037/CSA B125.37, and shall be installed in accordance with Section 603.5.1. No manually controlled flushometer valve shall be used to flush more than one urinal, and each such urinal flushometer valve shall be an approved, self-closing type discharging a predetermined quantity of water. Flushometers shall be installed so that they will be accessible for repair. Flushometer valves shall not be used where the water pressure is insufficient to operate them properly. Where the valve is operated, it shall complete the cycle of operation automatically, opening fully, and closing positively under the line water pressure. Each flushometer shall be provided with a means for regulating the flow through it.

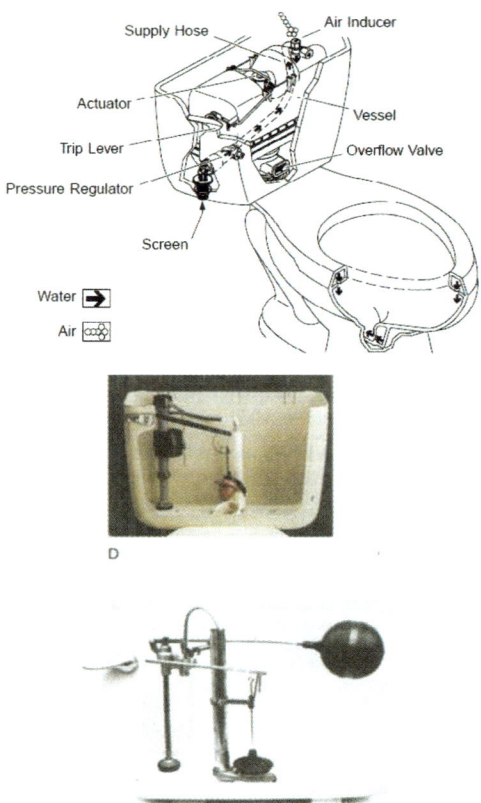

FIGURE 413.1
FLUSHOMETER AND FLUSH-TYPE WATER CLOSETS

413.3 Flush Tanks. Flush tanks for manual flushing shall be equipped with a flush valve that complies with ASME A112.19.5/CSA B45.15 and an antisiphon fill valve (ballcock) that complies with ASSE 1002/ASME A112.1002/CSA B125.12 and installed in accordance with Section 603.5.2.

413.4 Water Supply for Flush Tanks. An adequate quantity of water shall be provided to flush and clean the fixture served. The water supply for flushing tanks and flushometer tanks equipped for manual flushing shall be controlled by a float valve or other automatic device designed to refill the tank after each discharge and to shut completely off the water flow to the tank where the tank is filled to operational capacity. Provision shall be made to automatically supply water to the fixture to refill the trap seal after each flushing.

413.5 Overflows in Flush Tanks. Flush tanks shall be provided with overflows discharging into the water closet or urinal connected thereto. Overflows supplied as original parts with the fixture shall be of sufficient size to prevent tank flooding at the maximum rate at which the tank is supplied with water under normal operating conditions and where installed in accordance with the manufacturer's installation instructions.

414.0 Dishwashing Machines.

414.1 Application. Domestic dishwashing machines shall comply with UL 749. Commercial dishwashing machines shall comply with NSF 3 and UL 921.

414.2 Backflow Protection. The water supply connection to a commercial dishwashing machine shall be protected by an air gap or a backflow prevention device in accordance with Section 603.3.2, Section 603.3.5, Section 603.3.6, or that complies with ASSE 1004.

Water supply connections to commercial dishwashers need to be protected against possible contamination from the chemicals used in conjunction with dishwasher operations. This protection can be accomplished by the use of backflow prevention devices suitable for the application and water temperature.

414.3 Drainage Connection. Domestic dishwashing machines shall discharge indirectly through an air gap fitting in accordance with Section 807.3 into a waste receptor, a wye branch fitting on the tailpiece of a kitchen sink, or dishwasher connection of a food waste disposer. Commercial dishwashing machines shall discharge indirectly through an air break or direct connection. The indirect discharge for commercial dishwashing machines shall be in accordance with Section 807.1, and the direct discharge shall be in accordance with Section 704.3.

Figure 807.3a and **Figure 807.3b** in Chapter 8 illustrates the proper drainage connections for domestic dishwashers. Commercial dishwashers need to discharge through an air break or direct connection. If directly connected to the waste system, the commercial dishwasher shall be provided with floor drain protection as shown in **Figure 704.3** in Chapter 7.

415.0 Drinking Fountains.

415.1 Application. Drinking fountains shall be self-closing and comply with ASME A112.19.1/CSA B45.2, ASME A112.19.2/CSA B45.1, or ASME A112.19.3/CSA B45.4. Drinking fountains shall also comply with NSF 61. Permanently installed electric water coolers shall also comply with UL 399.

415.2 Drinking Fountain Alternatives. Where food is consumed indoors, water stations shall be permitted to be substituted for drinking fountains. Bottle filling stations shall be permitted to be substituted for drinking fountains up to 50 percent of the requirements for drinking fountains. Drinking fountains shall not be required for an occupant load of 30 or less.

415.3 Drainage Connection. Drinking fountains shall be permitted to discharge directly into the drainage system or indirectly through an air break in accordance with Section 809.1.

A drinking fountain is the only plumbing fixture that may be installed using either a direct or indirect connection chosen at the discretion of the installer. If the drinking fountain discharges by means of an indirect waste pipe, all of the provisions of Chapter 8 apply to this installation. Because the drinking fountain is supplied with a water distribution airgap, it may be installed with an indirect waste discharging through an airbreak. For building occupancy loads greater than 30, Table 422.1 guides the designer or installer on the number of drinking fountains required per type of occupancy.

415.4 Location. Drinking fountains shall not be installed in toilet rooms.

416.0 Emergency Eyewash and Shower Equipment.

416.1 Application. Emergency eyewash and shower equipment shall comply with ISEA Z358.1.

These showers are designed as safety features and act as a water deluge to wash away contaminants or toxins from a worker's body (see **Figure 416.1**). Therefore, they are not subject to water conservation measures. The International Safety Equipment Association (ISEA) Z358.1, Emergency Eye Wash and Shower Equipment, was revised in 2014. This standard helps the user in selecting and installing emergency equipment to meet OSHA requirements. Some manufactures have developed a compliance checklist to assist the installer in meeting the many requirements of this standard. Drains are not required for this emergency equipment making the installation more accessible to the hazard locations. When drains are provided the requirements in Section 811.0 must be followed since chemical or industrial liquid wastes are likely to damage or increase maintenance costs on the sanitary sewer system.

**FIGURE 416.1
EMERGENCY EYEWASH AND SHOWER STATION**

416.2 Water Supply. Emergency eyewash and shower equipment shall not be limited in the water supply flow rates. Where hot and cold water is supplied to an emergency shower or eyewash station, the temperature of the water supply shall be controlled by a temperature actuated mixing valve complying with ASSE 1071. The flow rate, discharge pattern, and temperature of flushing fluids shall be provided in accordance with ISEA Z358.1.

Where hot and cold water are supplied to this equipment, a temperature actuated mixing valve complying with ASSE 1071 must be installed. Good plumbing practice would also dictate that the hot water supply be on a circulating system to provide optimum water temperatures to the emergency equipment when in use.

Care should be taken when installing these types of safety equipment to insure water temperatures will not cause additional damage to parts of the body affected by chemical contamination. Section 4.5.6 of ISEA Z358 maintains that water temperatures that may accelerate chemical reaction should be tempered to lessen the effects of the chemicals on affected body parts. Additionally, the standard recommends that the safety/health advisor for the facility to be consulted to establish the optimum water temperature for the specific types of chemical or chemicals that may be found in areas near the emergency eyewash or shower equipment. Specific water temperature guidelines are not found in the standard, but a temperature range of not less than 60°F (15.5°C) to prevent hypothermia to not more than 100°F (38°C) to prevent additional damage to eyes is recommended.

416.3 Installation. Emergency eyewash and shower equipment shall be installed in accordance with the manufacturer's installation instructions.

416.4 Location. Emergency eyewash and shower equipment shall be located on the same level as the hazard and accessible for immediate use. The path of travel shall be free of obstructions and shall be clearly identified with signage.

416.5 Drain. A drain shall not be required for emergency eyewash or shower equipment. Where a drain is provided, the discharge shall be in accordance with Section 811.0.

417.0 Faucets and Fixture Fittings.

417.1 Application. Faucets and fixture fittings shall comply with ASME A112.18.1/CSA B125.1. Fixture fittings covered under the scope of NSF 61 shall comply with the requirements of NSF 61.

A fixture fitting is a device that controls the flow of water to a fixture. Most often it refers to a faucet or control valve such as a shower valve. It can either be attached as a lavatory faucet or be separate but accessible such as a mop sink faucet or spray valve (see **Figure 417.1**).

NSF 61 is a standard created by The National Sanitation Foundation, a standards development organization dedicated to public health (see www.nsf.org for more information about NSF), to cover products that are specific to and come in contact with drinking water. To be certified as compliant with NSF 61, the product, whether it be pipe, fixture, or fixture fitting, must be tested and found not to contain contaminants or toxins and that using the product will sustain the integrity of the drinking water. All materials used in the potable water system must be compliant to NSF 61.

FIGURE 417.1
FIXTURE FITTINGS

417.2 Deck Mounted Bath/Shower Valves. Deck mounted bath/shower transfer valves with integral backflow protection shall comply with ASME A112.18.1/CSA B125.1. This shall include handheld showers, and other bathing appliances mounted on the deck of bathtubs or other bathing appliances that incorporate a hose or pull out feature.

417.3 Handheld Showers. Handheld showers shall comply with ASME A112.18.1/CSA B125.1. Handheld showers with integral backflow protection shall comply with ASME A112.18.1/CSA B125.1 or shall have a backflow prevention device that complies with ASME A112.18.3 or ASSE 1014.

With the many different variations of bath/shower valves, handheld showers and faucets with hose connections (see Section 417.4), it is imperative that the water supply is protected from possible contamination. These devises are required to comply with the ASME or ASSE standard when manufactured with integral backflow protection or protected by the use of a backflow device.

417.4 Faucets and Fixture Fittings with Hose Connected Outlets. Faucets and fixture fittings with pull out spout shall comply with ASME A112.18.1/CSA B125.1. Faucets and fixture fittings with pull out spouts with integral backflow protection shall comply with ASME A112.18.1/CSA B125.1 or shall have a backflow preventer device that complies with ASME A112.18.3.

417.5 Separate Controls for Hot and Cold Water. Where two separate handles control the hot and cold water, the left-hand control of the faucet where facing the fixture fitting outlet shall control the hot water. Faucets and diverters shall be connected to the water distribution system so that hot water corresponds to the left side of the fixture fitting.

Single-handle mixing valves installed in showers and tub-shower combinations shall have the flow of hot water corresponding to the markings on the fixture fitting.

This requirement of hot on the left hand side makes the water supply uniform to prevent accidental use of the hot water and possible scalding if the temperature is too high (see **Figure 417.5**). "Left side hot, right side cold" has been the traditional position for water controls on plumbing fixtures. The introduction of single-handle faucets or fixture fittings is a separate issue. The ability to reverse cartridges or use different markings on the faucets to accommodate the correct relationship of hot water to cold has made the use of back-to-back tub and shower installations much easier. The markings on the mixing valve for showers and tub-shower combinations, however, must correspond to the actual direction for hot water.

FIGURE 417.5
FAUCET HOT AT LEFT

PLUMBING FIXTURES AND FIXTURE FITTINGS

418.0 Floor Drains.

418.1 Application. Floor drains shall comply with ASME A112.3.1, ASME A112.6.3, or CSA B79.

418.2 Strainer. Floor drains shall be considered plumbing fixtures and each such drain shall be provided with an approved-type strainer having a waterway equivalent to the area of the tailpiece. Floor drains shall be of an approved type and shall provide a watertight joint on the floor.

🔧 Floor drains are plumbing fixtures and are intended to collect accidental spills, overflows or water utilized for washing the floor surface. They are not intended as secondary fixtures that receive waste from primary fixtures that require either an airgap or an airbreak. That fixture is the floor receptor or floor sink. Each has a strainer and a flange to provide for the collection of leakage, which passes by the strainer and may cause damage to the floor below.

418.3 Location of Floor Drains. Floor drains shall be installed in the following areas:

(1) Toilet rooms containing two or more water closets or a combination of one water closet and one urinal, except in a dwelling unit.
(2) Commercial kitchens and in accordance with Section 704.3.
(3) Laundry rooms in commercial buildings and common laundry facilities in multi-family dwelling buildings.
(4) Boiler rooms.

🔧 As stated above, the floor drain is intended for emergency use. Public toilet rooms, commercial kitchens and laundry rooms are susceptible to overflow of the fixtures in the room and require a floor drain to contain that spillage. Boiler rooms require a floor drain for disposing of the accumulation of liquid wastes incident to cleaning, recharging, and routine maintenance. Single-dwelling units (homes, apartments or condos) are not required to have a floor drain.

418.4 Food Storage Areas. Where drains are provided in storerooms, walk-in freezers, walk-in coolers, refrigerated equipment, or other locations where food is stored, such drains shall have indirect waste piping. Separate waste pipes shall be run from each food storage area, each with an indirect connection to the building sanitary drainage system. Traps shall be provided in accordance with Section 801.3.2 of this code and shall be vented.

Indirect drains shall be permitted to be located in freezers or other spaces where freezing temperatures are maintained, provided that traps, where supplied, shall be located where the seal will not freeze. Otherwise, the floor of the freezer shall be sloped to a floor drain located outside of the storage compartment.

🔧 The floor drain located in food storage areas is required to drain to the drainage system by means of an indirect drain. This ensures that if there is backflow in the system, it will not enter the food storage area and contaminate the food. See Chapter 8 for further requirements.

418.5 Floor Slope. Floors shall be sloped to floor drains.

🔧 An important part of plumbing system installation is the protection of the system while the concrete slab is poured. A floor drain is installed with grade in mind and the strainer top should not be raised level to the floor. Care should be taken to ensure that the floor drain maintains proper elevation for drainage of the floor area.

419.0 Food Waste Disposers.

419.1 Application. Food waste disposal units shall comply with UL 430. Residential food waste disposers shall also comply with ASSE 1008.

419.2 Drainage Connection. Approved wye or other directional-type branch fittings shall be installed in continuous wastes connecting or receiving the discharge from a food waste disposer. No dishwasher drain shall be connected to a sink tailpiece, continuous waste, or trap on the discharge side of a food waste disposer.

🔧 Any type of force or pumped drain unit such as food waste disposers must enter the system through a wye type or a directional fitting (see **Figure 419.2a**). This will provide a smooth path for the waste rather than a tee connection, which would cause the waste to flow against the back of the receiving pipe and flow both upward and downward because of the force of the pumping action.

Dishwasher drains are not permitted to be connected to a sink tailpiece, continuous waste or trap on the discharge side of a food waste disposal unit. This is because of the possibility of the pressured waste from either the disposal or the dishwasher backing up into the other. If a disposal is present, then the dishwasher waste must connect to the disposal inlet supplied with the disposal (see **Figure 419.2b**).

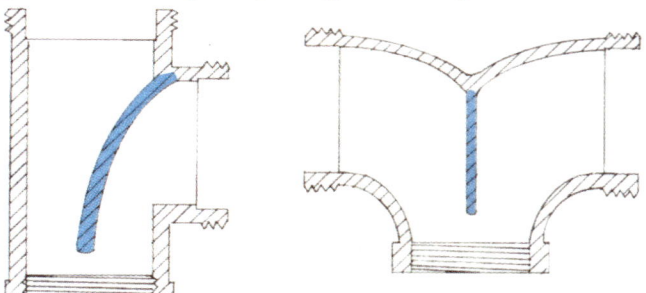

**FIGURE 419.2A
DIRECTIONAL FITTING**

419.3 Water Supply. A cold water supply shall be provided for food waste disposers. Such connection to the water supply shall be protected by an air gap or backflow prevention device in accordance with Section 603.2.

420.0 Sinks.

420.1 Application. Sinks shall comply with ASME A112.19.1/CSA B45.2, ASME A112.19.2/CSA B45.1, ASME A112.19.3/CSA B45.4, CSA B45.5/IAPMO Z124, CSA B45.8/IAPMO Z403, or CSA B45.12/IAPMO Z402. Moveable sink systems shall comply with ASME A112.19.12.

PLUMBING FIXTURES AND FIXTURE FITTINGS

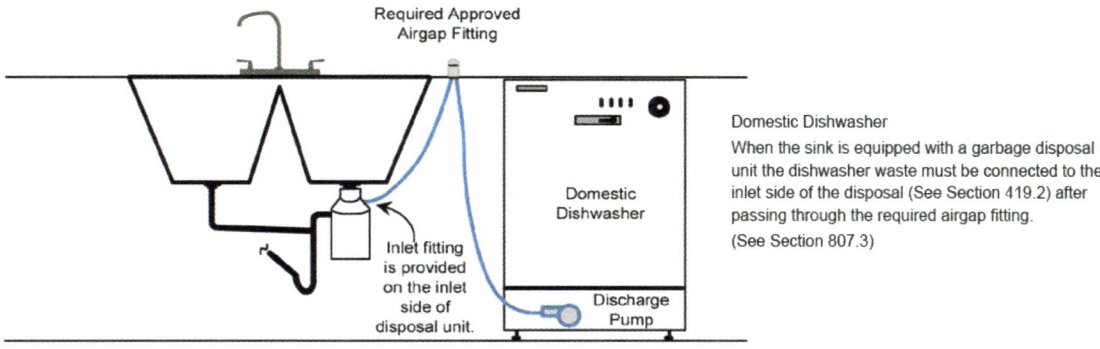

FIGURE 419.2B
CORRECT INSTALLATION OF DISHWASHER AND GARBAGE DISPOSAL

420.2 Water Consumption. Sink faucets shall have a maximum flow rate of not more than 2.2 gpm at 60 psi (8.3 L/m at 414 kPa).

Exceptions:

(1) Clinical sinks

(2) Laundry trays

(3) Service sinks

420.3 Pre-Rinse Spray Valve. Commercial food service pre-rinse spray valves shall have a maximum flow rate of 1.6 gallons per minute (gpm) at 60 pounds-force per square inch (psi) (6.0 L/m at 414 kPa) and shall be equipped with an integral automatic shutoff.

A pre-rinse spray valve is a handheld device that uses a spray of water to remove food waste from dishes prior to cleaning in a commercial dishwasher (see **Figure 420.3**). Pre-rinse spray valves typically include a nozzle, squeeze lever, dish guard bumper and usually have a clip to lock the handle in the on position. Prior to January 1, 2006 pre-rinse spray valves had flow rates ranging from one to five gallons per minute. With a typical restaurant consuming over two-thirds of their total water usage at the dishwashing station and in some cases half of that consumed by the pre-rinse valve, regulating the maximum flow from the valve will have substantial water and energy savings.

FIGURE 420.3
PRE-RINSE SPRAY VALVE

420.4 Waste Outlet. Kitchen and laundry sinks shall have a waste outlet and fixture tailpiece not less than 1½ inches (40 mm) in diameter. Service sinks shall have a waste outlet and fixture tailpiece not less than 2 inches (50 mm) in diameter. Fixture tailpieces shall be constructed from the materials specified in Section 701.2 for drainage piping. Waste outlets shall be provided with an approved strainer.

421.0 Floor Sinks.

421.1 Application. Floor sinks shall comply with ASME A112.6.7.

421.2 Strainers. The waste outlet of a floor sink shall be provided with an approved strainer or grate that is removable and accessible.

See **Figure 702.1b**

422.0 Minimum Number of Required Fixtures.

422.1 Fixture Count. Plumbing fixtures shall be provided for the type of building occupancy and in the minimum number shown in Table 422.1. The total occupant load and occupancy classification shall be determined in accordance with the building code. Occupancy classification not shown in Table 422.1 shall be considered separately by the Authority Having Jurisdiction.

The minimum number of fixtures shall be calculated at 50 percent male and 50 percent female based on the total occupant load. Where information submitted indicates a difference in the distribution of the sexes such information shall be used to determine the number of fixtures for each sex. Once the occupancy load and occupancy are determined, Table 422.1 shall be applied to determine the minimum number of plumbing fixtures required. Where applying the fixture ratios in Table 422.1 results in fractional numbers, such numbers shall be rounded to the next whole number. For multiple occupancies, fractional numbers shall be first summed and then rounded to the next whole number.

Before you can use Table 422.1 as it is intended, it is crucial to note the second sentence in this section that states "The total occupant load and occupancy classification shall be determined in accordance with the building code". You must have that information before you attempt to use Table 422.1. The Authority Having Jurisdiction will determine the occupant load for the building or space.

The revised Table 422.1 is based on research conducted by the American Society of Plumbing Engineers (ASPE) and the Stevens Institute. The research is based on the "Queuing Theory" for plumbing design. This theory provides answers to the following questions:

- How long might a user wait?
- How many people will form in the line?

- How many service equipment items are needed?

The theory provides a tool for determining the number of plumbing fixtures for a preferred level of service expressed in:

- Waiting times during peak periods of use.
- Fixture utilization.
- The probability of finding a vacant fixture.

Fixture counts in the past tended to underestimate the needs for small occupancies and to overestimate the needs for larger occupancies. The "Queuing Theory" tends to provide a better analysis of user service distribution to determine the minimum number of fixtures. The revisions correlate with the classifications of building codes along with recommendations from the American Restroom Association and the Journal of Planning Literature, Gender and Family Issues in Planning and Designing Public Restrooms. The revisions also address issues such as potty parity, employee and customer facilities, occupancies with peak usage, and waiting lines for women in various occupancies.

422.1.1 Family or Assisted-Use Toilet and Bathing Facilities. Where family or assisted-use toilet and bathing rooms are required, in applicable building regulations, the facilities shall be installed in accordance with those regulations.

422.2 Separate Facilities. Separate toilet facilities shall be provided for each sex.

Exceptions:

(1) Residential installations.
(2) In occupancies with a total occupant load of 10 or less, including customers and employees, one toilet facility, designed for use by no more than one person at a time, shall be permitted for use by both sexes.
(3) In business and mercantile occupancies with a total occupant load of 50 or less including customers and employees, one toilet facility, designed for use by no more than one person at a time, shall be permitted for use by both sexes.

422.2.1 Single Use Toilet Facilities. Single use toilet facilities and family or assisted use toilet facilities shall be identified with signage indicating use by either sex.

422.2.2 Family or Assisted-Use Toilet Facilities. Where a separate toilet facility is required for each sex, and each toilet facility is required to have only one water closet, two family or assisted-use toilet facilities shall be permitted in place of the required separate toilet facilities.

422.3 Fixture Requirements for Special Occupancies. Additional fixtures shall be permitted to be required where unusual environmental conditions or referenced activities are encountered. In food preparation areas, fixture requirements shall be permitted to be dictated by health codes.

422.4 Toilet Facilities Serving Employees and Customers. Each building or structure shall be provided with toilet facilities for employees and customers. Requirements for customers and employees shall be permitted to be met with a single set of restrooms accessible to both groups.

Required toilet facilities for employees and customers located in shopping malls or centers shall be permitted to be met by providing a centrally located toilet facility accessible to several stores. The maximum travel distance from entry to any store to the toilet facility shall not exceed 300 feet (91 440 mm).

Required toilet facilities for employees and customers in other than shopping malls or centers shall have a maximum travel distance not to exceed 500 feet (152 m).

422.4.1 Access to Toilet Facilities. In multi-story buildings, accessibility to the required toilet facilities shall not exceed one vertical story. Access to the required toilet facilities for customers shall not pass through areas designated as for employee use only such as kitchens, food preparation areas, storage rooms, closets, or similar spaces. Toilet facilities accessible only to private offices shall not be counted to determine compliance with this section.

422.5 Toilet Facilities for Workers. Toilet facilities shall be provided and maintained in a sanitary condition for the use of workers during construction.

PLUMBING FIXTURES AND FIXTURE FITTINGS

TABLE 422.1
MINIMUM PLUMBING FACILITIES[1]

Each building shall be provided with sanitary facilities, including provisions for persons with disabilities as prescribed by the Department Having Jurisdiction. Table 422.1 applies to new buildings, additions to a building, and changes of occupancy or type in an existing building resulting in increased occupant load.

TYPE OF OCCUPANCY[2]	WATER CLOSETS (FIXTURES PER PERSON)[3]		URINALS (FIXTURES PER PERSON)[4]	LAVATORIES (FIXTURES PER PERSON)[5,6]		BATHTUBS OR SHOWERS (FIXTURES PER PERSON)	DRINKING FOUNTAINS/ FACILITIES (FIXTURES PER PERSON)	OTHER
A-1 Assembly occupancy (fixed or permanent seating)- theatres, concert halls and auditoriums	Male 1: 1-100 2: 101-200 3: 201-400	Female 1: 1-25 2: 26-50 3: 51-100 4: 101-200 6: 201-300 8: 301-400	Male 1: 1-200 2: 201-300 3: 301-400 4: 401-600	Male 1: 1-200 2: 201-400 3: 401-600 4: 601-750	Female 1: 1-100 2: 101-200 3: 201-300 4: 301-500 5: 301-500 6: 501-750	—	1: 1-250 2: 251-500 3: 501-750	1 service sink or laundry tray
	Over 400, add 1 fixture for each additional 500 males and 1 fixture for each additional 125 females.		Over 600, add 1 fixture for each additional 300 males.	Over 750, add 1 fixture for each additional 250 males and 1 fixture for each additional 200 females.			Over 750, add 1 fixture for each additional 500 persons.	
A-2 Assembly occupancy- restaurants, pubs, lounges, night clubs and banquet halls	Male 1: 1-50 2: 51-150 3: 151-300 4: 301-400	Female 1: 1-25 2: 26-50 3: 51-100 4: 101-200 6: 201-300 8: 301-400	Male 1: 1-200 2: 201-300 3: 301-400 4: 401-600	Male 1: 1-150 2: 151-200 3: 201-400	Female 1: 1-150 2: 151-200 4: 201-400	—	1: 1-250 2: 251-500 3: 501-750	1 service sink or laundry tray
	Over 400, add 1 fixture for each additional 250 males and 1 fixture for each 125 females.		Over 600, add 1 fixture for each additional 300 males.	Over 400, add 1 fixture for each additional 250 males and 1 fixture for each additional 200 females.			Over 750, add 1 fixture for each additional 500 persons.	
A-3 Assembly occupancy (typical without fixed or permanent seating)- arcades, places of worship, museums, libraries, lecture halls, gymnasiums (without spectator seating), indoor pools (without spectator seating)	Male 1: 1-100 2: 101-200 3: 201-400	Female 1: 1-25 2: 26-50 3: 51-100 4: 101-200 6: 201-300 8: 301-400	Male 1: 1-100 2: 101-200 3: 201-400 4: 401-600	Male 1: 1-200 2: 201-400 3: 401-600 4: 601-750	Female 1: 1-100 2: 101-200 4: 201-300 5: 301-500 6: 501-750	—	1: 1-250 2: 251-500 3: 501-750	1 service sink or laundry tray
	Over 400, add 1 fixture for each additional 500 males and 1 fixture for each additional 125 females.		Over 600, add 1 fixture for each additional 300 males.	Over 750, add 1 fixture for each additional 250 males and 1 fixture for each additional 200 females.			Over 750, add 1 fixture for each additional 500 persons.	
A-4 Assembly occupancy (indoor activities or sporting events with spectator seating)- swimming pools, skating rinks, arenas and gymnasiums	Male 1: 1-100 2: 101-200 3: 201-400	Female 1: 1-25 2: 26-50 3: 51-100 4: 101-200 6: 201-300 8: 301-400	Male 1: 1-100 2: 101-200 3: 201-400 4: 401-600	Male 1: 1-200 2: 201-400 3: 401-750	Female 1: 1-100 2: 101-200 4: 201-300 5: 301-500 6: 501-750	—	1: 1-250 2: 251-500 3: 501-750	1 service sink or laundry tray
	Over 400, add 1 fixture for each additional 500 males and 1 fixture for each additional 125 females.		Over 600, add 1 fixture for each additional 300 males.	Over 750, add 1 fixture for each additional 250 males and 1 fixture for each additional 200 females.			Over 750, add 1 fixture for each additional 500 persons.	

PLUMBING FIXTURES AND FIXTURE FITTINGS

TABLE 422.1
MINIMUM PLUMBING FACILITIES[1] (continued)

TYPE OF OCCUPANCY[2]	WATER CLOSETS (FIXTURES PER PERSON)[3]		URINALS (FIXTURES PER PERSON)[4]	LAVATORIES (FIXTURES PER PERSON)[5, 6]		BATHTUBS OR SHOWERS (FIXTURES PER PERSON)	DRINKING FOUNTAINS/ FACILITIES (FIXTURES PER PERSON)	OTHER
A-5 Assembly occupancy (outdoor activities or sporting events)- amusement parks, grandstands and stadiums	Male 1: 1-100 2: 101-200 3: 201-400	Female 1: 1-25 2: 26-50 3: 51-100 4: 101-200 6: 201-300 8: 301-400	Male 1: 1-100 2: 101-200 3: 201-400 4: 401-600	Male 1: 1-200 2: 201-400 3: 401-750	Female 1: 1-100 2: 101-200 4: 201-300 5: 301-500 6: 501-750	—	1: 1-250 2: 251-500 3: 501-750	1 service sink or laundry tray
	Over 400, add 1 fixture for each additional 500 males and 1 fixture for each additional 125 females.		Over 600, add 1 fixture for each additional 300 males.	Over 750, add 1 fixture for each additional 250 males and 1 fixture for each additional 200 females.			Over 750, add 1 fixture for each additional 500 persons.	
B Business occupancy (office, professional or service type transactions)- banks, vet clinics, hospitals, car wash, banks, beauty salons, ambulatory health care facilities, laundries and dry cleaning, educational institutions (above high school), or training facilities not located within school, post offices and printing shops	Male 1: 1-50 2: 51-100 3: 101-200 4: 201-400	Female 1: 1-15 2: 16-30 3: 31-50 4: 51-100 8: 101-200 11: 201-400	Male 1: 1-100 2: 101-200 3: 201-400 4: 401-600	Male 1: 1-75 2: 76-150 3: 151-200 4: 201-300 5: 301-400	Female 1: 1-50 2: 51-100 3: 101-150 4: 151-200 5: 201-300 6: 301-400	—	1 per 150	1 service sink or laundry tray
	Over 400, add 1 fixture for each additional 500 males and 1 fixture for each additional 150 females.		Over 600, add 1 fixture for each additional 300 males.	Over 400, add 1 fixture for each additional 250 males and 1 fixture for each additional 200 females.				
E Educational occupancy-private or public schools	Male 1 per 50	Female 1 per 30	Male 1 per 100	Male 1 per 40	Female 1 per 40	—	1 per 150	1 service sink or laundry tray
F1, F2 Factory or Industrial occupancy-fabricating or assembly work	Male 1: 1-50 2: 51-75 3: 76-100	Female 1: 1-50 2: 51-75 3: 76-100	—	Male 1: 1-50 2: 51-75 3: 76-100	Female 1: 1-50 2: 51-75 3: 76-100	1 shower for each 15 persons exposed to excessive heat or to skin contamination with poisonous, infectious or irritating material.	1: 1-250 2: 251-500 3: 501-750	1 service sink or laundry tray
	Over 100, add 1 fixture for each additional 40 persons.		—	Over 100, add 1 fixture for each additional 40 persons.			Over 750, add 1 fixture for each additional 500 persons.	
I-1 Institutional occupancy (houses more than 16 persons on a 24-hour basis)- substance abuse centers, assisted living, group homes, or residential facilities	Male 1 per 15	Female 1 per 15	—	Male 1 per 15	Female 1 per 15	1 per 8	1 per 150	1 service sink or laundry tray

PLUMBING FIXTURES AND FIXTURE FITTINGS

TABLE 422.1
MINIMUM PLUMBING FACILITIES[1] (continued)

TYPE OF OCCUPANCY[2]		WATER CLOSETS (FIXTURES PER PERSON)[3]		URINALS (FIXTURES PER PERSON)[4]	LAVATORIES (FIXTURES PER PERSON)[5,6]		BATHTUBS OR SHOWERS (FIXTURES PER PERSON)	DRINKING FOUNTAINS/ FACILITIES (FIXTURES PER PERSON)	OTHER
I-2 Institutional occupancy-medical, psychiatric, surgical or nursing homes	Hospitals and nursing homes-individual rooms and ward room	1 per room		—	1 per room		1 per room	1 per 150	1 service sink or laundry tray
		1 per 8 patients		—	1 per 10 patients		1 per 20 patients		
	Hospital Waiting or Visitor Rooms	1 per room		—	1 per room		—	1 per room	—
	Employee Use	Male 1: 1-15 2: 16-35 3: 36-55	Female 1: 1-15 3: 16-35 4: 36-55	—	Male 1 per 40	Female 1 per 40	—	—	—
		Over 55, add 1 fixture for each additional 40 persons.							
I-3 Institutional occupancy (houses more than 5 people)	Prisons	1 per cell		—	1 per cell		1 per 20	1 per cell block/floor	—
	Correctional facilities or juvenile center	1 per 8		—	1 per 10		1 per 8	1 per floor	1 service sink or laundry tray
	Employee Use	Male 1: 1-15 2: 16-35 3: 36-55	Female 1: 1-15 3: 16-35 4: 36-55		Male 1 per 40	Female 1 per 40	—	1 per 150	—
		Over 55, add 1 fixture for each additional 40 persons.							
I-4 Institutional occupancy (any age that receives care for less than 24 hours)		Male 1: 1-15 2: 16-35 3: 36-55	Female 1: 1-15 3: 16-35 4: 36-55		Male 1 per 40	Female 1 per 40		1 per 150	1 service sink or laundry tray
		Over 55, add 1 fixture for each additional 40 persons.							
M Mercantile occupancy (the sale of merchandise and accessible to the public)		Male 1: 1-100 2: 101-200 3: 201-400	Female 1: 1-100 2: 101-200 4: 201-300 6: 301-400	Male 0: 1-200 1: 201-400	Male 1: 1-200 2: 201-400	Female 1: 1-200 2: 201-300 3: 301-400	—	1: 1-250 2: 251-500 3: 501-750	1 service sink or laundry tray
		Over 400, add 1 fixture for each additional 500 males and 1 fixture for each 200 females.		Over 400, add 1 fixture for each additional 500 males.	Over 400, add 1 fixture for each additional 500 males and 1 fixture for each 400 females.		—	Over 750, add 1 fixture for each additional 500 persons.	—
R-1 Residential occupancy (minimal stay)-hotels, motels, bed and breakfast homes		1 per sleeping room		—	1 per sleeping room		1 per sleeping room	—	1 service sink or laundry tray

2018 UNIFORM PLUMBING CODE ILLUSTRATED TRAINING MANUAL

TABLE 422.1
MINIMUM PLUMBING FACILITIES[1] (continued)

TYPE OF OCCUPANCY[2]		WATER CLOSETS (FIXTURES PER PERSON)[3]		URINALS (FIXTURES PER PERSON)[4]	LAVATORIES (FIXTURES PER PERSON)[5,6]		BATHTUBS OR SHOWERS (FIXTURES PER PERSON)	DRINKING FOUNTAINS/ FACILITIES (FIXTURES PER PERSON)	OTHER
R-2 Residential occupancy (long-term or permanent)	Dormitories	Male 1 per 10	Female 1 per 8	1 per 25	Male 1 per 12	Female 1 per 12	1 per 8	1 per 150	1 service sink or laundry tray
		Add 1 fixture for each additional 25 males and 1 fixture for each additional 20 females.		Over 150, add 1 fixture for each additional 50 males.	Add 1 fixture for each additional 20 males and 1 fixture for each additional 15 females.				
	Employee Use	Male 1: 1-15 2: 16-35 3: 36-55	Female 1: 1-15 3: 16-35 4: 36-55		Male 1 per 40	Female 1 per 40	—	—	
		Over 55, add 1 fixture for each additional 40 persons							
	Apartment house/unit	1 per apartment		—	1 per apartment		1 per apartment	—	1 kitchen sink per apartment. 1 laundry tray or 1 automatic clothes washer connection per unit or 1 laundry tray or 1 automatic clothes washer connection for each 12 units
R-3 Residential occupancy (long-term or permanent in nature) for more than 5 but does not exceed 16 occupants)		Male 1 per 10	Female 1 per 8	—	Male 1 per 12	Female 1 per 12	1 per 8	1 per 150	1 service sink or laundry tray
		Add 1 fixture for each additional 25 males and 1 fixture for each additional 20 females.			Add 1 fixture for each additional 20 males and 1 fixture for each additional 15 females.				
R-3 Residential occupancy (one and two family dwellings)		1 per one and two family dwelling		—	1 per one and two family dwelling		1 per one and two family dwelling	—	1 kitchen sink and 1 automatic clothes washer connection per one and two family dwelling
R-4 Residential occupancy (residential care or assisted living)		Male 1 per 10	Female 1 per 8	—	Male 1 per 12	Female 1 per 12	1 per 8	1 per 150	1 service sink or laundry tray
		Add 1 fixture for each additional 25 males and 1 fixture for each additional 20 females.			Add 1 fixture for each additional 20 males and 1 fixture for each additional 15 females.				
S-1, S-2 Storage occupancy-storage of goods, warehouse, aircraft hanger, food products, appliances		Male 1: 1-100 2: 101-200 3: 201-400	Female 1: 1-100 2: 101-200 3: 201-400	—	Male 1: 1-200 2: 201-400 3: 401-750	Female 1: 1-200 2: 201-400 3: 401-750	—	1: 1-250 2: 251-500 3: 501-750	1 service sink or laundry tray
		Over 400, add 1 fixture for each additional 500 males and 1 fixture for each additional 150 females.			Over 750, add 1 fixture for each additional 500 persons.			Over 750, add 1 fixture for each additional 500 persons.	

Notes:
[1] The figures shown are based upon one fixture being the minimum required for the number of persons indicated or any fraction thereof.
[2] A restaurant is defined as a business that sells food to be consumed on the premises.
   a. The number of occupants for a drive-in restaurant shall be considered as equal to the number of parking stalls.
   b. Hand-washing facilities shall be available in the kitchen for employees.
[3] The total number of required water closets for females shall be not less than the total number of required water closets and urinals for males.
[4] For each urinal added in excess of the minimum required, one water closet shall be permitted to be deducted. The number of water closets shall not be reduced to less than two-thirds of the minimum requirement.
[5] Group lavatories that are 24 lineal inches (610 mm) of wash sink or 18 inches (457 mm) of a circular basin, where provided with water outlets for such space, shall be considered equivalent to one lavatory.
[6] Metering or self closing faucets shall be installed on lavatories intended to serve the transient public.

CHAPTER 5

WATER HEATERS

501.0 General.

501.1 Applicability. The regulations of this chapter shall govern the construction, location, and installation of fuel-burning and other types of water heaters heating potable water, together with chimneys, vents, and their connectors. The minimum capacity for storage water heaters shall be in accordance with the first-hour rating listed in Table 501.1(2). No water heater shall be hereinafter installed that does not comply with the manufacturer's installation instructions and the type and model of each size thereof approved by the Authority Having Jurisdiction. A list of accepted water heater appliance standards is referenced in Table 501.1(1). Listed appliances shall be installed in accordance with the manufacturer's installation instructions. Unlisted water heaters shall be permitted in accordance with Section 504.3.2.

This chapter and Chapter 12, Fuel Piping, are an extraction of the requirements contained in NFPA 54, National Fuel Gas Code. They are not a full extraction of the entire NFPA 54 nor are they solely extractions from that code. These chapters extract the sections and subsections that pertain to the subject matter of the chapter. Here in Chapter 5, the material extracted relates to the installation of water heaters. The material extracted can be identified by the identification of the corresponding section number in NFPA 54, such as this example: [NFPA 54: 3.3.17.2]

The types of water heaters regulated by this chapter are fuel-gas burning, oil burning and electric, all of which heat potable water. Chapter 5 also regulates the air supply and venting systems associated with these water heaters. These appliances should not be confused with boilers, which are regulated by the Uniform Mechanical Code (UMC).

Water heaters are constructed to comply with either Canadian Standards Association (CSA) or Underwriters Laboratory (UL) design standards. Conversely, a water boiler is considered to be a "pressure vessel," and it must be constructed in compliance with the American Society of Mechanical Engineers (ASME), which is considerably more stringent than those applicable to water heaters. There are differences between a water heater and a boiler. There are a variety of performance values that distinguish water heaters from water boilers (pressure vessels), even though there may be considerable similarity in the appearance of one when compared with the other.

A water-heating device that exceeds any one of the following shall be classified as a water boiler:
- 120 gallons (454.2 liters) nominal water storage capacity;
- 160 psi (1103.2 kPa) operating pressure;
- 210°F (99°C) operating temperature; or
- 200,000 Btu/hr. (58,620 kW) heat input.

Most typically, it is elevated heat input that results in larger water-heating devices becoming classified as water boilers. The distinction between water heaters and water boilers takes on special significance in many jurisdictions. A number of state or local regulations require a National Board inspector to approve all boiler installations before those devices can be fired or have their power source(s) energized. It is not uncommon for there to be additional regulatory considerations with which the installer must become familiar.

One of the sections not contained in NFPA 54 is the reference to Table 501.1(1). The minimum capacity of residential water heaters are to be in accordance with the provisions of Table 501.1(2). The water heater installed in a house, apartment or condo must be sized to meet the applicable "First Hour Rating." The first hour rating represents how much hot water the unit can supply in a 1-hour period if it starts with a full tank of hot water. For example, in **Figure 501.1** the Energy Guide Label illustrates a water heater with a first hour rating of 57 gallons. This water heater would be able to be installed in a residence with up to three bedrooms and one and one-half bathrooms or a residence of two bedrooms and two to two and one-half bathrooms. The note at the bottom of the table requires solar water heaters to be sized this way.

The manufacturer's instructions will give the proper design parameters not only for the physical installation of the water heater but also for the size and installation of the venting system unless it is an electric water heater. This chapter is specific to water heaters; other appliances should be installed per the UMC.

TABLE 501.1(1)
WATER HEATERS

TYPE	STANDARD
Electric, Household	UL 174
Oil-Fired Storage Tank	UL 732
Gas, 75 000 Btu/h or less	CSA Z21.10.1
Gas, Above 75 000 Btu/h	CSA Z21.10.3
Electric, Commercial	UL 1453
Solid Fuel	UL 2523

For SI units: 1000 British thermal units per hour = 0.293 kW

502.0 Permits.

502.1 General. It shall be unlawful for a person to install, remove, or replace or cause to be installed, removed, or replaced a water heater without first obtaining a permit from the Authority Having Jurisdiction to do so.

The replacement or rearrangement of valves and fittings, including water heaters, must be accompanied by a permit approved by the Authority Having Jurisdiction (AHJ). This will then require the proper inspections for the appliance, vents and possibly the piping if altered.

WATER HEATERS

TABLE 501.1(2)
FIRST HOUR RATING[1]

Number of Bathrooms	1 to 1.5			2 to 2.5				3 to 3.5			
Number of Bedrooms	1	2	3	2	3	4	5	3	4	5	6
First Hour Rating,[2] Gallons	38	49	49	49	62	62	74	62	74	74	74

For SI units: 1 gallon = 3.785 L

Notes:
[1] The first hour rating is found on the "Energy Guide" label.
[2] Solar water heaters shall be sized to meet the appropriate first hour rating as shown in the table.

FIGURE 501.1
EXAMPLE OF ENERGY GUIDE LABEL

503.0 Inspection.

503.1 Inspection of Chimneys or Vents. This inspection shall be made after chimneys, vents, or parts thereof, authorized by the permit, have been installed and before such vent or part thereof has been covered or concealed.

503.2 Final Water Heater Inspection. This inspection shall be made after work authorized by the permit has been installed. The Authority Having Jurisdiction will make such inspection as deemed necessary to be assured that the work has been installed in accordance with the intent of this code. No appliance or part thereof shall be covered or concealed until the same has been inspected and approved by the Authority Having Jurisdiction.

Two inspections are required for the water heater installation. The first "rough" inspection will encompass the vent system and piping while the second "final" inspection will be of the water heater itself. All necessary components, such as the gas connection, venting, combustion air and, in Seismic design categories C, D, E, and F (see Section 507.2), anchors and supports, should also be installed before requesting the final inspection.

504.0 Water Heater Requirements.

504.1 Location. Water heater installations in bedrooms and bathrooms shall comply with one of the following [NFPA 54:10.27.1]:

(1) Fuel-burning water heaters shall be permitted to be installed in a closet located in the bedroom or bathroom provided the closet is equipped with a listed, gasketed door assembly and a listed self-closing device. The self-closing door assembly shall meet the requirements of Section 504.1.1. The door assembly shall be installed with a threshold and bottom door seal and shall meet the requirements of Section 504.1.2. Combustion air for such installations shall be obtained from the outdoors in accordance with Section 506.4. The closet shall be for the exclusive use of the water heater.

(2) Water heater shall be of the direct vent type. [NFPA 54: 10.27.1(2)]

The installation of a water heater in a closet opening into a bedroom or bathroom has been previously prohibited. Earlier water heaters were not as efficient as today's water heaters and produced quite a bit of lethal gases. Many deaths had occurred because of the accumulation of exhaust gases, especially in bedrooms because of poorly vented systems. Recently, there has been the desire to utilize as much space as possible in buildings, especially manufactured homes, apartments and condos; therefore, the above requirements were placed into the code.

The occupants of a bedroom or bathroom are protected by the requirement for gasketing the door and having a self-closing device on the door. Direct-vent water heaters, which pipe the combustion air and the exhaust to the outside, have been allowed in these areas for some time.

504.1.1 Self-Closing Doors. Self-closing doors shall swing easily and freely, and shall be equipped with a self-closing device to cause the door to close and latch each time it is opened. The closing mechanism shall not have a hold-open feature.

504.1.2 Gasketing. Gasketing on gasketed doors or frames shall be furnished in accordance with the published listings of the door, frame, or gasketing material manufacturer.

Exception: Where acceptable to the Authority Having Jurisdiction, gasketing of non-combustible or limited-combustible material shall be permitted to be applied to the frame, provided closing and latching of the door are not inhibited.

504.2 Vent. Water heaters of other than the direct-vent type shall be located as close as practical to the chimney or gas vent.

The requirement to keep the water heater as close to the vent or chimney as possible is meant to keep the installation from having long horizontal runs of vent. This requirement, along with the requirements below, should keep the installation compact, neat and free from interference with other installations.

504.3 Clearance. The clearance requirements for water heaters shall comply with Section 504.3.1 or Section 504.3.2.

WATER HEATERS

504.3.1 Listed Water Heaters. The clearances shall not be such as to interfere with combustion air, draft hood clearance and relief, and accessibility for servicing. Listed water heaters shall be installed in accordance with their listings and the manufacturer's installation instructions.

504.3.2 Unlisted Water Heaters. Unlisted water heaters shall be installed with a clearance of 12 inches (305 mm) on all sides and rear. Combustible floors under unlisted water heaters shall be protected in an approved manner. [NFPA 54:10.27.2.2]

504.4 Pressure-Limiting Devices. A water heater installation shall be provided with overpressure protection using an approved, listed device installed in accordance with the terms of its listing and the manufacturer's installation instructions.

504.5 Temperature-Limiting Devices. A water heater installation or a hot water storage vessel installation shall be provided with overtemperature protection by means of an approved, listed device installed in accordance with the terms of its listing and the manufacturer's installation instructions.

504.6 Temperature, Pressure, and Vacuum Relief Devices. Temperature, pressure, and vacuum relief devices or combinations thereof, and automatic gas shutoff devices shall be installed in accordance with the terms of their listings and the manufacturer's installation instructions. A shutoff valve shall not be placed between the relief valve and the water heater or on discharge pipes between such valves and the atmosphere. The hourly British thermal units (Btu) (kW•h) discharge capacity or the rated steam relief capacity of the device shall be not less than the input rating of the water heater. Discharge piping shall be installed in accordance with Section 608.5.

Sections 504.4, 504.5 and 504.6 provide the primary safety features of the water heater installation. Pressure-relief valves first appeared during the mid-1800s in response to a number of major mishaps involving steam-powered equipment. These devices were simply designed to protect against explosive overpressure conditions caused by "runaway" boilers. At that time, there were few domestic water heaters; consequently, overheated storage water had not become a major concern. As time passed, domestic water heaters became fashionable. Water heated to intended water temperatures represented no threat; however, unintentional overheating of stored water soon became a major cause of domestic tank explosions. In rare instances they still do.

In the early 1930s temperature-relief valves were developed to protect against contained water reaching temperatures in excess of 212°F (100°C). Temperatures below 212°F (100°C) may cause scalding and other personal injury to individuals, but at temperatures above 212°F (100°C) contained water becomes explosive when it flashes into uncontained steam. The transformation of water into steam can be catastrophically explosive when it flashes as it is suddenly released from a containment vessel, such as a water heater storage tank to the free atmosphere.

Later, the pressure and temperature relief functions were combined into a single valve that would allow one valve to protect against both overpressure and over-temperature. Data plates attached to these valves became more comprehensive and complex as a reflection of this dual purpose. The data plate information reflects each of these independent safety functions.

There are three levels of protection required for a water heater. The first level is a temperature-limiting device that must be incorporated with the water heater. This will limit the water heater from becoming a steam boiler. This control is integral to the water heater and is included with the heater's primary controls (see **Figure 504.6a**). They are installed directly on the water heater by the manufacturer. At a temperature not to exceed 210°F (99°C), this factory-installed device will activate shutting down the energy source to the water heater or boiler. Activation of the high-limit switch [210°F (99°C)] requires manual resetting of the circuit breakers or relighting of the standing pilot, when applicable. The presumption here is that an equipment failure has occurred that requires repair or replacement of the water heater or boiler.

The second level of protection is protection against an increase of pressure caused by the heating of the water. This is not accomplished by expansion tanks. A pressure-relief valve is required to ensure that the pressure is dissipated quickly. Section 608.3 addresses protection from thermal expansion and requires either a listed expansion tank or other device designed for thermal expansion control as an acceptable method of preventing excessive pressure within the system. This expansion device is in addition to the pressure relief-valve.

The third level of protection requires the water heater to be protected from vacuum situations that may occur in the piping system due to its installation. This is discussed in Section 608.7, Vacuum Relief Valves, of Chapter 6 which explains the installation of the vacuum-relief valve. Also see **Figure 608.7**.

Section 504.6 allows the use of a combination valve to accomplish these safety measures, usually the temperature and pressure protection. Section 608.3 of Chapter 6 also requires that a temperature- and pressure-relief valve (T&P valve) be installed on all water heaters and hot water boilers in addition to the factory-installed over-temperature safety devices. Thus, no matter what protection is used that is integral to the water heater, it must be protected by a T&P valve. The CSA Z21.22 standard requires markings for relief valves and is found on the relief valve tag plate (see **Figure 504.6b**). Three of the most important markings required for the combination temperature-and-pressure relief valve (T&P) are the set pressure, temperature limit, and the CSA rating. The set opening pressure is the minimum pressure the valve will open. The temperature limit is the temperature the valve will discharge. The CSA rating is the temperature steam rate and thermal expansion water rate in terms of Btu/hr.

The temperature steam rate and thermal expansion water rate is used to size the relief valve for its discharge capacity. This CSA discharge rating in terms of Btu/hr must exceed the Btu input indicated on the water heater label. The valve must be capable of discharging more Btu/hr than the heater is capable of.

Another marking shown on T&P data plates may include the ASME Pressure Steam Rating. Frequently, this rating is simply ignored. Why? Because it normally reflects a larger Btu input rating (based upon ASME standards) than the CSA rating. In all cases, when selecting a T&P relief valve, one must select the lowest rating shown on the data plate (the CSA rating) regardless of the water-heating method (whether gas or electric). The Btu input from the heat source must never be allowed to exceed the CSA Temperature Steam Rating as indicated on the data plate. The ASME rating reflects the maximum amount of heat input that a given pressure-relief valve is able to vent as it strives to maintain the maximum pressure (psi) for which it is rated. This Btu input limitation is worthwhile information to know; however, it is never the determining factor when selecting a T&P valve.

Another matter for consideration by the design professional is temperature and pressure protection of stand-alone hot water storage tanks. Protection must address both temperature and pressure within these tanks. The pressure limitations of these tanks may be well below the common 150 psi (1,034 kPa) design standard applied to most water heaters. Protection is frequently not possible with use of a single valve. Storage pressure tanks are designed and listed as ASME pressure vessels and, as such, they are limited operationally to the pressures designated on their data plates. Frequently, this pressure limitation will be considerably less than the pressure rating of a standard T&P valve. T&P valves can be special-ordered with activation pressures set as low as 75 psi (517.11 kPa). However, the more common T&P pressure ranges of 125 psi (861.84 kPa) to 150 psi (1,034.21 kPa) may be all that is available to the installer on short notice. In that case, it may be necessary to install two independent valves in the shell of the tank: one specifically to relieve pressure before it reaches the design limit of the tank (possibly as low as 75 psi) and the other to maintain water temperature at 210°F (99°C) or less.

The installation of the T&P valve is covered in Section 608.3. It is also discussed in the UA Water Supply Manual. In summary, it is important to remember the following:

- Both gas and electric water heaters require T&P valves when they are connected to storage tanks with a dimension greater than three inches (80 mm) in diameter.
- When two or more Btu ratings are referenced on the nameplate of a combination T&P valve, always use the lowest ratings given for temperature control purposes.
- Electric water heaters generally have a relatively smaller Btu input than that of a comparable gas-fired water heater. Nonetheless, electric heaters still require installation of T&P valves of appropriate capacity. To convert electrical energy input into equivalent Btu per hour, multiply kilowatt hours by a factor of 3.413.

505.0 Oil-Burning and Other Water Heaters.

These sections concerning water heaters using fuels other than gas, such as oil-fired water heaters, reiterate the protection requirements in the previous sections. An external combination T&P valve is required in addition to the integral primary temperature controls and integral over-temperature safety controls for these water heaters.

505.1 Water Heaters. Water heaters deriving heat from fuels or types of energy other than gas shall comply with the standards referenced in Table 501.1(1), Section 505.3, or Section 505.4. Vents or chimneys for such appliances shall be of approved types. An adequate supply of air for combustion and for adequate ventilation of heater rooms or compartments shall be provided. Each such appliance shall be installed in a location approved by the Authority Having Jurisdiction and local and state fire-prevention agencies.

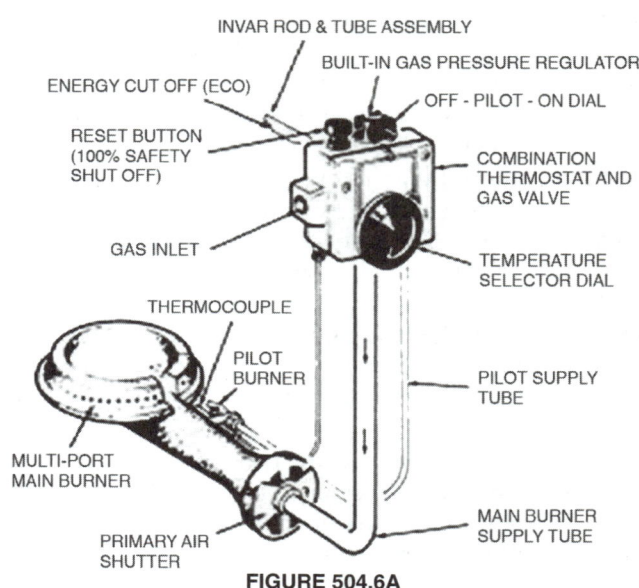

FIGURE 504.6A
PRIMARY WATER HEATER CONTROLS

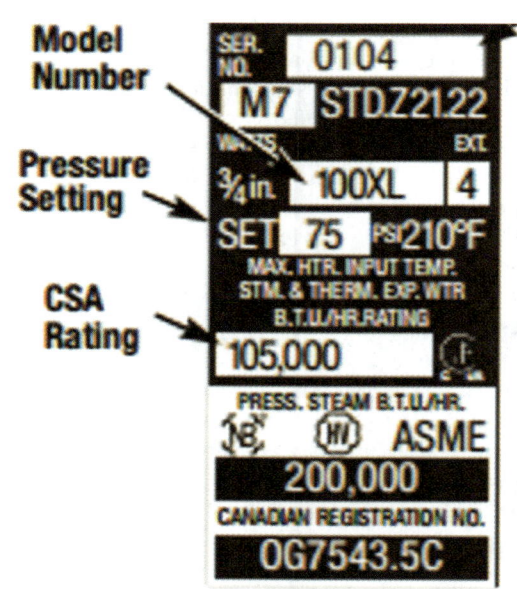

FIGURE 504.6B
EXAMPLE OF T&P VALVE RATINGS

505.2 Safety Devices. Storage-type water heaters and hot water boilers deriving heat from fuels or types of energy other than gas, shall be provided with, in addition to the primary temperature controls, an over-temperature safety protection device that complies with and is installed in accordance with nationally recognized applicable standards for such devices and a combination temperature and pressure-relief valve.

505.3 Oil-Fired Water Heaters. Oil-fired water heaters shall be installed in accordance with NFPA 31.

505.4 Indirect-Fired Water Heaters. Indirect-fired water heaters shall be in accordance with the applicable sections of the ASME Boiler and Pressure Vessel Code or shall comply with one of the other applicable standards shown in Table 501.1(1). Each water heater shall bear a label in accordance with ASME requirements, or an approved testing agency, certifying and attesting that such an appliance has been tested, inspected and meets the requirements of the applicable standards or code.

The indirect water heater is basically a heat exchanger and storage device where the heating of the water takes place in another device, such as a boiler (see **Figure 505.4**). The indirect water heater can be either a dual- or single-walled exchanger. The single wall is the more controversial device due to the possibility that if a leak developed in a single-wall device, it would create a cross connection in the potable water system. Therefore, the medium being circulated between the heating element of the system and the indirect-fired water heater must be potable water or another medium that meets FDA standards as "food grade." If the heating medium is food-grade liquid, then a cross-connection condition is avoided.

505.4.1 Single-Wall Heat Exchanger. An indirect-fired water heater that incorporates a single-wall heat exchanger shall be in accordance with the following requirements:

(1) The heat transfer medium shall be either potable water or contain fluids recognized as safe by the Food and Drug Administration (FDA) as food grade.

(2) Bear a label with the word "Caution," followed by the following statements:

(a) The heat-transfer medium shall be potable water or other nontoxic fluid recognized as safe by the FDA.

(b) The maximum operating pressure of the heat exchanger shall not exceed the maximum operating pressure of the potable water supply.

(3) The word "Caution" and the statements in letters shall have an uppercase height of not less than 0.120 of an inch (3.048 mm). The vertical spacing between lines of type shall be not less than 0.046 of an inch (1.168 mm). Lowercase letters shall be compatible with the uppercase letter size specification.

506.0 Air for Combustion and Ventilation.

In order for fuels to burn properly and safely, there must be a sufficient amount of fresh air available at the appliance location to support combustion, for dilution of the draft hood and to have equal pressure throughout the area of the appliance location. A relatively standard proportion of oxygen molecules to gas molecules are required in order to ensure a complete burn (oxidation) of the gas molecules. In a laboratory setting, this ratio is two oxygen molecules to one gas vapor molecule.

The composition of air is approximately 80-percent nitrogen, 19-percent oxygen and 1-percent trace gases (argon, helium, hydrogen, etc). Consequently, it requires 10 cubic feet of air to provide two cubic feet of oxygen (O_2) at the temperature and pressure of a "standard" atmosphere (pressure 29.92 inches of mercury at 59°F). When either atmospheric pressure or temperature is different from this standard, the density of the air will vary. Consequently, the number of molecules per cubic foot will also vary.

Because a "clean" burn (noncarbonizing) requires a ratio of two cubic feet of O_2 to one cubic foot of gas, any variation in air density will affect the burn. Therefore, it is sometimes necessary to adjust the quantity of gas molecules to match the available oxygen molecules. High altitude (less pressure, less density) is an example of when a smaller orifice in the fuel-gas supply is needed in order to match the available supply of oxygen molecules per cubic foot of air.

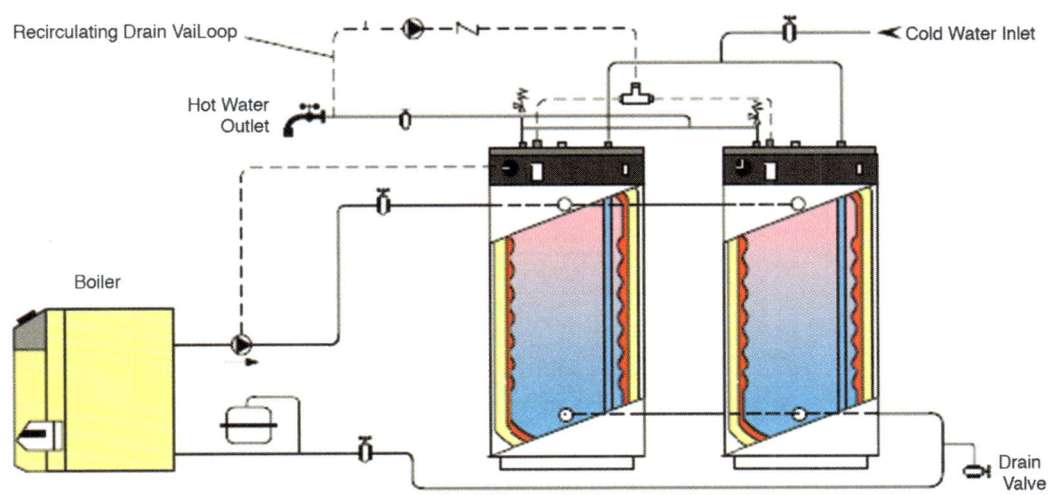

FIGURE 505.4
EXAMPLE OF COMMERCIAL INDIRECT-FIRED WATER HEATER APPLICATION

WATER HEATERS

Undersized combustion air openings will result in the deficiency of available air (O_2); thus, there will be insufficient air for the primary and secondary burn, dilution air (at the draft hood) and ventilation air (high and low opening for air intake and outflow). Proper design of combustion air openings is critical to safe and efficient operation of a fuel-burning appliance or device.

There are five methods used to supply combustion air to the appliance:

1. 100-percent air from indoors.
2. 100-percent air from the outdoors.
3. A combination of indoor and outdoor air.
4. Air supplied by engineered systems.
5. Air supplied by mechanical systems.

The "Combustion Air Chart" in **Figure 506.0** is a good visual guide to the requirements and options for supplying combustion air. The plumber oftentimes is not responsible for the method used to provide this supply of air. It is usually designed by an engineer and the system is either provided for by openings in wall systems of the building or duct work installed by sheet metal workers. There are times, however, especially when remodeling or replacing appliances, that combustion air should be considered. It is important to remember that this chapter is for the installation of water heaters and the installation of its vent or its connection to a vent or chimney system. The UMC or the full NFPA 54 contains provisions for appliances other than water heaters.

506.1 General. Air for combustion, ventilation, and dilution of flue gases for appliances installed in buildings shall be obtained by application of one of the methods covered in Section 506.2 through Section 506.7.3. Where the requirements of Section 506.2 are not met, outdoor air shall be introduced in accordance with methods covered in Section 506.4 through Section 506.7.3.

Exception: This provision shall not apply to direct-vent appliances. [NFPA 54:9.3.1.1]

506.1.1 Other Types of Appliances. Appliances of other than natural draft design appliances not designated as Category I vented appliances, and appliances equipped with power burners shall be provided with combustion, ventilation, and dilution air in accordance with the appliance manufacturer's instructions. [NFPA 54:9.3.1.2]

506.1.2 Draft Hood and Regulators. Where used, a draft hood or a barometric draft regulator shall be installed in the same room or enclosure as the appliance served so as to prevent a difference in pressure between the hood or regulator and the combustion air supply. [NFPA 54:9.3.1.4]

506.1.3 Makeup Air. Where exhaust fans, clothes dryers, and kitchen ventilation systems interfere with the operation of appliances, makeup air shall be provided. [NFPA 54:9.3.1.5]

506.2 Indoor Combustion Air. The required volume of indoor air shall be determined in accordance with the method in Section 506.2.1 or Section 506.2.2, except that where the air infiltration rate is known to be less than 0.40 ACH (air

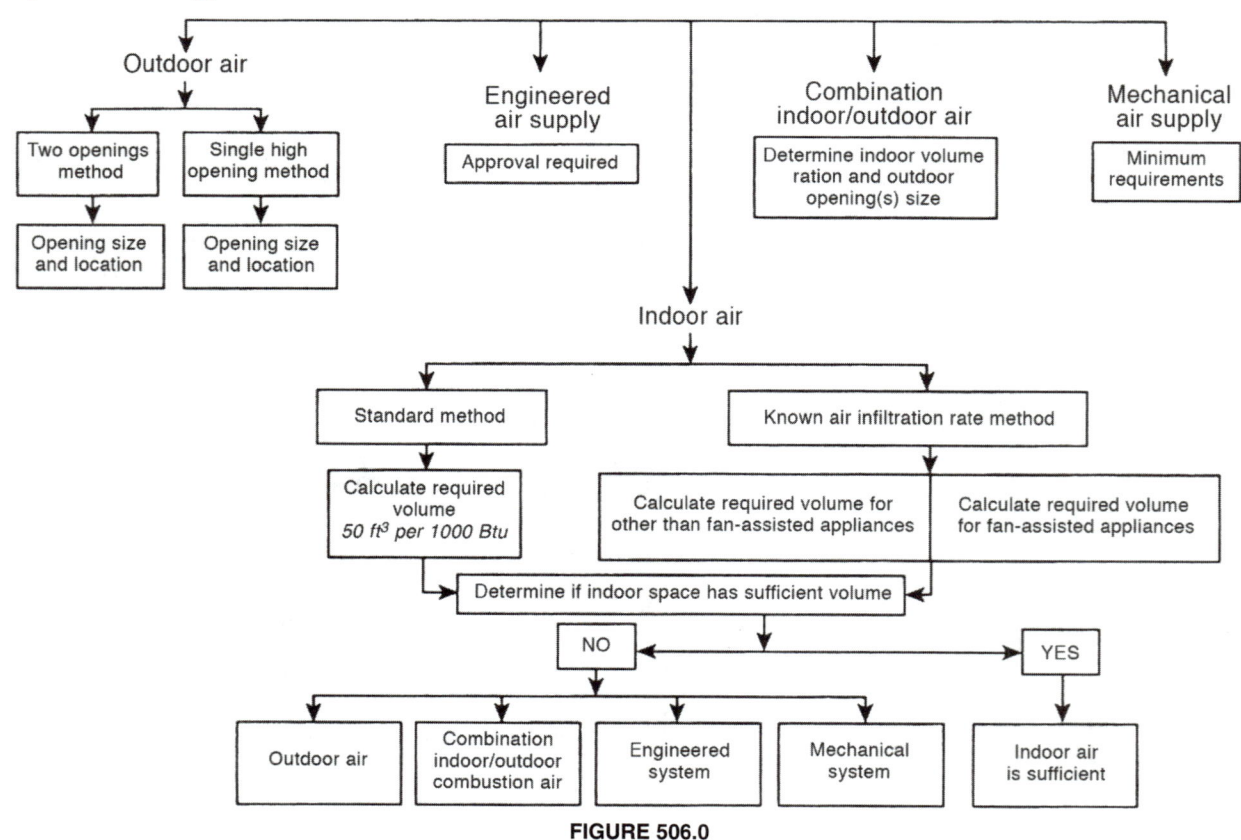

FIGURE 506.0
COMBUSTION AIR OPTIONS

WATER HEATERS

change per hour), the method in Section 506.2.2 shall be used. The total required volume shall be the sum of the required volume calculated for all appliances located within the space. Rooms communicating directly with the space in which the appliances are installed through openings not furnished with doors, and through combustion air openings sized and located in accordance with Section 506.3 are considered a part of the required volume. [NFPA 54:9.3.2]

There are two ways to determine the amount of indoor combustion air required from indoors. One is the Standard Method calculated by a ratio 50cf:1000 Btu/hr. However, if the air infiltration rate is known to be 0.4 ACH or less, the Known Air Infiltration Rate (KAIR) Method must be used. This method is used to calculate the required volume in newer built homes with low air infiltration rates. Air infiltration rates may be found by using the ASHRAE Method, a blower door test, or by a method of establishing a ventilation rate adopted by a local jurisdiction based on its experience.

506.2.1 Standard Method. The required volume shall be not less than 50 cubic feet per 1000 British thermal units per hour (Btu/h) (4.83 m³/kW). [NFPA 54:9.3.2.1]

Example using the Standard Method:

Problem: A 50,000 Btu/hr water heater with a draft hood is installed in a room 20 feet by 16 feet by 8 feet (See **Figure 506.2.1**). Find the minimum volume of indoor air required.

Solution: The Standard Method requires 50 cubic feet per 1,000 Btu/hr of all appliances in the space to be used to figure the volume.

50,000 Btu/hr ÷ 1000 Btu/hr = 50 Btu/hr

50 Btu/hr x 50 ft³ = 2,500 ft³ of indoor combustion air required

Available room volume:

20 x 16 x 8 = 2,560 ft³ per hour

Combustion air may be supplied by indoor air.

506.2.2 Known Air Infiltration Rate Method. Where the air infiltration rate of a structure is known, the minimum required volume shall be determined as follows: [NFPA 54:9.3.2.2]:

(1) For appliances other than fan-assisted, calculate using the following Equation 506.2.2(1). [NFPA 54:9.3.2.2(1)]

[Equation 506.2.2(1)]

Required volume $_{other}$ ≥ (21 ft³/ACH) x (I_{other}/1000 Btu/h)

(2) For fan-assisted appliances, calculate using the following Equation 506.2.2(2). [NFPA 54:9.3.2.2(2)]

[Equation 506.2.2(2)]

Required volume $_{fan}$ ≥ (15 ft³/ACH) x (I_{fan}/1000 Btu/h)

Where:

I_{other} = All appliances other than fan-assisted input in (Btu/h)

I_{fan} = Fan-assisted appliance input in (Btu/h)

ACH = Air change per hour (percent of volume of space exchanged per hour, expressed as a decimal)

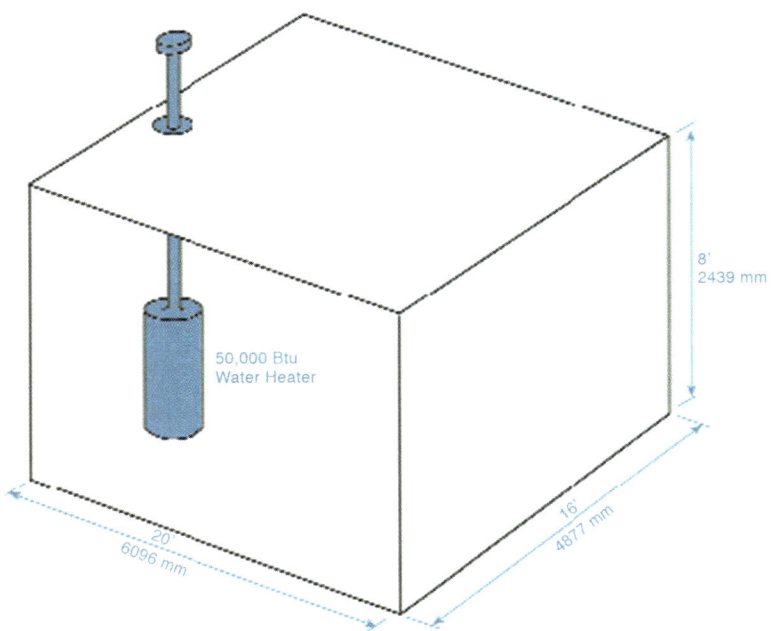

50 cubic ft. per 1,000 Btu Input

50 x $\frac{50,000}{1000}$ = 50 x 50 = 2,500 cu. ft. (required)

20 ft. x 16 ft. = 320 ft² x 8 ft = 2560 ft³

FIGURE 506.2.1
THE STANDARD METHOD FOR DETERMINING THE REQUIRED VOLUME OF INDOOR AIR

For SI units: 1 cubic foot = 0.0283 m³, 1000 British thermal units per hour = 0.293 kW

(3) For purposes of these calculations, an infiltration rate greater than 0.60 ACH shall not be used in the equations in Section 506.2.2(1) and Section 506.2.2(2). [NFPA 54:9.3.2.2(3)]

🔧 The KAIR Method requires a calculation of the required air volume based on two appliance types: fan-assisted and other than fan-assisted as shown above. The formula for each appliance type is based on the total combustion air needs of each type of appliance, which differ due to the amount of dilution air required. If installations include both types of appliance, then a separate calculation is done for each type of appliance. These calculations are combined to determine the total required air volume. These calculations determine the amount of air for combustion and ventilation required by the appliance for complete combustion.

Combustion air includes three components:

1. The amount of air required for complete combustion, 10 cubic feet per cubic foot gas burned;
2. Excess air to help ensure complete combustion, typically 5 cubic feet per cubic foot of gas;
3. Dilution air for proper venting, typically 6 cubic feet per cubic foot of gas.

The historical assumptions for excess and dilution air are not for fan-assisted appliances because they have neither a draft hood nor a flue collar, and therefore dilution air is limited to leakage from vent fittings. (Research by Battelle, published as Combustion-Air Issues Related to Residential Gas Appliances in Confined Spaces and Unusually Tight Construction, January 2001, has verified this.)

Example using the KAIR Method:

Problem: An 80,000 Btu/hr. fan-assisted furnace and a 50,000 Btu/hr water heater with a draft hood are installed in a room 35 feet by 35 feet by eight feet. By using a blower door test, the air infiltration rate was determined to be 0.65. Find the minimum volume of indoor air required.

Solution: When using the KAIR Method, the maximum air infiltration rate that can be used is 0.6, so for this example 0.6 shall be used. Since there is a draft hood appliance and a fan-assisted appliance, the required volume must be figured separately.

Required volume for the furnace using Equation 506.2.2(2):

(15 ft³ ÷ .60) x (80,000 ÷ 1,000) = 2,000 ft³

Required volume of the water heater using Equation 506.2.2(1):

(21 ft³ ÷ .60) x (50,000 ÷ 1,000) = 1,750 ft³

Total volume required for the enclosure would be:

2,000 ft³ + 1,750 ft³ = 3,750 ft³

Available volume of the room:

35 x 35 x 8 = 9,800 ft³

No additional combustion would be required.

506.3 Indoor Opening Size and Location. Openings used to connect indoor spaces shall be sized and located in accordance with the following:

(1) Each opening shall have a free area of not less than 1 square inch per 1000 Btu/h (0.002 m²/kW) of the total input rating of appliances in the space, but not less than 100 square inches (0.065 m²). One opening shall commence within 12 inches (305 mm) of the top of the enclosure, and one opening shall commence within 12 inches (305 mm) of the bottom of the enclosure (see **Figure 506.3**). The dimension of air openings shall be not less than 3 inches (76 mm).

🔧 See **Figure 506.3a.**

(2) The volumes of spaces in different stories shall be considered as communicating spaces where such spaces are connected by one or more openings in doors or floors having a total free area of not less than 2 square inches per 1000 Btu/h (0.004 m²/kW) of total input rating of appliances. [NFPA 54:9.3.2.3]

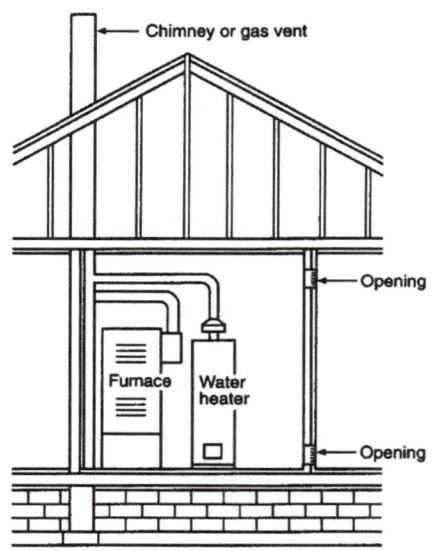

FIGURE 506.3
COMBUSTION AIR FROM ADJACENT INDOOR SPACES THROUGH INDOOR COMBUSTION AIR OPENINGS
[NFPA 54: FIGURE A.9.3.2.3(1)]

506.4 Outdoor Combustion Air. Outdoor combustion air shall be provided through opening(s) to the outdoors in accordance with methods in Section 506.4.1 or Section 506.4.2. The dimension of air openings shall be not less than 3 inches (76 mm). [NFPA 54:9.3.3]

506.4.1 Two Permanent Openings Method. Two permanent openings, one commencing within 12 inches (305 mm) of the top of the enclosure and one commencing within 12 inches (305 mm) of the bottom of the enclosure, shall be provided. The openings shall communicate directly, or by ducts, with the outdoors or spaces that freely communicate with the outdoors, as follows:

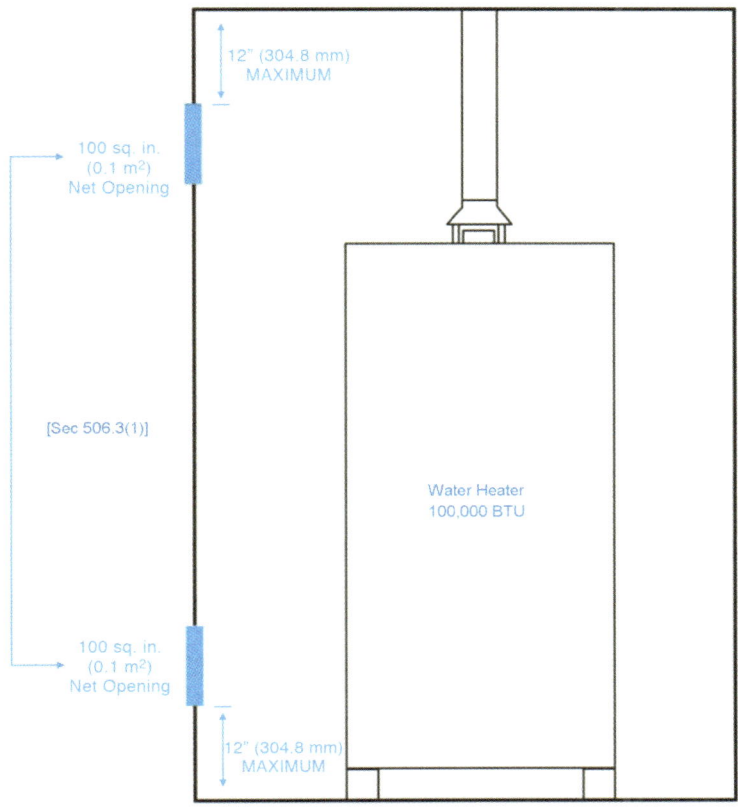

FIGURE 506.3A
COMBINING SPACES ON THE SAME STORY

(1) Where directly communicating with the outdoors or where communicating to the outdoors through vertical ducts, each opening shall have a minimum free area of 1 square inch per 4000 Btu/h (0.0005 m²/kW) of total input rating of all appliances in the enclosure. [See **Figure 506.4.1(1)** and **Figure 506.4.1(2)**]

(2) Where communicating with the outdoors through horizontal ducts, each opening shall have a minimum free area of 1 square inch per 2000 Btu/h (0.001 m²/kW) of total input rating of all appliances in the enclosure. [See **Figure 506.4.1(3)**] [NFPA 54:9.3.3.1]

🔧 The combustion air openings are to be within 12 inches of the floor or ceiling. Any ducts used shall be installed directly to the outside or to spaces communicating directly with the outdoors; ducts for upper and lower combustion air openings shall not be combined.

The reason for locating these openings, as defined above, is to provide natural ventilation of the enclosure. Heat rises; therefore, heat generated by fuel-burning appliances will rise and flow out of the enclosure. Cooler air will be drawn into the enclosure through the lower opening to replace the departing hot air. This natural circulation stabilizes the air temperature in the enclosure; thus, the air density is also stabilized, resulting in the best air density possible. Consequently, the air/gas ratio is closer to standard than would be the case if this enclosure ventilation were not occurring.

506.4.2 One Permanent Opening Method. One permanent opening, commencing within 12 inches (305 mm) of the top of the enclosure, shall be provided. The appliance shall have clearances of at least 1 inch (25.4 mm) from the sides and back and 6 inches (152 mm) from the front of the appliance. The opening shall directly communicate with the outdoors or shall communicate through a vertical or horizontal duct to the outdoors or spaces that freely communicate with the outdoors (see **Figure 506.4.2**) and shall have a minimum free area of the following:

(1) One square inch per 3000 Btu/h (0.0007 m²/kW) of the total input rating of all appliances located in the enclosure.

(2) Not less than the sum of the areas of all vent connectors in the space. [NFPA 54:9.3.3.2]

🔧 Problem: A 140,000 Btu/hr fan-assisted furnace and two 50,000 Btu/hr water heaters are installed in a room 35 feet by 35 feet by eight feet. All air will be from the outside. Find the minimum free area required for one permanent opening.

Solution:

140,000 Btu/hr + 50,000 Btu/hr + 50,000 Btu/hr = 240,000 Btu/hr

240,000 Btu/hr ÷ 3,000 Btu/hr = 80 inches² of combustion air required.

An 11-inch round duct ($\pi r^2 = \pi 5.5^2 = 95$ square inches) or nine-inch by nine-inch rectangular duct (81 square inches)

WATER HEATERS

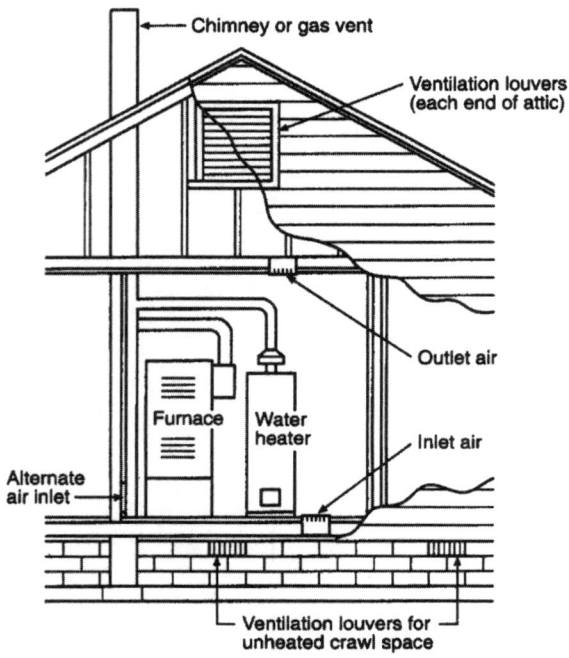

**FIGURE 506.4(1)
COMBUSTION AIR FROM OUTDOORS
INLET AIR FROM VENTILATED CRAWL SPACE AND
OUTLET AIR TO VENTILATED ATTIC
[NFPA 54: FIGURE A.9.3.3.1(1)(a)]**

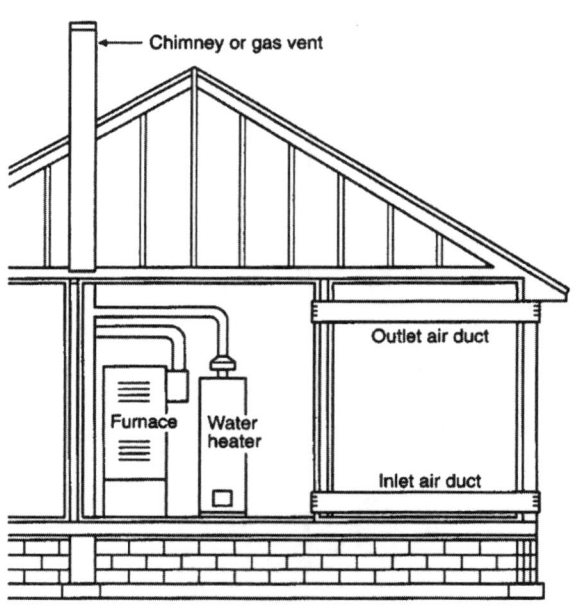

**FIGURE 506.4(3)
COMBUSTION AIR FROM
OUTDOORS THROUGH HORIZONTAL DUCTS
[NFPA 54: FIGURE A.9.3.3.1(2)]**

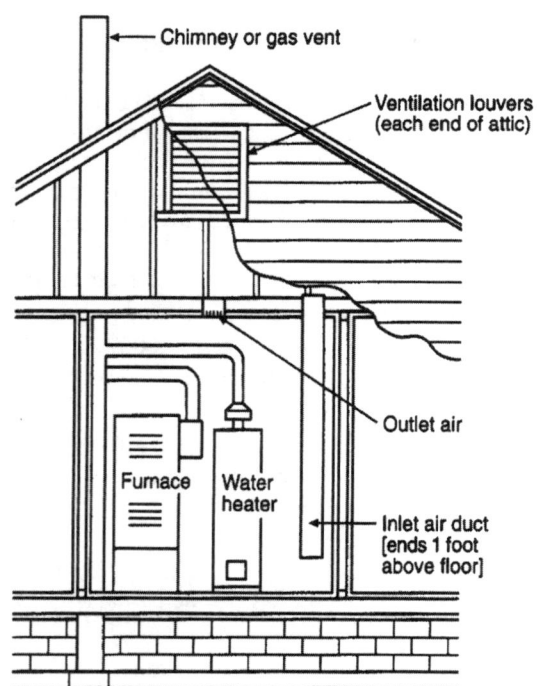

For SI units: 1 foot = 304.8 mm

**FIGURE 506.4(2)
COMBUSTION AIR FROM OUTDOORS
THROUGH VENTILATED ATTIC
[NFPA 54: FIGURE A.9.3.3.1(1)(B)]**

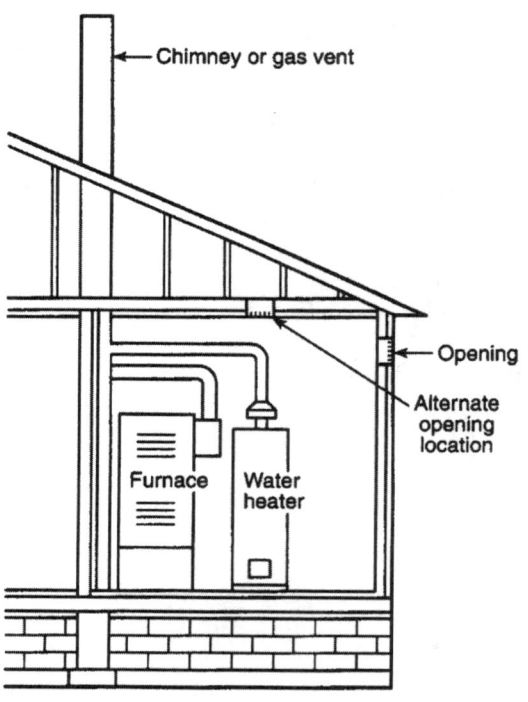

**FIGURE 506.4.2
COMBUSTION AIR FROM OUTDOORS THROUGH
SINGLE COMBUSTION AIR OPENING
[NFPA 54: FIGURE A.9.3.3.2]**

would be required. The area of all the vent connectors combined shall not be less than 80 square inches.

506.5 Combination Indoor and Outdoor Combustion Air. The use of a combination of indoor and outdoor combustion air shall be in accordance with Section 506.5.1 through Section 506.5.3. [NFPA 54:9.3.4] (See Appendix J for example calculations).

🔧 Using a combination of indoor and outdoor air helps to reduce the cost of heating and air conditioning by reducing the amount of unconditioned air leaking into the building and conditioned air leaking out. By obtaining part of the combustion air from inside the building, the opening to the exterior may be reduced.

506.5.1 Indoor Openings. Where used, openings connecting the interior spaces shall be in accordance with Section 506.3. [NFPA 54:9.3.4(1)]

506.5.2 Outdoor Openings. Outdoor openings shall be located in accordance with Section 506.4. [NFPA 54:9.3.4(2)]

506.5.3 Outdoor Opening(s) Size. The outdoor opening(s) size shall be calculated in accordance with the following:

(1) The ratio of interior spaces shall be the volume of the communicating spaces divided by the required volume.

(2) The outdoor size reduction factor shall be one minus the ratio of interior spaces.

(3) The size of outdoor openings shall be not less than the full size of outdoor openings calculated in accordance with Section 506.4, multiplied by the reduction factor. The dimension of air openings shall be not less than 3 inches (76 mm). [NFPA 54:9.3.4(3)]

🔧 Problem: A 140,000 Btu/hr fan-assisted furnace and two 50,000 Btu/hr water heaters are installed in a room 35 feet by 35 feet by 8 feet. Find the minimum size of the outdoor opening.

Solution:
Using the Standard Method, the required volume of the room would be 50 cubic feet per 1,000 Btu/hr.

140,000 + 50,000 + 50,000 = 240,000 Btu/hr total

50 ft^3 x (240,000 Btu/hr ÷ 1,000 Btu/hr) = 12,000 ft^3 of indoor combustion air required

Available room volume: 35 x 35 x 8 = 9,800 ft^3

12,000 ft^3 required – 9,800 ft^3 available = 2,200 ft^3 additional combustion air needed.

In this case, the volume of indoor air would not be enough to supply all the appliances and the following three steps are needed to calculate the size for the outdoor opening for additional combustion air.

Step (1) Find the ratio between the interior space and the required volume:

9,800 ÷ 12,000 = 0.82

Step (2) Find the reduction factor:

1.0 - 0.82 = 0.18

Step (3) Find the size of the outdoor opening for one permanent opening:

Section 506.4.2 (1) requires the free area of the opening to be one square inch per 3,000 Btu/hr. of the total input rating of all the appliances.

240,000 total Btu/hr. ÷ 3,000 Btu/hr. = 80

Multiply 80 by the reduction factor 0.18. 80 x 0.18 = 14.4 in^2

Therefore, a single pipe directly to the outside shall have 14.4 square inches.

506.6 Engineered Installations. Engineered combustion air installations shall provide an adequate supply of combustion, ventilation, and dilution air and shall be approved by the Authority Having Jurisdiction. [NFPA 54:9.3.5]

506.7 Mechanical Combustion Air Supply. Where combustion air is provided by a mechanical air supply system, the combustion air shall be supplied from outdoors at the minimum rate of 0.35 cubic feet per minute per 1000 Btu/h [0.034 (m^3/min)/kW] for appliances located within the space. [NFPA 54:9.3.6]

506.7.1 Exhaust Fans. Where exhaust fans are installed, additional air shall be provided to replace the exhausted air. [NFPA 54:9.3.6.1]

506.7.2 Interlock. Each of the appliances served shall be interlocked to the mechanical air supply system to prevent main burner operation where the mechanical air supply system is not in operation. [NFPA 54:9.3.6.2]

506.7.3 Specified Combustion Air. Where combustion air is provided by the building's mechanical ventilation system, the system shall provide the specified combustion air rate in addition to the required ventilation air. [NFPA 54:9.3.6.3]

506.8 Louvers, Grilles, and Screens. The required size of openings for combustion, ventilation, and dilution air shall be based on the net free area of each opening. Where the free area through a design of louver, grille, or screen is known, it shall be used in calculating the size opening required to provide the free area specified. Where the louver and grille design and free area are not known, it shall be assumed that wood louvers have 25 percent free area and metal louvers and grilles have 75 percent free area. Nonmotorized louvers and grilles shall be fixed in the open position. [NFPA 54:9.3.7.1]

🔧 Problem: If 100 square inches of net free opening are required, and the grill is metal (75-percent free area), what size grill would be required? See **Figure 506.8**.

Solution:

100 in^2 ÷ .75 = 133 in^2 minimum grill size

Grill size: 10" x 14" = 144 in^2

506.8.1 Minimum Screen Mesh Size. Screens shall be not less than ¼ of an inch (6.4 mm) mesh. [NFPA 54:9.3.7.2]

506.8.2 Motorized Louvers. Motorized louvers shall be interlocked with the appliance so they are proven in the full open position prior to main burner ignition and during main burner operation. Means shall be provided to prevent the main burner from igniting should the louver fail to open during burner startup and to shut down the main burner if the louvers close during burner operation. [NFPA 54:9.3.7.3]

WATER HEATERS

  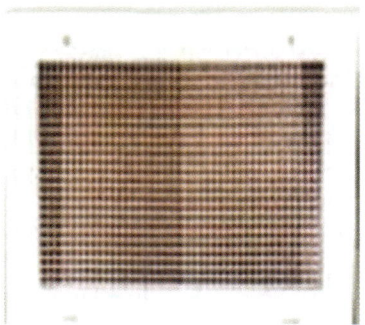

**FIGURE 506.8
METAL, WOOD, AND MESH GRILLS**

506.9 Combustion Air Ducts. Combustion air ducts shall comply with the following [NFPA 54:9.3.8]:

(1) Ducts shall be constructed of galvanized steel or a material having equivalent corrosion resistance, strength, and rigidity.

Exception: Within dwelling units, unobstructed stud and joist spaces shall not be prohibited from conveying combustion air, provided that not more than one fireblock is removed. [NFPA 54:9.3.8.1]

(2) Ducts shall terminate in an unobstructed space, allowing free movement of combustion air to the appliances. [NFPA 54:9.3.8.2]

(3) Ducts shall serve a single space. [NFPA 54:9.3.8.3]

(4) Ducts shall not serve both upper and lower combustion air openings where both such openings are used. The separation between ducts serving upper and lower combustion air openings shall be maintained to the source of combustion air. [NFPA 54:9.3.8.4]

(5) Ducts shall not be screened where terminating in an attic space. [NFPA 54:9.3.8.5]

(6) Combustion air intake openings located on the exterior of the building shall have the lowest side of the combustion air intake openings located at least 12 inches (305 mm) vertically from the adjoining finished ground level. [NFPA 54:9.3.8.8]

(7) Horizontal upper combustion air ducts shall not slope downward toward the source of combustion air. [NFPA 54:9.3.8.6]

(8) The remaining space surrounding a chimney liner, gas vent, special gas vent, or plastic piping installed within a masonry, metal, or factory-built chimney shall not be used to supply combustion air.

Exception: Direct-vent appliances designed for installation in a solid-fuel-burning fireplace where installed in accordance with the manufacturer's installation instructions. [NFPA 54:9.3.8.7]

507.0 Appliance and Equipment Installation Requirements.

507.1 Dielectric Insulator. The Authority Having Jurisdiction shall have the authority to require the use of an approved dielectric insulator on the water piping connections of water heaters and related water heating appliances.

🔧 Dielectric unions are not required by the UPC; however, this section allows the AHJ to require them if it so chooses.

507.2 Seismic Provisions. In seismic design categories C, D, E, and F, water heaters shall be anchored or strapped to resist horizontal displacement due to earthquake motion. Strapping shall be at points within the upper one-third and lower one-third of its vertical dimensions. At the lower point, a distance of not less than 4 inches (102 mm) shall be maintained from the controls with the strapping.

🔧 Seismic activity can cause the water heater to be displaced, creating a potentially dangerous condition. The gas connector may break (which could allow gas to flow freely into the space), or the vent connection may separate, allowing products of combustion to enter the space. Consideration must also be given to the possibility of a fire started by the water heater from damage to either the fuel-gas or electrical connections to the tank. In the event of a prolonged interruption of utilities, water heaters may provide an emergency source of drinking water, particularly if they are properly restrained and not destroyed during the earthquake. See **Figure 507.13a** for an example of water heater restraints.

507.3 Appliance Support. Appliances and equipment shall be furnished either with load-distributing bases or with a sufficient number of supports to prevent damage to either the building structure or the appliance and the equipment. [NFPA 54:9.1.8.1]

507.3.1 Structural Capacity. At the locations selected for installation of appliances and equipment, the dynamic and static load carrying capacities of the building structure shall be checked to determine whether they are adequate to carry the additional loads. The appliances and equipment shall be supported and shall be connected to the piping so as not to exert undue stress on the connections. [NFPA 54:9.1.8.2]

507.4 Ground Support. A water heater supported from the earth shall rest on level concrete or other approved base extending not less than 3 inches (76 mm) above the adjoining ground level.

507.5 Drainage Pan. Where a water heater is located in an attic, in or on an attic ceiling assembly, floor-ceiling assem-

WATER HEATERS

bly, or floor-subfloor assembly where damage results from a leaking water heater, a watertight pan of corrosion-resistant materials shall be installed beneath the water heater with not less than ¾ of an inch (20 mm) diameter drain to an approved location. Such pan shall be not less than 1½ inches (38 mm) in depth.

🔧 The point of termination for the water heater pan drain should be visible. It should be in a location that will draw attention to the need for investigation to determine the cause of the leak (see **Figure 507.5**). See Section 608.5 (7) that prohibits the discharge from a relief valve into a water heater pan.

**FIGURE 507.5
STANDARD WATER HEATER SAFETY PAN**

507.6 Added or Converted Equipment or Appliances. When additional or replacement appliances or equipment is installed or an appliance is converted to gas from another fuel, the location in which the appliances or equipment is to be operated shall be checked to verify the following:

(1) Air for combustion and ventilation is provided where required, in accordance with the provisions of Section 506.0. Where existing facilities are not adequate, they shall be upgraded to meet Section 506.0 specifications.

(2) The installation components and appliances meet the clearances to combustible material provisions of Section 507.27. It shall be determined that the installation and operation of the additional or replacement appliances do not render the remaining appliances unsafe for continued operation.

(3) The venting system is constructed and sized in accordance with the provisions of Section 509.0. Where the existing venting system is not adequate, it shall be upgraded to comply with Section 509.0. [NFPA 54:9.1.2]

🔧 The replacement or addition of water-heating equipment requires a permit (see Section 104.0) and must conform to this code. In addition to all the water, gas (or electric), and flue pipe connections to the water heater, this section requires that the location of the water heating equipment shall meet the provisions in this code for combustion, ventilation, clearances, and venting whenever the heating equipment is replaced or added.

507.7 Types of Gases. The appliance shall be connected to the fuel gas for which it was designed. No attempt shall be made to convert the appliance from the gas specified on the rating plate for use with a different gas without consulting the installation instructions, the serving gas supplier, or the appliance manufacturer for complete instructions. [NFPA 54:9.1.3]

507.8 Safety Shutoff Devices for Unlisted LP-Gas Appliance Used Indoors. Unlisted appliances for use with undiluted liquefied petroleum gases and installed indoors, except attended laboratory equipment, shall be equipped with safety shutoff devices of the complete shutoff type. [NFPA 54:9.1.4]

507.9 Use of Air or Oxygen Under Pressure. Where air or oxygen under pressure `is used in connection with the gas supply, effective means such as a backpressure regulator and relief valve shall be provided to prevent air or oxygen from passing back into the gas piping. Where oxygen is used, installation shall be in accordance with NFPA 51. [NFPA 54:9.1.5]

507.10 Protection of Gas Appliances from Fumes or Gases other than Products of Combustion. Non-direct-vent appliances installed in beauty shops, barber shops, or other facilities where chemicals that generate corrosive or flammable products such as aerosol sprays are routinely used shall be located in a mechanical room separate or partitioned off from other areas with provisions for combustion and dilution air from outdoors. Direct-vent appliances in such facilities shall be in accordance with the appliance manufacturer's installation instructions. [NFPA 54:9.1.6.2]

507.11 Process Air. In addition to air needed for combustion in commercial or industrial processes, process air shall be provided as required for cooling of appliances, equipment, or material; for controlling dew point, heating, drying, oxidation, dilution, safety exhaust, odor control, air for compressors; and for comfort and proper working conditions for personnel. [NFPA 54:9.1.7]

507.12 Flammable Vapors. Appliances shall not be installed in areas where the open use, handling, or dispensing of flammable liquids occurs, unless the design, operation, or installation reduces the potential of ignition of the flammable vapors. Appliances installed in compliance with Section 507.13 through Section 507.15 shall be considered to comply with the intent of this provision. [NFPA 54:9.1.9]

🔧 The prohibition of gas appliances from installation in areas where flammable vapors may occur is different than the requirement in the next section for residential garages. The intent here is to keep the appliance away from where these vapors will consistently occur or from where they are likely to occur because of the work being done. In this case, the appliance must be installed in a separate area away from these flammable vapors. The water heaters manufactured today are flammable vapor ignition resistant and are meant to protect from only accidental spills that may occur in a residential garage rather than during continuous exposure to these vapors.

507.13 Installation in Residential Garages. Appliances in residential garages and in adjacent spaces that open to the garage and are not part of the living space of a dwelling unit shall be installed so that all burners and burner-ignition devices are located not less than 18 inches (457 mm) above the floor unless listed as flammable vapor ignition resistant. [NFPA 54:9.1.10.1]

🔧 This requirement for installation of the water heater controls at 18 inches above the floor is to minimize the possibility of igniting heavier-than-air flammable vapors that may be present in a garage, such as caused by spilled gasoline or solvents, unless the water heater is listed as flammable vapor ignition resistant (see **Figure 507.13a**).

Much concern has been voiced over the past several years regarding the ignition of flammable vapors by gas appliances, specifically residential water heaters. The standard for water heaters, ANSI Z21.10.1, *Gas Water Heaters, Volume I—Storage, Water Heaters with Input Ratings of 75,000 Btu per Hour or Less*, require that all residential water heaters pass a flammable vapor ignition-resistance test in order to be listed.

All water heaters manufactured and listed to ANSI Z21.10.1 will be resistant to the ignition of flammable vapors. The design used to enable the new water heaters to meet the flammable-resistance ignition test is based on the principle first used in the mine safety lamp. In a mine safety lamp, a fine mesh screen encloses an oil lamp flame. If flammable gas enters the screened area, the gas ignites but the flame does not propagate outside the screen. The flame is kept within the combustion chamber of the water heater and eventually causes a sensor to trip, disabling the water heater (see F**igure 507.13b**).

Some models are being built that allow the homeowner to reignite the pilot after the condition causing the outage is resolved. Others require a replacement of the water heater.

Note that direct-vent water heaters produced prior to June 30, 2005, are not tested as resistant to the ignition of flammable vapors and cannot be installed on the floor of a residential garage unless specifically listed for garage floor installation. Direct-vent heaters are not sealed tightly enough to prevent flame propagation outside the firing chamber. Direct-vent sealed combustion chamber water heaters are allowed to be installed directly on the floor in a garage.

507.13.1 Physical Damage. Appliances installed in garages, warehouses, or other areas subject to mechanical damage shall be guarded against such damage by being installed behind protective barriers or by being elevated or located out of the normal path of vehicles.

507.13.2 Access from the Outside. Where appliances are installed in a separate, enclosed space having access only from outside of the garage, such appliances shall be permitted to be installed at floor level, providing the required combustion air is taken from the exterior of the garage. [NFPA 54:9.1.10.3]

507.14 Installation in Commercial Garages. Appliances installed in commercial garages shall comply with Section 507.14.1 and Section 507.14.2.

507.14.1 Parking Structures. Appliances installed in enclosed, basement, and underground parking structures shall be installed in accordance with NFPA 88A. [NFPA 54:9.1.11.1]

507.14.2 Repair Garages. Appliances installed in repair garages shall be installed in accordance with NFPA 30A. [NFPA 54:9.1.11.2]

507.15 Installation in Aircraft Hangars. Heaters in aircraft hangars shall be installed in accordance with NFPA 409. [NFPA 54:9.1.12]

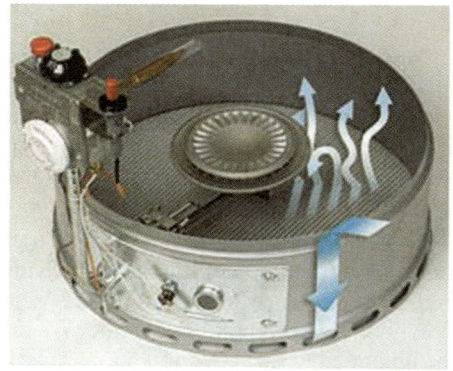

FIGURE 507.13B
FLAMMABLE VAPOR-RESISTANT WATER HEATER COMBUSTION CHAMBER

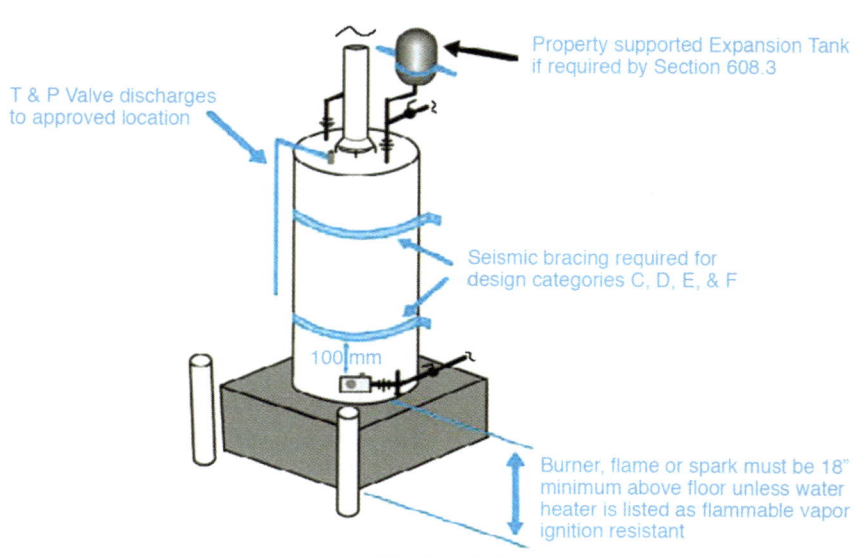

FIGURE 507.13A
RESIDENTIAL GARAGE WATER HEATER INSTALLATION

507.16 Venting of Flue Gases. Appliances shall be vented in accordance with the provisions of Section 509.0. [NFPA 54:9.1.14]

507.17 Extra Device or Attachment. No device or attachment shall be installed on an appliance that is capable of impairing the combustion of gas. [NFPA 54:9.1.15]

507.18 Adequate Capacity of Piping. When additional appliances are being connected to a gas piping system, the existing piping shall be checked to determine whether it has adequate capacity. Where the capacity is inadequate, the existing system shall be enlarged as necessary, or separate gas piping of adequate capacity shall be run from the point of delivery to the appliance. [NFPA 54:9.1.16]

507.19 Avoiding Strain on Gas Piping. Appliances shall be supported and connected to the piping so as not to exert undue strain on the connections. [NFPA 54:9.1.17]

507.20 Gas Appliance Pressure Regulators. Where the gas supply pressure is higher than that at which the appliance is designed to operate or varies beyond the design pressure limits of the appliance, a gas appliance pressure regulator shall be installed. [NFPA 54:9.1.18]

507.21 Venting of Gas Appliance Pressure Regulators. Venting of gas appliance pressure regulators shall comply with the following requirements:

(1) Appliance pressure regulators requiring access to the atmosphere for successful operation shall be equipped with vent piping leading outdoors or, if the regulator vent is an integral part of the appliance, into the combustion chamber adjacent to a continuous pilot, unless constructed or equipped with a vent limiting means to limit the escape of gas from the vent opening in the event of diaphragm failure.

(2) Vent limiting means shall be employed on listed appliance pressure regulators only.

(3) In the case of vents leading outdoors, means shall be employed to prevent water from entering this piping and also to prevent blockage of vents by insects and foreign matter.

(4) Under no circumstances shall a regulator be vented to the appliance flue or exhaust system.

(5) In the case of vents entering the combustion chamber, the vent shall be located so the escaping gas is readily ignited by the pilot and the heat liberated thereby does not adversely affect the normal operation of the safety shutoff system. The terminus of the vent shall be securely held in a fixed position relative to the pilot. For manufactured gas, the need for a flame arrester in the vent piping shall be determined.

(6) A vent line(s) from an appliance pressure regulator and a bleed line(s) from a diaphragm-type valve shall not be connected to a common manifold terminating in a combustion chamber. Vent lines shall not terminate in positive-pressure-type combustion chambers. [NFPA 54:9.1.19]

507.22 Bleed Lines for Diaphragm-Type Valves. Bleed lines shall comply with the following requirements:

(1) Diaphragm-type valves shall be equipped to convey bleed gas to the outdoors or into the combustion chamber adjacent to a continuous pilot.

(2) In the case of bleed lines leading outdoors, means shall be employed to prevent water from entering this piping and also to prevent blockage of vents by insects and foreign matter.

(3) Bleed lines shall not terminate in the appliance flue or exhaust system.

(4) In the case of bleed lines entering the combustion chamber, the bleed line shall be located so the bleed gas is readily ignited by the pilot and the heat liberated thereby does not adversely affect the normal operation of the safety shutoff system. The terminus of the bleed line shall be securely held in a fixed position relative to the pilot. For manufactured gas, the need for a flame arrester in the bleed line piping shall be determined.

(5) A bleed line(s) from a diaphragm-type valve and a vent line(s) from an appliance pressure regulator shall not be connected to a common manifold terminating in a combustion chamber. Bleed lines shall not terminate in positive-pressure-type combustion chambers. [NFPA 54:9.1.20]

507.23 Combination of Appliances and Equipment. A combination of appliances, equipment, attachments, or devices used together in a manner shall be in accordance with the standards that apply to the individual appliance and equipment. [NFPA 54:9.1.21]

507.24 Installation Instructions. The installing agency shall comply with the appliance and equipment manufacturer's installation instructions in completing an installation. The installing agency shall leave the manufacturer's installation, operating, and maintenance instructions in a location on the premises where they will be readily available for reference and guidance for the Authority Having Jurisdiction, service personnel, and the owner or operator. [NFPA 54:9.1.22]

The requirement to conform to the manufacturer's installation instructions abounds in this code and for good reason. The manufacturer designs a product to a specific standard for a specific installation. If those standards are not met or the installation requirements not followed, not only will the warranty for the equipment be voided but any liability from a failure or an accident will be on the shoulders of the installer, which can include the contractor and the individual plumber. Always consult and follow the manufacturer's installation instructions. If there are conflicts between the code and the instructions, the installer should contact the AHJ and the manufacturer for clarification; he or she should never decide the issue on his or her own (see **Figure 507.24**).

507.25 Protection of Outdoor Appliances. Appliances not listed for outdoor installation but installed outdoors shall be provided with protection to the degree that the environment requires. Appliances listed for outdoor installation shall be permitted to be installed without protection in accordance with the provisions of its listing and the manufacturer's installation instructions.

WATER HEATERS

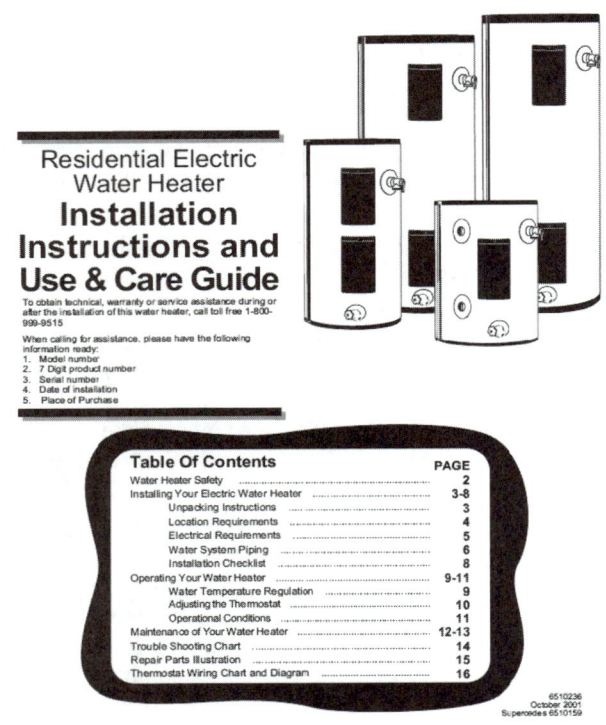

**FIGURE 507.24
INSTALLATION INSTRUCTIONS**

507.26 Accessibility for Service. Appliances shall be located with respect to building construction and other equipment so as to permit access to the appliance. Sufficient clearance shall be maintained to permit cleaning of heating surfaces; the replacement of filters, blowers, motors, burners, controls, and vent connections; the lubrication of moving parts where necessary; the adjustment and cleaning of burners and pilots; and the proper functioning of explosion vents, where provided. For attic installation, the passageway and servicing area adjacent to the appliance shall be floored. [NFPA 54:9.2.1]

Normally, water heaters are installed where there is easy access to allow the replacement of the water heater since it will eventually need to be replaced. The design of the installation should take into consideration that a large and heavy object will need to be removed and another large and heavy object put in its place. However, building designers may not place the maintenance of the plumbing system as their first priority. Hence, the plumber may find him or herself lugging a water heater down the ladders or passageways required in this section.

507.27 Clearance to Combustible Materials. Appliances and their vent connectors shall be installed with clearances from combustible material so their operation does not create a hazard to persons or property. Minimum clearances between combustible walls and the back and sides of various conventional types of appliances and their vent connectors are specified in Section 509.0. [NFPA 54:9.2.2]

For reference, the definitions for combustible, noncombustible and limited combustible in the UMC are:

Combustible material pertains to materials adjacent to or in contact with heat-producing appliances, vent connectors, gas vents, chimneys, steam and hot water pipes, and warm air ducts, materials made of or surfaced with wood, compressed paper, plant fibers, or other materials that are capable of being ignited and burned. Such material shall be considered combustible even though flame-proofed, fire retardant treated, or plastered. [NFPA 54:3.3.67.1]

Noncombustible material as applied to building construction material, means a material that in the form in which it is used is either one of the following: (1) A material that, in the form in which it is used and under the conditions anticipated, will not ignite, burn, support combustion, or release flammable vapors when subjected to fire or heat. Materials that are reported as passing ASTM E136 are considered noncombustible material. (2) Material having a structural base of noncombustible material as defined in item 1 above, with a surfacing material not over 1/8 of an inch (3.2 mm) thick that has a flame-spread index not higher than 50. Noncombustible does not apply to surface finish materials. Material required to be noncombustible for reduced clearances to flues, heating appliances, or other sources of high temperature shall refer to material in accordance with item 1 above. No material shall be classed as noncombustible that is subject to increase in combustibility or flame-spread index beyond the limits herein established, through the effects of age, moisture, or other atmospheric condition.

Limited-Combustible material refers to a building construction material that does not comply with the definition of noncombustible material that, in the form in which it is used, has a potential heat value not exceeding 3500 British thermal units per pound-force (Btu/lb) (8141 kJ/kg), where tested in accordance with NFPA 259, and includes either of the following:

(1) Materials having a structural base of noncombustible material, with a surfacing not exceeding a thickness of 1/8 of an inch (3.2 mm), that has a flame-spread index not greater than 50.

(2) Materials, in the form and thickness used, having neither a flame-spread index greater than 25 nor evidence of continued progressive combustion, and of such composition that surfaces that would be exposed by cutting through the material on any plane would have neither a flame-spread index greater than 25 nor evidence of continued progressive combustion, where tested in accordance with ASTM E84.

508.0 Appliances on Roofs.

508.1 General. Appliances on roofs shall be designed or enclosed so as to withstand climatic conditions in the area in which they are installed. Where enclosures are provided, each enclosure shall permit easy entry and movement, shall be of reasonable height, and shall have at least a 30 inch (762 mm) clearance between the entire service access panel(s) of the appliance, and the wall of the enclosure. [NFPA 54:9.4.1.1]

508.1.1 Load Capacity. Roofs on which appliances are to be installed shall be capable of supporting the additional load or shall be reinforced to support the additional load. [NFPA 54:9.4.1.2]

508.1.2 Fasteners. Access locks, screws, and bolts shall be of corrosion-resistant material. [NFPA 54:9.4.1.3]

508.2 Installation of Appliances on Roofs. Appliances shall be installed in accordance with the manufacturer's installation instructions. [NFPA 54:9.4.2.1]

508.2.1 Clearance. Appliances shall be installed on a well-drained surface of the roof. At least 6 feet (1829 mm) of clearance shall be available between any part of the appliance, and the edge of a roof or similar hazard, or rigidly fixed rails, guards, parapets, or other building structures at least 42 inches (1067 mm) in height shall be provided on the exposed side. [NFPA 54:9.4.2.2]

508.2.2 Electrical Power. All Appliances requiring an external source of electrical power for its operation shall be provided with the following:

(1) A readily accessible electrical disconnecting means within sight of the appliance that completely de-energizes the appliance.

(2) A 120 V-ac grounding-type receptacle outlet on the roof adjacent to the appliance on the supply side of the disconnect switch. [NFPA 54:9.4.2.3]

508.2.3 Platform or Walkway. Where water stands on the roof at the appliance or in the passageways to the appliance, or where the roof is of a design having a water seal, a suitable platform, walkway, or both shall be provided above the waterline. Such platform(s) or walkway(s) shall be located adjacent to the appliance and control panels so that the appliance can be safely serviced where water stands on the roof. [NFPA 54:9.4.2.4]

508.3 Appliances on Roofs. Appliances located on roofs or other elevated locations shall be accessible. [NFPA 54:9.4.3.1]

508.3.1 Access. Buildings exceeding 15 feet (4572 mm) in height shall have an inside means of access to the roof unless other means acceptable to the Authority Having Jurisdiction are used. [NFPA 54:9.4.3.2]

508.3.2 Access Type. The inside means of access shall be a permanent or fold away inside stairway or ladder, terminating in an enclosure, scuttle, or trap door. Such scuttles or trap doors shall be at least 22 inches by 24 inches (559 mm by 610 mm) in size, shall open easily and safely under all conditions, especially snow; and shall be constructed so as to permit access from the roof side unless deliberately locked on the inside.

At least 6 feet (1829 mm) of clearance shall be available between the access opening and the edge of the roof or similar hazard, or rigidly fixed rails or guards a minimum of 42 inches (1067 mm) in height shall be provided on the exposed side. Where parapets or other building structures are utilized in lieu of guards or rails, they shall be a minimum of 42 inches (1067 mm) in height. [NFPA 54:9.4.3.3]

See **Figure 508.3.2**.

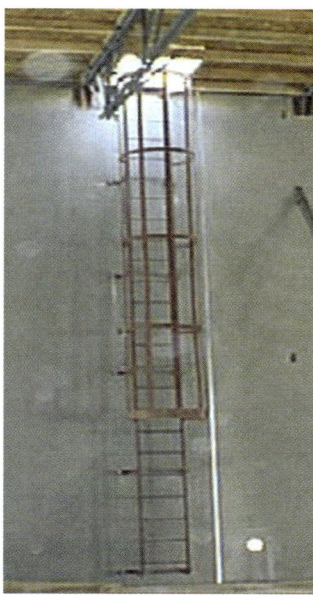

**FIGURE 508.3.2
LADDER ACCESS TO ROOF**

508.3.3 Permanent Lighting. Permanent lighting shall be provided at the roof access. The switch for such lighting shall be located inside the building near the access means leading to the roof. [NFPA 54:9.4.3.4]

508.4 Appliances in Attics and Under-Floor Spaces. An attic or under-floor space in which an appliance is installed shall be accessible through an opening and passageway, not less than as large as the largest component of the appliance, and not less than 22 inches by 30 inches (559 mm by 762 mm).

See **Figure 508.4**.

508.4.1 Length of Passageway. Where the height of the passageway is less than 6 feet (1829 mm), the distance from the passageway access to the appliance shall not exceed 20 feet (6096 mm) measured along the centerline of the passageway. [NFPA 54:9.5.1.1]

508.4.2 Width of Passageway. The passageway shall be unobstructed and shall have solid flooring not less than 24 inches (610 mm) wide from the entrance opening to the appliance. [NFPA 54:9.5.1.2]

508.4.3 Work Platform. A level working platform not less than 30 inches by 30 inches (762 mm by 762 mm) shall be provided in front of the service side of the appliance. [NFPA 54:9.5.2]

508.4.4 Lighting and Convenience Outlet. A permanent 120 V receptacle outlet and a lighting fixture shall be installed near the appliance. The switch controlling the lighting fixture shall be located at the entrance to the passageway. [NFPA 54:9.5.3]

509.0 Venting of Appliances.

Although gas is a clean-burning fuel, the products of combustion must not be allowed to accumulate inside of a building (see **Figure 509.0** for the products of combustion

WATER HEATERS

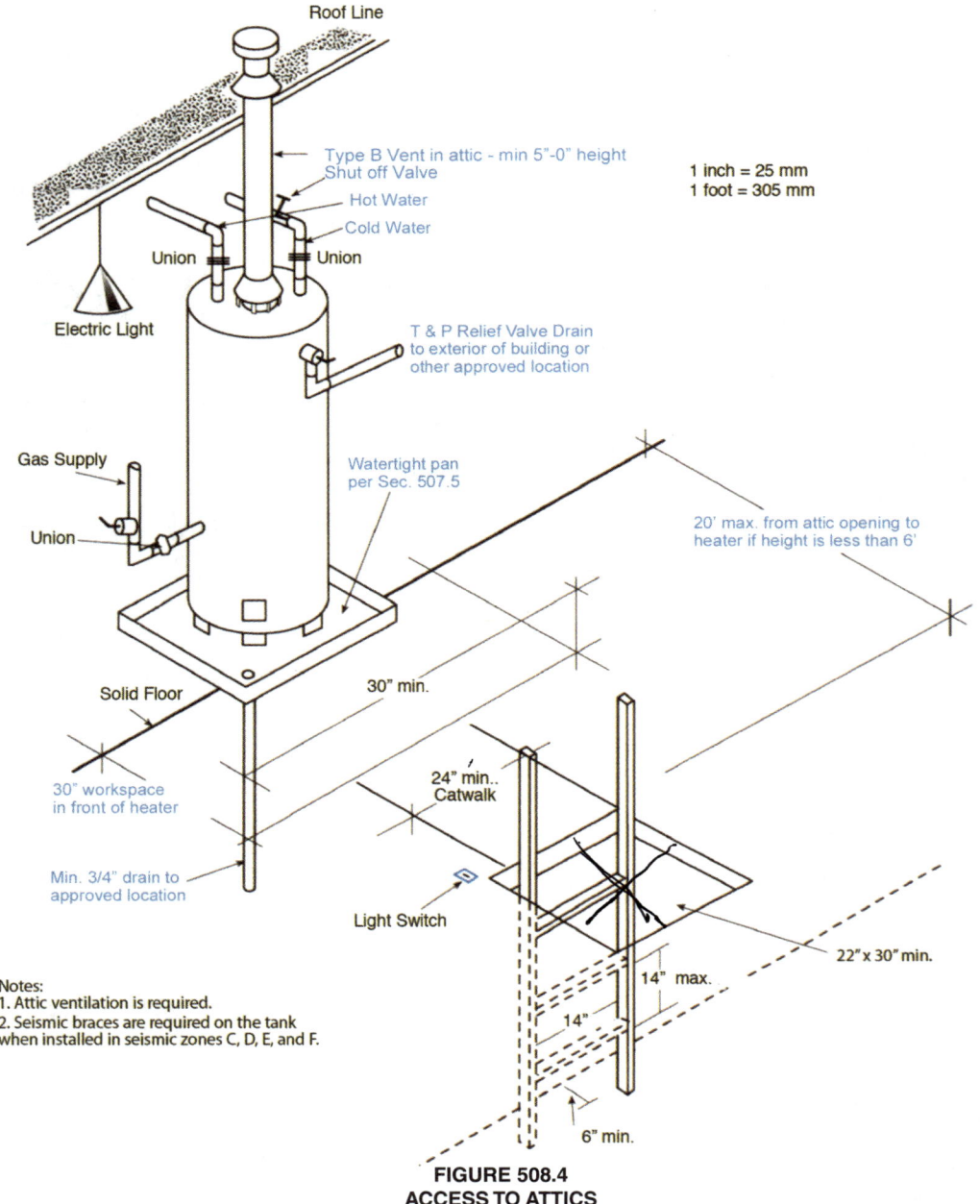

**FIGURE 508.4
ACCESS TO ATTICS**

formed under ideal conditions). Therefore, venting of most gas-utilization equipment is required. A properly installed and maintained venting system will perform the following functions to provide for proper appliance function and the safety of building occupants:

- Convey all of the combustion products to the outside atmosphere.
- Prevent damage to the gas equipment, vent building and furnishings from water vapor condensation in the flue gases.
- Prevent overheating of walls, building structure and other combustible materials that are installed with required clearance to the appliance and venting system.
- Provide fast priming of natural draft-venting systems to minimize spillage of combustion products into the building.

The operation of a venting system may appear complicated, but remember that natural draft venting works because hot air rises. When gas is burned, the products of combustion are hotter than the ambient air and, therefore, rise. The venting system's job is to channel these combustion products out of the building. To do so, the venting system must keep the combustion products as warm as possible in the vent. Heat retention maximizes the draft produced.

Prior to the 1980s, when the Department of Energy's minimum appliance efficiency regulation mandated higher-efficiency appliances, gas furnaces had a seasonal efficiency of about 60 percent. This meant that vent gases were hotter, providing ample heat for venting and for controlling condensation. With higher-efficiency space-heating appliances, the old safety margin no longer applies, and extra care must be taken so that the vents operate properly.

The Category I (mid-efficiency) appliances available today have a seasonal efficiency of about 80 percent. Some use a fan to assist the flow of combustion products through the appliance and are not equipped with a draft hood. Other appliances are equipped with a draft hood but reduce their total flue losses by using an automatic damper to cut off the flow in the vent during the off-cycle. Problems with early models and vent system failures were traced to condensation and resulted in the development of new venting tables, shown in Section 510.0, to provide proper sizing of vents for these appliances.

Water heaters must also meet minimum efficiency requirements, but they are still typically draft hood equipped. Vent sizing for draft hood-equipped water heaters (and other appliances) connected to a dedicated venting system did not require an extensive reevaluation. However, the size of common vents serving water heaters and Category I fan-assisted appliances has changed; sizing is to be done in accordance with the venting tables in Section 510.0.

High-efficiency condensing appliances have a seasonal efficiency of 90 percent or higher, which reduces vent gas temperatures to a point where the water vapor produced as a product of combustion condenses to liquid water in the appliance or in the vent. These condensing appliances carry a vented appliance category of Category IV. This type of appliance produces much cooler vent gases, resulting in water condensing in the vent. Venting must be accomplished with a fan because the vent gases are not hot enough to operate the natural draft vent. Water will condense in the vent and will dissolve some of the gases produced during combustion, which are slightly acidic. The vent materials used with these appliances must be able to resist the acidic condensate.

The advantage of high-efficiency appliances is that they reduce the amount of gas consumed with no loss in output. A mid-efficiency appliance uses one-third less gas than a conventional appliance, and a condensing appliance uses only one-half of the gas of a conventional appliance. The savings in fuel are offset by higher first cost and the higher maintenance requirements of high-efficiency appliances.

The following section concerns venting equipment, including the construction and installation of venting equipment, but not the sizing of the vents. Section 510.0 covers sizing of venting systems. Section 509.0 covers installation requirements. However, safe installation of venting systems requires engineering calculations to ensure that sufficient draft is present. The tables in Section 510.0 perform these calculations. They do not cover all vent sizes and configurations, and other methods can be used in lieu of the tables.

The sub sections in Section 510.0 generally follow the procedures an installer would use to design or evaluate a venting system for use with a Category I mid-efficiency water heater. Remember, this chapter is intended for the installation of water heaters. A water heater vent may be combined with other gas appliance vents but the other appliances should be installed per the UMC or NFPA 54. The other category appliances should be vented per their manufacturer's installation instructions only. The appliance category for most listed appliances can be found on the appliance nameplate. Some appliances equipped with draft hoods might not have a category on the nameplate. These appliances should be installed as Category I appliances.

The venting system is roughly divided into vertical and horizontal components. The main, vertical section of the venting system is generally a chimney or Type B vent. The main vent is attached to the appliance(s) by the vent connector, which often has a horizontal component. Each portion of the venting system has different requirements, which are detailed in each portion's corresponding section.

There are three basic types of venting systems used for conventional venting systems:

1. Masonry, metal and factory-built chimneys.
2. Gas vents.
3. Single-wall metal pipe.

The methods of constructing, installing and terminating these systems are discussed in this section. Other methods of venting are also discussed. They are normally used in industrial and commercial applications and rarely for water heater vents. They were included in the chapter due to the extraction process from NFPA 54; however, they should be discussed under the UMC or NFPA 54 and not here.

509.1 Listing. Type B and Type B-W gas vents shall comply with UL 441, and Type L gas vents shall comply with UL 641.

509.1.1 Installation. Listed vents shall be installed in accordance with Section 509.0 and the manufacturer's installation instructions. [NFPA 54:12.2.1]

509.1.2 Prohibited Discharge. Appliance vents shall not discharge into a space enclosed by screens having openings less than ¼ of an inch (6.4 mm) mesh.

509.2 Connection to Venting Systems. Except as permitted in Section 509.2.1 through Section 509.2.7, all appliances shall be connected to venting systems. [NFPA 54:12.3.1]

509.2.1 Appliances Not Required to be Vented. The following appliances shall not be required to be vented:

(1) A single listed booster-type (automatic instantaneous) water heater, when designed and used solely for the sanitizing rinse requirements of a dishwashing machine, provided that the appliance is installed with the draft hood in place and unaltered, if a draft hood is required, in a commercial kitchen having a mechanical exhaust system. Where installed in this manner, the draft hood outlet shall not be less than 36 inches (914 mm) vertically and 6 inches (152 mm) horizontally from any surface other than the appliance. [NFPA 54:12.3.2(5)]

(2) Other appliances listed for unvented use and not provided with flue collars. [NFPA 54:12.3.2(10)]

509.2.2 Maximum Input Rating. Where any or all of the appliances in Section 509.2.1(1) and Section 509.2.1(2) are

WATER HEATERS

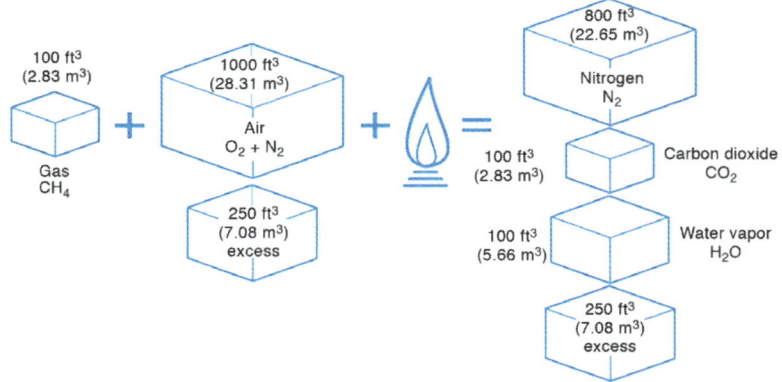

**FIGURE 509.0
THE PRODUCTS OF COMBUSTION OF NATURAL GAS**

installed so the aggregate input rating exceeds 20 (Btu/h)/ft^3 (207 W/m^3) of room or space in which it is installed, one or more shall be provided with venting systems or other approved means for conveying the vent gases to the outdoors so that the aggregate input rating of the remaining unvented appliance does not exceed 20 (Btu/h)/ft^3 (207 W/m^3). [NFPA 54:12.3.2.1]

509.2.3 Adjacent Room or Space. Where the calculation includes the volume of an adjacent room or space, the room or space in which the appliances are installed shall be directly connected to the adjacent room or space by a doorway, archway, or other opening of comparable size that cannot be closed. [NFPA 54:12.3.2.2]

509.2.4 Ventilating Hoods. Ventilating hoods and exhaust systems shall be permitted to be used to vent appliances installed in commercial applications and to vent industrial appliances, particularly where the process itself requires fume disposal. [NFPA 54:12.3.3]

🔧 This method is used frequently in restaurant kitchens to vent booster water heaters used to provide the very hot water required for sanitation. Information on the design and installation of ventilating hoods in industrial plants can be obtained from NFPA 91, *Standard for Exhaust Systems for Air Conveying of Vapors, Gases, Mists, and Noncombustible Particulate Solids.*

509.2.5 Well-Ventilated Spaces. The operation of industrial appliances such that its flue gases are discharged directly into a large and well-ventilated space shall be permitted. [NFPA 54:12.3.4]

509.2.6 Direct-Vent Appliances. Listed direct-vent appliances shall be installed in accordance with the manufacturer's installation instructions and Section 509.8.2. [NFPA 54:12.3.5]

🔧 The venting system of a direct-vent appliance is considered to be part of its listing. Listed appliances are tested with inlet and outlet configurations taken from their installation directions. Therefore, the installation instructions must be followed carefully. **Figure 509.2.6** shows one type of a direct-vent appliance that uses a concentric vent and air supply. Some direct-vent appliances use independent ducts for the inlet and outlets. Section 509.8.2 provides requirements for the location of the vent terminations of direct-vent appliances.

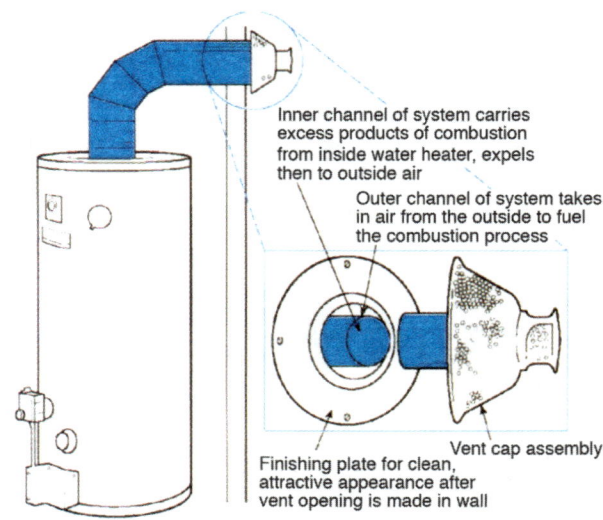

**FIGURE 509.2.6
DIRECT VENT WATER HEATER USING A CONCENTRIC VENT AND SUPPLY**

509.2.7 Appliances with Integral Vents. Appliances incorporating integral venting means shall be installed in accordance with the manufacturer's installation instructions, Section 509.8 and Section 509.8.1. [NFPA 54:12.3.6]

509.3 Design and Construction. Venting systems shall be designed and constructed to convey flue, vent, or both gases to the outdoors. [NFPA 54:12.1]

509.3.1 Appliance Draft Requirements. A venting system shall satisfy the draft requirements of the appliance in accordance with the manufacturer's instructions. [NFPA 54:12.4.1]

🔧 This section provides an overall performance requirement for a natural draft vent system. The vent system must remove all the products of combustion from gas appliances from a building. The principle on which natural draft vents operate is simple. Reduced to fundamentals, heat is the power that operates a natural draft vent or chimney, as shown in **Figure 509.3.1a**. Combustion gases rise in the chimney or vent only because they are hotter and, therefore, lighter than the surrounding air. The hotter the gases and the higher the vent, the more swiftly and powerfully the gases will rise. Conversely, the cooler the gases and the shorter the vent, the more sluggish the gases' movement will be. The flue gases in the vent must remain hot enough over the length of the vent to provide a strong draft. If cooled enough, the upward motion of the gases stops altogether and combustion gases can spill into the building through the relief opening of the appliance draft hood.

When the combustion products and dilution air rise in the vent, this volume must be replaced by air from outside the building. This replacement air can be supplied through normal air infiltration (through small openings in the building walls), through outdoor openings purposely installed in outside walls or by a mechanical air system. Thus, to ensure proper vent operation, the building's air tightness and other devices exhausting air from the building must be taken into account.

Fan-assisted combustion appliances cause some confusion because they are listed as Category I appliances, meaning that their vents operate by natural draft due to the heat of the vent gas. However, these appliances have a fan to assist the flow of the combustion product through the appliance. The fan is necessary because modern, higher-efficiency appliances have a higher pressure drop through their heat exchangers. The pressure provided by the induced draft blower is matched carefully to the resistance of the heat exchanger. Once the combustion products exit the appliance, the natural buoyancy takes over in the vent. If the vent is designed using the tables of Section 510.0, the pressure in the vent system will be negative.

Some furnaces are listed as Category I or III appliances, depending on the method used to vent the appliance. The fan in these furnaces is sized to result in negative vent pressure when connected to a properly sized vertical vent (Category I) or in a positive vent pressure when vented horizontally (Category III). Category III installation requires using venting materials recommended by the manufacturer's installation instructions.

Blockage of a natural draft vent may cause flue gases to spill into the building through the relief opening of the draft hood. Spillage can also be caused if the pressure in the vicinity of the appliance is much lower than the pressure outside the building (depressurization). Depressurization may be caused by mechanical exhausts, fireplaces, wind — anything that removes air from the building. Consequently, periodically verifying the performance of appliance venting systems is wise. Verification can be done readily with a natural draft system, described as follows:

- Operate the appliance for at least 5 minutes to allow the flue gases to heat the vent.
- Move a lighted match across the entire width of the draft hood relief opening (see **Figure 509.3.1b**).
- If the match flame is blown downward or extinguished, have the venting system or chimney checked by a qualified agency.
- If the appliance is not equipped with a draft hood, the match test may not be applicable. In this case, the plumber should consult the appliance instruction manual.

Good air circulation in adequate amounts is also vital for good venting and effective appliance operation. Signs of improper operation include a yellow or wavering flame, discoloration around access doors, a pungent odor in the building, excessive humidity, condensation or mold, corrosion of the vent material and soot near the burner or vent area. If any of these conditions are observed, a qualified agency should be contacted to have the installation inspected.

Under extremely adverse conditions, carbon monoxide can be produced as a result of either improper venting of combustion products, insufficient fresh air to support the proper burning of gas, or improperly adjusted appliances. If symptoms of carbon monoxide poisoning are experienced — headache, yawning, ringing in ears, weariness, vomiting, heart fluttering or throbbing — fresh air is needed promptly. Open windows or go outdoors. Any gas appliance suspected of operating improperly should be shut off and immediately checked by a qualified agency.

509.3.2 Appliance Venting Requirements. Appliances required to be vented shall be connected to a venting system designed and installed in accordance with the provisions of Section 509.4 through Section 509.15. [NFPA 54:12.4.2]

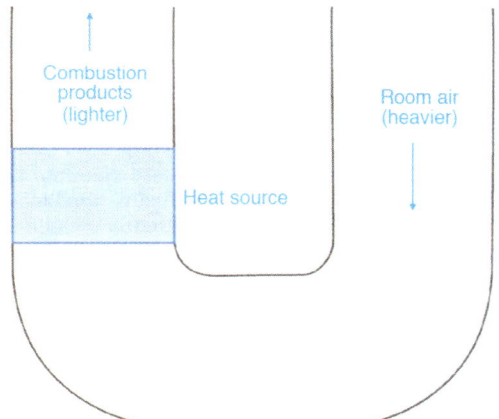

FIGURE 509.3.1A
PRINCIPLES OF VENT OPERATION

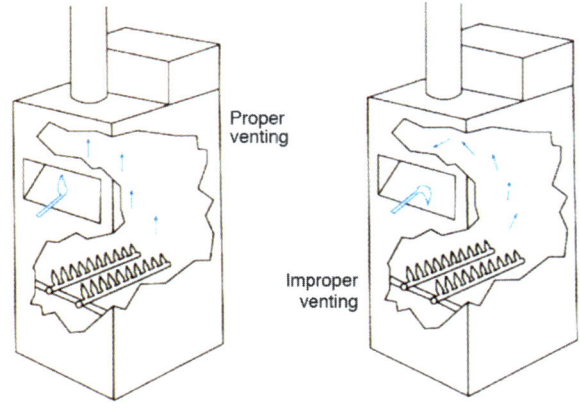

FIGURE 509.3.1B
MATCH TEST TO DETERMINE PROPER AND IMPROPER VENTING

WATER HEATERS

509.3.3 Mechanical Draft Systems. Mechanical draft systems shall be listed and installed in accordance with both the appliance and the mechanical draft system manufacturer's installation instructions. [NFPA 54:12.4.3.1]

509.3.3.1 Venting. Appliances requiring venting shall be permitted to be vented by means of mechanical draft systems of either forced or induced draft design. [NFPA 54:12.4.3.2]

509.3.3.2 Leakage. Forced draft systems and portions of induced draft systems under positive pressure during operation shall be designed and installed so as to prevent leakage of flue or vent gases into a building. [NFPA 54:12.4.3.3]

The key term here is "positive pressure." Vents serving fan-assisted appliances operate under negative pressure when vented into a chimney, even though the appliance uses a mechanical fan. Fan-assisted appliances can be common-vented with other fan-assisted appliances, draft hood-equipped appliances and all other Category I vented appliances.

509.3.3.3 Vent Connectors. Vent connectors serving appliances vented by natural draft shall not be connected into any portion of mechanical draft systems operating under positive pressure. [NFPA 54:12.4.3.4]

509.3.3.4 Operation. Where a mechanical draft system is employed, provision shall be made to prevent the flow of gas to the main burners where the draft system is not performing so as to satisfy the operating requirements of the appliance for safe performance. [NFPA 54:12.4.3.5]

509.3.3.5 Exit Terminals. The exit terminals of mechanical draft systems shall be not less than 7 feet (2134 mm) above finished ground level where located adjacent to public walkways and shall be located as specified in Section 509.8 and Section 509.8.1 of this code. [NFPA 54:12.4.3.6]

This section provides the installer with guidance on the sizing of mechanical draft systems. This information is needed because of the increased use of power vent kits to vent appliances. The manufacturer of a listed power vent kit provides sizing information in the instructions provided with the kit.

A minimum elevation of seven feet above grade is required so that combustion gases do not create a nuisance for pedestrians. Potential nuisances include the high temperature of the vent and the possibility of condensation of water in the vent gases forming a pool of water, ice or snow on the pavement. The definition of a "public walkway" is not specified in the code. The dictionary definition and common sense should be used. Section 509.8 specifies the location of vent terminals in relation to air inlets, doors and windows.

509.3.4 Ventilating Hoods and Exhaust Systems. Ventilating hoods and exhaust systems shall be permitted to be used to vent appliances installed in commercial applications. [NFPA 54:12.4.4.1]

509.3.4.1 Automatically Operated Appliances. Where automatically operated appliances, other than commercial cooking appliances, are vented through a ventilating hood or exhaust system equipped with a damper or with a power means of exhaust, provisions shall be made to allow the flow of gas to the main burners where the damper is open to a position to properly vent the appliance and where the power means of exhaust is in operation. [NFPA 54:12.4.4.2]

This subsection recognizes a safe, alternative method for venting of gas utilization equipment in commercial applications. For example, in restaurant kitchens a "booster" water heater is used to provide very hot water for sanitation. In this application, the booster water heater is operated only while the restaurant kitchen is in operation, and the range hood is used to vent both the range and the water heater.

Range hoods are covered under the UMC, which requires an interlock that operates when the airflow falls below a predetermined value. This circuit can be interlocked with the main gas valve of the water heater (or other gas appliance vented via the range hood) so that gas will not flow to the main burners unless a minimum airflow is achieved in the vent system.

509.3.5 Circulating Air Ducts and Furnace Plenums. Venting systems shall not extend into or pass through a fabricated air duct or furnace plenum. [NFPA 54:12.4.5.1]

509.3.6 Above-ceiling or Nonducted Air Handling System. Where a venting system passes through an above-ceiling air space or other nonducted portion of an air-handling system, it shall conform to one of the following requirements:

(1) The venting system shall be a listed special gas vent, other system serving a Category III or Category IV appliance, or other positive pressure vent, with joints sealed in accordance with the appliance or vent manufacturer's instructions.

(2) The vent system shall be installed such that no fittings or joints between sections are installed in the above-ceiling space.

(3) The venting system shall be installed in a conduit or enclosure with joints between the interior of the enclosure and the ceiling space sealed. [NFPA 54:12.4.5.2]

509.4 Type of Venting System to be Used. The type of venting system to be used shall be in accordance with Table 509.4. [NFPA 54:12.5.1]

Table 509.4 refers to appliance Category I through IV. The categories are based on vent temperature and pressure. A specific temperature is not provided because it is not the same for all appliances. The ANSI Z21 standards can be referenced for the manufacture and testing of appliances for this information.

To use Table 509.4, locate in the left-hand column the type of gas utilization equipment to be vented and read across the row to find the type(s) of venting system(s) that is permitted. For example, listed Category I equipment can be vented using a Type B gas vent, chimney, single-wall metal pipe, chimney lining system that is listed for gas venting or a special gas vent listed for the equipment.

509.4.1 Plastic Piping. Where plastic piping is used to vent an appliance, the appliance shall be listed for use with such venting materials and the appliance manufacturer's installation instruction shall identify the specific plastic piping material. [NFPA 54:12.5.2]

WATER HEATERS

TABLE 509.4
TYPE OF VENTING SYSTEM TO BE USED
[NFPA 54: TABLE 12.5.1]

APPLIANCES	TYPE OF VENTING SYSTEM	LOCATION OF REQUIREMENTS
Listed Category I appliances	Type B gas vent	Section 509.6
Listed appliances equipped with draft hood	Chimney	Section 509.5
Appliances listed for use with Type B gas vent	Single-wall metal pipe	Section 509.7
	Listed chimney lining system for gas venting	Section 509.5.3
	Special gas vent listed for appliances	Section 509.4.3
Listed vented wall furnaces	Type B-W gas vent	Section 509.6(2), Section 509.6.1.2
Category II appliances Category III appliances Category IV appliances	As specified or furnished by manufacturers of listed appliances	Section 509.4.1 and Section 509.4.3
Appliances that can be converted to use solid fuel Unlisted combination gas- and oil-burning appliances Combination gas- and solid-fuel-burning appliances Appliance listed for use with chimneys only Unlisted appliances	Chimney	Section 509.5
Listed combination gas- and oil-burning appliances	Type L vent	Section 509.6
	Chimney	Section 509.5
Decorative appliances in vented fireplace	Chimney	UMC Section 911.2
Gas-fired toilets	Single-wall metal pipe	Section 509.7
Direct-vent appliances	—	Section 509.2.6
Appliances with integral vents	—	Section 509.2.7

509.4.2 Plastic Vent Joints.
Plastic pipe and fittings used to vent appliances shall be installed in accordance with the appliance manufacturer's installation instructions. Where primer is required, it shall be of a contrasting color. [NFPA 54:12.5.3]

509.4.3 Special Gas Vents.
Special gas vents shall be listed and installed in accordance with the special gas vent manufacturer's installation instructions. [NFPA 54:12.5.4]

All special gas vents are listed vent materials in accordance with UL 1738, Standard for Venting Systems for Gas-Burning Appliances, Categories II, III and IV. Installation instructions for special gas vents include limitations on operating temperature, categories of appliance to be used with each vent, clearance to combustible materials, types of fittings and joint sealant to be used and vent termination requirements.

Special attention should be given to the following areas:

- Proper support for the special vent to prevent sagging and to allow for expansion, contraction and condensate drainage.
- Proper cutting and cleaning of joints and fittings, and the use of recommended joint sealants (substitutes are not usually permitted).
- Construction of a condensate trap (see appliance manufacturer's instruction for special requirements).
- Wall penetrations (the pipe should not be secured at a thimble, as the pipe must be allowed to move to accommodate expansion and contraction).
- The vent pipe or the fittings of the inside of a wall thimble must not be insulated when polymeric (nonmetallic) vent materials are used.

509.5 Masonry, Metal, and Factory-Built Chimneys.
Chimneys shall be installed in accordance with Section 509.5.1 through Section 509.5.3.

509.5.1 Factory-Built Chimneys.
Factory-built chimneys shall be installed in accordance with the manufacturer's installation instructions. Factory-built chimneys used to vent appliances that operate at positive vent pressure shall be listed for such application. [NFPA 54:12.6.1.1]

Factory-built chimneys shown in **Figures 509.5.1a through 509.5.1d** consist entirely of factory-built parts, such as chimney sections, supports, thimbles, flashings, caps and other parts required to complete a particular installation. The parts are designed to be assembled with other parts of the same model without requiring field alteration or construction. All of the parts for a particular model as described in the chimney manufacturer's instructions are to be used. **Figure 509.5.1a** shows the hardware for a positive-pressure factory-built chimney.

The following three types of factory-built chimneys are available:

- Residential-type and building heating appliance chimneys that are intended for venting flue gases at a temperature not exceeding 1,000°F (538°C) under continuous operating conditions
- 1,400°F (760°C) chimneys for venting flue gases at a temperature not exceeding 1,400°F (760°C) under continuous operating conditions
- Medium-heat appliance chimneys for venting flue gases at a temperature not exceeding 1,800°F (982°C).

The installation of chimneys serving appliances with vent temperatures over 1,000°F (538°C) is not covered in this code.

WATER HEATERS

See NFPA 211, *Standard for Chimneys, Fireplaces, Vents, and Solid Fuel-Burning Appliances*, which also describes the various types of heating appliances mentioned here.

Factory-built chimneys must be installed in accordance with the terms of their listing and the manufacturers' instructions, which are furnished with each chimney assembly and have been verified by the listing agency. The product standard for chimneys, UL 103, S*tandard for Factory-Built Chimneys for Residential Type and Building Heating Appliances*, includes separate tests to verify the proper operation of factory-built chimneys that are designed for positive combustion-product pressure.

FIGURE 509.5.1A
FACTORY-BUILT METAL POSITIVE PRESSURE CHIMNEY

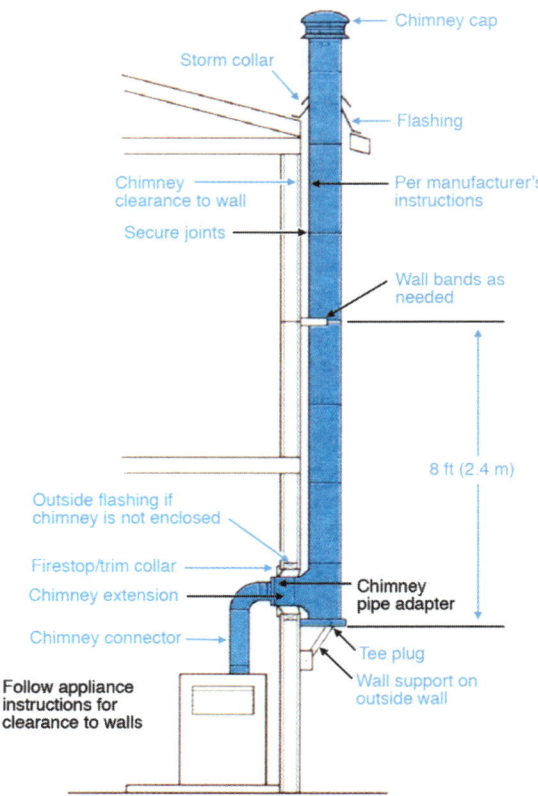

FIGURE 509.5.1C
ALTERNATIVE INSTALLATION OF RESIDENTIAL FACTORY-BUILT CHIMNEY OUTSIDE OF BUILDING

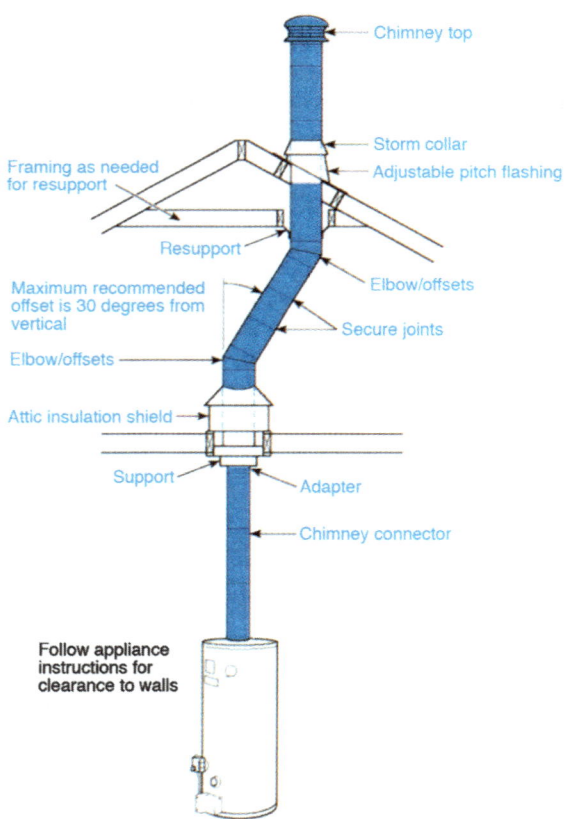

FIGURE 509.5.1B
TYPICAL INSTALLATION OF RESIDENTIAL FACTORY-BUILT CHIMNEY

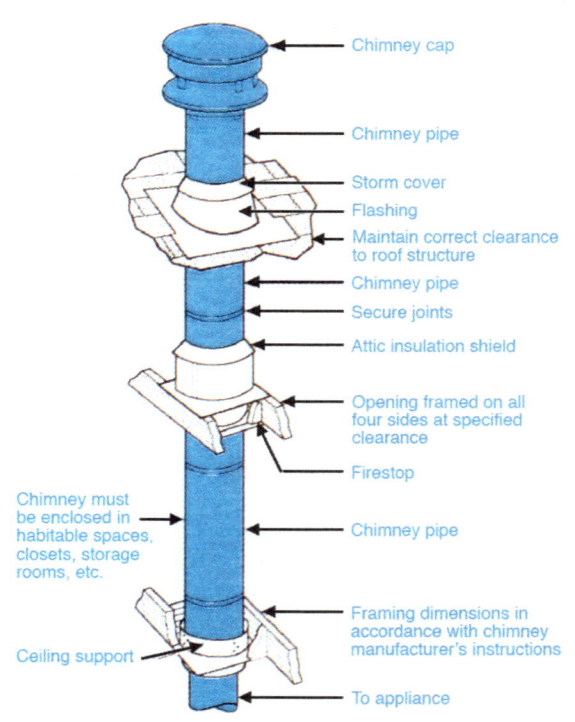

FIGURE 509.5.1D
ALTERNATIVE INSTALLATION OF RESIDENTIAL FACTORY-BUILT CHIMNEY INSIDE A BUILDING - ROOF PENETRATION

Category II, III and IV appliances are not suitable for venting by the conventional method of natural draft venting. These appliances must not be connected to chimneys. These appliances should be vented in accordance with the appliance manufacturer's instructions, which cover the type, size and length of material to be used; how the venting system is to be installed; and the minimum clearance to combustible material. These instructions have been verified by the listing agency.

509.5.1.1 Decorative Shrouds. Decorative shrouds addressed in Section 509.5.4.3 shall comply with UL 103 for factory-built residential chimneys.

509.5.1.2 Listing Requirements. Factory-built chimneys shall comply with the requirements of UL 103 or UL 959. Factory-built chimneys for use with wood-burning appliances shall comply with the Type HT requirements of UL 103. [NFPA 211:6.1.3.1]

509.5.2 Metal Chimneys. Metal chimneys shall be in accordance with NFPA 211. [NFPA 54:12.6.1.2]

Metal chimneys are field-constructed chimneys made of sheet steel and intended for industrial applications. These chimneys differ greatly from factory-built chimneys in that they are made of metal. The term "metal chimneys" is sometimes misapplied to describe factory-built chimneys. The construction and installation of metal chimneys is covered in NFPA 211, *Standard for Chimneys, Fireplaces, Vents and Solid Fuel-Burning Appliances*, which prohibits the installation of metal chimneys in residential structures.

509.5.3 Masonry Chimneys. Masonry chimneys shall be in accordance with NFPA 211 and lined with approved clay flue lining, a listed chimney lining system, or other approved material that resists corrosion, erosion, softening, or cracking from vent gases at temperatures not exceeding 1800°F (982°C).

Exception: Masonry chimney flues lined with a chimney lining system specifically listed for use with listed appliances with draft hoods, Category I appliances, and other appliances listed for use with Type B vents shall be permitted. The liner shall be installed in accordance with the liner manufacturer's installation instructions. A permanent identifying label shall be attached at the point where the connection is to be made to the liner. The label shall read: "This chimney liner is for appliances that burn gas only. Do not connect to solid- or liquid-fuel-burning appliances or incinerators." [NFPA 54:12.6.1.3]

Masonry chimneys can be used for all appliances permitted in Section 509.4 that can be vented using chimneys. The exception to Section 509.5.3 permits installation of liners in masonry chimneys that may not be covered in NFPA 211. The labeling requirement notifies future installers of the vent liner's limitations. Notification is important if a change in appliance fuels is considered (see **Figure 509.5.3**).

509.5.4 Termination. A chimney for a residential-type or low-heat appliance shall extend not less than 3 feet (914 mm) above the highest point where it passes through the roof of a building and not less than 2 feet (610 mm) higher than a portion of a building within a horizontal distance of 10 feet (3048 mm). (See **Figure 509.5.4**) [NFPA 54:12.6.2.1]

509.5.4.1 Medium-Heat Gas Appliances. A chimney for medium-heat appliances shall extend at least 10 feet (3048 mm) higher than any portion of any building within 25 feet (7620 mm). [NFPA 54:12.6.2.2]

509.5.4.2 Chimney Height. A chimney shall extend not less than 5 feet (1524 mm) above the highest connected appliance draft hood outlet or flue collar. [NFPA 54:12.6.2.3]

509.5.4.3 Decorative Shrouds. Decorative shrouds shall not be installed at the termination of factory-built chimneys except where such shrouds are listed and labeled for use with the specific factory-built chimney system and are installed in accordance with the manufacturer's installation instructions. [NFPA 54:12.6.2.4]

This requirement prohibits the installation of unlisted shrouds around factory-built chimneys that could affect the flow of flue gases and trap hot flue gases in close proximity to the combustible roof. Shrouds are used to hide chimneys. Shrouds can interfere with vent operation if they extend above the top of the vent.

509.5.5 Size of Chimneys. The effective area of a chimney venting system serving listed appliances with draft hoods, Category I appliances, and other appliances listed for use with Type B vents shall be in accordance with one of the following methods:

(1) Those listed in Section 510.0.

(2) For sizing an individual chimney venting system for a single appliance with a draft hood, the effective areas of the vent connector and chimney flue shall be not less than the area of the appliance flue collar or draft hood outlet or greater than seven times the draft hood outlet area.

(3) For sizing a chimney venting system connected to two appliances with draft hoods, the effective area of the chimney flue shall be not less than the area of the larger draft hood outlet plus 50 percent of the area of the smaller draft hood outlet or greater than seven times the smaller draft hood outlet area.

(4) Chimney venting systems using mechanical draft shall be sized in accordance with approved engineering methods.

(5) Other approved engineering methods. [NFPA 54:12.6.3.1]

The sizing of chimneys is provided in Section 510.0 in the form of tables and additional requirements. These venting requirements were developed by a research project funded by the Gas Research Institute (GRI) and conducted by Battelle.

The requirements in Section 510.1.10 recognize that there is a significant difference between the operation of interior and exterior chimneys. An exterior chimney has one or more sides exposed to the outdoors below the roofline, whereas an interior chimney is not exposed to the outdoors below the building's roofline. (Chimneys that pass through unheated attics or garages are not considered to be outdoors.) An interior chimney is largely isolated from weather changes. An exterior chimney is affected by the outside ambient temperature; therefore, the requirements for exterior chimneys

WATER HEATERS

are keyed to the lowest temperature expected in different parts of the country.

The requirements for sizing chimneys were developed using a flue loss of 17 percent for the heating appliance. Most real furnaces and boilers operate at a higher flue loss; therefore, the code requirements are conservative. Several manufacturers offer customized vent kits, which provide venting requirements for masonry chimneys that are keyed to the exact performance of their appliances. In this case, their recommendations should be used.

The sizing methods in Sections 509.5.5(2) and 509.5.5(3) permit a simple, alternative, chimney-sizing method that limits the vent to a maximum size of seven times the smallest draft hood outlet area and a minimum size based on the draft hood outlet areas. These methods limit the maximum size of the vent connector and gas vent to minimize condensation in the gas vent caused by insufficient flow of hot vent gases to heat an excessively large gas vent. Excessive condensation can cause premature failure of a chimney.

Note that Section 509.5.5(3) limits the use of this alternative to chimney venting systems serving only two appliances. Sample calculations using the venting tables in Section 510.0 with the alternative method resulted in vent sizes that were too small when four or more appliances are involved, and for some tall vents. The alternative always provides acceptable vent sizes for two appliances. Venting systems serving more than two appliances must use the vent sizing tables in Section 510.0 or other engineering methods.

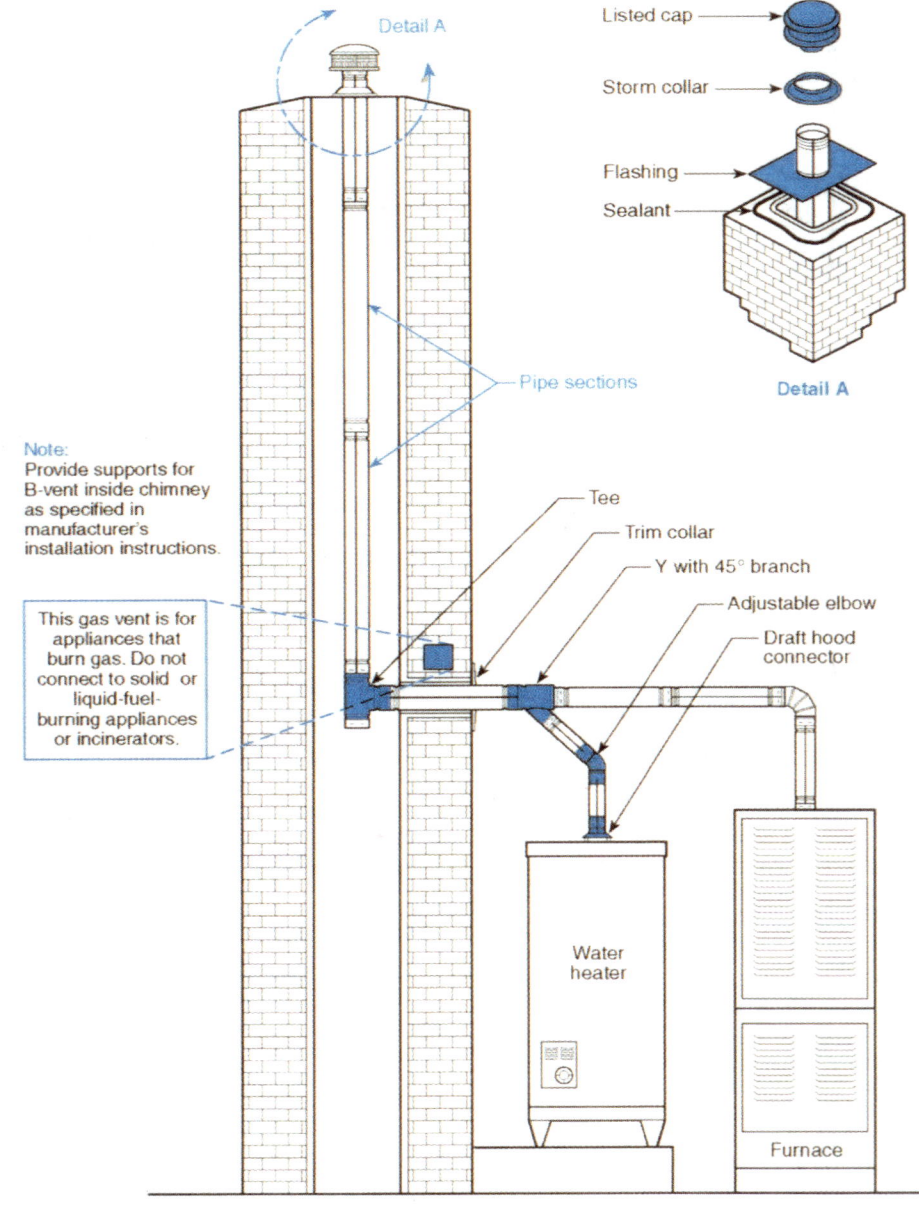

FIGURE 509.5.3
TYPICAL INSTALLATION USING TYPE B VENT TO LINE MASONRY CHIMNEY

WATER HEATERS

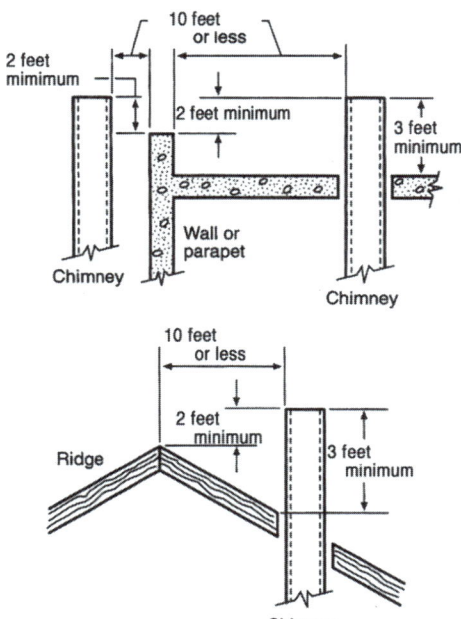

(a) Termination 10 feet or Less from Ridge, Wall, or Parapet

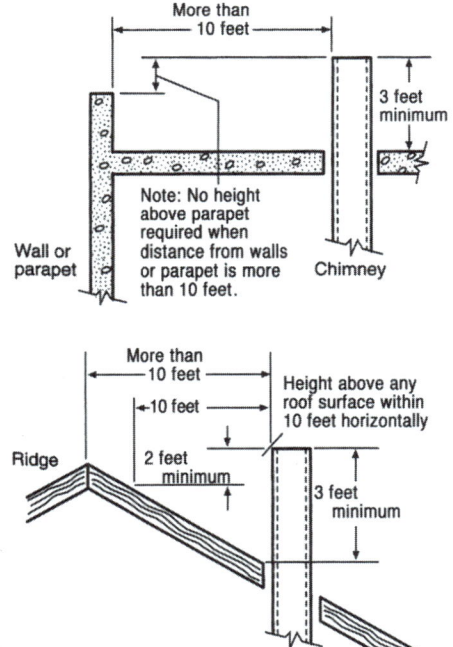

(b) Termination More Than 10 feet from Ridge, Wall, or Parapet

For SI units: 1 foot = 304.8 mm

FIGURE 509.5.4
TYPICAL TERMINATION LOCATIONS FOR CHIMNEYS AND SINGLE-WALL METAL PIPES SERVING RESIDENTIAL-TYPE AND LOW-HEAT APPLIANCE
[NFPA 54: FIGURE A.12.6.2.1]

509.5.6 Inspection of Chimneys. Before replacing an existing appliance or connecting a vent connector to a chimney, the chimney passageway shall be examined to ascertain that it is clear and free of obstructions and shall be cleaned where previously used for venting solid- or liquid-fuel-burning appliances or fireplaces. [NFPA 54:12.6.4.1]

Inspection prior to the connection of a new appliance to a chimney that previously served appliances using fuels other than gas fuels is needed because a buildup of soot or creosote that may have been caused by oil- or solid-fuel-burning appliances can be softened by the products of gas combustion. These deposits, if softened, can break away and block the chimney.

Oil- and solid-fuel-burning appliances can deposit sulfur residue in the chimney. When this sulfur is exposed to water vapor from gas appliances, it can form sulfuric acid that leads to rapid chimney deterioration. The requirement to clean the chimney will help minimize, but not eliminate this problem.

Some common chimney troubles and ways to detect and remedy them are shown in **Figure 509.5.6**.

509.5.6.1 Standard. Chimneys shall be lined in accordance with NFPA 211.

Exception: Existing chimneys shall be permitted to have their use continued when an appliance is replaced by an appliance of similar type, input rating, and efficiency, where the chimney complies with Section 509.5.6 through Section 509.5.6.3 and the sizing of the chimney is in accordance with Section 509.5.5. [NFPA 54:12.6.4.2]

509.5.6.2 Cleanouts. Cleanouts shall be examined to determine that they will remain tightly closed where not in use. [NFPA 54:12.6.4.3]

509.5.6.3 Existing Chimney. Where inspection reveals that an existing chimney is not safe for the intended application, it shall be repaired, rebuilt, lined, relined, or replaced with a vent or chimney in accordance with NFPA 211, and shall be approved for the appliances to be attached. [NFPA 54:12.6.4.4]

509.5.7 Chimney Serving Appliances Burning Other Fuels. An appliance shall not be connected to a chimney flue serving a separate appliance designed to burn solid fuel. [NFPA 54:12.6.5.1]

509.5.7.1 Gas and Liquid Fuel-Burning Appliances. Where one chimney serves gas appliances and liquid fuel-burning appliances, the appliances shall be connected through separate openings or shall be connected through a single opening where joined by a fitting located as close as practical to the chimney. Where two or more openings are provided into one chimney flue, they shall be at different levels. Where the gas appliance is automatically controlled, it shall be equipped with a safety shutoff device. [NFPA 54:12.6.5.2]

509.5.7.2 Gas and Solid Fuel-Burning Appliances. A listed combination gas- and solid-fuel-burning appliance connected to a single chimney flue shall be equipped with a manual reset device to shut off gas to the main burner in the event of sustained backdraft or flue gas spillage. The chimney flue shall be sized to properly vent the appliance. [NFPA 54:12.6.5.3]

WATER HEATERS

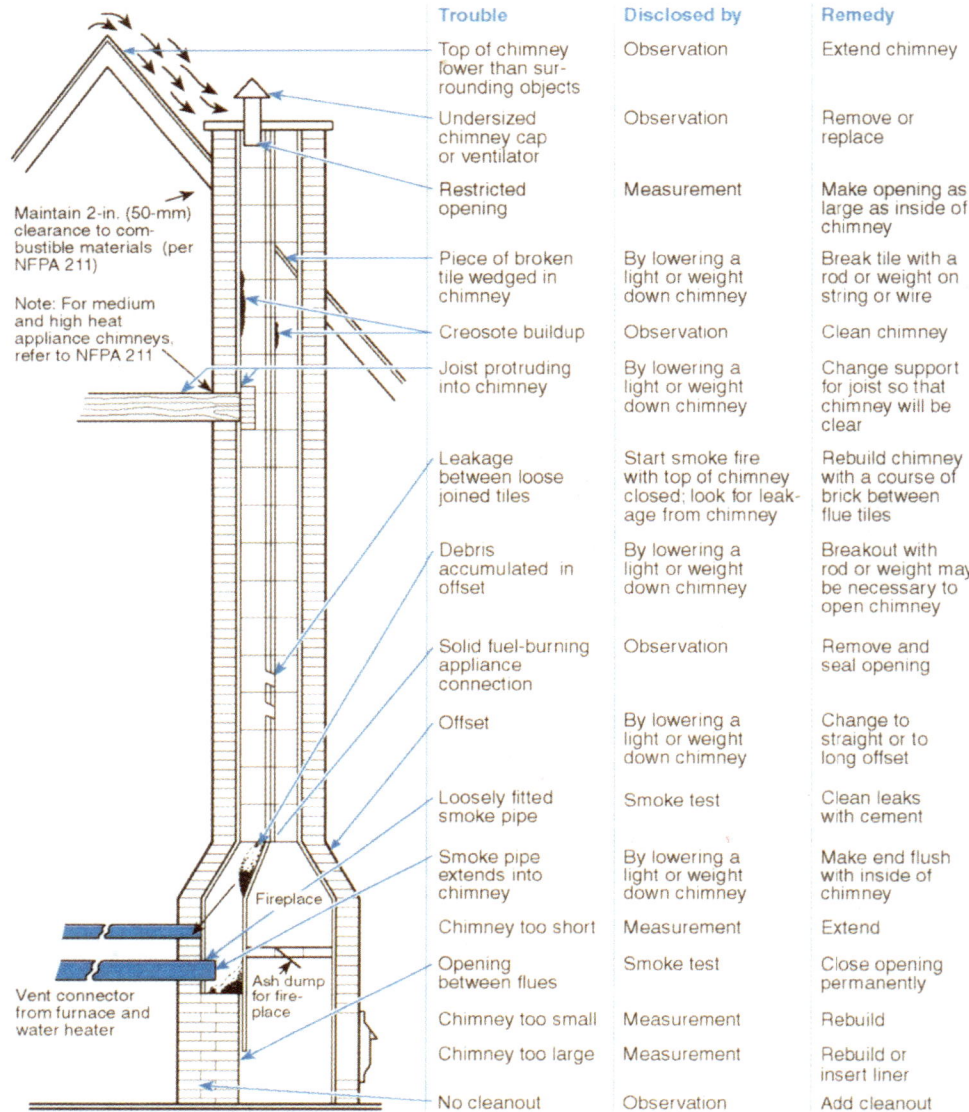

**FIGURE 509.5.6
COMMON CHIMNEY PROBLEMS AND REMEDIES**

509.5.7.3 Combination Gas- and Oil-Burning Appliances. A single chimney flue serving a listed combination gas- and oil-burning appliance shall be sized to properly vent the appliance. [NFPA 54:12.6.5.4]

509.5.8 Support of Chimneys. All portions of chimneys shall be supported for the design and weight of the materials employed. Listed factory-built chimneys shall be supported and spaced in accordance with the manufacturer's installation instructions. [NFPA 54:12.6.6]

509.5.9 Cleanouts. Where a chimney that formerly carried flue products from liquid- or solid-fuel-burning appliances is used with an appliance using fuel gas, an accessible cleanout shall be provided. The cleanout shall have a tight-fitting cover and be installed, so its upper edge is not less than 6 inches (152 mm) below the lower edge of the lowest chimney inlet opening. [NFPA 54:12.6.7]

When a gas-heating unit or conversion burner is connected to a chimney that previously served for venting another fuel, after a period of operation on natural gas, carbon deposits or mortar materials can become dislodged from the chimney wall and drop to the base of the chimney. The accumulation of this debris, if great enough, can result in blockage of the venting system, sometimes with fatal results.

A six-inch drop leg and inspection/maintenance cleanout is required. The cleanout provides a sump to collect a considerable amount of debris before a potential accident occurs. The cleanout also enables homeowners and service personnel to clean debris easily and to inspect conditions.

509.5.10 Space Surrounding Lining or Vent. The remaining space surrounding a chimney liner, gas vent, special gas vent, or plastic piping installed within a masonry chimney shall not be used to vent another appliance.

Exception: The insertion of another liner or vent within the chimney as provided in this code and the liner or vent manufacturer's instructions. [NFPA 54:12.6.8.1]

🔧 The use of a masonry chimney as a chase for vents is a practical solution to an existing unlined, oversized or failed chimney. When installing vents in a chimney space, the manufacturer's installation requirements for all of the vent material must be followed. If required by the vent manufacturer, penetrating the masonry chimney to install supports might be necessary.

When using an abandoned chimney as a chase for multiple vents, care should be taken to ensure that the vent materials are compatible and that vent failure does not result inadvertently. Category I or III gas appliance vent gas temperatures can exceed 300°F (150°C). Contact between the high-temperature Category I or III vent material and the PVC vent pipe can cause the PVC pipe to degrade or soften to the extent of failure. With the chimney opening sealed around the vent pipes, the products of combustion, which can escape from the failed PVC pipe, will be trapped in the chimney. These gases could eventually leak back into the living space or corrode the other nearby vents.

509.5.10.1 Combustion Air. The remaining space surrounding a chimney liner, gas vent, special gas vent, or plastic piping installed within masonry, metal, or factory-built chimney flue shall not be used to supply combustion air.

Exception: Direct-vent appliances designed for installation in a solid-fuel-burning fireplace where installed in accordance with the manufacturer's installation instructions. [NFPA 54:12.6.8.2]

509.6 Gas Vents. The installation of gas vents shall meet the following requirements:

(1) Gas vents shall be installed in accordance with the manufacturer's installation instructions.

(2) A Type B-W gas vent shall have a listed capacity not less than that of the listed vented wall furnace to which it is connected.

(3) Gas vents installed within masonry chimneys shall be installed in accordance with the manufacturer's installation instructions. Gas vents installed within masonry chimneys shall be identified with a permanent label installed at the point where the vent enters the chimney. The label shall contain the following language: "This gas vent is for appliances that burn gas. Do not connect to solid or liquid fuel–burning appliances or incinerators."

(4) Screws, rivets, and other fasteners shall not penetrate the inner wall of double-wall gas vents, except at the transition from the appliance draft hood outlet, flue collar, or single-wall metal connector to a double-wall vent. [NFPA 54:12.7.1]

🔧 This section applies to gas vents of all types. Not included are chimneys, which are covered in Section 509.5, and single-wall metal pipe, which is covered in Section 509.7. A gas vent is defined as a listed factory-made vent pipe and fittings (see Section 224.0). Therefore, this section requires listed gas vents to be installed in accordance with the manufacturer's installation instructions.

Type B and B-W gas vents that currently are listed are of double-wall construction: basically, a pipe within a pipe, with airspace between the two walls. This construction reduces the heat loss from the vent gases in much the same way that an insulated coffee mug works (see **Figure 509.6a**).

Usually, the inner wall is made of aluminum to resist corrosion and the outer wall is made of galvanized steel for strength. The parts of the vent — pipe sections, supports, spacers, caps and flashings — are furnished for field erection to form a continuous passageway from the gas appliance to the terminus of the roof, including the cap or roof assembly. These parts are manufactured by several companies that furnish the vent pipe and other related parts needed to erect a complete vent. The piping and parts produced by the different manufacturers are not interchangeable without the use of special, listed adapters.

Type B gas vents can be round or oval (see **Figure 509.6b**). Type B-W gas vents are always oval because they are designed to be installed within a stud space. Type L vents are similar in construction to Type B gas vents, except that the inner pipe of a Type L vent is stainless steel. Type L vents are intended for venting flue gases at higher temperatures than are normally produced by gas appliances, such as for oil furnaces listed for use with Type L vents. Typical Type B-W and B gas vent installations are diagrammed in F**igures 509.6c** and **509.6d**, respectively.

509.6.1 Termination Requirements. A gas vents shall terminate in accordance with one of the following:

(1) Gas vents that are 12 inches (300 mm) or less in size and located not less than 8 feet (2438 mm) from a vertical wall or similar obstruction shall terminate above the roof in accordance with **Figure 509.6.1** and Table 509.6.1.

(2) Gas vents that are over 12 inches (300 mm) in size or are located less than 8 feet (2438 mm) from a vertical wall or similar obstruction shall terminate not less than 2 feet (610 mm) above the highest point where they pass through the roof and not less than 2 feet (610 mm) above any portion of a building within 10 feet (3048 mm) horizontally.

(3) Industrial appliances as provided in Section 509.2.5.

🔧 Section 509.2.5 is included here to clarify that vent termination may not apply to spaces in which industrial gas equipment is installed if they are well-ventilated.

(4) Direct-vent systems as provided in Section 509.2.6.

(5) Appliances with integral vents as provided in Section 509.2.7.

(6) Mechanical draft systems as provided in Section 509.3.3 through Section 509.3.3.5.

🔧 Sections 509.6.1(2) through 509.6.1(6) are exceptions to the requirement of terminating a vent above the roof. Such appliances are not required to connect to a venting systems according to the sections referenced.

(7) Ventilating hoods and exhaust systems as provided in Section 509.3.4 and Section 509.3.4.1. [NFPA 54:12.7.2(1)]

WATER HEATERS

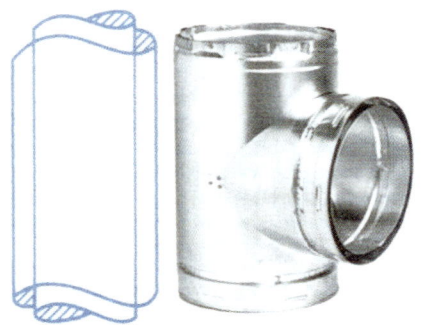

FIGURE 509.6A
CONSTRUCTION OF DOUBLE WALL TYPE B GAS VENT

FIGURE 509.6B
TYPE B AND B-W DOUBLE WALL GAS VENT

509.6.1.1 Type B and L Vents. A Type B or a Type L gas vent shall terminate at least 5 feet (1524 mm) in vertical height above the highest connected appliance draft hood or flue collar. [NFPA 54:12.7.2(2)]

509.6.1.2 Type B-W Vents. A Type B-W gas vent shall terminate at least 12 feet (3658 mm) in vertical height above the bottom of the wall furnace. [NFPA 54:12.7.2(3)]

509.6.1.3 Exterior Wall Termination. A gas vent extending through an exterior wall shall not terminate adjacent to the wall or below eaves or parapets, except as provided in Section 509.2.6 and Section 509.3.3 through Section 509.3.3.5. [NFPA 54:12.7.2(4)]

509.6.1.4 Decorative Shrouds. Decorative shrouds shall not be installed at the termination of gas vents except where such shrouds are listed for use with the specific gas venting system and are installed in accordance with the manufacturer's installation instructions. [NFPA 54:12.7.2(5)]

509.6.1.5 Termination Cap. All gas vents shall extend through the roof flashings roof jack, or roof thimble and terminate with a listed cap or listed roof assembly. [NFPA 54:12.7.2(6)]

Listed vent caps for Type B and B-W gas vents are designed and tested to serve the following four purposes:

(1) To reduce the chance of downdraft caused by wind effect;
(2) To provide proper venting under wind conditions;
(3) To permit vent termination at minimum height above the roof surface; and
(4) To prevent rain and debris from entering the vent.

Notice that a listed vent cap allows the vent termination to be closer to the nearby roofline, because the terminations are tested to operate properly using the clearances shown in **Figure 509.6.1**.

509.6.1.6 Forced Air Inlet. A gas vent shall terminate at least 3 feet (914 mm) above a forced air inlet located within 10 feet (3048 mm). [NFPA 54:12.7.2(7)]

509.6.1.7 Insulation Shield. Where a vent passes through an insulated assembly, an approved metal shield shall be installed between the vent and insulation. The shield shall extend not less than 2 inches (51 mm) above the insulation and be secured to the structure in accordance with the manufacturer's installation instructions.

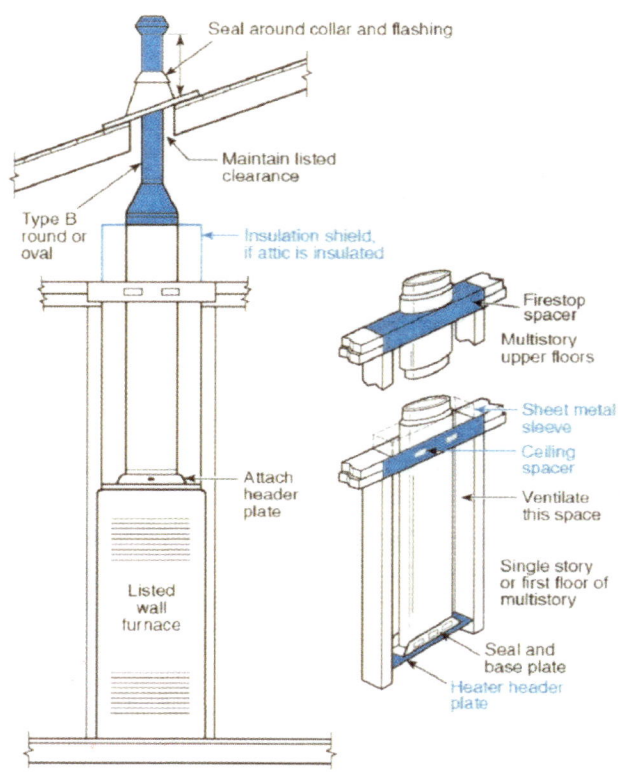

FIGURE 509.6C
TYPICAL TYPE B-W GAS VENT INSTALLATION

509.6.2 Size of Gas Vents. Venting systems shall be sized and constructed in accordance with Section 510.0 or other approved engineering methods and the gas vent and the appliance manufacturer's instructions. [NFPA 54:12.7.3]

Category I equipment is permitted to be installed in accordance with the Section 510.0 tables, which incorporate Category I vent-sizing tables. These tables provide three columns for each set of vent horizontal and vertical runs: one column for draft hood-equipped appliances and two columns that specify both minimum and maximum capacities for fan-assisted appliances. Additional information on use of the tables can be found in Section 510.0. In lieu of using the tables in Section 510.0 for vent sizing, approved engineering methods are permitted.

WATER HEATERS

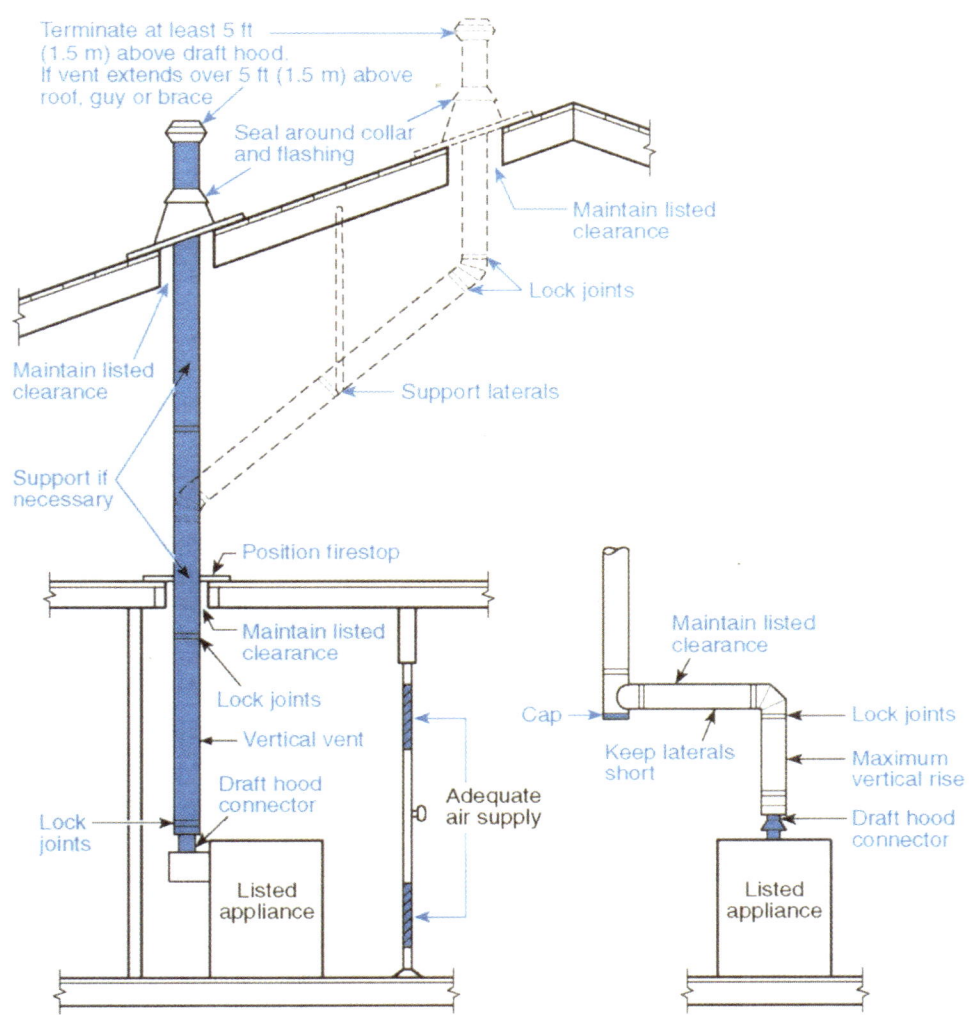

FIGURE 509.6D
TYPE B GAS VENT SINGLE APPLIANCE INSTALLATION

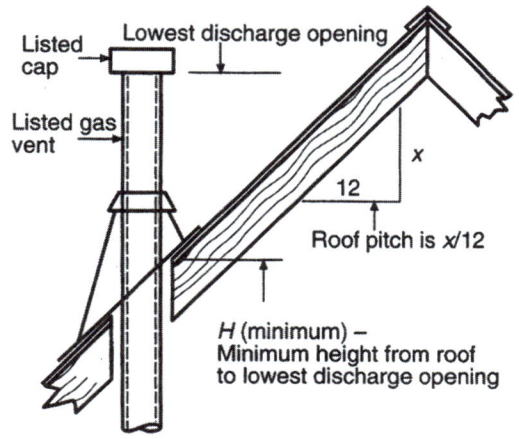

FIGURE 509.6.1
TERMINATION LOCATIONS FOR GAS VENTS WITH LISTED CAPS 12 INCHES OR LESS IN SIZE AT LEAST 8 FEET FROM A VERTICAL WALL
[NFPA 54: FIGURE 12.7.2]

TABLE 509.6.1
ROOF SLOPE HEIGHTS
[NFPA 54: TABLE 12.7.2]

ROOF PITCH	H (minimum) (feet)
Flat to 6/12	1.0
Over 6/12 to 7/12	1.25
Over 7/12 to 8/12	1.5
Over 8/12 to 9/12	2.0
Over 9/12 to 10/12	2.5
Over 10/12 to 11/12	3.25
Over 11/12 to 12/12	4.0
Over 12/12 to 14/12	5.0
Over 14/12 to 16/12	6.0
Over 16/12 to 18/12	7.0
Over 18/12 to 20/12	7.5
Over 20/12 to 21/12	8.0

For SI Units: 1 inch = 25.4 mm, 1 foot = 304.8 mm

WATER HEATERS

509.6.2.1 Category I Appliances. The sizing of natural draft venting systems serving one or more listed appliances equipped with a draft hood or appliances listed for use with Type B gas vent, installed in a single story of a building, shall be in accordance with one of the following:

(1) The provisions of Section 510.0.

(2) Vents serving fan-assisted combustion system appliances, or combinations of fan-assisted combustion systems and draft-hood-equipped appliances, shall be sized in accordance with Section 510.0 or other approved engineering methods.

(3) For sizing an individual gas vent for a single, draft-hood-equipped appliance, the effective area of the vent connector and the gas vent shall be not less than the area of the appliance draft hood outlet or exceeding seven times the draft hood outlet area.

(4) For sizing a gas vent connected to two appliances with draft hoods, the effective area of the vent shall be not less than the area of the larger draft hood outlet plus 50 percent of the area of the smaller draft hood outlet or exceeding seven times the smaller draft hood outlet area.

(5) Approved engineering practices. [NFPA 54:12.7.3.1]

The tables in Section 510.0 provide a safe method for sizing vents for all Category I and draft hood-equipped appliances. The alternative methods permitted in Section 509.6.2.1(3) and (4) offer simple techniques for installers to follow for commonly encountered vent arrangements. This simpler method is applicable only to one- and two-appliance venting situations that have a draft hood. These methods provide a maximum size of seven times the smallest draft hood outlet area and a minimum size based on the draft hood outlet area. This provision limits the maximum size of the vent connector and gas vent to minimize gas vent condensation caused by insufficient flow of hot vent gases to heat an excessively large gas vent. Excessive condensation can cause premature failure of a gas vent.

These alternative methods cannot be used for fan-assisted combustion appliances. The tables in Section 510.0 (or other engineering methods) must be used for all installations of fan-assisted combustion appliances. Vents for Category II, III and IV appliances must be in accordance with the equipment manufacturers' instructions. These appliances are listed by definition, and the manufacturers' instructions reflect the test results determined by the listing laboratory.

Venting systems serving more than two appliances must use the vent sizing tables in Section 510.0 or other engineering methods. Refer to the commentary following Section 509.5.5 for more information.

509.6.2.2 Vent Offsets. Type B and Type L vents sized in accordance with Section 509.6.2.1(3) or Section 509.6.2.1(4) shall extend in a generally vertical direction with offsets not exceeding 45 degrees (0.79 rad), except that a vent system having not more than one 60 degree (1.05 rad) offset shall be permitted. Any angle greater than 45 degrees (0.79 rad) from the vertical is considered horizontal. The total horizontal distance of a vent plus the horizontal vent connector serving draft hood-equipped appliances shall not be greater than 75 percent of the vertical height of the vent. [NFPA 54:12.7.3.2]

A Type B or L gas vent is limited to one 60-degree offset as shown in **Figure 509.6d**. The offset is also limited to 75 percent of the total vertical height of the vent. If the system is designed and sized in accordance with the tables in Section 510.0, a Type B gas vent can include two 90-degree turns and a horizontal run greater than 75 percent of the vertical portion of the vent. For systems with more than two 90-degree turns, Sections 510.2.4 and 510.2.5 permit the use of Tables 510.2(1) through 510.2(9) with a derating of the maximum capacity values on five percent for each turn less than or equal to 45 degrees, and 10 percent for each 90-degree turn over two 90-degree turns.

509.6.2.3 Category II, Category III, and Category IV Appliances. The sizing of gas vents for Category II, Category III, and Category IV appliances shall be in accordance with the appliance manufacturer's instructions. The sizing of plastic pipe specified by the appliance manufacturer as a venting material for Category II, Category III, and Category IV appliances shall be in accordance with the appliance manufacturers' instructions. [NFPA 54:12.7.3.3]

509.6.2.4 Sizing. Chimney venting systems using mechanical draft shall be sized in accordance with approved engineering methods. [NFPA 54:12.7.3.4]

509.6.3 Gas Vents Serving Appliances on more than One Floor. A common vent shall be permitted in multistory installations to vent Category I appliances located on more than one-floor level, provided the venting system is designed and installed in accordance with approved engineering methods.

For the purpose of this section, crawl spaces, basements, and attics shall be considered floor levels. [NFPA 54:12.7.4.1]

509.6.3.1 Occupiable Space. All appliances connected to the common vent shall be located in rooms separated from an occupiable space. Each of these rooms shall have provisions for an adequate supply of combustion, ventilation, and dilution air that is not supplied from an occupiable space. (See **Figure 509.6.3.1**) [NFPA 54:12.7.4.2]

Safety can be achieved in high-rise, multilevel appliance installations by separating the appliance from any occupiable space. The separation resolves the question of safety that arises when intercommunication of vents exists between various floors of a building. Separation ensures that no vent gases will enter the occupied space from the appliance room if the common vent becomes obstructed at any level or if the outlet is blocked. When such stoppage occurs, all vent gases from appliances operating below the obstruction will exit through the upper draft hood relief opening rather than through the vent outlet. Large quantities of vent gases will be dumped into the space containing those appliances immediately below the obstruction, while at the same time the appliances located at lower levels will appear to be operating normally.

One practical plan to separate or isolate appliances, if installation can be made adjacent to an outside wall, is shown by **Figure 509.6.3.1** and **Figure 509.6.3.1a**. Access to the appliance room is through a door that opens onto an outdoor balcony. A panel separates the appliance room from the inside of the building. If the appliance is a central furnace, the cold air return and outlet ducts are attached to the furnace.

No requirements are given in the code for installation of Category II, III or IV appliances using a common vent in multistory buildings. A Category II, III or IV appliance must be installed in accordance with the terms of its listing and the manufacturer's instructions. Category I appliance common vents are not usually suitable for venting Category II, III or IV appliances.

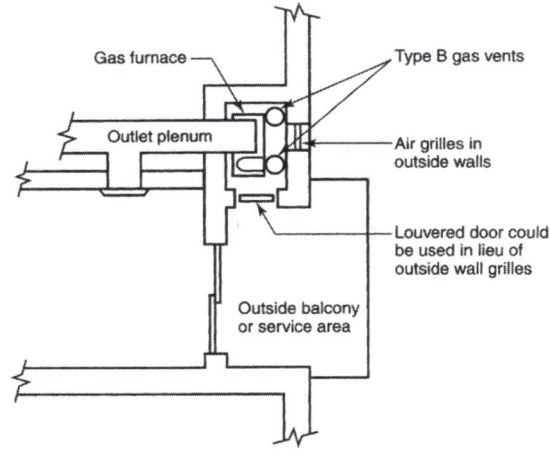

FIGURE 509.6.3.1
PLAN VIEW OF PRACTICAL SEPARATION METHOD FOR MULTISTORY GAS VENTING
[NFPA 54: FIGURE A.12.7.4.2]

509.6.3.2 Multistory Venting System. The size of the connectors and common segments of multistory venting systems for appliances listed for use with a Type B double-wall gas vent shall be in accordance with Table 510.2(1), provided all of the following apply:

(1) The available total height (H) for each segment of a multistory venting system is the vertical distance between the level of the highest draft hood outlet or flue collar on that floor and the centerline of the next highest interconnection tee.

(2) The size of the connector for a segment is determined from the appliance's gas input rate and available connector rise, and shall not be smaller than the draft hood outlet or flue collar size.

(3) The size of the common vertical vent segment, and of the interconnection tee at the base of that segment, is based on the total appliance's gas input rate entering that segment and its available total height. [NFPA 54:12.7.4.3]

509.6.4 Support of Gas Vents. Gas vents shall be supported and spaced in accordance with the manufacturer's installation instructions. [NFPA 54:12.7.5]

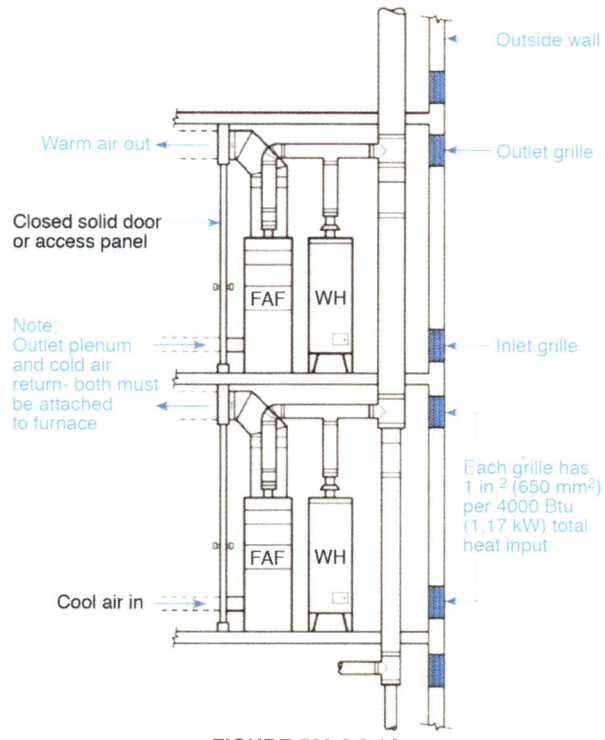

FIGURE 509.6.3.1A
OUTSIDE WALL AIR SUPPLY IN MULTISTORY BUILDING

509.6.5 Marking. In those localities where solid and liquid fuels are used extensively, gas vents shall be permanently identified by a label attached to the wall or ceiling at a point where the vent connector enters the gas vent. The label shall read: "This gas vent is for appliances that burn gas. Do not connect to solid- or liquid-fuel-burning appliances or incinerators." The Authority Having Jurisdiction shall determine whether its area constitutes such a locality. [NFPA 54:12.7.6]

509.7 Single-Wall Metal Pipe. Single-wall metal pipe shall be constructed of galvanized sheet steel not less than 0.0304 of an inch (0.7722 mm) thick or of other approved, noncombustible, corrosion-resistant material. [NFPA 54:12.8.1]

509.7.1 Cold Climate. Uninsulated single-wall metal pipe shall not be used outdoors for venting appliances in regions where the 99 percent winter design temperature is below 32°F (0°C). [NFPA 54:12.8.2]

The use of single-wall metal pipe (e.g., stovepipe) as a vent for gas utilization equipment is limited by the code, especially in dwellings, for several reasons, including the following:

- The surface of the pipe is usually hot enough to burn a person who accidentally contacts the pipe.
- The high surface temperature can cause overheating of adjacent combustible material.
- The heat loss from a long length of pipe in a cool area can result in reduced draft and the condensation of water vapor in the flue gases.

WATER HEATERS

The use of single-wall metal pipe as a vent, where permitted in Section 509.7, is limited to Category I appliances. Note that some building codes do not permit single-wall metal chimneys and unlisted metal chimneys inside one- and two-family dwellings. See Section 509.7.3

Single-wall metal pipe is a venting material that has limitations due to its inherently higher heat loss than chimneys and gas vents. **Figure 509.7.1** shows the limitations of single-wall gas vents when compared with characteristics of double-wall gas vents.

The outdoor use of single-wall metal pipe in cold climates is restricted to prevent premature corrosion of the vent. In cold climates, the low ambient temperature can cool vent gases to the point at which water vapor condenses in the vent and causes corrosion.

can ignite combustible material placed in contact with it. When this type of pipe passes through a combustible roof, protection of the roof is required, and a thimble (see **Figure 509.7.2**) is specified.

**FIGURE 509.7.2
WALL THIMBLE**

509.7.3 Installation with Appliances Permitted by Section 509.4. Single-wall metal pipe shall not be used as a vent in dwellings and residential occupancies. [NFPA 54:12.8.4.1]

509.7.3.1 Limitations. Single-wall metal pipe shall be used for runs directly from the space in which the appliance is located through the roof or exterior wall to the outer air. A pipe passing through a roof shall extend without interruption through the roof flashing, roof jacket, or roof thimble. [NFPA 54:12.8.4.2]

509.7.3.2 Attic or Concealed Space. Single-wall metal pipe shall not originate in an unoccupied attic or concealed space and shall not pass through an attic, inside wall, concealed space, or floor. [NFPA 54:12.8.4.3]

509.7.3.3 Incinerator. Single-wall metal pipe used for venting an incinerator shall be exposed and readily examinable for its full length and shall have required clearances maintained.

509.7.3.4 Clearances. Minimum clearances from single-wall metal pipe to combustible material shall be in accordance with Table 509.7.3.4(1). Reduced clearances from single-wall metal pipe to combustible material shall be as specified for vent connectors in Table 509.7.3.4(2). [NFPA 54:12.8.4.4]

509.7.3.5 Combustible Exterior Wall. Single-wall metal pipe shall not pass through a combustible exterior wall unless guarded at the point of passage by a ventilated metal thimble not smaller than the following:

(1) For listed appliances with draft hoods and appliances listed for use with Type B gas vents, the thimble shall be a minimum of 4 inches (100 mm) larger in diameter than the metal pipe. Where there is a run of not less than 6 feet (1829 mm) of metal pipe in the opening between the draft hood outlet and the thimble, the thimble shall be a minimum of 2 inches (50 mm) larger in diameter than the metal pipe.

(2) For unlisted appliances having draft hoods, the thimble shall be a minimum of 6 inches (150 mm) larger in diameter than the metal pipe.

(3) For residential and low-heat appliances, the thimble shall be a minimum of 12 inches (300 mm) larger in diameter than the metal pipe.

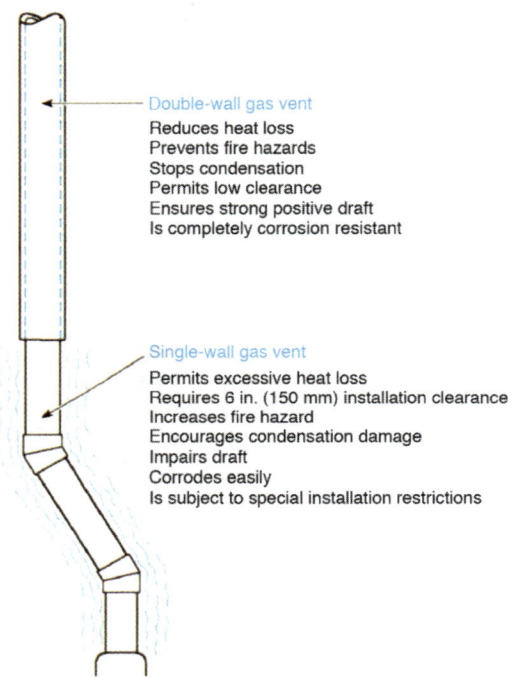

**FIGURE 509.7.1
COMPARISON OF SINGLE WALL AND DOUBLE WALL
GAS VENT CHARACTERISTICS**

509.7.2 Termination. The termination of single-wall metal pipe shall meet the following requirements:

(1) Single-wall metal pipe shall terminate at least 5 feet (1524 mm) in vertical height above the highest connected appliance draft hood outlet or flue collar.

(2) Single-wall metal pipe shall extend at least 2 feet (610 mm) above the highest point where it passes through a roof of a building and at least 2 feet (610 mm) higher than any portion of a building within a horizontal distance of 10 feet (3048 mm). (See **Figure 509.5.4**)

(3) An approved cap or roof assembly shall be attached to the terminus of a single-wall metal pipe. [NFPA 54:12.8.3]

These limitations are based on the fact that single-wall metal pipe can operate at a high temperature that easily

WATER HEATERS

TABLE 509.7.3.4(1)
CLEARANCE FOR CONNECTORS
[NFPA 54: TABLE 12.8.4.4]*

APPLIANCE	MINIMUM DISTANCE FROM COMBUSTIBLE MATERIAL (inches)			
	LISTED TYPE B GAS VENT MATERIAL	LISTED TYPE L VENT MATERIAL	SINGLE-WALL METAL PIPE	FACTORY-BUILT CHIMNEY SECTIONS
Listed appliances with draft hoods and appliances listed for use with Type B gas vents	As listed	As listed	6	As listed
Residential boilers and furnaces with listed gas conversion burner and with draft hood	6	6	9	As listed
Residential appliances listed for use with Type L vents	Not permitted	As listed	9	As listed
Unlisted residential appliances with draft hood	Not permitted	6	9	As listed
Residential and low-heat appliances other than those above	Not permitted	9	18	As listed
Medium-heat appliances	Not permitted	Not permitted	36	As listed

For SI units: 1 inch = 25.4 mm
* These clearances shall apply unless the installation instructions of a listed appliance or connector specify different clearances, in which case the listed clearances shall apply.

Exception: In lieu of thimble protection, all combustible material in the wall shall be removed a sufficient distance from the metal pipe to provide the specified clearance from such metal pipe to combustible material. Any material used to close up such opening shall be noncombustible. [NFPA 54:12.8.4.6]

509.7.3.6 Roof Thimble. Where a single-wall metal pipe passes through a roof constructed of combustible material, a noncombustible, nonventilating thimble shall be used at the point of passage. The thimble shall extend not less than 18 inches (457 mm) above and 6 inches (152 mm) below the roof with the annular space open at the bottom and closed at the top. The thimble shall be sized in accordance with Section 509.7.3.5. [NFPA 54:12.8.4.5]

509.7.4 Size of Single-Wall Metal Pipe. Single-wall metal piping shall comply with the Section 509.7.4.1 through Section 509.7.4.3. [NFPA 54:12.8.5]

509.7.4.1 Sizing of Venting System. A venting system of a single-wall metal pipe shall be sized in accordance with one of the following methods and the appliance manufacturer's instructions:

(1) For a draft hood-equipped appliance, in accordance with Section 510.0.

(2) For a venting system for a single appliance with a draft hood, the areas of the connector and the pipe each shall be not less than the area of the appliance flue collar or draft hood outlet, whichever is smaller. The vent area shall not exceed seven times the draft hood outlet area.

(3) Other approved engineering methods. [NFPA 54:12.8.5(1)]

The vent sizing tables in Section 510.0 must be used to size single-wall metal pipe used to vent draft hood-equipped appliances. Fan-assisted appliances may not be vented with single-wall pipe due to the heat loss in the single-wall metal pipe, which will quickly lead to condensation and corrosion. The alternative methods allowed for chimneys and gas vents are not permitted to be used where single-wall metal pipe is used as a vent material.

509.7.4.2 Non-Round Metal Pipe. Where a single-wall metal pipe is used and has a shape other than round, it shall have an effective area equal to the effective area of the round pipe for which it is substituted, and the internal dimension of the pipe shall be not less than 2 inches (50 mm). [NFPA 54:12.8.5(2)]

509.7.4.3 Venting Capacity. The vent cap or a roof assembly shall have a venting capacity not less than that of the pipe to which it is attached. [NFPA 54:12.8.5(3)]

509.7.5 Support of Single-Wall Metal Pipe. Portions of single-wall metal pipe shall be supported for the design and weight of the material employed. [NFPA 54:12.8.6]

509.7.6 Marking. Single-wall metal pipe shall comply with the marking provisions of Section 509.6.5. [NFPA 54:12.8.7]

509.8 Through-the-Wall Vent Termination. A mechanical draft venting system shall terminate at least 3 feet (914 mm) above any forced air inlet located within 10 feet (3048 mm) (See **Figure 509.8**).

Exceptions:

(1) This provision shall not apply to the combustion air intake of a direct-vent appliance.

(2) This provision shall not apply to the separation of the integral outdoor air inlet and flue gas discharge of listed outdoor appliances. [NFPA 54:12.9.1]

The intent here is to prevent appliance combustion products from being drawn into a building through fresh air inlets, including operable windows. This requirement recognizes that vent gases are lighter than air. Exception 1 recognizes that direct-vent appliance inlets do not communicate with the air in a building. Exception 2 prevents confusion in the installation of outdoor gas equipment. Some authorities have misinterpreted the code to prohibit such equipment or to require it to be modified in the field, which is not the intent of Section 509.8. An example of this type of equipment is a packaged rooftop air conditioner, which incorporates a gas vent and a circulation air inlet used for the building air supply.

WATER HEATERS

TABLE 509.7.3.4(2)
REDUCTION OF CLEARANCES WITH SPECIFIED FORMS OF PROTECTION[1, 2, 3, 4, 5, 6, 7, 8, 9, 10, 11]
[NFPA 54: TABLE 10.2.3]

TYPE OF PROTECTION APPLIED TO AND COVERING SURFACES OF COMBUSTIBLE MATERIAL WITHIN THE DISTANCE SPECIFIED AS THE REQUIRED CLEARANCE WITH NO PROTECTION [SEE FIGURE 509.7.3.4(1) THROUGH FIGURE 509.7.3.4(3)]	WHERE THE REQUIRED CLEARANCE WITH NO PROTECTION FROM APPLIANCE, VENT CONNECTOR, OR SINGLE-WALL METAL PIPE IS:									
	36 (inches)		18 (inches)		12 (inches)		9 (inches)		6 (inches)	
	ALLOWABLE CLEARANCES WITH SPECIFIED PROTECTION (inches)									
	USE COLUMN 1 FOR CLEARANCES ABOVE APPLIANCE OR HORIZONTAL CONNECTOR. USE COLUMN 2 FOR CLEARANCES FROM APPLIANCES, VERTICAL CONNECTOR, AND SINGLE-WALL METAL PIPE.									
	ABOVE COLUMN 1	SIDES AND REAR COLUMN 2	ABOVE COLUMN 1	SIDES AND REAR COLUMN 2	ABOVE COLUMN 1	SIDES AND REAR COLUMN 2	ABOVE COLUMN 1	SIDES AND REAR COLUMN 2	ABOVE COLUMN 1	SIDES AND REAR COLUMN 2
(1) 3½ inch thick masonry wall without ventilated air space	—	24	—	12	—	9	—	6	—	5
(2) ½ of an inch insulation board over 1 inch glass fiber or mineral wool batts	24	18	12	9	9	6	6	5	4	3
(3) 0.024 inch sheet metal over 1 inch glass fiber or mineral wool batts reinforced with wire on rear face with ventilated air space	18	12	9	6	6	4	5	3	3	3
(4) 3½ inch thick masonry wall with ventilated air space	—	12	—	6	—	6	—	6	—	6
(5) 0.024 inch sheet metal with ventilated air space	18	12	9	6	6	4	5	3	3	2
(6) ½ of an inch thick insulation board with ventilated air space	18	12	9	6	6	4	5	3	3	3
(7) 0.024 inch sheet metal with ventilated air space over 0.024 inch sheet metal with ventilated air space	18	12	9	6	6	4	5	3	3	3
(8) 1 inch glass fiber or mineral wool batts sandwiched between two sheets 0.024 inch sheet metal with ventilated air space	18	12	9	6	6	4	5	3	3	3

For SI units: 1 inch = 25.4 mm, °C = (°F-32)/1.8

Notes:

[1] Reduction of clearances from combustible materials shall not interfere with combustion air, draft hood clearance and relief, and accessibility of servicing.

[2] Clearances shall be measured from the outer surface of the combustible material to the nearest point on the surface of the appliance, disregarding an intervening protection applied to the combustible material.

[3] Spacers and ties shall be of noncombustible material. No spacer or tie shall be used directly opposite the appliance or connector.

[4] Where clearance reduction systems use a ventilated air space, adequate provision for air circulation shall be provided as described. [See Figure 509.7.3.4(2) and Figure 509.7.3.4(3)]

[5] There shall be not less than 1 inch (25.4 mm) between clearance reduction systems and combustible walls and ceilings for reduction systems using a ventilated air space.

[6] Where a wall protector is mounted on a single flat wall away from corners, it shall have a minimum 1 inch (25.4 mm) air gap. To provide air circulation, the bottom and top edges, or only the side and top edges, or edges shall be left open.

[7] Mineral wool batts (blanket or board) shall have a minimum density of 8 pounds per cubic foot (lb/ft^3) (128 kg/m^3) and a minimum melting point of 1500°F (816°C).

[8] Insulation material used as part of a clearance reduction system shall have a thermal conductivity of 1 British thermal unit inch per hour square foot degree Fahrenheit [Btu•in/(h•ft^2•°F)] [0.1 W/(m•K)] or less.

[9] There shall be not less than 1 inch (25.4 mm) between the appliance and the protector. In no case shall the clearance between the appliance and the combustible surface be reduced below that allowed in this table.

[10] Clearances and thicknesses are minimum; larger clearances and thicknesses are acceptable.

[11] Listed single-wall connectors shall be installed in accordance with the terms of their listing and the manufacturer's installation instructions.

WATER HEATERS

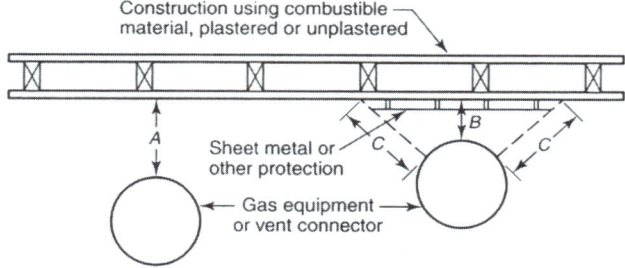

Notes:

1. A – Equals the clearance with no protection specified in Table 509.7.3.4(1) and Table 509.7.3.4(2) and in the sections applying to various types of appliances.
2. B – Equals the reduced clearance permitted in accordance with Table 509.7.3.4(2).
3. The protection applied to the construction using combustible material shall extend far enough in each direction to make C equal to A.

FIGURE 509.7.3.4(1)[1, 2, 3]
EXTENT OF PROTECTION NECESSARY TO REDUCE CLEARANCES FROM GAS APPLIANCES OR VENT CONNECTORS
[NFPA 54: FIGURE 10.3.2.3(a)]

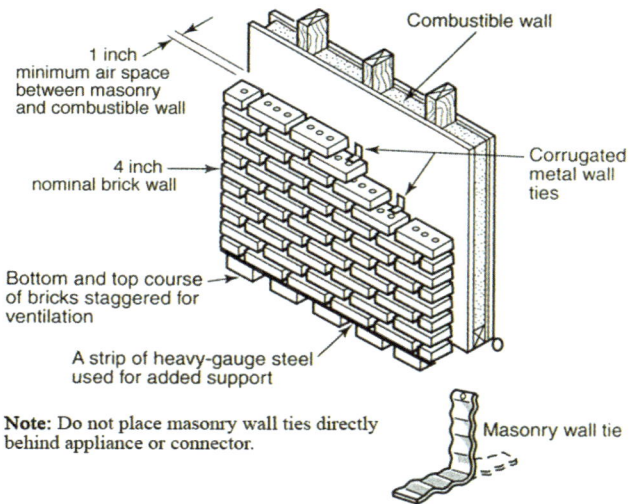

For SI units: 1 inch = 25.4 mm

FIGURE 509.7.3.4(3)
MASONRY CLEARANCE REDUCTION SYSTEM
[NFPA 54: FIGURE 10.3.2.3(c)]

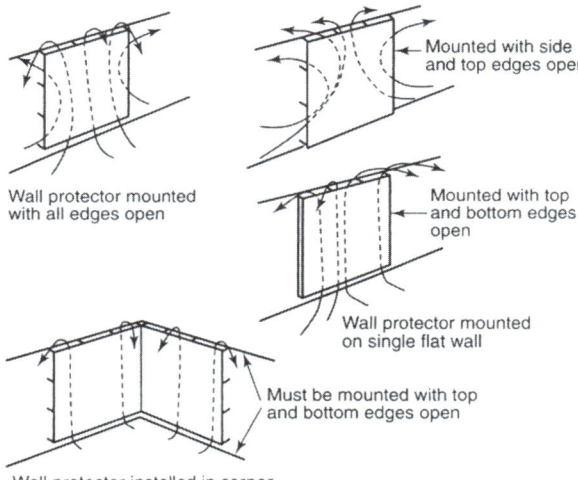

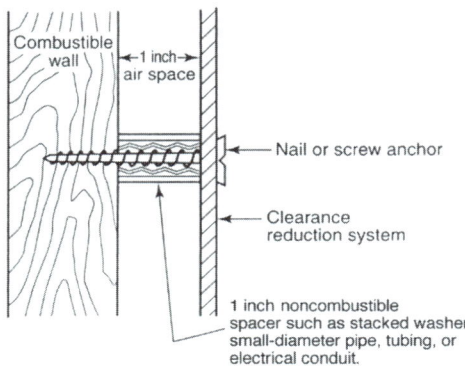

For SI units: 1 inch = 25.4 mm

Note: Masonry walls shall be attached to combustible walls using wall ties. Spacers shall not be used directly behind appliance or connector.

FIGURE 509.7.3.4(2)
WALL PROTECTION REDUCTION SYSTEM
[NFPA 54: FIGURE 10.3.2.3(b)]

509.8.1 Mechanical Draft Venting System. A mechanical draft venting system of other than direct-vent type shall terminate not less than 4 feet (1219 mm) below, 4 feet (1219 mm) horizontally from, or 1 foot (305 mm) above a door, operable window, or gravity air inlet into a building. The bottom of the vent terminal shall be located not less than 12 inches (305 mm) above finished ground level. [NFPA 54:12.9.2]

509.8.2 Direct-Vent Appliance. The vent terminal of a direct-vent appliance with an input of 10 000 Btu/h (2.93 kW) or less shall be located at least 6 inches (152 mm) from any air opening into a building, an appliance with an input over 10 000 Btu/h (2.9.3 kW) but not over 50 000 Btu/h (14.7 kW) shall be installed with a 9 inch (229 mm) vent termination clearance, and an appliance with an input over 50 000 Btu/h (14.7 kW) shall have at least a 12 inch (305 mm) vent termination clearance. The bottom of the vent terminal and the air intake shall be located at least 12 inches (305 mm) above finished ground level. [NFPA 54:12.9.3]

This section permits the vent terminals of direct-vent appliances to be located much closer to air inlets than permitted for nondirect-vent appliances. Studies have shown that vent gases from direct-vent appliances disperse rapidly upon leaving the vent terminal, even when the terminal is located under an open window; however, a window is unlikely to be open when heat is needed. The section also specifies the location of the exit terminal of direct-vent appliances. All of these locations are shown in **Figure 509.8**.

509.8.3 Category I through Category IV and Noncategorized Appliances. Through-the-wall vents for Category II and Category IV appliances and noncategorized condensing appliances shall not terminate over public walkways or over an area where condensate or vapor could create a nuisance or hazard or could be detrimental to the operation of regulators,

WATER HEATERS

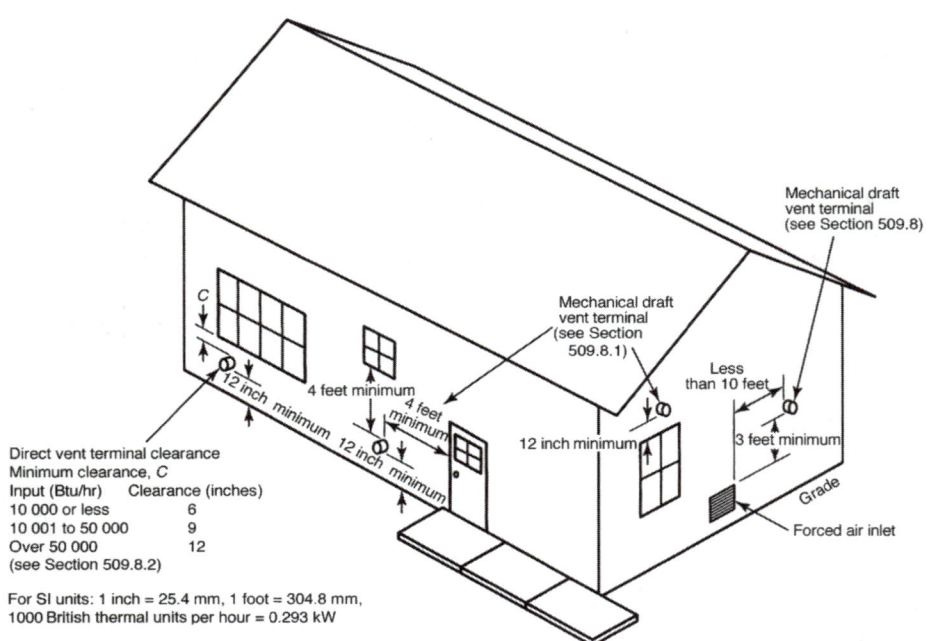

FIGURE 509.8
EXIT TERMINALS OF MECHANICAL DRAFT AND DIRECT-VENT VENTING SYSTEMS
[NFPA 54: FIGURE A.12.9]

relief valves, or other equipment. Where local experience indicates that condensate is a problem with Category I and Category III appliances, this provision shall also apply.

Drains for condensate shall be installed in accordance with the appliance and the vent manufacturer's installation instructions. [NFPA 54:12.9.4]

509.8.4 Annular Spaces. Where vents, including those for direct-vent appliances or combustion air intake pipes, penetrate outside walls of buildings, the annular spaces around such penetrations shall be permanently sealed using approved materials to prevent entry of combustion products into the building. [NFPA 54:12.9.5]

509.8.5 Vent Terminals. Vent systems for Category IV appliances that terminate through an outside wall of a building and discharge flue gases perpendicular to the adjacent wall shall be located not less than 10 feet (3048 mm) horizontally from an operable opening in an adjacent building.

Exception: This shall not apply to vent terminals that are 2 feet (610 mm) or more above or 25 feet (7620 mm) or more below operable openings. [NFPA 54:12.9.6]

509.9 Condensation Drain. Provision shall be made to collect and dispose of condensate from venting systems serving Category II and Category IV appliances and noncategorized condensing appliances in accordance with Section 509.8.3. [NFPA 54:12.10.1]

509.9.1 Local Experience. Where local experience indicates that condensation is a problem, provision shall be made to drain off and dispose of condensate from venting systems serving Category I and Category III appliances in accordance with Section 509.8.3. [NFPA 54:12.10.2]

Condensate drains should be installed per Section 814.0, Condensate Wastes and Control, in Chapter 8.

509.10 Vent Connectors for Category I Appliances. A vent connector shall be used to connect an appliance to a gas vent, chimney, or single-wall metal pipe, except where the gas vent, chimney, or single-wall metal pipe is directly connected to the appliance. [NFPA 54:12.11.1]

Vent connectors are relatively short runs of single-wall metal pipe, Type B vent or other vent materials that are used to connect an appliance to a chimney or vent. When two or more appliances are connected to a chimney or vent, at least one connector is used. When one appliance is connected to a vent with no change in size or venting material, there is no vent connector.

The vent connector has an important effect on the overall operation of the venting system. GRI-sponsored research has proven that a vent primes better and has a stronger draft if there is a connector rise directly over the appliance. Retaining as much heat as possible in the flue gases as they pass through the connector is also very important. For more information, refer to the GRI report by S. M. Ricci, et. al., '*Vent Oversizing and the "Seven-Times" Rule: Analysis and Recommendations*,' GRI-98/0285.

509.10.1 Materials. A vent connector shall be made of noncombustible, corrosion-resistant material capable of withstanding the vent gas temperature produced by the appliance and of a thickness to withstand physical damage. [NFPA 54:12.11.2.1]

509.10.1.1 Unconditioned Area. Where the vent connector used for an appliance having a draft hood or a Category I appliance is located in or passes through an unconditioned area, attic or crawl space, that portion of the vent connector shall be listed Type B, Type L, or listed vent material having equivalent insulation qualities.

Exception: Single-wall metal pipe located within the exterior walls of the building and located in an unconditioned area other than an attic or a crawl space having a local 99 percent winter design temperature of 5°F (-15°C) or higher. [NFPA 54:12.11.2.2]

Attics and unconditioned areas of buildings, such as garages and some basements, are commonly used for storage. Consequently, Type B and L vent material, addressed in Section 509.10.1.1, is specified for vent connectors of appliances installed in attics, because fires occur as a result of the ignition of combustible material that has been carelessly placed on or near single-wall metal vent connectors. Type B or equivalent materials are also needed to make sure that excessive condensation does not occur inside the vent connector [see **Figure G 101.2(1)** in Appendix G].

509.10.1.2 Residential-Type Appliances. Vent connectors for residential-type appliances shall comply with the following:

(1) Vent connectors for listed appliances having draft hoods, appliances having draft hoods and equipped with listed conversion burners, and Category I appliances that are not installed in attics, crawl spaces, or other unconditioned areas shall be one of the following:

(a) Type B or Type L vent material.

(b) Galvanized sheet steel not less than 0.018 of an inch (0.457 mm) thick.

(c) Aluminum (1100 or 3003 alloy or equivalent) sheet not less than 0.027 of an inch (0.686 mm) thick.

(d) Stainless steel sheet not less than 0.012 of an inch (0.305 mm) thick.

(e) Smooth interior wall metal pipe having resistance to heat and corrosion equal to or greater than that of Section 509.10.1.2(1)(b), Section 509.10.1.2(1)(c), or Section 509.10.1.2(1)(d).

(f) A listed vent connector.

(2) Vent connectors shall not be covered with insulation.

Exception: Listed insulated vent connectors shall be installed in accordance with the manufacturer's installation instructions. [NFPA 54:12.11.2.3]

509.10.1.3 Non-Residential Low-Heat Appliances. A vent connector for a non-residential low-heat appliance shall be a factory-built chimney section or steel pipe having resistance to heat and corrosion equivalent to that for the appropriate galvanized pipe as specified in Table 509.10.1.3. Factory-built chimney sections shall be joined together in accordance with the chimney manufacturer's instructions. [NFPA 54:12.11.2.4]

509.10.1.4 Medium-Heat Appliances. Vent connectors for medium-heat appliances shall be constructed of factory-built, medium-heat chimney sections or steel of a thickness not less than that specified in Table 509.10.1.4 and shall comply with the following:

(1) A steel vent connector for an appliance with a vent gas temperature in excess of 1000°F (538°C) measured at the entrance to the connector shall be lined with medium-duty fire brick or the equivalent.

(2) The lining shall be at least 2½ inches (64 mm) thick for a vent connector having a diameter or greatest cross-sectional dimension of 18 inches (457 mm) or less.

(3) The lining shall be at least 4½ inches (114 mm) thick laid on the 4½ inches (114 mm) bed for a vent connector having a diameter or greatest cross-sectional dimension greater than 18 inches (457 mm).

(4) Factory-built chimney sections, if employed, shall be joined together in accordance with the chimney manufacturer's instructions. [NFPA 54:12.11.2.5]

TABLE 509.10.1.3
MINIMUM THICKNESS FOR GALVANIZED STEEL VENT CONNECTORS FOR LOW-HEAT APPLIANCES
[NFPA 54: TABLE 12.11.2.4]

DIAMETER OF CONNECTOR (inches)	MINIMUM THICKNESS (inches)
Less than 6	0.019
6 to less than 10	0.023
10 to 12 inclusive	0.029
14 to 16 inclusive	0.034
Over 16	0.056

For SI units: 1 inch = 25.4 mm, 1 square inch = 0.000645 m^2

TABLE 509.10.1.4
MINIMUM THICKNESS FOR STEEL VENT CONNECTORS FOR MEDIUM-HEAT APPLIANCES
[NFPA 54: TABLE 12.11.2.5]

VENT CONNECTOR SIZE		
DIAMETER (inches)	AREA (square inches)	MINIMUM THICKNESS (inches)
Up to 14	Up to 154	0.053
Over 14 to 16	154 to 201	0.067
Over 16 to 18	201 to 254	0.093
Over 18	Exceeding 254	0.123

For SI units: 1 inch = 25.4 mm, 1 square inch = 0.000645 m^2

509.10.2 Size of Vent Connector. A vent connector for an appliance with a single draft hood or for a Category I fan-assisted combustion system appliance shall be sized and installed in accordance with Section 510.0 or other approved engineering methods. [NFPA 54:12.11.3.1]

When sizing connectors serving fan-assisted Category I appliances, the user will note that the selection of single-wall metal pipe connectors is limited. This limit is not included in Section 509.0 but becomes evident when the tables in Section 510.0 are used to size the vent and connector. The limited selection is due to the lower vent-operating temperature of these appliances, which reduces the allowable heat loss in the connector compared to nonfan-assisted combustion appliances. Single-wall connectors are not permitted when Tables 510.2(6) through 510.2(9) are being used to size connectors to exterior masonry chimneys.

509.10.2.1 Manifold. For a single appliance having more than one draft hood outlet or flue collar, the manifold shall be constructed according to the instructions of the appliance manufacturer. Where there are no instructions, the manifold shall be designed and constructed in accordance with approved engineering practices. As an alternative method, the effective area of the manifold shall equal the combined area of the flue collars or draft hood outlets, and the vent connectors shall have a minimum 1 foot (305 mm) rise. [NFPA 54:12.11.3.2]

In the absence of specific instructions, it is important to provide an adequate rise above the draft hoods. This rise is needed to ensure proper flue priming on startup. The minimum one foot connector rise required minimizes spillage.

509.10.2.2 Size. Where two or more appliances are connected to a common vent or chimney, each vent connector shall be sized in accordance with Section 510.0 or other approved engineering methods. [NFPA 54:12.11.3.3]

As an alternative method applicable where the appliances are draft-hood-equipped, each vent connector shall have an effective area not less than the area of the draft hood outlet of the appliance to which it is connected. [NFPA 54:12.11.3.4]

509.10.2.3 Height. Where two or more appliances are vented through a common vent connector or vent manifold, the common vent connector or vent manifold shall be located at the highest level consistent with available headroom and clearance to combustible material and sized in accordance with Section 510.0 or other approved engineering methods. [NFPA 54:12.11.3.5]

As an alternative method applicable only where there are two draft hood-equipped appliances, the effective area of the common vent connector or vent manifold and all junction fittings shall be not less than the area of the larger vent connector plus 50 percent of the area of the smaller flue collar outlet. [NFPA 54:12.11.3.6]

509.10.2.4 Size Increase. Where the size of a vent connector is increased to overcome installation limitations and obtain connector capacity equal to the appliance input, the size increase shall be made at the appliance draft hood outlet. [NFPA 54:12.11.3.7]

509.10.3 Two or More Appliances Connected to a Single Vent. Where two or more openings are provided into one chimney flue or vent, either of the following shall apply:

(1) The openings shall be at different levels.
(2) The connectors shall be attached to the vertical portion of the chimney or vent at an angle of 45 degrees (0.79 rad) or less relative to the vertical. [NFPA 54:12.11.4.1]

509.10.3.1 Height of Connector. Where two or more vent connectors enter a common vent, chimney flue, or single-wall metal pipe, the smaller connector shall enter at the highest level consistent with the available headroom or clearance to combustible material. [NFPA 54:12.11.4.2]

509.10.3.2 Pressure. Vent connectors serving Category I appliances shall not be connected to a portion of a mechanical draft system operating under positive static pressure, such as those serving Category III or Category IV appliances. [NFPA 54:12.11.4.3]

509.10.4 Clearance. Minimum clearances from vent connectors to combustible material shall comply with Table 509.7.3.4(1).

Exception: The clearance between a vent connector and combustible material shall be permitted to be reduced where the combustible material is protected as specified for vent connectors in Table 509.7.3.4(2). [NFPA 54:12.11.5]

509.10.5 Joints. Joints between sections of connector piping and connections to flue collars or draft hood outlets shall be fastened in accordance with one of the following methods:

(1) Sheet metal screws.
(2) Vent connectors of listed vent material assembled and connected to flue collars or draft hood outlets in accordance with the manufacturer's instructions.
(3) Other approved means. [NFPA 54:12.11.6]

Each bend or turn (elbow or tee) in a venting system reduces its capacity, especially in a natural draft vent. The capacities shown in the tables in Section 510.0 make an allowance for two 90-degree turns between the draft hood and the top of the venting system.

If the venting system contains more than two 90-degree turns, a capacity reduction of five percent should be made for each additional turn that is less than or equal to 45 degrees, and a capacity reduction of 10 percent should be taken for each elbow between 45 and 90 degrees. For example, the capacity of a venting system containing four 90-degree turns is 80 percent of that indicated in the tables, and a system with two 90-degree turns and two 45-degree turns is 90 percent of the table values (see Sections 510.1.2 and 510.2.5).

509.10.6 Slope. A vent connector shall be installed without any dips or sags and shall slope upward toward the vent or chimney at least ¼ inch per foot (20.8 mm/m).

Exception: Vent connectors attached to a mechanical draft system installed in accordance with appliance and draft system manufacturer's instructions. [NFPA 54:12.11.7]

509.10.7 Length of Vent Connector. The length of vent connectors shall comply with Section 509.10.7.1 or Section 509.10.7.2.

509.10.7.1 Single Wall Connector. The maximum horizontal length of a single-wall connector shall be 75 percent of the height of the chimney or vent except for engineered systems. [NFPA 54:12.11.8.1]

This section contains limitations on horizontal vent connectors. A single-wall vent connector for a single appliance is limited to 75 percent of the height in the horizontal position. If a vent is 10 feet (3048 mm) high, measured from the top of the appliance vent collar to the bottom of the rain cap, then 7.5 feet (2286 mm) of the total run of venting may be in a horizontal (greater than 45 degrees from the vertical) position. A single-wall vent connector radiates heat much faster than a Type B vent, and cooling flue gases present problems for the venting system. Condensate and loss of buoyancy may result in poor flue function and possible failure of the venting system. A Type B vent, when used as a connector, may be 100 percent of the height in the horizontal position (see Section 509.10.7.2).

509.10.7.2 Type B Double Wall Connector. The maximum horizontal length of a Type B double-wall connector shall be 100 percent of the height of the chimney or vent, except for engineered systems. The maximum length of an individual connector for a chimney or vent system serving multiple appliances, from the appliance outlet to the junction with the common vent or another connector, shall be 100 percent of the height of the chimney or vent. [NFPA 54:12.11.8.2]

509.10.8 Support. A vent connector shall be supported for the design and weight of the material employed to maintain clearances and prevent physical damage and separation of joints. [NFPA 54:12.11.9]

509.10.9 Chimney Connection. Where entering a flue in a masonry or metal chimney, the vent connector shall be installed above the extreme bottom to avoid stoppage. Where a thimble or slip joint is used to facilitate removal of the connector, the connector shall be attached to or inserted into the thimble or slip joint to prevent the connector from falling out. Means shall be employed to prevent the connector from entering so far as to restrict the space between its end and the opposite wall of the chimney flue. [NFPA 54:12.11.10]

509.10.10 Inspection. The entire length of a vent connector shall be readily accessible for inspection, cleaning, and replacement. [NFPA 54:12.11.11]

509.10.11 Fireplaces. A vent connector shall not be connected to a chimney flue serving a fireplace unless the fireplace flue opening is permanently sealed. [NFPA 54:12.11.12]

509.10.12 Passage through Ceilings, Floors, or Walls. A vent connector shall not pass through a ceiling, floor, or fire-resistance-rated wall. A single-wall metal pipe connector shall not pass through an interior wall.

Exception: Vent connectors made of listed Type B or Type L vent material and serving listed appliances with draft hoods and other appliances listed for use with Type B gas vents that pass through walls or partitions constructed of combustible material shall be installed with not less than the listed clearance to combustible material.

In penetrating an interior wall, floor or ceiling, a single-wall metal connector (e.g., stove pipe) would be entering a space or room of the building other than that in which the appliance is located. Such space could be a storeroom or other part of the building that is not normally occupied. In such cases, at least some portion of the connector would be out of sight; therefore, any damage to the connector, such as separation of joints, perforation by corrosion with consequent leakage of flue gases into the building or placement of combustible material near or on the connector, would escape early detection, creating a potentially hazardous situation.

509.10.12.1 Medium-Heat Appliances. Vent connectors for medium-heat appliances shall not pass through walls or partitions constructed of combustible material. [NFPA 54:12.11.13.2]

509.11 Vent Connectors for Category II, Category III, and Category IV Appliances. The vent connectors for Category II, Category III, and Category IV appliances shall be in accordance with Section 509.4 through Section 509.4.3. [NFPA 54:12.12]

509.12 Draft Hoods and Draft Controls. Vented appliances shall be installed with draft hoods.

Exception: Dual oven-type combination ranges; incinerators; direct-vent appliances; fan-assisted combustion system appliances; appliances requiring chimney draft for operation; single firebox boilers equipped with conversion burners with inputs exceeding 400 000 Btu/h (117 kW); appliances equipped with blast, power, or pressure burners that are not listed for use with draft hoods; and appliances designed for forced venting.

See **Figure 509.12a**.

Draft hoods on vent systems perform the following three functions:

1. The negative pressure in the vent system created by the hot exhaust gases draws in dilution air at the draft-hood opening. This dilution air is taken from the room in which the draft hood is located. This room air is much cooler than the exhaust gases, thereby lowering the net stack temperature and reducing fire hazards. Dilution air is also much drier than the exhaust gases, thereby raising the dew point and reducing any condensation.

2. The draft hood acts as a break between the vent system and the appliance and eliminates stack action. Appliance manufacturers design their equipment to operate with a specific range of airflow through the appliance. If there were no separation between the appliance and the vent system, excessive drafts created by tall chimneys would affect the combustion process and flame stability, possibly even pilot outage. Excessive drafts would also lower efficiency by moving the products of combustion through the heat exchanger before optimal heat transfer. Wind effects can also create temporary downdrafts.

3. Finally, a draft hood provides a relief opening in the event of a downdraft. Vent systems may temporarily experience poor venting at startup (before the vent heats up) or during windy conditions. Under these conditions, some of the products of combustion may ''spill out'' at the draft hood. The principal products of combustion from a properly burning appliance are carbon dioxide and water vapor and should cause no immediate harm. Once draft is established (or reestablished when the wind subsides), all of the combustion products are vented safely up the vent.

WATER HEATERS

During a sustained downdraft, such as in a blockage, all of the combustion products may spill into the living space and may eventually displace the oxygen in the room, potentially leading to incomplete combustion and the formation of carbon monoxide. New central heating appliances equipped with draft hoods must have safety switches, such as the spill switches, which will shut off the burner in the event of a sustained downdraft.

Draft hoods are an integral part of the equipment design and should never be altered. The height of the draft hood above the flue collar will affect combustion and the airflow through the appliance. If a draft hood were removed entirely, even a temporary downdraft would immediately affect the combustion process, potentially creating carbon monoxide.

Barometric draft regulators perform the same functions as a draft hood, but are generally used in connections with power burners and conversion burners. Where power burners are used, the gas input, combustion air, flame pattern and draft all must be carefully set to match the equipment they serve (see **Figure 509.12b**).

Barometric draft regulators are usually adjustable so that the amount of draft can be set for maximum efficiency and safe burner operation. Barometric draft regulators, when used with gas appliances, are double-acting so that they will act as a relief opening in the event of a downdraft. Safety shutoff devices are required on all conversion burners installed after 1990.

509.12.1 Installation. A draft hood supplied with or forming a part of listed vented appliances shall be installed without alteration, exactly as furnished and specified by the appliance manufacturer. [NFPA 54:12.13.2]

Where a draft hood is not supplied by the appliance manufacturer where one is required, a draft hood shall be installed, be of a listed or approved type, and, in the absence of other instructions, be of the same size as the appliance flue collar. Where a draft hood is required with a conversion burner, it shall be of a listed or approved type. [NFPA 54:12.13.2.1]

Where a draft hood of special design is needed or preferable, the installation shall be approved and in accordance with the recommendations of the appliance manufacturer. [NFPA 54:12.13.2.2]

509.12.2 Draft Control Devices. Where a draft control device is part of the appliance or is supplied by the appliance manufacturer, it shall be installed in accordance with the manufacturer's installation instructions. In the absence of manufacturer's installation instructions, the device shall be attached to the flue collar of the appliance or as near to the appliance as practical. [NFPA 54:12.13.3]

509.12.3 Additional Devices. Appliances requiring controlled chimney draft shall be permitted to be equipped with listed double-acting barometric draft regulators installed and adjusted in accordance with the manufacturer's installation instructions. [NFPA 54:12.13.4]

509.12.4 Location. Draft hoods and barometric draft regulators shall be installed in the same room or enclosure as the appliance in such a manner as to prevent a difference in pressure between the hood or regulator and the combustion air supply. [NFPA 54:12.13.5]

509.12.5 Positioning. Draft hoods and draft regulators shall be installed in the position for which they were designed with reference to the horizontal and vertical planes and shall be located so that the relief opening is not obstructed by a part of the appliance or adjacent construction. The appliance and its draft hood shall be located so that the relief opening is accessible for checking vent operation. [NFPA 54:12.13.6]

509.12.6 Clearance. A draft hood shall be located so that its relief opening is not less than 6 inches (152 mm) from a surface except that of the appliance it serves and the venting system to which the draft hood is connected. Where a greater or lesser clearance is indicated on the appliance label, the clearance shall not be less than that specified on the label. Such clearances shall not be reduced. [NFPA 54:12.13.7]

509.13 Manually Operated Dampers. A manually operated damper shall not be placed in an appliance vent connector. Fixed baffles shall not be classified as manually operated dampers. [NFPA 54:12.14]

509.14 Automatically Operated Vent Dampers. An automatically operated vent damper shall be of a listed type. [NFPA 54:12.15]

Preferably, automatically operated vent dampers, such as the one shown in **Figure 509.14**, should be included as part of the listed gas utilization equipment. Manually operated dampers are not permitted in vent connectors, and the requirement clarifies that fixed baffles, which are sometimes used to balance system draft on startup, are allowed.

509.14.1 Listing. Automatically operated vent dampers for oil fired appliances shall comply with UL 17. The automatic damper control shall comply with UL 378.

**FIGURE 509.12A
DRAFT HOOD**

**FIGURE 509.12B
BAROMETRIC DRAFT REGULATOR**

**FIGURE 509.14
AUTOMATICALLY OPERATED VENT DAMPERS**

509.15 Obstructions. Devices that retard the flow of vent gases shall not be installed in a vent connector, chimney, or vent. The following shall not be considered as obstructions:

(1) Draft regulators and safety controls specifically listed for installation in venting systems and installed in accordance with the manufacturer's installation instructions.

(2) Approved draft regulators and safety controls designed and installed in accordance with approved engineering methods.

(3) Listed heat reclaimers and automatically operated vent dampers installed in accordance with the manufacturer's installation instructions.

(4) Vent dampers serving listed appliances installed in accordance with Section 510.1 or Section 510.2 or other approved engineering methods.

(5) Approved economizers, heat reclaimers, and recuperators installed in venting systems of appliances not required to be equipped with draft hoods provided the appliance manufacturer's installation instructions cover the installation of such a device in the venting system and performance in accordance with Section 509.3 and Section 509.3.1 is obtained. [NFPA 54:12.16]

510.0 Sizing of Category I Venting Systems.

This section contains 15 code tables and explanatory text for calculating the required vent size for a variety of Category I venting systems. This chapter was first created for the 1992 edition of NFPA 54, by relocating the vent sizing tables and related notes from its Appendix G and extensively modifying them to cover fan-assisted combustion furnaces designed to meet U.S. energy minimums beginning in 1990. These furnaces provide higher efficiency, which is achieved by reducing vent temperature without jeopardizing vent operation.

The tables in Section 510.0 provide sizes for Category I appliances serving natural draft (NAT) appliances, fan-assisted (FAN) draft appliances, and combinations of both in one venting system. The tables incorporate the FAN Min and FAN Max columns to provide minimum and maximum capacities for each connector and vent serving fan-assisted combustion appliances. The tables "Appliance Input Rating" columns are described as follows [see Tables 510.1.2(1) through 510.2(9)].

- The NAT Max columns provide maximum capacities for each connector and vent serving natural draft appliances.
- The FAN + FAN column provides maximum capacities for each connector and vent serving multiple fan-assisted appliances.
- The FAN + NAT columns provide maximum capacities for each connector and vent serving combinations of natural draft and fan-assisted appliances.
- The NAT + NAT columns provide maximum capacities for each connector and vent serving multiple natural draft appliances.

Category I appliances are defined as appliances that operate with a nonpositive vent static pressure and with a vent gas temperature that avoids excessive condensate in the vent. Venting systems serving Category I appliances are conventional venting systems, in which the heat of the flue gases is the force that operates the vent. Section 510.0 includes the following:

- Six tables [Tables 510.1.2(1) to 510.1.2(6)] are listed for single-appliance installations and are listed in **Commentary Table – 1** for ease of use.
- Additional requirements for single-appliance vents are given. These qualifications must be considered when using single-appliance vent Tables 510.1.2(1) through 510.1.2(6).
- Nine vent tables are listed for multiple-appliance venting installations [Tables 510.2(1) to 510.2(9)].
- The five of the nine tables, listed in **Commentary Table – 2** are used for interior chimney or vent installations.
- Four tables covering exterior masonry chimneys are listed in **Commentary Table – 3**. The tables are organized by geographical regions of the United States based on the lowest anticipated winter design temperature. The minimum values shown in the tables are calculated to reduce the likelihood of condensation in the chimney.
- Additional requirements for multiple-appliance vents are provided. These qualifications must be considered when using multiple-appliance vent Tables 510.2(1) through 510.2(9).
- Tables 510.2(6) through 510.2(9) provide sizing information for exterior masonry chimneys serving heating appliances. The tables provide sizing options based on local 99-percent winter design temperatures in six temperature ranges. A map showing the boundaries of these temperature ranges in the United States is included in Figure 510.1.2(6).

Tables 510.1.2(1) through 510.1.2(6) were developed with the assumption that there are no restrictions in the path of the vent gas flow. An additional assumption is that there is no delib-

WATER HEATERS

erate attempt to remove heat; therefore, devices such as heat economizers may not be used in conjunction with the tables. Vent dampers installed as part of a listed appliance are to be installed in accordance with the manufacturer's instructions.

Further information on vent sizing and the sizing tables can be obtained from the UMC, NFPA 54 and the National Fuel Gas Code Handbook.

COMMENTARY TABLE – 1
Vent Table for Single Appliance Installations

Table No.	Vent	Connector
510.1.2(1)	Type B	None
510.1.2(2)	Type B	Single wall
510.1.2(3)	Masonry chimney	Type B
510.1.2(4)	Masonry chimney	Single wall
510.1.2(5)	Single-wall metal	None
510.1.2(6)	Exterior masonry chimney	Type B

COMMENTARY TABLE – 2
Vent Tables for Multiple-Appliance Venting Installations

Table No.	Vent	Connector
510.2(1)	Type B	Type B
510.2(2)	Type B	Single wall
510.2(3)	Masonry chimney	Type B
510.2(4)	Masonry chimney	Single wall
510.2(5)	Single wall	Single wall

COMMENTARY TABLE – 3
Exterior Masonry Chimney Tables

Table No.	Vent	Connector
510.2(6)	Exterior masonry chimney	NAT NAT
510.2(7)	Exterior masonry chimney	NAT NAT
510.2(8)	Exterior masonry chimney	FAN NAT
510.2(9)	Exterior masonry chimney	FAN NAT

510.1 Single Appliance Vent Table 510.1.2(1) through Table 510.1.2(6). Table 510.1.2(1) through Table 510.1.2(6) shall not be used where obstructions are installed in the venting system. The installation of vents serving listed appliances with vent dampers shall be in accordance with the appliance manufacturer's installation instructions or in accordance with the following:

(1) The maximum capacity of the vent system shall be determined using the NAT Max column.

(2) The minimum capacity shall be determined as though the appliance were a fan-assisted appliance, using the FAN Min column to determine the minimum capacity of the vent system. Where the corresponding "FAN Min" is "NA", the vent configuration shall not be permitted, and an alternative venting configuration shall be utilized. [NFPA 54:13.1.1]

This section provides guidance for using the venting tables with draft hood appliances that are equipped with a vent damper. An example of such appliance is a boiler that uses a vent damper to obtain higher efficiencies. The maximum capacity of the vent is found using the NAT Max column in these tables. This column treats the vent as one serving a fan-assisted appliance when the appliance is not operating. The reason for this unusual combination is that the appliance operates as a natural draft appliance, but when the appliance is not operating and the vent damper is closed, it is similar to a fan-assisted appliance in its propensity to condense water.

510.1.1 Vent Downsizing. Where the vent size determined from the tables is smaller than the appliance draft hood outlet or flue collar, the use of the smaller size shall be permitted provided that the installation is in accordance with the following requirements:

(1) The total vent height (H) is not less than 10 feet (3048 mm).

(2) Vents for appliance draft hood outlets or flue collars 12 inches (300 mm) in diameter or smaller are not reduced more than one table size.

(3) Vents for appliance draft hood outlets or flue collars exceeding 12 inches (300 mm) in diameter are not reduced more than two table sizes.

(4) The maximum capacity listed in the tables for a fan-assisted appliance is reduced by 10 percent (0.90 x maximum table capacity).

(5) The draft hood outlet exceeds 4 inches (100 mm) in diameter. A 3 inch (80 mm) diameter vent shall not be connected to a 4 inch (100 mm) diameter draft hood outlet. This provision shall not apply to fan-assisted appliances. [NFPA 54:13.1.2]

If the vent is smaller than the draft hood or flue collar, venting problems can occur. The restrictions in Section 510.1.1 recognize and avoid these venting problems by placing limits on "downsizing." In particular, note that a 4-inch draft hood outlet may not be reduced to 3 inches.

510.1.2 Elbows. Single-appliance venting configurations with zero lateral lengths in Table 510.1.2(1), Table 510.1.2(2), and Table 510.1.2(5) shall not have elbows in the venting system. Single-appliance venting with lateral lengths includes two 90 degree (1.57 rad) elbows. For each additional elbow up to and including 45 degrees (0.79 rad), the maximum capacity listed in the venting tables shall be reduced by 5 percent. For each additional elbow greater than 45 degrees (0.79 rad) up to and including 90 degrees (1.57 rad), the maximum capacity listed in the venting tables shall be reduced by 10 percent. Where multiple offsets occur in a vent, the total lateral length of offsets combined shall not exceed that specified in Table 510.1.2(1) through Table 510.1.2(5). [NFPA 54:13.1.3]

The sizing tables were designed with the assumption that up to two 90-degree turns were part of the venting system, except for the zero lateral length case, addressed in Section 510.1.3. The zero lateral case was assumed to extend straight up from the appliance outlet to the vent termination. Adding additional elbows to the system is possible.

The new derating factors for additional elbows are shown in **Commentary Table – 4**. Previously, a 10-percent reduction was required for each elbow, regardless of its angle. This revision, which was based on input from the Gas Technology Institute's contractor, Battelle, provides flexibility to installers forced to use vent offsets and encourages the use of elbows of less than 90 degrees.

When sizing a vent system with elbows, please note the following important factors:
- The tables include two 90-degree elbows. The table values should not be derated for venting systems with one or two elbows.
- For each additional elbow over two, (see **Figure 510.1.2**) derate the table values by 5 or 10 percent, depending on the angle of the elbow.

COMMENTARY TABLE – 4	
Table Capacity Derating for Elbows of Less Than 90 Degrees	
Elbow	Vent Table Capacity Reduction, Per Elbow (%)
0–45	5
>45–90	10

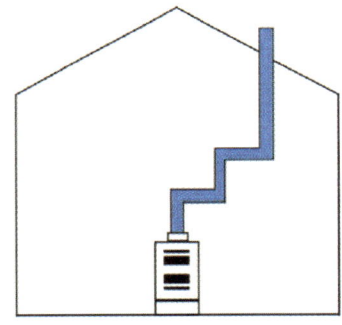

FIGURE 510.1.2
USE OF MORE THAN TWO ELBOWS

510.1.3 Zero Lateral. Zero lateral (L) shall apply to a straight vertical vent attached to a top outlet draft hood or flue collar. [NFPA 54:13.1.4]

🔧 This section does not permit the use of elbows in a venting system where the zero lateral is used (see **Figure 510.1.3**). If elbows are needed, the table rows for 2 feet (0.6 m) lateral length must be used. Elbows in a vent system with zero offset are not common but may be needed to route the vent to avoid a building obstruction, such as a beam.

510.1.4 High-Altitude Installations. Sea level input ratings shall be used where determining maximum capacity for high-altitude installation. Actual input (derated for altitude) shall be used for determining minimum capacity for high-altitude installation. [NFPA 54:13.1.5]

🔧 Using the sea level input rating for the maximum capacity, as required here, is a conservative measure because

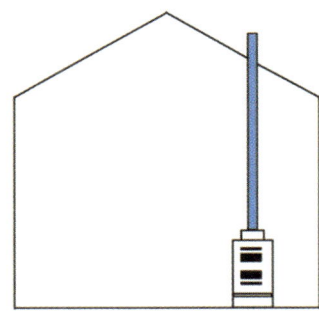

FIGURE 510.1.3
ZERO LATERAL USED

less draft is produced at high altitudes. The derating process will also make condensation more likely. The reduced input rate should be used for minimum capacity.

510.1.5 Multiple Input Ratings. For appliances with more than one input rate, the minimum vent capacity (FAN Min) determined from the tables shall be less than the lowest appliance input rating, and the maximum vent capacity (FAN Max/NAT Max) determined from the tables shall exceed the highest appliance rating input. [NFPA 54:13.1.6]

🔧 If the appliance has multiple input ratings, the minimum capacity is determined with the minimum input rate while the maximum is determined with the maximum input rate.

510.1.6 Corrugated Chimney Liner Reduction. Listed corrugated metallic chimney liner systems in masonry chimneys shall be sized by using Table 510.1.2(1) or Table 510.1.2(2) for Type B vents with the maximum capacity reduced by 20 percent (0.80 x maximum capacity) and the minimum capacity as shown in Table 510.1.2(1) or Table 510.1.2(2).

Corrugated metallic liner systems installed with bends or offsets shall have their maximum capacity further reduced in accordance with Section 510.1.2. The 20 percent reduction for corrugated metallic chimney liner systems includes an allowance for one long radius 90 degree (1.57 rad) turn at the bottom of the liner. [NFPA 54:13.1.7]

🔧 Because properly installed corrugated chimney liners have a heat loss similar to a Type B vent, they are sized using Table 510.1.2(1) or 510.1.2(2). However, such liners' corrugations and their tendency to spiral in the chimney require a 20-percent maximum capacity reduction. Many liners begin at the breaching and then bend up vertically. This 90-degree elbow at the beginning of the liner is included in the table capacity (see **Figure 509.5.1c**).

510.1.7 Connection to Chimney Liners. Connections between chimney liners and listed double-wall connectors shall be made with listed adapters designed for such purposes. [NFPA 54:13.1.8]

510.1.8 Vertical Vent Upsizing Using 7 x Rule. Where the vertical vent has a larger diameter than the vent connector, the vertical vent diameter shall be used to determine the minimum vent capacity, and the connector diameter shall be used to determine the maximum vent capacity. The flow area of the vertical vent shall not exceed seven times the flow area of the listed appliance categorized vent area, flue collar area, or draft hood outlet area unless designed in accordance with approved engineering methods. [NFPA 54:13.1.9]

In a vent system with a larger vent than the vent connector, the draft is limited by the small diameter of the connector (see **Figure 510.1.8**). Condensation will begin in the larger diameter. The maximum and minimum capacities must be determined accordingly. Practical limits exist as to how large the vertical vent may be relative to its source of vent gas flow; therefore, the flow area of the vent may not be more than seven times the flow area of the outlet of the appliance or draft hood.

A sudden large expansion of the vent system diameter at the vertical portion creates a pressure drop that may limit the draft and encourage condensation. This is the purpose of the limitations on vertical vent size versus vent connector size.

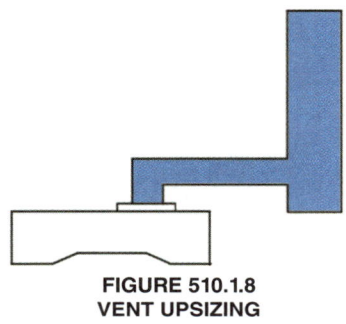

**FIGURE 510.1.8
VENT UPSIZING**

510.1.9 Draft Hood Conversion Accessories. Draft hood conversion accessories for use with masonry chimneys venting listed Category I fan-assisted appliances shall be listed and installed in accordance with the listed accessory manufacturer's installation instructions. [NFPA 54:13.1.10]

These are add-on kits that convert a fan-assisted appliance to a draft hood appliance by providing an opening for dilution air from the room to enter the venting system. A draft hood conversion accessory is a listed component that is usually provided by the appliance manufacturer. The accessory must be installed according to the instructions included with the kit. The addition of the draft hood kit allows the appliance to be sized as a draft hood appliance, which may be beneficial when venting into a masonry chimney, especially exterior masonry chimneys.

510.1.10 Chimney and Vent Locations. Table 510.1.2(1) through Table 510.1.2(5) shall be used only for chimneys and vents not exposed to the outdoors below the roof line. A Type B vent or listed chimney lining system passing through an unused masonry chimney flue shall not be considered to be exposed to the outdoors. Where vents extend outdoors above the roof more than 5 feet (1524 mm) higher than required by Table 509.6.1, and where vents terminate in accordance with Section 509.6.1(2), the outdoor portion of the vent shall be enclosed as required by this paragraph for vents not considered to be exposed to the outdoors, or such venting system shall be engineered. A Type B vent passing through an unventilated enclosure or chase insulated to a value of not less than R-8 shall not be considered to be exposed to the outdoors. Table 510.1.2(3) in combination with Table 510.1.2(6) shall be used for clay-tile-lined exterior masonry chimneys, provided all of the following requirements are met:

(1) The vent connector is Type B double wall.

(2) The vent connector length is limited to 18 inches per inch (18 mm/mm) of vent connector diameter.

(3) The appliance is draft hood-equipped.

(4) The input rating is less than the maximum capacity given in Table 510.1.2(3).

(5) For a water heater, the outdoor design temperature shall be not less than 5°F (-15°C).

(6) For a space-heating appliance, the input rating is greater than the minimum capacity given by Table 510.1.2(6). [NFPA 54:13.1.11]

This section specifies which tables to use when sizing chimneys and vents not exposed to the outdoors below the roof line. Special attention is paid to Type B vents or listed chimney lining systems passing through a masonry chimney flu or a chase insulated to a value of at least R8. Such installations are not considered as being exposed to the outdoors (see **Figure 510.1.10a**).

When sizing a clay-tile-lined exterior masonry chimney, Table 510.1.2(3) in combination with Table 510.1.2(6) shall be used, provided the six conditions are met. To size chimneys and vents serving only water heaters, the outdoor design temperature must be greater than 5°F. See **Figure 510.1.10** for a map of the United States showing the temperature zones. In colder climates, either the chimney must be lined with a metallic liner or another vent must be used. To size chimneys and vents serving space heating appliances, Table 510.1.2(6) for the minimum capacity and Table 510.1.2(3) for the maximum vent capacity will be used.

Note that only Type B vent connectors are allowed to minimize heat loss through vent connectors. Water heaters are treated differently from heating appliances because their total operating hours are not sufficient to keep a chimney warm in cold periods, resulting in condensation of water in the chimney vent.

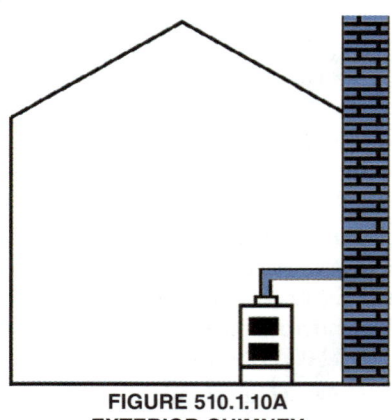

**FIGURE 510.1.10A
EXTERIOR CHIMNEY**

510.1.11 Corrugated Vent Connector Size. Corrugated vent connectors shall not be smaller than the listed appliance categorized vent diameter, flue collar diameter, or draft hood outlet diameter. [NFPA 54:13.1.12]

🔧 There is no capacity reduction required when using corrugated vent connectors as there is when using corrugated chimney liners.

510.1.12 Upsizing. Vent connectors shall not be upsized more than two sizes exceeding the listed appliance categorized vent diameter, flue collar diameter, or draft hood outlet diameter. [NFPA 54:13.1.13]

🔧 A sudden large expansion of the vent system diameter at the vertical portion creates a pressure drop that may limit the draft and encourage condensation. Therefore, there is a limit placed on upsizing.

510.1.13 Single Run of Vent. In a single run of vent or vent connector, more than one diameter and type shall be permitted to be used, provided that the sizes and types are permitted by the tables. [NFPA 54:13.1.14]

510.1.14 Interpolation. Interpolation shall be permitted in calculating capacities for vent dimensions that fall between table entries. [NFPA 54:13.1.15]

🔧 If the installation dimensions fall between two table entries for which there are defined values, the designer may calculate the "in between" value. This is called "interpolation." An example follows.

Table 510.1.2(1) provides the capacity of a three-inch diameter, 15-foot-high vent with zero lateral serving a natural draft appliance of 58,000 Btu/hr, and of a three-inch diameter, 20-foot-high vent with zero lateral of 61,000 Btu/hr. The capacity of a 3-inch diameter, 18-foot-high vent with zero lateral can be interpolated by taking three-fifths of the difference between 58,000 and 61,000 and adding it to the capacity of the 15-foot-high vent. In this case, the difference is 3,000 Btu/hr. Three-fifths of 3,000 is 1,800, making the capacity of the 18-foot vent is 59,800 Btu/hr.

510.1.15 Extrapolation. Extrapolation beyond the table entries shall not be permitted. [NFPA 54:13.1.16]

🔧 Extrapolation is estimating a value outside of the table. For example, Table 510.1.2(2) provides vent capacities for vents up to 100 feet (30 m). The reader is not permitted to use the numbers in Table 510.1.2(2) to estimate the capacity of a vent that is 110 feet (33 m) high.

510.1.16 Engineering Methods. For vent heights lower than 6 feet (1829 mm) and higher than shown in the tables, engineering methods shall be used to calculate vent capacities. [NFPA 54:13.1.17]

🔧 This section reinforces the requirement that vent heights outside the parameters of the tables in Section 510.0 must be calculated and that the tables cannot be used.

510.1.17 Height Entries. Where the actual height of a vent falls between entries in the height column of the applicable table in Table 510.1.2(1) through Table 510.1.2(6), one of the following shall be used:

(1) Interpolation.

(2) The lower appliance input rating shown in the table entries for FAN Max and NAT Max column values; and the higher appliance input rating for the FAN Min column values. [NFPA 54:13.1.18]

510.2 Multiple Appliance Vent Table 510.2(1) through Table 510.2(9). Venting Table 510.2(1) through Table 510.2(9) shall not be used where obstructions are installed in the venting system. The installation of vents serving listed appliances with vent dampers shall be in accordance with the appliance manufacturer's instructions or in accordance with the following:

(1) The maximum capacity of the vent connector shall be determined using the NAT Max column.

(2) The maximum capacity of the vertical vent or chimney shall be determined using the FAN + NAT column when the second appliance is a fan-assisted appliance, or the NAT + NAT column when the second appliance is equipped with a draft hood.

(3) The minimum capacity shall be determined as if the appliance were a fan-assisted appliance, as follows:

(a) The minimum capacity of the vent connector shall be determined using the FAN Min column.

(b) The FAN + FAN column shall be used when the second appliance is a fan-assisted appliance, and the FAN + NAT column shall be used when the second appliance is equipped with a draft hood, to determine whether the vertical vent or chimney configuration is not permitted (NA). Where the vent configuration is NA, the vent configuration shall not be permitted and an alternative venting configuration shall be utilized. [NFPA 54:13.2.1]

🔧 Tables 510.2(1) through 510.2(9) were developed with the assumption that there are no restrictions in the path of the vent gas flow. An additional assumption is that there is no deliberate attempt to remove heat; therefore, devices such as heat economizers may not be used in conjunction with the tables. Vent dampers installed as part of a listed appliance are to be installed in accordance with the manufacturers' instructions or the general requirements of this section.

Guidance for using the venting tables with draft hood appliances equipped with a vent damper is provided here. An example of such an appliance is a boiler, which uses a vent damper to obtain higher efficiencies as follows:

- The maximum capacity of the vent connector is found in the NAT Max column of the "Vent Connector Capacity" sections of Tables 510.2(1) through 510.2(4).

- The maximum capacity of the vent is found using the FAN Max column of the "Common Vent Capacity" sections of Tables 510.2(1) through 510.2(5).

The following treats the vent as one serving a natural draft appliance when the appliance is operating:

- The minimum capacity of the vent connector is found using the FAN Min column of the "Vent Connector Capacity" sections of Tables 510.2(1) through 510.2(4).

WATER HEATERS

The minimum capacity of the vent depends on the venting category of the second appliance (the first appliance is the boiler with the vent damper or other obstruction):

- If the second appliance is a fan-assisted appliance, the FAN + FAN column of the "Common Vent Capacity" portion of Tables 510.2(1) through 510.2(4) is used. If the second appliance is equipped with a draft hood, the FAN + NAT column of the "Common Vent Capacity" of Tables 510.2(1) through 510.2(4) is used.

This treats the vent as one serving a fan-assisted (increased efficiency) appliance when the appliance is not operating. The reason for this unusual combination is that the appliance operates as a natural draft appliance, but when the appliance is not operating and the vent damper is closed, it is similar to a fan-assisted (energy-saving) appliance in its propensity to condensate water.

510.2.1 Vent Connector Maximum Length. The maximum vent connector horizontal length shall be 18 inches per inch (18 mm/mm) of connector diameter as shown in Table 510.2.1, or as permitted by Section 510.2.2. [NFPA 54:13.2.2]

The multiple appliance venting tables do not contain lateral length rows like those found in the single appliance tables. This section limits the vent connector length to 18 inches per inch of connector diameter as shown in Table 510.2.1. If these lengths are exceeded, Section 510.2.2 provides guidance on derating the tables accordingly.

**TABLE 510.2.1
VENT CONNECTOR MAXIMUM LENGTH
[NFPA 54: TABLE 13.2.2]**

CONNECTOR DIAMETER (inches)	MAXIMUM CONNECTOR HORIZONTAL LENGTH (feet)
3	4½
4	6
5	7½
6	9
7	10½
8	12
9	13½
10	15
12	18
14	21
16	24
18	27
20	30
22	33
24	36

For SI units: 1 inch = 25.4 mm, 1 foot = 304.8 mm

510.2.2 Vent Connector Exceeding Maximum Length. The vent connector shall be routed to the vent utilizing the shortest possible route. Connectors with longer horizontal lengths than those listed in Table 510.2.1 are permitted under the following conditions:

(1) The maximum capacity (FAN Max or NAT Max) of the vent connector shall be reduced 10 percent for each additional multiple of the length listed in Table 510.2.1. For example, the maximum length listed for a 4 inch (100 mm) connector is 6 feet (1829 mm). With a connector length greater than 6 feet (1829 mm) but not exceeding 12 feet (3658 mm), the maximum capacity must be reduced by 10 percent (0.90 x maximum vent connector capacity). With a connector length greater than 12 feet (3658 mm) but not exceeding 18 feet (5486 mm), the maximum capacity shall be reduced by 20 percent (0.80 x maximum vent capacity).

(2) For a connector serving a fan-assisted appliance, the minimum capacity (FAN Min) of the connector shall be determined by referring to the corresponding single appliance table. For Type B double-wall connectors, Table 510.1.2(1) shall be used. For single-wall connectors, Table 510.1.2(2) shall be used. The height (H) and lateral (L) shall be measured according to the procedures for a single appliance vent, as if the other appliances were not present. [NFPA 54:13.2.3]

The common venting tables (the lower portion of Tables 510.2(1) through 510.2(5) are designed with the assumption that the vent connector is no more than 18 inches long for each inch of diameter. Sections 510.2.1 and 510.2.2, summarized as follows, provides guidance on how to handle connectors that are longer than this:

- The maximum capacity for vent connectors that exceed 18 inches per inch of diameter is reduced by 10 percent for each length of 18 inches per inch of vent diameter.
- The minimum capacity for connectors serving fan-assisted appliances is determined differently for single-wall and Type B double-wall connectors, as described in Section 510.2.2.

510.2.3 Vent Connector Manifold. Where the vent connectors are combined prior to entering the vertical portion of the common vent to form a common vent manifold, the size of the common vent manifold and the common vent shall be determined by applying a 10 percent reduction (0.90 x maximum common vent capacity) to the common vent capacity part of the common vent tables. The length of the common vent manifold (LM) shall not exceed 18 inches per inch (18 mm/mm) of common vent diameter (D). [NFPA 54:13.2.4]

Section 510.2.3 considers the "common vent" to be the vertical vent or chimney only. A combination of vent connections prior to entering a vertical chimney or vent is considered to be a vent manifold [see **Figure G 101.2(11)** in Appendix G], and guidance for sizing such is located here. Note that a chimney liner system that comes out through the breaching, as many do, should be considered a manifold. The capacity of the vent manifold is reduced by 10 percent from the table value. Also note that the length of a vent manifold is limited to 18 inches per inch of vent manifold diameter.

510.2.4 Vent Offset. Where the common vertical vent is offset, the maximum capacity of the common vent shall be reduced in accordance with Section 510.2.5, and the horizontal length of the common vent offset shall not exceed 18 inches per inch (18 mm/mm) of common vent diameter (D). Where multiple offsets occur in a common vent, the total horizontal length of offsets combined shall not exceed 18 inches per inch (18 mm/mm) of the common vent diameter. [NFPA 54:13.2.5]

🔧 Vent offsets, illustrated in **Figure 510.2.4**, typically occur high in the venting system and tend to reduce the draft produced because they slow the flow of vent gases. This reduction in draft is offset by the requirement for a 10-percent capacity reduction required in Section 510.2.5.

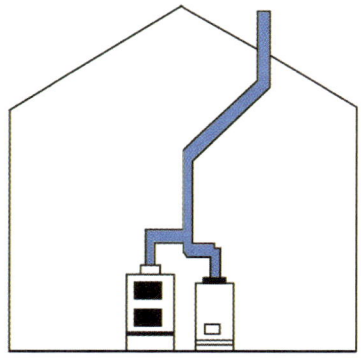

FIGURE 510.2.4
VENT OFFSETS

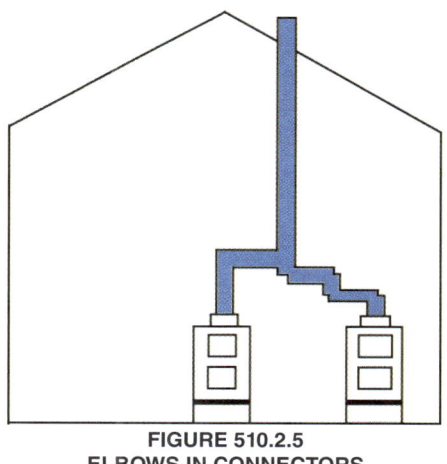

FIGURE 510.2.5
ELBOWS IN CONNECTORS

510.2.5 Elbow Reduction. For each elbow up to and including 45 degrees (0.79 rad) in the common vent, the maximum common vent capacity listed in the venting tables shall be reduced by 5 percent. For each elbow exceeding 45 degrees (0.79 rad) up to and including 90 degrees (1.57 rad), the maximum common vent capacity listed in the venting tables shall be reduced by 10 percent. [NFPA 54:13.2.6]

🔧 The sizing requirements were designed with the assumption that up to two 90-degree turns were part of the venting system, except for the zero lateral length case. The zero lateral case was assumed to extend straight from the appliance outlet to the vent termination. Adding additional elbows to the system is possible.

The requirements for derating the table capacities for elbows were revised to recognize elbows of less than 90 degrees. The new derating factors are shown in **Commentary Table – 4**. Previously, a 10-percent reduction was required for each elbow, regardless of its angle. This revision, which was based on input from the Gas Technology Institute's contractor, Battelle, provides flexibility to installers forced to use vent offsets and encourages the use of elbows of less than 90 degrees. See **Figure 510.2.5**.

When sizing a common vent with elbows, please note the following important factors:

• The tables include two 90-degree elbows. Do not derate the common vent table values for venting systems with one or two elbows.
• For each additional elbow over two in the common vent, derate the common vent table value by 5 or 10 percent, depending on the angle of the elbow.

510.2.6 Elbows in Connectors. The vent connector capacities listed in the common vent sizing tables include allowance for two 90 degree (1.57 rad) elbows. For each additional elbow up to and including 45 degrees (0.79 rad), the maximum vent connector capacity listed in the venting tables shall be reduced by 5 percent. For each elbow greater than 45 degrees (0.79 rad) up to and including 90 degrees (1.57 rad), the maximum vent connector capacity listed in the venting tables shall be reduced by 10 percent. [NFPA 54:13.2.7]

🔧 Please see commentary in Section 510.2.5.

510.2.7 Common Vent Minimum Size. The cross-sectional area of the common vent shall be equal to or greater than the cross-sectional area of the largest connector. [NFPA 54:13.2.8]

510.2.8 Tee and Wye Fittings. Tee and wye fittings connected to a common gas vent shall be considered as part of the common gas vent and constructed of materials consistent with that of the common gas vent. [NFPA 54:13.2.9]

510.2.9 Size of Fittings. At the point where tee or wye fittings connect to a common gas vent, the opening size of the fitting shall be equal to the size of the common vent. Such fittings shall not be prohibited from having reduced size openings at the point of connection of appliance gas vent connectors. [NFPA 54:13.2.10]

510.2.10 High-Altitude Installations. Sea level input ratings shall be used where determining maximum capacity for high-altitude installation. Actual input (derated for altitude) shall be used for determining minimum capacity for high-altitude installation. [NFPA 54:13.2.11]

🔧 Using the sea level input rating required by this section for the maximum capacity is a conservative measure because less draft is produced at high altitudes. The derating process will also make condensation more likely. The reduced input rate should be used for the minimum capacity.

510.2.11 Connector Rise. The connector rise (R) for each appliance connector shall be measured from the draft hood outlet or flue collar to the centerline where the vent gas streams come together. [NFPA 54:13.2.12]

510.2.12 Vent Height. For multiple appliances located on one floor, the total height (H) shall be measured from the highest draft hood outlet or flue collar up to the level of the outlet of the common vent. [NFPA 54:13.2.13]

510.2.13 Multistory Installations. For multistory installations, the total height (H) for each segment of the system shall be the vertical distance between the highest draft hood outlet or flue collar entering that segment and the centerline of the next higher interconnection tee. [NFPA 54:13.2.14]

510.2.14 Size of Vents for Multistory Installations. The size of the lowest connector and of the vertical vent leading to the lowest interconnection of a multistory system shall be in accordance with Table 510.1.2(1) or Table 510.1.2(2) for available total height (H) up to the lowest interconnection. [NFPA 54:13.2.15]

510.2.15 Vent Type Multistory Installations. Where used in multistory systems, vertical common vents shall be Type B double-wall and shall be installed with a listed vent cap. [NFPA 54:13.2.16]

510.2.16 Offsets in Multistory Installations. Offsets in multistory common vent systems shall be limited to a single offset in each system, and systems with an offset shall comply with the following:

(1) The offset angle shall not exceed 45 degrees (0.79 rad) from vertical.

(2) The horizontal length of the offset shall not exceed 18 inches per inch (18 mm/mm) of common vent diameter of the segment in which the offset is located.

(3) For the segment of the common vertical vent containing the offset, the common vent capacity listed in the common venting tables shall be reduced by 20 percent (0.80 x maximum common vent capacity).

(4) A multistory common vent shall not be reduced in size above the offset. [NFPA 54:13.2.17]

510.2.17 Vertical Vent Size Limitation. Where two or more appliances are connected to a vertical vent or chimney, the flow area of the largest section of vertical vent or chimney shall not exceed seven times the smallest listed appliance categorized vent areas, flue collar area, or draft hood outlet area unless designed in accordance with approved engineering methods. [NFPA 54:13.2.18]

510.2.18 Multiple Input Ratings. For appliances with more than one input rate, the minimum vent connector capacity (FAN Min) determined from the tables shall be less than the lowest appliance input rating, and the maximum vent connector capacity (FAN Max or NAT Max) determined from the table shall exceed the highest appliance input rating. [NFPA 54:13.2.19]

If the appliance has multiple input rates, the minimum capacity is determined with the minimum input rate while the maximum capacity is determined with the maximum input rate.

510.2.19 Corrugated Metallic Chimney Liner Reduction. Listed corrugated metallic chimney liner systems in masonry chimneys shall be sized by using Table 510.2(1) or Table 510.2(2) for Type B vents, with the maximum capacity reduced by 20 percent (0.80 x maximum capacity) and the minimum capacity as shown in Table 510.2(1) or Table 510.2(2). Corrugated metallic liner systems installed with bends or offsets shall have their maximum capacity further reduced in accordance with Section 510.2.5 and Section 510.2.6. The 20 percent reduction for corrugated metallic chimney liner systems includes an allowance for one long radius 90 degree (1.57 rad) turn at the bottom of the liner. [NFPA 54:13.2.20]

Because properly installed corrugated liners have a heat loss similar to that of a Type B vent, they are sized using Table 510.2(1) or 510.2(2), as stated in Section 510.2.17. However, such liners' corrugations and their tendency to spiral in the chimney require a 20-percent maximum capacity reduction.

Many liners begin at the breaching and then bend up vertically. This 90-degree elbow at the beginning of the liner is included in the table capacity.

510.2.20 Chimneys and Vents. Table 510.2(1) through Table 510.2(5) shall be used only for chimneys and vents not exposed to the outdoors below the roof line. A Type B vent or listed chimney lining system passing through an unused masonry chimney flue shall not be considered to be exposed to the outdoors. A Type B vent passing through an unventilated enclosure or chase insulated to a value of not less than R-8 shall not be considered to be exposed to the outdoors. Where vents extend outdoors above the roof more than 5 feet (1524 mm) higher than required by Table 509.6.1, and where vents terminate in accordance with Section 509.6.1(2), the outdoor portion of the vent shall be enclosed as required by this section for vents not considered to be exposed to the outdoors, or such venting system shall be engineered. Table 510.2(6) through Table 510.2(9) shall be used for clay-tile-lined exterior masonry chimneys, provided all the following conditions are met:

(1) The vent connector is Type B double-wall.

(2) At least one appliance is draft hood-equipped.

(3) The combined appliance input rating is less than the maximum capacity given by Table 510.2(6) (for NAT + NAT) or Table 510.2(8) (for FAN + NAT).

(4) The input rating of each space-heating appliance is greater than the minimum input rating given by Table 510.2(7) (for NAT + NAT) or Table 510.2(9) (for FAN + NAT).

(5) The vent connector sizing is in accordance with Table 510.2(3). [NFPA 54:13.2.22]

Guidance is provided in this section for using exterior masonry chimneys for heating appliances. These limitations are needed because of the high heat loss in chimneys exposed to the outdoors below the roofline in cold climates. Note that these restrictions do not apply to chimneys and vents not exposed to the outdoors below the roofline because these chimneys are considered to be interior chimneys.

510.2.21 Vent Connector Sizing. Vent connectors shall not be increased more than two sizes greater than the listed appliance categorized vent diameter, flue collar diameter, or draft hood outlet diameter. Vent connectors for draft hood-equipped appliances shall not be smaller than the draft hood outlet diameter. Where a vent connector size(s) determined from the tables for fan-assisted appliance(s) is smaller than the flue collar diameter, the use of the smaller size(s) shall be permitted provided that the installation complies with all of the following conditions:

(1) Vent connectors for fan-assisted appliance flue collars 12 inches (300 mm) in diameter or smaller are not reduced by more than one table size [e.g., 12 inches to 10 inches (300 mm to 250 mm) is a one size reduction] and those larger than 12 inches (300 mm) in diameter are not reduced more than two table sizes [e.g., 24 inches to 20 inches (600 mm to 500 mm) is a two size reduction].

(2) The fan-assisted appliance(s) is common vented with a draft hood-equipped appliance(s).

(3) The vent connector has a smooth interior wall. [NFPA 54:13.2.24]

A sudden, large expansion of the vent connector diameter creates a pressure drop that may limit the draft and encourage condensation. Therefore, this section places a limit on upsizing, limiting the vent connector diameter to two sizes larger than one of the following, whichever is applicable:

- Appliance-categorized vent connector diameter;
- Flue collar diameter; or
- Draft hood outlet diameter.

Note that an appliance includes either a draft hood or flue collar for connection to the vent or vent connector. The term "appliance-categorized vent diameter/area" is defined as follows:

The minimum vent area/diameter permissible for Category I appliances to maintain a nonpositive vent static pressure when tested in accordance with nationally recognized standards.

Therefore, where connecting an appliance to a vent connector, use the applicable tables in Section 510.0 to find the minimum size vent connector. Note that only smooth wall vent connectors may be downsized. It would also be advisable to check with the manufacturer to make sure it allows a downsizing.

510.2.22 Combination of Pipe Types and Sizes. All combinations of pipe sizes, single-wall metal pipe, and double-wall metal pipe shall be allowed within any connector run(s) or within the common vent, provided ALL of the appropriate tables permit ALL of the desired sizes and types of pipe, as if they were used for the entire length of the subject connector or vent. Where single-wall and Type B double-wall metal pipes are used for vent connectors within the same venting system, the common vent shall be sized using Table 510.2(2) or Table 510.2(4) as appropriate. [NFPA 54:13.2.25]

This section effectively requires the installer to check the minimum and maximum capacities for the vent section for each vent type, as if the entire vent section were made of that size. For example, suppose a vent connector was half single-wall and half Type B vent. The installer must show that a vent connector of that length would be allowed if it was all single-wall and also if it was all Type B vent.

510.2.23 Multiple Connector and Vent Sizes. Where a table permits more than one diameter of pipe to be used for a connector or vent, the permitted sizes shall be permitted to be used. [NFPA 54:13.2.26]

510.2.24 Interpolation. Interpolation shall be permitted in calculating capacities for vent dimensions that fall between table entries. [NFPA 54:13.2.27]

If the installation dimensions fall between two table entries for which there are defined values, the installer may calculate the "in between" value. This is called "interpolation." See Section 510.1.14 for an example of interpolation.

510.2.25 Extrapolation. Extrapolation beyond the table entries shall not be permitted. [NFPA 54:13.2.28]

Extrapolation is estimating the values outside of the tables. For example, Table 510.2(2) provides common vent capacities for vents up to 100 feet (30 m). The installer is not allowed to estimate the capacity of a vent 110 feet (33 m) high.

510.2.26 Engineering Methods. For vent heights lower than 6 feet (1829 mm) and higher than shown in the tables, engineering methods shall be used to calculate vent capacities. [NFPA 54:13.2.29]

This Section reinforces the requirement that vent heights outside the parameters of the tables must be calculated and that the tables cannot be used.

510.2.27 Height Entries. Where the actual height of a vent falls between entries in the height column of the applicable table in Table 510.2(1) through Table 510.2(9), one of the following shall be used:

(1) Interpolation.

(2) The lower appliance input rating shown in the table entries for FAN Max and NAT Max column values; and the higher appliance input rating for the FAN Min column values. [NFPA 54:13.2.30]

WATER HEATERS

99% Winter Design Temperatures for the Contiguous United States

This map is a necessarily generalized guide to temperatures in the contiguous United States. Temperatures shown for areas such as mountainous regions and large urban centers may not be accurate. The data used to develop this map are from the 1993 ASHRAE Handbook — Fundamentals (Chapter 24, Table 1: Climate Conditions for the United States).

For 99% winter design temperatures in Alaska, consult the ASHRAE Handbook — Fundamentals.

99% winter design temperatures for Hawaii are greater than 37°F.

For SI units: °C = (°F-32)/1.8

FIGURE 510.1.10
RANGE OF WINTER DESIGN TEMPERATURES USED IN ANALYZING EXTERIOR MASONRY CHIMNEYS IN THE UNITED STATES
[NFPA 54: FIGURE F.2.4]

TABLE 510.1.2(1)
TYPE B DOUBLE-WALL GAS VENT [NFPA 54: TABLE 13.1(a)]*

		NUMBER OF APPLIANCES:			SINGLE								
		APPLIANCE TYPE:			CATEGORY I								
		APPLIANCE VENT CONNECTION:			CONNECTED DIRECTLY TO VENT								

		VENT DIAMETER – D (inch)											
		3			4			5			6		7
		APPLIANCE INPUT RATING IN THOUSANDS OF BTU PER HOUR											
HEIGHT H (feet)	LATERAL L (feet)	FAN		NAT	FAN		NAT	FAN		NAT	FAN	NAT	FAN	NAT
		Min	Max	Max	Min	Max	Max	Min	Max	Max	Min	Max	Max	Min	Max	Max

H	L	FAN Min	FAN Max	NAT Max	FAN Min	FAN Max	NAT Max	FAN Min	FAN Max	NAT Max	FAN Min	FAN Max	NAT Max	FAN Min	FAN Max	NAT Max
6	0	0	78	46	0	152	86	0	251	141	0	375	205	0	524	285
	2	13	51	36	18	97	67	27	157	105	32	232	157	44	321	217
	4	21	49	34	30	94	64	39	153	103	50	227	153	66	316	211
	6	25	46	32	36	91	61	47	149	100	59	223	149	78	310	205
8	0	0	84	50	0	165	94	0	276	155	0	415	235	0	583	320
	2	12	57	40	16	109	75	25	178	120	28	263	180	42	365	247
	5	23	53	38	32	103	71	42	171	115	53	255	173	70	356	237
	8	28	49	35	39	98	66	51	164	109	64	247	165	84	347	227
10	0	0	88	53	0	175	100	0	295	166	0	447	255	0	631	345
	2	12	61	42	17	118	81	23	194	129	26	289	195	40	402	273
	5	23	57	40	32	113	77	41	187	124	52	280	188	68	392	263
	10	30	51	36	41	104	70	54	176	115	67	267	175	88	376	245
15	0	0	94	58	0	191	112	0	327	187	0	502	285	0	716	390
	2	11	69	48	15	136	93	20	226	150	22	339	225	38	475	316
	5	22	65	45	30	130	87	39	219	142	49	330	217	64	463	300
	10	29	59	41	40	121	82	51	206	135	64	315	208	84	445	288
	15	35	53	37	48	112	76	61	195	128	76	301	198	98	429	275
20	0	0	97	61	0	202	119	0	349	202	0	540	307	0	776	430
	2	10	75	51	14	149	100	18	250	166	20	377	249	33	531	346
	5	21	71	48	29	143	96	38	242	160	47	367	241	62	519	337
	10	28	64	44	38	133	89	50	229	150	62	351	228	81	499	321
	15	34	58	40	46	124	84	59	217	142	73	337	217	94	481	308
	20	48	52	35	55	116	78	69	206	134	84	322	206	107	464	295
30	0	0	100	64	0	213	128	0	374	220	0	587	336	0	853	475
	2	9	81	56	13	166	112	14	283	185	18	432	280	27	613	394
	5	21	77	54	28	160	108	36	275	176	45	421	273	58	600	385
	10	27	70	50	37	150	102	48	262	171	59	405	261	77	580	371
	15	33	64	NA	44	141	96	57	249	163	70	389	249	90	560	357
	20	56	58	NA	53	132	90	66	237	154	80	374	237	102	542	343
	30	NA	NA	NA	73	113	NA	88	214	NA	104	346	219	131	507	321
50	0	0	101	67	0	216	134	0	397	232	0	633	363	0	932	518
	2	8	86	61	11	183	122	14	320	206	15	497	314	22	715	445
	5	20	82	NA	27	177	119	35	312	200	43	487	308	55	702	438
	10	26	76	NA	35	168	114	45	299	190	56	471	298	73	681	426
	15	59	70	NA	42	158	NA	54	287	180	66	455	288	85	662	413
	20	NA	NA	NA	50	149	NA	63	275	169	76	440	278	97	642	401
	30	NA	NA	NA	69	131	NA	84	250	NA	99	410	259	123	605	376
100	0	NA	NA	NA	0	218	NA	0	407	NA	0	665	400	0	997	560
	2	NA	NA	NA	10	194	NA	12	354	NA	13	566	375	18	831	510
	5	NA	NA	NA	26	189	NA	33	347	NA	40	557	369	52	820	504
	10	NA	NA	NA	33	182	NA	43	335	NA	53	542	361	68	801	493
	15	NA	NA	NA	40	174	NA	50	321	NA	62	528	353	80	782	482
	20	NA	NA	NA	47	166	NA	59	311	NA	71	513	344	90	763	471
	30	NA	NA	NA	NA	NA	NA	78	290	NA	92	483	NA	115	726	449
	50	NA	NA	NA	NA	NA	NA	NA	NA	NA	147	428	NA	180	651	405

For SI units: 1 inch = 25.4 mm, 1 foot = 304.8 mm, 1000 British thermal units per hour = 0.293 kW, 1 square inch = 0.000645 m^2
* NA: Not applicable.

WATER HEATERS

TABLE 510.1.2(1)
TYPE B DOUBLE-WALL GAS VENT [NFPA 54: TABLE 13.1(a)] (continued)

							NUMBER OF APPLIANCES:	SINGLE						
							APPLIANCE TYPE:	CATEGORY I						
							APPLIANCE VENT CONNECTION:	CONNECTED DIRECTLY TO VENT						
		VENT DIAMETER – D (inch)												
		8		9			10			12		14		
		APPLIANCE INPUT RATING IN THOUSANDS OF BTU PER HOUR												
HEIGHT H (feet)	LATERAL L (feet)	FAN		NAT	FAN		NAT	FAN		NAT	FAN		NAT	FAN	NAT
		Min	Max	Max	Min	Max	Max	Min	Max	Max	Min	Max	Max	Min	Max	Max

HEIGHT H (feet)	LATERAL L (feet)	FAN Min	FAN Max	NAT Max	FAN Min	FAN Max	NAT Max	FAN Min	FAN Max	NAT Max	FAN Min	FAN Max	NAT Max	FAN Min	FAN Max	NAT Max
6	0	0	698	370	0	897	470	0	1121	570	0	1645	850	0	2267	1170
	2	53	425	285	63	543	370	75	675	455	103	982	650	138	1346	890
	4	79	419	279	93	536	362	110	668	445	147	975	640	191	1338	880
	6	93	413	273	110	530	354	128	661	435	171	967	630	219	1330	870
8	0	0	780	415	0	1006	537	0	1261	660	0	1858	970	0	2571	1320
	2	50	483	322	60	619	418	71	770	515	98	1124	745	130	1543	1020
	5	83	473	313	99	607	407	115	758	503	154	1110	733	199	1528	1010
	8	99	463	303	117	596	396	137	746	490	180	1097	720	231	1514	1000
10	0	0	847	450	0	1096	585	0	1377	720	0	2036	1060	0	2825	1450
	2	48	533	355	57	684	457	68	852	560	93	1244	850	124	1713	1130
	5	81	522	346	95	671	446	112	839	547	149	1229	829	192	1696	1105
	10	104	504	330	122	651	427	142	817	525	187	1204	795	238	1669	1080
15	0	0	970	525	0	1263	682	0	1596	840	0	2380	1240	0	3323	1720
	2	45	633	414	53	815	544	63	1019	675	86	1495	985	114	2062	1350
	5	76	620	403	90	800	529	105	1003	660	140	1476	967	182	2041	1327
	10	99	600	386	116	777	507	135	977	635	177	1446	936	227	2009	1289
	15	115	580	373	134	755	491	155	953	610	202	1418	905	257	1976	1250
20	0	0	1057	575	0	1384	752	0	1756	930	0	2637	1350	0	3701	1900
	2	41	711	470	50	917	612	59	1150	755	81	1694	1100	107	2343	1520
	5	73	697	460	86	902	599	101	1133	738	135	1674	1079	174	2320	1498
	10	95	675	443	112	877	576	130	1105	710	172	1641	1045	220	2282	1460
	15	111	654	427	129	853	557	150	1078	688	195	1609	1018	248	2245	1425
	20	125	634	410	145	830	537	167	1052	665	217	1578	990	273	2210	1390
30	0	0	1173	650	0	1548	855	0	1977	1060	0	3004	1550	0	4252	2170
	2	33	826	535	42	1072	700	54	1351	865	74	2004	1310	98	2786	1800
	5	69	811	524	82	1055	688	96	1332	851	127	1981	1289	164	2759	1775
	10	91	788	507	107	1028	668	125	1301	829	164	1944	1254	209	2716	1733
	15	105	765	490	124	1002	648	143	1272	807	187	1908	1220	237	2674	1692
	20	119	743	473	139	977	628	160	1243	784	207	1873	1185	260	2633	1650
	30	149	702	444	171	929	594	195	1189	745	246	1807	1130	305	2555	1585
50	0	0	1297	708	0	1730	952	0	2231	1195	0	3441	1825	0	4934	2550
	2	26	975	615	33	1276	813	41	1620	1010	66	2431	1513	86	3409	2125
	5	65	960	605	77	1259	798	90	1600	996	118	2406	1495	151	3380	2102
	10	86	935	589	101	1230	773	118	1567	972	154	2366	1466	196	3332	2064
	15	100	911	572	117	1203	747	136	1536	948	177	2327	1437	222	3285	2026
	20	113	888	556	131	1176	722	151	1505	924	195	2288	1408	244	3239	1987
	30	141	844	522	161	1125	670	183	1446	876	232	2214	1349	287	3150	1910
100	0	0	1411	770	0	1908	1040	0	2491	1310	0	3925	2050	0	5729	2950
	2	21	1155	700	25	1536	935	30	1975	1170	44	3027	1820	72	4313	2550
	5	60	1141	692	71	1519	926	82	1955	1159	107	3002	1803	136	4282	2531
	10	80	1118	679	94	1492	910	108	1923	1142	142	2961	1775	180	4231	2500
	15	93	1095	666	109	1465	895	126	1892	1124	163	2920	1747	206	4182	2469
	20	105	1073	653	122	1438	880	141	1861	1107	181	2880	1719	226	4133	2438
	30	131	1029	627	149	1387	849	170	1802	1071	215	2803	1663	265	4037	2375
	50	197	944	575	217	1288	787	241	1688	1000	292	2657	1550	350	3856	2250

For SI units: 1 inch = 25.4 mm, 1 foot = 304.8 mm, 1000 British thermal units per hour = 0.293 kW, 1 square inch = 0.000645 m^2

TABLE 510.1.2(1)
TYPE B DOUBLE-WALL GAS VENT [NFPA 54: TABLE 13.1(a)] (continued)

		NUMBER OF APPLIANCES:				SINGLE						
		APPLIANCE TYPE:				CATEGORY I						
		APPLIANCE VENT CONNECTION:				CONNECTED DIRECTLY TO VENT						
		VENT DIAMETER – D (inch)										
		16			18			20		22		24

HEIGHT H (feet)	LATERAL L (feet)	FAN Min	FAN Max	NAT Max	FAN Min	FAN Max	NAT Max	FAN Min	FAN Max	NAT Max	FAN Min	FAN Max	NAT Max	FAN Min	FAN Max	NAT Max
6	0	0	2983	1530	0	3802	1960	0	4721	2430	0	5737	2950	0	6853	3520
	2	178	1769	1170	225	2250	1480	296	2782	1850	360	3377	2220	426	4030	2670
	4	242	1761	1160	300	2242	1475	390	2774	1835	469	3370	2215	555	4023	2660
	6	276	1753	1150	341	2235	1470	437	2767	1820	523	3363	2210	618	4017	2650
8	0	0	3399	1740	0	4333	2220	0	5387	2750	0	6555	3360	0	7838	4010
	2	168	2030	1340	212	2584	1700	278	3196	2110	336	3882	2560	401	4634	3050
	5	251	2013	1330	311	2563	1685	398	3180	2090	476	3863	2545	562	4612	3040
	8	289	2000	1320	354	2552	1670	450	3163	2070	537	3850	2530	630	4602	3030
10	0	0	3742	1925	0	4782	2450	0	5955	3050	0	7254	3710	0	8682	4450
	2	161	2256	1480	202	2868	1890	264	3556	2340	319	4322	2840	378	5153	3390
	5	243	2238	1461	300	2849	1871	382	3536	2318	458	4301	2818	540	5132	3371
	10	298	2209	1430	364	2818	1840	459	3504	2280	546	4268	2780	641	5099	3340
15	0	0	4423	2270	0	5678	2900	0	7099	3620	0	8665	4410	0	10 393	5300
	2	147	2719	1770	186	3467	2260	239	4304	2800	290	5232	3410	346	6251	4080
	5	229	2696	1748	283	3442	2235	355	4278	2777	426	5204	3385	501	6222	4057
	10	283	2659	1712	346	3402	2193	432	4234	2739	510	5159	3343	599	6175	4019
	15	318	2623	1675	385	3363	2150	479	4192	2700	564	5115	3300	665	6129	3980
20	0	0	4948	2520	0	6376	3250	0	7988	4060	0	9785	4980	0	11 753	6000
	2	139	3097	2000	175	3955	2570	220	4916	3200	269	5983	3910	321	7154	4700
	5	219	3071	1978	270	3926	2544	337	4885	3174	403	5950	3880	475	7119	4662
	10	273	3029	1940	334	3880	2500	413	4835	3130	489	5896	3830	573	7063	4600
	15	306	2988	1910	372	3835	2465	459	4786	3090	541	5844	3795	631	7007	4575
	20	335	2948	1880	404	3791	2430	495	4737	3050	585	5792	3760	689	6953	4550
30	0	0	5725	2920	0	7420	3770	0	9341	4750	0	11 483	5850	0	13 848	7060
	2	127	3696	2380	159	4734	3050	199	5900	3810	241	7194	4650	285	8617	5600
	5	206	3666	2350	252	4701	3020	312	5863	3783	373	7155	4622	439	8574	5552
	10	259	3617	2300	316	4647	2970	386	5803	3739	456	7090	4574	535	8505	5471
	15	292	3570	2250	354	4594	2920	431	5744	3695	507	7026	4527	590	8437	5391
	20	319	3523	2200	384	4542	2870	467	5686	3650	548	6964	4480	639	8370	5310
	30	369	3433	2130	440	4442	2785	540	5574	3565	635	6842	4375	739	8239	5225
50	0	0	6711	3440	0	8774	4460	0	11 129	5635	0	13 767	6940	0	16 694	8430
	2	113	4554	2840	141	5864	3670	171	7339	4630	209	8980	5695	251	10 788	6860
	5	191	4520	2813	234	5826	3639	283	7295	4597	336	8933	5654	394	10 737	6818
	10	243	4464	2767	295	5763	3585	355	7224	4542	419	8855	5585	491	10 652	6749
	15	274	4409	2721	330	5701	3534	396	7155	4511	465	8779	5546	542	10 570	6710
	20	300	4356	2675	361	5641	3481	433	7086	4479	506	8704	5506	586	10 488	6670
	30	347	4253	2631	412	5523	3431	494	6953	4421	577	8557	5444	672	10 328	6603
100	0	0	7914	4050	0	10 485	5300	0	13 454	6700	0	16 817	8600	0	20 578	10 300
	2	95	5834	3500	120	7591	4600	138	9577	5800	169	11 803	7200	204	14 264	8800
	5	172	5797	3475	208	7548	4566	245	9528	5769	293	11 748	7162	341	14 204	8756
	10	223	5737	3434	268	7478	4509	318	9447	5717	374	11 658	7100	436	14 105	8683
	15	252	5678	3392	304	7409	4451	358	9367	5665	418	11 569	7037	487	14 007	8610
	20	277	5619	3351	330	7341	4394	387	9289	5613	452	11 482	6975	523	13 910	8537
	30	319	5505	3267	378	7209	4279	446	9136	5509	514	11 310	6850	592	13 720	8391
	50	415	5289	3100	486	6956	4050	572	8841	5300	659	10 979	6600	752	13 354	8100

For SI units: 1 inch = 25.4 mm, 1 foot = 304.8 mm, 1000 British thermal units per hour = 0.293 kW, 1 square inch = 0.000645 m^2

WATER HEATERS

TABLE 510.1.2(2)
TYPE B DOUBLE-WALL GAS VENT [NFPA 54: TABLE 13.1(b)]*

		NUMBER OF APPLIANCES:	SINGLE
		APPLIANCE TYPE:	CATEGORY I
		APPLIANCE VENT CONNECTION:	SINGLE-WALL METAL CONNECTOR

VENT DIAMETER – D (inch)														
		3			4			5			6			7		

| APPLIANCE INPUT RATING IN THOUSANDS OF BTU PER HOUR | | | | | | | | | | | | | | |

HEIGHT H (feet)	LATERAL L (feet)	FAN Min	FAN Max	NAT Max	FAN Min	FAN Max	NAT Max	FAN Min	FAN Max	NAT Max	FAN Min	FAN Max	NAT Max	FAN Min	FAN Max	NAT Max
6	0	38	77	45	59	151	85	85	249	140	126	373	204	165	522	284
	2	39	51	36	60	96	66	85	156	104	123	231	156	159	320	213
	4	NA	NA	33	74	92	63	102	152	102	146	225	152	187	313	208
	6	NA	NA	31	83	89	60	114	147	99	163	220	148	207	307	203
8	0	37	83	50	58	164	93	83	273	154	123	412	234	161	580	319
	2	39	56	39	59	108	75	83	176	119	121	261	179	155	363	246
	5	NA	NA	37	77	102	69	107	168	114	151	252	171	193	352	235
	8	NA	NA	33	90	95	64	122	161	107	175	243	163	223	342	225
10	0	37	87	53	57	174	99	82	293	165	120	444	254	158	628	344
	2	39	61	41	59	117	80	82	193	128	119	287	194	153	400	272
	5	52	56	39	76	111	76	105	185	122	148	277	186	190	388	261
	10	NA	NA	34	97	100	68	132	171	112	188	261	171	237	369	241
15	0	36	93	57	56	190	111	80	325	186	116	499	283	153	713	388
	2	38	69	47	57	136	93	80	225	149	115	337	224	148	473	314
	5	51	63	44	75	128	86	102	216	140	144	326	217	182	459	298
	10	NA	NA	39	95	116	79	128	201	131	182	308	203	228	438	284
	15	NA	NA	NA	NA	NA	72	158	186	124	220	290	192	272	418	269
20	0	35	96	60	54	200	118	78	346	201	114	537	306	149	772	428
	2	37	74	50	56	148	99	78	248	165	113	375	248	144	528	344
	5	50	68	47	73	140	94	100	239	158	141	363	239	178	514	334
	10	NA	NA	41	93	129	86	125	223	146	177	344	224	222	491	316
	15	NA	NA	NA	NA	NA	80	155	208	136	216	325	210	264	469	301
	20	NA	NA	NA	NA	NA	NA	186	192	126	254	306	196	309	448	285
30	0	34	99	63	53	211	127	76	372	219	110	584	334	144	849	472
	2	37	80	56	55	164	111	76	281	183	109	429	279	139	610	392
	5	49	74	52	72	157	106	98	271	173	136	417	271	171	595	382
	10	NA	NA	NA	91	144	98	122	255	168	171	397	257	213	570	367
	15	NA	NA	NA	115	131	NA	151	239	157	208	377	242	255	547	349
	20	NA	NA	NA	NA	NA	NA	181	223	NA	246	357	228	298	524	333
	30	NA	NA	NA	NA	NA	NA	NA	NA	NA	NA	NA	NA	389	477	305
50	0	33	99	66	51	213	133	73	394	230	105	629	361	138	928	515
	2	36	84	61	53	181	121	73	318	205	104	495	312	133	712	443
	5	48	80	NA	70	174	117	94	308	198	131	482	305	164	696	435
	10	NA	NA	NA	89	160	NA	118	292	186	162	461	292	203	671	420
	15	NA	NA	NA	112	148	NA	145	275	174	199	441	280	244	646	405
	20	NA	NA	NA	NA	NA	NA	176	257	NA	236	420	267	285	622	389
	30	NA	NA	NA	NA	NA	NA	NA	NA	NA	315	376	NA	373	573	NA
100	0	NA	NA	NA	49	214	NA	69	403	NA	100	659	395	131	991	555
	2	NA	NA	NA	51	192	NA	70	351	NA	98	563	373	125	828	508
	5	NA	NA	NA	67	186	NA	90	342	NA	125	551	366	156	813	501
	10	NA	NA	NA	85	175	NA	113	324	NA	153	532	354	191	789	486
	15	NA	NA	NA	132	162	NA	138	310	NA	188	511	343	230	764	473
	20	NA	NA	NA	NA	NA	NA	168	295	NA	224	487	NA	270	739	458
	30	NA	NA	NA	NA	NA	NA	231	264	NA	301	448	NA	355	685	NA
	50	NA	NA	NA	NA	NA	NA	NA	NA	NA	NA	NA	NA	540	584	NA

For SI units: 1 inch = 25.4 mm, 1 foot = 304.8 mm, 1000 British thermal units per hour = 0.293 kW, 1 square inch = 0.000645 m²
* NA: Not applicable.

TABLE 510.1.2(2)
TYPE B DOUBLE-WALL GAS VENT [NFPA 54: TABLE 13.1(b)] (continued)*

		NUMBER OF APPLIANCES:			SINGLE						
		APPLIANCE TYPE:			CATEGORY I						
		APPLIANCE VENT CONNECTION:			SINGLE-WALL METAL CONNECTOR						
		VENT DIAMETER – D (inch)									
		8			9			10		12	
		APPLIANCE INPUT RATING IN THOUSANDS OF BTU PER HOUR									
HEIGHT H (feet)	LATERAL L (feet)	FAN		NAT	FAN		NAT	FAN		NAT	FAN		NAT
		Min	Max	Max	Min	Max	Max	Min	Max	Max	Min	Max	Max
6	0	211	695	369	267	894	469	371	1118	569	537	1639	849
	2	201	423	284	251	541	368	347	673	453	498	979	648
	4	237	416	277	295	533	360	409	664	443	584	971	638
	6	263	409	271	327	526	352	449	656	433	638	962	627
8	0	206	777	414	258	1002	536	360	1257	658	521	1852	967
	2	197	482	321	246	617	417	339	768	513	486	1120	743
	5	245	470	311	305	604	404	418	754	500	598	1104	730
	8	280	458	300	344	591	392	470	740	486	665	1089	715
10	0	202	844	449	253	1093	584	351	1373	718	507	2031	1057
	2	193	531	354	242	681	456	332	849	559	475	1242	848
	5	241	518	344	299	667	443	409	834	544	584	1224	825
	10	296	497	325	363	643	423	492	808	520	688	1194	788
15	0	195	966	523	244	1259	681	336	1591	838	488	2374	1237
	2	187	631	413	232	812	543	319	1015	673	457	1491	983
	5	231	616	400	287	795	526	392	997	657	562	1469	963
	10	284	592	381	349	768	501	470	966	628	664	1433	928
	15	334	568	367	404	742	484	540	937	601	750	1399	894
20	0	190	1053	573	238	1379	750	326	1751	927	473	2631	1346
	2	182	708	468	227	914	611	309	1146	754	443	1689	1098
	5	224	692	457	279	896	596	381	1126	734	547	1665	1074
	10	277	666	437	339	866	570	457	1092	702	646	1626	1037
	15	325	640	419	393	838	549	526	1060	677	730	1587	1005
	20	374	616	400	448	810	526	592	1028	651	808	1550	973
30	0	184	1168	647	229	1542	852	312	1971	1056	454	2996	1545
	2	175	823	533	219	1069	698	296	1346	863	424	1999	1308
	5	215	806	521	269	1049	684	366	1324	846	524	1971	1283
	10	265	777	501	327	1017	662	440	1287	821	620	1927	1243
	15	312	750	481	379	985	638	507	1251	794	702	1884	1205
	20	360	723	461	433	955	615	570	1216	768	780	1841	1166
	30	461	670	426	541	895	574	704	1147	720	937	1759	1101
50	0	176	1292	704	220	1724	948	295	2223	1189	428	3432	1818
	2	168	971	613	209	1273	811	280	1615	1007	401	2426	1509
	5	204	953	602	257	1252	795	347	1591	991	496	2396	1490
	10	253	923	583	313	1217	765	418	1551	963	589	2347	1455
	15	299	894	562	363	1183	736	481	1512	934	668	2299	1421
	20	345	866	543	415	1150	708	544	1473	906	741	2251	1387
	30	442	809	502	521	1086	649	674	1399	848	892	2159	1318
100	0	166	1404	765	207	1900	1033	273	2479	1300	395	3912	2042
	2	158	1152	698	196	1532	933	259	1970	1168	371	3021	1817
	5	194	1134	688	240	1511	921	322	1945	1153	460	2990	1796
	10	238	1104	672	293	1477	902	389	1905	1133	547	2938	1763
	15	281	1075	656	342	1443	884	447	1865	1110	618	2888	1730
	20	325	1046	639	391	1410	864	507	1825	1087	690	2838	1696
	30	418	988	NA	491	1343	824	631	1747	1041	834	2739	1627
	50	617	866	NA	711	1205	NA	895	1591	NA	1138	2547	1489

For SI units: 1 inch = 25.4 mm, 1 foot = 304.8 mm, 1000 British thermal units per hour = 0.293 kW, 1 square inch = 0.000645 m^2
* NA: Not applicable.

WATER HEATERS

TABLE 510.1.2(3)
MASONRY CHIMNEY [NFPA 54: TABLE 13.1(c)]*

		NUMBER OF APPLIANCES:	SINGLE
		APPLIANCE TYPE:	CATEGORY I
		APPLIANCE VENT CONNECTION:	TYPE B DOUBLE-WALL CONNECTOR

TYPE B DOUBLE-WALL CONNECTOR DIAMETER – D (inch)
TO BE USED WITH CHIMNEY AREAS WITHIN THE SIZE LIMITS AT BOTTOM

HEIGHT H (feet)	LATERAL L (feet)	3 FAN Min	3 FAN Max	3 NAT Max	4 FAN Min	4 FAN Max	4 NAT Max	5 FAN Min	5 FAN Max	5 NAT Max	6 FAN Min	6 FAN Max	6 NAT Max	7 FAN Min	7 FAN Max	7 NAT Max
6	2	NA	NA	28	NA	NA	52	NA	NA	86	NA	NA	130	NA	NA	180
	5	NA	NA	25	NA	NA	49	NA	NA	82	NA	NA	117	NA	NA	165
8	2	NA	NA	29	NA	NA	55	NA	NA	93	NA	NA	145	NA	NA	198
	5	NA	NA	26	NA	NA	52	NA	NA	88	NA	NA	134	NA	NA	183
	8	NA	NA	24	NA	NA	48	NA	NA	83	NA	NA	127	NA	NA	175
10	2	NA	NA	31	NA	NA	61	NA	NA	103	NA	NA	162	NA	NA	221
	5	NA	NA	28	NA	NA	57	NA	NA	96	NA	NA	148	NA	NA	204
	10	NA	NA	25	NA	NA	50	NA	NA	87	NA	NA	139	NA	NA	191
15	2	NA	NA	35	NA	NA	67	NA	NA	114	NA	NA	179	53	475	250
	5	NA	NA	35	NA	NA	62	NA	NA	107	NA	NA	164	NA	NA	231
	10	NA	NA	28	NA	NA	55	NA	NA	97	NA	NA	153	NA	NA	216
	15	NA	NA	NA	NA	NA	48	NA	NA	89	NA	NA	141	NA	NA	201
20	2	NA	NA	38	NA	NA	74	NA	NA	124	NA	NA	201	51	522	274
	5	NA	NA	36	NA	NA	68	NA	NA	116	NA	NA	184	80	503	254
	10	NA	NA	NA	NA	NA	60	NA	NA	107	NA	NA	172	NA	NA	237
	15	NA	NA	NA	NA	NA	NA	NA	NA	97	NA	NA	159	NA	NA	220
	20	NA	NA	NA	NA	NA	NA	NA	NA	83	NA	NA	148	NA	NA	206
30	2	NA	NA	41	NA	NA	82	NA	NA	137	NA	NA	216	47	581	303
	5	NA	NA	NA	NA	NA	76	NA	NA	128	NA	NA	198	75	561	281
	10	NA	NA	NA	NA	NA	67	NA	NA	115	NA	NA	184	NA	NA	263
	15	NA	NA	NA	NA	NA	NA	NA	NA	107	NA	NA	171	NA	NA	243
	20	NA	NA	NA	NA	NA	NA	NA	NA	91	NA	NA	159	NA	NA	227
	30	NA	NA	NA	NA	NA	NA	NA	NA	NA	NA	NA	NA	NA	NA	188
50	2	NA	NA	NA	NA	NA	92	NA	NA	161	NA	NA	251	NA	NA	351
	5	NA	NA	NA	NA	NA	NA	NA	NA	151	NA	NA	230	NA	NA	323
	10	NA	NA	NA	NA	NA	NA	NA	NA	138	NA	NA	215	NA	NA	304
	15	NA	NA	NA	NA	NA	NA	NA	NA	127	NA	NA	199	NA	NA	282
	20	NA	NA	NA	NA	NA	NA	NA	NA	NA	NA	NA	185	NA	NA	264
	30	NA	NA	NA	NA	NA	NA	NA	NA	NA	NA	NA	NA	NA	NA	NA
Minimum internal area of chimney (square inches)		12			19			28			38			50		
Maximum internal area of chimney (square inches)		Seven times the listed appliance categorized vent area, flue collar area, or draft hood outlet areas.														

For SI units: 1 inch = 25.4 mm, 1 foot = 304.8 mm, 1000 British thermal units per hour = 0.293 kW, 1 square inch = 0.000645 m^2

* NA: Not applicable.

WATER HEATERS

TABLE 510.1.2(3)
MASONRY CHIMNEY [NFPA 54: TABLE 13.1(c)] (continued)*

		NUMBER OF APPLIANCES:	SINGLE
		APPLIANCE TYPE:	CATEGORY I
		APPLIANCE VENT CONNECTION:	TYPE B DOUBLE-WALL CONNECTOR

		TYPE B DOUBLE-WALL CONNECTOR DIAMETER – D (inch) TO BE USED WITH CHIMNEY AREAS WITHIN THE SIZE LIMITS AT BOTTOM									
		8			9			10			12		
		APPLIANCE INPUT RATING IN THOUSANDS OF BTU PER HOUR									
HEIGHT H (feet)	LATERAL L (feet)	FAN		NAT	FAN		NAT	FAN		NAT	FAN		NAT
		Min	Max	Max	Min	Max	Max	Min	Max	Max	Min	Max	Max
6	2	NA	NA	247	NA	NA	320	NA	NA	401	NA	NA	581
	5	NA	NA	231	NA	NA	298	NA	NA	376	NA	NA	561
8	2	NA	NA	266	84	590	350	100	728	446	139	1024	651
	5	NA	NA	247	NA	NA	328	149	711	423	201	1007	640
	8	NA	NA	239	NA	NA	318	173	695	410	231	990	623
10	2	68	519	298	82	655	388	98	810	491	136	1144	724
	5	NA	NA	277	124	638	365	146	791	466	196	1124	712
	10	NA	NA	263	155	610	347	182	762	444	240	1093	668
15	2	64	613	336	77	779	441	92	968	562	127	1376	841
	5	99	594	313	118	759	416	139	946	533	186	1352	828
	10	126	565	296	148	727	394	173	912	567	229	1315	777
	15	NA	NA	281	171	698	375	198	880	485	259	1280	742
20	2	61	678	375	73	867	491	87	1083	627	121	1548	953
	5	95	658	350	113	845	463	133	1059	597	179	1523	933
	10	122	627	332	143	811	440	167	1022	566	221	1482	879
	15	NA	NA	314	165	780	418	191	987	541	251	1443	840
	20	NA	NA	296	186	750	397	214	955	513	277	1406	807
30	2	57	762	421	68	985	558	81	1240	717	111	1793	1112
	5	90	741	393	106	962	526	125	1216	683	169	1766	1094
	10	115	709	373	135	927	500	158	1176	648	210	1721	1025
	15	NA	NA	353	156	893	476	181	1139	621	239	1679	981
	20	NA	NA	332	176	860	450	203	1103	592	264	1638	940
	30	NA	NA	288	NA	NA	416	249	1035	555	318	1560	877
50	2	51	840	477	61	1106	633	72	1413	812	99	2080	1243
	5	83	819	445	98	1083	596	116	1387	774	155	2052	1225
	10	NA	NA	424	126	1047	567	147	1347	733	195	2006	1147
	15	NA	NA	400	146	1010	539	170	1307	702	222	1961	1099
	20	NA	NA	376	165	977	511	190	1269	669	246	1916	1050
	30	NA	NA	327	NA	NA	468	233	1196	623	295	1832	984
Minimum internal area of chimney (square inches)		63			78			95			132		
Maximum internal area of chimney (square inches)		Seven times the listed appliance categorized vent area, flue collar area, or draft hood outlet areas.										

For SI units: 1 inch = 25.4 mm, 1 foot = 304.8 mm, 1000 British thermal units per hour = 0.293 kW, 1 square inch = 0.000645 m^2
* NA: Not applicable.

WATER HEATERS

TABLE 510.1.2(4)
MASONRY CHIMNEY [NFPA 54: TABLE 13.1(d)]*

		NUMBER OF APPLIANCES:			SINGLE								
		APPLIANCE TYPE:			CATEGORY I								
		APPLIANCE VENT CONNECTION:			SINGLE-WALL METAL CONNECTOR								

		SINGLE-WALL METAL CONNECTOR DIAMETER – D (inch) TO BE USED WITH CHIMNEY AREAS WITHIN THE SIZE LIMITS AT BOTTOM												
		3			4			5			6			7		
		APPLIANCE INPUT RATING IN THOUSANDS OF BTU PER HOUR												
HEIGHT H (feet)	LATERAL L (feet)	FAN		NAT	FAN		NAT	FAN		NAT	FAN		NAT	FAN		NAT
		Min	Max	Max	Min	Max	Max	Min	Max	Max	Min	Max	Max	Min	Max	Max
6	2	NA	NA	28	NA	NA	52	NA	NA	86	NA	NA	130	NA	NA	180
	5	NA	NA	25	NA	NA	48	NA	NA	81	NA	NA	116	NA	NA	164
8	2	NA	NA	29	NA	NA	55	NA	NA	93	NA	NA	145	NA	NA	197
	5	NA	NA	26	NA	NA	51	NA	NA	87	NA	NA	133	NA	NA	182
	8	NA	NA	23	NA	NA	47	NA	NA	82	NA	NA	126	NA	NA	174
10	2	NA	NA	31	NA	NA	61	NA	NA	102	NA	NA	161	NA	NA	220
	5	NA	NA	28	NA	NA	56	NA	NA	95	NA	NA	147	NA	NA	203
	10	NA	NA	24	NA	NA	49	NA	NA	86	NA	NA	137	NA	NA	189
15	2	NA	NA	35	NA	NA	67	NA	NA	113	NA	NA	178	166	473	249
	5	NA	NA	32	NA	NA	61	NA	NA	106	NA	NA	163	NA	NA	230
	10	NA	NA	27	NA	NA	54	NA	NA	96	NA	NA	151	NA	NA	214
	15	NA	NA	NA	NA	NA	46	NA	NA	87	NA	NA	138	NA	NA	198
20	2	NA	NA	38	NA	NA	73	NA	NA	123	NA	NA	200	163	520	273
	5	NA	NA	35	NA	NA	67	NA	NA	115	NA	NA	183	NA	NA	252
	10	NA	NA	NA	NA	NA	59	NA	NA	105	NA	NA	170	NA	NA	235
	15	NA	NA	NA	NA	NA	NA	NA	NA	95	NA	NA	156	NA	NA	217
	20	NA	NA	NA	NA	NA	NA	NA	NA	80	NA	NA	144	NA	NA	202
30	2	NA	NA	41	NA	NA	81	NA	NA	136	NA	NA	215	158	578	302
	5	NA	NA	NA	NA	NA	75	NA	NA	127	NA	NA	196	NA	NA	279
	10	NA	NA	NA	NA	NA	66	NA	NA	113	NA	NA	182	NA	NA	260
	15	NA	NA	NA	NA	NA	NA	NA	NA	105	NA	NA	168	NA	NA	240
	20	NA	NA	NA	NA	NA	NA	NA	NA	88	NA	NA	155	NA	NA	223
	30	NA	NA	NA	NA	NA	NA	NA	NA	NA	NA	NA	NA	NA	NA	182
50	2	NA	NA	NA	NA	NA	91	NA	NA	160	NA	NA	250	NA	NA	350
	5	NA	NA	NA	NA	NA	NA	NA	NA	149	NA	NA	228	NA	NA	321
	10	NA	NA	NA	NA	NA	NA	NA	NA	136	NA	NA	212	NA	NA	301
	15	NA	NA	NA	NA	NA	NA	NA	NA	124	NA	NA	195	NA	NA	278
	20	NA	NA	NA	NA	NA	NA	NA	NA	NA	NA	NA	180	NA	NA	258
	30	NA	NA	NA	NA	NA	NA	NA	NA	NA	NA	NA	NA	NA	NA	NA
Minimum internal area of chimney (square inches)		12			19			28			38			50		
Maximum internal area of chimney (square inches)		Seven times the listed appliance categorized vent area, flue collar area, or draft hood outlet areas.														

For SI units: 1 inch = 25.4 mm, 1 foot = 304.8 mm, 1000 British thermal units per hour = 0.293 kW, 1 square inch = 0.000645 m^2
* NA: Not applicable.

WATER HEATERS

TABLE 510.1.2(4)
MASONRY CHIMNEY [NFPA 54: TABLE 13.1(d)] (continued)*

		NUMBER OF APPLIANCES:	SINGLE
		APPLIANCE TYPE:	CATEGORY I
		APPLIANCE VENT CONNECTION:	SINGLE-WALL METAL CONNECTOR

		SINGLE-WALL METAL CONNECTOR DIAMETER – D (inch) TO BE USED WITH CHIMNEY AREAS WITHIN THE SIZE LIMITS AT BOTTOM									
		8			9			10			12		
		APPLIANCE INPUT RATING IN THOUSANDS OF BTU PER HOUR											
HEIGHT H (feet)	LATERAL L (feet)	FAN		NAT	FAN		NAT	FAN		NAT	FAN		NAT
		Min	Max	Max	Min	Max	Max	Min	Max	Max	Min	Max	Max
6	2	NA	NA	247	NA	NA	319	NA	NA	400	NA	NA	580
	5	NA	NA	230	NA	NA	297	NA	NA	375	NA	NA	560
8	2	NA	NA	265	NA	NA	349	382	725	445	549	1021	650
	5	NA	NA	246	NA	NA	327	NA	NA	422	673	1003	638
	8	NA	NA	237	NA	NA	317	NA	NA	408	747	985	621
10	2	216	518	297	271	654	387	373	808	490	536	1142	722
	5	NA	NA	276	334	635	364	459	789	465	657	1121	710
	10	NA	NA	261	NA	NA	345	547	758	441	771	1088	665
15	2	211	611	335	264	776	440	362	965	560	520	1373	840
	5	261	591	312	325	755	414	444	942	531	637	1348	825
	10	NA	NA	294	392	722	392	531	907	504	749	1309	774
	15	NA	NA	278	452	692	372	606	873	481	841	1272	738
20	2	206	675	374	258	864	490	252	1079	625	508	1544	950
	5	255	655	348	317	842	461	433	1055	594	623	1518	930
	10	312	622	330	382	806	437	517	1016	562	733	1475	875
	15	NA	NA	311	442	773	414	591	979	539	823	1434	835
	20	NA	NA	292	NA	NA	392	663	944	510	911	1394	800
30	2	200	759	420	249	982	556	340	1237	715	489	1789	1110
	5	245	737	391	306	958	524	417	1210	680	600	1760	1090
	10	300	703	370	370	920	496	500	1168	644	708	1713	1020
	15	NA	NA	349	428	884	471	572	1128	615	798	1668	975
	20	NA	NA	327	NA	NA	445	643	1089	585	883	1624	932
	30	NA	NA	281	NA	NA	408	NA	NA	544	1055	1539	865
50	2	191	837	475	238	1103	631	323	1408	810	463	2076	1240
	5	NA	NA	442	293	1078	593	398	1381	770	571	2044	1220
	10	NA	NA	420	355	1038	562	447	1337	728	674	1994	1140
	15	NA	NA	395	NA	NA	533	546	1294	695	761	1945	1090
	20	NA	NA	370	NA	NA	504	616	1251	660	844	1898	1040
	30	NA	NA	318	NA	NA	458	NA	NA	610	1009	1805	970
Minimum internal area of chimney (square inches)		63			78			95			132		
Maximum internal area of chimney (square inches)		Seven times the listed appliance categorized vent area, flue collar area, or draft hood outlet areas.											

For SI units: 1 inch = 25.4 mm, 1 foot = 304.8 mm, 1000 British thermal units per hour = 0.293 kW, 1 square inch = 0.000645 m²
* NA: Not applicable.

WATER HEATERS

TABLE 510.1.2(5)
SINGLE-WALL METAL PIPE OR TYPE B ASBESTOS-CEMENT VENT [NFPA 54: TABLE 13.1(e)]*

					NUMBER OF APPLIANCES:	SINGLE			
					APPLIANCE TYPE:	DRAFT HOOD-EQUIPPED			
					APPLIANCE VENT CONNECTION:	CONNECTED DIRECTLY TO PIPE OR VENT			
		DIAMETER – D (inch) TO BE USED WITH CHIMNEY AREAS WITHIN THE SIZE LIMITS AT BOTTOM							
		3	4	5	6	7	8	10	12
HEIGHT H (feet)	LATERAL L (feet)	APPLIANCE INPUT RATING IN THOUSANDS OF BTU PER HOUR							
		MAXIMUM APPLIANCE INPUT RATING IN THOUSANDS OF BTU PER HOUR							
6	0	39	70	116	170	232	312	500	750
	2	31	55	94	141	194	260	415	620
	5	28	51	88	128	177	242	390	600
8	0	42	76	126	185	252	340	542	815
	2	32	61	102	154	210	284	451	680
	5	29	56	95	141	194	264	430	648
	10	24	49	86	131	180	250	406	625
10	0	45	84	138	202	279	372	606	912
	2	35	67	111	168	233	311	505	760
	5	32	61	104	153	215	289	480	724
	10	27	54	94	143	200	274	455	700
	15	NA	46	84	130	186	258	432	666
15	0	49	91	151	223	312	420	684	1040
	2	39	72	122	186	260	350	570	865
	5	35	67	110	170	240	325	540	825
	10	30	58	103	158	223	308	514	795
	15	NA	50	93	144	207	291	488	760
	20	NA	NA	82	132	195	273	466	726
20	0	53	101	163	252	342	470	770	1190
	2	42	80	136	210	286	392	641	990
	5	38	74	123	192	264	364	610	945
	10	32	65	115	178	246	345	571	910
	15	NA	55	104	163	228	326	550	870
	20	NA	NA	91	149	214	306	525	832
30	0	56	108	183	276	384	529	878	1370
	2	44	84	148	230	320	441	730	1140
	5	NA	78	137	210	296	410	694	1080
	10	NA	68	125	196	274	388	656	1050
	15	NA	NA	113	177	258	366	625	1000
	20	NA	NA	99	163	240	344	596	960
	30	NA	NA	NA	NA	192	295	540	890
50	0	NA	120	210	310	443	590	980	1550
	2	NA	95	171	260	370	492	820	1290
	5	NA	NA	159	234	342	474	780	1230
	10	NA	NA	146	221	318	456	730	1190
	15	NA	NA	NA	200	292	407	705	1130
	20	NA	NA	NA	185	276	384	670	1080
	30	NA	NA	NA	NA	222	330	605	1010

For SI units: 1 inch = 25.4 mm, 1 foot = 304.8 mm, 1000 British thermal units per hour = 0.293 kW, 1 square inch = 0.000645 m²
* NA: Not applicable.

WATER HEATERS

TABLE 510.1.2(6)
EXTERIOR MASONRY CHIMNEY [NFPA 54: TABLE 13.1(f)][1,2]

	NUMBER OF APPLIANCES:	SINGLE
	APPLIANCE TYPE:	NAT
	APPLIANCE VENT CONNECTION:	TYPE B DOUBLE-WALL CONNECTOR

MINIMUM ALLOWABLE INPUT RATING OF SPACE-HEATING APPLIANCE IN THOUSANDS OF BTU PER HOUR

VENT HEIGHT H (feet)	INTERNAL AREA OF CHIMNEY (square inches)							
	12	19	28	38	50	63	78	113
	Local 99% winter design temperature: 37°F or greater							
6	0	0	0	0	0	0	0	0
8	0	0	0	0	0	0	0	0
10	0	0	0	0	0	0	0	0
15	NA	0	0	0	0	0	0	0
20	NA	NA	123	190	249	184	0	0
30	NA	NA	NA	NA	NA	393	334	0
50	NA	NA	NA	NA	NA	NA	NA	579
	Local 99% winter design temperature: 27°F to 36°F							
6	0	0	68	116	156	180	212	266
8	0	0	82	127	167	187	214	263
10	0	51	97	141	183	201	225	265
15	NA	NA	NA	NA	233	253	274	305
20	NA	NA	NA	NA	NA	307	330	362
30	NA	NA	NA	NA	NA	419	445	485
50	NA	NA	NA	NA	NA	NA	NA	763
	Local 99% winter design temperature: 17°F to 26°F							
6	NA	NA	NA	NA	NA	215	259	349
8	NA	NA	NA	NA	197	226	264	352
10	NA	NA	NA	NA	214	245	278	358
15	NA	NA	NA	NA	NA	296	331	398
20	NA	NA	NA	NA	NA	352	387	457
30	NA	NA	NA	NA	NA	NA	507	581
50	NA	NA	NA	NA	NA	NA	NA	NA
	Local 99% winter design temperature: 5°F to 16°F							
6	NA	NA	NA	NA	NA	NA	NA	416
8	NA	NA	NA	NA	NA	NA	312	423
10	NA	NA	NA	NA	NA	289	331	430
15	NA	NA	NA	NA	NA	NA	393	485
20	NA	NA	NA	NA	NA	NA	450	547
30	NA	NA	NA	NA	NA	NA	NA	682
50	NA	NA	NA	NA	NA	NA	NA	972
	Local 99% winter design temperature: -10°F to 4°F							
6	NA	NA	NA	NA	NA	NA	NA	484
8	NA	NA	NA	NA	NA	NA	NA	494
10	NA	NA	NA	NA	NA	NA	NA	513
15	NA	NA	NA	NA	NA	NA	NA	586
20	NA	NA	NA	NA	NA	NA	NA	650
30	NA	NA	NA	NA	NA	NA	NA	805
50	NA	NA	NA	NA	NA	NA	NA	1003
	Local 99% winter design temperature: -11°F or lower Not recommended for any vent configurations							

For SI units: 1 inch = 25.4 mm, 1 foot = 304.8 mm, 1000 British thermal units per hour = 0.293 kW, 1 square inch = 0.000645 m^2, °C = (°F-32)/1.8

Notes:
[1] See Figure 510.1.2(6) for a map showing local 99 percent winter design temperatures in the United States.
[2] NA: Not applicable.

WATER HEATERS

TABLE 510.2(1)
TYPE B DOUBLE-WALL VENT [NFPA 54: TABLE 13.2(a)]*

		NUMBER OF APPLIANCES:											TWO OR MORE	
		APPLIANCE TYPE:											CATEGORY I	
		APPLIANCE VENT CONNECTION:											TYPE B DOUBLE-WALL CONNECTOR	
		VENT CONNECTOR CAPACITY												
		TYPE B DOUBLE-WALL VENT AND CONNECTOR DIAMETER – D (inch)												
		3			4			5			6		7	
		APPLIANCE INPUT RATING LIMITS IN THOUSANDS OF BTU PER HOUR												
VENT HEIGHT H (feet)	CONNECTOR RISE R (feet)	FAN		NAT	FAN		NAT	FAN		NAT	FAN		FAN	NAT
		Min	Max	Max	Min	Max	Max	Min	Max	Max	Min	Max	Max	Min	Max	Max

VENT HEIGHT H (feet)	CONNECTOR RISE R (feet)	FAN Min	FAN Max	NAT Max	FAN Min	FAN Max	NAT Max	FAN Min	FAN Max	NAT Max	FAN Min	FAN Max	NAT Max	FAN Min	FAN Max	NAT Max
6	1	22	37	26	35	66	46	46	106	72	58	164	104	77	225	142
	2	23	41	31	37	75	55	48	121	86	60	183	124	79	253	168
	3	24	44	35	38	81	62	49	132	96	62	199	139	82	275	189
8	1	22	40	27	35	72	48	49	114	76	64	176	109	84	243	148
	2	23	44	32	36	80	57	51	128	90	66	195	129	86	269	175
	3	24	47	36	37	87	64	53	139	101	67	210	145	88	290	198
10	1	22	43	28	34	78	50	49	123	78	65	189	113	89	257	154
	2	23	47	33	36	86	59	51	136	93	67	206	134	91	282	182
	3	24	50	37	37	92	67	52	146	104	69	220	150	94	303	205
15	1	21	50	30	33	89	53	47	142	83	64	220	120	88	298	163
	2	22	53	35	35	96	63	49	153	99	66	235	142	91	320	193
	3	24	55	40	36	102	71	51	163	111	68	248	160	93	339	218
20	1	21	54	31	33	99	56	46	157	87	62	246	125	86	334	171
	2	22	57	37	34	105	66	48	167	104	64	259	149	89	354	202
	3	23	60	42	35	110	74	50	176	116	66	271	168	91	371	228
30	1	20	62	33	31	113	59	45	181	93	60	288	134	83	391	182
	2	21	64	39	33	118	70	47	190	110	62	299	158	85	408	215
	3	22	66	44	34	123	79	48	198	124	64	309	178	88	423	242
50	1	19	71	36	30	133	64	43	216	101	57	349	145	78	477	197
	2	21	73	43	32	137	76	45	223	119	59	358	172	81	490	234
	3	22	75	48	33	141	86	46	229	134	61	366	194	83	502	263
100	1	18	82	37	28	158	66	40	262	104	53	442	150	73	611	204
	2	19	83	44	30	161	79	42	267	123	55	447	178	75	619	242
	3	20	84	50	31	163	89	44	272	138	57	452	200	78	627	272

COMMON VENT CAPACITY

TYPE B DOUBLE-WALL COMMON VENT DIAMETER – D (inch)

COMBINED APPLIANCE INPUT RATING IN THOUSANDS OF BTU PER HOUR

VENT HEIGHT H (feet)	4 FAN+FAN	4 FAN+NAT	4 NAT+NAT	5 FAN+FAN	5 FAN+NAT	5 NAT+NAT	6 FAN+FAN	6 FAN+NAT	6 NAT+NAT	7 FAN+FAN	7 FAN+NAT	7 NAT+NAT
6	92	81	65	140	116	103	204	161	147	309	248	200
8	101	90	73	155	129	114	224	178	163	339	275	223
10	110	97	79	169	141	124	243	194	178	367	299	242
15	125	112	91	195	164	144	283	228	206	427	352	280
20	136	123	102	215	183	160	314	255	229	475	394	310
30	152	138	118	244	210	185	361	297	266	547	459	360
50	167	153	134	279	244	214	421	353	310	641	547	423
100	175	163	NA	311	277	NA	489	421	NA	751	658	479

For SI units: 1 inch = 25.4 mm, 1 foot = 304.8 mm, 1000 British thermal units per hour = 0.293 kW, 1 square inch = 0.000645 m^2

* NA: Not applicable.

WATER HEATERS

TABLE 510.2(1)
TYPE B DOUBLE-WALL VENT [NFPA 54: TABLE 13.2(a)] (continued)

		NUMBER OF APPLIANCES:		TWO OR MORE				
		APPLIANCE TYPE:		CATEGORY I				
		APPLIANCE VENT CONNECTION:		TYPE B DOUBLE-WALL CONNECTOR				
		VENT CONNECTOR CAPACITY						

		TYPE B DOUBLE-WALL VENT AND CONNECTOR DIAMETER – D (inch)						
		8			9			10		
		APPLIANCE INPUT RATING LIMITS IN THOUSANDS OF BTU PER HOUR						
VENT HEIGHT H (feet)	CONNECTOR RISE R (feet)	FAN		NAT	FAN		NAT	FAN		NAT
		Min	Max	Max	Min	Max	Max	Min	Max	Max
6	1	92	296	185	109	376	237	128	466	289
	2	95	333	220	112	424	282	131	526	345
	3	97	363	248	114	463	317	134	575	386
8	1	100	320	194	118	408	248	138	507	303
	2	103	356	230	121	454	294	141	564	358
	3	105	384	258	123	492	330	143	612	402
10	1	106	341	200	125	436	257	146	542	314
	2	109	374	238	128	479	305	149	596	372
	3	111	402	268	131	515	342	152	642	417
15	1	110	389	214	134	493	273	162	609	333
	2	112	419	253	137	532	323	165	658	394
	3	115	445	286	140	565	365	167	700	444
20	1	107	436	224	131	552	285	158	681	347
	2	110	463	265	134	587	339	161	725	414
	3	113	486	300	137	618	383	164	764	466
30	1	103	512	238	125	649	305	151	802	372
	2	105	535	282	129	679	360	155	840	439
	3	108	555	317	132	706	405	158	874	494
50	1	97	627	257	120	797	330	144	984	403
	2	100	645	306	123	820	392	148	1014	478
	3	103	661	343	126	842	441	151	1043	538
100	1	91	810	266	112	1038	341	135	1285	417
	2	94	822	316	115	1054	405	139	1306	494
	3	97	834	355	118	1069	455	142	1327	555

	COMMON VENT CAPACITY								
	TYPE B DOUBLE-WALL COMMON VENT DIAMETER – D (inch)								
	8			9			10		
	COMBINED APPLIANCE INPUT RATING IN THOUSANDS OF BTU PER HOUR								
VENT HEIGHT H (feet)	FAN +FAN	FAN +NAT	NAT +NAT	FAN +FAN	FAN +NAT	NAT +NAT	FAN +FAN	FAN +NAT	NAT +NAT
6	404	314	260	547	434	335	672	520	410
8	444	348	290	602	480	378	740	577	465
10	477	377	315	649	522	405	800	627	495
15	556	444	365	753	612	465	924	733	565
20	621	499	405	842	688	523	1035	826	640
30	720	585	470	979	808	605	1209	975	740
50	854	706	550	1164	977	705	1451	1188	860
100	1025	873	625	1408	1215	800	1784	1502	975

For SI units: 1 inch = 25.4 mm, 1 foot = 304.8 mm, 1000 British thermal units per hour = 0.293 kW, 1 square inch = 0.000645 m^2

WATER HEATERS

TABLE 510.2(1)
TYPE B DOUBLE-WALL VENT [NFPA 54: TABLE 13.2(a)] (continued)*

		NUMBER OF APPLIANCES:				TWO OR MORE					
		APPLIANCE TYPE:				CATEGORY I					
		APPLIANCE VENT CONNECTION:				TYPE B DOUBLE-WALL CONNECTOR					
		VENT CONNECTOR CAPACITY									
		TYPE B DOUBLE-WALL VENT AND CONNECTOR DIAMETER – D (inch)									
		12			14			16		18	
		APPLIANCE INPUT RATING LIMITS IN THOUSANDS OF BTU PER HOUR									
VENT HEIGHT H (feet)	CONNECTOR RISE R (feet)	FAN		NAT	FAN		NAT	FAN		NAT	FAN		NAT
		Min	Max	Max	Min	Max	Max	Min	Max	Max	Min	Max	Max
6	2	174	764	496	223	1046	653	281	1371	853	346	1772	1080
	4	180	897	616	230	1231	827	287	1617	1081	352	2069	1370
	6	NA	NA	NA	NA	NA	NA	NA	NA	NA	NA	NA	NA
8	2	186	822	516	238	1126	696	298	1478	910	365	1920	1150
	4	192	952	644	244	1307	884	305	1719	1150	372	2211	1460
	6	198	1050	772	252	1445	1072	313	1902	1390	380	2434	1770
10	2	196	870	536	249	1195	730	311	1570	955	379	2049	1205
	4	201	997	664	256	1371	924	318	1804	1205	387	2332	1535
	6	207	1095	792	263	1509	1118	325	1989	1455	395	2556	1865
15	2	214	967	568	272	1334	790	336	1760	1030	408	2317	1305
	4	221	1085	712	279	1499	1006	344	1978	1320	416	2579	1665
	6	228	1181	856	286	1632	1222	351	2157	1610	424	2796	2025
20	2	223	1051	596	291	1443	840	357	1911	1095	430	2533	1385
	4	230	1162	748	298	1597	1064	365	2116	1395	438	2778	1765
	6	237	1253	900	307	1726	1288	373	2287	1695	450	2984	2145
30	2	216	1217	632	286	1664	910	367	2183	1190	461	2891	1540
	4	223	1316	792	294	1802	1160	376	2366	1510	474	3110	1920
	6	231	1400	952	303	1920	1410	384	2524	1830	485	3299	2340
50	2	206	1479	689	273	2023	1007	350	2659	1315	435	3548	1665
	4	213	1561	860	281	2139	1291	359	2814	1685	447	3730	2135
	6	221	1631	1031	290	2242	1575	369	2951	2055	461	3893	2605
100	2	192	1923	712	254	2644	1050	326	3490	1370	402	4707	1740
	4	200	1984	888	263	2731	1346	336	3606	1760	414	4842	2220
	6	208	2035	1064	272	2811	1642	346	3714	2150	426	4968	2700

	COMMON VENT CAPACITY											
	TYPE B DOUBLE-WALL COMMON VENT DIAMETER – D (inch)											
	12			14			16			18		
	COMBINED APPLIANCE INPUT RATING IN THOUSANDS OF BTU PER HOUR											
VENT HEIGHT H (feet)	FAN +FAN	FAN +NAT	NAT +NAT	FAN +FAN	FAN +NAT	NAT +NAT	FAN +FAN	FAN +NAT	NAT +NAT	FAN +FAN	FAN +NAT	NAT +NAT
6	900	696	588	1284	990	815	1735	1336	1065	2253	1732	1345
8	994	773	652	1423	1103	912	1927	1491	1190	2507	1936	1510
10	1076	841	712	1542	1200	995	2093	1625	1300	2727	2113	1645
15	1247	986	825	1794	1410	1158	2440	1910	1510	3184	2484	1910
20	1405	1116	916	2006	1588	1290	2722	2147	1690	3561	2798	2140
30	1658	1327	1025	2373	1892	1525	3220	2558	1990	4197	3326	2520
50	2024	1640	1280	2911	2347	1863	3964	3183	2430	5184	4149	3075
100	2569	2131	1670	3732	3076	2450	5125	4202	3200	6749	5509	4050

For SI units: 1 inch = 25.4 mm, 1 foot = 304.8 mm, 1000 British thermal units per hour = 0.293 kW, 1 square inch = 0.000645 m²
* NA: Not applicable.

WATER HEATERS

TABLE 510.2(1)
TYPE B DOUBLE-WALL VENT [NFPA 54: TABLE 13.2(a)] (continued)*

		NUMBER OF APPLIANCES:			TWO OR MORE				
		APPLIANCE TYPE:			CATEGORY I				
		APPLIANCE VENT CONNECTION:			TYPE B DOUBLE-WALL CONNECTOR				
		VENT CONNECTOR CAPACITY							
		TYPE B DOUBLE-WALL VENT AND CONNECTOR DIAMETER – D (inch)							
		20			22			24	
		APPLIANCE INPUT RATING LIMITS IN THOUSANDS OF BTU PER HOUR							
VENT HEIGHT H (feet)	CONNECTOR RISE R (feet)	FAN		NAT	FAN		NAT	FAN		NAT
		Min	Max	Max	Min	Max	Max	Min	Max	Max
6	2	NA	NA	NA	NA	NA	NA	NA	NA	NA
	4	NA	NA	NA	NA	NA	NA	NA	NA	NA
	6	NA	NA	NA	NA	NA	NA	NA	NA	NA
8	2	NA	NA	NA	NA	NA	NA	NA	NA	NA
	4	471	2737	1800	560	3319	2180	662	3957	2590
	6	478	3018	2180	568	3665	2640	669	4373	3130
10	2	NA	NA	NA	NA	NA	NA	NA	NA	NA
	4	486	2887	1890	581	3502	2280	686	4175	2710
	6	494	3169	2290	589	3849	2760	694	4593	3270
15	2	NA	NA	NA	NA	NA	NA	NA	NA	NA
	4	523	3197	2060	624	3881	2490	734	4631	2960
	6	533	3470	2510	634	4216	3030	743	5035	3600
20	2	NA	NA	NA	NA	NA	NA	NA	NA	NA
	4	554	3447	2180	661	4190	2630	772	5005	3130
	6	567	3708	2650	671	4511	3190	785	5392	3790
30	2	NA	NA	NA	NA	NA	NA	NA	NA	NA
	4	619	3840	2365	728	4861	2860	847	5606	3410
	6	632	4080	2875	741	4976	3480	860	5961	4150
50	2	NA	NA	NA	NA	NA	NA	NA	NA	NA
	4	580	4601	2633	709	5569	3185	851	6633	3790
	6	594	4808	3208	724	5826	3885	867	6943	4620
100	2	NA	NA	NA	NA	NA	NA	NA	NA	NA
	4	523	5982	2750	639	7254	3330	769	8650	3950
	6	539	6143	3350	654	7453	4070	786	8892	4810

	COMMON VENT CAPACITY								
	TYPE B DOUBLE-WALL COMMON VENT DIAMETER – D (inch)								
	20			22			24		
	COMBINED APPLIANCE INPUT RATING IN THOUSANDS OF BTU PER HOUR								
VENT HEIGHT H (feet)	FAN +FAN	FAN +NAT	NAT +NAT	FAN +FAN	FAN +NAT	NAT +NAT	FAN +FAN	FAN +NAT	NAT +NAT
6	2838	2180	1660	3488	2677	1970	4206	3226	2390
8	3162	2439	1860	3890	2998	2200	4695	3616	2680
10	3444	2665	2030	4241	3278	2400	5123	3957	2920
15	4026	3133	2360	4971	3862	2790	6016	4670	3400
20	4548	3552	2640	5573	4352	3120	6749	5261	3800
30	5303	4193	3110	6539	5157	3680	7940	6247	4480
50	6567	5240	3800	8116	6458	4500	9837	7813	5475
100	8597	6986	5000	10 681	8648	5920	13 004	10 499	7200

For SI units: 1 inch = 25.4 mm, 1 foot = 304.8 mm, 1000 British thermal units per hour = 0.293 kW, 1 square inch = 0.000645 m^2

* NA: Not applicable.

WATER HEATERS

TABLE 510.2(2)
TYPE B DOUBLE-WALL VENT [NFPA 54: TABLE 13.2(b)]*

		NUMBER OF APPLIANCES:	TWO OR MORE
		APPLIANCE TYPE:	CATEGORY I
		APPLIANCE VENT CONNECTION:	SINGLE-WALL METAL CONNECTOR

VENT CONNECTOR CAPACITY

SINGLE-WALL METAL VENT CONNECTOR DIAMETER – D (inch)

VENT HEIGHT H (feet)	CONNECTOR RISE R (feet)	3 FAN Min	3 FAN Max	3 NAT Max	4 FAN Min	4 FAN Max	4 NAT Max	5 FAN Min	5 FAN Max	5 NAT Max	6 FAN Min	6 FAN Max	6 NAT Max	7 FAN Min	7 FAN Max	7 NAT Max
6	1	NA	NA	26	NA	NA	46	NA	NA	71	NA	NA	102	207	223	140
	2	NA	NA	31	NA	NA	55	NA	NA	85	168	182	123	215	251	167
	3	NA	NA	34	NA	NA	62	121	131	95	175	198	138	222	273	188
8	1	NA	NA	27	NA	NA	48	NA	NA	75	NA	NA	106	226	240	145
	2	NA	NA	32	NA	NA	57	125	126	89	184	193	127	234	266	173
	3	NA	NA	35	NA	NA	64	130	138	100	191	208	144	241	287	197
10	1	NA	NA	28	NA	NA	50	119	121	77	182	186	110	240	253	150
	2	NA	NA	33	84	85	59	124	134	91	189	203	132	248	278	183
	3	NA	NA	36	89	91	67	129	144	102	197	217	148	257	299	203
15	1	NA	NA	29	79	87	52	116	138	81	177	214	116	238	291	158
	2	NA	NA	34	83	94	62	121	150	97	185	230	138	246	314	189
	3	NA	NA	39	87	100	70	127	160	109	193	243	157	255	333	215
20	1	49	56	30	78	97	54	115	152	84	175	238	120	233	325	165
	2	52	59	36	82	103	64	120	163	101	182	252	144	243	346	197
	3	55	62	40	87	107	72	125	172	113	190	264	164	252	363	223
30	1	47	60	31	77	110	57	112	175	89	169	278	129	226	380	175
	2	51	62	37	81	115	67	117	185	106	177	290	152	236	397	208
	3	54	64	42	85	119	76	122	193	120	185	300	172	244	412	235
50	1	46	69	34	75	128	60	109	207	96	162	336	137	217	460	188
	2	49	71	40	79	132	72	114	215	113	170	345	164	226	473	223
	3	52	72	45	83	136	82	119	221	123	178	353	186	235	486	252
100	1	45	79	34	71	150	61	104	249	98	153	424	140	205	585	192
	2	48	80	41	75	153	73	110	255	115	160	428	167	212	593	228
	3	51	81	46	79	157	85	114	260	129	168	433	190	222	603	256

COMMON VENT CAPACITY

TYPE B DOUBLE-WALL COMMON VENT DIAMETER – D (inch)

COMBINED APPLIANCE INPUT RATING IN THOUSANDS OF BTU PER HOUR

VENT HEIGHT H (feet)	4 FAN+FAN	4 FAN+NAT	4 NAT+NAT	5 FAN+FAN	5 FAN+NAT	5 NAT+NAT	6 FAN+FAN	6 FAN+NAT	6 NAT+NAT	7 FAN+FAN	7 FAN+NAT	7 NAT+NAT
6	NA	78	64	NA	113	99	200	158	144	304	244	196
8	NA	87	71	NA	126	111	218	173	159	331	269	218
10	NA	94	76	163	137	120	237	189	174	357	292	236
15	121	108	88	189	159	140	275	221	200	416	343	274
20	131	118	98	208	177	156	305	247	223	463	383	302
30	145	132	113	236	202	180	350	286	257	533	446	349
50	159	145	128	268	233	208	406	337	296	622	529	410
100	166	153	NA	297	263	NA	469	398	NA	726	633	464

For SI units: 1 inch = 25.4 mm, 1 foot = 304.8 mm, 1000 British thermal units per hour = 0.293 kW, 1 square inch = 0.000645 m²
* NA: Not applicable.

TABLE 510.2(2)
TYPE B DOUBLE-WALL VENT [NFPA 54: TABLE 13.2(b)] (continued)

		NUMBER OF APPLIANCES:			TWO OR MORE				
		APPLIANCE TYPE:			CATEGORY I				
		APPLIANCE VENT CONNECTION:			SINGLE-WALL METAL CONNECTOR				
		VENT CONNECTOR CAPACITY							
		SINGLE-WALL METAL VENT CONNECTOR DIAMETER – *D* (inch)							
		8			9			10	
		APPLIANCE INPUT RATING LIMITS IN THOUSANDS OF BTU PER HOUR							
VENT HEIGHT *H* (feet)	CONNECTOR RISE *R* (feet)	FAN		NAT	FAN		NAT	FAN		NAT
		Min	Max	Max	Min	Max	Max	Min	Max	Max
6	1	262	293	183	325	373	234	447	463	286
	2	271	331	219	334	422	281	458	524	344
	3	279	361	247	344	462	316	468	574	385
8	1	285	316	191	352	403	244	481	502	299
	2	293	353	228	360	450	292	492	560	355
	3	302	381	256	370	489	328	501	609	400
10	1	302	335	196	372	429	252	506	534	308
	2	311	369	235	381	473	302	517	589	368
	3	320	398	265	391	511	339	528	637	413
15	1	312	380	208	397	482	266	556	596	324
	2	321	411	248	407	522	317	568	646	387
	3	331	438	281	418	557	360	579	690	437
20	1	306	425	217	390	538	276	546	664	336
	2	317	453	259	400	574	331	558	709	403
	3	326	476	294	412	607	375	570	750	457
30	1	296	497	230	378	630	294	528	779	358
	2	307	521	274	389	662	349	541	819	425
	3	316	542	309	400	690	394	555	855	482
50	1	284	604	245	364	768	314	507	951	384
	2	294	623	293	376	793	375	520	983	458
	3	304	640	331	387	816	423	535	1013	518
100	1	269	774	249	345	993	321	476	1236	393
	2	279	788	299	358	1011	383	490	1259	469
	3	289	801	339	368	1027	431	506	1280	527

	COMMON VENT CAPACITY								
	TYPE B DOUBLE-WALL COMMON VENT DIAMETER – *D* (inch)								
	8			9			10		
	COMBINED APPLIANCE INPUT RATING IN THOUSANDS OF BTU PER HOUR								
VENT HEIGHT *H* (feet)	FAN +FAN	FAN +NAT	NAT +NAT	FAN +FAN	FAN +NAT	NAT +NAT	FAN +FAN	FAN +NAT	NAT +NAT
6	398	310	257	541	429	332	665	515	407
8	436	342	285	592	473	373	730	569	460
10	467	369	309	638	512	398	787	617	487
15	544	434	357	738	599	456	905	718	553
20	606	487	395	824	673	512	1013	808	626
30	703	570	459	958	790	593	1183	952	723
50	833	686	535	1139	954	689	1418	1157	838
100	999	846	606	1378	1185	780	1741	1459	948

For SI units: 1 inch = 25.4 mm, 1 foot = 304.8 mm, 1000 British thermal units per hour = 0.293 kW, 1 square inch = 0.000645 m^2

WATER HEATERS

TABLE 510.2(3)
MASONRY CHIMNEY [NFPA 54: TABLE 13.2(c)]*

		NUMBER OF APPLIANCES:	TWO OR MORE
		APPLIANCE TYPE:	CATEGORY I
		APPLIANCE VENT CONNECTION:	TYPE B DOUBLE-WALL CONNECTOR

VENT CONNECTOR CAPACITY

VENT HEIGHT H (feet)	CONNECTOR RISE R (feet)	TYPE B DOUBLE-WALL VENT CONNECTOR DIAMETER – D (inch)													
		3			4			5			6			7		
		APPLIANCE INPUT RATING LIMITS IN THOUSANDS OF BTU PER HOUR														
		FAN		NAT	FAN		NAT	FAN		NAT	FAN		NAT	FAN		NAT
		Min	Max	Max	Min	Max	Max	Min	Max	Max	Min	Max	Max	Min	Max	Max
6	1	24	33	21	39	62	40	52	106	67	65	194	101	87	274	141
	2	26	43	28	41	79	52	53	133	85	67	230	124	89	324	173
	3	27	49	34	42	92	61	55	155	97	69	262	143	91	369	203
8	1	24	39	22	39	72	41	55	117	69	71	213	105	94	304	148
	2	26	47	29	40	87	53	57	140	86	73	246	127	97	350	179
	3	27	52	34	42	97	62	59	159	98	75	269	145	99	383	206
10	1	24	42	22	38	80	42	55	130	71	74	232	108	101	324	153
	2	26	50	29	40	93	54	57	153	87	76	261	129	103	366	184
	3	27	55	35	41	105	63	58	170	100	78	284	148	106	397	209
15	1	24	48	23	38	93	44	54	154	74	72	277	114	100	384	164
	2	25	55	31	39	105	55	56	174	89	74	299	134	103	419	192
	3	26	59	35	41	115	64	57	189	102	76	319	153	105	448	215
20	1	24	52	24	37	102	46	53	172	77	71	313	119	98	437	173
	2	25	58	31	39	114	56	55	190	91	73	335	138	101	467	199
	3	26	63	35	40	123	65	57	204	104	75	353	157	104	493	222
30	1	24	54	25	37	111	48	52	192	82	69	357	127	96	504	187
	2	25	60	32	38	122	58	54	208	95	72	376	145	99	531	209
	3	26	64	36	40	131	66	56	221	107	74	392	163	101	554	233
50	1	23	51	25	36	116	51	51	209	89	67	405	143	92	582	213
	2	24	59	32	37	127	61	53	225	102	70	421	161	95	604	235
	3	26	64	36	39	135	69	55	237	115	72	435	180	98	624	260
100	1	23	46	24	35	108	50	49	208	92	65	428	155	88	640	237
	2	24	53	31	37	120	60	51	224	105	67	444	174	92	660	260
	3	25	59	35	38	130	68	53	237	118	69	458	193	94	679	285

COMMON VENT CAPACITY

VENT HEIGHT H (feet)	MINIMUM INTERNAL AREA OF MASONRY CHIMNEY FLUE (square inches)														
	12			19			28			38			50		
	COMBINED APPLIANCE INPUT RATING IN THOUSANDS OF BTU PER HOUR														
	FAN +FAN	FAN +NAT	NAT +NAT	FAN +FAN	FAN +NAT	NAT +NAT	FAN +FAN	FAN +NAT	NAT +NAT	FAN +FAN	FAN +NAT	NAT +NAT	FAN +FAN	FAN +NAT	NAT +NAT
6	NA	74	25	NA	119	46	NA	178	71	NA	257	103	NA	351	143
8	NA	80	28	NA	130	53	NA	193	82	NA	279	119	NA	384	163
10	NA	84	31	NA	138	56	NA	207	90	NA	299	131	NA	409	177
15	NA	NA	36	NA	152	67	NA	233	106	NA	334	152	523	467	212
20	NA	NA	41	NA	NA	75	NA	250	122	NA	368	172	565	508	243
30	NA	NA	NA	NA	NA	NA	NA	270	137	NA	404	198	615	564	278
50	NA	NA	NA	NA	NA	NA	NA	NA	NA	NA	NA	NA	NA	620	328
100	NA	NA	NA	NA	NA	NA	NA	NA	NA	NA	NA	NA	NA	NA	348

For SI units: 1 inch = 25.4 mm, 1 foot = 304.8 mm, 1000 British thermal units per hour = 0.293 kW, 1 square inch = 0.000645 m^2

* NA: Not applicable.

WATER HEATERS

TABLE 510.2(3)
MASONRY CHIMNEY [NFPA 54: TABLE 13.2(c)] (continued)*

		NUMBER OF APPLIANCES:			TWO OR MORE				
		APPLIANCE TYPE:			CATEGORY I				
		APPLIANCE VENT CONNECTION:			TYPE B DOUBLE-WALL CONNECTOR				
		VENT CONNECTOR CAPACITY							
		TYPE B DOUBLE-WALL VENT CONNECTOR DIAMETER – D (inch)							
		8			9			10	
		APPLIANCE INPUT RATING LIMITS IN THOUSANDS OF BTU PER HOUR							
VENT HEIGHT H (feet)	CONNECTOR RISE R (feet)	FAN		NAT	FAN		NAT	FAN		NAT
		Min	Max	Max	Min	Max	Max	Min	Max	Max
6	1	104	370	201	124	479	253	145	599	319
	2	107	436	232	127	562	300	148	694	378
	3	109	491	270	129	633	349	151	795	439
8	1	113	414	210	134	539	267	156	682	335
	2	116	473	240	137	615	311	160	776	394
	3	119	517	276	139	672	358	163	848	452
10	1	120	444	216	142	582	277	165	739	348
	2	123	498	247	145	652	321	168	825	407
	3	126	540	281	147	705	366	171	893	463
15	1	125	511	229	153	658	297	184	824	375
	2	128	558	260	156	718	339	187	900	432
	3	131	597	292	159	760	382	190	960	486
20	1	123	584	239	150	752	312	180	943	397
	2	126	625	270	153	805	354	184	1011	452
	3	129	661	301	156	851	396	187	1067	505
30	1	119	680	255	145	883	337	175	1115	432
	2	122	715	287	149	928	378	179	1171	484
	3	125	746	317	152	968	418	182	1220	535
50	1	115	798	294	140	1049	392	168	1334	506
	2	118	827	326	143	1085	433	172	1379	558
	3	121	854	357	147	1118	474	176	1421	611
100	1	109	907	334	134	1222	454	161	1589	596
	2	113	933	368	138	1253	497	165	1626	651
	3	116	956	399	141	1282	540	169	1661	705

	COMMON VENT CAPACITY								
	MINIMUM INTERNAL AREA OF MASONRY CHIMNEY FLUE (square inches)								
	63			78			113		
	COMBINED APPLIANCE INPUT RATING IN THOUSANDS OF BTU PER HOUR								
VENT HEIGHT H (feet)	FAN +FAN	FAN +NAT	NAT +NAT	FAN +FAN	FAN +NAT	NAT +NAT	FAN +FAN	FAN +NAT	NAT +NAT
6	NA	458	188	NA	582	246	1041	853	NA
8	NA	501	218	724	636	278	1144	937	408
10	606	538	236	776	686	302	1226	1010	454
15	682	611	283	874	781	365	1374	1156	546
20	742	668	325	955	858	419	1513	1286	648
30	816	747	381	1062	969	496	1702	1473	749
50	879	831	461	1165	1089	606	1905	1692	922
100	NA	NA	499	NA	NA	669	2053	1921	1058

For SI units: 1 inch = 25.4 mm, 1 foot = 304.8 mm, 1000 British thermal units per hour = 0.293 kW, 1 square inch = 0.000645 m²
* NA: Not applicable.

WATER HEATERS

TABLE 510.2(4)
MASONRY CHIMNEY [NFPA 54: TABLE 13.2(d)]*

		NUMBER OF APPLIANCES:	TWO OR MORE
		APPLIANCE TYPE:	CATEGORY I
		APPLIANCE VENT CONNECTION:	SINGLE-WALL METAL CONNECTOR

VENT CONNECTOR CAPACITY

		SINGLE-WALL METAL VENT CONNECTOR DIAMETER – D (inch)												
		3			4			5			6		7	

APPLIANCE INPUT RATING LIMITS IN THOUSANDS OF BTU PER HOUR

VENT HEIGHT H (feet)	CONNECTOR RISE R (feet)	FAN Min	FAN Max	NAT Max	FAN Min	FAN Max	NAT Max	FAN Min	FAN Max	NAT Max	FAN Min	FAN Max	NAT Max	FAN Min	FAN Max	NAT Max
6	1	NA	NA	21	NA	NA	39	NA	NA	66	179	191	100	231	271	140
	2	NA	NA	28	NA	NA	52	NA	NA	84	186	227	123	239	321	172
	3	NA	NA	34	NA	NA	61	134	153	97	193	258	142	247	365	202
8	1	NA	NA	21	NA	NA	40	NA	NA	68	195	208	103	250	298	146
	2	NA	NA	28	NA	NA	52	137	139	85	202	240	125	258	343	177
	3	NA	NA	34	NA	NA	62	143	156	98	210	264	145	266	376	205
10	1	NA	NA	22	NA	NA	41	130	151	70	202	225	106	267	316	151
	2	NA	NA	29	NA	NA	53	136	150	86	210	255	128	276	358	181
	3	NA	NA	34	97	102	62	143	166	99	217	277	147	284	389	207
15	1	NA	NA	23	NA	NA	43	129	151	73	199	271	112	268	376	161
	2	NA	NA	30	92	103	54	135	170	88	207	295	132	277	411	189
	3	NA	NA	34	96	112	63	141	185	101	215	315	151	286	439	213
20	1	NA	NA	23	87	99	45	128	167	76	197	303	117	265	425	169
	2	NA	NA	30	91	111	55	134	185	90	205	325	136	274	455	195
	3	NA	NA	35	96	119	64	140	199	103	213	343	154	282	481	219
30	1	NA	NA	24	86	108	47	126	187	80	193	347	124	259	492	183
	2	NA	NA	31	91	119	57	132	203	93	201	366	142	269	518	205
	3	NA	NA	35	95	127	65	138	216	105	209	381	160	277	540	229
50	1	NA	NA	24	85	113	50	124	204	87	188	392	139	252	567	208
	2	NA	NA	31	89	123	60	130	218	100	196	408	158	262	588	230
	3	NA	NA	35	94	131	68	136	231	112	205	422	176	271	607	255
100	1	NA	NA	23	84	104	49	122	200	89	182	410	151	243	617	232
	2	NA	NA	30	88	115	59	127	215	102	190	425	169	253	636	254
	3	NA	NA	34	93	124	67	133	228	115	199	438	188	262	654	279

COMMON VENT CAPACITY

MINIMUM INTERNAL AREA OF MASONRY CHIMNEY FLUE (square inches)

	12			19			28			38			50		

COMBINED APPLIANCE INPUT RATING IN THOUSANDS OF BTU PER HOUR

VENT HEIGHT H (feet)	FAN+FAN	FAN+NAT	NAT+NAT	FAN+FAN	FAN+NAT	NAT+NAT	FAN+FAN	FAN+NAT	NAT+NAT	FAN+FAN	FAN+NAT	NAT+NAT	FAN+FAN	FAN+NAT	NAT+NAT
6	NA	NA	25	NA	118	45	NA	176	71	NA	255	102	NA	348	142
8	NA	NA	28	NA	128	52	NA	190	81	NA	276	118	NA	380	162
10	NA	NA	31	NA	136	56	NA	205	89	NA	295	129	NA	405	175
15	NA	NA	36	NA	NA	66	NA	230	105	NA	335	150	NA	400	210
20	NA	NA	NA	NA	NA	74	NA	247	120	NA	362	170	NA	503	240
30	NA	NA	NA	NA	NA	NA	NA	NA	135	NA	398	195	NA	558	275
50	NA	NA	NA	NA	NA	NA	NA	NA	NA	NA	NA	NA	NA	612	325
100	NA	NA	NA	NA	NA	NA	NA	NA	NA	NA	NA	NA	NA	NA	NA

For SI units: 1 inch = 25.4 mm, 1 foot = 304.8 mm, 1000 British thermal units per hour = 0.293 kW, 1 square inch = 0.000645 m^2
* NA: Not applicable.

TABLE 510.2(4)
MASONRY CHIMNEY [NFPA 54: TABLE 13.2(d)] (continued)*

		NUMBER OF APPLIANCES:		TWO OR MORE					
		APPLIANCE TYPE:		CATEGORY I					
		APPLIANCE VENT CONNECTION:		SINGLE-WALL METAL CONNECTOR					
		VENT CONNECTOR CAPACITY							
		SINGLE-WALL METAL VENT CONNECTOR DIAMETER – D (inch)							
		8			9			10	
		APPLIANCE INPUT RATING LIMITS IN THOUSANDS OF BTU PER HOUR							
VENT HEIGHT H (feet)	CONNECTOR RISE R (feet)	FAN		NAT	FAN		NAT	FAN		NAT
		Min	Max	Max	Min	Max	Max	Min	Max	Max
6	1	292	366	200	362	474	252	499	594	316
	2	301	432	231	373	557	299	509	696	376
	3	309	491	269	381	634	348	519	793	437
8	1	313	407	207	387	530	263	529	672	331
	2	323	465	238	397	607	309	540	766	391
	3	332	509	274	407	663	356	551	838	450
10	1	333	434	213	410	571	273	558	727	343
	2	343	489	244	420	640	317	569	813	403
	3	352	530	279	430	694	363	580	880	459
15	1	349	502	225	445	646	291	623	808	366
	2	359	548	256	456	706	334	634	884	424
	3	368	586	289	466	755	378	646	945	479
20	1	345	569	235	439	734	306	614	921	387
	2	355	610	266	450	787	348	627	986	443
	3	365	644	298	461	831	391	639	1042	496
30	1	338	665	250	430	864	330	600	1089	421
	2	348	699	282	442	908	372	613	1145	473
	3	358	729	312	452	946	412	626	1193	524
50	1	328	778	287	417	1022	383	582	1302	492
	2	339	806	320	429	1058	425	596	1346	545
	3	349	831	351	440	1090	466	610	1386	597
100	1	315	875	328	402	1181	444	560	1537	580
	2	326	899	361	415	1210	488	575	1570	634
	3	337	921	392	427	1238	529	589	1604	687

	COMMON VENT CAPACITY								
	MINIMUM INTERNAL AREA OF MASONRY CHIMNEY FLUE (square inches)								
	63			78			113		
	COMBINED APPLIANCE INPUT RATING IN THOUSANDS OF BTU PER HOUR								
VENT HEIGHT H (feet)	FAN +FAN	FAN +NAT	NAT +NAT	FAN +FAN	FAN +NAT	NAT +NAT	FAN +FAN	FAN +NAT	NAT +NAT
6	NA	455	187	NA	579	245	NA	846	NA
8	NA	497	217	NA	633	277	1136	928	405
10	NA	532	234	771	680	300	1216	1000	450
15	677	602	280	866	772	360	1359	1139	540
20	765	661	321	947	849	415	1495	1264	640
30	808	739	377	1052	957	490	1682	1447	740
50	NA	821	456	1152	1076	600	1879	1672	910
100	NA	NA	494	NA	NA	663	2006	1885	1046

For SI units: 1 inch = 25.4 mm, 1 foot = 304.8 mm, 1000 British thermal units per hour = 0.293 kW, 1 square inch = 0.000645 m^2

* NA: Not applicable.

WATER HEATERS

TABLE 510.2(5)
SINGLE-WALL METAL PIPE OR TYPE B ASBESTOS-CEMENT VENT [NFPA 54: TABLE 13.2(e)]*

		NUMBER OF APPLIANCES:			TWO OR MORE		
		APPLIANCE TYPE:			DRAFT HOOD-EQUIPMENT		
		APPLIANCE VENT CONNECTION:			DIRECT TO PIPE OR VENT		
		VENT CONNECTOR CAPACITY					
		VENT CONNECTOR DIAMETER – D (inch)					
TOTAL VENT HEIGHT H (feet)	CONNECTOR RISE R (feet)	3	4	5	6	7	8
		APPLIANCE INPUT RATING IN THOUSANDS OF BTU PER HOUR					
6-8	1	21	40	68	102	146	205
	2	28	53	86	124	178	235
	3	34	61	98	147	204	275
15	1	23	44	77	117	179	240
	2	30	56	92	134	194	265
	3	35	64	102	155	216	298
30 and up	1	25	49	84	129	190	270
	2	31	58	97	145	211	295
	3	36	68	107	164	232	321

COMMON VENT CAPACITY						
COMMON VENT DIAMETER – D (inch)						
TOTAL VENT HEIGHT H (feet)	4	5	6	7	8	10	12
COMBINED APPLIANCE INPUT RATING IN THOUSANDS OF BTU PER HOUR						
6	48	78	111	155	205	320	NA
8	55	89	128	175	234	365	505
10	59	95	136	190	250	395	560
15	71	115	168	228	305	480	690
20	80	129	186	260	340	550	790
30	NA	147	215	300	400	650	940
50	NA	NA	NA	360	490	810	1190

For SI units: 1 inch = 25.4 mm, 1 foot = 304.8 mm, 1000 British thermal units per hour = 0.293 kW, 1 square inch = 0.000645 m^2
* NA: Not applicable.

TABLE 510.2(6)
EXTERIOR MASONRY CHIMNEY [NFPA 54-12: TABLE 13.2(f)]*

	NUMBER OF APPLIANCES:				TWO OR MORE			
	APPLIANCE TYPE:				NAT + NAT			
	APPLIANCE VENT CONNECTION:				TYPE B DOUBLE-WALL CONNECTOR			
COMBINED APPLIANCE MAXIMUM INPUT RATING IN THOUSANDS OF BTU PER HOUR								
VENT HEIGHT H (feet)	INTERNAL AREA OF CHIMNEY (square inches)							
	12	19	28	38	50	63	78	113
6	25	46	71	103	143	188	246	NA
8	28	53	82	119	163	218	278	408
10	31	56	90	131	177	236	302	454
15	NA	67	106	152	212	283	365	546
20	NA	NA	NA	NA	NA	325	419	648
30	NA	NA	NA	NA	NA	NA	496	749
50	NA	NA	NA	NA	NA	NA	NA	922
100	NA	NA	NA	NA	NA	NA	NA	NA

For SI units: 1 inch = 25.4 mm, 1 foot = 304.8 mm, 1000 British thermal units per hour = 0.293 kW, 1 square inch = 0.000645 m^2
* NA: Not applicable.

WATER HEATERS

TABLE 510.2(7)
EXTERIOR MASONRY CHIMNEY [NFPA 54: TABLE 13.2(g)][1,2]

			NUMBER OF APPLIANCES:	TWO OR MORE
			APPLIANCE TYPE:	NAT + NAT
			APPLIANCE VENT CONNECTION:	TYPE B DOUBLE-WALL CONNECTOR

MINIMUM ALLOWABLE INPUT RATING OF SPACE-HEATING APPLIANCE IN THOUSANDS OF BTU PER HOUR

VENT HEIGHT H (feet)	INTERNAL AREA OF CHIMNEY (square inches)							
	12	19	28	38	50	63	78	113
Local 99% winter design temperature: 37°F or greater								
6	0	0	0	0	0	0	0	NA
8	0	0	0	0	0	0	0	0
10	0	0	0	0	0	0	0	0
15	NA	0	0	0	0	0	0	0
20	NA	NA	NA	NA	NA	184	0	0
30	NA	NA	NA	NA	NA	393	334	0
50	NA	NA	NA	NA	NA	NA	NA	579
100	NA	NA	NA	NA	NA	NA	NA	NA
Local 99% winter design temperature: 27°F to 36°F								
6	0	0	68	NA	NA	180	212	NA
8	0	0	82	NA	NA	187	214	263
10	0	51	NA	NA	NA	201	225	265
15	NA	NA	NA	NA	NA	253	274	305
20	NA	NA	NA	NA	NA	307	330	362
30	NA	NA	NA	NA	NA	NA	445	485
50	NA	NA	NA	NA	NA	NA	NA	763
100	NA	NA	NA	NA	NA	NA	NA	NA
Local 99% winter design temperature: 17°F to 26°F								
6	NA	NA	NA	NA	NA	NA	NA	NA
8	NA	NA	NA	NA	NA	NA	264	352
10	NA	NA	NA	NA	NA	NA	278	358
15	NA	NA	NA	NA	NA	NA	331	398
20	NA	NA	NA	NA	NA	NA	387	457
30	NA	NA	NA	NA	NA	NA	NA	581
50	NA	NA	NA	NA	NA	NA	NA	862
100	NA	NA	NA	NA	NA	NA	NA	NA
Local 99% winter design temperature: 5°F to 16°F								
6	NA	NA	NA	NA	NA	NA	NA	NA
8	NA	NA	NA	NA	NA	NA	NA	NA
10	NA	NA	NA	NA	NA	NA	NA	430
15	NA	NA	NA	NA	NA	NA	NA	485
20	NA	NA	NA	NA	NA	NA	NA	547
30	NA	NA	NA	NA	NA	NA	NA	682
50	NA	NA	NA	NA	NA	NA	NA	NA
100	NA	NA	NA	NA	NA	NA	NA	NA
Local 99% winter design temperature: 4°F or lower *Not recommended for any vent configurations*								

For SI units: 1 inch = 25.4 mm, 1 foot = 304.8 mm, 1000 British thermal units per hour = 0.293 kW, 1 square inch = 0.000645 m^2, °C = (°F-32)/1.8

Notes:
[1] See Figure 510.1.2(6) for a map showing local 99 percent winter design temperatures in the United States.
[2] NA: Not applicable.

WATER HEATERS

TABLE 510.2(8)
EXTERIOR MASONRY CHIMNEY [NFPA 54: TABLE 13.2(h)]*

	NUMBER OF APPLIANCES:	TWO OR MORE
	APPLIANCE TYPE:	FAN + NAT
	APPLIANCE VENT CONNECTION:	TYPE B DOUBLE-WALL CONNECTOR

COMBINED APPLIANCE MAXIMUM INPUT RATING IN THOUSANDS OF BTU PER HOUR								
VENT HEIGHT H (feet)	INTERNAL AREA OF CHIMNEY (square inches)								
	12	19	28	38	50	63	78	113
6	74	119	178	257	351	458	582	853
8	80	130	193	279	384	501	636	937
10	84	138	207	299	409	538	686	1010
15	NA	152	233	334	467	611	781	1156
20	NA	NA	250	368	508	668	858	1286
30	NA	NA	NA	404	564	747	969	1473
50	NA	NA	NA	NA	NA	831	1089	1692
100	NA	NA	NA	NA	NA	NA	NA	1921

For SI units: 1 inch = 25.4 mm, 1 foot = 304.8 mm, 1000 British thermal units per hour = 0.293 kW, 1 square inch = 0.000645 m^2

* NA: Not applicable.

TABLE 510.2(9)
EXTERIOR MASONRY CHIMNEY [NFPA 54: TABLE 13.2(i)][1,2]

	NUMBER OF APPLIANCES:	TWO OR MORE
	APPLIANCE TYPE:	FAN + NAT
	APPLIANCE VENT CONNECTION:	TYPE B DOUBLE-WALL CONNECTOR

MINIMUM ALLOWABLE INPUT RATING OF SPACE-HEATING APPLIANCE IN THOUSANDS OF BTU PER HOUR

VENT HEIGHT H (feet)	INTERNAL AREA OF CHIMNEY (square inches)							
	12	19	28	38	50	63	78	113
Local 99% winter design temperature: 37°F or greater								
6	0	0	0	0	0	0	0	0
8	0	0	0	0	0	0	0	0
10	0	0	0	0	0	0	0	0
15	NA	0	0	0	0	0	0	0
20	NA	NA	123	190	249	184	0	0
30	NA	NA	NA	334	398	393	334	0
50	NA	NA	NA	NA	NA	714	707	579
100	NA	NA	NA	NA	NA	NA	NA	1600
Local 99% winter design temperature: 27°F to 36°F								
6	0	0	68	116	156	180	212	266
8	0	0	82	127	167	187	214	263
10	0	51	97	141	183	201	225	265
15	NA	111	142	183	233	253	274	305
20	NA	NA	187	230	284	307	330	362
30	NA	NA	NA	330	319	419	445	485
50	NA	NA	NA	NA	NA	672	705	763
100	NA	NA	NA	NA	NA	NA	NA	1554
Local 99% winter design temperature: 17°F to 26°F								
6	0	55	99	141	182	215	259	349
8	52	74	111	154	197	226	264	352
10	NA	90	125	169	214	245	278	358
15	NA	NA	167	212	263	296	331	398
20	NA	NA	212	258	316	352	387	457
30	NA	NA	NA	362	429	470	507	581
50	NA	NA	NA	NA	NA	723	766	862
100	NA	NA	NA	NA	NA	NA	NA	1669
Local 99% winter design temperature: 5°F to 16°F								
6	NA	78	121	166	214	252	301	416
8	NA	94	135	182	230	269	312	423
10	NA	111	149	198	250	289	331	430
15	NA	NA	193	247	305	346	393	485
20	NA	NA	NA	293	360	408	450	547
30	NA	NA	NA	377	450	531	580	682
50	NA	NA	NA	NA	NA	797	853	972
100	NA	NA	NA	NA	NA	NA	NA	1833
Local 99% winter design temperature: -10°F to 4°F								
6	NA	NA	145	196	249	296	349	484
8	NA	NA	159	213	269	320	371	494
10	NA	NA	175	231	292	339	397	513
15	NA	NA	NA	283	351	404	457	586
20	NA	NA	NA	333	408	468	528	650
30	NA	NA	NA	NA	NA	603	667	805
50	NA	NA	NA	NA	NA	NA	955	1003
100	NA	NA	NA	NA	NA	NA	NA	NA
Local 99% winter design temperature: -11°F or lower Not recommended for any vent configurations								

For SI units: 1 inch = 25.4 mm, 1 foot = 304.8 mm, 1000 British thermal units per hour = 0.293 kW, 1 square inch = 0.000645 m², °C = (°F-32)/1.8

Notes:
[1] See Figure 510.1.2(6) for a map showing local 99 percent winter design temperatures in the United States.
[2] NA: Not applicable.

CHAPTER 6
WATER SUPPLY AND DISTRIBUTION

601.0 General.

601.1 Applicability. This chapter shall govern the materials, design, and installation of water supply systems, including methods and devices used for backflow prevention.

601.2 Hot and Cold Water Required. Except where not deemed necessary for safety or sanitation by the Authority Having Jurisdiction, each plumbing fixture shall be provided with an adequate supply of potable running water piped thereto in an approved manner, so arranged as to flush and keep it in a clean and sanitary condition without danger of backflow or cross-connection. Water closets and urinals shall be flushed using an approved flush tank or flushometer valve.

Exception: Listed fixtures that do not require water for their operation and are not connected to the water supply.

In occupancies where plumbing fixtures are installed for private use, hot water shall be required for bathing, washing, laundry, cooking purposes, dishwashing or maintenance. In occupancies where plumbing fixtures are installed for public use, hot water shall be required for bathing and washing purposes. This requirement shall not supersede the requirements for individual temperature control limitations for public lavatories and public and private bidets, bathtubs, whirlpool bathtubs, and shower control valves.

For safety and sanitation reasons, each plumbing fixture that requires water for its operation is required to be provided with potable water to wash the side walls of the fixture and to replenish the trap seal. The exception to this requirement is for reclaimed water systems that may be installed if Chapter 15 is adopted or the installation is approved per Section 301.3, Alternate Materials and Methods of Construction Equivalency. See Chapter 15 for more information.

Fixtures such as floor drains, receptors, floor sinks, incinerators, gas-fired water closets, chemical-treated toilets, composting toilets, and nonwater urinals do not require water for their operation and are not connected to the water supply.

601.3 Identification of a Potable and Nonpotable Water System. In buildings where potable water and nonpotable water systems are installed, each system shall be clearly identified in accordance with Section 601.3.1 through Section 601.3.5.

Identification of the water supply system is critical to the safe functioning of the building and the protection of the occupants of that building. The system cannot be compromised in any fashion. The first step in the protection of a safe and pure water supply is the correct labeling of the various water systems in the building. This is important not only during construction but especially after the building is occupied when it is subject to maintenance and possibly when added to or altered to later on. Therefore the requirements above must be adhered to on every installation where potable and nonpotable water systems are present.

601.3.1 Potable Water. Green background with white lettering.

601.3.2 Color and Information. Each system shall be identified with a colored pipe or band and coded with paints, wraps, and materials compatible with the piping.

Except as required by Section 601.3.3, nonpotable water systems shall have a yellow background with black uppercase lettering, with the words "CAUTION: NONPOTABLE WATER, DO NOT DRINK." Each nonpotable system shall be identified to designate the liquid being conveyed, and the direction of normal flow shall be clearly shown. The minimum size of the letters and length of the color field shall comply with Table 601.3.2.

The background color and required information shall be indicated every 20 feet (6096 mm) but not less than once per room, and shall be visible from the floor level.

TABLE 601.3.2
MINIMUM LENGTH OF COLOR FIELD AND SIZE OF LETTERS

OUTSIDE DIAMETER OF PIPE OR COVERING (inches)	MINIMUM LENGTH OF COLOR FIELD (inches)	MINIMUM SIZE OF LETTERS (inches)
½ to 1¼	8	½
1½ to 2	8	¾
2½ to 6	12	1¼
8 to 10	24	2½
Over 10	32	3½

For SI units: 1 inch = 25.4 mm

601.3.3 Alternate Water Sources. Alternate water source systems shall have a purple (Pantone color No. 512, 522C, or equivalent) background with uppercase lettering and shall be field or factory marked as follows:

(1) Gray water systems shall be marked in accordance with this section with the words "CAUTION: NONPOTABLE GRAY WATER, DO NOT DRINK" in black letters.

(2) Reclaimed (recycled) water systems shall be marked in accordance with this section with the words: "CAUTION: NONPOTABLE RECLAIMED (RECYCLED) WATER, DO NOT DRINK" in black letters.

(3) On-site treated water systems shall be marked in accordance with this section with the words: "CAUTION: ON-SITE TREATED NONPOTABLE WATER, DO NOT DRINK" in black letters.

(4) Rainwater catchment systems shall be marked in accordance with this section with the words: "CAUTION: NONPOTABLE RAINWATER WATER, DO NOT DRINK" in black letters.

WATER SUPPLY AND DISTRIBUTION

601.3.4 Fixtures. Where vacuum breakers or backflow preventers are installed with fixtures listed in Table 1701.1, identification of the discharge side shall be permitted to be omitted.

601.3.5 Outlets. Each outlet on the nonpotable water line that is used for special purposes shall be posted with black uppercase lettering as follows: "CAUTION: NON-POTABLE WATER, DO NOT DRINK."

602.0 Unlawful Connections.

602.1 Prohibited Installation. No installation of potable water supply piping, or part thereof, shall be made in such a manner that it will be possible for used, unclean, polluted, or contaminated water, mixtures, or substances to enter a portion of such piping from a tank, receptor, equipment, or plumbing fixture by reason of backsiphonage, suction, or other cause, either during normal use and operation thereof, or where such tank, receptor, equipment, or plumbing fixture is flooded or subject to pressure exceeding the operating pressure in the hot or cold water piping.

The building potable water supply installation must not be subjected to contamination or pollution from any source. Chemicals that could produce toxic conditions in the systems are prohibited from being introduced into any part of the potable water piping system. Similarly, piping materials that could produce toxic conditions and any piping that has been used for conveying fluids other than potable water are prohibited from use in the potable water supply system.

Potable water supply piping, water outlets and equipment must be protected from any contaminated or polluted liquid or substance, thereby avoiding potential contamination from such sources. The system must be protected from any type of cross-connection – back siphonage or back pressure – at all times and in all instances. This protection must occur at the installation of the system.

602.2 Cross-Contamination. No person shall make a connection or allow one to exist between pipes or conduits carrying domestic water supplied by a public or private building supply system, and pipes, conduits, or fixtures containing or carrying water from any other source or containing or carrying water that has been used for any purpose whatsoever, or piping carrying chemicals, liquids, gases, or substances whatsoever, unless there is provided a backflow prevention device approved for the potential hazard and maintained in accordance with this code. Each point of use shall be separately protected where potential cross-contamination of individual units exists.

No person, during installation or after the building is occupied and maintained, may make any connection that might introduce used, contaminated or polluted water or any other substance into the potable water supply by any means. Different sources of water should be kept separate to ensure that if one source is contaminated the other source will not be contaminated. Where contamination can occur, backflow prevention devices or methods shall be used to protect the system. See Section 603.3. Each fixture supply or outlet must be protected.

602.3 Backflow Prevention. No plumbing fixture, device, or construction shall be installed or maintained, or shall be connected to a domestic water supply, where such installation or connection provides a possibility of polluting such water supply or cross-connection between a distributing system of water for drinking and domestic purposes and water that becomes contaminated by such plumbing fixture, device, or construction unless there is provided a backflow prevention device approved for the potential hazard.

No installation of a fixture, device or an arrangement of the piping system that may cause a cross-connection shall be allowed unless it is protected by a backflow prevention device or method. Nor shall any maintenance or repair of the piping system cause a cross-connection.

602.4 Approval by Authority. No water piping supplied by a private water supply system shall be connected to any other source of supply without the approval of the Authority Having Jurisdiction, Health Department, or other department having jurisdiction.

Direct connection between any private water supply system and a public potable water supply system, except as approved by the applicable authority, is prohibited. Once the water has passed the well or the meter, water purveyors consider it to be used. The utility supplying this water can ask for protection at the meter (containment backflow protection or premise isolation) to prevent water from an unapproved source or any other foreign substance from entering the public water supply system.

603.0 Cross-Connection Control.

603.1 General. Cross-connection control shall be provided in accordance with the provisions of this chapter.

No person shall install a water-operated equipment or mechanism, or use a water-treating chemical or substance, where it is found that such equipment, mechanism, chemical, or substance causes pollution or contamination of the domestic water supply. Such equipment or mechanism shall be permitted where equipped with an approved backflow prevention device or assembly.

The prevention of cross-connections is a primary responsibility of the plumber, maintenance or service worker. A thorough knowledge of backflow prevention and installation and testing of backflow prevention devices is needed to ensure the protection of the water system. IAPMO's Backflow Reference Manual contains comprehensive information on installation, testing, cross-connection control, repair, and inspections.

Cross-connection means any actual or potential connection or structural arrangement between a public or private potable water system and any other source or system through which it is possible to introduce into any part of the potable system any used water, industrial fluid, gas or substance other than the intended potable water with which the system is supplied. Bypass arrangements, jumper connections, removable sections, swivel or changeover devices and other temporary devices through which backflow can occur are considered to be potential unintended cross-connections.

There are two types of cross-connections – direct and indirect cross-connections. There are two types of backflow caused by these cross-connections – backsiphonage and backpressure backflow – and there are two types of backflow conditions – pollution (low hazard) or contamination (high hazard).

Polluted water may have an undesirable taste, color, or odor, but it is not considered unfit for human consumption and will not cause sickness. Contaminated water is water that is not safe for human consumption and will cause sickness or even death. Contaminated water is always considered to be nonpotable and requires a more absolute method or device to prevent backflow. It must be understood that a device or method that is suitable for only isolating potable water from a potential pollutant may be unacceptable as a means of protection against contamination. For example, a double check valve backflow assembly will be acceptable for a pollutant (low hazard) but not for a contaminant (high hazard).

For a cross-connection to occur, a link or channel between a polluted water source and pipes carrying potable water must exist. Two factors are essential for backflow to occur in a cross-connection. First, there must be a link between the two systems; this consists of either a physical connection or an arrangement or situation where backflow can occur. Second, an unbalanced force of liquid pressure must act toward the potable water supply.

In a direct cross-connection there must be a direct physical connection existing between the potable and nonpotable system. An increase in pressure must also exist on the downstream side of this connection causing a reversal of flow from the original intended direction of flow. Backflow occurs when the pressure within a polluted system exceeds the pressure in the potable water system. This is a backpressure cross-connection. A typical example of this type of cross-connection is the quickfill line for a boiler. The treated water in the boiler and connected piping represent the hazard, in this case a high hazard. This is because the pump circulating the fluid in the piping system could cause an increase of pressure above the quickfill line supply pressure and cause a backflow of treated water into the potable supply. Therefore, Section 603.5.10 requires a reduced pressure principle backflow prevention assembly where chemicals are introduced into the system (see **Figure 603.1a**).

The other type of cross-connection is an indirect cross-connection. There are two types of indirect cross-connections – under-rim or submerged connections, and over-rim connections. The under-rim or submerged cross-connection is where the water inlet comes into the bottom or side of a receptacle and is immersed in a polluted or contaminated substance (see D in **Figure 603.1c**). Without some form of protection just the filling of the lavatory to its rim could cause backflow into the potable water supply.

An over-rim connection is one where the water supply terminates above the flood level rim of a fixture but has a hose fitting or connection that creates a potential for an under-rim termination (see A, B and F in **Figure 603.1c**). In this instance, the over-rim supply line may not be continuously submerged unless a hose is permanently attached. Backflow may occur if the hose is left in the sink and something happens to cause a negative or lower pressure in the fixture supply, such as a break in the line or draining the system while leaving the fixture valve on. This will cause a siphon to occur drawing the possibly contaminated water into the potable water supply. The type of backflow that occurs in these instances is backsiphonage. See **Figure 603.1c** for other examples of backsiphonage.

The control of backflow requires removal of one of the two essential factors that can cause the backflow; namely, the physical link or the cause of the negative or low pressure. Removal of the physical link or cross-connection, such as in **Figure 603.1b** is a positive means of preventing backflow and is the only true means of preventing bacterial contamination. This can also be accomplished with a backflow prevention device or an air gap as specified in the following sections. The appropriate selection, installation and testing of the devices are functions essential to the process of isolating pollutants and contaminants from the potable water system.

There are two basic areas where backflow protection is required. The first is the protection of the public or private water supply. The water purveyor may require all connections to the public water supply to have some form of protection. In some areas this may mean that every water service to a public, private or residential building may require a backflow prevention device to protect the system. This is called containment protection. The public water supply is contained and protected from contamination.

FIGURE 603.1A
BOILER QUICK FILL VALVE WITH BACKFLOW PROTECTION

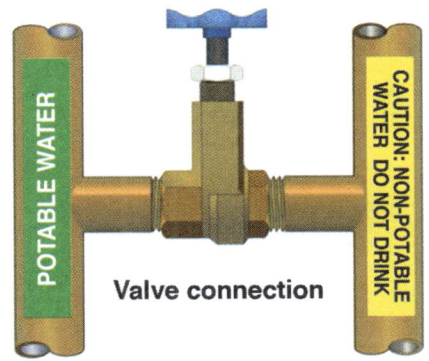

FIGURE 603.1B
DIRECT LINK CROSS-CONNECTION

WATER SUPPLY AND DISTRIBUTION

The other area of protection is inside or outside the building downstream of the containment protection. There may be connections to equipment or systems that could contaminate the building water supply. Backflow protection is required at this connection. This is called isolation protection. The contaminated fluid is isolated downstream of the connection to the building water supply. An atmospheric vacuum breaker on a urinal is a case of isolation protection.

603.2 Approval of Devices or Assemblies. Before a device or an assembly is installed for the prevention of backflow, it shall have first been approved by the Authority Having Jurisdiction. Devices or assemblies shall be tested in accordance with recognized standards or other standards acceptable to the Authority Having Jurisdiction. Backflow prevention devices and assemblies shall comply with Table 603.2, except for specific applications and provisions as stated in Section 603.5.1 through Section 603.5.21.

Devices or assemblies installed in a potable water supply system for protection against backflow shall be maintained in good working condition by the person or persons having control of such devices or assemblies. Such devices or assemblies shall be tested at the time of installation, repair, or relocation and not less than on an annual schedule thereafter, or more often where required by the Authority Having Jurisdiction. Where found to be defective or inoperative, the device or assembly shall be repaired or replaced. No device or assembly shall be removed from use or relocated or other device or assembly substituted, without the approval of the Authority Having Jurisdiction.

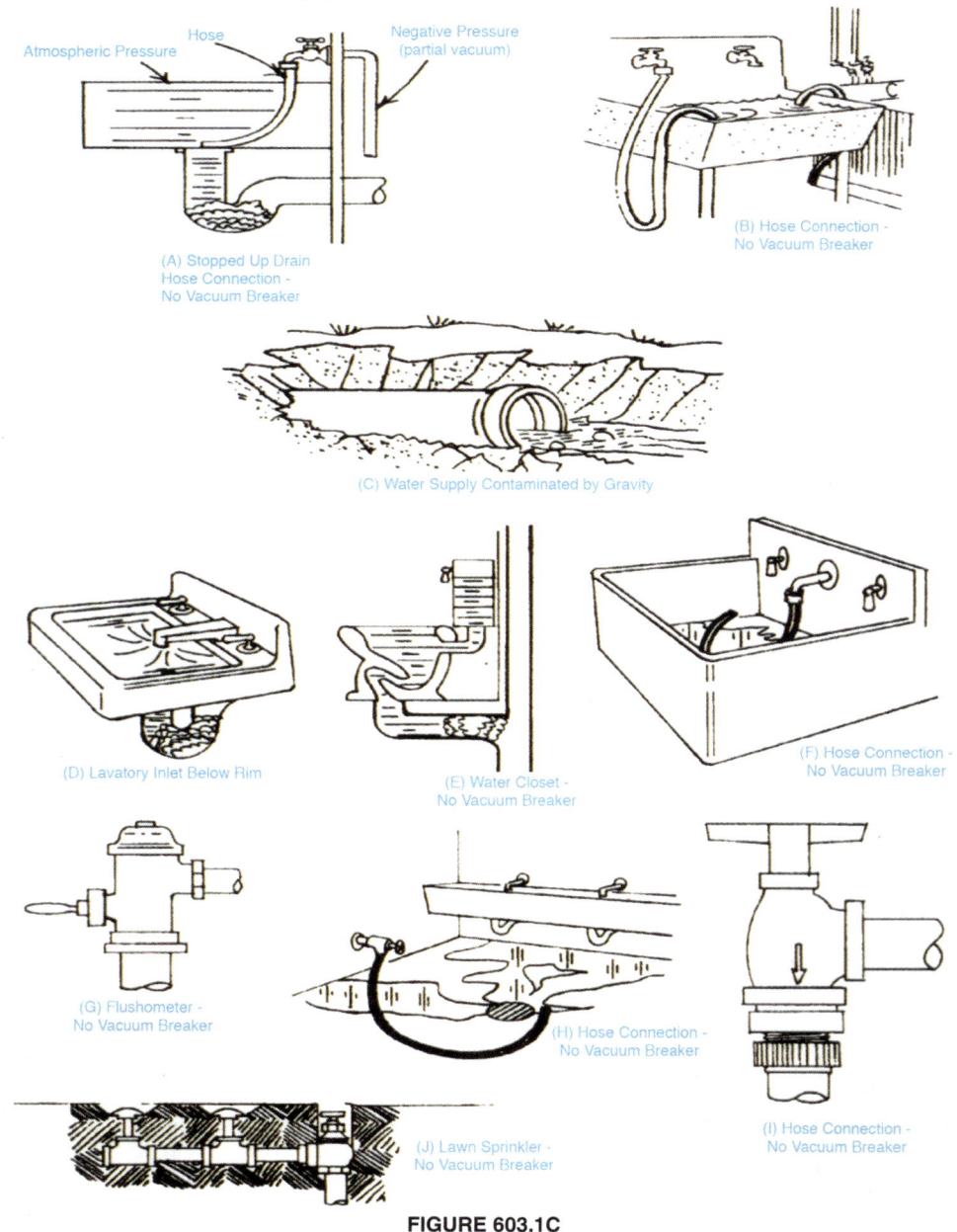

FIGURE 603.1C
CROSS-CONNECTIONS

WATER SUPPLY AND DISTRIBUTION

TABLE 603.2
BACKFLOW PREVENTION DEVICES, ASSEMBLIES, AND METHODS

DEVICE, ASSEMBLY, OR METHOD[1]	APPLICABLE STANDARDS	DEGREE OF HAZARD				INSTALLATION[2,3]
		POLLUTION (LOW HAZARD)		CONTAMINATION (HIGH HAZARD)		
		BACK-SIPHONAGE	BACK-PRESSURE	BACK-SIPHONAGE	BACK-PRESSURE	
Air gap	ASME A112.1.2	X	—	X	—	See Table 603.3.1 in this chapter.
Air gap fittings for use with plumbing fixtures, appliances, and appurtenances	ASME A112.1.3	X	—	X	—	Air gap fitting is a device with an internal air gap, and typical installation includes plumbing fixtures, appliances, and appurtenances. The critical level shall not be installed below the flood level rim.
Atmospheric vacuum breaker (consists of a body, checking member and atmospheric port)	ASSE 1001 or CSA B64.1.1	X	—	X	—	Upright position. No valve downstream. Minimum of 6 inches or listed distance above all downstream piping and flood level rim of receptor.[4,5]
Antisiphon fill valve (ballcocks) for gravity water closet flush tanks and urinal tanks	ASSE 1002/ ASME A112.1002/ CSA B125.12	X	—	X	—	Installation of gravity water closet flush tank and urinal tanks with the fill valve installed with the critical level not less than 1 inch above the opening of the overflow pipe.[4,5]
Vacuum breaker wall hydrants, hose bibbs, freeze resistant, automatic draining type	ASSE 1019 or CSA B64.2.1.1	X	—	X	—	Installation includes wall hydrants and hose bibbs. Such devices are not for use under continuous pressure conditions (means of shutoff downstream of device is prohibited).[4,5]
Hose connection vacuum breakers	ASSE 1011	X	—	X	—	Such devices are not for use under continuous pressure conditions. No valve downstream.[4,6]
Hose connection backflow preventers	ASSE 1052	X	—	X	—	Such devices are not for use under continuous pressure conditions.[4,6]
Dual check backflow preventer wall hydrants, freeze resistant	ASSE 1053	X	—	X	—	Such devices are not for use under continuous pressure conditions.[4]
Freeze resistant sanitary yard hydrants	ASSE 1057	X	—	X	—	Such devices are not for use under continuous pressure conditions.[4]
Backflow preventer for Carbonated Beverage Dispensers (two independent check valves with a vent to the atmosphere)	ASSE 1022	X	—	—	—	Installation includes carbonated beverage machines or dispensers. These devices operate under intermittent or continuous pressure conditions.
Spill-Resistant Pressure Vacuum Breaker (single check valve with air inlet vent and means of field testing)	ASSE 1056	X	—	X	—	Upright position. Minimum of 12 inches or listed distance above all downstream piping and flood-level rim of receptor.[5]
Double Check Valve Backflow Prevention Assembly (two independent check valves and means of field testing)	ASSE 1015; AWWA C510; CSA B64.5 or CSA B64.5.1	X	X	—	—	Horizontal unless otherwise listed. Access and clearance shall be in accordance with the manufacturer's instructions, and not less than a 12 inch clearance at the bottom for maintenance. May need platform/ladder for test and repair. Does not discharge water.

WATER SUPPLY AND DISTRIBUTION

TABLE 603.2
BACKFLOW PREVENTION DEVICES, ASSEMBLIES, AND METHODS (continued)

DEVICE, ASSEMBLY, OR METHOD[1]	APPLICABLE STANDARDS	DEGREE OF HAZARD				INSTALLATION[2,3]
		POLLUTION (LOW HAZARD)		CONTAMINATION (HIGH HAZARD)		
		BACK-SIPHONAGE	BACK-PRESSURE	BACK-SIPHONAGE	BACK-PRESSURE	
Double Check Detector Fire Protection Backflow Prevention Assembly (two independent check valves with a parallel detector assembly consisting of a water meter and a double check valve backflow prevention assembly and means for field testing)	ASSE 1048	X	X	—	—	Horizontal unless otherwise listed. Access and clearance shall be in accordance with the manufacturer's instructions, and not less than a 12 inch clearance at the bottom for maintenance. May need platform/ladder for test and repair. Does not discharge water. Installation includes a fire protection system and is designed to operate under continuous pressure conditions.
Pressure Vacuum Breaker Backflow Prevention Assembly (loaded air inlet valve, internally loaded check valve and means for field testing)	ASSE 1020 or CSA B64.1.2	X	—	X	—	Upright position. May have valves downstream. Minimum of 12 inches above all downstream piping and flood-level rim of the receptor. May discharge water.
Reduced Pressure Principle Backflow Prevention Assembly (two independently acting loaded check valves, a differential pressure relief valve and means for field testing)	ASSE 1013; AWWA C511; CSA B64.4 or CSA B64.4.1	X	X	X	X	Horizontal unless otherwise listed. Access and clearance shall be in accordance with the manufacturer's instructions, and not less than a 12 inch clearance at the bottom for maintenance. May need platform/ladder for test and repair. May discharge water.
Reduced Pressure Detector Fire Protection Backflow Prevention Assembly (two independently acting loaded check valves, a differential pressure relief valve, with a parallel detector assembly consisting of a water meter and a reduced-pressure principle backflow prevention assembly, and means for field testing)	ASSE 1047	X	X	X	X	Horizontal unless otherwise listed. Access and clearance shall be in accordance with the manufacturer's instructions, and not less than a 12 inch clearance at the bottom for maintenance. May need platform/ladder for test and repair. May discharge water. Installation includes a fire protection system and is designed to operate under continuous pressure conditions.

For SI units: 1 inch = 25.4 mm

Notes:
[1] See the description of devices and assemblies in this chapter.
[2] Installation in pit or vault requires previous approval by the Authority Having Jurisdiction.
[3] Refer to the general and specific requirement for installation.
[4] Not to be subjected to operating pressure for more than 12 hours in a 24 hour period.
[5] For deck-mounted and equipment-mounted vacuum breaker, see Section 603.5.13.
[6] Shall be installed in accordance with Section 603.5.7.

WATER SUPPLY AND DISTRIBUTION

Testing or maintenance shall be performed by a certified backflow assembly tester or repairer in accordance with ASSE Series 5000 or otherwise approved by the Authority Having Jurisdiction.

Several organizations list approved backflow prevention devices. IAPMO, American Society of Sanitary Engineering (ASSE) and the University of Southern California Foundation for Cross-Connection Control are just a few. One of the easiest methods to ensure that the device is listed and approved is to acquire a device with the IAPMO shield.

The requirement for testing and maintaining the backflow prevention device is contained in this section. The device must be tested for proper working conditions upon installation and at least on an annual basis. In many instances the water authority or the health department may be the Authority Having Jurisdiction (AHJ) over cross-connection control. Requirements for installation may differ from jurisdiction to jurisdiction. Be sure to contact the local AHJ for those requirements.

603.3 Backflow Prevention Devices, Assemblies, and Methods. Backflow prevention devices, assemblies, and methods shall comply with Section 603.3.1 through Section 603.3.9.

603.3.1 Air Gap. The minimum air gap to afford backflow protection shall be in accordance with Table 603.3.1.

The airgap as a means of backflow protection is the primary method of protecting the water supply at the fixture (see **Figure 603.3.1a**). When the appropriate airgap separation is provided at each fixture between its flood level rim and the water supply outlet, primary protection will have been provided for the building water supply as well as for the public water supply. This section requires all installations utilizing an airgap to be in accordance with Table 603.3.1.

Table 603.3.1 establishes the minimum vertical distance above the flood level rim of the fixture to the water supply outlet. In most instances the distance of a fixture faucet, tub spout or pipe opening to the rim of the fixture is being discussed. There are other installations of airgaps such as a drainage system airgap. This type of installation also must adhere to the requirements contained in Table 603.3.1.

Table 603.3.1 applies mostly to fixture and fixture fitting manufacturers. The table refers to two types of fixtures – fixtures without sidewalls or ribs (see **Figure 603.3.1a**) and fixtures with side walls (see **Figure 603.3.1b**). The fixture manufacturer will use this table to design the fixture with the proper height of the faucet above the rim for the proper airgap. Although most plumbers will properly install the specified and required fixture and faucet and be done with it, Table 603.3.1 does have very important information for them.

The minimum airgap required for any installation is contained in the first row of the table. For effective openings not greater than one-half inch, a minimum airgap of one inch is required. Therefore, all airgaps must be at least one inch in vertical distance above the flood level rim. As one goes down the table, the effective opening increases in size and the resulting airgap also increases, culminating in the requirement for any effective opening greater than one inch which requires an airgap of two times the diameter of the opening. These are two of the most important factors for the protection of the water supply and they will be repeated again in the code.

As noted above, in determining the proper size of the airgap, the effective opening is used. This must not be confused with the faucet or spout opening. Note 3 at the bottom of Table 603.3.1 defines the "effective opening" as being the "minimum cross-sectional area at the seat of the control valve or supply pipe." This location is usually at the point in the valve where the two supply lines (hot and cold) intersect with the supply to the faucet spout. The opening

TABLE 603.3.1
MINIMUM AIR GAPS FOR WATER DISTRIBUTION[4]

FIXTURES	WHERE NOT AFFECTED BY SIDEWALLS[1] (inches)	WHERE AFFECTED BY SIDEWALLS[2] (inches)
Effective openings[3] not greater than ½ of an inch in diameter	1	1½
Effective openings[3] not greater than ¾ of an inch in diameter	1½	2¼
Effective openings[3] not greater than 1 inch in diameter	2	3
Effective openings[3] greater than 1 inch in diameter	Two times the diameter of effective opening	Three times the diameter of effective opening

For SI units: 1 inch = 25.4 mm

Notes:
[1] Sidewalls, ribs, or similar obstructions do not affect air gaps where spaced from the inside edge of the spout opening a distance exceeding three times the diameter of the effective opening for a single wall, or a distance exceeding four times the effective opening for two intersecting walls.
[2] Vertical walls, ribs, or similar obstructions extending from the water surface to or above the horizontal plane of the spout opening other than specified in Footnote 1 above. The effect of three or more such vertical walls or ribs has not been determined. In such cases, the air gap shall be measured from the top of the wall.
[3] The effective opening shall be the minimum cross-sectional area at the seat of the control valve or the supply pipe or tubing that feeds the device or outlet. Where two or more lines supply one outlet, the effective opening shall be the sum of the cross-sectional areas of the individual supply lines or the area of the single outlet, whichever is smaller.
[4] Air gaps less than 1 inch (25.4 mm) shall be approved as a permanent part of a listed assembly that has been tested under actual backflow conditions with vacuums of 0 to 25 inches of mercury (85 kPa).

WATER SUPPLY AND DISTRIBUTION

or seat to the faucet or valve spout is usually the smallest opening and is thus the effective opening (see A in **Figure 603.3.1c**). In most faucets and valves this opening (See C in **Figure 603.3.1c**) is approximately three-eighth inch and the airgap required is one inch.

In cases where the end of a pipe is the effective opening, such as a boiler blowoff, the minimum airgap is then two times the diameter of the end of the pipe. For example, the boiler blowoff is 2 inches in diameter the airgap above the floor sink receiving the blowoff would be 4 inches.

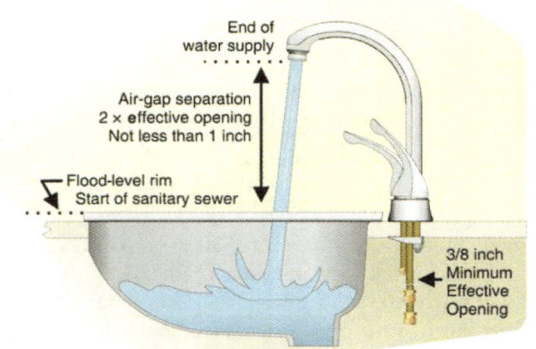

FIGURE 603.3.1A
AIR GAP

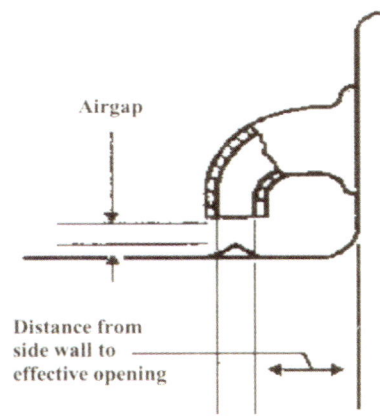

FIGURE 603.3.1B
AIR GAP AFFECTED BY SIDE WALL

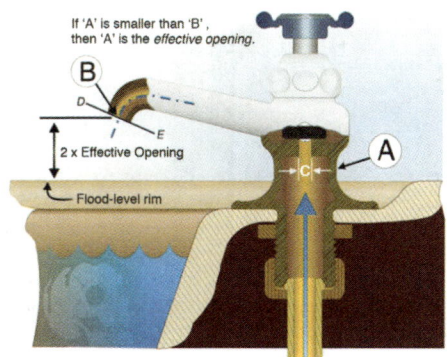

FIGURE 603.3.1C
EFFECTIVE OPENING

603.3.2 Atmospheric Vacuum Breaker (AVB). An atmospheric vacuum breaker consists of a body, a checking member, and an atmospheric port.

The purpose of a vacuum breaker is to stop back-siphonage. The atmospheric vacuum breaker (AVB) consists of a check valve member and an air vent that is normally closed when the device is pressurized. The air vent allows air into the piping system for preventing a siphon at the point of use (see **Figure 603.3.2a**). A common design element is that the air- inlet valve and check valve are nearly always the same mechanical component.

Although simple in its design, the AVB has several installation requirements to function properly:

- Because the AVB provides back-siphonage protection only, it is considered isolation protection only. These devices are located on individual plumbing fixtures and appliances. They will not be installed at the service line to a building, for example, where the device may be subject to backpressure.

- All vacuum breakers are considered high-hazard protection. These devices protect the potable system from toxic materials. Remember that any backflow preventer that provides high-hazard protection will be acceptable for low- hazard applications.

- The AVB must be installed with its critical level a minimum of six inches above the highest downstream piping and flood level rim of receptor (see **Figure 603.3.2b**). A deck mounted AVB must be installed with its critical level no less than one-inch above the flood level rim of the fixture.

- The AVB must not be subjected to more than twelve hours of continuous water pressure. Continuous water pressure is where upstream pressure is applied continuously to a device or assembly. If the AVB is used with continuous pressure, the valve may become stuck or sealed closed and not open on demand to atmosphere. The reasons for this may include mineral deposits or chemical bonding from water quality conditions, or a spider's web holding the valve closed to atmosphere.

- Valves are not permitted downstream of an AVB. Components such as check valves, gate valves, solenoid valves or pressure-regulating devices can trap or suspend pressure in the AVB, thereby allowing continuous pressure.

- An AVB must not be subjected to backpressure. This valve allows air into the system to stop a siphon with a water pressure loss. The device cannot determine the water's direction of flow, as long as water pressure holds the valve closed to atmosphere. Therefore, elevated piping, auxiliary sources of water and pressure pumps, for example, are not permitted downstream.

- AVBs must be installed upright. The valve that allows air into the downstream piping relies on gravity to function properly. If the device is installed out of plumb, the valve may not fully open.

603.3.3 Hose Connection Backflow Preventer. A hose connection backflow preventer consists of two inde-

WATER SUPPLY AND DISTRIBUTION

pendent check valves with an independent atmospheric vent between and a means of field testing and draining.

A hose connection vacuum breaker (HVB) should be installed on each faucet or hose bibb that is connected to the potable water supply to prevent backflow into the water supply. HVB's must be installed at least six inches above the ground surface. The HVB is an AVB and must follow the installation parameters of Table 603.2. Most HVBs have a set screw that prevents them from being easily removed once they are installed. This prevents them from being removed when the garden hose is removed. Once the HVB is installed, no further adjustments are required (see **Figure 603.3.3**).

A HVB prevents backflow to the water supply by venting water to the atmosphere (onto the ground) when backflow conditions exist. A spring-loaded check valve is opened by the water supply pressure when outflow occurs through the valve. When pressure is sufficient to open the check valve, flow is directed into the garden hose. When the supply pressure is interrupted or when the pressure in the hose becomes greater than the supply pressure, outflow stops and the spring-loaded check valve closes, simultaneously opening a vent to the atmosphere. In this mode of operation, any water that flows backwards through the HVB is vented onto the ground.

The spring-loaded check valve does not allow drainage of water from between the hose bibb and the upper part of the HVB; thus, freeze protection must be provided, just as all outdoor plumbing would need to be protected under freezing conditions.

HVBs should be inspected and tested periodically to ensure that they are working properly. Fortunately, their mode of operation permits inspections to be easily made. Verify that the check valve closes and the atmospheric vent opens reliably whenever the water supply is shut off. It is very simple to do this if a nozzle that can be shut off is used on the end of the hose. When the nozzle shuts off, turn on the faucet and allow the hose to pressurize. Then, shut off the faucet while the hose is pressurized. After a few seconds, the hose pressure should be released in a small spray as the atmospheric vent suddenly opens.

603.3.4 Double Check Valve Backflow Prevention Assembly (DC). A double check valve backflow prevention assembly consists of two independently acting internally loaded check valves, four properly located test cocks, and two isolation valves.

The double check valve backflow prevention assembly (DC) is just what it says – two check valves in series to prevent backflow (see **Figure 603.3.4**). The first check will close in a backflow condition. Usually one psi of backflow pressure will close the check. The second check is incorporated as a backup if the first check fails; however, there is no way to know if the second check has failed until it is tested. If the device fails, there is no means for removal of the polluted or contaminated water, which may then travel into the potable line. Therefore, the DC may only be used on low-hazard applications. Any liquid connected to the piping downstream of a DC should be a pollutant and not a contaminant.

The DC provides backpressure and backsiphonage protection. A DC may be used as meter-service protection as a containment, or on individual fixtures and appliances as isolation protection. A field test is required on installation, at least annually thereafter, after any repair and when relocated. The device must be installed a minimum of 12 inches above the surrounding ground or floor.

603.3.5 Pressure Vacuum Breaker Backflow Prevention Assembly (PVB). A pressure vacuum breaker

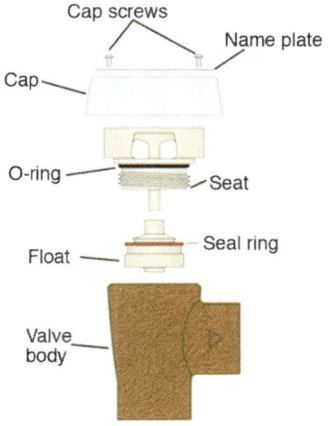

**FIGURE 603.3.2A
ATMOSPHERIC VACUUM BREAKER (AVB)**

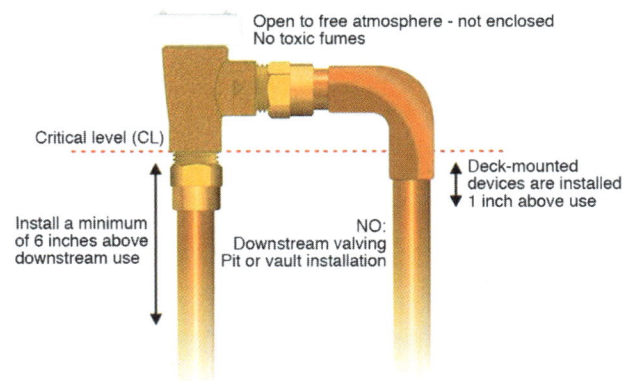

**FIGURE 603.3.2B
INSTALLATION OF AVB**

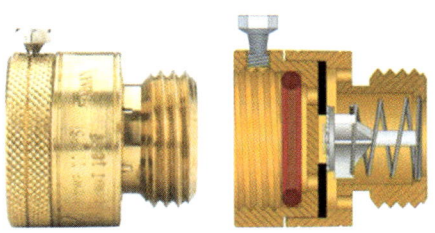

**FIGURE 603.3.3
HOSE CONNECTION VACUUM BREAKER**

WATER SUPPLY AND DISTRIBUTION

backflow prevention assembly consists of a loaded air inlet valve, an internally loaded check valve, two properly located test cocks, and two isolation valves. This device shall be permitted to be installed indoors where provisions for spillage are provided.

🔧 The pressure type vacuum breaker (PVB) evolved from the AVB. There are a number of AVB installation restrictions due to the limits of the design. One of the AVB's limitations is the restriction of no continuous water pressure. The concern of the inlet valve sticking closed is serious, as it would render the AVB useless. The PVB was designed to overcome this problem.

The AVB has one moving part that works as an air inlet or port and it serves as a check valve to the supply piping when no water pressure is present. The PVB check valve is similar but spring loaded. When the piping-system pressure is reduced, the spring forces the check open to the atmosphere, allowing air into the system and breaking the downstream siphon.

PVBs, therefore, are designed to operate under pressure for long periods of time without becoming inoperative. The internal check valves are spring-loaded so that any tendency for the valve to stick closed due to long periods of applied pressure and consequent fouling will be counteracted.

The continuous pressure restriction would no longer apply and would allow valves downstream and water pressure 24 hours per day. This type of assembly may be used only where outlet pressure will never exceed inlet pressure and is not subject to backpressure. This device may be installed indoors only if provisions for spillage from the dome are provided. They must be installed a minimum of 12 inches above the highest piping downstream of the PVB (see **Figure 603.3.5**).

603.3.6 Spill-Resistant Pressure Vacuum Breaker (SVB). A pressure-type vacuum breaker backflow prevention assembly consists of one check valve force-loaded closed and an air inlet vent valve force-loaded open to atmosphere, positioned downstream of the check valve and located between and including two tightly closing shutoff valves and test cocks.

🔧 The spill-resistant type vacuum breaker (SVB) is essentially a next-generation PVB. This assembly evolved because of the spillage encountered with the PVB. The function of the SVB is to eliminate the water spillage of the PVB.

The pressure vacuum breaker requires that both the check valve and air inlet act independently, but the components of the SVB are not required to be independent of each other. During operation (flow), the diaphragm at the bottom of the valve chamber allows the poppet to close before the

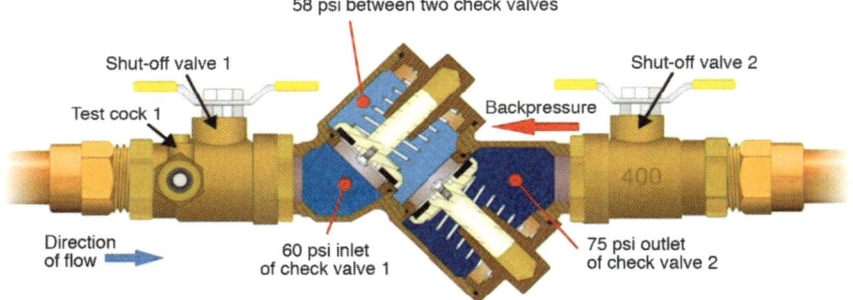

Double check valve assembly - static condition with checks closed - downstream back pressure.
(Pressure values shown for example only)

FIGURE 603.3.4
DOUBLE CHECK VALVE BACKFLOW PREVENTION ASSEMBLY

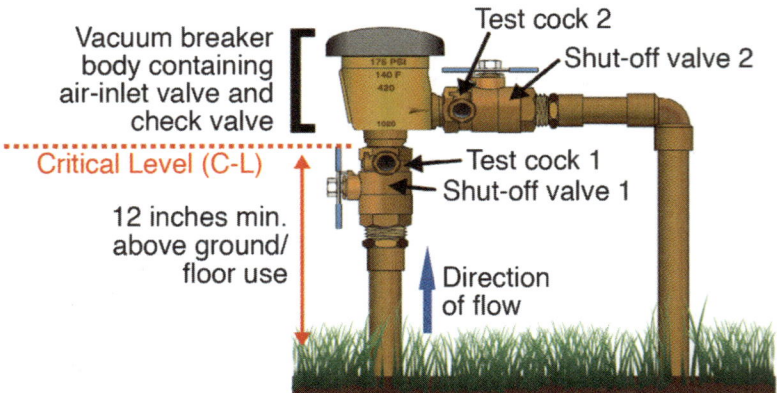

FIGURE 603.3.5
PRESSURE VACUUM BREAKER BACKFLOW PREVENTION ASSEMBLY INSTALLATION

WATER SUPPLY AND DISTRIBUTION

internal check valve opens, preventing most of the spillage that normally occurs with a PVB. In this way, the air inlet can be closed as the check is closed again, eliminating the spill (see **Figure 603.3.6**).

The SVB provides the same type of protection as the PVB and can only be used where the PVB can be used.

603.3.7 Reduced-Pressure Principle Backflow Prevention Assembly (RP).
A reduced-pressure principle backflow prevention assembly consists of two independently acting internally loaded check valves, a differential pressure relief valve, four properly located test cocks, and two isolation valves.

🔧 The reduced pressure principle backflow prevention assembly is referred to as the RP, RPPZ, RPZ as well as other acronyms. The RP is the best mechanical backflow preventer available today. It is considered as proper protection for high- or low-hazard applications, backpressure or backsiphonage backflow and containment or isolation protection.

The name of this assembly is derived from a zone of reduced pressure needed for the operation of the relief or vent valve. This device is an assembly of two internally loaded, specially designed and independently operating check valves that have a relief valve installed between the primary valves, specifically designed to maintain a zone of relative differential pressure between the two check valves unless one or both of the check valves are fouled. This assembly has a tightly closing upstream and a tightly closing downstream shutoff valve (see **Figure 603.3.7**).

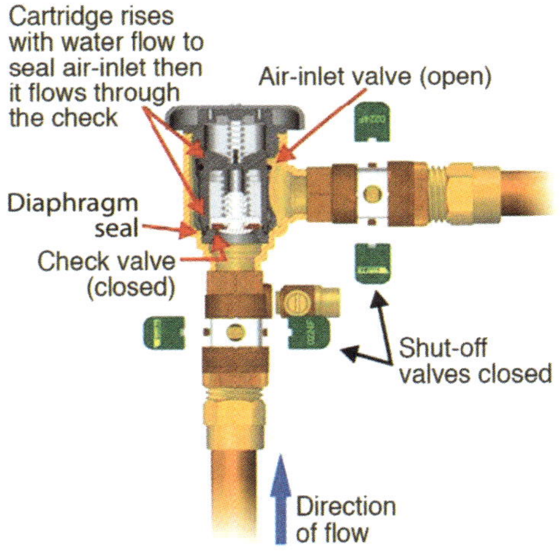

As water flows through the first check valve, a pressure drop is created. The assembly's relief valve determines if the supply pressure is greater and will remain closed; if the difference in pressure between the supply and zone is reduced, the relief valve will open. With a loss of supply pressure and both checks closing, the relief valve will open to maintain a minimum pressure of two psi less than the supply pressure of the fluid in the zone and downstream line. If there is a complete loss of supply pressure, the pressure between the two check valves will become atmospheric because the spring has opened the relief valve.

The RP may be used at actual or potential high- or low-hazard applications. As previously mentioned, the assembly provides backpressure and backsiphonage protection.

An RP may be used as meter-service containment protection or on individual fixtures and appliance isolation protection.

A field test of an RP is required on installation, at least annually thereafter, after any repair or when relocated. The assembly must be installed a minimum of 12 inches above the surrounding ground or floor.

An RP's relief valve has the potential to discharge a significant amount of water. Care should be taken as to its location and provision for indirect drainage. The drainage piping should be sized adequately to take the full discharge of the system. Always refer to the manufacturer's installation requirements.

603.3.8 Double Check Detector Fire Protection Backflow Prevention Assembly.
A double check valve backflow prevention assembly with a parallel detector assembly consisting of a water meter and a double check valve backflow prevention assembly (DC).

603.3.9 Reduced Pressure Detector Fire Protection Backflow Prevention Assembly.
A reduced-pressure principle backflow prevention assembly with a parallel

FIGURE 603.3.6
SPILL RESISTANT PRESSURE VACUUM BREAKER

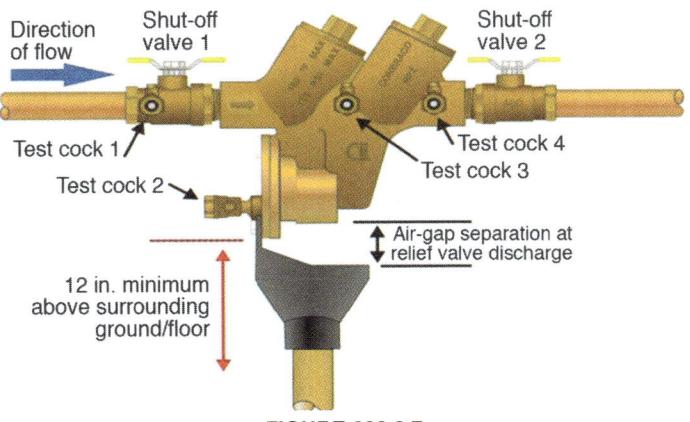

FIGURE 603.3.7
REDUCED PRESSURE PRINCIPLE BACKFLOW ASSEMBLY

detector assembly consisting of a water meter and a reduced-pressure principle backflow prevention assembly (RP).

603.4 General Requirements. Assemblies shall comply with listed standards and be acceptable to the Authority Having Jurisdiction, with jurisdiction over the selection and installation of backflow prevention assemblies.

🔧 Any backflow prevention assembly or device must be a listed and approved device. Again, many organizations provide the standard the devices should be listed to and may also provide a list of approved devices. IAPMO accomplishes this with Table 1701.1 and is referenced in Table 603.2.

603.4.1 Backflow Prevention Valve. Where more than one backflow prevention valve is installed on a single premise, and the valves are installed in one location, each separate valve shall be permanently identified by the permittee in a manner satisfactory to the Authority Having Jurisdiction.

🔧 As with any multiple valve system, each valve or device in a multiple device installation must be labeled. The label should identify which system or location the device serves. This is often accomplished with brass disk tagging with the lettering stamped into the tag.

603.4.2 Testing. The premise owner or responsible person shall have the backflow prevention assembly tested by a certified backflow assembly tester at the time of installation, repair, or relocation and not less than on an annual schedule thereafter, or more often where required by the Authority Having Jurisdiction. The periodic testing shall be performed in accordance with the procedures referenced in ASSE Series 5000 by a tester qualified in accordance with those standards.

🔧 This section reaffirms the requirement for testing each device by a certified tester. It also places the responsibility for ensuring the device is tested squarely on the shoulders of the owner of the building. If the owner employs a building manager or superintendent then he or she assumes that responsibility.

603.4.3 Access and Clearance. Access and clearance shall be provided for the required testing, maintenance, and repair. Access and clearance shall be in accordance with the manufacturer's instructions, and not less than 12 inches (305 mm) between the lowest portion of the assembly and grade, floor, or platform. Installations elevated that exceed 5 feet (1524 mm) above the floor or grade shall be provided with a platform capable of supporting a tester or maintenance person.

🔧 Any installation of a backflow prevention device must take into consideration that at some point in time the device may need to be replaced or repaired. Testable devices will need clearance for testing; meaning that the tester must be able to access the test cocks and set up the test kit appropriately. AVBs, PVBs and SVBs must also have free area around them so that air flow in and out of the valves will not be compromised. Space needs to be given for each of these needs, not just the 1 foot clearance from the bottom of the device.

The code takes seriously the requirement for testing and maintenance of the device by requiring any installation above 5 feet to include a platform to work on the device. The measurement is usually taken from the center line of the device to the floor. Testing or repairing these devices takes space and sometimes involves bulky equipment. The requirement for the platform provides some safety for the plumber working on the device. It is also recognized that if a device is not easily accessible, a tester or maintenance worker might be reluctant to do a proper test, repair or even inspect the device

603.4.4 Connections. Direct connections between potable water piping and sewer-connected wastes shall not be permitted to exist under any condition with or without backflow protection. Where potable water is discharged to the drainage system, it shall be by means of an approved air gap of two pipe diameters of the supply inlet, but in no case shall the gap be less than 1 inch (25.4 mm). Connection shall be permitted to be made to the inlet side of a trap provided that an approved vacuum breaker is installed not less than 6 inches (152 mm), or the distance according to the device's listing, above the flood-level rim of such trapped fixture, so that at no time will such device be subjected to backpressure.

🔧 Under no circumstance should there be a direct connection of the potable water supply to the drainage system. Normally this is not an issue. A plumber should understand the reasons for this and never intentionally tie the two systems together. But sometimes it is forgotten that a relief line from a pressure relief valve or temperature pressure relief valve or the discharge from a water treatment system is a direct connection to the potable system. Lines such as this shall discharge to the drainage system through an air-gap two times the diameter of the pipe or a minimum of one inch (see **Figure 603.4.4**).

The discharge line of the AVB can connect to the inlet side of a trap, which is the tailpiece of a fixture above the trap seal. The AVB is required to be installed a minimum of six inches above the flood rim of the fixture. The measurement is from the critical level of the device to the flood rim.

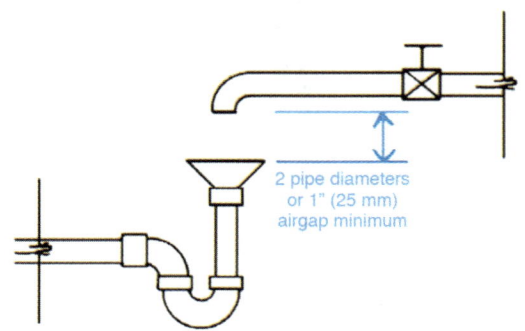

FIGURE 603.4.4
POTABLE WATER AND SEWER CONNECTED WASTE AIRGAP

603.4.5 Hot Water Backflow Preventers. Backflow preventers for hot water exceeding 110°F (43°C) shall be a type designed to operate at temperatures exceeding 110°F (43°C) without rendering a portion of the assembly inoperative.

Installations for water over 110 degrees must be designed for that temperature and the device used must be designed and listed for the maximum temperature of the water. The device should be insulated, taking care to have the test cocks accessible and the drain ports open.

603.4.6 Integral Backflow Preventers. Fixtures, appliances, or appurtenances with integral backflow preventers or integral air gaps manufactured as a unit shall be installed in accordance with their listing requirements and the manufacturer's installation instructions.

There may be some fixture or appliances that have backflow devices integrally installed, such as some carbonated beverage dispensers or dialysis machines. Always follow the manufacturer's instructions with these items. Sometimes there may be a conflict with local health standards and these devices. For example, in some areas the AHJ may require a carbonated soda fountain to be protected by an RP rather than the installed vented double check. Always check with the AHJ before installation.

603.4.7 Freeze Protection. In cold climate areas, backflow assemblies and devices shall be protected from freezing with an outdoor enclosure that complies with ASSE 1060 or by a method acceptable to the Authority Having Jurisdiction.

603.4.8 Drain Lines. Drain lines serving backflow devices or assemblies shall be sized in accordance with the discharge rates of the manufacturer's flow charts of such devices or assemblies.

Disregarding the flow rates of the discharge from an RP is one of the most common mistakes in backflow installations. There have been many instances where an RP is installed and the discharge line is piped to a floor sink that is too small for the gallons-per-minute flow of the discharge. This will of course lead to flooding of the area because the drainage system cannot keep up with the flow from the RP. This is not just the responsibility of the plumbing design professional but of everyone involved with the installation. The plumber must make sure that the flow rate of the device is checked against the proposed size of the receptacle and the proper size is selected.

603.4.9 Prohibited Locations. Backflow prevention devices with atmospheric vents or ports shall not be installed in pits, underground, or submerged locations. Backflow preventers shall not be located in an area containing fumes that are toxic, poisonous, or corrosive.

603.5 Specific Requirements. Specific requirements for backflow prevention shall comply with Section 603.5.1 through Section 603.5.21.

603.5.1 Atmospheric Vacuum Breaker. Water closet and urinal flushometer valves shall be protected against backflow by an approved backflow prevention assembly, device, or method. Where the valves are equipped with an atmospheric vacuum breaker, the vacuum breaker shall be installed on the discharge side of the flushometer valve with the critical level not less than 6 inches (152 mm), or the distance according to its listing, above the overflow rim of a water closet bowl or the highest part of a urinal.

The highest part of the urinal is the top of the urinal; it is not the overflow rim of the bowl of the urinal. The measurement to the critical level is taken from the top of the urinal (see **Figure 603.5.1**).

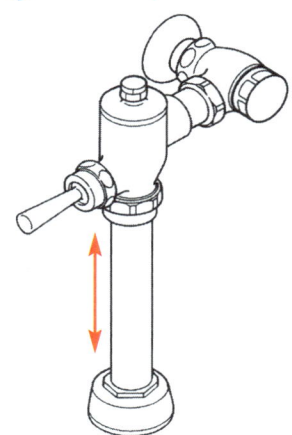

FIGURE 603.5.1
FLUSHOMETER VALVE FROM HIGHEST POINT OF URINAL TO CRITICAL LEVEL

603.5.2 Ballcock. Water closet and urinal tanks shall be equipped with a ballcock. The ballcock shall be installed with the critical level not less than 1 inch (25.4 mm) above the full opening of the overflow pipe. In cases where the ballcock has no hush tube, the bottom of the water supply inlet shall be installed 1 inch (25.4 mm) above the full opening of the overflow pipe.

603.5.3 Backflow Prevention. Water closet flushometer tanks shall be protected against backflow by an approved backflow prevention assembly, device, or method.

603.5.4 Heat Exchangers. Heat exchangers used for heat transfer, heat recovery, or solar heating shall protect the potable water system from being contaminated by the heat-transfer medium. Single-wall heat exchangers used in indirect-fired water heaters shall meet the requirements of Section 505.4.1. Double-wall heat exchangers shall separate the potable water from the heat-transfer medium by providing a space between the two walls that are vented to the atmosphere.

See **Figure 603.5.4**.

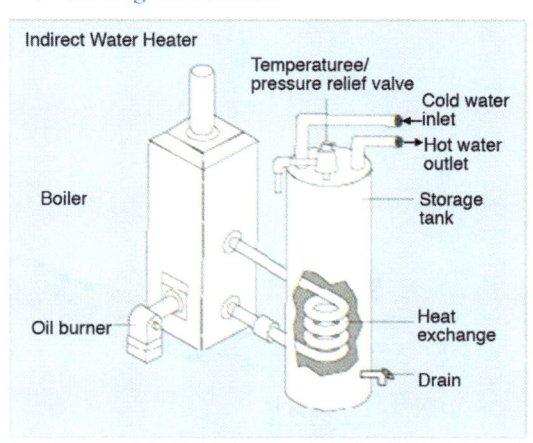

FIGURE 603.5.4
INDIRECT FIRED WATER HEATER

WATER SUPPLY AND DISTRIBUTION

603.5.5 Water Supply Inlets. Water supply inlets to tanks, vats, sumps, swimming pools, and other receptors shall be protected by one of the following means:

(1) An approved air gap.

(2) A listed vacuum breaker installed on the discharge side of the last valve with the critical level not less than 6 inches (152 mm) or in accordance with its listing.

(3) A backflow preventer suitable for the degree of hazard, installed in accordance with the requirements for that type of device or assembly as set forth in this chapter.

603.5.6 Protection from Lawn Sprinklers and Irrigation Systems. Potable water supplies to systems having no pumps or connections for pumping equipment, and no chemical injection or provisions for chemical injection, shall be protected from backflow by one of the following devices:

(1) Atmospheric vacuum breaker (AVB)

(2) Pressure vacuum breaker backflow prevention assembly (PVB)

(3) Spill-resistant pressure vacuum breaker (SVB)

(4) Reduced-pressure principle backflow prevention assembly (RP)

See **Figure 603.5.6**.

603.5.6.1 Systems with Pumps. Where sprinkler and irrigation systems have pumps, connections for pumping equipment, or auxiliary air tanks, or are otherwise capable of creating backpressure, the potable water supply shall be protected by the following type of device where the backflow device is located upstream from the source of backpressure:

(1) Reduced-pressure principle backflow prevention assembly (RP)

603.5.6.2 Systems with Backflow Devices. Where systems have a backflow device installed downstream from a potable water supply pump or a potable water supply pump connection, the device shall be one of the following:

(1) Atmospheric vacuum breaker (AVB)

(2) Pressure vacuum breaker backflow prevention assembly (PVB)

(3) Spill-resistant pressure vacuum breaker (SVB)

(4) Reduced-pressure principle backflow prevention assembly (RP)

603.5.6.3 Systems with Chemical Injectors. Where systems include a chemical injector or provisions for chemical injection, the potable water supply shall be protected by a reduced-pressure principle backflow prevention assembly (RP).

603.5.7 Outlets with Hose Attachments. Potable water outlets with hose attachments, other than water heater drains, boiler drains, and clothes washer connections, shall be protected by a nonremovable hose bibb-type backflow preventer, a nonremovable hose bibb-type vacuum breaker, or by an atmospheric vacuum breaker installed not less than 6 inches (152 mm) above the highest point of usage located on the discharge side of the last valve. In climates where freezing temperatures occur, a listed self-draining frost-proof hose bibb with an integral backflow preventer or vacuum breaker shall be used.

603.5.8 Water-Cooled Equipment. Water-cooled compressors, degreasers, or other water-cooled equipment shall be protected by a backflow preventer installed in accordance with the requirements of this chapter. Water-cooled equipment that produces backpressure shall be equipped with the appropriate protection.

603.5.9 Aspirators. Water inlets to water-supplied aspirators shall be equipped with a vacuum breaker installed in accordance with its listing requirements and this chapter. The discharge shall drain through an air gap. Where the tailpiece of a fixture to receive the discharge of an aspirator is used, the air gap shall be located above the flood-level rim of the fixture.

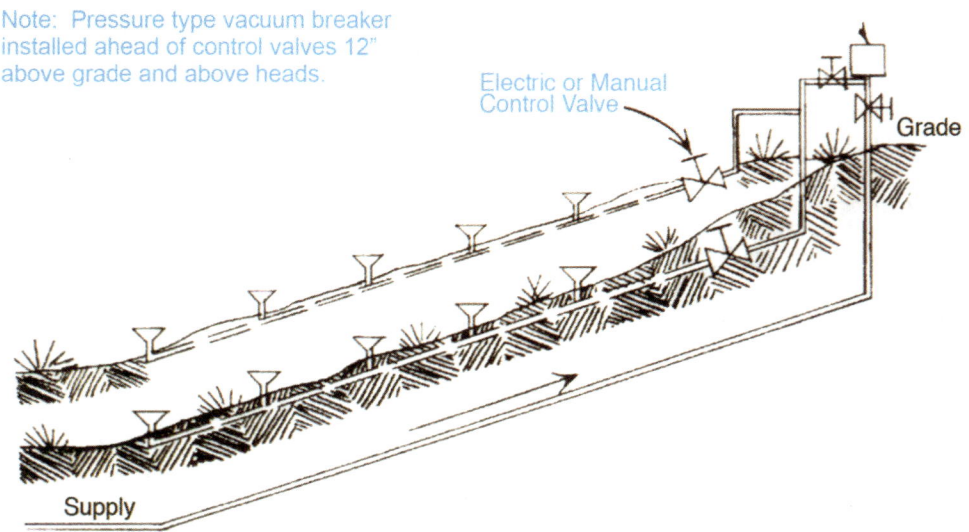

FIGURE 603.5.6
LAWN IRRIGATION SYSTEM WITH PVB INSTALLATION

WATER SUPPLY AND DISTRIBUTION

603.5.10 Steam or Hot Water Boilers. Potable water connections to steam or hot water boilers shall be protected from backflow by a double check valve backflow prevention assembly or reduced pressure principle backflow prevention assembly in accordance with Table 603.2. Where chemicals are introduced into the system a reduced pressure principle backflow prevention assembly shall be provided in accordance with Table 603.2.

603.5.11 Nonpotable Water Piping. In cases where it is impractical to correct individual cross-connections on the domestic waterline, the line supplying such outlets shall be considered a nonpotable water line. No drinking or domestic water outlets shall be connected to the nonpotable waterline. Where possible, portions of the nonpotable waterline shall be exposed, and exposed portions shall be properly identified in a manner satisfactory to the Authority Having Jurisdiction. Each outlet on the nonpotable waterline that is permitted to be used for drinking or domestic purposes shall be posted: "CAUTION: NONPOTABLE WATER, DO NOT DRINK."

This reference to a nonpotable water line refers to the piping downstream of the backflow prevention device. The possibility of backflow occurring in the line from the device causes the line to be designated as nonpotable. In many installations a water line is installed leading to fixture or appliance connections. In order to save costs only one backflow device is installed and the fixture or appliances are connected to this line. The entire line from the backflow device is nonpotable.

603.5.12 Beverage Dispensers. Potable water supply to beverage dispensers carbonated beverage dispensers, or coffee machines shall be protected by an air gap or a vented backflow preventer that complies with ASSE 1022. For carbonated beverage dispensers, piping material installed downstream of the backflow preventer shall not be affected by carbon dioxide gas.

In some jurisdictions the carbonated beverage dispenser is required to have an RP installed rather than the vented double check device. Some AHJs, usually health departments, believe the possible contamination caused by the copper carbonate created by the mixing of copper pipe and carbon dioxide from the carbonator is too high a hazard for just the vented double check device. Plumbers should check with the AHJ in their area for the specific requirements. Also remember that the piping downstream from the backflow device on a carbonator should not be copper or brass materials. Another factor is that there are three standards for devices to these carbonators. Be sure to use the correct device required by the manufacturers of the carbonator (see **Figure 603.5.12**).

603.5.13 Deck-Mounted and Equipment-Mounted Vacuum Breakers. Deck-mounted or equipment-mounted vacuum breakers shall be installed in accordance with their listing and the manufacturer's installation instructions, with the critical level not less than 1 inch (25.4 mm) above the flood-level rim.

603.5.14 Protection from Fire Systems. Except as provided in Section 603.5.14.1 and Section 603.5.14.2,

Standard Carbonator with ASSE 1022 Ventable Check Valve

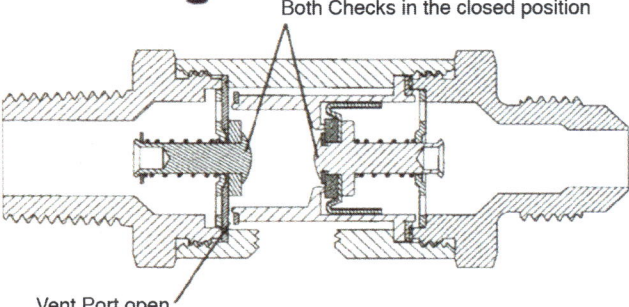

Both Checks in the closed position

Vent Port open

FIGURE 603.5.12
CARBONATED BEVERAGE DISPENSER AND DOUBLE CHECK WITH VENTED PORT - CUTAWAY VIEW

potable water supplies to fire protection systems that are normally under pressure, including but not limited to standpipes and automatic sprinkler systems, except in one- or two-family or townhouse residential sprinkler systems, piped in materials approved for potable water distribution systems shall be protected from backpressure and backsiphonage by one of the following testable devices:

(1) Double check valve backflow prevention assembly (DC)

(2) Double check detector fire protection backflow prevention assembly

(3) Reduced pressure principle backflow prevention assembly (RP)

(4) Reduced pressure detector fire protection backflow prevention assembly

Potable water supplies to fire protection systems that are not normally under pressure shall be protected from backflow and shall be in accordance with the requirements of the appropriate standards referenced in Table 1701.1.

603.5.14.1 Fire Department Connection. Where fire protection systems supplied from a potable water system include a fire department (siamese) connection that is located less than 1700 feet (518.2 m) from a nonpotable water source that is capable of being used by the fire department as a secondary water supply, the potable water supply shall be protected by one of the following:

(1) Reduced pressure principle backflow prevention assembly (RP)

(2) Reduced pressure detector fire protection backflow prevention assembly

Nonpotable water sources include fire department vehicles carrying water of questionable quality or water that is treated with antifreeze, corrosion inhibitors, or extinguishing agents.

See **Figure 603.5.14.1**.

603.5.14.2 Chemicals. Where antifreeze, corrosion inhibitors, or other chemicals are added to a fire protection system supplied from a potable water supply, the potable water system shall be protected by one of the following:

(1) Reduced pressure principle backflow prevention assembly (RP)

(2) Reduced pressure detector fire protection backflow prevention assembly

603.5.14.3 Hydraulic Design. Where a backflow device is installed in the potable water supply to a fire protection system, the hydraulic design of the system shall account for the pressure drop through the backflow device. Where such devices are retrofitted for an existing fire protection system, the hydraulics of the sprinkler system design shall be checked to verify that there will be sufficient water pressure available for satisfactory operation of the fire sprinklers.

The installation of backflow prevention devices in a fire sprinkler system has always been controversial. Most sprinkler system professionals believe that the pressure in the system should not be compromised by the backflow device and that the incidence of backflow is smaller than the possibility for fire. Most also believe that any possibility of contamination, in this case from a system of stagnant water, requires a backflow device. In any case, if a backflow device is installed in a fire sprinkler system, the pressure drop through the device must be taken into account in the design of the fire sprinkler system.

603.5.15 Health Care or Laboratory Areas. Vacuum breakers for washer-hose bedpans shall be located not less than 5 feet (1524 mm) above the floor. Hose connections in health care or laboratory areas shall be not less than 6 feet (1829 mm) above the floor.

603.5.16 Special Equipment. Portable cleaning equipment and dental vacuum pumps shall be protected from backflow by an air gap, an atmospheric vacuum breaker, a spill-resistant vacuum breaker, or a reduced pressure principle backflow preventer.

603.5.17 Potable Water Outlets and Valves. Potable water outlets, freeze-proof yard hydrants, combination stop-and-waste valves, or other fixtures that incorporate a stop and waste feature that drains into the ground shall not be installed underground.

603.5.18 Pure Water Process Systems. The water supply to a pure water process system, such as dialysis water systems, semiconductor washing systems, and similar process piping systems, shall be protected from backpressure and backsiphonage by a reduced-pressure principle backflow preventer.

603.5.18.1 Dialysis Water Systems. The individual connections of the dialysis related equipment to the dialysis pure water system shall not require additional backflow protection.

Over the years, manufacturers have become increasingly aware of the importance of water purity and its effect on the quality of the final product. Pure water process systems are used for a variety of uses, including but not limited to industrial, food and beverage, machinery and equipment, medical, power generation and research. Pure water process systems must include protection from backpressure and back-siphonage by a reduced pressure principle backflow preventer installed between the potable water supply and the water process system. There are several standard methods used by operators to disinfect their pure water systems such as the use of chemicals, shock sterilization with steam, ultraviolet radiation and sterile filtration. If the pure water created by these systems flows back into the water supply, the piping system could be damaged. Pure water will attack metallic pipe and corrode it in a short amount of time.

603.5.19 Plumbing Fixture Fittings. Plumbing fixture fittings with integral backflow protection shall comply with ASME A112.18.1/CSA B125.1.

603.5.20 Swimming Pools, Spas, and Hot Tubs. Potable water supply to swimming pools, spas, and hot tubs shall be protected by an air gap or a reduced pressure principle backflow preventer in accordance with the following:

(1) The unit is equipped with a submerged fill line.

(2) The potable water supply is directly connected to the unit circulation system.

The plumber should always be aware of pool or Jacuzzi installations and provide adequate protection of the potable water supply. An AVB on a hose bibb used to fill a pool or Jacuzzi would be considered adequate protection.

603.5.21 Chemical Dispensers. The water supply to chemical dispensers shall be protected against backflow. The chemical dispenser shall comply with ASSE 1055 or the water supply shall be protected by one of the following methods:

(1) Air gap

(2) Atmospheric vacuum breaker (AVB)

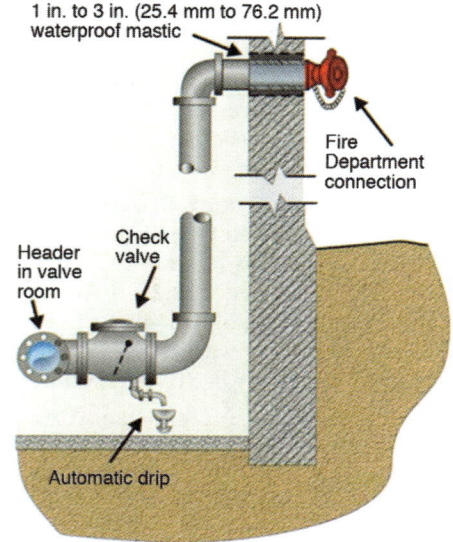

**FIGURE 603.5.14.1
FIRE DEPARTMENT CONNECTION**

(3) Pressure vacuum breaker backflow prevention assembly (PVB)
(4) Spill-resistant pressure vacuum breaker (SVB)
(5) Reduced-pressure principle backflow prevention assembly (RP)

🔧 ASSE 1055-09 is the standard for chemical dispensers that include two types of devices (**Figure 603.5.21a**). Those that include chemicals pressurized above atmospheric pressure (Type A), and those that do not pressurize chemicals above atmospheric pressure (Type B) (**Figure 603.5.21b**). Connections of these devices to existing code compliant appliances or fixtures shall not compromise potable water supplies.

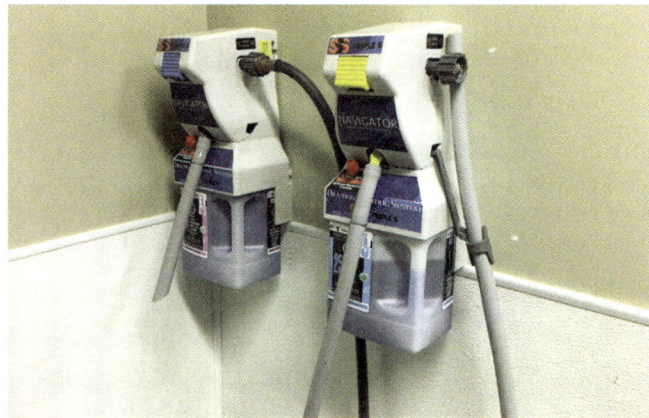

**FIGURE 603.5.21A
CHEMICAL DISPENSER**

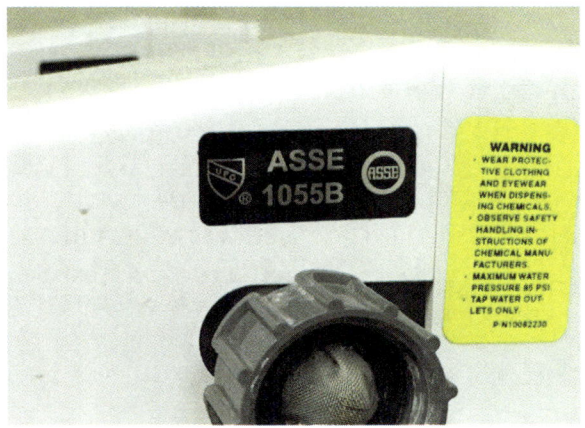

**FIGURE 603.5.21B
CHEMICAL DISPENSER TYPE B**

604.0 Materials.

604.1 Pipe, Tube, and Fittings. Pipe, tube, fittings, solvent cement, thread sealants, solders, and flux used in potable water systems intended to supply drinking water shall comply with NSF 61. Where fittings and valves are made from copper alloys containing more than 15 percent zinc by weight and are used in plastic piping systems, they shall be resistant to dezincification and stress corrosion cracking in compliance with NSF 14.

Materials used in the water supply system, except valves and similar devices, shall be of a like material, except where otherwise approved by the Authority Having Jurisdiction.

Materials for building water piping and building supply piping shall comply with the applicable standards referenced in Table 604.1.

🔧 The NSF standard applies to all pipe, pipe fittings and valve materials that come in contact with potable water. All must comply with the water safe measures contained in the standard.

When installing water supply systems there should be no interconnection of piping materials except where transitions are required. In other words, if an installer is using copper pipe, he or she would not put lengths of plastic pipe in the system just because the copper pipe ran out. As stated, the only time different materials should be intermixed is for valves or fittings designed to be used with the material being used or at transition points. At these transitions, the proper transition fittings must be used to make that transition joint.

Table 604.1 is a chart designed to guide the user to the correct usage of materials for the water supply system. The table identifies if the material is used for water distribution pipe and fittings (mains, risers and branches within the building) and whether it can be used for hot or cold water or both. It also identifies whether the material is used for building supply pipe and fittings (water or yard service pipe and fittings). For example, PE or polyethylene pipe is not for use in the water distribution system but it can be used for the building supply piping. Also included in this version of Table 604.1 are the standards that apply to the piping material for the water supply system.

The ASTM standards for these products allow for several different alloys. Some of these alloys allow for more zinc and less copper. Red brass has the least amount of zinc. The less zinc in the brass the greater the resistance to dezincification.

604.2 Lead Content. The maximum allowable lead content in pipes, pipe fittings, plumbing fittings, and fixtures intended to convey or dispense water for human consumption shall be not more than a weighted average of 0.25 percent with respect to the wetted surfaces of pipes, pipe fittings, plumbing fittings, and fixtures. For solder and flux, the lead content shall be not more than 0.2 percent where used in piping systems that convey or dispense water for human consumption.

Exceptions:

(1) Pipes, pipe fittings, plumbing fittings, fixtures, or backflow preventers used for nonpotable services such as manufacturing, industrial processing, irrigation, outdoor watering, or any other uses where the water is not used for human consumption.

(2) Flush valves, fill valves, flushometer valves, tub fillers, shower valves, service saddles, or water distribution main gate valves that are 2 inches (50 mm) in diameter or larger.

WATER SUPPLY AND DISTRIBUTION

TABLE 604.1
MATERIALS FOR BUILDING SUPPLY AND WATER DISTRIBUTION PIPING AND FITTINGS

MATERIAL	BUILDING SUPPLY PIPE AND FITTINGS	WATER DISTRIBUTION PIPE AND FITTINGS	REFERENCED STANDARD(S) PIPE	REFERENCED STANDARD(S) FITTINGS
Copper and Copper Alloys	X	X	ASTM B42, ASTM B43, ASTM B75, ASTM B88, ASTM B135, ASTM B251, ASTM B302, ASTM B447	ASME B16.15, ASME B16.18, ASME B16.22, ASME B16.26, ASME B16.50^2, ASME B16.51, ASSE 1061
CPVC	X	X	ASTM D2846, ASTM F441, ASTM F442, CSA B137.6	ASSE 1061, ASTM D2846, ASTM F437, ASTM F438, ASTM F439, ASTM F1970, CSA B137.6
CPVC-AL-CPVC	X	X	ASTM F2855	ASTM D2846
Ductile-Iron	X	X	AWWA C151	ASME B16.4, AWWA C110, AWWA C153
Galvanized Steel	X	X	ASTM A53	—
Malleable Iron	X	X	—	ASME B16.3
PE	X^1	—	ASTM D2239, ASTM D2737, ASTM D3035, AWWA C901, CSA B137.1	ASTM D2609, ASTM D2683, ASTM D3261, ASTM F1055, CSA B137.1
PE-AL-PE	X	X	ASTM F1282, CSA B137.9	ASTM F1282, ASTM F1974, CSA B137.9
PE-AL-PEX	X	X	ASTM F1986	ASTM F1986
PE-RT	X	X	ASTM F2769, CSA B137.18	ASTM D3261, ASTM F1055, ASSE 1061, ASTM F1807, ASTM F2098, ASTM F2159, ASTM F2735, ASTM F2769, CSA B137.18
PEX	X	X	ASTM F876, ASTM F877, CSA B137.5, AWWA C904^1	ASSE 1061, ASTM F877, ASTM F1807, ASTM F1960, ASTM F1961, ASTM F2080, ASTM F2159, ASTM F2735, CSA B137.5
PEX-AL-PEX	X	X	ASTM F1281, CSA B137.10, ASTM F2262,	ASTM F1281, ASTM F1974, ASTM F2434, CSA B137.10,
PP	X	X	ASTM F2389, CSA B137.11	ASTM F2389, CSA B137.11
PVC	X^1	—	ASTM D1785, ASTM D2241, AWWA C900	ASTM D2464, ASTM D2466, ASTM D2467, ASTM F1970, AWWA C907
Stainless Steel	X	X	ASTM A269, ASTM A312	—

Notes:
^{1} For building supply or exterior cold-water applications, not for water distribution piping.
^{2} For brazed fittings only.

604.2.1 Lead Content of Water Supply Pipe and Fittings. Pipes, pipe fittings, valves, and faucets utilized in the water supply system for non-drinking water applications shall have a maximum of 8 percent lead content.

604.3 Copper or Copper Alloy Tube. Copper or copper alloy tube for water piping shall have a weight of not less than Type L.

Exception: Type M copper or copper alloy tubing shall be permitted to be used for water piping where piping is aboveground in, or on, a building or underground outside of structures.

604.4 Hard-Drawn Copper or Copper Alloy Tubing. Hard-drawn copper or copper alloy tubing for water supply and distribution in addition to the required incised marking shall be marked in accordance with ASTM B88. The colors shall be: Type K, green; Type L, blue; and Type M, red.

See Figure 604.4 and Learning Links goo.gl/q3BR68 and goo.gl/TxPUH2.

WATER SUPPLY AND DISTRIBUTION

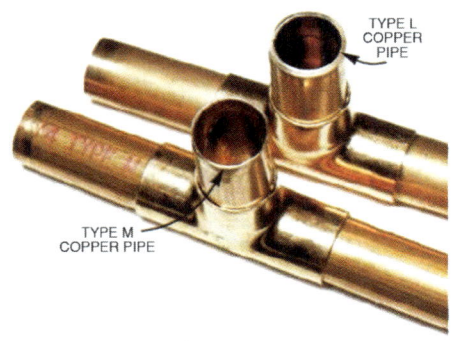

**FIGURE 604.4
COPPER PIPING MARKING**

604.5 Flexible Connectors. Flexible water connectors shall be installed in readily accessible locations, and where under continuous pressure shall comply with ASME A112.18.6/CSA B125.6. Flexible water connectors with an excess flow shutoff device shall comply with CSA B125.5/IAPMO Z600.

604.6 Cast-Iron Fittings. Cast-iron fittings up to and including 2 inches (50 mm) in size, where used in connection with potable water piping, shall be galvanized.

604.7 Malleable Iron Fittings. Malleable iron water fittings shall be galvanized.

See **Figure 604.7.**

**FIGURE 604.7
GALVANIZED FITTINGS**

604.8 Previously Used Piping and Tubing. Piping and tubing that has previously been used for a purpose other than for potable water systems shall not be used.

604.9 Epoxy Coating. The epoxy coating used on existing, underground steel building supply piping shall comply with NSF 61 and AWWA C210.

604.10 Plastic Materials. Approved plastic materials shall be permitted to be used in building supply piping, provided that where metal building supply piping is used for electrical grounding purposes, replacement piping, therefore, shall be of like materials.

Exception: Where a grounding system acceptable to the Authority Having Jurisdiction is installed, inspected, and approved, the metallic pipe shall be permitted to be replaced with nonmetallic pipe.

One of the factors to take into consideration when using plastic piping is how the grounding of the building is handled. Often the building is grounded using a grounding clamp on the metallic building supply. The underground metallic pipe acts as the grounding rod for the electrical system. If plastic pipe is used or if metallic piping is replaced with plastic pipe, another method must be used for grounding purposes. Not doing so would lead to a building without grounding protection.

Another provision for plastic piping is the installation of a tracer wire. Locating the pipe without the tracer is much more difficult and can result in loss of time.

604.10.1 Tracer Wire. Plastic materials for building supply piping outside underground shall have an electrically continuous corrosion-resistant blue insulated copper tracer wire, or other approved conductor installed adjacent to the piping. Access shall be provided to the tracer wire, or the tracer wire shall terminate aboveground at each end of the nonmetallic piping. The tracer wire size shall be not less than 18 AWG, and the insulation type shall be suitable for direct burial.

604.11 Solder. Solder shall comply with the requirements of Section 604.2.

604.12 Flexible Corrugated Connectors. Flexible corrugated connectors of copper, copper alloy, or stainless steel shall be limited to the following connector lengths:

(1) Fixture Connectors – 30 inches (762 mm)

(2) Washing Machine Connectors – 72 inches (1829 mm)

(3) Dishwasher and Icemaker Connectors – 120 inches (3048 mm)

The measurements above are from the fixture or appliance supply valve to the fixture or appliance inlet connection.

604.13 Water Heater Connectors. Flexible metallic (copper and stainless steel), reinforced flexible, braided stainless steel, or polymer braided with EPDM core connectors that connect a water heater to the piping system shall comply with ASME A112.18.6/CSA B125.6. Copper, copper alloy, or stainless steel flexible connectors shall not exceed 24 inches (610 mm). PEX, PEX-AL-PEX, PE-AL-PE, or PE-RT tubing shall not be installed within the first 18 inches (457 mm) of piping connected to a water heater.

605.0 Joints and Connections.

605.1 Copper or Copper Alloy Pipe, Tubing, and Joints. Joining methods for copper or copper alloy pipe, tubing, and fittings shall be installed in accordance with the manufacturer's installation instructions and shall comply with Section 605.1.1 through Section 605.1.5.

605.1.1 Brazed Joints. Brazed joints between copper or copper alloy pipe or tubing and fittings shall be made with brazing alloys having a liquid temperature above 1000°F (538°C). The joint surfaces to be brazed shall be cleaned bright by either manual or mechanical means. Tubing shall be cut square and reamed to full inside diameter. Brazing flux shall be applied to the joint surfaces where required by manufacturer's recommendation. Brazing filler metal shall conform to AWS A5.8 and shall be applied at the point where the pipe or tubing enters the socket of the fitting.

605.1.2 Flared Joints. Flared joints for soft copper or copper alloy water tubing shall be made with fittings that comply with the applicable standards referenced in Table 604.1. Pipe or tubing shall be cut square using an appropriate tubing

WATER SUPPLY AND DISTRIBUTION

cutter. The tubing shall be reamed to full inside diameter, resized to round, and expanded with a proper flaring tool.

605.1.3 Mechanical Joints. Mechanical joints shall include, but are not limited to, compression, flanged, grooved, pressed, and push fit fittings.

605.1.3.1 Mechanically Formed Tee Fittings. Mechanically formed tee fittings shall have extracted collars that shall be formed in a continuous operation consisting of drilling a pilot hole and drawing out the pipe or tube surface to form a collar having a height not less than three times the thickness of the branch tube wall. The branch pipe or tube shall be notched to conform to the inner curve of the run pipe or tube and shall have two dimple depth stops to ensure that penetration of the branch pipe or tube into the collar is of a depth for brazing and that the branch pipe or tube does not obstruct the flow in the main line pipe or tube. Dimple depth stops shall be in line with the run of the pipe or tube. The second dimple shall be ¼ of an inch (6.4 mm) above the first and shall serve as a visual point of inspection. Fittings and joints shall be made by brazing. Soldered joints shall not be permitted.

605.1.3.2 Press-Connect Fittings. Press-connect fittings for copper or copper alloy pipe or tubing shall have an elastomeric o-ring that forms the joint. The pipe or tubing shall be fully inserted into the fitting, and the pipe or tubing marked at the shoulder of the fitting. Pipe or tubing shall be cut square, chamfered, and reamed to full inside diameter. The fitting alignment shall be checked against the mark on the pipe or tubing to ensure the pipe or tubing is inserted into the fitting. The joint shall be pressed using the tool recommended by the manufacturer.

The pressed fitting is a type of fitting that is directly attached to a copper tube by the mechanical deformation or pressing of the fitting onto the tube. A pressing tool creates a seal and a restrained connection. These fittings include an O-ring seal and may incorporate a corrosion-resistant mechanical grip ring. Press fittings are used with copper tubing only (see **Figure 605.1.3.2** and Learning Link **goo.gl/hBrJr4**).

605.1.3.3 Push Fit Fittings. Removable and nonremovable push fit fittings for copper or copper alloy tubing or pipe that employ quick assembly push fit connectors shall comply with ASSE 1061. Push fit fittings for copper or copper alloy pipe or tubing shall have an approved elastomeric o-ring that forms the joint. Pipe or tubing shall be cut square, chamfered, and reamed to full inside diameter. The tubing shall be fully inserted into the fitting, and the tubing marked at the shoulder of the fitting. The fitting alignment shall be checked against the mark on the tubing to ensure the tubing is inserted into the fitting and gripping mechanism has engaged on the pipe.

This is a type of joining method that joins a pipe to a fitting by means of hand assembly with no tools required. The pipe is inserted and then pushed into one end of the fitting, resulting in a positive connection. These fittings are typically comprised of elastomeric seals (a type of O-ring) and corrosion-resistant tube grippers (a friction ring). Such joints are permanent but, depending on the design, the joint may be disassembled. Fittings should be installed in accordance with the manufacturer's installation instructions. See Learning Link **goo.gl/hBrJr4**.

605.1.4 Soldered Joints. Soldered joints between copper or copper alloy pipe or tubing and fittings shall be made in accordance with ASTM B828 with the following sequence of joint preparation and operation as follows: measuring and cutting, reaming, cleaning, fluxing, assembly and support, heating, applying the solder, cooling and cleaning. Pipe or tubing shall be cut square and reamed to the full inside diameter including the removal of burrs on the outside of the pipe or tubing. Surfaces to be joined shall be cleaned bright by manual or mechanical means. Flux shall be applied to pipe or tubing and fittings and shall conform to ASTM B813, and shall become noncorrosive and nontoxic after soldering. Insert pipe or tubing into the base of the fitting and remove excess flux. Pipe or tubing and fitting shall be supported to ensure a uniform capillary space around the joint. Heat shall be applied using an air or fuel torch with the flame perpendicular to the pipe or tubing using acetylene or an LP gas. Preheating shall depend on the size of the joint. The flame shall be moved to the fitting cup and alternate between the pipe or tubing and fitting. Solder conforming to ASTM B32 shall be applied to the joint surfaces until capillary action draws the molten solder into the cup. Solder and fluxes with a lead content that exceeds 0.2 percent shall be prohibited in piping systems conveying potable water. Joint surfaces shall not be disturbed until cool and any remaining flux residue shall be cleaned.

605.1.5 Threaded Joints. Threaded joints for copper or copper alloy pipe shall be made with pipe threads that comply with ASME B1.20.1. Thread sealant tape or compound shall be applied only on male threads, and such material shall be of approved types, insoluble in water, and nontoxic.

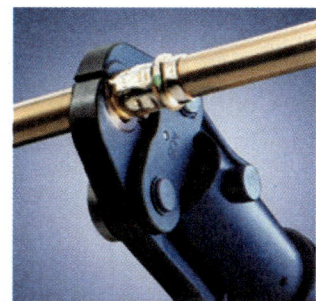

ProPress Fittings (1/2" to 2" Connections)

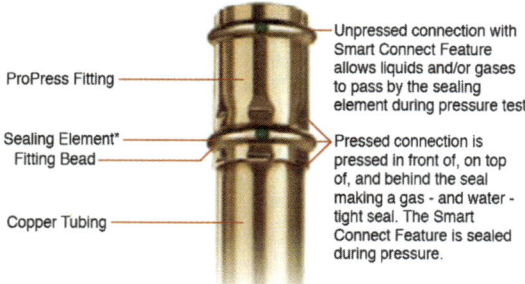

**FIGURE 605.1.3.2
PRESSED FITTING**

WATER SUPPLY AND DISTRIBUTION

605.2 CPVC Plastic Pipe and Joints. CPVC plastic pipe and fitting joining methods shall be installed in accordance with the manufacturer's installation instructions and shall comply with Section 605.2.1 through Section 605.2.3.

605.2.1 Mechanical Joints. Mechanical joints shall include compression, flanged, grooved and push fit fittings.

605.2.1.1 Push Fit Fittings. Removable and nonremovable push fit fittings that employ a quick assembly push fit connector shall comply with ASSE 1061.

605.2.2 Solvent Cement Joints. Solvent cement joints for CPVC pipe and fittings shall be clean from dirt and moisture. Solvent cements shall comply with ASTM F493, requiring the use of a primer shall be orange in color. The primer shall be colored and shall comply with ASTM F656. Listed solvent cement that complies with ASTM F493 and that does not require the use of primers, yellow or red in color, shall be permitted for pipe and fittings that comply with ASTM D2846, ½ of an inch (15 mm) through 2 inches (50 mm) in diameter or ASTM F442, ½ of an inch (15 mm) through 3 inches (80 mm) in diameter. Apply primer where required inside the fitting and to the depth of the fitting on pipe. Apply liberal coat of cement to the outside surface of pipe to depth of fitting and inside of fitting. Place pipe inside fitting to forcefully bottom the pipe in the socket and hold together until joint is set.

605.2.3 Threaded Joints. Threads shall comply with ASME B1.20.1. A minimum of Schedule 80 shall be permitted to be threaded; however, the pressure rating shall be reduced by 50 percent. The use of molded fittings shall not result in a 50 percent reduction in the pressure rating of the pipe provided that the molded fittings shall be fabricated so that the wall thickness of the material is maintained at the threads. Thread sealant compound that is compatible with the pipe and fitting, insoluble in water, and nontoxic shall be applied to male threads. Caution shall be used during assembly to prevent over tightening of the CPVC components once the thread sealant has been applied. Female CPVC threaded fittings shall be used with plastic male threads only.

605.3 CPVC/AL/CPVC Plastic Pipe and Joints. Chlorinated polyvinyl chloride/aluminum/chlorinated polyvinyl chloride (CPVC/AL/CPVC) plastic pipe and fitting joining methods shall be installed in accordance with the manufacturer's installation instructions and shall comply with Section 605.3.1 or Section 605.3.2.

605.3.1 Solvent Cement Joints. Solvent cement joints for CPVC/AL/CPVC pipe and fittings shall be clean from dirt and moisture. Solvent cements that comply with ASTM F493, requiring the use of a primer shall be orange in color. The primer shall be colored and shall comply with ASTM F656. Listed solvent cement that complies with ASTM F493 and that does not require the use of primers, yellow in color, shall be permitted to join pipe that comply with ASTM F2855 and fittings that comply with ASTM D2846, ½ of an inch (15 mm) through 2 inches (50 mm) in diameter. Apply primer where required inside the fitting and to the depth of the fitting on pipe. Apply liberal coat of cement to the outside surface of pipe to depth of fitting and inside of fitting. Place pipe inside fitting to forcefully bottom the pipe in the socket and hold together until joint is set.

605.3.2 Mechanical Joints. Mechanical joints shall include flanged, grooved, and push fit fittings.

605.3.2.1 Push Fit Fittings. Removable and nonremovable push fit fittings that employ a quick assembly push fit connector shall comply with ASSE 1061.

605.4 Ductile Iron Pipe and Joints. Ductile iron pipe and fitting joining methods shall be installed in accordance with the manufacturer's installation instructions and shall comply with Section 605.4.1 or Section 605.4.2.

605.4.1 Mechanical Joints. Mechanical joints for ductile iron pipe and fittings shall consist of a bell that is cast integrally with the pipe or fitting and provided with an exterior flange having bolt holes and a socket with annular recesses for the sealing gasket and the plain end of the pipe or fitting. The elastomeric gasket shall comply with AWWA C111. Lubricant recommended for potable water application by the pipe manufacturer shall be applied to the gasket and plain end of the pipe.

605.4.2 Push-On Joints. Push-on joints for ductile iron pipe and fittings shall consist of a single elastomeric gasket that shall be assembled by positioning the elastomeric gasket in an annular recess in the pipe or fitting socket and forcing the plain end of the pipe or fitting into the socket. The plain end shall compress the elastomeric gasket to form a positive seal and shall be designed so that the elastomeric gasket shall be locked in place against displacement. The elastomeric gasket shall comply with AWWA C111. Lubricant recommended for potable water application by the pipe manufacturer shall be applied to the gasket and plain end of the pipe.

605.5 Galvanized Steel Pipe and Joints. Galvanized steel pipe and fitting joining methods shall be installed in accordance with the manufacturer's installation instructions and shall comply with Section 605.5.1 or Section 605.5.2.

605.5.1 Mechanical Joints. Mechanical joints shall be made with an approved and listed elastomeric gasket.

605.5.2 Threaded Joints. Threaded joints shall be made with pipe threads that comply with ASME B1.20.1. Thread sealant tape or compound shall be applied only on male threads, and such material shall be of approved types, insoluble in water, and nontoxic.

605.6 PE Plastic Pipe/Tubing and Joints. PE plastic pipe or tubing and fitting joining methods shall be installed in accordance with the manufacturer's installation instructions and shall comply with Section 605.6.1 or Section 605.6.2.

605.6.1 Heat-Fusion Joints. Heat-fusion joints between PE pipe or tubing and fittings shall be assembled in accordance with Section 605.6.1.1 through Section 605.6.1.3 using butt, socket, and electro-fusion heat methods.

See Figure 605.6.1.

605.6.1.1 Butt-Fusion Joints. Butt-fusion joints shall be made in accordance with ASTM F2620. Joints shall be made by heating the squared ends of two pipes, pipe and fitting, or two fittings by holding ends against a heated element. The heated element shall be removed where the

WATER SUPPLY AND DISTRIBUTION

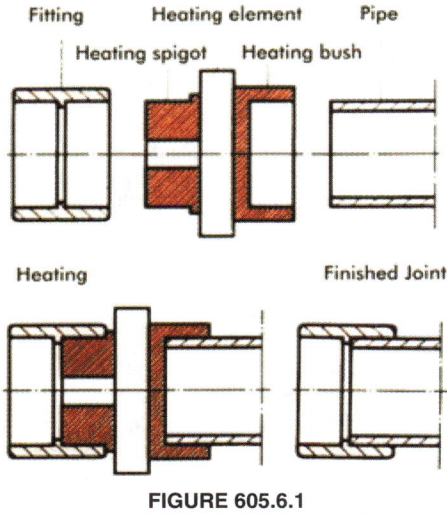

FIGURE 605.6.1
HEAT FUSION WELD JOINTS

proper melt is obtained and joined ends shall be placed together with applied force.

605.6.1.2 Electro-Fusion Joints. Electro-fusion joints shall be heated internally by a conductor at the interface of the joint. Align and restrain fitting to pipe to prevent movement and apply electric current to the fitting. Turn off the current when the proper time has elapsed to heat the joint. The joint shall fuse together and remain undisturbed until cool.

605.6.1.3 Socket-Fusion Joints. Socket-fusion joints shall be made in accordance with ASTM F2620. Joints shall be made by simultaneously heating the outside surface of a pipe end and the inside of a fitting socket. Where the proper melt is obtained, the pipe and fitting shall be joined by inserting one into the other with applied force. The joint shall fuse together and remain undisturbed until cool.

605.6.2 Mechanical Joints. Mechanical joints between PE pipe or tubing and fittings shall include insert and mechanical compression fittings that provide a pressure seal resistance to pullout. Joints for insert fittings shall be made by cutting the pipe square, using a cutter designed for plastic piping, and removal of sharp edges. Two stainless steel clamps shall be placed over the end of the pipe. Fittings shall be checked for proper size based on the diameter of the pipe. The end of pipe shall be placed over the barbed insert fitting, making contact with the fitting shoulder. Clamps shall be positioned equal to 180 degrees (3.14 rad) apart and shall be tightened to provide a leak tight joint. Compression type couplings and fittings shall be permitted for use in joining PE piping and tubing. Stiffeners that extend beyond the clamp or nut shall be prohibited. Bends shall be not less than 30 pipe diameters, or the coil radius where bending with the coil. Bends shall not be permitted closer than 10 pipe diameters of a fitting or valve. Mechanical joints shall be designed for their intended use.

605.7 PE-AL-PE Plastic Pipe/Tubing and Joints. PE-AL-PE plastic pipe or tubing and fitting joining methods shall be installed in accordance with the manufacturer's installation instructions and shall comply with Section 605.7.1 and Section 605.7.1.1.

PE-AL-PE pipe is multilayer polyethylene and aluminum bonded together, with the aluminum "sandwiched" between the plastic layers. This provides the benefits of a smooth interior plastic pipe and a semirigid pipe much like soft copper pipe. The pipe can be laid out and remain rigid without recoil. PE-AL-PE is manufactured to ASTM F1282, Standard Specification for Polyethylene/Aluminum/Polyethylene (PE-AL-PE) Composite Pressure Pipe.

Joining methods for this layered pipe use crimp, compression or press fit fittings. The fittings usually incorporate O-rings to provide water and gas-tight joints. Because of this the end of the pipe must be reamed or beveled so the O-ring is not damaged by the end of the pipe.

Use only the PE-AL-PE pipe manufacturer's recommended fittings and tools for installation. Consult the manufacturer's installation guide for information on the proper sizing methods to be used for the installation.

605.7.1 Mechanical Joints. Mechanical joints for PE-AL-PE pipe or tubing and fittings shall be either of the metal insert fittings with a split ring and compression nut or metal insert fittings with copper crimp rings. Metal insert fittings shall comply with ASTM F1974. Crimp insert fittings shall be joined to the pipe by placing the copper crimp ring around the outer circumference of the pipe, forcing the pipe material into the space formed by the ribs on the fitting until the pipe contacts the shoulder of the fitting. The crimp ring shall then be positioned on the pipe so the edge of the crimp ring is $\frac{1}{8}$ of an inch (3.2 mm) to $\frac{1}{4}$ of an inch (6.4 mm) from the end of the pipe. The jaws of the crimping tool shall be centered over the crimp ring and tool perpendicular to the barb. The jaws shall be closed around the crimp ring and shall not be crimped more than once.

605.7.1.1 Compression Joints. Compression joints for PE-AL-PE pipe or tubing and fittings shall be joined through the compression of a split ring, by a compression nut around the circumference of the pipe. The compression nut and split ring shall be placed around the pipe. The ribbed end of the fitting shall be inserted into the pipe until the pipe contacts the shoulder of the fitting. Position and compress the split ring by tightening the compression nut onto the insert fitting.

605.8 PE-RT. Polyethylene of raised temperature (PE-RT) tubing and fitting joining methods and shall comply with Section 605.8.1.

605.8.1 Mechanical Joints. Fittings for PE-RT tubing shall comply with the applicable standards listed in Table 604.1. Mechanical joints for PE-RT tubing shall be installed in accordance with the manufacturer's installation instructions.

605.9 PEX Plastic Tubing and Joints. PEX plastic tubing and fitting joining methods shall be installed in accordance with the manufacturer's installation instructions and shall comply with Section 605.9.1 through Section 605.9.3.

PEX piping used in the water distribution system should be marked with ASTM F 876, Standard Specification for Cross-Linked Polyethylene (PEX) Tubing, or ASTM F 877, Standard Specification for Cross-Linked Polyethylene (PEX) Plastic Hot and Cold Water Distribution Systems.

WATER SUPPLY AND DISTRIBUTION

There are several types of joining methods or fittings used with PEX plumbing systems. All are mechanical fittings that are either directional or transitional. PEX piping cannot be joined by solvent cementing. Most PEX piping manufacturers have their own mechanical fitting system. The method of connection should comply with the manufacturer's recommendations and instructions.

The four main types of joining methods for PEX piping are:

1. Cold Expansion Fittings with PEX Reinforced Rings (see **Figure 605.9a**).
2. Cold Expansion Fittings With Metal Compression Sleeves (see **Figure 605.9b**).
3. Metal or Plastic Insert Fittings
- Using Copper Crimp Ring (see **Figure 605.9c**)
- Using Stainless Steel Clamp
- Using Stainless Steel Sleeve
4. Push Type Fittings (see **Figure 605.9d**).

Cold expansion joining methods require that the tubing is expanded before it is placed over the oversized fitting. This requires a special tool to expand the tubing. Be sure to have the correct tool from the manufacturer of the tubing to correctly perform the expansion (see **Figure 605.9e**).

For joining methods that use a metal compression sleeve, crimp ring or clamp use the appropriate tool per the PEX tubing manufacturer's instructions. Intermixing different brands of PEX tubing and fittings is not recommended and may not provide a water tight or testable joint. Always consult the PEX tubing manufacturer's installation guide for information on the proper fittings, tools and sizing methods to be used for the installation.

605.9.1 Fittings. Fittings for PEX tubing shall comply with the applicable standards referenced in Table 604.1. PEX tubing that complies with ASTM F876 shall be marked with the applicable standard designation for the fittings, specified by the tubing manufacturer for use with the tubing.

605.9.2 Mechanical Joints. Mechanical joints shall be installed in accordance with the manufacturer's installation instructions.

605.9.3 Push Fit Fittings. Removable and nonremovable push fit fittings that employ a quick assembly push fit connector shall comply with ASSE 1061.

605.10 PEX-AL-PEX Plastic Tubing and Joints. PEX-AL-PEX plastic pipe or tubing and fitting joining methods shall be installed in accordance with the manufacturer's installation instructions and shall comply with Section 605.10.1 and Section 605.10.1.1.

PEX-AL-PEX pipe is multilayer crosslinked polyethylene and aluminum bonded together, with the aluminum "sandwiched" between the plastic layers. This provides the benefits of a smooth interior plastic pipe and a semirigid pipe much like soft copper pipe. The pipe can be laid out and remain rigid without recoil. PEX-AL-PEX is manufactured to ASTM F1281, Standard Specification for Crosslinked Polyethylene/Aluminum/Crosslinked Polyethylene (PEX-AL-PEX) Pressure Pipe.

Joining methods for this layered pipe use crimp, compression or press fit fittings (see **Figure 605.10**). The fittings

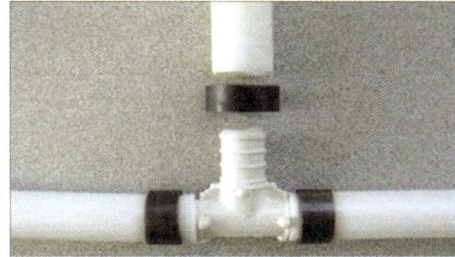

FIGURE 605.9C
PEX INSERT FITTINGS WITH COPPER CRIMP RINGS

FIGURE 605.9A
PEX FITTINGS WITH PEX REINFORCED RINGS

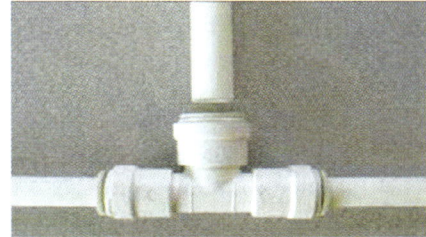

FIGURE 605.9D
PEX PUSH FIT FITTINGS

FIGURE 605.9B
PEX FITTINGS WITH METAL COMPRESSION SLEEVES

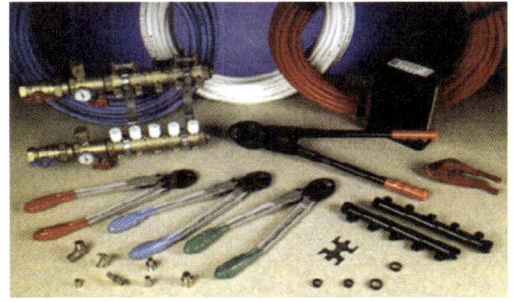

FIGURE 605.9E
PEX TOOLS

usually incorporate O-rings to provide water and gas-tight joints. Because of this the end of the pipe must be reamed or beveled so the O-ring is not damaged by the end of the pipe.

Use only the PEX-AL-PEX pipe manufacturer's recommended fittings and tools for installation. Consult the manufacturer's installation guide for information on the proper sizing methods to be used for the installation.

FIGURE 605.10
PEX-AL-PEX COMPRESSION FITTINGS

605.10.1 Mechanical Joints. Mechanical joints between PEX-AL-PEX tubing and fittings shall include mechanical and compression type fittings and insert fittings with a crimping ring. Insert fittings utilizing a crimping ring shall comply with ASTM F1974 or ASTM F2434. Crimp joints for crimp insert fittings shall be joined to PEX-AL-PEX pipe by the compression of a crimp ring around the outer circumference of the pipe, forcing the pipe material into annular spaces formed by ribs on the fitting.

605.10.1.1 Compression Joints. Compression joints shall include compression insert fittings and shall be joined to PEX-AL-PEX pipe through the compression of a split ring or compression nut around the outer circumference of the pipe, forcing the pipe material into the annular space formed by the ribs on the fitting.

605.11 Polypropylene (PP) Piping and Joints. PP pipe and fittings shall be installed in accordance with the manufacturer's installation instructions and shall comply with Section 605.11.1 through Section 605.11.3.

605.11.1 Heat-Fusion Joints. Heat-fusion joints for polypropylene (PP) pipe and fitting joints shall be installed with socket-type heat-fused polypropylene fittings, fusion outlets, butt-fusion polypropylene fittings or pipe, or electro-fusion polypropylene fittings. Joint surfaces shall be clean and free from moisture. The joint shall be undisturbed until cool. Joints shall be made in accordance with ASTM F2389 or CSA B137.11.

605.11.2 Mechanical and Compression Sleeve Joints. Mechanical and compression sleeve joints shall be installed in accordance with the manufacturer's installation instructions.

Unlike other plastic piping that can be solvent-welded, polypropylene must be joined by heat fusion or mechanical joints. Socket fusion type pipe and fittings are joined by heat fusing the polypropylene material with a thermostatically controlled heat tool.

605.11.3 Threaded Joints. PP pipe shall not be threaded. PP transition fittings for connection to other piping materials shall only be threaded by use of copper alloy or stainless steel inserts molded in the fitting.

605.12 PVC Plastic Pipe and Joints. PVC plastic pipe and fitting joining methods shall be installed in accordance with the manufacturer's installation instructions and shall comply with Section 605.12.1 through Section 605.12.3.

PVC piping shall not be exposed to direct sunlight unless the piping does not exceed 24 inches (610 mm) and is wrapped with not less than 0.04 of an inch (1.02 mm) thick tape or otherwise protected from UV degradation.

605.12.1 Mechanical Joints. Mechanical joints shall be designed to provide a permanent seal and shall be of the mechanical or push-on joint. The mechanical joint shall include a pipe spigot that has a wall thickness to withstand without deformation or collapse; the compressive force exerted where the fitting is tightened. The push-on joint shall have a minimum wall thickness of the bell at any point between the ring and the pipe barrel. The elastomeric gasket shall comply with ASTM D3139, and be of such size and shape as to provide a compressive force against the spigot and socket after assembly to provide a positive seal.

605.12.2 Solvent Cement Joints. Solvent cement joints for PVC pipe and fittings shall be clean from dirt and moisture. Pipe shall be cut square and pipe shall be deburred. Where surfaces to be joined are cleaned and free of dirt, moisture, oil, and other foreign material, apply primer purple in color that complies with ASTM F656. Primer shall be applied to the surface of the pipe and fitting is softened. Solvent cement that complies with ASTM D2564 shall be applied to all joint surfaces. Joints shall be made while both the inside socket surface and outside surface of pipe are wet with solvent cement. Hold joint in place and undisturbed for 1 minute after assembly.

605.12.3 Threaded Joints. Threads shall comply with ASME B1.20.1. A minimum of Schedule 80 shall be permitted to be threaded; however, the pressure rating shall be reduced by 50 percent. The use of molded fittings shall not result in a 50 percent reduction in the pressure rating of the pipe provided that the molded fittings shall be fabricated so that the wall thickness of the material is maintained at the threads. Thread sealant compound that is compatible with the pipe and fitting, insoluble in water and nontoxic shall be applied to male threads. Caution shall be used during assembly to prevent over tightening of the PVC components once the thread sealant has been applied. Female PVC threaded fittings shall be used with plastic male threads only.

605.13 Stainless Steel Pipe and Joints. Joining methods for stainless steel pipe and fittings shall be installed in accordance with the manufacturer's installation instructions and shall comply with Section 605.13.1 or Section 605.13.2.

605.13.1 Mechanical Joints. Mechanical joints shall be designed for their intended use. Such joints shall include compression, flanged, grooved, press-connect, and threaded.

605.13.2 Welded Joints. Welded joints shall be either

WATER SUPPLY AND DISTRIBUTION

fusion or resistance welded based on the selection of the base metal. The chemical composition of the filler metal shall comply with AWS A5.9 based on the alloy content of the piping material.

605.14 Slip Joints. In water piping, slip joints shall be permitted to be used only on the exposed fixture supply.

605.15 Dielectric Unions. Dielectric unions where installed at points of connection where there is a dissimilarity of metals shall be in accordance with ASSE 1079.

605.16 Joints Between Various Materials. Joints between various materials shall be installed in accordance with the manufacturer's installation instructions and shall comply with Section 605.16.1 through Section 605.16.3.

605.16.1 Copper or Copper Alloy Pipe or Tubing to Threaded Pipe Joints. Joints from copper or copper alloy pipe or tubing to threaded pipe shall be made using copper alloy adapter, copper alloy nipple [minimum 6 inches (152 mm)], dielectric fitting, or dielectric union in accordance with ASSE 1079. The joint between the copper or copper alloy pipe or tubing and the fitting shall be a soldered, brazed, flared, or press-connect joint and the connection between the threaded pipe and the fitting shall be made with a standard pipe size threaded joint.

605.16.2 Plastic Pipe to Other Materials. Where connecting plastic pipe to other types of piping, approved types of adapter or transition fittings designed for the specific transition intended shall be used.

PVC, CPVC, polyethylene and PEX are the plastics most commonly used as piping materials. The use of these materials is restricted to certain applications and locations within the plumbing system. Some are preferred and required by design for certain applications and locations. Therefore, it has become necessary to design special joints and special adapters to join one material to another.

Adapting plastic pipe to other materials requires the use of proper adapters. There are many types of adapters. They are used specifically for what their names indicate. They are specially designed for a specific job when the connection of different types of materials is necessary. When an adapter is necessary for a specific material, see the manufacturer's catalog on fittings and adapters compatible to the materials being used. Also, check the fitting and material's listings before using an adapter. Some adapters may appear as though they may work in various situations but, in many cases, are limited to a specific use. For example, Section 605.12.3 restricts female PVC threaded fittings to be used only with plastic male threads. Therefore, a male PVC fitting shall be used to transition to a female copper fitting.

605.16.3 Stainless Steel to Other Materials. Where connecting stainless steel pipe to other types of piping, mechanical joints of the compression type, dielectric fitting, or dielectric union in accordance with ASSE 1079 and designed for the specific transition intended shall be used.

606.0 Valves.

606.1 General. Valves up to and including 2 inches (50 mm) in size shall be copper alloy or other approved material. Sizes exceeding 2 inches (50 mm) shall be permitted to have cast iron or copper alloy bodies. Each gate or ball valve shall be a fullway or full-port type with working parts of the non-corrosive material. Valves carrying water used in potable water systems intended to supply drinking water shall comply with the requirements of NSF 61 and ASME A112.4.14, ASME B16.34, ASTM F1970, ASTM F2389, AWWA C500, AWWA C504, AWWA C507, IAPMO Z1157, MSS SP-67, MSS SP-70, MSS SP-71, MSS SP-72, MSS SP-78, MSS SP-80, MSS SP-110, MSS SP-122, or NSF 359.

Fullway gate (or ball) valves are intended to be fully opened or fully closed when in service and are for the sole purpose of providing isolation of a complete piping system or isolation of an individual zone within a system. These valves are not designed or intended to be utilized for regulation or modulation of water flow. Conversely, globe valves (regulating valves) are not designed or intended to be utilized as "control" valves. As applied in Section 606.0, the word "control" means to "check or restrain" the flow of water; consequently, fullway valves are designed to always be in one of two possible positions—full restraint or no restraint. These two operating positions (full open or full closed) are consistent with the listing standards for gate valves and with the manufacturer's recommended operating instructions (see **Figure 606.1**).

Gate valves should never be left partially open or partially closed. A valve gate that extends into the water stream will lose material through the process of erosion and, eventually, will be useless as a means of stopping water flow.

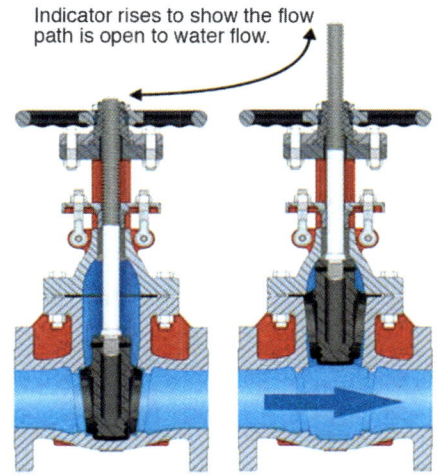

Outside stem and yoke (OS&Y) is a common indicator valve used in fire-sprinkler systems.

FIGURE 606.1
ONE TYPE OF FULLWAY VALVE

606.2 Fullway Valve. A fullway valve controlling outlets shall be installed on the discharge side of each water meter and each unmetered water supply. Water piping supplying more than one building on one premise shall be equipped with a separate fullway valve to each building, so arranged that the water supply can be turned on or off to an individual or separate building provided; however, that supply piping to a single-family residence and building accessory

WATER SUPPLY AND DISTRIBUTION

thereto shall be permitted to be controlled by one valve. Such shutoff valves shall be accessible. A fullway valve shall be installed on the discharge piping from water supply tanks at or near the tank. A fullway valve shall be installed on the cold water supply pipe to each water heater at or near the water heater.

🔧 Fullway control valves are used primarily in the water supply system to isolate separate systems, buildings, apartments or dwelling units and fixtures. This is done so that a repair, alteration or addition can be made without disrupting or shutting down entire systems and leaving occupants without running water.

606.3 Multidwelling Units. In multidwelling units, one or more shutoff valves shall be provided in each dwelling unit so that the water supply to a plumbing fixture or group of fixtures in that dwelling unit can be shut off without stopping water supply to fixtures in other dwelling units. These valves shall be accessible in the dwelling unit that they control.

606.4 Multiple Openings. Valves used to control two or more openings shall be fullway gate valves, ball valves, or other approved valves designed and approved for the service intended.

606.5 Control Valve. A control valve shall be installed immediately ahead of each water-supplied appliance and immediately ahead of each slip joint or appliance supply.

Parallel water distribution systems shall provide a control valve either immediately ahead of each fixture being supplied or installed at the manifold, and shall be identified with the fixture being supplied. Where parallel water distribution system manifolds are located in attics, crawl spaces, or other locations not readily accessible, a separate shutoff valve shall be required immediately ahead of each individual fixture or appliance served.

🔧 In this section the requirement for a control valve for each appliance or slip joint means that each fixture will need a control valve. Most fixtures and fixture fittings connect with slip joints; therefore, each fixture will require a control valve, which is normally an angle stop (see **Figure 606.5a**).

The conventional method of water distribution is a main and branch system. This system utilizes a main to service several branches, which in turn service individual fixtures. A parallel water distribution system usually refers to plastic pipe systems, usually PEX or PEX-AL-PEX systems that use a manifold in the system. There are two types of manifold systems:

- The "home run" system utilizes a centrally located manifold to individually distribute supply lines to each fixture (see **Figure 606.5b**).
- The "remote manifold" system utilizes a trunk or main, which services several small manifolds that in turn service a group of individual fixtures.

The manifold, as long as it incorporates a fullway shutoff valve for each fixture and is not located in an attic, crawl space, or other locations not readily accessible, can be used in lieu of a control valve at the fixture itself. This will entail shutting off the supply to a fixture from the manifold, not an angle stop at the fixture. The ports on the manifold must be properly identified as to the fixture and location it controls. For manifolds located in attics, crawl spaces, or other locations not readily accessible, a shut off valve must be ahead of the fixture or appliance.

All valves and control manifolds within the water supply system shall be accessible no matter where they are located on or about the premises.

606.6 Accessible. Required shutoff or control valves shall be accessible.

606.7 Multiple Fixtures. A single control valve shall be installed on a water supply line ahead of an automatic metering valve that supplies a battery of fixtures.

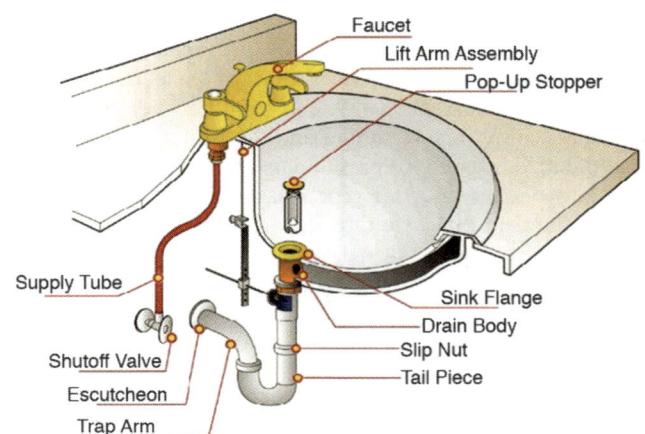

**FIGURE 606.5A
SHUTOFF VALVE AND SLIP NUT**

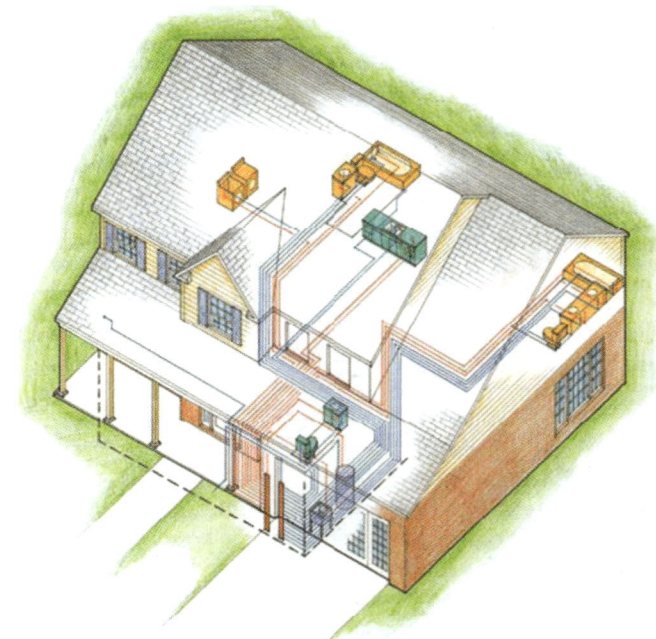

**FIGURE 606.5B
PARALLEL WATER DISTRIBUTION PEX HOME RUN
MANIFOLD SYSTEM**

607.0 Potable Water Supply Tanks.

607.1 General. Potable water supply tanks shall be installed in accordance with the manufacturer's installation instructions and supported in accordance with the building code.

607.2 Potable Water Tanks. Potable water supply tanks, interior tank coatings, or tank liners intended to supply drinking water shall comply with NSF 61.

607.3 Venting. Tanks used for potable water shall be tightly covered and vented in accordance with the manufacturer's installation instructions. Such vent shall be screened with a corrosion-resistant material of not less than number 24 mesh.

607.4 Overflow. Tanks shall have not less than a 16 square inch (0.01 m^2) overflow that is screened with a corrosion-resistant material of not less than number 24 mesh.

607.5 Valves. Pressurized tanks shall be provided with a listed pressure-relief valve installed in accordance with the manufacturer's installation instructions. The relief valve shall be discharged in accordance with Section 608.5. Where a potable water supply tank is located above the fixtures, appliances, or system components it serves, it shall be equipped with a vacuum relief valve that complies with CSA Z21.22.

608.0 Water Pressure, Pressure Regulators, Pressure Relief Valves, and Vacuum Relief Valves.

608.1 Inadequate Water Pressure. Where the water pressure in the main or other source of supply will not provide a residual water pressure of not less than 15 pounds force per square inch (psi) (103 kPa), after allowing for friction and other pressure losses, a tank and a pump or other means that will provide said 15 psi (103 kPa) pressure shall be installed. Where fixtures, fixture fittings, or both are installed that, require residual pressure exceeding 15 psi (103 kPa), that minimum residual pressure shall be provided.

The residual water pressure required for any water supply system is 15 psi. By definition, Residual Pressure is the pressure available at the fixture or water outlet after allowance is made for pressure drop due to friction loss, head, meters, and other losses in the system during maximum demand periods (See Section 218.0). There may be certain fixtures or appliances in an installation that may require a higher minimum pressure above 15 psi. In this case, that minimum pressure will be the required minimum pressure. If the main line pressure cannot provide this minimum, the system must be designed using pumps, tanks or both to achieve the minimum required.

608.2 Excessive Water Pressure. Where static water pressure in the water supply piping is exceeding 80 psi (552 kPa), an approved-type pressure regulator preceded by an adequate strainer shall be installed and the static pressure reduced to 80 psi (552 kPa) or less. Pressure regulator(s) equal to or exceeding 1½ inches (40 mm) shall not require a strainer. Such regulator(s) shall control the pressure to water outlets in the building unless otherwise approved by the Authority Having Jurisdiction. Each such regulator and strainer shall be accessibly located aboveground or in a vault equipped with a properly sized and sloped boresighted drain to daylight, shall be protected from freezing, and shall have the strainer readily accessible for cleaning without removing the regulator or strainer body or disconnecting the supply piping.

Pipe size determinations shall be based on 80 percent of the reduced pressure where using Table 610.4.

An approved expansion tank shall be installed in the cold water distribution piping downstream of each such regulator to prevent excessive pressure from developing due to thermal expansion and to maintain the pressure setting of the regulator. Expansion tanks used in potable water systems intended to supply drinking water shall comply with NSF 61. The expansion tank shall be properly sized and installed in accordance with the manufacturer's installation instructions and listing. Systems designed by registered design professionals shall be permitted to use approved pressure relief valves in lieu of expansion tanks provided such relief valves have a maximum pressure relief setting of 100 psi (689 kPa) or less.

A limit of 80 psi (551.6 kPa) is the maximum static pressure of any water supply system. The reason for this is to reduce water hammer, unnecessary use of water, splashing, excessive discharge of pressure relief valves and to protect appliance and fixture valves and mechanisms from pressure that exceeds their design limits. Any installation with pressures above 80 psi will require a pressure regulating valve to limit the pressure to 80 psi or below (see **Figure 608.2**).

By design, pressure regulating or reducing valves are modulating valves, which have a high level of flow resistance and consequent pressure drop through them even when fully open. Therefore, pipe sizing downstream of the pressure regulator must be based on "worst-case" pressure loss during a maximum demand water flow. Worst-case pressure loss through a listed pressure regulator is presumed to be no greater than 20 percent; therefore, the water system is sized based upon 80 percent of the pressure regulator's "set" pressure, this being a selected static pressure that is presumed not to exceed 80 psi. Therefore, all pipe size determinations downstream of the regulator must be based on 80 percent of this reduced pressure when using Table 610.4.

For example, a water system has a pressure of 100 psi. A pressure regulator will be installed and set at 80 psi. For sizing purposes using Table 610.4, the maximum pressure would be 64 psi, which is 80 percent of 80 psi.

608.3 Expansion Tanks, and Combination Temperature and Pressure-Relief Valves. A water system provided with a check valve, backflow preventer, or other normally closed device that prevents dissipation of building pressure back into the water main, independent of the type of water heater used, shall be provided with an approved, listed, and adequately sized expansion tank or other approved device having a similar function to control thermal expansion. Such expansion tank or other approved device shall be installed on the building side of the check valve, backflow preventer, or other device and shall be

WATER SUPPLY AND DISTRIBUTION

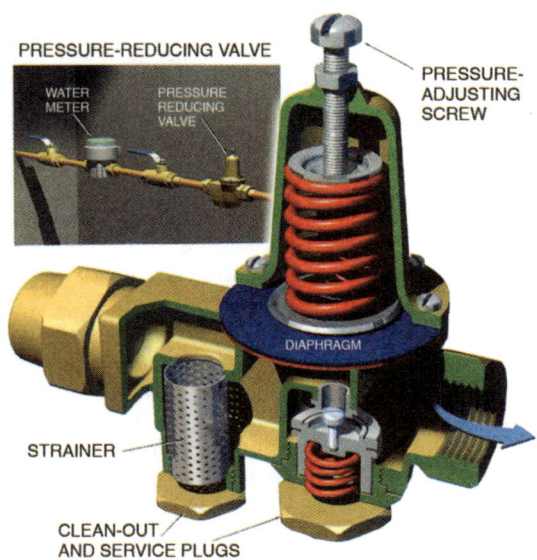

**FIGURE 608.2
PRESSURE REDUCING VALVE WITH STRAINER**

sized and installed in accordance with the manufacturer's installation instructions.

A water system containing storage water heating equipment shall be provided with an approved, listed, adequately sized combination temperature and pressure-relief valve, except for listed nonstorage instantaneous heaters having an inside diameter of not more than 3 inches (80 mm). Each such approved combination temperature and pressure-relief valve shall be installed on the water-heating device in an approved location based on its listing requirements and the manufacturer's installation instructions. Each such combination temperature and pressure-relief valve shall be provided with a drain in accordance with Section 608.5.

Water is, for all practical purposes, an incompressible liquid. Because it is incompressible, water placed in a closed container can build up high pressure when heated, even when heated from 40°F to room temperature. This is thermal expansion.

A building water distribution system having a check valve, pressure regulator, backflow preventer or other device that prevents pressure buildup in the building from being able to dissipate back into the source or water supply main regardless of the type of water heater used, must be equipped with an expansion tank or other means to control the thermal expansion pressures. The expansion tank will absorb excess pressure within the tank (see **Figure 608.3a**). The pressure relief valve relieves the excess pressure by discharging until the pressure drops below the set point (see **Figure 608.3b**).

When the water system is equipped with a storage water heater, pressures rise even more because of the increase in temperature and the high volume of water contained in the tank. All storage water heaters are required to be equipped with a combination temperature and pressure relief valve (T&P valve). Note in **Figure 608.3b**, the combination T&P valve has a temperature sensing tube. The pressure relief valve does not.

Storage water heaters must be manufactured to the ANSI Z21.10 series of standards. These water heaters are protected in three stages. The primary stage is the thermostat. Should the thermostat fail, the secondary stage or high-limit switch will turn off the source of energy to the heater. If the high-limit switch fails, the combination T&P valve opens to prevent a catastrophic failure of the water heater.

Instantaneous water heaters having an inside dimensional width of three inches or less are exempt from having to meet the requirement for a T&P valve. An example of this exemption is the small single point of use heater for a hand sink. The typical dimension for these units are 11" x 5" x 2.8" and they have no storage capacity other than in the pipe itself to cause over temperature and pressure build up. When the need for hot water has been met the instantaneous water heater shuts off. If the water supply to this small heater has a backflow preventer, an adequately sized expansion tank shall be provided (see **Figure 608.3d**) even though a T&P valve is not required.

When installing a T&P valve it is critical that the installer check the rating plate on the combination T&P valve before installation to make sure that the Btu input rating of the water heater does not exceed the maximum Btu rating of the valve. The use of an undersized combination T&P valve could result in a catastrophic water heater failure should both the water heater thermostat and the water heater high-limit switch fail. In instances where there are two separate Btu ratings on the valve plate, the smaller of the two is used in making this determination. For more information, see Chapter 5.

Manufacturers of T&P valves require that the valve be installed with its temperature-sensing element immersed within the top 6 inches of the tank since this is where the hottest water in the tank is located (see **Figure 608.3c**). Manufacturers recommend that they be manually opened at least once a year to ensure that they are functioning. In areas that have a high mineral content in the water, scale can form around the valve seat and render the valve inoperable. In such areas, it may be necessary to manually open the valve every three months or less. A water heater is a potential bomb that is prevented from exploding by three separate safety devices. The combination T&P valve is a lifesaver and is treated accordingly.

It must be emphasized that water heaters operating at pressures above 150 psi are classified as boilers. In fact, water heaters that have a storage volume greater than 120 gallons, have a Btu input rating greater than 200,000, or operate at temperatures higher than 210°F, are also classified as boilers and are manufactured to the ASME *Pressure Vessel Standard*.

608.4 Pressure Relief Valves. Each pressure relief valve shall be an approved automatic type with drain, and each such relief valve shall be set at a pressure of not more than 150 psi (1034 kPa). No shutoff valve shall be installed between the relief valve and the system.

608.5 Discharge Piping. The discharge piping serving a temperature relief valve, pressure relief valve, or combina-

WATER SUPPLY AND DISTRIBUTION

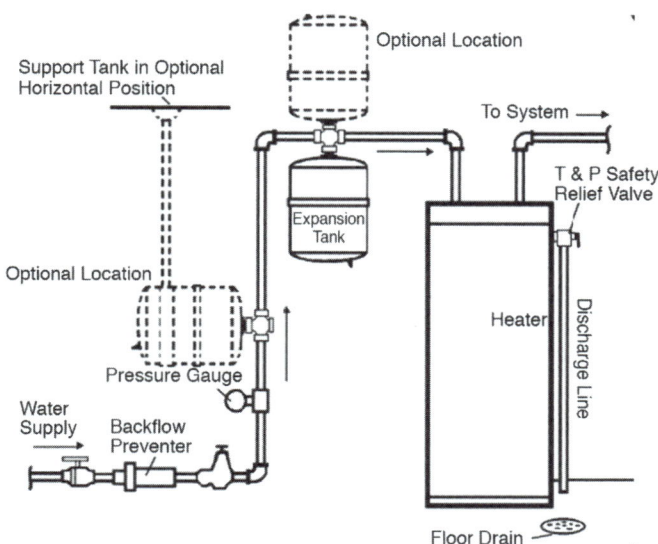

FIGURE 608.3A
INSTALLATION OF THERMAL EXPANSION TANK

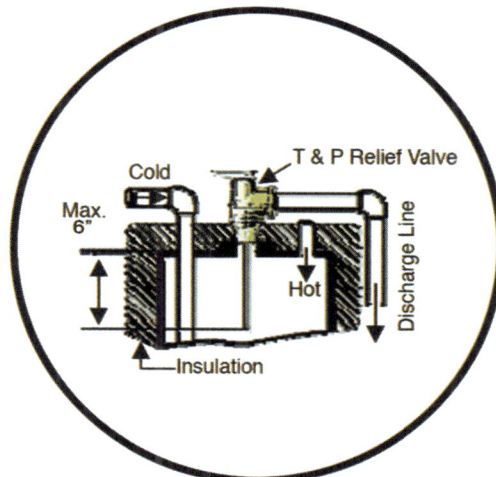

FIGURE 608.3C
PROPER INSTALLATION OF T&P VALVE

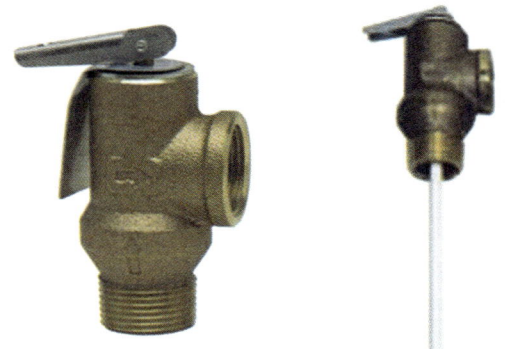

FIGURE 608.3B
PRESSURE RELIEF VALVE AND COMBINATION TEMPERATURE PRESSURE RELIEF VALVE

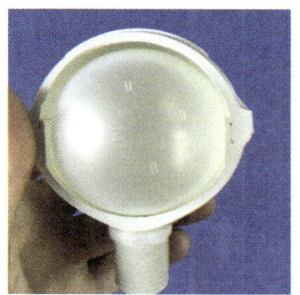 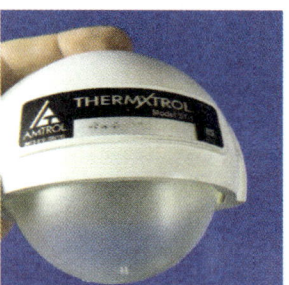

FIGURE 608.3D
INSTANTANEOUS WATER HEATER EXPANSION TANK

tion of both shall have no valves, obstructions, or means of isolation and be provided with the following:

(1) Equal to the size of the valve outlet and shall discharge full size to the flood level of the area receiving the discharge and pointing down.

(2) Materials shall be rated at not less than the operating temperature of the system and approved for such use or shall comply with ASME A112.4.1.

(3) Discharge pipe shall discharge independently by gravity through an air gap into the drainage system or outside of the building with the end of the pipe not exceeding 2 feet (610 mm) and not less than 6 inches (152 mm) above the ground and pointing downwards.

(4) Discharge in such a manner that does not cause personal injury or structural damage.

(5) No part of such discharge pipe shall be trapped or subject to freezing.

(6) The terminal end of the pipe shall not be threaded.

(7) Discharge from a relief valve into a water heater pan shall be prohibited.

🔧 The pan is not an approved receptacle and is not intended for a pressurized discharge from relief valves. The pan is only intended to contain tank leakage and dispose it through a gravity drain.

608.6 Water-Heating Devices. A water-heating device connected to a separate storage tank and having valves between said heater and tank shall be provided with an approved water pressure relief valve.

🔧 Any time there is a possibility for pressure buildup or thermal expansion because of a piping or valve arrangement, the pressure must be relieved through a pressure regulator.

608.7 Vacuum Relief Valves. Where a hot-water storage tank or an indirect water heater is located at an elevation above the fixture outlets in the hot-water system, a vacuum relief valve that complies with CSA Z21.22 shall be installed on the storage tank or heater.

🔧 If the water heater is installed above fixtures, there is a possibility of siphonage from the water heater. If the water supply is turned off or pressure drops as fixtures are being used, there is a possibility of a siphon being created because of the fixture water supply opening being lower than the tank. This could result in the tank emptying and possible steam being created in the tank. There have been instances where this siphonic action has been so pow-

WATER SUPPLY AND DISTRIBUTION

erful as to collapse the tank. For these reasons, a vacuum relief valve must be installed (see **Figure 608.7**).

609.0 Installation, Testing, Unions, and Location.

609.1 Installation. Water piping shall be adequately supported in accordance with Table 313.3. Burred ends shall be reamed to the full bore of the pipe or tube. Changes in direction shall be made by the appropriate use of fittings, except that changes in direction in copper or copper alloy tubing shall be permitted to be made with bends, provided that such bends are made with bending equipment that does not deform or create a loss in the cross-sectional area of the tubing. Changes in direction are allowed with flexible pipe and tubing without fittings in accordance with the manufacturer's instructions. Provisions shall be made for expansion in hot-water piping. Piping, equipment, appurtenances, and devices shall be installed in a workmanlike manner in accordance with the provisions and intent of the code. Building supply yard piping shall be not less than 12 inches (305 mm) below the average local frost depth. The cover shall be not less than 12 inches (305 mm) below finish grade.

These installation requirements of the water supply piping reiterate many of the requirements for installation of piping in Chapter 3. The limitation on the bending of plastic piping is to prevent crimping the pipe, thus preventing a decrease in the diameter of the water supply pipe. Consult the piping manufacturer's installation requirements for these bend limitations.

The minimum depth of any water service yard piping (building supply pipe) is 12 inches below finished grade. Where there is a possibility of the ground cover freezing, the minimum depth is 12 inches below the average local frost depth. Be sure to check with the AHJ to determine the frost depth at the installation site.

609.2 Trenches. Water pipes shall not be run or laid in the same trench as building sewer or drainage piping constructed of clay or materials that are not approved for use within a building unless both of the following conditions are met:

(1) The bottom of the water pipe shall be not less than 12 inches (305 mm) above the top of the sewer or drain line.

(2) The water pipe shall be placed on a solid shelf excavated at one side of the common trench with a clear horizontal distance of not less than 12 inches (305 mm) from the sewer or drain line.

Water pipes crossing sewer or drainage piping constructed of clay or materials that are not approved for use within a building shall be laid not less than 12 inches (305 mm) above the sewer or drain pipe.

For a variety of reasons, either water or drainage piping may develop leaks over a period of time. The fill material (backfill) around these pipes will become saturated when a leak occurs. Saturated soil becomes a bridge for bacterial travel between pipes. When one of these pipes is carrying potable water and the other is transporting waste water, there is a considerable threat to the potable water system. Therefore, it is essential that potable water and waste piping not be allowed to share a common trench unless the building sewer is constructed of materials approved for use within the building.

The requirement of a "solid shelf" that is 12 inches above and 12 inches horizontal from the top of a sewer pipe is to provide an independent "trough" for each pipe (see **Figure 609.2**). A failure in the sewer piping will eventually result in saturation and settlement of the fill material above the sewer. However, because the water piping is supported on an independent trench base and is off to one side of the common trench, it will be unaffected by this event.

Water pipe is permitted to be laid in the same trench as the building sewer, and at the same elevation, or to cross over the sewer with little separation, provided the building sewer is constructed using materials (for its entire length) that are approved for use within the building. Materials for use under and within the building are much more rigid and stronger than materials that are used only outside of the building foundation. For example, cast iron drainage pipe is sturdier than PVC Pipe SDR 25. The joints in the former are also less susceptible to leaking or separating than the latter. Therefore, there is less chance of contamination occurring using drainage materials for use under and within the building.

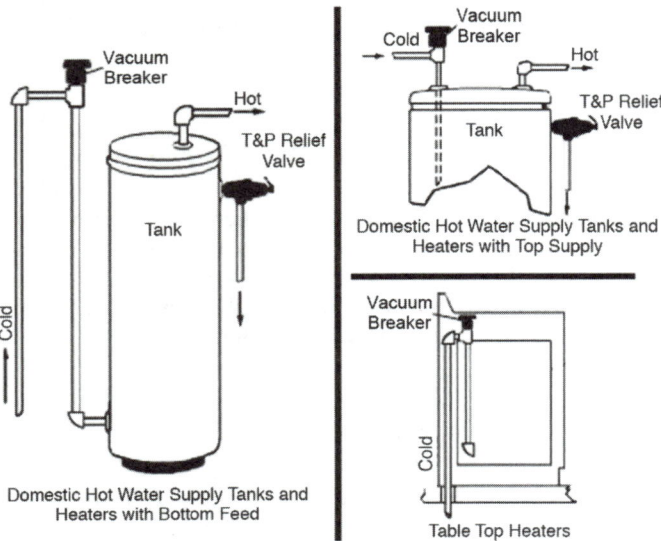

FIGURE 608.7
VACUUM RELIEF VALVE INSTALLATION

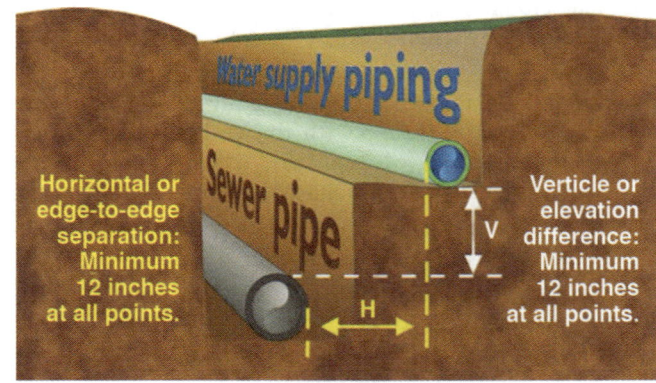

FIGURE 609.2
WATER AND SEWER LINE SEPARATION

609.3 Under Concrete Slab. Water piping installed within a building and in or under a concrete floor slab resting on the ground shall be installed in accordance with the following requirements:

(1) Ferrous piping shall have a protective coating of an approved type; machine applied and in accordance with recognized standards. Field wrapping shall provide equivalent protection and shall be restricted to those short sections and fittings necessarily stripped for threading. Zinc coating (galvanizing) shall not be deemed adequate protection for piping or fittings. Approved nonferrous piping shall not be required to be wrapped.

(2) Copper or copper alloy tubing shall be installed without joints where possible. Where joints are permitted, they shall be brazed, and fittings shall be wrought copper.

For the purpose of this section, "within a building" shall mean within the fixed limits of the building foundation.

Knowledge of the mineral content of the soil in which piping is being laid is extremely important. This knowledge will dictate what type of piping material to use. For example, soil with high iron content is unsuitable for copper piping unless the copper is sleeved continuously for its entire length underground. Even if the copper is sleeved and there is a pinhole in the protective covering, the corrosion caused by galvanic action will occur even quicker than if the pipe was not sleeved at all. This is due to the concentration of the corrosion – transfer of ions – in one spot and will wear the pipe wall at a faster rate than if the transfer occurred along the entire length of pipe.

Of course copper pipe is not the only material that can be eroded. The point being made is, know the soils where the installation will occur and know the limitations of the materials that are being used.

609.4 Testing. Upon completion of a section or of the entire hot and cold water supply system, the system shall be tested with water or air. The potable water test pressure shall be greater than or equal to the working pressure under which the system is to be used. The air pressure shall be a minimum of 50 psi (345 kPa). Plastic pipe shall not be tested with air. The piping system shall withstand the test pressure without showing evidence of leakage for a period of not less than 15 minutes.

Exception: PEX, PP or PE-RT tube shall be permitted to be tested with air where permitted by the manufacturer's instructions.

One of the prime tenets of plumbing installation is to have the water system pressure tested, inspected and approved before the system or any portion thereof is covered or concealed. The hot and cold water piping must be subjected to a pressure test of not less than its working pressure without leaking. Except for plastic piping, a 50 psi air pressure test may be used. Due to the possibility of plastic piping, especially PVC, shattering under air pressure plastic piping must be tested with water only. For an instructive video follow the Learning Link **goo.gl/dsywgJ**.

When specifically authorized by the manufacturer's instructions, pipe and fittings such as PEX, PP or PE-RT can be tested by air when not otherwise prohibited because it is unlikely to shatter under pressure. There are climate conditions that would prevent water from being utilized in test and in addition, excessive use of water resources could be decreased by testing with air as an alternative.

609.5 Unions. Unions shall be installed in the water supply piping not more than 12 inches (305 mm) of regulating equipment, water heating, conditioning tanks, and similar equipment that requires service by removal or replacement in a manner that will facilitate its ready removal.

609.6 Location. Except as provided in Section 609.7, no building supply shall be located in a lot other than the lot that is the site of the building or structure served by such building supply.

609.7 Abutting Lot. Nothing contained in this code shall be construed to prohibit the use of an abutting lot to:

(1) Provide access to connect a building supply to an available public water service where proper cause and legal easement not in violation of other requirements have been first established to the satisfaction of the Authority Having Jurisdiction.

(2) Provide additional space for a building supply where the proper cause, transfer of ownership, or change of boundary not in violation of other requirements have been first established to the satisfaction of the Authority Having Jurisdiction. The instrument recording such action shall constitute an agreement with the Authority Having Jurisdiction, which shall clearly state and show that the areas so joined or used shall be maintained as a unit during the time they are so used. Such an agreement shall be recorded in the office of the County Recorder as a part of the conditions of ownership of said properties, and shall be binding on heirs, successors, and assigns to such properties. A copy of the instrument recording such proceedings shall be filed with the Authority Having Jurisdiction.

609.8 Low-Pressure Cutoff Required on Booster Pumps for Water Distribution Systems. Where a booster pump (excluding a fire pump) is connected to a building supply or underground water pipe, a low-pressure cutoff switch on the inlet side of the pump shall be installed not more than 5 feet (1524 mm) of the inlet. The cutoff switch shall be set for not less than 10 psi (69 kPa). A pressure gauge shall be installed between the shutoff valve and the pump.

The low pressure cutoff is required because of the possibility of an excessive pressure drop resulting in negative pressure on the suction side (inlet) of the pump creating a phenomenon called cavitation. When cavitation occurs, cavities or voids form in the pump creating turbulence that causes damage to the impellers and eventually to pump failure. This failure could also lead to backflow contamination.

609.9 Disinfection of Potable Water System. New or repaired potable water systems shall be disinfected prior to use where required by the Authority Having Jurisdiction.

The method to be followed shall be that prescribed by the Health Authority or, in case no method is prescribed by it, the following:

(1) The pipe system shall be flushed with clean, potable water until potable water appears at the points of the outlet.

(2) The system or parts thereof shall be filled with a water-chlorine solution containing not less than 50 parts per million of chlorine, and the system or part thereof shall be valved-off and allowed to stand for 24 hours; or, the system or part thereof shall be filled with a water-chlorine solution containing not less than 200 parts per million of chlorine and allowed to stand for 3 hours.

(3) Following the allowed standing time, the system shall be flushed with clean, potable water until the chlorine residual in the water coming from the system does not exceed the chlorine residual in the flushing water.

(4) The procedure shall be repeated where it is shown by a bacteriological examination made by an approved agency that contamination persists in the system.

Disinfection of the water distribution system is vitally important to the health and safety of the building occupants. Piping manufacturers, supply distribution centers and the installer should all take care to protect pipe before it is installed. However, contaminants can enter the pipe during storage or installation, or by reason of breakage or leakage. The pipe must be returned to safe drinking water limits. Usually the health department or water purveyor has jurisdiction over water supply disinfections and has its own requirements. If not, this section may be used. Once again be sure to consult the AHJ as to the proper requirements for disinfection.

609.10 Water Hammer. Building water supply systems where quick-acting valves are installed shall be provided with water hammer arrester(s) to absorb high pressures resulting from the quick closing of these valves. Water hammer arresters shall be approved mechanical devices that comply with ASSE 1010 or PDI-WH 201 and shall be installed as close as possible to quick-acting valves.

609.10.1 Mechanical Devices. Where listed mechanical devices are used, the manufacturer's specifications as to location and method of installation shall be followed.

Although water hammer is a subject usually left up to plumbing engineers, the effects of water hammer must be dealt with every day by plumbing contractors everywhere. Water hammer is easily recognized by the banging or thumping noise that is heard when valves are shut off. Although this is an easy way to recognize the problem, water hammer does not always make these telltale noises. Water hammer occurs when the flow of moving water is suddenly stopped by a closing valve. This sudden stop results in a tremendous spike of pressure behind the valve, which acts like a tiny explosion inside the pipe. This pressure spike reverberates throughout the plumbing system, rattling and shaking pipes, until it is absorbed. Normally, a sufficient pocket of air will absorb such a pressure spike, but if no pocket of air is present, expensive fixtures and appliances within the plumbing system will be damaged as they are left to absorb this pressure spike.

It used to be thought that an air chamber, or capped stand pipe, was an effective solution to controlling water hammer. However, within an air chamber, nothing separates the air from the water. It only takes a few short weeks before the air is absorbed into the water, leaving the air chamber waterlogged and completely ineffective against water hammer. Laboratory tests confirm that the air is depleted by simple air permeation and by interaction between static pressure and flow pressure. In the diagram shown in **Figure 609.10.1a**, notice the difference in water level between "Static Line Pressure" and "Postcycle Static Level." One can easily see that this method of protecting against water hammer is ineffective.

The most effective means of controlling water hammer is a measured, compressible cushion of air which is permanently separated from the water system. The arrester model shown in **Figure 609.10.1b** employs a pressurized cushion of air and a two O-ring piston, which permanently separates this air cushion from the water system. When the valve closes and the water flow is suddenly stopped, the pressure spike pushes the piston up the arrester chamber against the pressurized cushion of air. The air cushion in the arrester reacts instantly, absorbing the pressure spike that causes water hammer.

Many other styles of water hammer arresters are manufactured. There are residential installation models and industrial models that are also represented in **Figure 609.10.1b**. Always consult the manufacturer's installation and product manuals for the proper size and installation location for a water hammer arrester.

609.11 Pipe Insulation. Insulation of domestic hot water piping shall be in accordance with Section 609.11.1 and Section 609.11.2.

The water in uninsulated ½-inch nominal pipe surrounded by room temperature air cools down from 120°F to 105°F in about 10 minutes; in ¾ inch nominal pipe it cools down in about 15 minutes. R-3 pipe insulation roughly doubles the cool down time to 20 minutes for ½ inch piping and roughly triples it to 45 minutes for ¾ inch piping. When the time between hot water events exceeds one hour, the water in the insulated pipes is likely to cool down back to ambient, minimizing the benefit of pipe insulation for spread out draws. Based on the Lawrence Berkeley National Laboratory research findings (2012), about 30 percent of hot water draws are within 10 and 60 minutes apart. Pipe insulation will eliminate most of the water and energy wasted while waiting for these hot water draws. See **Figure 609.11**.

609.11.1 Insulation Requirements. Domestic hot water piping shall be insulated.

609.11.2 Pipe Insulation Wall Thickness. Hot water pipe insulation shall have a minimum wall thickness of not less than the diameter of the pipe for a pipe up to 2 inches (50 mm) in diameter. Insulation wall thickness shall be not less than 2 inches (51 mm) for a pipe of 2 inches (50 mm) or more in diameter.

WATER SUPPLY AND DISTRIBUTION

Exceptions:

(1) Piping that penetrates framing members shall not be required to have pipe insulation for the distance of the framing penetration.

(2) Hot water piping between the fixture control valve or supply stop and the fixture or appliance shall not be required to be insulated.

610.0 Size of Potable Water Piping.

610.1 Size. The size of each water meter and each potable water supply pipe from the meter or other source of supply to the fixture supply branches, risers, fixtures, connections, outlets, or other uses shall be based on the total demand and shall be determined according to the methods and procedures outlined in this section. Water piping systems shall be designed to ensure that the maximum velocities allowed by the code and the applicable standard are not exceeded.

🔧 Water pipe sizing has two primary objectives—maintaining flow velocities at a level that is appropriate for the type of pipe being installed and matching pipe sizes to the residual pressure of the system. Matching these two principles together will allow for a water supply distribution system that will deliver the appropriate amount of water volume at the appropriate pressures to allow the fixtures or appliances to perform their functions without causing damage to the system, fixture or appliance.

Each type of piping material has limitations on its ability to withstand the erosive effects of water flow at high velocities. Maintaining the proper velocity of water flowing through the system, therefore, ensures a long lasting piping system. For instance, the corrosive resistance of copper pipe as well as other metals is due to an oxide film that naturally forms on the walls of the copper pipe as it comes in contact with flowing water. If the velocity of water is too great, then impingement, also referred to as erosion corrosion, occurs, stripping the protective oxide coating. Impingement damage occurs when water velocity in copper pipe exceeds eight feet per second for cold water and five feet per second for hot water. Hot water, especially in a pumped circulation system, is required to maintain lower velocities because of hot water's greater tendency to release oxygen bubbles, which increases with velocity, causing acceleration in corrosion.

When installing, the failure to ream pipe ends after cutting and threading also creates its own erosive action at the pipe wall and greatly exacerbates the effects associated with high velocities, turbulence, and cavitation, which become an additional source of pressure loss as well as causing physical damage to piping materials.

The selection of piping material and how it is sized and installed in the building are critical elements in the delivery of a potable water supply that satisfies the total demand load imposed upon the system. This total demand is estab-

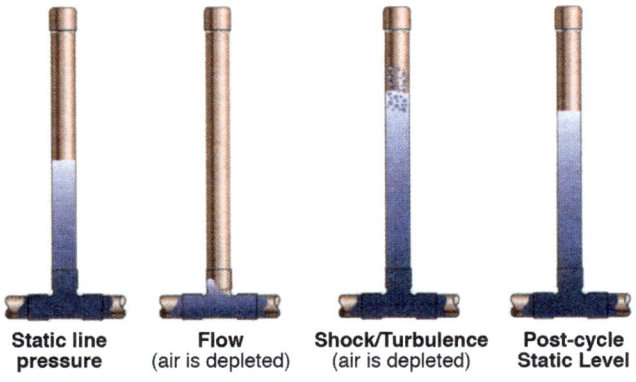

FIGURE 609.10.1A
CHAMBER AIR CUSHION DEPLETION

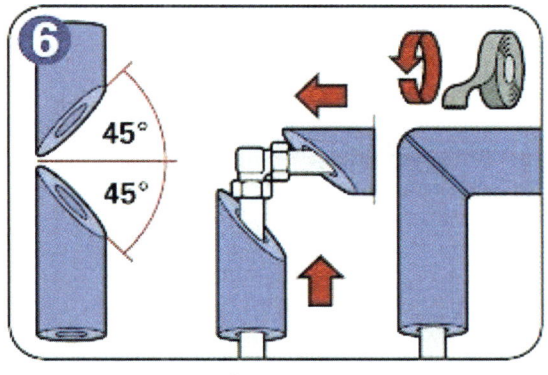

FIGURE 609.11
PIPE INSULATION

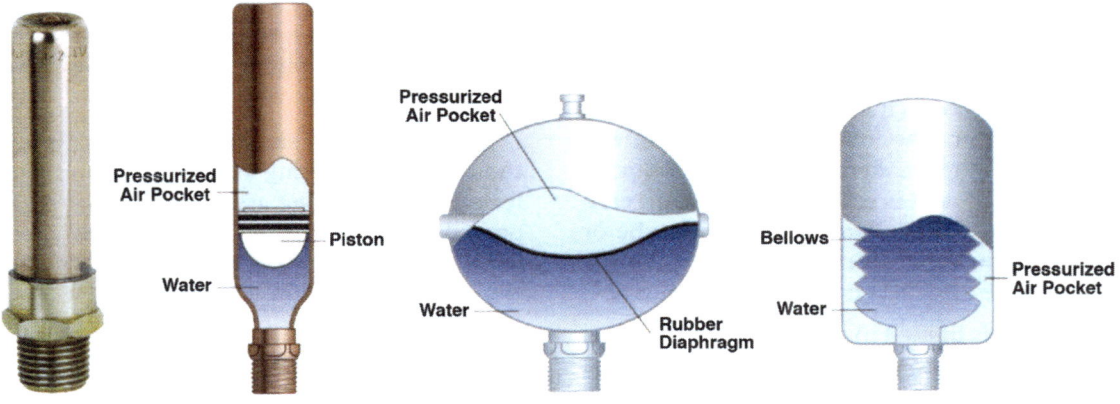

FIGURE 609.10.1B
TYPES OF WATER HAMMER ARRESTERS

lished by the choice of type and number of fixtures and equipment, their location in the building and the minimum water pressure requirements at each point of service within the system. The following sections will guide the user in providing for that total demand.

610.2 Pressure Loss. Where a water filter, water softener, backflow prevention device, tankless water heater, or similar device is installed in a water supply line, the pressure loss through such devices shall be included in the pressure loss calculations of the system, and the water supply pipe and meter shall be adequately sized to provide for such a pressure loss.

No water filter, water softener, backflow prevention device, or similar device regulated by this code shall be installed in a potable water supply piping where the installation of such device produces an excessive pressure drop in such water supply piping. In the absence of specific pressure drop information, the diameter of the inlet or outlet of such device or its connecting piping shall be not less than the diameter of such water distribution piping to the fixtures served by the device.

Such devices shall be of a type approved by the Authority Having Jurisdiction and shall be tested for flow rating and pressure loss by an approved laboratory or recognized testing agency to standards consistent with the intent of this chapter.

Devices such as those noted above will cause pressure drops in the system and this resultant loss must be included in the pressure loss calculations when sizing the system. The size of the water supply piping and meter is then designed to offset the pressure drops caused by the frictional losses through these devices. The manufacturer of the device will provide the pressure loss factor in its installation material.

It is possible that after pressure loss calculations are completed and gallons-per-minute flow is taken into consideration, the installer could have a larger size of pipe than the inlet and outlet of the device. This is allowable and not considered a prohibited reduction of flow as long as the calculations are provided to the AHJ. When the pressure loss factors are unknown, the inlet and outlet of the device causing the pressure loss must have the same diameter as the supply piping.

610.3 Quantity of Water. The quantity of water required to be supplied to every plumbing fixture shall be represented by fixture units, as shown in Table 610.3. Equivalent fixture values shown in Table 610.3 include both hot and cold water demand.

Table 610.3 lists the water supply fixture unit equivalents for common plumbing fixtures installed in private, public and assembly occupancies. A water supply fixture unit is a numerical load factor used to compare the relative load-producing effects of one type of fixture to another type of fixture. When the concept of water supply fixture units was originally developed in Methods of Estimating Loads in Plumbing Systems (BMS65), an advanced method of probability and a select comparative scale of 1 to 10 were chosen. As a result, a lavatory had a numerical load factor of 2 fixture units relative to a Flushometer valve water closet having a numerical load factor of 10 fixture units. All other fixtures had relative values in between. Since the 1992 Energy Policy Act (EPAct), fixture unit values have been reduced to reflect the reduction of flow rates mandated by the act.

Water supply fixture units are not to be confused with drainage fixture units that were based on a different scale and a different probability algorithm (see commentary in Chapter 7). A drainage fixture unit is based on the relative discharge of one type of fixture as compared to the discharge of another type of fixture. For example, a bathtub installed in a private-use occupancy is assigned four water supply fixture units. This same bathtub is assigned only two drainage fixture units.

Another factor affecting the number of water supply fixture units assigned to a specific fixture is the probability of simultaneous use. Where three 1.6 gallon per flush (gpf) flush tank water closets are installed in a private-use or even a public-use occupancy, the probability that all three will be flushed at the same time is quite low; however, where installed in an assembly occupancy, the probability of simultaneous or near simultaneous flushing is quite high. This factor is the reason that the water supply fixture unit rating assigned to this fixture is 2.5 for private and public use and 3.5 for assembly use.

The water supply fixture unit values shown in Table 610.3 represent the load on the meter and building supply. When sizing the building supply pipe according to Section 610.8, utilize the listed fixture unit values in Table 610.3. When sizing for hot and cold water branches, it is permissible to use ¾ the listed fixture unit value, only for fixtures having both hot and cold water connections. For example, the fixture unit value for a lavatory is one fixture unit. This value represents the load when sizing the meter and building supply. When sizing the separate hot and cold water branches, the fixture unit value may be .75 for each branch. This may not be significant for small systems, but for larger systems this may result in reduced pipe size and cost savings.

Table 610.3 also establishes the minimum pipe size of a given fixture branch. For example, a clothes washer will have a minimum fixture branch pipe size of ½-inch. It will be assigned a fixture unit value of four water supply fixture units (WSFU) in private use and four WSFU in public use. There is no value for assembly use because it would not apply to a clothes washer.

As with any table in the code, the note section at the bottom of the table is very important as to the use of the table, and in this case to sizing the water supply system. Following is a review of these notes:

1. The minimum fixture branch pipe size applies to the cold branch pipe for a cold-only supplied fixture. It also applies to both the hot and cold fixture branch size for a fixture supplied with hot and cold water. For example, a clothes washer fixture branch hot water supply is ½-inch and the cold water supply is ½-inch.

2. For fixtures or appliances not listed in Table 610.3 one must acquire the gallons per minute flow and establish

TABLE 610.3
WATER SUPPLY FIXTURE UNITS (WSFU) AND MINIMUM FIXTURE BRANCH PIPE SIZES[3]

APPLIANCES, APPURTENANCES OR FIXTURES[2]	MINIMUM FIXTURE BRANCH PIPE SIZE[1,4] (inches)	PRIVATE	PUBLIC	ASSEMBLY[6]
Bathtub or Combination Bath/Shower (fill)	½	4.0	4.0	—
¾ inch Bathtub Fill Valve	¾	10.0	10.0	—
Bidet	½	1.0	—	—
Clothes Washer	½	4.0	4.0	—
Dental Unit, cuspidor	½	—	1.0	—
Dishwasher, domestic	½	1.5	1.5	—
Drinking Fountain or Water Cooler	½	0.5	0.5	0.75
Hose Bibb	½	2.5	2.5	—
Hose Bibb, each additional[8]	½	1.0	1.0	—
Lavatory	½	1.0	1.0	1.0
Lawn Sprinkler, each head[5]	—	1.0	1.0	—
Mobile Home, each (minimum)	—	12.0	—	—
Sinks	—	—	—	—
Bar	½	1.0	2.0	—
Clinical Faucet	½	—	3.0	—
Clinical Flushometer Valve with or without faucet	1	—	8.0	—
Kitchen, domestic with or without dishwasher	½	1.5	1.5	—
Laundry	½	1.5	1.5	—
Service or Mop Basin	½	1.5	3.0	—
Washup, each set of faucets	½	—	2.0	—
Shower, per head	½	2.0	2.0	—
Urinal, 1.0 GPF Flushometer Valve	¾	See Footnote[7]		—
Urinal, greater than 1.0 GPF Flushometer Valve	¾	See Footnote[7]		—
Urinal, flush tank	½	2.0	2.0	3.0
Urinal, Hybrid	½	1.0	1.0	1.0
Wash Fountain, circular spray	¾	—	4.0	—
Water Closet, 1.6 GPF Gravity Tank	½	2.5	2.5	3.5
Water Closet, 1.6 GPF Flushometer Tank	½	2.5	2.5	3.5
Water Closet, 1.6 GPF Flushometer Valve	1	See Footnote[7]		—
Water Closet, greater than 1.6 GPF Gravity Tank	½	3.0	5.5	7.0
Water Closet, greater than 1.6 GPF Flushometer Valve	1	See Footnote[7]		—

For SI units: 1 inch = 25 mm

Notes:
[1] Size of the cold branch pipe, or both the hot and cold branch pipes.
[2] Appliances, appurtenances, or fixtures not referenced in this table shall be permitted to be sized by reference to fixtures having a similar flow rate and frequency of use.
[3] The listed fixture unit values represent their load on the cold water building supply. The separate cold water and hot water fixture unit value for fixtures having both hot and cold water connections shall be permitted to be each taken as three-quarter of the listed total value of the fixture.
[4] The listed minimum supply branch pipe sizes for individual fixtures are the nominal (I.D.) pipe size.
[5] For fixtures or supply connections likely to impose continuous flow demands, determine the required flow in gallons per minute (gpm) (L/s), and add it separately to the demand in gpm (L/s) for the distribution system or portions thereof.
[6] Assembly [Public Use (See Table 422.1)].
[7] Where sizing flushometer systems, see Section 610.10.
[8] Reduced fixture unit loading for additional hose bibbs is to be used where sizing total building demand and for pipe sizing where more than one hose bibb is supplied by a segment of water distribution pipe. The fixture branch to each hose bibb shall be sized on the basis of 2.5 fixture units.

the frequency of use from the fixture manufacturer. If there is a comparable fixture, the installer may use that fixture value for the new fixture.

3. This note allows the system designer to apply a 75 percent method to design the water supply system. In calculations determining the total demand of the cold water supply, the fixture would be assigned the total fixture unit value. A clothes washer would be assigned four WSFU when determining the meter and building supply line. Where the branch servicing the fixture tees off from the main, the branch may be sized using 75 percent of that WSFU value. The branch serving the clothes washer would be assigned three fixture units. This method will provide benefits mostly in larger systems allowing a reduction in branch sizing, material cost and water usage. However, the most common water distribution system will likely be a small system that may not benefit from this method. See **Figure 610.3a** for an example of the 75 percent method of sizing branches.

4. The size listed in the minimum fixture branch size is an interior diameter or nominal size. An assignment of ½-inch copper pipe would actually be five-eighths of an inch outside diameter pipe, not a one-half of an inch outside diameter.

5. As stated earlier, the table was designed for intermittent flow and not a continuous flow or use. For sprinkler heads that have a continuous flow, the gallons per minute flow must be determined and applied to the system demand where it occurs.

6. The term "assembly" is further defined in Table 422.1 as Assembly occupancy –A-1 through A-5.

7. Section 610.10 applies to these fixtures when supplied with flushometers. Thus, the fixture unit value of Table 610.10 and method of sizing contained within the table will be used for these fixtures.

8. Table 610.3 assigns a hose bibb 2.5 WSFU. When there are multiple hose bibbs on the system, the user must follow Note 8. For example, assume there are three hose bibbs on a system. When determining the total demand as per Section 610.7, they would be assigned fixture units in this fashion, one hose bibb at 2.5 WSFU, one additional hose bibb at one WSFU and one more additional hose bibb at one WSFU for a total of 4.5 WSFU. This also applies if a branch or line segment in the system serves the three hose bibbs. If there were two hose bibbs on a common branch or segment, one hose bibb would be assigned 2.5 WSFU and the other hose bibb would be assigned one WSFU for a total of 3.5 WSFU. It makes no difference which hose bibb is assigned: one fixture unit or 2.5 fixture units; the installer is sizing the demand on a segment of pipe and not on the hose bibbs themselves. If there is one hose bibb on a branch, it is assigned 2.5 WSFU (see **Figure 610.3b**).

610.4 Sizing Water Supply and Distribution Systems. Systems within the range of Table 610.4 shall be permitted to be sized from that table or by the method in accordance with Section 610.5.

Listed parallel water distribution systems shall be installed in accordance with their listing, but at no time shall a portion of the system exceed the maximum velocities allowed by the code.

610.5 Sizing per Appendices A and C. Except as provided in Section 610.4, the size of each water piping system shall be determined in accordance with the procedure set forth in Appendix A. For alternate methods of sizing water supply systems, see Appendix C.

The Uniform Plumbing Code (UPC) provides different ways to size the water supply system for the plumber or design professional. They are:

1. Systems with fixture unit values and pipe lengths within the range of Table 610.4 may be sized using Tables 610.3 and 610.4 and Sections 610.7 through 610.10. When sizing for flushometers using Table 610.4, Table 610.10 must be used. The maxiumum range of Table 610.4 are 654 WSFU and 1,000 feet in developed length.

2. Usually used for very large systems, Appendix A, Recommended Rules for Sizing the Water Supply System, may be used to size the water distribution system. Appendix A explains how to compensate for aging of pipe due to lime and calcium accumulations and the resultant decrease in pipe capacity. All friction-loss data is calculated using the "Fairly Rough" or "Rough" charts, unless the pipe used has no accumulation characteristics. Pressure losses due to the installation of meters, pressure-reducing valves, backflow preventers, fittings, water-treating devices or other flow-restricting devices are also addressed in Appendix A. Consideration for other devices with significant pressure loss must also be taken into account.

3. Appendix C, Alternate Plumbing Systems, where adopted, may be used to size water distribution systems. This appendix provides for the grouping of fixtures for sizing purposes and may result in a lower total demand for the water supply and consequently a reduction in total cost for the system.

4. The last method is contained in Note 3 of Table 610.3. This is the 75 percent method explained earlier.

610.6 Friction and Pressure Loss. Except where the type of pipe used and the water characteristics are such that no decrease in capacity due to the length of service (age of system) is expected, friction-loss data shall be obtained from the "Fairly Rough" or "Rough" charts in Appendix A of this code. Friction or pressure losses in a water meter, valve, and fittings shall be obtained from the same sources. Pressure losses through water-treating equipment, backflow prevention devices, or other flow-restricting devices shall be computed in accordance with Section 610.2.

610.7 Conditions for Using Table 610.4. On a proposed water piping installation sized using Table 610.4, the following conditions shall be determined:

(1) Total number of fixture units as determined from Table 610.3, Equivalent Fixture Units, for the fixtures to be installed.

WATER SUPPLY AND DISTRIBUTION

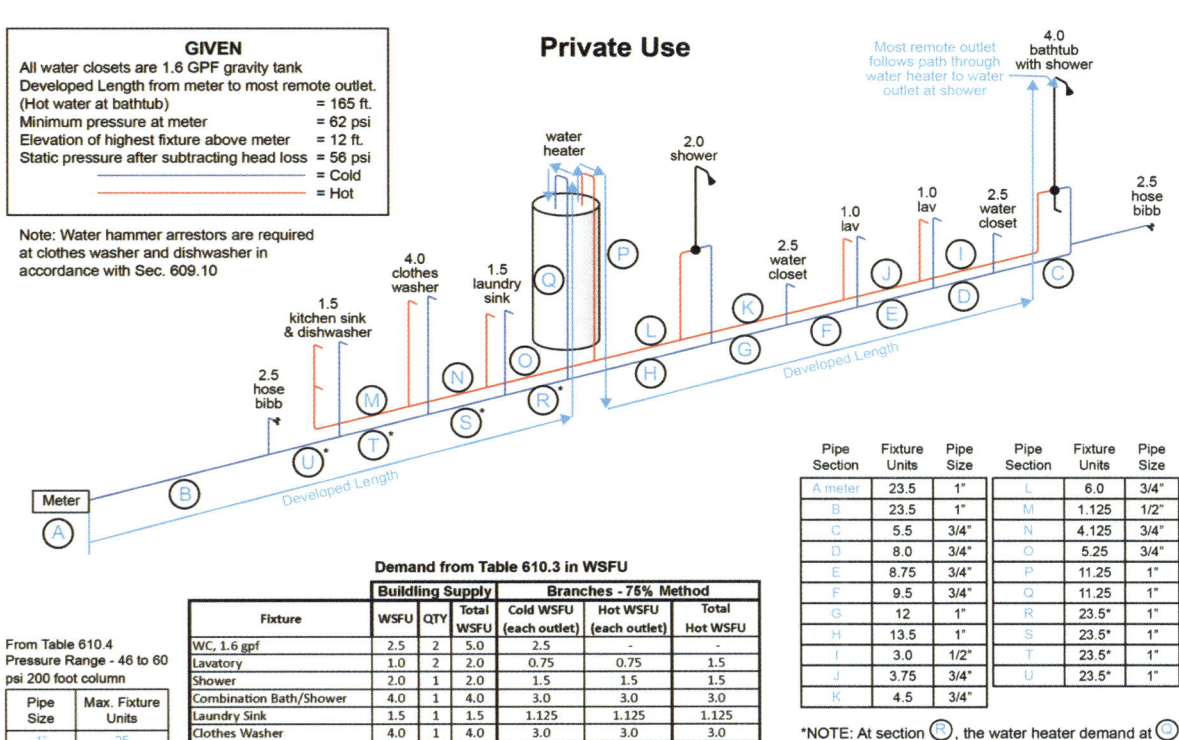

FIGURE 610.3A
SIZING EXAMPLE 1 USING THE 75% METHOD

Hose Bibb Sizing

Table 610.3 contains a listing for "Hose Bibb 0 each Additional" with a reduced water supply fixture unit value of 1.0. This reduced water supply fixture unit value is assigned only to hose pipe sections serving more than one hose bibb, hence the term "additional". Each pipe section serving only on hose bibb would be assigned a water supply fixture unit value of 2.5. A pipe section serving two or more hose bibbs would be assigned a water supply fixture unit value of 2.5 for the first hose bibb plus 1 w.s.f.u. for each additional hose bibb. The method accounts for the fact that multiple hose bibs are seldom used at the same time in the same building. Please see the illustration and explanation below

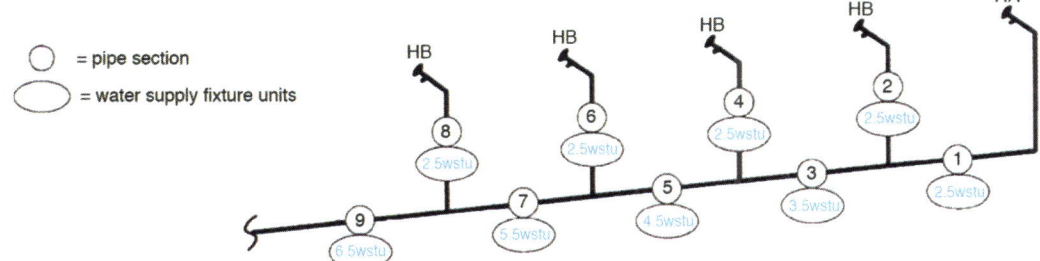

Sizing Notes

1) Pipe sections 1, 2, 4, 6, and 8 serve one hose bibb. Each of these sections is assigned a water supply fixture unit value of 2.5
2) Pipe section 3 serves one hose bibb (2.5 w.s.f.u.) + one "additional" hose bibb (1.0.f.u.) for a total of 3.5 water supply fixture until.
3) Pipe section 5 serves one hose bibb (2.5 w.s.f.u.) + two "additional" hose bibb (2.0.f.u.) for a total of 4.5 water supply fixture until.
4) Pipe section 7 serves one hose bibb (2.5 w.s.f.u.) + three "additional" hose bibb (3.0.f.u.) for a total of 5.5 water supply fixture until.
5) Pipe section 9 serves one hose bibb (2.5 w.s.f.u.) + four "additional" hose bibb (4.0.f.u.) for a total of 6.5 water supply fixture until.

FIGURE 610.3B
HOSE BIBB SIZING USING NOTE 8 FROM TABLE 610.3

TABLE 610.4
FIXTURE UNIT TABLE FOR DETERMINING WATER PIPE AND METER SIZES

METER AND STREET SERVICE (inches)	BUILDING SUPPLY AND BRANCHES (inches)	MAXIMUM ALLOWABLE LENGTH (feet)													
		40	60	80	100	150	200	250	300	400	500	600	700	800	900	1000
PRESSURE RANGE – 30 to 45 psi[1]																
¾	½[2]	6	5	4	3	2	1	1	1	0	0	0	0	0	0	0
¾	¾	16	16	14	12	9	6	5	5	4	4	3	2	2	2	1
¾	1	29	25	23	21	17	15	13	12	10	8	6	6	6	6	6
1	1	36	31	27	25	20	17	15	13	12	10	8	6	6	6	6
¾	1¼	36	33	31	28	24	23	21	19	17	16	13	12	12	11	11
1	1¼	54	47	42	38	32	28	25	23	19	17	14	12	12	11	11
1½	1¼	78	68	57	48	38	32	28	25	21	18	15	12	12	11	11
1	1½	85	84	79	65	56	48	43	38	32	28	26	22	21	20	20
1½	1½	150	124	105	91	70	57	49	45	36	31	26	23	21	20	20
2	1½	151	129	129	110	80	64	53	46	38	32	27	23	21	20	20
1	2	85	85	85	85	85	85	82	80	66	61	57	52	49	46	43
1½	2	220	205	190	176	155	138	127	120	104	85	70	61	57	54	51
2	2	370	327	292	265	217	185	164	147	124	96	70	61	57	54	51
2	2½	445	418	390	370	330	300	280	265	240	220	198	175	158	143	133
PRESSURE RANGE – 46 to 60 psi[1]																
¾	½[2]	7	7	6	5	4	3	2	2	1	1	1	0	0	0	0
¾	¾	20	20	19	17	14	11	9	8	6	5	4	4	3	3	3
¾	1	39	39	36	33	28	23	21	19	17	14	12	10	9	8	8
1	1	39	39	39	36	30	25	23	20	18	15	12	10	9	8	8
¾	1¼	39	39	39	39	39	39	34	32	27	25	22	19	19	17	16
1	1¼	78	78	76	67	52	44	39	36	30	27	24	20	19	17	16
1½	1¼	78	78	78	78	66	52	44	39	33	29	24	20	19	17	16
1	1½	85	85	85	85	85	85	80	67	55	49	41	37	34	32	30
1½	1½	151	151	151	151	128	105	90	78	62	52	42	38	35	32	30
2	1½	151	151	151	151	150	117	98	84	67	55	42	38	35	32	30
1	2	85	85	85	85	85	85	85	85	85	85	85	85	85	83	80
1½	2	370	370	340	318	272	240	220	198	170	150	135	123	110	102	94
2	2	370	370	370	370	368	318	280	250	205	165	142	123	110	102	94
2	2½	654	640	610	580	535	500	470	440	400	365	335	315	285	267	250
PRESSURE RANGE – Over 60 psi[1]																
¾	½[2]	7	7	7	6	5	4	3	3	2	1	1	1	1	1	0
¾	¾	20	20	20	20	17	13	11	10	8	7	6	6	5	4	4
¾	1	39	39	39	39	35	30	27	24	21	17	14	13	12	12	11
1	1	39	39	39	39	38	32	29	26	22	18	14	13	12	12	11
¾	1¼	39	39	39	39	39	39	39	39	34	28	26	25	23	22	21
1	1¼	78	78	78	78	74	62	53	47	39	31	26	25	23	22	21
1½	1¼	78	78	78	78	78	74	65	54	43	34	26	25	23	22	21
1	1½	85	85	85	85	85	85	85	85	81	64	51	48	46	43	40
1½	1½	151	151	151	151	151	151	130	113	88	73	51	51	46	43	40
2	1½	151	151	151	151	151	151	142	122	98	82	64	51	46	43	40
1	2	85	85	85	85	85	85	85	85	85	85	85	85	85	85	85
1½	2	370	370	370	370	360	335	305	282	244	212	187	172	153	141	129
2	2	370	370	370	370	370	370	370	340	288	245	204	172	153	141	129
2	2½	654	654	654	654	654	650	610	570	510	460	430	404	380	356	329

For SI units: 1 inch = 25 mm, 1 foot = 304.8 mm, 1 pound-force per square inch = 6.8947 kPa

Notes:
[1] Available static pressure after head loss.
[2] Building supply, not less than ¾ of an inch (20 mm) nominal size.

WATER SUPPLY AND DISTRIBUTION

🔧 This will be the total of all fixture units using Table 610.3. The easiest way to determine the total demand is to create your own demand chart similar to the one shown in **Figure 610.3a**. The individual fixtures are listed with their WSFU values times the number of same fixtures, which equals the total for that fixture. These are then added, giving the total demand for the system. In **Figure 610.3a** the total is 23.5 WSFU. The total demand is used for sizing the meter and the building supply.

Determining the hot water demand can be calculated by adding all the WSFU values of the fixtures supplied with hot water. Using the 75 percent method (see footnote 3 of Table 610.3) in **Figure 610.3a** the line segment "Q" leading to the water heater is given a value of 11.25 WSFU.

(2) Developed length of supply pipe from meter to the most remote outlet.

🔧 The most remote outlet is the furthest fixture branch outlet from the water meter or other supply source. The most remote outlet may be either a cold or hot water outlet, provided that the water flowing to this outlet follows the longest possible path in the system (see **Figure 610.7**). This reasoning comes from the definition of "remote outlet" in Chapter 2: "When used for sizing water piping, it is the furthest outlet dimension, measuring from the meter, either the developed length of the cold-water piping or through the water heater to the furthest outlet on the hot-water piping." It is important to note that the location of the water heater within the system may have a significant effect on the determination of the most remote outlet and, consequently, on the column in Table 610.4 used to size the system. See **Figure 610.9b** for examples on how the water heater affects the size of the piping and the fixture unit demand on individual pipe segments.

The measured distance can now be used to determine which column to use in Table 610.4 for the maximum allowable length.

(3) Difference in elevation between the meter or other source of supply and the highest fixture or outlet.

🔧 The difference in elevation between the highest outlet in the water distribution system and the meter will be higher, lower or the same as the meter. For example, if the highest outlet is 10 feet higher than the meter, the difference is 10 feet. This value will be used in item 2 of Section 610.8.

(4) Pressure in the street main or another source of supply at the locality where the installation is to be made.

(5) In localities where there is a fluctuation of pressure in the main throughout the day, the water piping system shall be designed on the basis of the minimum pressure available.

🔧 The pressure at the service supply to the meter must be determined. Most systems will have pressure fluctuations in the system. The minimum pressure of these fluctuations must be used. This information may be obtained by taking a pressure reading using a gauge that will register the minimum pressure as measured over a 24-hour period or it may be obtained from the water purveyor.

610.8 Size of Meter and Building Supply Pipe Using Table 610.4.
The size of the meter and the building supply pipe shall be determined as follows:

(1) Determine the available pressure at the water meter or other source of supply.

(2) Add or subtract depending on positive or negative elevation change, ½ psi (3.4 kPa) for each foot (305 mm) of difference in elevation between such source of supply and the highest water supply outlet in the building or on the premises.

(3) Use the "pressure range" group within which this pressure will fall using Table 610.4.

(4) Select the "length" column that is equal to or longer than the required length.

(5) Follow down the column to a fixture unit value equal to or exceeding the total number of fixture units required by the installation.

(6) Having located the proper fixture unit value for the required length, sizes of meter and building supply pipe as found in the two left-hand columns shall be applied.

No building supply pipe shall be less than ¾ of an inch (20 mm) in diameter.

🔧 The following steps are an example of how to size the meter and building supply pipe using Table 610.4.

1. The available pressure at the meter or service pipe may be obtained by either measuring with a gauge or from the water purveyor. This is a static pressure (when there is no use) and will typically vary within a range of minimum to maximum depending on time of day

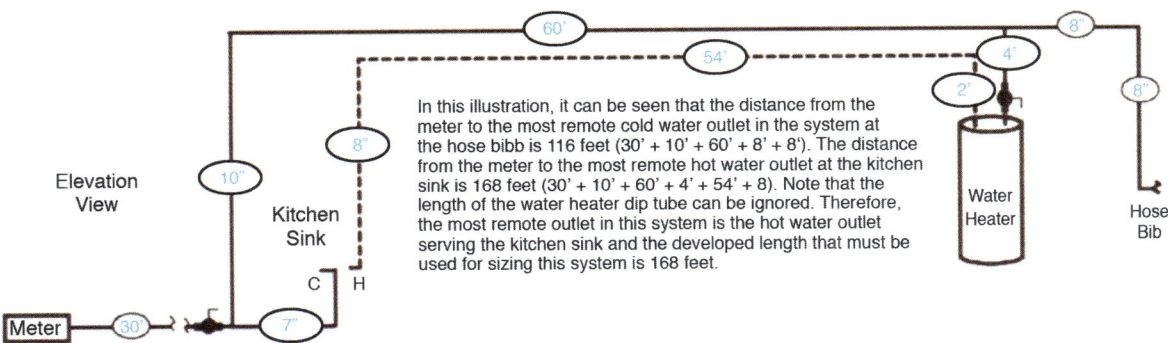

FIGURE 610.7
CALCULATING THE DEVELOPED LENGTH OF THE MOST REMOTE OUTLET

WATER SUPPLY AND DISTRIBUTION

and the amount of water stored for use. When there is a variation in pressure, sizing is always based on minimum pressure.

2. An adjustment must be made for an increase or a decrease in head pressure caused by the elevation differential between the source of supply (water meter) and the highest water supply outlet in the system. Adding or subtracting 0.5 psi to or from the minimum pressure for each foot of difference in elevation between the two provides this adjustment. If the highest outlet is above the meter, the calculation is subtracted from the minimum pressure, resulting in less pressure. If the highest outlet is below the meter, the calculation is added to the minimum pressure, resulting in a higher pressure. If there is a 10-foot differential, then 5 psi will be either added or subtracted from the minimum pressure available. The corrected value will be used to determine which pressure range to use. **Figure 610.8a** illustrates these conditions.

3. Once the minimum available pressure is determined, use the pressure range that the minimum pressure falls within to size the system. For example, with a 48 psi minimum pressure available, use the 46 to 60 psi pressure range of Table 610.4 to calculate the sizing of the system.

4. Notice Table 610.4 contains columns designated as "Maximum Allowable Length" in feet. Once the developed length from the meter to the most remote outlet has been measured, use that measurement to find the appropriate column in the correct pressure range. For example, if the developed length to the most remote outlet is 87 feet, use the Maximum Allowable Length column designated 100 feet. Use this column within the 46 to 60 psi range according to our example.

5. Let's assume the plumbing system has a total amount of 57 WSFU. Knowing the pressure range (46 to 60 psi) and length column (100 feet) find the first value equal to or exceeding 57 WSFU. Under these conditions 67 WSFU is the correct value.

6. The value in the same row in the first column, Meter and Street Service, will determine the size of the service pipe to the meter and the meter itself. Within these parameters the size is one-inch.

The second column, Building Supply and Branches, determines the size of the building supply (the pipe segment from the meter to the first tee) and all other segments of the system. Continuing the example, the building supply size found in the second column, from the same row as the meter size, is 1-1/4 inches. See **Figure 610.8b**.

Both Section 610.8(6) and Table 610.4 footnote 2 require the minimum size of the building supply to be ¾-in.

610.9 Size of Branches.
Where Table 610.4 is used, the minimum size of each branch shall be determined by the total fixture units served by that branch and then following the steps in Section 610.8. No branch piping shall exceed the total demand in fixture units for the system computed from Table 610.3.

When using Table 610.4 for the sizing of branches, the Maximum Allowable Length column will determine the maximum number of fixture units permitted on a given pipe size. Using **Figure 610.3a** as an example, the total developed length to the most remote outlet is 165 feet. Select the column under The Maximum Allowable Length that has the next greater length, which is 200 feet (see **Figure 610.9a**). **Figure 610.3a** also indicates the total demand is 23.5 WSFU, which requires a one-inch meter and street service and a one-inch building supply within the pressure range that includes 56 psi (see **Figure 610.9a**).

The branch sizes can now be determined. Staying within the 200-foot column the following chart can be created (see **Figure 610.9a**).

½" Branch – WSFU ≤ 3
¾" Branch – WSFU > 3 and ≤ 11
1" Branch – WSFU > 11 and ≤ 25

We can now apply this chart to Figure 610.3A to every branch segment using the optional 75 percent method (consider Table 610.3 footnote 3, only fixtures with both hot and cold water connections may use the 75 percent method). Beginning with the most remote cold fixture branch, segment C has a total of 5.5 WSFU [2.5 for the HB and 3 (75% of 4) for the shower] requiring a pipe size of 3/4 inch. Continuing toward the meter, segment D supplies 8 WSFU, segment E supplies 8.75 WSFU and segment F supplies 9.5 WSFU all of which are in the range to be supplied by ¾ pipe. Segment G is supplying 12 WSFU requiring 1 inch and so on. The same procedure is then applied to the hot fixture segments or branches. Whether the water heater is in the middle of the piping segments or at an end, begin with a remote end from the water heater and add the demand at each segment. In this example the total demand for hot water is 11.25 WSFU. That means the supply into (Q) and out of (P) the water heater will be 11.25 WSFU. That demand is added to the cold demand where they connect at R.

Notice branch segment R in **Figure 610.3a**. Branch segment R includes both the cold WSFU upstream and the hot WSFU supplied in to and out of the water heater. The cumulative cold WSFU at H is 13.5 and the total hot WSFU is 11.25 having a combined total of 24.75 WSFU, which exceeds the total demand originally computed for the system. When this occurs, Section 610.9 establishes a cap on fixture units for branch piping that will not exceed the total demand for the system, which in this case is 23.5 WSFU. Therefore, branch segments R, S, T and U shall be assigned 23.5 WSFU. Table 610.4 requires a branch supplying 23.5 WSFU to have a minimum pipe size of 1¼ inch.

There is another element of Table 610.4 that will likely affect the sizing of branches. Using the same **Figure 610.3a** as an example, but this time the utility within the jurisdiction installed a ¾-inch meter and street service. Using the factors already established Table 610.4 only permits 23 WSFU on a ¾-inch meter and street service with a one-inch building supply. However, if we move down the column to the next ¾-inch meter and street service, the building supply is now 1¼ inch, which can supply 39

WATER SUPPLY AND DISTRIBUTION

Pressure (pounds per square inch) [kpa] = 0.5 x Head (feet)]mm]

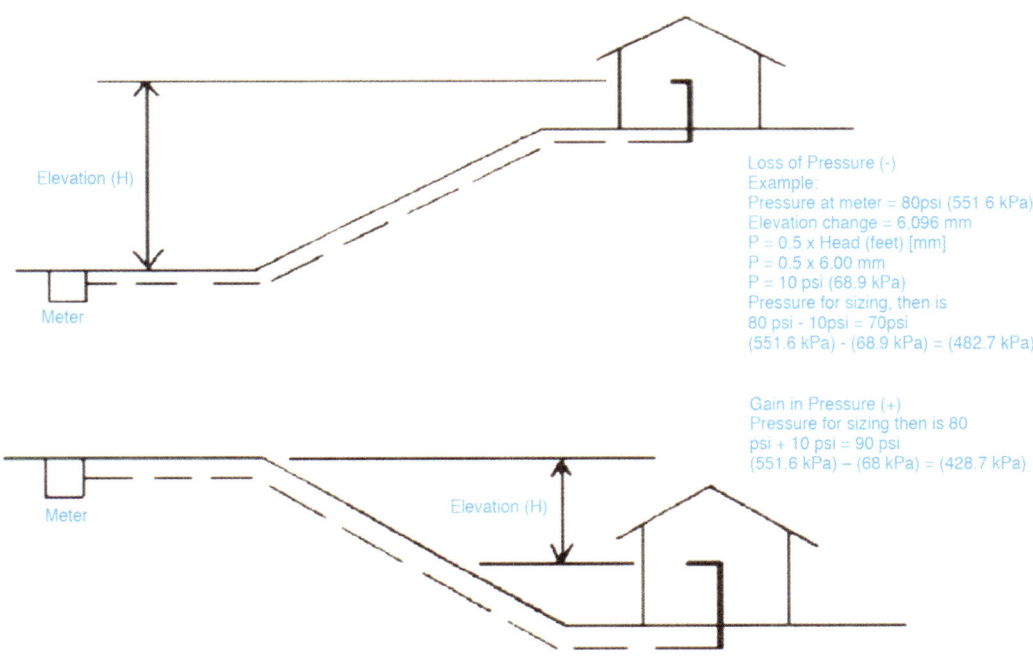

Loss of Pressure (-)
Example:
Pressure at meter = 80psi (551.6 kPa)
Elevation change = 6,096 mm
P = 0.5 x Head (feet) [mm]
P = 0.5 x 6.00 mm
P = 10 psi (68.9 kPa)
Pressure for sizing, then is
80 psi - 10psi = 70psi
(551.6 kPa) - (68.9 kPa) = (482.7 kPa)

Gain in Pressure (+)
Pressure for sizing then is 80
psi + 10 psi = 90 psi
(551.6 kPa) – (68 kPa) = (428.7 kPa)

FIGURE 610.8A
CALCULATING PRESSURE DIFFERENTIAL

TABLE 610.4
FIXTURE UNIT TABLE FOR DETERMINING WATER PIPE AND METER SIZES

METER AND STREET SERVICE (inches)	BUILDING SUPPLY AND BRANCHES (inches)	MAXIMUM ALLOWABLE LENGTH (feet)												
		40	60	80	100	150	200	250	300	400	500	600	700	800	900	1000
PRESSURE RANGE – 30 to 45 psi[1]																
¾	½[2]	6	5	4	3	2	1	1	1	0	0	0	0	0	0	0
¾	¾	16	16	14	12	9	6	5	5	4	4	3	2	2	2	1
¾	1	29	25	23	21	17	15	13	12	10	8	6	6	6	6	6
1	1	36	31	27	25	20	17	15	13	12	10	8	6	6	6	6
¾	1¼	36	33	31	28	24	23	21	19	17	16	13	12	12	11	11
1	1¼	54	47	42	38	32	28	25	23	19	17	14	12	12	11	11
1½	1¼	78	68	57	48	38	32	28	25	21	18	15	12	12	11	11
1	1½	85	84	79	65	56	48	43	38	32	28	26	22	21	20	20
1½	1½	150	124	105	91	70	57	49	45	36	31	26	23	21	20	20
2	1½	151	129	129	110	80	64	53	46	38	32	27	23	21	20	20
1	2	85	85	85	85	85	85	82	80	66	61	57	52	49	46	43
1½	2	220	205	190	176	155	138	127	120	104	85	70	61	57	54	51
2	2	370	327	292	265	217	185	164	147	124	96	70	61	57	54	51
2	2½	445	418	390	370	330	300	280	265	240	220	198	175	158	143	133
PRESSURE RANGE – 46 to 60 psi[1]																
¾	½[2]	7	7	6	5	4	3	2	2	1	1	1	0	0	0	0
¾	¾	20	20	19	17	14	11	9	8	6	5	4	4	3	3	3
¾	1	39	39	36	33	28	23	21	19	17	14	12	10	9	8	8
1	1	39	39	39	36	30	25	23	20	18	15	12	10	9	8	8
¾	1¼	39	39	39	39	39	39	34	32	27	25	22	19	19	17	16
1	1¼	78	78	76	67	52	44	39	36	30	27	24	20	19	17	16
1½	1¼	78	78	78	78	66	52	44	39	33	29	24	20	19	17	16
1	1½	85	85	85	85	85	85	80	67	55	49	41	37	34	32	30
1½	1½	151	151	151	151	128	105	90	78	62	52	42	38	35	32	30

FIGURE 610.8B
SIZING THE METER AND BUILDING SUPPLY USING TABLE 610.4

WATER SUPPLY AND DISTRIBUTION

WSFU (See **Figure 610.9a-a**). You must now revise your previous chart as follows:

½" Branch – WSFU ≤ 3
¾" Branch – WSFU > 3 and ≤ 11
1" Branch – WSFU > 11 and ≤ 23
1 ¼" Branch – WSFU > 23 and ≤ 39

This can affect branch sizing since the building supply is now 1¼ inch. Branch segment R S, T and U each supplying 23.5 WSFU are increased to 1¼".

Table 610.4 allows greater utility for easy pipe sizing by accounting for different meter sizing jurisdictions may adopt. Review the other sizing examples, **Figures 610.9b through 610.9e** at the end of the chapter (Note: the optional 75 percent sizing method is not used for these Figures).

610.10 Sizing for Flushometer Valves. Where using Table 610.4 to size water supply systems serving flushometer valves, the number of flushometer fixture units assigned to every section of pipe, whether branch or main, shall be determined by the number and category of flushometer valves served by that section of pipe, in accordance with Table 610.10. Piping supplying a flushometer valve shall be not less in size than the valve inlet.

Where using Table 610.10 to size water piping, care shall be exercised to assign flushometer fixture units based on the number and category of fixtures served.

TABLE 610.10
FLUSHOMETER FIXTURE UNITS FOR WATER SIZING USING TABLE 610.3

FIXTURE CATEGORY: WATER CLOSET WITH FLUSHOMETER VALVES		
NUMBER OF FLUSHOMETER VALVES	INDIVIDUAL FIXTURE UNITS ASSIGNED IN DECREASING VALUE	FIXTURE UNITS ASSIGNED FOR WATER CLOSETS AND SIMILAR 10-UNIT FIXTURES IN ACCUMULATIVE VALUES
1	40	40
2	30	70
3	20	90
4	15	105
5 or more	10 each	115 plus 10 for each additional fixture in excess of 5
FIXTURE CATEGORY: URINALS WITH FLUSHOMETER VALVES		
NUMBER OF FLUSHOMETER VALVES	INDIVIDUAL FIXTURE UNITS ASSIGNED IN DECREASING VALUE	FIXTURE UNITS ASSIGNED FOR URINALS AND SIMILAR 5-UNIT FIXTURES IN ACCUMULATIVE VALUES
1	20	20
2	15	35
3	10	45
4	8	53
5 or more	5 each	58 plus 5 for each additional fixture in excess of 5

In the example below, fixture units assigned to each section of pipe are computed. Each capital letter refers to the section of pipe above it unless otherwise shown.

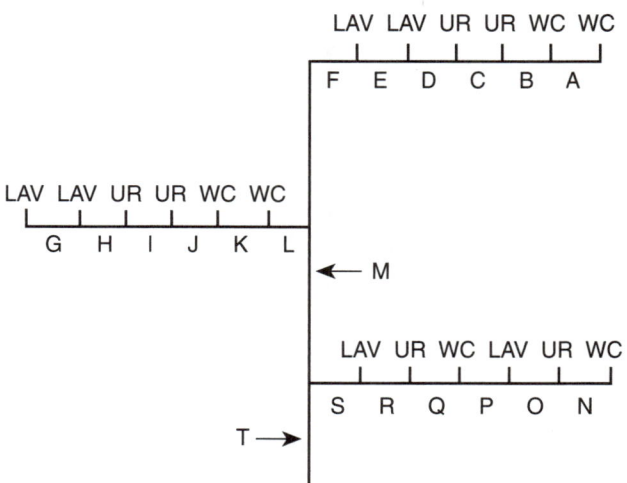

A: 1 WC = 40 F.U.
B: 2 WC = 70 F.U.
C: 2 WC (70) + 1 UR (20) = 90 F.U.
D: 2 WC (70) + 2 UR (35) = 105 F.U.
E: 2 WC (70) + 2 UR (35) + 1 LAV (1) = 106 F.U.
F: 2 WC (70) + 2 UR (35) + 2 LAV (2) = 107 F.U.
G: 1 LAV = 1 F.U.
H: 2 LAV = 2 F.U.
I: 2 LAV (2) + 1 UR (20) = 22 F.U.
J: 2 LAV (2) + 2 UR (35) = 37 F.U.
K: 2 LAV (2) + 2 UR (35) + 1 WC (40) = 77 F.U.
L: 2 LAV (2) + 2 UR (35) + 2 WC (70) = 107 F.U.
M: 4 WC (105) + 4 UR (53) + 4 LAV (4) = 162 F.U.
N: 1 WC = 40 F.U.
O: 1 WC (40) + 1 UR (20) = 60 F.U.
P: 1 WC (40) + 1 UR (20) + 1 LAV (1) = 61 F.U.
Q: 2 WC (70) + 1 UR (20) + 1 LAV (1) = 91 F.U.
R: 2 WC (70) + 2 UR (35) + 1 LAV (1) = 106 F.U.
S: 2 WC (70) + 2 UR (35) + 2 LAV (2) = 107 F.U.
T: 6 WC (125) + 6 UR (63) + 6 LAV (6) = 194 F.U.

EXAMPLE 610.10

SIZING METHOD FOR PUBLIC USE FIXTURES USING TABLE 610.10

Fixture unit values are based on Hunter's curves for estimating the water supply demand (Dr. Roy Hunter was a physicist at the National Bureau of Standards from 1920 to 1945). See Appendix A, Chart A 103.1(1) and Chart A 103.1(2) for an illustration of these curves. Notice that there are two separate curves that merge together at some point near 1000 fixture units. The lower curve is used for sizing systems with tank-type toilets and the upper curve is used for sizing systems with Flushometer valves.

WATER SUPPLY AND DISTRIBUTION

TABLE 610.4
FIXTURE UNIT TABLE FOR DETERMINING WATER PIPE AND METER SIZES

METER AND STREET SERVICE (inches)	BUILDING SUPPLY AND BRANCHES (inches)	MAXIMUM ALLOWABLE LENGTH (feet)													
		40	60	80	100	150	200	250	300	400	500	600	700	800	900	1000
PRESSURE RANGE – 30 to 45 psi[1]																
¾	½[2]	6	5	4	3	2	1	1	1	0	0	0	0	0	0	0
¾	¾	16	16	14	12	9	6	5	5	4	4	3	2	2	2	1
¾	1	29	25	23	21	17	15	13	12	10	8	6	6	6	6	6
1	1	36	31	27	25	20	17	15	13	12	10	8	6	6	6	6
¾	1¼	36	33	31	28	24	23	21	19	17	16	13	12	12	11	11
1	1¼	54	47	42	38	32	28	25	23	19	17	14	12	12	11	11
1½	1¼	78	68	57	48	38	32	28	25	21	18	15	12	12	11	11
1	1½	85	84	79	65	56	48	43	38	32	28	26	22	21	20	20
1½	1½	150	124	105	91	70	57	49	45	36	31	26	23	21	20	20
2	1½	151	129	129	110	80	64	53	46	38	32	27	23	21	20	20
1	2	85	85	85	85	85	85	82	80	66	61	57	52	49	46	43
1½	2	220	205	190	176	155	138	127	120	104	85	70	61	57	54	51
2	2	370	327	292	265	217	185	164	147	124	96	70	61	57	54	51
2	2½	445	418	390	370	330	300	280	265	240	220	198	175	158	143	133
PRESSURE RANGE – 46 to 60 psi[1]																
¾	½[2]	7	7	6	5	4	3	2	2	1	1	1	0	0	0	0
¾	¾	20	20	19	17	14	11	9					4	3	3	3
¾	1	39	39	36	33	28	23	21					10	9	8	8
1	1	39	39	39	36	30	25	23	20	18	15	12	10	9	8	8
¾	1¼	39	39	39	39	39	39	34	32	27	25	22	19	19	17	16
1	1¼	78	78	76	67	52	44	39	36	30	27	24	20	19	17	16
1½	1¼	78	78	78	78	66	52	44	39	33	29	24	20	19	17	16

½" Branch – WSFU ≤ 3
¾" Branch – WSFU > 3 and ≤ 11
1" Branch – WSFU > 11 and ≤ 25

1-inch Meter and Street Service with 1-inch Building Supply

FIGURE 610.9A
SIZING THE BRANCHES USING TABLE 610.4

TABLE 610.4
FIXTURE UNIT TABLE FOR DETERMINING WATER PIPE AND METER SIZES

METER AND STREET SERVICE (inches)	BUILDING SUPPLY AND BRANCHES (inches)	MAXIMUM ALLOWABLE LENGTH (feet)													
		40	60	80	100	150	200	250	300	400	500	600	700	800	900	1000
PRESSURE RANGE – 30 to 45 psi[1]																
¾	½[2]	6	5	4	3	2	1	1	1	0	0	0	0	0	0	0
¾	¾	16	16	14	12	9	6	5	5	4	4	3	2	2	2	1
¾	1	29	25	23	21	17	15	13	12	10	8	6	6	6	6	6
1	1	36	31	27	25	20	17	15	13	12	10	8	6	6	6	6
¾	1¼	36	33	31	28	24	23	21	19	17	16	13	12	12	11	11
1	1¼	54	47	42	38	32	28	25	23	19	17	14	12	12	11	11
1½	1¼	78	68	57	48	38	32	28	25	21	18	15	12	12	11	11
1	1½	85	84	79	65	56	48	43	38	32	28	26	22	21	20	20
1½	1½	150	124	105	91	70	57	49	45	36	31	26	23	21	20	20
2	1½	151	129	129	110	80	64	53	46	38	32	27	23	21	20	20
1	2	85	85	85	85	85	85	82	80	66	61	57	52	49	46	43
1½	2	220	205	190	176	155	138	127	120	104	85	70	61	57	54	51
2	2	370	327	292	265	217	185	164	147	124	96	70	61	57	54	51
2	2½	445	418	390	370	330	300	280	265	240	220	198	175	158	143	133
PRESSURE RANGE – 46 to 60 psi[1]																
¾	½[2]	7	7	6	5	4	3	2					0	0	0	0
¾	¾	20	20	19	17	14	11	9					3	3	3	3
¾	1	39	39	36	33	28	23	21					9	9	8	8
1	1	39	39	39	36	30	25	23					10	9	8	8
¾	1¼	39	39	39	39	39	39	34	32	27	25	22	19	19	17	16
1	1¼	78	78	76	67	52	44	39	36	30	27	24	20	19	17	16
1½	1¼	78	78	78	78	66	52	44	39	33	29	24	20	19	17	16

½" Branch – WSFU ≤ 3
¾" Branch – WSFU > 3 and ≤ 11
1" Branch – WSFU > 11 and ≤ 23
1¼" Branch – WSFU > 23 and ≤ 39

¾-inch Meter and Street Service with 1¼-inch Building Supply

FIGURE 610.9A-A
SIZING WITH A SMALLER STREET SERVICE USING TABLE 610.4

WATER SUPPLY AND DISTRIBUTION

Table 610.4 was developed based only on the lower curve for tank-type toilets. So an adjustment needs to be made for Flushometer-type toilets fixture unit values when using this table. This adjustment was made by increasing the fixture unit value for the Flushometer-type toilet to match the value of the lower curve. For example, the fixture unit value for

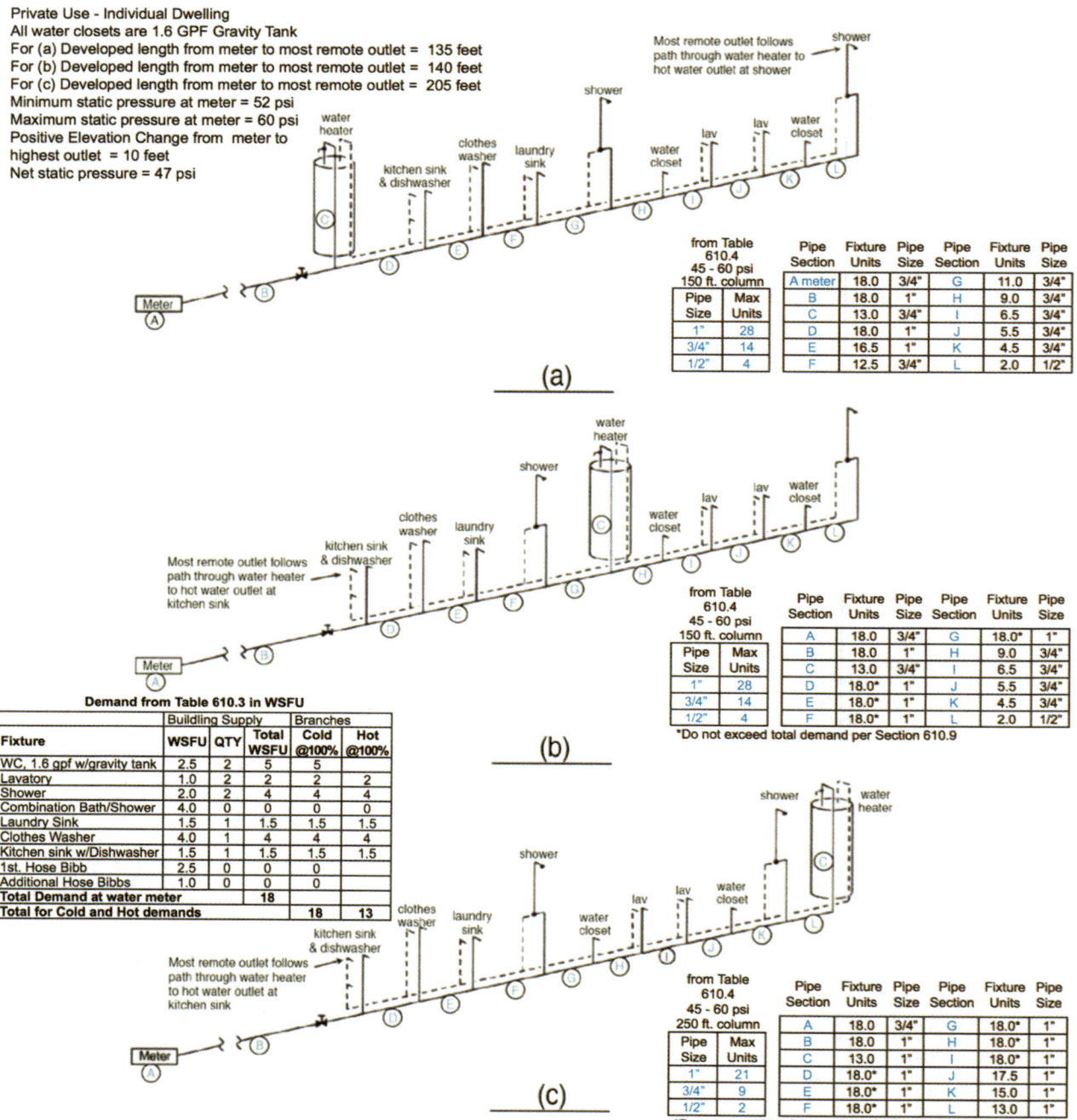

Pipe sizes are affected by two factors in these drawings. First, the hot fixture units are added to the cold water units where the two connect as you work your way from the most remote outlet toward the building supply. When adding those two demands it is possible to exceed the total WSFU demand for the system. This is a result of sizing the total demand for the meter and building supply, and then separating the distribution system into hot and cold branches. To accommodate this Section 610.9 states "no branch piping shall exceed the total demand in fixture units for the system computed from Table 610.3." (For each drawing on this page the total WSFU or demand is 18.) Succinctly, stop adding fixture unit when you reach the total demand previously computed for the water meter.

The second factor is the developed length of the system which will likely increase depending on the location of the water heater as in drawing (c). Longer developed lengths results in fewer fixture units carried by a given pipe size, per Table 610.4.

FIGURE 610.9B
SIZING EXAMPLE 2 - WATER HEATER PLACEMENT

WATER SUPPLY AND DISTRIBUTION

each public 1.6 GPF Flushometer Valve is 8.0 in Appendix A, Table A 103.1. Table 610.10 requires the first Flushometer Valve to have 40 fixture units, the second to have 30 fixture units, the third to have 20 fixture units, etc. until the fifth and following, which are then to have a value of 10 each.

So how do these adjusted fixture unit values match the lower curve? Refer to Chart A 103.1(2) in Appendix A. Suppose there are five Flushometer valves in the system, each having a fixture unit value of eight according to Appendix Table A 103.1. The total value of 40 WSFU requires a demand of approximately 47gpm according to the upper curve in Chart A 103.1(2). Using Table 610.10, five Flushometer valves have a total of 115 WSFU. Now using the lower curve in Chart A 103.1(2), 115 WSFU requires the same demand of approximately 47gpm. Therefore, the adjusted fixture unit values for the Flushometer valves match the same demand required by the upper curve when using the lower curve.

This code section also provides an example on how to apply the adjusted fixture unit values to a plumbing system. First, each branch must be treated independently. Therefore pipe sections A, K and N each require a value of 40 WSFU. The values are then decreased in each branch based on the decreasing value in Table 610.10. Secondly, the main is sized accumulatively. When sizing pipe section M, the accumulative value is used in Table 610.10. Pipe section M serves four Flushometer valve toilets having an accumulative value of 105 WSFU rather than 140 WSFU if the branch values were added. Pipe Section T is calculated the same as M but now includes the last branch. Pipe section T serves six Flushometer valve toilets with an accumulated total of 125 WSFU. See also the example in Example 610.10.

610.11 Sizing Systems for Flushometer Tanks. The size of branches and mains serving flushometer tanks shall be consistent with the sizing procedures for flush tank water closets.

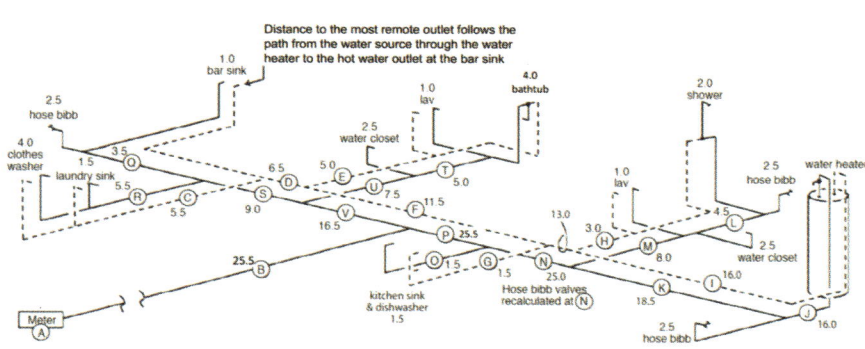

FIGURE 610.9C
SIZING EXAMPLE 3

WATER SUPPLY AND DISTRIBUTION

🔧 Although water closets with flushometer tanks are a separate line item in Table 610.3, there is no difference in WSFU value from gravity tanks. There is no other consideration to be made when sizing systems with water closets with flushometer tanks, so they may be sized the same as water closets with flush tanks.

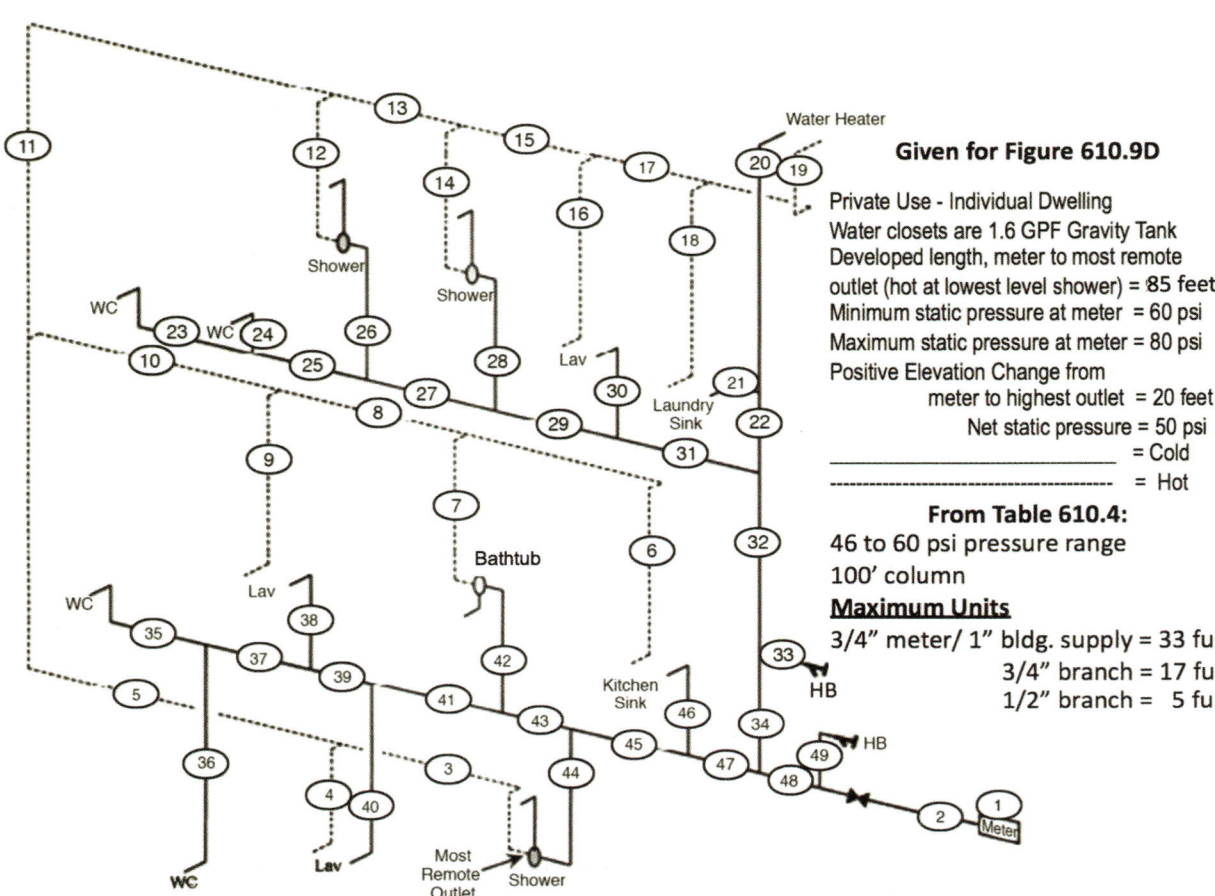

FIGURE 610.9D
SIZING EXAMPLE 4

WATER SUPPLY AND DISTRIBUTION

610.12 Sizing for Velocity. Water piping systems shall not exceed the maximum velocities listed in this section or Appendix A.

610.12.1 Copper Tube Systems. Maximum velocities in copper and copper alloy tube and fitting systems shall not exceed 8 feet per second (ft/s) (2.4 m/s) in cold water and 5 ft/s (1.5 m/s) in hot water.

610.12.2 Tubing Systems Using Copper Fittings. Maximum velocities through copper fittings in tubing other than copper shall not exceed 8 ft/s (2.4 m/s) in cold water and 5 ft/s (1.5 m/s) in hot water.

Previous sections discussed velocities and the affect that high velocities have on the piping system. These sections reiterate the importance of staying within the velocity ratings of the materials used. Under normal conditions using Tables 610.3, 610.4 and 610.10 will keep velocities within those limits. Water distribution systems using copper alloy fittings can be either copper pipe systems or plastic pipe or tubing systems utilizing copper alloy fit-

Table 610.4

Pipe Size	Maximum w.s.f.u. Loading
2"	368
1-1/2"	150
1-1/4"	66
1"	30
3/4"	14

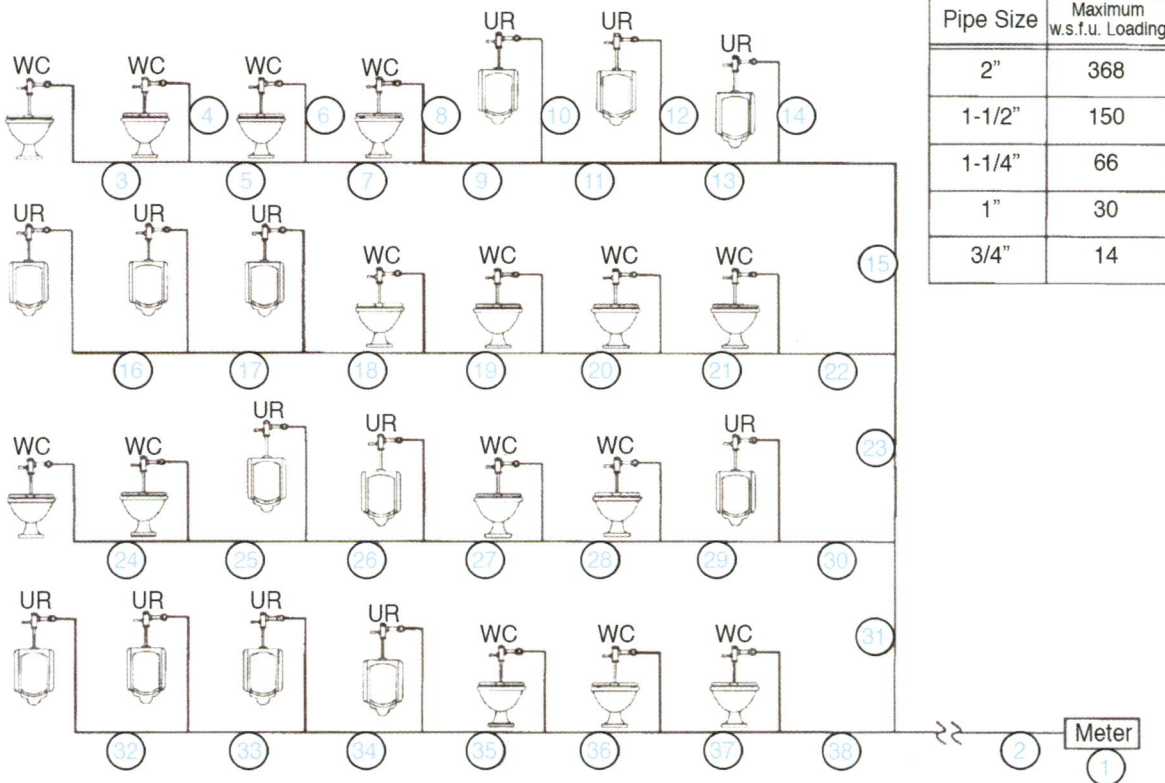

Pipe Section	w.s.f.u.	Pipe Size	Pipe Section	w.s.f.u.	Pipe Size	Pipe Section	w.s.f.u.	Pipe Size
1 meter	313	2"	14	20	1"	27	105	1-1/2"
2	313	2"	15	150	1-1/2"	28	125	1-1/2"
3	40	1-1/4"	16	20	1"	29	140	1-1/2"
4	40	1-1/4"	17	35	1-1/4"	30	150	2"
5	70	1-1/2"	18	45	1-1/4"	31	263	1"
6	40	1-1/4"	19	85	1-1/2"	32	20	1-1/4"
7	90	1-1/2"	20	115	1-1/2"	33	35	1-1/4"
8	40	1-1/4"	21	135	1-1/2"	34	45	1-1/4"
9	105	1-1/2"	22	150	1-1/2"	35	53	1-1/2"
10	20	1"	23	208	2"	36	93	1-1/2"
11	125	1-1/2"	24	40	1-1/4"	37	123	1-1/2"
12	20	1"	25	70	1-1/2"	38	143	
13	140	1-1/2"	26	90	1-1/2"			

Given
1) WC = Water Closet
2) UR = Urinal
3) All WCs are 1.6 GPF
4) All URs are 1.0 GPF
5) Static pressure at meter = 74 psi
6) Distance from meter to most remote outlet = 150 ft
7) Elevation of highest outlet above meter = 40 ft

FIGURE 610.9E
FLUSHOMETER SIZING EXAMPLE

WATER SUPPLY AND DISTRIBUTION

tings. The same velocities apply to the copper alloy fitting as it does to the copper pipe. There have been reports of copper alloy fittings failing due to high velocities in plastic tubing systems. When installing and sizing these systems be sure to follow the manufacturer's installation instructions and/or Appendix A.

610.13 Exceptions. The provisions of this section relative to the size of water piping shall not apply to the following:

(1) Water supply piping systems designed in accordance with recognized engineering procedures acceptable to the Authority Having Jurisdiction.

(2) Alteration of or minor additions to existing installations provided the Authority Having Jurisdiction finds that there will be an adequate supply of water to operate fixtures.

(3) Replacement of existing fixtures or appliances.

(4) Piping that is part of fixture equipment.

(5) Unusual conditions where, in the judgment of the Authority Having Jurisdiction, an adequate supply of water is provided to operate fixtures and equipment.

(6) The size and material of irrigation water piping installed outside of a building or structure and separated from the potable water supply by means of an approved air gap or backflow prevention device are not regulated by this code. The potable water piping system supplying each such irrigation system shall be adequately sized as required elsewhere in this chapter to deliver the full connected demand of both the domestic use and the irrigation systems.

🔧 Irrigation systems or lawn sprinkling systems are not subject to the UPC. The supply piping and backflow prevention device leading to the irrigation system are covered under the UPC. The supply piping is sized per the demand of the irrigation system but the installation and sizing of the irrigation system is dealt with by the system's piping and sprinkler manufacturer.

611.0 Drinking Water Treatment Units.

611.1 Application. Drinking water treatment units shall comply with NSF 42 or NSF 53. Water softeners shall comply with NSF 44. Ultraviolet water treatment systems shall comply with NSF 55. Reverse osmosis drinking water treatment systems shall comply with NSF 58. Drinking water distillation systems shall comply with NSF 62.

611.2 Air Gap Discharge. Discharge from drinking water treatment units shall enter the drainage system through an air gap in accordance with Table 603.3.1 or an air gap device that complies with Table 603.2, NSF 58, or IAPMO PS 65.

🔧 Most types of water treatment or soft water treatment systems have a discharge to the drainage system. This is normally a small diameter plastic tube that is sometimes placed in the standpipe of the automatic clothes washer – without an airgap. This of course is a cross connection; if the washer backs up there is a possibility of the waste reversing in the discharge tube and contaminating the water supply. This discharge line must always discharge through an airgap. There are many devices on the market that make this an easy install such as that shown in **Figure 611.2a**.

Any type of water treatment device that connects directly to the potable water supply shall be protected by the appropriate backflow protection device. If the water treatment device utilizes a discharge line to the drainage system, it shall discharge through an airgap (see **Figure 611.2b**).

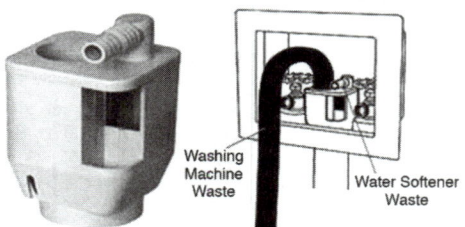

FIGURE 611.2A
WATER SOFTENER AIRGAP

611.3 Connection Tubing. The tubing to and from drinking water treatment units shall be of a size and material as recommended by the manufacturer.

611.4 Sizing of Residential Softeners. Residential-use water softeners shall be sized in accordance with Table 611.4.

TABLE 611.4
SIZING OF RESIDENTIAL WATER SOFTENERS[4]

REQUIRED SIZE OF SOFTENER CONNECTION (inches)	NUMBER OF BATHROOM GROUPS SERVED[1]
¾	up to 2[2]
1	up to 4[3]

For SI units: 1 inch = 25 mm

Notes:
[1] Installation of a kitchen sink and dishwasher, laundry tray, and automatic clothes washer permitted without additional size increase.
[2] An additional water closet and lavatory permitted.
[3] Over four bathroom groups, the softener size shall be engineered for the specific installation.
[4] See also Appendix A, Recommended Rules for Sizing the Water Supply System, and Appendix C, Alternate Plumbing Systems, for alternate methods of sizing water supply systems.

🔧 For example, a residence has two bedrooms and an additional water closet and lavatory – a two and one-half bath house. The size of the water softener connection shall be three-fourths of an inch.

612.0 Residential Fire Sprinkler Systems.

🔧 The Standard for residential sprinkler systems is NFPA 13D, Installation of Sprinkler Systems in One- and Two-Family Dwellings and Manufactured Homes. Annex A in this standard contains explanatory material to help further understand and apply these provisions for residential sprinkler systems.

612.1 Where Required. Where residential sprinkler systems are required in one and two-family dwellings or townhouses, the systems shall be installed by personnel, installer, or both, certified in accordance with ASSE Series 7000 in accordance with this section or NFPA 13D. This

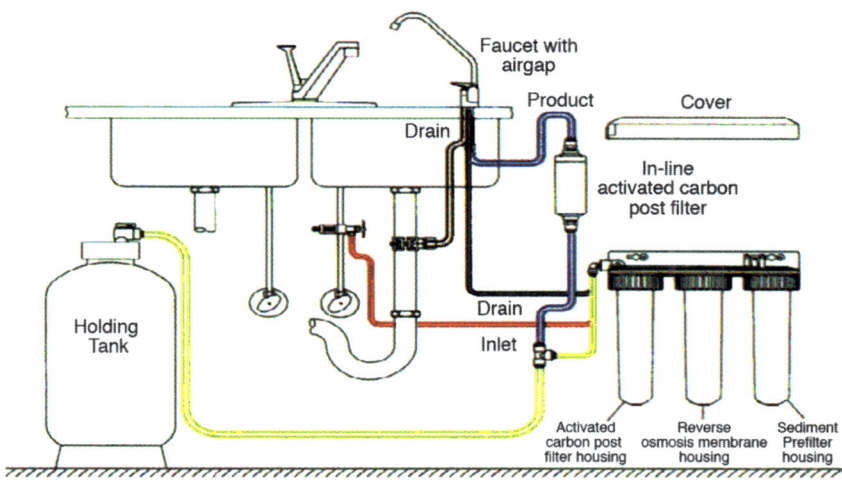

FIGURE 611.2B
INSTALLATION OF DRINKING WATER TREATMENT DEVICE WITH AIR GAP

section shall be considered equivalent to NFPA 13D. Partial residential sprinkler systems shall be permitted to be installed in buildings not required to be equipped with a residential sprinkler system.

🔧 When sizing a multipurpose water distribution/fire suppression system, the largest pipe size required based on either the sprinkler piping calculations or the water distribution piping calculations in Section 610.4 (see **Figure 612.1**).

Care should be taken during the design and installation process to prevent dead legs. Dead legs are sections of piping or plumbing that have been altered or terminated so that water cannot flow through them. The US Department of Labor recommends elimination of dead legs to prevent the growth of Legionaries disease bacteria (LDB).

612.2 Types of Systems. This section shall apply to stand-alone and multipurpose wet-pipe sprinkler systems that do not include the use of antifreeze. A multipurpose fire sprinkler system shall provide potable water to both fire sprinklers and plumbing fixtures. A stand-alone sprinkler system shall be separate and independent from the potable water distribution system. A backflow preventer shall not be required to separate a stand-alone sprinkler system from the water distribution system where the sprinkler system material is in accordance with the requirements of Section 604.0.

612.3 Sprinklers. Sprinklers shall be installed in accordance with Section 612.3.1 through Section 612.3.7.

612.3.1 Required Sprinkler Locations. Sprinklers shall be installed to protect all floor areas of a dwelling unit in one and two-family dwellings or townhouses.

Exceptions:

(1) Attics, crawl spaces, and normally unoccupied concealed spaces that do not contain fuel-fired appliances do not require sprinklers. In attics, crawl spaces, and normally unoccupied concealed spaces that contain fuel-fired equipment, a sprinkler shall be provided to protect the equipment; however, sprinklers shall not be required in the remainder of the space.

(2) Clothes closets, linen closets, and pantries that do not exceed 24 square feet (2.2 m^2) in area, with the smallest dimension not exceeding 3 feet (914 mm) and having wall and ceiling surfaces of gypsum board.

(3) Bathrooms and toilet rooms that do not exceed 55 square feet (5.1 m^2) in area.

(4) Garages; carports; exterior porches; unheated entry areas, such as mud rooms, that are adjacent to an exterior door; and similar areas.

(5) Covered unheated projections of the building at entrances/exits provided it is not the only means of egress from the dwelling unit.

(6) Ceiling pockets that meet the following requirements:

(a) The total volume of an unprotected ceiling pocket does not exceed 100 cubic feet (2.83 m^3).

(b) The entire floor under the unprotected ceiling pocket is protected by the sprinklers at the lower ceiling elevation.

(c) Each unprotected ceiling pocket is separated from an adjacent unprotected ceiling pocket by not less than a 10 feet (3048 mm) horizontal distance.

(d) The interior finish of the unprotected ceiling pocket is noncombustible material.

(e) Skylights not exceeding 32 square feet (2.97 m^2).

612.3.2 Sprinkler Installation. Sprinklers shall be listed residential sprinklers and shall be installed in accordance with the sprinkler manufacturer's installation instructions.

612.3.3 Temperature Rating and Separation from Heat Sources. Sprinklers shall have a temperature rating of not less than 135°F (57°C) and not more than 170°F (77°C). Sprinklers shall be separated from heat sources in accordance with the sprinkler manufacturer's installation instructions.

Exception: Sprinklers located close to a heat source in accordance with Section 612.3.3.1 shall be intermediate temperature sprinklers.

612.3.3.1 Intermediate Temperature Sprinklers. Sprinklers shall have an intermediate temperature rating of

WATER SUPPLY AND DISTRIBUTION

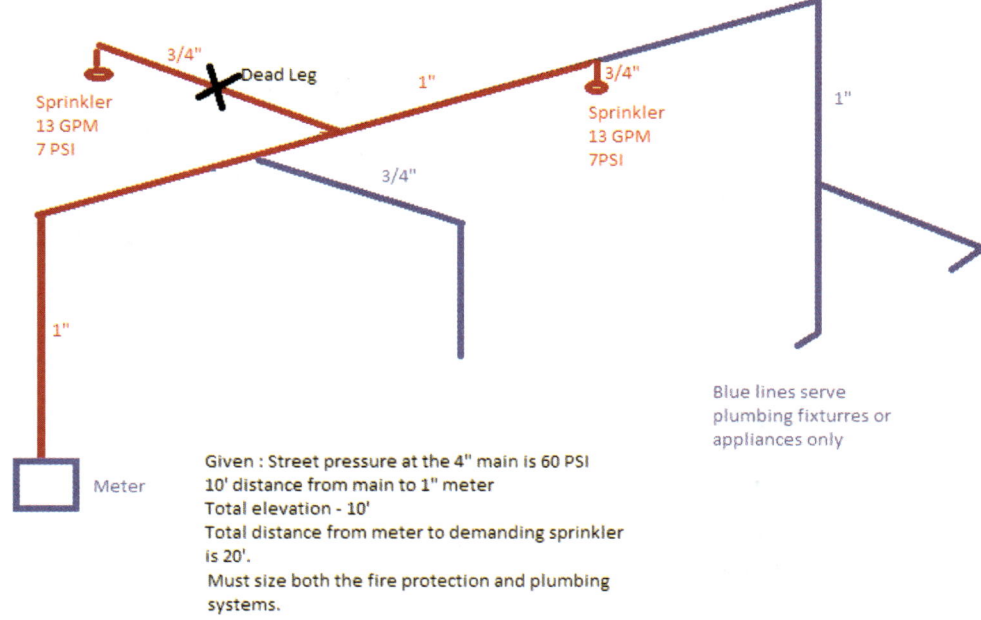

FIGURE 612.1
MULTI-PURPOSE SYSTEM

not less than 175°F (79°C) and not more than 225°F (107°C) where installed in the following locations:

(1) Directly under skylights, where the sprinkler is exposed to direct sunlight.

(2) In attics and concealed spaces located directly beneath a roof.

(3) Within the distance to a heat source in accordance with Table 612.3.3.1.

612.3.4 Freezing Areas. The piping system shall be protected in accordance with the requirements of Chapter 3. Where sprinklers are required in areas that are subject to freezing, dry-sidewall or dry-pendent sprinklers extending from a non-freezing area into a freezing area shall be installed.

612.3.5 Coverage Area Limit. The area of coverage of a single sprinkler shall be based on the sprinkler listing and the sprinkler manufacturer's installation instructions. The area of coverage of a single sprinkler shall not exceed 400 square feet (37.16 m^2).

612.3.6 Obstructions to Sprinkler Coverage. The water discharge from a sprinkler shall not be blocked by obstructions unless additional sprinklers are installed to protect the obstructed area. Additional sprinklers shall not be required where sprinkler separation from obstructions is in accordance with the requirements of Table 612.3.6, or the minimum distances specified in the sprinkler manufacturer's installation instructions.

612.3.6.1 Additional Requirements for Pendent Sprinklers. Pendent sprinklers located within 3 feet (914 mm) of the center of a ceiling fan, surface-mounted ceiling luminaire, or similar object shall be considered to be obstructed, and additional sprinklers shall be provided.

612.3.6.2 Additional Requirements for Sidewall Sprinklers. Sidewall sprinklers located within 5 feet (1524 mm) of the center of a ceiling fan, surface-mounted

TABLE 612.3.3.1
LOCATIONS WHERE INTERMEDIATE TEMPERATURE SPRINKLERS ARE REQUIRED

HEAT SOURCE	DISTANCE FROM HEAT SOURCE[1]	
	MINIMUM DISTANCE[2] (inches)	MAXIMUM DISTANCE (inches)
Fireplace, Side of Open or Recessed Fireplace	12	36
Fireplace, Front of Recessed Fireplace	36	60
Coal and Wood Burning Stove	12	42
Kitchen Range Top	9	18
Oven	9	18
Vent Connector or Chimney Connector	9	18
Heating Duct, Not Insulated	9	18
Hot Water Pipe, Not Insulated	6	12
Side of Ceiling or Wall Warm Air Register	12	24
Front of Wall Mounted Warm Air Register	18	36
Water Heater, Furnace, or Boiler	3	6
Luminaire up to 250 Watts	3	6
Luminaire 250 Watts up to 499 Watts	6	12

For SI units: 1 inch = 25.4 mm

Notes:
[1] Distances shall be measured in a straight line from the nearest edge of the heat source to the nearest edge of the sprinkler.
[2] Sprinklers shall not be located at distances less than the minimum table distance unless the sprinkler listing allows a lesser distance.

TABLE 612.3.6
MINIMUM SEPARATION FROM OBSTRUCTION

PENDENT SPRINKLERS

DISTANCE FROM DEFLECTOR TO PLANE AT BOTTOM OF OBSTRUCTION (A) (inches)	MINIMUM DISTANCE TO OBSTRUCTION (B) (feet)
1	1½
3	3
5	4
7	4½
9	6
11	6½
14	7

SIDEWALL SPRINKLER SIDE OBSTRUCTION

DISTANCE FROM DEFLECTOR TO PLANE AT BOTTOM OF OBSTRUCTION (A) (inches)	MINIMUM DISTANCE TO OBSTRUCTION (B) (feet)
1	1½
3	3
5	4
7	4½
9	6
11	6½
14	7

SIDEWALL SPRINKLER FORWARD OBSTRUCTION

DISTANCE FROM DEFLECTOR TO PLANE AT BOTTOM OF OBSTRUCTION (A) (inches)	MINIMUM DISTANCE TO OBSTRUCTION (B) (feet)
1	8
2	10
3	11
4	12
6	13
7	14
9	15
11	16
14	17

For SI units: 1 inch = 25.4 mm, 1 foot = 304.8 mm

ceiling luminaire, or similar object shall be considered to be obstructed, and additional sprinklers shall be provided.

612.3.7 Sprinkler Modifications Prohibited. Sprinklers shall not be painted, caulked, or modified. A sprinkler that has been painted, caulked, modified, or damaged shall be replaced with a new sprinkler.

612.4 Sprinkler Piping System. Sprinkler piping systems shall be installed in accordance with Section 612.4.1 through Section 612.4.5.

612.4.1 General. Sprinkler piping shall be installed in accordance with the requirements for water distribution piping. Sprinkler piping shall comply with the material requirements for cold water distribution piping. For multipurpose piping systems, the sprinkler piping shall connect to and be a part of the cold water distribution piping system.

612.4.2 Nonmetallic Pipe and Tubing. Nonmetallic pipe and tubing, such as CPVC, PEX-AL-PEX, PE-RT, and PEX, shall be certified for residential sprinkler installations and shall have a pressure rating of not less than 130 psi (896 kPa) at 120°F (49°C).

612.4.2.1 Nonmetallic Pipe Protection. Nonmetallic pipe and tubing systems shall be protected from exposure to the occupied space by a layer of not less than ⅜ of an inch (9.5 mm) thick gypsum wallboard, ½ of an inch (12.7 mm) thick plywood, or other material having a 15-minute fire rating.

Exceptions:
(1) Pipe protection shall not be required in areas that are not required to be protected with sprinklers in accordance with Section 612.3.1.
(2) Pipe protection shall not be required where exposed piping is permitted by the pipe third party listing.

612.4.2.2 Sprinkler Installation on Systems Assembled with Solvent Cement. The solvent cementing of fittings shall be completed, and threaded adapters for sprinklers shall be verified as being clear of excess cement before the installation of sprinklers on systems assembled with solvent cement.

612.4.3 Shutoff Valves Prohibited. Shutoff valves shall not be installed in a location where the valve would isolate piping serving one or more sprinklers. Shutoff valves shall only be permitted for the entire water distribution system.

612.4.4 Single Dwelling Limit. The sprinkler piping beyond the service valve located at the beginning of the water distribution system shall serve only one dwelling unit.

612.4.5 Drain. A ½ inch (15 mm) drain for the sprinkler system shall be provided on the system side of the water distribution shutoff valve.

612.5 Sprinkler Piping Design. Sprinkler piping systems shall be sized in accordance with Section 612.5.1 through Section 612.5.3.2.2.

612.5.1 Determining System Design Flow. The sizing of the sprinkler piping system shall be based on the flow rate and pressure of each sprinkler in accordance with Section 612.5.1.1 and the number of sprinklers in accordance with Section 612.5.1.2.

612.5.1.1 Determining Required Flow Rate for Each Sprinkler. The minimum flow rate and pressure for each residential sprinkler shall be in accordance with the manufacturer's published data for the specific sprinkler model based on the following:
(1) The area of coverage.
(2) The ceiling configuration.
(3) The temperature rating.
(4) Additional conditions specified by the sprinkler manufacturer.

612.5.1.2 System Flow Rate. The flow rate used for sizing the sprinkler piping system shall be based on the following:
(1) The flow rate for a room having only one sprinkler shall be the flow rate required for the sprinkler in accordance with Section 612.5.1.1.
(2) The flow rate for a room having two or more sprinklers shall be determined by identifying the sprinkler in the room with the highest required flow rate in accordance with Section 612.5.1.1 and multiplying that flow rate by 2.
(3) Where the sprinkler manufacturer specifies different criteria for ceiling configurations that are not smooth, flat, and horizontal the required flow rate for that room shall be in accordance with the sprinkler manufacturer's instructions.
(4) The flow rate used for sizing the sprinkler system shall be the flow required by the room with the largest flow rate in accordance with Section 612.5.1.2(1), Section 612.5.1.2(2), and Section 612.5.1.2(3).
(5) For the purpose of this section, it shall be permissible to reduce the flow rate for a room by subdividing the space into two or more rooms, where each room is evaluated separately on the required design flow rate. Each room shall be bounded by walls and a ceiling. Openings in walls shall have a lintel not less than 8 inches (203 mm) in depth, and each lintel shall form a solid barrier between the ceiling and the top of the opening.

612.5.2 Sprinkler Pipe Water Supply. The water supply for a multipurpose or stand-alone sprinkler system shall be provided by the public water main, private water main, private well system, or storage tank. The water supply required shall be determined in accordance with Section 612.5.1.2 at a pressure not less than that used in accordance with Section 612.5.3.

612.5.2.1 Water Pressure from Individual Sources. Where a dwelling unit water supply is from a tank system, a private well system, or a combination of these, the available water pressure shall be based on the minimum pressure control setting of the pump.

612.5.2.2 Required Capacity. The water supply shall have the capacity to provide the required flow rate to the sprinklers for a period of time as follows:

(1) Seven minutes for one story dwelling units less than 2000 square feet (185.8 m²) in area.

(2) Ten minutes for multi-level dwelling units and one story dwelling units not less than 2000 square feet (185.8 m²) in the area.

Where a well system, a water supply tank system, or a combination thereof is used a combination of well capacity and tank storage shall be permitted to meet the capacity requirement.

612.5.3 Sprinkler Pipe Sizing. The sprinkler piping shall be sized for the flow rate in accordance with Section 612.5.1. The flow rate required to supply the plumbing fixtures shall not be required to be added to the sprinkler design flow for multipurpose or stand alone piping systems. The sizing of the water supply to the plumbing fixtures shall be determined in accordance with this chapter. For multipurpose piping systems, the largest pipe size required based on either the sprinkler piping calculations or the water distribution piping calculations shall be installed.

612.5.3.1 Sprinkler Pipe Sizing Method. The sprinkler pipe shall be sized using the prescriptive method in Section 612.5.3.2 or by hydraulic calculation in accordance with NFPA 13D. The sprinkler pipe size from the water supply source to a sprinkler shall be not less than ¾ of an inch (20 mm) in diameter. Threaded adapter fittings at the point where sprinklers are attached to the piping shall be not less than ½ of an inch (15 mm) in diameter.

612.5.3.2 Prescriptive Pipe Sizing Method. The sprinkler pipe shall be sized by determining the available pressure to offset friction loss in piping and based on the piping material, diameter and length using the equation in Section 612.5.3.2.1 and the procedure in Section 612.5.3.2.2.

612.5.3.2.1 Available Pressure Equation. The available system pressure (P_t) for sizing the sprinkler piping shall be determined in accordance with the Equation 612.5.3.2.1.

(Equation 612.5.3.2.1)
$$P_t = P_{sup} - PL_{ws} - PL_m - PL_d - PL_e - P_{sp}$$

Where:
- P_t = Pressure used for sizing the system in Table 612.5.3.2(4) through Table 612.5.3.2(9)
- P_{sup} = Pressure available from the water supply source
- PL_{ws} = Pressure loss in the water service pipe
- PL_m = Pressure loss through the water meter
- PL_d = Pressure loss from devices other than the water meter
- PL_e = Pressure loss associated with changes in elevation
- P_{sp} = Maximum pressure required by a sprinkler

612.5.3.2.2 Calculation Procedure. The following procedure shall be used to determine the minimum size of the residential sprinkler piping:

Step 1 - Determine P_{sup}

Obtain the supply pressure available from the water main from the water purveyor, or for an individual source; the available supply pressure shall be in accordance with Section 612.5.2.1. The pressure shall be the flowing pressure available at the flow rate used when applying Table 612.5.3.2(1).

Step 2 – Determine PL_{ws}

Use Table 612.5.3.2(1) to determine the pressure loss in the water service pipe based on the size of the water service. Where the water service supplies more than one dwelling unit, 5 gpm (0.3 L/s) shall be added to the sprinkler flow rate.

Step 3 – Determine PL_m

Use Table 612.5.3.2(2) to determine the pressure loss from the water meter based on the water meter size.

Step 4 – Determine PL_d

Determine the pressure loss from devices, other than the water meter, installed in the piping system supplying sprinklers such as pressure-reducing valves, backflow preventers, water softeners, or water filters. Device pressure losses shall be based on the device manufacturer's specifications. The flow rate used to determine pressure loss shall be the sprinkler flow rate from Section 612.5.1.2. As an alternative to deducting pressure loss for a device, an automatic bypass valve shall be installed to divert flow around the device when a sprinkler activates.

Step 5 – Determine PL_e

Use Table 612.5.3.2(3) to determine the pressure loss associated with changes in elevation. The elevation used in applying the table shall be the difference between the elevation where the water source pressure was measured and the elevation of the highest sprinkler.

Step 6 – Determine P_{sp}

Determine the maximum pressure required by an individual sprinkler based on the flow rate from Section 612.5.1.1. The minimum pressure required is specified in the sprinkler manufacturer's published data for the specific sprinkler model based on the selected flow rate.

Step 7 – Calculate P_t

Using Equation 612.5.3.2.1, calculate the available system pressure for sizing the sprinkler piping.

Step 8 – Determine the maximum allowable pipe length

Use Table 612.5.3.2(4) through Table 612.5.3.2(9) to select a material and size for the residential sprinkler piping. The piping material and size shall be acceptable where the developed length of pipe between the inside water service valve and the most remote sprinkler does not exceed the maximum allowable length specified by the applicable table. Interpolation of P_t between the tabular values shall be permitted.

The maximum allowable length of piping in Table 612.5.3.2(4) through Table 612.5.3.2(9) incorporates an adjustment for pipe fittings, and no additional consideration

WATER SUPPLY AND DISTRIBUTION

TABLE 612.5.3.2(1)
WATER SERVICE PRESSURE LOSS (PL_{ws})[1, 2, 3]

FLOW RATE (gpm)	¾ INCH WATER SERVICE PRESSURE LOSS (psi)				1 INCH WATER SERVICE PRESSURE LOSS (psi)				1¼ INCH WATER SERVICE PRESSURE LOSS (psi)			
	40 FEET OR LESS	41 FEET TO 75 FEET	76 FEET TO 100 FEET	101 FEET TO 150 FEET	40 FEET OR LESS	41 FEET TO 75 FEET	76 FEET TO 100 FEET	101 FEET TO 150 FEET	40 FEET OR LESS	41 FEET TO 75 FEET	76 FEET TO 100 FEET	101 FEET TO 150 FEET
8	5.1	8.7	11.8	17.4	1.5	2.5	3.4	5.1	0.6	1.0	1.3	1.9
10	7.7	13.1	17.8	26.3	2.3	3.8	5.2	7.7	0.8	1.4	2.0	2.9
12	10.8	18.4	24.9	NP	3.2	5.4	7.3	10.7	1.2	2.0	2.7	4.0
14	14.4	24.5	NP	NP	4.2	7.1	9.6	14.3	1.6	2.7	3.6	5.4
16	18.4	NP	NP	NP	5.4	9.1	12.4	18.3	2.0	3.4	4.7	6.9
18	22.9	NP	NP	NP	6.7	11.4	15.4	22.7	2.5	4.3	5.8	8.6
20	27.8	NP	NP	NP	8.1	13.8	18.7	27.6	3.1	5.2	7.0	10.4
22	NP	NP	NP	NP	9.7	16.5	22.3	NP	3.7	6.2	8.4	12.4
24	NP	NP	NP	NP	11.4	19.3	26.2	NP	4.3	7.3	9.9	14.6
26	NP	NP	NP	NP	13.2	22.4	NP	NP	5.0	8.5	11.4	16.9
28	NP	NP	NP	NP	15.1	25.7	NP	NP	5.7	9.7	13.1	19.4
30	NP	NP	NP	NP	17.2	NP	NP	NP	6.5	11.0	14.9	22.0
32	NP	NP	NP	NP	19.4	NP	NP	NP	7.3	12.4	16.8	24.8
34	NP	NP	NP	NP	21.7	NP	NP	NP	8.2	13.9	18.8	NP
36	NP	NP	NP	NP	24.1	NP	NP	NP	9.1	15.4	20.9	NP

For SI units: 1 gallon per minute = 0.06 L/s, 1 pound-force per square inch = 6.89 kPa, 1 inch = 25 mm, 1 foot = 304.8 mm

Notes:
[1] Values are applicable for underground piping materials and are based on polyethylene pipe having an SDR of 11 and a Hazen Williams C Factor of 150.
[2] Values include the following length allowances for fittings: 25 percent length increase for actual lengths up to 100 feet (30 480 mm) and 15 percent length increase for actual lengths over 100 feet (30 480 mm).
[3] NP – Means not permitted.

TABLE 612.5.3.2(2)
MINIMUM WATER METER PRESSURE LOSS (PL_m)[1, 2]

FLOW RATE (gpm)	⅝ INCH METER PRESSURE LOSS (psi)	¾ INCH METER PRESSURE LOSS (psi)	1 INCH METER PRESSURE LOSS (psi)
8	2	1	1
10	3	1	1
12	4	1	1
14	5	2	1
16	7	3	1
18	9	4	1
20	11	4	2
22	NP	5	2
24	NP	5	2
26	NP	6	2
28	NP	6	2
30	NP	7	2
32	NP	7	3
34	NP	8	3
36	NP	8	3

For SI units: 1 gallon per minute = 0.06 L/s, 1 pound-force per square inch = 6.89 kPa, 1 inch = 25 mm

Notes:
[1] Table 612.5.3.2(2) establishes conservative values for water meter pressure loss for installations where the water meter loss is unknown. Where the actual water meter pressure loss is known, PL_m shall be the pressure loss as specified by the meter manufacturer.
[2] NP – Means not permitted.

of friction losses associated with pipe fittings shall be required.

612.6 Instructions and Signs. An owner's manual for the fire sprinkler system shall be provided to the owner. A sign or valve tag shall be installed at the main shutoff valve to the water distribution system stating the following: *"Warning, the water system for this home supplies fire sprinklers that require certain flow and pressure to fight a fire. Devices that restrict the flow decrease the pressure, or automatically shut-off the water to the fire sprinkler system, such as water softeners, filtration systems, and automatic shutoff valves shall not be added to this system without a review of the fire sprinkler system by a fire protection specialist. Do not remove this sign."*

612.7 Inspection and Testing. The inspection and testing of sprinkler systems shall be in accordance with Section 612.7.1 and Section 612.7.2.

612.7.1 Pre-Concealment Inspection. The following shall be verified prior to the concealment of any sprinkler system piping:

(1) Sprinklers are installed in all areas in accordance with Section 612.3.1.

(2) Where sprinkler water spray patterns are obstructed by construction features, luminaires or ceiling fans, additional sprinklers are installed in accordance with Section 612.3.6.

(3) Sprinklers are the correct temperature rating and are installed at or beyond the required separation distances

WATER SUPPLY AND DISTRIBUTION

TABLE 612.5.3.2(3)
ELEVATION LOSS (*PL*ₑ)

ELEVATION (feet)	PRESSURE LOSS (psi)
5	2.2
10	4.4
15	6.5
20	8.7
25	10.9
30	13.0
35	15.2
40	17.4

For SI units: 1 foot = 304.8 mm, 1 pound-force per square inch = 6.89 kPa

from heat sources in accordance with Section 612.3.3 and Section 612.3.3.1.

(4) The minimum pipe size in accordance with the requirements of Table 612.5.3.2(4) through Table 612.5.3.2(9) or, where the piping system was hydraulically calculated in accordance with Section 612.5.3.1, the size used in the hydraulic calculation.

(5) The pipe length does not exceed the length permitted by Table 612.5.3.2(4) through Table 612.5.3.2(9) or, where the piping system was hydraulically calculated in accordance with Section 612.5.3.1, pipe lengths and fittings shall not exceed those used in the hydraulic calculation.

(6) Nonmetallic piping that conveys water to sprinklers is

TABLE 612.5.3.2(4)
ALLOWABLE PIPE LENGTH FOR ¾ INCH TYPE M COPPER WATER TUBING*

SPRINKLER FLOW RATE (gpm)	WATER DISTRIBUTION SIZE (inch)	AVAILABLE PRESSURE - P_t (psi)									
		15	20	25	30	35	40	45	50	55	60
		ALLOWABLE LENGTH OF PIPE FROM SERVICE VALVE TO FARTHEST SPRINKLER (feet)									
8	¾	217	289	361	434	506	578	650	723	795	867
9	¾	174	232	291	349	407	465	523	581	639	697
10	¾	143	191	239	287	335	383	430	478	526	574
11	¾	120	160	200	241	281	321	361	401	441	481
12	¾	102	137	171	205	239	273	307	341	375	410
13	¾	88	118	147	177	206	235	265	294	324	353
14	¾	77	103	128	154	180	205	231	257	282	308
15	¾	68	90	113	136	158	181	203	226	248	271
16	¾	60	80	100	120	140	160	180	200	220	241
17	¾	54	72	90	108	125	143	161	179	197	215
18	¾	48	64	81	97	113	129	145	161	177	193
19	¾	44	58	73	88	102	117	131	146	160	175
20	¾	40	53	66	80	93	106	119	133	146	159
21	¾	36	48	61	73	85	97	109	121	133	145
22	¾	33	44	56	67	78	89	100	111	122	133
23	¾	31	41	51	61	72	82	92	102	113	123
24	¾	28	38	47	57	66	76	85	95	104	114
25	¾	26	35	44	53	61	70	79	88	97	105
26	¾	24	33	41	49	57	65	73	82	90	98
27	¾	23	30	38	46	53	61	69	76	84	91
28	¾	21	28	36	43	50	57	64	71	78	85
29	¾	20	27	33	40	47	53	60	67	73	80
30	¾	19	25	31	38	44	50	56	63	69	75
31	¾	18	24	29	35	41	47	53	59	65	71
32	¾	17	22	28	33	39	44	50	56	61	67
33	¾	16	21	26	32	37	42	47	53	58	63
34	¾	NP	20	25	30	35	40	45	50	55	60
35	¾	NP	19	24	28	33	38	42	47	52	57
36	¾	NP	18	22	27	31	36	40	45	49	54
37	¾	NP	17	21	26	30	34	38	43	47	51
38	¾	NP	16	20	24	28	32	36	40	45	49
39	¾	NP	15	19	23	27	31	35	39	42	46
40	¾	NP	NP	18	22	26	29	33	37	40	44

For SI units: 1 pound-force per square inch = 6.89 kPa, 1 gallon per minute = 0.06 L/s, 1 inch = 25 mm, 1 foot = 304.8 mm
* NP – Means not permitted.

2018 UNIFORM PLUMBING CODE ILLUSTRATED TRAINING MANUAL

certified as having a pressure rating of not less than 130 psi (896 kPa) at 120°F (49°C).

(7) Piping is properly supported.

(8) The piping system is tested in accordance with Section 609.4.

612.7.2 Final Inspection. Upon completion of the residential sprinkler system, the system shall be inspected. The following shall be verified during the final inspection:

(1) Sprinklers are not painted, damaged, or otherwise hindered from the operation.

(2) Where a pump is required to provide water to the system, the pump starts automatically upon system water demand.

(3) Pressure reducing valves, water softeners, water filters, or other impairments to water flow that were not part of the original design has not been installed.

(4) The sign or valve tag in accordance with Section 612.6 is installed, and the owner's manual for the system is present.

TABLE 612.5.3.2(5)
ALLOWABLE PIPE LENGTH FOR 1 INCH TYPE M COPPER WATER TUBING

SPRINKLER FLOW RATE (gpm)	WATER DISTRIBUTION SIZE (inch)	AVAILABLE PRESSURE - P_t (psi)									
		15	20	25	30	35	40	45	50	55	60
		ALLOWABLE LENGTH OF PIPE FROM SERVICE VALVE TO FARTHEST SPRINKLER (feet)									
8	1	806	1075	1343	1612	1881	2149	2418	2687	2955	3224
9	1	648	864	1080	1296	1512	1728	1945	2161	2377	2593
10	1	533	711	889	1067	1245	1422	1600	1778	1956	2134
11	1	447	596	745	894	1043	1192	1341	1491	1640	1789
12	1	381	508	634	761	888	1015	1142	1269	1396	1523
13	1	328	438	547	657	766	875	985	1094	1204	1313
14	1	286	382	477	572	668	763	859	954	1049	1145
15	1	252	336	420	504	588	672	756	840	924	1008
16	1	224	298	373	447	522	596	671	745	820	894
17	1	200	266	333	400	466	533	600	666	733	799
18	1	180	240	300	360	420	479	539	599	659	719
19	1	163	217	271	325	380	434	488	542	597	651
20	1	148	197	247	296	345	395	444	493	543	592
21	1	135	180	225	270	315	360	406	451	496	541
22	1	124	165	207	248	289	331	372	413	455	496
23	1	114	152	190	228	267	305	343	381	419	457
24	1	106	141	176	211	246	282	317	352	387	422
25	1	98	131	163	196	228	261	294	326	359	392
26	1	91	121	152	182	212	243	273	304	334	364
27	1	85	113	142	170	198	226	255	283	311	340
28	1	79	106	132	159	185	212	238	265	291	318
29	1	74	99	124	149	174	198	223	248	273	298
30	1	70	93	116	140	163	186	210	233	256	280
31	1	66	88	110	132	153	175	197	219	241	263
32	1	62	83	103	124	145	165	186	207	227	248
33	1	59	78	98	117	137	156	176	195	215	234
34	1	55	74	92	111	129	148	166	185	203	222
35	1	53	70	88	105	123	140	158	175	193	210
36	1	50	66	83	100	116	133	150	166	183	199
37	1	47	63	79	95	111	126	142	158	174	190
38	1	45	60	75	90	105	120	135	150	165	181
39	1	43	57	72	86	100	115	129	143	158	172
40	1	41	55	68	82	96	109	123	137	150	164

For SI units: 1 pound-force per square inch = 6.89 kPa, 1 gallon per minute = 0.06 L/s, 1 inch = 25 mm, 1 foot = 304.8 mm

WATER SUPPLY AND DISTRIBUTION

TABLE 612.5.3.2(6)
ALLOWABLE PIPE LENGTH FOR ¾ INCH IPS CPVC PIPE

SPRINKLER FLOW RATE (gpm)	WATER DISTRIBUTION SIZE (inch)	AVAILABLE PRESSURE - P_t (psi)									
		15	20	25	30	35	40	45	50	55	60
		ALLOWABLE LENGTH OF PIPE FROM SERVICE VALVE TO FARTHEST SPRINKLER (feet)									
8	¾	348	465	581	697	813	929	1045	1161	1278	1394
9	¾	280	374	467	560	654	747	841	934	1027	1121
10	¾	231	307	384	461	538	615	692	769	845	922
11	¾	193	258	322	387	451	515	580	644	709	773
12	¾	165	219	274	329	384	439	494	549	603	658
13	¾	142	189	237	284	331	378	426	473	520	568
14	¾	124	165	206	247	289	330	371	412	454	495
15	¾	109	145	182	218	254	290	327	363	399	436
16	¾	97	129	161	193	226	258	290	322	354	387
17	¾	86	115	144	173	202	230	259	288	317	346
18	¾	78	104	130	155	181	207	233	259	285	311
19	¾	70	94	117	141	164	188	211	234	258	281
20	¾	64	85	107	128	149	171	192	213	235	256
21	¾	58	78	97	117	136	156	175	195	214	234
22	¾	54	71	89	107	125	143	161	179	197	214
23	¾	49	66	82	99	115	132	148	165	181	198
24	¾	46	61	76	91	107	122	137	152	167	183
25	¾	42	56	71	85	99	113	127	141	155	169
26	¾	39	52	66	79	92	105	118	131	144	157
27	¾	37	49	61	73	86	98	110	122	135	147
28	¾	34	46	57	69	80	92	103	114	126	137
29	¾	32	43	54	64	75	86	96	107	118	129
30	¾	30	40	50	60	70	81	91	101	111	121
31	¾	28	38	47	57	66	76	85	95	104	114
32	¾	27	36	45	54	63	71	80	89	98	107
33	¾	25	34	42	51	59	68	76	84	93	101
34	¾	24	32	40	48	56	64	72	80	88	96
35	¾	23	30	38	45	53	61	68	76	83	91
36	¾	22	29	36	43	50	57	65	72	79	86
37	¾	20	27	34	41	48	55	61	68	75	82
38	¾	20	26	33	39	46	52	59	65	72	78
39	¾	19	25	31	37	43	50	56	62	68	74
40	¾	18	24	30	35	41	47	53	59	65	71

For SI units: 1 pound-force per square inch = 6.89 kPa, 1 gallon per minute = 0.06 L/s, 1 inch = 25 mm, 1 foot = 304.8 mm

WATER SUPPLY AND DISTRIBUTION

TABLE 612.5.3.2(7)
ALLOWABLE PIPE LENGTH FOR 1 INCH IPS CPVC PIPE

SPRINKLER FLOW RATE (gpm)	WATER DISTRIBUTION SIZE (inch)	AVAILABLE PRESSURE - P_t (psi)									
		15	20	25	30	35	40	45	50	55	60
		ALLOWABLE LENGTH OF PIPE FROM SERVICE VALVE TO FARTHEST SPRINKLER (feet)									
8	1	1049	1398	1748	2098	2447	2797	3146	3496	3845	4195
9	1	843	1125	1406	1687	1968	2249	2530	2811	3093	3374
10	1	694	925	1157	1388	1619	1851	2082	2314	2545	2776
11	1	582	776	970	1164	1358	1552	1746	1940	2133	2327
12	1	495	660	826	991	1156	1321	1486	1651	1816	1981
13	1	427	570	712	854	997	1139	1281	1424	1566	1709
14	1	372	497	621	745	869	993	1117	1241	1366	1490
15	1	328	437	546	656	765	874	983	1093	1202	1311
16	1	291	388	485	582	679	776	873	970	1067	1164
17	1	260	347	433	520	607	693	780	867	954	1040
18	1	234	312	390	468	546	624	702	780	858	936
19	1	212	282	353	423	494	565	635	706	776	847
20	1	193	257	321	385	449	513	578	642	706	770
21	1	176	235	293	352	410	469	528	586	645	704
22	1	161	215	269	323	377	430	484	538	592	646
23	1	149	198	248	297	347	396	446	496	545	595
24	1	137	183	229	275	321	366	412	458	504	550
25	1	127	170	212	255	297	340	382	425	467	510
26	1	118	158	197	237	276	316	355	395	434	474
27	1	111	147	184	221	258	295	332	368	405	442
28	1	103	138	172	207	241	275	310	344	379	413
29	1	97	129	161	194	226	258	290	323	355	387
30	1	91	121	152	182	212	242	273	303	333	364
31	1	86	114	143	171	200	228	257	285	314	342
32	1	81	108	134	161	188	215	242	269	296	323
33	1	76	102	127	152	178	203	229	254	280	305
34	1	72	96	120	144	168	192	216	240	265	289
35	1	68	91	114	137	160	182	205	228	251	273
36	1	65	87	108	130	151	173	195	216	238	260
37	1	62	82	103	123	144	165	185	206	226	247
38	1	59	78	98	117	137	157	176	196	215	235
39	1	56	75	93	112	131	149	168	187	205	224
40	1	53	71	89	107	125	142	160	178	196	214

For SI units: 1 pound-force per square inch = 6.89 kPa, 1 gallon per minute = 0.06 L/s, 1 inch = 25 mm, 1 foot = 304.8 mm

TABLE 612.5.3.2(8)
ALLOWABLE PIPE LENGTH FOR ¾ INCH PEX TUBING*

SPRINKLER FLOW RATE (gpm)	WATER DISTRIBUTION SIZE (inch)	AVAILABLE PRESSURE - P_t (psi)									
		15	20	25	30	35	40	45	50	55	60
		ALLOWABLE LENGTH OF PIPE FROM SERVICE VALVE TO FARTHEST SPRINKLER (feet)									
8	¾	93	123	154	185	216	247	278	309	339	370
9	¾	74	99	124	149	174	199	223	248	273	298
10	¾	61	82	102	123	143	163	184	204	225	245
11	¾	51	68	86	103	120	137	154	171	188	205
12	¾	44	58	73	87	102	117	131	146	160	175
13	¾	38	50	63	75	88	101	113	126	138	151
14	¾	33	44	55	66	77	88	99	110	121	132
15	¾	29	39	48	58	68	77	87	96	106	116
16	¾	26	34	43	51	60	68	77	86	94	103
17	¾	23	31	38	46	54	61	69	77	84	92
18	¾	21	28	34	41	48	55	62	69	76	83
19	¾	19	25	31	37	44	50	56	62	69	75
20	¾	17	23	28	34	40	45	51	57	62	68
21	¾	16	21	26	31	36	41	47	52	57	62
22	¾	NP	19	24	28	33	38	43	47	52	57
23	¾	NP	17	22	26	31	35	39	44	48	52
24	¾	NP	16	20	24	28	32	36	40	44	49
25	¾	NP	NP	19	22	26	30	34	37	41	45
26	¾	NP	NP	17	21	24	28	31	35	38	42
27	¾	NP	NP	16	20	23	26	29	33	36	39
28	¾	NP	NP	15	18	21	24	27	30	33	36
29	¾	NP	NP	NP	17	20	23	26	28	31	34
30	¾	NP	NP	NP	16	19	21	24	27	29	32
31	¾	NP	NP	NP	15	18	20	23	25	28	30
32	¾	NP	NP	NP	NP	17	19	21	24	26	28
33	¾	NP	NP	NP	NP	16	18	20	22	25	27
34	¾	NP	NP	NP	NP	NP	17	19	21	23	25
35	¾	NP	NP	NP	NP	NP	16	18	20	22	24
36	¾	NP	NP	NP	NP	NP	15	17	19	21	23
37	¾	NP	NP	NP	NP	NP	NP	16	18	20	22
38	¾	NP	NP	NP	NP	NP	NP	16	17	19	21
39	¾	NP	NP	NP	NP	NP	NP	NP	16	18	20
40	¾	NP	NP	NP	NP	NP	NP	NP	16	17	19

For SI units: 1 pound-force per square inch = 6.89 kPa, 1 gallon per minute = 0.06 L/s, 1 inch = 25 mm, 1 foot = 304.8 mm

* NP – Means not permitted.

WATER SUPPLY AND DISTRIBUTION

TABLE 612.5.3.2(9)
ALLOWABLE PIPE LENGTH FOR 1 INCH PEX TUBING

SPRINKLER FLOW RATE (gpm)	WATER DISTRIBUTION SIZE (inch)	AVAILABLE PRESSURE - P_t (psi)									
		15	20	25	30	35	40	45	50	55	60
		ALLOWABLE LENGTH OF PIPE FROM SERVICE VALVE TO FARTHEST SPRINKLER (feet)									
8	1	314	418	523	628	732	837	941	1046	1151	1255
9	1	252	336	421	505	589	673	757	841	925	1009
10	1	208	277	346	415	485	554	623	692	761	831
11	1	174	232	290	348	406	464	522	580	638	696
12	1	148	198	247	296	346	395	445	494	543	593
13	1	128	170	213	256	298	341	383	426	469	511
14	1	111	149	186	223	260	297	334	371	409	446
15	1	98	131	163	196	229	262	294	327	360	392
16	1	87	116	145	174	203	232	261	290	319	348
17	1	78	104	130	156	182	208	233	259	285	311
18	1	70	93	117	140	163	187	210	233	257	280
19	1	63	84	106	127	148	169	190	211	232	253
20	1	58	77	96	115	134	154	173	192	211	230
21	1	53	70	88	105	123	140	158	175	193	211
22	1	48	64	80	97	113	129	145	161	177	193
23	1	44	59	74	89	104	119	133	148	163	178
24	1	41	55	69	82	96	110	123	137	151	164
25	1	38	51	64	76	89	102	114	127	140	152
26	1	35	47	59	71	83	95	106	118	130	142
27	1	33	44	55	66	77	88	99	110	121	132
28	1	31	41	52	62	72	82	93	103	113	124
29	1	29	39	48	58	68	77	87	97	106	116
30	1	27	36	45	54	63	73	82	91	100	109
31	1	26	34	43	51	60	68	77	85	94	102
32	1	24	32	40	48	56	64	72	80	89	97
33	1	23	30	38	46	53	61	68	76	84	91
34	1	22	29	36	43	50	58	65	72	79	86
35	1	20	27	34	41	48	55	61	68	75	82
36	1	19	26	32	39	45	52	58	65	71	78
37	1	18	25	31	37	43	49	55	62	68	74
38	1	18	23	29	35	41	47	53	59	64	70
39	1	17	22	28	33	39	45	50	56	61	67
40	1	16	21	27	32	37	43	48	53	59	64

For SI units: 1 pound-force per square inch = 6.89 kPa, 1 gallon per minute = 0.06 L/s, 1 inch = 25 mm, 1 foot = 304.8 mm

CHAPTER 7
SANITARY DRAINAGE

Part I – Drainage Systems.
701.0 General.

701.1 Applicability. This chapter shall govern the materials, design, and installation of sanitary drainage systems and building sewers.

701.2 Drainage Piping. Materials for drainage piping shall be in accordance with one of the referenced standards in Table 701.2 except that:

🔧 Requirements for drainage and sewer piping will also apply to vent pipe materials. Table 701.2 is a good reference for determining which material is allowed for building drainage or sewer piping. The drainage waste and vent columns refer to pipe for use within the building either above ground or underground. Building sewer applications are for use outside of the foundation of the building only.

(1) No galvanized wrought-iron or galvanized steel pipe shall be used underground and shall be kept not less than 6 inches (152 mm) aboveground.

🔧 Due to the possibility of wrought iron or steel piping rusting, whether galvanized or not, it is not for use underground. The use of galvanized pipe in the drainage system is increasingly rare except for pipe nipple applications. This is because of the increased cost of the material and labor versus other available materials.

(2) ABS and PVC DWV piping installations shall be installed in accordance with applicable standards referenced in Table 701.2 and Chapter 14 "Firestop Protection." Except for individual single-family dwelling units, materials exposed within ducts or plenums shall have a flame-spread index of not more than 25 and a smoke-developed index of not more than 50, where tested in accordance with ASTM E84 or UL 723. These tests shall comply with all requirements of the standards to include the sample size, both for width and length. Plastic pipe shall not be tested filled with water.

🔧 The use of plastic materials for drainage and vent piping within the building has been limited in past years because of the high flammability and smoke-creating potential of the material. It would either be prohibited or limited to three or four stories in residential buildings. Plastic pipe ABS and PVC for drain waste and vent (DWV) applications are prohibited within ducts or plenums unless they meet the above specifications for flame spread and smoke-developed indexes according to the requirements of Chapter 14. There are still some areas of the country that prohibit plastics within the building, but most building codes will allow its use when the requirements for fire protection and wall penetration protection are adhered to. Plastic DWV fittings are illustrated in **Figure 701.2(2)**.

(3) No vitrified clay pipe or fittings shall be used above-

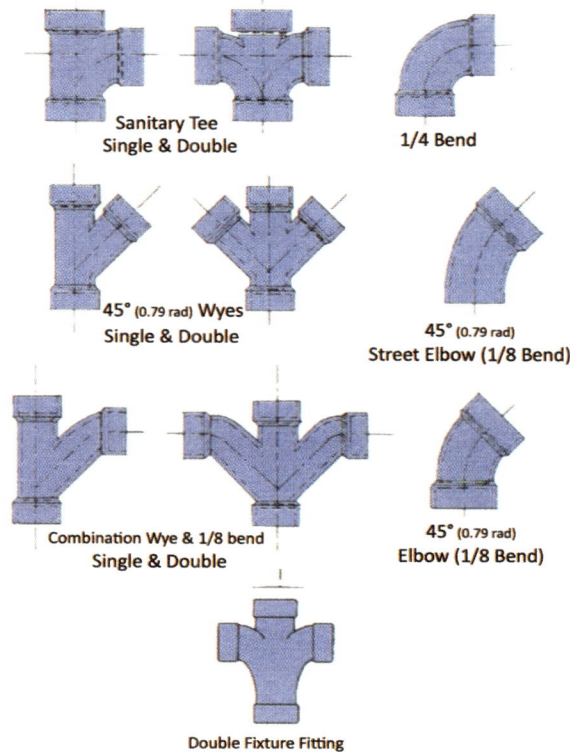

**FIGURE 701.2(2)
PLASTIC DWV FITTINGS**

ground or where pressurized by a pump or ejector. They shall be kept not less than 12 inches (305 mm) belowground.

(4) Copper or copper alloy tube for drainage and vent piping shall have a weight of not less than that of copper or copper alloy drainage tube type DWV.

(5) Stainless steel 304 pipe and fittings shall not be installed underground and shall be kept not less than 6 inches (152 mm) aboveground.

🔧 Type 304 stainless steel is approved for aboveground applications and Type 316L is approved for above- or below-ground applications. Type 316L has superior corrosion-resistant properties compared to Type 304 stainless steel. All stainless steel systems must be installed per the manufacturer's installation instructions.

Stainless steel has properties that far exceed many materials on the market today, and selecting the correct grade ensures that the product will have a maintenance-free and cost-effective installation. Type 304 stainless steel is an austenitic grade that can be severely drawn, spun, rolled, machined or worked in the cold state. Type 304L is the low carbon version of Type 304. It is used in heavy gauge components for improved welding ability. This type of

SANITARY DRAINAGE

stainless steel has an excellent corrosion resistance in a wide variety of environments including chlorides.

Type 316L stainless steel contains low carbon and high nickel content and an addition of molybdenum that gives it improved corrosion resistance. The molybdenum tends to help minimize pitting or pin-hole corrosion under certain conditions.

ASME A112.3.1 contains installation procedures for stainless steel drainage systems for sanitary, storm and chemical applications, above and below ground. This standard covers requirements for socket-type, seam-welded stainless steel pipe, fittings, joints and drains for use in sanitary and storm, DWV and chemical waste systems [see **Figure 701.2(5)**].

**FIGURE 701.2(5)
STAINLESS STEEL PIPE AND FITTINGS**

(6) Cast-iron soil pipe and fittings and the stainless steel couplings used to join these products shall be listed and tested in accordance with standards referenced in Table 701.2. Such pipe and fittings shall be marked with the country of origin, manufacturer's name or registered trademark as defined in the product standards, the third party certifier's mark, and the class of the pipe or fitting.

701.3 Drainage Fittings. Materials for drainage fittings shall comply with the applicable standards referenced in Table 701.2 of the same diameter as the piping served, and such fittings shall be compatible with the type of pipe used.

701.3.1 Screwed Pipe. Fittings on screwed pipe shall be of the recessed drainage type. Burred ends shall be reamed to the full bore of the pipe.

701.3.2 Threads. The threads of drainage fittings shall be tapped to allow ¼ inch per foot (20.8 mm/m) grade.

701.3.3 Type. Fittings used for drainage shall be of the drainage type, have a smooth interior water-way, and be constructed to allow ¼ inch per foot (20.8 mm/m) grade.

🔧 The drainage system relies on gravity, not on any pressure system, to move the effluent through the system. In order for the waste to travel down the system the installation must be uniformly smooth without sharp turns or snags. To accomplish this, pipe and fittings are designed and manufactured with smooth bores and transitions between pipe and fittings to provide the best pathway of least resistance (see **Figure 701.3.3**). Solids are entrained in the effluent and must be transported to their destination without plugging the system. Research and design of pipe and fittings are accomplished for a reason; therefore, the system must be installed correctly or it will not accomplish its designated task. The pipe must be cut, deburred and joined with fittings correctly.

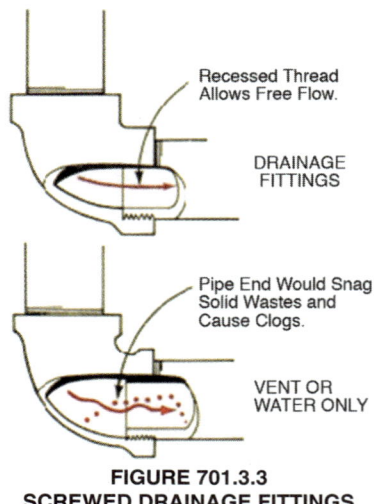

**FIGURE 701.3.3
SCREWED DRAINAGE FITTINGS**

701.4 Continuous Wastes. Continuous wastes and fixture tailpieces shall be constructed from the materials specified in Section 701.2 for drainage piping, provided, however, that such connections where exposed or accessible shall be permitted to be of seamless drawn brass not less than No. 20 B & S Gauge (0.032 inches) (0.8 mm).

🔧 Continuous wastes and tailpieces, if concealed, shall consist of materials used in the drainage system. For example, a floor sink tailpiece would be required to be a cast iron or Schedule 40 ABS or PVC pipe if it is installed below the floor in the ground. Where exposed, as in a kitchen sink tailpiece and continuous waste, the piping material shall be a minimum of No. 20 Brown and Sharp sheet metal gauge. This equals 0.0320 inches (see **Figure 701.4**).

701.5 Lead. (See Table 1701.1) Sheet lead shall comply with the following:

(1) For safe pans – not less than 4 pounds per square foot (lb/ft^2) (19 kg/m^2) or ¹⁄₁₆ of an inch (1.6 mm) thick.

🔧 A safe pan is the material placed under or around a fixture, in this case a floor drain or shower drain, to

SANITARY DRAINAGE

TABLE 701.2
MATERIALS FOR DRAIN, WASTE, VENT PIPE AND FITTINGS

MATERIAL	UNDERGROUND DRAIN, WASTE, VENT PIPE AND FITTINGS	ABOVEGROUND DRAIN, WASTE, VENT PIPE AND FITTINGS	BUILDING SEWER PIPE AND FITTINGS	REFERENCED STANDARD(S) PIPE	REFERENCED STANDARD(S) FITTINGS
ABS (Schedule 40)	X	X	X	ASTM D2661, ASTM D2680*	ASTM D2661, ASTM D2680*
Cast-Iron	X	X	X	ASTM A74, ASTM A888, CISPI 301	ASME B16.12, ASTM A74, ASTM A888, CISPI 301
Co-Extruded ABS (Schedule 40)	X	X	X	ASTM F628	ASTM D2661, ASTM D2680*
Co-Extruded Composite (Schedule 40)	X	X	X	ASTM F1488	ASTM D2661, ASTM D2665, ASTM F794*, ASTM F1866
Co-Extruded PVC (Schedule 40)	X	X	X	ASTM F891, ASTM F1760	ASTM D2665, ASTM F794*, ASTM F1336*, ASTM F1866
Copper and Copper Alloys (Type DWV)	X	X	X	ASTM B43, ASTM B75, ASTM B251, ASTM B302, ASTM B306	ASME B16.23, ASME B16.29
Galvanized Malleable Iron	—	X	—	—	ASME B16.3
Galvanized Steel	—	X	—	ASTM A53	—
Polyethylene	—	—	X	ASTM F714, ASTM F894	—
PVC (Schedule 40)	X	X	X	ASTM D1785, ASTM D2665, ASTM F794*	ASTM D2665, ASTM F794*, ASTM F1866
PVC (Sewer and Drain)	—	—	X	ASTM D2729	ASTM D2729
PVC PSM	—	—	X	ASTM D3034	ASTM D3034
Stainless Steel 304	—	X	—	ASME A112.3.1	ASME A112.3.1
Stainless Steel 316L	X	X	X	ASME A112.3.1	ASME A112.3.1
Vitrified Clay (Extra strength)	—	—	X	ASTM C700	ASTM C700

* For building sewer applications.

ensure all the waste empties into the drain and does not spill over or empty away from the drain. Sheet lead (see **Figure 701.5**) may be used for this purpose, and if it is, it must be at least one-sixteenth of an inch thick. It is placed between the flange and the clamping ring of the drain. Tightening down the clamping ring creates a seal and any waste will then drain to the weep holes in the drain.

(2) For flashings or vent terminals – not less than 3 lb/ft^2 (15 kg/m^2) or 0.0472 of an inch (1.2 mm) thick.

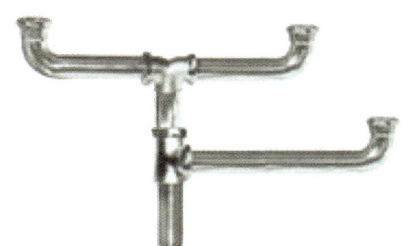

FIGURE 701.4
CONTINUOUS WASTE FOR TRIPLE SINK

SANITARY DRAINAGE

(3) Lead bends and lead traps shall be not less than ⅛ of an inch (3.2 mm) in wall thickness.

701.6 Caulking Ferrules. Caulking ferrules shall be manufactured from copper or copper alloy and shall be in accordance with Table 701.6.

🔧 Caulking ferrules are for use in the wiping of lead joints between lead pipe and cast iron, steel or wrought iron. Although not used in many areas anymore, plumbers may still come across a repair or acid waste application where lead might be used. See **Figure 701.6** for an example of a brass caulking ferrule used to make the transition from lead to cast iron.

**TABLE 701.6
CAULKING FERRULES**

PIPE SIZE (inches)	INSIDE DIAMETER (inches)	LENGTH (inches)	MINIMUM WEIGHT EACH	
			pounds	ounces
2	2¼	4½	1	0
3	3¼	4½	1	12
4	4¼	4½	2	8

For SI units: 1 inch = 25 mm, 1 pound = 0.453 kg, 1 ounce = 0.02834 kg

701.7 Soldering Bushings. Soldering bushings shall be of copper or copper alloy and shall be in accordance with Table 701.7.

**TABLE 701.7
SOLDERING BUSHINGS**

PIPE SIZE (inches)	MINIMUM WEIGHT EACH		PIPE SIZE (inches)	MINIMUM WEIGHT EACH	
	pounds	ounces		pounds	ounces
1¼	0	6	2½	1	6
1½	0	8	3	2	0
2	0	14	4	3	8

For SI units: 1 inch = 25 mm, 1 pound = 0.453 kg, 1 ounce = 0.02834 kg

702.0 Fixture Unit Equivalents.

702.1 Trap Size. The unit equivalent of plumbing fixtures shown in Table 702.1 shall be based on the size of the trap required, and the unit equivalent of fixtures and devices not shown in Table 702.1 shall be based on the size of trap or trap arm.

Maximum drainage fixture units for a fixture trap and trap arm loadings for sizes up to 4 inches (100 mm) shall be

**TABLE 702.2(1)
MAXIMUM DRAINAGE FIXTURE UNITS FOR A
TRAP AND TRAP ARM***

SIZE OF TRAP AND TRAP ARM (inches)	DRAINAGE FIXTURE UNIT VALUES (DFU)
1¼	1 unit
1½	3 units
2	4 units
3	6 units
4	8 units

For SI Units: 1 inch = 25 mm
* **Exception:** On self-service laundries.

in accordance with Table 702.2(1).

**FIGURE 701.5
SHEET LEAD**

**FIGURE 701.6
BRASS CAULKING FERRULE**

🔧 A fixture unit is a numerical load factor which measures on some select scale the load-producing effects of a single plumbing fixture of a given kind. The fixture unit idea originated with Dr. Roy Hunter, physicist of the National Bureau of Standards, in the early 1920s. In the laboratory, the discharge rate of fixtures most commonly used in residential homes was measured. The fixture having the least discharge flow rate was the lavatory (an average of 7.5 gpm) and the fixture having the greatest discharge rate was the toilet (an average of 45 gpm). The ratio between the two is 6:1. Therefore, the fixtures were weighted on a scale of one to six. The lavatory with the least discharge rate was assigned a fixture unit value of one and the toilet was assigned the greatest fixture unit value of six. All other fixtures were weighted in between. The bathtub with a 1½" trap with a discharge rate of 15 gpm was assigned two fixture units, a kitchen sink with a discharge rate of 11.3 gpm was assigned 1.5 fixture units and a laundry sink with a discharge rate of 22.5 gpm was assigned three fixture units. Over the years, fixture unit values have been modified with many more plumbing fixtures being assigned fixture unit values.

SANITARY DRAINAGE

TABLE 702.1
DRAINAGE FIXTURE UNIT VALUES (DFU)

PLUMBING APPLIANCES, APPURTENANCES, OR FIXTURES	MINIMUM SIZE TRAP AND TRAP ARM[7] (inches)	PRIVATE	PUBLIC	ASSEMBLY[8]
Bathtub or Combination Bath/Shower	1½	2.0	2.0	—
Bidet	1¼	1.0	—	—
Bidet	1½	2.0	—	—
Clothes Washer, domestic, standpipe[5]	2	3.0	3.0	3.0
Dental Unit, cuspidor	1¼	—	1.0	1.0
Dishwasher, domestic, with independent drain[2]	1½	2.0	2.0	2.0
Drinking Fountain or Water Cooler	1¼	0.5	0.5	1.0
Food Waste Disposer, commercial	2	—	3.0	3.0
Floor Drain, emergency	2	—	0.0	0.0
Floor Drain (for additional sizes see Section 702.0)	2	2.0	2.0	2.0
Shower, single-head trap	2	2.0	2.0	2.0
Multi-head, each additional	2	1.0	1.0	1.0
Lavatory	1¼	1.0	1.0	1.0
Lavatories in sets	1½	2.0	2.0	2.0
Washfountain	1½	—	2.0	2.0
Washfountain	2	—	3.0	3.0
Mobile Home, trap	3	12.0	—	—
Receptor, indirect waste[1,3]	1½		See footnote[1,3]	
Receptor, indirect waste[1,4]	2		See footnote[1,4]	
Receptor, indirect waste[1]	3		See footnote[1]	
Sinks	—	—	—	—
Bar	1½	1.0	—	—
Bar[2]	1½	—	2.0	2.0
Clinical	3	—	6.0	6.0
Commercial with food waste[2]	1½	—	3.0	3.0
Exam Room	1½	—	1.0	—
Special Purpose[2]	1½	2.0	3.0	3.0
Special Purpose	2	3.0	4.0	4.0
Special Purpose	3	—	6.0	6.0
Kitchen, domestic[2] (with or without food waste disposer, dishwasher, or both)	1½	2.0	2.0	—
Laundry[2] (with or without discharge from a clothes washer)	1½	2.0	2.0	2.0
Service or Mop Basin	2	—	3.0	3.0
Service or Mop Basin	3	—	3.0	3.0
Service, flushing rim	3	—	6.0	6.0
Wash, each set of faucets	—	—	2.0	2.0
Urinal, integral trap 1.0 GPF[2]	2	2.0	2.0	5.0
Urinal, integral trap greater than 1.0 GPF	2	2.0	2.0	6.0
Urinal, exposed trap[2]	1½	2.0	2.0	5.0
Water Closet, 1.6 GPF Gravity Tank[6]	3	3.0	4.0	6.0
Water Closet, 1.6 GPF Flushometer Tank[6]	3	3.0	4.0	6.0
Water Closet, 1.6 GPF Flushometer Valve[6]	3	3.0	4.0	6.0
Water Closet, greater than 1.6 GPF Gravity Tank[6]	3	4.0	6.0	8.0
Water Closet, greater than 1.6 GPF Flushometer Valve[6]	3	4.0	6.0	8.0

For SI units: 1 inch = 25 mm

Notes:

[1] Indirect waste receptors shall be sized based on the total drainage capacity of the fixtures that drain therein to, in accordance with Table 702.2(2).
[2] Provide a 2 inch (50 mm) minimum drain.
[3] For refrigerators, coffee urns, water stations, and similar low demands.
[4] For commercial sinks, dishwashers, and similar moderate or heavy demands.
[5] Buildings having a clothes-washing area with clothes washers in a battery of three or more clothes washers shall be rated at 6 fixture units each for purposes of sizing common horizontal and vertical drainage piping.
[6] Water closets shall be computed as 6 fixture units where determining septic tank sizes based on Appendix H of this code.
[7] Trap sizes shall not be increased to the point where the fixture discharge is capable of being inadequate to maintain their self-scouring properties.
[8] Assembly [Public Use (see Table 422.1)].

SANITARY DRAINAGE

Fixture units when applied individually to plumbing fixtures can represent equivalence to a discharge flow rate of 7.5gpm per fixture unit. However, when applied to a plumbing system, fixture units represent a probable load factor and not equivalence of 7.5 gpm per fixture unit. For example, Table 703.2 allows a maximum loading of 216 fixture units for a four-inch horizontal drain pipe. This cannot mean 7.5 gpm per fixture unit, because that would indicate 1,620 gpm flow rate in a four-inch pipe. Using Manning's formula, at one-fourth of an inch per foot slope, a four-inch pipe will flow full bore at 181 gpm. So what do the 216 fixture units represent? They represent a load factor based upon the laws of probability, because not all discharges will occur simultaneously nor will all discharging flows be coincident. Only a portion of the discharges will overlap in the piping system. Furthermore, as the discharges approach the building drain the flow rates will decrease toward terminal velocity and tend toward a uniform flow.

So what can the expected flow rate be for a four-inch building drain when there are 216 fixture units on the system? The algorithms provided by the National Bureau of Standards are published in Recommended Minimum Requirements for Plumbing, BH13 and Methods of Estimating Loads in Plumbing Systems, BMS65. Based on these algorithms, by the time the probable discharges reach the four-inch building drain, there will be an expected uniform average flow rate of approximately 30gpm, which is about one-third the flow capacity of a four-inch diameter cast iron pipe at one-fourth of an inch per foot slope.

Drainage fixture units are not to be confused with water supply fixture units that were based on a different scale and a modified algorithm (see commentary in Section 610.3).

Table 702.2(1) determines the maximum drainage fixture units for a given trap up to four inches. For traps 1¼ inches in size, one fixture unit is the maximum allowed on that trap. For traps two inches in size four fixture units are allowed on that trap and so on. The exception is for self-service laundries, which will be explained in the commentary for Table 702.1 under Note 5.

Using this table will help determine the fixture unit rating of a fixture not listed in Table 702.1. For example, if a plumber is installing a custom fixture not listed in Table 702.1, the trap size required for the fixture will determine the fixture unit rating according to Table 702.2(1)

The simplest and most used method of assigning fixture units to fixtures is by using Table 702.1. Most of the common types of plumbing fixtures, with their assigned fixture unit values, minimum trap and trap arm sizes, are listed in this table.

Table 702.1 contains separate columns for fixtures, trap and trap arm sizes, and drainage fixture unit (DFU) ratings for three different use classifications – private, public and assembly. Reading the table is straightforward. A single lavatory (a lavatory by itself) requires a 1¼-inch minimum sized trap and trap arm, and is assigned one fixture unit in all three use classifications. The fixtures list in this table is more extensive compared to Table 610.3 because of the varying characteristics in drainage. Fixtures can drain individually or in sets, as with lavatories. There are different sizes of traps for the same fixture, and there are fixtures without a water supply yet receive discharging wastes, such as an indirect waste receptor and floor drain. There is even a fixture that has no fixture unit rating — floor drains for emergency use. The emergency use floor drain is not assigned a fixture unit value because it does not add a drainage load on the drainage system. This floor drain is installed to capture overflowing waste from a fixture or fixtures whose drains are plugged. The system has already been designed for these drainage loads, so there is no need to add more fixture units and possibly oversize the pipe for floor drains that add no load of their own (see **Figure 702.1a**).

The notes to Table 702.1 are one of the most important features of this table. They are as follows:

1. This note refers to indirect waste receptors, which is most often a floor sink (see **Figure 702.1b**). The size of the receptor and its trap, and the fixture unit rating will be based on the discharge flow rate of the equipment, fixtures, or devices draining into them. This is accomplished by adding the flow rate of the drains discharging into the receptor and using Table 702.2(2) to determine the equivalent fixture units. Then using Table 702.2(1), the fixture units will determine the proper receptor and trap size for the installation.

2. Dishwashers, certain sinks and urinals are referenced by this note. These footnoted fixtures must be supplied with a two-inch drain although the trap and trap arms may be 1½ inches (see **Figure 702.1c**). This footnote implicitly prohibits the use of a 1½ inch diameter vertical drain with these specific fixtures having 2 DFU, which is permitted by Table 703.2. Table 703.2, footnote 2 corresponds to this requirement.

3. The 1½-inch size receptor is for these low flow fixtures or devices.

4. The two-inch receptor is for moderate or heavy flow demands.

5. Clothes washers are normally rated at three DFU each. Clothes washers in a battery of three or more shall be rated at 6 fixture units only when sizing the common horizontal and vertical waste pipe sizing. The increase in the fixture unit value takes into account the greater probability of simultaneous discharge of the clothes washers and the suds- producing effect in the drainage pipe. The increase in fixture unit value will occur only on the piping common to three or more clothes washers. It is also only for drainage piping; vents will be sized using three DFU per washer according to Table 703.2 (see **Figure 702.1d** for sizing example). Suds will be further discussed in Section 711.0.

6. If Appendix H is used in conjunction with Tables 702.1 and 703.2 for determining the size of a septic tank for a private installation, the water closets shall be rated at six DFU.

7. Table 702.1 establishes the trap and trap arm size for the fixtures in that table. The fixtures are designed with this established trap size to accommodate the flow from the fixture and to keep the trap free from sediment by providing a scouring action when it drains. Over-sizing the trap will diminish this effect. However, Section 1003.3 of Chapter 10 allows an increase of one pipe size for the trap and trap arm.

8. Assembly is considered a specific Public Use classification in Table 422.1 having five sub-categories. The use of toilet rooms in these types of occupancies is typically higher than other types and has a greater load-producing effect upon the plumbing system. Therefore, urinals and toilets are rated with higher fixture units.

702.2 Intermittent Flow. Drainage fixture units for intermittent flow into the drainage system shall be computed on the rated discharge capacity in gallons per minute (gpm) (L/s) in accordance with Table 702.2(2).

TABLE 702.2(2)
DISCHARGE CAPACITY IN GALLONS PER MINUTE FOR INTERMITTENT FLOW ONLY*

GPM	FIXTURE UNITS
Up to 7½	Equals 1 Fixture Unit
Greater than 7½ to 15	Equals 2 Fixture Units
Greater than 15 to 30	Equals 4 Fixture Units
Greater than 30 to 50	Equals 6 Fixture Units

For SI units: 1 gallon per minute = 0.06 L/s

* Discharge capacity exceeding 50 gallons per minute (3.15 L/s) shall be determined by the Authority Having Jurisdiction.

Intermittent flows differ from continuous flows. Intermittent flows derive from fixtures discharging into the drainage system that are subject to the predictability of being on or off during a period of time. All the fixtures in Table 702.1 have intermittent flows and therefore are assigned fixture units as discussed in the commentary for Section 702.1. With respect to fixture discharge, each fixture unit is equivalent to 7.5 gallons per minute (GPM). Table 702.2(2) shows discharge capacities in gallons per minute computed to fixture units up to 50 gpm. This table could be used for fixtures or appliances not listed in Table 702.1 that have a known discharge rate. For example, there are many types and sizes of commercial dishwashers, glassware washers, pots and pans washers, etc that are not listed in Table 702.1. Fixture unit values may be determined from the discharge rate (gpm) of the appliance. Notice that Table 702.2(2) states "For Intermittent Flow Only."

Discharges from pumps are not applicable to this table because the flow is not derived from fixtures or appliances discharging immediately into the drainage system. The discharges are first collected into a sump and when reaching the sump's capacity, a pump produces a continuous discharge flow into the drainage system. There is no element of predictability when the pump will be on or off during a period of time. Also, the rates of discharge of most pumps exceed Table 702.2(2). These discharges are called continuous and fall under the requirements of Section 702.3. The fixture unit values in Table 702.1 are applied in Table 703.2 by taking into account intermittent use and the probability of the fixture being on or off. For flows over 50 gpm, the Authority Having Jurisdiction (AHJ) will determine the fixture unit value of the fixture.

702.3 Continuous Flow. For a continuous flow into a drainage system, such as from a pump, sump ejector, air conditioning equipment, or similar device, 2 fixture units shall be equal to each gallon per minute (gpm) (L/s) of flow.

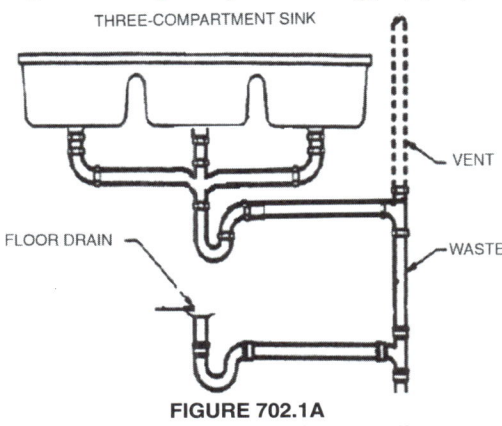

FIGURE 702.1A
EMERGENCY-USE FLOOR DRAIN

FIGURE 702.1B
RECEPTOR (FLOOR SINK)

FIGURE 702.1C
KITCHEN SINK ROUGH IN - 1 1/2 INCH TRAP ARM TO 2 INCH DRAIN

SANITARY DRAINAGE

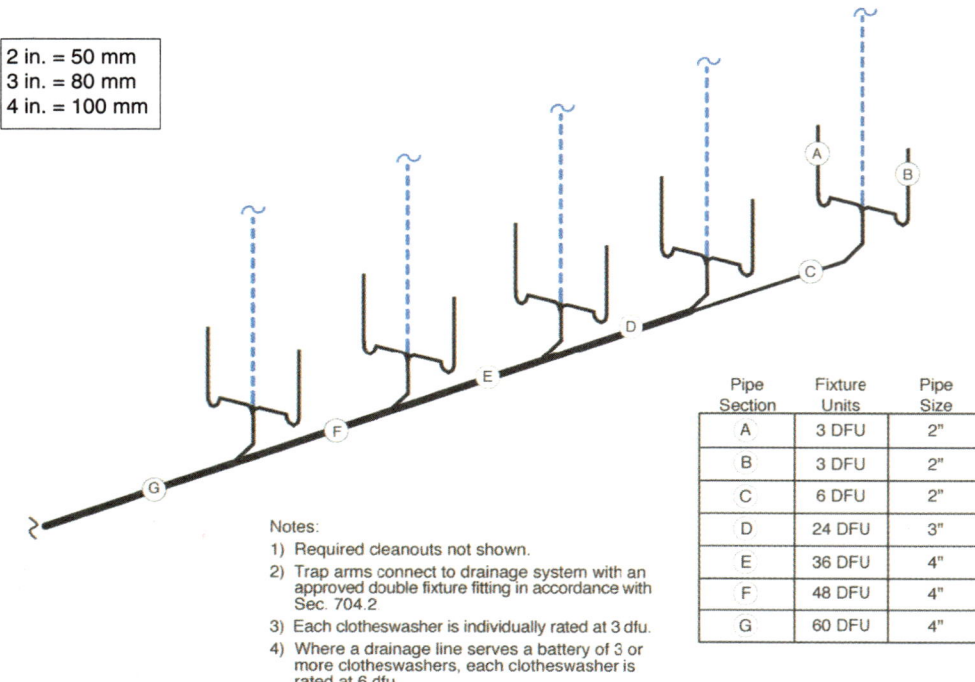

FIGURE 702.1D
ILLUSTRATION OF CLOTHES WASHER SIZING PER NOTE 5 OF TABLE 702.1

Continuous flows do not derive from individual plumbing fixtures discharging immediately into the drainage system, and are not subject to the predictability of being on or off as discussed in Section 702.2. Continuous flows will typically have higher discharge rates than most plumbing fixtures. This type of flow, such as from a sewage discharge pump, has a greater load on the system than does an intermittent flow from a plumbing fixture since the pump flows continuous for a longer period of time at a higher discharge flow rates.

The flow rate from the continuously flowing pump or device is assigned two fixture units for each gallon per minute of discharge. This provision was first added to the UPC in 1952 (albeit in another section on Drainage Below Curb), which corresponds with the *Plumbing Manual* (BMS66) published by the National Bureau of Standards in 1940. The *Plumbing Manual* (p.15) assigned 50 fixture units to sewage ejector or sump pumps for each 25 gallons per minute (the same as two fixture units per gallon per minute). Pump sizes were typically incremented in discharge flow rates of approximately 25 gpm. Possibly, the following two steps may have been used to derive the 50 fixture units per 25 gpm. First, if 25 gpm was converted to fixture units by dividing 25 by 7.5 (1 fixture unit equal to 7.5 gpm), then the result is 3.3 fixture units. Secondly, if the duration of the pump was contrasted to the duration of a toilet flush, which approximated six seconds, and if the duration of a pump was considered 90 seconds during peak loads, then the duration of the pump is 15 times longer than a toilet flush. The factor of 15 multiplied to the 3.3 fixture units renders a total of 49.5 fixture units, or 50 fixture units rounded. So for an ejector pump with a discharge capacity of 75 gpm, divide 75 by 7.5 and multiply by the factor of 15. The result is 150 fixture units, or two fixture units per gallon per minute. The installation of sump pumps is discussed further in Section 710.0.

703.0 Size of Drainage Piping.

703.1 Minimum Size. The minimum sizes of vertical, horizontal, or both drainage piping shall be determined from the total of fixture units connected thereto, and additionally, in the case of vertical drainage pipes, in accordance with their length.

703.2 Maximum Number of Fixture Units. Table 703.2 shows the maximum number of fixture units allowed on a vertical or horizontal drainage pipe, building drain, or building sewer of a given size; the maximum number of fixture units allowed on a branch interval of a given size; and the maximum length (in feet and meters) of a vertical drainage pipe of a given size.

Drainage sizing is accomplished by these steps:

1. Assign the proper fixture unit value, trap and trap arm sizes to each fixture.

2. Assign the proper fixture unit value to each pipe segment of the drainage system for the total number of fixtures served by that segment. Always start at the highest part of the drainage system and work down to the end of the system, totaling each pipe segment along the way.

3. Assign the proper pipe size to each segment as determined by Table 703.2.

Table 703.2 contains the maximum drainage fixture

SANITARY DRAINAGE

TABLE 703.2
MAXIMUM UNIT LOADING AND MAXIMUM LENGTH OF DRAINAGE AND VENT PIPING

SIZE OF PIPE (inches)	1¼	1½	2	3	4	5	6	8	10	12
Maximum Units Drainage Piping[1]										
Vertical	1	2[2]	16[3]	48[4]	256	600	1380	3600	5600	8400
Horizontal	1	1	8[3]	35[4]	216[5]	428[5]	720[5]	2640[5]	4680[5]	8200[5]
Maximum Length Drainage Piping										
Vertical, (feet)	45	65	85	212	300	390	510	750	—	—
Horizontal (unlimited)										
Vent Piping Horizontal and Vertical[6]										
Maximum Units	1	8[3]	24	84	256	600	1380	3600	—	—
Maximum Lengths, (feet)	45	60	120	212	300	390	510	750		

For SI units: 1 inch = 25 mm, 1 foot = 304.8 mm

Notes:
[1] Excluding trap arm.
[2] Except sinks, urinals, and dishwashers – exceeding 1 fixture unit.
[3] Except six-unit traps or water closets.
[4] Only four water closets or six-unit traps allowed on a vertical pipe or stack; and not to exceed three water closets or six-unit traps on a horizontal branch or drain.
[5] Based on ¼ inch per foot (20.8 mm/m) slope. For ⅛ of an inch per foot (10.4 mm/m) slope, multiply horizontal fixture units by a factor of 0.8.
[6] The diameter of an individual vent shall be not less than 1¼ inches (32 mm) nor less than one-half the diameter of the drain to which it is connected. Fixture unit load values for drainage and vent piping shall be computed from Table 702.1 and Table 702.2(2). Not to exceed one-third of the total permitted length of a vent shall be permitted to be installed in a horizontal position. Where vents are increased one pipe size for their entire length, the maximum length limitations specified in this table do not apply. This table is in accordance with the requirements of Section 901.3.

unit value per pipe size for horizontal and vertical drainage piping as well as for vent piping. Here we shall only discuss sizing for vertical and drainage piping. See Chapter 9 for discussion on using Table 703.2 for vent pipe sizing.

Table 703.2 is a very simplified table based on the research efforts of the National Bureau of Standards originally published in 1923 called *Recommended Minimum Requirements for Plumbing* (BH13). Many of the values for pipe sizes three-inch and greater still reflect the original recommendations of BH13. For example, the maximum units and the maximum lengths for vertical drainage piping are extracted from Table 14 and Table 16 respectively found in BH13. The maximum units for horizontal drainage piping was originally extracted from Table 15 in BH13, but have been modified since 1973.

The difference in maximum fixture unit values between vertical and horizontal drainage piping is because of the differing flow characteristics between the two. A vertical pipe has fifteen to eighteen times the carrying capacity of a horizontal pipe laid with a one-half inch per foot slope. However, the carrying capacity of a vertical stack is limited to one-third of the flow that it can carry full at maximum velocity. Since the controlling of air pressures is a greater problem in vertical pipes than in horizontal sloping pipes, and because of the greater velocities of flow attained in vertical pipes, this limitation allows for the separation and circulation of air. Even with this limitation, the carrying capacity of vertical pipes is still moderately greater than horizontal pipes. The maximum fixture unit values in Table 703.2 for vertical pipes are the probable demand loads that will not exceed this one-third limitation.

The carrying capacity of horizontal pipes varies with slope. As the slope decreases, so also does the velocity of the flow in the pipe. This indicates that capacity is directly related to velocity. There are two common pipe flow formulas that relate velocity to capacity (in gallons per minute), namely the Darcy and Manning's formula. Using the right coefficients, both formulas have approximate results. So for example, using Darcy's formula, a cast-iron four-inch horizontal pipe flowing full with a slope of one-fourth of an inch per foot has a capacity of 110 gpm with a velocity of 2.8 feet per second. Changing the slope to one-eighth of an inch per foot the capacity is reduced to 77 gpm with a velocity of 1.98 feet per second.

Notice that the maximum length for horizontal drainage piping is unlimited. The effect of gravity alone upon the hydraulic gradient of a sloping pipe will result in uniform flow at a constant velocity as the length increases. Therefore, there is no limitation to the length of horizontal drains. However, there is a length limitation for vertical drainage piping. Length limitation is not based upon excessive velocity precautions. In fact, experiments have determined that terminal velocity occurs in a relatively short distance based upon the maximum allowable fixture unit loading for vertical pipes. Due to frictional resistance within the pipe, acceleration does not increase continually, but will terminate, allowing velocity to remain constant. For example, a three-inch vertical pipe flowing with a capacity of 100 gpm will reach a terminal velocity of 16.8 feet per second at approximately 30 feet. So whether the length is

30 feet or 100 feet, the velocity will not exceed 16.8 feet per second. Therefore, there is no length limitation based upon velocity precautions (see also discussion in Section 907.1).

Limits for lengths of vertical pipes were set on the assumption that a certain amount of sewer ventilation through the stack is necessary. These limits were based on the recommendations in BH13 where it was determined that the venting capabilities of a three-inch vent stack in a two-pipe system (where a vent stack parallels a soil stack) limited the height of the system to 150 feet given the maximum allowable fixture unit loading for the soil stack. Height limitations were determined for vertical drainage pipes of other diameters having equivalent venting capabilities to allow sewer ventilation. The requirements for sewer ventilation are discussed further in Chapter 9.

The notes at the bottom of Table 703.2 provide other requirements for horizontal and vertical drainage piping.

1. Trap and trap arms are not defined as drainage piping. Trap arm sizes are sized by Table 702.1.

2. Most fixtures with 1½-inch traps and an assigned value of 2 fixture units by Table 702.1 will usually have a 1½-inch vertical drainage pipe. The exception is for sinks, dishwashers and urinals, which are required to have a minimum size drainage pipe of two inches in diameter per Note 2 in Table 702.1 (see **Figure 702.1c**).

3. Note 3 refers to 2-inch drainage piping. Although the maximum unit loading would seem to allow six-unit traps [a 3-inch trap and trap arm according to Table 702.2(a)] or a toilet to drain into a 2-inch pipe, the flow of these larger output fixtures are too great for this size pipe. Six-unit traps and toilets will require a 3-inch drain line.

4. Due to the volume of flow in water closets or six-unit traps, a 3-inch vertical pipe is limited to four water closets or four six-unit traps, and a 3-inch horizontal pipe is limited to three water closets or three six-unit traps. This would also require an increase in pipe size if there was an installation that had four water closets on a three-inch vertical stack, which then had to turn horizontally. The horizontal drainage pipe could not be a 3-inch pipe. The pipe would have to increase in size to 4 inches. See **Figure 703.2a**.

5. Table 703.2 is assumed to be based on drainage piping having a slope of one-fourth of an inch per foot. The flows in piping installed at one-eighth of an inch per foot will have less velocity, which reduces the carrying capacity as mentioned above. Therefore the maximum unit loading must be adjusted accordingly. This is accomplished by multiplying the maximum fixture units allowed for that pipe size by 0.8. The reduction allows only 80 percent of the maximum allowable fixture units at one fourth of an inch per foot slope. For example, a 4-inch horizontal drainage piped sloped at 1/4 of an inch per foot may have a total fixture unit load of 216 fixture units. The same pipe sloped at 1/8 inch per foot would have a total fixture unit value of 216 x 0.8 or 173 fixture units. This applies only to 4-inch and larger drainage piping. Three-inch and smaller pipe sizes may not be laid at 1/8 of an inch slope (see Section 708.1).

6. This note refers to vent sizing, which will be discussed in Chapter 9.

An example of drainage sizing is illustrated in **Figure 703.2b**.

703.3 Sizing per Appendix C. For alternate method of sizing drainage piping, see Appendix C.

704.0 Fixture Connections (Drainage).

704.1 Inlet Fittings. Drainage piping shall be provided with approved inlet fittings for fixture connections, correctly located according to the size and type of fixture proposed to be connected.

The fixture connection to the drainage piping is the transition from the trap arm to the drainage pipe. The approved fitting to receive drainage from fixtures is usually in the form of a sanitary tee and is referred to as a fixture fitting. The wye fitting or the combination wye and one-eighth bend fitting are not used for this purpose since the vent opening will be below the weir of the trap, which is prohibited in Section 1002.4. However, they may be used to connect a trap arm to the drainage pipe when the fixture has an integral trap, such as a water closet.

704.2 Single Vertical Drainage Pipe. Two fixtures set back-to-back, or side-by-side, within the distance allowed between a trap and its vent, shall be permitted to be served by a single vertical drainage pipe provided that each fixture wastes separately into an approved double-fixture fitting having inlet openings at the same level.

Grouping plumbing fixtures in certain areas of buildings can simplify the design of a plumbing system. In back-to-back and side-by-side installations, one drainage pipe serves two fixtures. The fixture fitting used is similar to the one used for single fixtures but with a dual opening. It is specially designed with more of an angle for the entry of flow into the barrel of the fitting. In this manner, the flow from one fixture cannot flow into the trap arm of the adjoining fixture. The proper term for this fitting is a "double-fixture fitting" (see **Figure 704.2a**). It also is sometimes referred to as a "Figure 5 fitting." This is due to the illustration number as it appears in the cast iron fitting catalog. **Figure 704.2b** indicates inlet fittings for single- and double-fixture fittings.

704.3 Commercial Sinks. Pot sinks, scullery sinks, dishwashing sinks, silverware sinks, and other similar fixtures shall be connected directly to the drainage system. A floor drain shall be provided adjacent to the fixture, and the fixture shall be connected on the sewer side of the floor drain trap, provided that no other drainage line is connected between the floor drain waste connection and the fixture drain. The fixture and floor drain shall be trapped and vented in accordance with this code.

Where direct connections exist for these types of kitchen equipment, a floor drain must be provided adjacent

SANITARY DRAINAGE

to the fixture and connected on the sewer side of the floor drain trap. No other drainage waste connection is permitted to connect between the floor drain and the fixture drain. This practice provides protection for the fixture drain of such sinks in the event of a blockage downstream of the drain. The floor drain provides a visual indication of stoppage since drainage will rise up onto the floor through the floor drain before ever contaminating the sink (see **Figure 704.3**).

705.0 Joints and Connections.

705.1 ABS and ABS Co-Extruded Plastic Pipe and Joints. Joining methods for ABS plastic pipe and fittings shall be installed in accordance with the manufacturer's installation instructions and shall comply with Section 705.1.1 through Section 705.1.3.

705.1.1 Mechanical Joints. Mechanical joints shall be designed to provide a permanent seal and shall be of the mechanical or push-on joint. The push-on joint shall include an elastomeric gasket that complies with ASTM D3212 and shall provide a compressive force against the spigot and socket after assembly to provide a permanent seal.

705.1.2 Solvent Cement Joints. Solvent cement joints for ABS pipe and fittings shall be clean from dirt and moisture. Pipe shall be cut square and shall be deburred. Where surfaces to be joined are cleaned, and free of dirt, moisture,

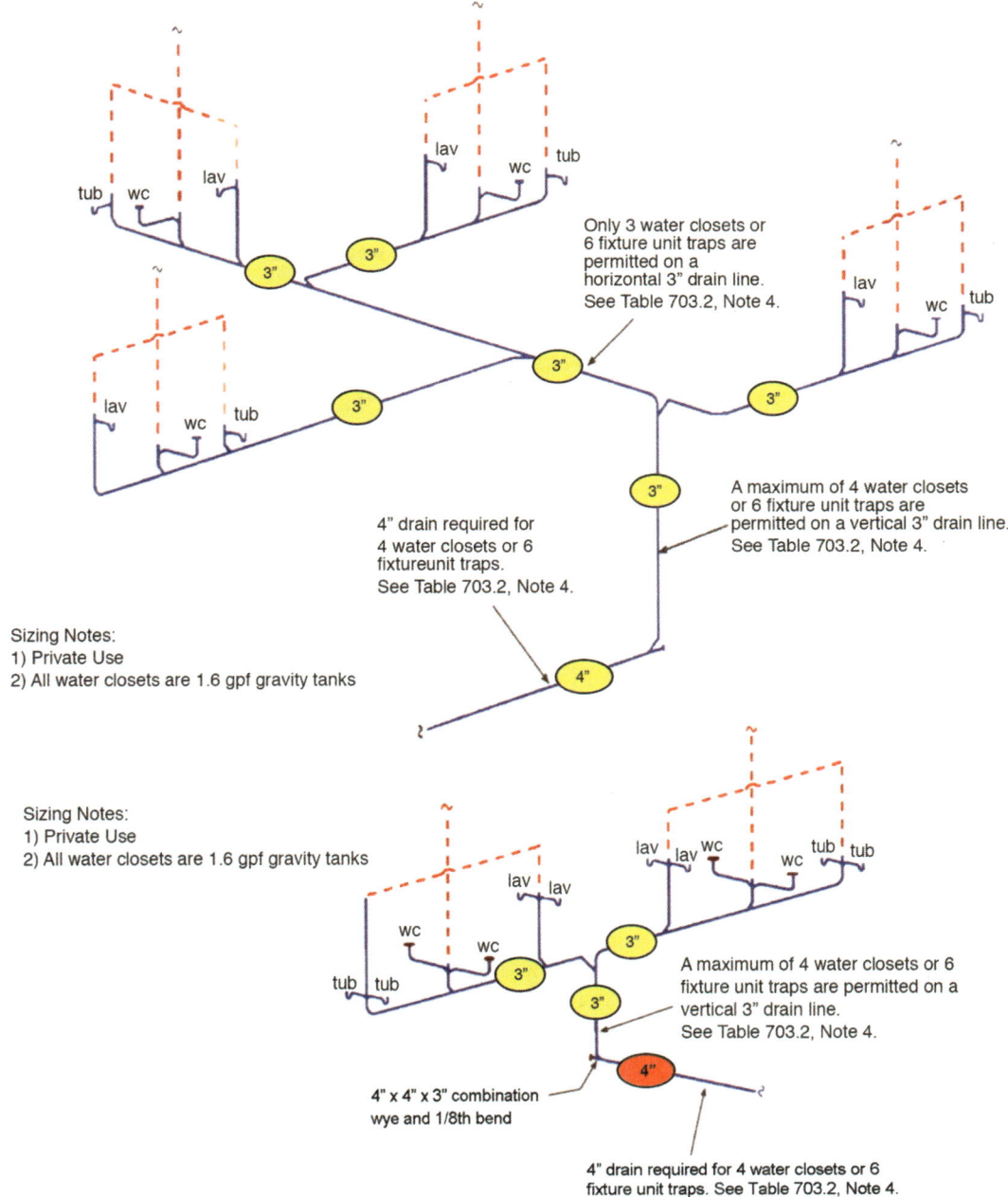

FIGURE 703.2A
ILLUSTRATION OF TABLE 703.2, NOTE 4 - WATER CLOSETS ON 3-INCH PIPE

SANITARY DRAINAGE

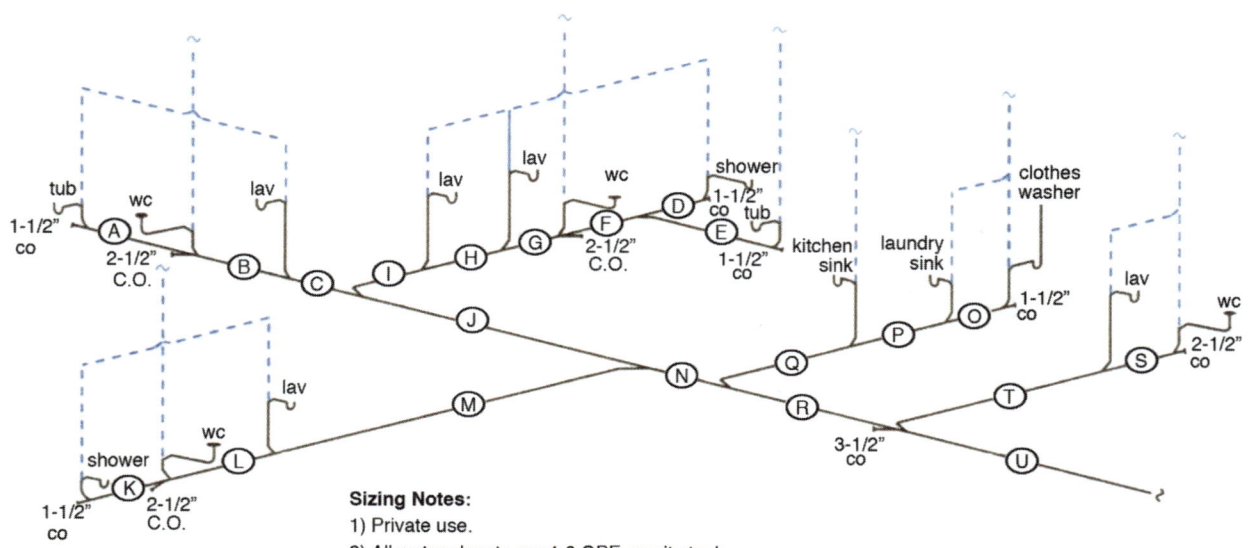

Sizing Notes:
1) Private use.
2) All water closets are 1.6 GPF gravity tank.
3) Horizontal drainage piping is below the floor level of the lowest floor.
4) All horizontal drainage pipe sections exceed 5 feet in length.
5) Cleanouts are sized in accordance with Table 707.1.
6) Section U serves 4 water closets and a total load of 32 drainage fixture units. Note 4 at the bottom of Table 703.2 limits a 3" horizontal drain to a maximum of 3 water closets or 6 fixture unit traps.

From Table 702.1

Fixture	DFU
Water Closet	= 3.0 dfu
Lavatory	= 1.0 dfu
Shower	= 2.0 dfu
Bathtub	= 2.0 dfu
Kitchen Sink	= 2.0 dfu
Laundry Sink	= 2.0 dfu
Clothes Washer	= 3.0 dfu

Pipe Section	Drainage Fixture Units	Pipe Size	Pipe Section	Drainage Fixture Units	Pipe Size
A	2.0	2"	L	5.0	3"
B	5.0	3"	M	6.0	3"
C	6.0	3"	N	21.0	3"
D	2.0	2"	O	3.0	2"
E	2.0	2"	P	5.0	2"
F	4.0	2"	Q	7.0	2"
G	7.0	3"	R	28.0	3"
H	8.0	3"	S	3.0	3"
I	9.0	3"	T	4.0	3"
J	15.0	3"	U	32.0	4"
K	2.0	2"			

Vertical Drainage Pipe Sizing
(not shown on drawing)

Water Closet	= 3"
Lavatory	= 1-1/4"
Shower	= 2"
Bathtub	= 1-1/2"
Kitchen Sink	= 2"
Laundry Sink	= 1-1/2"
Clothes Washer	= 2"

**FIGURE 703.2B
SIZING THE DRAINAGE SYSTEM**

oil, and other foreign material, the solvent cement that complies with ASTM D2235 shall be applied to all joint surfaces. Joints shall be made while both the inside socket surface and outside surface of pipe are wet with solvent cement. Hold joint in place and undisturbed for 1 minute after assembly.

705.1.3 Threaded Joints. Threads shall comply with ASME B1.20.1. A minimum of Schedule 80 shall be permitted to be threaded. Molded threads on adapter fittings for the transition to threaded joints shall be permitted. Thread sealant compound shall be applied to male threads, insoluble in water, and nontoxic. The joint between the pipe and transition fitting shall be of the solvent cement type. Caution shall be used during assembly to prevent over tightening of the ABS components once the thread sealant compound has been applied.

705.2 Cast-Iron Pipe and Joints. Joining methods for cast-iron pipe and fittings shall be installed in accordance with the manufacturer's installation instructions and shall comply with Section 705.2.1 or Section 705.2.2.

705.2.1 Caulked Joints. Caulked joints shall be firmly packed with oakum or hemp and filled with molten lead to a depth of not less than 1 inch (25.4 mm) in one continuous pour. The lead shall be caulked thoroughly at the inside and outside edges of the joint. After caulking, the finished joint shall not exceed 1/8 of an inch (3.2 mm) below the rim of the hub. No paint, varnish, or other coatings shall be permitted on the joining material until after the joint has been tested and approved.

See **Figure 705.2.1**.

705.2.2 Mechanical Joints and Compression Joints. Mechanical joints for cast-iron pipe and fittings shall be of the elastomeric compression type or mechanical joint couplings. Compression type joints with an elastomeric gasket for cast-iron hub and spigot pipe shall comply with ASTM C564 and be tested in accordance with ASTM C1563. Hub and spigot shall be clean and free of dirt, mud, sand, and foreign materials. Cut pipe shall be free from sharp edges. Fold and insert gasket into the hub. Lubricate the joint following manufacturer's instructions. Insert spigot into hub until the spigot end of the pipe bottom out in the hub. Use the same procedure for the installation of fittings.

A mechanical joint shielded coupling type for hubless cast-iron pipe and fittings shall have a metallic shield that complies with ASTM A1056, ASTM C1277, ASTM C1540, or CISPI 310. The elastomeric gasket shall comply with ASTM C564. Hubless cast-iron pipe and fittings shall be clean and free of dirt, mud, sand, and foreign materials. Cut pipe shall be free from sharp edges. Gasket shall be placed on the end of the pipe or fitting and the stainless steel shield and clamp assembly on the end of the other pipe or fitting. Pipe or fittings shall be seated against the center stop inside the elastomeric sleeve. Slide the stainless steel shield and clamp assembly into a position centered over the gasket and tighten. Bands shall be tightened using an approved calibrated torque wrench specifically set by the manufacturer of the couplings.

705.3 Copper or Copper Alloy Pipe (DWV) and Joints. Joining methods for copper or copper alloy pipe and fittings shall be installed in accordance with the manufacturer's installation instructions and shall comply with Section 705.3.1 through Section 705.3.4.

705.3.1 Brazed Joints. Brazed joints between copper or copper alloy pipe and fittings shall be made with brazing alloys having a liquid temperature above 1000°F (538°C). The joint surfaces to be brazed shall be cleaned bright by either manual or mechanical means. Piping shall be cut square and reamed to full inside diameter. Brazing flux shall be applied to the joint surfaces where required by manufacturer's recommendation. Brazing filler metal shall conform to AWS A5.8 and shall be applied at the point where the pipe or tubing enters the socket of the fitting.

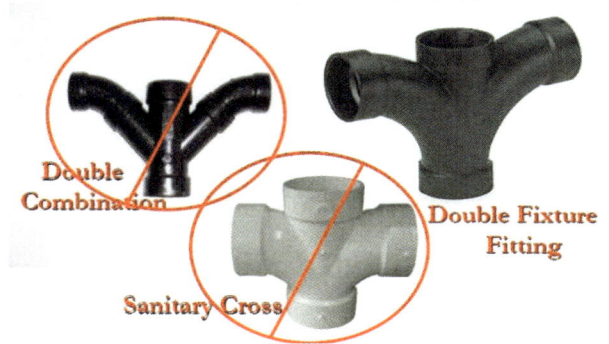

FIGURE 704.2A
DOUBLE FIXTURE FITTING

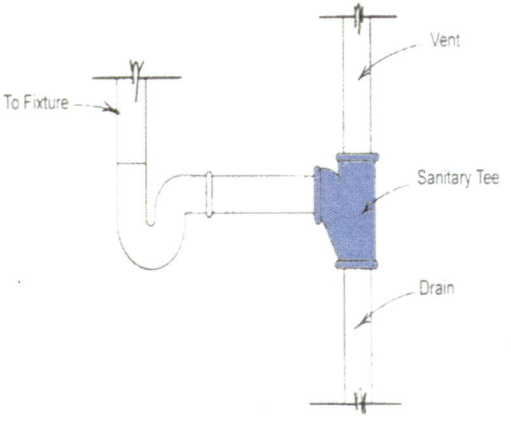

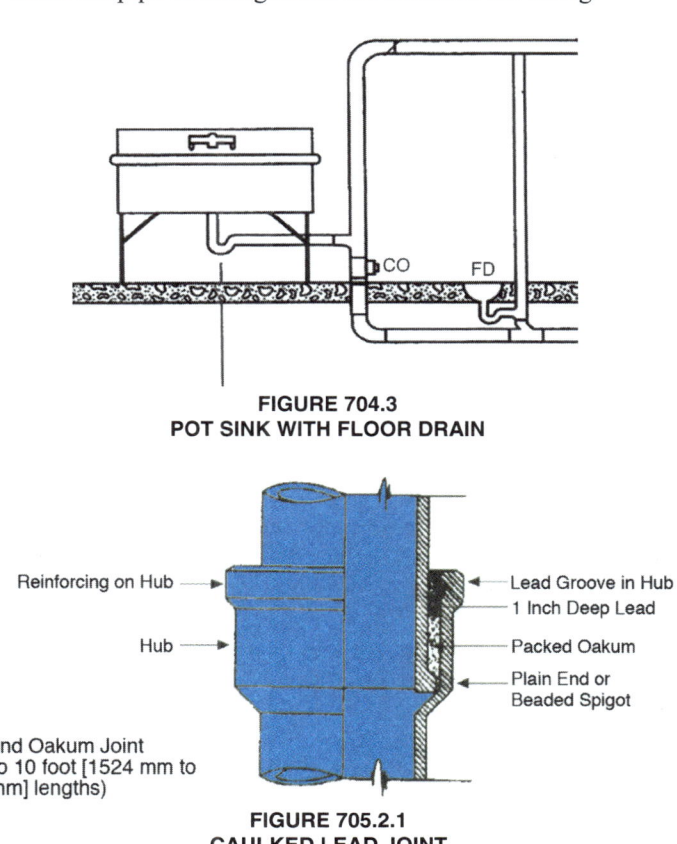

FIGURE 704.3
POT SINK WITH FLOOR DRAIN

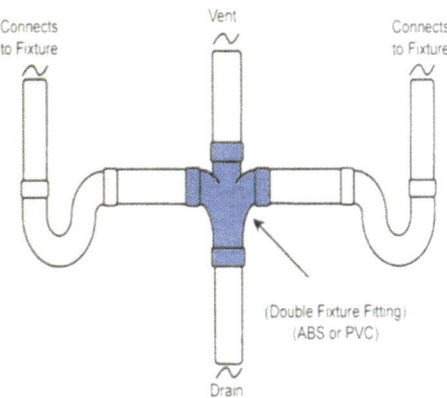

FIGURE 704.2B
SANITARY TEE AND DOUBLE FIXTURE FITTING

FIGURE 705.2.1
CAULKED LEAD JOINT

705.3.2 Mechanical Joints. Mechanical joints in copper or copper alloy piping shall be made with a mechanical coupling with grooved end piping or approved joint designed for the specific application.

705.3.3 Soldered Joints. Soldered joints between copper or copper alloy pipe and fittings shall be made in accordance with ASTM B828 with the following sequence of joint preparation and operation as follows: measuring and cutting, reaming, cleaning, fluxing, assembly and support, heating, applying the solder, cooling, and cleaning. Pipe shall be cut square and reamed to the full inside diameter including the removal of burrs on the outside of the pipe. Surfaces to be joined shall be cleaned bright by manual or mechanical means. Flux shall be applied to pipe and fittings and shall conform to ASTM B813, and shall become noncorrosive and nontoxic after soldering. Insert pipe into the base of the fitting and remove excess flux. Pipe and fitting shall be supported to ensure a uniform capillary space around the joint. Heat shall be applied using air or fuel torch with the flame perpendicular to the pipe using acetylene or an LP gas. Preheating shall depend on the size of the joint. The flame shall be moved to the fitting cup and alternate between the pipe and fitting. Solder conforming to ASTM B32 shall be applied to the joint surfaces until capillary action draws the molten solder into the cup. Joint surfaces shall not be disturbed until cool, and any remaining flux residue shall be cleaned.

705.3.4 Threaded Joints. Threaded joints for copper or copper alloy pipe shall be made with pipe threads that comply with ASME B1.20.1. Thread sealant tape or compound shall be applied only to male threads, and such material shall be approved types, insoluble in water, and nontoxic.

705.4 Galvanized Steel Pipe and Joints. Joining methods for galvanized steel pipe and fittings shall be installed in accordance with the manufacturer's installation instructions and shall comply with Section 705.4.1 or Section 705.4.2.

705.4.1 Mechanical Joints. Mechanical joints shall be made with an elastomeric gasket.

705.4.2 Threaded Joints. Threaded joints shall be made with pipe threads that comply with ASME B1.20.1. Thread sealant tape or compound shall be applied only to male threads, and such material shall be of approved types, insoluble in water, and nontoxic.

705.5 Polyethylene (PE) Sewer Pipe. Polyethylene (PE) sewer pipe or tubing and fitting joining methods shall be installed in accordance with the manufacturer's installation instructions and shall comply with Section 705.5.1 through Section 705.5.1.3.

705.5.1 Heat-Fusion Joints. Heat-fusion joints between PE sewer pipe or tubing and fittings shall be assembled in accordance with Section 705.5.1.1 through Section 705.5.1.3 using butt-fusion, electro-fusion, or socket-fusion heat methods. Do not disturb the joint until cooled to ambient temperature.

705.5.1.1 Butt-Fusion Joints. Butt-fusion joints for PE pipe shall be installed in accordance with ASTM F2620 and shall be made by heating the prepared ends of two pipes, pipe and fitting, or two fittings by holding ends against a heated element. The heated element shall be removed when the required melt or times are obtained and heated ends shall be placed together with applied force. Do not disturb the joint until cooled to ambient temperature.

705.5.1.2 Electro-Fusion Joints. Electro-fusion joints shall be heated internally by a conductor at the interface of the joint. Fittings shall comply with ASTM F1055 for the performance requirements of polyethylene electro-fusion fittings. The specified electro-fusion cycle used to form the joint requires consideration of the properties of the materials being joined, the design of the fitting being used, and the environmental conditions. Align and restrain fitting to pipe to prevent movement and apply electric current to the fitting. Turn off the current when the required time has elapsed to heat the joint. Do not disturb the joint until cooled to ambient temperature.

705.5.1.3 Socket-Fusion Joints. Socket fusion joints shall be installed in accordance with ASTM F2620 and shall be made by simultaneously heating the outside surface of a pipe end and the inside of a fitting socket. Where the required melt is obtained, the pipe and fitting shall be joined by inserting one into the other with applied force. Do not disturb the joint until cooled to ambient temperature.

705.6 PVC and PVC Co-Extruded Plastic Pipe and Joining Methods. Joining methods for PVC plastic pipe and fittings shall be installed in accordance with the manufacturer's installation instructions and shall comply with Section 705.6.1 through Section 705.6.3.

705.6.1 Mechanical Joints. Mechanical joints shall be designed to provide a permanent seal and shall be of the mechanical or push-on joint type. The push-on joint shall include an elastomeric gasket that complies with ASTM D3212 and shall provide a compressive force against the spigot and socket after assembly to provide a permanent seal.

705.6.2 Solvent Cement Joints. Solvent cement joints for PVC pipe and fittings shall be clean from dirt and moisture. Pipe shall be cut square, and pipe shall be deburred. Where surfaces to be joined are cleaned and free of dirt, moisture, oil, and other foreign material, apply primer purple in color that complies with ASTM F656. Primer shall be applied to the surface of the pipe and fitting is softened. Solvent cement that comply with ASTM D2564 shall be applied to all joint surfaces. Joints shall be made while both the inside socket surface and outside surface of pipe are wet with solvent cement. Hold joint in place and undisturbed for 1 minute after assembly.

705.6.3 Threaded Joints. Threads shall comply with ASME B1.20.1. A minimum of Schedule 80 shall be permitted to be threaded. Molded threads on adapter fittings for the transition to threaded joints shall be permitted. Thread sealant compound that is compatible with the pipe and fitting, insoluble in water and nontoxic shall be applied to male threads. The joint between the pipe and transition fitting shall be of the solvent cement type. Caution shall be used during assembly to prevent over tightening of the PVC components once the thread sealant has been applied. Female PVC threaded fittings shall be used with plastic male threads only.

705.7 Stainless Steel Pipe and Joints. Joining methods for stainless steel pipe and fittings shall be installed in accordance with the manufacturer's installation instructions and shall comply with Section 705.7.1 or Section 705.7.2.

705.7.1 Mechanical Joints. Mechanical joints between stainless steel pipe and fittings shall be of the compression, grooved coupling, hydraulic press-connect fittings, or flanged.

705.7.2 Welded Joints. Welded joints between stainless steel pipe and fittings shall be made in accordance with ASME A112.3.1 and shall be welded autogenously. Pipe shall be cleaned, free of scale and contaminating particles. Pipe shall be cut with a combination cutting and beveling tool that provides a square cut, and free of burrs. Mineral oil lubricant shall be used during the cutting and beveling process.

705.8 Vitrified Clay Pipe and Joints. Joining methods for vitrified clay pipe and fittings shall be installed in accordance with the manufacturer's installation instructions and shall comply with Section 705.8.1.

705.8.1 Mechanical Joints. Mechanical joints shall be designed to provide a permanent seal and shall be of the mechanical or push-on joint type. The push-on joint shall include an elastomeric gasket that complies with ASTM C425 and shall provide a compressive force against the spigot and socket after assembly to provide a permanent seal.

705.9 Special Joints. Special joints shall comply with Section 705.9.1 through Section 705.9.4.

705.9.1 Slip Joints. In fixture drains and traps, slip joints of approved materials shall be permitted to be used in accordance with their approvals.

A drainage slip joint is made of three pieces: a friction ring, a gasket and a slip joint nut. The gasket is made of a compressible pliable fiber, rubber or plastic material. The nut is tightened on the fitting and the gasket is compressed around the pipe being joined by the slip joint (see **Figure 705.9.1**). The use of slip joints is restricted to accessible or exposed locations (see Section 402.10). In waste piping, their use is permissible on the inlet side of the fixture trap. One slip joint on the outlet side of the fixture trap is permitted (see Section 1003.2). Ground joints and ferrule type connections are not considered slip joints (Section 705.9.3).

705.9.2 Expansion Joints. Expansion joints shall be accessible, except where in vent piping or drainage stacks, and shall be permitted to be used where necessary to provide for expansion and contraction of the pipes.

By its very name, it is easy to visualize the use of this joint. All piping materials are subject to expansion and contraction to varying degrees. Expansion joints must be

**FIGURE 705.9.1
DRAINAGE SLIP JOINTS**

compatible with the pipe and installed in accordance with the manufacturer's installation instructions. Many manufacturers require expansion joints due to the change in temperature (either water or air), length of run and piping material.

705.9.3 Ground Joint, Flared, or Ferrule Connections. Copper or copper alloy ground joint flared, or ferrule-type connections that allow adjustment of tubing, but provide a rigid joint where made up, shall not be considered as slip joints.

The ferrule connection is similar to the slip joint, except that instead of the flexible compressive ring, a machine brass ferrule, which matches the chamfered inner edge of the fitting, is used. This provides a rigid water-tight joint that is not considered a slip joint and therefore does not require accessibility.

705.9.4 Transition Joint. A solvent cement transition joint between ABS and PVC building drain and building sewer shall be made using listed transition solvent cement in accordance with ASTM D3138.

705.10 Joints Between Various Materials. Joints between various materials shall be installed in accordance with the manufacturer's installation instructions and with Section 705.10.1 through Section 705.10.4. Mechanical couplings used to join different materials shall comply with ASTM C1173 for belowground use, ASTM C1460 for aboveground use, or ASTM C1461 for aboveground and belowground use.

705.10.1 Copper or Copper Alloy Pipe to Cast-Iron Pipe. Joints from copper or copper alloy pipe or tubing to cast-iron pipe shall be made with a listed compression-type joint or copper alloy ferrule. The copper or copper alloy pipe or tubing shall be soldered or brazed to the ferrule, and the ferrule shall be joined to the cast-iron hub by a compression or caulked joint.

705.10.2 Copper or Copper Alloy Pipe to Threaded Pipe Joints. Joints from copper or copper alloy pipe or tubing to threaded pipe shall be made by the use of a listed copper alloy adapter or dielectric fitting. The joint between the copper or copper alloy pipe and the fitting shall be a soldered or brazed, and the connection between the

SANITARY DRAINAGE

threaded and the fittings shall be made with a standard pipe size threaded joint.

705.10.3 Plastic Pipe to Other Materials. Where connecting plastic pipe to other types of plastic or other types of piping material; approved listed adapter or transition fittings and listed for the specific transition intended shall be used.

705.10.4 Stainless Steel Pipe to Other Materials. Where connecting stainless steel pipe to other types of piping, listed mechanical joints of the compression type and listed for the specific transition intended shall be used.

706.0 Changes in Direction of Drainage Flow.

706.1 Approved Fittings. Changes in the direction of drainage piping shall be made by the appropriate use of approved fittings and shall be of the angles presented by a one-sixteenth bend, one-eighth bend, or one-sixth bend, or other approved fittings of equivalent sweep.

🔧 In a drainage system, the flow of liquid waste and sewage is constantly changing direction on its way to the public sewer or private sewage disposal system. Due to the laws of gravity, the flow in vertical piping will be much faster than the flow in horizontal piping of the same size. For efficient operation, it is necessary to make changes of direction in such a manner as to limit disruption of flow. There are provisions for three possible conditions:

1. Flow from a vertical pipe into a horizontal pipe.
2. Flow from a horizontal pipe into a vertical pipe.
3. Flow from a horizontal pipe into another horizontal pipe.

Figures 706.1a and **706.1b** identify the proper fittings to use in these conditions.

706.2 Horizontal to Vertical. Horizontal drainage lines, connecting with a vertical stack, shall enter through 45 degree (0.79 rad) wye branches, 60 degree (1.05 rad) wye branches, combination wye and one-eighth bend branches, sanitary tee or sanitary tapped tee branches, or other approved fittings of equivalent sweep. No fitting having more than one inlet at the same level shall be used unless such fitting is constructed so that the discharge from one inlet cannot readily enter any other inlet. Double sanitary tees shall be permitted to be used where the barrel of the fitting is not less than two pipe sizes larger than the largest inlet, (pipe sizes recognized for this purpose are 2 inches, 2½ inches, 3 inches, 3½ inches, 4 inches, 4½ inches, 5 inches, 6 inches, etc.) (50 mm, 65 mm, 80 mm, 90 mm, 100 mm, 115 mm, 125 mm, 150 mm, etc.).

🔧 A change in direction of drainage flows from horizontal to vertical occasions an increase of velocity, which also increases the carrying capacity of the pipe at the point of transition. Therefore, short pattern fittings such as sanitary tees and quarter bends are permitted. However, there is a difference between the carrying capacity of short pattern and long pattern fittings. Tests have determined that long pattern fittings used to change flow direction from horizontal to vertical, such as a wye or combination wye and one-eighth bend, have twice the capacity of short pattern fittings. The downward sweep of a long pattern fitting reduces obstruction of air space and presents a smoother transition to annular flow (see Chapter 9, Section 907.1 for further discussion on vertical stack flows and pressure transients).

For two horizontal branches connecting to the vertical stack at the same level, a double wye or a double combination wye and one-eighth bend should be used. The downward sweep of the wye pattern directs the flow to the pipe wall below the opposite outlet opening. A double sanitary tee would direct the flow into the outlet opening on the other side of the fitting (see **Figure 706.2**). However, a double sanitary tee could be used if the barrel of the sani-

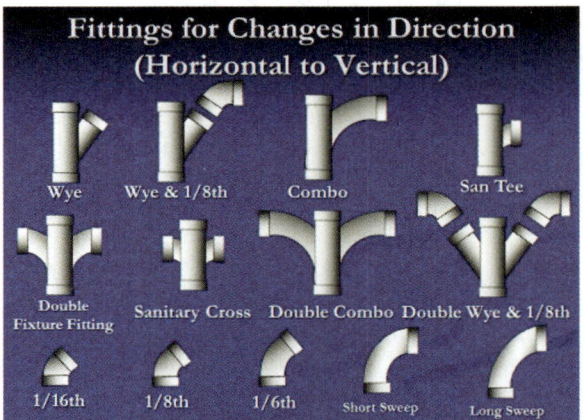

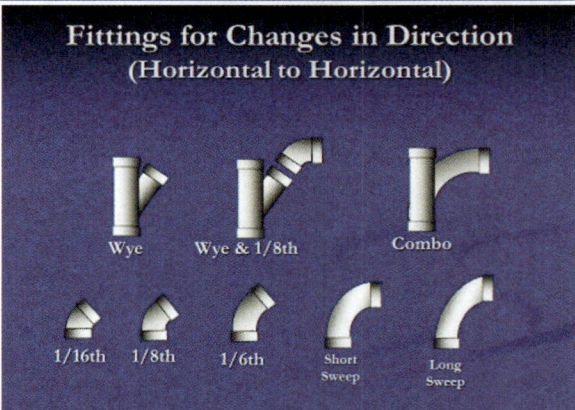

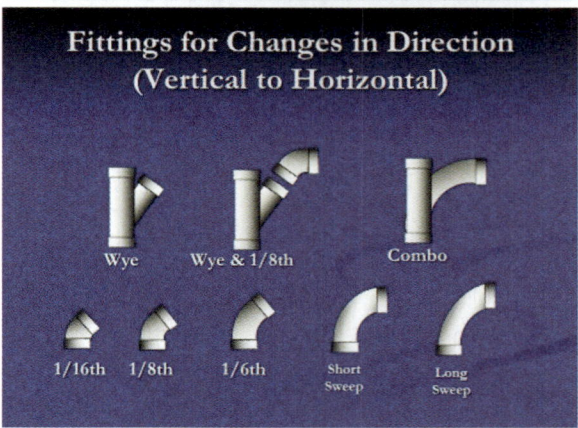

FIGURE 706.1A
FITTINGS FOR CHANGES IN DIRECTION OF FLOW

SANITARY DRAINAGE

Hubless Cast Iron	1-1/2"	2"	3"	4"	5"	6	Uses
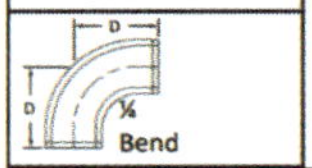 ¼ Bend	4-1/4"	4-1/2"	5"	5-1/2"	6-1/2"	7"	Minimum fitting for H to V Not for H to H or V to H
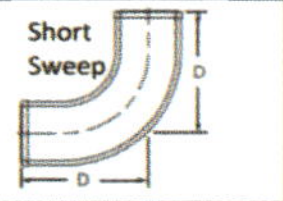 Short Sweep	*	6-1/2"	7"	7-1/2"	8-1/2"	9"	Minimum fitting for H to H Minimum fitting for V to H May be used anywhere
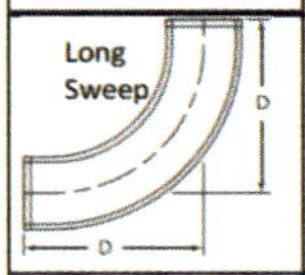 Long Sweep	9-1/4"	9-1/2"	10"	10-1/2"	11-1/2"	12"	Not required in UPC May be used anywhere

Service Cast Iron		2"	3"	4"	5"	6	
¼ Bend	X= D=	3-1/4" 6"	4" 7"	4-1/2" 8"	5" 8-1/2"	5-1/2" 9"	Minimum fitting for H to V Not for H to H or V to H
Short Sweep	X= D=	5-1/4" 8"	6" 9"	6-1/2" 10"	7" 10-1/2"	7-1/2" 11"	Minimum fitting for H to H Minimum fitting for V to H May be used anywhere
Long Sweep	X= D=	8-1/4" 11"	9" 12"	9-1/2" 13"	10" 13-1/2"	10-1/2" 14"	Not required in UPC May be used anywhere

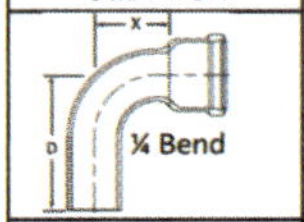

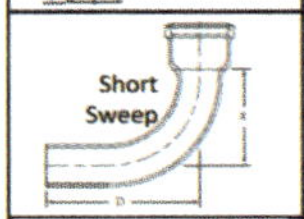

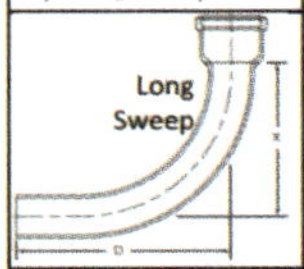

PVC/ABS	1-1/2"	2"	3"	4"			
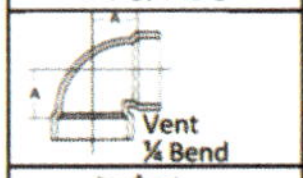 Vent ¼ Bend	1-3/16"	1-1/2"	1-7/8"				DWV Vents only
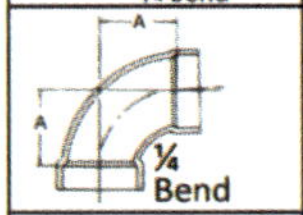 ¼ Bend	1-3/4"	2-15/16"	3-1/16"	3-7/8"			Minimum fitting for H to V Not for H to H or V to H
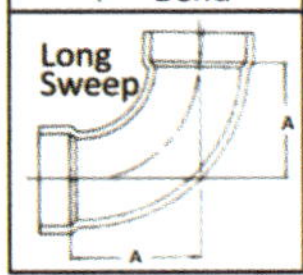 Long Sweep	2-3/4"	3-1/4"	4-1/16"	4-15/16"			Minimum fitting for H to H Minimum fitting for V to H May be used anywhere

DWV Copper	1-1/2"	2"	3"	4"			
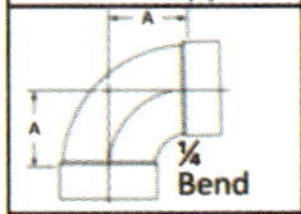 ¼ Bend	1-7/16"	1-15/16"	2-25/32"	3-25/32"			Minimum fitting for H to V Not for H to H or V to H
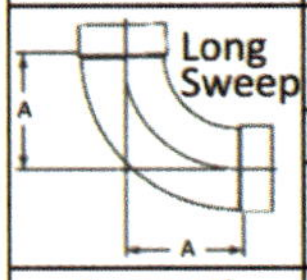 Long Sweep	2-5/16"	2-3/4"	3-1/16"	3-7/8"			Minimum fitting for H to H Minimum fitting for V to H May be used anywhere

Recessed Drainage Fitting	1-1/2"	2"	3"	4"			
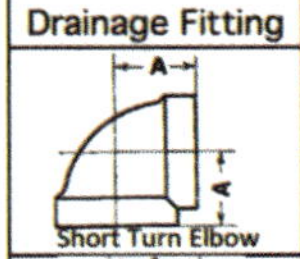 Short Turn Elbow	2"	2-5/8"	3-3/8"	3-15/16"			DWV Vents only
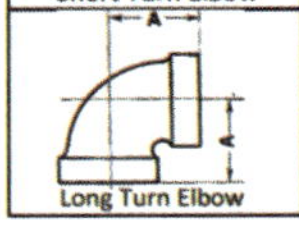 Long Turn Elbow	2-11/16"	2-7/16"	5-7/8"	6-3/4"			Minimum fitting for H to V Not for H to H or V to H

FIGURE 706.1B
DIMENSIONAL DRAWINGS OF FITTING PATTERNS

SANITARY DRAINAGE

tary tee is two pipe sizes larger than the largest of the two inlet pipes. The larger diameter of the stack allows a longer distance for the horizontal flow to arc across the diameter of the vertical pipe and to fall below the opposite outlet opening. Notice that one-half sizes are used for this purpose. Also remember that these are not fixture or trap arm connections. They are branch connections. Therefore, when Section 706.2 allows double sanitary tees to be installed, it is referencing branch-piping connections, not fixture fitting connections. For back-to-back or side-by-side fixture connections, see Section 704.2.

**FIGURE 706.2
DOUBLE SANITARY TEE**

706.3 Horizontal to Horizontal. Horizontal drainage lines connecting with other horizontal drainage lines shall enter through 45 degree (0.79 rad) wye branches, combination wye and one-eighth bend branches, or other approved fittings of equivalent sweep.

Unlike horizontal to vertical changes in direction, a horizontal to horizontal change of direction does not occasion an increase in velocity, but tends to obstruct and diminish velocity. Hence, horizontal drainage lines connecting with other horizontal drainage lines shall enter through 45-degree fittings with as smooth and wide a turn as possible. Any interruption to the flow by short or abrupt changes of direction will cause turbulence and reduce the velocity of that flow. The adverse effects that could follow are pipes filling upstream from the point of transition, increased pressure changes within the drainage system, and insufficient velocity to carry away solids.

706.4 Vertical to Horizontal. Vertical drainage lines connecting with horizontal drainage lines shall enter through 45 degree (0.79 rad) wye branches, combination wye and one-eighth bend branches, or other approved fittings of equivalent sweep. Branches or offsets of 60 degrees (1.05 rad) shall be permitted to be used where installed in a true vertical position.

One of the most critical changes of direction from vertical to horizontal flows in the drainage system is at the base of a stack. As the vertical flow makes a forced entry into a horizontal pipe, the flow becomes violently turbulent and abruptly increases in depth downstream the vertical stack causing what is called a "hydraulic jump" (see also commentary in Section 901.2). This jump can cause backpressures upstream of the jump and affect trap seals in close proximity to the base of the stack. The proper transitional fitting with the proper sweep will reduce the affect of the hydraulic jump on the system. These fittings will be with 45 degree angles (wyes and combination wye and one-eighth bend) or sweeps (long and short). The short sweep fitting is not to be confused with a quarter bend fitting, which has a short ninety-degree radial bend. The short sweep radius approximates the radius of two one-eighth bends put together.

707.0 Cleanouts.

707.1 Plug. Each cleanout fitting for cast-iron pipe shall consist of a cast-iron or copper alloy body and an approved plug. Each cleanout for galvanized wrought iron, galvanized steel, copper, or copper alloy pipe shall consist of a plug as specified in Table 707.1, or a standard weight copper alloy cap, or an approved ABS or PVC plastic plug, or an approved stainless steel cleanout or plug. Plugs shall have raised square heads or approved countersunk rectangular slots.

**TABLE 707.1
CLEANOUTS**

SIZE OF PIPE (inches)	SIZE OF CLEANOUT (inches)	THREADS (per inches)
1½	1½	11½
2	1½	11½
2½	2½	8
3	2½	8
4 & larger	3½	8

For SI units: 1 inch = 25 mm

Horizontal drainage piping is required to have cleanouts at specific locations. In the event of a stoppage, a point of entry is provided where special tools may be used to remove stoppages.

The size of the cleanout plug, not the body of the cleanout fitting, is stated in Table 707.1. The smallest cleanout plug is 1½ inches (40 mm). The largest required is 3½ inches (90 mm). For large drainage piping, this is still the largest opening required. This is because most of the equipment used for clearing stoppages can fit through this opening (see **Figure 707.1**).

**FIGURE 707.1
MINIMUM SIZE CLEANOUT - 3 1/2 INCHES**

SANITARY DRAINAGE

707.2 Approved. Each cleanout fitting and each cleanout plug or cap shall be of an approved type.

707.3 Watertight and Gastight. Cleanouts shall be designed to be watertight and gastight.

707.4 Location. Each horizontal drainage pipe shall be provided with a cleanout at its upper terminal, and each run of piping, that is more than 100 feet (30 480 mm) in total developed length, shall be provided with a cleanout for each 100 feet (30 480 mm), or fraction thereof, in length of such piping. An additional cleanout shall be provided in a drainage line for each aggregate horizontal change in direction exceeding 135 degrees (2.36 rad). A cleanout shall be installed above the fixture connection fitting, serving each urinal, regardless of the location of the urinal in the building.

Exceptions:

(1) Cleanouts shall be permitted to be omitted on a horizontal drain line less than 5 feet (1524 mm) in length unless such line is serving sinks or urinals.

(2) Cleanouts shall be permitted to be omitted on a horizontal drainage pipe installed on a slope of 72 degrees (1.26 rad) or less from the vertical angle (one-fifth bend).

(3) Excepting the building drain, its horizontal branches, and urinals, a cleanout shall not be required on a pipe or piping that is above the floor level of the lowest floor of the building.

(4) An approved type of two-way cleanout fitting, installed inside the building wall near the connection between the building drain and the building sewer or installed outside of a building at the lower end of a building drain and extended to grade, shall be permitted to be substituted for an upper terminal cleanout.

Cleanouts are required for horizontal drainage piping (see **Figure 707.4**). For long runs of piping, a cleanout is required every 100 feet. This measurement is taken from the highest end of the horizontal drainage piping to the point of connection with the building sewer. Installing a cleanout above the fixture connection fitting serving each urinal will provide access for mechanically cleaning the drain line without having to remove the urinal from the wall. Of course, there are exceptions to this requirement:

1. Except for sinks and urinals, a horizontal drain line less than five feet does not require a cleanout.

2. A direct projection from the building drain in a vertical direction does not require a cleanout. The angle of less than 72 degrees from the vertical does not meet the definition of "vertical" until it reaches a 45-degree angle, but this is referring to pipe with a significant slope from the drain.

3. This is a significant exception. A cleanout is only required on the building drain as it is defined in Chapter 2: Building Drain – "That part of lowest piping of a drainage system which receives the discharge from soil, waste and other drainage pipes inside the walls of the building and conveys it to the building sewer beginning two feet outside the building wall." The building drain and its horizontal branches, along with urinals regardless of their location in the building, are required to have cleanouts. Piping above the first floor of the building is not required to have a cleanout.

4. A two-way cleanout at the transition of the building drain and sewer can substitute for an upper terminal cleanout only if the distance to the furthest horizontal drainage piping is within 100 feet.

707.5 Cleaning. Each cleanout shall be installed so that it opens to allow cleaning in the direction of flow of the soil or waste or at right angles thereto and, except in the case of wye branch and end-of-line cleanouts, shall be installed vertically above the flow line of the pipe.

707.6 Extension. Each cleanout extension shall be considered as drainage piping and each 90 degree (1.57 rad) cleanout extension shall be extended from a wye-type fitting or other approved fitting of equivalent sweep.

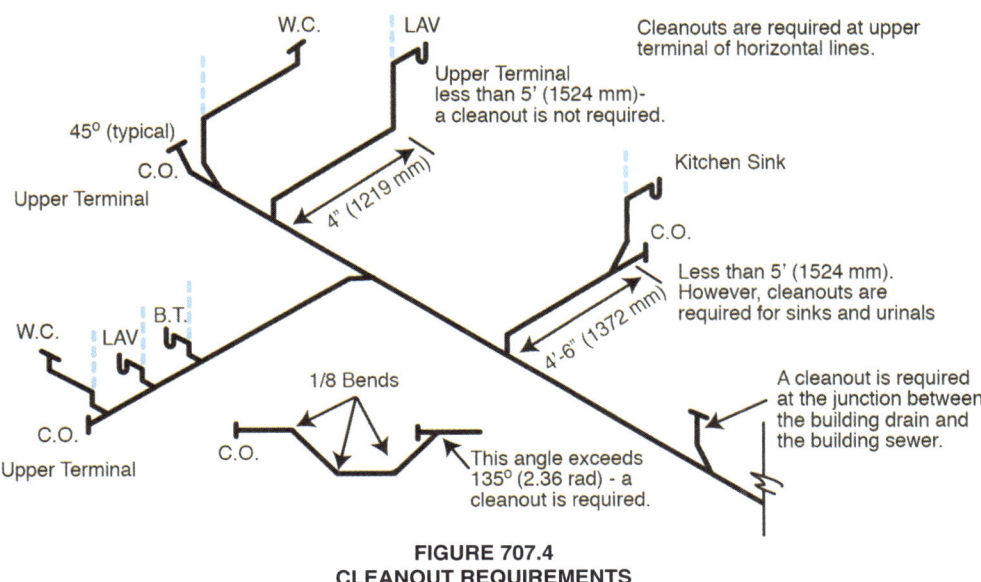

**FIGURE 707.4
CLEANOUT REQUIREMENTS**

SANITARY DRAINAGE

707.7 Interceptor. Each cleanout for an interceptor shall be outside of such interceptor.

707.8 Access. Each cleanout, unless installed under an approved cover plate, shall be above grade, readily accessible, and so located as to serve the purpose for which it is intended. Cleanouts located under cover plates shall be so installed as to provide the clearances and accessibility required by this section.

707.9 Clearance. Each cleanout in piping 2 inches (50 mm) or less in size shall be so installed that there is a clearance of not less than 18 inches (457 mm) by 18 inches (457 mm) in front of the cleanout. Cleanouts in piping exceeding 2 inches (50 mm) shall have a clearance of not less than 24 inches (610 mm) by 24 inches (610 mm) in front of the cleanout. Cleanouts in under-floor piping shall be extended to or above the finished floor or shall be extended outside the building where there is less than 18 inches (457 mm) vertical overall, allowing for obstructions such as ducts, beams, and piping, and 30 inches of (762 mm) horizontal clearance from the means of access to such cleanout. No under-floor cleanout shall be located exceeding 5 feet (1524 mm) from an access door, trap door, or crawl hole.

For practical reasons, a cleanout shall have sufficient clear space in front of the cleanout or around it so that it can be accessed to serve the purpose for which it is intended. The above clearances are for the benefit of the service plumber. They allow enough room for a repair to be completed without too much disruption to the occupants in the building and so the job can be done quickly and efficiently (see **Figure 707.9**).

707.10 Fittings. Cleanout fittings shall be not less in size than those given in Table 707.1.

707.11 Pressure Drainage Systems. Cleanouts shall be provided for pressure drainage systems as classified under Section 710.7.

707.12 Countersunk Cleanout Plugs. Countersunk cleanout plugs shall be installed where raised heads cause a hazard.

707.13 Hubless Blind Plugs. Where a hubless blind plug is used for a required cleanout, the complete coupling and plug shall be accessible for removal or replacement.

707.14 Trap Arms. Cleanouts for trap arms shall be installed in accordance with Section 1002.3.

708.0 Grade of Horizontal Drainage Piping.

708.1 General. Horizontal drainage piping shall be run in practical alignment and a uniform slope of not less than ¼ inch per foot (20.8 mm/m) or 2 percent toward the point of disposal provided that, where it is impractical due to the depth of the street sewer, to the structural features, or to the arrangement of a building or structure to obtain a slope of ¼ inch per foot (20.8 mm/m) or 2 percent, such pipe or piping 4 inches (100 mm) or larger in diameter shall be permitted to have a slope of not less than ⅛ inch per foot (10.4 mm/m) or 1 percent, where first approved by the Authority Having Jurisdiction.

Horizontal drains that depend on gravity alone for movement of flow require slope. The slope should be sufficient to produce a scouring effect in the pipe so that the drainage system essentially is self-cleansing. A scouring effect requires a velocity of at least two feet per second. For smaller size pipes, a minimum of two percent slope (one-fourth of an inch per foot) is required to produce a velocity of at least two feet per second. For larger size pipes (four-inch and larger), a velocity of two feet per second can be achieved with a one percent slope (one-eighth of an inch per foot). This corresponds with what is stated in Table 703.2 and its notes. Pipe sizes less than 4 inches may not be run at a one-eighth of an inch per foot slope. Pipe sizes four inches and larger may be run at a one-eighth of an inch per foot slope but only if job-site conditions preclude the installation of drainage piping at one-fourth of an inch per foot slope and if the AHJ approves. If the drainage piping is installed at one-eighth of an inch per foot slope, Note 5 of Table 703.2 must now be adhered to.

709.0 Gravity Drainage Required.

709.1 General. Where practicable, plumbing fixtures shall be drained to the public sewer or private sewage disposal system by gravity.

There are three methods or conditions that apply to the flow of drainage piping and the transport of the effluent to the public or private sewer or private sewage disposal system: 1) gravity drainage, 2) drainage of fixtures located below the level of the next upstream manhole, and 3) drainage of fixtures located below the level of the main sewer. Each of them has separate requirements for installation.

This section requires the drainage system shall only be by gravity drainage. The two other conditions mentioned above indicate where gravity drainage is not practicable. The previous sections of the code have given the requirements for gravity drainage – proper slope, the proper fittings to use in flow transitions and the proper joining methods for the pipe and fittings (see **Figure 709.1**). The following sections will give the requirements for the other two conditions.

710.0 Drainage of Fixtures Located Below the Next Upstream Manhole or Below the Main Sewer Level.

710.1 Backflow Protection. Fixtures installed on a floor level that is lower than the next upstream manhole cover of the public, or private sewer shall be protected from backflow of sewage by installing an approved type of backwater valve. Fixtures on such floor level that are not below the next upstream manhole cover shall not be required to be protected by a backwater valve. Fixtures on floor levels above such elevation shall not discharge through the backwater valve. Cleanouts for drains that pass through a backwater valve shall be clearly identified with a permanent label stating "backwater valve downstream."

In this situation, the flow of the drainage piping is still accomplished by gravity but some of the fixtures within

SANITARY DRAINAGE

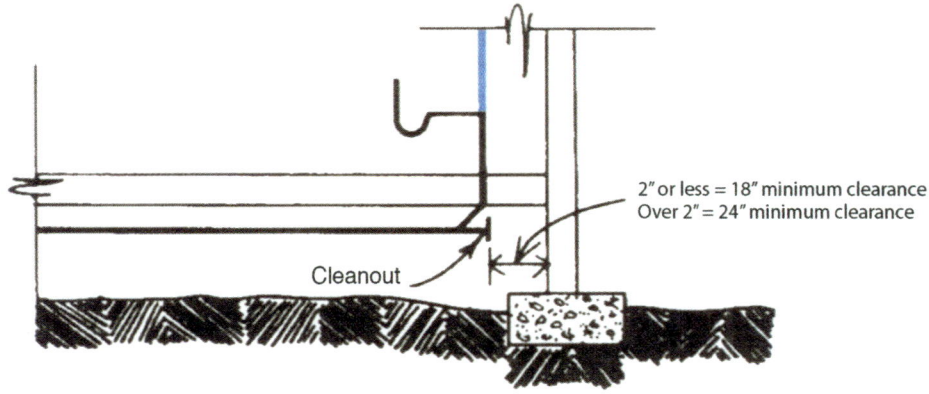

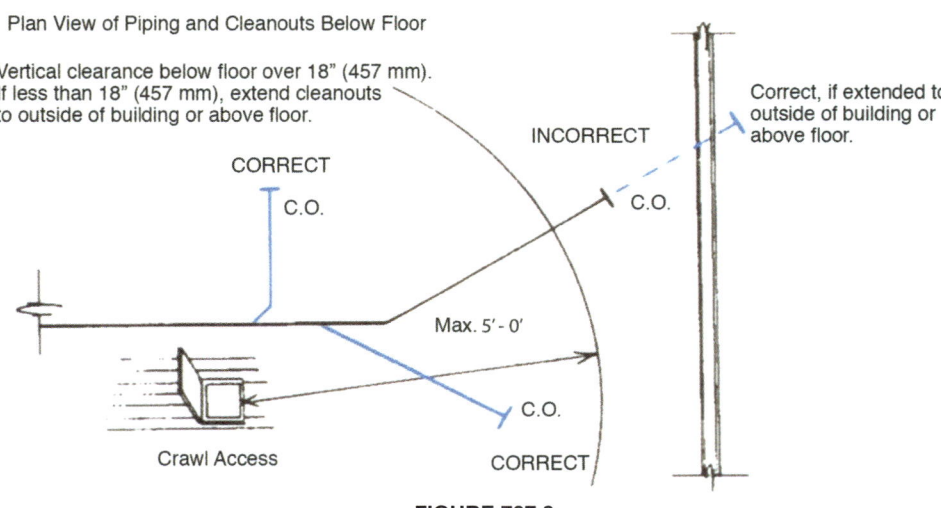

**FIGURE 707.9
CLEANOUT ACCESS**

the building have an elevation that is below the upstream manhole cover. This means that if the sewer becomes plugged or overloaded and backs up, then the effluent would back up into the fixtures below the elevation of the top of the upstream manhole and overflow into the building. In order to protect the occupants of the building and prevent the backflow of sewage or liquid waste, an approved-type backwater valve must be installed (see **Figure 710.1**).

Backwater valves shall be installed only in that branch or section of the drainage system that receives the discharge from fixtures installed on a floor level that is below the next upstream manhole. The waste from the fixtures located above that level is prohibited from flowing through the backwater valve. This is to provide an unobstructed path to the atmosphere for venting purposes. If all the waste flowed through the backwater valve flapper, which is normally shut, it would prevent the free flow of air within the system as required in Chapter 9.

710.2 Sewage Discharge. Drainage piping serving fixtures that are located below the crown level of the main sewer shall discharge into an approved watertight sump or receiving tank, so located as to receive the sewage or wastes by gravity. From such sump or receiving tank, the sewage or other liquid wastes shall be lifted and discharged into the building drain or building sewer by approved ejectors, pumps, or other equally efficient approved mechanical devices.

The effluent from piping and fixtures located below the crown level of the sewer (level of waste flow in the sewer pipe) must be pumped above and then drained into the gravity system (see **Figure 710.2**). The following sections provide the requirements for the installation of sewage pumps, sumps and receiving tanks, and sizing for drainage and vent.

710.3 Sewage Ejector and Pumps. A sewage ejector or sewage pump receiving the discharge of water closets or urinals:

(1) Shall have a discharge capacity of not less than 20 gpm (1.26 L/s).

(2) In single dwelling units, the ejector or pump shall be capable of passing an 1½ inch (38 mm) diameter solid ball, and the discharge piping of each ejector or pump shall have a backwater valve and gate valve, and be not less than 2 inches (50 mm) in diameter.

(3) In other than single-dwelling units, the ejector or pump

SANITARY DRAINAGE

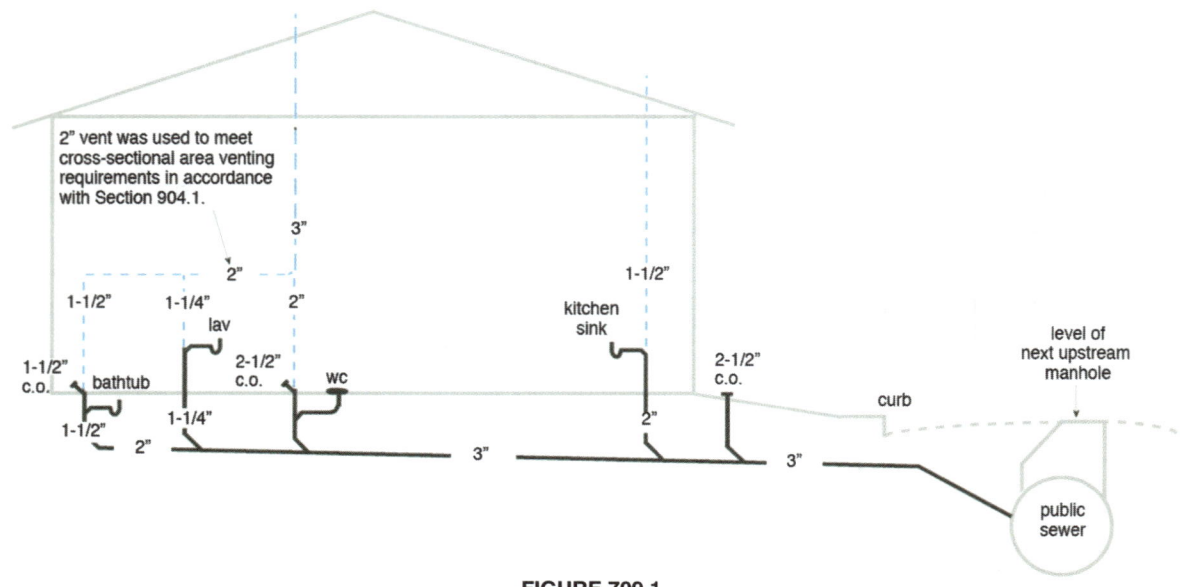

**FIGURE 709.1
GRAVITY FLOW TO SEWER**

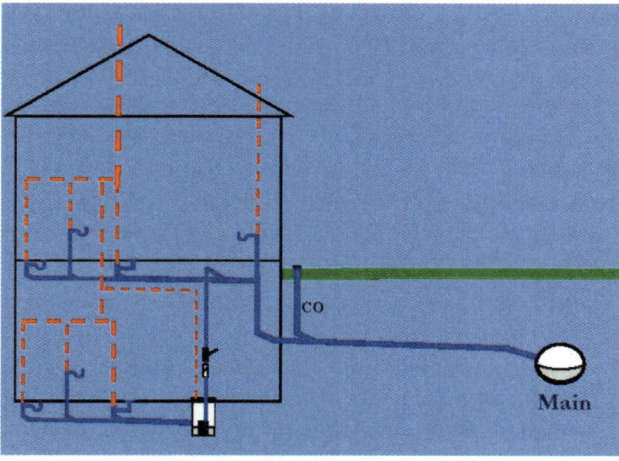

**FIGURE 710.1
BACKWATER VALVE**

**FIGURE 710.2
SEWAGE EJECTOR SYSTEM**

shall be capable of passing a 2 inch (51 mm) diameter solid ball, and the discharge piping of each ejector or pump shall have a backwater valve and gate valve, and be not less than 3 inches (80 mm) in diameter.

🔧 To design this pumped waste system and correctly size the sump pump depends on the flow rate of the fixtures flowing into the sump. This is usually accomplished by the plumbing design professional. The discharge capacity of the pump is determined by the pump's performance curve showing the relation between pumped gallons per minute and the total dynamic head in feet (see **Figure 710.3a**). The greater the head is in height, the less the performance is in pumped gallons per minute. Therefore, the minimum discharge capacity of 20 gpm depends on the head height. For example, if the performance curve shows a discharge capacity of 20 gpm at 15 feet of head, then any height exceeding 15 feet does not meet the minimum discharge capacity of the pump. There is also a difference in design requirements between single-dwelling units (private) with one pump and a discharge pipe size of two inches in diameter, and other occupancies (public) with two pumps and a discharge pipe size of three inches in diameter (see **Figure 710.3b**).

710.4 Discharge Line. The discharge line from such ejector, pump, or another mechanical device shall be of approved pressure rated material and be provided with an accessible backwater or swing check valve and gate or ball valve. Where the gravity drainage line to which such discharge line connects is horizontal, the method of connection shall be from the top through a wye branch fitting. The gate or ball valve shall be located on the discharge side of the backwater or check valve.

Gate or ball valves, where installed in drainage piping, shall be fullway type with working parts of corrosion-resistant metal. Sizes 4 inches (100 mm) or more in diameter

SANITARY DRAINAGE

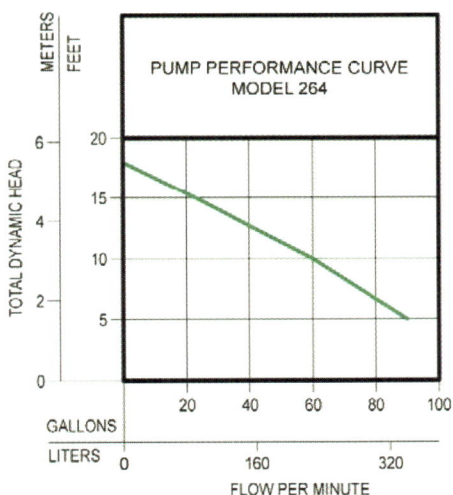

**FIGURE 710.3A
PUMP PERFORMANCE CURVE**
Reprinted with Permission from Zoeller Pump Co.

tional to the square of the flow rate ($F=KV^2$). With two pumps operating in parallel, the flow rate doubles, whereas the friction pressure loss is exponential. Therefore, the output discharge is a proportion of the combined flow rates. A system curve shows the relationship between flow rate and pressure loss and determines the flow rate when two pumps operate in parallel. **Figure 710.5** shows a system curve intersecting a single operating pump at 320 gpm and two pumps in parallel at 400 gpm.

In all cases, the load imposed at the point of pump discharge connection to the gravity system will be added to the existing gravity fixture unit load created by fixtures installed upstream from the point of connection.

The aggregate total will seldom represent the maximum fixture unit capacity of the pipe to which the pump system is being connected. Sizing tables that illustrate maximum capacity (fixture unit loading) for various pipe sizes reflect enormous incremental changes with each increase in pipe size. There would nearly always be surplus capacity to take up the excess load when both pumps go into action during an emergency condition. However, designers should increase the gravity pipe size if the drainage system loading appears to be near its capacity and when there is concern that activation of both pumps might pressurize the gravity system.

710.5 Size of Building Drains and Sewers. Building drains or building sewers receiving a discharge from a pump or ejector shall be adequately sized to prevent overloading. Two fixture units shall be allowed for each gallon per minute (L/s) of flow.

Discharges from a sewage pump system will eventually connect to the building's gravity drainage system. In order to prevent this high-volume, high-velocity discharge from overloading the gravity drain lines, it is necessary for a fixture unit rating to be established that provides for this "continuous flow" (see commentary in Section 702.3 for more discussion on continuous flow). Both this section and 702.3 stipulate that all pump discharges shall be assigned two fixture units for each gallon per minute of discharge. Therefore, a pump that is designed to have a 20 gpm discharge rate would impose a 40-fixture unit load on the gravity drain line at the point where the pump discharge piping is connected to the gravity system.

Where more than one pump is installed (as required in public-use buildings according to Section 710.9) the fixture unit loading is based upon the discharge from one pump rather than both due to redundancy. Section 710.9 requires an arrangement of alternate function between the two pumps and simultaneity will only occur in an overload condition where an alarm will sound. Hence, a properly designed system will seldom, if ever, have both pumps discharging.

In the event of an emergency condition that requires both pumps to operate simultaneously, the resulting combined discharge will normally be short-lived. The increased loading, resulting from the combined output of both pumps, will represent only a proportion of the combined discharge flow rates. The combined discharge will not represent an amount that is twice their separate capacity. This is because friction pressure loss is propor-

710.6 Backwater Valves. Backwater valves, gate valves, fullway ball valves, unions, motors, compressors, air tanks, and other mechanical devices required by this section shall be located where they will be accessible for inspection and repair and, unless continuously exposed, shall be enclosed in a masonry pit fitted with an adequately sized removable cover.

Backwater valves shall comply with ASME A112.14.1, and have bodies of cast-iron, plastic, copper alloy, or other approved materials; shall have noncorrosive bearings, seats, and self-aligning discs; and shall be constructed to ensure a positive mechanical seal. Such backwater valves shall remain open during periods of low flows to avoid screening of solids and shall not restrict capacities or cause excessive turbulence during peak loads. Unless otherwise listed, valve access covers shall be bolted type with gasket, and each valve shall bear the manufacturer's name cast into the body and the cover.

Some backwater valves do not require a pit for installation but are designed to provide access via a riser pipe and fitting (see **Figure 710.6**).

710.7 Drainage and Venting Systems. The drainage and venting systems, in connection with fixtures, sumps, receiving tanks, and mechanical waste-lifting devices shall be installed under the same requirements as provided for in this code for gravity systems.

This provision establishes the requirements for the installation of this pumped or pressurized system. Simply put, it must be installed as any other drainage and vent system is installed under this code. All cleanout and vent sizing conditions or other provisions of the code that

SANITARY DRAINAGE

apply to drainage and vent piping also apply to this installation except where specific direction is otherwise given.

710.8 Sump and Receiving Tank Construction. Sumps and receiving tanks shall be watertight and shall be constructed of concrete, metal, or other approved materials. Where constructed of poured concrete, the walls and bottom shall be adequately reinforced and designed to recognized acceptable standards. Metal sumps or tanks shall be of such thickness as to serve their intended purpose and shall be treated internally and externally to resist corrosion.

710.9 Alarm. Such sumps and receiving tanks shall be automatically discharged and, wherein a "public use" occupancy, shall be provided with dual pumps or ejectors arranged to function alternately in normal use and independently in case of overload or mechanical failure. The pumps shall have an audio and visual alarm, readily accessible, that signals pump failure or an overload condition. The lowest inlet shall have a clearance of not less than 2 inches (51 mm) from the high-water or "starting" level of the sump.

There is a basic purpose for providing dual pumps in public-use occupancies. It is essential for public health reasons that waste removal systems always function and that the building is never deemed to have become "insanitary" due to lack of waste removal. This condition would legally require the building to be taken out of service. Obviously, it is essential that failure of either pump be made

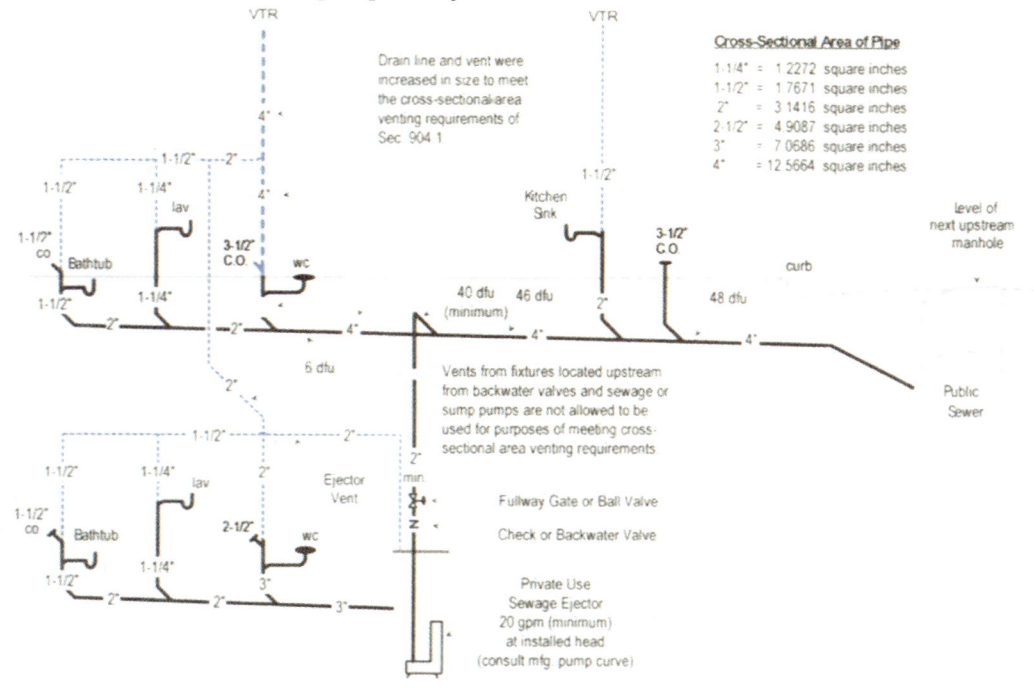

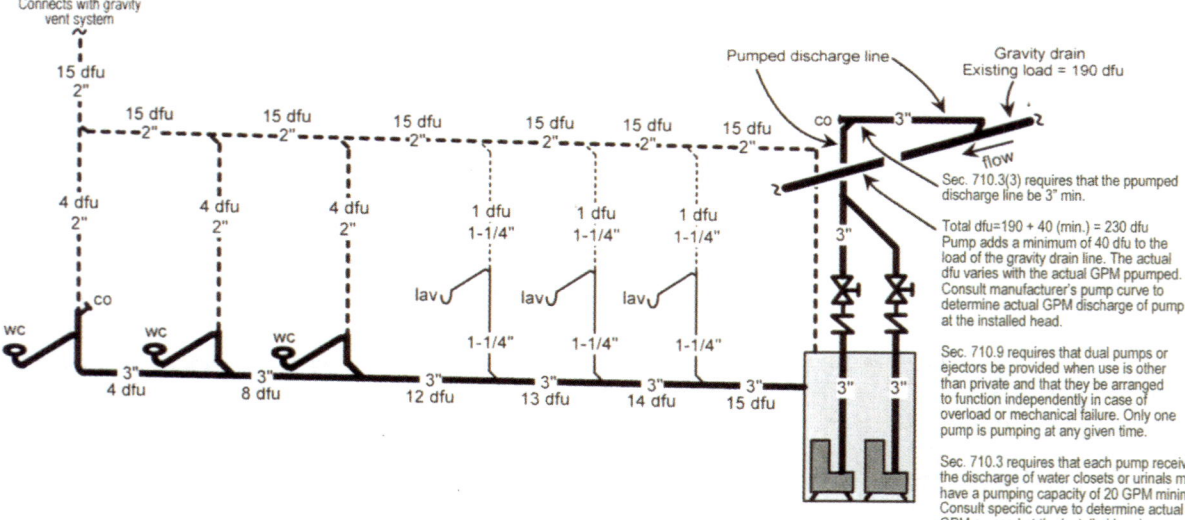

FIGURE 710.3B
FIXTURES LOCATED BELOW THE SEWER LEVEL AND DUAL SUMP PUMP INSTALLATION

SANITARY DRAINAGE

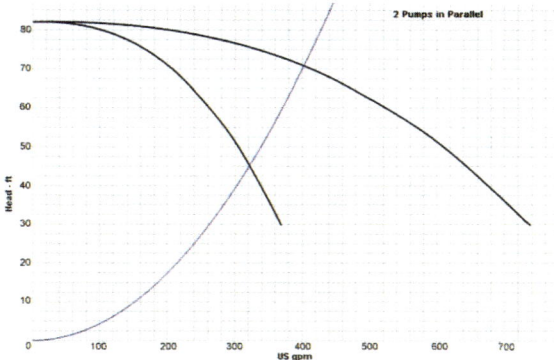

FIGURE 710.5
SYSTEM CURVE FOR PARALLEL PUMPS

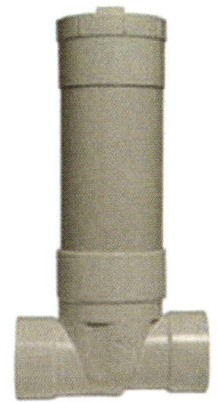

FIGURE 710.6
BACKWATER VALVE

evident through a fail-proof alarm system so that timely repairs may be made and the building kept in service.

In this dual-pump installation it is necessary that a "cycling" switch be included as a basic element of these installations so that each pump is regularly activated on every other cycle. Every time pumping action is called for, the pump that was not active on the previous cycle will now become the active pump. The objective in doing this is twofold: Alternate cycles

1. Exercise each pump as a test to confirm that both pumps are operational, and
2. Working both pumps by having them on for approximately equal periods of time.

An inactive pump, as an emergency standby pump, will often become useless over a period of time simply because of the deleterious effect water will have on a submerged dormant pump. The second pump in a dual-pump system is required by code, not because it is considered to be an "emergency" pump, but rather, to prevent an emergency condition. It is required as an operational pump that is intended to share the load equally with its counterpart (see **Figure 710.9**).

In addition to switching pumps alternately, there must be an alarm system installed (both visual and audio) to provide notification when either of these pumps fails to activate. The cycling switch will sense pump failure and will cycle back to the pump that was last active, ensuring that waste discharge will always occur.

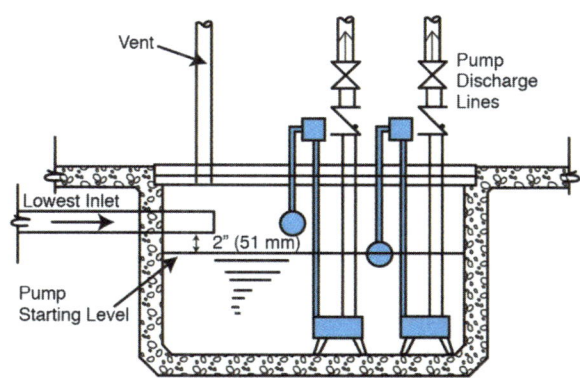

FIGURE 710.9
DUAL SUMP PUMP INSTALLATION

710.10 Sump and Receiving Tank Covers and Vents. Sumps and receiving tanks shall be provided with substantial covers having a bolt-and-gasket-type manhole or equivalent opening to permit access for inspection, repairs, and cleaning. The top shall be provided with a vent pipe that shall extend separately through the roof or, where permitted, be combined with other vent pipes. Such vent shall be large enough to maintain atmospheric pressure within the sump under normal operating conditions and, in no case, shall be less in size than that required by Table 703.2 for the number and type of fixtures discharging into the sump, nor less than 1½ inches (40 mm) in diameter. Where the preceding requirements are met and the vent, after leaving the sump, is combined with vents from fixtures discharging into the sump, the size of the combined vent need not exceed that required for the total number of fixtures discharging into the sump. No vent from an air-operating sewage ejector shall combine with other vents.

The statement that the vent from the sump, when combined with the fixture vents from the fixtures draining into the sump, need not exceed the required vent for the total number of fixtures discharging into the sump can be confusing. What is stated here is that the size of the combined vent, i.e. the common vent serving the sump and fixtures discharging into the sump, needs to be sized only according to Table 703.2 for the number and type of fixtures discharging into the sump and does not need to increase when adding the sump vent.

710.11 Air Tanks. Air tanks shall be so proportioned as to be of equal cubical capacity to the ejectors connected in addition to that in which there shall be maintained an air pressure of not less than 2 pounds per foot (lb/ft) (3 kg/m) of height the sewage is to be raised. No water-operated ejectors shall be permitted.

710.12 Grinder Pump Ejector. Grinder pumps shall be permitted to be used.

SANITARY DRAINAGE

🔧 Grinder pumps shred solids and sewage discharge into a finely ground slurry that can be discharged through a small diameter pipe. These grinder pumps have fittings with special cutting ring systems that are mounted to the pump's motor shaft. When the wastewater collects and reaches a certain level, the sewage grinder pump turns on to grind the waste and force the wastewater through the pressure pipe to the sewage system. This submersible pump includes a heavy-duty shredding unit with a centrifugal design and an open impeller. A grinder pump typically consists of corrosive-resistant stainless steel and powder-coated cast-iron parts to withstand the processing of corrosive materials.

710.12.1 Discharge Piping. The discharge piping shall be sized in accordance with the manufacturer's installation instructions and shall be not less than 1¼ inches (32 mm) in diameter. A check valve and fullway-type shutoff valve shall be located on the discharge line.

710.13 Macerating Toilet Systems and Pumped Waste Systems. Fixtures shall be permitted to discharge to a macerating toilet system, or pumped waste system shall be permitted as an alternate to a sewage pump system where approved by the Authority Having Jurisdiction. Such systems shall comply with ASME A112.3.4/CSA B45.9 and shall be installed in accordance with the manufacturer's installation instructions.

🔧 Many macerating toilet systems incorporate a grinder pump that grinds the waste and discharges the slurry into the sanitary drainage system. These systems are intended to collect waste from a single water closet, lavatory and bathtub located in the same room and grind and pump these wastes to the sanitary drainage system.

The macerating system consists of three major components: (a) a container that houses the operating mechanisms; (b) a pressure chamber that automatically or manually activates and deactivates the induction motor; and (c) an induction motor that drives the shredder blades and pump assembly, which is permitted to be combined into a single unit (see **Figure 710.13a**).

A macerating toilet looks very much like a conventional water closet; however, the wastewater discharges through a rear-spigot outlet in the water closet bowl to a macerator pump contained in a small plastic box that is positioned on or inside the wall behind the water closet (see **Figure 710.13b**). The macerator uses a fast rotating cutting blade to liquefy human waste in the flush water. Within seconds, this slurry discharges under pressure through a minimum three-fourths of an inch pipe into the sanitary drainage system. These systems are typically installed above the floor where it is impossible or cost-prohibitive to install conventional plumbing fixtures that require below-floor drainage piping.

710.13.1 Sumps. The sump shall be watertight and gastight.

710.13.2 Discharge Piping. The discharge piping shall be sized in accordance with manufacturer's instructions and

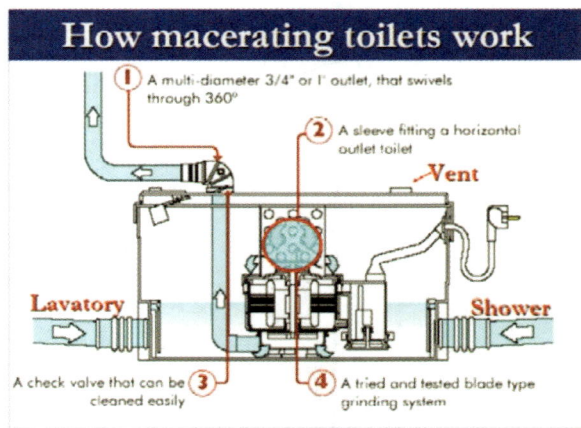

FIGURE 710.13A
MACERATING TOILET

FIGURE 710.13B
MACERATING TOILET SYSTEM

shall be not less than ¾ of an inch (20 mm) in diameter. The developed length of the discharge piping shall not exceed the manufacturer's instructions. A check valve and fullway-type shutoff valve shall be located within the discharge line or internally within the device.

710.13.3 Venting. The plumbing fixtures that discharge into the macerating device shall be vented in accordance with this code. The sump shall be vented in accordance with the manufacturer's instructions, and such vent shall be permitted to connect to the fixture venting.

711.0 Suds Relief.

711.1 General. Drainage connections shall not be made into a drainage piping system within 8 feet (2438 mm) of a vertical to horizontal change of direction of a stack containing suds-producing fixtures. Bathtubs, laundries, washing machine standpipes, kitchen sinks, and dishwashers shall be considered suds-producing fixtures. Where parallel vent stacks are required, they shall connect to the drainage stack at a point 8 feet (2438 mm) above the lowest point of the drainage stack.

Exceptions:

(1) Single-family residences.

(2) Stacks receiving the discharge from less than three stories of plumbing fixtures.

The above-named suds producing fixtures discharge high-sudsing detergents that, when mixed with other liquid wastes (in addition to air in a stack), create areas of increased turbulence in lower portions of soil stacks. The liquid wastes, being heavier than suds, will flow through the suds-loaded piping without carrying the suds discharge along with the flow (see **Figure 711.1a**). However, the suds become compressed, forcing them to move through any available path of relief. The relief path may be any of the branches connected to the building drain or a vent stack or branch vents. High suds pressure will blow trap seals, leaving residue in the fixtures and allowing sewer gas to enter the building (see **Figure 711.1b**). Any vertical-to-horizontal change of direction of a stack receiving the discharge of suds-producing fixtures has the potential to create a sudsing problem.

712.0 Testing.

Per Section 105.2.5, Responsibility, of Chapter 1, the equipment, material and labor necessary for inspection and tests shall be furnished by the person to whom the permit is issued. This person also has the responsibility to ensure that the work being inspected will stand the test.

712.1 Media. The piping of the plumbing, drainage, and venting systems shall be tested with water or air except that plastic pipe shall not be tested with air. The Authority Having Jurisdiction shall be permitted to require the removal of cleanouts, etc., to ascertain whether the pressure has reached all parts of the system. After the plumbing fixtures have been set and their traps filled with water, they shall be submitted to a final test.

712.2 Water Test. The water test shall be applied to the drainage and vent systems either in its entirety or in sections. Where the test is applied to the entire system, openings in the piping shall be tightly closed, except the highest opening, and the system filled with water to the point of overflow. Where the system is tested in sections, each opening shall be tightly plugged, except the highest opening of the section under test, and each section shall be filled with water, but no section shall be tested with less than a 10 foot head of water (30 kPa). In testing successive sections, not less than the upper 10 feet (3048 mm) of the next preceding section shall be tested, so that no joint or pipe in the building (except the uppermost 10 feet (3048 mm) of the system) shall have been submitted to a test of less than a 10 foot head of water (30 kPa). The water shall be kept in the system, or in the portion under test, for not less than 15 minutes before inspection starts. The system shall then be tight at all points.

712.3 Air Test. The air test shall be made by attaching an air compressor testing apparatus to a suitable opening and, after closing all other inlets and outlets to the system, forcing air into the system until there is a uniform gauge pressure of 5 pounds-force per square inch (psi) (34 kPa) or sufficient to balance a column of mercury 10 inches (34 kPa) in height. The pressure shall be held without the introduction of additional air for a period of not less than 15 minutes.

Part II – Building Sewers.

Part I of this chapter provides the requirements for transporting sewage waste from a fixture located in a building to the building sewer. Part II provides the requirements for transporting sewage waste from the building sewer to the public or private sewer. In some installations, the sewage may discharge to a private sewage disposal system (see **Figure 713.2**). The installation of a private sewage disposal system is addressed in Appendix H.

Many of the requirements for the building sewer are the same as for the building drainage system. Most of them have already been discussed in either this chapter or in Chapters 1 and 3.

713.0 Sewer Required.

713.1 Where Required. A building in which plumbing fixtures are installed and premises having drainage piping thereon shall have a connection to a public or private sewer, except as provided in Section 713.2, and Section 713.4.

See **Figure 713.1**.

713.2 Private Sewage Disposal System. Where no public sewer intended to serve a lot or premises is available in a thoroughfare or right of way abutting such lot or premises, drainage piping from a building or works shall be connected to an approved private sewage disposal system.

See **Figure 713.2**.

713.3 Public Sewer. Within the limits prescribed by Section 713.4 hereof, the rearrangement or subdivision into smaller parcels of a lot that abuts and is served by a public sewer shall not be deemed cause to permit the construction of a private sewage disposal system, and plumbing or drainage systems on a smaller parcel or parcels shall connect to the public sewer.

713.4 Public Sewer Availability. The public sewer shall be permitted to be considered as not being available where such public sewer or a building or an exterior drainage facility connected thereto is located more than 200 feet (60 960 mm) from a proposed building or exterior drainage facility on a lot or premises that abut and is served by such public sewer.

The installation of a private sewage disposal system is allowed when the public sewer is further than 200 feet of a building that is located on the lot that abuts the sewer.

713.5 Permit. No permit shall be issued for the installation, alteration, or repair of a private sewage disposal system, or part thereof, on a lot for which a connection with a public sewer is available.

If the public sewer is available when a permit to repair or replace a private sewage disposal system is requested, the permit will not be issued and the owner will have to connect to the public sewer. Section 713.7, Installation, makes an exception to this when there is insufficient depth for the owner to connect to the public sewer by gravity.

SANITARY DRAINAGE

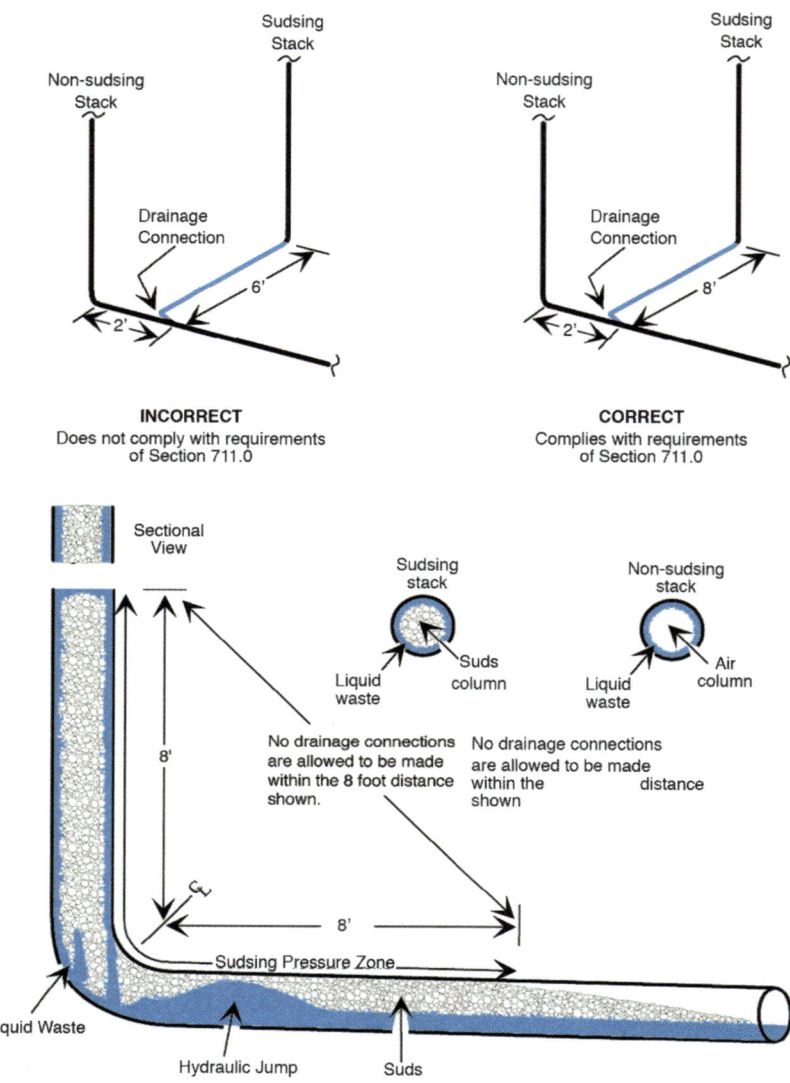

FIGURE 711.1A
SUDS RELIEF IN THE DRAINAGE SYSTEM

FIGURE 711.1B
SUDS LOADING

713.6 Lot. On every lot or premises hereafter connected to a public sewer, plumbing, and drainage systems or parts thereof on such lot or premises shall be connected with such public sewer.

713.7 Installation. In cities, counties, or both where the installation of building sewers is under the jurisdiction of a department other than the Authority Having Jurisdiction, the provisions of this code relating to building sewers need not apply.

Exception: Single-family dwellings and buildings or structures accessory thereto, existing and connected to an approved private sewage disposal system prior to the time of connecting the premises to the public sewer shall be permitted, where no hazard, nuisance, or insanitary condition is evidenced, and written permission has been obtained from the Authority Having Jurisdiction, remain connected to such properly maintained private sewage disposal system where there is insufficient grade or fall to permit drainage to the sewer by gravity.

714.0 Damage to Public Sewer or Private Sewage Disposal System.

714.1 Unlawful Practices. It shall be unlawful for a

SANITARY DRAINAGE

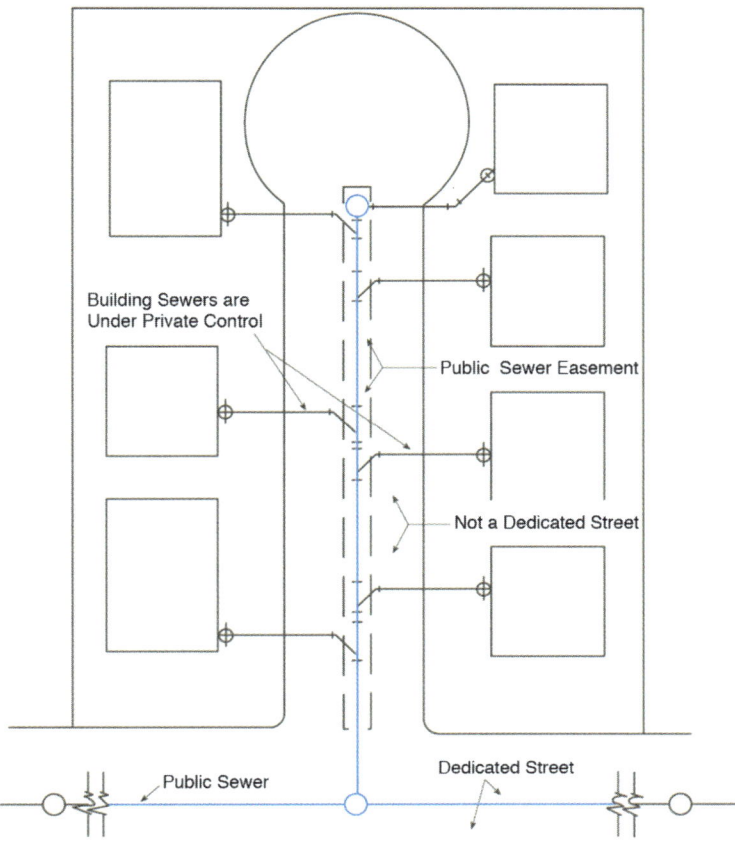

FIGURE 713.1
PUBLIC SEWER SYSTEM - INDIVIDUAL CONNECTIONS

person to deposit, by means whatsoever, into a plumbing fixture, floor drain, interceptor, sump, receptor, or device which is connected to a drainage system, public sewer, private sewer, septic tank, or cesspool, ashes; cinders; solids; rags; flammable, poisonous, or explosive liquids or gases; oils; grease; and whatsoever that is capable of causing damage to the public sewer, private sewer, or private sewage disposal system.

714.2 Prohibited Water Discharge. No rain, surface, or subsurface water shall be connected to or discharged into a drainage system unless first approved by the Authority Having Jurisdiction.

714.3 Prohibited Sewer Connection. No cesspool, septic tank, seepage pit, or drain field shall be connected to a public sewer or to a building sewer leading to such public sewer.

🔧 These parts of the private sewage disposal system are not meant to discharge to the building sewer. If a connection has to be made to the building sewer, then these installations shall be removed and the piping from the building connected directly to the building sewer.

714.4 Commercial Food Waste Disposer. The Authority Having Jurisdiction shall review before approval, the installation of a commercial food waste disposer connecting to a private sewage disposal system.

714.5 Tanks. An approved type, watertight sewage or wastewater holding tank, the contents of which, due to their character, shall be periodically removed and disposed of at some approved off-site location, shall be installed where required by the Authority Having Jurisdiction or the Health Officer to prevent anticipated surface or subsurface contamination or pollution, damage to the public sewer, or other hazardous or nuisance conditions.

🔧 A holding tank is an unacceptable device when utilized in a plumbing system. It is an expedient method of collecting and holding waste products for short-term special applications, but even then only after specific approval by the AHJ and the health authority. In most cases, approval is time-limited and is based upon a specified frequency of maintenance/service to ensure that this installation remains safe and functional.

715.0 Building Sewer Materials.

715.1 Materials. The building sewer, beginning 2 feet (610 mm) from a building or structure, shall be of such materials as prescribed in this code.

🔧 Refer to Table 701.2 for the materials allowed for the building sewer.

715.2 Joining Methods and Materials. Joining methods and materials shall be as prescribed in this code.

715.3 Existing Sewers. Replacement of existing

SANITARY DRAINAGE

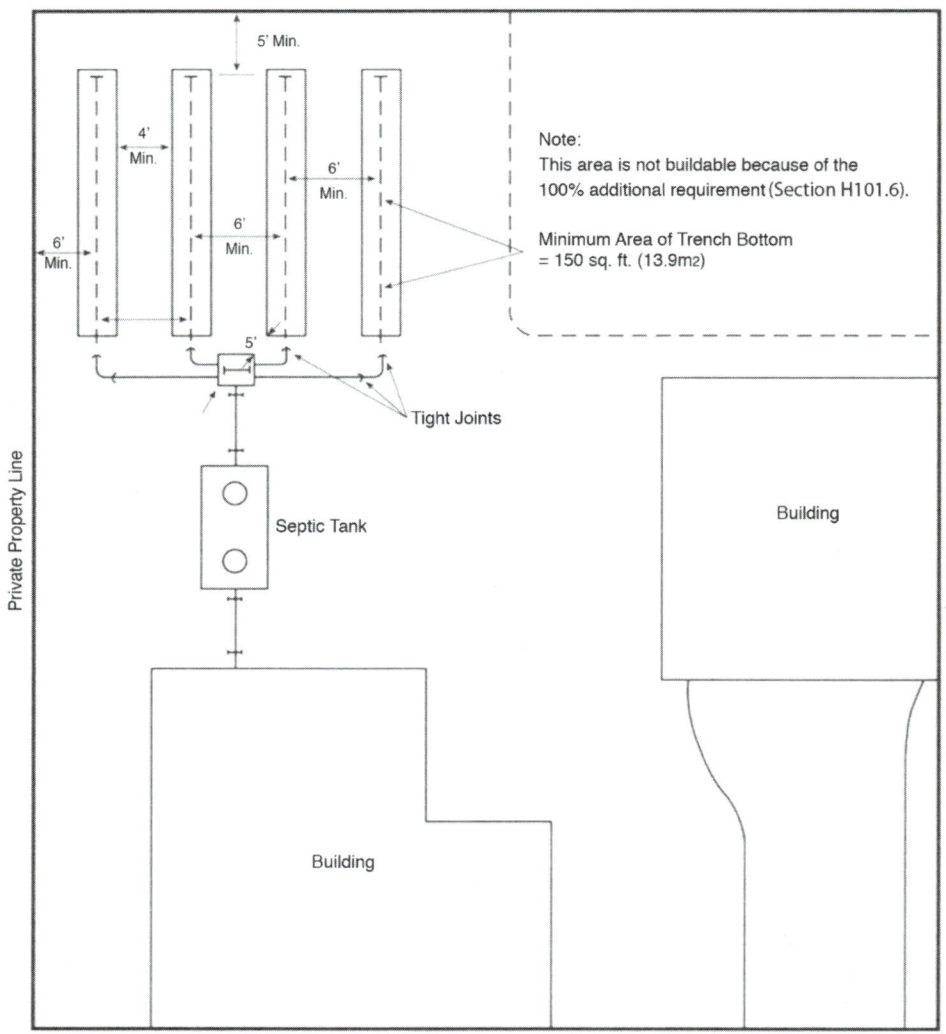

**FIGURE 713.2
PRIVATE SEWAGE DISPOSAL SYSTEM**

building sewer and building storm sewers using trenchless methodology and materials shall be installed in accordance with ASTM F1216. Cast-iron soil pipes and fittings shall not be repaired or replaced by using this method aboveground or belowground. Replacement using cured-in-place pipe liners shall not be used on collapsed piping or when the existing piping is compromised.

Trenchless methodologies have been in use in the United States since early 2000. Trenchless methodologies for replacing existing sewers do not require continuous open trenches, but rather utilizes subsurface techniques such as tunneling, directional boring, pipe bursting (see **Figure 715.3a**), and cured-in-place pipe (**Figure 715.3b**). Figure below is an example of a pipe bursting trenchless methodology.

Cured in place pipe rehabilitation (CIPP) involves a linear insertion of a resin-saturated liner into the host pipe. The liner is either pulled or inverted through the host pipe. When inverted, the liner is turned inside-out as air is forced into one end of the liner. After the resin liner expands to the inside dimensions of the pipe, a curing process using heated water or steam polymerizes the resin liner to form a pipe within a pipe. ASTM F1216 provides a more detailed explanation of the CIPP procedures.

This standard applies specifically to the installation of a resin-impregnated, flexible tube that is cured in place of the pipe. It does not apply to a pipe bursting technique. The standard provides installation requirements and inspection practices. Inspection requirements include cut samples for flexural and tensile testing, leakage testing, pressure pipe testing, delamination test, wall thickness, and a visual or closed-circuit television inspection.

Trenchless methodology in accordance with ASTM F1216 is not allowed where the pipe is collapsed or compromised. This means that the cross-sectional area of the pipe is more than 40% obstructed preventing the insertion of the resin-impregnated tube. Piping of cast iron material is not permitted to be repaired by this method because it conflicts with the manufacturer's product standards.

SANITARY DRAINAGE

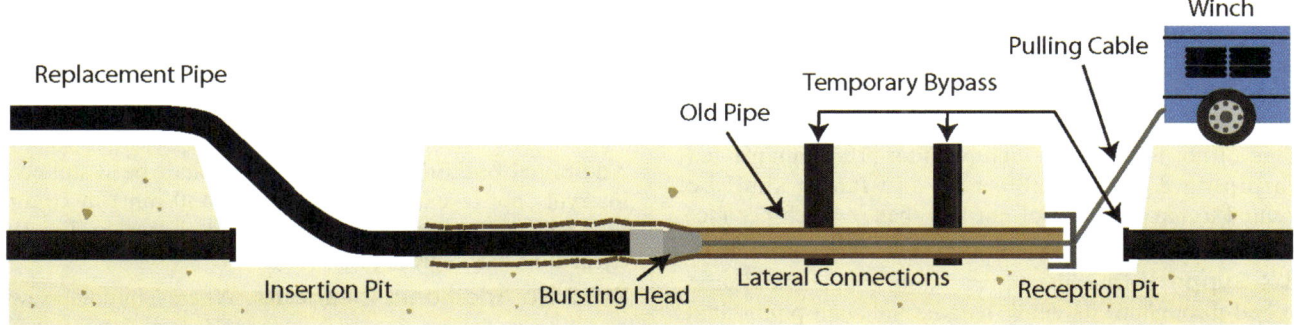

**FIGURE 715.3A
PIPE BURSTING**

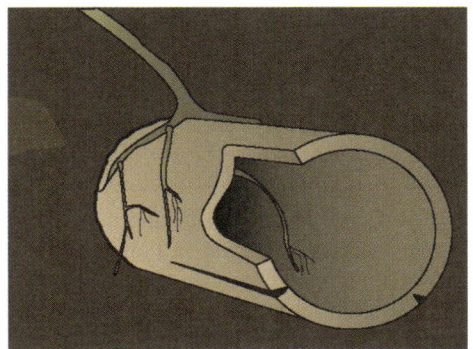 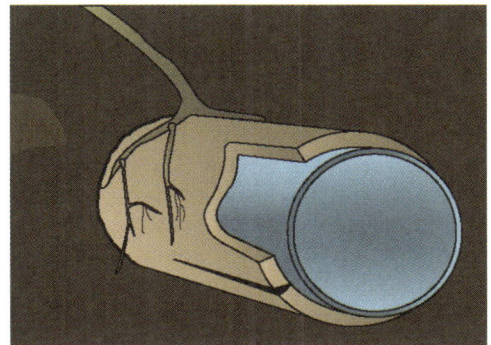

**FIGURE 715.3B
CURED IN PLACE PIPE**

716.0 Markings.

716.1 General. Pipe, brick, block, prefabricated septic tanks, prefabricated septic tank or seepage pit covers, or other parts or appurtenances incidental to the installation of building sewers or private sewage disposal systems shall be in accordance with the approval requirements of Chapter 3 of this code.

717.0 Size of Building Sewers.

717.1 General. The minimum size of a building sewer shall be determined on the basis of the total number of fixture units drained by such sewer, in accordance with Table 717.1. No building sewer shall be smaller than the building drain.

For alternate methods of sizing building sewers, see Appendix C.

Table 717.1 stipulates minimum fixture unit loading as well as maximum fixture unit loading for sewer sizes eight inches and larger. **Figure 717.1** contains an example of sewer sizing. As pipe size increases, the depth of the "spring line" (flowing water depth) decreases if all else remains constant. Shallow water will not support suspended solids at an elevation sufficiently high enough to prevent them from dragging along the bottom of the pipe. As these solids drag along, they slow down and, eventually, the water flows past, leaving the solids in the bottom of the pipe. An accumulation will eventually cause a stoppage, resulting in frequent service disruption and high maintenance costs. By requiring minimum fixture unit loading, there is an attempt to increase the likelihood that there will be adequate liquid depth and scouring to remove all solids on a daily basis.

Table 717.1 also provides for pipe sizes eight inches and larger to be installed at slopes less than one-eighth of an inch per foot. Table 703.2 makes allowance for the building drain to be installed at a slope of one-eighth of an inch per foot for four-inch and above pipe with a restriction on the total fixture unit rating of the pipe. Table 717.1 also takes this into consideration. As grade diminishes, the maximum fixture unit rating of the pipe decreases. At the same time, the minimum loading of the pipe increases. This provides a proper liquid level in the pipe as discussed above. Section 718.1 has further slope allowances.

The difficulty in maintaining a constant slope of the sewage pipe is greatly exacerbated as the angle of the slope decreases. Failure to maintain a constant shallow slope will result in low areas that become another collection point for solids and, potentially, another source of functional weakness and maintenance misery. Systems at these grades will be installed using a laser level to ensure proper slope of the pipe.

718.0 Grade, Support, and Protection of Building Sewers.

718.1 Slope. Building sewers shall be run in practical alignment and at a uniform slope of not less than ¼ inch per foot (20.8 mm/m) toward the point of disposal.

Exception: Where approved by the Authority Having

SANITARY DRAINAGE

Jurisdiction and where it is impractical, due to the depth of the street sewer or to the structural features or the arrangement of a building or structure, to obtain a slope of ¼ inch per foot (20.8 mm/m), such pipe or piping 4 inches (100 mm) through 6 inches (150 mm) shall be permitted to have a slope of not less than ⅛ inch per foot (10.4 mm/m) and such piping 8 inches (200 mm) and larger shall be permitted to have a slope of not less than ¹⁄₁₆ inch per foot (5.2 mm/m).

718.2 Support. Building sewer piping shall be laid on a firm bed throughout its entire length, and such piping laid in made or filled-in ground shall be laid on a bed of approved materials and shall be properly supported as required by the Authority Having Jurisdiction.

718.3 Protection from Damage. No building sewer or other drainage piping or part thereof, which is constructed of materials other than those approved for use under or within a building, shall be installed under or within 2 feet (610 mm) of a building or structure, or part thereof, nor less than 1 foot (305 mm) below the surface of the ground. The provisions of this subsection include structures such as porches and steps, whether covered or uncovered; breezeways; roofed porte cochere; roofed patios; carports; covered walks; covered driveways; and similar structures or appurtenances.

TABLE 717.1
MAXIMUM/MINIMUM FIXTURE UNIT LOADING ON BUILDING SEWER PIPING*

SIZE OF PIPE (inches)	SLOPE, (inches per foot)		
	¹⁄₁₆	⅛	¼
6 and smaller	(As specified in Table 703.2/ No minimum loading)		
8	1950/1500	2800/625	3900/275
10	3400/1600	4900/675	6800/300
12	5600/1700	8000/725	11 200/325

For SI units: 1 inch = 25 mm, 1 inch per foot = 83.3 mm/m

* See also Appendix H, Private Sewage Disposal Systems. For alternate methods of sizing drainage piping, see Appendix C.

719.0 Cleanouts.

719.1 Locations. Cleanouts shall be placed inside the building near the connection between the building drain and the building sewer or installed outside the building at the lower end of the building drain and extended to grade.

Additional building sewer cleanouts shall be installed at intervals not to exceed 100 feet (30 480 mm) in straight runs and for each aggregate horizontal change in direction exceeding 135 degrees (2.36 rad).

719.2 No Additional Cleanouts. Where a building sewer or a branch thereof does not exceed 10 feet (3048 mm) in length and is a straight-line projection from a building drain that is provided with a cleanout, no cleanout will be required at its point of connection to the building drain.

719.3 Building Sewer Cleanouts. Required building sewer cleanouts shall be extended to grade and shall be in accordance with the appropriate sections of cleanouts, Section 707.0, for sizing, construction, and materials. Where building sewers are located under buildings, the cleanout requirements of Section 707.0 shall apply.

719.4 Cleaning. Each cleanout shall be installed so that it opens to allow cleaning in the direction of flow of the soil or waste or at right angles thereto and, except in the case of wye branch and end-of-line cleanouts, shall be installed vertically above the flow line of the pipe.

719.5 Access. Cleanouts installed under concrete or asphalt paving shall be made accessible by yard boxes or by extending flush with paving with approved materials and shall be adequately protected.

The cleanouts for sewer piping are somewhat different than for building drains. Most cleanouts are installed in parking or lot areas. Either the cleanout or the yard box covering the cleanout must be brought to grade. The cleanout plug size is according to Table 707.1, and is typically 3½ inches because the building sewer is usually four inches or larger.

1/4" per foot		1/8" per foot	
A	4"	A	4"
B	3"	B	3"*
C	4"	C	4"
D	3"	D	4"
E	3"	E	3"*
F	4"	F	4"
G	4"	G	4"
H	4"	H	4"
J	3"	J	4"
K	4"	K	4"
L	4"	L	4"
M	4"	M	4"
N	4"	N	5"
O	4"	O	5"
P	5"	P	5"
Q	5"	Q	5"
R	5"	R	5"

See Table 703.2, footnote 4, for limitations on the number of water closets on 3" (75 mm) horizontal or vertical piping

This job has been sized for 1/4 inch per foot (21 mm/m) slope and also 1/8 inch per foot (10 mm/m) slope. See the sizing chart at the left for correct pipe size, and compare.

*1/8" per foot (11 mm/m) not permitted on 3" (76 mm) pipe. See section 708.0

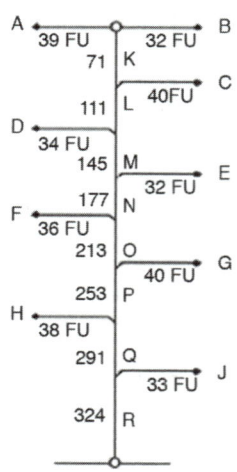

FIGURE 717.1
EXAMPLE OF SEWER SIZING

SANITARY DRAINAGE

719.6 Manholes. Approved manholes shall be permitted to be installed in lieu of cleanouts, where first approved by the Authority Having Jurisdiction. The maximum distance between manholes shall not exceed 300 feet (91 440 mm).

The inlet and outlet connections shall be made by the use of a flexible compression joint not less than 12 inches (305 mm) and not exceeding 3 feet (914 mm) from the manhole. No flexible compression joints shall be embedded in the manhole base.

Manholes may be used in lieu of cleanouts because of the clear area where they are installed. The manhole is a more complicated installation than a normal cleanout. The base of the cleanout and the inside of the waste line within the manhole are usually formed in concrete. This has to be done with a smooth interior waterway so the flow will be uninterrupted. **Figure 719.6a** illustrates a typical manhole found on civil drawings. **Figure 719.6b** shows the interior of a manhole.

720.0 Sewer and Water Pipes.

720.1 General. Building sewers or drainage piping of clay or materials that are not approved for use within a building shall not be run or laid in the same trench as the water pipes unless the following requirements are met:

(1) The bottom of the water pipe, at points, shall be not less than 12 inches (305 mm) above the top of the sewer or drain line.

(2) The water pipe shall be placed on a solid shelf excavated at one side of the common trench with a clear horizontal distance of not less than 12 inches (305 mm) from the sewer or drain line.

(3) Water pipes crossing sewer or drainage piping constructed of clay or materials that are not approved for use within a building shall be laid not less than 12 inches (305 mm) above the sewer or drain pipe.

For the purpose of this section, "within a building" shall mean within the fixed limits of the building foundation.

See **Figure 720.1**.

721.0 Location.

721.1 Building Sewer. Except as provided in Section 721.2, no building sewer shall be located in a lot other than the lot that is the site of the building or structure served by such sewer nor shall a building sewer be located at a point having less than the minimum distances referenced in Table 721.1.

721.2 Abutting Lot. Nothing contained in this code shall be construed to prohibit the use of all or part of an abutting lot to:

(1) Provide access to connect a building sewer to an available public sewer where proper cause and legal easement, not in violation of other requirements, has been first established to the satisfaction of the Authority Having Jurisdiction.

(2) Provide additional space for a building sewer where the proper cause, transfer of ownership, or change of boundary, not in violation of other requirements, has been first established to the satisfaction of the

FIGURE 719.6B
INTERIOR OF MANHOLE
(NOTE: THE WATER WAY IS NOT TO CODE)

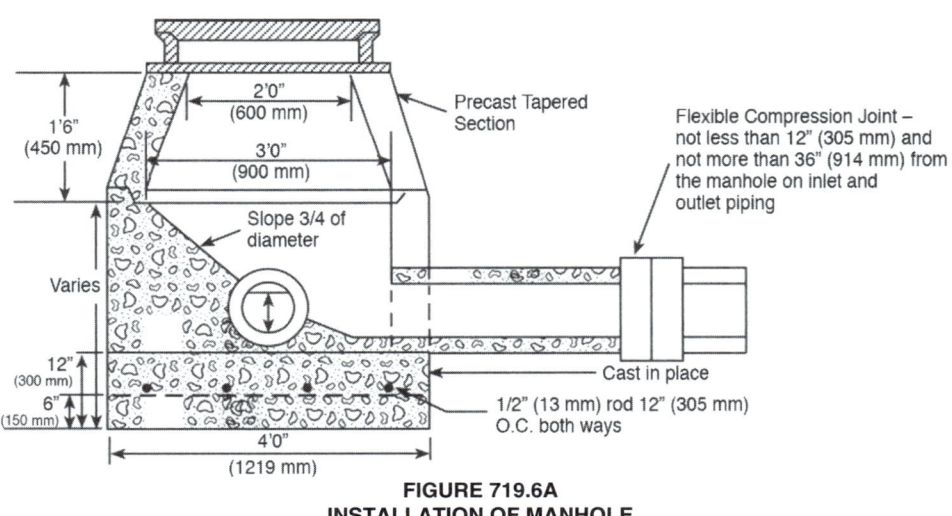

FIGURE 719.6A
INSTALLATION OF MANHOLE

SANITARY DRAINAGE

TABLE 721.1
MINIMUM HORIZONTAL DISTANCE REQUIRED FROM BUILDING SEWER (feet)

Buildings or structures[1]	2
Property line adjoining private property	Clear[2]
Water supply wells	50[3]
Streams	50
On-site domestic water service line	1[4]
Public water main	10[5, 6]

For SI units: 1 foot = 304.8 mm

Notes:
[1] Including porches and steps, whether covered or uncovered; breezeways; roofed porte-cochere; roofed patios; carports; covered walks; covered driveways; and similar structures or appurtenances.
[2] See also Section 312.3.
[3] Drainage piping shall clear domestic water supply wells by not less than 50 feet (15 240 mm). This distance shall be permitted to be reduced to not less than 25 feet (7620 mm) where the drainage piping is constructed of materials approved for use within a building.
[4] See Section 720.0.
[5] For parallel construction.
[6] For crossings, approval by the Health Department or the Authority Having Jurisdiction shall be required.

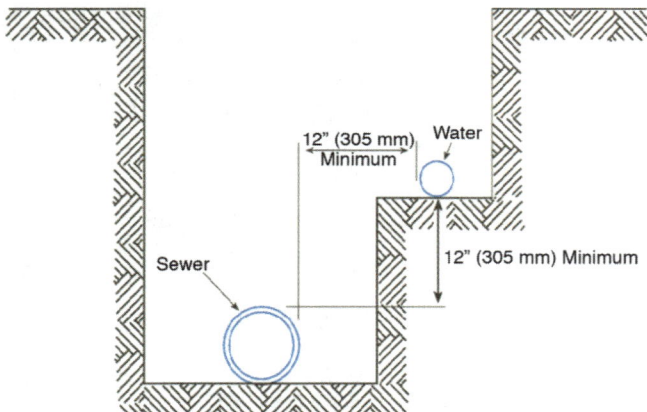

FIGURE 720.1
SEPARATION OF SEWER AND WATER LINES

Authority Having Jurisdiction. The instrument recording such action shall constitute an agreement with the Authority Having Jurisdiction and shall clearly state and show that the areas so joined or used shall be maintained as a unit during the time they are so used. Such an agreement shall be recorded in the office of the County Recorder as part of the conditions of ownership of said properties, and shall be binding on heirs, successors, and assigns to such properties. A copy of the instrument recording such proceedings shall be filed with the Authority Having Jurisdiction.

722.0 Abandoned Sewers and Sewage Disposal Facilities.

722.1 Building (House) Sewer. An abandoned building (house) sewer, or part thereof, shall be plugged or capped in an approved manner within 5 feet (1524 mm) of the property line.

All openings into a drainage or vent system, except those openings to which plumbing fixtures are properly connected or which constitute vent terminals, must be permanently plugged or capped in an approved manner using appropriate materials.

722.2 Cesspools, Septic Tanks, and Seepage Pits. A cesspool, septic tank, and seepage pit that has been abandoned or has been discontinued otherwise from further use, or to which no waste or soil pipe from a plumbing fixture is connected, shall have the sewage removed therefrom and be completely filled with earth, sand, gravel, concrete, or other approved material.

The abandonment of private sewage disposal facilities without filling the large underground areas of these facilities is a dangerous practice. The possibility of someone, especially children, getting into or falling into these deep installations is reason enough to fill them, but they also represent health hazards due to the continued festering of bacteria and mold. Therefore, they must be covered and filled within 30 days of abandonment (see Section 722.5).

722.3 Filling. The top cover or arch over the cesspool, septic tank, or seepage pit shall be removed before filling, and the filling shall not extend above the top of the vertical portions of the sidewalls or above the level of the outlet pipe until inspection has been called and the cesspool, septic tank, or seepage pit has been inspected. After such inspection, the cesspool, septic tank, or seepage pit shall be filled to the level of the top of the ground.

722.4 Ownership. No person owning or controlling a cesspool, septic tank, or seepage pit on the premises of such person or in that portion of a public street, alley, or other public property abutting such premises, shall fail, refuse, or neglect to comply with the provisions of this section or upon receipt of notice so to comply from the Authority Having Jurisdiction.

722.5 Disposal Facilities. Where disposal facilities are abandoned consequent to connecting premises with the public sewer, the permittee making the connection shall fill abandoned facilities in accordance with the Authority Having Jurisdiction within 30 days from the time of connecting to the public sewer.

723.0 Building Sewer Test.

723.1 General. Building sewers shall be tested by plugging the end of the building sewer at its points of connection to the public sewer or private sewage disposal system and completely filling the building sewer with water from the lowest to the highest point thereof, or by approved equivalent low-pressure air test. Plastic DWV piping systems shall not be tested by the air test method. The building sewer shall be watertight.

Building sewers are tested by plugging the end of the building sewer at its point of connection with the public sewer or private sewage disposal system. The building sewer is then completely filled with water from the lowest point to the highest point (see **Figure 723.1a**). There is no requirement for any specific head pressure. An air-pressure test may be used in lieu of the water test, except on plastic DWV piping systems due to safety issues. See **Figure 723.1b** of a lethal plastic shard exploding from a compressed air test on plastic pipe. The requirements for using air as testing media are found in Section 712.3.

See Learning Link **goo.gl/d1sywgJ** about testing plastic DWV systems with air.

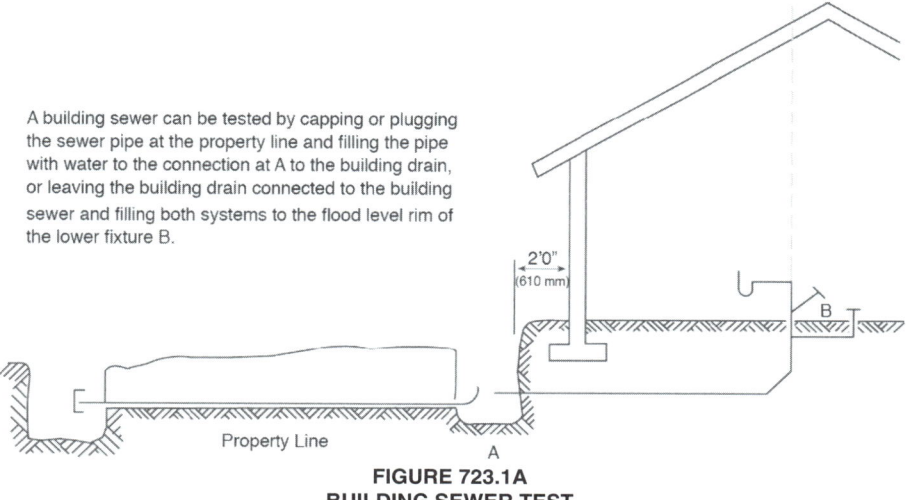

**FIGURE 723.1A
BUILDING SEWER TEST**

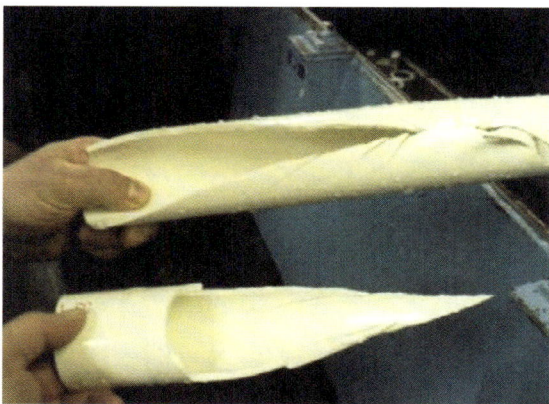

**FIGURE 723.1B
PLASTIC SHARD FROM COMPRESSED AIR TEST**

CHAPTER 8

INDIRECT WASTES

801.0 General.

801.1 Applicability. This chapter shall govern the materials, design, and installation of indirect waste piping, receptors, and connections; and provisions for discharge and disposal of condensate wastes, chemical wastes, industrial wastes, and clear water wastes.

801.2 Air Gap or Air Break Required. Indirect waste piping shall discharge into the building drainage system through an air gap or air break as set forth in this code. Where a drainage air gap is required by this code, the minimum vertical distance as measured from the lowest point of the indirect waste pipe or the fixture outlet to the flood-level rim of the receptor shall be not less than 1 inch (25.4 mm).

An indirect waste pipe does not connect directly to the drainage system, but conveys liquid waste by discharging it into a plumbing fixture, receptor or receptacle, which in turn is directly connected to the drainage system. Various types of fixtures, appliances and equipment are required to discharge their wastes to the drainage system by indirect waste.

There are two types of indirect waste piping:

1. Indirect waste that discharges to the receptor through an air break
2. Indirect waste that discharges to the receptor through an air gap. There are two types of air gaps:
a. Water distribution air gap.
b. Drainage air gap.

An air break provides a physical separation between the fixture indirect waste pipe and the receptor. If allowed by the code, the fixture indirect waste pipe may terminate below the flood level rim of the receptor, creating a low inlet into the receptor. An air break simply requires that there be no direct physical connection between the fixture indirect waste pipe and the receptor (see **Figure 801.2a**).

The drainage air gap is not the same as the water distribution air gap. The water distribution air gap is defined in Chapter 6 and determined by Table 603.3.1 (see **Figure 801.2b**). The water distribution air gap minimum size is also one inch but also requires that the air gap height be a minimum of two times the diameter of the indirect waste pipe if the size is over one-half of an inch.

The drainage air gap (see **Figure 801.2c**) is defined as the vertical distance as measured from the lowest point of the indirect waste pipe or fixture outlet to the overflow rim of the receptor. This distance is a minimum of one inch. There are no further requirements, unlike those that exist for the water distribution air gap.

Where both air break and air gap are permitted (e.g. Sections 801.3.2 and 801.4) an air break is typically used. However, if only a drainage air gap is required (e.g. Sections 801.3.3 and 801.6) there is a greater need to protect the discharging fixture from contamination.

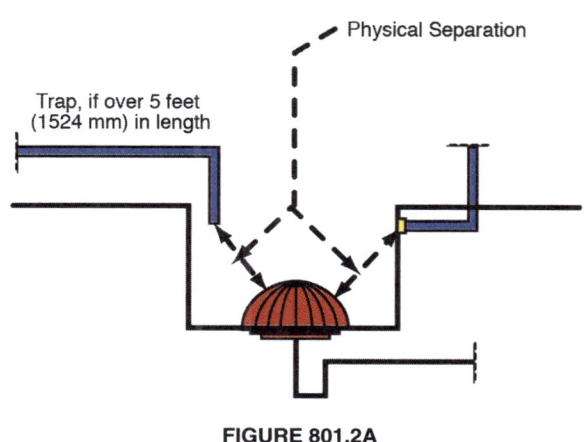

FIGURE 801.2A
AIR BREAK

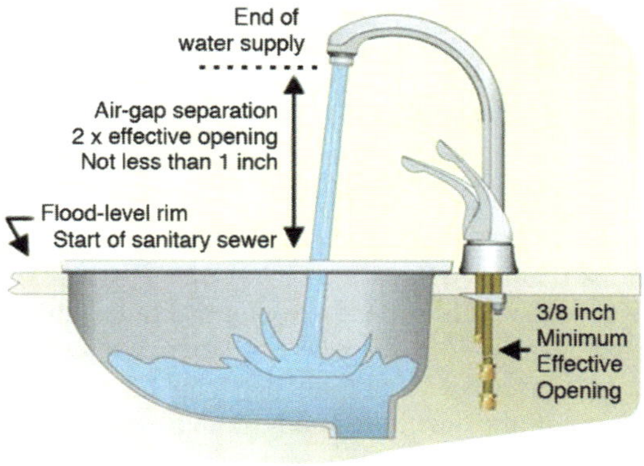

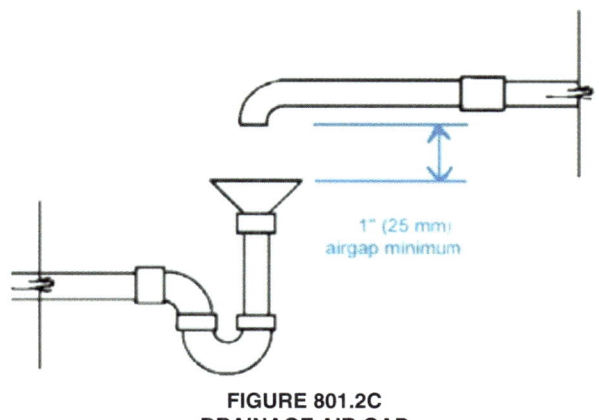

FIGURE 801.2B
WATER DISTRIBUTION AIR GAP

FIGURE 801.2C
DRAINAGE AIR GAP

INDIRECT WASTES

801.3 Food and Beverage Handling Establishments. Establishments engaged in the storage, preparation, selling, serving, processing, or other handling of food and beverage involving the following equipment that requires drainage shall provide indirect waste piping for refrigerators, refrigeration coils, freezers, walk-in coolers, iceboxes, ice-making machines, steam tables, egg boilers, coffee urns and brewers, hot-and-cold drink dispensers, and similar equipment.

🔧 An indirect waste is required for any type of fixture or equipment that in any way may contain or come into contact with food. An indirect waste is also required for drains from refrigeration equipment or coils. The purpose of the indirect waste is to isolate the fixture or equipment from drainage system waste that, in a receptor overflow situation, could contaminate the contents of the equipment, such as in **Figure 801.3**.

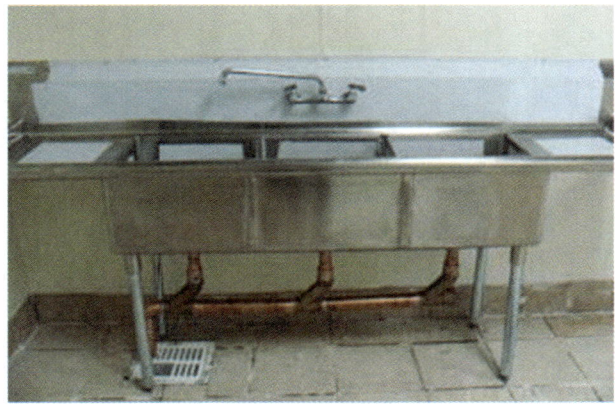

**FIGURE 801.3
INDIRECT WASTE PIPING**

801.3.1 Size of Indirect Waste Pipes. Except for refrigeration coils and ice-making machines, the size of the indirect waste pipe shall be not smaller than the drain on the unit, but shall be not smaller than 1 inch (25 mm), and the maximum developed length shall not exceed 15 feet (4572 mm). Indirect waste pipe for ice-making machines shall be not less than the drain on the unit and in no case less than ¾ of an inch (20 mm).

🔧 Indirect waste piping shall be a minimum of one inch in size but not smaller than the drain of the equipment or fixture. For example, if a food preparation sink outlet is 1½ inches, the indirect waste size would be a minimum of 1½ inches. This indirect waste pipe would be limited in length to a maximum of 15 feet.

An exception to the length requirement is made for refrigeration coils and ice-making machines, which shall be a minimum of three-fourths of an inch in size but not smaller than the drain connection. There is no limitation on the length of the indirect waste piping from this type of equipment. The difference here is that the piping referred to is condensate or condensate overflow piping from the refrigeration unit that poses very little danger of a contamination. This equipment is also usually placed in high areas or mounted on tall equipment, making it difficult to limit the piping to 15 feet as well as for any contamination to reach the coils or airflow of the equipment where it could spread the contamination.

801.3.2 Walk-In Coolers. For walk-in coolers, floor drains shall be permitted to be connected to a separate drainage line discharging into an outside receptor. The flood-level rim of the receptor shall be not less than 6 inches (152 mm) lower than the lowest floor drain. Such floor drains shall be trapped and individually vented. Cleanouts shall be provided at 90 degree (1.57 rad) turns and shall be accessibly located. Such waste shall discharge through an air gap or air break into a trapped and vented receptor, except that a full-size air gap is required where the indirect waste pipe is under vacuum.

🔧 Walk-in coolers are usually erected separately inside the building structure. They are distinct room areas with walls, floors and ceilings designed to refrigerate its contents to a set temperature (see **Figure 801.3.2a**). The contents are most often food and beverage either in open or sealed containers. If the cooler is provided with floor drains, they cannot be directly connected to the drainage system.

The possibility of contaminating the cooler contents in the event of the drainage system backing up, requires that the floor drains in the cooler indirectly discharge into a receptor through an air break or an air gap (an air gap is required where there are negative vacuum conditions). The drains must be trapped and vented as normal, except they may not connect with other vents of the drainage system (see Section 803.3). This is to prevent any contamination that may occur because of airflow between the drainage system and the cooler vents if the traps become dry (see **Figure 801.3.2b**).

In this installation there is one other difference from a normal drainage system. This is the requirement for cleanouts at every 90-degree change in direction. The piping installation under a structure is not easily removed and its proximity to food or beverages requires more care than the normal situation; thus the need for more cleanouts to ensure any blockage can be removed.

One other consideration is the cooler's operating temperature. If it is below freezing, the floor drain traps must be protected from freezing. This can usually be accomplished by applying heat tracing to the trap.

801.3.3 Food-Handling Fixtures. Food-preparation sinks, steam kettles, potato peelers, ice cream dipper wells, and similar equipment shall be indirectly connected to the drainage system by means of an air gap. Bins, sinks, and other equipment having drainage connections and used for the storage of unpackaged ice used for human ingestion, or used in direct contact with ready-to-eat food, shall be indirectly connected to the drainage system by means of an air gap. Each indirect waste pipe from food-handling fixtures or equipment shall be separately piped to the indirect waste receptor and shall not combine with other indirect waste pipes. The piping from the equipment to the receptor shall

INDIRECT WASTES

be not less than the drain on the unit and in no case less than ½ of an inch (15 mm).

🔧 Any fixture, appliance or piece of equipment, such as an ice cream dipper well (see **Figure 801.3.3**), that can contain or may come in contact with food is required to be indirectly wasted through an air gap only. This is to keep any possible contaminates from coming into contact with food, items that may contain food, or any utensils used to eat food. Not doing so could cause a severe health hazard, so the installer should be well aware of the regulations governing such installations.

Ice-maker storage bins or storage containers for ice are also considered food storage equipment. The drain must discharge its waste through an air gap. The size of the indirect waste pipe shall be no smaller than the drain of the container but in no case smaller than one-half of an inch. The indirect waste pipes from this equipment may not combine with other indirect wastes; they must be individually piped to the receptor.

801.4 Bar and Fountain Sink Traps. Where the sink in a bar, soda fountain, or counter is so located that the trap serving the sink cannot be vented, the sink drain shall discharge through an air gap or air break (see Section 801.3.3) into an approved receptor that is vented. The developed length from the fixture outlet to the receptor shall not exceed 5 feet (1524 mm).

🔧 Installing equipment and fixtures in a bar area can be a challenge. There is limited space to place all the equipment and fixtures needed for this installation. In most cases, only a back wall is available to vent the drainage system. This is no help when the sinks are in the counter across from the wall. Therefore, this section was included in the code to allow sinks in this type of installation to be installed with indirect wastes.

Although other sections of this chapter apply to this installation, the length requirement of indirect waste piping is excluded. In this installation it is limited to five feet. This is due to the fact that the items placed in a bar sink are prone to cause more odors than normal, so limiting the length will help to eliminate this problem. Great care should be taken at the time of fixture lay out and rough-in to prevent the later possibility of having to adjust the position of any sinks or plumbing to comply with this requirement.

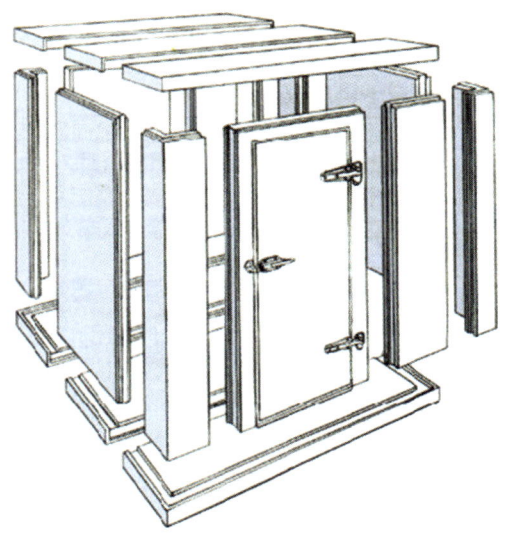

FIGURE 801.3.2A
WALK-IN COOLER INSTALLATION

FIGURE 801.3.3
ICE CREAM DIPPER WELL - ATTACHES TO ICE CREAM COUNTER

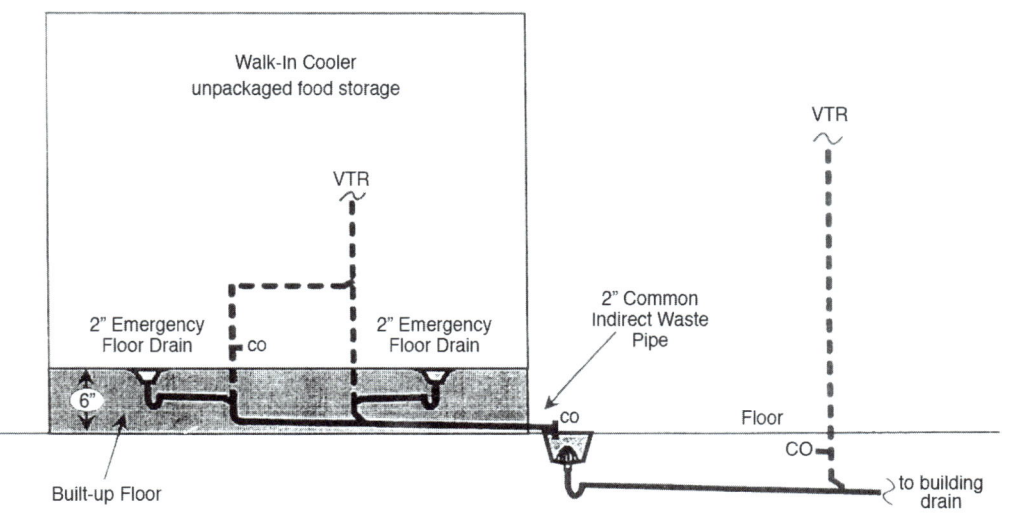

FIGURE 801.3.2B
WALK-IN COOLER FLOOR DRAIN PIPING

2018 UNIFORM PLUMBING CODE ILLUSTRATED TRAINING MANUAL

INDIRECT WASTES

801.5 Connections from Water Distribution System. Indirect waste connections shall be provided for drains, overflows, or relief pipes from potable water pressure tanks, water heaters, boilers, and similar equipment that is connected to the potable water distribution system. Such indirect waste connections shall be made using a water-distribution air gap constructed in accordance with Table 603.3.1.

🔧 The water distribution air gap referred to here is contained in the bottom row of Table 603.3.1. For effective openings of one inch or greater, the required air gap is two times the diameter of the opening. For example, if there is a combination temperature pressure relief valve drain of three-fourths of an inch, the required air gap would be 1½ inches (see **Figure 801.5**).

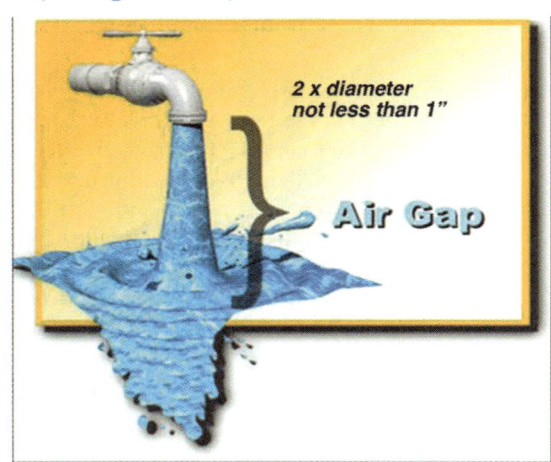

**FIGURE 801.5
WATER DISTRIBUTION PIPE AIR GAP**

801.6 Sterilizers. Lines, devices, or apparatus such as stills, sterilizers, and similar equipment requiring waste connections and used for sterile materials shall be indirectly connected using an air gap. Each such indirect waste pipe shall be separately piped to the receptor and shall not exceed 15 feet (4572 mm). Such receptors shall be located in the same room.

🔧 See **Figure 806.1**.

801.7 Drip or Drainage Outlets. Appliances, devices, or apparatus not regularly classified as plumbing fixtures, but which have a drip or drainage outlets, shall be permitted to be drained by indirect waste pipes discharging into an open receptor through either an air gap or air break (see Section 801.3.1).

🔧 There are many types of devices that have water supplied to them and consequently must have a drain. An example might be a coffee maker with an automatic water fill (see **Figure 801.7**). These devices may be drained by indirect drains discharging to an approved receptor. The size of the drain, its length and need for an air gap or air break are dealt with in other sections of this chapter and will also apply to these installations.

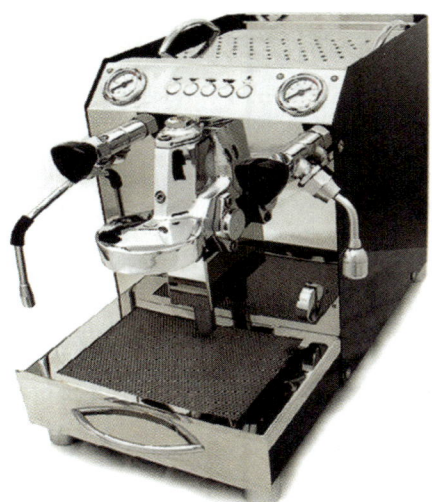

**FIGURE 801.7
ESPRESSO MACHINE WITH DRIP OUTLET**

802.0 Approvals.

802.1 General. No plumbing fixtures served by indirect waste pipes or receiving discharge therefrom shall be installed until first approved by the Authority Having Jurisdiction.

🔧 Before any plumbing fixture that requires an indirect waste pipe or any receptor that receives the discharge from any indirect waste pipe is installed, approval from the Authority Having Jurisdiction (AHJ) is required. Section 104.3.1, Construction Documents, requires that plans and specifications be submitted for approval prior to the installation. In some areas of the country, it is also required that the health department review and approve the plans and specifications prior to submitting them to the AHJ for its approval.

803.0 Indirect Waste Piping.

803.1 Materials. Pipe, tube, and fittings conveying indirect waste shall be of such materials and design as to perform their intended function to the satisfaction of the Authority Having Jurisdiction.

803.2 Copper and Copper Alloys. Joints and connections in copper and copper alloy pipe and tube shall be installed in accordance with Section 705.3.

803.3 Pipe Size and Length. Except as hereinafter provided, the size of indirect waste piping shall be in accordance with other sections of this code applicable to drainage and vent piping. No vent from indirect waste piping shall combine with a sewer-connected vent, but shall extend separately to the outside air. Indirect waste pipes exceeding 5 feet (1524 mm), but less than 15 feet (4572 mm) in length shall be directly trapped, but such traps need not be vented.

Indirect waste pipes less than 15 feet (4572 mm) in length shall be not less than the diameter of the drain outlet or tailpiece of the fixture, appliance, or equipment served, and in no case less than ½ of an inch (15 mm). Angles and changes of direction in such indirect waste pipes shall be provided with cleanouts to permit flushing and cleaning.

INDIRECT WASTES

🔧 The installation of indirect waste piping shall be governed as is other drainage and vent piping and all chapters and sections that apply to drainage and vent piping shall also apply to indirect waste piping. However, the vents from indirect waste shall not intersect with other drainage system vents so as to prevent possible contamination from contaminants emitted in the airflow from the drainage system.

There are two other requirements for indirect waste that are different than normal drainage piping. The first requirement deals with the length of pipe. If the indirect waste piping is longer than five feet, the indirect waste must be trapped but not vented (see **Figure 803.3**). This is to prevent odors from materials that might lie in the piping from entering the fixtures or equipment. Notice that there is no stated requirement for venting of indirect waste over 15 feet. This is because indirect waste lines, except from refrigeration coils and ice machines, are limited to a maximum of 15 feet in length. If an indirect waste line over 15 feet were approved, it would need to be vented as described in Chapter 9.

The second requirement is for cleanouts at all angles and changes of direction. This implies a cleanout is needed at every 45- or 90-degree fitting. This could entail quite a bit of work unless the installation is accomplished with the least amount of piping and fittings as possible.

804.0 Indirect Waste Receptors.

804.1 Standpipe Receptors. Plumbing fixtures or other receptors receiving the discharge of indirect waste pipes shall be approved for the use proposed and shall be of such shape and capacity as to prevent splashing or flooding and shall be located where they are readily accessible for inspection and cleaning. No standpipe receptor for a clothes washer shall extend more than 30 inches (762 mm), or not less than 18 inches (457 mm) above its trap. No trap for a clothes washer standpipe receptor shall be installed below the floor, but shall be roughed in not less than 6 inches (152 mm) and not more than 18 inches (457 mm) above the floor. No indirect waste receptor shall be installed in a toilet room, closet, cupboard, or storeroom, or in a portion of a building not in general use by the occupants thereof; except standpipes for clothes washers shall be permitted to be installed in toilet and bathroom areas where the clothes washer is installed in the same room.

🔧 One of the primary duties of the design professional and the plumber concerning indirect waste is to ensure that the receptor is large enough in physical dimensions and in pipe size to accept the waste flowing into it. The receptor must be deep and wide enough so that there will be no overloading or splashing on the floor that could cause a slip hazard. The size of the receptor should be determined using the parameters set forth in Section 702.0, Fixture Unit Equivalents, Table 702.1 and Table 702.2(2).

The clothes washer standpipe is considered an indirect waste receptor, not a fixture tailpiece, and, as such, has specific requirements for its use. The minimum and maximum elevations of the trap and the standpipe allow for flexibility in installation resulting in a total rise above the floor between 24 and 48 inches, the two possible extremes.

A 24-inch flexibility in the height of the standpipe is helpful, for example, when installing a standpipe with a wash basin (see **Figure 804.1b**). In this example, it would be good practice to elevate the height of the standpipe above the flood level height of the wash basin. In the event of a blockage downstream, the waste would back up in the wash basin before it would overflow through the standpipe. The allowable range of height for the trap and standpipe will allow the clothes washer standpipe to terminate above an adjacent sink (see **Figure 804.1a**). Also, the 18-inch minimum height of the standpipe intends to prevent the possibility of the clothes washer's pumped waste from overflowing the standpipe.

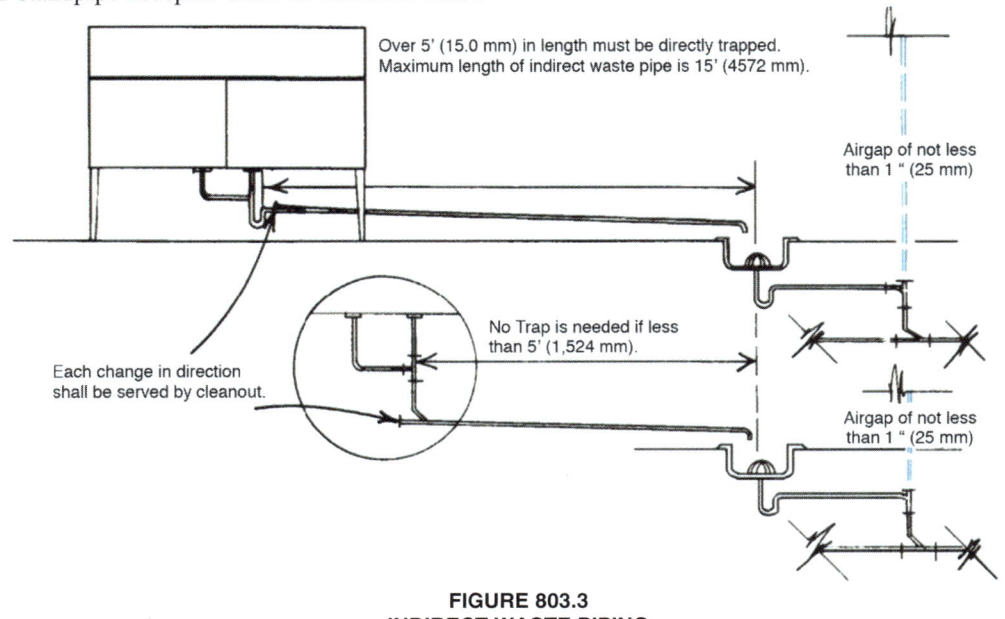

FIGURE 803.3
INDIRECT WASTE PIPING

INDIRECT WASTES

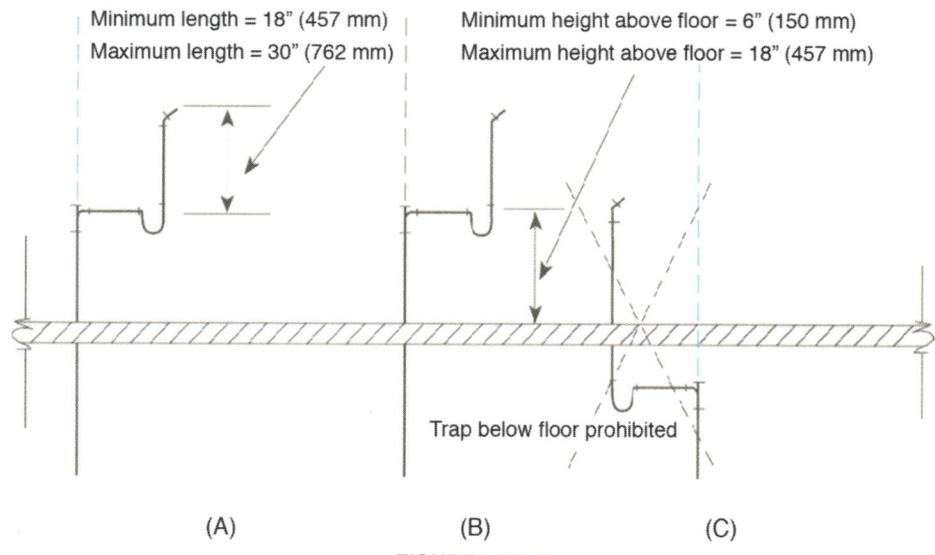

FIGURE 804.1A
CLOTHES WASHER STANDPIPE ELEVATIONS

FIGURE 804.1B
CLOTHES WASHER STANDPIPE INSTALLATION

805.0 Pressure Drainage Connections.

805.1 General. Indirect waste connections shall be provided for drains, overflows, or relief vents from the water supply system, and no piping or equipment carrying wastes or producing wastes or other discharges under pressure shall be directly connected to a part of the drainage system.

The preceding shall not apply to an approved sump pump or to an approved pressure-wasting plumbing fixture or device where the Authority Having Jurisdiction has been satisfied that the drainage system is adequately sized to accommodate the anticipated discharge thereof.

🔧 Since water supply systems are pressurized, any discharges from such systems to the drainage system shall be indirectly wasted. Examples are T&P relief valves from a water heater and pressure-reducing valves (see also Section 801.5).

Pressurized waste is prohibited from direct connection to the drainage system because of the potential to overload or pressurize the system. The drainage system flows by gravity so the introduction of an unaccounted pressurized flow will disrupt that flow.

This provision does not apply to any approved sump pump or pressure-wasting plumbing fixture such as a commercial dishwasher (see Section 414.3), provided the drainage system is adequately sized to receive the anticipated discharge and has previously been approved by the AHJ. Sizing for continuous flow is provided for in Section 710.5.

806.0 Sterile Equipment.

806.1 General. Appliances, devices, or apparatus such as stills, sterilizers, and similar equipment requiring water and waste and used for sterile materials shall be drained through an air gap.

🔧 The plumber should take extreme care to ensure that an air gap and not an air break is used when installing the discharge line from sterilization equipment. **Figure 806.1** is an illustration of an indirectly wasted sterilizer which clearly shows how sterilization equipment should be installed.

807.0 Appliances.

807.1 Non-Classed Apparatus. Commercial dishwashing machines, silverware washing machines, and other appliances, devices, equipment, or other apparatus not regularly classed as plumbing fixtures, which are equipped with pumps, drips, or drainage outlets, shall be permitted to

INDIRECT WASTES

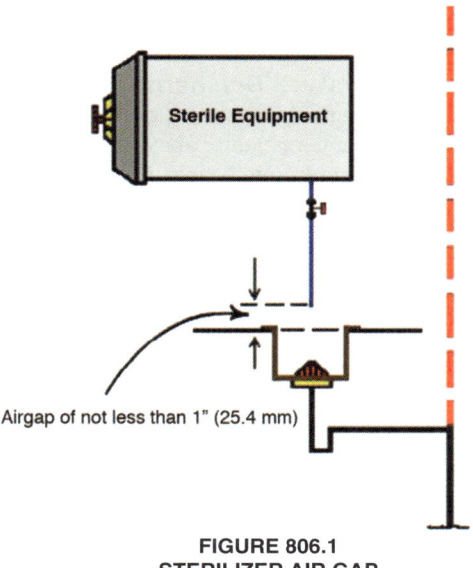

FIGURE 806.1
STERILIZER AIR GAP

be drained by indirect waste pipes discharging into an approved type of open receptor.

807.2 Undiluted Condensate Waste. Where undiluted condensate waste from a fuel-burning condensing appliance is discharged into the drainage system, the material in the drainage system shall be cast-iron, galvanized iron, plastic, or other materials approved for this use.

Exceptions:

(1) Where the above condensate is discharged to an exposed fixture tailpiece and trap, such tailpiece and trap shall be permitted to be a copper alloy.

(2) Materials approved in Section 701.0 shall be permitted to be used where data is provided that the condensate waste is adequately diluted.

807.3 Domestic Dishwashing Machine. No domestic dishwashing machine shall be directly connected to a drainage system or food waste disposer without the use of an approved dishwasher air gap fitting on the discharge side of the dishwashing machine. Listed air gaps shall be installed with the flood-level (FL) marking at or above the flood level of the sink or drainboard, whichever is higher.

ASSE 1021 and the IAPMO material and property standard PS 23 provide for approved listings for domestic dishwasher drain airgaps. Dishwasher drain airgaps are constructed to prevent the backflow of liquid waste when installed with the critical level at/or above the flood level rim of the fixture. The drain air gap will have the marking CL or c/l (ASSE 1021), or FL or F/L (IAPMO PS 23) located on the device to indicate the position of the dishwasher drain airgap in relation to the flood level of the receptacle (see **Figures 807.3a** and **807.3b**).

808.0 Cooling Water.

808.1 General. Where permitted by the Authority Having Jurisdiction, clean running water used exclusively as a cooling medium in an appliance, device, or apparatus shall be permitted to discharge into the drainage system through the inlet side of a fixture trap in the event that a suitable fixture is not available to receive such discharge. Such trap connection shall be by means of a pipe connected to the inlet side of an approved fixture trap, the upper end terminating in a funnel-shaped receptacle set adjacent, and not less than 6 inches (152 mm) above the overflow rim of the fixture.

This method of indirect waste discharge utilizes a funnel connecting to a fixture tailpiece placed 6 inches above the overflow rim of the fixture. Again, note that this only applies to clean water waste used as a cooking medium (see **Figure 808.1**).

809.0 Drinking Fountains.

809.1 General. Drinking fountains shall be permitted to be installed with indirect wastes.

A drinking fountain may be installed using either

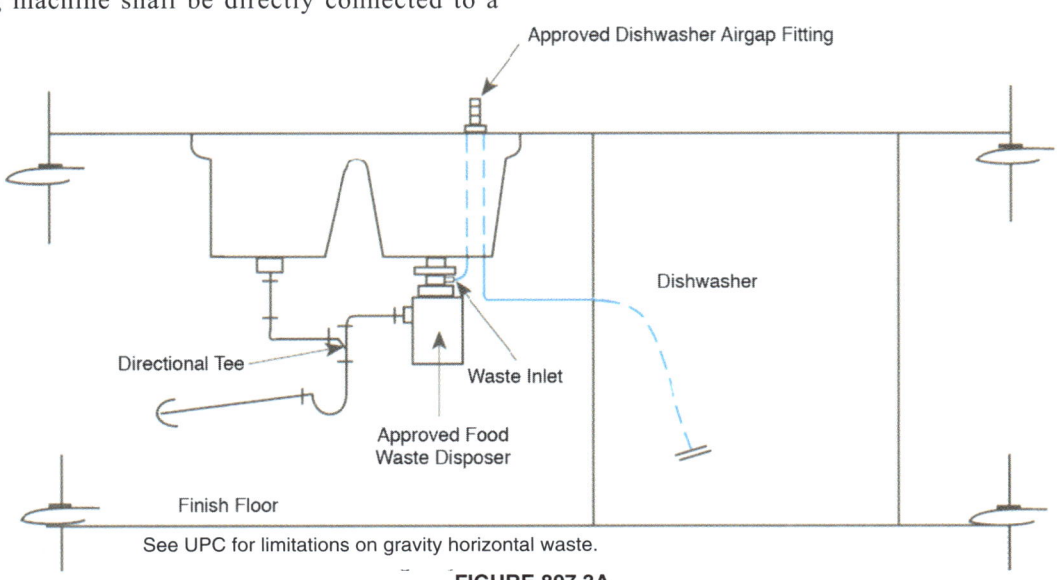

FIGURE 807.3A
DISHWASHER AIR GAP CONNECTION THROUGH DISPOSAL

INDIRECT WASTES

a direct or indirect connection chosen at the discretion of the installer. If the drinking fountain discharges by means of an indirect waste pipe, all of the provisions of Chapter 8 apply to this installation. Because the drinking fountain is supplied with a water distribution air gap, it may be installed with an indirect waste discharging through an air break.

810.0 Steam and Hot Water Drainage Condensers and Sumps.

810.1 High-Temperature Discharge. No steam pipe shall be directly connected to plumbing or drainage system, nor shall water having a temperature above 140°F (60°C) be discharged under pressure directly into a drainage system.

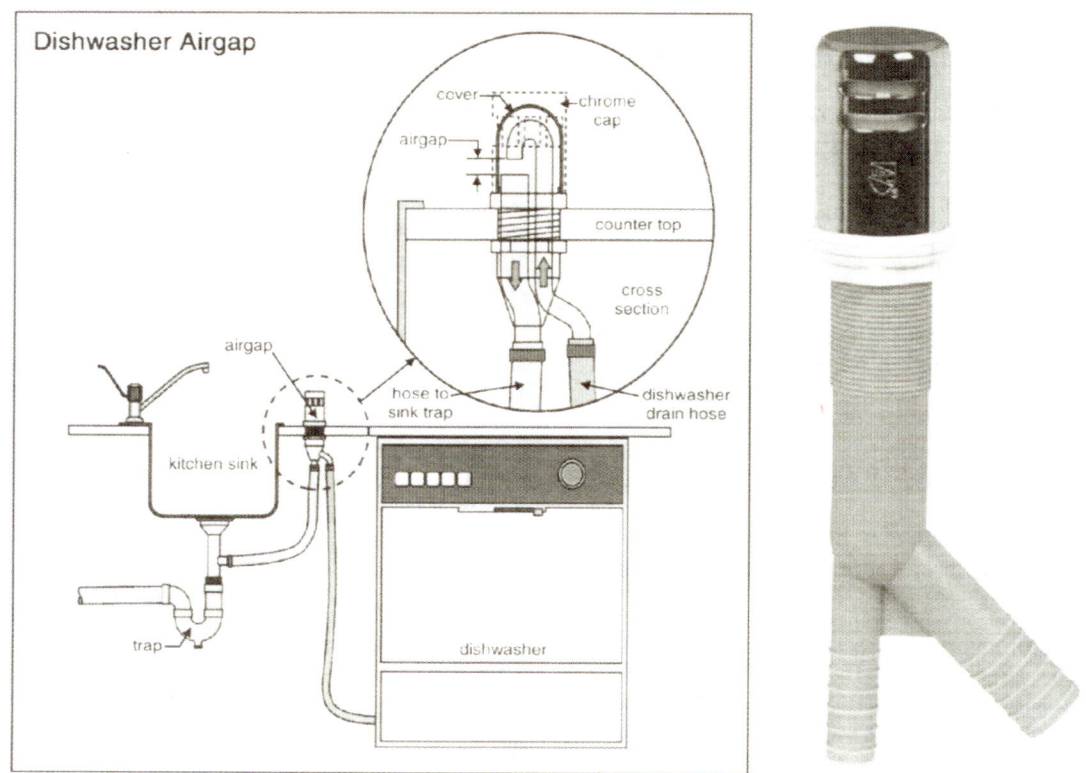

FIGURE 807.3B
DISHWASHER AIR GAP CONNECTION THROUGH SINK TAILPIECE AND AIR GAP FITTING

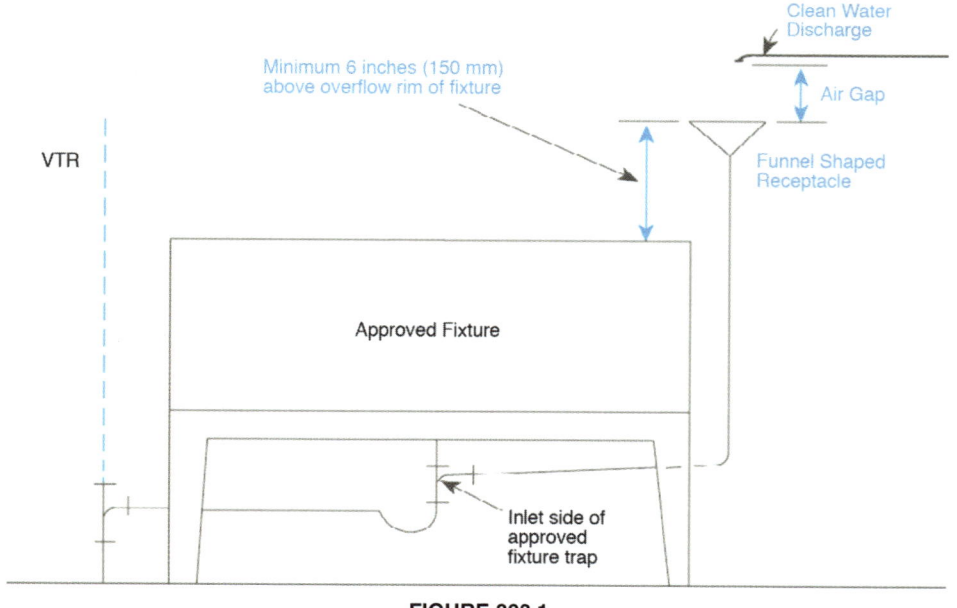

FIGURE 808.1
CLEAR WATER WASTE

INDIRECT WASTES

Pipes from boilers shall discharge by means of indirect waste piping as determined by the Authority Having Jurisdiction or the boiler manufacturer's recommendations. Such pipes shall be permitted to be indirectly connected by discharging into an open or closed condenser or an intercepting sump of an approved type that will prevent the entrance of steam or such water under pressure into the drainage system. Closed condensers or sumps shall be provided with a vent that shall be taken off the top and extended separately, full size above the roof. Condensers and sumps shall be properly trapped at the outlet with a deep seal trap extending to within 6 inches (152 mm) of the bottom of the tank. The top of the deep seal trap shall have a ¾ of an inch (19.1 mm) opening located at the highest point of the trap to serve as a siphon breaker. Outlets shall be taken off from the side in such a manner as to allow a waterline to be maintained that will permanently occupy not less than one-half the capacity of the condenser or sump. Inlets shall enter above the waterline. Wearing plates or baffles shall be installed in the tank to protect the shell. The sizes of the blowoff line inlet, the water outlets, and the vent shall be as shown in Table 810.1. The contents of condensers receiving steam or hot water under pressure shall pass through an open sump before entering the drainage system.

The object here is to eliminate the possibility of steam or hot water 140°F or above from directly entering the drainage system and pressurizing the system. This does not mean that water 140°F or above may not enter the drainage system; rather it may not enter the system by means of a direct connection. The steam or hot water is directed to a containment vessel and then discharged by indirect waste to the drainage system. **Figure 810.1** incorporates all requirements of this section and illustrates the use of the cylindrical metal closed condenser.

TABLE 810.1
PIPE CONNECTIONS IN BLOWOFF CONDENSERS AND SUMPS
(inches)

BOILER BLOWOFF	WATER OUTLET	VENT
¾*	¾*	2
1	1	2½
1¼	1¼	3
1½	1½	4
2	2	5
2½	2½	6

For SI units: 1 inch = 25 mm

* To be used only with boilers of 100 square feet (9.29 m²) of heating surface or less.

Table 810.1 can be used for boilers; however, the plumber should once again be sure to consult the manufacturer's instructions and requirements for the boiler blowoff, outlet and vent.

810.2 Sumps, Condensers, and Intercepting Tanks. Sumps, condensers, or intercepting tanks that are constructed of concrete shall have walls and bottom, not less than 4 inches (102 mm) in thickness, and the inside shall be cement plastered not less than ½ of an inch (12.7 mm) in thickness. Condensers constructed of metal shall be not less than No. 12 U.S. standard gauge (0.109 inch) (2.77 mm), and such metal condensers shall be protected from external corrosion by an approved bituminous coating.

Sumps are constructed either on site or are a package unit and should be installed by the plumber. Installers should always follow the manufacturer's recommendations and instructions when working with these items.

810.3 Cleaning. Sumps and condensers shall be provided with suitable means of access for cleaning and shall contain

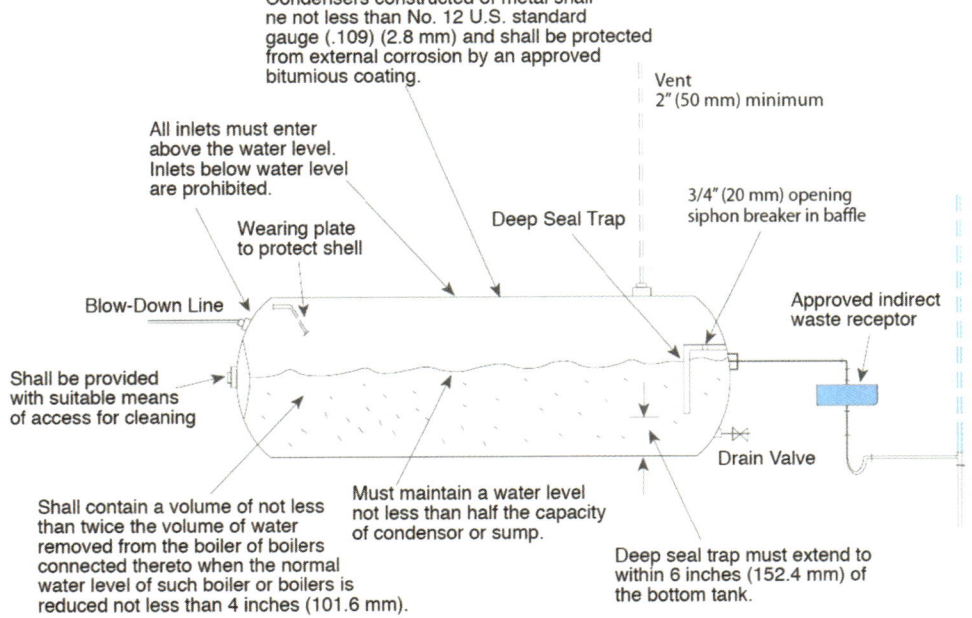

FIGURE 810.1
HOT WATER CONDENSER

INDIRECT WASTES

a volume of not less than twice the volume of water removed from the boiler or boilers connected to it where the normal water level of such boiler or boilers is reduced not less than 4 inches (102 mm).

810.4 Strainers. An indirect waste interceptor is receiving discharge-containing particles that would clog the receptor drain shall have a readily removable beehive strainer.

811.0 Chemical Wastes.

Sections 811.1 through 811.8 mirror or expand on those sections applying to chemical waste in Chapters 3 and 7. They should also be taken into consideration when installing the chemical waste systems described here (see **Figure 811.0**).

811.1 Pretreatment. Chemical or liquid industrial wastes that are likely to damage or increase maintenance costs on the sanitary sewer system, detrimentally affect sewage treatment or contaminate surface or subsurface waters shall be pretreated to render them innocuous before discharge into a drainage system. Detailed construction documents of the pretreatment facilities shall be required by the Authority Having Jurisdiction.

Piping conveying industrial, chemical, or process wastes from their point of origin to sewer-connected pretreatment facilities shall be of such material and design as to adequately perform its intended function to the satisfaction of the Authority Having Jurisdiction. Drainage discharge piping from pretreatment facilities or interceptors shall be in accordance with standard drainage installation procedures.

Copper or copper alloy tube shall not be used for chemical or industrial wastes as defined in this section.

811.2 Waste and Vent Pipes. Each waste pipe receiving or intended to receive the discharge of a fixture into which acid or corrosive chemical is placed, and each vent pipe connected thereto, shall be constructed of chlorinated polyvinyl chloride (CPVC), polypropylene (PP), polyvinylidene fluoride (PVDF), chemical-resistant glass, high-silicon iron pipe, or lead pipe with a wall thickness of not less than $\frac{1}{8}$ of an inch (3.2 mm); an approved type of ceramic glazed or unglazed vitrified clay; or other approved corrosion-resistant materials. CPVC pipe and fittings shall comply with ASTM F2618. PP pipe and fittings shall comply with ASTM F1412 or CSA B181.3. PVDF pipe and fittings shall comply with ASTM F1673 or CSA B181.3. Chemical-resistant glass pipe and fittings shall comply with ASTM C1053. High-silicon iron pipe and fittings shall comply with ASTM A861.

811.3 Joining Materials. Joining materials shall be of approved type and quality.

811.4 Access. Where practicable, the piping shall be readily accessible and installed with the maximum of clearance from other services.

811.5 Permanent Record. The owner shall make and keep a permanent record of the location of piping and venting carrying chemical waste.

811.6 Chemical Vent. No chemical vent shall intersect vents for other services.

811.7 Discharge. Chemical wastes shall be discharged in a manner approved by the Authority Having Jurisdiction.

811.8 Diluted Chemicals. The provisions of this section about materials and methods of construction shall not apply to installations such as photographic or x-ray darkrooms or

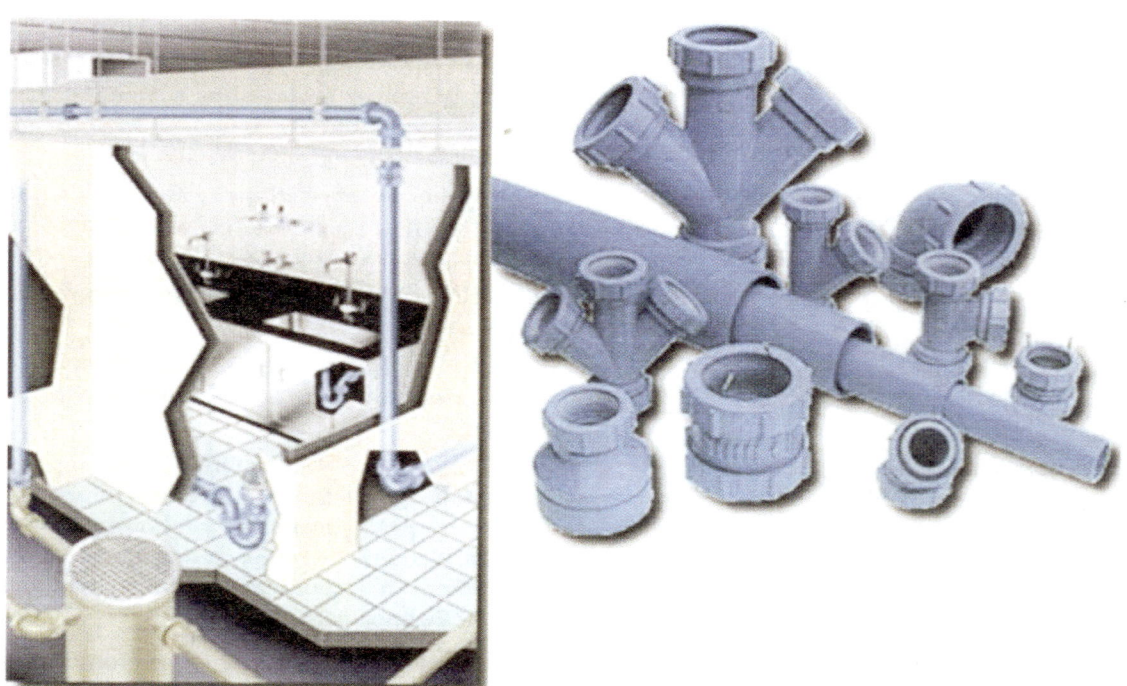

FIGURE 811.0
CHEMICAL WASTE PIPING AND FITTINGS

research or control laboratories where minor amounts of adequately diluted chemicals are discharged.

812.0 Clear Water Wastes.

812.1 General. Water lifts, expansion tanks, cooling jackets, sprinkler systems, drip or overflow pans, or similar devices that discharge clear wastewater into the building drainage system shall discharge through an indirect waste.

813.0 Swimming Pools.

813.1 General. Pipes carrying wastewater from swimming or wading pools, including pool drainage and backwash from filters, shall be installed as an indirect waste. Where a pump is used to discharge pool waste water to the drainage system, the pump discharge shall be installed as an indirect waste.

See **Figure 813.1**.

814.0 Condensate Waste and Control.

814.1 Condensate Disposal. Condensate from air washers, air-cooling coils, condensing appliances, and the overflow from evaporative coolers and similar water-supplied equipment or similar air-conditioning equipment shall be collected and discharged to an approved plumbing fixture or disposal area. Where discharged into the drainage system, equipment shall drain using an indirect waste pipe. The waste pipe shall have a slope of not less than ⅛ inch per foot (10.4 mm/m) or 1 percent slope and shall be of an approved corrosion-resistant material not smaller than the outlet size in accordance with Section 814.3 or Section 814.4 for air-cooling coils or condensing appliances, respectively. Condensate or wastewater shall not drain over a public way.

Condensate waste water is produced from air- or water-cooled appliances and must be discharged to the drainage system (see **Figure 814.1**). Because there are no solids in the condensate waste, it may be run at one-percent slope. This will also be of value for longer runs of piping in ceiling installations.

814.1.1 Condensate Pumps. Where approved by the Authority Having Jurisdiction, condensate pumps shall be installed in accordance with the manufacturer's installation instructions. Pump discharge shall rise vertically to a point where it is possible to connect to a gravity condensate drain and discharged to an approved disposal point. Each condensing unit shall be provided with a separate sump and interlocked with the equipment to prevent the equipment from operating during a failure. Separate pumps shall be permitted to connect to a single gravity indirect waste where equipped with check valves and approved by the Authority Having Jurisdiction.

814.2 Condensate Control. Where an equipment or appliance is installed in a space where damage is capable of resulting from condensate overflow, other than damage to replaceable lay-in ceiling tiles, a drain line shall be provided and shall be drained in accordance with Section 814.1. An additional protection method for condensate overflow shall be provided in accordance with one of the following:

(1) A water level detecting device that will shut off the equipment or appliance in the event the primary drain is blocked.

(2) An additional watertight pan of corrosion-resistant material, with a separate drain line, installed beneath the cooling coil, unit, or the appliance to catch the overflow condensate due to a clogged primary condensate drain.

(3) An additional drain line at a level that is higher than the primary drain line connection of the drain pan.

(4) An additional watertight pan of corrosion-resistant material with a water level detection device installed beneath the cooling coil, unit, or the appliance to catch the overflow condensate due to a clogged primary condensate drain and to shut off the equipment.

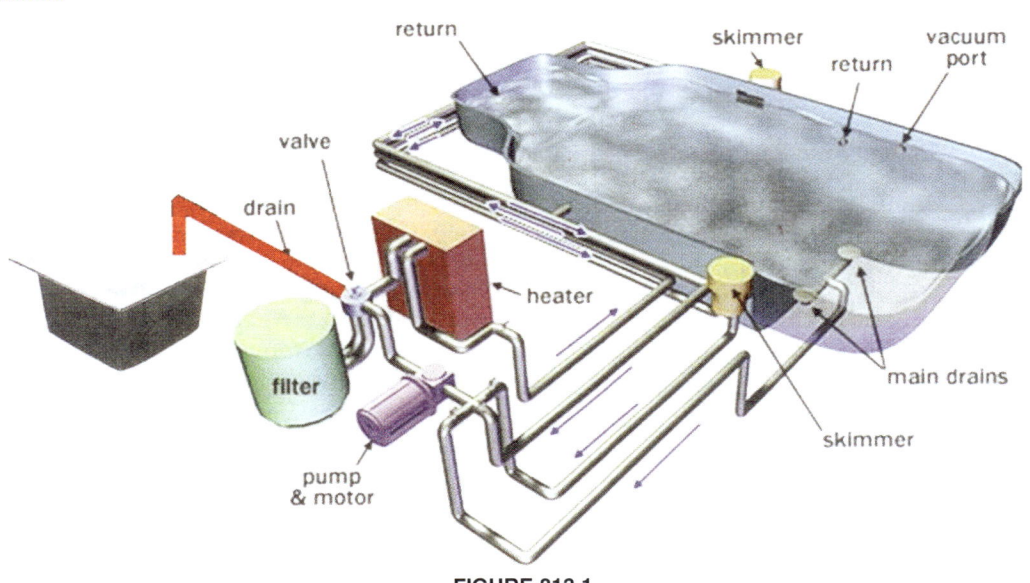

**FIGURE 813.1
SWIMMING POOL INDIRECT WASTE**

INDIRECT WASTES

**FIGURE 814.1
CONDENSATE WASTE**

The additional pan or the additional drain line connection shall be provided with a drain pipe of not less than ¾ of an inch (20 mm) nominal pipe size, discharging at a point that is readily observed.

🔧 When equipment or appliances are installed in a space where damage may result from a blocked primary condensate drain, an overflow or secondary drain must be provided. There are several ways to meet this requirement and the appropriate method is mostly dependent on the design of the equipment or appliance. If the equipment or appliance is equipped with an overflow (secondary) connection, it is to be piped to discharge to a conspicuous location so the public would be alerted that the primary drain is stopped. Approved locations could vary with jurisdictions but would include a point above a showerhead in a bathtub or shower or outdoors above an exterior window. If no secondary drain outlet is provided, a water-tight, corrosion-resistant pan is to be installed under the equipment or appliance to collect the overflow. The pan should be piped similarly to the secondary drain pipe. The pan drain pipe shall be a minimum ¾ inch, discharging at a point that can be readily observed.

814.2.1 Protection of Appurtenances. Where insulation or appurtenances are installed where damage is capable of resulting from a condensate drain pan overfill, such installations shall occur above the rim of the drain pan with supports. Where the supports are in contact with the condensate waste, the supports shall be of approved corrosion-resistant material.

814.3 Condensate Waste Pipe Material and Sizing. Condensate waste pipes from air-cooling coils shall be sized in accordance with the equipment capacity as specified in Table 814.3. The material of the piping shall comply with the pressure and temperature rating of the appliance or equipment and shall be approved for use with the liquid being discharged.

**TABLE 814.3
MINIMUM CONDENSATE PIPE SIZE**

EQUIPMENT CAPACITY IN TONS OF REFRIGERATION	MINIMUM CONDENSATE PIPE DIAMETER (inches)
Up to 20	¾
21 – 40	1
41 – 90	1¼
91 – 125	1½
126 – 250	2

For SI units: 1 ton of refrigerant = 3.52 kW, 1 inch = 25 mm

The size of condensate waste pipes is for one unit or a combination of units, or as recommended by the manufacturer. The capacity of waste pipes assumes a ⅛ inch per foot (10.4 mm/m) or 1 percent slope, with the pipe running three-quarters full at the following pipe conditions:

Outside Air – 20%		Room Air – 80%	
DB	WB	DB	WB
90°F	73°F	75°F	62.5°F

For SI units: °C = (°F-32)/1.8

Condensate drain sizing for other slopes or other conditions shall be approved by the Authority Having Jurisdiction.

Air-conditioning waste pipes shall be constructed of materials specified in Chapter 7.

🔧 Always consult the equipment manufacturer's installation instructions for proper sizing of the condensate drain. When there are no requirements Table 814.3 can be used to size the drain. Drains may be combined using Table 814.3 for sizing. For example, if there are two units at 20 tons of refrigeration each, they would both require a ¾ inch drain. When the wastes are combined in an indirect waste line it would be sized for 40 tons of refrigeration, which would require a minimum of 1 inch indirect waste line.

The sizing table in this section could be used in most areas, except for those areas with excessively high humidity weather patterns. In determining the size of the pipe, convert the Btu/hr capacity of the cooling equipment into tons of refrigeration. 12,000 Btu/h of cooling capacity = 1 ton of refrigeration.

814.3.1 Cleanouts. Condensate drain lines shall be configured or provided with a cleanout to permit the clearing of blockages and for maintenance without requiring the drain line to be cut.

814.4 Appliance Condensate Drains. Condensate drain lines from individual condensing appliances shall be sized as required by the manufacturer's instructions. Condensate drain lines serving more than one appliance shall be approved by the Authority Having Jurisdiction prior to installation.

Condensing appliances should have means of disposal of the condensation similar to cooling coils. Usually, the manufacturer provides detailed instructions on how to dispose of the condensation, including the minimum size of condensate piping (based on the predicted amounts of condensation).

Since the condensation tends to be corrosive in nature, the installation instructions should be followed closely as to the type of condensate piping material to be used and also what the suitability is of drainage piping material receiving the corrosive discharge.

814.5 Point of Discharge. Air-conditioning condensate waste pipes shall connect indirectly, except where permitted in Section 814.6, to the drainage system through an air gap or air break to trapped and vented receptors, dry wells, leach pits, or the tailpiece of plumbing fixtures. A condensate drain shall be trapped in accordance with the appliance manufacturer's instructions or as approved.

The installation for this type of equipment may result in outdoor installations located far from a receptor. Therefore, the discharge may drain to a dry well, leach pit or other type of receptor, but only if approved by the AHJ beforehand. This type of discharge may be continuous, depending on the appliance and humidity of the local area. The drain should never discharge to a sidewalk or walkway, which may cause a hazardous condition.

814.6 Condensate Waste From Air-Conditioning Coils. Where the condensate waste from air-conditioning coils discharges by direct connection to a lavatory tailpiece or to an approved accessible inlet on a bathtub overflow, the connection shall be located in the area controlled by the same person controlling the air-conditioned space.

A condensate indirect waste line may discharge and connect to the tailpiece of a lavatory (see **Figure 814.6a**) or the overflow of a bathtub (see **Figure 814.6b**). Both areas of connection must be accessible and both should be wye-type connections. This only applies to condensate waste. The user of this Code should consult the AHJ prior to installing commercial appliances to determine whether a direct or indirect connection in required on a specific appliance.

If the condensate is discharging in a space you control, you also need to be able control when condensate can drain in that location by controlling the thermostat. Imagine if there was a stoppage where condensate was discharging and you don't have access to shut-off the AC equipment.

814.7 Plastic Fittings. Female plastic screwed fittings shall be used with male plastic fittings and plastic threads.

Since metal pipe threads are usually tapered (IPS), the code does not allow female threaded plastic fittings from joining male metal fitting as the plastic fitting would crack if the joint is tightened beyond the allowable tensile strength of the plastic.

FIGURE 814.6A
LAVATORY TAILPIECE CONNECTION

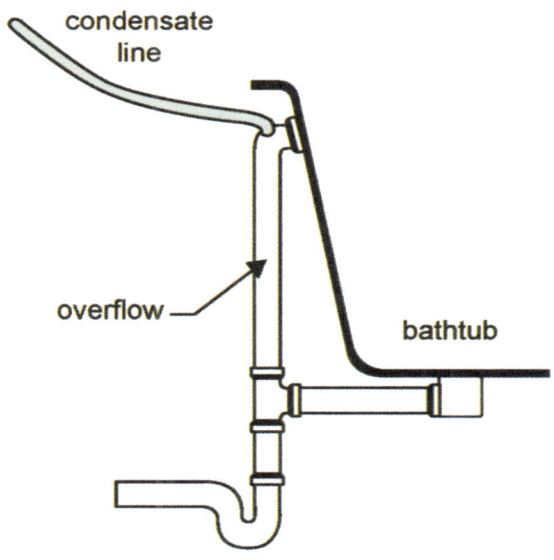

FIGURE 814.6B
BATHTUB OVERFLOW CONNECTION

CHAPTER 9

VENTS

901.0 General.

901.1 Applicability. This chapter shall govern the materials, design, and installation of plumbing vent systems.

901.2 Vents Required. Each plumbing fixture trap, except as otherwise provided in this code, shall be protected against siphonage and backpressure, and air circulation shall be ensured throughout all parts of the drainage system by means of vent pipes installed in accordance with the requirements of this chapter and as otherwise required by this code.

This section requires that the fixture trap be protected against siphonage and backpressure by means of a vent system. The principle of ventilating the drainage system to retain trap seals has been the basis of modern plumbing systems since the end of the 19th century. Prior to the 20th century, plumbing fixtures were rarely used inside the building. The protection of the trap and its trap seal (see **Figure 901.2a**) by a vent has allowed the plumbing fixture to be placed safely inside the building (see also commentary for traps and trap protection in Sections 1001.2 and 1002.1). Without a vent, the trap seal is subject to siphonage or backpressure, which would allow sewer gas to enter the building or room. The consequences of sewer gas escaping the drainage system and entering the room can be lethal, as was proven in the SARS epidemic (Severe Acute Respiratory Syndrome) reported in Hong Kong, China in 2003, where 65 deaths and 321 infected people resulted from the virus being transmitted in part through dry traps.

siphon will occur. Also, a negative pressure occurs in the drain when a fixture is discharged. The discharging fixture may siphon an adjacent fixture trap that is not vented. For these reasons, venting is required to relieve any aspiration that may affect a trap seal.

Siphonage may also occur by a fixture siphoning itself at the end of its discharge. This is called self-siphonage. According to *Recommended Minimum Requirements for Plumbing* (BH13) published by the National Bureau of Standards (NBS), self-siphonage occurs when there is an unbalanced water column in an inverted U shape caused by the P-trap as seen in the common siphon (see **Figure 901.2b**). Self-siphonage is also caused by the fluid velocity, or the momentum pull exerted by the moving column of water, whether the pipe is horizontal, vertical, or inclined. In a horizontal drain from the trap, the momentum pull in its full intensity is exerted at the end of the discharge (see **Figure 901.2c**). Therefore, such configurations are prohibited without the fixture traps properly vented to prevent self-siphonage.

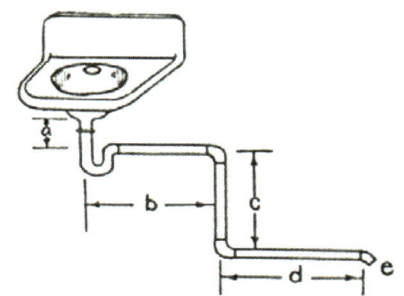

FIGURE 901.2B
SELF SIPHONAGE CAUSED BY INVERTED TRAP

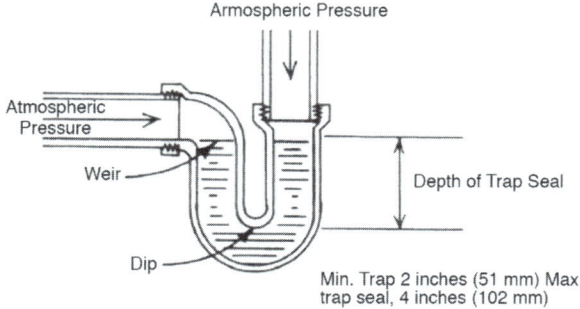

FIGURE 901.2A
THE TRAP AND TRAP SEAL

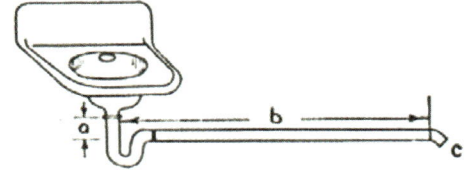

FIGURE 901.2C
SELF SIPHONAGE CAUSED BY MOMENTUM PULL
(Reprinted from *Recommended Minimum Requirements for Plumbing*, BH13)

Siphonage

Siphonage can occur in two ways. It occurs when there is sufficient negative pressure (aspiration or negative pressure below atmospheric) to draw water out of the fixture trap. A negative pressure occurs in a drainage stack as water rapidly flows downward, drawing air into itself (entrainment). If the sheet of water flows past an unvented branch connecting to the stack, a negative pressure will also be produced in the branch affecting any trap seals in the branch. If the negative pressure is great enough, then a

Backpressure

Backpressure occurs when there is positive air pressure (increase or positive pressure above atmospheric) in the drainage system. As water flows down a stack drawing air along with it, it also pushes air ahead of itself. The air volume will flow through the building drain and dissipate into the sewer drainage system unless there is a blockage in the air path, such as when the building drain is full, or when there is a building trap or a backwater valve. Such block-

Vents

ages in the air path generate positive pressures and may have an adverse effect on trap seals. This back pressure on the outlet side of the trap could push air bubbles through the trap way releasing sewer gas into the living space and, if significant enough, could even blow the water out of the trap way (see **Figure 901.2d**).

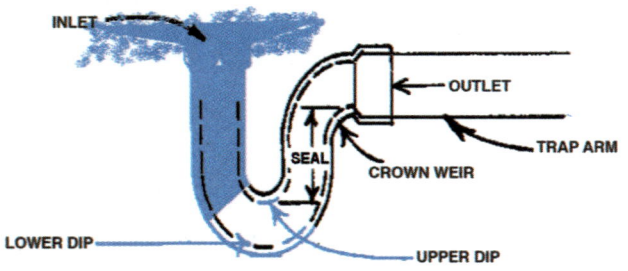

FIGURE 901.2D
TRAP SEAL BLOWOUT CAUSED BY BACKPRESSURE

A particular blockage or closure in the air path is the hydraulic jump. A hydraulic jump is a flow condition where the flow rate becomes greater than the drain capacity causing the depth of the flow to be greater than the pipe diameter. This sudden rise is called a jump and may completely block air flow (see **Figure 901.2e**). Hydraulic jumps may occur at horizontal junctions where one branch flows into another and at the base of drainage stacks (see also commentary in Section 706.4). When the hydraulic jump completely fills the drain, positive air pressure waves will be generated, and if they last long enough then traps seals in close proximity may be affected. Backpressures generated by hydraulic jumps must be relieved by the venting system to protect trap seals.

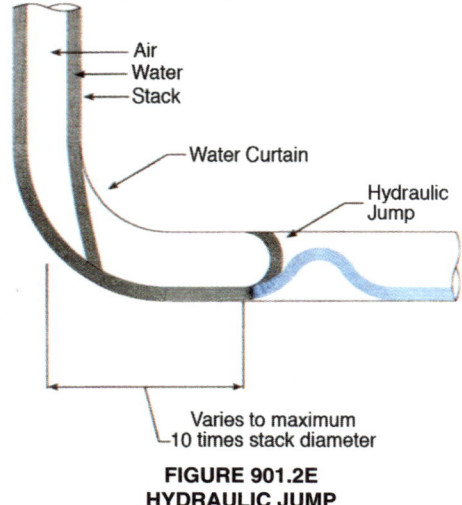

FIGURE 901.2E
HYDRAULIC JUMP

Air Circulation

Section 901.2 also requires air circulation shall be ensured throughout all parts of the drainage system by means of vent pipes. This statement means that air must circulate throughout the drainage system including the building drain, building sewer, and public sewer. The basic principles of venting ensures that the equalization of atmospheric pressure on both sides of the trap seal within plus or minus one-inch water column, will be provided at all times and in all parts of the drainage system.

As already mentioned, building traps (see **Figure 901.2f**) block air circulation to the building drain and sewer and therefore are prohibited in Section 1008.1 except where allowed by the Authority Having Jurisdiction (AHJ). See Figure 1101.15b for an example of where building traps are used in the combined sewer and storm drainage systems. The reason the building trap is not used extensively in the Uniform Plumbing Code (UPC) is that the storm system is intended to be separate from the sewer system. Sewer manholes are sealed unlike storm system manholes and do not provide for air circulation in the sewer system. Therefore, the circulation of air for the public sewer, the building sewer and the building drainage system must come from the vents in the building alone. In Section 1008.1, the building trap is allowed but will require a relieving vent to provide for the circulation of air upstream of the building trap, and in Section 904.1 it will require another vent downstream of the building trap to allow circulation of air to the public sewer or private sewage disposal system.

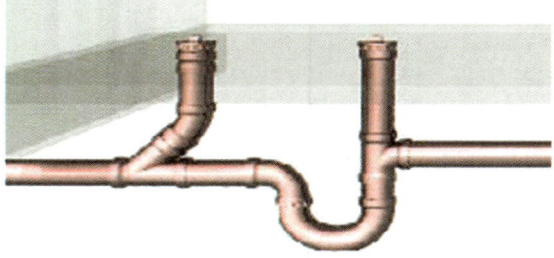

FIGURE 901.2F
ONE TYPE OF BUILDING TRAP

901.3 Trap Seal Protection. The vent system shall be designed to prevent a trap seal from being exposed to a pressure differential that exceeds 1 inch water column (0.24 kPa) on the outlet side of the trap.

The commentary in Section 901.2 explained that positive and negative air pressures within the drainage system may affect fixture trap seals. At what point will the pressures produced be detrimental to the trap seal? This was the question raised by the subcommittee on plumbing under the direction of the National Bureau of Standards in the 1920s [see *Recommended Minimum Requirements for Plumbing* (BH13) p. 198]. The subcommittee agreed that any trap having a full seal depth of two inches should not be reduced to less than one inch. This is equivalent to a variation of plus or minus 2-inches of water column (.07 psi). A factor of safety was introduced into the testing criterion whereby the pressure variations within the branch lines were not to exceed plus or minus 1-inch of water column (.04 psi), which in a trap with a two-inch seal would leave a 1½ inch seal under suction and approximately a full seal under back pressure. This factor of safety has determined vent sizing in all plumbing codes.

902.0 Vents Not Required.

902.1 Interceptor. Vent piping shall be permitted to be omitted on an interceptor where such interceptor acts as a primary settling tank and discharges through a horizontal indirect waste pipe into a secondary interceptor. The second interceptor shall be properly trapped and vented.

See **Figure 902.1** for an illustration of this installation. Section 1009.0, Interceptors (Clarifiers) and Separators, and Section 1016.0, Sand Interceptors, are the areas affected by this exception.

902.2 Bars, Soda Fountains, and Counter. Traps serving sinks that are part of the equipment of bars, soda fountains, and counters need not be vented where the location and construction of such bars, soda fountains, and counters are such as to make it impossible to do so. Where such conditions exist, said sinks shall discharge using approved indirect waste pipes into a floor sink or other approved type of receptor.

This section allows indirect waste for special cases where venting is impossible. An example would be an oval bar with seating around it where an adjacent wall is not available for a vent. Traps serving soda fountains and counter sinks that are part of the bar equipment may be indirectly wasted into an approved receptor. Therefore, Chapter 8, Indirect Wastes will apply. See **Figure 902.2** for an example of this installation.

903.0 Materials.

903.1 Applicable Standards. Vent pipe and fittings shall comply with the applicable standards referenced in Table 701.2, except that:

(1) No galvanized steel or 304 stainless steel pipe shall be installed underground and shall be not less than 6 inches (152 mm) aboveground.

(2) ABS and PVC DWV piping installations shall be in accordance with Chapter 14 "Firestop Protection." Except for individual single-family dwelling units, materials exposed within ducts or plenums shall have a flame-spread index of not more than 25 and a smoke-developed index of not more than 50 where tested in accordance with ASTM E84 or UL 723. These tests shall comply with all requirements of the standards to include the sample size, both for width and length. Plastic pipe shall not be tested filled with water.

903.2 Use of Copper or Copper Alloy Tubing. Copper or copper alloy tube for underground drainage and vent piping shall have a weight of not less than that of copper or copper alloy drainage tube type DWV.

903.2.1 Aboveground. Copper or copper alloy tube for aboveground drainage and vent piping shall have a weight of not less than that of copper or copper alloy drainage tube type DWV.

903.2.2 Prohibited Use. Copper or copper alloy tube

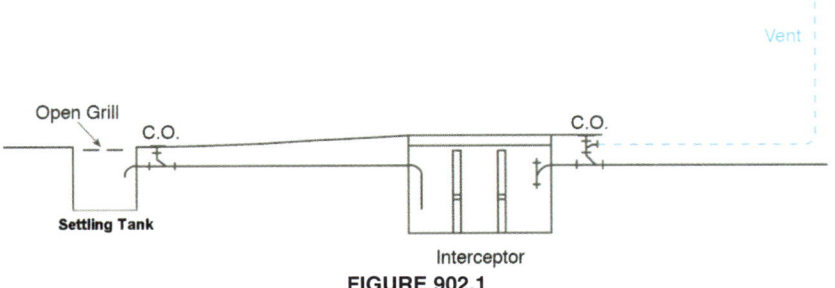

FIGURE 902.1
VENT NOT REQUIRED ON SETTLING TANK

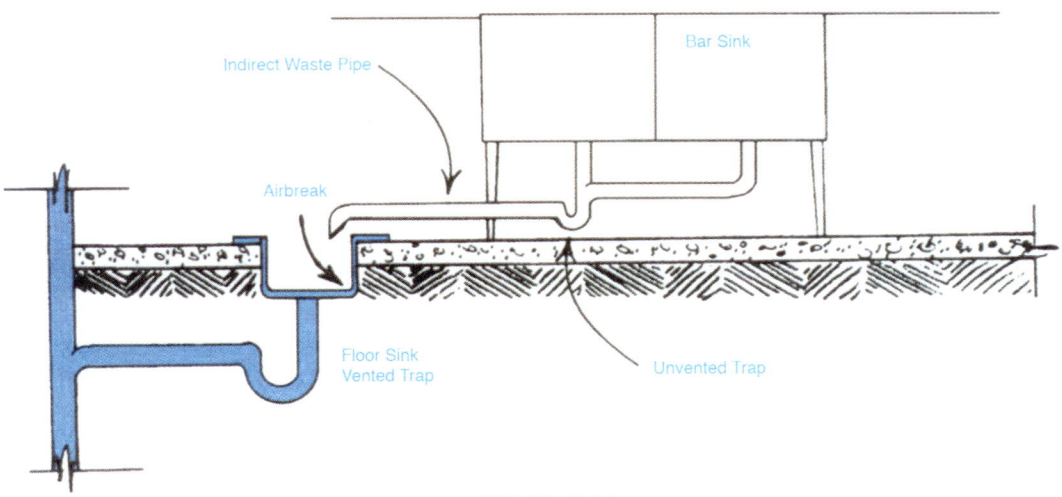

FIGURE 902.2
INDIRECT WASTE - VENT NOT REQUIRED

shall not be used for chemical or industrial wastes as defined in Section 811.0.

903.2.3 Marking. Copper or copper alloy tubing, in addition to the required incised marking, shall be marked in accordance with either ASTM B306 or ASTM B88. The colors shall be Type K, green; Type L, blue; Type M, red; and Type DWV, yellow.

903.3 Changes in Direction. Changes in the direction of vent piping shall be made by the appropriate use of approved fittings, and no such pipe shall be strained or bent. Burred ends shall be reamed to the full bore of the pipe.

🔧 Approved fittings used for changes of direction of vent piping are practically the same for drainage piping. One difference is the vent 90 fitting (see **Figure 903.3a**), which can be used anywhere in the vent system except below the overflow rim of the fixture (see Section 905.3). Above the overflow rim of the fixture any type of drainage fitting can be used for venting purposes.

As with drainage fittings, vent fittings must also be installed in the direction of flow. The direction of air flow in the venting system has a different orientation than drainage flows. For example, vent branch-fittings are usually turned in the opposite orientation than drainage fittings to enhance air flow. Not only for airflow, but also a path must be provided for condensate to flow to the fixture drain without it being trapped in the horizontal piping and blocking the flow of air. The sanitary tee is a good example of a fitting used in one position in drainage piping and reversed in vent piping (see **Figure 903.3b**).

FIGURE 903.3A
ONE-FORTH BEND AND VENT 90

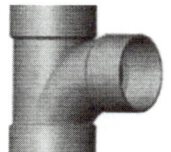 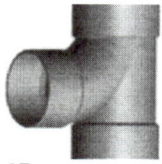

FIGURE 903.3B
SANITARY TEE FOR DRAINAGE AND VENTING

904.0 Size of Vents.

904.1 Size. The size of vent piping shall be determined from its length and the total number of fixture units connected thereto, in accordance with Table 703.2. The diameter of an individual vent shall be not less than 1¼ inches (32 mm) nor less than one-half the diameter of the drain to which it is connected. In addition, the drainage piping of each building and each connection to a public sewer or a private sewage disposal system shall be vented by means of one or more vent pipes, the aggregate cross-sectional area of which shall be not less than that of the largest required building sewer as determined from Table 703.2. Vent pipes from fixtures located upstream from pumps, ejectors, backwater valves, or other devices that obstruct the free flow of air and other gases between the building sewer and the outside atmosphere shall not be used for meeting the cross-sectional area venting requirements of this section.

Exception: Where connected to a common building sewer, the drainage piping of two or more buildings located on the same lot and under one ownership shall be permitted to be vented by means of piping sized in accordance with Table 703.2, provided the aggregate cross-sectional area of vents is not less than that of the largest required common building sewer.

🔧 This code section contains three important elements that determine vent pipe sizing. They each address a separate area of sizing the vent system that may also have implications on drainage pipe sizing. These three elements that determine the sizing of vents are summarized in the following points and will be expanded upon afterward.

1. The length of the vent pipe and the total number of fixture units connected to it as determined by Table 703.2, with the caveat of meeting the minimum size requirements.

2. The requirement to provide aggregate ventilation for the building and public sewer with vent piping equivalent to the diameter of the required building sewer, with the caveat that certain vent pipes are unable to provide aggregate ventilation.

3. A common building sewer may be vented by the aggregate venting systems of two or more buildings connected to it.

Length and Fixture Units

The first important element that determines vent pipe sizing is the length of the vent pipe and the total number of fixture units connected to it as determined by Table 703.2. Since the vent system is directly related to the drainage system by providing airflow to it, the same sizing table for drain piping in Chapter 7, Sanitary Drainage, is also used for vent piping. The last row in Table 703.2 is specific for vent piping and determines the diameter of the vent pipe according to the number of fixture units connected to it and the length of the vent pipe. Vent piping is assigned drainage fixture units (DFUs) from Table 702.1 in the same manner as drainage piping. Each pipe size is limited to the maximum fixture unit value listed in the column beneath it. For example, two-inch vent piping is limited to 24 total fixture units.

The diameter of the vent pipe is also influenced by its total length. The reason for this is that air is a gaseous fluid and is subject to friction loss as water is. Air has capacity, or volume, in cubic feet or gallons and has velocity in cubic feet per minute or gallons per minute. As air moves through the piping system, the physical contact with the pipe wall creates a resistance to flow (friction), which varies as the square of the velocity. In other words, when the air flow velocity doubles, the drop of pressure due to friction (or the frictional resistance) increases by a factor of four. If the length of vent is too long to the source of air, frictional resistance will eventually diminish the capacity of the air needed to relieve negative and positive pressures to protect trap seals. For this reason, vent piping is limited to a length that only allows a pressure drop of plus or minus one-inch of water column as factor of safety required for trap seal protection (see commentary for Section 901.3). Referring to Table 703.2 in the "Vent Piping" row and "Maximum Lengths", a two-inch vent is limited to 120 feet of total developed length.

There are two footnote references to Table 703.2 that apply to vent pipes. Footnote 6 reiterates the beginning of Section 904.1 regarding minimum vent size requirements. The minimum vent size of 1¼ inch was considered the minimum practical size for installation purposes when this recommendation was given in the early 1900s. Pipe sizes can actually be further reduced using gas flow formulas when applied in an alternate engineered vent system (Section 912.0).

Footnote 6 also references Tables 702.1 and 702.2(2) for the determination of drainage fixture units (see commentary in Chapter 7 for these sections), and limits vent piping installed in a horizontal position to one-third of the total permitted length. The reason for this limitation dates back to the late 1940s when the American Standards Association was developing the *American Standard National Plumbing Code* (known as the A40 Code). This association participated in harmonizing the requirements of several plumbing codes then in existence. Here in the A40 Code is the first appearance of a limitation on the horizontal portion of the vent pipe. Under *Size of Vent Piping*, it states that "Twenty per cent of the total length may be installed in a horizontal position." Most likely, this was borrowed from the City of Detroit Code. This twenty percent rule was later modified by the Uniform Plumbing Code (UPC) Committee to one-third the total permitted developed length of the vent and introduced in the 1952 edition of the UPC. Although gas flow formulas do not make a distinction in frictional resistance whether air flows are vertical or horizontal, nor did the early research of the National Bureau of Standards indicate the need to limit the lengths of horizontal vent pipes, the limitation was an accepted idea. For example, a 2-inch vent is limited to a maximum of 40 feet in a horizontal position.

Lastly, footnote 6 eliminates the length restriction when vent piping is increased one pipe size. An increase of one pipe size will increase the capacity of the vent pipe 50 to 75 percent. This increase in capacity will permit an equivalent volume of air to flow with less friction loss, making the length limitation unnecessary. This exemption is applicable to vertical and horizontal vent piping. A 2-inch vent will serve a maximum 24 fixture units at a maximum length of 120 feet and limited to 40 feet in the horizontal position. When increased to 3 inches, there are no length limitations whether vertical or horizontal. This principle is also restated in Section 904.2.

Footnote 3 affects the maximum fixture unit loading of 1½-inch vent pipe. One and one-half inch vent pipe may serve eight fixture units but it may not serve six-unit traps [i.e. 3-inch traps per Table 702.2(1)] or water closets. If a water closet or a six-unit trap is included in these eight-fixture units, the vent pipe must be increased in size from 1½ inches to 2 inches. Therefore, a water closet will require a 2-inch minimum sized vent.

Aggregate Ventilation

The second important element that determines vent pipe sizing is the requirement to provide aggregate ventilation for the building and public sewer with vent piping equivalent to the diameter of the required building sewer. Section 904.1 states: "the drainage piping of each building and each connection to a public sewer or private sewage disposal system shall be vented by means of one or more vent pipes, the aggregate cross-sectional area of which shall not be less than that of the largest required building sewer as determined from Table 703.2." We shall consider this requirement in three parts.

First, the drainage piping of each building refers to the building drain and the connection to the public or private sewer refers to the building sewer (see Chapter 2 for definitions of each), both of which are sized according to Table 703.2. In order for the building sewer to be properly ventilated, a vent pipe of the same diameter as the building sewer needs to connect to the drainage piping and extend through the roof undiminished in size (see **Figure 904.1a**). Notice that the Code requires the vent pipe to be the same diameter as the building sewer *as determined by Table 703.2*. A local utility may install a building sewer larger than what is determined by Table 703.2, yet the required diameter for the vent pipe does not need to increase likewise.

Secondly, this code section allows more than one vent pipe to ventilate the building sewer. When using this option, the cumulative sum capacity of the vent pipes communicated from the vent terminals to the building sewer must be equal to or greater than the capacity of the building sewer. The sum capacity is calculated by means of aggregating (summing) the cross-sectional area of each vent pipe used to ventilate the building drain. The cross-sectional area is simply the area of a circle (πr^2 or $\pi d^2/4$) represented by the radius or diameter of the pipe (see **Figure 904.1b**). For example, a building sewer with a diameter of four inches has a cross-sectional area that would equal the area of a 4-inch circle, which is 12.57 in^2 ($\pi 2^2$ or $\pi 4^2/4$). The vent or vents terminating through the roof must also equal 12.57 in^2. This may be accomplished by one 4-inch vent or any combination of sizes and numbers that equal 12.57 in^2. See **Figure 904.1c** for different aggregate options equal to or greater than 12.57 in^2.

Furthermore, the cross-sectional area of each vent must be conveyed through the drainage pipe of at least the same size as the cumulated aggregate total. For example, in **Figure 904.1d** three vents in one bathroom group are chosen to ventilate a 3-inch building sewer having a cross-sectional area of 7.07 in^2 (Pipe Section C). The toilet vent is the required minimum of 2 inches, and the lavatory and shower vents are increased to 1½ inches and 2 inch respectively to meet the minimum aggregate sum of 7.07 in^2. The aggregate cross-sectional area of all three vents is 8.05 in^2,

VENTS

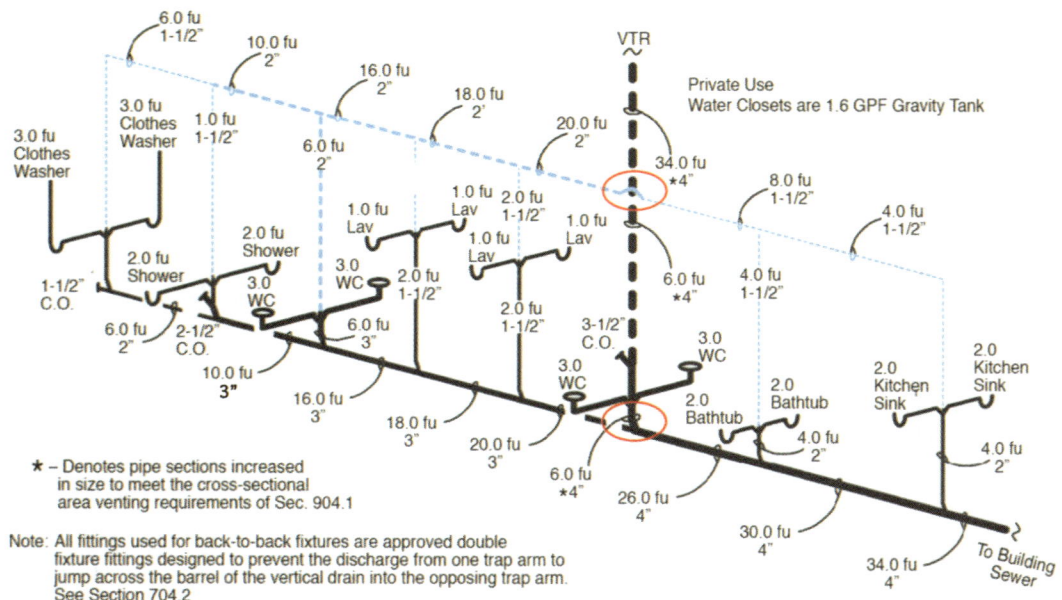

**FIGURE 904.1A
SIZE INCREASE DUE TO CROSS-SECTIONAL AREA**

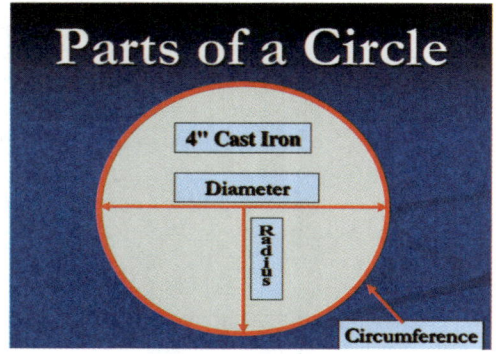

**FIGURE 904.1B
PARTS OF A CIRCLE**

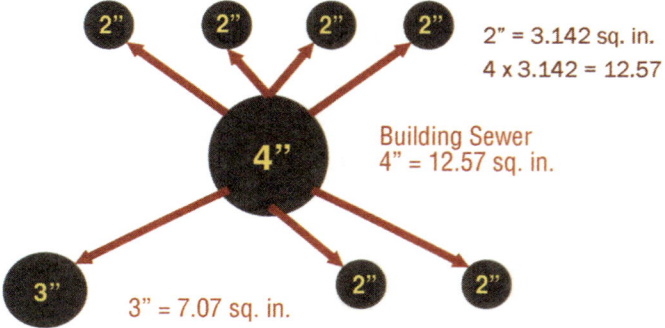

**FIGURE 904.1C
AGGREGATE CROSS-SECTIONAL AREA**

which exceeds the minimum required cross-sectional area of 7.07 in^2. According to Table 702.1 and Table 703.2, pipe section B is permitted to be 2 inches. However, the 2-inch drain does not convey the cumulative aggregate cross-sectional area of both the lavatory and shower vents to the three-inch building sewer. Notice Chart 1 in **Figure 904.1d**. The cumulative cross-sectional area of the lavatory and shower vents is 4.91 in^2. This cross-sectional area exceeds the cross-sectional area of the 2-inch drain, which is only 3.14 in^2. Therefore, only 3.14 in^2 can be aggregated with the 2-inch toilet vent, which totals 6.28 in^2 for venting the building sewer. This is less than the required 7.07 in^2. Therefore, pipe section B is required to increase to 3 inches with a cross-sectional area of 7.07 in^2. In Chart 2, pipe section B is increased to 3 inches, allowing the aggregate cross-sectional area of the lavatory and shower vents (4.91 in^2) communicate to Pipe Section C. The toilet vent adds to the aggregate sum (8.05 in^2) meeting the minimum required cross sectional area of the building sewer of 7.07 in^2.

Figure 904.1e is another example where either the vent pipe needs to increase one pipe size (Option 1) or the drain pipe needs to increase one pipe size (Option 2).

In complying with the aggregate cross-sectional area principle, the vents may individually terminate through the roof or they may be combined together so only one vent terminates through the roof. If the vents are combined so only one vent terminates through the roof, then each vent must connect to the vent terminal whose diameter is equal to or greater than the required diameter for the building drain. Using the example in **Figure 904.1d**, the three vents must connect to a vent terminal whose cross-sectional area is equal to or greater than 7.07 in^2 (see **Figure 904.1f**). See also **Figure 904.1g** as another method of connecting multiple vents to a single 3-inch vent stack through the roof.

Thirdly, certain vent pipes are unable to provide aggregate ventilation to the building sewer. Section 904.1 further states, "Vent pipes from fixtures located upstream from pumps, ejectors, backwater valves, or other devices that

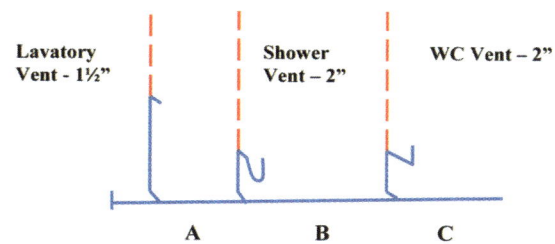

Chart 1: Aggregate Venting						
Pipe Section	DFU	Drain Size	Vent Size	Cross-sectional Area of Drain	Cross-sectional Area of Vent	Cumulative Aggregate Total
A	1	1 ½"	1 ½"	1.77 in^2	1.77 in^2	1.77 in^2
B	3	2"	2"	3.14 in^2	3.14 in^2	4.91 in^2
C	6	3"	2"	7.07 in^2	3.14 in^2	6.28 in^2

Chart 2: Aggregate Venting						
Pipe Section	DFU	Drain Size	Vent Size	Cross-sectional Area of Drain	Cross-sectional Area of Vent	Cumulative Aggregate Total
A	1	1 ½"	1 ½"	1.77 in^2	1.77 in^2	1.77 in^2
B	3	3"	2"	7.07 in^2	3.14 in^2	4.91 in^2
C	6	3"	2"	7.07 in^2	3.14 in^2	8.05 in^2

FIGURE 904.1D
AGGREGATE VENTING THROUGH DRAINAGE PIPES

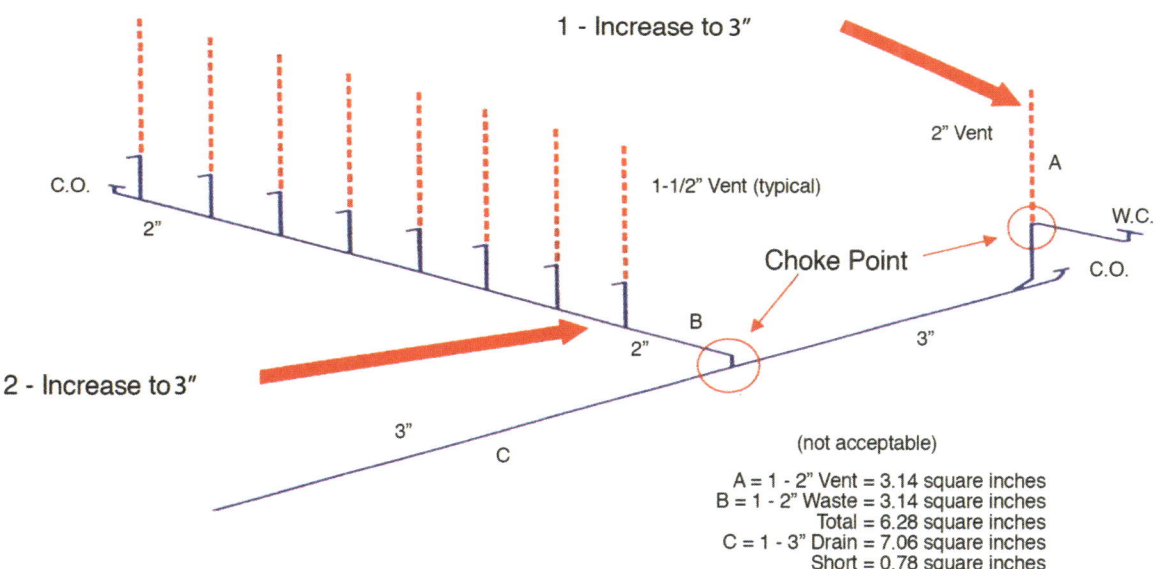

FIGURE 904.1E
SIZE INCREASE DUE TO CROSS-SECTIONAL AREA

obstruct the free flow of air and other gases between the building sewer and the outside atmosphere shall not be used for meeting the cross-sectional area venting requirements of this section." These are conditions where the flow of air in the drainage system is diminished or blocked from entering the building drain and sewer. For example, the flapper of the backwater valve is at rest or closed in normal position where air cannot flow through the valve in either direction. This effectively prevents any air flow through the vent and drainage piping to ventilate the building sewer. Therefore, vent pipes upstream of the backwater valve may not contribute to any of the cross-sectional area needed to satisfy the requirements of aggregate venting (see **Figure 904.1h**). Also, the vent from a sump or vents from fixtures flowing into a sump cannot contribute to the aggregate venting requirements either. Such vents are venting fixture traps below the elevation of the building drain and sewer. The drains that these vents serve are not physically connected to the drainage system in a way that will allow the air flow in the system. **Figure 904.1i** provides an example where all the venting principles discussed so far are displayed.

VENTS

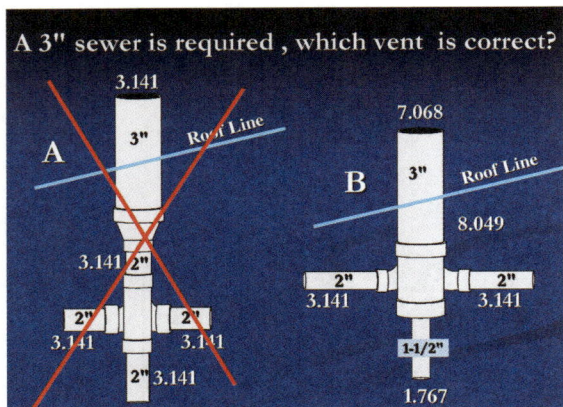

FIGURE 904.1F
AGGREGATE VENTING COMBINED AT VENT TERMINAL

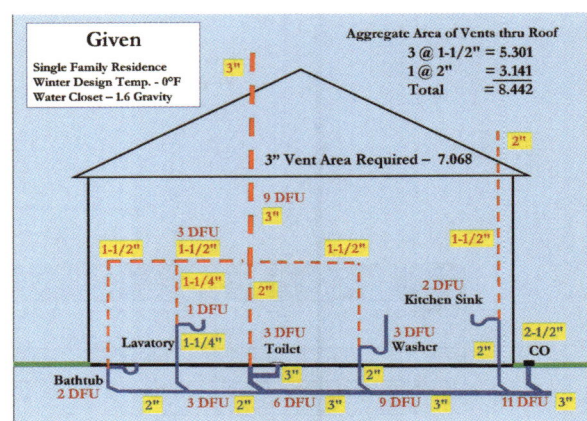

FIGURE 904.1G
AGGREGATE VENTING COMBINED AT VENT STACK

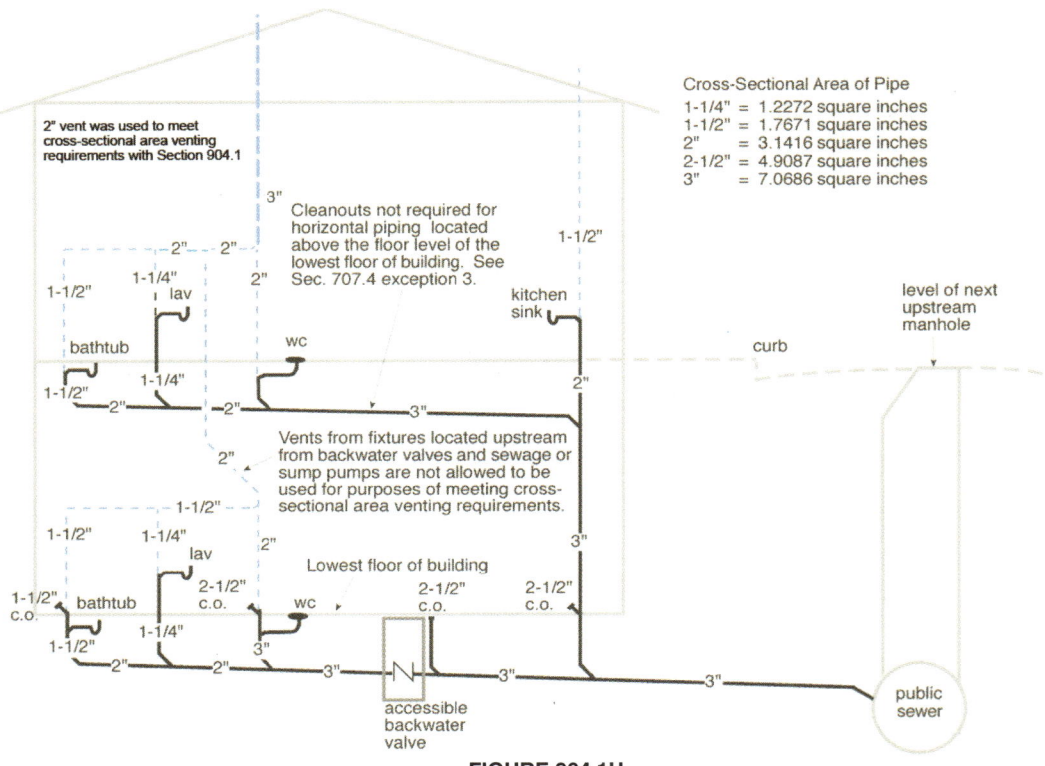

FIGURE 904.1H
FIXTURES LOCATED BELOW UPSTREAM MANHOLE COVER

Common Building Sewer

The third important element that determines vent pipe sizing is when a common building sewer may be vented by the aggregate venting systems of two or more buildings connected to it. The exception of Section 904.1 states that if there is more than one building on a lot, under a single ownership and the buildings are connected to the same building sewer, the vents for the buildings can be combined to satisfy the cross-sectional area requirement of the common building sewer. **Figure 311.1** illustrates two buildings on one lot.

In summary, the steps to follow for sizing the vent system are:

1. Assign fixture units and trap and trap arm sizes to each fixture.

2. Assign the fixture unit loading to each pipe segment in the system. In the drainage system, begin at the uppermost fixture on the system and work down the system, totaling fixture units as each pipe segment is assigned. In the vent system it is just the opposite. Begin at the lowest fixture on the system and work upwards, totaling fixture units as each pipe segment is assigned.

3. Using Table 703.2, now assign pipe sizes to the system based on the fixture units that were assigned to each pipe segment.

Check for the aggregate cross-sectional area as explained above.

VENTS

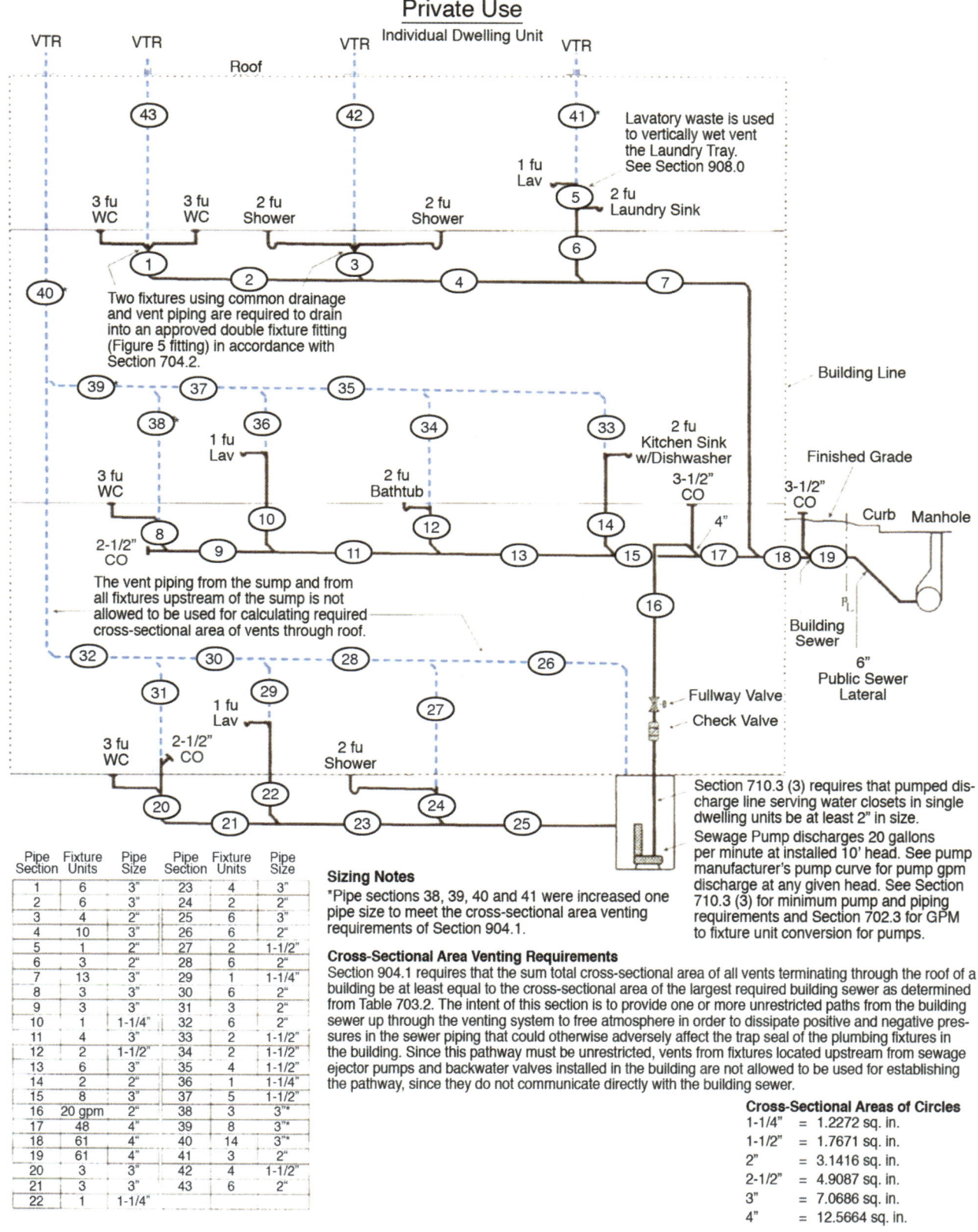

**FIGURE 904.1I
SIZING EXAMPLE**

2018 UNIFORM PLUMBING CODE ILLUSTRATED TRAINING MANUAL

VENTS

904.2 Length. Not more than one-third of the total permitted length, in accordance with Table 703.2, of a minimum-sized vent shall be installed in a horizontal position.

Exception: Where a minimum-sized vent is increased one pipe size for its entire length, the maximum length limitation shall not apply.

905.0 Vent Pipe Grades and Connections.

905.1 Grade. Vent and branch vent pipes shall be free from drops or sags, and each such vent shall be level or shall be so graded and connected as to drip back by gravity to the drainage pipe it serves.

🔧 The atmosphere inside the drainage and vent system is very humid. There will be conditions when condensation is produced inside the vent system and it must be allowed to drain to the drainage piping. When connecting individual vents, branch vents or interconnecting vent stacks or other vents, there should be no sags or back pitch of the vent pipe, which will trap condensate in the pipe. This would reduce the area of that portion of the venting system and diminish the efficiency of the vent system. Therefore, vents must be level or graded back to the fixture drain (see **Figure 905.1**).

FIGURE 905.1
VENT PIPING LEVEL OR GRADED

905.2 Horizontal Drainage Pipe. Where vents connect to a horizontal drainage pipe, each vent pipe shall have its invert taken off above the drainage centerline of such pipe downstream of the trap being served.

🔧 Although vent piping may be installed in a horizontal position, vent connections to the drainage piping they serve may not be connected horizontally. The "invert" (interior bottom surface of a pipe) of the vent connection must be above the centerline of the drainage piping to which it is connected (see **Figure 905.2a**). The centerline of the drainage pipe is the elevation within the horizontal drain line that would normally not be exceeded by the flow of effluent except during surge conditions. The objective here is to keep the vent opening above the flow of effluent. This will prevent the effluent flow through the drain line from blocking the flow of air through the vent and from flowing into the vent piping. **Figure 905.2b** illustrates how this is accomplished. The vent will remain "dry" unless a drain line stoppage creates a flooded condition.

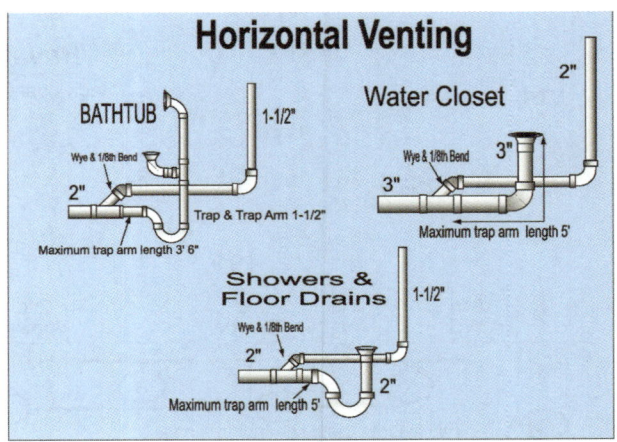

FIGURE 905.2A
HORIZONTAL VENTING

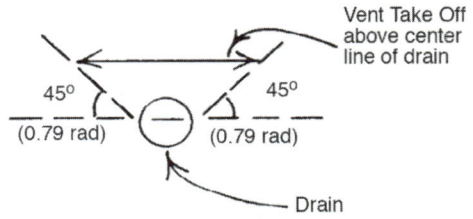

FIGURE 905.2B
VENT CONNECTION TO HORIZONTAL DRAIN LINE

905.3 Vent Pipe Rise. Unless prohibited by structural conditions, each vent shall rise vertically to a point not less than 6 inches (152 mm) above the flood-level rim of the fixture served before offsetting horizontally, and where two or more vent pipes converge, each such vent pipe shall rise to a point not less than 6 inches (152 mm) in height above the flood-level rim of the plumbing fixture it serves before being connected to any other vent. Vents less than 6 inches (152 mm) above the flood-level rim of the fixture shall be installed with approved drainage fittings, material, and grade to the drain.

🔧 The ideal installation for a plumbing fixture would be a vertical fixture drain connecting with the fixture trap arm by a sanitary tee with the vent rising vertically and connecting with a branch vent or going through the roof. This is defined as a "continuous vent" (see **Figure 905.3a**). This continuous vent must remain vertical until it is at least six inches above the overflow rim of the fixture. It may then turn horizontally or connect to other vents in the system. This ensures that if there is a stoppage and an overflow condition, the waste from one fixture will not flow into the other fixtures via the vent.

The vent connection is always below the overflow rim of the fixture but normally there is only a short vertical piece that is below the fixture rim; however, this is not always possible. Sometimes there will be obstructions in the structure of the building that will prevent this ideal condition. Significant portions of the vent may have to be installed below the flood-level rim of the fixture. For example, there

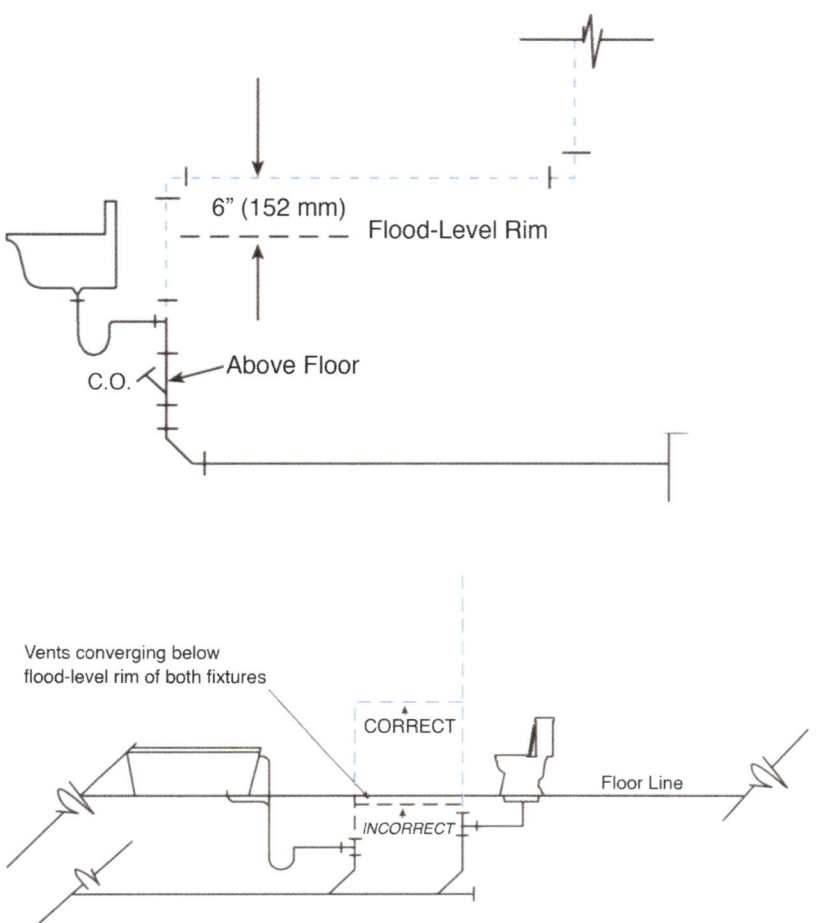

FIGURE 905.3A
VENTS MUST RISE 6 INCHES ABOVE THE OVERFLOW RIM OF THE FIXTURE BEFORE CONNECTING WITH OTHER VENTS

may be an installation of a floor drain where the distance from a wall that might be used to place the drain's vent exceeds the trap arm distance. In this installation the vent would have to be placed in a horizontal position until it could turn vertically in the wall (see **Figure 905.3b**).

This vent, until it rises 6 inches above the overflow rim of the fixture, is in fact a drain. If there is an overflow condition, the vent would be filled to the rim of the fixture. When the stoppage is cleared, the waste in the vent has to flow to the drain and it must do this using drainage fittings. The horizontal portion of the vent must also be graded back to the drain to allow the overflow effluent to recede.

905.4 Roof Termination. Vent pipes shall extend undiminished in size above the roof, or shall be reconnected with soil or waste vent of the proper size.

There shall be no reduction in the size of a vent as this would of course reduce the cross-sectional area of the vent. The individual vent may be connected with other vents using the sizing methods discussed earlier. The vent stack must terminate above the roof of the building to eliminate the possibility of gases from the plumbing system to enter the building. See Section 906.0 for additional requirements for vent terminals.

FIGURE 905.3B
VENT INSTALLED IN HORIZONTAL POSITION

905.5 Location of Opening. The vent pipe opening from soil or waste pipe, except for water closets and similar fixtures, shall not be below the weir of the trap.

This section is identical to Section 1002.4. A vent

pipe connection that is below the weir of the trap (see Definitions, Crown Weir (Trap Weir), and **Figure 1001.2a**) allows the potential for the fixture to induce self-siphonage (see commentary for Section 1002.2 for further discussion). Therefore, lengths of trap arms are limited and slopes are restricted to prevent the vent pipe connection to be below the weir of the trap (see Table 1002.2). If the vent connection is below the weir of the trap, this indicates that either the length or the slope of the trap arm is not code compliant.

This will determine what type of fitting may be used to connect the trap arm, vent and fixture drain. For example, in vertical installations, the sanitary tee is the proper fitting to use for this connection. A combination wye and one-eighth bend or a wye will not be allowed here since the opening of the vent would be below the weir of the trap (see **Figure 905.5a** and commentary for Section 1002.2 for further discussion).

The above requirement does not apply to water closets and similar siphon-waste fixtures. In reality, these fixtures are intentionally S trapped so that siphonage will occur, thereby emptying the entire trap seal and its contents with every flush. These fixtures rely on a mechanical refill device (ballcock) to seal the empty S trap. Consequently, the primary function of vents serving these fixtures is to provide a source of air to follow behind the surge of water being discharged from water closets and other below-the-weir waste/vent fixture connections (see **Figure 905.5b**).

905.6 Common Vertical Pipe. Two fixtures shall be permitted to be served by a common vertical pipe where each such fixture wastes separately into an approved double fitting having inlet openings at the same level.

Fixtures may be set back to back or side by side and use a common vent sized by combining the drainage fixture units of both fixtures to obtain the correct vent pipe size. The fitting used must be a double fixture fitting (see **Figure 905.6**). The double fixture fitting provides an opening that directs the flow from the two trap arms down the fixture drain. A double sanitary tee does not have this feature and could allow the effluent from one fixture to flow into the trap arm of the other fixture. The double combination or double wye and one-eighth bend could not be used

(except for water closets or similar fixtures) because the vent opening would be below the weir of the trap.

906.0 Vent Termination.

The following sections provide requirements for placement of the vent termination through the roof. The purpose of these requirements is to provide enough separation from the termination of the vent to allow for the diffusion of

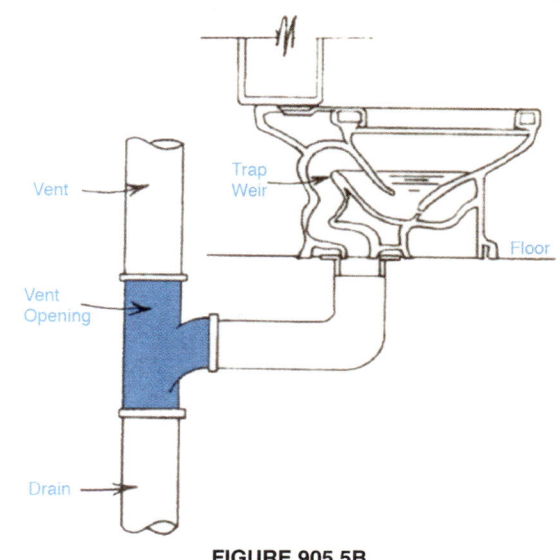

FIGURE 905.5B
WATER CLOSET VENT

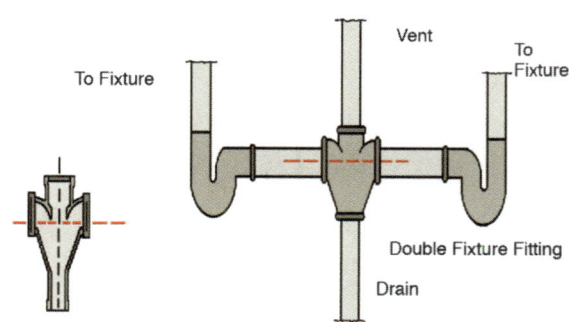

FIGURE 905.6
BACK-TO-BACK FIXTURES INSTALLED WITH DOUBLE FIXTURE FITTING

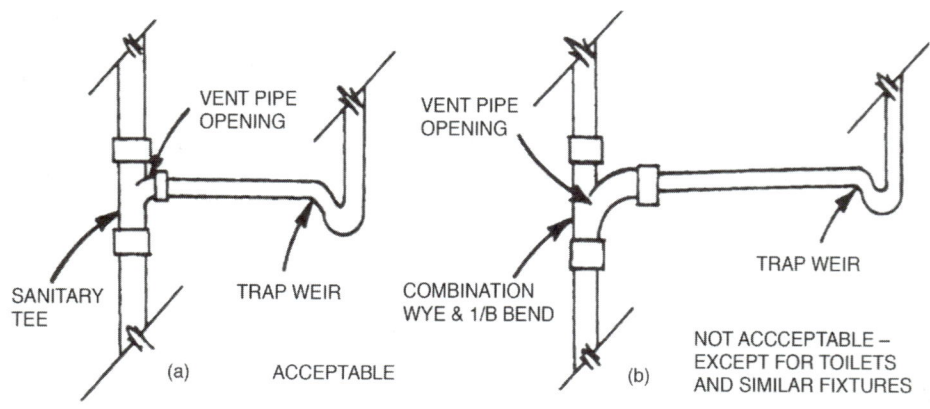

FIGURE 905.5A
THE PROPER FITTING FOR VERTICAL VENT

the atmosphere inside the vent to the outside air, preventing it from entering the building. Sufficient forethought should be given to vent terminations in planning the path of the drainage and vent system. Be aware of all equipment locations, air intakes, openings into the building and lot lines. See Figure 906.0 for an illustration of these requirements.

906.1 Roof Termination. Each vent pipe or stack shall extend through its flashing and shall terminate vertically not less than 6 inches (152 mm) above the roof nor less than 1 foot (305 mm) from a vertical surface. ABS and PVC piping exposed to sunlight shall be protected by water based synthetic latex paints.

906.2 Clearance. Each vent shall terminate not less than 10 feet (3048 mm) from, or not less than 3 feet (914 mm) above, an openable window, door, opening, air intake, or vent shaft, or not less than 3 feet (914 mm) in every direction from a lot line, alley and street excepted.

906.3 Use of Roof. Vent pipes shall be extended separately or combined, of full required size, not less than 6 inches (152 mm) above the roof or firewall. Flagpoling of vents shall be prohibited except where the roof is used for assembly purposes or parking. Vents within 10 feet (3048 mm) of a part of the roof that is used for assembly purposes or parking shall extend not less than 7 feet (2134 mm) above such roof and shall securely stay.

Vent pipes above roofs used for parking or public assembly require a prescriptive height that would prevent sewer gases from polluting the occupied area. The purpose of this section is to protect persons within ten feet horizontally of the plumbing vent from harmful sewer gases.

906.4 Outdoor Installations. Vent pipes for outdoor installations shall extend not less than 10 feet (3048 mm) above the surrounding ground and shall be securely supported.

906.5 Joints. Joints at the roof around vent pipes shall be made watertight by the use of approved flashings or flashing material.

906.6 Lead. (See Table 1701.1) Sheet lead shall comply with the following:

(1) For safe pans – not less than 4 pounds per square foot (lb/ft^2) (19 kg/m^2) or $1/16$ of an inch (1.6 mm) thick.

(2) For flashings or vent terminals – not less than 3 lb/ft^2 (15 kg/m^2) or 0.0472 of an inch (1.2 mm) thick.

(3) Lead bends and lead traps shall be not less than $1/8$ of an inch (3.2 mm) in wall thickness.

906.7 Frost or Snow Closure. Where frost or snow closure is likely to occur in locations having minimum design temperature below 0°F (-17.8°C), vent terminals shall be not less than 2 inches (50 mm) in diameter, but in no event smaller than the required vent pipe. The change in diameter shall be made inside the building not less than 1 foot (305 mm) below the roof in an insulated space and terminate not less than 10 inches (254 mm) above the roof, or in accordance with the Authority Having Jurisdiction.

The phenomenon of frost closure of roof vents was investigated by the National Bureau of Standards (NBS) and the results were published in *Frost Closure of Roof Vents in Plumbing Systems* BMS 142 (1954). The report explained the phenomenon as follows. Partial or complete closures of roof vents occurred during very cold weather typical in the northern states. When temperatures reached below freezing, ice or frost buildup could occur and close or lessen the opening of the vent. This formation of ice or frost results from condensation and freezing of the moisture carried by the relatively warm air that rises in the stack during periods of cold weather at times when there is little or no flow of waste water down the stack. The upward

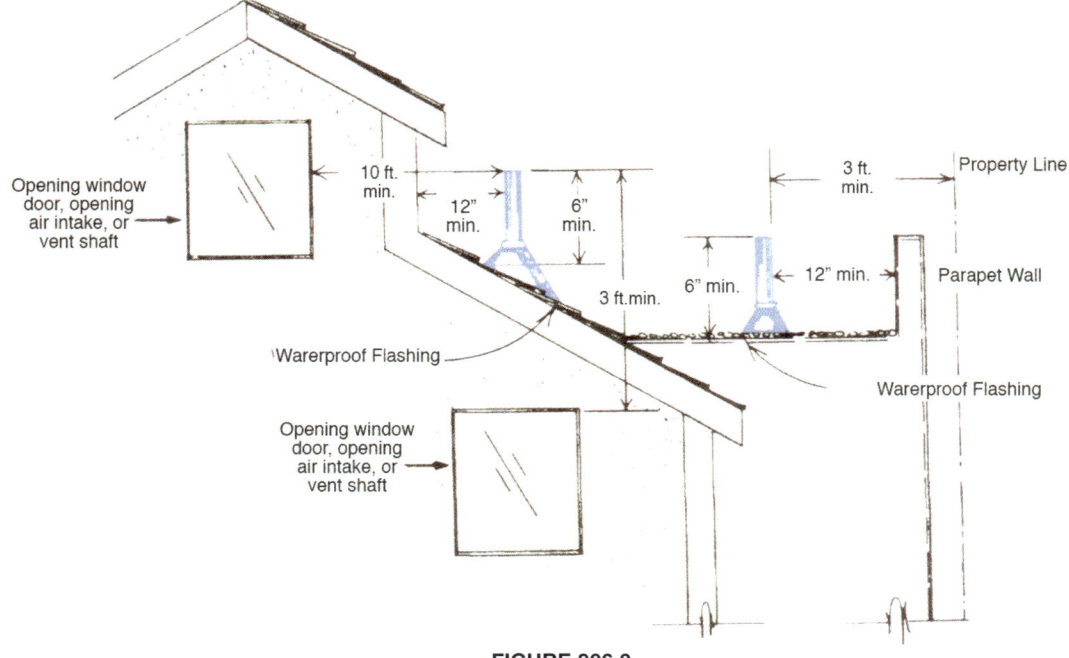

FIGURE 906.0
VENT TERMINATION THROUGH THE ROOF

current of air in the stack or vent is due to temperature differences between the outside atmosphere and the air in the stack and sewer causing convection. Moist water vapor produced whenever fixtures are discharged into the stack, especially when the discharge is hot water, tends to diffuse through the stack and vent system and contact the cold interior surface of the vent pipe above the roof where it may freeze. The cold wind gusts blowing down on the opening of the vent as well as the tendency of cold air forcing its way into the vent also contribute to the freezing of moisture arising out of the pipe. The variableness of partial to complete closure depends upon the concentration of moisture, the steadiness of convection and the consistency of freezing temperatures.

To abate the process of frost closure, the simplest recommendations were to limit the height of the exposed length of vent above the roof and to enlarge the vent just before it passes through the roof. Obviously, the larger the diameter of pipe, the less likely it is to close completely. However, there is no rule of thumb stating the correct diameter of pipe to be used to prevent frost closure. NBS testing revealed the average rate of closure for different diameters of pipe in order to predict to what extent a vent will freeze up, but not how fast. Field reports disclosed complete frost closure for even four-inch vents given the right conditions.

The UPC requires a minimum two-inch vent terminating no less than ten inches above the roof and the change in diameter to be made at least twelve inches below the roof as the minimum starting point to lessen the possibility of frost closure. Because of all the variable conditions surrounding frost closure, installers should check with the local AHJ for the required minimum size in their area. In very cold regions a larger pipe size may be needed.

Even though the condition of heavy snowfall on the roof could block vent openings, it was not considered a mitigating factor in reducing vent terminal lengths to ten-inches above the roof because of the porous nature of snow permitting some passage of air, and the fact that coverage is short-lived. Again, installers should check with the local AHJ for the proper height of vent terminals in their area.

907.0 Vent Stacks and Relief Vents.

907.1 Drainage Stack. Each drainage stack that extends 10 or more stories shall be served by a parallel vent stack, which shall extend undiminished in size from its upper terminal and connect to the drainage stack at or immediately below the lowest fixture drain. Each such vent stack shall also be connected to the drainage stack at each fifth floor, counting down from the uppermost fixture drain, using a yoke vent, the size of which shall be not less in diameter than either the drainage or the vent stack, whichever is smaller.

Pressure transients that can cause siphonage and backpressure that affect trap seals, and the need to provide air circulation throughout the drainage system was discussed in Section 901.2. This section introduces special venting requirements specific to multistory drainage stacks of 10 stories or more (see **Figure 907.1a**). We shall first discuss flow characteristics in a stack that effect pneumatic changes developing within the drainage stack between various floor levels to better understand the code requirements.

Flow Characteristics

The drainage stack receives the waste flow from horizontal branches by means of sanitary tee or combination fittings. The type of branch connection fitting to the stack and the size of the horizontal branch affects the amount of airflow movement down the stack (see **Figures 907.1b** and **907.1c**). Long pattern fittings used to change flow direction from horizontal to vertical allow twice the capacity of short pattern fittings and greater air movement (see commentary for Section 706.2). The waste flowing from the horizontal branch into the stack crosses the pipe diameter to the opposite pipe wall and initiates a downward, spiraling flow, forming a sheet of water down the wall of the pipe. This sheet of water continues a downward flow against the stack wall, increasing in velocity, while leaving a core of air in the center of the pipe that allows air to flow freely (see **Figure 907.1d**). However, this core of air is impinged when a branch flow deflects the sheet as it enters the stack (see **Figure 907.1e**). This impingement causes a substantial increase in negative pressure within the stack as well as causing backpressure in the horizontal branch. This flow characteristic and effect is significant especially in tall stacks with multiple branch connections.

As the flow transitions from vertical to horizontal at the base of the stack, a "hydraulic jump" occurs (see commentary Section 901.2 and **Figure 901.2e**) tending to fill the horizontal pipe (sometimes completely), blocking the flow of air in the drain. When the hydraulic jump completely fills the drain, the entrained air from the stack will compress and send shock waves in reverse direction equal to the flow velocity. Although the shock waves are relatively mild since air is compressible unlike water, and if they last long enough, then traps seals in close proximity may be affected. The hydraulic jump will abate after approximately 10 pipe diameters from the stack and the waste will return to normal flow. The transitional flow from vertical to horizontal sometimes creates a curtain of water across the diameter of the bend or wye, stopping the flow of air through the system. Offsets in the drainage stack also can cause these conditions. Using long sweeps, two one-eighth bends or wye connections can decrease both of these conditions. However, pressure fluctuations will still occur, depending upon flow volumes.

Terminal velocity

It was commonly believed that the waste flowing down the stack would continue to increase in velocity until reaching relatively high speeds, creating enough force to "blow out" its piping connection to the building drain. With the investigations of stack flow characteristics at the National Bureau of Standards in the early 1920s, Dr. Roy Hunter has been accredited with contributing to hydraulic science on the whole subject of vertical velocity in plumbing stacks. He discovered that the height of stacks does not need to be limited because of the fear of excessive downward velocities. Acceleration of falling water under the influence of gravity and the friction of water on the sides of the pipe

VENTS

reaches a maximum, which allows velocity to remain constant. This is known as terminal velocity. Maximum velocity, i.e. terminal velocity, occurs in a comparatively short fall and increases as the volume of flow increases. This is a significant factor when tall stacks having multiple branches contribute increasing volumes of flow to the stack. Terminal velocity will increase causing greater pressure transients in the stack.

One way to control excessive pressure is to limit the carrying capacity of the stack, which will also reduce terminal velocity. The maximum fixture unit values in Table 703.2 for vertical pipes are the probable demand loads that will not exceed one-third of the flow that a vertical stack can carry full at maximum velocity (see also commentary for Section 703.2). If these flow parameters are exceeded, the waste flow will begin to flatten out as friction and air resistance increase, forming a plug across the pipe diameter, thus closing the air core (see **Figure 907.1f**). This plug, if persistent creates a piston-like action down the stack. This can cause severe pressure fluctuations within the drainage system, causing trap seals to blowout below the flow due to high pressures as air is pushed ahead of the flow. Excessive flow rates in a stack could also cause trap seal siphonage due to negative pressures occurring behind the flow as air rushes to fill the void left by the rapidly flowing effluent. Even with limiting the carrying capacity of the stack to one-third to control terminal velocity, pneumatic pressure changes of appreciable magnitude may still develop within a tall stack between branch connections.

Pressure changes

The best way to visualize pressure changes that occur in a tall stack is the use of pressure curve charts, three of which have been provided to facilitate the explanation. The greatest aspiration, or negative pressure, occurs in a stack a short distance below the discharging branch. As the flow approaches the base of the stack, the pressure approaches atmospheric pressure and then transitions to positive pressure due to back pressure at the base of the stack as explained above (see **Figure 907.1g** Graph Figure 29). If the flow in the stack passes a vented branch, then notice how the pressure curves collapse toward atmospheric pressure (see **Figure 907.1g** Graph Figure 30). The vented branch relieves the negative pressure. If the flow continues and is interrupted by a discharging branch, then the negative pressures significantly resume because of the impingement of the core of air when the sheet of water is deflected by branch flows (see **Figure 907.1h** Graph Figure 32 and **Figure 907.1i**). This phenomenon occurs throughout a tall stack with multiple branch connections with varying degrees of pressure transients. The code requirements of

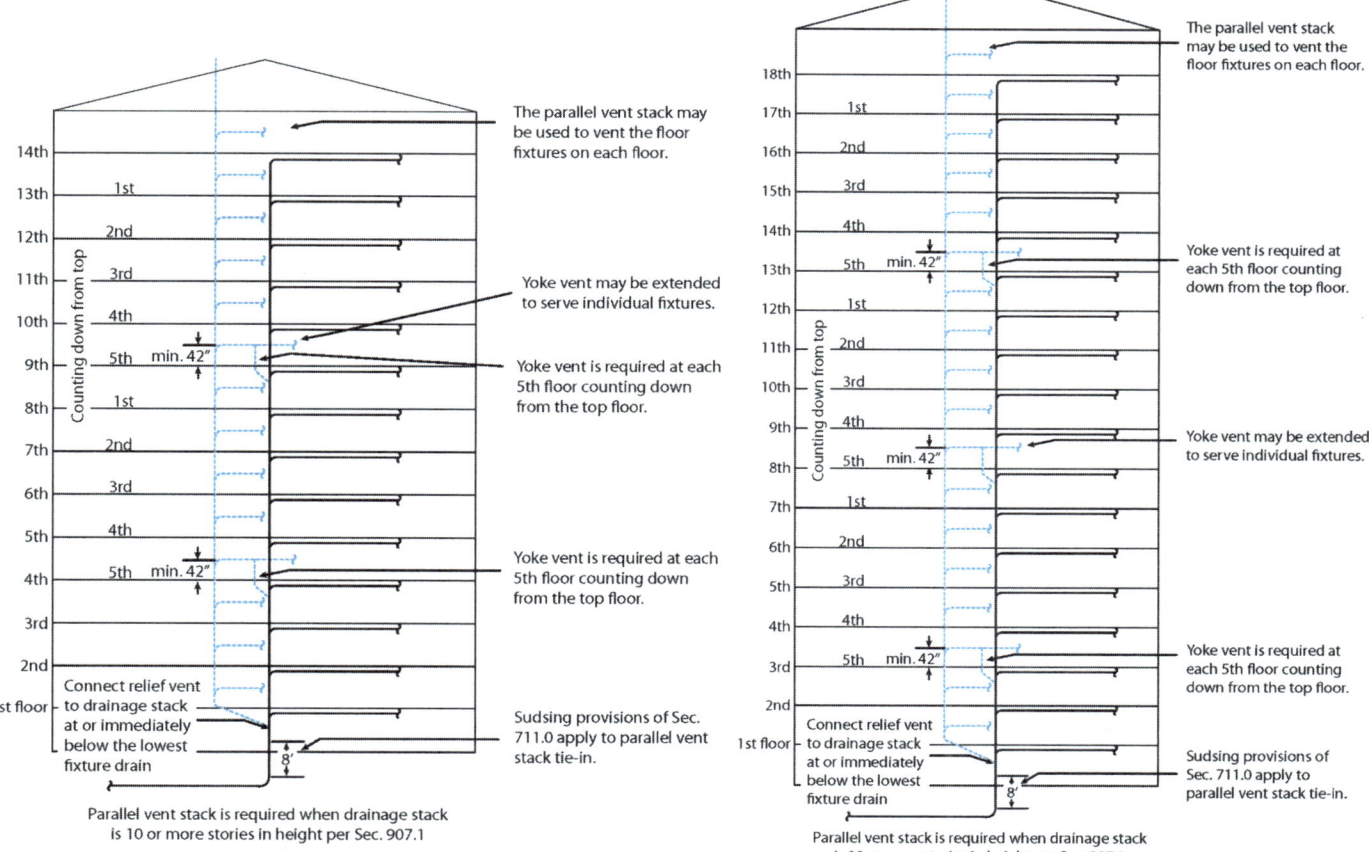

FIGURE 907.1A
PARALLEL VENT STACK SYSTEM

VENTS

this section provide a means for controlling these variations.

The Code Requirements

As Table 703.2 controls excessive pressures by limiting the carrying capacity of the stack, Section 907.1 additionally controls excessive pressure transients that occur in tall stacks as previously discussed. Typically, multistory buildings less than ten stories will also have a parallel vent stack to provide branch vents at each floor for venting fixture traps and may diminish in size as it reaches the lower level. However, for ten or more stories the required parallel vent stack must remain undiminished in size for its entire length, from the termination through the roof to its connection below the lowest fixture drain connected to the drainage stack. This parallel stack not only provides branch venting for individual fixtures at each story, but also vents the drainage stack to relieve the pressure transients that occur in a tall stack.

The parallel vent stack performs this function by the use of yoke vents. A yoke vent is a pipe connection between the drainage and vent stack to relieve pressures within the drainage stack (see **Figure 907.1j**). The term coined "yoke vent" originally derived its name from the appearance of the piping as being similar to a yoke used on oxen. Although the appearance has been modified, the term has persisted in plumbing vernacular. An alternate term is a relief vent.

The yoke vent connection is to be made at every fifth floor. This is determined by counting down from the uppermost horizontal branch connection to the stack, allowing five horizontal branches between the yoke vent connections (see **Figure 907.1a**). The yoke vent accomplishes the same thing as the vented branch vent seen in **Figure 907.1g**, Graph Figure 30. At the point of the yoke vent connection, the pressure in the stack will quickly approach atmospheric pressure. Negative pressures will resume when lower

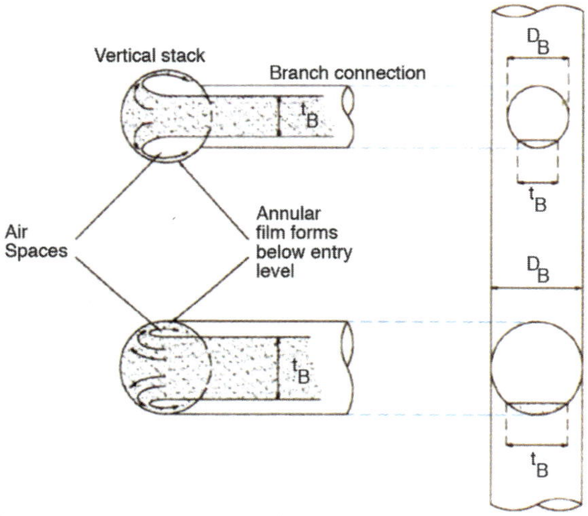

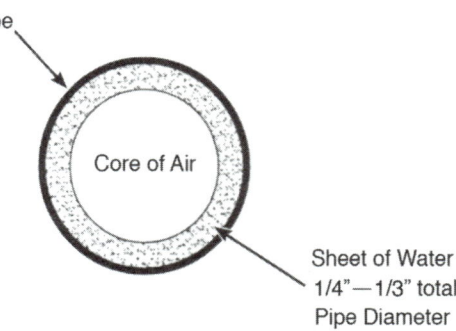

Flow in Drainage Stack
FIGURE 907.1D
FLOW IN DRAINAGE STACK

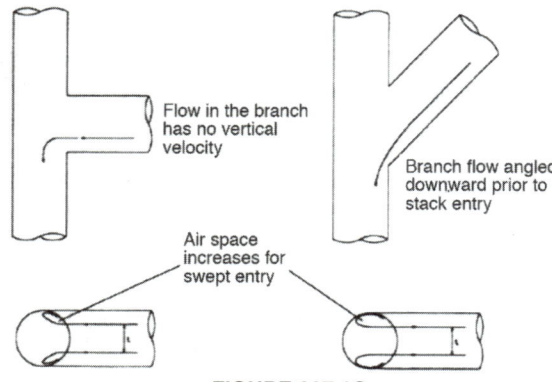

FIGURE 907.1B
BRANCH FLOW INTO STACK - EFFECT ON STACK OF LARGE AND SMALL BRANCHES
(Reprinted by Permission of the Publishers, In Engineered Design of Building Drainage Systems by J.A. Swaffield and L.S. Galowin [Farnham: Ashgate 1992], P. 81. Copyright © 1992)

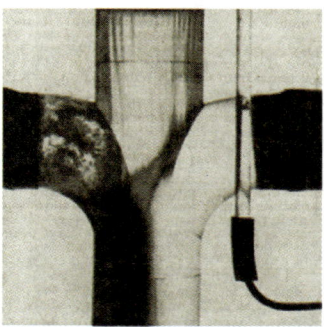

FIGURE 907.1E
BRANCH FLOW DEFLECTING THE VERTICAL SHEET OF WATER
(Reprinted from NBS Monograph 31)

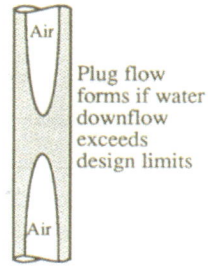

FIGURE 907.1F
PLUG BLOCKING AIRFLOW
(Reprinted by Permission of the Publishers, In Engineered Design of Building Drainage Systems by J.A. Swaffield and L.S. Galowin [Farnham: Ashgate, 1992], p 83. Copyright © 1992)

FIGURE 907.1C
EFFECT OF FITTINGS ON BRANCH FLOW
(Reprinted by Permission of the Publishers, In Engineered Design of Building Drainage Systems by J.A. Swaffield and L.S. Galowin [Farnham: Ashgate 1992], P. 83. Copyright © 1992)

VENTS

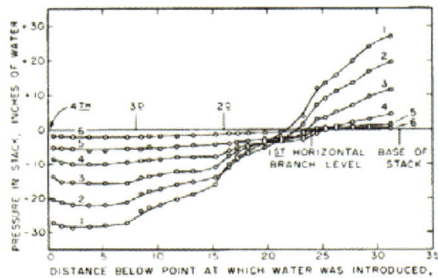

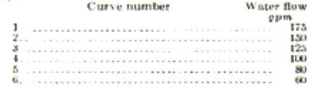

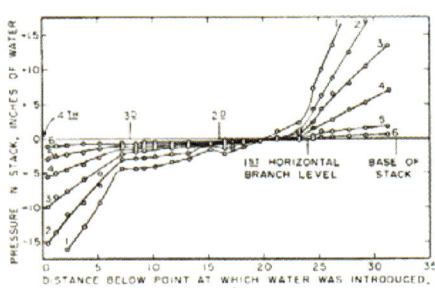

FIGURE 907.1G
COMPARISON OF PRESSURE CHANGES BETWEEN AN UNVENTED AND VENTED STACK
(Reprinted from NBS Monograph 31)

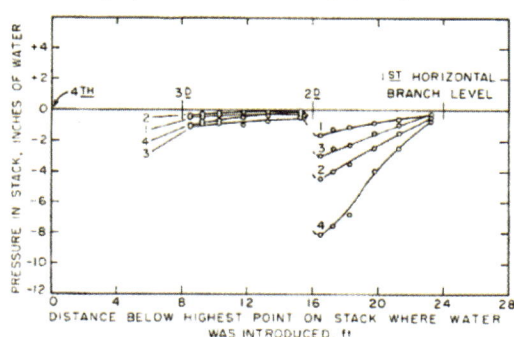

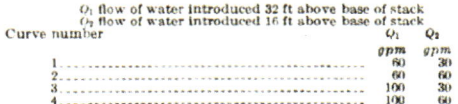

FIGURE 907.1H
PRESSURE CHANGES FROM INTERRUPTED BRANCH DISCHARGE
(Reprinted from NBS Monograph 31)

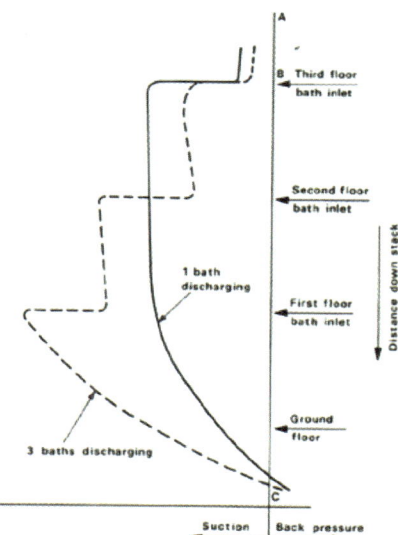

FIGURE 907.1I
PRESSURE CHANGES FROM MULTIPLE INTERRUPTED BRANCH DISCHARGES
(Reprinted from CIB W062, 1973)

branch flows enter the stack until the flow reaches another yoke vent, which brings the pressure close to atmospheric again, and so on down the stack. Once the flow reaches the base of the stack where the hydraulic jump occurs, the positive backpressure will be relieved since the parallel vent stack connects below the lowest horizontal branch, hence protecting the lower branch interval trap seals from excessive positive pressure. This is why the parallel vent stack must remain undiminished in size. The size of the vent pipe at the base of the stack must be determined by the maximum fixture units allowed on the stack to relieve the positive back pressure.

The size of the yoke vent connection between the two stacks is determined by the diameter of the smallest stack. Typically, the parallel vent stack will be larger than drain stack at the upper floor levels. The diameter of the drain stack will then determine the size of the yoke vent connection. As the drainage stack increases in size, so then must the yoke vent connection.

907.2 Yoke Vent. The yoke vent connection to the vent stack shall be placed not less than 42 inches (1067 mm) above the floor level, and the yoke vent connection to the drainage stack shall be using a wye-branch fitting placed below the lowest drainage branch connection serving that floor.

See **Figure 907.1j**.

908.0 Wet Venting.

908.1 Vertical Wet Venting. Wet venting is limited to vertical drainage piping receiving the discharge from the trap arm of one and two fixture unit fixtures that also serves as a vent not exceeding four fixtures. Wet-vented fixtures shall be within the same story; provided, further, that fixtures with a continuous vent discharging into a wet vent shall be within the same story as the wet-vented fixtures. No wet vent shall exceed 6 feet (1829 mm) in developed length.

The vertical wet vent is a vertical drain that acts as a vent for the trap arms located below it. Hence, a wetted vent (see **Figure 908.1a** where pipe sections A, B and C is the vertical wet vented portion). It serves two purposes; one as a drain and one as a vent. This fact is very important. In the vertical wet vent only trap arms are connected to the wet vent, not waste branches. This effectively limits the connection to the wet vent section to single fixture connections.

Fixture trap arm connections discharging into the vertical wet vent are limited to one and two drainage fixtures units. This limits the type of fixtures that can discharge into the vertical wet vent. For example, no water closet or clothes washer may discharge into the vertical wet vent because of the large volume of water discharge and high flow rates that could siphon the other fixture traps. However, the toilet and clothes washer may be vented by the vertical wet vent (see **Figures 908.1a** and **908.1b**).

The vertical wet vent cannot serve more than four fixtures. In application, the maximum drainage fixture units served by any vertical wet vent will be equal to four sets of two drainage fixtures units, for a total of eight DFUs. This will effectively limit the size of the wet vent. The minimum size for the wet vent is two inches, and eight DFUs require a minimum two-inch waste.

Fixtures that connect to the vertical wet vent must be on the same floor as well as the fixture serving as the wet vent. In other words, the fixture serving as the vertical wet vent is the uppermost fixture with the continuous vent. This uppermost fixture may not be on an upper floor level wet venting fixtures on the lower level.

908.1.1 Size. The vertical piping between two consecutive inlet levels shall be considered a wet-vented section. Each wet-vented section shall be not less than one pipe size exceeding the required minimum waste pipe size of the upper fixture or shall be one pipe size exceeding the

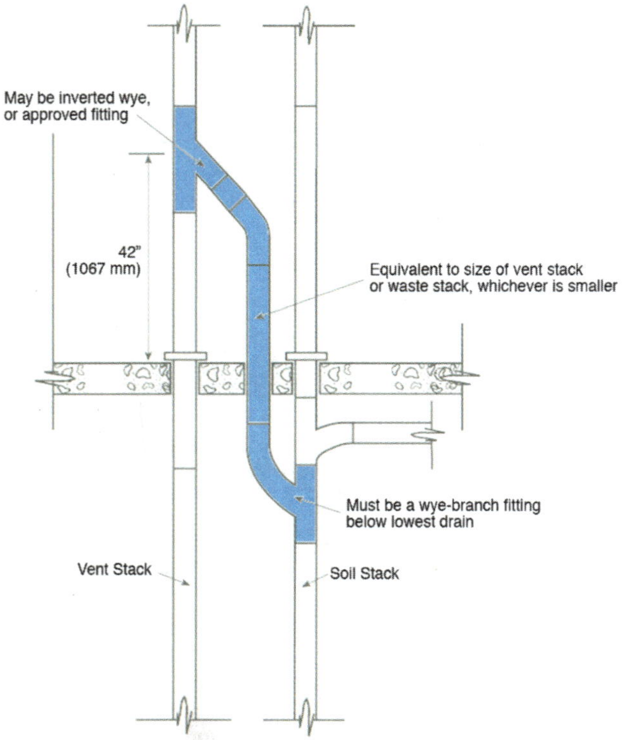

FIGURE 907.1J
YOKE VENT INSTALLATION

Pipe Section	Fixture Units	Pipe Size
A	1	2"
B	3	2-1/2"
C	5	3"
D	8	3"
E	8	2"

Isometric Compass

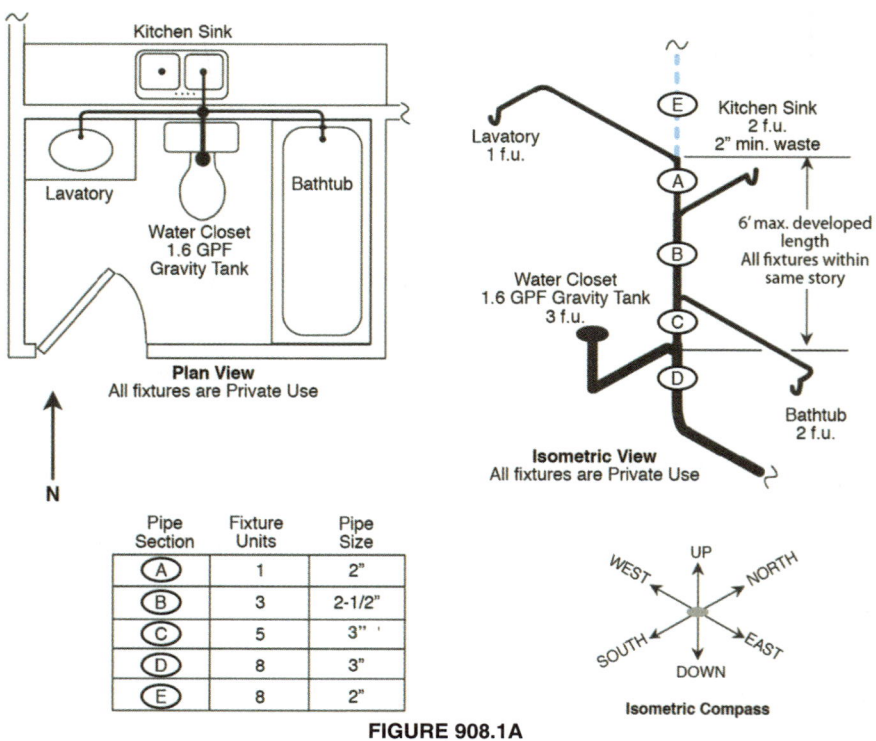

FIGURE 908.1A
WET VENT SIZING EXAMPLE 1

required minimum pipe size for the sum of the fixture units served by such wet-vented section, whichever is larger, but in no case less than 2 inches (50 mm) in diameter.

🔧 Section 908.1.1 requires each wet-vented section to be increased one pipe size larger than the minimum required pipe size, yet must be a minimum of 2 inches. For example, a lavatory having a 1¼ inches trap arm serving as a vertical wet vent is required to be increased one pipe size. However, the next pipe size is 1½ inches, which is not permitted. Therefore, the pipe diameter needs to increase again to the minimum two inches (see **Figure 908.1b**). Therefore, the largest size required for the wet vent will be 3 inches. So the choices for the size of the wet vent section are either 2 inches or 3 inches.

908.1.2 Vent Connection. Common vent sizing shall be the sum of the fixture units served but, in no case, smaller than the minimum vent pipe size required for a fixture served, or by Section 904.0.

🔧 The common vent serving a lavatory that wet vents a kitchen may be 1½ inch according to Table 703.2, but must increase to 2 inches if the lavatory is wet venting a toilet, even though the wet vent portion is a minimum 2 inches, since the toilet requires a minimum 2 inch vent (see **Figure 908.1a**). The common vent may also need to increase if it is used for aggregate venting according to Section 904.0. Notice in **Figures 908.1.2a** and **908.1.2b** the reduction in the amount of vent piping when vertical wet venting is applied as well as examples for sizing.

908.2 Horizontal Wet Venting for a Bathroom Group. A bathroom group located on the same floor level shall be permitted to be vented by a horizontal wet vent where all of the conditions of Section 908.2.1 through Section 908.2.5 are met.

🔧 The idea of *group venting* rather than individual venting of fixtures within a bathroom group is the basic idea of wet venting. Originally, the design permitted the use of only one lavatory vent to provide the needed airflow to protect the trap seal for a bathtub. It did this through the piping of the lavatory's vertical drain connection to the horizontal tub drain. The vent for the bathtub became wetted by the discharge of the lavatory, hence a wet vent. The design also worked suitably for back-to-back bathroom groups with only one common vent for the lavatories. Typically, the toilet was directly connected to and vented by a stack vent and required additional venting where the bath-

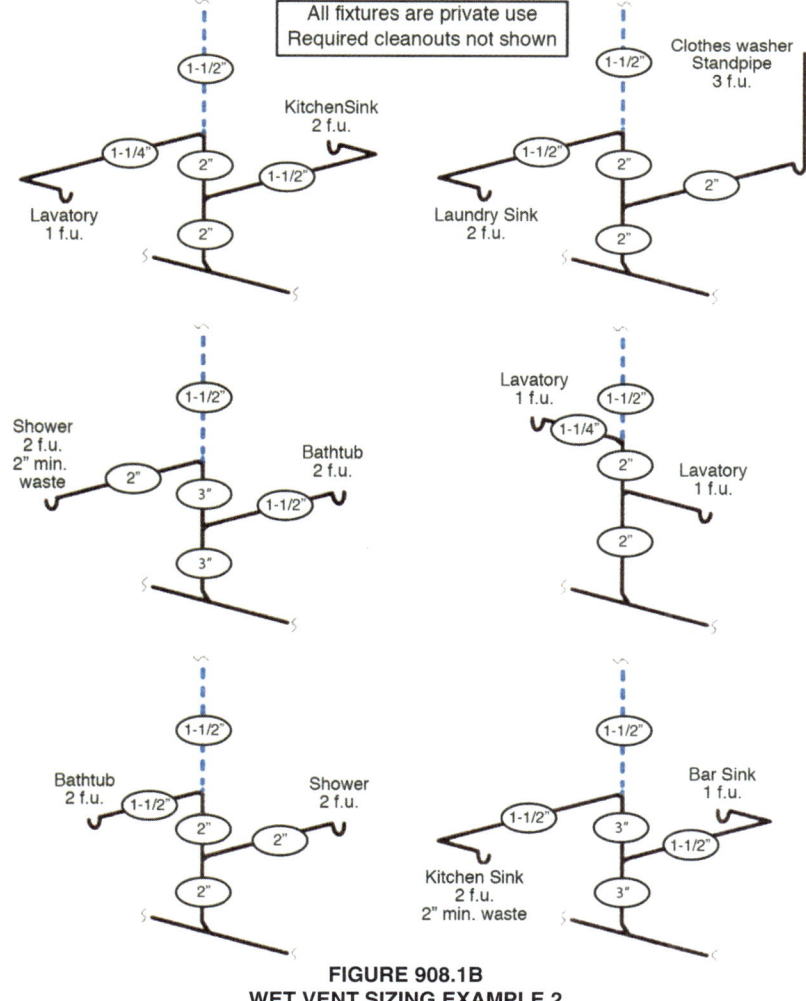

**FIGURE 908.1B
WET VENT SIZING EXAMPLE 2**

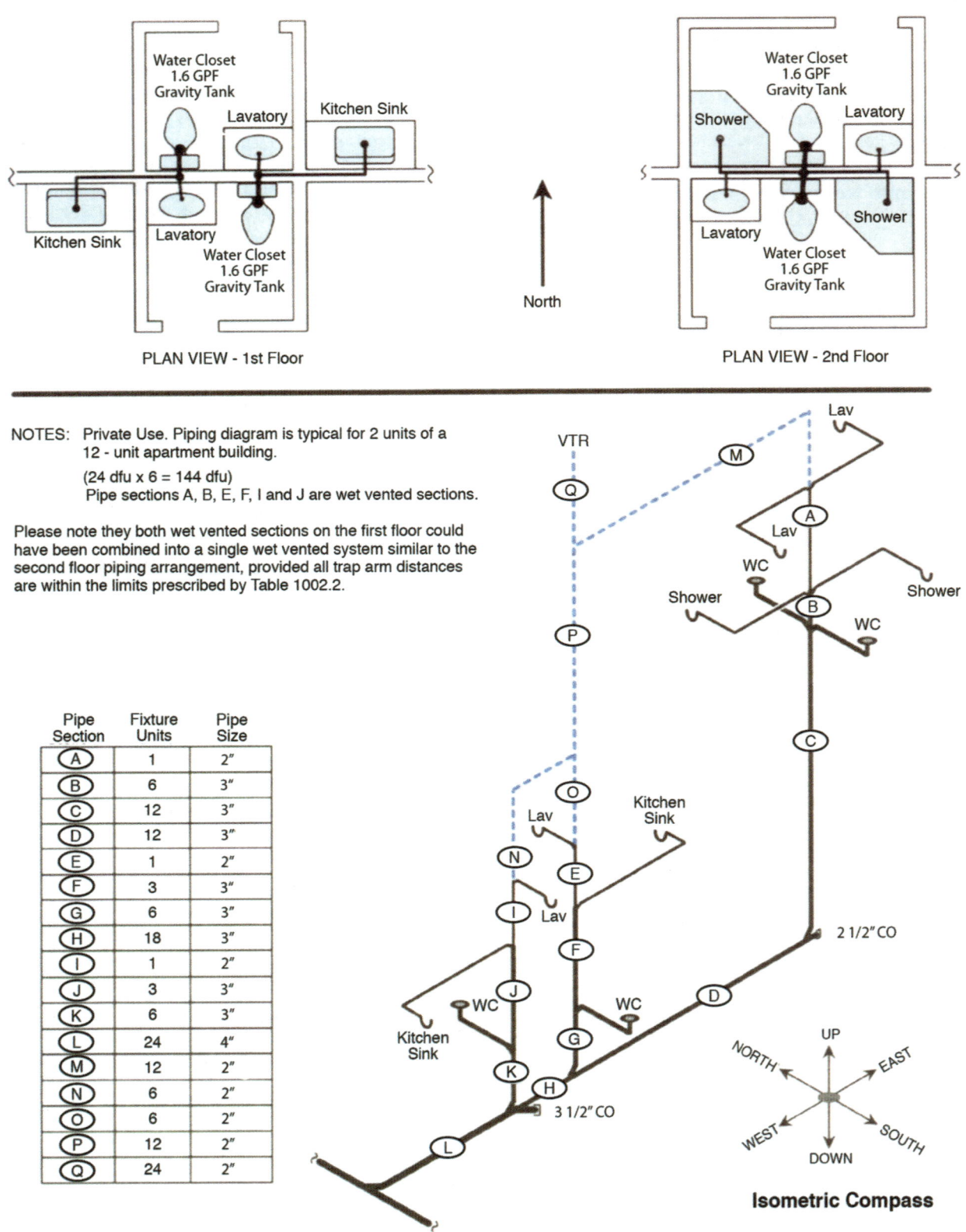

FIGURE 908.1.2A
WET VENT SIZING EXAMPLE 3

VENTS

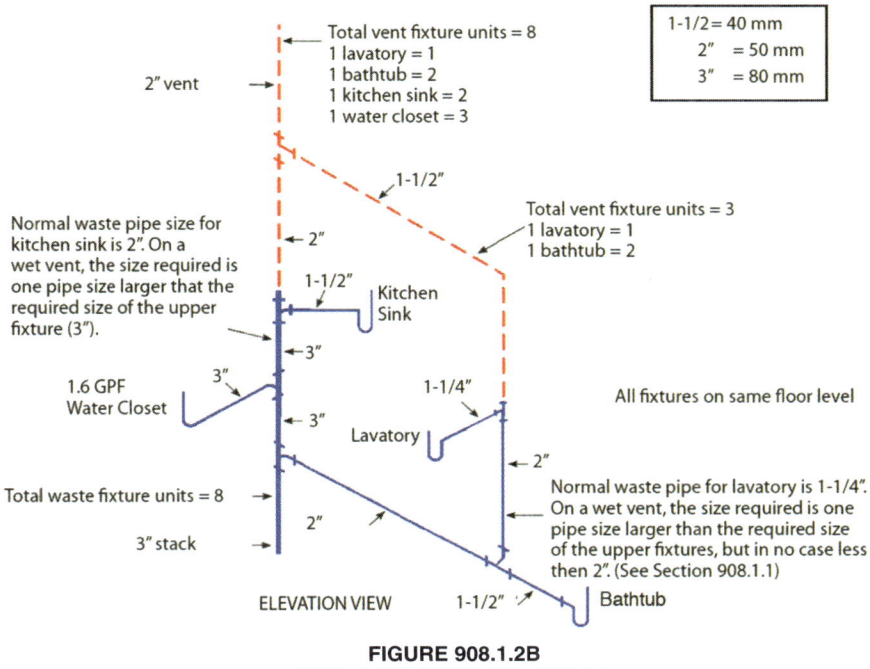

**FIGURE 908.1.2B
WET VENT SIZING EXAMPLE 4**

room groups were located on lower levels of a multistory building and subject to stack pressures caused from fixtures discharging above. This particular group venting had very specific design parameters that needed to be followed, such as the diameter, length and slope of the tub drain, the size of the lavatory drain that vented the bathroom group, and the diameter of the horizontal branch and where it was to connect to the stack relative to the toilet.

The idea of *horizontal* wet venting was a later development and significantly differs from the group venting just explained. Horizontal wet venting uses the concept of the combination waste and vent system (see Section 910.0) to eliminate the need to individually vent every fixture by relying on the existence of a continuous air space above the mean water surface in the horizontal drain, and the absence of excessive surges that would crest above the mean water surface.

The fixtures allowed to be vented by a wet vent either have low discharge surges or last for only a short duration that do not significantly impact the air flow in the horizontal drain. That is why the water closet is distinguished from the other fixtures with the requirement to connect downstream of all the fixture drain connections to the horizontal wet vent (see Section 908.2.4). Although having a short surge duration, the toilet typically has a high discharge rate that could significantly impinge the air space in the horizontal drain and affect trap seals.

The design of the horizontal wet vent requires that each fixture within the bathroom group (see definition for Bathroom Group in Chapter 2) connects independently and laterally to the horizontal wet vent pipe and not exceed the lengths from the trap to the lateral connection as determined by Table 1002.2 (see **Figure 908.2a**). This reduces the potential to interrupt the airflow above the mean water surface, and also prevents the vented upper portion of the horizontal wet vent pipe from being below the weir of the trap. Surge flows of relatively short durations follow discharges from the lavatory and water closet, while tubs and showers could tend toward steady flows when the drainage is extended over a period of time (such as extended shower use and tubs draining when full). The condition of flows in the horizontal wet vent branch could be steady, surge, or no flow. These lateral connections are therefore the critical points of the system.

See Figure **908.2b** and **Figure 908.2c** for examples of layout options and sizing requirements.

908.2.1 Vent Connection. The dry vent connection to the wet vent shall be an individual vent for the bidet, shower, or bathtub. One or two vented lavatory(s) shall be permitted to serve as a wet vent for a bathroom group. Only one wet-vented fixture drain or trap arm shall discharge upstream of the dry-vented fixture drain connection. Dry vent connections to the horizontal wet vent shall be in accordance with Section 905.2 and Section 905.3.

The only fixtures that may serve as the wet venting fixture for the bathroom group are the bidet, shower, bathtub, and lavatory(s). This excludes the water closet, floor drain, or any fixtures outside the bathroom. This also excludes a back-to-back lavatory since the lavatories would be in separate bathroom groups. Two lavatories serving as a wet vent are only permitted if they are within the same bathroom group as Section 908.2.5 requires.

The code allows only one wet-vented fixture drain or trap arm to discharge upstream of the dry-vented fixture

VENTS

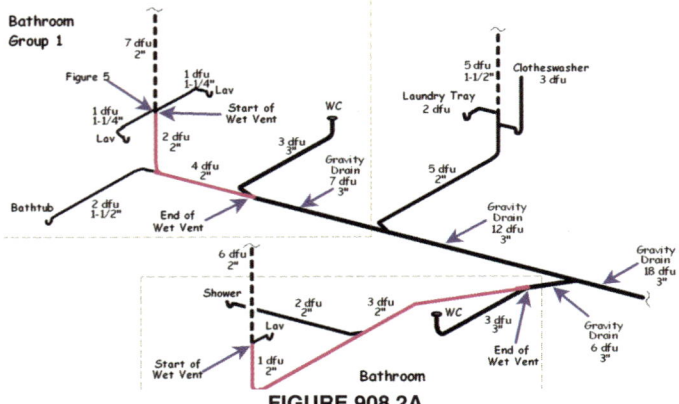

FIGURE 908.2A
LATERAL CONNECTIONS TO THE HORIZONTAL WET VENT

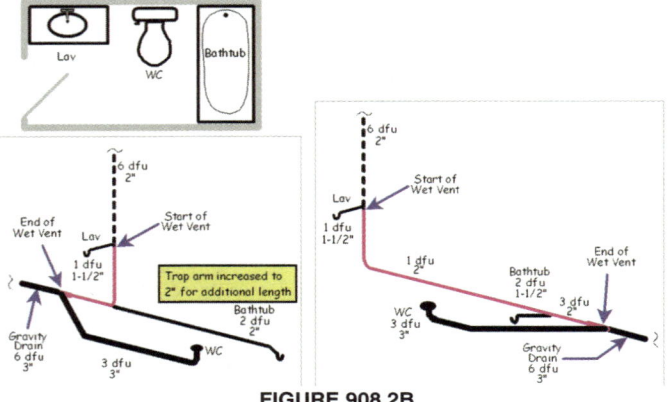

FIGURE 908.2B
EXAMPLE 1: LAYOUT OPTIONS

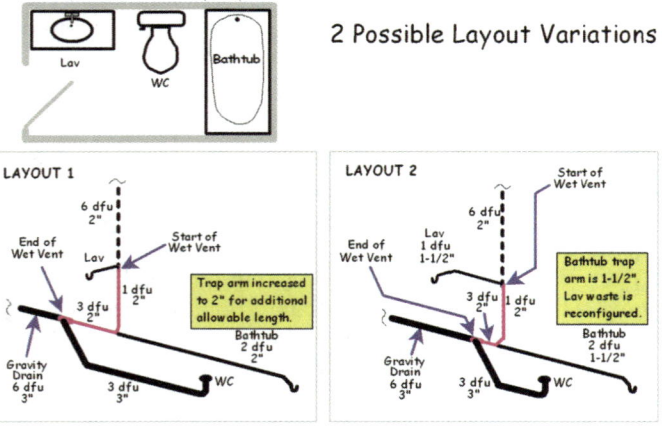

FIGURE 908.2C
EXAMPLE 2: LAYOUT OPTIONS

drain connection. To avoid confusion of what it means for the fixture drain or trap arm to be upstream the dry vented fixture drain connection, consider **Figure 908.2.1a**. Always begin the point of reference at the dry-vented fixture drain. The lavatory is the dry-vented fixture that is wet venting the water closet. Even though the water closet is positioned laterally upstream the lavatory, the water closet drain connection is downstream the lavatory dry-vented drain connection. **Figure 908.2.1b** shows a bathtub that is laterally upstream the lavatory fixture, but it is not upstream the dry-vented fixture drain connection where the arrow shows the start of the wet vent. This is the point where the dry vent connects to the fixture drain that is wet-venting the bathroom group. To be upstream of the start of the wet vent, the bathtub would have to take off from the point where the vertical drain of the lavatory transitions horizontally as shown in **Figure 908.2.1c**. Therefore, **Figure 908.2.1b** would allow another fixture, such as a shower, to be upstream of the start of the wet vent. This would also require the sizing of the horizontal wet vent to increase.

908.2.2 Size. The wet vent shall be sized based on the fixture unit discharge into the wet vent. The wet vent shall be not less than 2 inches (50 mm) in diameter for 4 drainage fixture units (dfu) or less, and not less than 3 inches (80 mm) in diameter for 5 dfu or more. The dry vent shall be sized in accordance with Table 702.1 and Table 703.2 based on the total fixture units discharging into the wet vent.

🔧 In order to prevent excessive interference at the junction of the lateral connections, the sizing requirements for the horizontal wet vent branch are more restrictive than what is allowed in Table 703.2. If combined flows at the junctions are excessive, then local flooding or large pneumatic pressure fluctuations may occur. Therefore, the minimum size for the horizontal wet vent pipe is two-inches and limited to four fixture units. A system having five or more fixture units must be increased to three inches. The reduced fixture units limit the practical flow capacity allowing the drain sufficient volume for continuous air space above the flow. Along with sufficient volume within the pipe, it is also critical to maintain a minimum uniform slope of ¼- inch per foot to prevent surges of water to crest the crown of the pipe (see Section 708.1).

908.2.3 Trap Arm. The length of the trap arm shall not exceed the limits in Table 1002.2. The trap size shall be in accordance with Section 1003.3. The vent pipe opening from the horizontal wet vent, except for water closets and similar fixtures, shall not be below the weir of the trap.

🔧 The last sentence prohibiting the vent pipe opening below the weir of the trap, would not allow an unvented lavatory to connect independently to the wet vented horizontal branch even though the distance of the lavatory trap arm to the wet vented horizontal branch is in accordance with Table 1002.2. An unvented lavatory trap arm would form an S-trap, which is in conflict with Section 1004.1 that prohibits S-traps. With this provision, the lavatory(s) may serve as the wet vent, or shall be individually vented when not serving as the wet vent (as shown in **Figure 908.2.3**).

908.2.4 Water Closet. The water closet fixture drain or trap arm connection to the wet vent shall be downstream of fixture drain or trap arm connections to the horizontal wet vent.

VENTS

908.2.5 Additional Fixtures. Additional fixtures shall discharge downstream of the wet vent system and be conventionally vented. Only the fixtures within the bathroom group shall connect to the wet-vented horizontal branch.

909.0 Special Venting for Island Fixtures.

909.1 General. Traps for island sinks and similar equipment shall be roughed in above the floor and shall be permitted to be vented by extending the vent as high as possible, but not less than the drainboard height and then returning it downward and connecting it to the horizontal sink drain immediately downstream from the vertical fixture drain. The return vent shall be connected to the horizontal drain through a wye-branch fitting and shall, in addition, be provided with a foot vent taken off the vertical fixture vent by means of a wye branch immediately below the floor and extending to the nearest partition and then through the roof to the open air, or shall be permitted to be connected to other vents at a point not less than 6 inches (152 mm) above the flood-level rim of the fixtures served. Drainage fittings shall be used on the vent below the floor

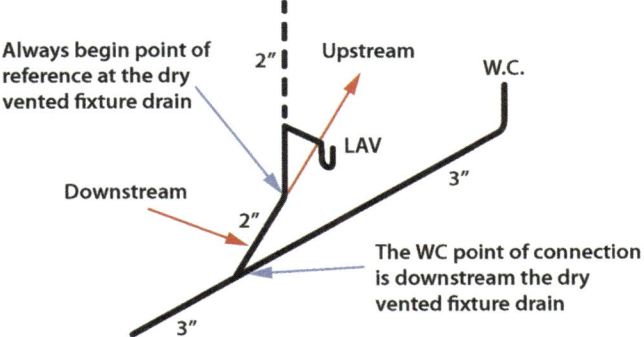

FIGURE 908.2.1A
UPSTREAM DOWNSTREAM DRY-VENT CONNECTION

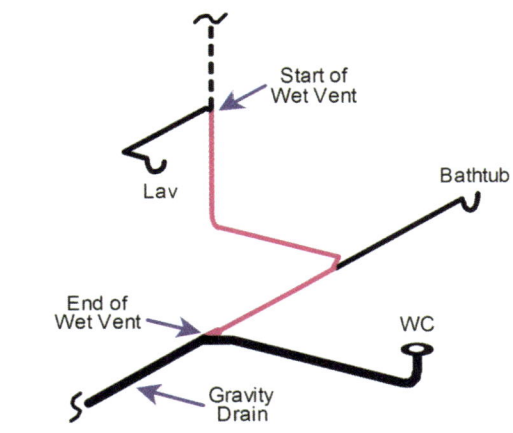

FIGURE 908.2.1B
BATHTUB IS NOT UPSTREAM DRY-VENT CONNECTION

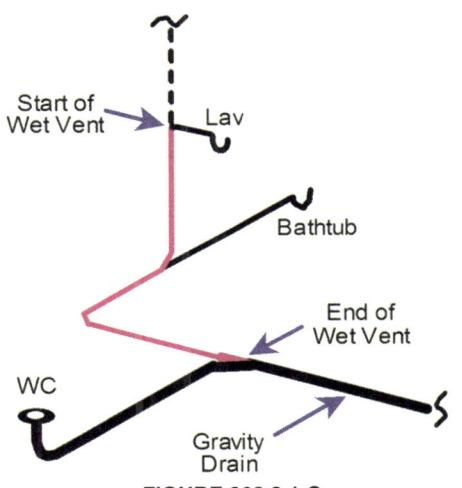

FIGURE 908.2.1.C
BATHTUB UPSTREAM DRY-VENT CONNECTION

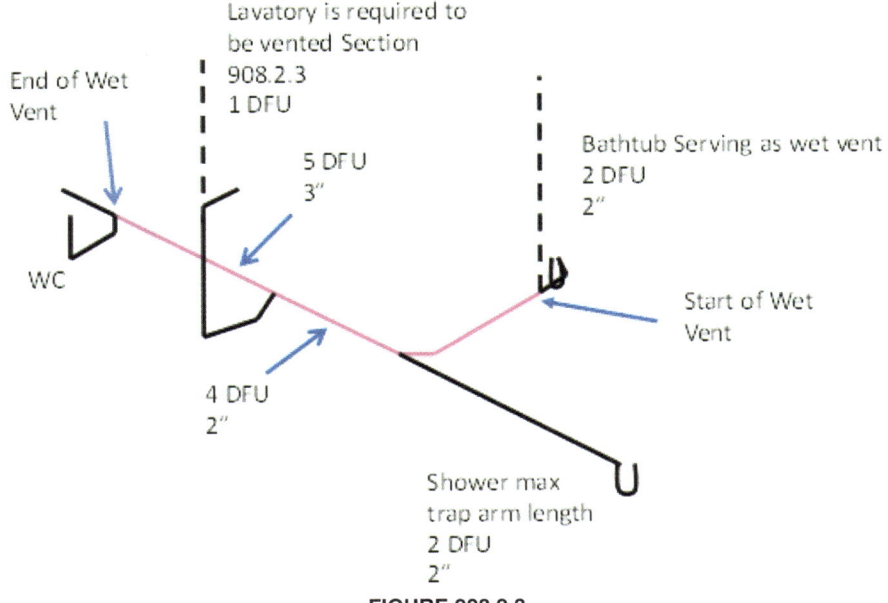

FIGURE 908.2.3
LAVATORY REQUIRING AN INDIVIDUAL VENT

level, and a slope of not less than ¼ inch per foot (20.8 mm/m) back to the drain shall be maintained. The return bend used under the drainboard shall be a one-piece fitting or an assembly of a 45 degree (0.79 rad), a 90 degree (1.57 rad), and a 45 degree (0.79 rad) elbow in the order named. Pipe sizing shall be as elsewhere required in this code. The island sink drain, upstream of the returned vent, shall serve no other fixtures. An accessible cleanout shall be installed in the vertical portion of the foot vent.

An island fixture is a fixture remote from a plumbing wall and requires special venting that differs from continuous, individual, common, and wet venting. An island fixture differs from a fixture mounted in cabinets that are attached to a wall forming a peninsula and must be vented according to Section 905.3 as shown in **Figure 909.1a**.

The unique feature of the island vent is that it returns downward to a horizontal vent (foot vent) below the floor before rising vertically above the spill rim of the sink (see **Figure 909.1b**). Other than floor-mounted fixtures and island fixtures, no other fixture is permitted to have its vent installed below the floor.

The vent from the trap arm connection is required to extend vertically as high as possible inside the cabinet, but not less than the drainboard height. The configuration of the return bend using 45 degree and 90 degree elbows purposes to maintain a vertical angle (see Section 224.0, Vertical Pipe). This is to protect the return bend from accumulating waste and to prevent waste from flowing down the return side of the bend if a blockage occurs.

Notice in **Figure 909.1b** how the return vent is connected to both the horizontal vent (foot vent) and the horizontal drain. Wye-branch fittings must be used for both connections and are considered drainage fittings. The wye-branch fitting connecting the return vent to the foot vent is to be in vertical position with the wye-branch sloping down toward the drain rather than up toward the vent, which would better assist air flow. Although frictional resistance of air flow would increase with a down-turned branch fitting, it is insignificant and still allows airflow up the vent to protect the trap seal. The critical reason for the down-turned wye fitting at the foot vent connection is the flooding potential of the foot vent in the event of a kitchen waste blockage. Envision a blockage occurring downstream the kitchen drain. The waste would begin to flow up the vertical piping and spill into the foot vent and continue vertically until it spills into the basin of the sink. When the blockage is released, then the waste in the foot vent must flow toward the horizontal drain with a slope of not less than one-fourth of an inch per foot. Therefore, the wye-branch fitting is required to slope down toward the drain and not up toward the vent.

The installation of a cleanout on the vertical component of the foot vent installation is also required. Although not required by code, it is a good idea to install this cleanout above the spill rim of the sink since a drain blockage would allow the vertical section of the foot vent to fill to the same level as the waste spilled into the sink basin. This would allow safe opening of the cleanout for rodding.

910.0 Combination Waste and Vent Systems.

910.1 Where Permitted. Combination waste and vent systems shall be permitted where structural conditions preclude the installation of conventional systems as otherwise prescribed by this code.

The structural condition referenced in this section is where there is an absence or lack of vertical walls to install vent pipes. This plumbing system design begins with the concept of an appropriately sized horizontal pipe that serves a dual purpose as a drain and a vent where the wetted perimeter is low enough to allow adequate air movement in the upper portion of the pipe.

The drain and vent area ratio in the horizontal pipe is controlled by sizing traps and horizontal branches two pipe sizes larger than the fixtures tailpiece it serves as well as prohibiting high discharge fixtures such as toilets and urinals. For purposes of this section, half sizes are included when determining two pipe sizes larger.

910.2 Approval. Construction documents for each combination waste and vent system shall first be approved by the

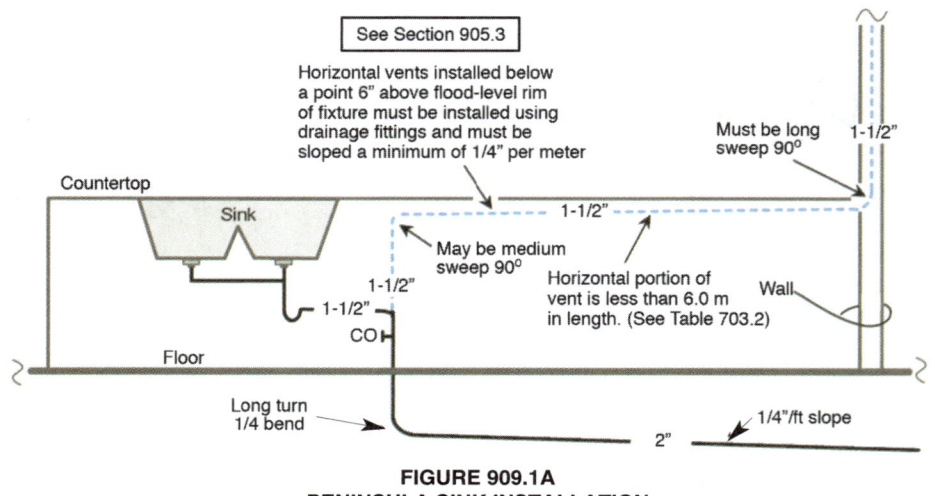

**FIGURE 909.1A
PENINSULA SINK INSTALLATION**

VENTS

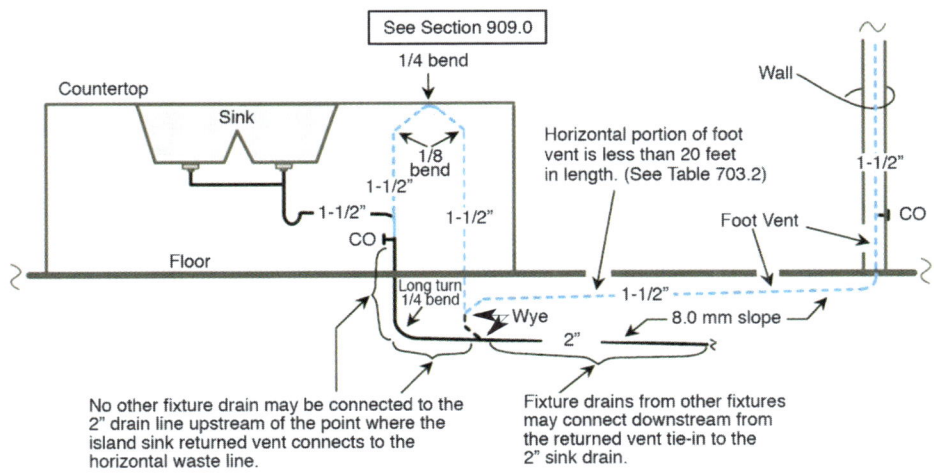

**FIGURE 909.1B
ISLAND SINK INSTALLATION**

Authority Having Jurisdiction before a portion of such system is installed.

910.3 Vents. Each combination waste and vent system, as defined in Chapter 2, shall be provided with a vent or vents adequate to ensure free circulation of air. A branch exceeding 15 feet (4572 mm) in length shall be separately vented in an approved manner. The area of a vent installed in a combination waste and vent system shall be not less than one-half the inside cross-sectional area of the drain pipe served. The vent connection shall be downstream of the uppermost fixture.

In **Figure 910.3a**, the drain pipe is four inches. The vent must be at least half the area of the drain pipe. A 4-inch pipe has an area of 12.566 in^2. Half of the area is 6.283 in^2. This would require that the terminal vent and any relief vents must be 3-inches.

This vent is sometimes called "the terminal vent." It must be placed downstream of the uppermost fixture (see **Figure 910.3a** and **910.3b**). This will provide a washing action to keep the base of the vent clear. The vent must take off above the center line of the waste. A low-use fixture, such as a drinking fountain, may be connected to the terminal vent. This will also help wash the vent clear.

910.4 Size. Each waste pipe and each trap in such a system shall be not less than two pipe sizes exceeding the sizes required by Chapter 7 of this code, and not less than two pipe sizes exceeding a fixture tailpiece or connection.

See Appendix B, Section B 101.6.1 for pipe size increments when increasing two pipe sizes. See also **Figure 910.4a** and **910.4b**.

910.5 Vertical Waste Pipe. No vertical waste pipe shall be used in such a system, except the tailpiece or connection between the outlet of a plumbing fixture and the trap. Such tailpieces or connections shall be as short as possible, and in no case shall exceed 2 feet (610 mm).

Exception: Branch lines shall be permitted to have 45 degree (0.79 rad) vertical offsets.

All drains are required to make horizontal entries where connecting to other drains. A top entry would introduce a curtain of water into that portion of the drain line that is dedicated to venting, thereby creating a disturbance and disruption of the venting function. A compromised venting system will have an adverse effect on flow velocity of the waste stream, the result being a non-scouring drainage system which is unable to remove self-induced settlement (see **Figures 910.5a** and **910.5b**).

910.6 Cleanouts. An accessible cleanout shall be installed in each vent for the combination waste and vent system. Cleanouts shall not be required on a wet-vented branch serving a single trap where the fixture tailpiece or connection is not less than 2 inches (50 mm) in diameter and provides ready access for cleaning through the trap.

910.7 Fixtures. No water closet or urinal shall be installed on such a system. Other one, two, or three unit fixtures remotely located from the sanitary system and adjacent to a combination waste and vent system shall be permitted to be connected to such system in the conventional manner by means of waste and vent pipes of regular sizes, providing that the two pipe size increase required in Section 910.4 is based on the total fixture unit load connected to the system.

See Appendix B of this code for explanatory notes on the design of combination waste and vent systems.

While water closets and urinals are prohibited on such systems, they may connect downstream the combination waste and vent system provided they are conventionally vented and sized, and that the two pipe-size increase requirements of Section 910.4 include the total fixture unit load of all connected fixtures. See **Figure 910.7**.

Review Appendix B for additional information and clarification about these systems.

911.0 Circuit Venting.

Circuit venting has been a common venting practice used in plumbing in the United States since the 1920's.

VENTS

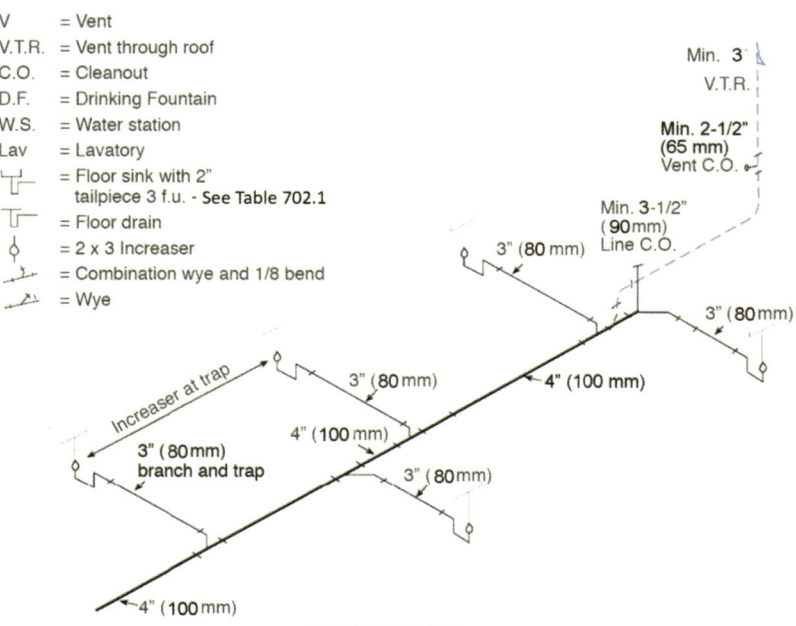

FIGURE 910.3A
COMBINATION WASTE AND VENT WITH SINGLE VENT

FIGURE 910.3B
UNDERGROUND INSTALLATION OF TERMINAL VENT

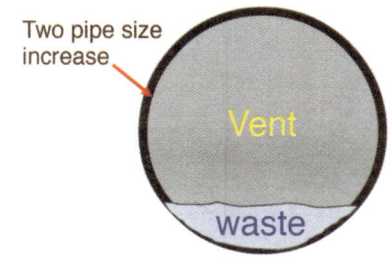

FIGURE 910.4A
VISUALIZATION OF FLOW IN COMBINATION WASTE AND VENT

FIGURE 910.4B
OVERSIZED TRAP WITH NORMAL SIZED TAILPIECE

Dr. Hunter included circuit venting in the Plumbing Manual BMS 66 published in 1940. By this time, circuit venting had become widely accepted and recognized throughout the United States. Circuit venting is an effective and efficient method of venting a battery of fixtures connected to a horizontal branch. Without the turbulent action of a vertical stack, maintaining the pressure within a horizontal branch becomes less demanding. As designed in BMS 66 circuit venting was intended solely for public restrooms with a battery of water closets and pedestal urinals with no more than 8 fixtures for a four-inch horizontal branch.

A circuit vent is a vent that connects to a horizontal branch (see definition in Chapter 2) and vents two traps to a maximum of eight traps connected into a battery of fixtures (see **Figure 911.0**). The vent for a circuit vent is typically 2 inch in diameter providing more than an adequate amount of air for venting the battery of fixtures. One vent for every 8 fixtures is all that is required to maintain the pressure within the branch to plus or minus 1-inch of a water column.

911.1 Circuit Vent Permitted. A maximum of eight fixtures connected to a horizontal branch drain shall be permitted to be circuit vented. Each fixture drain shall connect horizontally to the horizontal branch being circuit vented. The horizontal branch drain shall be classified as a

VENTS

vent from the most downstream fixture drain connection to the most upstream fixture drain connection to the horizontal branch.

🔧 When installing circuit venting, the fixture drain or trap arm is the portion of pipe from the fixture connection to the circuit vented horizontal branch. Although it is not addressed in Section 911.1, it is generally recognized that fixture drains or trap arms are limited to the lengths prescribed in Table 1002.2.

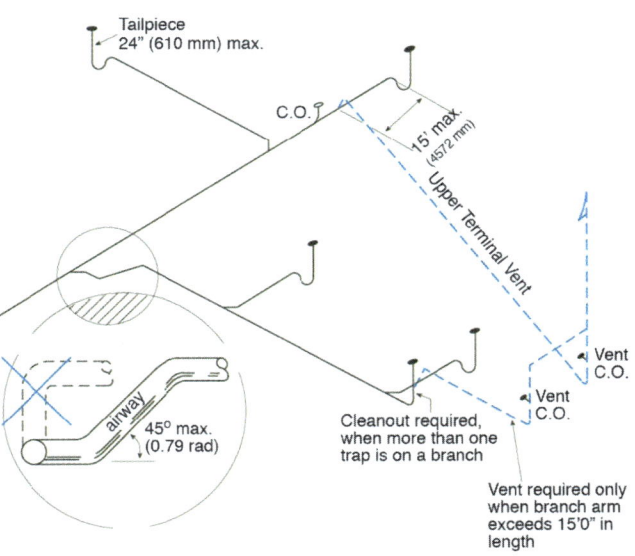

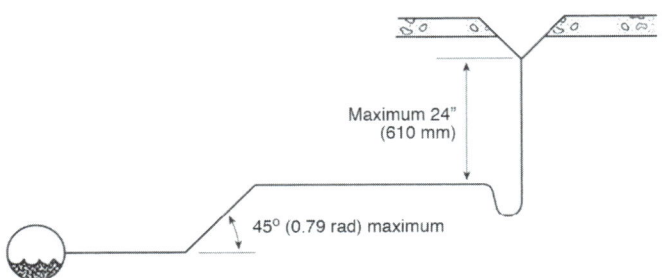

FIGURE 910.5A
VERTICAL TAILPIECE AND HORIZONTAL ENTRY INTO MAIN

FIGURE 910.5B
45 DEGREE OFFSETS FOR BRANCHES

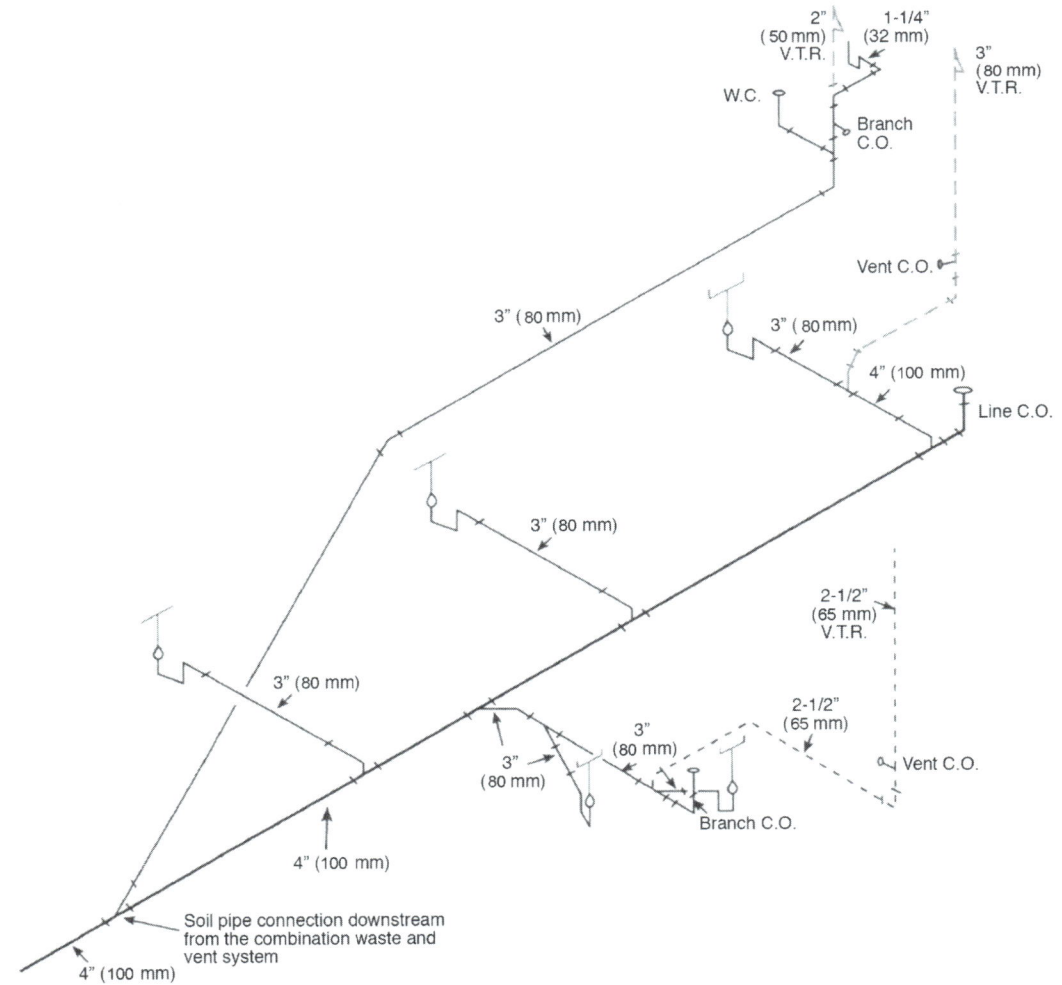

FIGURE 910.7
WATER CLOSET MAY NOT BE CONNECTED TO SYSTEM AND CLEANOUT PLACEMENT

911.1.1 Multiple Circuit-Vented Branches. Circuit-vented horizontal branch drains are permitted to be connected together. Each group of a maximum of eight fixtures shall be considered a separate circuit vent and shall be in accordance with the requirements of this section.

🔧 See **Figure 911.0** first floor example and Section 911.3.1 for additional provisions.

911.2 Vent Size and Connection. The circuit vent shall be not less than 2 inches (50 mm) in diameter, and the connection shall be located between the two most upstream fixture drains. The vent shall connect to the horizontal branch on the vertical. The circuit vent pipe shall not receive the discharge of soil or waste.

🔧 The vent shall connect to the horizontal branch on the vertical between the two most upstream trap arms.

911.3 Slope and Size of Horizontal Branch. The slope of the vent section of the horizontal branch drain shall be not more than 1 inch per foot (83.3 mm/m). The entire length of the vented section of the horizontal branch drain shall be sized for the total drainage discharge to the branch.

911.3.1 Size of Multiple Circuit Vent. Multiple circuit vented branches shall be permitted to connect on the same floor level. Each separate circuit-vented horizontal branch that is interconnected shall be sized independently in accordance with Section 911.3. The downstream circuit-vented horizontal branch shall be sized for the total discharge into the branch, including the upstream branches and the fixtures within the branch.

911.4 Relief Vent. A 2 inch (50 mm) relief vent shall be provided for circuit-vented horizontal branches receiving the discharge of four or more water closets and connecting to a drainage stack that receives the discharge of soil or waste from upper horizontal branches.

🔧 There are two conditions that mandate the installation of a relief vent for a circuit-vented horizontal branch:

1) when the circuit-vented horizontal branch receives the discharge of four or more water closets, and

2) when the circuit-vented horizontal branch connects to a stack that receives discharges from upper floor fixtures.

See **Figure 911.4**.

911.4.1 Connection and Installation. The relief vent shall connect to the horizontal branch drain between the stack and the most downstream fixture drain of the circuit vent. The relief vent shall be installed on the vertical to the horizontal branch.

911.4.2 Fixture Drain or Branch. The relief vent is permitted to be a fixture drain or fixture branch for a fixture

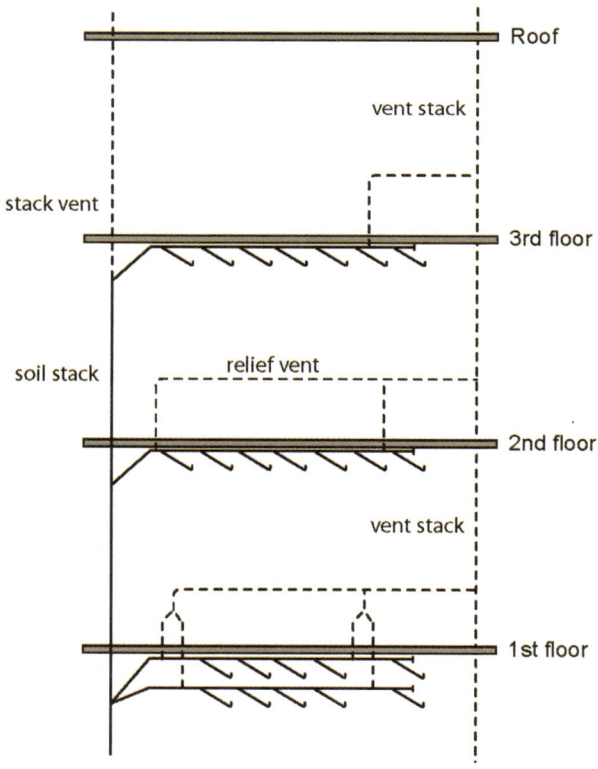

3rd Floor without relief vent
2nd Floor with one relief vent
1st Floor with multiple circuit-vented branches

**FIGURE 911.0
CIRCUIT VENTING**

located within the same branch interval as the circuit-vented horizontal branch. The discharge to a relief vent shall not exceed 4 fixture units.

While a branch interval is not defined in Chapter 2 Definitions, there is a definition found in Appendix C. A branch interval is defined as a length of soil or waste stack corresponding in general to a story height, but in no case less than 8 feet (2438 mm), within which the horizontal branches from one floor or story of the building are connected to the stack.

911.5 Additional Fixtures. Fixtures, other than the circuit-vented fixtures, are permitted to discharge to the horizontal branch drain. Such fixtures shall be located on the same floor as the circuit-vented fixtures and shall be either individually or common vented.

912.0 Engineered Vent System.

912.1 General. The design and sizing of a vent system shall be permitted to be determined by accepted engineering practices. The system shall be designed by a registered design professional and approved in accordance with Section 301.5.

912.2 Minimum Requirements. An engineered vent system shall provide protection of the trap seal in accordance with Section 901.3.

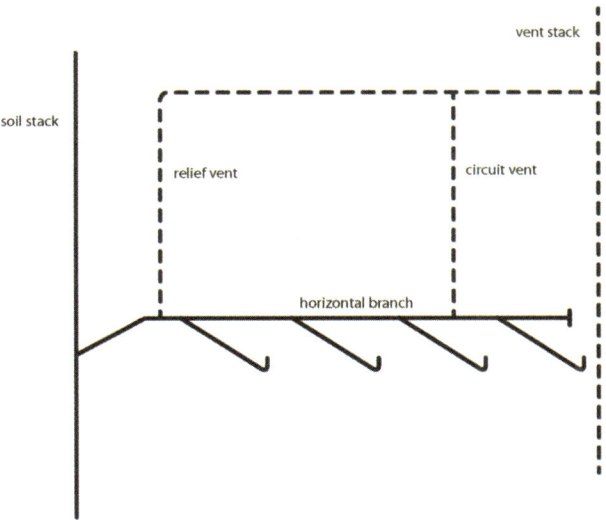

**FIGURE 911.4
RELIEF VENT**

CHAPTER 10
TRAPS AND INTERCEPTORS

1001.0 General.

1001.1 Applicability. This chapter shall govern the materials, design, and installation of traps and interceptors.

1001.2 Where Required. Each plumbing fixture shall be separately trapped by an approved type of liquid seal trap. This section shall not apply to fixtures with integral traps. Not more than one trap shall be permitted on a trap arm. Food waste disposers installed with a set of restaurant, commercial, or industrial sinks shall be connected to a separate trap. Each domestic clothes washer and each laundry tub shall be connected to a separate and independent trap, except that a trap serving a laundry tub shall also be permitted to receive the waste from a clothes washer set adjacent to it. The vertical distance between a fixture outlet and the trap weir shall be as short as practicable, but in no case shall the tailpiece from a fixture exceed 24 inches (610 mm) in length. One trap shall be permitted to serve a set of not more than three single compartment sinks or laundry tubs of the same depth or three lavatories immediately adjacent to each other and in the same room where the waste outlets are not more than 30 inches (762 mm) apart, and the trap is centrally located where three compartments are installed.

The requirement for fixtures within the building to be trapped by a water seal (see **Figure 1001.2a**) has been in plumbing codes for almost 100 years. The trap's main purpose is to eliminate the possibility of sewer gas from entering the confines of the room or building. The water seal, or trap seal, effectively accomplishes this by creating a barrier of 2 to 4 inches of water that prevents sewer gas from entering the room. A secondary benefit is that it prevents access of pests or vermin to drains, as well as access to the building through the fixture by way of the sewer system.

Each fixture must be separately trapped. Some fixtures such as water closets and urinals have integral traps and do not require an additional trap since the water seal is built in its structure (see **Figure 1001.2c**).

The trap arm, the piping from the end of the trap to the inner edge of the vent connection, may only serve one trap (see **Figures 1001.2b** and **1001.2d**). This eliminates the possibility of one trap being siphoned as the other fixture is used. The flow along the trap arm passing the other trap connection could siphon the unused trap. This is the reason why each trap must be protected by a separate vent as is required in Section 1002.2 and in Section 901.2, Vents Required.

The continuous waste provides a method of draining multiple compartments to a single trap. The continuous waste is limited to up to three compartments, sinks or lavatories, but only if they are within the 30 inch limitation. In a three-sink configuration the trap must be located under the center sink. This is to provide uniform drain sizing and to shorten the distance to the trap and its drain (see **Figures 1001.2e, 1001.2f** and **1001.2g**).

The installation of a food waste disposal in a commercial kitchen is normally under the jurisdiction of a health department. In some areas, this installation may even be prohibited. Commercial kitchen sinks will often be connected to a grease interceptor or hydromechanical device so depositing food waste into these devices will negatively affect their function. When they are allowed to be connected to the drainage system they must be installed on a separate trap. This also means that they will have a separate trap arm, vent, and fixture drain. This will effectively separate the sink from the disposal, and if the disposal line becomes plugged, the sink will still function (see **Figure 1001.2h**).

If a clothes washer and laundry tub are both to be installed in the same area, they must each have their own drainage connection. If a laundry tub, sometimes called a laundry tray or sink, is installed but there are no connec-

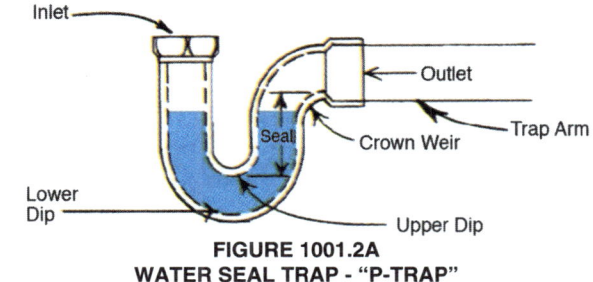

FIGURE 1001.2A
WATER SEAL TRAP - "P-TRAP"

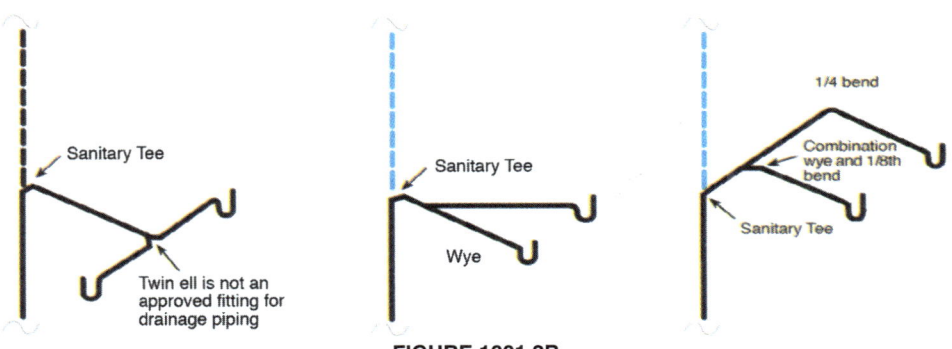

FIGURE 1001.2B
TWO TRAPS ON ONE TRAP ARM - PROHIBITED

tions for a clothes washer, then the clothes washer may be set next to the sink and drain into it. The laundry sink compartment will be able to handle the pumped waste from the clothes washer because of its volume.

The fixture tailpiece should be as short as possible to limit the fall of waste into the trap. Excessive vertical lengths of the tailpiece will increase the effective head pressure above the inlet of the trap by increasing the height of the column of water, which may be sufficient to induce self-siphonage. As the length of the column increases, so also the velocity of the moving column of water, increasing the momentum pull on the trap seal. Research has proven that limiting the tailpiece length to 24 inches will help to eliminate the possibility of the trap to self-siphon. The tailpiece may offset but the developed length of the tailpiece must not exceed 24 inches (see **Figure 1001.2i**).

Other methods developed to eliminate self-siphonage of the trap are the limitation of trap arm lengths (see Section 1002.2) and the provision of a vent for the trap (see Section 1002.1).

The clothes washer standpipe is not the tailpiece for a clothes washer. The standpipe for the clothes washer is considered to be an indirect waste receptor rather than a tailpiece and, therefore, is limited by Section 804.1, Standpipe Receptors, and not by this section of the code.

1002.0 Traps Protected by Vent Pipes.

1002.1 Vent Pipes. Each plumbing fixture trap, except as otherwise provided in this code, shall be protected against siphonage, backpressure, and air circulation shall be assured throughout the drainage system using a vent pipe installed in accordance with the requirements of this code.

The section (see also Section 901.2) requires the trap to be protected from siphonage and backpressure by a vent (see **Figure 1002.1**), and that air circulation is assured throughout the system. Most of the measures to accomplish

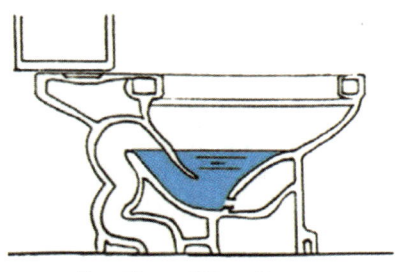
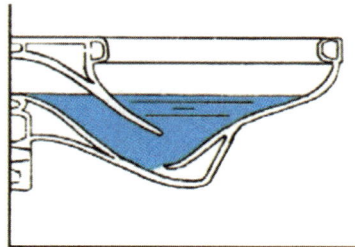
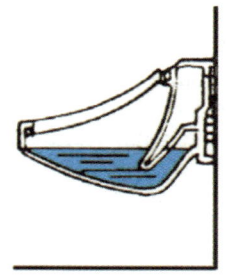

FIGURE 1001.2C
WATER CLOSET AND URINAL INTEGRAL TRAPS

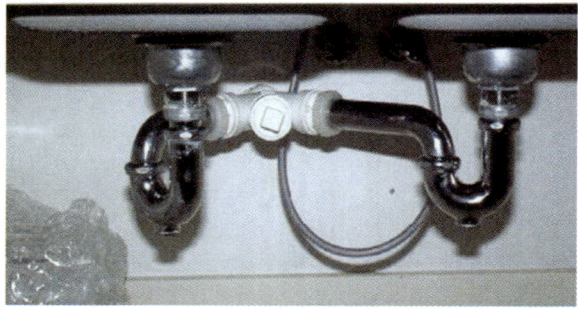

FIGURE 1001.2D
TWO TRAPS ON ONE TRAP ARM - PROHIBITED

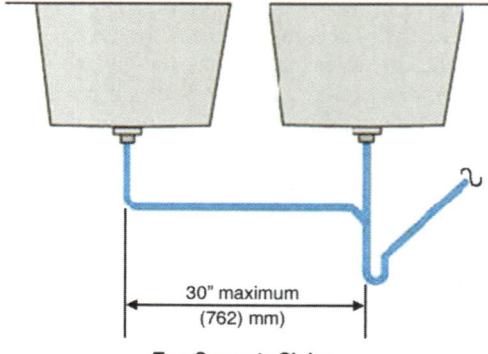

FIGURE 1001.2F
TWO FIXTURES ON ONE TRAP - PROHIBITED IF OVER 30 INCH CENTER TO CENTER OR IF OVERFLOW RIM NOT AT SAME LEVEL

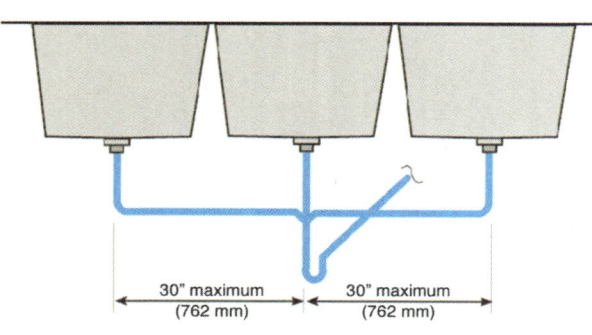

FIGURE 1001.2E
THREE COMPARTMENT SINK ON ONE TRAP

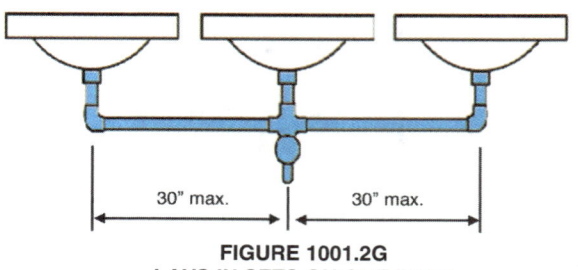

FIGURE 1001.2G
LAVS IN SETS ON ONE TRAP

this are discussed in Chapter 9 and should be reviewed along with this chapter.

The protection of the waterseal trap is assured upon a simple pneumatic principle, namely, equal atmospheric pressure must be maintained on both sides of the trap. When a trap is filled, the atmospheric pressure on the fixture side of the trap water seal is equal to the atmospheric pressure on the outlet side of the trap water seal. The vent connection to the trap arm maintains that equal pressure on the outlet side of the trap. As is discussed in Chapter 9, the vent system is designed to maintain pressure variation within the system of plus or minus one-inch water column (.25 kPa). Pressures above or below that design pressure will cause a trap to lose its water seal. A positive pressure on the downstream side of the trap will cause air to blow through the trap seal, pushing its contents into the fixture. A negative pressure will siphon the trap seal into the trap arm and the drain.

This allowable pressure variation of ± one-inch water column is a considerable factor of safety for a trap seal of 2 inches or greater. For example, if a trap seal has a depth of 2 inches (see **Figure 1005.1**) and is subjected to a negative pressure of one-inch water column, then the remaining trap seal depth would be 1½ inches. If the trap seal is 4 inches, then the remaining seal depth would be 3½ inches. Hence, it would take a negative pressure of a four-inch water column to siphon the water from a trap with a 2-inch seal below the dip of the trap, and eight-inches of water column to siphon a 4-inch seal below the dip of the trap. Designing a venting system according to Chapter 9 ensures considerable safety margins by maintaining the pressure variations within ± one-inch water column.

1002.2 Fixture Traps. Each fixture trap shall have a protecting vent so located that the developed length of the trap arm from the trap weir to the inner edge of the vent shall be within the distance given in Table 1002.2 but in no case less than two times the diameter of the trap arm.

The effect of length of unvented fixture drains on self-siphonage was the subject of much research at the National Bureau of Standards [see Recommended Minimum Requirements for Plumbing BH13 (1928) and Self-Siphonage of Fixture Traps BMS126 (1951)]. One of the objectives of the research was to find the maximum length of a fixture drain without self-siphoning.

This research demonstrated that the slope should be confined between the limits of ¼ inch and ½ inch per foot and that the outlet end of the fixture drain should not extend below the weir of the trap. Hence, Table 1002.2

TABLE 1002.2
HORIZONTAL LENGTHS OF TRAP ARMS
(EXCEPT FOR WATER CLOSETS AND SIMILAR FIXTURES)[1, 2]

TRAP ARM PIPE DIAMETER (inches)	DISTANCE TRAP TO VENT MINIMUM (inches)	LENGTH MAXIMUM (inches)
1¼	2½	30
1½	3	42
2	4	60
3	6	72
4	8	120
Exceeding 4	2 x Diameter	120

For SI units: 1 inch = 25.4 mm
Notes:
[1] Maintain ¼ inch per foot slope (20.8 mm/m).
[2] The developed length between the trap of a water closet or similar fixture (measured from the top of the closet flange to the inner edge of the vent) and its vent shall not exceed 6 feet (1829 mm).

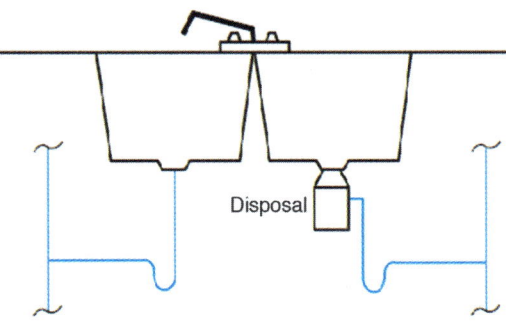

FIGURE 1001.2H
SEPARATE DRAINAGE CONNECTION FOR COMMERCIAL GARBAGE DISPOSAL

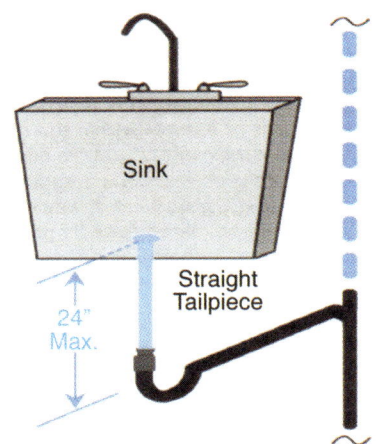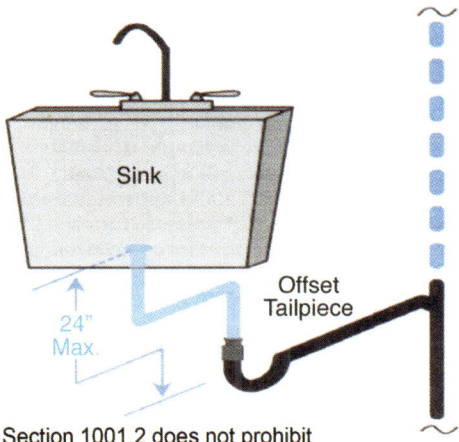

Section 1001.2 does not prohibit an offset in a tailpiece.

FIGURE 1001.2I
LENGTH OF TAILPIECE

TRAPS AND INTERCEPTORS

shows maximum horizontal lengths of trap arms calculated using a grade of ¼ inch per foot to an extent where the trap arm outlet will not be below the weir of the trap. For example, a 1¼ inch trap arm may have a maximum horizontal distance of 30 inches without producing the effect of self-siphonage (see **Figures 1002.2a** and **1002.2b**).

The research also revealed the variations between short and long turn pattern fittings (sanitary tee and combination wye-eighth bend), the short turn fitting permitting the use of longer unvented lengths of fixture drains. Thus, the short turn fitting has better characteristics to prevent self-siphonage and is preferred when connecting a fixture drain to a vertical vent. The sanitary tee also allows the vent opening to remain above the trap weir. This better allows air into the trap arm while the fixture is draining and retards any momentum pull that could cause siphonage.

Because the water closet and similar integrally trapped fixtures function as self-siphonage fixtures, their traps are an S-type. In this case, the trap arm length is calculated along the developed length of the piping from the top of the closet flange to the inner edge of the vent. Footnote 2 in Table 1002.2 stipulates for the water closet trap arm a maximum length of 6 feet. This limitation is required whether the water closet trap arm is 3 inches in diameter or 4 inches.

The same research discussed above also discovered that the vent could be placed too close to the weir of the trap. A style of venting the trap called the "crown vent" had been used but it was found to be subject to clogging. When the fixture discharges, there is an upward vertical momentum of flow on the outlet side of the trap which would enter the vent if installed on the crown of the trap causing the vent to eventually foul and clog. Also, the vent at the crown of the trap or close to the crown is covered by the initial full flow into the trap arm and is useless to prohibit siphonage. Therefore, there is a minimum trap arm length requirement. The minimum length is two times the diameter of the trap arm. For example, the 1¼ inch trap arm must have a minimum distance of 2½ inches from the weir of the trap to the inner edge of the vent. Two times the diameter of the pipe effectively moves the inner edge of the vent away from the area of full flow in the trap arm and allows the vent to function properly (see **Figure 1002.2c**).

FIGURE 1002.2B
THREE-INCH TRAP ARM

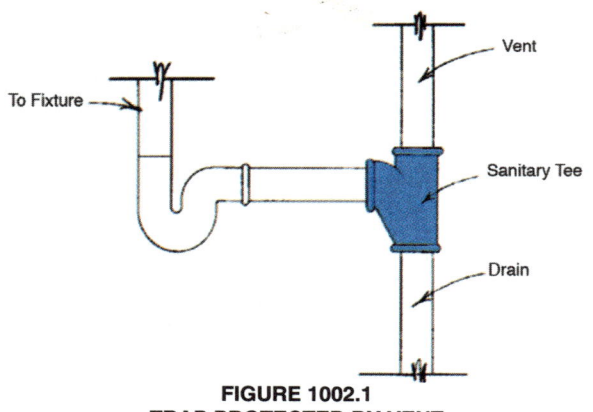

FIGURE 1002.1
TRAP PROTECTED BY VENT

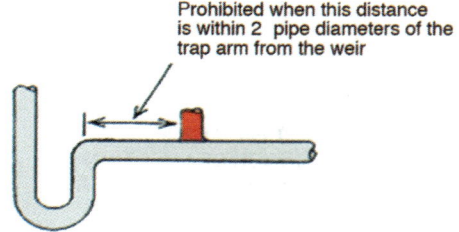

FIGURE 1002.2C
MINIMUM TRAP ARM DISTANCE

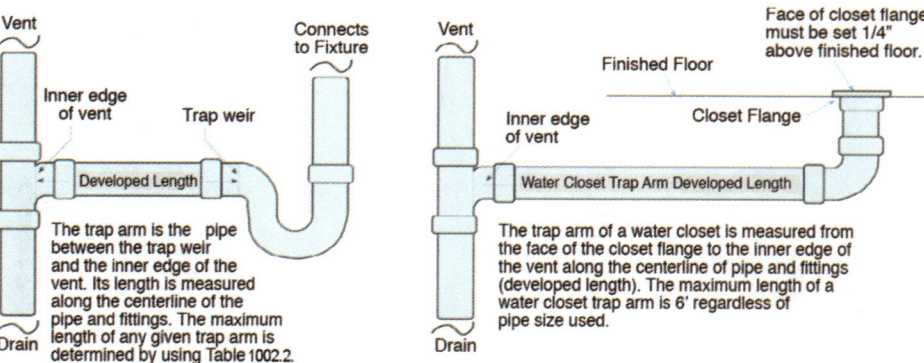

FIGURE 1002.2A
DEVELOPED LENGTH OF TRAP ARM

1002.3 Change of Direction. A trap arm shall be permitted to change direction without the use of a cleanout where such change of direction does not exceed 90 degrees (1.57 rad). Horizontal changes in the direction of trap arms shall be in accordance with Section 706.3.

Exception: For trap arms, 3 inches (80 mm) in diameter and larger, the change of direction shall not exceed 135 degrees (2.36 rad) without the use of a cleanout.

🔧 A very common installation of the trap arm is a 90-degree turn in the trap arm between the trap and the connection of the trap arm to the waste and vent (see **Figure 1002.3**). This is commonly called a "dirty arm." Research has determined that a 90-degree elbow in a horizontal trap arm at a distance no greater than 18 to 20 inches from the vertical inlet of the trap actually reduces the self-siphonage effect (NBS, Recommended Minimum Requirements for Plumbing BH13). The 90-degree turn is allowable if the fitting used complies with Section 706.3, which states which fittings may connect horizontal lines with other horizontal lines. This essentially limits the 90-degree change of direction to a long sweep or two 45-degree fittings. However, in most areas a Durham 90-degree fitting is allowed. Note that only the Durham or drainage fitting can be used here.

1002.4 Vent Pipe Opening. The vent pipe opening from soil or waste pipe, except for water closets and similar fixtures, shall not be below the weir of the trap.

🔧 This requirement is identical to Section 905.5 and has been discussed in Section 1002.2. Caution is needed in using the proper fitting in connecting the trap arm to the waste and vent connection. In the vertical position the sanitary tee will be the proper fitting for most installations. The use of a combination wye and one-eighth bend will not be allowed as it places the inner edge of the vent below the weir of the trap. See **Figure 1002.4** for an improper use of a double wye and the improper length of clothes washer standpipes.

1003.0 Traps — Described.

1003.1 General Requirements. Each trap, except for traps within an interceptor or similar device shall be self-cleaning. Traps for bathtubs, showers, lavatories, sinks, laundry tubs, floor drains, urinals, drinking fountains, dental units, and similar fixtures shall be of standard design, weight and shall be of ABS, cast-brass, cast-iron, lead, PP, PVC, or other approved material. An exposed and readily accessible drawn-copper alloy tubing trap, not less than 17 B & S Gauge (0.045 inch) (1.143 mm), shall be permitted to be used on fixtures discharging domestic sewage.

Exception: Drawn-copper alloy tubing traps shall not be used for urinals. Each trap shall have the manufacturer's name stamped legibly in the metal of the trap, and each tubing trap shall have the gauge of the tubing in addition to the manufacturer's name. A trap shall have a smooth and uniform interior waterway.

🔧 The requirement for self-cleaning traps does not refer to any internal mechanism within the trap itself. It simply refers to the self-scouring potential that the flow rate of discharge has within the trap itself to prevent fouling. The P

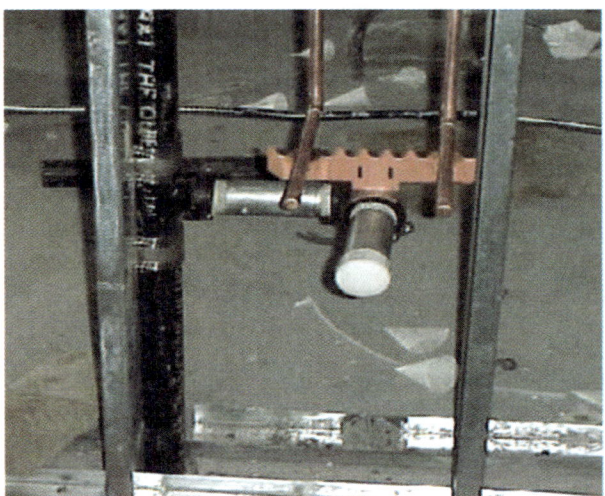

FIGURE 1002.3
"DIRTY ARM" - USE ONLY DURHAM-STYLE FITTING

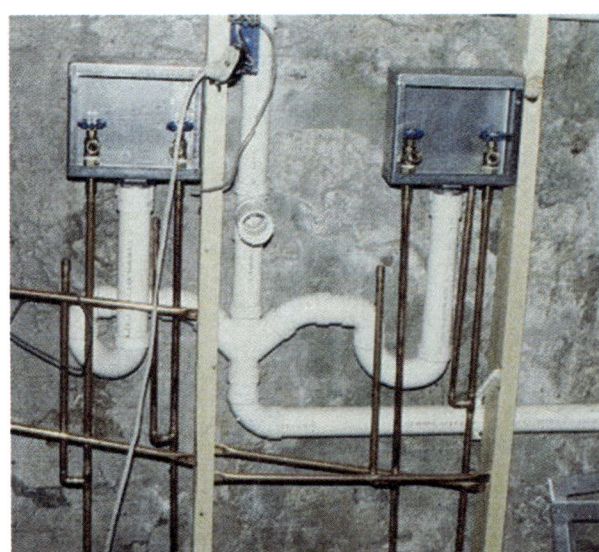

FIGURE 1002.4
IMPROPER FITTINGS USED FOR VENT CONNECTION

trap is already designed to be self-cleaning or self-scouring if sized correctly. Although an increase of trap seal depth reduces the potential for self-siphonage, it will also decrease the scouring properties by reducing the rate of flow through the trap. A drum trap is an example of a trap that is not self-cleaning (see commentary for Section 1004.1).

Traps are part of the drainage system and must be approved for drainage materials listed. The brass tubing trap of not less than 17 Brown and Sharp (B&S) sheet metal gauge must be readily accessible. The trap gauge is thicker than the allowable continuous waste gauge of 20 B&S gauge because the trap will hold waste water continuously and must be more resistant to corrosion than the continuous waste or tailpiece.

The use of drawn brass traps for urinals is not allowed because of the corrosive contents (urine) of the liquid waste in the urinal trap. Each trap must also be readily identifiable not only for the installer but for the inspector and maintenance personnel.

1003.2 Slip Joint Fittings. A maximum of one approved slip joint fitting shall be permitted to be used on the outlet side of a trap, and no tubing trap shall be installed without a listed tubing trap adapter. Listed plastic trap adapters shall be permitted to be used to connect listed metal tubing traps.

🔧 This would allow the slip joint on a trap arm extension or a 45-degree offset. Remodeling a sink cabinet is an application where such a slip joint may be needed. The new cabinet may offset the fixture trap from the waste connection. To reconnect the trap to the existing trap adapter, a slip extension piece or a 45-degree offset slip joint may be needed.

An approved trap adapter must be used at the connection of the trap arm to the drainage fitting. Whether it is a plastic pipe end or a galvanized nipple, a trap adapter must be used. The adapter prevents the trap arm from extending too far down the drainage pipe or, if placed close to the waste and vent connection, from passing into the drainage fitting and blocking the waste or airflow (see **Figure 1003.2**).

**FIGURE 1003.2
TRAP ADAPTER - NOTICE STOP FOR P-TRAP ARM**

1003.3 Size. The size (nominal diameter) of a trap for a given fixture shall be sufficient to drain the fixture rapidly but in no case less than nor more than one pipe size larger than given in Table 702.1. The trap shall be the same size as the trap arm to which it is connected.

🔧 Trap sizes are generally based on the size of the fixture outlet. The size of the trap and trap arm should correspond to the size of the fixture outlet to provide proper discharge flow rate and prevent overflowing or sluggish drainage. If the trap is smaller than what is required in Table 702.1 Drainage Fixture Unit Valves (DFU), then the effect of self-siphonage would increase. If it is oversized more than one pipe size, then the flow from the fixture would not be enough to provide the proper scouring or cleaning effect for the trap. However, allowing the trap to increase only one size larger than the minimum requirement will not significantly affect the scouring ability of the trap and will also decrease the effect of self-siphonage.

For example, a kitchen sink trap that is a minimum of 1½ inches can be increased to 2 inches. This would allow a longer trap arm developed length. Notice that this section requires the trap to be the same size as the trap arm. If the option to use a 2-inch trap arm is selected, this code section also mandates that a 2-inch trap be installed. A 2-inch trap could not be used with a 1½-inch trap arm as this would be an obstruction to flow prohibited in Chapter 3.

1004.0 Traps.

1004.1 Prohibited. No form of trap that depends for its seal upon the action of movable parts shall be used. No trap that has concealed interior partitions, except those of plastic, glass, or similar corrosion-resisting material, shall be used. "S" traps, bell traps, and crown-vented traps shall be prohibited. No fixture shall be double trapped. Drum and bottle traps shall be installed for special conditions. No trap shall be installed without a vent, except as otherwise provided in this code.

🔧 Manufacturers have produced various kinds of traps in the past and continue to do so with innovative designs. However, not all designs have been found favorable for a safe plumbing system. Certain anti-siphon traps have been designed with movable parts or mechanical means to maintain its trap seal (see **Figure 1004.1a**). The problem with these types of traps is that they may foul and malfunction over time since they are not self-cleansing. Traps with interior partitions are permitted only if the partition is of noncorrosive material (see **Figure 1004.1b**). Interior partitions are not visible and if the partition corrodes, the trap seal will be lost without notice.

The unsuitability of crown vented traps to be used in the drainage system has already been discussed in Section 1002.2, and the S trap is ineffective in protecting the trap seal because the arrangement of the S trap makes the trap seal too susceptible to siphonage (see **Figure 1004.1c**).

The bell trap (see **Figure 1004.1d**) consists of a cup with a standpipe in the middle over which is a bell that dips into the water contained in the cup to form a seal. Also, this trap is not self-cleansing, the seal depth is very shallow (often less than one-half inch), and it is known to be very sluggish in draining.

Drum and bottle traps (see **Figures 1004.1e** and **1004.1f**) may be installed only when permitted by the Authority Having Jurisdiction (AHJ) for use under special conditions. Drum traps are used as sediment traps, separating out small solids before they enter the drainage system. The bottle trap is similar to the P trap, but with a slightly different configuration and is somewhat more prone to siphonage than the P trap. It is smaller than the P trap and used where space saving is necessary. These traps may still be used upon approval but there should be a valid reason.

When a fixture is double trapped, the piping between the two traps could become air bound, resulting in a slow drain and increasing the possibility of a stoppage. There is also the possibility of one or both of the traps being siphoned by the flow through them; therefore, no fixture may be double trapped.

TRAPS AND INTERCEPTORS

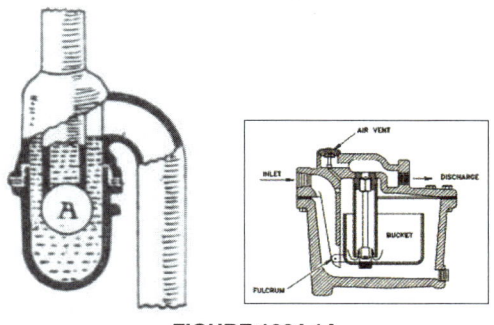

FIGURE 1004.1A
ANTI-SIPHON TRAPS

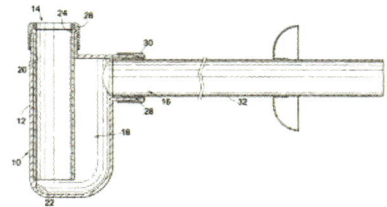

FIGURE 1004.1B
TRAP WITH INTERIOR PARTITION

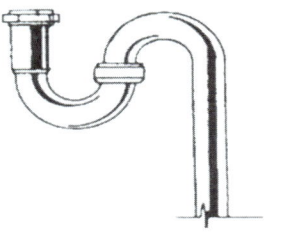

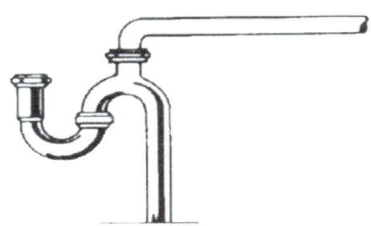

FIGURE 1004.1C
S-TYPE AND CROWN VENTED TRAPS - PROHIBITED

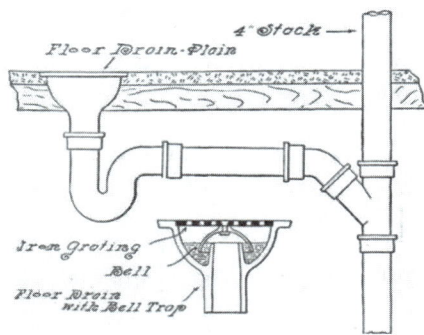

FIGURE 1004.1D
BELL TRAP

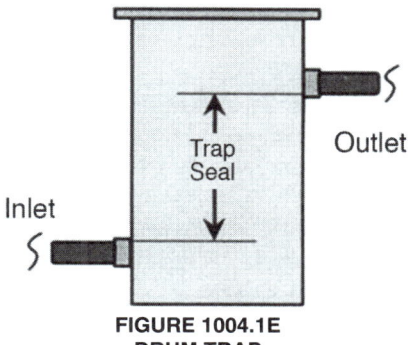

FIGURE 1004.1E
DRUM TRAP

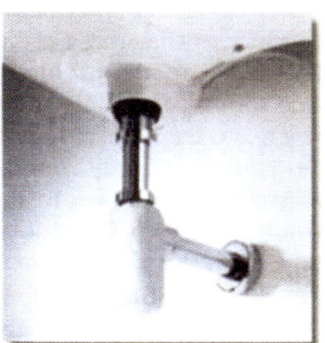

FIGURE 1004.1F
BOTTLE TRAP

1004.2 Movable Parts. Bladders, check valves or another type of devices with moveable parts shall be prohibited to serve as a trap.

🔧 This section prohibits movable parts that maintain a trap seal. Some traps have been manufactured to function without liquid seals. These traps utilize a bladder or diaphragm mechanism that is typically closed to seal off sewer gas while allowing expansion of the diaphragm when there is discharge of water. Similarly, check valves do not rely on trapped water to prevent the emission of sewer gas. Although these products may be initially effective, there still exist conditions where fouling may cause the movable parts to no longer seal tight and allow the emission of gases.

1005.0 Trap Seals.

1005.1 General. Each fixture trap shall have a liquid seal of not less than 2 inches (51 mm) and not more than 4 inches (102 mm), except where a deeper seal is found necessary by the Authority Having Jurisdiction. Traps shall be set true with respect to their liquid seals and, where necessary, they shall be protected from freezing.

🔧 The water seal that forms the trap barrier is the maximum vertical depth of liquid that a trap will retain. It is measured between the crown weir and the top of the dip of the trap. A water seal of not less than two inches and not more than four inches was found to be the optimum depth to provide the necessary protection against sewer gas and still provide the self-scouring effect needed to keep the trap from fouling (see **Figure 1005.1**). A trap with less than two-inch-

TRAPS AND INTERCEPTORS

es of seal depth offers less resistance to self-siphonage and more readily loses its seal by evaporation. Obviously the greater the seal depth, the opposite is proportionately true.

Traps for special equipment or circumstances, such as a grease interceptor, may have a deeper trap seal but these are usually for commercial or industrial applications that require a greater depth. Also, in most of these cases the trap seal is not in a P-trap but in the depth of water in a tank.

The trap must be set true (level) with respect to the liquid bound in the trap. This means that the water columns on both sides of the trap must be level with each other on the same horizontal plane of the trap weir. If not, the depth of the trap seal will be compromised with a lesser depth, reducing protection from backpressure or siphonage.

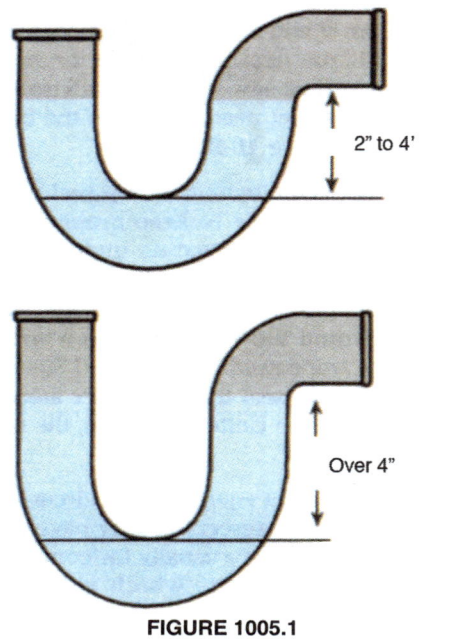

**FIGURE 1005.1
STANDARD TRAP SEAL AND DEEP SEAL TRAP**

1006.0 Floor Drain Traps.

1006.1 General. Floor drains shall connect into a trap so constructed that it can be readily cleaned and of a size to serve efficiently the purpose for which it is intended. The drain inlet shall be so located that it is in full view. Where subject to the reverse flow of sewage or liquid waste, such drains shall be equipped with an approved backwater valve.

🔧 Floor drains are usually an emergency device placed in the floor to catch the overflow from sinks or equipment. Because they are at floor level they could eventually collect sediment or other debris in the trap. If there is any flow at all into most floor drain traps, it will be a very low or intermittent flow and not enough to provide a scouring action in the trap. Sediment may build up and eventually plug the trap; therefore, they must be installed in open sight for the ability to service them.

1007.0 Trap Seal Protection.

1007.1 General. Floor drain or similar traps directly connected to the drainage system and subject to infrequent use shall be protected with a trap seal primer, except where not deemed necessary for safety or sanitation by the Authority Having Jurisdiction. Trap seal primers shall be accessible for maintenance.

🔧 Floor drain traps that do not connect to a water-supplied fixture will tend to evaporate. Other traps that are not periodically used or are installed in overly dry areas will lose their seal by simple evaporation. The result of this loss of trap seal allows known health hazards such as hydrogen sulfide, methane, bacteria, viruses and mold to enter the living space. The SARS epidemic (Severe Acute Respiratory Syndrome) reported in Hong Kong, China in 2003 caused in part by a dry and leaking trap, has raised the awareness of the dangers of an unprotected sanitary drainage system.

One of the first lines of defense against trap failure is the trap primer, which is a piping device that helps to maintain a water seal in a remote trap. A trap primer connects to a cold water inlet or other water supply. Every time that inlet or supply is used, a small amount of water flows down to the trap to keep the water seal from drying up. If the primer is electronic, it is set on a timer that daily releases enough water to maintain the trap seal (see **Figures 1007.1a** and **1007.1b**).

It is clear that traps can easily dry up in rooms where floor drains are seldom used, but it may not be obvious that traps dry up in rooms that are used daily. This is a common problem in mechanical and service rooms, which are often dry and overly warm. Years ago, traps would be regularly replenished on the janitor's nightly rounds. Enough of the soapy water from those old fashioned heavy mops would go down the floor drain to replenish the trap. Today, with the popularity of dry chemicals for cleaning, not much water is used on the floors, at least not enough to keep the traps seal full. By providing a trickle of water every time water is used, trap seal primers refill the seal of a P-trap.

Making the correct choice of which trap primer to use should take some thought. There are three main kinds of trap seal primers:

1. Pressure drop activated.
2. Flush valve operated.
3. Multiple and electronic activated.

Pressure-drop-activated trap seal primers are usually brass, have a one-half-inch Mail Iron Pipe (MIP) inlet connection, and one-half-inch Female Iron Pipe (FIP) outlet connection. An interior cartridge seals when the line pressure is in a static state, but when the line pressure drops just three psi (20.68 kPa) by the flushing of a toilet, a faucet opened, etc., the cartridge will rise. It rises because a pressure differential has occurred within the primer. At this point, a metered amount of water is discharged under pressure into the line connected to the floor drain trap.

Pressure drop activated trap seal primers are designed to minimize the problem of clogging. These trap seal primers are constructed with a fine mesh filter, keeping line debris at bay.

TRAPS AND INTERCEPTORS

The flush-valve-operated trap seal primer gets its name from its installation (see **Figure 1007.1c**). It is placed below the flush valve on a water closet or urinal and connects to the floor drain by a tube so that the tube can catch some of the water discharged from the flush valve. The water runs down the tube to the floor drain and the water seal is maintained. This trap seal primer has two positive features: 1) low water consumption since it only uses a portion of the discharged water from the flush valve operation, and 2) no moving parts. However, this trap seal primer also introduces some restrictions. Its use may not be practical because the flush-valve may be situated too far from the floor drains. In this case, a pressure-drop-activated trap primer is more practical since it can be installed close to the drain that needs to be primed.

When a number of pressure-drop-activated trap seal primers are located close to one another, they can be primed via a multiple trap primer (see **Figure 1007.1d**). In this case, a metered amount of water from the trap primer is evenly distributed to more than one floor drain trap by means of a distribution unit or an electronic primer manifold. For every 20 feet of floor drain trap make-up water line, the primer must be a minimum of 1 foot elevation from the finished floor. It should generally be mounted level with the water supply line.

These multiple units can be installed in a variety of situations. For example, such a unit is useful when a men's restroom is located next to a women's restroom. In this instance, one trap seal primer can serve both floor drain traps, omitting the need for additional trap primers, extra shutoff valves, another access door and air gap fittings, which saves time, money, and labor. They can also be installed with an electronic timer to provide automatic filling of the trap seal. In this installation there is no worry about having enough pressure drop or use of the fixtures to keep the trap filled.

The trap primer must be accessible. The access door needs to be large enough to allow the plumber to reach the trap seal primer, the line shutoff valve and any other accessories present. Installing a line shutoff valve will make repair and maintenance that much easier.

1007.2 Trap Seal Primers. Potable water supply trap seal primer valves shall comply with ASSE 1018. Drainage and electronic design type trap seal primer devices shall comply with ASSE 1044.

See **Figure 1007.2**.

1008.0 Building Traps.

1008.1 General. Building traps shall not be installed except where required by the Authority Having

FIGURE 1007.1B
PRIMER CONNECTION TO FLOOR DRAIN PREPARED FOR CONCRETE POUR

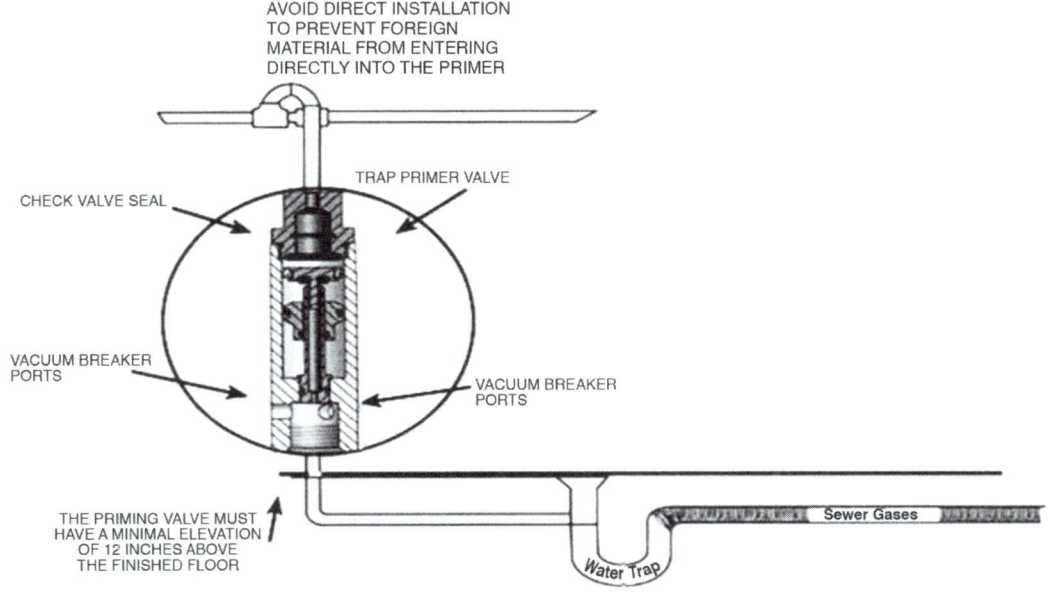

FIGURE 1007.1A
TRAP SEAL PRIMER RECOMMENDED INSTALLATION

TRAPS AND INTERCEPTORS

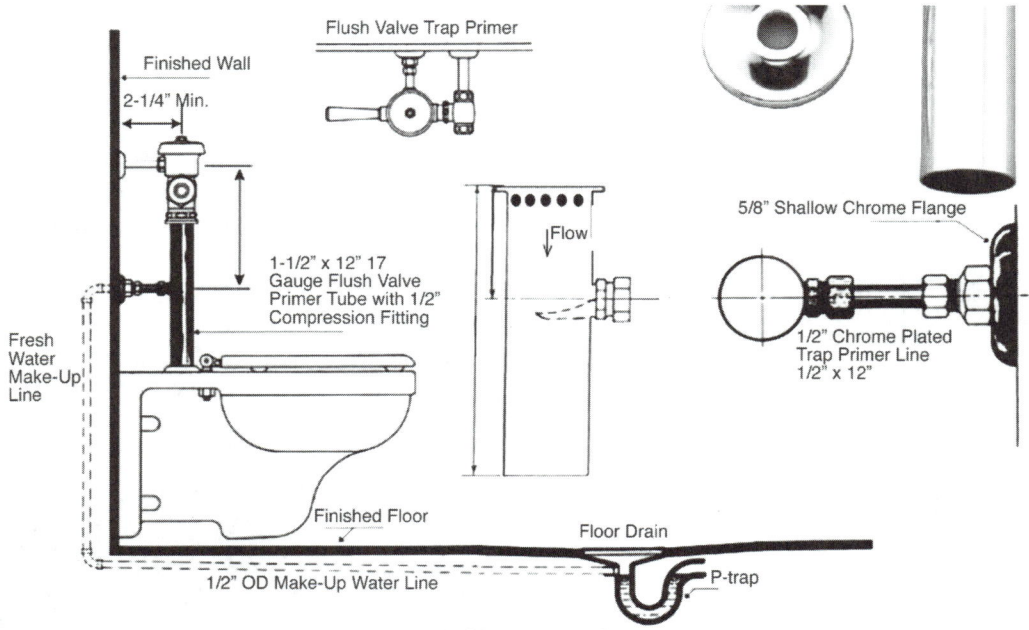

FIGURE 1007.1C
FLUSH VALVE TRAP PRIMER

FIGURE 1007.1D
MULTIPLE TRAP PRIMER INSTALLATION

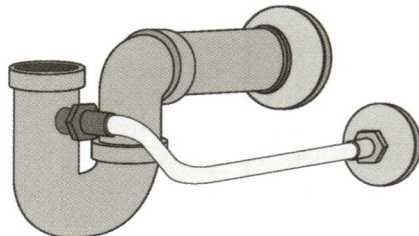

FIGURE 1007.2
DRAINAGE TRAP SEAL PRIMER

Building traps (see **Figure 901.2f**) are generally not installed although there may be reasons for their inclusion in the building sewer. Therefore the code allows the AHJ to make the determination when they are necessary. Some of the reasons for prohibiting building traps are:

1) it interferes with the flow of sewage and air movements in the plumbing system;
2) it reduces the carrying capacity of the building drain due to the added resistance;
3) it increases the potential of back pressure in the soil stack; and
4) it reduces the ventilation of the street sewers.

Some of the reasons for permitting the use of building traps are:

1) it affords an additional barrier against admission of sewer gas in case of trap evaporation or leakage;
2) it is needed in very cold climates to prevent frost closure of a vent terminal by reducing the circulation of air communicated from the street sewer through the plumbing system;

Jurisdiction. Each building trap where installed shall be provided with a cleanout and with a relieving vent or fresh-air intake on the inlet side of the trap, which needs not be larger than one-half the diameter of the drain to which it connects. Such relieving vent or fresh-air intake shall be carried above grade and terminate in a screened outlet located outside the building.

3) it prevents exposing neighboring occupants from foul odors communicated through a vent terminal where buildings are in close proximity to each other.

The building trap is also used where there is a combined sewer and storm system (see Section 1101.15). Sewer gas is not permitted into the storm drainage system by the use of a building trap installed on the main storm drain before the wye connection to the combined building sewer. The Code does not require a fresh air inlet for a building storm drain trap, but a cleanout has to be provided on the outlet side of the trap.

1009.0 Interceptors (Clarifiers) and Separators.

1009.1 Where Required. Interceptors (clarifiers) (including grease, oil, sand, solid interceptors, etc.) shall be required by the Authority Having Jurisdiction where they are necessary for the proper handling of liquid wastes containing grease, flammable wastes, sand, solids, acid or alkaline substances, or other ingredients harmful to the building drainage system, the public or private sewer, or to public or private sewage disposal.

Interceptors and separators serve a different purpose than fixture traps in that they protect the drainage and sewer systems from the introduction of substances that might be harmful to those systems. An interceptor or separator is a device designed and installed to separate, retain and treat deleterious, hazardous and undesirable matter before they enter the sewer system (see **Figure 1009.1**). Normal sewage or liquid wastes bypass these devices and discharge into the sewer system via the building drain and sewer.

If large amounts of grease, soaps and oils enter a sewer system, the materials will coagulate, solidify and adhere to the inside wall of the pipe, eventually blocking or partially blocking the pipe. Similarly, solids such as sand, plaster, metal chips and stones will cause stoppages or partial stoppages if permitted to enter the system. Oils, kerosene, gasoline, naphtha and paraffin present the danger of fire and explosion in sewer piping. Acids, alkalis and other chemical wastes in concentrated amounts could attack the piping through corrosion. In addition, all of the above are detrimental to proper functioning of sewage treatment plants, both public and private, and must be collected and treated before they discharge to the sewer system.

There are three principal methods of intercepting and treating these unwanted wastes. They are:

1. *Floatation*. Gasoline, oil, grease and other volatile liquids are lighter than water and are not readily soluble in water. They tend to separate out of the liquid waste and float to the top of the container where they can then be collected and removed. Hydromechanical grease interceptors, gravity grease interceptors and oil or gas interceptors are examples of devices that use floatation as a primary method of separation.

2. *Sedimentation*. Sand and other solids, which are heavier than water and also are not soluble in water, tend to settle out of the waste water to the bottom of the container. The solids can then be pumped or vacuumed out of the interceptor. Sand interceptors, septic tanks and drum traps for smaller installations are examples of devices that use sedimentation as the primary method of separation.

3. *Neutralization*. Chemical wastes that could be detrimental to the sewage system in concentrated amounts must be neutralized before entering the sewer system. Chemical waste is collected before entry into the sewer and treated or neutralized before being discharged to the sewer. Acid neutralizing tanks and other such devices are examples of systems that use neutralization as the primary method of separation.

Occasionally, waste with high concentrations of grease or emulsified grease may require further treatment, such as additional settling, rock filtration and chemical or mechanical treatment to remove the grease. There are many instances where one of the three methods or a combination of methods for treating these wastes is needed.

Examples of the use of these devices are: restaurants, hotels, cafeterias, schools, hospitals and institutional or commercial buildings where food is served in great quantity may produce grease in sufficient amounts to warrant the installation of grease interceptors; dairies, slaughterhouses and commercial food-processing plants have grease and fats as byproducts and require interceptors; gasoline service stations, garages, automobile repair shops, laundries, dry-cleaning plants, machine shops and industries using chemical processing are sources of flammable and volatile wastes that must be removed and treated; machine shops, garages, service stations, hospitals, medical clinics, dental laboratories and fish preparation areas contribute solids such as metals, plaster, sand and fish scales; hair and lint intercep-

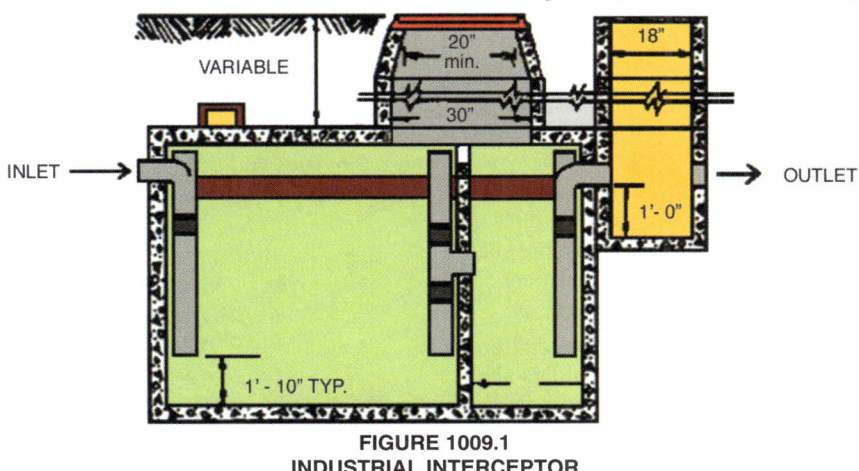

**FIGURE 1009.1
INDUSTRIAL INTERCEPTOR**

tors may be necessary in beauty salons, barbershops, surgical preparation rooms and veterinary facilities; and dental laboratories and jewelry manufacturers may require interceptors for the recovery of precious metals.

Separators and interceptors will be designed by the manufacturer and the installation location designated by a plumbing engineer or design professional. The plumber's or plumbing contractor's responsibility is to install the system properly. In addition to the manufacturer's installation instructions, the following sections provide guidance for the installation of the interceptor or separator.

1009.2 Approval. The size, type, and location of each interceptor (clarifier) or separator shall be approved by the Authority Having Jurisdiction. Except where otherwise specifically permitted, no wastes other than those requiring treatment or separation shall be discharged into an interceptor (clarifier).

1009.3 Design. Interceptors (clarifiers) for sand and similar heavy solids shall be so designed and located as to be readily accessible for cleaning and shall have a water seal of not less than 6 inches (152 mm).

1009.4 Relief Vent. Interceptors (clarifiers) shall be so designed that they will not become air-bound where closed covers are used. Each interceptor (clarifier) shall be properly vented.

1009.5 Location. Each interceptor (clarifier) cover shall be readily accessible for servicing and maintaining the interceptor (clarifier) in working and operating condition. The use of ladders or the removal of bulky equipment to service interceptors (clarifiers) shall constitute a violation of accessibility. Location of interceptors (clarifiers) shall be shown on the approved building plan.

1009.6 Maintenance of Interceptors. Interceptors shall be maintained in efficient operating condition by periodic removal of accumulated grease, scum, oil, or other floating substances and solids deposited in the interceptor.

1009.7 Discharge. The waste pipe from oil and sand interceptors shall discharge as approved by the Authority Having Jurisdiction.

1010.0 Slaughterhouses, Packing Establishments, etc.
1010.1 General. A fish, fowl, and animal slaughterhouse or establishment; a fish, fowl, and meat packing or curing establishment; a soap factory, tallow-rendering, fat-rendering, and a hide-curing establishment shall be connected to and shall drain or discharge into an approved grease interceptor (clarifier).

1011.0 Minimum Requirements for Auto Wash Racks.
1011.1 General. A private or public wash rack or floor or slab used for cleaning machinery or machine parts shall be adequately protected against storm or surface water and shall drain or discharge into an approved interceptor (clarifier).

1012.0 Commercial and Industrial Laundries.
1012.1 General. Laundry equipment in commercial and industrial buildings that do not have integral strainers shall discharge into an interceptor having a wire basket or similar device that is removable for cleaning and that will prevent passage into the drainage system of solids ½ of an inch (12.7 mm) or larger in maximum dimensions, such as string, rags, buttons, or other solid materials detrimental to the public sewerage system.

1013.0 Bottling Establishments.
1013.1 General. Bottling plants shall discharge their process wastes into an interceptor that will provide for the separation of broken glass or other solids, before discharging liquid wastes into the drainage system.

1014.0 Grease Interceptors.
Prior to the 2000 UPC, the term "grease trap" was used to identify almost any installation that provided fats, oils or grease (FOG) removal. The grease trap was actually used to provide the fixture trap for one fixture if certain requirements were met. However, research discovered that the fixture trap did a better job of protecting the trap seal so the use of the grease trap as a fixture trap was eliminated. Because of the fact that the grease trap is not a fixture trap and in order to create a uniform description of these devices, the 2003 UPC termed them "grease interceptors."

The UPC identifies four separate types of devices used in the collection, retention and disposal of FOG. They may be used separately or in combination to create a complete FOG removal and disposal system. For convenience, they are listed here with their definition from Chapter 2.

1. *Hydromechanical Grease Interceptor – A plumbing appurtenance or appliance that is installed in a sanitary drainage system to intercept nonpetroleum fats, oil, and grease (FOG) from a wastewater discharge and is identified by flow rate, and separation and retention efficiency. The design incorporates air entrainment, hydromechanical separation, interior baffling, or barriers in combination or separately, and one of the following:*

A – External flow control, with air intake (vent), directly connected

B – External flow control, without air intake (vent), directly connected

C – Without external flow control, directly connected

D – Without external flow control, indirectly connected these interceptors comply with the requirements of Table 1014.2.1. Hydromechanical grease interceptors are generally installed inside.

The hydromechanical grease interceptor (HGI) is a smaller device and is basically what was referred to as the "grease trap," as earlier models did in fact function as the fixture trap (see **Figure 1014.0a**). The term "hydromechanical" refers to the use of air-injecting flow controls, counter flow baffles and barrier baffles along with the natural buoyancy of FOG to accomplish its function of FOG separation.

2. *Grease Removal Device (GRD) – A hydromechanical grease interceptor that automatically, mechanically removes non-petroleum fats, oils and grease (FOG) from the interceptor, the control of which are either automatic or manually initiated.*

The grease removal device (GRD) (see **Figure 1014.0b**) not only separates the FOG from the wastewater but also removes the FOG from the interceptor. Normally, a hydromechanical grease interceptor (HGI) design is used in conjunction with the method of removing the FOG.

There are two basic types of GRDs:

a. *Timer controlled.* This typically utilizes a disk or belt that passes through the FOG layer and a squeegee device to wipe the accumulated FOG from the disk or belt into a drain trough and into a FOG receptacle. Other means of removing the FOG include a pump or gravity flow activated by the timer.

b. *Sensor-controlled devices.* These sense the presence of FOG and initiate the removal process only when necessary and as often as necessary. The GRD can always keep the retained FOG below the rated capacity of the device. They use valving and gravity or pump-assisted FOG removal.

3. *Gravity Grease Interceptor – A plumbing appurtenance or appliance that is installed in a sanitary drainage system to intercept nonpetroleum fats, oils, and greases (FOG) from a wastewater discharge and is identified by volume, 30-minute retention time, baffle(s) not less than two compartments, a total volume of not less than 300 gallons (1135 L), and gravity separation. [These interceptors comply with the requirements of Chapter 10 or are designed by a registered professional engineer.] Gravity grease interceptors are generally installed outside.*

The gravity grease interceptor is normally a large volume tank constructed of steel, concrete, fiberglass or plastic. It separates FOG from wastewater by gravity alone. The FOG-entrained wastewater enters the tank using tee baffles to direct the flow below the floating FOG level in order to leave the floating FOG undisturbed. Retention time is used to separate the FOG from the wastewater by floatation (see **Figures 1009.1** and **1014.3**).

4. *FOGG Disposal System – A grease interceptor that reduces nonpetroleum fats, oils, and grease (FOG) in effluent by separation, mass, and volume reduction.*

The FOG disposal system usually employs the HGI method for separating out the FOG. Disposal or remediation of FOG is accomplished by thermal, chemical, electrical or biochemical means. Each method attempts to change the molecular composition of the FOG to render it harmless or to enhance the digestion of the FOG as it is retained in the interceptor.

1014.1 General. Where it is determined by the Authority Having Jurisdiction that waste pretreatment is required, an approved type of grease interceptor(s) complies with ASME A112.14.3, ASME A112.14.4, CSA B481, PDI G-101, or PDI G-102, and sized in accordance with Section 1014.2.1 or Section 1014.3.6, shall be installed in accordance with the manufacturer's installation instructions to receive the drainage from fixtures or equipment that produce grease-laden waste located in areas of establishments where food is prepared, or other establishments where grease is introduced into the drainage or sewage system in quantities that can effect line stoppage or hinder sewage treatment or private sewage disposal systems. A combination of hydromechanical, gravity grease interceptors and engineered systems shall be allowed to meet this code and other applicable requirements of the Authority Having Jurisdiction where space or existing physical constraints of existing buildings necessitate such installations. A grease interceptor shall not be required for individual dwelling units or private living quarters. Water closets, urinals, and other plumbing fixtures conveying human waste shall not drain into or through the grease interceptor.

Although much FOG is created in residential kitchens and disposed of by draining to the sanitary drainage system, there is no requirement for individual residential grease interceptors. Human waste from the drainage system must bypass the interceptor. Adding this waste to the interceptor will render it useless by overfilling it with solids.

1014.1.1 Trapped and Vented. Each fixture discharging into a grease interceptor shall be individually trapped and vented in an approved manner.

1014.1.2 Maintenance. Grease interceptors shall be maintained in efficient operating condition by periodic removal of the accumulated grease and latent material. No such collected grease shall be introduced into drainage piping or a public or private sewer. Where the Authority Having Jurisdiction determines that a grease interceptor is not being properly cleaned or maintained, the Authority Having Jurisdiction shall have the authority to mandate the

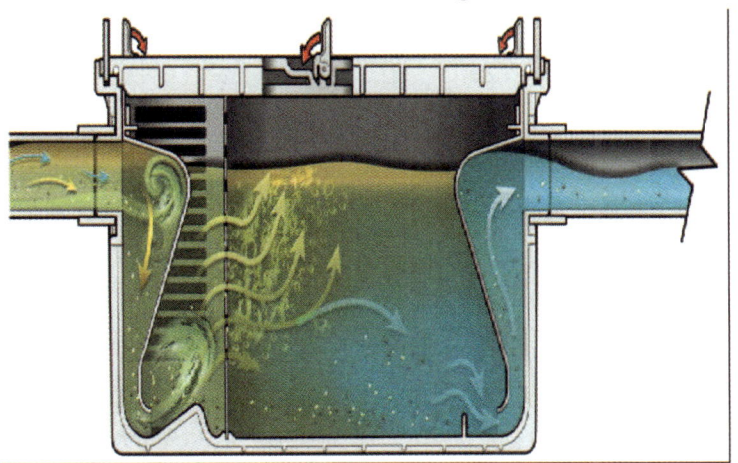

FIGURE 1014.0A
HYDROMECHANICAL GREASE INTERCEPTOR DESIGN AND OPERATION

TRAPS AND INTERCEPTORS

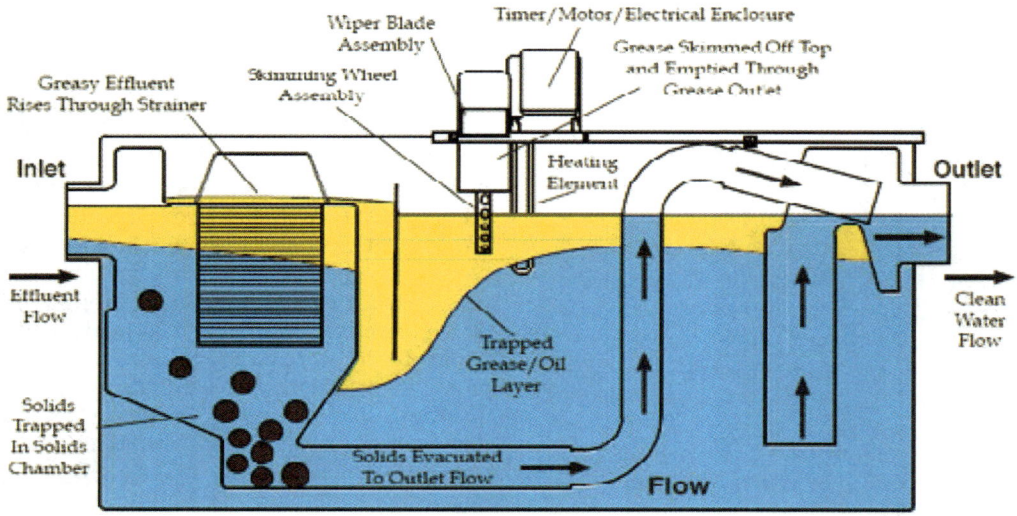

**FIGURE 1014.0B
GREASE REMOVAL DEVICE**

installation of additional equipment or devices and to mandate a maintenance program.

🔧 Maintenance is the most difficult part of the FOG removal system. The installation of these systems is mandated by codes and it is relatively easy to check compliance. However, ensuring that the maintenance of these systems is accomplished is another matter. Although their maintenance is mandated, it is difficult, if not impossible, to ensure that it is being done. A poorly maintained system is as useful as no system at all.

When maintaining these systems two factors must be addressed — service interval and FOG disposal. If the service interval is too long, the FOG will either simply pass through the overloaded interceptor or begin to breakdown into other harmful compounds. Long-term storage of FOG in a large interceptor can create a dangerous condition. The creation of hydrogen sulfide is a byproduct of FOG decomposition and could lead to the corrosion of piping and the interceptor itself, not to mention that it is a lethal gas if it escapes the system.

The development of the GRD and the FOG disposal system is a result of this maintenance problem. They can at least eliminate some of the maintenance problems by removal or remediation of the FOG. However, these devices do not solve all the maintenance issues. The devices will still need to be cleaned periodically and solids that have entered the interceptor removed.

1014.1.3 Food Waste Disposers and Dishwashers.
No food waste disposer or dishwasher shall be connected to or discharge into a grease interceptor. Commercial food waste disposers shall be permitted to discharge directly into the building's drainage system.

Exception: Food waste disposers shall be permitted to discharge to grease interceptors that are designed to receive the discharge of food waste.

🔧 The only time a food waste disposer should discharge to a grease interceptor is when the grease interceptor is specifically designed to accept food waste. Otherwise, a food waste disposer should never discharge to a grease interceptor as such a discharge can cause the grease interceptor to fail. The solids from the disposal will become entrained in the interceptor and begin to fill it unless cleaned out. If a disposal is connected to the grease interceptor not designed to receive food waste, a solids interceptor must precede the grease interceptor. The entrance of soapy water from a dishwasher into the interceptor will also be harmful to the system. The chemicals in dishwater will adversely affect the function of the interceptor and must be kept from the system.

1014.2 Hydromechanical Grease Interceptors.
Plumbing fixtures or equipment connected to a Type A and B hydromechanical grease interceptor shall discharge through an approved type of vented flow control installed in a readily accessible and visible location. Flow control devices shall be designed and installed so that the total flow through such device or devices shall at no time be greater than the rated flow of the connected grease interceptor. No flow control device having adjustable or removable parts shall be approved. The vented flow control device shall be located such that no system vent shall be between the flow control and the grease interceptor inlet. The vent or air inlet of the flow control device shall connect with the sanitary drainage vent system, as elsewhere required by this code, or shall terminate through the roof of the building, and shall not terminate to the free atmosphere inside the building.

Exception: Listed grease interceptors with integral flow controls or restricting devices shall be installed in an accessible location in accordance with the manufacturer's installation instructions.

🔧 The hydromechanical grease interceptor (HGI) uses the principles of fluid dynamics by taking advantage of air entrained in the effluent through the use of the vented flow control (see **Figure 1014.2a**). The FOG-laden wastewater passes through the flow control on its way into the interceptor. As the effluent passes through the orifice of the flow control, air is introduced to the flow from the air vent. Upon entering the interceptor, the effluent is directed through the separation chamber. The entrained air will separate from the effluent quickly and then accomplishes two tasks. The escaping air accelerates the separation of FOG as it rises rapidly to the surface of the water in the separation chamber. The rising air burbles literally pull the FOG globules to the top of the water. The released air then provides a small amount of positive pressure above the contents of the separation chamber to regulate the internal running water level of the grease interceptor, keeping it from becoming air bound.

A flow control shall be installed on each fixture discharging to the HGI. A single flow control is allowed for multiple fixtures as long as the resulting total flow from the fixtures meets the capacity requirements of Section 1014.2.1 (see **Figure 1014.2b**). There should also be no fixture vent between the flow control and the interceptor. This would allow too much air to be entrained into the flow.

The placement of the flow control is vitally important to the functioning of the interceptor. It should be placed as close as possible to the fixture. There should be as little vertical height as possible between the fixture outlet and the flow control. If the flow control is placed well below the fixture, the added head pressure above the flow control will render the flow control useless as it will increase the flow through the system (see **Figure 1014.2c**).

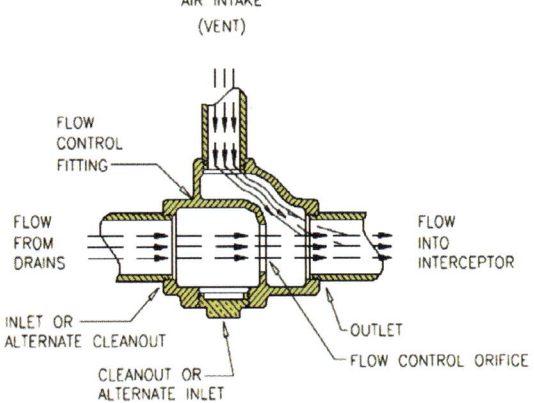

**FIGURE 1014.2A
FLOW CONTROL DEVICE**

1014.2.1 Capacity. The total capacity in gallons (gal) (L) of fixtures discharging into a hydromechanical grease interceptor shall not exceed two and one-half times the certified gallon per minute (gpm) (L/s) flow rate of the interceptor in accordance with Table 1014.2.1.

For this section, the term "fixture" shall mean and include each plumbing fixture, appliance, apparatus, or other equipment required to be connected to or discharged into a grease interceptor by a provision of this section.

When sizing for a hydromechanical grease interceptor (HGI) using fixture capacity, determine the compartment sizes of the fixtures and calculate the volume in cubic inches (LxWxD). Divide the volume by 231 to calculate the volume of liquid in gallons. Per Example 1014.2.1, multiply the gallons by the fill factor .75. This fill factor takes into

**TABLE 1014.2.1
HYDROMECHANICAL GREASE INTERCEPTOR SIZING
USING GRAVITY FLOW RATES[1]**

DIAMETER OF GREASE WASTE PIPE (inches)	MAXIMUM FULL PIPE FLOW (gpm)[2]	SIZE OF GREASE INTERCEPTOR	
		ONE-MINUTE DRAINAGE PERIOD (gpm)	TWO-MINUTE DRAINAGE PERIOD (gpm)
2	20	20	10
3	60	75	35
4	125	150	75
5	230	250	125
6	375	400	200

For SI units: 1 inch = 25 mm, 1 gallon per minute = 0.06 L/s
Notes:
[1] For interceptor sizing by the fixture capacity see the example below.
[2] ¼ inch slope per foot (20.8 mm/m) based on Manning's formula with friction factor N =.012.

**EXAMPLE 1014.2.1
SIZING HYDROMECHANICAL GREASE INTERCEPTOR(S)
USING FIXTURE CAPACITY**

Step 1: Determine the flow rate from each fixture.

[Length] X [Width] X [Depth] / [231] = Gallons X [.75 fill factor] / [Drain Period (1 minute or 2 minutes)]

Step 2: Calculate the total load from fixtures that discharge into the interceptor.

FIXTURES	COMPART-MENTS	LOAD (gallons)	SIZE OF GREASE INTERCEPTOR ONE-MINUTE DRAINAGE PERIOD (gpm)	TWO-MINUTE DRAINAGE PERIOD (gpm)
Compartment size	—	—	—	—
24 inches x 24 inches x 12 inches	2	44.9	—	—
Hydrant	—	3	—	—
Rated Appliance	—	2	—	—
—	—	49.9	50	25

For SI units: 1 inch = 25.4 mm, 1 gallon per minute = 0.06 L/s, 1 gallon = 3.785 L

account that the fixture is normally filled to about 75 percent of capacity, with the other 25 percent being displaced by the items placed in the fixture for washing. Therefore, the actual drainage load is 75 percent of the fixture capacity.

Once the actual drainage load is calculated, determine the flow rate per minute using a one minute or two minute drainage period. Typically, sinks will drain within a one-minute period, which determines the drainage load in gallons per minute.

When would the two-minute drainage period be used? When the capacity of the fixtures (in gallons) discharging into the interceptor is 2½ times the flow rate capacity of the

TRAPS AND INTERCEPTORS

HGI, a two-minute drainage period would be used. The code allows the capacity of the fixtures to be 2½ times the flow rate of the HGI to give latitude to space constraints for an interceptor. A smaller interceptor may be used, but the consequence is a longer drainage period.

For example, a four-compartment sink is to discharge into a HGI. Each compartment size is 24x24x12. Using the method described in the first paragraph, the actual drainage load is 89.8 gallons. Using the drainage period of one minute, a HGI equal to or greater than 89.8 gpm would be required (a typical size is 100 gpm). Suppose a 100 gpm interceptor would not fit due to space limitations. The Code would allow a 50 gpm interceptor with the acceptance that the fixtures would take longer to drain. Instead of one minute, the drainage period would be extended to two minutes. Remember, the flow control device will not allow the discharge rate to exceed 50 gpm. The drainage load of 89.8 gallons is within 2½ times the flow rate of 50 gpm. If the capacity of the fixtures had exceeded 125 gallons, then a 50 gpm interceptor would not be permitted.

1014.2.2 Vent. A vent shall be installed downstream of hydromechanical grease interceptors in accordance with the requirements of this code.

🔧 The HGI itself must be vented to keep it from becoming air bound (see **Figure 1014.2a**).

1014.3 Gravity Grease Interceptors. Required gravity grease interceptors shall comply with the provisions of Section 1014.3.1 through Section 1014.3.7.

🔧 The gravity grease interceptor, as stated earlier, is

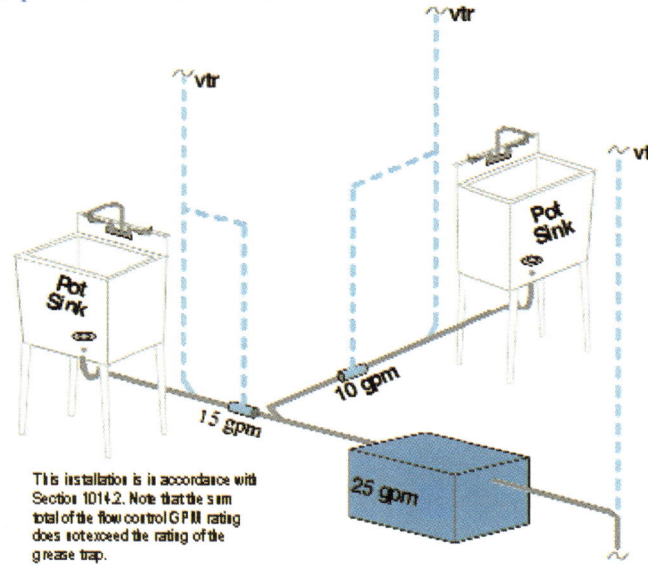

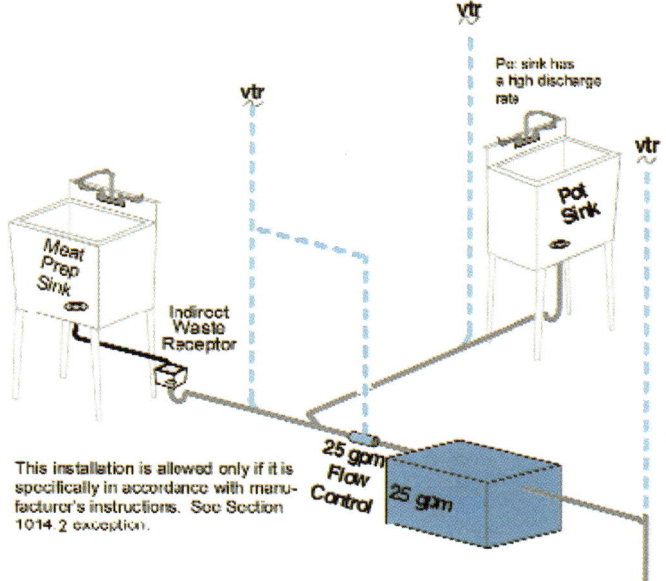

FIGURE 1014.2B
FLOW CONTROL INSTALLATION

a large volume tank with two sections. Its function is relatively simple. The effluent enters the tank through a baffled tee so as to not disturb the accumulated FOG floating at the top of the water level. The effluent is retained in the initial chamber until flow forces it into the next chamber and eventually out to the sewer. The FOG is separated by gravity – the FOG floats to the top of the water level. The tank is sized to allow enough retention time to permit the FOG to separate.

A sampling box is installed in the outflow line to enable maintenance personnel to check on the functioning of the interceptor. The following sections give guidance to the design, location, and installation of the gravity grease interceptor (see **Figure 1014.3**).

1014.3.1 General. The provisions of this section shall apply to the design, construction, installation, and testing of commercial kitchen gravity grease interceptors.

1014.3.2 Waste Discharge Requirements. Waste discharge in establishments from fixtures and equipment which contain grease, including but not limited to, scullery sinks, pot and pan sinks, dishwashers, soup kettles, and floor drains located in areas where grease-containing materials exist, shall be permitted to be drained into the sanitary waste through the interceptor where approved by the Authority Having Jurisdiction.

1014.3.2.1 Toilets and Urinals. Toilets, urinals, and other similar fixtures shall not drain through the interceptor.

1014.3.2.2 Inlet Pipe. Waste shall enter the interceptor through the inlet pipe.

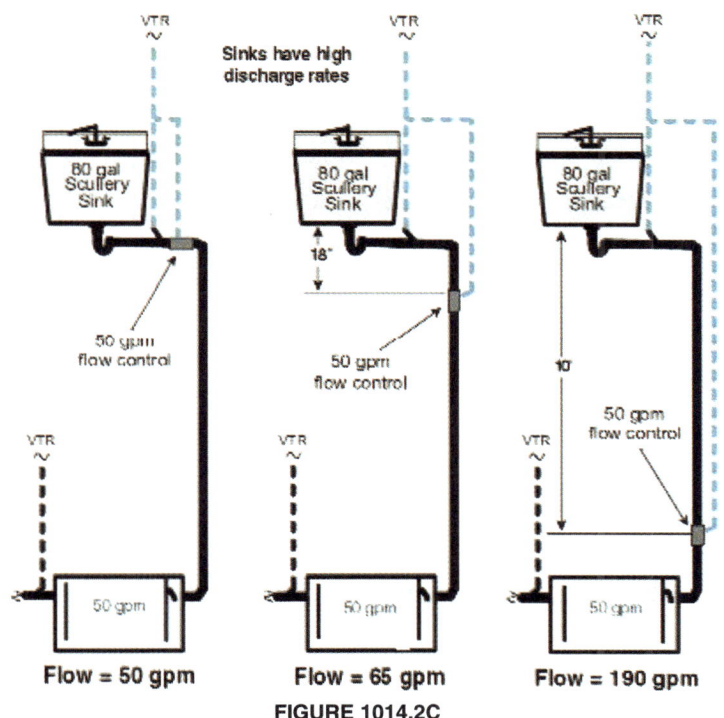

FIGURE 1014.2C
PLACEMENT OF FLOW CONTROL DEVICE

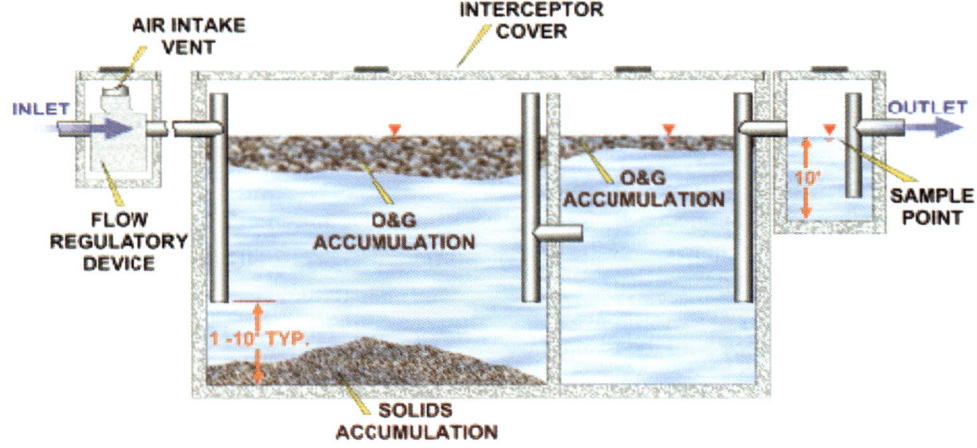

FIGURE 1014.3
GRAVITY GREASE INTERCEPTOR WITH FLOW CONTROL (IF NEEDED) AND SAMPLE BOX

1014.3.3 Design. Gravity interceptors shall be constructed in accordance with the applicable standard in Table 1701.1 or the design approved by the Authority Having Jurisdiction.

1014.3.4 Location. Each grease interceptor shall be so installed and connected that it shall be easily accessible for inspection, cleaning, and removal of the intercepted grease. A gravity grease interceptor that complies with IAPMO Z1001 shall not be installed in a building where food is handled. Location of the grease interceptor shall meet the approval of the Authority Having Jurisdiction.

1014.3.4.1 Interceptors. Interceptors shall be placed as close as practical to the fixtures they serve.

1014.3.4.2 Business Establishment. Each business establishment for which a gravity grease interceptor is required shall have an interceptor which shall serve that establishment unless otherwise approved by the Authority Having Jurisdiction.

1014.3.4.3 Access. Each gravity grease interceptor shall be located to be readily accessible to the equipment required for maintenance.

1014.3.5 Construction Requirements. Gravity grease interceptors shall be designed to remove grease from effluent and shall be sized in accordance with this section. Gravity grease interceptors shall also be designed to retain grease until accumulations can be removed by pumping the interceptor. It is recommended that a sample box is located at the outlet end of gravity grease interceptors so that the Authority Having Jurisdiction can periodically sample effluent quality.

1014.3.6 Sizing Criteria. The volume of the interceptor shall be determined by using Table 1014.3.6. Where drainage fixture units (DFUs) are not known, the interceptor shall be sized based on the maximum DFUs allowed for the pipe size connected to the inlet of the interceptor. Refer to Table 703.2, Drainage Piping, Horizontal.

1014.3.7 Abandoned Gravity Grease Interceptors. Abandoned grease interceptors shall be pumped and filled as required for abandoned sewers and sewage disposal facilities in Section 722.0.

1015.0 FOG (Fats, Oils, and Greases) Disposal System.

1015.1 Purpose. The purpose of this section is to provide the necessary criteria for the sizing, application, and installation of FOG disposal systems designated as a pretreatment or discharge water quality compliance strategy.

1015.2 Components, Materials, and Equipment. FOG disposal systems, including components, materials, and equipment necessary for the proper function of the system, shall comply with ASME A112.14.6.

1015.3 Sizing and Installation. FOG disposal systems shall be sized and installed in accordance with the manufacturer's installation instructions.

1015.4 Performance. FOG disposal systems shall produce an effluent quality not to exceed 5.84 grains per gallon (gr/gal) (100 mg/L) FOG.

1016.0 Sand Interceptors.

A sand interceptor is a device designed and installed to separate and retain deleterious, hazardous or undesirable matter from normal wastes and permit normal sewage or liquid wastes to discharge into the sewer by

TABLE 1014.3.6
GRAVITY GREASE INTERCEPTOR SIZING

DRAINAGE FIXTURE UNITS[1, 3] (DFUs)	INTERCEPTOR VOLUME[2] (gallons)
8	500
21	750
35	1000
90	1250
172	1500
216	2000
307	2500
342	3000
428	4000
576	5000
720	7500
2112	10 000
2640	15 000

For SI units: 1 gallon = 3.785 L

Notes:
[1] The maximum allowable DFUs plumbed to the kitchen drain lines that will be connected to the grease interceptor.
[2] This size is based on DFUs, the pipe size from this code; Table 703.2; Useful Tables for flow in half-full pipes (ref: Mohinder Nayyar Piping Handbook, 3rd Edition, 1992). Based on 30-minute retention time (ref.: George Tchobanoglous and Metcalf & Eddy. Wastewater Engineering Treatment, Disposal, and Reuse, 3rd Ed. 1991 & Ronald Crites and George Tchobanoglous. Small and Decentralized Wastewater Management Systems, 1998. Rounded up to nominal interceptor volume.
[3] Where the flow rate of directly connected fixture(s) or appliance(s) have no assigned DFU values, the additional grease interceptor volume shall be based on the known flow rate (gpm) (L/s) multiplied by 30 minutes.

EXAMPLE 1014.3.6
GRAVITY GREASE INTERCEPTOR SIZING EXAMPLE

Given: A restaurant with the following fixtures and equipment.

One food preparation sink; three-floor drains - one in the food prep area, one in the grill area, and one receiving the indirect waste from the ice machine and a mop sink.

Kitchen Drain Line DFU Count (from Table 702.1):

3 floor drains at 2 DFUs each = 6 DFUs
Mop sink at 3 DFUs each = 3 DFUs
Food prep sink at 3 DFUs each = 3 DFUs
Total = 12 DFUs

Using Table 1014.3.6, the grease interceptor will be sized at 750 gallons (2389 L).

gravity (see **Figure 1016.0**). The sand interceptor is often used in garage or car wash installations to collect deposited sand before entry to the public sewer. Multiple floor drains are used and may drain to a single interceptor. The floor drains from these installations need not be trapped as they are not connected directly to the sanitary system and could become plugged with accumulated sand.

1016.1 Discharge. Where the discharge of a fixture or drain contains solids or semi-solids heavier than water that would be harmful to a drainage system or cause a stoppage within the system, the discharge shall be through a sand interceptor. Multiple floor drains shall be permitted to discharge into one sand interceptor.

1016.2 Authority Having Jurisdiction. Sand interceptors are required where the Authority Having Jurisdiction deems it advisable to have a sand interceptor to protect the drainage system.

1016.3 Construction and Size. Sand interceptors shall be built of brick or concrete, prefabricated coated steel, or other watertight material. The interceptor shall have an interior baffle for full separation of the interceptor into two sections. The outlet pipe shall be the same size as the inlet pipe of the sand interceptor, the minimum being 3 inches (80 mm), and the baffle shall have two openings of the same diameter as the outlet pipe and at the same invert as the outlet pipe. These openings shall be staggered so that there cannot be a straight line flow between the inlet pipe and the outlet pipe. The invert of the inlet pipe shall be no lower than the invert of the outlet pipe.

The sand interceptor shall have a minimum dimension of 2 square feet (0.2 m^2) for the net free opening of the inlet section and a minimum depth under the invert of the outlet pipe of 2 feet (610 mm).

For each 5 gpm (0.3 L/s) flow or fraction thereof over 20 gpm (1.26 L/s), the area of the sand interceptor inlet section is to be increased by 1 square foot (0.09 m^2). The outlet section shall at all times have a minimum area of 50 percent of the inlet section.

The outlet section shall be covered by a solid removable cover, set flush with the finished floor, and the inlet section shall have an open grating, set flush with the finished floor and suitable for the traffic in the area in which it is located.

1016.4 Separate Use. Sand and similar interceptors for every solid shall be so designed and located as to be readily accessible for cleaning, shall have a water seal of not less than 6 inches (152 mm), and shall be vented.

1017.0 Oil and Flammable Liquid Interceptors.

The oil separator provides for the separation and removal of oil or flammable liquids. See **Figure 1017.0** for an illustration of this interceptor.

1017.1 Interceptors Required. Repair garages and gasoline stations with grease racks or grease pits, and factories that have oily, flammable, or both types of wastes as a result of manufacturing, storage, maintenance, repair, or testing processes, shall be provided with an oil or flammable liquid interceptor that shall be connected to necessary floor drains. The separation or vapor compartment shall be independently vented to the outer air. Where two or more separation or vapor compartments are used, each shall be vented to the outer air or shall be permitted to connect to a header that is installed at a minimum of 6 inches (152 mm) above the spill line of the lowest floor drain and vented independently to the outer air. The minimum size of a flammable vapor vent shall be not less than 2 inches (50 mm), and, where vented through a sidewall, the vent shall be not less

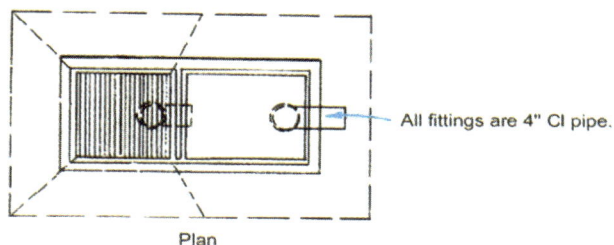
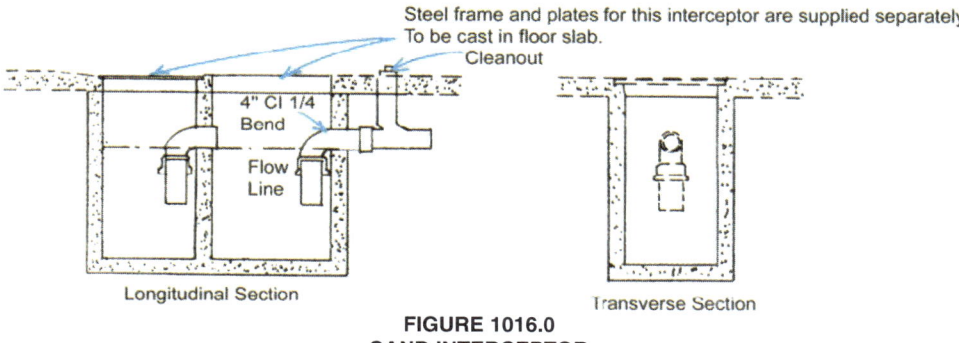

**FIGURE 1016.0
SAND INTERCEPTOR**

than 10 feet (3048 mm) above the adjacent level at an approved location. The interceptor shall be vented on the sewer side and shall not connect to a flammable vapor vent. Oil and flammable interceptors shall be provided with gastight cleanout covers that shall be readily accessible. The waste line shall be not less than 3 inches (80 mm) in diameter with a full-size cleanout to grade. Where an interceptor is provided with an overflow, it shall be provided with an overflow line [not less than 2 inches (50 mm) in diameter] to an approved waste oil tank having a minimum capacity of 550 gallons (2082 L) and meeting the requirements of the Authority Having Jurisdiction. The waste oil from the separator shall flow by gravity or shall be pumped to a higher elevation by an automatic pump. Pumps shall be adequately sized and accessible. Waste oil tanks shall have a 2 inch (50 mm) minimum pump-out connection at grade and an 1½ inch (40 mm) minimum vent to atmosphere at an approved location not less than 10 feet (3048 mm) above grade.

1017.2 Design of Interceptors. Each manufactured interceptor that is rated shall be stamped or labeled by the manufacturer with an indication of its full discharge rate in gpm (L/s). The full discharge rate to such an interceptor shall be determined at full flow. Each interceptor shall be rated equal to or greater than the incoming flow and shall be provided with an overflow line to an underground tank.

Interceptors not rated by the manufacturer shall have a depth of not less than 2 feet (610 mm) below the invert of the discharge drain. The outlet opening shall have not less than an 18 inch (457 mm) water seal and shall have a minimum capacity as follows: Where not more than three motor vehicles are serviced, stored, or both, interceptors shall have a minimum capacity of 6 cubic feet (0.2 m^3), and 1 cubic foot (0.03 m^3) of capacity shall be added for each vehicle up to 10 vehicles. Above 10 vehicles, the Authority Having Jurisdiction shall determine the size of the interceptor required. Where vehicles are serviced and not stored, interceptor capacity shall be based on a net capacity of 1 cubic foot (0.03 m^3) for each 100 square feet (9.29 m^2) of the surface to be drained into the interceptor, with a minimum of 6 cubic feet (0.2 m^3).

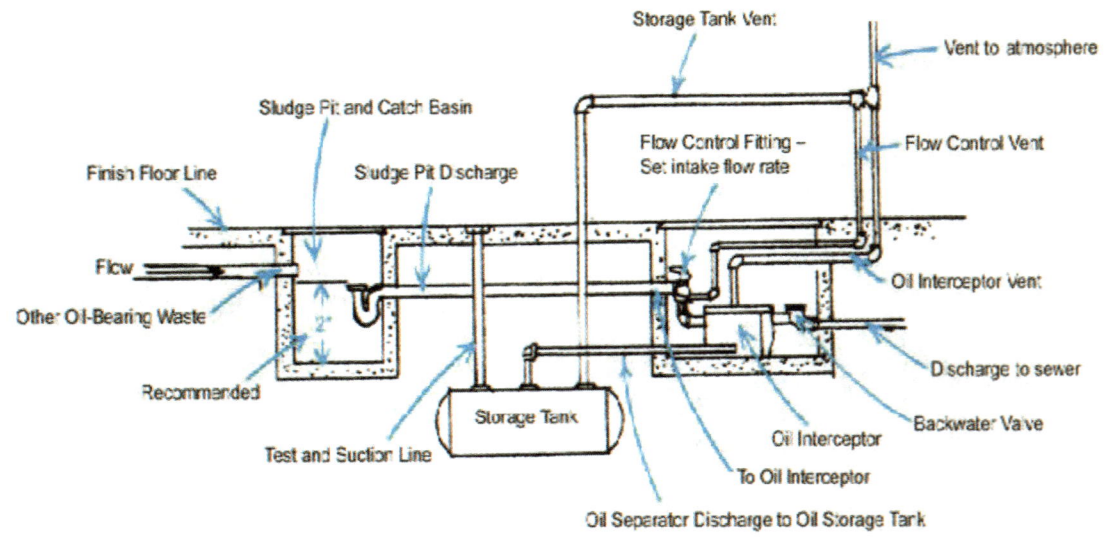

FIGURE 1017.0
OIL INTERCEPTOR

CHAPTER 11
STORM DRAINAGE

1101.0 General.

1101.1 Applicability. This chapter shall govern the materials, design, and installation of storm water drainage systems.

1101.2 Where Required. Roofs, paved areas, yards, courts, courtyards, vent shafts, light wells, or similar areas having rainwater, shall be drained into a separate storm sewer system, or into a combined sewer system where a separate storm sewer system is not available, or to some other place of disposal satisfactory to the Authority Having Jurisdiction. In the case of one- and two-family dwellings, storm water shall be permitted to be discharged on flat areas, such as streets or lawns, so long as the storm water shall flow away from the building and away from adjoining property, and shall not create a nuisance.

There are three basic methods used to deal with storm water that falls upon the roof or in and around a building. Which method is used will depend on how the building is designed and/or the type of public sewage system used in the area. The methods are:

1. Gravity Drainage – Storm water simply drains off the roof and other areas to the ground. Storm water flows away from the building naturally to storm drains or, in some cases, to the street. For most single-family residences, the roof is pitched, allowing rainwater to flow by gravity to a gutter system and to discharge to grade.

2. Separate Storm and Sanitary Drainage Systems – Water may accumulate on the roof because of building design and a building storm drainage system is installed, which will consist of a primary and a secondary system. Storm water is then carried away to either a collection point or to the graded area around the building and flows naturally away from the building. The building and public sewer system is only a sanitary system and may not collect storm water. There may be a separate storm sewer system but it is independent of the sanitary system (see **Figure 1101.2** and **1101.15a**).

3. Combined Storm and Sanitary Systems – Water may accumulate on the roof because of the building design. The public sewer is a combined storm and sanitary sewer system. Therefore the building will have a primary and secondary storm water drainage system that will collect the storm water and convey it to the combined sanitary/storm drainage sewer (see **Figure 1101.15b**).

1101.3 Storm Water Drainage to Sanitary Sewer Prohibited. Storm water shall not be drained into sewers intended for sanitary drainage.

Sanitary-only sewers are designed and sized to carry only waste and waste water from fixtures within the buildings they serve. The addition of storm water to the sanitary sewer could cause the sewer system to overload

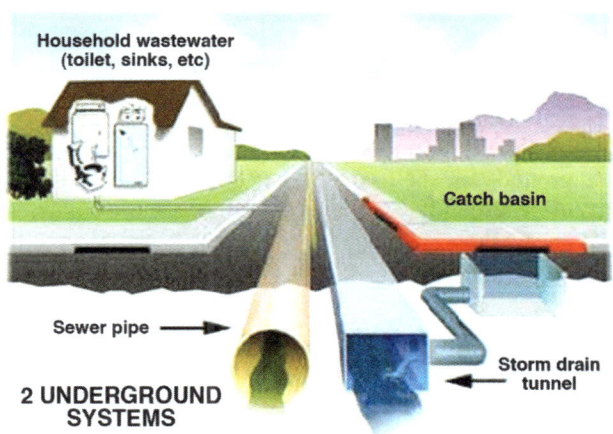

FIGURE 1101.2
SEPARATE DRAINAGE SYSTEMS - SANITARY AND STORM

and back up into the street or even into the building. The sewage treatment plant is not sized to accommodate and treat storm water and could be severely compromised if storm water were allowed to reach the plant.

1101.4 Material Uses. Pipe, tube, and fittings conveying rainwater shall be of such materials and design as to perform their intended function to the satisfaction of the Authority Having Jurisdiction. Conductors within a vent or shaft shall be of cast-iron, galvanized steel, wrought iron, copper, copper alloy, lead, Schedule 40 ABS DWV, Schedule 40 PVC DWV, stainless steel 304 or 316L [stainless steel 304 pipe and fittings shall not be installed underground and shall be kept not less than 6 inches (152 mm) aboveground], or other approved materials, and changes in direction shall be in accordance with the requirements of Section 706.0. ABS and PVC DWV piping installations shall be installed in accordance with applicable standards referenced in Table 1701.1 and Chapter 14 "Firestop Protection." Except for individual single-family dwelling units, materials exposed within ducts or plenums shall have a flame-spread index of not more than 25 and a smoke-developed index of not more than 50, where tested in accordance with ASTM E84 or UL 723. These tests shall comply with all requirements of the standards to include the sample size, both for width and length. Plastic pipe shall not be tested filled with water.

All rainwater piping and fittings located within a building must be made of the same materials required for sanitary drainage piping located within a building per Chapter 7.

The storm drainage piping system shall be under the same parameters as sanitary drainage piping per Chapter 7.

1101.4.1 Copper and Copper Alloys. Joints and connections in copper and copper alloy pipe and tube shall be installed in accordance with Section 705.3.

STORM DRAINAGE

1101.4.2 Conductors. Conductors installed aboveground in buildings shall comply with the applicable standards referenced in Table 701.2 for aboveground drain, waste, and vent pipe. Conductors installed aboveground level shall be of seamless copper water tube, Type K, L, or M; Schedule 40 copper pipe or Schedule 40 copper alloy pipe; Type DWV copper drainage tube; service weight cast-iron soil pipe or hubless cast-iron soil pipe; standard weight galvanized steel pipe; stainless steel 304 or 316L [stainless steel 304 pipe and fittings shall not be installed underground and shall be kept not less than 6 inches (152 mm) aboveground], or Schedule 40 ABS or Schedule 40 PVC plastic pipe.

1101.4.3 Leaders. Leaders installed outside shall comply with the applicable standards referenced in Table 701.2 for aboveground drain, waste, and vent pipe; aluminum sheet metal; galvanized steel sheet metal; or copper sheet metal.

1101.4.4 Underground Building Storm Drains. Underground building storm drains shall comply with the applicable standards referenced in Table 701.2 for underground drain, waste, and vent pipe.

1101.4.5 Building Storm Sewers. Building storm sewers shall comply with the applicable standards referenced in Table 701.2 for building sewer pipe.

1101.4.6 Subsoil Drains. Subsoil drains shall be open jointed, perforated, or both and constructed of materials in conformance with Table 1101.4.6.

**TABLE 1101.4.6
MATERIALS FOR SUBSOIL DRAIN PIPE AND FITTINGS**

MATERIAL	REFERENCED STANDARD(S)
PE	ASTM F667
PVC	ASTM D2729
Vitrified Clay (Extra strength)	ASTM C4, ASTM C700

1101.5 Expansion Joints Required. Expansion joints or sleeves shall be provided where warranted by temperature variations or physical conditions.

Expansion joints or sleeves must be installed to protect the storm water piping from expansion, contraction, and building movement (see **Figure 1101.5**). A typical location for the installation of an expansion joint is the connection between a roof drain body attached to a plywood decked roof and a vertical conductor. Movement of the wood roof and expansion of the piping would otherwise either damage the piping or damage the water-tight joint between the roof and the roof or overflow drain. Expansion joints are also frequently needed to compensate for expansion and contraction of conductors located in attic spaces that are subject to widely fluctuating temperatures during different times of the day.

1101.6 Subsoil Drains. Subsoil drains shall be provided around the perimeter of buildings having basements, cellars, crawl spaces, or floors below grade. Such subsoil drains shall be permitted to be positioned inside or outside

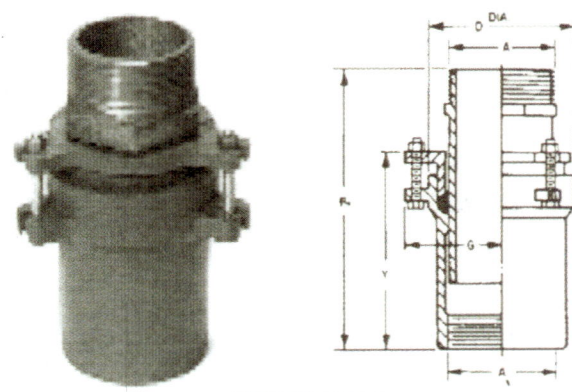

**FIGURE 1101.5
EXPANSION JOINT**

of the footing, shall be of perforated or open-jointed approved drain tile or pipe, not less than 3 inches (80 mm) in diameter, and shall be laid in gravel, slag, crushed rock, approved ¾ of an inch (19.1 mm) crushed, recycled glass aggregate, or other approved porous material with not less than 4 inches (102 mm) surrounding the pipe. Filter media shall be provided for exterior subsoil piping.

Besides storm water that falls upon a roof or areaway, there is another form of storm water that must be dealt with – subsoil water. Subsoil water can either be rainwater that gathers under or around the building during storm activity and must be collected or drained away, or it can be water that seeps from an underground water source either because of rising water levels from storms or because of shallow groundwater. In either case, a method to collect the water and convey it away from the building must be installed or the building could be severely damaged. Because these systems are below ground level they are termed "subsoil drains" (see **Figure 1101.6a**).

Subsoil drains are required to be installed when the building has floors below grade – a basement, cellar or crawl space. Most foundations are constructed of concrete and may develop cracks over time, which will then allow groundwater to enter the building and cause damage. Concrete slabs below grade are typically poured after the footings and foundation walls are in place. This method results in a "cold" joint at the juncture of the two surfaces. As the concrete cures and shrinks, openings form that allow the entrance of groundwater into the building. During rainy periods of the year, the ground may become saturated with water. In many types of soils, groundwater will develop a head pressure around the foundation walls that will force water into the building through any cracks or openings.

Subsoil drains will carry away groundwater to an approved point of disposal. These drains may be installed either inside or outside of the building foundation. They are usually constructed of perforated pipe (see **Figure 1101.6b**). Open-joint drain pipe or tile is rarely used nowadays. Care must be taken that the perforations of the drain tiles or pipe are not larger than the granular material around the pipe. Sand and fine gravel must not be used, as they

STORM DRAINAGE

will be quickly washed into the subsoil drain and over time will cause a stoppage in the pipe. When these subsoil drains are installed outside of the building, filter fabric or media must be installed over the pipe to prevent silt, soil, and sand from infiltrating the piping and causing stoppages (see **Figure 1101.6c**).

1101.6.1 Discharge. Subsoil drains shall be piped to a storm drain, to an approved water course, to the front street curb or gutter, to an alley, or the discharge from the subsoil drains shall be conveyed to the alley by a concrete gutter. Where a continuously flowing spring or groundwater is encountered, subsoil drains shall be piped to a storm drain or an approved water course.

🔧 As with storm water systems, the subsoil water must be removed without causing a nuisance. It must be piped to a naturally draining waterway or the area surrounding the building must be designed to carry the storm water using a method such as guttering in a parking lot as in **Figure 1101.6.1**. It should never be allowed to run across a sidewalk or walkway but drain to a curb or gutter. If the water is continuously flowing, then it should be piped to a storm drain or waterway rather than constantly running over open and traveled areas.

1101.6.2 Sump. Where it is not possible to convey the drainage by gravity, subsoil drains shall discharge to an accessible sump provided with an approved automatic electric pump. The sump shall be not less than 15 inches (381 mm) in diameter, 18 inches (457 mm) in depth, and provided with a fitted cover. The sump pump shall have an adequate capacity to discharge water coming into the sump as it accumulates to the required discharge point, and the capacity of the pump shall be not less than 15 gallons per minute (gpm) (0.95 L/s). The discharge piping from the sump pump shall be not less than 1½ inches (40 mm) in diameter and have a union or other approved quick-disconnect assembly to make the pump accessible for servicing.

🔧 The requirements for pumping water out of a collection sump for the disposal of subsoil water below

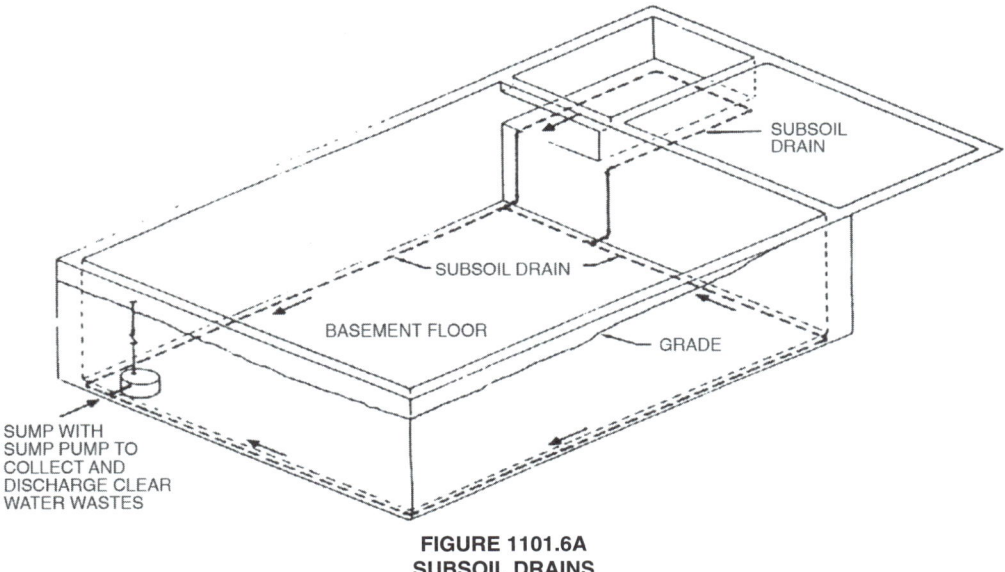

FIGURE 1101.6A
SUBSOIL DRAINS

FIGURE 1101.6B
PERFORATED DRAIN PIPE

FIGURE 1101.6C
INSTALLATION OF SUBSOIL DRAINS

STORM DRAINAGE

gravity flow are similar in principle to sewer ejector systems. One difference is that the sump does not need to be air tight and vented. However, it shall meet the minimum size requirements with a fitted cover. Also, the pump for ground water has a smaller minimum gallon-per-minute requirement, 15 gpm instead of 20 gpm (see **Figure 1101.6.2**).

1101.6.3 Splash Blocks. For separate dwellings not serving continuously flowing springs or groundwater, the sump discharge pipe shall be permitted to discharge onto a concrete splash block with a minimum length of 24 inches (610 mm). This pipe shall be within 4 inches (102 mm) of the splash block and positioned to direct the flow parallel to the recessed line of the splash block.

🔧 Splash blocks are not allowed to be used for pumped discharge if the pump receives water from continuously flowing springs or from a continuously high water table because the discharged water may drain back to the source. This continuously flowing sump discharge should be piped to a collection area.

The splash block is required to protect the ground beneath the block from erosion. The splash block must direct the water away from the building and the ground surface around the building must slope away from the building (see **Figure 1101.6.3**).

1101.6.4 Backwater Valve. Subsoil drains subject to backflow where discharging into a storm drain shall be provided with a backwater valve in the drain line so located as to be accessible for inspection and maintenance.

🔧 As with the sanitary system, if there is a possibility of backflow into the subsoil system then a backwater valve must be installed. This ensures that a stoppage in the building storm sewer will not result in water from gutters, downspouts, roof drains and even parking lot catch basins to back up into the subsoil system, causing damage to the building.

1101.6.5 Open Area. Nothing in Section 1101.6 shall prevent drains that serve either subsoil drains or areaways of a detached building from discharging to a properly graded open area, provided that:

(1) They do not serve continuously flowing springs or groundwater.

(2) The point of discharge is not less than 10 feet (3048 mm) from a property line.

(3) It is impracticable to discharge such drains to a storm drain, to an approved water course, to the front street curb or gutter, or to an alley.

🔧 Discharging subsoil water from a pump to an open area is allowed, provided the above three provisions are met. Care must be taken to ensure that the soil type or pavement at the discharge point will be capable of draining the anticipated discharge so as not to create a nuisance and also that the location is not in conflict with local requirements. See **Figure 1101.6.5** for examples of discharging to a curb and graded area.

1101.7 Building Subdrains. Building subdrains located below the public sewer level shall discharge into a sump or

FIGURE 1101.6.1
CONCRETE CONDUIT OR GUTTER

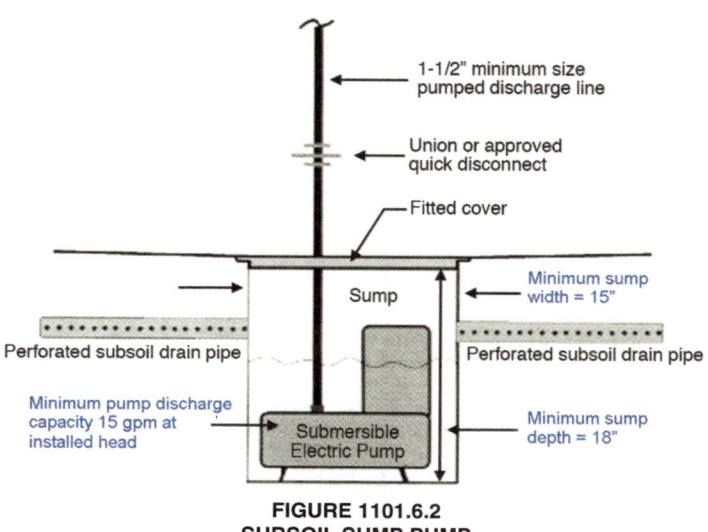

FIGURE 1101.6.2
SUBSOIL SUMP PUMP

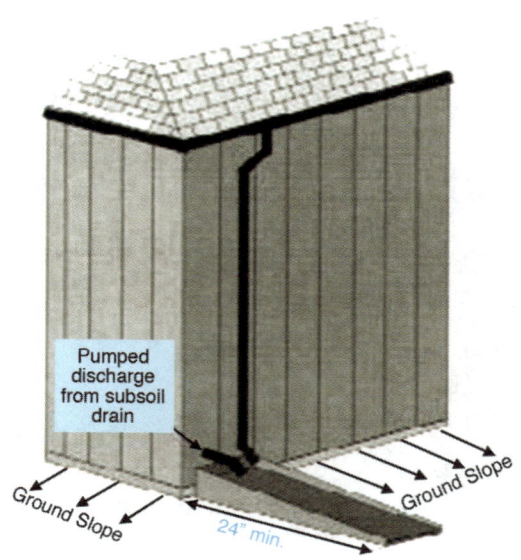

FIGURE 1101.6.3
SPLASH BLOCK

STORM DRAINAGE

**FIGURE 1101.6.5
DISCHARGE TO CURB OR PAVED AREA**

receiving tank, the contents of which shall be automatically lifted and discharged into the drainage system as required for building sumps.

🔧 A building subdrain is that portion of a sanitary drainage system that does not drain by gravity into the building sewer and must be distinguished from subsoil drains. The building subdrain shall not discharge into the storm drainage system or to any water course, but only into the building sanitary drain (see Section 710.2).

1101.8 Areaway Drains. Open subsurface space adjacent to a building, serving as an entrance to the basement or cellar of a building, shall be provided with a drain or drains. The areaway drains shall be not less than 2 inches (50 mm) in diameter for areaways at a maximum of 100 square feet (9.29 m^2) in area, and shall be discharged in the manner provided for subsoil drains not serving continuously flowing springs or groundwater (see Section 1101.6.1). Areaways exceeding 100 square feet (9.29 m^2) shall not drain into subsoil drains. The drains for areaways exceeding 100 square feet (9.29 m^2) shall be sized in accordance with Table 1103.2.

🔧 Area drains or areaway drains are receptors designed to collect surface or storm water from an open area usually set below grade level. An example of this is a landing located at the bottom of an exterior stairwell that leads to a basement entrance to a building. Failure to provide adequate drainage for these locations could lead to flooding of the building (see **Figure 1101.8**).

Small areas (under 100 ft^2) may drain to the subsoil drain and be discharged from the building through the subsoil pump. Larger areas (above 100 ft^2) could conceivably receive enough rainfall to overload a subsoil drainage system and cause damage to the building. For this reason, these types of areas must connect to a building storm drainage system or be separately pumped out of the building, terminating to an approved location. These larger drains shall be sized using Table 1103.2, Sizing of Horizontal Rainwater Piping. See Section 1103.2 and the sizing examples at the end of this chapter for sizing information.

1101.9 Window Areaway Drains. Window areaways at a maximum of 10 square feet (0.93 m^2) in area shall be permitted to discharge to the subsoil drains through a 2 inch (50 mm) diameter pipe. However, window areaways exceeding 10 square feet (0.93 m^2) in area shall be handled in the manner provided for entrance areaways (see Section 1101.8).

🔧 Window areaways are located at windows situated below the surrounding ground level. Storm water or flooding of the surrounding areas could flow into these lower areas and then into the building, potentially causing damage. Sizing for areas greater than 10 ft^2 shall be sized using the previous section, Section 1101.8 (see **Figure 1101.8**).

1101.10 Filling Stations and Motor Vehicle Washing Establishments. Public filling stations and motor vehicle washing establishments shall have the paved area sloped toward sumps or gratings within the property lines. Curbs not less than 6 inches (152 mm) high shall be placed where required to direct water to gratings or sumps.

🔧 Gas stations can experience large quantities of petroleum product runoff. Spillages resulting from overfilling gas tanks, gas pumps, oil filling, transmission fluid filling and antifreeze must not be allowed to flow into the storm drainage system. Car-washing facilities can also receive large quantities of these hazardous chemicals, as well as large quantities of sediments. The paved areas of these types of establishments must be sloped from the property line to collecting sumps or trench drain systems in a manner that will not allow the surface runoff from these areas to enter the public storm drainage system without pretreatment. Many jurisdictions will require the installation of sand and oil/water separators to serve these types of drains prior to allowing them to connect to the public system. The requirement for curbing will ensure that the waste water will flow to drainage grates or sumps and then to the interceptor (see Chapter 10 for installation of interceptors).

1101.11 Paved Areas. Where the occupant creates surface water drainage, the sumps, gratings, or floor drains shall be piped to a storm drain or an approved water course.

🔧 Rainwater falling on unimproved areas is disposed of by a combination of surface runoff to a water course and absorption into the soil. When areas are

STORM DRAINAGE

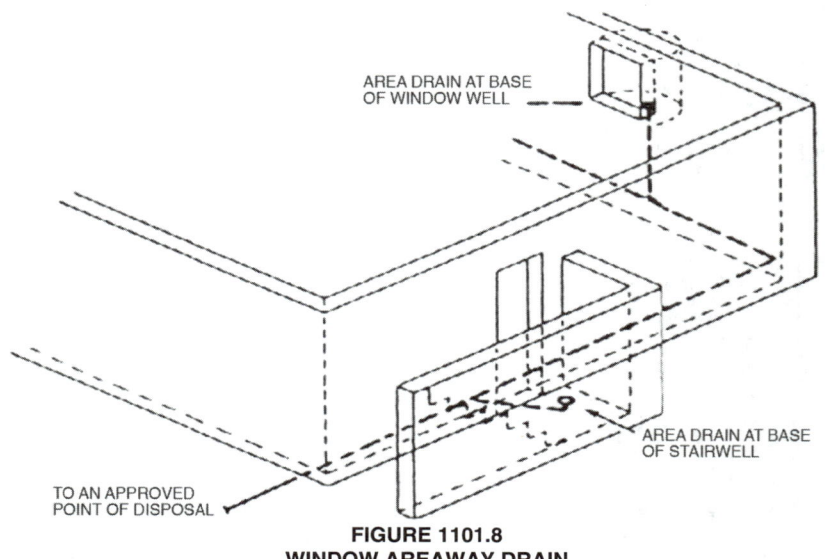

FIGURE 1101.8
WINDOW AREAWAY DRAIN

improved or building sites are paved, rainwater can no longer be absorbed into the soil. The paved area will cause surface water to collect during a storm, which must be drained and discharged properly.

1101.12 Roof Drainage. Roof drainage shall comply with Section 1101.12.1 and Section 1101.12.2.

1101.12.1 Primary Roof Drainage. Roof areas of a building shall be drained by roof drains or gutters. The location and sizing of drains and gutters shall be coordinated with the structural design and pitch of the roof. Unless otherwise required by the Authority Having Jurisdiction, roof drains, gutters, vertical conductors or leaders, and horizontal storm drains for primary drainage shall be sized based on a storm of 60 minutes duration and 100 year return period. Refer to Table D 101.1 (in Appendix D) for 100 years, 60-minute storms at various locations.

🔧 Paramount to eliminating the threat of roof collapse due to water accumulation is the installation of roof drains or gutters. Roof drains are generally installed on flat roofs or roofs surrounded by parapet walls. Pitched roofs not surrounded by parapet walls are generally drained by gutters. Close coordination with structural engineering relative to location and roof pitch as well as proper sizing in accordance with local climatic conditions are the basis for the design of the roof drainage system.

Rates for rainfall are established by the U.S. Weather Bureau based on 100-year storm occurrence with a 60-minute duration and can be located in Table D 101.1 of Appendix D.

The Table of Recurrence Intervals and Probabilities of Occurrences, shown in **Figure 1101.12.1** from the U.S. Geological Service, provide a measure of the probability and frequency of these storms. This table explains that the size of a storm represented by a 100-year storm has a probability of 1 in 100 of occurring in any given year and that there is a 1 percent chance of a 100-year storm occurring in any given year. This type of storm is used as the basis for sizing roof drainage systems to ensure that the system can handle the largest estimated storm that could occur in any given year.

When using Appendix D, make sure that local area requirements are also observed. Many times the amount of rainfall shown in Appendix D is increased locally for a greater measure of safety.

Recurrence Intervals and Probabilities of Occurrences		
Recurrence interval, in years	Probability of occurrence in any given year	Percent chance of occurrence in any given year
100	1 in 100	1
50	1 in 50	2
25	1 in 25	4
10	1 in 10	10
5	1 in 5	20
2	1 in 2	50

FIGURE 1101.12.1
RECURRENCE INTERVALS AND PROBABILITIES OF OCCURRENCES

1101.12.2 Secondary Drainage. Secondary (emergency) roof drainage shall be provided by one of the methods specified in Section 1101.12.2.1 or Section 1101.12.2.2.

🔧 To ensure that rainwater accumulation on a roof is drained away, two independent systems of roof drainage are required. They are the primary roof drainage system and the secondary roof drainage system, commonly referred to as an "overflow system." The primary system consists of roof drains, piping serving those drains and the discharge method used – either surface or gravity drainage or pumped drainage. The secondary system ensures that if the primary system or primary drains are plugged or overloaded, the secondary system will handle the rainwater and drain it away, protecting the building and its occupants from harm (see **Figure 1101.12.2**).

The secondary roof drainage system can be of two methods – roof scuppers, open sides of parapet walls, or secondary roof drains. The secondary roof drains can be of two types – roof drains with an independent piping system or roof drains that combine with the primary roof drainage piping, which will require an increase in piping size.

Once the required rainfall data has been determined, the primary and secondary roof drainage system must be sized in accordance with Section 1103.0 using Table 1103.1 (vertical drains and pipe), Table 1103.2 (horizontal drain and pipe) and Table 1103.3 (gutters).

1101.12.2.1 Roof Scuppers or Open Side. Secondary roof drainage shall be provided by an open-sided roof or scuppers where the roof perimeter construction extends above the roof in such a manner that water will be entrapped. An open-sided roof or scuppers shall be sized to prevent the depth of ponding water from exceeding that for which the roof was designed as determined by Section 1101.12.1. Scupper openings shall be not less than 4 inches (102 mm) high and have a width equal to the circumference of the roof drain required for the area served, sized in accordance with Table 1103.1.

An open-sided roof area used for secondary roof drainage can be just a simple opening in the parapet wall surrounding the roof, as in **Figure 1101.12.2.1a**. A scupper can be a piped penetration in the parapet wall, as in **Figures 1101.12.2.1b** and **1101.12.2.1c**. When either method is used, they must be sized by Table 1101.12. The width of the opening must be equal to the circumference of the drain required. For example, if the roof drain is 4 inches, the opening of the scupper or the open-sided roof opening must be 12.56 inches in length. If it is a piped scupper, it must be a 4-inch or larger pipe size.

1101.12.2.2 Secondary Roof Drain. Secondary roof drains shall be provided. The secondary roof drains shall be located not less than 2 inches (51 mm) above the roof surface. The maximum height of the roof drains shall be a height to prevent the depth of ponding water from exceeding that for which the roof was designed as determined by Section 1101.12.1. The secondary roof drains shall connect to a piping system in accordance with Section 1101.12.2.2.1 or Section 1101.12.2.2.2.

If roof drains are provided for the secondary system, they must be located so that they will be a minimum of 2 inches higher than the inlet of the primary roof drain (see **Figures 1101.12.2** and **1101.12.2.2**). The drain must also be placed so that it will serve the area of the primary drain. Care must also be taken that the ponding of water, which will occur if the primary drain does not drain, does not exceed the structural capacity of the roof. The roof must be designed with the weight of that standing water in mind.

1101.12.2.2.1 Separate Piping System. The secondary roof drainage system shall be a separate system of piping, independent of the primary roof drainage system. The discharge shall be above grade, in a location observable by the building occupants or maintenance personnel. Secondary roof drain systems shall be sized in accordance with Section 1101.12.1 based on the rainfall rate for which the primary system is sized.

1101.12.2.2.2 Combined System. The secondary roof drains shall connect to the vertical piping of the primary storm drainage conductor downstream of the last horizontal offset located below the roof. The primary storm drainage system shall connect to the building storm water that connects to an underground public storm sewer. The combined secondary and primary roof drain systems shall be sized in accordance with Section 1103.0 based on double the rainfall rate for the local area.

The secondary roof drain piping can be installed using the method in either Section 1101.12.2.2.1 or Section 1101.12.2.2.2. The secondary piping may be an entirely independent system consisting of roof drains, conduits and leaders, which independently discharge to the outside of the building. This secondary system is a mirror of the primary system, sized and piped identical to the primary system. If this method is used, the primary drainage piping must be piped to the curb, gutter or other approved area. The secondary system, basically an emergency overflow system, can be piped to an open area even if it is a walkway or other traveled outside area. The reason for this is that the secondary system should seldom be used, and if it is used, it would give a warning that something is wrong with the primary system. This secondary piped system gives the

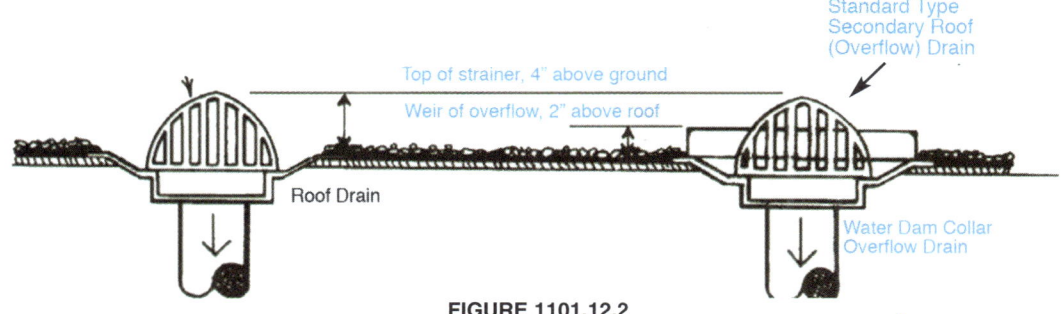

FIGURE 1101.12.2
PRIMARY AND SECONDARY ROOF DRAINS - SECONDARY DRAIN WEIR 2 INCH ABOVE PRIMARY - STRAINER 4 INCH ABOVE THE ROOF

STORM DRAINAGE

FIGURE 1101.12.2.1A
OPEN-SIDED SCUPPER AT PIPE TRADES TRAINING CENTER, UA LOCAL 525 LAS VEGAS, NV

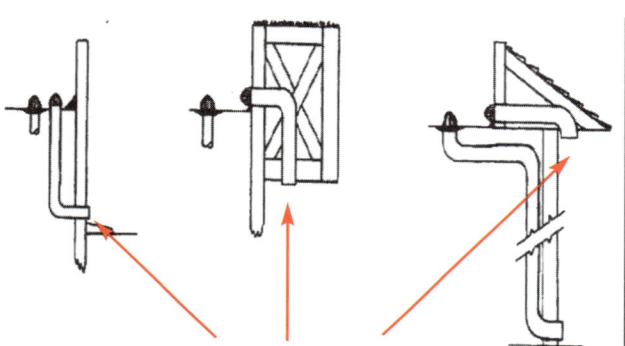

FIGURE 1101.12.2.1B
EXAMPLES OF PIPED SCUPPERS

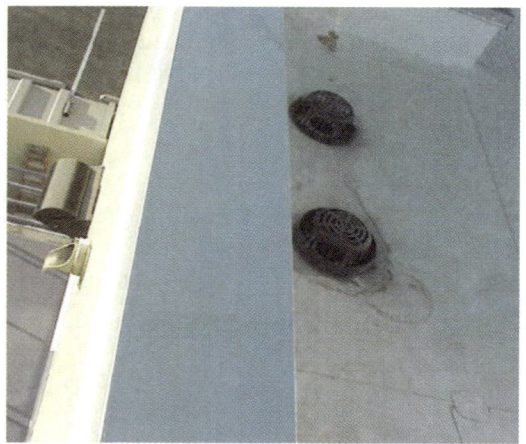

FIGURE 1101.12.2.1C
PRIMARY AND SECONDARY ROOF DRAINS WITH PIPED SCUPPER OVERFLOW

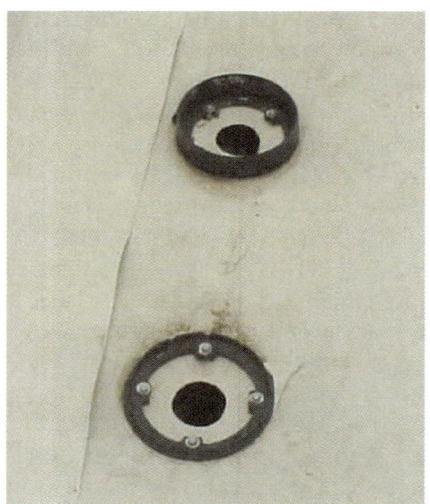

FIGURE 1101.12.2.2
PRIMARY AND SECONDARY ROOF DRAINS - SECONDARY WITH 2 INCH WEIR

owner of the building the largest margin of safety for the building and its occupants.

The other method of secondary roof drainage piping is to connect the secondary roof drains to the piping serving the primary roof drains (see **Figures 1101.12.2.2.2a, 1101.12.2.2.2b** and **1101.12.2.2.2c**). The connection of the secondary drain must be in a vertical section of piping downstream of any horizontal piping below the roof. If this method is used the size of the conduits, leaders and downspouts downstream of the connection must be increased by using twice the rainfall rate required for the primary piping. For example, if the rainfall rate for the area is 3 inches per hour the size of the combined piping system must be calculated using 6 inches of rainfall per hour.

1101.13 Cleanouts. Cleanouts for building storm drains shall comply with the requirements of Section 719.0 of this code.

🔧 Cleanouts are required to be installed in the building storm sewer in the same manner as those required for building sanitary sewers in Chapter 7.

1101.13.1 Rain Leaders and Conductors. Rain leaders and conductors connected to a building storm sewer shall have a cleanout installed at the base of the outside leader or outside conductor before it connects to the horizontal drain.

🔧 One specific requirement for cleanouts not contained in Chapter 7 but required here, is the requirement for a cleanout in the vertical piping outside of the building if the leader connects to a building storm sewer.

1101.14 Rainwater Sumps. Rainwater sumps serving "public use" occupancy buildings shall be provided with dual pumps arranged to function alternately in the case of overload or mechanical failure. Pumps rated 600 V or less shall comply with UL 778 and shall be installed in accordance with the manufacturer's installation instructions.

STORM DRAINAGE

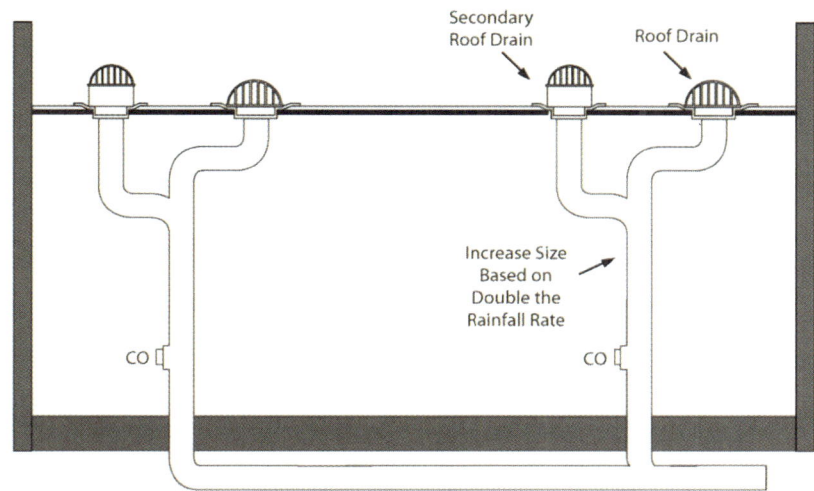

FIGURE 1101.12.2.2.2A
COMBINED ROOF DRAINAGE SYSTEM

FIGURE 1101.12.2.2.2B
SECONDARY DRAINAGE CONNECTION TO THE VERTICAL CONDUCTOR

🔧 A roof drainage system will rarely be used in a private-use building; however, if it is, it will require only a single-pump system. As in Chapter 7 for sanitary systems, public-use buildings will require a dual-pump system for any sump installation. Public use as defined in Section 218.0 is any fixtures "not defined as private or private use". This includes business, multifamily, mercantile occupancies, etc. When dual-pump systems are installed the pit or sump shall be designed to facilitate the removal of the pump without pulling on the cords or on the discharge pipe.

1101.15 Traps on Storm Drains and Leaders.
Leaders and storm drains, where connected to a combined sewer, shall be trapped. Floor and area drains connected to a storm drain shall be trapped.

Exception: Traps shall not be required where roof drains, rain leaders, and other inlets are at locations permitted under Section 906.0, Vent Termination.

🔧 When a combined sewer system (sanitary drainage and storm drainage) is used, traps are required on storm drains and leaders and also on any floor drains or area drains connected to the storm drain system. This is of

FIGURE 1101.12.2.2.2C
SYSTEM SIZED BASED ON DOUBLE THE RAINFALL RATE

course to prevent sewer gas in the combined system from entering the building. Traps will not be required for areas such as roofs where vents would be installed open ended. For example, a roof drain would not be required to be trapped because it would be open to the outside of the building on the roof. See **Figures 1101.15a** and **1101.15b**.

1101.15.1 Where Not Required.
No trap shall be required for leaders or conductors that are connected to a sewer carrying storm water exclusively.

STORM DRAINAGE

1101.15.2 Trap Size. Traps, where installed for individual conductors, shall be the same size as the horizontal drain to which they are connected.

1101.15.3 Method of Installation of Combined Sewer. Individual storm-water traps shall be installed on the stormwater drain branch serving each storm-water inlet, or a single trap shall be installed in the main storm drain just before its connection with the combined building sewer. Such traps shall be provided with an accessible cleanout on the outlet side of the trap.

🔧 At some point in a combined storm and sanitary system the storm-only portion must connect to the combined or the sanitary system. At this point, a trap must be installed to prevent sewer gas from entering the building. This installation will be similar to the building trap installation (see Chapter 7). See "D" in **Figure 1101.15b**.

1101.16 Leaders, Conductors, and Connections. Leaders or conductors shall not be used as soil, waste, or vent pipes nor shall soil, waste, or vent pipes be used as leaders or conductors.

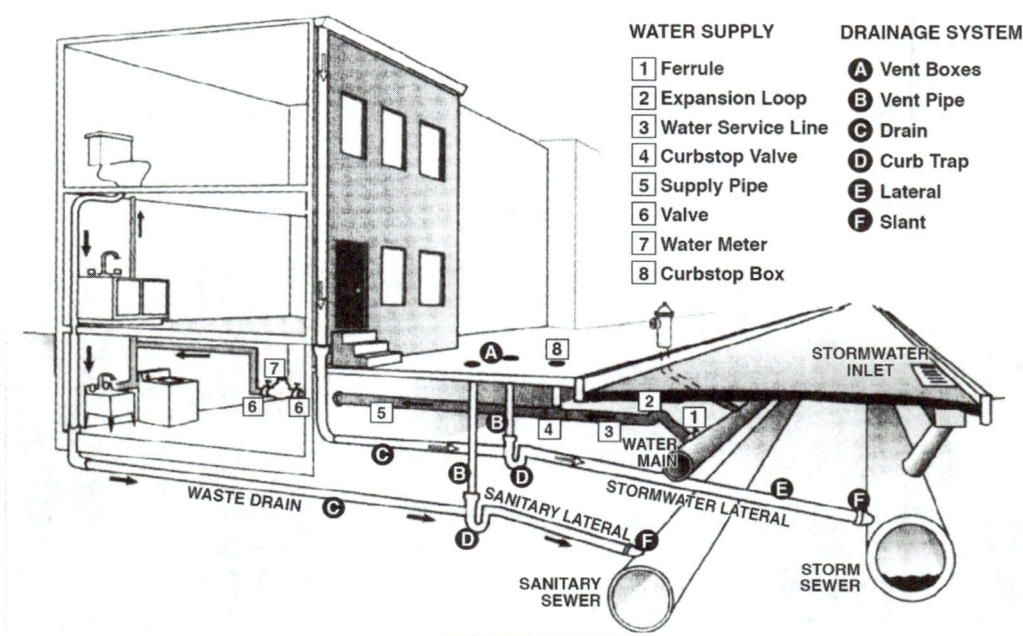

FIGURE 1101.15A
SEPARATE SANITARY AND STORM DRAINAGE SYSTEMS

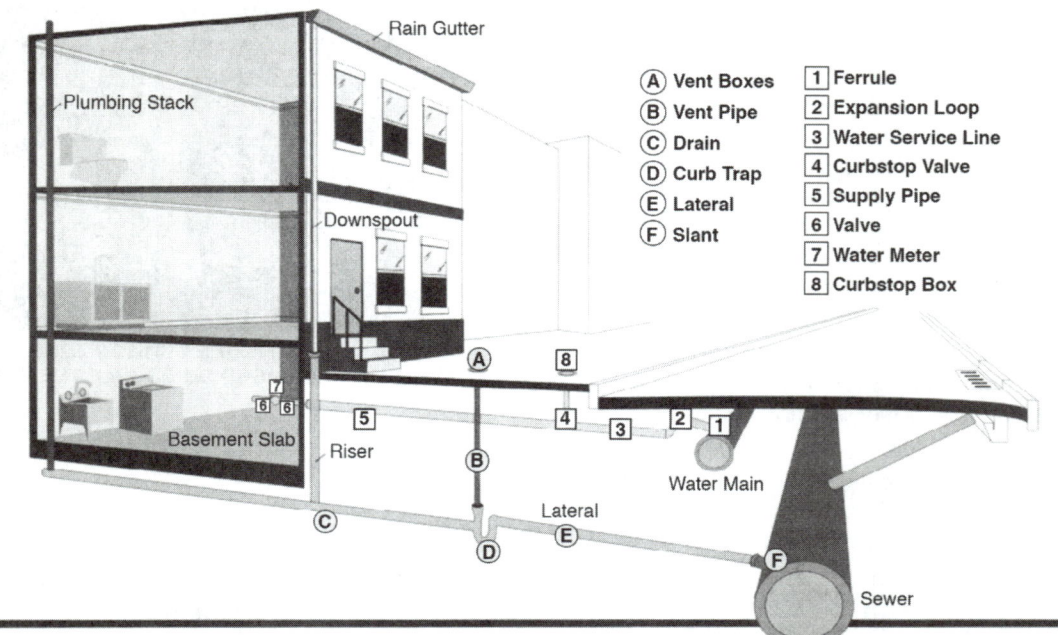

FIGURE 1101.15B
COMBINED SEWER SYSTEM - SANITARY AND STORM SYSTEM

STORM DRAINAGE

🔧 The following sub-sections are "common-sense" requirements for the installation of roof drainage systems. Roof drainage piping shall not be used for sanitary or waste water drainage or vice versa. Leaders on the outside of buildings need to be protected from damage, usually by bollards, as shown in **Figure 1101.16**. The requirement to keep the connection of the roof drain piping downstream of the combined or sanitary stack by at least 10 feet is to keep this connection away from the hydraulic jump area of the stack, where the piping tends to be full.

1101.16.1 Protection of Leaders. Leaders installed along alleyways, driveways, or other locations where exposed to damage shall be protected by metal guards, recessed into the wall, or constructed from the ferrous pipe.

1101.16.2 Combining Storm with Sanitary Drainage. The sanitary and storm drainage system of a building shall be entirely separate, except where a combined sewer is used, in which case the building storm drain shall be connected in the same horizontal plane through a single wye fitting to the combined building sewer not less than 10 feet (3048 mm) downstream from a soil stack.

1102.0 Roof Drains.

1102.1 Applications. Roof drains shall be constructed of aluminum, cast-iron, copper alloy of not more than 15 percent zinc, leaded nickel bronze, stainless steel, ABS, PVC, polypropylene, polyethylene, or nylon and shall comply with ASME A112.3.1 or ASME A112.6.4.

🔧 Roof drains and the materials they are made of are normally specified by the design engineer and should be verified that they are constructed of materials in accordance with the ASME standards listed here.

1102.2 Dome Strainers Required. Roof drains shall have domed strainers.

Exception: Roof drain strainers for use on sun decks, parking decks, and similar areas that are normally serviced and maintained, shall be permitted to be of the flat surface type. Such roof drain strainers shall be level with the deck.

🔧 The strainer for roof drains is constructed so that debris such as leaves and small papers could be caught against the dome strainer and the drain will still function (see **Figures 1102.2a** and **1102.2b**).

1102.3 Roof Drain Flashings. The connection between the roof and roof drains that pass through the roof and into the interior of the building shall be made watertight by the use of proper flashing material.

🔧 The most important element in the installation of the roof drain is its attachment to the roof itself. If this connection is not watertight and leakproof, the water damage that can occur could be almost as damaging as a roof collapse, and indeed a leaking roof drain could lead to the eventual failure of the roof. The roof drain to roof installation will connect and seal much like the floor drain to the shower safe pan installation. The main difference will be the under deck clamping ring or system used to tighten and seal the roof drain to the roof (see **Figure 1102.3**).

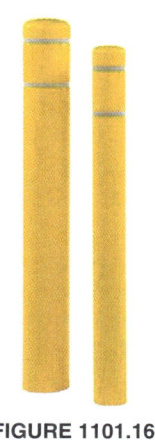

**FIGURE 1101.16
BOLLARDS**

**FIGURE 1102.2A
ROOF DRAIN WITH 4 INCH DOME STRAINER**

**FIGURE 1102.2B
FLAT DECK OR PROMENADE DECK ROOF DRAIN**

**FIGURE 1102.3
UNDER DECK CLAMPING RING**

STORM DRAINAGE

1102.3.1 Lead Flashing. Where lead flashing material is used, it shall be not less than 4 pounds per square foot (lb/ft^2) (19 kg/m^2).

1102.3.2 Copper Flashing. Where copper flashing material is used, it shall be not less than 12 ounces per square foot (oz/ft^2) (3.7 kg/m^2).

1103.0 Size of Leaders, Conductors, and Storm Drains.

🔧 The following sections provide the guidelines to size the storm drainage system. There are four specific items that need to be sized in this system:

1. Roof drain size, which will require the use of Table 1103.1, Sizing Roof Drains, Leaders and Vertical Rainwater Piping.
2. Vertical pipe size, which will require the use of Table 1103.1, Sizing Roof Drains, Leaders and Vertical Rainwater Piping.
3. Horizontal pipe size, which will require the use of Table 1103.2, Sizing of Horizontal Rainwater Piping, whether inside or outside of the building or whether it is storm drain or storm sewer piping.
4. Roof gutter size, which will require the use of Table 1103.3, Size of Gutters.

There are two factors that need to be determined in order to size the above items:

1. The maximum projected roof area in square feet, which drains rainwater to each individual roof drain or area drain. Three steps are needed for this:
a) Calculate the "flat" area of the roof.
b) Calculate the contributing side wall area.
c) Calculate any continuous flow conditions.
2. Determine rainfall rate in the area using Appendix D or by consulting the Authority Having Jurisdiction (AHJ) for the rainfall rate used in your area.

Each item will be discussed as they occur in the sections below.

1103.1 Vertical Conductors and Leaders. Vertical conductors and leaders shall be sized by the maximum projected roof area and Table 1103.1.

1103.2 Size of Horizontal Storm Drains and Sewers. The size of building storm drains, or building storm sewers or their horizontal branches shall be based on the maximum projected roof or paved area to be handled and Table 1103.2.

1103.3 Size of Roof Gutters. The size of semi-circular gutters shall be based on the maximum projected roof area and Table 1103.3.

🔧 Sections 1103.1, 1103.2 and 1103.3 require sizing the storm drainage system based on the "maximum projected roof area." This term refers to the amount of roof area that will collect or contribute to the collection of storm water on the roof. In Table 1103.1 the drain or pipe size is determined by using the "Maximum Allowable Horizontal Projected Roof Areas." This will include the area of the roof itself unless there is part of the roof that is sloped and deposits the rainwater naturally to the ground, thus not contributing to the collection of storm water on the roof. Simply put, the area of the roof in square feet that a roof drain serves is the calculation to determine the horizontal projected area.

Roofs are constructed in many different ways. Some may be flat roofs with a slight slope to one side so that the rainwater drains to the side where roof drains are installed (see **Figure 1103.3a**), while others may contain many such areas throughout the roof that slope (or cant) the roof several times to areas where roof drains are to be installed (see **Figure 1103.3b**). Whichever the case may be, the size of each roof drain is determined by the square foot area of the roof that is sloped to the roof drain. For example, **Figure 1103.3a** illustrates a flat roof with dimensions of 20 feet by 60 feet. The maximum projected roof area sloping to the roof drain is therefore 1,200 ft^2.

In **Figure 1103.3b** a multiple roof drain system with a canted roof creating many individual roof drain areas is

TABLE 1103.1
SIZING ROOF DRAINS, LEADERS, AND VERTICAL RAINWATER PIPING[2, 3]

SIZE OF DRAIN, LEADER, OR PIPE	FLOW	MAXIMUM ALLOWABLE HORIZONTAL PROJECTED ROOF AREAS AT VARIOUS RAINFALL RATES (square feet)											
inches	gpm[1]	1 (in/h)	2 (in/h)	3 (in/h)	4 (in/h)	5 (in/h)	6 (in/h)	7 (in/h)	8 (in/h)	9 (in/h)	10 (in/h)	11 (in/h)	12 (in/h)
2	30	2880	1440	960	720	575	480	410	360	320	290	260	240
3	92	8800	4400	2930	2200	1760	1470	1260	1100	980	880	800	730
4	192	18 400	9200	6130	4600	3680	3070	2630	2300	2045	1840	1675	1530
5	360	34 600	17 300	11 530	8650	6920	5765	4945	4325	3845	3460	3145	2880
6	563	54 000	27 000	17 995	13 500	10 800	9000	7715	6750	6000	5400	4910	4500
8	1208	116 000	58 000	38 660	29 000	23 200	19 315	16 570	14 500	12 890	11 600	10 545	9600

For SI units: 1 inch = 25 mm, 1 gallon per minute = 0.06 L/s, 1 inch per hour = 25.4 mm/h, 1 square foot = 0.0929 m^2

Notes:
[1] Maximum discharge capacity, gpm (L/s) with approximately 1¾ inch (44 mm) head of water at the drain.
[2] For rainfall rates other than those listed, determine the allowable roof area by dividing the area given in the 1 inch per hour (25.4 mm/h) column by the desired rainfall rate.
[3] Vertical piping shall be round, square, or rectangular. Square pipe shall be sized to enclose its equivalent round pipe. Rectangular pipe shall have not less than the same cross-sectional area as its equivalent round pipe, except that the ratio of its side dimensions shall not exceed 3 to 1.

illustrated. The red outlined section of the roof represents the area of the roof sloping to the roof drain at the column line B-1. To find the maximum projected roof area, multiply the length of the red outlined area by its width. Obtaining the maximum projected roof area requires the user to find the dimensions of the roof no matter how complicated a roof might be, determine the portion of the roof sloping to each roof drain, and then calculate that area in square feet.

1103.4 Side Walls Draining onto a Roof. Where vertical walls project above a roof to permit storm water to drain into the roof area below, the adjacent roof area shall be permitted to be computed from Table 1103.1 as follows:

🔧 In determining this maximum horizontal projected roof area the user is calculating the area of roof that rain may fall upon and the amount of rainwater demand, calculated in square feet, it creates. But this is not the only area that contributes to the amount of rainwater on the roof. As rain falls it rarely falls in a direct vertical line to the ground. Wind normally accompanies rain and causes the rain to fall at an angle to the ground. If there are side walls projecting vertically up from the roof, they will block the rainfall and cause it to drain onto the roof. This causes an increase in the amount of rainwater that can collect on the roof. These contributing, or "tributary," areas must be

TABLE 1103.2
SIZING OF HORIZONTAL RAINWATER PIPING[1, 2]

SIZE OF PIPE	FLOW (⅛ inch per foot slope)	MAXIMUM ALLOWABLE HORIZONTAL PROJECTED ROOF AREAS AT VARIOUS RAINFALL RATES (square feet)					
inches	gpm	1 (in/h)	2 (in/h)	3 (in/h)	4 (in/h)	5 (in/h)	6 (in/h)
3	34	3288	1644	1096	822	657	548
4	78	7520	3760	2506	1880	1504	1253
5	139	13 360	6680	4453	3340	2672	2227
6	222	21 400	10 700	7133	5350	4280	3566
8	478	46 000	23 000	15 330	11 500	9200	7670
10	860	82 800	41 400	27 600	20 700	16 580	13 800
12	1384	133 200	66 600	44 400	33 300	26 650	22 200
15	2473	238 000	119 000	79 333	59 500	47 600	39 650

SIZE OF PIPE	FLOW (¼ inch per foot slope)	MAXIMUM ALLOWABLE HORIZONTAL PROJECTED ROOF AREAS AT VARIOUS RAINFALL RATES (square feet)					
inches	gpm	1 (in/h)	2 (in/h)	3 (in/h)	4 (in/h)	5 (in/h)	6 (in/h)
3	48	4640	2320	1546	1160	928	773
4	110	10 600	5300	3533	2650	2120	1766
5	196	18 880	9440	6293	4720	3776	3146
6	314	30 200	15 100	10 066	7550	6040	5033
8	677	65 200	32 600	21 733	16 300	13 040	10 866
10	1214	116 800	58 400	38 950	29 200	23 350	19 450
12	1953	188 000	94 000	62 600	47 000	37 600	31 350
15	3491	336 000	168 000	112 000	84 000	67 250	56 000

SIZE OF PIPE	FLOW (½ inch per foot slope)	MAXIMUM ALLOWABLE HORIZONTAL PROJECTED ROOF AREAS AT VARIOUS RAINFALL RATES (square feet)					
inches	gpm	1 (in/h)	2 (in/h)	3 (in/h)	4 (in/h)	5 (in/h)	6 (in/h)
3	68	6576	3288	2192	1644	1310	1096
4	156	15 040	7520	5010	3760	3010	2500
5	278	26 720	13 360	8900	6680	5320	4450
6	445	42 800	21 400	14 267	10 700	8580	7140
8	956	92 000	46 000	30 650	23 000	18 400	15 320
10	1721	165 600	82 800	55 200	41 400	33 150	27 600
12	2768	266 400	133 200	88 800	66 600	53 200	44 400
15	4946	476 000	238 000	158 700	119 000	95 200	79 300

For SI units: 1 inch = 25 mm, 1 gallon per minute = 0.06 L/s, ⅛ inch per foot = 10.4 mm/m, 1 inch per hour = 25.4 mm/h, 1 square foot = 0.0929 m^2

Notes:
[1] The sizing data for horizontal piping are based on the pipes flowing full.
[2] For rainfall rates other than those listed, determine the allowable roof area by dividing the area given in the 1 inch per hour (25.4 mm/h) column by the desired rainfall rate.

STORM DRAINAGE

TABLE 1103.3
SIZE OF GUTTERS

DIAMETER OF GUTTER (1/16 inch per foot slope)	MAXIMUM RAINFALL RATES BASED ON ROOF AREA (square feet)				
inches	2 (in/h)	3 (in/h)	4 (in/h)	5 (in/h)	6 (in/h)
3	340	226	170	136	113
4	720	480	360	288	240
5	1250	834	625	500	416
6	1920	1280	960	768	640
7	2760	1840	1380	1100	918
8	3980	2655	1990	1590	1325
10	7200	4800	3600	2880	2400

DIAMETER OF GUTTER (1/8 inch per foot slope)	MAXIMUM RAINFALL RATES BASED ON ROOF AREA (square feet)				
inches	2 (in/h)	3 (in/h)	4 (in/h)	5 (in/h)	6 (in/h)
3	480	320	240	192	160
4	1020	681	510	408	340
5	1760	1172	880	704	587
6	2720	1815	1360	1085	905
7	3900	2600	1950	1560	1300
8	5600	3740	2800	2240	1870
10	10 200	6800	5100	4080	3400

DIAMETER OF GUTTER (1/4 inch per foot slope)	MAXIMUM RAINFALL RATES BASED ON ROOF AREA (square feet)				
inches	2 (in/h)	3 (in/h)	4 (in/h)	5 (in/h)	6 (in/h)
3	680	454	340	272	226
4	1440	960	720	576	480
5	2500	1668	1250	1000	834
6	3840	2560	1920	1536	1280
7	5520	3680	2760	2205	1840
8	7960	5310	3980	3180	2655
10	14 400	9600	7200	5750	4800

DIAMETER OF GUTTER (1/2 inch per foot slope)	MAXIMUM RAINFALL RATES BASED ON ROOF AREA (square feet)				
inches	2 (in/h)	3 (in/h)	4 (in/h)	5 (in/h)	6 (in/h)
3	960	640	480	384	320
4	2040	1360	1020	816	680
5	3540	2360	1770	1415	1180
6	5540	3695	2770	2220	1850
7	7800	5200	3900	3120	2600
8	11 200	7460	5600	4480	3730
10	20 000	13 330	10 000	8000	6660

For SI units: 1 inch = 25 mm, 1/16 inch per foot = 5.2 mm/m, 1 inch per hour = 25.4 mm/h, 1 square foot = 0.0929 m^2

STORM DRAINAGE

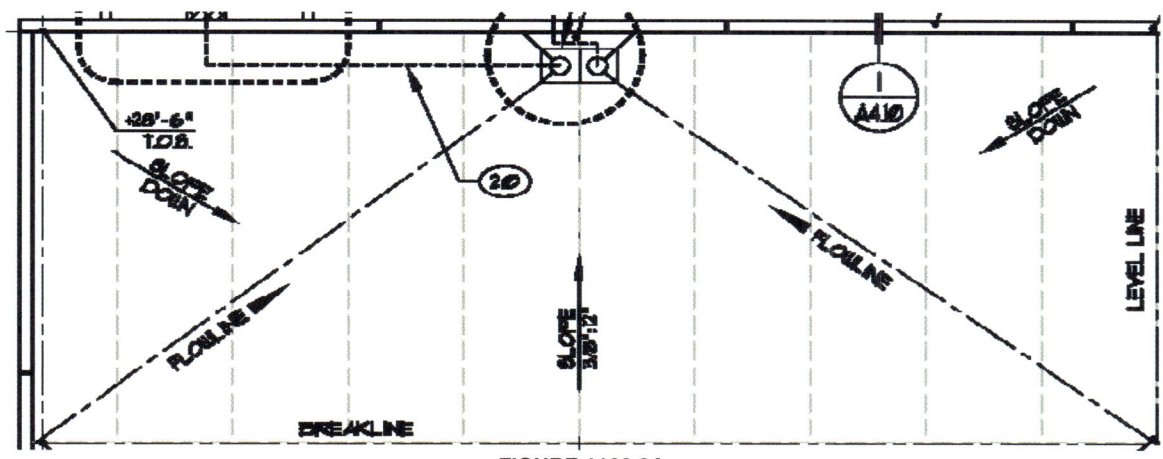

FIGURE 1103.3A
FLAT ROOF WITH PRIMARY AND SECONDARY DRAIN - NO SIDE WALLS
(DIMENSIONS - 20' WIDE BY 60' LONG = 1200 FT.² OF PROJECTED ROOF AREA)

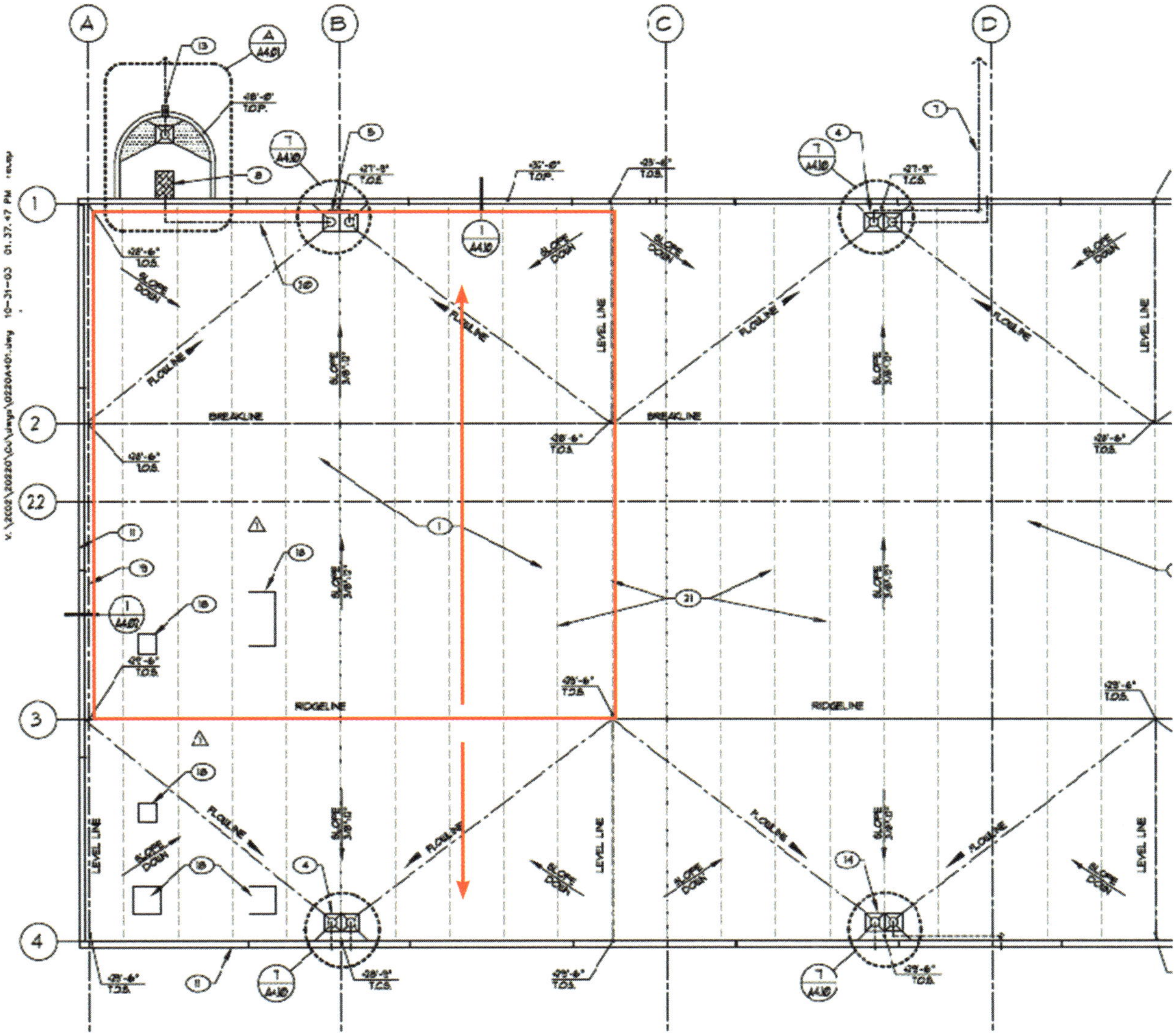

FIGURE 1103.3B
PARTIAL ROOF LAYOUT OF MULTIPLE ROOF DRAINS - RED BOX REPRESENTS THE PROJECTED ROOF AREA FOR ROOF DRAIN AT COLUMN LINE B-1

added to the total projected roof area and be represented in square feet so that one can easily determine the size of the roof drain and piping. To accomplish this, Section 1103.4 contains the parameters to determine the square foot area to be added to the maximum horizontal projected roof area for the seven different side-wall conditions that may occur.

Although there are seven conditions discussed here, there are only two percentages that must be used to calculate the total roof area. The first percentage is for a wall that is above other walls on a roof, or a wall that stands alone and is not adjacent or opposite other walls. Fifty percent of this wall area is added to the flat area of the roof. This factor takes into consideration that the wall does not collect rainfall as a flat area would; thus, only half of the area of the wall needs to be added to the total area of the roof.

The second percentage is for adjacent or connecting side walls. Thirty-five percent of the total wall area of both walls is added to the projected roof area. This factor takes into consideration the fact that one connecting wall will block a portion of the rainfall from falling on the other wall. Thus, only 35 percent of both wall areas are added to the total area of the roof.

For the following examples, the flat area of the roof will be given 500 ft^2, and the typical wall area represented by each wall segment will be given a height of 10 feet with a width of 20 feet, which is 200 ft^2.

(1) For one wall – add 50 percent of the wall area to the roof area figures.

🔧 In **Example 1**, a single wall projecting above the roof would add 50 percent of the total area of the wall or 100 ft^2 to the flat area of the roof. The total projected roof area would then be 600 ft^2. This amount would then be used this to determine the size of the horizontal drain, conductor, or gutter.

(2) For two adjacent walls of equal height – add 35 percent of the total wall areas.

🔧 In **Example 2**, using the same square footage as above, there are two walls of equal height with the sum of 400 ft^2. Because the walls are adjacent – next to each other rather than opposite each other – only 35 percent of the 400, which is 140 ft^2 will be used in the total square feet. The total projected roof area of 640 ft^2 would then be used to determine the size of the horizontal drain, conductor or gutter.

In computing the area of adjacent walls of equal height, it has been determined that all of the wall area of the adjacent walls, whether one wall is longer than the other or not, should be used and computed at 35 percent. For example, if one of the walls above was 30 feet wide, the computation would be: 10 x 30 = 300 + (10x20) = 500 x 35% = 175 ft^2.

(3) For two adjacent walls of unequal height – add 35 percent of the total common height and add 50 percent of the remaining height of the highest wall.

🔧 In computing the area of adjacent walls of unequal height, use the method used in **Example 2** for the adjacent wall area that is of equal height. The remaining wall area above the area of equal height will be calculated as in **Example 1**.

In **Example 3**, the two adjacent areas of equal height (wall sections 1a and 2) are calculated as follows: (10 x 20) + (10 x 20) = 400 ft^2 x 35% = 140 ft^2. The area above the equal height walls (wall section 1b), 10 x 20 = 200 ft^2, would be calculated at 50 percent, which is 100 ft^2. Add this to 140 ft^2 for a total of 240 ft^2. The total projected roof area of 740 ft^2 would then be used to determine the size of the horizontal drain, conductor, or gutter.

(4) Two opposite walls of same height – add no additional area.

🔧 In **Example 4**, two opposite walls of the same height will "cancel" each other. One wall will catch rain adding to the rainwater on the roof, while the other wall will block the same amount of rain from falling on the roof. Therefore, no area is added to the 500 square feet of flat roof area.

(5) Two opposite walls of differing heights – add 50 percent of the wall area above the top of the lower wall.

🔧 As in **Example 4** so also in **Example 5** the two opposite walls of the same height will cancel each other (wall sections 1 and 2a). Wall section 2b is higher than the other wall, and is calculated as if it was a wall by itself – no wall opposite and no wall adjacent. Therefore, 50 percent of the area of wall section 2b is added to the flat roof area. Using the same numbers (10 x 20 x 50%) would add 100 ft^2 to the total flat roof area of 500 ft^2 for a total of 600 ft^2 of projected roof area. This amount would then be used to determine the size of the horizontal drain, conductor, or gutter.

(6) Walls on three sides – add 50 percent of the area of the inner wall below the top of the lowest wall, plus an allowance for the area of the wall above the top of the lowest wall, in accordance with Section 1103.4(3) and Section 1103.4(5) above.

🔧 For **Example 6**, in order to determine the computation of the square footage that is added to the total projected roof area for three walls of differing heights, use the same criteria as in the previous examples. Per **Example 4**, the two opposite walls (wall sections 1 and 3a) cancel each other. Per **Example 1**, the inner wall (wall section 2a) will be calculated as one wall (since wall sections 1 and 3a have been cancelled) at 50 percent, or 100 ft^2 (as in **Example 1** above). Wall sections 2b and 3b are adjacent walls of equal height and will be calculated as **Example 3** at 35 percent, or 140 ft^2. Wall section 3c will be calculated as **Example 3** as a wall section above adjacent walls at 50 percent, or 100 ft^2. A total of 340 ft^2 must be added to the 500 ft^2 flat roof projected area for a total of 840 ft^2. This amount would then be used to determine the size of the horizontal drain, conductor, or gutter.

(7) Walls on four sides – no allowance for wall areas below the top of the lowest wall – add for areas above

the top of the lowest wall in accordance with Section 1103.4(1), Section 1103.4(3), Section 1103.4(5), and Section 1103.4(6) above.

🔧 For **Example 7**, in order to determine the computation of the square footage to be added to the total projected roof area for four walls of differing heights, use the same criteria as in the previous examples. Wall sections 1, 2a, 3a, and 4a are of the same height and opposite each other, and therefore will cancel each other as in **Example 4**. For wall sections 2b and 4b, the two are opposite walls and cancel each other. Wall section 3b will be computed as one wall standing alone at 50 percent, or 100 ft^2 per **Example 1**. Wall sections 3c and 4c are adjacent walls of equal height and will be computed at 35 percent, or 140 ft^2, as **Example 2**. Wall section 4d will be computed at 50 percent, or 100 ft^2, as **Example 3**. Therefore, a total of 340 ft^2 will be added to the 500 ft^2 "flat" roof projected area for a total of 840 ft^2. This amount would then be used to determine the size of the horizontal drain, conductor, or gutter.

Although roof drainage sizing may seem complicated it can be very simple. Just remember that if the wall is standing alone or above the other walls add 50 percent of the wall area to the "flat" area of the roof. If the walls are adjacent add 35 percent of the wall area to the "flat" area of the roof. And if the walls are opposite and of equal height they cancel each other.

1104.0 Values for Continuous Flow.

1104.1 General. Where there is a continuous or semi-continuous discharge into the building storm drain or building storm sewer, as from a pump, ejector, air-conditioning plant, or similar device, 1 gpm (0.06 L/s) of such discharge shall be computed as being equivalent to 24 square feet (2.2 m^2) of roof area, based upon a rate of rainfall of 4 inches per hour (in/h) (102 mm/h).

🔧 Values for continuous flow discharge, such as from a storm system sump pump, have to be computed in a fashion that will represent the rainwater gallon-per-minute (gpm) flow into a drain and then into the piping system. The requirement here is to first find the flow rate of the pump and then multiply this number by 24 ft^2. For example, if a sump pump has a flow rate of 15 gpm, add 360 ft^2 to the projected roof area flowing to that drain. Then use, at a minimum, the 4 inches/hour rainfall column in Tables 1103.1, 1103.2 and 1103.3, whichever apply, to size the system. That column would also have to be used for the piping from the drain accepting the waste from the sump and all the piping downstream.

The storm drainage system can now be sized. Specific steps and examples of this will be at the end of the chapter.

1105.0 Controlled-Flow Roof Drainage.

1105.1 Application. Instead of sizing the storm drainage system in accordance with Section 1103.0, the roof drainage shall be permitted to be sized by controlled flow and storage of the storm water on the roof, provided the following conditions are met:

(1) The water from a 25-year frequency storm shall not be stored on the roof exceeding 24 hours.

(2) During the storm, the water depth on the roof shall not exceed the depths specified in Table 1105.1(1).

(3) Not less than two drains shall be installed in roof areas of 10 000 square feet (929 m^2) or less, and not less than one additional drain shall be installed for each 10 000 square feet (929 m^2) of roof area exceeding 10 000 square feet (929 m^2).

(4) Each roof drain shall have a precalibrated, fixed (non-adjustable), and proportional weir (notched) in a standing water collar inside the strainer. No mechanical devices or valves shall be permitted.

(5) Pipe sizing shall be based on the pre-calibrated rate of flow (gpm) (L/s) of the pre-calibrated weir for the maximum allowable water depth, and Table 1103.1 and Table 1103.2.

(6) The height of stones or other granular material above the waterproofed surface shall not be considered in water depth measurement, and the roof surface in the vicinity of the drain shall not be recessed to create a reservoir.

(7) Roof design, where controlled-flow roof drainage is used, shall be such that the design roof live load is not less than 30 lb/ft^2 (146 kg/m^2) to provide a safety factor exceeding the 15 lb/ft^2 (73 kg/m^2) represented by the depth of water stored on the roof in accordance with Table 1105.1(1).

(8) Scuppers shall be provided in parapet walls. The distance of scupper bottoms above the roof level at the drains shall not exceed the maximum distances specified in Table 1105.1(2).

(9) Scupper openings shall be not less than 4 inches (102 mm) high and have a width equal to the circumference of the roof drain required for the area served, sized in accordance with Table 1103.1.

(10) Flashings shall extend above the top of the scuppers.

(11) At a wall or parapet, 45 degree (0.79 rad) cants shall be installed.

(12) Separate storm and sanitary drainage systems shall be provided within the building.

(13) Calculations for the roof drainage system shall be submitted along with the plans to the Authority Having Jurisdiction for approval.

🔧 Controlled flow roof drainage is normally an engineer-designed system. The average journeyman plumber will not design this system. There are too many parameters that the journeyman will not have access to or control of to design this system. This is an alternative sizing method for storm drains that utilizes the roof to retain water and precalibrated weirs on the roof drains to control the flow into the storm drainage system (see **Figure 1105.1a**). There are several conditions enumerated above that must be met before utilizing this sizing method. **Figure 1105.1b** illustrates scupper provisions in a parapet wall.

STORM DRAINAGE

Example (1)

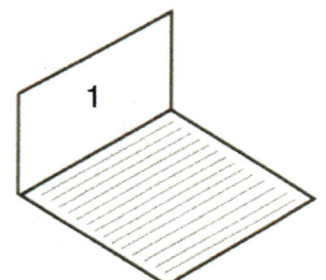

1   50% Tributary Area

Add to 100% of roof area.

Example (2)

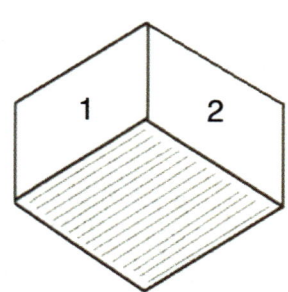

35% Tributary Areas

Add to 100% of roof area.

Example (3)

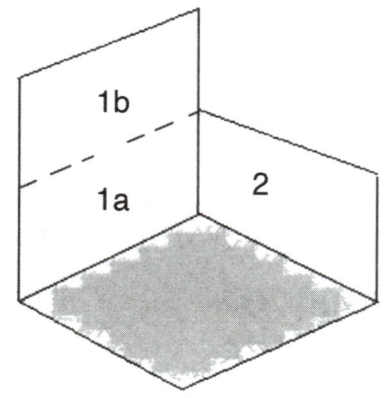
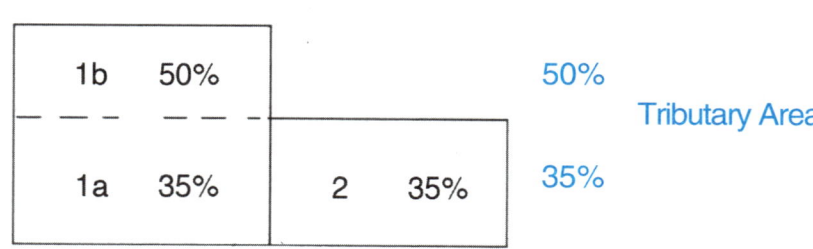

50%
Tributary Areas
35%

Add to 100% of roof area.

Example (4)

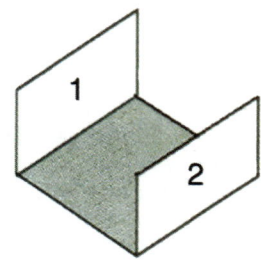
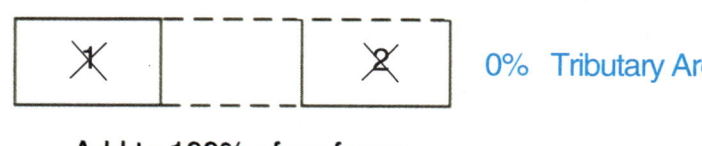

0% Tributary Areas

Add to 100% of roof area.

STORM DRAINAGE

Example (5)

Example (6)

Example (7)

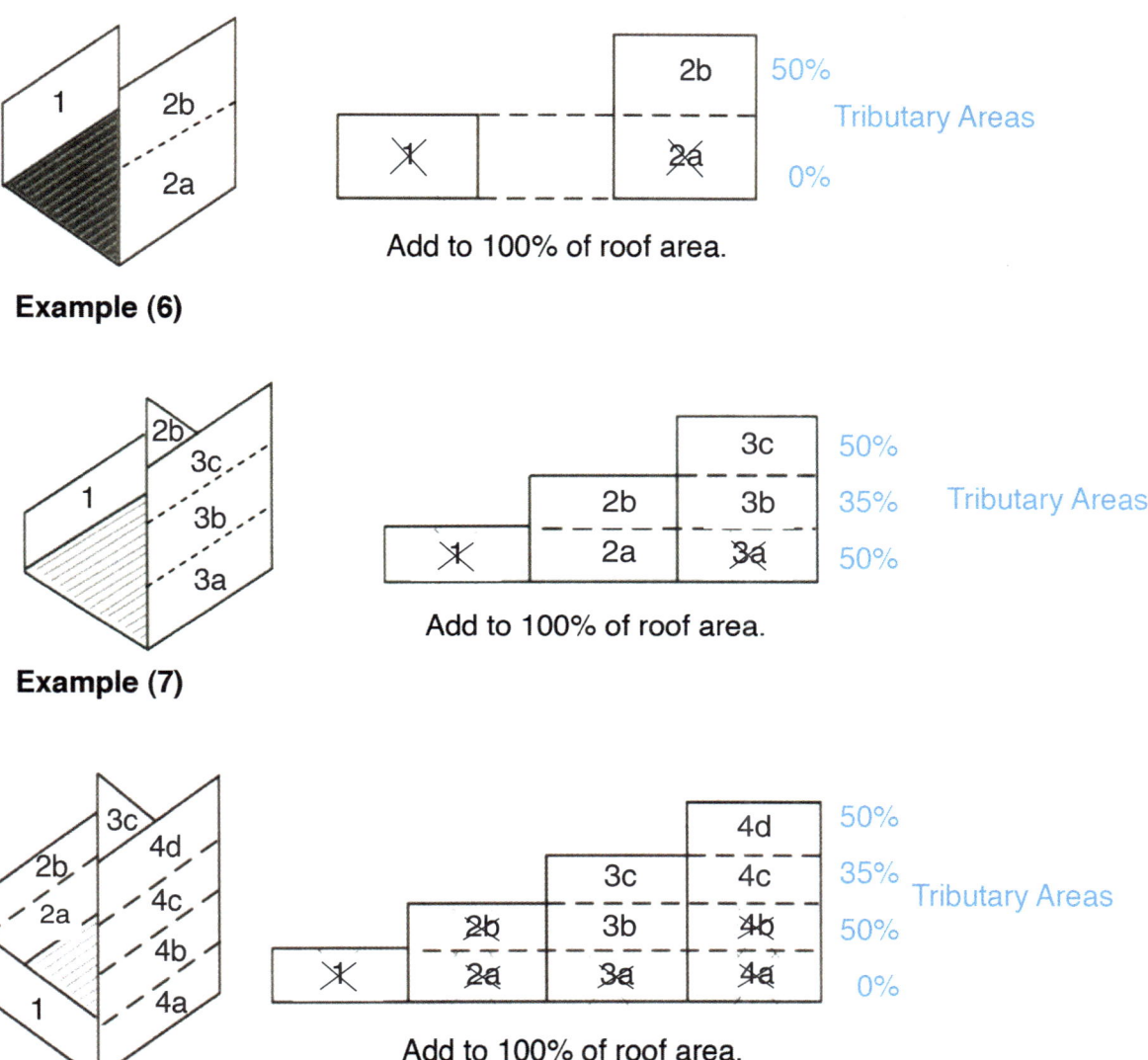

1105.2 Setback Roofs. Drains on setback roofs shall be permitted to be connected to the controlled-flow drainage systems provided:

(1) The setback is designed for storing water, or

(2) The square footage of the setback drainage area is converted as outlined in Section 1105.0 to gpm, and the storm-water pipe sizes in the controlled-flow system are based on the sum of the loads.

(3) The branch from each of the roof drains that are not provided with controlled flow shall be sized in accordance with Table 1103.1.

1106.0 Engineered Storm Drainage System.

1106.1 General. The design and sizing of a storm drainage system shall be permitted to be determined by accepted engineering practices. The system shall be designed by a registered design professional and approved in accordance with Section 301.5.

1106.2 Siphonic Roof Drainage Systems. The design of a siphonic roof drainage system shall comply with ASPE 45.

1106.3 Siphonic Roof Drains. Siphonic roof drains shall comply with ASME A112.6.9.

Siphonic roof drains are designed with a specially engineered and tested roof drain baffle to allow negative atmospheric pressure to be sustained in the piping by inhibiting the admission of air (see **Figure 1106.3a**). This allows a full-bore flow with higher volumes and velocities in the piping system.

STORM DRAINAGE

A registered design professional is required to employ hydraulic calculations to ensure hydraulic balance is achieved and that the piping system will automatically fill to induce a siphon. Since the flow is induced by negative pressure, the system allows for a level horizontal installation serving multiple roof drains (**Figure 1106.3b**).

As the siphonic roof drainage discharges from the vertical stack to a storm sewer, the influence of gravity takes over to sustain the flow within the storm sewer system.

1107.0 Testing.

Storm drainage systems shall be tested in the same manner as outlined in Section 712.0 Testing, for sanitary drainage and vent systems. Refer to this section for further explanation of the requirements stated below.

1107.1 Testing Required. New building storm drainage systems and parts of existing systems that have been altered, extended, or repaired shall be tested in accordance with Section 1107.2.1 or Section 1107.2.2 to disclose leaks and defects.

1107.2 Methods of Testing Storm Drainage Systems. Except for outside leaders and perforated or open-jointed drain tile, the piping of storm drain systems shall be tested upon completion of the rough piping installation by water or air, except that plastic pipe shall not be tested with air, and proved tight. The Authority Having Jurisdiction shall be permitted to require the removal of cleanout plugs to ascertain whether the pressure has reached parts of the system. One of the following test

TABLE 1105.1(1)
CONTROLLED-FLOW MAXIMUM ROOF WATER DEPTH

ROOF RISE* (inches)	MAXIMUM WATER DEPTH AT DRAIN (inches)
Flat	3
2	4
4	5
6	6

For SI units: 1 inch = 25.4 mm

* Vertical measurement from the roof surface at the drain to the highest point of the roof surface served by the drain, ignoring a local depression immediately adjacent to the drain.

TABLE 1105.1(2)
DISTANCE OF SCUPPER BOTTOMS ABOVE ROOF

ROOF RISE* (inches)	ABOVE ROOF LEVEL AT DRAIN (inches)
Flat	3
2	4
4	5
6	6

For SI units: 1 inch = 25.4 mm

* Vertical measurement from the roof surface at the drain to the highest point of the roof surface served by the drain, ignoring a local depression immediately adjacent to the drain.

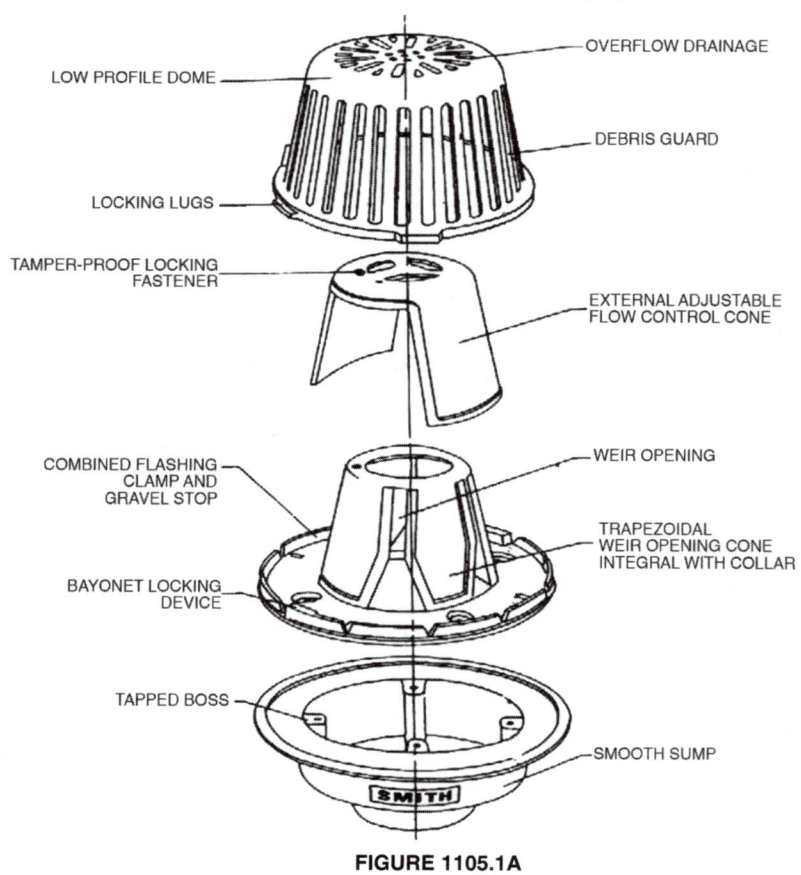

FIGURE 1105.1A
CONTROLLED FLOW ROOF DRAIN

STORM DRAINAGE

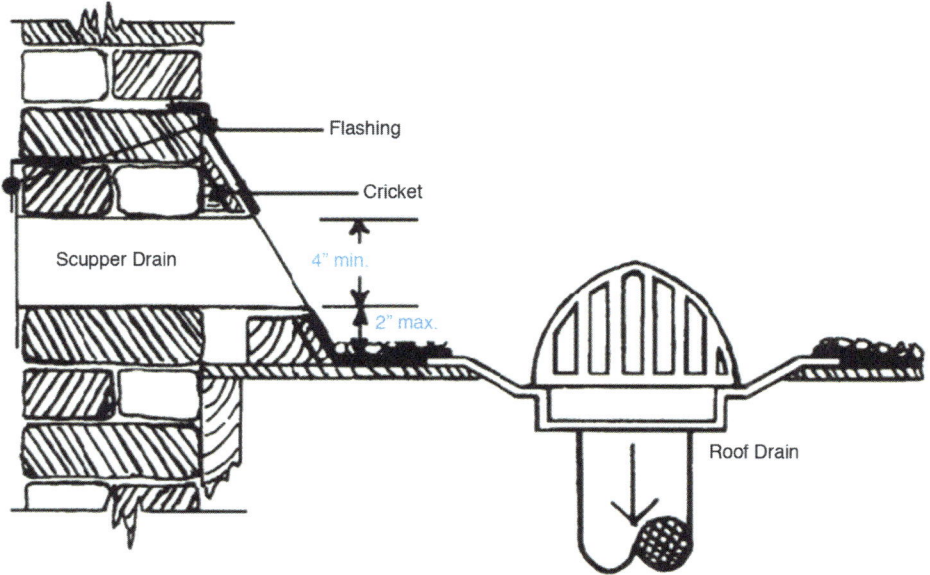

**FIGURE 1105.1B
SCUPPER OPENING**

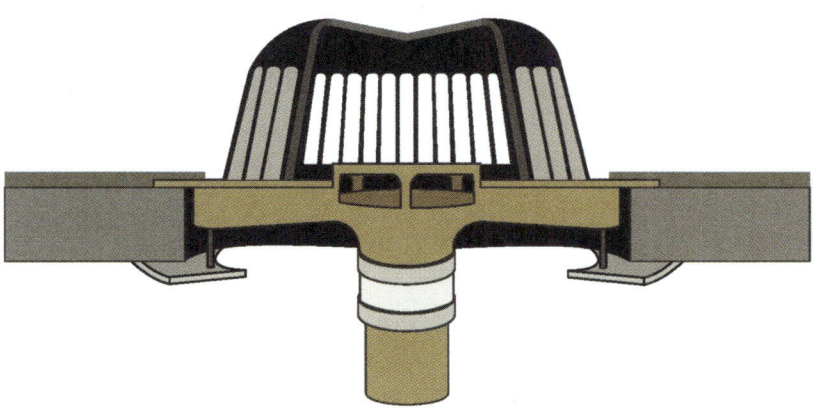

**FIGURE 1106.3A
SIPHONIC ROOF DRAIN**

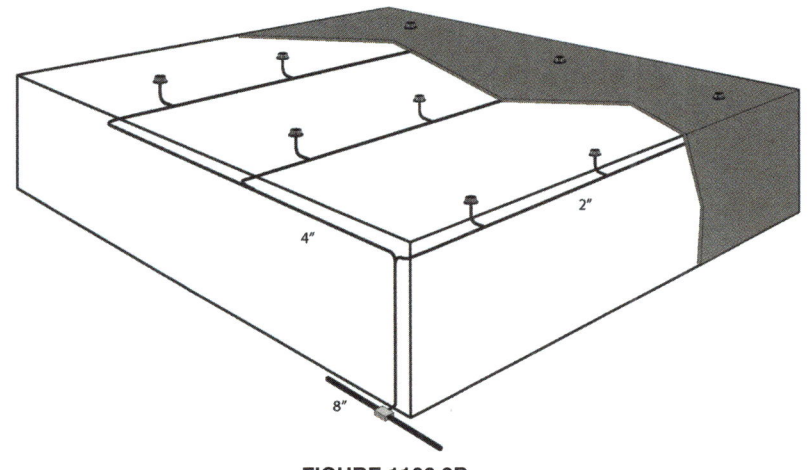

**FIGURE 1106.3B
SIPHONIC ROOF DRAIN SYSTEM**

methods shall be used in accordance with Section 1107.2.1 through Section 1107.2.3.

1107.2.1 Water Test. After piping has been installed, the water test shall be applied to the drainage system, either to the entire system or sections. Where the test is applied to the entire system, all openings in the piping shall be tightly closed except for the highest opening, and the system shall be filled with water to the point of overflow. Where the system is tested in sections, each opening shall be tightly plugged except for the highest opening of the section under test, and each section shall be filled with water, but no section shall be tested with less than a 10 foot (3048 mm) head of water. In testing successive sections, not less than the upper 10 feet (3048 mm) of the next preceding section shall be tested so that no joint of pipe in the building except the uppermost 10 feet (3048 mm) of a roof drainage system, which shall be filled with water to the flood level of the uppermost roof drain, shall have been submitted to a test of less than 10 foot (3048 mm) head of water. The water shall be kept in the system or the portion of the test for not less than 15 minutes before inspection starts; the system shall then be tight.

1107.2.2 Air Test. The air test shall be made by attaching an air compressor testing apparatus to a suitable opening after closing other inlets and outlets to the system, forcing air into the system until there is a uniform gauge pressure of 5 pounds-force per square inch (psi) (34 kPa) or sufficient pressure to balance a column of mercury 10 inches (34 kPa) in height. This pressure shall be held without the introduction of additional air for not less than 15 minutes.

1107.2.3 Exceptions. Where circumstances exist that make air and water tests described in Section 1107.2.1 and Section 1107.2.2 impractical, see Section 105.3.

Sizing the Storm Drainage System

The four sections of this chapter that contain the requirements to size the storm drainage system are:

1. Section 1101.12 requires the use of rainfall rates in inches per hour from Table D 101.1 of Appendix D or the AHJ's recommended rainfall rate to size the system.
2. Section 1103.0 requires the system be sized based on the maximum horizontal projected roof area serving each drain and each pipe segment. This section's subsections then require the use of Tables 1103.1, 1103.2, and 1103.3 to find the size of drains, vertical and horizontal piping and gutters.
3. Section 1103.4 provides the information needed to determine side wall tributary areas
4. Section 1104.0 provides the information for continuous flow calculations to reach the maximum projected roof area needed to size each drain or pipe segment.

There are three basic steps to sizing the storm drainage system:

1. Determine the maximum projected roof area including tributary side wall and continuous flow areas for each drain or gutter and each pipe segment.
2. Determine the maximum amount of rainfall per hour in the local area. Use Table D 101.1 of Appendix D for the rain fall rate in the area, or the maximum rate established locally by the AHJ, if higher than Table D 101.1.
3. Determine sizes from the following tables:
a. Use Table 1103.1 to determine drain and vertical leader size.
b. Use Table 1103.2 for horizontal rainwater piping.
i) Determine the grade used for the horizontal piping.
c. Use Table 1103.3 for gutter sizing:
i) Determine the grade used for the gutters.

The following example illustrates the above steps.

A building in Tucson, Arizona, is under construction and has a 2,000 ft^2 roof. Horizontal piping is at 1/8-inch per foot grade. Determine the roof drain, vertical pipe, horizontal pipe and gutter size.

Step 1 – Table D 101.1 of Appendix D states that Tucson, Arizona, has a rainfall rate of 3 inches per hour for a 100-year storm with 60-minute duration.

Step 2 – Determine the maximum roof area. The maximum projected roof area is given as 2,000 ft^2.

Step 3 – Determine sizes.

a) Determine the Roof Drain size in Table 1103.1. Find the 3-inch-per-hour rainfall column in Table 1103.1. Find the row with not less than 2,000 ft^2 or above maximum allowable area. Go to the far left column for the pipe size – it is 3 inches. The required size for one roof drain for the 2,000 ft^2 roof at 3-inch-per-hour rainfall is 3 inches. Notice that a 2-inch roof drain will serve 960 ft^2. The use of three 2-inch roof drains instead of one 3-inch drain would be allowable.

b) Find the vertical pipe size also in Table 1103.1. The vertical pipe size is also 3 inches. If other drains connect to the vertical pipe, each segment would have to be sized based on the total projected area served by the pipe. See **Figure 2** for a multiple roof drain sizing example.

c) Find the horizontal pipe grade and then the size in Table 1103.2.

The pipe grade is given as 1/8-inch per foot. In Table 1103.2 find the section of the table for 1/8-inch per foot piping. Find the 3-inch-per-hour rainfall column. Find the row with not less than 2,000 ft^2 or above maximum allowable area. Go to the far left hand column for the pipe size – it is 4 inches. If other drains connect to the horizontal pipe, each segment would have to be sized based on the total projected area served by the pipe (see **Figure 2** for a multiple roof drain sizing example).

d) Find the grade and gutter size in Table 1103.3. The pipe grade is given as 1/8-inch per foot. In Table 1103.3 find the section of the table for 1/8-inch per foot piping. Find the 3-inch-per-hour rainfall column. Find the row with not less than 2,000 ft^2 or above maximum allowable area. Go to the far left hand

STORM DRAINAGE

column for the gutter size – it is 7 inches. If other drains or roof sections drain to the gutter, each segment would have to be sized based on the total projected area served by the gutter.

Tables 1103.2 and 1103.1 provide footnotes that furnish the assumptions that the tables are based upon or give further qualifying instructions.

Note 1 in Table 1103.1 states that the flow rate in the second column is based on approximately 1¾-inches head of water at the roof drain. This assumes that a minimal ponding occurs around the roof drain as the water is draining in the leader. The flow rate capacity of the leader sizes are based on this static head.

Note 1 in Table 1103.2 states that horizontal piping is designed to flow full. Again there are no traps to "blow out" and the system is usually open ended to the discharge area. As a result of this, the pressure buildup will be dissipated at the end of the pipe with no problems in the system.

Note 2 in Tables 1103.2 and 1103.1 provide a method to determine the allowable roof area for a listed size pipe at a listed slope for amounts of rain that are not included in the table. To calculate this, divide the allowable square feet of roof area allowed for a 1-inch rainfall rate by the rainfall rate for the local area. For example, for the 3.2-inch-per-hour rainfall rate for a 6-inch vertical pipe, divide the 1-inch-per-hour rainfall allowable area for 6 inches in Table 1103.1 by 3.2. This would be 54,000/ 3.2 = 16,875 ft^2. Therefore, a 6-foot pipe at 3.2-inch-per-hour rainfall rate could serve up to 16,875 ft^2. If the pipe size to be used is not known, this process would have to be done for each pipe size to find the correct size for the maximum projected roof area. The same method is used for Table 1103.2.

Note 3 in Table 1103.1 refers to the fact that round, square or rectangular rainwater pipe may be used in the system. Square or rectangular pipe is considered equivalent to round pipe if the same size round pipe would be able to be enclosed within the square or rectangular pipe. Rectangular pipe can also be used if it is equal to the area of the round pipe as long as the rectangle sides do not exceed the ratio of 3-1. For example a 12-inch by 3-inch rectangle would not be allowed.

The secondary or overflow system will be sized in the same manner. If the secondary system is a separate system as referred to in Section 1101.12.2.2.1, it should be identical to the primary system. If the secondary system is a combined system, as referred to in Section 1101.12.2.2.2, then the primary and secondary drain must be sized using twice the rainfall rate for the local area. The rest of the piping downstream must be sized using this rate also.

One other factor to take into consideration while sizing the system is illustrated in **Figure 1**. Suppose this is a single roof drain system for a 4,000 ft^2 roof with no tributary side wall areas and a rainfall rate of 4-inch-per-hour. The horizontal piping is at 1/8 per foot grade.

The drain size for both A and B in **Figure 1** as determined from Table 1103.1, is 4 inches. In A the horizontal conductor, as determined from Table 1103.2, is 6 inches and the vertical leader, as determined from Table 1103.1, is 4 inches. However, Section 310.5 in Chapter 3 prohibits a reduction in size that would retard flow or create a restriction. Increasing size from 4 to 6 inches and then back to 4 inches creates a restriction. Therefore, the vertical leader in A must be 6 inches in size. Care must be given to take this "problem" into consideration when sizing from horizontal to vertical. This problem does not occur in B. The drain and leader are 4 inches and the horizontal drain is 6 inches.

It is advisable to follow step by step the sizing example in **Figure 2**. The situation noted in **Figure 1** also occurs in this example.

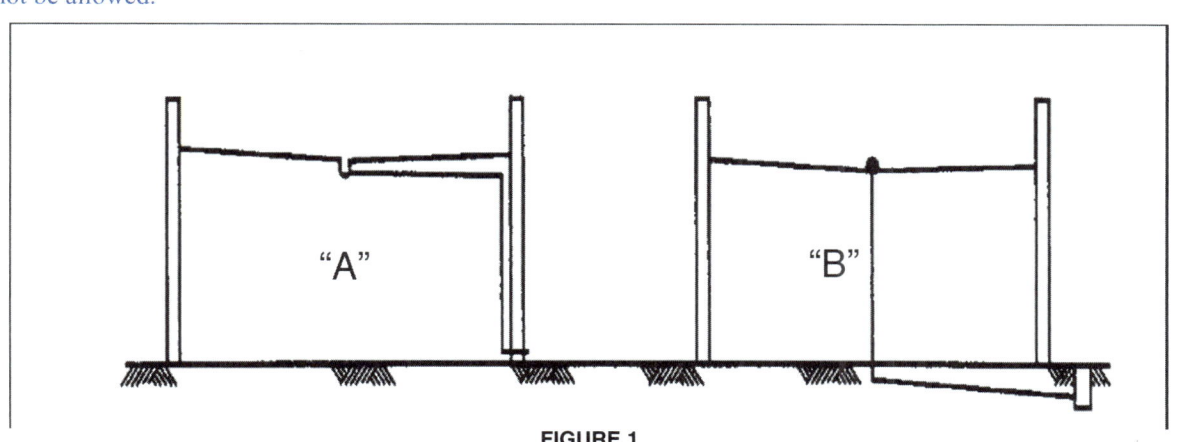

FIGURE 1
VERTICAL TO HORIZONTAL AND HORIZONTAL TO VERTICAL SIZING

STORM DRAINAGE

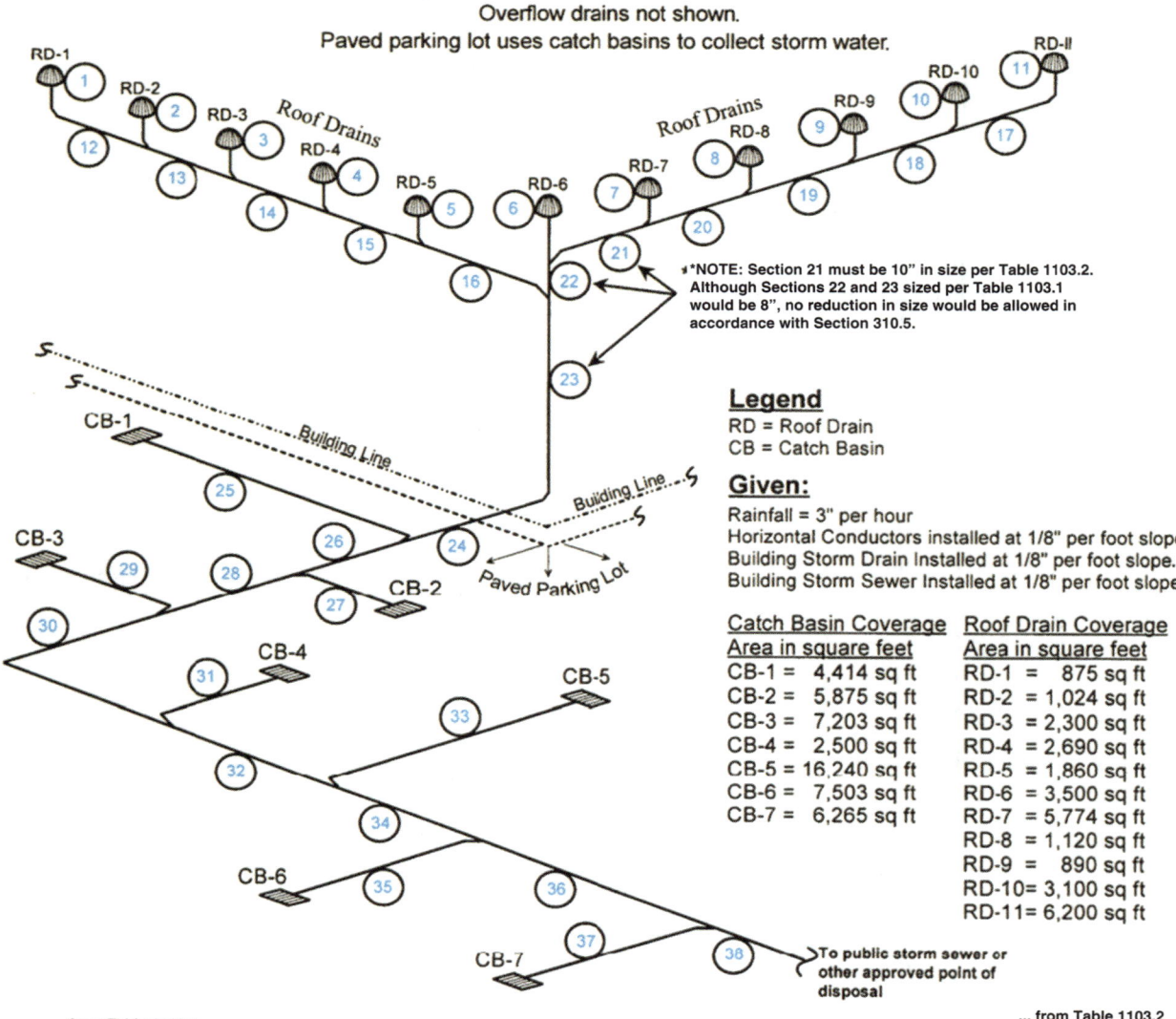

FIGURE 2
MULTIPLE ROOF DRAIN AND AREA WAY DRAIN SIZING EXAMPLE

CHAPTER 12

FUEL GAS PIPING

1201.0 General.

1201.1 Applicability. The regulations of this chapter shall govern the installation of fuel gas piping in or in connection with a building, structure or within the property lines of premises up to 5 pounds-force per square inch (psi) (34 kPa) for natural gas and 10 psi (69 kPa) for undiluted propane, other than service pipe. Fuel oil piping systems shall be installed in accordance with NFPA 31.

This code covers the installation of fuel gas systems (natural gas and LP-gas systems) within the building or within the property line of the premises. The service piping supplied by the public utility and fuel oil systems are not covered under this code. Visit www.nfpa.org for a complete list of NFPA codes.

1202.0 Coverage of Piping System.

1202.1 General. Coverage of piping systems shall extend from the point of delivery to the appliance connections. For other than undiluted liquefied petroleum gas (LP-Gas) systems, the point of delivery shall be the outlet of the service meter assembly or the outlet of the service regulator or service shutoff valve where no meter is provided. For undiluted LP-Gas systems, the point of delivery shall be considered to be the outlet of the final pressure regulator, exclusive of line gas regulators where no meter is installed. Where a meter is installed, the point of delivery shall be the outlet of the meter. [NFPA 54:1.1.1.1(A)]

Fuel gas systems covered under Chapter 12 are natural gas systems, liquefied petroleum gas (LP-gas) vapor only systems and manufactured gas systems. The vast majority of systems installed today are natural gas systems; therefore, most of the information covered here will pertain to natural gas systems.

The system component covered by this chapter is the piping from the shutoff valve at the meter (natural gas) or at the final regulator (LP-gas) to the gas appliance. This includes all pipe, fittings, valves, regulators and connectors used in the system to deliver fuel gas to the appliance (see **Figure 1202.1**).

Historically, fuel gas systems were installed to provide either heat for cooking and room heating or lighting. Today there are literally hundreds of different types of devices that utilize fuel gas for their operation. Some are appliances such as water heaters, furnaces and ovens. Others are not defined as appliances, such as tiki torches, decorative gas logs and even man-made volcanoes (such as the Mirage Hotel in Las Vegas, Nevada).

1202.2 Piping System Requirements. Requirements for piping systems shall include design, materials, components, fabrication, assembly, installation, testing, inspection, operation, and maintenance. [NFPA 54:1.1.1.1(C)]

The scope of this chapter will include all parts of the fuel gas system, from the design of the piping system, to its installation, inspection, use and maintenance. The scope of the chapter will not apply to the following enumerated systems and their components.

1202.3 Applications. This code shall not apply to the following items (reference standards for some of which appear in Chapter 17):

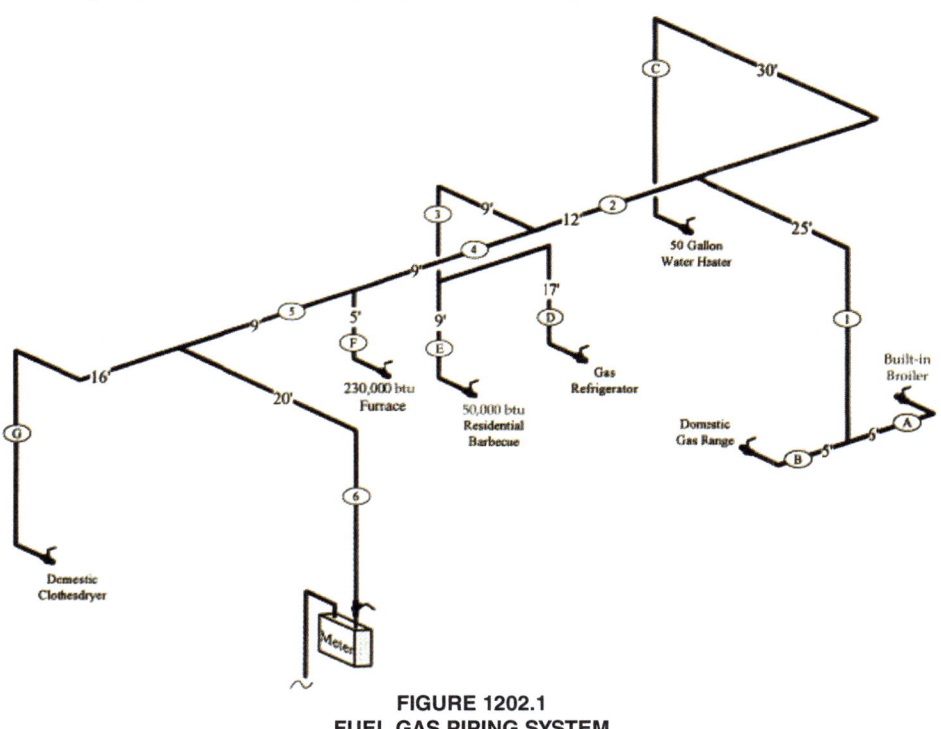

FIGURE 1202.1
FUEL GAS PIPING SYSTEM

FUEL GAS PIPING

(1) Portable LP-Gas appliances and equipment of all types that are not connected to a fixed fuel piping system.
(2) Installation of appliances such as brooders, dehydrators, dryers, and irrigation equipment used for agricultural purposes.
(3) Raw material (feedstock) applications except for piping to special atmosphere generators.
(4) Oxygen-fuel gas cutting and welding systems.
(5) Industrial gas applications using such gases as acetylene and acetylenic compounds, hydrogen, ammonia, carbon monoxide, oxygen, and nitrogen.
(6) Petroleum refineries, pipeline compressor or pumping stations, loading terminals, compounding plants, refinery tank farms, and natural gas processing plants.
(7) Large integrated chemical plants or portions of such plants where flammable or combustible liquids or gases are produced by chemical reactions or used in chemical reactions.
(8) LP-Gas installations at utility gas plants.
(9) Liquefied natural gas (LNG) installations.
(10) Fuel gas piping in electric utility power plants.
(11) Proprietary items of equipment, apparatus, or instruments such as gas-generating sets, compressors, and calorimeters.
(12) LP-Gas equipment for vaporization, gas mixing, and gas manufacturing.
(13) LP-Gas piping for buildings under construction or renovations that is not to become part of the permanent building piping system—that is, temporary fixed piping for building heat.
(14) Installation of LP-Gas systems for railroad switch heating.
(15) Installation of LP-Gas and compressed natural gas (CNG) systems on vehicles.
(16) Gas piping, meters, gas-pressure regulators, and other appurtenances used by the serving gas supplier in distribution of gas, other than undiluted LP-Gas.
(17) Building design and construction, except as specified herein.
(18) Fuel gas systems on recreational vehicles manufactured in accordance with NFPA 1192.
(19) Fuel gas systems using hydrogen as a fuel.
(20) Construction of appliances. [NFPA 54:1.1.1.2]

1203.0 Inspection.

1203.1 Inspection Notification. Upon completion of the installation, alteration, or repair of gas piping, and prior to the use thereof, the Authority Having Jurisdiction shall be notified that such gas piping is ready for inspection.

1203.2 Excavation. Excavations required for the installation of underground piping shall be kept open until the piping has been inspected and approved. Where such piping is covered or concealed before such approval, it shall be exposed upon the direction of the Authority Having Jurisdiction.

1203.3 Type of Inspections. The Authority Having Jurisdiction shall make the following inspections and either shall approve that portion of the work as completed or shall notify the permit holder wherein the same fails to be in accordance with this code.

1203.3.1 Rough Piping Inspection. This inspection shall be made after gas piping authorized by the permit has been installed and before such piping has been covered or concealed or fixture or appliance has been attached thereto. This inspection shall include a determination that the gas piping size, material, and installation meet the requirements of this code.

Although a pressure test is not specifically required in the rough inspection, it is necessary that the plumber tests the system for leaks before any work is covered or concealed. Otherwise, a great amount of time and money will be wasted to repair a leak once walls are covered and painted.

1203.3.2 Final Piping Inspection. This inspection shall be made after piping authorized by the permit has been installed, and after portions, thereof that are to be covered or concealed are so concealed, and before fixture, appliance, or shutoff valve has been attached thereto. This inspection shall comply with Section 1213.1. Test gauges used in conducting tests shall be in accordance with Section 318.0.

These tests shall be made using only air, carbon dioxide (CO_2), or nitrogen pressure in accordance with Section 1213.3 and shall be made in the presence of the Authority Having Jurisdiction (AHJ). All necessary apparatus for conducting tests shall be furnished by the permit holder.

It is necessary that a thoroughly accurate determination of line tightness be made within a reasonable period. This can only be done if the gauge recording the line test pressure is sensitive, accurate and of such graduations that small leaks can be detected quickly. Be sure to consult Section 318.0, Test Gauges, for the proper gauge requirement. Detailed pressure testing and inspection guidelines are contained in Section 1213.0.

1203.4 Inspection Waived. In cases where the work authorized by the permit consists of a minor installation of additional piping to piping already connected to a gas meter, the preceding inspections shall be permitted to be waived at the discretion of the Authority Having Jurisdiction. In this event, the Authority Having Jurisdiction shall make such inspection as deemed advisable to be assured that the work has been performed in accordance with the intent of this code.

Minor changes to an existing gas piping installation, such as extending an existing line for the relocation of a gas appliance, may be performed without requiring a pressure test of the entire system if approved by the AHJ. The AHJ may require an approved gas detector or a leak detection fluid to test for leaks (see Section 1213.4.1).

1204.0 Certificate of Inspection.

1204.1 Issuance. Whereupon final piping inspection, the installation is found to be in accordance with the provisions of this code, a certificate of inspection shall be permitted to be issued by the Authority Having Jurisdiction.

1204.2 Gas Supplier. A copy of the certificate of such final piping inspection shall be issued to the serving gas supplier supplying gas to the premises.

1204.3 Unlawful. It shall be unlawful for a serving gas supplier, or person is furnishing gas, to turn on or cause to be turned on, a fuel gas or a gas meter or meters, until such certificate of final inspection, as herein provided, has been issued.

🔧 Because of the hazards associated with fuel gas, it is extremely important to ensure that the gas system has been inspected and tested and that it is safe to turn on the gas supply to the premises. The final inspection ensures that the gas piping has not been damaged during construction and that all openings have been capped or plugged. After the certificate of inspection is approved, some jurisdictions will install a gas tag at the gas meter. This tag indicates that the gas piping system has been inspected, tested and complies with the Uniform Plumbing Code (UPC). The gas supplier would then be allowed to install the gas service line (if not already installed) and the gas meter and begin gas service.

1205.0 Authority to Render Gas Service.

1205.1 Authorized Personnel. It shall be unlawful for a person, firm, or corporation, excepting an authorized agent or employee of a person, firm, or corporation engaged in the business of furnishing or supplying gas and whose service pipes supply or connect with the particular premises, to turn on or reconnect gas service in or on a premises where and when gas service is, at the time, not being rendered.

1205.2 Outlets. It shall be unlawful to turn on or connect gas in or on the premises unless outlets are securely connected to gas appliances or capped or plugged with screw joint fittings.

🔧 Only the serving gas supplier or public utility has the right to turn on the gas service. Before the gas is turned on or reconnected to the building, all the gas outlets shall be connected to appliances and any unused outlets shall be capped or plugged. A gas valve cannot be used as a plug or cap due to the potential of leaking or accidentally being turned on. All venting from the appliances, if applicable, shall also be installed. Following the requirements in this section will provide a safe environment for the appliances to function properly without causing any danger to the occupants or the building.

1206.0 Authority to Disconnect.

1206.1 Disconnection. The Authority Having Jurisdiction or the serving gas supplier is hereby authorized to disconnect gas piping or appliance or both that shall be found not to be in accordance with the requirements of this code or that are found defective and in such condition as to endanger life or property.

🔧 When a hazard is discovered, the AHJ or the gas supplier has the authority to disconnect the faulty gas appliance, the gas piping or both. Examples of faulty appliances might be a safety control valve that is leaking or a crack in the heat exchanger of a forced-air furnace that could allow products of combustion (carbon monoxide) to spill into the building.

1206.2 Notice. Where such disconnection has been made, a notice shall be attached to such gas piping or appliance or both that shall state the same has been disconnected, together with the reasons thereof.

1206.3 Capped Outlets. It shall be unlawful to remove or disconnect gas piping or gas appliance without capping or plugging with a screw joint fitting, the outlet from which said pipe or appliance was removed. Outlets to which gas appliances are not connected shall be left capped and gastight on a piping system that has been installed, altered, or repaired.

Exception: Where an approved listed quick-disconnect device is used.

1207.0 Temporary Use of Gas.

1207.1 General. Where temporary use of gas is desired, and the Authority Having Jurisdiction deems the use necessary, a permit shall be permitted to be issued for such use for a period not to exceed that designated by the Authority Having Jurisdiction, provided that such gas piping system otherwise is in accordance with the requirements of this code regarding material, sizing, and safety.

🔧 There are times when a temporary gas system is needed, such as to supply heat during a remodel of an existing space or when a business is going to occupy a space on a temporary basis. The AHJ may allow a gas system to be used for a specific amount of time when deemed necessary, provided that all requirements regarding material, sizing, installation and safety are enforced.

1208.0 Gas Piping System Design, Materials, and Components.

1208.1 Installation of Piping System. Where required by the Authority Having Jurisdiction, a piping sketch or plan shall be prepared before proceeding with the installation. The plan shall show the proposed location of piping, the size of different branches, the various load demands, and the location of the point of delivery. [NFPA 54:5.1.1]

1208.1.1 Addition to Existing System. When additional appliances are being connected to a gas piping system, the existing piping shall be checked to determine whether it has adequate capacity. If inadequate, the existing system shall be enlarged as required, or separate gas piping of adequate capacity shall be provided. [NFPA 54:5.1.2.1 – 5.1.2.2]

🔧 The installation of any fuel gas system or an addition to a fuel gas system, large or small, will require a permit. To obtain the permit, a plan or design must be submitted to the building department. The plan must show the system layout, sizing parameters and, of course, the appliances to be installed. In large systems, the design will be accomplished by a plumbing design professional; however, for smaller applications, a journeyman or foreman may sketch the system. **Figure 1202.1** represents one such sketch. The plans check personnel of the building department will then review the plans and verify that the fuel gas system is code compliant.

1208.2 Provision for Location of Point of Delivery. The location of the point of delivery shall be acceptable to the serving gas supplier. [NFPA 54:5.2]

🔧 The service gas supplier will determine the point of delivery based upon company policies, the regulations of the U.S. Department of Transportation or the state public utility commission. State law may also govern the point of delivery for LP-gas systems. The installer must know the point of delivery so that the length of the piping system can be properly determined in order to provide a sufficient volume of gas for proper appliance operation. Failure to do so may result in an inadequate piping system or one that does not coincide with the point of delivery. This may require a re-piping of the system. **Figure 1208.2** is an example of the service gas supplier's notation on a set of plans.

FUEL GAS PIPING

1208.3 Interconnections Between Gas Piping Systems. Where two or more meters, or two or more service regulators where meters are not provided, are located on the same premises and supply separate users, the gas piping systems shall not be interconnected on the outlet side of the meters or service regulators. [NFPA 54:5.3.1]

🔧 Each separate fuel gas system must be served by a separate meter or service. Interconnected gas systems can cause a potential safety problem. Maintenance, emergency and utility personnel would have no way of knowing which services must be turned off to take the system safely out of service if these systems were interconnected.

Interconnected gas systems are not to be confused with a natural gas supplier providing multiple meters or service regulators for the purpose of capacity to one user. This situation can occur when the supplier has a limitation in the fuel supply or the meter or regulator capacity will be exceeded, especially for large volume users. In this case, the outlet piping systems are interconnected at the immediate outlet of the meter or service regulator assembly. Such an interconnection falls under the meter or service regulator assembly, which is governed by the U.S. Department of Transportation.

1208.3.1 Interconnections for Standby Fuels. Where a supplementary gas for standby use is connected downstream from a meter or a service regulator where a meter is not provided, equipment to prevent backflow shall be installed. A three-way valve installed to admit the standby supply, and at the same time shut off the regular supply, shall be permitted to be used for this purpose. [NFPA 54:5.3.2.1 – 5.3.2.2]

🔧 As with water supply backflow, there should be no situation where a cross connection or a backflow of gasses can occur. Different types of gas supplies must not be mixed. If a building is served by a natural gas pipeline and uses a propane system for standby, it is most important that propane not be fed into the natural gas main and that natural gas not be fed into the propane storage tanks. This would destroy the integrity of the fuel supply and could cause combustion problems. A properly approved three-way valve as shown in **Figure 1208.3.1** may be used for this purpose instead of single valves with backflow devices.

1208.4 Sizing of Gas Piping Systems. Gas piping systems shall be of such size and so installed as to provide a supply gas sufficient to meet the maximum demand and supply gas to each appliance inlet at not less than the minimum supply pressure required by the appliance. [NFPA 54:5.4.1]

1208.4.1 Maximum Gas Demand. The volumetric flow rate of gas to be provided shall be the sum of the maximum inputs of the appliances served. The volumetric flow rate of gas to be provided shall be adjusted for altitude where the installation is above 2000 feet (610 m). [NFPA 54:5.4.2.1 – 5.4.2.2]. Where the input rating is not indicated, the gas supplier, appliance manufacturer, or a qualified agency shall be contacted, or the rating from Table 1208.4.1 shall be used for estimating the volumetric flow rate of gas to be supplied.

The total connected hourly load shall be used as the basis for piping sizing, assuming all the appliances are operating at full capacity simultaneously.

FIGURE 1208.3.1
THREE-WAY GAS VALVE

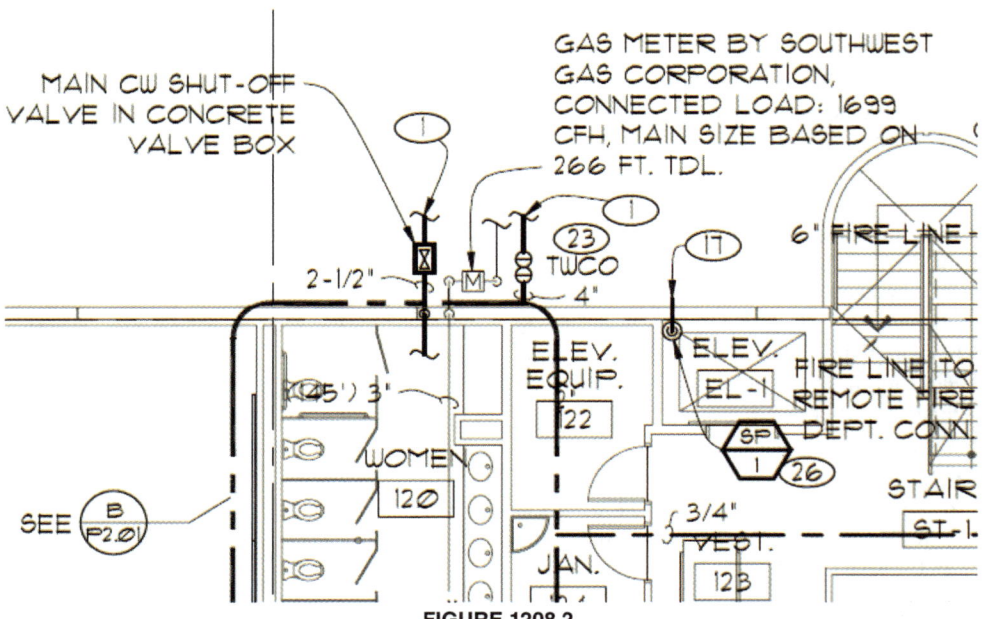

FIGURE 1208.2
LOCATION OF POINT OF DELIVERY DESIGNATED BY GAS SUPPLIER

Exception: Sizing shall be permitted to be based upon established load diversity factors. [NFPA 54:5.4.2.3]

1208.4.2 Sizing Methods. Gas piping shall be sized in accordance with one of the following:

(1) Pipe sizing tables or sizing equations in this chapter.

(2) Other approved engineering methods acceptable to the Authority Having Jurisdiction.

(3) Sizing tables included in a listed piping system manufacturer's installation instructions. [NFPA 54:5.4.3]

🔧 Sizing methods are described in Section 1215.0. The statement "other approved engineering methods" refers to calculations for sizing the system completed by a plumbing design engineer using equations other than those contained in Section 1215.3. These calculations must be approved by the AHJ.

TABLE 1208.4.1
APPROXIMATE GAS INPUT FOR TYPICAL APPLIANCES
[NFPA 54: TABLE A 5.4.2.1]

APPLIANCE	INPUT (Btu/h approx.)
Space Heating Units	
Warm air furnace	
Single family	100 000
Multifamily, per unit	60 000
Hydronic boiler	
Single family	100 000
Multifamily, per unit	60 000
Space and Water Heating Units	
Hydronic boiler	
Single-family	120 000
Multifamily, per unit	75 000
Water Heating Appliances	
Water heater, automatic storage	
30 to 40 gallon tank	35 000
Water heater, automatic storage	
50 gallon tank	50 000
Water heater, automatic instantaneous	
Capacity at 2 gallons per minute	142 800
Capacity at 4 gallons per minute	285 000
Capacity at 6 gallons per minute	428 400
Water heater, domestic, circulating or side-arm	35 000
Cooking Appliances	
Range, freestanding, domestic	65 000
Built-in oven or broiler unit, domestic	25 000
Built-in top unit, domestic	40 000
Other Appliances	
Refrigerator	3000
Clothes dryer, Type 1 (domestic)	35 000
Gas fireplace direct vent	40 000
Gas log	80 000
Barbecue	40 000
Gaslight	2500

For SI units: 1000 British thermal units per hour = 0.293 kW

The statement "sizing tables from a listed piping manufacturer's installation instructions" refers to piping materials that are not included in sizing Tables 1215.2(1) through 1215.2(36). This includes corrugated stainless steel tubing (CSST). The tables included in this chapter for CSST refer to the equivalent hydraulic diameter (EHD) rather than the pipe size. The CSST manufacturer's sizing tables will refer to a specific pipe size. This is due to the fact that the different manufacturers of CSST use slightly different diameters and fittings for their pipe. This will be further discussed under Section 1208.6.4.4.

1208.4.3 Allowable Pressure Drop. The design pressure loss in any piping system under maximum probable flow conditions, from the point of delivery to the inlet connection of the appliance, shall be such that the supply pressure at the appliance is greater than or equal to the minimum pressure required by the appliance. [NFPA 54:5.4.4]

🔧 Gas piping systems must supply the volume of gas required by each appliance at a pressure within the design range established by the appliance manufacturer. The sizing tables take pressure loss into consideration and most systems will fall within the pressure loss values of the tables. If the system does not fall within the range of the tables, then engineered calculations must be used or the use of a higher inlet pressure should be considered.

The gas inlet pressure to the connected equipment must also remain within the manufacturer's design limits when the equipment "cycles off." Gas piping systems that operate at pressures exceeding the connected gas appliance's rated inlet pressure or that deliver gas at a pressure that varies outside the equipment manufacturer's design inlet pressure limits must incorporate pressure regulators into the system so that gas delivered to the equipment will be within the design pressure range of the appliance. See Section 1208.8 for line gas pressure regulator requirements.

1208.5 Maximum Design Operating Pressure. The maximum design operating pressure for piping systems located inside buildings shall not exceed 5 psi (34 kPa) unless one or more of the following conditions are met:

(1) The piping system is welded.

(2) The piping is located in a ventilated chase or otherwise enclosed for protection against accidental gas accumulation.

(3) The piping is located inside buildings or separate areas of buildings used exclusively for one of the following:

(a) Industrial processing or heating

(b) Research

(c) Warehousing

(d) Boiler or mechanical rooms

(4) The piping is a temporary installation for buildings under construction.

(5) The piping serves appliances or equipment used for agricultural purposes.

(6) The piping system is an LP-Gas piping system with a design operating pressure greater than 20 psi (138 kPa) and complies with NFPA 58. [NFPA 54:5.5.1]

🔧 See **Figure 1208.5**.

FUEL GAS PIPING

**FIGURE 1208.5
STEEL GAS PIPING THREADED AND WELDED**

1208.5.1 LP-Gas Systems. LP-Gas systems designed to operate below -5°F (-21°C) or with butane or a propane-butane mix shall be designed to either accommodate liquid LP-Gas or to prevent LP-Gas vapor from condensing back into a liquid. [NFPA 54:5.5.2]

1208.6 Acceptable Piping Materials and Joining Methods. Materials used for piping systems shall either comply with the requirements of this chapter or be acceptable to the Authority Having Jurisdiction. [NFPA 54:5.6.1.1]

1208.6.1 Used Materials. Pipe, fittings, valves, or other materials shall not be used again unless they are free of foreign materials and have been ascertained to be approved for the service intended. [NFPA 54:5.6.1.2]

1208.6.2 Other Materials. Material not covered by the standards specifications listed herein shall meet the following criteria:

(1) Be investigated and tested to determine that it is safe and suitable for the proposed service.

(2) Be recommended for that service by the manufacturer.

(3) Be acceptable to the Authority Having Jurisdiction. [NFPA 54:5.6.1.3]

Piping materials for fuel gas, new or used, shall be compatible with the gas used and listed for that use. Used materials should not be used unless absolutely necessary. Those wishing to use materials not covered here should follow the requirements for alternate materials in Chapter 3 for approval to use the material.

1208.6.3 Metallic Pipe. Cast-iron pipe shall not be used. [NFPA 54:5.6.2.1]

1208.6.3.1 Steel and Wrought-Iron Pipe. Steel and wrought-iron pipe shall be at least of standard weight (Schedule 40) and shall comply with one of the following standards:

(1) ASME B36.10

(2) ASTM A53

(3) ASTM A106 [NFPA 54:5.6.2.2]

1208.6.3.2 Copper and Copper Alloy Pipe. Copper and copper alloy pipe shall not be used if the gas contains more than an average of 0.3 grains of hydrogen sulfide per 100 standard cubic feet (scf) of gas (0.7 mg/100 L).

Threaded copper, copper alloy, or aluminum alloy pipe shall not be used with gases corrosive to such material. [NFPA 54:5.6.2.3 – 5.6.2.4]

Copper for natural gas systems has been used in North America and throughout the world for many years (see **Figure 1208.6.3.2a**); however, its use in the western part of the United States was prohibited in the 1950s. This was due to the fact that systems had been failing in this region because of corrosion caused by the "sulfidation" of the copper pipe and copper components of gas meters. Studies were conducted and found that hydrogen sulfide, occurring naturally in the natural gas supplied, was in high concentrations in these areas and was the cause of this corrosion.

Natural gas is a mixture of many hydrocarbon gases and other elements. One of those elements is hydrogen sulfide (H_2S). Hydrogen sulfide will react with copper, creating a coating of black or dark brown "dust," referred to as "sulfidation" (see **Figure 1208.6.3.2b**). This sulfidation is perceived to cause two possible corrosion problems in copper natural gas systems:

- Continual flaking of the copper sulfide, thinning the pipe and eventually causing pinholes and, thus, leaks; and
- Continual flaking of the copper sulfide, causing the flakes themselves to fall and be carried into the appliance and possibly block burners or be deposited into gas valves, causing the valves to foul.

Two extensive studies have been done on the sulfidation of copper in natural gas installations that have shed light on these two problems. The first was a study completed in 1996 in the United Kingdom entitled, Safety Aspects of the Effects of Hydrogen Sulfide Concentrations in Natural Gas, prepared by WS Atkins Environment for the British government. The second was a study completed in 2002 entitled, Copper Tubing in a Natural Gas Environment, work performed by Wayne T. Yuen, engineer for a joint effort by the Copper Development Association and the Southern California Gas Company. The studies found that the first problem, the thinning of the copper wall and eventual failure or pin holing of the pipe, is unlikely ever to occur. Although the copper sulfide coating may flake or fall off the wall of the pipe and could continually do so, "the corrosion rates suggest failure of the copper pipe due to sulfidation should not represent a concern. Even at high concentrations [of H_2S] the pipe would have a life expectancy of 100 years."

**FIGURE 1208.6.3.2A
COPPER TUBE FOR NATURAL GAS USE**

**FIGURE 1208.6.3.2B
SULFIDATION OF COPPER PIPE**

The second problem with sulfidation, the possible continual flaking and thus the plugging of burners and the fouling of valves, is the serious concern with these copper systems. Both studies found evidence of flaking. The UK study stated that "Approximately 21,000 properties in Great Britain are estimated to be affected by the presence of hydrogen sulfide in gas. [Ninety percent] 90% are estimated to be associated with blockage of burners and about 5 percent with central heating boiler gas valve failure." However, they also stated, "Only one accident has been reported due to gas valve failure caused by sulfidation."

Both studies also concluded that the elimination of the high content of H2S would eliminate the problems in the use of copper pipe for natural gas installations. The requirement in this section for the average of 0.3 grains of H_2S per 100 scf of gas (0.7 mg/100 L) falls within the safe limits for natural gas. In recent years the natural gas industry has been removing H_2S from natural gas along with water vapor, which can also cause problems in the system. To be sure of the H_2S content of the natural gas used in his or her area, a plumber should consult the serving gas supplier. It can provide the content of all elements contained in the natural gas supplied. This also applies to Section 1208.6.4.2 below. See **Figure 1208.6.3.2c** for a copper natural gas installation.

1208.6.3.3 Aluminum Alloy Pipe. Aluminum alloy pipe shall comply with ASTM B241 (except that the use of alloy 5456 is prohibited) and shall be marked at each end of each length indicating compliance. Aluminum alloy pipe shall be coated to protect against external corrosion where it is in contact with masonry, plaster, insulation or is subject to repeated wettings by such liquids as water, detergents, or sewage. [NFPA 54:5.6.2.5]

Aluminum alloy pipe shall not be used in exterior locations or underground. [NFPA 54:5.6.2.6]

Strong acids and strong bases readily corrode aluminum and its alloys. The assistance of a metallurgist or qualified corrosion specialist is strongly recommended in selecting an aluminum alloy because the corrosion behavior of different aluminum alloys varies significantly. Aluminum pipe and tubing are not used extensively for gas piping in the United States, but they are used in some other countries. Aluminum tubing is used for piping that is an integral part of appliances; however, such piping is not covered by this code (see **Figure 1208.6.3.3**).

1208.6.4 Metallic Tubing. Seamless copper, aluminum alloy, or steel tubing shall not be used with gases corrosive to such material. [NFPA 54:5.6.3]

1208.6.4.1 Steel Tubing. Steel tubing shall comply with ASTM A254. [NFPA 54:5.6.3.1]

1208.6.4.2 Copper and Copper Alloy Tubing. Copper and copper alloy tubing shall not be used where the gas contains more than an average of 0.3 grains of hydrogen sulfide per 100 scf of gas (0.7 mg/100 L). Copper tubing shall comply with standard Type K or L of ASTM B88 or ASTM B280. [NFPA 54:5.6.3.2]

1208.6.4.3 Aluminum Alloy Tubing. Aluminum alloy tubing shall comply with ASTM B210 or ASTM B241. Aluminum alloy tubing shall be coated to protect against external corrosion where it is in contact with masonry, plaster, insulation, or is subject to repeated wettings by such liquids as water, detergent, or sewage. Aluminum alloy tubing shall not be used in exterior locations or underground. [NFPA 54:5.6.3.3]

1208.6.4.4 Corrugated Stainless Steel Tubing. Corrugated stainless steel tubing shall be listed in accordance with CSA LC-1. [NFPA 54:5.6.3.4]

CSST consists of a continuous, flexible, stainless steel pipe with an exterior PVC covering. The piping is produced in coils that are air-tested for leaks (see **Figure 1208.6.4.4**). It is most often installed in a central manifold configuration (also called "parallel configuration") with "home run" lines that extend to gas appliances. Flexible gas piping is lightweight and requires fewer connections than traditional gas piping because it can be easily bent and routed around obstacles.

There are several CSST manufacturers. Each has a slightly different Equivalent Hydraulic Diameter noted in the sizing charts as 'flow designation' or different joining method. Be sure to consult the specific manufacturer's installation instructions. Always use the fittings recommended by the manufacturer. Never substitute fittings from one manufacturer to another unless specifically allowed by the manufacturer.

FIGURE 1208.6.3.2C
COPPER NATURAL GAS SYSTEM IN CANADA

FIGURE 1208.6.3.3
ALUMINUM TUBING

FUEL GAS PIPING

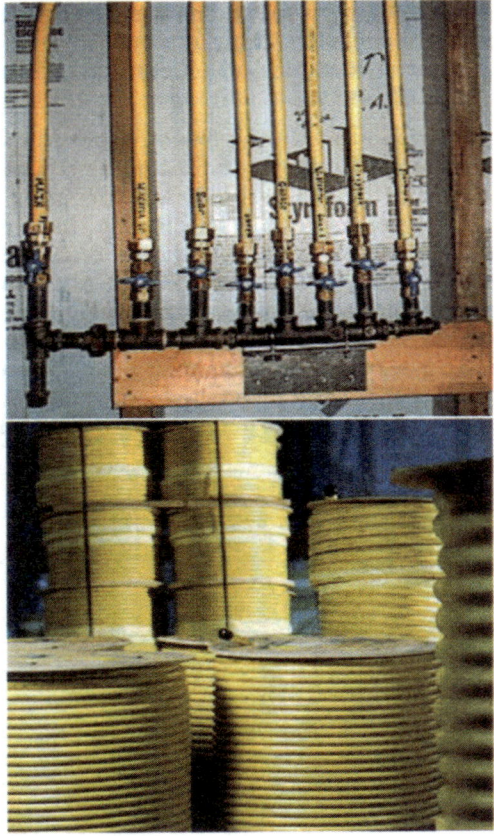

**FIGURE 1208.6.4.4
CORRUGATED STAINLESS STEEL
BY GASTITE DIVISION TITEFLEX CORPORATION**

1208.6.5 Plastic Pipe, Tubing, and Fittings. Polyethylene plastic pipe, tubing, and fittings used to supply fuel gas shall conform to ASTM D2513. Pipe to be used shall be marked "gas" and "ASTM D2513." Polyvinyl chloride (PVC) and chlorinated polyvinyl chloride (CPVC) plastic pipe, tubing, and fittings shall not be used to supply fuel gas. [NFPA 54:5.6.4.1.1 – 5.6.4.1.3]

🔧 As stated, plastic gas piping may only be used outside of the building foundation and in underground installations. Plastic piping normally used for gas systems is polyethylene pipe. Polyethylene pipe is by far the most extensively used pipe for gas piping systems. Methods of joining are discussed in Section 1208.6.12 (see **Figure 1208.6.5**).

1208.6.6 Regulator Vent Piping. Plastic pipe and fittings used to connect regulator vents to remote vent terminations shall be PVC conforming to UL 651(Schedule 40 and 80). PVC vent piping shall not be installed indoors. [NFPA 54:5.6.4.2]

1208.6.7 Anodeless Risers. Anodeless risers shall comply with the following:

(1) Factory-assembled anodeless risers shall be recommended by the manufacturer for the gas used and shall be leak-tested by the manufacturer in accordance with written procedures.

🔧 Anodeless risers are used to make the transition between underground PE pipe or tubing and metal pipe aboveground. As PE pipe must be installed below ground, risers are commonly used to connect the underground PE to above-ground piping materials. Anodeless risers are available as factory-assembled units and field-assembled kits. Anodeless risers are made from PE pipe inside a protective metal sheath, usually Schedule 40 steel pipe. The metal is protected from corrosion by a factory-applied coating and a separate anode is not required; hence the name "anodeless" (see **Figure 1208.6.7a**). Factory-assembled risers usually have a 90-degree bend at the PE connection end and come in several lengths depending on the burial depth of the PE pipe.

(2) Service head adapters and field-assembled anodeless risers incorporating service head adapters shall be recommended by the manufacturer for the gas used and shall be design-certified to be in accordance with the requirements of Category I of ASTM D2513. The manufacturer shall provide the user qualified installation instructions.

🔧 The service head adapter is a transition riser, normally flexible, that will allow easy connection of the service meter or other transition needs. They will come with UV protective sheathing and a swivel head for connection to the metal piping (see **Figure 1208.6.7b**).

(3) The use of plastic pipe, tubing, and fittings in undiluted LP-Gas piping systems shall be in accordance with NFPA 58. [NFPA 54:5.6.4.3(3)]

1208.6.8 Workmanship and Defects. Gas pipe, tubing, and fittings shall be clear and free from cutting burrs and defects in structure or threading, and shall be thoroughly brushed and chip and scale blown. Defects in pipe, tubing, and fittings shall not be repaired. Defective pipe, tubing, and fittings shall be replaced. [NFPA 54:5.6.5]

🔧 This is a restatement of Section 309.0, Workmanship, which also applies to gas installations. Installers must inspect and clean piping materials before assembly to ensure compliance with the code. Small pieces of metal can clog the small clearances between moving parts and in orifices. Repairing defective components is not an appropriate remedy.

1208.6.9 Protective Coating. Where in contact with material or atmosphere exerting a corrosive action, metallic piping and fittings coated with a corrosion-resistant material shall be used. External or internal coatings or linings used on piping or components shall not be considered as adding strength. [NFPA 54:5.6.6]

🔧 Materials commonly used for protective coating include paint, polyethylene jacketing over a mastic primer and fusion-bonded epoxy. Generally, paint is used on above-grade applications and the others on pipe that is installed underground.

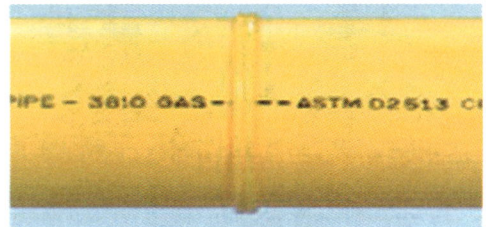

**FIGURE 1208.6.5
PLASTIC HIGH-DENSITY POLYETHYLENE PIPE
FOR NATURAL GAS**

1208.6.10 Metallic Pipe Threads. Metallic pipe and fitting threads shall be taper pipe threads and shall comply with ASME B1.20.1. [NFPA 54:5.6.7.1]

1208.6.10.1 Damaged Threads. Pipe with threads that are stripped, chipped, corroded, or otherwise damaged shall not be used. Where a weld opens during the operation of cutting or threading, that portion of the pipe shall not be used. [NFPA 54:5.6.7.2]

1208.6.10.2 Number of Threads. Field threading of metallic pipe shall be in accordance with Table 1208.6.10.2. [NFPA 54:5.6.7.3]

1208.6.10.3 Thread Joint Compounds. Thread joint compounds shall be resistant to the action of LP-Gas or to any other chemical constituents of the gases to be conducted through the piping. [NFPA 54:5.6.7.4]

1208.6.11 Metallic Piping Joints and Fittings. The type of piping joint used shall be suitable for the pressure and temperature conditions and shall be selected giving consideration to joint tightness and mechanical strength under the service conditions. The joint shall be able to sustain the maximum end force due to the internal pressure and any additional forces due to temperature expansion or contraction, vibration, fatigue, or the weight of the pipe and its contents. [NFPA 54:5.6.8]

1208.6.11.1 Pipe Joints. Pipe joints shall be threaded, flanged, brazed, welded, or press-connect fittings that comply with CSA LC-4. Where nonferrous pipe is brazed, the brazing materials shall have a melting point in excess of 1000°F (538°C). Brazing alloys shall not contain more than 0.05 percent phosphorus. [NFPA 54:5.6.8.1]

1208.6.11.2 Tubing Joints. Tubing joints shall be made with approved gas tubing fittings, be brazed with a material having a melting point in excess of 1000°F (538°C), or made by press-connect fittings that comply with CSA LC-4. Brazing alloys shall not contain more than 0.05 percent phosphorus. [NFPA 54:5.6.8.2]

1208.6.11.3 Flared Joints. Flared joints shall be used only in systems constructed from nonferrous pipe and tubing where experience or tests have demonstrated that the joint is approved for the conditions and where provisions are made in the design to prevent separation of the joints. [NFPA 54:5.6.8.3]

TABLE 1208.6.10.2
SPECIFICATIONS FOR THREADING METALLIC PIPE
[NFPA 54: TABLE 5.6.7.3]

IRON PIPE SIZE (inches)	APPROXIMATE LENGTH OF THREADED PORTION (inches)	APPROXIMATE NUMBER OF THREADS TO BE CUT
½	¾	10
¾	¾	10
1	⅞	10
1¼	1	11
1½	1	11
2	1	11
2½	1½	12
3	1½	12
4	1⅝	13

For SI units: 1 inch = 25.4 mm

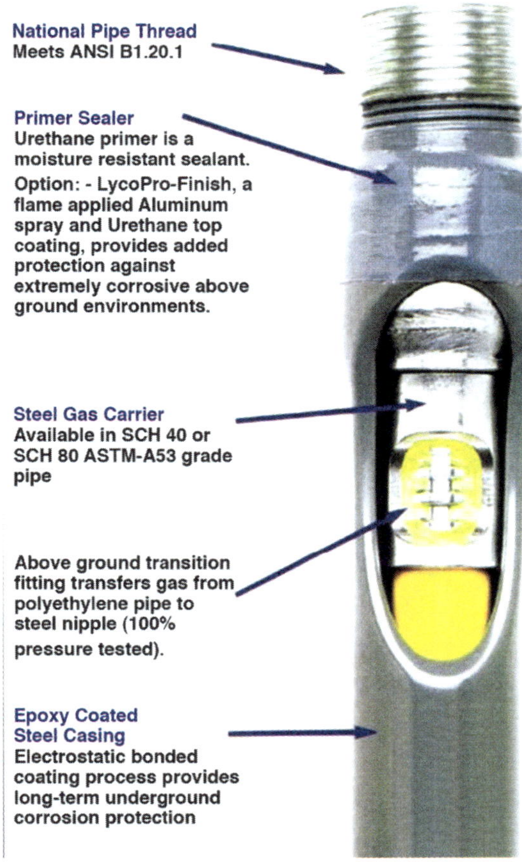

FIGURE 1208.6.7A
ANODELESS RISER

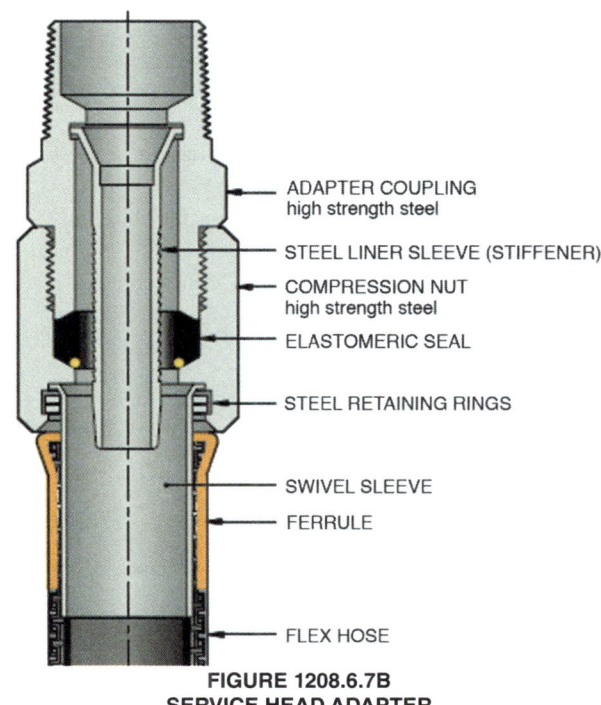

FIGURE 1208.6.7B
SERVICE HEAD ADAPTER

FUEL GAS PIPING

1208.6.11.4 Metallic Pipe Fittings (Including Valves, Strainers, Filters). Metallic pipe fittings shall comply with the following:

(1) Threaded fittings in sizes larger than 4 inches (100 mm) shall not be used unless approved by the Authority Having Jurisdiction.

(2) Fittings used with steel or wrought-iron pipe shall be steel, copper alloy, malleable iron, or cast-iron.

(3) Fittings used with copper or copper alloy pipe shall be copper or copper alloy.

(4) Fittings used with aluminum alloy pipe shall be of aluminum alloy.

(5) Cast-iron fittings shall comply with the following:

(a) Flanges shall be permitted.

(b) Bushings shall not be used.

(c) Fittings shall not be used in systems containing flammable gas-air mixtures.

(d) Fittings in sizes 4 inches (100 mm) and larger shall not be used indoors unless approved by the Authority Having Jurisdiction.

(e) Fittings in sizes 6 inches (150 mm) and larger shall not be used unless approved by the Authority Having Jurisdiction.

(6) Aluminum alloy fitting threads shall not form the joint seal.

(7) Zinc-aluminum alloy fittings shall not be used in systems containing flammable gas-air mixtures.

(8) Special fittings such as couplings; proprietary-type joints; saddle tees; gland-type compression fittings; and flared, flareless, or compression-type tubing fittings shall be as follows:

(a) Used within the fitting manufacturer's pressure-temperature recommendations.

(b) Used within the service conditions anticipated with respect to vibration, fatigue, thermal expansion, or contraction.

(c) Installed or braced to prevent separation of the joint by gas pressure or external physical damage.

(d) Approved by the Authority Having Jurisdiction. [NFPA 54:5.6.8.4 (1-8)]

The use of different materials for fittings than the material for the piping shall not be allowed. For example, steel fittings shall not be used for aluminum alloy pipe and so on. Not only would this intermingling of materials look like shoddy workmanship but, in most cases, it will cause corrosion or at least a possible reduction in pipe size, depending on the materials used. Only properly listed and approved transition fittings may be used to join together different materials.

1208.6.12 Plastic Piping, Joints, and Fittings. Plastic pipe, tubing, and fittings shall be installed in accordance with the manufacturer's installation instructions. Section 1208.6.12.1 through Section 1208.6.12.4 shall be observed where making such joints. [NFPA 54:5.6.9]

The following description of heat fusion welding of polyethylene pipe is from Oxford Plastics Inc. (see **Figure 1208.6.12**).

The fundamental of heat fusion welding is to heat two HDPE pipe surfaces to an appropriate temperature, changing the resin's molecular structure to a pliable state, and then fuse them together by application of prescribed force until cooling occurs, returning the material to a crystalline state and creating one homogeneous pipe.

When fusion pressure is applied at the designated temperature and prescribed force, the molecules from each pipe surface end mix. As the joint cools, the molecules return to their crystalline form, the original interfaces have been removed, and the two pipes have become one continuous length. The end result is a fusion joint that is as strong as or stronger than the pipe itself, and this creates the leak-free joint that is one of the amazing strengths of HDPE pipe.

Generally speaking, HDPE pipe is Butt fused together using a "fusion welder." Welding machines vary depending on the Outside Diameter (OD) of the pipe to be welded. The pipe pieces are held axially by a clamping device to allow subsequent operations to take place. Large diameter pipes may require hoisting assistance such as an excavator or crane. Once the pipe is clamped, the pipe ends are "faced" with a machining tool to establish clean, parallel mating surfaces, perpendicular to the centerline of each pipe. A heating element or heating plate is inserted in between the faced ends, and the pipe is drawn together against the heating plate. A melt pattern that penetrates into the pipe ends is formed around both pipe ends. Once the correct melt temperature is reached, the heating plate is quickly removed, and the melt ends are drawn together with a specified force. The specified force on the joint must be continuous, and held until the joint cools. A small melt bead forms at the joint. At completion, the fused pipe is removed from the welding machine.

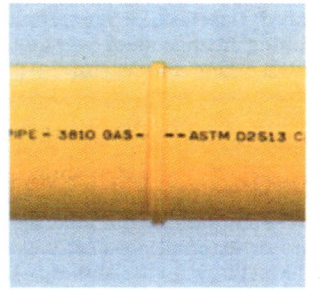 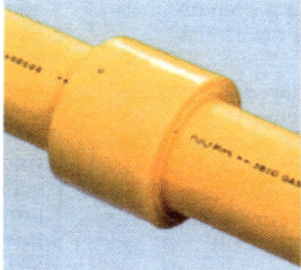 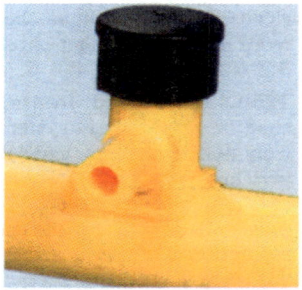

FIGURE 1208.6.12
POLYETHYLENE HEAT FUSION JOINTS (FROM LEFT TO RIGHT) BUTT FUSION JOINT, SOCKET FUSION JOINT, AND SADDLE FUSION JOINT

1208.6.12.1 Joint Design. The joint shall be designed and installed so that the longitudinal pullout resistance of the joint will be at least equal to the tensile strength of the plastic piping material. [NFPA 54:5.6.9(1)]

1208.6.12.2 Heat-Fusion Joint. Heat-fusion joints shall be made in accordance with qualified procedures that have been established and proven by test to produce gastight joints as strong as the pipe or tubing being joined. Joints shall be made with the joining method recommended by the pipe manufacturer. Heat-fusion fittings shall be marked "ASTM D2513." [NFPA 54:5.6.9(2)]

1208.6.12.3 Compression-Type Mechanical Joints. Where compression-type mechanical joints are used, the gasket material in the fitting shall be compatible with the plastic piping and with the gas distributed by the system. An internal tubular rigid stiffener shall be used in conjunction with the fitting. The stiffener shall be flush with the end of the pipe or tubing and shall extend not less than the outside end of the compression fitting where installed. The stiffener shall be free of rough or sharp edges and shall not be a forced fit in the plastic. Split tubular stiffeners shall not be used. [NFPA 54:5.6.9(3)]

Compression-type mechanical joints for plastic piping come in many forms and materials. The components are generally a body; a threaded compression nut; an elastomer seal ring or O-ring; a stiffener and, in some cases, a grip ring (see **Figure 1208.6.12.3a**). The seal and grip rings, when compressed, grip the outside of the pipe, affecting a pressure-tight seal and, in most designs, providing pullout resistance that exceeds the yield strength of the polyethylene pipe. It is important that the inside of the pipe wall be supported by the stiffener under the seal ring and under the gripping ring (if incorporated in the design) to avoid deflection of the pipe. A lack of this support could result in a loss of the seal or the gripping of the pipe for pullout resistance. This fitting style is normally used in service lines for gas or water pipe two inches (50 mm) Iron Pipe Size (IPS) and smaller. It is also important to consider that three categories of this type of joining device are available. The compression type is recommended to provide a seal only, the stab type provides a seal and some restraint from pullout and the mechanical bolt type provides a seal plus full pipe restraint against pullout.

Stab-type mechanical fittings also have many styles; however, the design concept, as illustrated in **Figure 1208.6.12.3b**, is similar in most styles. Internally, there are specially designed components, including an elastomer seal, such as an O-ring, and a gripping device to affect pressure sealing and pullout resistance capabilities. Self-contained stiffeners are included in this design. With this style fitting, the operator prepares the pipe ends, marks the stab depth on the pipe and "stabs" the pipe into the depth prescribed for the fitting being used. These fittings are available in sizes from one-half-inch Copper Tubing Size (CTS) through two inches IPS and are all of ASTM D 2513(2) Category I design, indicating seal and full restraint against pullout.

Mechanical bolt-type couplings for large diameter pipes are available to join polyethylene to polyethylene or other types of pipe such as PVC, steel and cast iron in sizes from 1¼ inches IPS and larger. Components for this style of fitting are shown in **Figure 1208.6.12.3c**. As with the mechanical compression fittings, these couplings work on the general principle of compressing an elastomeric gasket around each pipe end to be joined in order to form a seal. The gasket, when compressed against the outside of the pipe by tightening the bolts, produces a pressure seal.

1208.6.12.4 Liquefied Petroleum Gas Piping Systems. Plastic piping joints and fittings for use in LP-Gas piping systems shall be in accordance with NFPA 58. [NFPA 54:5.6.9(4)]

1208.6.13 Flange Specification. Flanges shall comply with Section 1208.6.13.1 through Section 1208.6.13.7.

1208.6.13.1 Cast Iron Flanges. Cast iron flanges shall be in accordance with ASME B16.1. [NFPA 54:5.6.10.1.1]

1208.6.13.2 Steel Flanges. Steel flanges shall be in accordance with the following:

(1) ASME B16.5 or

(2) ASME B16.47. [NFPA 54:5.6.10.1.2]

1208.6.13.3 Non-Ferrous Flanges. Non-ferrous flanges shall be in accordance with ASME B16.24. [NFPA 54:5.6.10.1.3]

1208.6.13.4 Ductile Iron Flanges. Ductile iron flanges shall be in accordance with ASME B16.42. [NFPA 54:5.6.10.1.4]

1208.6.13.5 Dissimilar Flange Connections. Raised-face flanges shall not be joined to flat-faced cast iron, ductile iron or nonferrous material flanges. [NFPA 54:5.6.10.2]

1208.6.13.6 Flange Facings. Standard facings shall be permitted for use under this code. Where 150 psi (1034 kPa) steel flanges are bolted to Class 125 cast-iron flanges, the raised face on the steel flange shall be removed. [NFPA 54:5.6.10.3]

1208.6.13.7 Lapped Flanges. Lapped flanges shall be used only aboveground or in exposed locations accessible for inspection. [NFPA 54:5.6.10.4]

1208.6.14 Flange Gaskets. The material for gaskets shall be capable of withstanding the design temperature and pressure of the piping system and the chemical constituents of the gas being conducted without change to its chemical and physical properties. The effects of fire exposure to the joint shall be considered in choosing the material. [NFPA 54:5.6.11]

1208.6.14.1 Flange Gasket Materials. Acceptable materials shall include the following:

(1) Metal (plain or corrugated)

(2) Composition

(3) Aluminum o-rings

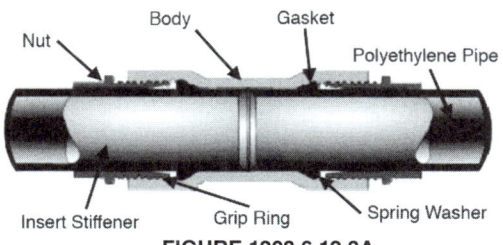

FIGURE 1208.6.12.3A
POLYETHYLENE PIPE COMPRESSION JOINT

FUEL GAS PIPING

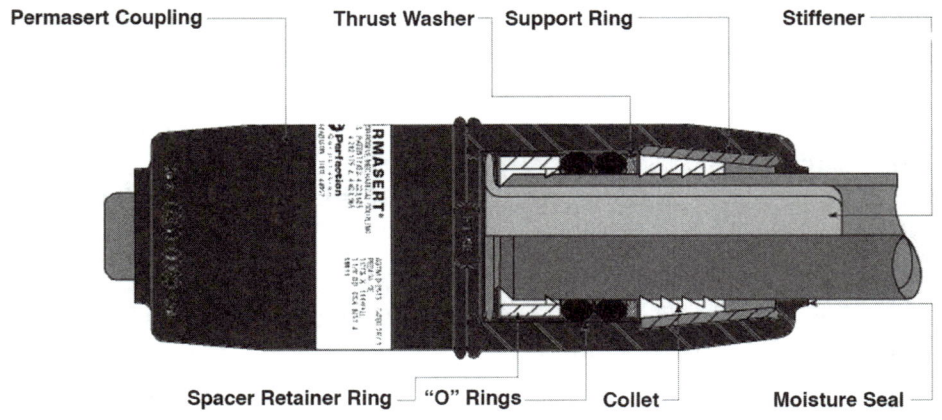

FIGURE 1208.6.12.3B
POLYETHYLENE PIPE STAB-TYPE JOINT

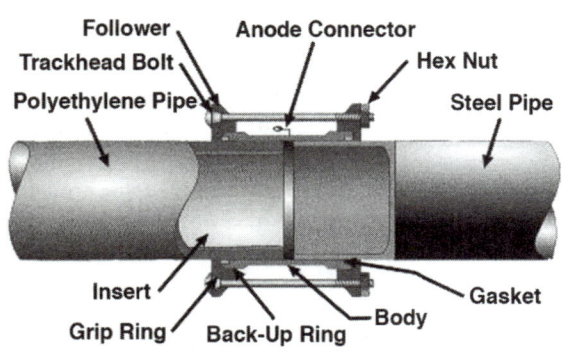

FIGURE 1208.6.12.3C
POLYETHYLENE PIPE MECHANICAL BOLT-TYPE JOINT

(4) Spiral-wound metal gaskets

(5) Rubber-faced phenolic

(6) Elastomeric [NFPA 54:5.6.11.1]

1208.6.14.2 Metallic Flange Gaskets. Metallic flange gaskets shall be in accordance with ASME B16.20. [NFPA 54:5.6.11.2.1]

1208.6.14.3 Non-Metallic Flange Gaskets. Non-metallic flange gaskets shall be in accordance with ASME B16.21. [NFPA 54:5.6.11.2.2]

1208.6.14.4 Full-Face Flange Gasket. Full-face flange gaskets shall be used with all non-steel flanges. [NFPA 54:5.6.11.3]

1208.6.14.5 Separated Flanges. When a flanged joint is separated, the gasket shall be replaced. [NFPA 54:5.6.11.2.4]

1208.7 Gas Meters. Gas meters shall be selected for the maximum expected pressure and permissible pressure drop. [NFPA 54:5.7.1]

The gas meter will almost always be installed by the gas service supplier or gas utility. As stated earlier, the supplier will designate the point of delivery and install the service piping. The gas supplier must also comply with the following requirements for gas meters. However, the apprentice or journeyman plumber will normally not install the meter or service piping unless working for the service company. The following requirements are very specific and need no comment.

1208.7.1 Location. Gas meters shall be located in ventilated spaces readily accessible for examination, reading, replacement, or necessary maintenance. [NFPA 54:5.7.2.1]

1208.7.1.1 Subject to Damage. Gas meters shall not be placed where they will be subjected to damage, such as adjacent to a driveway; under a fire escape; in public passages, halls, or where they will be subject to excessive corrosion or vibration. [NFPA 54:5.7.2.2]

1208.7.1.2 Extreme Temperatures. Gas meters shall not be located where they will be subjected to extreme temperatures or sudden extreme changes in temperature or in areas where they are subjected to temperatures beyond those recommended by the manufacturer. [NFPA 54:5.7.2.3]

See **Figure 1208.7.1.2**.

FIGURE 1208.7.1.2
INAPPROPRIATE LOCATION OF METER

1208.7.2 Supports. Gas meters shall be supported or connected to rigid piping so as not to exert a strain on the meters. Where flexible connectors are used to connect a gas meter to downstream piping at mobile homes in mobile home parks, the meter shall be supported by a post or bracket placed in a firm footing or by other means providing equivalent support. [NFPA 54:5.7.3]

1208.7.3 Meter Protection. Meters shall be protected against overpressure, backpressure, and vacuum. [NFPA 54:5.7.4]

1208.7.4 Identification. Gas piping at multiple meter installations shall be marked by a metal tag or other permanent means designating the building or the part of the building being supplied and attached by the installing agency. [NFPA 54:5.7.5]

🔧 See **Figure 1208.7.4**.

1208.8 Gas Pressure Regulators. A line pressure regulator or gas appliance pressure regulator, as applicable, shall be installed where the gas supply pressure exceeds that at which the branch supply line or appliances are designed to operate or vary beyond design pressure limits. [NFPA 54:5.8.1]

🔧 Gas pressure regulators are designed to maintain constant downstream gas pressure, regardless of change in the gas flow or in the conditions of the upstream pressure. Ideally, they restrict the flow from the inlet pressure side to balance exactly what the downstream gas pressure should be (see **Figure 1208.8a**). There are four basic areas where regulators are used:

1. A service pressure regulator is installed to reduce the main gas service pressure to the design pressure of the piping system. In **Figure 1208.7.4** the service pressure regulator is shown upstream of the meters.

2. A line gas pressure regulator may be used to reduce the gas piping main pressure to a lower pressure. A typical example would be in a hybrid pressure system – for example, a 2 psi (13.8 kPa) main pressure to an 11 inch (2.7kPa) water column pressure to the branches serving the appliances.

3. An appliance regulator may be needed at the outlet to the appliance which will reduce the branch pressure to the specific working pressure of the appliance (see **Figures 1208.8b** and **1208.8c**).

4. There may be a pressure regulator in the appliance itself, which will reduce the outlet pressure to the working pressure within the appliance.

Gas pressure regulators have a vent that allows the air above the regulator diaphragm to be displaced so the diaphragm can move. This air is vented out of the regulator. This air may also contain gas products and, if in large enough quantities or if the diaphragm breaks, could cause an explosion within the building. The regulator, if required, must be vented per the requirements of this section.

FIGURE 1208.7.4
MULTIPLE METER INSTALLATION AND METER TAG

1208.8.1 Listing. Line pressure regulators shall be listed in accordance with CSA Z21.80. [NFPA 54:5.8.2]

1208.8.2 Location. The gas pressure regulator shall be accessible for servicing. [NFPA 54:5.8.3]

1208.8.3 Regulator Protection. Pressure regulators shall be protected against physical damage. [NFPA 54:5.8.4]

1208.8.4 Venting of Line Pressure Regulators. Line pressure regulators shall comply with all of the following:

(1) An independent vent to the exterior of the building, sized in accordance with the regulator manufacturer's instructions, shall be provided where the location of a regulator is such that a ruptured diaphragm will cause a hazard. Where more than one regulator is at a location, each regulator shall have a separate vent to the outdoors or, if approved by the Authority Having Jurisdiction, the vent lines shall be permitted to be manifolded in accordance with accepted engineering practices to minimize backpressure in the event of diaphragm failure. Materials for vent piping shall be in accordance with Section 1208.6 through Section 1208.6.12.3.

Exception: A regulator and vent limiting means combination listed as complying with CSA Z21.80 shall be permitted to be used without a vent to the outdoors.

(2) The vent shall be designed to prevent the entry of water, insects, or other foreign materials that could cause a blockage.

(3) The regulator vent shall terminate at least 3 feet (914 mm) from a source of ignition.

(4) At locations where regulators might be submerged during floods, a special antiflood-type breather vent fitting shall be installed, or the vent line shall be extended above the height of the expected flood waters.

(5) A regulator shall not be vented to the appliance flue or exhaust system. [NFPA 54:5.8.5.1]

🔧 See **Figure 1208.8.4**.

1208.8.5 Venting of Gas Appliance Pressure Regulators. For venting of gas appliance pressure regulators see Section 507.21. [NFPA 54:5.8.5.2]

1208.8.6 Discharge of Vents. The discharge of vents shall be in accordance with the following requirements:

(1) The discharge stacks, vents, or outlet parts of all pressure-relieving and pressure-limiting devices shall be located so that gas is safely discharged to the outdoors. Discharge stacks or vents shall be designed to prevent the entry of water, insects, or other foreign material that could cause a blockage.

(2) The discharge stack or vent line shall be at least the same size as the outlet of the pressure-relieving device. [NFPA 54:5.9.8.1 – 5.9.8.2]

1208.8.7 Bypass Piping. Valved and regulated bypasses shall be permitted to be placed around gas line pressure regulators where continuity of service is imperative. [NFPA 54:5.8.6]

1208.8.8 Identification. Line pressure regulators at multiple regulator installations shall be marked by a metal tag or other permanent means designating the building or the part of the building being supplied. [NFPA 54:5.8.7]

FUEL GAS PIPING

1208.9 Overpressure Protection Devices. Where the serving gas supplier delivers gas at a pressure greater than 2 psi for piping systems serving appliances designed to operate at a gas pressure of 14 inches water column (3.5 kPa) or less, overpressure protection devices shall be installed. Piping systems serving equipment designed to operate at inlet pressures greater than 14 inches water column (3.5 kPa) shall be equipped with overpressure protection devices as required by the appliance manufacturer's installation instructions. [NFPA 54:5.9.1]

Sections 1208.9 through 1208.10.4 are direct extractions from the latest version of NFPA 54 dealing with overpressure protection devices (OPD) for fuel gas systems exceeding 2 psi pressure. The requirements for these sections are based, in part, on the provisions found in standard ANSI Z21.80/Can 6.22 Line Pressure Regulators, Section 3.8 (see **Figure 1208.9**). This section of the standard applies to overpressure protection devices which are integral parts, or provided with Class I or Class II regulators with inlet pressures exceeding 2 psi pressure and outlet pressures limited to less than 1/2 psi pressure. Class I regulators can be certified for inlet pressures of 2, 5 and 10 psi with maximum outlet pressure settings of 1/2 psi. Class II regulators can be certified for inlet pressures of 5 or 10 psi with maximum outlet pressure of 2 psi. Per Section 1208.10.1, overpressure protection devices

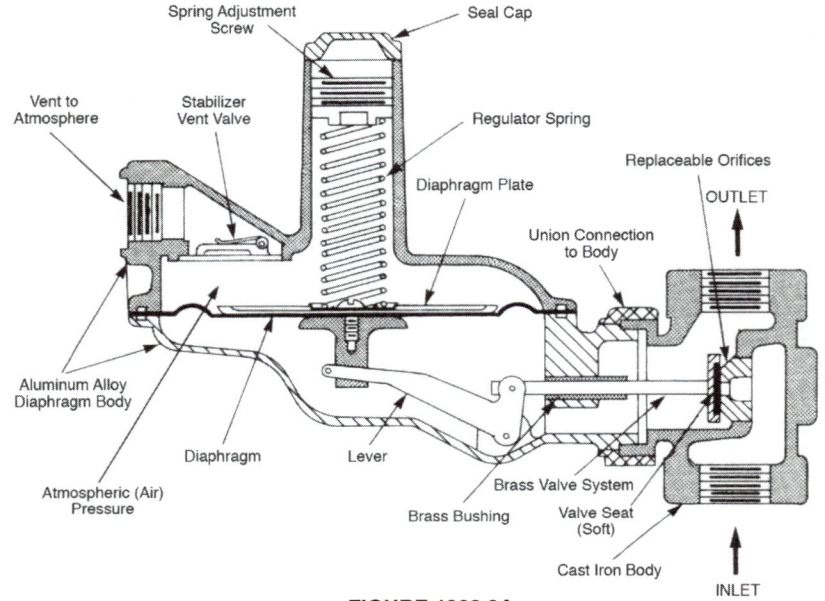

FIGURE 1208.8A
TYPICAL GAS PRESSURE REGULATOR (VENT TO ATMOSPHERE MAY NEED TO BE PIPED)

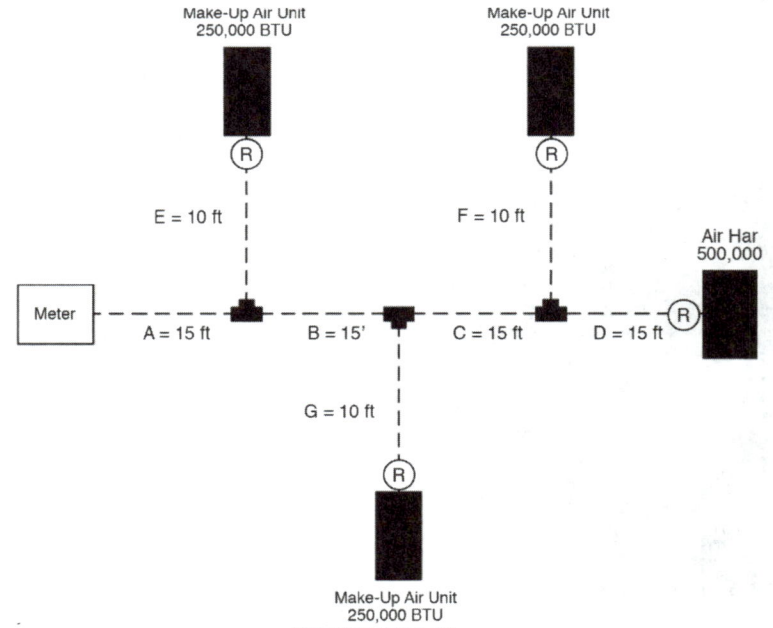

FIGURE 1208.8B
TYPICAL 2PSI GAS SYSTEM WITH INDIVIDUAL APPLIANCE REGULATORS

must also be provided for Class II regulators that have outlet pressures that are higher than 1/2 psi but less than 2 psi. Line pressure regulators with separate overpressure protection devices shall be factory preassembled, tested and supplied to the field as a unit. In lieu of an overpressure protection device, a pressure relief valve may be provided downstream of the regulator and discharge to the exterior of the structure. In the event that the line pressure regulator fails wide open the relief valve must provide overpressure protection that does not exceed the regulator's operating inlet pressure.

1208.10 Pressure Limitation Requirements. Where piping systems serving appliances designed to operate with a gas supply pressure of 14 inches water column (3.5 kPa) or less are required to be equipped with overpressure protection by Section 1208.9, each overpressure protection device shall be adjusted to limit the gas pressure to each connected appliance to 2 psi (14 kPa) or less upon a failure of the line pressure regulator. [NFPA 54:5.9.2.1]

1208.10.1 Overpressure Protection Required. Where piping systems serving appliances designed to operate with a gas supply pressure greater than 14 inches water column (3.5 kPa) are required to be equipped with overpressure protection by Section 1208.9, each overpressure protection device shall be adjusted to limit the gas pressure to each connected appliance as required by the appliance manufacturer's installation instructions. [NFPA 54:5.9.2.2]

1208.10.2 Overpressure Protection Devices. Each overpressure protection device installed to meet the requirements of this section shall be capable of limiting the pressure to its connected appliance(s) as required by this section independently of any other pressure control equipment in the piping system. [NFPA 54:5.9.2.3]

1208.10.3 Detection of Failure. Each gas piping system for which an overpressure protection device is required by this section shall be designed and installed so that a failure of the primary pressure control device(s) is detectable. [NFPA 54:5.9.2.4]

1208.10.4 Flow Capacity. If a pressure relief valve is used to meet the requirements of this section, it shall have a flow capacity such that the pressure in the protected system is maintained at or below the limits specified in Section 1208.10 under the following conditions:

(1) The line pressure regulator for which the relief valve is providing overpressure protection has failed wide open.

(2) The gas pressure at the inlet of the line pressure regulator for which the relief valve is providing overpressure protection is not less than the regulator's normal operating inlet pressure. [NFPA 54:5.9.2.5]

1208.11 Backpressure Protection. Protective devices shall be installed as close to the equipment as practical where the design of the equipment connected is such that air, oxygen, or standby gases are capable of being forced into the gas supply system.

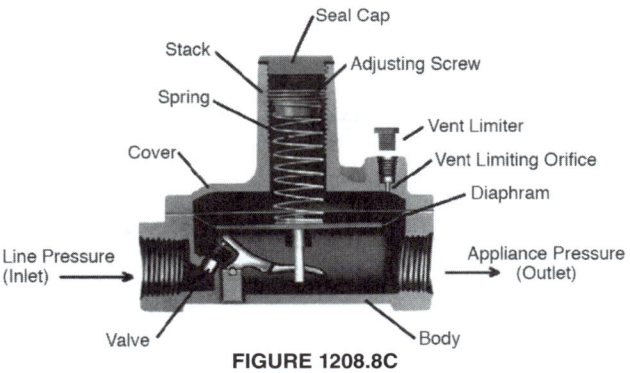

FIGURE 1208.8C
TYPICAL APPLIANCE REGULATOR

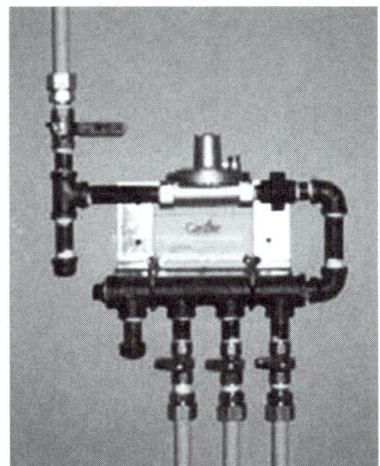

FIGURE 1208.8.4
LINE PRESSURE REGULATOR IN CSST SYSTEM

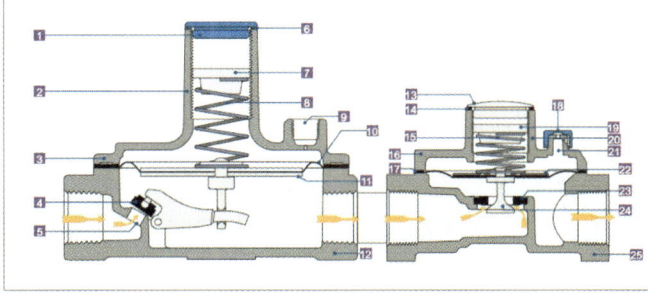

FIGURE 1208.9
LINE PRESSURE REGULATOR
(courtesy of Maxitrol Company)

FUEL GAS PIPING

Gas and air combustion mixers incorporating double diaphragm "zero" or "atmosphere" governors or regulators shall require no further protection unless connected directly to compressed air or oxygen at pressures of 5 psi (34 kPa) or more. [NFPA 54:5.10.1.1 – 5.10.1.2]

1208.11.1 Protective Devices. Protective devices shall include, but not be limited to the following:

(1) Check valves.

(2) Three-way valves (of the type that completely closes one side before starting to open the other side).

(3) Reverse flow indicators controlling positive shutoff valves.

(4) Normally closed air-actuated positive shutoff pressure regulators. [NFPA 54:5.10.2]

Some gas systems or appliances are designed to function at a higher efficiency by adding air or other gases to the gas supply. These other gases are usually in high pressure tanks or pipe lines. The supplemental gasses must not be allowed to enter the fuel gas supply and will require backpressure protection.

1208.12 Low-Pressure Protection. A protective device shall be installed between the meter and the appliance or equipment if the operation of the appliance or equipment is such that it could produce a vacuum or a dangerous reduction in gas pressure at the meter. Such protective devices include, but are not limited to, mechanical, diaphragm-operated, or electrically operated low-pressure shutoff valves. [NFPA 54:5.11]

If a low-pressure shutoff valve is installed, a manual reset device is recommended. This device ensures that the gas will not come back on unexpectedly. It is very important to protect the gas supply system from low pressure and vacuum. If the equipment could cause low pressure or vacuum in the supply system, it could adversely affect the supply pressure of other customers on the line. This low-pressure situation could create a dangerous condition for an unsuspecting user, and it would be difficult to determine the source if caused by another customer.

1208.13 Shutoff Valves. Shutoff valves shall be approved and shall be selected giving consideration to pressure drop, service involved, emergency use, and reliability of operation. Shutoff valves of size 1 inch (25 mm) National Pipe Thread and smaller shall be listed. [NFPA 54:5.12]

Shutoff valves are required at the meter (see **Figure 1208.7.4**), before a pressure regulator (see **Figure 1208.8.4**) and at the appliance itself. Always ensure that the valve is for use with the gas in service and for the temperatures and pressures required for the system.

1208.14 Expansion and Flexibility. Piping systems shall be designed to prevent failure from thermal expansion or contraction. [NFPA 54:5.14.1]

Expansion and contraction in the gas piping system is usually provided for in the design of the system itself. Drops and risers are normally piped with elbows and, simply because of the nature of this piping, create swing joints that will provide the needed relief for the system. Expansion joints may be needed if the runs are very long and if there are large changes in the ambient temperature. Anchoring the piping in a manner that allows expansion to take place where it is planned to should also be considered.

1208.14.1 Special Local Conditions. Where local conditions include earthquake, tornado, unstable ground, or flood hazards, special consideration shall be given to increased strength and flexibility of piping supports and connections. [NFPA 54:5.14.2]

1209.0 Excess Flow Valve.

1209.1 General. Where automatic excess flow valves are installed, they shall be listed to CSA Z21.93 and shall be sized and installed in accordance with the manufacturer's instructions. [NFPA 54:5.13]

Excess flow valves automatically shut off the gas when the flow to a residence or commercial facility exceeds design limits. This excess in gas flow can be caused by a break in the service line from ground movement, natural disasters or third-party damage. Areas of the country where earthquakes occur require an excess flow valve on the main service line or at the appliance outlet (see **Figure 1209.1**).

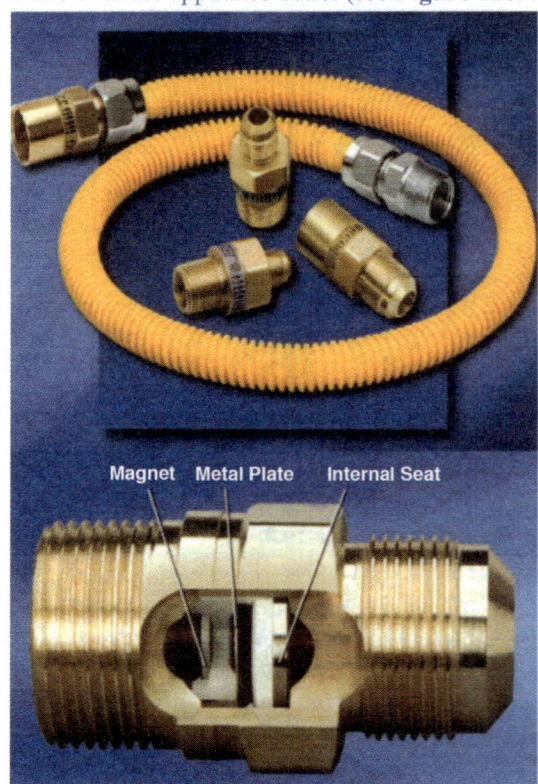

**FIGURE 1209.1
EXCESS FLOW VALVE**

1210.0 Gas Piping Installation.

1210.1 Piping Underground. Underground gas piping shall be installed with sufficient clearance from any other underground structure to avoid contact therewith, to allow maintenance, and to protect against damage from proximity to other structures. In addition, underground plastic piping shall be installed with sufficient clearance or shall be insulated from sources of heat to prevent the heat from impairing the serviceability of the pipe. [NFPA 54:7.1.1]

1210.1.1 Cover Requirements. Underground piping systems shall be installed with a minimum of 12 inches (305 mm) of cover. The minimum cover shall be increased to 18 inches (457 mm) if external damage to the pipe or tubing from external forces is likely to result. Where a minimum of 12 inches (305 mm) of cover cannot be provided, the pipe shall be installed in conduit or bridged (shielded). [NFPA 54:7.1.2.1 (A) (B)]

1210.1.2 Trenches. The trench shall be graded so that the pipe has a firm, substantially continuous bearing on the bottom of the trench. [NFPA 54:7.1.2.2]

1210.1.2.1 Backfilling. Where flooding of the trench is done to consolidate the backfill, care shall be exercised to see that the pipe is not floated from its firm bearing on the trench bottom. [NFPA 54:7.1.2.3]

The intent of these requirements for underground piping is to provide protection for the installed piping from corrosion or physical damage. Damage to pipes is not limited to any one material of pipe. Plastic or copper tubing is severed easily with a shovel, and digging tools can damage the coating on steel pipe, leading to corrosion. Digging with powered digging equipment, such as a backhoe, can damage any type of pipe.

There are several ways to protect the pipe from physical damage. One is to protect the pipe from damage caused by digging, so the protection from digging tools must be of sufficient strength to protect the pipe. Such protection can be provided by installing the gas pipe inside a larger pipe, which becomes a shield, or by locating a protective material, such as a steel plate with corrosion protection, above the pipe. In most cases, the pipe is buried at a sufficient depth so that protective devices are not needed.

The second means is to provide excavators with an indicator that a pipe is buried below. Some jurisdictions also require a yellow plastic tape printed with a warning that gas pipe is buried below to be placed a foot above the pipe to warn future diggers that a gas pipe is buried below. The tape is an indicator and not a protective device.

1210.1.3 Protection Against Corrosion. Steel pipe and steel tubing installed underground shall be installed in accordance with Section 1210.1.3.1 through Section 1210.1.3.9. [NFPA 54:7.1.3]

1210.1.3.1 Zinc Coating. Zinc coating (galvanizing) shall not be deemed adequate protection for underground gas piping. [NFPA 54:7.1.3.1]

1210.1.3.2 Underground Piping. Underground piping shall comply with one or more of the following unless approved technical justification is provided to demonstrate that protection is unnecessary:

(1) The piping shall be made of a corrosion-resistant material that is suitable for the environment in which it will be installed.

(2) Pipe shall have a factory-applied, electrically insulating coating. Fittings and joints between sections of coated pipe shall be coated in accordance with the coating manufacturer's instructions.

(3) The piping shall have a cathodic protection system installed, and the system shall be maintained in accordance with Section 1210.1.3.3 or Section 1210.1.3.6. [NFPA 54:7.1.3.2]

1210.1.3.3 Cathodic Protection. Cathodic protection systems shall be monitored by testing, and the results shall be documented. The test results shall demonstrate one of the following:

(1) A pipe-to-soil voltage of −0.85 volts or more negative is produced, with reference to a saturated copper-copper sulfate half cell.

(2) A pipe-to-soil voltage of −0.78 volts or more negative is produced, with reference to a saturated KCl calomel half cell.

(3) A pipe-to-soil voltage of −0.80 volts or more negative is produced, with reference to a silver-silver chloride half cell.

(4) Compliance with a method described in Appendix D of Title 49 of the code of Federal Regulations, Part 192. [NFPA 54:7.1.3.3]

Steel pipe installed underground or in contact with water deteriorates through a process called galvanic corrosion. When steel pipe is used, the galvanic action is avoided by coating the pipe with an electrically insulating material. Factory-applied coatings include polyethylene over a mastic primer and fusion-bonded epoxy. Field-applied coatings usually consist of either hot- or cold-applied tapes over a complementary primer.

Another step in properly protecting a steel pipe underground is to maintain an electrical charge on the pipe relative to the surrounding soil and adequate to ensure that the corrosion- causing transfer of electrons will not occur. Maintaining this electrical charge is called cathodic protection and is accomplished by an impressed voltage supplied by either a passive sacrificial anode system or by an active rectifier system. Sizing and spacing of either the anodes or the rectifiers are important. Manufacturers of these components are a good source of guidance.

Field coating steel pipe but not using cathodic protection is usually worse than unprotected pipe. Uncoated pipe will corrode over its entire surface area, whereas corrosion of coated pipe lacking cathodic protection will be concentrated at any pinhole or other imperfection in the coating. Because obtaining and maintaining a perfect coating application is virtually impossible, protection of underground steel pipe should include cathodic protection. Any cathodically protected pipe should be electrically isolated from upstream and downstream components to prevent the loss of the required electrical charge. This isolation is accomplished with the use of dielectric unions or flange-insulating kits.

Piping carrying hazardous materials is often protected by a manufactured coating supplemented with cathodic protection. Cathodic protection is a means of preventing metal structures such as pipelines from reacting with the environment and corroding. Carbon steel piping exposed to the elements will break down electrochemically and ultimately fail. Cathodic protection systems prevent the oxidation process from occurring by creating a current flow from the cathodic protection system to the piping. There are two types of cathodic protection systems: galvanic and impressed current.

FUEL GAS PIPING

Galvanic cathodic protection utilizes an electrical-chemical process where one metal is more susceptible to corrosion than another when both metals are linked electrically. Sacrificial anodes, many times made of magnesium, are used to protect steel piping. (see **Figure 1210.1.3.3a**).

In many applications, the difference between the sacrificial anode and the steel piping is not enough to generate sufficient current for cathodic protection to occur. In these cases, a power supply, or rectifier, is used to generate larger potential differences, enabling more current to flow to the piping being protected. This is referred to as impressed current cathodic protection. In these systems DC current with an output of up to 50 amperes and 50 volts is used to protect the piping system. The voltage depends on several factors, such as size of the pipeline, coating quality and environmental conditions. The typical configuration of the wiring for this system is connecting cables from the positive DC terminal of the rectifier to the anode group and the negative terminal to the pipe. Anodes can be installed in a ground bed consisting of vertical holes backfilled with a material that improves the performance and life of the anodes called coke. They may also be laid in a trench surrounded by coke and backfilled (see **Figure 1210.1.3.3b**).

It is important that these systems be designed by professionals possessing expertise in cathodic protection and knowledge of the piping systems to be protected.

Galvanic Cathodic Protection

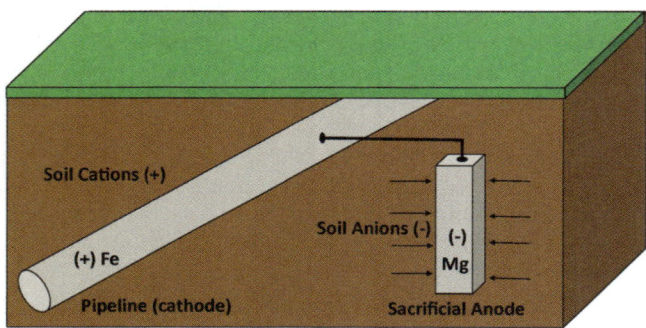

The magnesium rod (sacrificial anode) will protect the steel pipeline from corrosion. Magnesium is more easily oxidized than iron therefore acting as an anode in a galvanic cell. The steel pipeline becomes the cathode and oxygen is reduced protecting the steel pipeline.

FIGURE 1210.1.3.3A
GALVANIC CATHODIC PROTECTION

Impressed Current Cathodic Protection

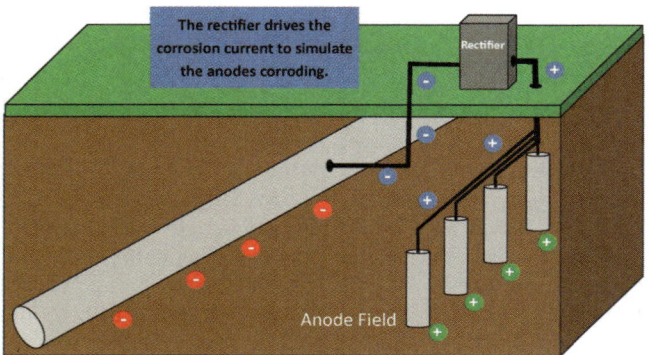

FIGURE 1210.1.3.3B
IMPRESSED CURRENT CATHODIC PROTECTION

1210.1.3.4 Sacrificial Anodes. Sacrificial anodes shall be tested in accordance with the following:

(1) Upon installation of the cathodic protection system, except where prohibited by climatic conditions, in which case the testing shall be performed not later than 180 days after the installation of the system.

(2) 12 to 18 months after the initial test.

(3) Upon successful verification testing in accordance with Section 1210.1.3.4(1) and Section 1210.1.3.4(2), periodic follow-up testing shall be performed at intervals not to exceed 36 months. [NFPA 54:7.1.3.4]

1210.1.3.5 System Failing Tests. Systems failing a test shall be repaired not more than 180 days after the date of the failed testing. The testing schedule shall be restarted as required in Section 1210.1.3.4(1) and Section 1210.1.3.4(2), and the results shall comply with Section 1210.1.3.3. [NFPA 54:7.1.3.5]

1210.1.3.6 Impressed Current Cathodic Protection. Impressed current cathodic protection systems shall be inspected and tested in accordance with the following schedule:

(1) The impressed current rectifier voltage output shall be checked at intervals not exceeding two months.

(2) The pipe-to-soil voltage shall be tested at least annually. [NFPA 54:7.1.3.6]

1210.1.3.7 Documentation. Documentation of the results of the two most recent tests shall be retained. [NFPA 54:7.1.3.7]

1210.1.3.8 Dissimilar Metals. Where dissimilar metals are joined underground, an insulating coupling or fitting shall be used. [NFPA 54:7.1.3.8]

1210.1.3.9 Steel Risers. Steel risers, other than anodeless risers, connected to plastic piping shall be cathodically protected by means of a welded anode. [NFPA 54:7.1.3.9]

1210.1.4 Protection Against Freezing. Where the formation of hydrates or ice is known to occur, the piping shall be protected against freezing. [NFPA 54:7.1.4]

Natural gas may contain water vapor that could cause ice buildup if installed in areas with colder temperatures. If the gas used is not dry gas then precautions must be taken to protect the gas from freezing. Ice problems are all but nonexistent with pipeline natural gas and propane as almost all manufacturers remove the water vapor inherent in these gases. It can still exist in gas from alternative sources, such as local wells, landfills or waste treatment plants. Consult the gas supplier about the content of the fuel gas and the possibility of ice.

1210.1.5 Piping Through Foundation Wall. Underground piping, where installed through the outer foundation or basement wall of a building shall be encased in a protective sleeve or protected by an approved device or method. The space between the gas piping and the sleeve and between the sleeve and the wall shall be sealed to prevent entry of gas and water. [NFPA 54:7.1.5]

See **Figure 1210.1.5**.

1210.1.6 Piping Underground Beneath Buildings. Where gas piping is installed underground beneath buildings, the piping shall be either of the following:

(1) Encased in an approved conduit designed to withstand

the imposed loads and installed in accordance with Section 1210.1.6.1 or Section 1210.1.6.2.

(2) A piping or encasement system listed for installation beneath buildings. [NFPA 54:7.1.6]

Gas leaks in underground piping can migrate along the trench and into the building instead of venting to the atmosphere above grade. This condition can be extremely hazardous. In some areas, there is also a requirement that the installation should be made so that the piping can be removed or replaced without removal of the flooring materials. The requirement to seal the casing against water entry is in keeping with the desire to exercise good workmanship. See **Figure 1210.1.6** for an illustration of a below-grade foundation wall penetration.

1210.1.6.1 Conduit with One End Terminating Outdoors. The conduit shall extend into an accessible portion of the building and, at the point where the conduit terminates in the building, the space between the conduit and the gas piping shall be sealed to prevent the possible entrance of a gas leakage. Where the end sealing is of a type that retains the full pressure of the pipe, the conduit shall be designed for the same pressure as the pipe. The conduit shall extend at least 4 inches (102 mm) outside the building, be vented outdoors above finished ground level, and be installed to prevent the entrance of water and insects. [NFPA 54:7.1.6.1]

1210.1.6.2 Conduit with Both Ends Terminating Indoors. Where the conduit originates and terminates within the same building, the conduit shall originate and terminate in an accessible portion of the building and shall not be sealed. [NFPA 54:7.1.6.2]

1210.1.7 Plastic Piping. Plastic piping shall be installed outdoors, underground only.

Plastic pipe and its tubing are more susceptible to inadvertent damage during installation than most metallic pipe. For this reason, special attention should be given to proper compaction below the pipe, to the elimination of shear points on connections during backfilling and to the materials in the backfill, making certain that angular and large materials are not used near the pipe.

Exceptions:

(1) Plastic piping shall be permitted to terminate aboveground where an anodeless riser is used.

The use of anode-type risers and transition fittings with anodes [see **Figure 1210.1.7(1)**] has been allowed in several editions of the code. In both of these applications the plastic-to-steel transition is made below grade. Anodeless risers were addressed in Section 1208.6.7.

(2) Plastic piping shall be permitted to terminate with a wall head adapter aboveground in buildings, including basements, where the plastic piping is inserted in a piping material permitted for use in buildings. [NFPA 54:7.1.7.1]

Wall head adapters [see **Figure 1210.1.7(2)**] often are used by utilities for insertion renewals of steel service lines. Within the scope of this code, wall head adapters can be used for renewals of steel pipes from meters at the property line or in renewals of existing underground steel pipes between buildings, as well as for making a clean-sleeved, below-grade foundation wall penetration.

1210.1.7.1 Connections Between Metallic and Plastic Piping. Connections made between metallic and plastic piping shall be made with fittings conforming to one of the following:

(1) ASTM D2513, Category I transition fittings

(2) ASTM F1973

(3) ASTM F2509 [NFPA 54:7.1.7.2]

1210.1.7.2 Tracer. An electrically continuous corrosion-resistant tracer shall be buried with the plastic pipe to facilitate locating. The tracer shall be one of the following:

(1) A product specifically designed for that purpose.

(2) Insulated copper conductor not less than 14 AWG.

Where tracer wire is used, access shall be provided from

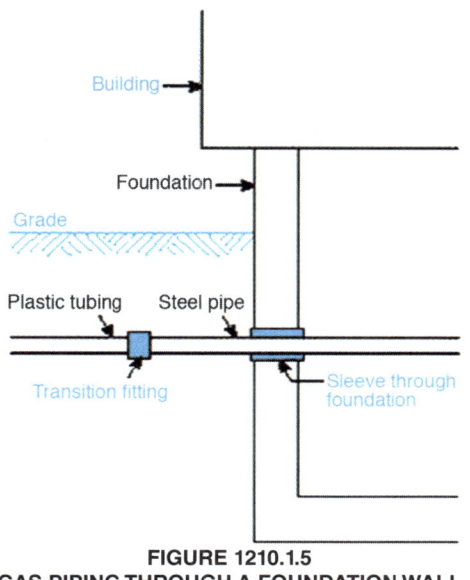

FIGURE 1210.1.5
GAS PIPING THROUGH A FOUNDATION WALL

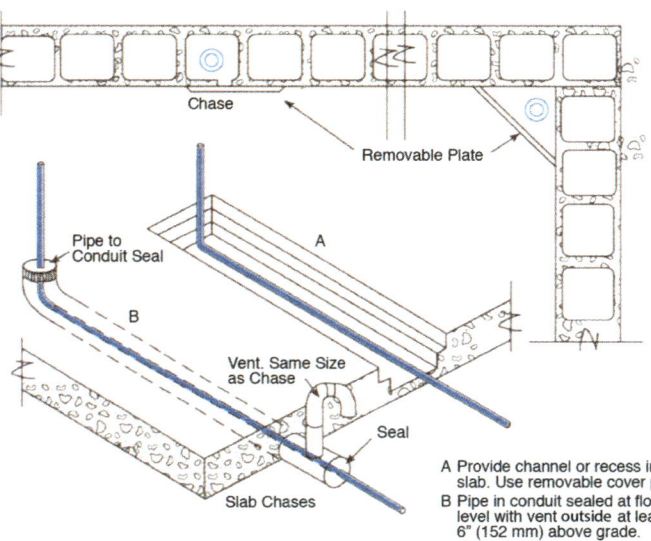

FIGURE 1210.1.6
GAS PIPING FLOOR INSTALLATIONS

FUEL GAS PIPING

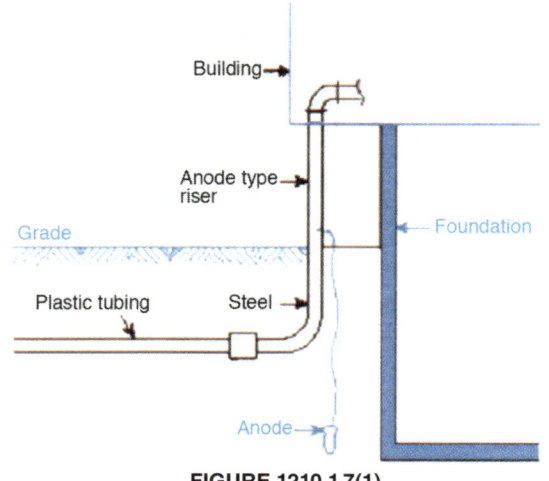

FIGURE 1210.1.7(1)
ANODE-TYPE RISER

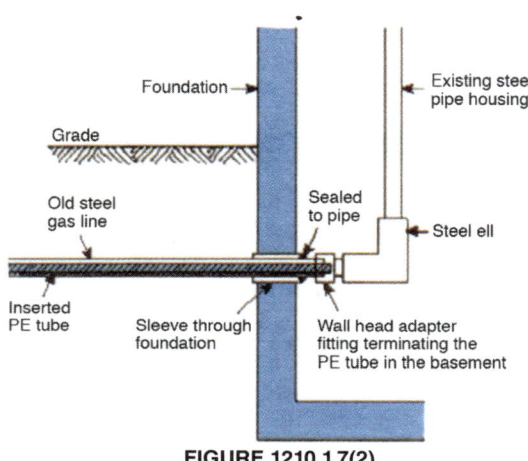

FIGURE 1210.1.7(2)
WALL HEAD ADAPTER

aboveground, or one end of the tracer wire or tape shall be brought aboveground at a building wall or riser. [NFPA 54:7.1.7.3 – 7.1.7.3.2]

🔧 A product specifically designed for that purpose is not the typical yellow plastic tape printed with a warning that gas pipe is buried below and used as an indicator. The product the UPC refers to is a tape that is manufactured with the capacity to identify the specific location of buried plastic gas piping.

1210.2 Installation of Piping. Piping installed aboveground shall be securely supported and located where it will be protected from physical damage. Where passing through an exterior wall, the piping shall also be protected from corrosion by coating or wrapping with an inert material approved for such applications. The piping shall be sealed around its circumference at the point of the exterior penetration to prevent the entry of water, insects, and rodents. Where piping is encased in a protective pipe sleeve, the annular spaces between the gas piping and the sleeve and between the sleeve and the wall opening shall be sealed. [NFPA 54:7.2.1]

🔧 In the previous sections, wall penetrations underground were discussed. The same care must be taken with piping above ground. The requirements of Section 312.10, Sleeves, apply to fuel gas as do the requirements of Chapter 14, Firestop Protection.

1210.2.1 Building Structure. The installation of gas piping shall not cause structural stresses within building components to exceed allowable design limits. Approval shall be obtained before any beams or joists are cut or notched. [NFPA 54:7.2.2.1 – 7.2.2.2]

🔧 The requirements of Section 312.11 shall also apply to fuel gas installations. Also consult the building code for cutting and notching requirements. Always try to bore the stud or joist rather than notch or cut as in **Figure 1210.2.1**.

FIGURE 1210.2.1
BORING IS ALWAYS PREFERABLE TO NOTCHING OR CUTTING

1210.2.2 Gas Piping to be Sloped. Piping for other than dry gas conditions shall be sloped not less than ¼ inch in 15 feet (6.4 mm in 4.6 m) to prevent traps. [NFPA 54:7.2.3]

🔧 As stated earlier, one of the elements in natural gas is water vapor. When it is in enough quantity that cold temperatures will cause it to condense and form water droplets that can be corrosive or might foul valves or orifices, the pipe must be graded in order to prevent water buildup in low spots or traps and to guide the water to drip legs required by Section 1210.6.

Dry gas is defined as a gas having a moisture and hydrocarbon dew point below any normal temperature to which the gas piping is exposed. When these conditions are met, the installation of drips is unnecessary. The gas service supplier will be able to verify if dry gas is used in a particular area.

1210.2.2.1 Ceiling Locations. Gas piping shall be permitted to be installed in accessible spaces between a fixed ceiling and a dropped ceiling, whether or not such spaces are used as a plenum. Valves shall not be located in such spaces.

Exception: Appliance or equipment shutoff valves required by this code shall be permitted to be installed in accessible spaces containing vented appliances.

🔧 Often, furnaces or other appliances are installed in ceiling or attic locations. This exception allows the shutoff valve to be placed near this equipment and requires the location to be accessible.

1210.2.3 Prohibited Locations. Gas piping inside any building shall not be installed in or through a clothes chute, chimney or gas vent, dumbwaiter, elevator shaft, or air duct, other than combustion air ducts. [NFPA 54:7.2.4]

Exception: Ducts used to provide ventilation air in accordance with Section 506.0 or to above-ceiling spaces in accordance with Section 1210.2.2.1.

🔧 The intent of this subsection is to prevent the mechanical transport of gas leakage throughout the building.

1210.2.4 Hangers, Supports, and Anchors. Piping shall be supported with metal pipe hooks, metal pipe straps, metal bands, metal brackets, metal hangers, or building structural components, approved for the size of piping; of adequate strength and quality; and located at intervals to prevent or damp out excessive vibration. Piping shall be anchored to prevent undue strains on connected appliances and equipment and shall not be supported by other piping. Pipe hangers and supports shall conform to the requirements of MSS SP-58. [NFPA 54:7.2.5.1]

1210.2.4.1 Spacing. Spacing of supports in gas piping installations shall not exceed the distance shown in Table 1210.2.4.1. Spacing of supports for CSST shall be in accordance with the CSST manufacturer's instructions. [NFPA 54:7.2.5.2]

TABLE 1210.2.4.1
SUPPORT OF PIPING
[NFPA 54: TABLE 7.2.5.2]

STEEL PIPE, NOMINAL SIZE OF PIPE (inches)	SPACING OF SUPPORTS (feet)	NOMINAL SIZE OF TUBING SMOOTH-WALL (inches O.D.)	SPACING OF SUPPORTS (feet)
½	6	½	4
¾ or 1	8	⅝ or ¾	6
1¼ or larger (horizontal)	10	⅞ or 1 (horizontal)	8
1¼ or larger (vertical)	Every floor level	1 or larger (vertical)	Every floor level

For SI units: 1 inch = 25 mm, 1 foot = 304.8 mm

1210.2.4.2 Expansion and Contraction. Supports, hangers, and anchors shall be installed so as not to interfere with the free expansion and contraction of the piping between anchors. All parts of the supporting system shall be designed and installed, so they are not disengaged by movement of the supported piping. [NFPA 54:7.2.5.3]

🔧 Table 1210.2.4.1 lists the minimum distances allowable for support of piping, including steel pipe and smooth-wall tubing. Table 313.3 also contains these requirements. The other requirements of Chapter 3 also apply to fuel gas piping and are mirrored with these provisions.

The installation of CSST piping should be per the manufacturer's installation instructions. Each manufacturer has design and installation guides. For example, **Figure 1210.2.4.2** is the support table from the Design and Installation Guide for Gastite CSST.

1210.2.4.3 Piping on Roof Tops. Gas piping installed on the roof surfaces shall be elevated above the roof surface and shall be supported in accordance with Table 1210.2.4.1. [NFPA 54:7.2.5.4]

1210.3 Concealed Piping in Buildings. Gas piping in concealed locations shall be installed in accordance with this section. [NFPA 54:7.3.1]

1210.3.1 Connections. Where gas piping is to be concealed, connections shall be of the following type:

(1) Pipe fittings such as elbows, tees, couplings, and right/left nipple/couplings.

(2) Joining tubing by brazing (see Section 1208.6.11.2).

(3) Fittings listed for use in concealed spaces or that have been demonstrated to sustain, without leakage, forces due to temperature expansion or contraction, vibration, or fatigue based on their geographic location, application, or operation.

(4) Where necessary to insert fittings in the gas pipe that has been installed in a concealed location, the pipe shall be reconnected by welding, flanges, or the use of a right/left nipple/coupling.

🔧 The list of approved fittings in this subsection consists of those that are least likely to leak in concealed installations. Left and right nipple and couplings therefore, are acceptable in this type of installation.

Fittings that are specifically listed and approved for use in concealed locations are a part of a listed system designed to meet the concerns that prohibit other tubing fittings in concealed locations.

1210.3.2 Piping in Partitions. Concealed gas piping shall not be located in solid partitions. [NFPA 54:7.3.3]

1210.3.3 Tubing in Partitions. This provision shall not apply to tubing that pierces walls, floors, or partitions. Tubing installed vertically and horizontally inside hollow walls or partitions without protection along its entire concealed length shall meet the following requirements:

(1) A steel striker barrier not less than 0.0508 of an inch (1.3 mm) thick, or equivalent, is installed between the tubing and the finished wall and extends at least 4 inches (102 mm) beyond concealed penetrations of plates, firestops, wall studs, and similar construction features.

(2) The tubing is installed in single runs and is rigidly secured. [NFPA 54:7.3.4]

🔧 A striker plate of 0.0508-inch thickness is approximately 16 gauge metal. These plates are required at tubing penetrations to protect the tubing from nail or screw penetrations. Nails and screws are more apt to be able to pin down and penetrate a tube within the wall plate because it is restrained by the penetration. This hazard is also the basis for the requirement that tubing located away from penetrations must not be secured within the wall. If the tube is not secured, it can move within the wall space. This protection is not only for initial construction but for the home or building owner

Table 4-2			
Support Spacing			
Gastite® Part No.	Size	EHD	Vertical or Horizontal
S93-6A	3/8"	13	4 Feet
S93-8A	1/2"	18	6 Feet
S93-11B	3/4"	23	8 feet (USA) 6 feet (Canada)
S93-16A	1"	31	8 feet (USA) 6 feet (Canada)
S93-20A	1-1/4"	37	8 feet (USA) 6 feet (Canada)
S93024A	1-1/2"	47	8 feet (USA) 6 feet (Canada)
S93-32A	2"	60	8 feet (USA) 6 feet (Canada)

FIGURE 1210.2.4.2
CCST SUPPORT TABLE FROM GASTITE
CSST DESIGN AND INSTALLATION GUIDE

FUEL GAS PIPING

who may want to attach something to the wall (see **Figure 1210.3.3**). Some jurisdictions may require the same protection for steel piping. Again, be sure to follow the manufacturer's requirements for CSST installations.

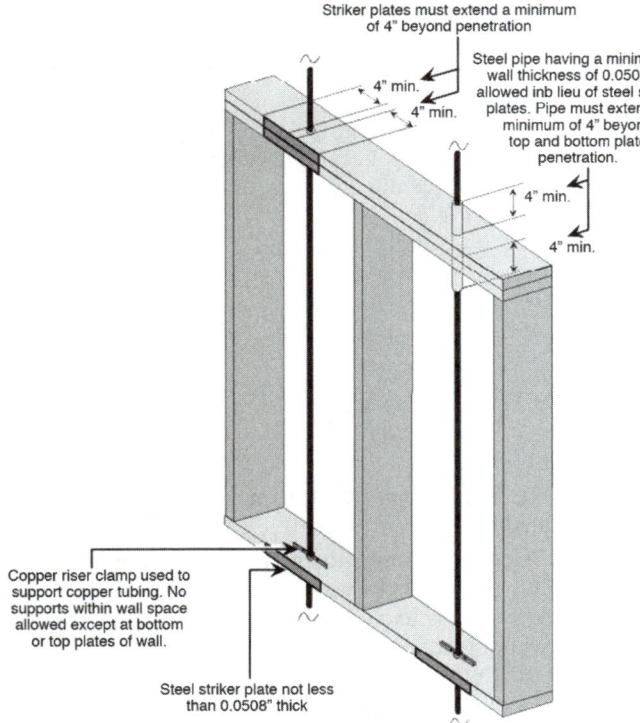

FIGURE 1210.3.3
TUBING INSTALLATION WITHIN PARTITIONS

1210.3.4 Piping in Floors. In industrial occupancies, gas piping in solid floors such as concrete shall be laid in channels in the floor and covered to permit access to the piping with minimum damage to the building. Where piping in floor channels is exposed to excessive moisture or corrosive substances, the piping shall be protected in an approved manner.

In other than industrial occupancies and where approved by the Authority Having Jurisdiction, gas piping embedded in concrete floor slabs constructed with portland cement shall be surrounded with a minimum of 1½ inches (38 mm) of concrete and shall not be in physical contact with other metallic structures such as reinforcing rods or electrically neutral conductors. All piping, fittings, and risers shall be protected against corrosion in accordance with Section 1208.6.9. Piping shall not be embedded in concrete slabs containing quick-set additives or cinder aggregate. [NFPA 54:7.3.5.1 – 7.3.5.2]

Concrete containing cinder aggregate or quickset additives, usually calcium chloride, is not permitted. This is due to the fact that such products attack the metals from which gas piping is made.

It is important to remember that different requirements apply to piping installed in floors in industrial occupancies than in other occupancies. In industrial occupancies, access to the gas piping is necessary because of the use of corrosive materials that can damage the piping.

1210.4 Piping in Vertical Chases. Where gas piping exceeding 5 psi (34 kPa) is located within vertical chases in accordance with Section 1208.5(2), the requirements of Section 1210.4.1 through Section 1210.4.3 shall apply. [NFPA 54:7.4]

Special requirements for fuel gas piping systems containing gas pressures above 5 psi are needed due to the dangers of what a leak in this system might do. A leak in this high pressure system would allow far greater quantities of gas into the building. For this reason, the piping must be welded rather than using another joining system. The piping must be located in a ventilated chase rather than in walls and through trusses where normal fuel gas systems are installed.

1210.4.1 Pressure Reduction. Where pressure reduction is required in branch connections for compliance with Section 1208.5, such reduction shall take place either inside the chase or immediately adjacent to the outside wall of the chase. Regulator venting and downstream overpressure protection shall comply with Section 1208.8.4 and Section 1208.9 through Section 1208.10.4. The regulator shall be accessible for service and repair, and vented in accordance with one of the following:

(1) Where the fuel gas is lighter than air, regulators equipped with a vent limiting means shall be permitted to be vented into the chase. Regulators not equipped with a vent limiting means shall be permitted to be vented either directly to the outdoors or to a point within the top 1 foot (305 mm) of the chase.

(2) Where the fuel gas is heavier than air, the regulator vent shall be vented only directly to the outdoors. [NFPA 54:7.4.1]

1210.4.2 Chase Construction. Chase construction shall comply with local building codes with respect to fire resistance and protection of horizontal and vertical openings. [NFPA 54:7.4.2]

1210.4.3 Ventilation. A chase shall be ventilated to the outdoors and only at the top. The opening(s) shall have a minimum free area [in square inches (square meters)] equal to the product of one-half of the maximum pressure in the piping [in pounds per square inch (kilopascals)] times the largest nominal diameter of that piping [in inches (millimeters)], or the cross-sectional area of the chase, whichever is smaller. Where more than one fuel gas piping system is present, the free area for each system shall be calculated, and the largest area used. [NFPA 54:7.4.3]

1210.5 Gas Pipe Turns. Changes in direction of gas pipe shall be made by the use of fittings, factory bends, or field bends. [NFPA 54:7.5]

1210.5.1 Metallic Pipe. Metallic pipe bends shall comply with the following:

(1) Bends shall be made with bending equipment and procedures intended for that purpose.

(2) Bends shall be smooth and free from buckling, cracks, or other evidence of mechanical damage.

(3) The longitudinal weld of the pipe shall be near the neutral axis of the bend.

(4) The pipe shall not be bent through an arc of more than 90 degrees (1.57 rad).

(5) The inside radius of a bend shall be not less than six times the outside diameter of the pipe. [NFPA 54:7.5.1]

1210.5.2 Plastic Pipe. Plastic pipe bends shall comply with the following:

(1) The pipe shall not be damaged, and the internal diameter of the pipe shall not be effectively reduced.

(2) Joints shall not be located in pipe bends.

(3) The radius of the inner curve of such bends shall not be less than 25 times the inside diameter of the pipe.

(4) Where the piping manufacturer specifies the use of special bending tools or procedures, such tools or procedures shall be used. [NFPA 54:7.5.2]

The practice of bending metallic pipe, especially steel pipe, is not used much these days; however, plastic pipe is bent on a regular basis. Gas piping mains and service pipe are often installed in polyethylene pipe. Turns made by bending the pipe are the norm and the manufacturer's installation instructions must always be followed.

1210.5.3 Elbows. Factory-made welding elbows or transverse segments cut therefrom shall have an arc length measured along the crotch at least 1 inch (25.4 mm) for pipe sizes 2 inches (50 mm) and larger. [NFPA 54:7.5.3]

See **Figure 1210.5.3**.

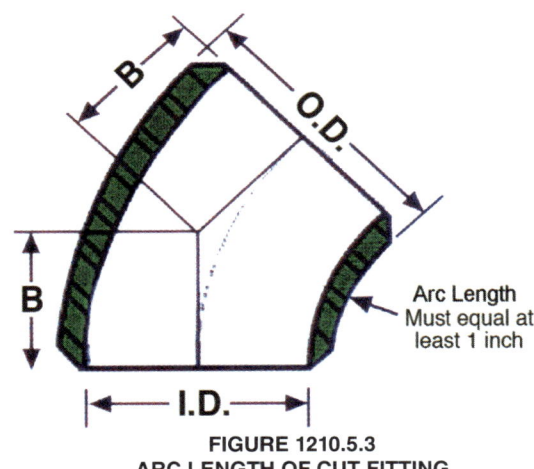

**FIGURE 1210.5.3
ARC LENGTH OF CUT FITTING**

1210.6 Drips and Sediment Traps. For other than dry gas conditions, a drip shall be provided at a point in the line of pipe where condensate could collect. Where required by the Authority Having Jurisdiction or the serving gas supplier, a drip shall also be provided at the outlet of the meter. This drip shall be so installed as to constitute a trap wherein an accumulation of condensate shuts off the flow of gas before it runs back into the meter. [NFPA 54:7.6.1]

Dry gas is defined as a gas having a moisture and hydrocarbon dew point below any normal temperature to which the gas piping is exposed. The phrase "for other than dry gas conditions" clarifies that drips are not necessary when only dry gas is expected to be present.

Drips collect condensate from gases that contain condensable products (usually water) or oil added in the gas transmission process. Usually, drips are located near the outlet of the meter or the service entrance and at other locations where condensate could collect, such as before the appliance inlet. The drip is installed at the bottom of a downward-flowing line by placing a tee at the bottom of the line, installing a nipple and cap in the run of the tee and continuing the pipe run out the side of the tee. Drips must be located so that they can be emptied to prevent liquid from causing an obstruction to the flow of gas. Drip locations also must be protected from freezing (see **Figures 1208.6.4.4** and **1208.8.4**).

1210.6.1 Location of Drips. All drips shall be installed only in such locations that they are readily accessible to permit cleaning or emptying. A drip shall not be located where the condensate is likely to freeze. [NFPA 54:7.6.2]

1210.6.2 Sediment Traps. The installation of sediment traps shall be in accordance with Section 1212.9. [NFPA 54:7.6.3]

Sediment traps, sometimes confused with drip traps, are installed to collect solid foreign particles to prevent such material from entering close-fitting parts or small passageways (e.g., valves and orifices). Sediment traps are identical to drip traps and are to be installed according to Section 1212.8 (see **Figure 1212.9**). Although some gases can contain foreign particles, this situation is not likely to be a problem in either utility gas or LP-gas because of the methods and equipment used in handling these products. Dirt and pipe material cuttings that are placed unavoidably in the system itself (usually during construction) and are present in limited amounts are the target of sediment traps. Thus, sediment traps seldom need to be opened for service or cleaning.

Many appliance manufacturers are incorporating sediment traps in their appliances, and a number of the ANSI Z21 standards for gas appliances require the installation of a sediment trap. Section 1212.9 requires the installation of a sediment trap at the time of installation of most appliances if they are not already equipped with one.

1210.7 Outlets. Outlets shall be located and installed in accordance with the following requirements:

(1) The outlet fittings or piping shall be securely fastened in place.

(2) Outlets shall not be located behind doors.

(3) Outlets shall be located far enough from floors, walls, patios, slabs, and ceilings to permit the use of wrenches without straining, bending, or damaging the piping.

(4) The unthreaded portion of gas piping outlets shall extend not less than 1 inch (25.4 mm) through finished ceilings or indoor or outdoor walls.

(5) The unthreaded portion of gas piping outlets shall extend not less than 2 inches (51 mm) above the surface of floors or outdoor patios or slabs.

(6) The provisions of Section 1210.7(4) and Section 1210.7(5) shall not apply to listed quick-disconnect devices of the flush-mounted type or listed gas convenience outlets. Such devices shall be installed in accordance with the manufacturer's installation instructions. [NFPA 54:7.7.1.1 – 7.7.1.6]

The provisions of this section are straightforward. Minimum projections of pipe ends are specified in the case of walls and ceilings to allow a wrench to grip the pipe. In the case of floors, the extra projection length will tend to protect the threads from flooding and mechanical damage.

1210.7.1 Cap Outlets. Each outlet, including a valve, shall be closed gastight with a threaded plug or cap immediately after installation and shall be left closed until the appliance or equipment is connected thereto. Where an appliance or equipment is disconnected from an outlet, and the outlet is not to be used again immediately, it shall be capped or plugged gastight.

Exceptions:

(1) Laboratory appliances installed in accordance with Section 1212.3.1 shall be permitted.

(2) The use of a listed quick-disconnect device with integral shutoff or listed gas convenience outlet shall be permitted. [NFPA 54:7.7.2.1]

A plug or cap is required for all gas pipe openings. Closing a valve is not enough to satisfy this requirement because the valve can be opened inadvertently or accidentally. No temporary or makeshift closure is permitted because anything except a properly tightened plug or cap could leak (see **Figure 1210.7.1**). Listed quick-disconnect devices and gas convenience outlets are permitted to go uncapped or unplugged because they are required by their listing to have valves that will automatically shut off the gas either prior to or during the disconnect.

FIGURE 1210.7.1
MULTIPLE OUTLETS CAPPED AND READY FOR TEST

1210.7.2 Appliance Shutoff Valves. Appliance shutoff valves installed in fireplaces shall be removed, and the piping capped gastight where the fireplace is used for solid-fuel burning. [NFPA 54:7.7.2.2]

When burning solid fuel, the removal of equipment shutoff valves from fireplaces is required. Experience has shown that these valves will leak if they are subjected to the heat of a wood fire. Whenever possible, all gas piping in a fireplace that is going to be used to burn solid fuel should also be removed as a preventive measure. Leaving the gas piping in the fireplace subjects the piping to extreme temperatures, which over time can cause the piping connections to leak.

1210.8 Branch Pipe Connection. Where a branch outlet is placed on the main supply line before it is known what size pipe will be connected to it, the outlet shall be of the same size as the line that supplies it. [NFPA 54:7.8]

This requirement is for those times during construction when an appliance is called for or added during installation of the piping system. Sometimes, the gas demand or size of an inlet is not known and the system has to be completed. When this happens the branch to the appliance shall be the same size as the line the branch is taken from. For example, if the line is 1 inch and an unknown appliance is added, the added branch would also have to be 1 inch in size.

1210.9 Manual Gas Shutoff Valves. An accessible gas shutoff valve shall be provided upstream of each gas pressure regulator. Where two gas pressure regulators are installed in series in a single gas line, a manual valve shall not be required at the second regulator. [NFPA 54:7.9.1]

See **Figures 1208.7.4** and **1208.8.4**.

1210.9.1 Valves Controlling Multiple Systems. Main gas shutoff valves controlling several gas piping systems shall be readily accessible for operation and installed to be protected from physical damage. They shall be marked with a metal tag or other permanent means attached by the installing agency so that the gas piping systems supplied through them are readily identified. [NFPA 54:7.9.2.1]

1210.9.2 Shutoff Valves for Multiple House Lines. In multiple-tenant buildings supplied through a master meter, through one service regulator where a meter is not provided, or where meters or service regulators are not readily accessible from the appliance or equipment location, an individual shutoff valve for each apartment or tenant line shall be provided at a convenient point of general accessibility. In a common system serving a number of individual buildings, shutoff valves shall be installed at each building. [NFPA 54:7.9.2.2]

1210.9.3 Emergency Shutoff Valves. An exterior shutoff valve to permit turning off the gas supply to each building in an emergency shall be provided. The emergency shutoff valves shall be plainly marked as such and their locations posted as required by the Authority Having Jurisdiction. [NFPA 54:7.9.2.3]

The reason for requiring a shutoff valve for each building's gas line is to allow for the system to be separated for repair or maintenance. In a case where only part of the system has to be shut down for repair, the remaining system can continue to operate. Shutoff valves must be plainly marked so that in the future service workers can determine which lines service which units or buildings.

The requirement for a shutoff valve at each building provides emergency responders with the ability to shut off the flow of gas to a building involved in fire (see **Figure 1210.9.3**).

In natural gas installations, this shutoff valve is part of the utility's service and meter installation. The natural gas supplier provides a shutoff valve for the supply to its gas meter, under the U.S. Department of Transportation regulations (49 CFR 192). The emergency shutoff valve required in this section can be the same valve that is required in Section 1210.9.2 for multiple house lines. If the natural gas supplier installs a shutoff valve and gas meter (or shutoff valve if no meter is provided) then the shutoff valve can also serve as the emergency shutoff valve. If the fuel gas line is run from this location to another building underground, the line must be brought up outside and another shutoff valve installed. This second shutoff will be the emergency shutoff valve and the valve controlling multiple systems. In any piping system, there must be an accessible exterior shutoff valve.

In LP-gas systems, the emergency shutoff valve is normally the propane tank shutoff valve.

**FIGURE 1210.9.3
MANUAL GAS SHUTOFF VALVE**

1210.9.4 Shutoff Valve for Laboratories. Each laboratory space containing two or more gas outlets installed on tables, benches, or in hoods in educational, research, commercial and industrial occupancies shall have a single shutoff valve through which all such gas outlets are supplied. The shutoff valve shall be accessible, located within the laboratory or adjacent to the laboratory's egress door, and identified. [NFPA 54:7.9.2.4]

1210.10 Prohibited Devices. No device shall be placed inside the gas piping or fittings that reduces the cross-sectional area or otherwise obstructs the free flow of gas, except where an allowance in the piping system design has been made for such a device and where approved by the Authority Having Jurisdiction. [NFPA 54:7.10]

1210.11 Systems Containing Gas-Air Mixtures Outside the Flammable Range. Where gas-air mixing machines are employed to produce mixtures above or below the flammable range, they shall be provided with stops to prevent adjustment of the mixture to within or approaching the flammable range. [NFPA 54:7.11]

1210.12 Systems Containing Flammable Gas-Air Mixtures. Systems containing flammable gas-air mixtures shall be in accordance with Section 1210.12.1 through Section 1210.15.6.

1210.12.1 Required Components. A central premix system with a flammable mixture in the blower or compressor shall consist of the following components:

(1) Gas-mixing machine in the form of an automatic gas-air proportioning device combined with a downstream blower or compressor.

(2) Flammable mixture piping, minimum Schedule 40.

(3) Automatic firecheck(s).

(4) Safety blowout(s) or backfire preventers for systems utilizing flammable mixture lines above 2½ inches (65 mm) nominal pipe size (NPS) or the equivalent. [NFPA 54:7.12.1]

1210.12.2 Optional Components. The following components shall also be permitted to be utilized in any type of central premix system:

(1) Flowmeter(s)

(2) Flame arrester(s) [NFPA 54:7.12.2]

1210.12.3 Additional Requirements. Gas-mixing machines shall have nonsparking blowers and shall be constructed so that a flashback does not rupture machine casings. [NFPA 54:7.12.3]

1210.12.4 Special Requirements for Mixing Blowers. A mixing blower system shall be limited to applications with minimum practical lengths of mixture piping, limited to a maximum mixture pressure of 10 inches water column (2.5 kPa) and limited to gases containing no more than 10 percent hydrogen. The blower shall be equipped with a gas-control valve at its air entrance arranged so that gas is admitted to the airstream, entering the blower in proper proportions for correct combustion by the type of burners employed, the said gas-control valve being of either the zero governor or mechanical ratio valve type that controls the gas and air adjustment simultaneously. No valves or other obstructions shall be installed between the blower discharge and the burner or burners. [NFPA 54:7.12.4]

1210.12.5 Installation of Gas-Mixing Machines. Installation of gas-mixing machines shall comply with the following:

(1) The gas-mixing machine shall be located in a well-ventilated area or in a detached building or cutoff room provided with room construction and explosion vents in accordance with sound engineering principles. Such rooms or below-grade installations shall have adequate positive ventilation.

(2) Where gas-mixing machines are installed in well-ventilated areas, the type of electrical equipment shall be in accordance with NFPA 70, for general service conditions unless other hazards in the area prevail. Where gas-mixing machines are installed in small detached buildings or cutoff rooms, the electrical equipment and wiring shall be installed in accordance with NFPA 70 for hazardous locations.

(3) Air intakes for gas-mixing machines using compressors or blowers shall be taken from outdoors whenever practical.

(4) Controls for gas-mixing machines shall include interlocks and a safety shutoff valve of the manual reset type in the gas supply connection to each machine arranged to automatically shut off the gas supply in the event of high or low gas pressure. Except for open-burner installations

FUEL GAS PIPING

only, the controls shall be interlocked so that the blower or compressor stops operating following a gas supply failure. Where a system employs pressurized air, means shall be provided to shut off the gas supply in the event of air failure.

(5) Centrifugal gas-mixing machines in parallel shall be reviewed by the user and equipment manufacturer before installation, and means or plans for minimizing the effects of downstream pulsation and equipment overload shall be prepared and utilized as needed. [NFPA 54:7.12.5.1 – 7.12.5.5]

1210.12.6 Use of Automatic Firechecks, Safety Blowouts, or Backfire Preventers.
Automatic firechecks and safety blowouts or backfire preventers shall be provided in piping systems distributing flammable air-gas mixtures from gas-mixing machines to protect the piping and the machines in the event of flashback, in accordance with the following:

(1) Approved automatic firechecks shall be installed upstream as close as practical to the burner inlets following the firecheck manufacturer's instructions.

(2) A separate manually operated gas valve shall be provided at each automatic firecheck for shutting off the flow of the gas-air mixture through the firecheck after a flashback has occurred. The valve shall be located upstream as close as practical to the inlet of the automatic firecheck.

Caution: These valves shall not be reopened after a flashback has occurred until the firecheck has cooled sufficiently to prevent re-ignition of the flammable mixture and has been reset properly.

(3) A safety blowout or backfiring preventer shall be provided in the mixture line near the outlet of each gas-mixing machine where the size of the piping is larger than 2½ inches (65 mm) NPS, or equivalent, to protect the mixing equipment in the event of an explosion passing through an automatic firecheck. The manufacturer's instructions shall be followed when installing these devices, particularly after a disc has burst. The discharge from the safety blowout or backfire preventer shall be located or shielded so that particles from the ruptured disc cannot be directed towards personnel. Wherever there are interconnected installations of gas-mixing machines with safety blowouts or backfire preventers; provision shall be made to keep the mixture from other machines from reaching any ruptured disc opening. Check valves shall not be used for this purpose.

(4) Large-capacity premix systems provided with explosion heads (rupture discs) to relieve excessive pressure in pipelines shall be located at and vented to a safe outdoor location. Provisions shall be provided for automatically shutting off the supply of the gas-air mixture in the event of a rupture. [NFPA 54:7.12.6]

1211.0 Electrical Bonding and Grounding.

1211.1 Pipe and Tubing other than CSST.
Each aboveground portion of a gas piping system other than CSST that is likely to become energized shall be electrically continuous and bonded to an effective ground-fault current path. Gas piping, other than CSST, shall be considered to be bonded where it is connected to appliances that are connected to the appliance grounding conductor of the circuit supplying that appliance. [NFPA 54:7.13.1]

There is normally one way that gas piping becomes electrically energized: through a failure of the appliance's wiring or electrical components that then energizes the appliance and potentially any gas piping connected to it. If this failure occurs, the grounding electrode provided to the appliance (where a three-wire circuit is provided) will protect the gas piping from becoming electrically energized. If there are no appliances with electrical components connected to the piping system, then the gas piping system is not likely to become energized.

The net result is that gas piping does not require a separate bonding connection unless the following situations occur:

1. There are gas appliances with electrical connections that are connected to ungrounded wiring systems (two-prong plugs).

2. There are reasons to believe that the gas piping system can become grounded by a source of electricity outside the piping system (this is highly unlikely).

1211.2 Bonding of CSST Gas Piping.
CSST gas piping systems, and gas piping systems containing one or more segments of CSST, shall be bonded to the electrical service grounding electrode system or, where provided, lightning protection grounding electrode system. [NFPA 54:7.13.2]

CSST systems are required by the manufacturer to be electrically bonded (see **Figure 1211.2a**).

Grounding vs Bonding

Grounding is a type of bonding, in which the conductive objects are connected to the earth using a conductor or metallic rod. The connection between an electrical circuit or instrument and the earth is known as Grounding. Grounding, sometimes called Earthing, ensures that all metal parts of an electrical circuit that an individual might contact are connected to the earth, thus ensuring zero voltage.

Bonding, by definition, is the joining of metallic pieces to form a conducting path which ensures safe electrical continuity. Bonding is generally done as protection from electrical shocks. Two or more conductive objects are required for a bonding connection, which is accomplished by connecting the metallic pieces together by means of a conductor (wire). Bonding provides the safety in the case of fault current. In the case of CSST (corrugated stainless steel tubing) used in fuel gas piping, bonding may reduce the risk of damage and fire from lighting strike by reducing the risk of arcing. According to the National Electrical Code (NEC), gas piping systems are considered to be direct-bonded if connected to the electrical service equipment enclosure, the grounded conductor at the electrical service, the grounding electrode conductor or one of more of the grounding electrodes used.

A common bonding installation for single or multi-family structures would be a single connection on the steel nipple downstream of the gas meter and prior to the first CSST connection (see **Figure 1211.2b**). The conductor should be no smaller than a 6 AWG copper wire and should be attached in accordance with the requirements found in the NEC. Bonding

FUEL GAS PIPING

clamps shall be installed in accordance with their listings per UL 467 and need to make metal-to-metal contact with a steel pipe component or the first CSST fitting (see **Figure 1211.2c**).

1211.2.1 Bonding Jumper Connection. The bonding jumper shall connect to a metallic pipe, pipe fitting, or CSST fitting. [NFPA 54:7.13.2.1]

1211.2.2 Bonding Jumper Size. The bonding jumper shall not be smaller than 6 AWG copper wire or equivalent. [NFPA 54:7.13.2.2]

1211.2.3 Bonding Jumper Length. The length of the jumper between the connection to the gas piping system and the grounding electrode system shall not exceed 75 feet (22 875 mm). Any additional electrodes shall be bonded to the electrical service grounding electrode system or, where provided, lightning protection grounding electrode system. [NFPA 54:7.13.2.3]

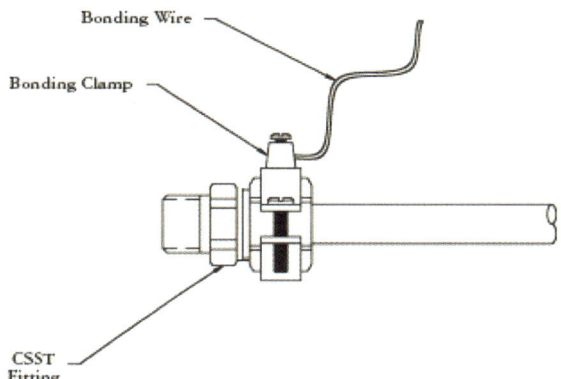

FIGURE 1211.2A
ELECTRICAL BONDING OF CSST SYSTEM

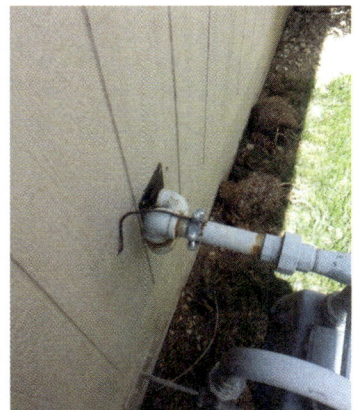

FIGURE 1211.2B
BONDING CONNECTION AT METER

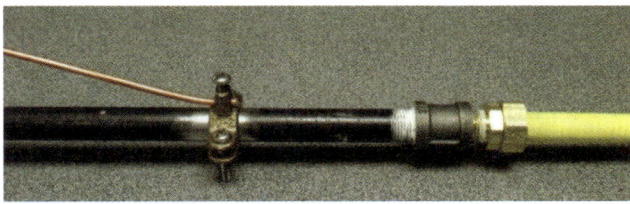

FIGURE 1211.2C
BONDING CLAMP AND CONDUCTOR

1211.2.4 Bonding Connections. Bonding connections shall be in accordance with NFPA 70. [NFPA 54:7.13.2.4]

1211.2.5 Devices Used for Bonding. Devices used for the bonding connection shall be listed for the application in accordance with UL 467. [NFPA 54:7.13.2.5]

1211.3 Prohibited Use. Gas piping shall not be used as a grounding conductor or electrode. [NFPA 54:7.13.3]

1211.4 Lightning Protection System. Where a lightning protection system is installed, the bonding of the gas piping shall be in accordance with NFPA 780. [NFPA 54:7.13.4]

1211.5 Electrical Circuits. Electrical circuits shall not utilize gas piping or components as conductors.

Exception: Low-voltage (50V or less) control circuits, ignition circuits, and electronic flame detection device circuits shall be permitted to make use of piping or components as a part of an electric circuit. [NFPA 54:7.14]

1211.6 Electrical Connections. All electrical connections between the wiring and electrically operated control devices in a piping system shall conform to the requirements of NFPA 70. [NFPA 54:7.15.1]

1211.6.1 Safety Control. Any essential safety control depending on electric current as the operating medium shall be of a type that will shut off (fail safe) the flow of gas in the event of current failure. [NFPA 54:7.15.2]

The safest way to ensure that the fuel gas system is free from becoming energized is to follow the manufacturer's installation requirements for the appliance, this code for installation of the piping and the National Electrical Code for installation and connection to electrical sources.

1212.0 Appliance and Equipment Connections to Building Piping.

1212.1 Connecting Appliances and Equipment. Appliances and equipment shall be connected to the building piping in compliance with Section 1212.5 through Section 1212.8 by one of the following:

The appliance shall be connected to the fuel gas piping with one of the following connection methods only after a manual shutoff valve is installed upstream of the connector (see Section 1212.6). A quick disconnect device, when used, must also be installed downstream from a manual shutoff valve (see Section 1212.7).

(1) Rigid metallic pipe and fittings.

See **Figure 1212.1(1)**.

(2) Semirigid metallic tubing and metallic fittings. Aluminum alloy tubing shall not be used in exterior locations.

(3) A listed connector in compliance with CSA Z21.24. The connector shall be used in accordance with the manufacturer's installation instructions and shall be in the same room as the appliance. Only one connector shall be used per appliance.

See **Figure 1212.1(3)**.

(4) A listed connector in compliance with CSA Z21.75. Only one connector shall be used per appliance.

(5) CSST where installed in accordance with the manufacturer's installation instructions.

FUEL GAS PIPING

(6) Listed nonmetallic gas hose connectors in accordance with Section 1212.3.

See **Figure 1212.1(6)**.

(7) Unlisted gas hose connectors for use in laboratories and educational facilities in accordance with Section 1212.4. [NFPA 54:9.6.1]

1212.1.1 Commercial Cooking Appliances. Connectors used with commercial cooking appliances that are moved for cleaning and sanitation purposes shall be installed in accordance with the connector manufacturer's installation instructions. Such connectors shall be listed in accordance with CSA Z21.69. [NFPA 54:9.6.1.3]

1212.1.2 Restraint. Movement of appliances with casters shall be limited by a restraining device installed in accordance with the connector and appliance manufacturer's installation instructions. [NFPA 54:9.6.1.4]

1212.2 Suspended Low-Intensity Infrared Tube Heaters. Suspended low-intensity infrared tube heaters shall be connected to the building piping system with a connector listed for the application in accordance with CSA Z21.24 as follows:

(1) The connector shall be installed in accordance with the tube heater installation instructions and shall be in the same room as the appliance.

(2) Only one connector shall be used per appliance. [NFPA 54:9.6.1.5]

1212.3 Use of Nonmetallic Gas Hose Connectors. Listed gas hose connectors shall be installed in accordance with the manufacturer's installation instructions and in accordance with Section 1212.3.1 and Section 1212.3.2. [NFPA 54:9.6.2]

1212.3.1 Indoor. Indoor gas hose connectors shall be used only to connect laboratory, shop, and ironing appliances requiring mobility during operation and installed in accordance with the following:

(1) An appliance shutoff valve shall be installed where the connector is attached to the building piping.

(2) The connector shall be of minimum length and shall not exceed 6 feet (1829 mm).

(3) The connector shall not be concealed and shall not extend from one room to another or pass through wall partitions, ceilings, or floors. [NFPA 54:9.6.2(1)]

1212.3.2 Outdoor. Where outdoor gas hose connectors are used to connect portable outdoor appliances, the connector shall be listed in accordance with CSA Z21.54 and installed in accordance with the following:

(1) An appliance shutoff valve, a listed quick-disconnect device, or a listed gas convenience outlet shall be installed where the connector is attached to the supply piping and in such a manner to prevent the accumulation of water or foreign matter.

(2) This connection shall be made only in the outdoor area where the appliance is to be used. [NFPA 54:9.6.2(2)]

The connector length shall not exceed 15 feet (4572 mm).

FIGURE 1212.1(1)
RIGID CONNECTION TO APPLIANCE

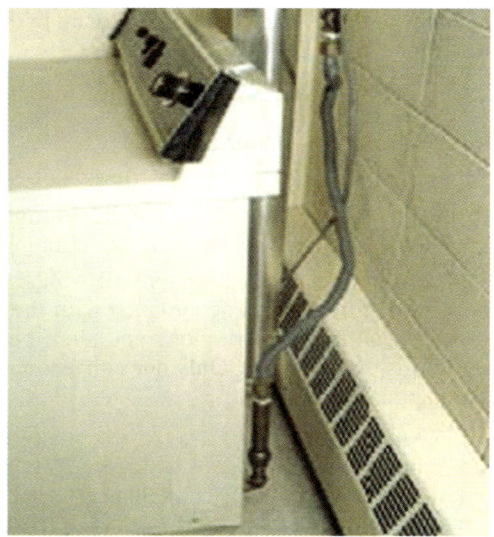

FIGURE 1212.1(3)
FLEXIBLE APPLIANCE CONNECTION

FIGURE 1212.1(6)
KITCHEN EQUIPMENT ON CASTERS WITH QUICK DISCONNECT AND SWIVEL

FUEL GAS PIPING

🔧 Two important requirements for maximum lengths of appliance connectors are contained in Sections 1212.3.1 and 1212.3.2. First, indoor appliance connectors, no matter of what material, shall be a maximum of 6 feet in length. Second, outdoor gas hose connections shall be a maximum of 15 feet in length (see **Figure 1212.3.2**). The following sections give no measurement as to the allowable length for flexible hoses for mobile equipment. The maximums above could be used for this equipment, but to be sure consult the AHJ.

FIGURE 1212.3.2
CONNECTION OF OUTDOOR EQUIPMENT

1212.4 Injection (Bunsen) Burners. Injection (Bunsen) burners used in laboratories and educational facilities shall be permitted to be connected to the gas supply by an unlisted hose. [NFPA 54:9.6.3]

1212.5 Connection of Portable and Mobile Industrial Gas Appliances. Where portable industrial appliances, or appliances requiring mobility or subject to vibration, are connected to the building gas piping system by the use of a flexible hose, the hose shall be suitable and safe for the conditions under which it can be used. [NFPA 54:9.6.4.1]

1212.5.1 Swivel Joints or Couplings. Where industrial appliances requiring mobility are connected to the rigid piping by the use of swivel joints or couplings, the swivel joints or couplings shall be suitable for the service required, and only the minimum number required shall be installed. [NFPA 54:9.6.4.2]

1212.5.2 Metal Flexible Connectors. Where industrial appliances subject to vibration are connected to the building piping system by the use of all metal flexible connectors, the connectors shall be suitable for the service required. [NFPA 54:9.6.4.3]

1212.5.3 Flexible Connectors. Where flexible connections are used, they shall be of the minimum practical length and shall not extend from one room to another or pass through any walls, partitions, ceilings, or floors. Flexible connections shall not be used in any concealed location. They shall be protected against physical or thermal damage and shall be provided with gas shutoff valves in readily accessible locations in rigid piping upstream from the flexible connections. [NFPA 54:9.6.4.4]

1212.6 Appliance Shutoff Valves and Connections. Each appliance connected to a piping system shall have an accessible, approved manual shutoff valve with a nondisplaceable valve member or a listed gas convenience outlet. Appliance shutoff valves and convenience outlets shall serve a single appliance only. The shutoff valve shall be located within 6 feet (1829 mm) of the appliance it serves. Where a connector is used, the valve shall be installed upstream of the connector. A union or flanged connection shall be provided downstream from the valve to permit removal of appliance controls. Shutoff valves serving decorative appliances shall be permitted to be installed in fireplaces if listed for such use. [NFPA 54:9.6.5, 9.6.5.1(A)(B)]

Exceptions:

(1) Shutoff valves shall be permitted to be accessibly located inside or under an appliance where such appliance is removed without removal of the shutoff valve.

(2) Shutoff valves shall be permitted to be accessibly located inside wall heaters and wall furnaces listed for recessed installation where necessary maintenance is performed without removal of the shutoff valve.

1212.7 Quick-Disconnect Devices. Quick-disconnect devices used to connect appliances to the building piping shall be listed to CSA Z21.41. Where installed indoors, an approved manual shutoff valve with a non-displaceable valve member shall be installed upstream of the quick-disconnect device. [NFPA 54:9.6.6.1 – 9.6.6.2]

1212.8 Gas Convenience Outlets. Appliances shall be permitted to be connected to the building piping by means of a listed gas convenience outlet, in conjunction with a listed appliance connector, installed in accordance with the manufacturer's installation instructions.

Gas convenience outlets shall be listed in accordance with CSA Z21.90 and installed in accordance with the manufacturer's installation instructions. [NFPA 54:9.6.7]

1212.9 Sediment Trap. Where a sediment trap is not incorporated as a part of the appliance, a sediment trap shall be installed downstream of the appliance shutoff valve as close to the inlet of the appliance as practical, before the flex connector, where used at the time of appliance installation. The sediment trap shall be either a tee fitting with a capped nipple in the bottom outlet, as illustrated in **Figure 1212.9**, or other device recognized as an effective sediment trap. Illuminating appliances, ranges, clothes dryers, decorative appliances for installation in vented fireplaces, gas fireplaces, and outdoor grills shall not be required to be so equipped. [NFPA 54:9.6.8]

🔧 The design of the sediment trap was discussed in Section 1210.6.2. Appliances other than those noted above

FUEL GAS PIPING

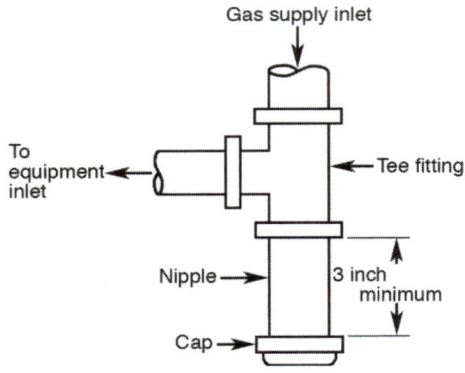

For SI units: 1 inch = 25.4 mm

FIGURE 1212.9
METHOD OF INSTALLING A TEE FITTING SEDIMENT TRAP
[NFPA 54: FIGURE 9.6.8]

shall be provided with a sediment trap. It is located prior to the connector allowing the sediment to settle into the leg of the trap and can be removed or serviced. A sediment trap after the flex connector serves no purpose as the sediment will already be trapped in the connector.

1212.10 Installation of Piping. Piping shall be installed in a manner not to interfere with inspection, maintenance, or servicing of the appliance. [NFPA 54:9.6.9]

When installing the fuel gas system care must be taken to ensure that the arrangement of the piping and connectors will allow the removal and maintenance of the appliance when necessary. This is also a requirement of Section 309.0, Workmanship.

1212.11 Liquefied Petroleum Gas Facilities and Piping. Liquefied petroleum gas facilities shall be in accordance with NFPA 58.

NFPA 58 regulates the installation of LP-gas supply to the gas regulator. This chapter of the code then regulates LP-gas piping from the regulator to the appliance.

1213.0 Pressure Testing, Inspection, and Purging.

1213.1 Piping Installations. Prior to acceptance and initial operation, all piping installations shall be visually inspected and pressure tested to determine that the materials, design, fabrication, and installation practices comply with the requirements of this code. [NFPA 54:8.1.1.1]

Before being put into operation, piping installations must be inspected and tested to determine that the total project complies with the code. The piping installation includes all fixed piping from the point of delivery to the equipment shutoff valves. Appliances and appliance connectors are not part of the fixed piping system and are not included in this test. The pressure test normally is conducted twice but only required once. The system should be tested at the rough inspection to ensure that any concealed piping does not leak. The second test is the final inspection prior to initial operation of the piping system. The system could be tested one more time if the piping system is modified or repaired.

The procedures contained in the following subsections are very specific and self explanatory. They should be followed in their entirety every time a fuel gas system is tested. Plumbers should always be mindful of the explosive nature of the contents of this piping. The protection of the lives of the occupants in the building in which this piping is installed is truly in the plumber's hands; it is a responsibility that should not be taken lightly.

1213.1.1 Inspection Requirements. Inspection shall consist of visual examination during or after manufacture, fabrication, assembly, or pressure tests. [NFPA 54:8.1.1.2]

1213.1.2 Repairs and Additions. Where repairs or additions are made following the pressure test, the affected piping shall be tested. Minor repairs and additions are not required to be pressure-tested provided that the work is inspected, and connections are tested with a noncorrosive leak-detecting fluid or other leak-detecting methods approved by the Authority Having Jurisdiction. [NFPA 54:8.1.1.3]

1213.1.3 New Branches. Where new branches are installed to new appliances the newly installed branch(es) shall be required to be pressure tested. Connections between the new piping and the existing piping shall be tested with a noncorrosive leak-detecting fluid or approved leak-detecting methods. [NFPA 54:8.1.1.4]

1213.1.4 Piping System. A piping system shall be tested as a complete unit or in sections. Under no circumstances shall a valve in a line be used as a bulkhead between gas in one section of the piping system and test medium in an adjacent section, unless a double block and bleed valve system is installed. A valve shall not be subjected to the test pressure unless it can be determined that the valve, including the valve closing mechanism, is designed to safely withstand the pressure. [NFPA 54:8.1.1.5]

1213.1.5 Regulators and Valves. Regulator and valve assemblies fabricated independently of the piping system in which they are to be installed shall be permitted to be tested with inert gas or air at the time of fabrication. [NFPA 54:8.1.1.6]

1213.1.6 Test Medium. The test medium shall be air, nitrogen, carbon dioxide, or an inert gas. OXYGEN SHALL NEVER BE USED. [NFPA 54:8.1.2]

1213.2 Test Preparation. Test preparation shall comply with Section 1213.2.1 through Section 1213.2.6.

1213.2.1 Pipe Joints. Pipe joints, including welds, shall be left exposed for examination during the test.

Exception: Covered or concealed pipe end joints that have been previously tested in accordance with this code. [NFPA 54:8.1.3.1]

1213.2.2 Expansion Joints. Expansion joints shall be provided with temporary restraints, where required, for the additional thrust load under test. [NFPA 54:8.1.3.2]

1213.2.3 Appliances and Equipment. Appliances and equipment that are not to be included in the test shall be either disconnected from the piping or isolated by blanks, blind flanges, or caps. Flanged joints at which blinds are inserted to blank off other equipment during the test shall not be required to be tested. [NFPA 54:8.1.3.3]

1213.2.4 Designed for Operating Pressures Less Than Test Pressure. Where the piping system is connected to appliances or equipment designed for operating pressures of less than the test pressure, such appliances or equipment shall be isolated from the piping system by disconnecting them and capping the outlet(s). [NFPA 54:8.1.3.4]

See **Figure 1210.7.1**.

1213.2.5 Designed for Operating Pressures Equal to or Greater Than Test Pressure. Where the piping system is connected to appliances or equipment designed for operating pressures equal to or greater than the test pressure, such appliances or equipment shall be isolated from the piping system by closing the individual appliance or equipment shutoff valve(s). [NFPA 54:8.1.3.5]

1213.2.6 Safety. All testing of piping systems shall be performed in a manner that protects the safety of employees and the public during the test. [NFPA 54:8.1.3.6]

1213.3 Test Pressure. This inspection shall include an air, CO_2, or nitrogen pressure test, at which time the gas piping shall stand a pressure of not less than 10 psi (69 kPa) gauge pressure. Test pressures shall be held for a length of time satisfactory to the Authority Having Jurisdiction but in no case less than 15 minutes with no perceptible drop in pressure. For welded piping, and for piping carrying gas at pressures in excess of 14 inches water column pressure (3.5 kPa), the test pressure shall be not less than 60 psi (414 kPa) and shall be continued for a length of time satisfactory to the Authority Having Jurisdiction, but in no case for less than 30 minutes. For CSST carrying gas at pressures in excess of 14 inches water column (3.5 kPa) pressure, the test pressure shall be not less than 30 psi (207 kPa) for 30 minutes. These tests shall be made using air, CO_2, or nitrogen pressure and shall be made in the presence of the Authority Having Jurisdiction. Necessary apparatus for conducting tests shall be furnished by the permit holder. Test gauges used in conducting tests shall be in accordance with Section 318.0.

1213.4 Detection of Leaks and Defects. The piping system shall withstand the test pressure specified without showing any evidence of leakage or other defects. Any reduction of test pressures as indicated by pressure gauges shall be deemed to indicate the presence of a leak unless such reduction can be readily attributed to some other cause. [NFPA 54:8.1.5.1]

1213.4.1 Detecting Leaks. The leakage shall be located by means of an approved gas detector, a noncorrosive leak detection fluid, or other approved leak detection methods. [NFPA 54:8.1.5.2]

See **Figure 1213.4.1**.

1213.4.2 Repair or Replace. Where leakage or other defects are located, the affected portion of the piping system shall be repaired or replaced and retested. [NFPA 54:8.1.5.3]

1213.5 Piping System Leak Test. Leak checks using fuel gas shall be permitted in piping systems that have been pressure-tested in accordance with Section 1213.0. [NFPA 54:8.2.1]

Once the fuel gas system has been pressure tested it must be leak checked before approval to turn on the fuel gas is given. The pressure test and leak check are often confused with each other. A pressure test is required for new piping installations and additions to piping installations, while a leak check is required whenever the gas system is initially placed into service or when the gas is turned back on after being turned off.

The test for leakage differs from the pressure test in Sections 1213.2 and 1213.3 in that it requires no special preparations. The medium used for a leak check is fuel gas at its normal supply pressure. The gas is applied to the total system (i.e., piping, equipment and equipment connections and valves).

Other test methods can be used. It is recommended that a written procedure for the method be developed and that steps are taken to ensure that all employees follow the method so that companies test every system identically.

1213.5.1 Turning Gas On. During the process of turning gas on into a system of new gas piping, the entire system shall be inspected to determine that there are no open fittings or ends and that valves at unused outlets are closed and plugged or capped. [NFPA 54:8.2.2]

1213.5.2 Leak Check. Immediately after the gas is turned on into a new system or into a system that has been initially restored after an interruption of service, the piping system shall be checked for leakage. Where leakage is indicated, the gas supply shall be shut off until the necessary repairs have been made. [NFPA 54:8.2.3]

1213.5.3 Placing Appliances and Equipment in Operation. Appliances and equipment shall not be placed in operation until after the piping system has been checked in accordance with Section 1213.5.2; connections to the appli-

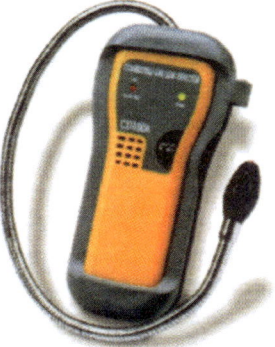

**FIGURE 1213.4.1
GAS LEAK DETECTION - LIQUID AND ELECTRONIC**

FUEL GAS PIPING

ance are checked for leakage and purged in accordance with Section 1213.6. [NFPA 54:8.2.4]

1213.6 Purging Requirements. The purging of piping shall be in accordance with Section 1213.6.1 through Section 1213.6.3. [NFPA 54:8.3]

The purging requirements for purging indoors and outdoors provide increased safety procedures including gas detection, monitoring, and the location of discharge points. Purging requirements are specified depending upon the size of the piping system in terms of pipe diameter/length and operating pressure. The pipe diameter/length and pressure criteria are selected to distinguish between large commercial systems (industrial systems) and small commercial systems (including residential systems). Large systems have pipe volumes or potential for higher flow rates that require procedures to ensure that large volumes of fuel gases are not released indoors and that flammable mixtures do not occur within the piping itself. Installers of these complex systems deal with considerably more variables that may result in a higher potential for discharge of large gas volumes during purging operations. All large systems are required to be purged outdoors and the use of a combustible gas indicator is mandated. For smaller systems, the purging requirements allow five options that have been shown to be effective and are widely used. These include purging directly to the outdoors; purging through the appliance's burner; purging through a standalone burner; purging to the indoors using a combustible gas detector; and purging in accordance with the written purging procedures of a gas supplier. The purging requirements recognize that the gas supplier has been successfully conducting purging operations with their trained personnel. Installers of smaller systems have familiarity with purging these systems and the potential for discharge of large gas volumes during purging operation is low.

1213.6.1 Piping Systems Required to be Purged Outdoors. The purging of piping systems shall be in accordance with the provisions of Section 1213.6.1.1 through Section 1213.6.1.5 where the piping system meets either of the following:

(1) The design operating gas pressure is greater than 2 psig (14 kPa).

(2) The piping being purged contains one or more sections of pipe or tubing meeting the size and length criteria of Table 1213.6.1. [NFPA 54:8.3.1]

TABLE 1213.6.1
SIZE AND LENGTH OF PIPING
[NFPA 54: TABLE 8.3.1]*

NOMINAL PIPING SIZE (inches)	LENGTH OF PIPING (feet)
≥ 2½ < 3	> 50
≥ 3 < 4	> 30
≥ 4 < 6	> 15
≥ 6 < 8	> 10
≥ 8	Any length

For SI units: 1 inch = 25 mm, 1 foot = 304.8 mm

* CSST EHD size of 62 is equivalent to nominal 2 inches (50 mm) pipe or tubing size.

1213.6.1.1 Removal from Service. Where existing gas piping is opened, the section that is opened shall be isolated from the gas supply and the line pressure vented in accordance with Section 1213.6.1.4. Where gas piping meeting the criteria of Table 1213.6.1 is removed from service, the residual fuel gas in the piping shall be displaced with an inert gas. [NFPA 54:8.3.1.1]

1213.6.1.2 Removal of Piping. Where piping containing gas is to be removed, the line shall be first disconnected from sources of gas and then thoroughly purged with air, water, or inert gas before cutting, or welding is done.

Residual gas fumes remaining in the fuel gas piping system must be removed when any type of repair or replacement occurs that will include fire or sparks. Although this subsection permits the use of air for purging, an inert gas or water may provide an increase in safety for larger volume systems. Cutting is not limited to flame cutting because mechanical methods, such as cutting with a chop saw, can produce sparks that can ignite a gas-air mixture.

1213.6.1.3 Placing in Operation. Where gas piping containing air and meeting the criteria of Table 1213.6.1 is placed in operation, the air in the piping shall first be displaced with an inert gas. The inert gas shall then be displaced with fuel gas in accordance with Section 1213.6.1.4. [NFPA 54:8.3.1.2]

1213.6.1.4 Outdoor Discharge of Purged Gases. The open end of a piping system being pressure vented or purged shall discharge directly to an outdoor location. Purging operations shall comply with the following requirements:

(1) The point of discharge shall be controlled with a shutoff valve.

(2) The point of discharge shall be located at least 10 feet (3048 mm) from sources of ignition, at least 10 feet (3048 mm) from building openings, and at least 25 feet (7620 mm) from mechanical air intake openings.

(3) During discharge, the open point of discharge shall be continuously attended and monitored with a combustible gas indicator that is in accordance with Section 1213.6.1.5.

(4) Purging operations introducing fuel gas shall be stopped where 90 percent fuel gas by volume is detected within the pipe.

(5) Persons not involved in the purging operations shall be evacuated from areas within 10 feet (3048 mm) of the point of discharge. [NFPA 54:8.3.1.3]

1213.6.1.5 Combustible Gas Indicator. Combustible gas indicators shall be listed and calibrated in accordance with the manufacturer's instructions. Combustible gas indicators shall numerically display a volume scale from 0 percent to 100 percent in 1 percent or smaller increments. [NFPA 54:8.3.1.4]

1213.6.2 Piping Systems Allowed to be Purged Indoors or Outdoors. The purging of piping systems shall be in accordance with the provisions of Section 1213.6.2.1 where the piping system meets both of the following:

(1) The design operating pressure is 2 psig (14 kPa) or less.

(2) The piping being purged is constructed entirely from

pipe or tubing not meeting the size and length criteria of Table 1213.6.1. [NFPA 54:8.3.2]

1213.6.2.1 Purging Procedure. The piping system shall be purged in accordance with one or more of the following:

(1) The piping shall be purged with fuel gas and shall discharge to the outdoors.

(2) The piping shall be purged with fuel gas and shall discharge to the indoors or outdoors through an appliance burner not located in a combustion chamber. Such burner shall be provided with a continuous source of ignition.

(3) The piping shall be purged with fuel gas and shall discharge to the indoors or outdoors through a burner that has a continuous source of ignition, and that is designed for such purpose.

(4) The piping shall be purged with fuel gas that is discharged to the indoors or outdoors, and the point of discharge shall be monitored with a listed combustible gas detector in accordance with Section 1213.6.2.2. Purging shall be stopped where fuel gas is detected.

(5) The piping shall be purged by the gas supplier in accordance with written procedures. [NFPA 54:8.3.2.1]

1213.6.2.2 Combustible Gas Detector. Combustible gas detectors shall be listed and calibrated or tested in accordance with the manufacturer's instructions. Combustible gas detectors shall be capable of indicating the presence of fuel gas. [NFPA 54:8.3.2.2]

1213.6.3 Purging Appliances and Equipment. After the piping system has been placed in operation, appliances and equipment shall be purged before being placed into operation. [NFPA 54:8.3.3]

1214.0 Required Gas Supply.

1214.1 General. The following regulations shall comply with this section and Section 1215.0, shall be the standard for the installation of gas piping. Natural gas regulations and tables are based on the use of a gas having a specific gravity of 0.60 and for undiluted liquefied petroleum gas, having a specific gravity of 1.50. Where gas of a different specific gravity is to be delivered, the specific gravity conversion factors provided by the serving gas supplier shall be used in sizing piping systems from the pipe sizing tables in this chapter.

The parameters of the sizing tables are described in this section. When using the sizing tables the installer should be sure to use the correct table for the gas, type of pipe, inlet pressure and specific gravity of gas that is being used. For example, Tables 1215.2(1) to 1215.2(23) require a specific gravity of 0.60. Tables 1215.2(24) to 1215.2(36) use a specific gravity of 1.50. Specific gravity is the ratio between the density, mass per unit volume, of natural gas and the density of air. The user should always double check the table for the correct parameters for the system he or she is sizing.

1214.2 Volume. The hourly volume of gas required at each piping outlet shall be taken as not less than the maximum hourly rating as specified by the manufacturer of the appliance or appliances to be connected to each such outlet.

Every appliance will have a Btu rating located on its information plate or label as shown in **Figure 1214.2**.

1214.3 Gas Appliances. Where the gas appliances to be installed have not been specified, Table 1208.4.1 shall be permitted to be used as a reference to estimate requirements of typical appliances.

To obtain the cubic feet per hour (m^3/h) of gas required, divide the input of the appliances by the average Btu (kW•h) heating value per cubic foot (m^3) of the gas. The average Btu (kW•h) per cubic foot (m^3) of the gas in the area of the installation shall be permitted to be obtained from the serving gas supplier.

If the appliance input rating is not known, Table 1208.4.1 may be used to estimate the rating. The table is very easy to use. It has two columns: one for common appliances and the other for the approximate input Btu/h rating of those appliances. For example, if the input rating for a 50-gallon water heater is not known, Table 1208.4.1 can be used. Referring to the table, a 50-gallon water heater has an approximate input rating of 50,000 Btu/h.

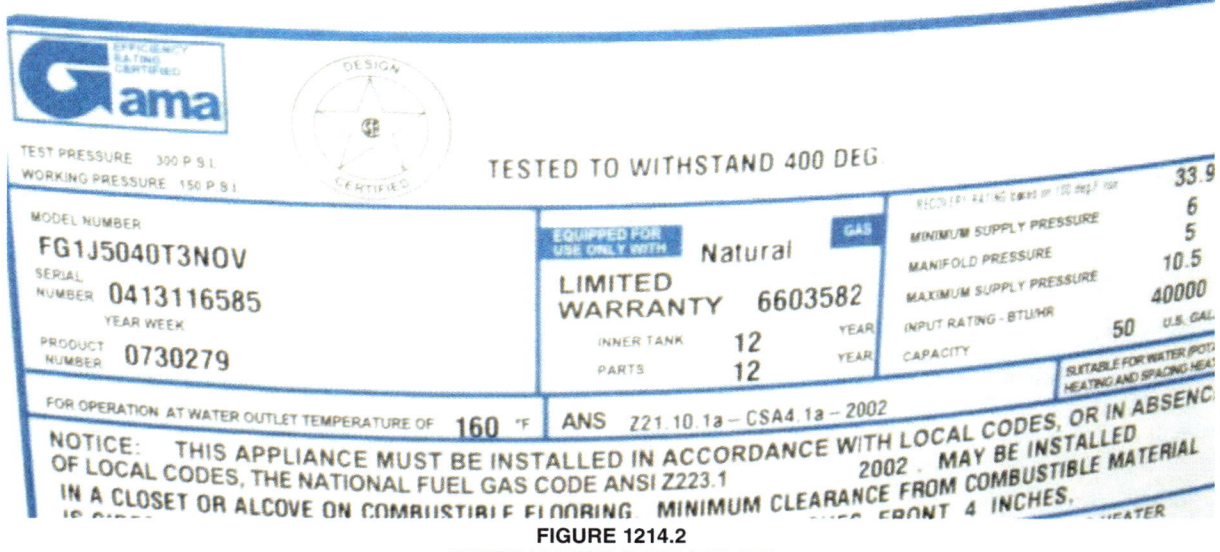

FIGURE 1214.2
WATER HEATER RATING PLATE

FUEL GAS PIPING

As seen on the label shown in **Figure 1214.2**, this gas water heater has a 50-gallon capacity and has an input of 40,000 Btu/h. If Table 1208.4.1 had been used, 50,000 Btu/h would be the input rating. To be precise and accurate, the user should have the true input rating of each appliance before sizing the system; however, this is sometimes impossible. Houses and buildings are designed and sometimes constructed before appliances are chosen. Therefore, the user should be conservative in his or her choice of appliance input ratings to ensure that the proper gas demands are met.

Natural gas meters typically record use in cubic feet (ft^3), the appliances demand is in British thermal units per hour (Btu/h), and the gas utility sells gas in a unit called a therm or 100,000 Btu. A therm factor is used by (Natural) gas suppliers to convert the volume of gas used to its heat equivalent (Btu/ft^3), and thus calculate the actual energy use.

To size the gas system for an adequate supply, the demand for each outlet must be known in Btu/h. When using the charts in Chapter 12, Btu/h must be converted to cubic foot per hour. To determine the cubic foot per hour value of the gas, the user will need to know the average Btu heating value per cubic foot of the gas being supplied. The local gas supplier can furnish this information. However, it may be easier to review a local gas bill looking for the term: billing factor, therm multiplier or an equivalent factor or multiplier. If, for example the therm multiplier or factor was 1.038, the Btu/ft^3 would be 1,038. If the factor was .947 the Btu/ft^3 would be 947.

To determine the cubic feet per hour of gas required, divide the input rating of the appliance by the average Btu heating value per cubic foot of the gas.

Example 1
The gas being supplied has a value of 1,000 Btu per cubic foot and the input rating of an appliance is 125,000 Btu/h. Divide 125,000 by 1,000. 125,000 ÷ 1,000 = 125 cubic feet per hour (ft^3/h).

Example 2
The gas being supplied has a value of 1,055 Btu per cubic foot and the input rating of an appliance is 125,000 Btu/h. Divide 125,000 by 1,055. 125,000 ÷ 1,055 = 118.48 cubic feet per hour or rounded to 119 ft^3/h.

Figure 1215.1.1 is an example using 1,100 Btu/ft^3 as the heat value for the gas. This is an example of how to convert Btu/h into cubic feet per hour. It is not a statement that all natural gas contains 1,100 Btu/ft^3.

Btu/ft^3 vary by source, geographic area, elevation and range from 800 to 1,200 Btu/ft^3.

Types of fuel:
Fuel gases are typically gaseous or begin as a liquid. Gaseous fuels most commonly encountered are natural gas, propane/air mix and butane. These are generally odorless and typically have a chemical called mercaptan (described as having the smell of rotting cabbage or malodorous socks) added to make it easier to detect leaks.

The heating value of these gases commonly vary from 800 to 1,200 Btu/ft^3 depending on their source. Some suppliers blend these gases attempting to supply gas at approximately 1,000 Btu/ft^3, however the blended gases are typically more but can also be less than 1,000 Btu/ft^3. The variations in the heating values per ft^3 is the reason gas sizing charts are in ft3/hr and not in Btu.

While gas meters measure gas use in cubic feet gas, utilities sell gas in therms. A therm is 100,000 Btu. To compute the actual heating value of the gas, the utility uses what is sometimes called a therm multiplier, therm factor, or therm billing.

The gas meter may register a month of use at 80 ft^3. When it comes time for billing, the gas supplier, knowing what the average Btu/ft^3 was that month, uses the therm multiplier to charge for every Btu. For example, if the average Btu/ft^3 for that month was 1023, the therm multiplier becomes 1.023. The actual use is:

80 (ft^3) x 100,000 (therm) x 1.023 (therm multiplier) = 8,184,000 Btu.

With that additional understanding you can now accurately calculate the ft^3/h required at each gas outlet.

Examples:
1. Demand of 40,000 Btu/h where the therm multiplier is 1.031

   40,000/1031 = 38.79 or rounded to 39 ft^3/h

2. Demand of 65,000 Btu/h where the therm multiplier is 1.045

   65,000/1045 = 62.2009 or rounded to 62 ft^3/h

Precautions when choosing the heating value:

1. What happens if the heating value of the gas changes next month or next year? You cannot account for changes in the future.

   Indeed, there is that possibility but how can you know how much the change might be and will it be a higher or lower heating value. Size the system accurately for the conditions that exist at the time of installation.

2. Using 1,000 Btu/ft^3 simplifies the math and leaves a margin for error. On the surface that sounds reasonable but what if the Btu/ft^3 is 867?

Many installers use 1,000 Btu/ft^3 typically over-sizing the system which is acceptable. Just understand what you are doing and why you are doing it.

1214.4 Size of Piping Outlets. The size of the supply piping outlet for a gas appliance shall be not less than ½ of an inch in diameter (15 mm).

The size of a piping outlet for a mobile home shall be not less than ¾ of an inch in diameter (20 mm).

🔧 Notice that the one-half inch requirement is for the minimum size of the supply piping to the appliance. It is not the minimum size of the appliance connector. The minimum size of the appliance connector should be the size of the appliance inlet. Often, there may be a three-eighths inch inlet to the appliance. The minimum pipe size, however, is still one-half inch.

1215.0 Required Gas Piping Size.

1215.1 Pipe Sizing Methods. Where the pipe size is to be determined using any of the methods in Section 1215.1.1 through Section 1215.1.3, the diameter of each pipe segment shall be obtained from the pipe sizing tables in Section 1215.2 or the sizing equations in Section 1215.3. [NFPA 54:6.1]

🔧 Section 1208.4.2, Sizing Methods, gives a clear direction in Item 1 of that section. It states that gas piping shall be sized in accordance with pipe sizing tables or sizing equations in this chapter. Systems falling within the range of the tables may be sized using those tables and the methods described earlier. It should be noted, however, that very large or long systems may benefit from the equation method (Section 1215.3), which could lead to smaller or more accurate pipe sizes.

These two methods of sizing are very similar. They both require the length of piping from the meter or point of delivery to reach the most remote outlet in the system. The objective of using this distance is to ensure that the required demand or amount of fuel gas needed for the last appliance along the developed length of system is delivered to that appliance. The sizing tables [Tables 1215.2(1) to 1215.2(36)] have incorporated the average pressure losses for each distance in the "Length" column for each pipe size in the "Capacity in Cubic Feet of Gas Per Hour" rows. Separate pressure loss calculations are not necessary if the system is designed with the average amount of fittings and length of pipe. This, by the way, would be the proper way to design the system.

1215.1.1 Longest Length Method. The pipe size of each section of gas piping shall be determined using the longest length of piping from the point of delivery to the most remote outlet and the load of the section (see calculation example in Figure 1215.1.1). [NFPA 54:6.1.1]

🔧 The Longest Length Method requires only one distance measurement – the length from the meter to the most remote outlet. That measurement is then used to size the entire system. In all of the sizing tables the far left column is the "Length" column. For example, if there is a system with 137 feet of pipe from the meter to the most remote outlet and Table 1215.2(1) is being used, you would use the 150 foot row to size all pipe segments in the system.

This is a very simple method for sizing a fuel gas system and, for relatively small systems, it yields very similar sizes of pipe compared to using the Branch Length Method. However, if the system is a large installation, the use of the Branch Length Method could give smaller sizes in branch segments, and therefore, provide cost savings for the owner.

1215.1.2 Branch Length Method. Pipe shall be sized as follows:

(1) The pipe size of each section of the longest pipe run from the point of delivery to the most remote outlet shall be determined using the longest run of piping and the load of the section.

(2) The pipe size of each section of branch piping not previously sized shall be determined using the length of piping from the point of delivery to the most remote outlet in each branch and the load of the section. [NFPA 54:6.1.2]

🔧 The Branch Length Method also requires the distance of the pipe to be from the meter to the most remote outlet but only for the purpose of sizing the system main. The piping from the remote outlet to the meter is also sized using the "Length" column in the sizing tables, but each branch off the main is measured from the meter to the furthest outlet on that branch. The piping from the remote outlet on that branch to the main is then sized from that "Length" column.

For example, if there is a distance from the meter to the most remote outlet of 137 feet, the 150 foot length row would be used again to size the system main. Next, each branch off of the main would have to be measured from the meter to the remote outlet on the branch, and then the piping from the remote outlet on the branch to the main would be sized. If there is a distance of 47 feet from the remote outlet on the branch to the meter, use the 50 foot row would be used to size the branch piping. Each branch would have to be sized in this manner.

1215.1.3 Hybrid Pressure. The pipe size for each section of higher pressure gas piping shall be determined using the longest length of piping from the point of delivery to the most remote line pressure regulator. The pipe size from the line pressure regulator to each outlet shall be determined using the length of piping from the regulator to the most remote outlet served by the regulator. [NFPA 54:6.1.3]

🔧 Large systems are sometimes designed to have a higher pressure than the normal fuel gas system. This inlet pressure is the pressure at the meter outlet, which is also the piping inlet. Using a higher inlet pressure will allow smaller main and branch pipe sizes. However, most appliances require a pressure regulator to bring the pressure down to the working pressure of the appliance. In a system such as this, with regulators at each appliance, it would be termed a "medium" or "high pressure" system, depending on the pressure of the gas. Sizing would be accomplished using only one sizing table.

The Hybrid Pressure Method is used to size fuel gas systems utilizing two or more different inlet pressure values. A hybrid pressure system is one that would require a line pressure regulator to step the higher main line pressure to a lower pressure, which is usually the appliance working pressure. In this design, only one line regulator is needed rather than several appliance regulators as in a high pressure system. Two sizing tables will have to be used to size the system – one to size the higher pressure portion of the system and another to size the reduced pressure portion of the system. Either sizing method, Branch or Longest Length, could be used in sizing the hybrid system. **Figure 1215.1.3** illustrates a complicated hybrid pressure system.

1215.2 Tables for Sizing Gas Piping Systems. Table 1215.2(1) through Table 1215.2(36) shall be used to size gas piping in conjunction with one of the methods described in Section 1215.1.1 through Section 1215.1.3. [NFPA 54:6.2]

🔧 Each sizing table is designed for a specific type of gas, a specific piping material, a specific inlet pressure, a specific pressure drop and a specific gravity of fuel gas. For example, Table 1215.2(1) is for natural gas with an inlet pressure less than two psi, for Schedule 40 Metallic Pipe, at a pressure drop of 0.5 inches of water column and for a gas with a specific gravity of 0.60.

Be sure to use the proper table when sizing the system. Not using the proper table will not only cause a plumber to fail a sizing exam but, if installed, the system could be undersized or oversized and thus might not provide the proper amount of fuel gas for the appliances used.

FUEL GAS PIPING

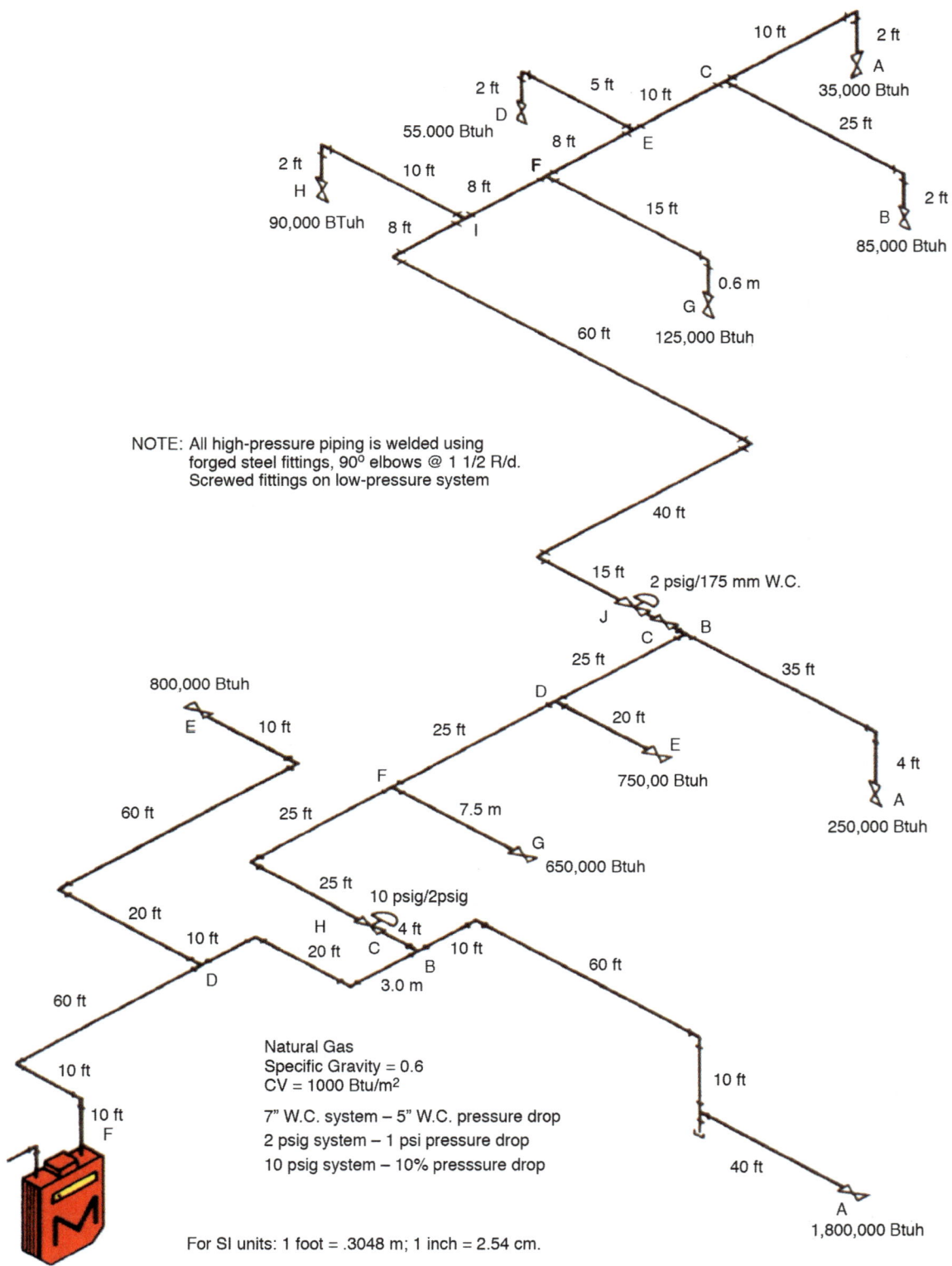

**FIGURE 1215.1.3
HYBRID PRESSURE GAS SYSTEM**

FUEL GAS PIPING

1215.3 Sizing Equations. The inside diameter of smooth-wall pipe or tubing shall be determined by Equation 1215.3(1), Equation 1215.3(2), Table 1215.3 and using the equivalent pipe length determined by the methods in Section 1215.1.1 through Section 1215.1.3. [NFPA 54:6.4]

EQUATION 1215.3(1)
LOW-PRESSURE GAS FORMULA (LESS THAN 1.5 psi)
[NFPA 54:6.4.1]

$$D = \frac{Q^{0.381}}{19.17 \left(\frac{\Delta H}{Cr \times L}\right)^{0.206}}$$

Where:
- D = inside diameter of pipe, inches
- Q = input rate appliance(s), cubic feet per hour at 60°F and 30 inch mercury column
- L = equivalent length of pipe, feet
- ΔH = pressure drop, inches water column
- Cr = in accordance with Table 1215.3

EQUATION 1215.3(2)
HIGH-PRESSURE GAS FORMULA (1.5 psi AND ABOVE)
[NFPA 54:6.4.2]

$$D = \frac{Q^{0.381}}{18.93 \left[\frac{(P_1^2 - P_2^2) \cdot Y}{Cr \times L}\right]^{0.206}}$$

Where:
- D = inside diameter of pipe, inches
- Q = input rate appliance(s), cubic feet per hour at 60°F and 30 inch mercury column
- P_1 = upstream pressure, psia (P_1 + 14.7)
- P_2 = downstream pressure, psia (P_2 + 14.7)
- L = equivalent length of pipe, feet
- Cr = in accordance with Table 1215.3
- Y = in accordance with Table 1215.3

For SI units: 1 cubic foot = 0.0283 m³, 1000 British thermal units per hour = 0.293 kW, 1 inch = 25 mm, 1 foot = 304.8 mm, 1 pound-force per square inch = 6.8947 kPa, °C = (°F-32)/1.8, 1 inch mercury column = 3.39 kPa, 1 inch water column = 0.249 kPa

TABLE 1215.3
Cr AND Y FOR NATURAL GAS AND UNDILUTED PROPANE AT STANDARD CONDITIONS
[NFPA 54: TABLE 6.4.2]

GAS	FORMULA FACTORS	
	Cr	Y
Natural Gas	0.6094	0.9992
Undiluted Propane	1.2462	0.9910

These are the equations that generated the Tables. The equations can be easily used once the givens are known. Suppose the natural gas is low pressure less than 2psi. The input rating of the appliance is 50 cubic feet per hour and the length is 100 feet. The pressure drop is 0.5 inches water column. Using Equation 1215.3(1), find the required diameter. The givens are,

$Q = 50$

$L = 100$

$\Delta H = 0.5$

$Cr = 0.6094$ (Table 1216.3, Natural Gas)

Therefore, the equation will look like this,

$$D = \frac{50^{0.381}}{19.17 \left(\frac{0.5}{0.6094 \times 100}\right)^{0.206}}$$

$D = 0.6229$

Refer to Table 1215.2(1) and find 100 feet in the first column and 50 cubic feet per hour in the second column. The required diameter is ½ inch or 0.622 inches ID, which is the same that the equation yielded. These equations can be utilized when exact lengths and input ratings are known.

1215.4 Sizing of Piping Sections. To determine the size of each section of pipe in a system within the range of Table 1215.2(1) through Table 1215.2(36), proceed as follows:

(1) Measure the length of the pipe from the gas meter location to the most remote outlet on the system.

(2) Select the length in feet column and row showing the distance, or the next longer distance where the table does not give the exact length.

(3) Starting at the most remote outlet, find in the row just selected the gas demand for that outlet. Where the exact figure of demand is not shown, choose the next larger figure in the row.

(4) At the top of this column will be found the correct size of pipe.

(5) Using this same row, proceed in a similar manner to each section of pipe serving this outlet. For each section of pipe, determine the total gas demand supplied by that section. Where gas piping sections serve both heating and cooling appliances and the installation prevents both units from operating simultaneously, the larger of the two demand loads needs to be used in sizing these sections.

(6) Size each section of branch piping not previously sized by measuring the distance from the gas meter location to the most remote outlet in that branch and follow the procedures of steps 2, 3, 4, and 5 above. Size branch piping in the order of their distance from the meter location, beginning with the most distant outlet not previously sized.

The steps enumerated here are for sizing by the Branch Length Method. For the Longest Length Method, (6) is eliminated and at (5) each outlet and pipe segment in the entire system is sized from the same length row.

FUEL GAS PIPING

1215.5 Engineering Methods. For conditions other than those covered by Section 1215.1, such as longer runs or greater gas demands, the size of each gas piping system shall be determined by standard engineering methods acceptable to the Authority Having Jurisdiction, and each such system shall be so designed that the total pressure drop between the meter or another point of supply and an outlet where full demand is being supplied to all outlets, shall be in accordance with the requirements of Section 1208.4.

1215.6 Variable Gas Pressure. Where the supply gas pressure exceeds 5 psi (34.6 kPa) for natural gas and 10 psi (69 kPa) for undiluted propane or is less than 6 inches (1.5 kPa) of water column, or where diversity demand factors are used, the design, pipe, sizing, materials, location, and use of such systems first shall be approved by the Authority Having Jurisdiction. Piping systems designed for pressures exceeding the serving gas supplier's standard delivery pressure shall have prior verification from the gas supplier of the availability of the design pressure.

Sizing the Gas System

Sizing the gas system can be very easy if the proper steps are followed. The following steps summarize all of the sizing criteria that have been discussed in this chapter.

1. Determine the type of gas piping material to be used, the inlet pressure of the gas and its specific gravity and the allowable pressure drop. Use this information to determine which sizing table will be used to size the system.
2. Identify the appliances on the system and their Btu input rating. Refer to the commentary following section 1214.3 when the appliance Btu input rating is unknown.
3. Determine the heating value of the gas supplied and convert the appliance Btu/h input rating to cubic feet per hour (ft3/h).
4. Note the cubic feet per hour at each appliance as in **Figure 1215.1.1** Example Illustrating Use of Tables 1208.4.1 and 1215.2(1).
5. Measure each appliance outlet from the meter to the appliance along the piping system and note the distance for each appliance on the print or paper. The branch with the longest distance will be the most remote outlet. The piping from this outlet back to the meter will be the "system main."
6. Note the total cubic feet per hour for each pipe segment of the branches and main in the system starting from the most remote outlet from the meter and working back to the meter, totaling the demand for each segment.

Knowing these steps use the example provided in **Figure 1215.1.1** when applying the following sizing methods below. The given piping material is Schedule 40 metallic pipe. The gas to be used is natural gas with a specific gravity of 0.60 and a heating value of 1,100 Btu/ft^3, delivered at less than 2 inches w.c. and with a pressure drop of 0.5 inches water column.

Based on these givens, Table 1215.2(1) will be used to size the system. The system is now ready to be sized using either of the two sizing methods: Longest Length Method or the Branch Length Method.

The Hybrid Pressure Method would not be used because there is only one pressure value within the system.

Longest Length Method

1. Measure the length of the pipe from the gas meter location to the most remote outlet on the system. In this example, the most remote outlet is A at 60 feet from the meter.
2. In Table 1215.2(1) select the distance in the "Length" column that is equal to or greater than 60 feet. This will determine the row used for pipe sizing (see highlighted row in Table 1215.2(1) extraction on following page).
3. Convert each appliance's Btu/h rating into cubic feet per hour by dividing the Btu/h rating by 1,100 and round to the closest whole number.

If the Btu/h rating is not known, use Table 1208.4.1.

30-gallon water heater:	32 ft^3/h
Gas Refrigerator:	3 ft^3/h
Range:	59 ft^3/h
Furnace:	136 ft^3/h

4. Starting at the most remote outlet, find in this same row the gas demand for that outlet. If the exact number for the demand is not shown, choose the next larger number in the row.

Outlet	Appliance	Cubic Feet of Gas/hr	Nominal Pipe Diameter
A	Water Heater, 30gal	32	½"
B	Gas Refrigerator	3	½"
C	Range	59	½"
D	Furnace	136	¾"

5. At the top of this column, the nominal pipe size will be found.

Using the same method in the previous step, proceed in a similar manner using the 60 feet length row for each section of pipe serving the outlets.

Pipe Section	Cubic Feet of Gas/hr	Nominal Pipe Diameter
1	35	½"
2	94	¾"
3	230	1"

Branch Length Method

1. Follow the same steps 1-3 as the Longest Length Method.
2. Follow the same step 5 as the Longest Length Method. In the Branch Length Method, the longest pipe run is sized first. Since the longest run is 60 feet, use the 60 feet Length row in Table 1215.2(1) to size each section of the

longest run. The results are the same as tabulated in step 5 of the previous method.

3. Size each branch as determined by the length from the point of delivery to the outlet. This is the differing step from the Longest Length Method. Determine the length of the pipe run from the beginning of the meter to each outlet. Then use that Length row in Table 1215.2(1) that is equal to or greater than the length to find the diameter of pipe.

Outlet	Length	Cubic Feet of Gas/hr	Nominal Pipe Diameter
A	60	32	½"
B	55	3	½"
C	55	59	½"
D	50	136	¾"

For outlets A, B, and C, the 60 feet Length row is used in Table 1215.2(1) (see the example on the next page). Outlet D uses the 50 feet Length row. The sizing results are the same using either method in this small example. Actual gas systems will be more complicated, and each method may yield differing results allowing the choice for the more economical method.

Table 1216.2(1)
Schedule 40 Metallic Pipe [NFPA 54: Table 6.2(b)][1, 2]

									Gas:	Natural			
									INLET PRESSURE:	Less than 2 psi			
									Pressure Drop:	0.5 in. w.c.			
									Specific Gravity:	0.60			

	Pipe Size (inch)												
Nominal:	1/2	3/4	1	1 1/4	1 1/2	2	2 1/2	3	4	5	6	8	10	12
Actual ID:	0.622	0.824	1.049	1.380	1.610	2.067	2.469	3.068	4.026	5.047	6.065	7.981	10.020	11.938
Length (feet)	Capacity in Cubic Feet of Gas per Hour												
10	172	360	678	1390	2090	4020	6400	11 300	23 100	41 800	67 600	139 000	252 000	399 000
20	118	247	466	957	1430	2760	4400	7780	15 900	28 700	46 500	95 500	173 000	275 000
30	95	199	374	768	1150	2220	3530	6250	12 700	23 000	37 300	76 700	139 000	220 000
40	81	170	320	657	985	1900	3020	5350	10 900	19 700	31 900	65 600	119 000	189 000
50	72	151	284	583	873	1680	2680	4740	9660	17 500	28 300	58 200	106 000	167 000
60	65	137	257	528	791	1520	2430	4290	8760	15 800	25 600	52 700	95 700	152 000
70	60	126	237	486	728	1400	2230	3950	8050	14 600	23 600	48 500	88 100	139 000
80	56	117	220	452	677	1300	2080	3670	7490	13 600	22 000	45 100	81 900	130 000
90	52	110	207	424	635	1220	1950	3450	7030	12 700	20 600	42 300	76 900	122 000
100	50	104	195	400	600	1160	1840	3260	6640	12 000	19 500	40 000	72 600	115 000

EXAMPLE EXTRACTED FROM TABLE 1216.2(1)

FUEL GAS PIPING

FIGURE 1215.1.1
EXAMPLE ILLUSTRATING USE OF TABLE 1208.4.1 AND TABLE 1215.2(1)

Problem: Determine the required pipe size of each section and outlet of the piping system shown in Figure 1215.1.1. Gas to be used has a specific gravity of 0.60 and 1100 British thermal units (Btu) per cubic foot (0.0114 kW·h/L), delivered at 8 inch water column (1.9 kPa) pressure.

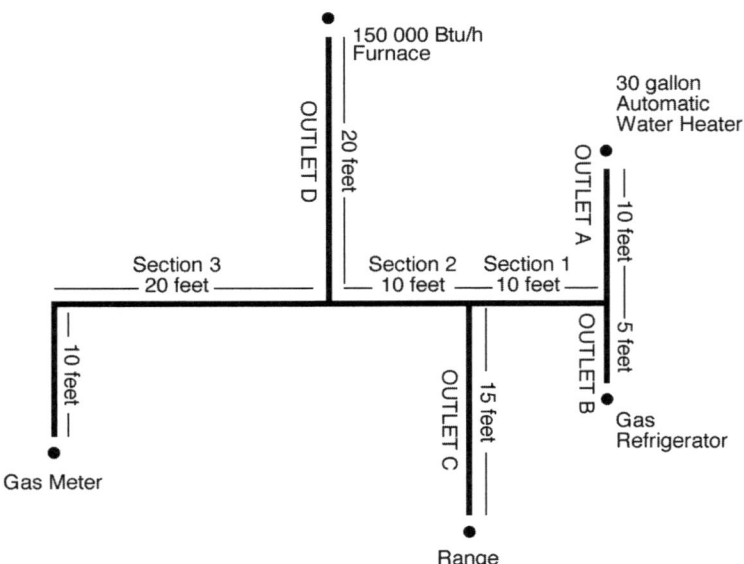

For SI units: 1 foot = 304.8 mm, 1 gallon = 3.785 L, 1000 British thermal units per hour = 0.293 kW, 1 cubic foot per hour = 0.0283 m³/h

Solution:

(1) Maximum gas demand of Outlet A —

    32 cubic feet per hour (0.91 m³/h) (from Table 1208.4.1).

Maximum gas demand of Outlet B —

    3 cubic feet per hour (0.08 m³/h) (from Table 1208.4.1).

Maximum gas demand of Outlet C —

    59 cubic feet per hour (1.67 m³/h) (from Table 1208.4.1).

Maximum gas demand of Outlet D —

    136 cubic feet per hour (3.85 m³/h) [150 000 Btu/hour (44 kW) divided by 1100 Btu per cubic foot (0.0114 kW·h/L)].

(2) The length of pipe from the gas meter to the most remote outlet (Outlet A) is 60 feet (18 288 mm).

(3) Using the length in feet column row marked 60 feet (18 288 mm) in Table 1215.2(1):

Outlet A, supplying 32 cubic feet per hour (0.91 m³/h), requires ½ of an inch (15 mm) pipe.

Section 1, supplying Outlets A and B, or 35 cubic feet per hour (0.99 m³/h) requires ½ of an inch (15 mm) pipe.

Section 2, supplying Outlets A, B, and C, or 94 cubic feet per hour (2.66 m³/h) requires ¾ of an inch (20 mm) pipe.

Section 3, supplying Outlets A, B, C, and D, or 230 cubic feet per hour (6.51 m³/h), requires 1 inch (25 mm) pipe.

(4) Using the column marked 60 feet (18 288 mm) in Table 1215.2(1) [no column for actual length of 55 feet (16 764 mm)]:

Outlet B supplying 3 cubic feet per hour (0.08 m³/h), requires ½ of an inch (15 mm) pipe.

Outlet C, supplying 59 cubic feet per hour (1.67 m³/h), requires ½ of an inch (15 mm) pipe.

(5) Using the column marked 60 feet (18 288 mm) in Table 1215.2(1):

Outlet D, supplying 136 cubic feet per hour (3.85 m³/h), requires ¾ of an inch (20 mm) pipe.

FUEL GAS PIPING

TABLE 1215.2(1)
SCHEDULE 40 METALLIC PIPE [NFPA 54: TABLE 6.2(b)][1, 2]

									GAS:	NATURAL			
									INLET PRESSURE:	LESS THAN 2 psi			
									PRESSURE DROP:	0.5 in. w.c.			
									SPECIFIC GRAVITY:	0.60			
	PIPE SIZE (inch)												
NOMINAL:	½	¾	1	1¼	1½	2	2½	3	4	5	6	8	10	12
ACTUAL ID:	0.622	0.824	1.049	1.380	1.610	2.067	2.469	3.068	4.026	5.047	6.065	7.981	10.020	11.938
LENGTH (feet)	CAPACITY IN CUBIC FEET OF GAS PER HOUR												
10	172	360	678	1390	2090	4020	6400	11 300	23 100	41 800	67 600	139 000	252 000	399 000
20	118	247	466	957	1430	2760	4400	7780	15 900	28 700	46 500	95 500	173 000	275 000
30	95	199	374	768	1150	2220	3530	6250	12 700	23 000	37 300	76 700	139 000	220 000
40	81	170	320	657	985	1900	3020	5350	10 900	19 700	31 900	65 600	119 000	189 000
50	72	151	284	583	873	1680	2680	4740	9660	17 500	28 300	58 200	106 000	167 000
60	65	137	257	528	791	1520	2430	4290	8760	15 800	25 600	52 700	95 700	152 000
70	60	126	237	486	728	1400	2230	3950	8050	14 600	23 600	48 500	88 100	139 000
80	56	117	220	452	677	1300	2080	3670	7490	13 600	22 000	45 100	81 900	130 000
90	52	110	207	424	635	1220	1950	3450	7030	12 700	20 600	42 300	76 900	122 000
100	50	104	195	400	600	1160	1840	3260	6640	12 000	19 500	40 000	72 600	115 000
125	44	92	173	355	532	1020	1630	2890	5890	10 600	17 200	35 400	64 300	102 000
150	40	83	157	322	482	928	1480	2610	5330	9650	15 600	32 100	58 300	92 300
175	37	77	144	296	443	854	1360	2410	4910	8880	14 400	29 500	53 600	84 900
200	34	71	134	275	412	794	1270	2240	4560	8260	13 400	27 500	49 900	79 000
250	30	63	119	244	366	704	1120	1980	4050	7320	11 900	24 300	44 200	70 000
300	27	57	108	221	331	638	1020	1800	3670	6630	10 700	22 100	40 100	63 400
350	25	53	99	203	305	587	935	1650	3370	6100	9880	20 300	36 900	58 400
400	23	49	92	189	283	546	870	1540	3140	5680	9190	18 900	34 300	54 300
450	22	46	86	177	266	512	816	1440	2940	5330	8620	17 700	32 200	50 900
500	21	43	82	168	251	484	771	1360	2780	5030	8150	16 700	30 400	48 100
550	20	41	78	159	239	459	732	1290	2640	4780	7740	15 900	28 900	45 700
600	19	39	74	152	228	438	699	1240	2520	4560	7380	15 200	27 500	43 600
650	18	38	71	145	218	420	669	1180	2410	4360	7070	14 500	26 400	41 800
700	17	36	68	140	209	403	643	1140	2320	4190	6790	14 000	25 300	40 100
750	17	35	66	135	202	389	619	1090	2230	4040	6540	13 400	24 400	38 600
800	16	34	63	130	195	375	598	1060	2160	3900	6320	13 000	23 600	37 300
850	16	33	61	126	189	363	579	1020	2090	3780	6110	12 600	22 800	36 100
900	15	32	59	122	183	352	561	992	2020	3660	5930	12 200	22 100	35 000
950	15	31	58	118	178	342	545	963	1960	3550	5760	11 800	21 500	34 000
1000	14	30	56	115	173	333	530	937	1910	3460	5600	11 500	20 900	33 100
1100	14	28	53	109	164	316	503	890	1810	3280	5320	10 900	19 800	31 400
1200	13	27	51	104	156	301	480	849	1730	3130	5070	10 400	18 900	30 000
1300	12	26	49	100	150	289	460	813	1660	3000	4860	9980	18 100	28 700
1400	12	25	47	96	144	277	442	781	1590	2880	4670	9590	17 400	27 600
1500	11	24	45	93	139	267	426	752	1530	2780	4500	9240	16 800	26 600
1600	11	23	44	89	134	258	411	727	1480	2680	4340	8920	16 200	25 600
1700	11	22	42	86	130	250	398	703	1430	2590	4200	8630	15 700	24 800
1800	10	22	41	84	126	242	386	682	1390	2520	4070	8370	15 200	24 100
1900	10	21	40	81	122	235	375	662	1350	2440	3960	8130	14 800	23 400
2000	NA	20	39	79	119	229	364	644	1310	2380	3850	7910	14 400	22 700

For SI units: 1 inch = 25 mm, 1 foot = 304.8 mm, 1 cubic foot per hour = 0.0283 m³/h, 1 pound-force per square inch = 6.8947 kPa, 1 inch water column = 0.249 kPa

Notes:
[1] Table entries are rounded to 3 significant digits.
[2] NA means a flow of less than 10 ft³/h (0.283 m³/h).

FUEL GAS PIPING

TABLE 1215.2(2)
SCHEDULE 40 METALLIC PIPE [NFPA 54: TABLE 6.2(c)]*

								GAS:	NATURAL
								INLET PRESSURE:	LESS THAN 2 psi
								PRESSURE DROP:	3.0 in. w.c.
								SPECIFIC GRAVITY:	0.60

INTENDED USE: INITIAL SUPPLY PRESSURE OF 8.0 IN. W.C. OR GREATER									
	PIPE SIZE (inch)								
NOMINAL:	½	¾	1	1¼	1½	2	2½	3	4
ACTUAL ID:	0.622	0.824	1.049	1.380	1.610	2.067	2.469	3.068	4.026
LENGTH (feet)	CAPACITY IN CUBIC FEET OF GAS PER HOUR								
10	454	949	1790	3670	5500	10 600	16 900	29 800	60 800
20	312	652	1230	2520	3780	7280	11 600	20 500	41 800
30	250	524	986	2030	3030	5840	9310	16 500	33 600
40	214	448	844	1730	2600	5000	7970	14 100	28 700
50	190	397	748	1540	2300	4430	7060	12 500	25 500
60	172	360	678	1390	2090	4020	6400	11 300	23 100
70	158	331	624	1280	1920	3690	5890	10 400	21 200
80	147	308	580	1190	1790	3440	5480	9690	19 800
90	138	289	544	1120	1670	3230	5140	9090	18 500
100	131	273	514	1060	1580	3050	4860	8580	17 500
125	116	242	456	936	1400	2700	4300	7610	15 500
150	105	219	413	848	1270	2450	3900	6890	14 100
175	96	202	380	780	1170	2250	3590	6340	12 900
200	90	188	353	726	1090	2090	3340	5900	12 000
250	80	166	313	643	964	1860	2960	5230	10 700
300	72	151	284	583	873	1680	2680	4740	9660
350	66	139	261	536	803	1550	2470	4360	8890
400	62	129	243	499	747	1440	2290	4050	8270
450	58	121	228	468	701	1350	2150	3800	7760
500	55	114	215	442	662	1280	2030	3590	7330
550	52	109	204	420	629	1210	1930	3410	6960
600	50	104	195	400	600	1160	1840	3260	6640
650	47	99	187	384	575	1110	1760	3120	6360
700	46	95	179	368	552	1060	1690	3000	6110
750	44	92	173	355	532	1020	1630	2890	5890
800	42	89	167	343	514	989	1580	2790	5680
850	41	86	162	332	497	957	1530	2700	5500
900	40	83	157	322	482	928	1480	2610	5330
950	39	81	152	312	468	901	1440	2540	5180
1000	38	79	148	304	455	877	1400	2470	5040
1100	36	75	141	289	432	833	1330	2350	4780
1200	34	71	134	275	412	794	1270	2240	4560
1300	33	68	128	264	395	761	1210	2140	4370
1400	31	65	123	253	379	731	1160	2060	4200
1500	30	63	119	244	366	704	1120	1980	4050
1600	29	61	115	236	353	680	1080	1920	3910
1700	28	59	111	228	342	658	1050	1850	3780
1800	27	57	108	221	331	638	1020	1800	3670
1900	27	56	105	215	322	619	987	1750	3560
2000	26	54	102	209	313	602	960	1700	3460

For SI units: 1 inch = 25 mm, 1 foot = 304.8 mm, 1 cubic foot per hour = 0.0283 m^3/h, 1 pound-force per square inch = 6.8947 kPa, 1 inch water column = 0.249 kPa
* Table entries are rounded to 3 significant digits.

TABLE 1215.2(3)
SCHEDULE 40 METALLIC PIPE [NFPA 54: TABLE 6.2(d)]*

							GAS:	NATURAL
							INLET PRESSURE:	LESS THAN 2 psi
							PRESSURE DROP:	6.0 in. w.c.
							SPECIFIC GRAVITY:	0.60

INTENDED USE: INITIAL SUPPLY PRESSURE OF 11.0 IN. W.C. OR GREATER										
	PIPE SIZE (inch)										
NOMINAL:	½	¾	1	1¼	1½	2	2½	3	4
ACTUAL ID:	0.622	0.824	1.049	1.38	1.61	2.067	2.469	3.068	4.026
LENGTH (feet)	CAPACITY IN CUBIC FEET OF GAS PER HOUR								
10	660	1380	2600	5340	8000	15 400	24 600	43 400	88 500
20	454	949	1790	3670	5500	10 600	16 900	29 800	60 800
30	364	762	1440	2950	4410	8500	13 600	24 000	48 900
40	312	652	1230	2520	3780	7280	11 600	20 500	41 800
50	276	578	1090	2240	3350	6450	10 300	18 200	37 100
60	250	524	986	2030	3030	5840	9310	16 500	33 600
70	230	482	907	1860	2790	5380	8570	15 100	30 900
80	214	448	844	1730	2600	5000	7970	14 100	28 700
90	201	420	792	1630	2440	4690	7480	13 200	27 000
100	190	397	748	1540	2300	4430	7060	12 500	25 500
125	168	352	663	1360	2040	3930	6260	11 100	22 600
150	153	319	601	1230	1850	3560	5670	10 000	20 500
175	140	293	553	1140	1700	3270	5220	9230	18 800
200	131	273	514	1056	1580	3050	4860	8580	17 500
250	116	242	456	936	1400	2700	4300	7610	15 500
300	105	219	413	848	1270	2450	3900	6890	14 100
350	96	202	380	780	1170	2250	3590	6340	12 900
400	90	188	353	726	1090	2090	3340	5900	12 000
450	84	176	332	681	1020	1960	3130	5540	11 300
500	80	166	313	643	964	1860	2960	5230	10 700
550	76	158	297	611	915	1760	2810	4970	10 100
600	72	151	284	583	873	1680	2680	4740	9660
650	69	144	272	558	836	1610	2570	4540	9250
700	66	139	261	536	803	1550	2470	4360	8890
750	64	134	252	516	774	1490	2380	4200	8560
800	62	129	243	499	747	1440	2290	4050	8270
850	60	125	235	483	723	1390	2220	3920	8000
900	58	121	228	468	701	1350	2150	3800	7760
950	56	118	221	454	681	1310	2090	3690	7540
1000	55	114	215	442	662	1280	2030	3590	7330
1100	52	109	204	420	629	1210	1930	3410	6960
1200	50	104	195	400	600	1160	1840	3260	6640
1300	47	99	187	384	575	1110	1760	3120	6360
1400	46	95	179	368	552	1060	1690	3000	6110
1500	44	92	173	355	532	1020	1630	2890	5890
1600	42	89	167	343	514	989	1580	2790	5680
1700	41	86	162	332	497	957	1530	2700	5500
1800	40	83	157	322	482	928	1480	2610	5330
1900	39	81	152	312	468	901	1440	2540	5180
2000	38	79	148	304	455	877	1400	2470	5040

For SI units: 1 inch = 25 mm, 1 foot = 304.8 mm, 1 cubic foot per hour = 0.0283 m^3/h, 1 pound-force per square inch = 6.8947 kPa, 1 inch water column = 0.249 kPa
* Table entries are rounded to 3 significant digits.

FUEL GAS PIPING

TABLE 1215.2(4)
SCHEDULE 40 METALLIC PIPE [NFPA 54: TABLE 6.2(e)]*

						GAS:	NATURAL		
						INLET PRESSURE:	2.0 psi		
						PRESSURE DROP:	1.0 psi		
						SPECIFIC GRAVITY:	0.60		

	PIPE SIZE (inch)								
NOMINAL:	½	¾	1	1¼	1½	2	2½	3	4
ACTUAL ID:	0.622	0.824	1.049	1.380	1.610	2.067	2.469	3.068	4.026
LENGTH (feet)	CAPACITY IN CUBIC FEET OF GAS PER HOUR								
10	1510	3040	5560	11 400	17 100	32 900	52 500	92 800	189 000
20	1070	2150	3930	8070	12 100	23 300	37 100	65 600	134 000
30	869	1760	3210	6590	9880	19 000	30 300	53 600	109 000
40	753	1520	2780	5710	8550	16 500	26 300	46 400	94 700
50	673	1360	2490	5110	7650	14 700	23 500	41 500	84 700
60	615	1240	2270	4660	6980	13 500	21 400	37 900	77 300
70	569	1150	2100	4320	6470	12 500	19 900	35 100	71 600
80	532	1080	1970	4040	6050	11 700	18 600	32 800	67 000
90	502	1010	1850	3810	5700	11 000	17 500	30 900	63 100
100	462	934	1710	3510	5260	10 100	16 100	28 500	58 200
125	414	836	1530	3140	4700	9060	14 400	25 500	52 100
150	372	751	1370	2820	4220	8130	13 000	22 900	46 700
175	344	695	1270	2601	3910	7530	12 000	21 200	43 300
200	318	642	1170	2410	3610	6960	11 100	19 600	40 000
250	279	583	1040	2140	3210	6180	9850	17 400	35 500
300	253	528	945	1940	2910	5600	8920	15 800	32 200
350	232	486	869	1790	2670	5150	8210	14 500	29 600
400	216	452	809	1660	2490	4790	7640	13 500	27 500
450	203	424	759	1560	2330	4500	7170	12 700	25 800
500	192	401	717	1470	2210	4250	6770	12 000	24 400
550	182	381	681	1400	2090	4030	6430	11 400	23 200
600	174	363	650	1330	2000	3850	6130	10 800	22 100
650	166	348	622	1280	1910	3680	5870	10 400	21 200
700	160	334	598	1230	1840	3540	5640	9970	20 300
750	154	322	576	1180	1770	3410	5440	9610	19 600
800	149	311	556	1140	1710	3290	5250	9280	18 900
850	144	301	538	1100	1650	3190	5080	8980	18 300
900	139	292	522	1070	1600	3090	4930	8710	17 800
950	135	283	507	1040	1560	3000	4780	8460	17 200
1000	132	275	493	1010	1520	2920	4650	8220	16 800
1100	125	262	468	960	1440	2770	4420	7810	15 900
1200	119	250	446	917	1370	2640	4220	7450	15 200
1300	114	239	427	878	1320	2530	4040	7140	14 600
1400	110	230	411	843	1260	2430	3880	6860	14 000
1500	106	221	396	812	1220	2340	3740	6600	13 500
1600	102	214	382	784	1180	2260	3610	6380	13 000
1700	99	207	370	759	1140	2190	3490	6170	12 600
1800	96	200	358	736	1100	2120	3390	5980	12 200
1900	93	195	348	715	1070	2060	3290	5810	11 900
2000	91	189	339	695	1040	2010	3200	5650	11 500

For SI units: 1 inch = 25 mm, 1 foot = 304.8 mm, 1 cubic foot per hour = 0.0283 m^3/h, 1 pound-force per square inch = 6.8947 kPa

* Table entries are rounded to 3 significant digits.

TABLE 1215.2(5)
SCHEDULE 40 METALLIC PIPE [NFPA 54: TABLE 6.2(f)]*

								GAS:	NATURAL
								INLET PRESSURE:	3.0 psi
								PRESSURE DROP:	2.0 psi
								SPECIFIC GRAVITY:	0.60

	PIPE SIZE (inch)								
NOMINAL:	½	¾	1	1¼	1½	2	2½	3	4
ACTUAL ID:	0.622	0.824	1.049	1.380	1.610	2.067	2.469	3.068	4.026
LENGTH (feet)	CAPACITY IN CUBIC FEET OF GAS PER HOUR								
10	2350	4920	9270	19 000	28 500	54 900	87 500	155 000	316 000
20	1620	3380	6370	13 100	19 600	37 700	60 100	106 000	217 000
30	1300	2720	5110	10 500	15 700	30 300	48 300	85 400	174 000
40	1110	2320	4380	8990	13 500	25 900	41 300	73 100	149 000
50	985	2060	3880	7970	11 900	23 000	36 600	64 800	132 000
60	892	1870	3520	7220	10 800	20 800	33 200	58 700	120 000
70	821	1720	3230	6640	9950	19 200	30 500	54 000	110 000
80	764	1600	3010	6180	9260	17 800	28 400	50 200	102 000
90	717	1500	2820	5800	8680	16 700	26 700	47 100	96 100
100	677	1420	2670	5470	8200	15 800	25 200	44 500	90 800
125	600	1250	2360	4850	7270	14 000	22 300	39 500	80 500
150	544	1140	2140	4400	6590	12 700	20 200	35 700	72 900
175	500	1050	1970	4040	6060	11 700	18 600	32 900	67 100
200	465	973	1830	3760	5640	10 900	17 300	30 600	62 400
250	412	862	1620	3330	5000	9620	15 300	27 100	55 300
300	374	781	1470	3020	4530	8720	13 900	24 600	50 100
350	344	719	1350	2780	4170	8020	12 800	22 600	46 100
400	320	669	1260	2590	3870	7460	11 900	21 000	42 900
450	300	627	1180	2430	3640	7000	11 200	19 700	40 200
500	283	593	1120	2290	3430	6610	10 500	18 600	38 000
550	269	563	1060	2180	3260	6280	10 000	17 700	36 100
600	257	537	1010	2080	3110	5990	9550	16 900	34 400
650	246	514	969	1990	2980	5740	9150	16 200	33 000
700	236	494	931	1910	2860	5510	8790	15 500	31 700
750	228	476	897	1840	2760	5310	8470	15 000	30 500
800	220	460	866	1780	2660	5130	8180	14 500	29 500
850	213	445	838	1720	2580	4960	7910	14 000	28 500
900	206	431	812	1670	2500	4810	7670	13 600	27 700
950	200	419	789	1620	2430	4670	7450	13 200	26 900
1000	195	407	767	1580	2360	4550	7240	12 800	26 100
1100	185	387	729	1500	2240	4320	6890	12 200	24 800
1200	177	369	695	1430	2140	4120	6570	11 600	23 700
1300	169	353	666	1370	2050	3940	6290	11 100	22 700
1400	162	340	640	1310	1970	3790	6040	10 700	21 800
1500	156	327	616	1270	1900	3650	5820	10 300	21 000
1600	151	316	595	1220	1830	3530	5620	10 000	20 300
1700	146	306	576	1180	1770	3410	5440	9610	19 600
1800	142	296	558	1150	1720	3310	5270	9320	19 000
1900	138	288	542	1110	1670	3210	5120	9050	18 400
2000	134	280	527	1080	1620	3120	4980	8800	18 000

For SI units: 1 inch = 25 mm, 1 foot = 304.8 mm, 1 cubic foot per hour = 0.0283 m³/h, 1 pound-force per square inch = 6.8947 kPa

* Table entries are rounded to 3 significant digits.

FUEL GAS PIPING

TABLE 1215.2(6)
SCHEDULE 40 METALLIC PIPE [NFPA 54: TABLE 6.2(g)]*

						GAS:	NATURAL		
						INLET PRESSURE:	5.0 psi		
						PRESSURE DROP:	3.5 psi		
						SPECIFIC GRAVITY:	0.60		
	PIPE SIZE (inch)								
NOMINAL:	½	¾	1	1¼	1½	2	2½	3	4
ACTUAL ID:	0.622	0.824	1.049	1.380	1.610	2.067	2.469	3.068	4.026
LENGTH (feet)	CAPACITY IN CUBIC FEET OF GAS PER HOUR								
10	3190	6430	11 800	24 200	36 200	69 700	111 000	196 000	401 000
20	2250	4550	8320	17 100	25 600	49 300	78 600	139 000	283 000
30	1840	3720	6790	14 000	20 900	40 300	64 200	113 000	231 000
40	1590	3220	5880	12 100	18 100	34 900	55 600	98 200	200 000
50	1430	2880	5260	10 800	16 200	31 200	49 700	87 900	179 000
60	1300	2630	4800	9860	14 800	28 500	45 400	80 200	164 000
70	1200	2430	4450	9130	13 700	26 400	42 000	74 300	151 000
80	1150	2330	4260	8540	12 800	24 700	39 300	69 500	142 000
90	1060	2150	3920	8050	12 100	23 200	37 000	65 500	134 000
100	979	1980	3620	7430	11 100	21 400	34 200	60 400	123 000
125	876	1770	3240	6640	9950	19 200	30 600	54 000	110 000
150	786	1590	2910	5960	8940	17 200	27 400	48 500	98 900
175	728	1470	2690	5520	8270	15 900	25 400	44 900	91 600
200	673	1360	2490	5100	7650	14 700	23 500	41 500	84 700
250	558	1170	2200	4510	6760	13 000	20 800	36 700	74 900
300	506	1060	1990	4090	6130	11 800	18 800	33 300	67 800
350	465	973	1830	3760	5640	10 900	17 300	30 600	62 400
400	433	905	1710	3500	5250	10 100	16 100	28 500	58 100
450	406	849	1600	3290	4920	9480	15 100	26 700	54 500
500	384	802	1510	3100	4650	8950	14 300	25 200	51 500
550	364	762	1440	2950	4420	8500	13 600	24 000	48 900
600	348	727	1370	2810	4210	8110	12 900	22 900	46 600
650	333	696	1310	2690	4030	7770	12 400	21 900	44 600
700	320	669	1260	2590	3880	7460	11 900	21 000	42 900
750	308	644	1210	2490	3730	7190	11 500	20 300	41 300
800	298	622	1170	2410	3610	6940	11 100	19 600	39 900
850	288	602	1130	2330	3490	6720	10 700	18 900	38 600
900	279	584	1100	2260	3380	6520	10 400	18 400	37 400
950	271	567	1070	2190	3290	6330	10 100	17 800	36 400
1000	264	551	1040	2130	3200	6150	9810	17 300	35 400
1100	250	524	987	2030	3030	5840	9320	16 500	33 600
1200	239	500	941	1930	2900	5580	8890	15 700	32 000
1300	229	478	901	1850	2770	5340	8510	15 000	30 700
1400	220	460	866	1780	2660	5130	8180	14 500	29 500
1500	212	443	834	1710	2570	4940	7880	13 900	28 400
1600	205	428	806	1650	2480	4770	7610	13 400	27 400
1700	198	414	780	1600	2400	4620	7360	13 000	26 500
1800	192	401	756	1550	2330	4480	7140	12 600	25 700
1900	186	390	734	1510	2260	4350	6930	12 300	25 000
2000	181	379	714	1470	2200	4230	6740	11 900	24 300

For SI units: 1 inch = 25 mm, 1 foot = 304.8 mm, 1 cubic foot per hour = 0.0283 m^3/h, 1 pound-force per square inch = 6.8947 kPa

* Table entries are rounded to 3 significant digits.

FUEL GAS PIPING

TABLE 1215.2(7)
SEMI-RIGID COPPER TUBING [NFPA 54: TABLE 6.2(h)][1,2]

							GAS:	NATURAL		
							INLET PRESSURE:	LESS THAN 2 psi		
							PRESSURE DROP:	0.3 in. w.c.		
							SPECIFIC GRAVITY:	0.60		

		TUBE SIZE (inch)								
NOMINAL:	K & L:	¼	⅜	½	⅝	¾	1	1¼	1½	2
	ACR:	⅜	½	⅝	¾	⅞	1⅛	1⅜	–	–
OUTSIDE:		0.375	0.500	0.625	0.750	0.875	1.125	1.375	1.625	2.125
INSIDE:[3]		0.305	0.402	0.527	0.652	0.745	0.995	1.245	1.481	1.959
LENGTH (feet)		CAPACITY IN CUBIC FEET OF GAS PER HOUR								
10		20	42	85	148	210	448	806	1270	2650
20		14	29	58	102	144	308	554	873	1820
30		11	23	47	82	116	247	445	701	1460
40		10	20	40	70	99	211	381	600	1250
50		NA	17	35	62	88	187	337	532	1110
60		NA	16	32	56	79	170	306	482	1000
70		NA	14	29	52	73	156	281	443	924
80		NA	13	27	48	68	145	262	413	859
90		NA	13	26	45	64	136	245	387	806
100		NA	12	24	43	60	129	232	366	761
125		NA	11	22	38	53	114	206	324	675
150		NA	10	20	34	48	103	186	294	612
175		NA	NA	18	31	45	95	171	270	563
200		NA	NA	17	29	41	89	159	251	523
250		NA	NA	15	26	37	78	141	223	464
300		NA	NA	13	23	33	71	128	202	420
350		NA	NA	12	22	31	65	118	186	387
400		NA	NA	11	20	28	61	110	173	360
450		NA	NA	11	19	27	57	103	162	338
500		NA	NA	10	18	25	54	97	153	319
550		NA	NA	NA	17	24	51	92	145	303
600		NA	NA	NA	16	23	49	88	139	289
650		NA	NA	NA	15	22	47	84	133	277
700		NA	NA	NA	15	21	45	81	128	266
750		NA	NA	NA	14	20	43	78	123	256
800		NA	NA	NA	14	20	42	75	119	247
850		NA	NA	NA	13	19	40	73	115	239
900		NA	NA	NA	13	18	39	71	111	232
950		NA	NA	NA	13	18	38	69	108	225
1000		NA	NA	NA	12	17	37	67	105	219
1100		NA	NA	NA	12	16	35	63	100	208
1200		NA	NA	NA	11	16	34	60	95	199
1300		NA	NA	NA	11	15	32	58	91	190
1400		NA	NA	NA	10	14	31	56	88	183
1500		NA	NA	NA	NA	14	30	54	84	176
1600		NA	NA	NA	NA	13	29	52	82	170
1700		NA	NA	NA	NA	13	28	50	79	164
1800		NA	NA	NA	NA	13	27	49	77	159
1900		NA	NA	NA	NA	12	26	47	74	155
2000		NA	NA	NA	NA	12	25	46	72	151

For SI units: 1 inch = 25 mm, 1 foot = 304.8 mm, 1 cubic foot per hour = 0.0283 m³/h, 1 pound-force per square inch = 6.8947 kPa, 1 inch water column = 0.249 kPa

Notes:
[1] Table entries are rounded to 3 significant digits.
[2] NA means a flow of less than 10 ft³/h (0.283 m³/h).
[3] Table capacities are based on Type K copper tubing inside diameter (shown), which has the smallest inside diameter of the copper tubing products.

FUEL GAS PIPING

TABLE 1215.2(8)
SEMI-RIGID COPPER TUBING [NFPA 54: TABLE 6.2(i)][1, 2]

								GAS:	NATURAL
							INLET PRESSURE:		LESS THAN 2 psi
							PRESSURE DROP:		0.5 in. w.c.
							SPECIFIC GRAVITY:		0.60

		TUBE SIZE (inch)								
NOMINAL:	K & L:	¼	⅜	½	⅝	¾	1	1¼	1½	2
	ACR:	⅜	½	⅝	¾	⅞	1⅛	1⅜	–	–
OUTSIDE:		0.375	0.500	0.625	0.750	0.875	1.125	1.375	1.625	2.125
INSIDE:[3]		0.305	0.402	0.527	0.652	0.745	0.995	1.245	1.481	1.959
LENGTH (feet)		CAPACITY IN CUBIC FEET OF GAS PER HOUR								
10		27	55	111	195	276	590	1060	1680	3490
20		18	38	77	134	190	406	730	1150	2400
30		15	30	61	107	152	326	586	925	1930
40		13	26	53	92	131	279	502	791	1650
50		11	23	47	82	116	247	445	701	1460
60		10	21	42	74	105	224	403	635	1320
70		NA	19	39	68	96	206	371	585	1220
80		NA	18	36	63	90	192	345	544	1130
90		NA	17	34	59	84	180	324	510	1060
100		NA	16	32	56	79	170	306	482	1000
125		NA	14	28	50	70	151	271	427	890
150		NA	13	26	45	64	136	245	387	806
175		NA	12	24	41	59	125	226	356	742
200		NA	11	22	39	55	117	210	331	690
250		NA	NA	20	34	48	103	186	294	612
300		NA	NA	18	31	44	94	169	266	554
350		NA	NA	16	28	40	86	155	245	510
400		NA	NA	15	26	38	80	144	228	474
450		NA	NA	14	25	35	75	135	214	445
500		NA	NA	13	23	33	71	128	202	420
550		NA	NA	13	22	32	68	122	192	399
600		NA	NA	12	21	30	64	116	183	381
650		NA	NA	12	20	29	62	111	175	365
700		NA	NA	11	20	28	59	107	168	350
750		NA	NA	11	19	27	57	103	162	338
800		NA	NA	10	18	26	55	99	156	326
850		NA	NA	10	18	25	53	96	151	315
900		NA	NA	NA	17	24	52	93	147	306
950		NA	NA	NA	17	24	50	90	143	297
1000		NA	NA	NA	16	23	49	88	139	289
1100		NA	NA	NA	15	22	46	84	132	274
1200		NA	NA	NA	15	21	44	80	126	262
1300		NA	NA	NA	14	20	42	76	120	251
1400		NA	NA	NA	13	19	41	73	116	241
1500		NA	NA	NA	13	18	39	71	111	232
1600		NA	NA	NA	13	18	38	68	108	224
1700		NA	NA	NA	12	17	37	66	104	217
1800		NA	NA	NA	12	17	36	64	101	210
1900		NA	NA	NA	11	16	35	62	98	204
2000		NA	NA	NA	11	16	34	60	95	199

For SI units: 1 inch = 25 mm, 1 foot = 304.8 mm, 1 cubic foot per hour = 0.0283 m^3/h, 1 pound-force per square inch = 6.8947 kPa, 1 inch water column = 0.249 kPa

Notes:
[1] Table entries are rounded to 3 significant digits.
[2] NA means a flow of less than 10 ft^3/h (0.283 m^3/h).
[3] Table capacities are based on Type K copper tubing inside diameter (shown), which has the smallest inside diameter of the copper tubing products.

FUEL GAS PIPING

TABLE 1215.2(9)
SEMI-RIGID COPPER TUBING [NFPA 54: TABLE 6.2(j)][1, 2]

						GAS:	NATURAL			
						INLET PRESSURE:	LESS THAN 2 psi			
						PRESSURE DROP:	1.0 in. w.c.			
						SPECIFIC GRAVITY:	0.60			

INTENDED USE: TUBE SIZING BETWEEN HOUSE LINE REGULATOR AND THE APPLIANCE

NOMINAL:		TUBE SIZE (inch)								
	K & L:	¼	⅜	½	⅝	¾	1	1¼	1½	2
	ACR:	⅜	½	⅝	¾	⅞	1⅛	1⅜	–	–
OUTSIDE:		0.375	0.500	0.625	0.750	0.875	1.125	1.375	1.625	2.125
INSIDE:[3]		0.305	0.402	0.527	0.652	0.745	0.995	1.245	1.481	1.959
LENGTH (feet)		CAPACITY IN CUBIC FEET OF GAS PER HOUR								
10		39	80	162	283	402	859	1550	2440	5080
20		27	55	111	195	276	590	1060	1680	3490
30		21	44	89	156	222	474	853	1350	2800
40		18	38	77	134	190	406	730	1150	2400
50		16	33	68	119	168	359	647	1020	2130
60		15	30	61	107	152	326	586	925	1930
70		13	28	57	99	140	300	539	851	1770
80		13	26	53	92	131	279	502	791	1650
90		12	24	49	86	122	262	471	742	1550
100		11	23	47	82	116	247	445	701	1460
125		NA	20	41	72	103	219	394	622	1290
150		NA	18	37	65	93	198	357	563	1170
175		NA	17	34	60	85	183	329	518	1080
200		NA	16	32	56	79	170	306	482	1000
250		NA	14	28	50	70	151	271	427	890
300		NA	13	26	45	64	136	245	387	806
350		NA	12	24	41	59	125	226	356	742
400		NA	11	22	39	55	117	210	331	690
450		NA	10	21	36	51	110	197	311	647
500		NA	NA	20	34	48	103	186	294	612
550		NA	NA	19	32	46	98	177	279	581
600		NA	NA	18	31	44	94	169	266	554
650		NA	NA	17	30	42	90	162	255	531
700		NA	NA	16	28	40	86	155	245	510
750		NA	NA	16	27	39	83	150	236	491
800		NA	NA	15	26	38	80	144	228	474
850		NA	NA	15	26	36	78	140	220	459
900		NA	NA	14	25	35	75	135	214	445
950		NA	NA	14	24	34	73	132	207	432
1000		NA	NA	13	23	33	71	128	202	420
1100		NA	NA	13	22	32	68	122	192	399
1200		NA	NA	12	21	30	64	116	183	381
1300		NA	NA	12	20	29	62	111	175	365
1400		NA	NA	11	20	28	59	107	168	350
1500		NA	NA	11	19	27	57	103	162	338
1600		NA	NA	10	18	26	55	99	156	326
1700		NA	NA	10	18	25	53	96	151	315
1800		NA	NA	NA	17	24	52	93	147	306
1900		NA	NA	NA	17	24	50	90	143	297
2000		NA	NA	NA	16	23	49	88	139	289

For SI units: 1 inch = 25 mm, 1 foot = 304.8 mm, 1 cubic foot per hour = 0.0283 m^3/h, 1 pound-force per square inch = 6.8947 kPa, 1 inch water column = 0.249 kPa

Notes:
[1] Table entries are rounded to 3 significant digits.
[2] NA means a flow of less than 10 ft^3/h (0.283 m^3/h).
[3] Table capacities are based on Type K copper tubing inside diameter (shown), which has the smallest inside diameter of the copper tubing products.

FUEL GAS PIPING

TABLE 1215.2(10)
SEMI-RIGID COPPER TUBING [NFPA 54: TABLE 6.2(k)][2]

							GAS:	NATURAL		
						INLET PRESSURE:	LESS THAN 2 psi			
						PRESSURE DROP:	17.0 in. w.c.			
						SPECIFIC GRAVITY:	0.60			
		TUBE SIZE (inch)								
NOMINAL:	K & L:	¼	⅜	½	⅝	¾	1	1¼	1½	2
	ACR:	⅜	½	⅝	¾	⅞	1⅛	1⅜	–	–
OUTSIDE:		0.375	0.500	0.625	0.750	0.875	1.125	1.375	1.625	2.125
INSIDE:[1]		0.305	0.402	0.527	0.652	0.745	0.995	1.245	1.481	1.959
LENGTH (feet)		CAPACITY IN CUBIC FEET OF GAS PER HOUR								
10		190	391	796	1390	1970	4220	7590	12 000	24 900
20		130	269	547	956	1360	2900	5220	8230	17 100
30		105	216	439	768	1090	2330	4190	6610	13 800
40		90	185	376	657	932	1990	3590	5650	11 800
50		79	164	333	582	826	1770	3180	5010	10 400
60		72	148	302	528	749	1600	2880	4540	9460
70		66	137	278	486	689	1470	2650	4180	8700
80		62	127	258	452	641	1370	2460	3890	8090
90		58	119	243	424	601	1280	2310	3650	7590
100		55	113	229	400	568	1210	2180	3440	7170
125		48	100	203	355	503	1080	1940	3050	6360
150		44	90	184	321	456	974	1750	2770	5760
175		40	83	169	296	420	896	1610	2540	5300
200		38	77	157	275	390	834	1500	2370	4930
250		33	69	140	244	346	739	1330	2100	4370
300		30	62	126	221	313	670	1210	1900	3960
350		28	57	116	203	288	616	1110	1750	3640
400		26	53	108	189	268	573	1030	1630	3390
450		24	50	102	177	252	538	968	1530	3180
500		23	47	96	168	238	508	914	1440	3000
550		22	45	91	159	226	482	868	1370	2850
600		21	43	87	152	215	460	829	1310	2720
650		20	41	83	145	206	441	793	1250	2610
700		19	39	80	140	198	423	762	1200	2500
750		18	38	77	135	191	408	734	1160	2410
800		18	37	74	130	184	394	709	1120	2330
850		17	35	72	126	178	381	686	1080	2250
900		17	34	70	122	173	370	665	1050	2180
950		16	33	68	118	168	359	646	1020	2120
1000		16	32	66	115	163	349	628	991	2060
1100		15	31	63	109	155	332	597	941	1960
1200		14	29	60	104	148	316	569	898	1870
1300		14	28	57	100	142	303	545	860	1790
1400		13	27	55	96	136	291	524	826	1720
1500		13	26	53	93	131	280	505	796	1660
1600		12	25	51	89	127	271	487	768	1600
1700		12	24	49	86	123	262	472	744	1550
1800		11	24	48	84	119	254	457	721	1500
1900		11	23	47	81	115	247	444	700	1460
2000		11	22	45	79	112	240	432	681	1420

For SI units: 1 inch = 25 mm, 1 foot = 304.8 mm, 1 cubic foot per hour = 0.0283 m^3/h, 1 pound-force per square inch = 6.8947 kPa, 1 inch water column = 0.249 kPa

Notes:
[1] Table capacities are based on Type K copper tubing inside diameter (shown), which has the smallest inside diameter of the copper tubing products.
[2] Table entries are rounded to 3 significant digits.

TABLE 1215.2(11)
SEMI-RIGID COPPER TUBING [NFPA 54: TABLE 6.2(I)][2]

		GAS:	NATURAL
		INLET PRESSURE:	2.0 psi
		PRESSURE DROP:	1.0 psi
		SPECIFIC GRAVITY:	0.60

NOMINAL:	K & L:	¼	⅜	½	⅝	¾	1	1¼	1½	2
	ACR:	⅜	½	⅝	¾	⅞	1⅛	1⅜	–	–
OUTSIDE:		0.375	0.500	0.625	0.750	0.875	1.125	1.375	1.625	2.125
INSIDE:[1]		0.305	0.402	0.527	0.652	0.745	0.995	1.245	1.481	1.959
LENGTH (feet)		CAPACITY IN CUBIC FEET OF GAS PER HOUR								
10		245	506	1030	1800	2550	5450	9820	15 500	32 200
20		169	348	708	1240	1760	3750	6750	10 600	22 200
30		135	279	568	993	1410	3010	5420	8550	17 800
40		116	239	486	850	1210	2580	4640	7310	15 200
50		103	212	431	754	1070	2280	4110	6480	13 500
60		93	192	391	683	969	2070	3730	5870	12 200
70		86	177	359	628	891	1900	3430	5400	11 300
80		80	164	334	584	829	1770	3190	5030	10 500
90		75	154	314	548	778	1660	2990	4720	9820
100		71	146	296	518	735	1570	2830	4450	9280
125		63	129	263	459	651	1390	2500	3950	8220
150		57	117	238	416	590	1260	2270	3580	7450
175		52	108	219	383	543	1160	2090	3290	6850
200		49	100	204	356	505	1080	1940	3060	6380
250		43	89	181	315	448	956	1720	2710	5650
300		39	80	164	286	406	866	1560	2460	5120
350		36	74	150	263	373	797	1430	2260	4710
400		33	69	140	245	347	741	1330	2100	4380
450		31	65	131	230	326	696	1250	1970	4110
500		30	61	124	217	308	657	1180	1870	3880
550		28	58	118	206	292	624	1120	1770	3690
600		27	55	112	196	279	595	1070	1690	3520
650		26	53	108	188	267	570	1030	1620	3370
700		25	51	103	181	256	548	986	1550	3240
750		24	49	100	174	247	528	950	1500	3120
800		23	47	96	168	239	510	917	1450	3010
850		22	46	93	163	231	493	888	1400	2920
900		22	44	90	158	224	478	861	1360	2830
950		21	43	88	153	217	464	836	1320	2740
1000		20	42	85	149	211	452	813	1280	2670
1100		19	40	81	142	201	429	772	1220	2540
1200		18	38	77	135	192	409	737	1160	2420
1300		18	36	74	129	183	392	705	1110	2320
1400		17	35	71	124	176	376	678	1070	2230
1500		16	34	68	120	170	363	653	1030	2140
1600		16	33	66	116	164	350	630	994	2070
1700		15	31	64	112	159	339	610	962	2000
1800		15	30	62	108	154	329	592	933	1940
1900		14	30	60	105	149	319	575	906	1890
2000		14	29	59	102	145	310	559	881	1830

For SI units: 1 inch = 25 mm, 1 foot = 304.8 mm, 1 cubic foot per hour = 0.0283 m^3/h, 1 pound-force per square inch = 6.8947 kPa

Notes:
[1] Table capacities are based on Type K copper tubing inside diameter (shown), which has the smallest inside diameter of the copper tubing products.
[2] Table entries are rounded to 3 significant digits.

FUEL GAS PIPING

TABLE 1215.2(12)
SEMI-RIGID COPPER TUBING [NFPA 54: TABLE 6.2(m)][3]

							GAS:	NATURAL		
							INLET PRESSURE:	2.0 psi		
							PRESSURE DROP:	1.5 psi		
							SPECIFIC GRAVITY:	0.60		

INTENDED USE: PIPE SIZING BETWEEN POINT OF DELIVERY AND THE HOUSE LINE REGULATOR. TOTAL LOAD SUPPLIED BY A SINGLE HOUSE LINE REGULATOR NOT EXCEEDING 150 CUBIC FEET PER HOUR.[2]

		TUBE SIZE (inch)								
NOMINAL:	K & L:	¼	⅜	½	⅝	¾	1	1¼	1½	2
	ACR:	⅜	½	⅝	¾	⅞	1⅛	1⅜	–	–
OUTSIDE:		0.375	0.500	0.625	0.750	0.875	1.125	1.375	1.625	2.125
INSIDE:[1]		0.305	0.402	0.527	0.652	0.745	0.995	1.245	1.481	1.959
LENGTH (feet)		CAPACITY IN CUBIC FEET OF GAS PER HOUR								
10		303	625	1270	2220	3150	6740	12 100	19 100	39 800
20		208	430	874	1530	2170	4630	8330	13 100	27 400
30		167	345	702	1230	1740	3720	6690	10 600	22 000
40		143	295	601	1050	1490	3180	5730	9030	18 800
50		127	262	532	931	1320	2820	5080	8000	16 700
60		115	237	482	843	1200	2560	4600	7250	15 100
70		106	218	444	776	1100	2350	4230	6670	13 900
80		98	203	413	722	1020	2190	3940	6210	12 900
90		92	190	387	677	961	2050	3690	5820	12 100
100		87	180	366	640	907	1940	3490	5500	11 500
125		77	159	324	567	804	1720	3090	4880	10 200
150		70	144	294	514	729	1560	2800	4420	9200
175		64	133	270	472	670	1430	2580	4060	8460
200		60	124	252	440	624	1330	2400	3780	7870
250		53	110	223	390	553	1180	2130	3350	6980
300		48	99	202	353	501	1070	1930	3040	6320
350		44	91	186	325	461	984	1770	2790	5820
400		41	85	173	302	429	916	1650	2600	5410
450		39	80	162	283	402	859	1550	2440	5080
500		36	75	153	268	380	811	1460	2300	4800
550		35	72	146	254	361	771	1390	2190	4560
600		33	68	139	243	344	735	1320	2090	4350
650		32	65	133	232	330	704	1270	2000	4160
700		30	63	128	223	317	676	1220	1920	4000
750		29	60	123	215	305	652	1170	1850	3850
800		28	58	119	208	295	629	1130	1790	3720
850		27	57	115	201	285	609	1100	1730	3600
900		27	55	111	195	276	590	1060	1680	3490
950		26	53	108	189	268	573	1030	1630	3390
1000		25	52	105	184	261	558	1000	1580	3300
1100		24	49	100	175	248	530	954	1500	3130
1200		23	47	95	167	237	505	910	1430	2990
1300		22	45	91	160	227	484	871	1370	2860
1400		21	43	88	153	218	465	837	1320	2750
1500		20	42	85	148	210	448	806	1270	2650
1600		19	40	82	143	202	432	779	1230	2560
1700		19	39	79	138	196	419	753	1190	2470
1800		18	38	77	134	190	406	731	1150	2400
1900		18	37	74	130	184	394	709	1120	2330
2000		17	36	72	126	179	383	690	1090	2270

For SI units: 1 inch = 25 mm, 1 foot = 304.8 mm, 1 cubic foot per hour = 0.0283 m^3/h, 1 pound-force per square inch = 6.8947 kPa

Notes:

[1] Table capacities are based on Type K copper tubing inside diameter (shown), which has the smallest inside diameter of the copper tubing products.

[2] Where this table is used to size the tubing upstream of a line pressure regulator, the pipe or tubing downstream of the line pressure regulator shall be sized using a pressure drop no greater than 1 inch water column (0.249 kPa).

[3] Table entries are rounded to 3 significant digits.

FUEL GAS PIPING

TABLE 1215.2(13)
SEMI-RIGID COPPER TUBING [NFPA 54: TABLE 6.2(n)][2]

							GAS:	NATURAL		
							INLET PRESSURE:	5.0 psi		
							PRESSURE DROP:	3.5 psi		
							SPECIFIC GRAVITY:	0.60		

		TUBE SIZE (inch)								
NOMINAL:	**K & L:**	¼	⅜	½	⅝	¾	1	1¼	1½	2
	ACR:	⅜	½	⅝	¾	⅞	1⅛	1⅜	–	–
OUTSIDE:		0.375	0.500	0.625	0.750	0.875	1.125	1.375	1.625	2.125
INSIDE:[1]		0.305	0.402	0.527	0.652	0.745	0.995	1.245	1.481	1.959
LENGTH (feet)		CAPACITY IN CUBIC FEET OF GAS PER HOUR								
10		511	1050	2140	3750	5320	11 400	20 400	32 200	67 100
20		351	724	1470	2580	3650	7800	14 000	22 200	46 100
30		282	582	1180	2070	2930	6270	11 300	17 800	37 000
40		241	498	1010	1770	2510	5360	9660	15 200	31 700
50		214	441	898	1570	2230	4750	8560	13 500	28 100
60		194	400	813	1420	2020	4310	7750	12 200	25 500
70		178	368	748	1310	1860	3960	7130	11 200	23 400
80		166	342	696	1220	1730	3690	6640	10 500	21 800
90		156	321	653	1140	1620	3460	6230	9820	20 400
100		147	303	617	1080	1530	3270	5880	9270	19 300
125		130	269	547	955	1360	2900	5210	8220	17 100
150		118	243	495	866	1230	2620	4720	7450	15 500
175		109	224	456	796	1130	2410	4350	6850	14 300
200		101	208	424	741	1050	2250	4040	6370	13 300
250		90	185	376	657	932	1990	3580	5650	11 800
300		81	167	340	595	844	1800	3250	5120	10 700
350		75	154	313	547	777	1660	2990	4710	9810
400		69	143	291	509	722	1540	2780	4380	9120
450		65	134	273	478	678	1450	2610	4110	8560
500		62	127	258	451	640	1370	2460	3880	8090
550		58	121	245	429	608	1300	2340	3690	7680
600		56	115	234	409	580	1240	2230	3520	7330
650		53	110	224	392	556	1190	2140	3370	7020
700		51	106	215	376	534	1140	2050	3240	6740
750		49	102	207	362	514	1100	1980	3120	6490
800		48	98	200	350	497	1060	1910	3010	6270
850		46	95	194	339	481	1030	1850	2910	6070
900		45	92	188	328	466	1000	1790	2820	5880
950		43	90	182	319	452	967	1740	2740	5710
1000		42	87	177	310	440	940	1690	2670	5560
1100		40	83	169	295	418	893	1610	2530	5280
1200		38	79	161	281	399	852	1530	2420	5040
1300		37	76	154	269	382	816	1470	2320	4820
1400		35	73	148	259	367	784	1410	2220	4630
1500		34	70	143	249	353	755	1360	2140	4460
1600		33	68	138	241	341	729	1310	2070	4310
1700		32	65	133	233	330	705	1270	2000	4170
1800		31	63	129	226	320	684	1230	1940	4040
1900		30	62	125	219	311	664	1200	1890	3930
2000		29	60	122	213	302	646	1160	1830	3820

For SI units: 1 inch = 25 mm, 1 foot = 304.8 mm, 1 cubic foot per hour = 0.0283 m³/h, 1 pound-force per square inch = 6.8947 kPa

Notes:
[1] Table capacities are based on Type K copper tubing inside diameter (shown), which has the smallest inside diameter of the copper tubing products.
[2] Table entries are rounded to 3 significant digits.

FUEL GAS PIPING

TABLE 1215.2(14)
CORRUGATED STAINLESS STEEL TUBING (CSST) [NFPA 54: TABLE 6.2(o)][1, 2]

									GAS:	NATURAL			
									INLET PRESSURE:	LESS THAN 2 psi			
									PRESSURE DROP:	0.5 in. w.c.			
									SPECIFIC GRAVITY:	0.60			

	TUBE SIZE (EHD)[3]												
FLOW DESIGNATION:	13	15	18	19	23	25	30	31	37	39	46	48	60	62
LENGTH (feet)	CAPACITY IN CUBIC FEET OF GAS PER HOUR												
5	46	63	115	134	225	270	471	546	895	1037	1790	2070	3660	4140
10	32	44	82	95	161	192	330	383	639	746	1260	1470	2600	2930
15	25	35	66	77	132	157	267	310	524	615	1030	1200	2140	2400
20	22	31	58	67	116	137	231	269	456	536	888	1050	1850	2080
25	19	27	52	60	104	122	206	240	409	482	793	936	1660	1860
30	18	25	47	55	96	112	188	218	374	442	723	856	1520	1700
40	15	21	41	47	83	97	162	188	325	386	625	742	1320	1470
50	13	19	37	42	75	87	144	168	292	347	559	665	1180	1320
60	12	17	34	38	68	80	131	153	267	318	509	608	1080	1200
70	11	16	31	36	63	74	121	141	248	295	471	563	1000	1110
80	10	15	29	33	60	69	113	132	232	277	440	527	940	1040
90	10	14	28	32	57	65	107	125	219	262	415	498	887	983
100	9	13	26	30	54	62	101	118	208	249	393	472	843	933
150	7	10	20	23	42	48	78	91	171	205	320	387	691	762
200	6	9	18	21	38	44	71	82	148	179	277	336	600	661
250	5	8	16	19	34	39	63	74	133	161	247	301	538	591
300	5	7	15	17	32	36	57	67	95	148	226	275	492	540

For SI units: 1 inch = 25 mm, 1 foot = 304.8 mm, 1 cubic foot per hour = 0.0283 m^3/h, 1 pound-force per square inch = 6.8947 kPa, 1 inch water column = 0.249 kPa

Notes:
[1] Table entries are rounded to 3 significant digits.
[2] Table includes losses for four 90 degree (1.57 rad) bends and two end fittings. Tubing runs with larger numbers of bends, fittings, or both shall be increased by an equivalent length of tubing to the following equation: $L = 1.3\,n$, where L is additional length (ft) of tubing and n is the number of additional fittings, bends, or both.
[3] EHD = Equivalent Hydraulic Diameter, which is a measure of the relative hydraulic efficiency between different tubing sizes. The greater the value of EHD, the greater the gas capacity of the tubing.

TABLE 1215.2(15)
CORRUGATED STAINLESS STEEL TUBING (CSST) [NFPA 54: TABLE 6.2(p)][1, 2]

		GAS:	NATURAL
		INLET PRESSURE:	LESS THAN 2 psi
		PRESSURE DROP:	3.0 in. w.c.
		SPECIFIC GRAVITY:	0.60

INTENDED USE: INITIAL SUPPLY PRESSURE OF 8.0 INCH WATER COLUMN OR GREATER

FLOW DESIGNATION:	13	15	18	19	23	25	30	31	37	46	48	60	62
LENGTH (feet)	\multicolumn{13}{c}{CAPACITY IN CUBIC FEET OF GAS PER HOUR}
5	120	160	277	327	529	649	1180	1370	2140	4430	5010	8800	10 100
10	83	112	197	231	380	462	828	958	1530	3200	3560	6270	7160
15	67	90	161	189	313	379	673	778	1250	2540	2910	5140	5850
20	57	78	140	164	273	329	580	672	1090	2200	2530	4460	5070
25	51	69	125	147	245	295	518	599	978	1960	2270	4000	4540
30	46	63	115	134	225	270	471	546	895	1790	2070	3660	4140
40	39	54	100	116	196	234	407	471	778	1550	1800	3180	3590
50	35	48	89	104	176	210	363	421	698	1380	1610	2850	3210
60	32	44	82	95	161	192	330	383	639	1260	1470	2600	2930
70	29	41	76	88	150	178	306	355	593	1170	1360	2420	2720
80	27	38	71	82	141	167	285	331	555	1090	1280	2260	2540
90	26	36	67	77	133	157	268	311	524	1030	1200	2140	2400
100	24	34	63	73	126	149	254	295	498	974	1140	2030	2280
150	19	27	52	60	104	122	206	240	409	793	936	1660	1860
200	17	23	45	52	91	106	178	207	355	686	812	1440	1610
250	15	21	40	46	82	95	159	184	319	613	728	1290	1440
300	13	19	37	42	75	87	144	168	234	559	665	1180	1320

For SI units: 1 foot = 304.8 mm, 1 cubic foot per hour = 0.0283 m^3/h, 1 pound-force per square inch = 6.8947 kPa, 1 inch water column = 0.249 kPa

Notes:

[1] Table entries are rounded to 3 significant digits.

[2] Table includes losses for four 90 degree (1.57 rad) bends and two end fittings. Tubing runs with larger numbers of bends, fittings, or both shall be increased by an equivalent length of tubing to the following equation: $L = 1.3\,n$, where L is additional length (ft) of tubing and n is the number of additional fittings, bends, or both.

[3] EHD = Equivalent Hydraulic Diameter, which is a measure of the relative hydraulic efficiency between different tubing sizes. The greater the value of EHD, the greater the gas capacity of the tubing.

FUEL GAS PIPING

TABLE 1215.2(16)
CORRUGATED STAINLESS STEEL TUBING (CSST) [NFPA 54: TABLE 6.2(q)][1, 2]

							GAS:	NATURAL					
							INLET PRESSURE:	LESS THAN 2 psi					
							PRESSURE DROP:	6.0 in. w.c.					
							SPECIFIC GRAVITY:	0.60					
INTENDED USE: INITIAL SUPPLY PRESSURE OF 11.0 INCH WATER COLUMN OR GREATER													
	TUBE SIZE (EHD)[3]												
FLOW DESIGNATION:	13	15	18	19	23	25	30	31	37	46	48	60	62
LENGTH (feet)	CAPACITY IN CUBIC FEET OF GAS PER HOUR												
5	173	229	389	461	737	911	1690	1950	3000	6280	7050	12 400	14 260
10	120	160	277	327	529	649	1180	1370	2140	4430	5010	8800	10 100
15	96	130	227	267	436	532	960	1110	1760	3610	4100	7210	8260
20	83	112	197	231	380	462	828	958	1530	3120	3560	6270	7160
25	74	99	176	207	342	414	739	855	1370	2790	3190	5620	6400
30	67	90	161	189	313	379	673	778	1250	2540	2910	5140	5850
40	57	78	140	164	273	329	580	672	1090	2200	2530	4460	5070
50	51	69	125	147	245	295	518	599	978	1960	2270	4000	4540
60	46	63	115	134	225	270	471	546	895	1790	2070	3660	4140
70	42	58	106	124	209	250	435	505	830	1660	1920	3390	3840
80	39	54	100	116	196	234	407	471	778	1550	1800	3180	3590
90	37	51	94	109	185	221	383	444	735	1460	1700	3000	3390
100	35	48	89	104	176	210	363	421	698	1380	1610	2850	3210
150	28	39	73	85	145	172	294	342	573	1130	1320	2340	2630
200	24	34	63	73	126	149	254	295	498	974	1140	2030	2280
250	21	30	57	66	114	134	226	263	447	870	1020	1820	2040
300	19	27	52	60	104	122	206	240	409	793	936	1660	1860

For SI units: 1 foot = 304.8 mm, 1 cubic foot per hour = 0.0283 m^3/h, 1 pound-force per square inch = 6.8947 kPa, 1 inch water column = 0.249 kPa

Notes:
[1] Table entries are rounded to 3 significant digits.
[2] Table includes losses for four 90 degree (1.57 rad) bends and two end fittings. Tubing runs with larger numbers of bends, fittings, or both shall be increased by an equivalent length of tubing to the following equation: $L = 1.3\,n$, where L is additional length (ft) of tubing and n is the number of additional fittings, bends, or both.
[3] EHD = Equivalent Hydraulic Diameter, which is a measure of the relative hydraulic efficiency between different tubing sizes. The greater the value of EHD, the greater the gas capacity of the tubing.

TABLE 1215.2(17)
CORRUGATED STAINLESS STEEL TUBING (CSST) [NFPA 54: TABLE 6.2(r)][1, 2, 3, 4]

								GAS:	NATURAL					
								INLET PRESSURE:	2.0 psi					
								PRESSURE DROP:	1.0 psi					
								SPECIFIC GRAVITY:	0.60					
	TUBE SIZE (EHD)[5]													
FLOW DESIGNATION:	13	15	18	19	23	25	30	31	37	39	46	48	60	62
LENGTH (feet)	CAPACITY IN CUBIC FEET OF GAS PER HOUR													
10	270	353	587	700	1100	1370	2590	2990	4510	5037	9600	10 700	18 600	21 600
25	166	220	374	444	709	876	1620	1870	2890	3258	6040	6780	11 900	13 700
30	151	200	342	405	650	801	1480	1700	2640	2987	5510	6200	10 900	12 500
40	129	172	297	351	567	696	1270	1470	2300	2605	4760	5380	9440	10 900
50	115	154	266	314	510	624	1140	1310	2060	2343	4260	4820	8470	9720
75	93	124	218	257	420	512	922	1070	1690	1932	3470	3950	6940	7940
80	89	120	211	249	407	496	892	1030	1640	1874	3360	3820	6730	7690
100	79	107	189	222	366	445	795	920	1470	1685	3000	3420	6030	6880
150	64	87	155	182	302	364	646	748	1210	1389	2440	2800	4940	5620
200	55	75	135	157	263	317	557	645	1050	1212	2110	2430	4290	4870
250	49	67	121	141	236	284	497	576	941	1090	1890	2180	3850	4360
300	44	61	110	129	217	260	453	525	862	999	1720	1990	3520	3980
400	38	52	96	111	189	225	390	453	749	871	1490	1730	3060	3450
500	34	46	86	100	170	202	348	404	552	783	1330	1550	2740	3090

For SI units: 1 foot = 304.8 mm, 1 cubic foot per hour = 0.0283 m³/h, 1 pound-force per square inch = 6.8947 kPa

Notes:

[1] Table does not include effect of pressure drop across the line regulator. Where regulator loss exceeds 0.75 psi (5.17 kPa), DO NOT USE THIS TABLE. Consult with regulator manufacturer for pressure drops and capacity factors. Pressure drops across a regulator are capable of varying with flow rate.

[2] CAUTION: Capacities shown in table are capable of exceeding maximum capacity for a selected regulator. Consult with regulator or tubing manufacturer for guidance.

[3] Table includes losses for four 90 degree (1.57 rad) bends and two end fittings. Tubing runs with larger numbers of bends, fittings, or both shall be increased by an equivalent length of tubing according to the following equation: $L = 1.3\,n$, where L is additional length (ft) of tubing and n is the number of additional fittings, bends, or both.

[4] Table entries are rounded to 3 significant digits.

[5] EHD = Equivalent Hydraulic Diameter, which is a measure of the relative hydraulic efficiency between different tubing sizes. The greater the value of EHD, the greater the gas capacity of the tubing.

FUEL GAS PIPING

TABLE 1215.2(18)
CORRUGATED STAINLESS STEEL TUBING (CSST) [NFPA 54: TABLE 6.2(s)][1, 2, 3, 4]

														GAS:	NATURAL
														INLET PRESSURE:	5.0 psi
														PRESSURE DROP:	3.5 psi
														SPECIFIC GRAVITY:	0.60

	TUBE SIZE (EHD)[5]													
FLOW DESIGNATION:	13	15	18	19	23	25	30	31	37	39	46	48	60	62
LENGTH (feet)	CAPACITY IN CUBIC FEET OF GAS PER HOUR													
10	523	674	1080	1300	2000	2530	4920	5660	8300	9140	18 100	19 800	34 400	40 400
25	322	420	691	827	1290	1620	3080	3540	5310	5911	11 400	12 600	22 000	25 600
30	292	382	632	755	1180	1480	2800	3230	4860	5420	10 400	11 500	20 100	23 400
40	251	329	549	654	1030	1280	2420	2790	4230	4727	8970	10 000	17 400	20 200
50	223	293	492	586	926	1150	2160	2490	3790	4251	8020	8930	15 600	18 100
75	180	238	403	479	763	944	1750	2020	3110	3506	6530	7320	12 800	14 800
80	174	230	391	463	740	915	1690	1960	3020	3400	6320	7090	12 400	14 300
100	154	205	350	415	665	820	1510	1740	2710	3057	5650	6350	11 100	12 800
150	124	166	287	339	548	672	1230	1420	2220	2521	4600	5200	9130	10 500
200	107	143	249	294	478	584	1060	1220	1930	2199	3980	4510	7930	9090
250	95	128	223	263	430	524	945	1090	1730	1977	3550	4040	7110	8140
300	86	116	204	240	394	479	860	995	1590	1813	3240	3690	6500	7430
400	74	100	177	208	343	416	742	858	1380	1581	2800	3210	5650	6440
500	66	89	159	186	309	373	662	766	1040	1422	2500	2870	5060	5760

For SI units: 1 foot = 304.8 mm, 1 cubic foot per hour = 0.0283 m^3/h, 1 pound-force per square inch = 6.8947 kPa

Notes:

[1] Table does not include effect of pressure drop across the line regulator. Where regulator loss exceeds 1 psi (7 kPa), DO NOT USE THIS TABLE. Consult with regulator manufacturer for pressure drops and capacity factors. Pressure drops across regulator are capable of varying with the flow rate.

[2] CAUTION: Capacities shown in table are capable of exceeding the maximum capacity of selected regulator. Consult tubing manufacturer for guidance.

[3] Table includes losses for four 90 degree (1.57 rad) bends and two end fittings. Tubing runs with larger numbers of bends, fittings, or both shall be increased by an equivalent length of tubing to the following equation: $L = 1.3\,n$, where L is additional length (feet) of tubing and n is the number of additional fittings, bends, or both.

[4] Table entries are rounded to 3 significant digits.

[5] EHD = Equivalent Hydraulic Diameter, which is a measure of the relative hydraulic efficiency between different tubing sizes. The greater the value of EHD, the greater the gas capacity of the tubing.

TABLE 1215.2(19)
POLYETHYLENE PLASTIC PIPE [NFPA 54: TABLE 6.2(t)]*

						GAS:	NATURAL	
						INLET PRESSURE:	LESS THAN 2 psi	
						PRESSURE DROP:	0.3 in. w.c.	
						SPECIFIC GRAVITY:	0.60	
	PIPE SIZE (inch)							
NOMINAL OD:	½	¾	1	1 ¼	1 ½	2	3	4
DESIGNATION:	SDR 9.3	SDR 11	SDR 11	SDR 10	SDR 11	SDR 11	SDR 11	SDR 11
ACTUAL ID:	0.660	0.860	1.077	1.328	1.554	1.943	2.864	3.682
LENGTH (feet)	CAPACITY IN CUBIC FEET OF GAS PER HOUR							
10	153	305	551	955	1440	2590	7170	13 900
20	105	210	379	656	991	1780	4920	9520
30	84	169	304	527	796	1430	3950	7640
40	72	144	260	451	681	1220	3380	6540
50	64	128	231	400	604	1080	3000	5800
60	58	116	209	362	547	983	2720	5250
70	53	107	192	333	503	904	2500	4830
80	50	99	179	310	468	841	2330	4500
90	46	93	168	291	439	789	2180	4220
100	44	88	159	275	415	745	2060	3990
125	39	78	141	243	368	661	1830	3530
150	35	71	127	221	333	598	1660	3200
175	32	65	117	203	306	551	1520	2940
200	30	60	109	189	285	512	1420	2740
250	27	54	97	167	253	454	1260	2430
300	24	48	88	152	229	411	1140	2200
350	22	45	81	139	211	378	1050	2020
400	21	42	75	130	196	352	974	1880
450	19	39	70	122	184	330	914	1770
500	18	37	66	115	174	312	863	1670

For SI units: 1 inch = 25 mm, 1 foot = 304.8 mm, 1 cubic foot per hour = 0.0283 m^3/h, 1 pound-force per square inch = 6.8947 kPa, 1 inch water column = 0.249 kPa
* Table entries are rounded to 3 significant digits.

FUEL GAS PIPING

TABLE 1215.2(20)
POLYETHYLENE PLASTIC PIPE [NFPA 54: TABLE 6.2(u)]*

					GAS:	NATURAL		
					INLET PRESSURE:	LESS THAN 2 psi		
					PRESSURE DROP:	0.5 in. w.c.		
					SPECIFIC GRAVITY:	0.60		
	PIPE SIZE (inch)							
NOMINAL OD:	½	¾	1	1 ¼	1 ½	2	3	4
DESIGNATION:	SDR 9.3	SDR 11	SDR 11	SDR 10	SDR 11	SDR 11	SDR 11	SDR 11
ACTUAL ID:	0.660	0.860	1.077	1.328	1.554	1.943	2.864	3.682
LENGTH (feet)	CAPACITY IN CUBIC FEET OF GAS PER HOUR							
10	201	403	726	1260	1900	3410	9450	18 260
20	138	277	499	865	1310	2350	6490	12 550
30	111	222	401	695	1050	1880	5210	10 080
40	95	190	343	594	898	1610	4460	8630
50	84	169	304	527	796	1430	3950	7640
60	76	153	276	477	721	1300	3580	6930
70	70	140	254	439	663	1190	3300	6370
80	65	131	236	409	617	1110	3070	5930
90	61	123	221	383	579	1040	2880	5560
100	58	116	209	362	547	983	2720	5250
125	51	103	185	321	485	871	2410	4660
150	46	93	168	291	439	789	2180	4220
175	43	86	154	268	404	726	2010	3880
200	40	80	144	249	376	675	1870	3610
250	35	71	127	221	333	598	1660	3200
300	32	64	115	200	302	542	1500	2900
350	29	59	106	184	278	499	1380	2670
400	27	55	99	171	258	464	1280	2480
450	26	51	93	160	242	435	1200	2330
500	24	48	88	152	229	411	1140	2200

For SI units: 1 inch = 25 mm, 1 foot = 304.8 mm, 1 cubic foot per hour = 0.0283 m^3/h, 1 pound-force per square inch = 6.8947 kPa, 1 inch water column = 0.249 kPa
* Table entries are rounded to 3 significant digits.

TABLE 1215.2(21)
POLYETHYLENE PLASTIC PIPE [NFPA 54: TABLE 6.2(v)]*

					GAS:	NATURAL		
					INLET PRESSURE:	2.0 psi		
					PRESSURE DROP:	1.0 psi		
					SPECIFIC GRAVITY:	0.60		
	PIPE SIZE (inch)							
NOMINAL OD:	½	¾	1	1 ¼	1 ½	2	3	4
DESIGNATION:	SDR 9.3	SDR 11	SDR 11	SDR 10	SDR 11	SDR 11	SDR 11	SDR 11
ACTUAL ID:	0.660	0.860	1.077	1.328	1.554	1.943	2.864	3.682
LENGTH (feet)	CAPACITY IN CUBIC FEET OF GAS PER HOUR							
10	1860	3720	6710	11 600	17 600	31 600	87 300	169 000
20	1280	2560	4610	7990	12 100	21 700	60 000	116 000
30	1030	2050	3710	6420	9690	17 400	48 200	93 200
40	878	1760	3170	5490	8300	14 900	41 200	79 700
50	778	1560	2810	4870	7350	13 200	36 600	70 700
60	705	1410	2550	4410	6660	12 000	33 100	64 000
70	649	1300	2340	4060	6130	11 000	30 500	58 900
80	603	1210	2180	3780	5700	10 200	28 300	54 800
90	566	1130	2050	3540	5350	9610	26 600	51 400
100	535	1070	1930	3350	5050	9080	25 100	48 600
125	474	949	1710	2970	4480	8050	22 300	43 000
150	429	860	1550	2690	4060	7290	20 200	39 000
175	395	791	1430	2470	3730	6710	18 600	35 900
200	368	736	1330	2300	3470	6240	17 300	33 400
250	326	652	1180	2040	3080	5530	15 300	29 600
300	295	591	1070	1850	2790	5010	13 900	26 800
350	272	544	981	1700	2570	4610	12 800	24 700
400	253	506	913	1580	2390	4290	11 900	22 900
450	237	475	856	1480	2240	4020	11 100	21 500
500	224	448	809	1400	2120	3800	10 500	20 300
550	213	426	768	1330	2010	3610	9990	19 300
600	203	406	733	1270	1920	3440	9530	18 400
650	194	389	702	1220	1840	3300	9130	17 600
700	187	374	674	1170	1760	3170	8770	16 900
750	180	360	649	1130	1700	3050	8450	16 300
800	174	348	627	1090	1640	2950	8160	15 800
850	168	336	607	1050	1590	2850	7890	15 300
900	163	326	588	1020	1540	2770	7650	14 800
950	158	317	572	990	1500	2690	7430	14 400
1000	154	308	556	963	1450	2610	7230	14 000
1100	146	293	528	915	1380	2480	6870	13 300
1200	139	279	504	873	1320	2370	6550	12 700
1300	134	267	482	836	1260	2270	6270	12 100
1400	128	257	463	803	1210	2180	6030	11 600
1500	124	247	446	773	1170	2100	5810	11 200
1600	119	239	431	747	1130	2030	5610	10 800
1700	115	231	417	723	1090	1960	5430	10 500
1800	112	224	404	701	1060	1900	5260	10 200
1900	109	218	393	680	1030	1850	5110	9900
2000	106	212	382	662	1000	1800	4970	9600

For SI units: 1 inch = 25 mm, 1 foot = 304.8 mm, 1 cubic foot per hour = 0.0283 m^3/h, 1 pound-force per square inch = 6.8947 kPa

* Table entries are rounded to 3 significant digits.

FUEL GAS PIPING

TABLE 1215.2(22)
POLYETHYLENE PLASTIC TUBING [NFPA 54: TABLE 6.2(w)][2, 3]

	GAS:	NATURAL
	INLET PRESSURE:	LESS THAN 2.0 psi
	PRESSURE DROP:	0.3 in. w.c.
	SPECIFIC GRAVITY:	0.60
	PLASTIC TUBING SIZE (CTS)[1] (inch)	
NOMINAL OD:	½	1
DESIGNATION:	SDR 7	SDR 11
ACTUAL ID:	0.445	0.927
LENGTH (feet)	CAPACITY IN CUBIC FEET OF GAS PER HOUR	
10	54	372
20	37	256
30	30	205
40	26	176
50	23	156
60	21	141
70	19	130
80	18	121
90	17	113
100	16	107
125	14	95
150	13	86
175	12	79
200	11	74
225	10	69
250	NA	65
275	NA	62
300	NA	59
350	NA	54
400	NA	51
450	NA	47
500	NA	45

For SI units: 1 inch = 25 mm, 1 foot = 304.8 mm, 1 cubic foot per hour = 0.0283m^3/h, 1 pound-force per square inch = 6.8947 kPa, 1 inch water column = 0.249 kPa

Notes:
[1] CTS = Copper tube size.
[2] Table entries are rounded to 3 significant digits.
[3] NA means a flow of less than 10 ft^3/h (0.283 m^3/h).

TABLE 1215.2(23)
POLYETHYLENE PLASTIC TUBING [NFPA 54: TABLE 6.2(x)][2, 3]

	GAS:	NATURAL
	INLET PRESSURE:	LESS THAN 2.0 psi
	PRESSURE DROP:	0.5 in. w.c.
	SPECIFIC GRAVITY:	0.60
	PLASTIC TUBING SIZE (CTS)[1] (inch)	
NOMINAL OD:	½	1
DESIGNATION:	SDR 7	SDR 11
ACTUAL ID:	0.445	0.927
LENGTH (feet)	CAPACITY IN CUBIC FEET OF GAS PER HOUR	
10	72	490
20	49	337
30	39	271
40	34	232
50	30	205
60	27	186
70	25	171
80	23	159
90	22	149
100	21	141
125	18	125
150	17	113
175	15	104
200	14	97
225	13	91
250	12	86
275	11	82
300	11	78
350	10	72
400	NA	67
450	NA	63
500	NA	59

For SI units: 1 inch = 25 mm, 1 foot = 304.8 mm, 1 cubic foot per hour = 0.0283m^3/h, 1 pound-force per square inch = 6.8947 kPa, 1 inch water column = 0.249 kPa

Notes:
[1] CTS = Copper tube size.
[2] Table entries are rounded to 3 significant digits.
[3] NA means a flow of less than 10 ft^3/h (0.283 m^3/h).

TABLE 1215.2(24)
SCHEDULE 40 METALLIC PIPE [NFPA 54: TABLE 6.3(a)]*

						GAS:	UNDILUTED PROPANE		
						INLET PRESSURE:	10.0 psi		
						PRESSURE DROP:	1.0 psi		
						SPECIFIC GRAVITY:	1.50		

INTENDED USE: PIPE SIZING BETWEEN FIRST STAGE (HIGH PRESSURE) REGULATOR AND SECOND STAGE (LOW PRESSURE) REGULATOR

NOMINAL INSIDE:	PIPE SIZE (inch)								
	½	¾	1	1¼	1½	2	2½	3	4
ACTUAL:	0.622	0.824	1.049	1.380	1.610	2.067	2.469	3.068	4.026
LENGTH (feet)	CAPACITY IN THOUSANDS OF BTU PER HOUR								
10	3320	6950	13 100	26 900	40 300	77 600	124 000	219 000	446 000
20	2280	4780	9000	18 500	27 700	53 300	85 000	150 000	306 000
30	1830	3840	7220	14 800	22 200	42 800	68 200	121 000	246 000
40	1570	3280	6180	12 700	19 000	36 600	58 400	103 000	211 000
50	1390	2910	5480	11 300	16 900	32 500	51 700	91 500	187 000
60	1260	2640	4970	10 200	15 300	29 400	46 900	82 900	169 000
70	1160	2430	4570	9380	14 100	27 100	43 100	76 300	156 000
80	1080	2260	4250	8730	13 100	25 200	40 100	70 900	145 000
90	1010	2120	3990	8190	12 300	23 600	37 700	66 600	136 000
100	956	2000	3770	7730	11 600	22 300	35 600	62 900	128 000
125	848	1770	3340	6850	10 300	19 800	31 500	55 700	114 000
150	768	1610	3020	6210	9300	17 900	28 600	50 500	103 000
175	706	1480	2780	5710	8560	16 500	26 300	46 500	94 700
200	657	1370	2590	5320	7960	15 300	24 400	43 200	88 100
250	582	1220	2290	4710	7060	13 600	21 700	38 300	78 100
300	528	1100	2080	4270	6400	12 300	19 600	34 700	70 800
350	486	1020	1910	3930	5880	11 300	18 100	31 900	65 100
400	452	945	1780	3650	5470	10 500	16 800	29 700	60 600
450	424	886	1670	3430	5140	9890	15 800	27 900	56 800
500	400	837	1580	3240	4850	9340	14 900	26 300	53 700
550	380	795	1500	3070	4610	8870	14 100	25 000	51 000
600	363	759	1430	2930	4400	8460	13 500	23 900	48 600
650	347	726	1370	2810	4210	8110	12 900	22 800	46 600
700	334	698	1310	2700	4040	7790	12 400	21 900	44 800
750	321	672	1270	2600	3900	7500	12 000	21 100	43 100
800	310	649	1220	2510	3760	7240	11 500	20 400	41 600
850	300	628	1180	2430	3640	7010	11 200	19 800	40 300
900	291	609	1150	2360	3530	6800	10 800	19 200	39 100
950	283	592	1110	2290	3430	6600	10 500	18 600	37 900
1000	275	575	1080	2230	3330	6420	10 200	18 100	36 900
1100	261	546	1030	2110	3170	6100	9720	17 200	35 000
1200	249	521	982	2020	3020	5820	9270	16 400	33 400
1300	239	499	940	1930	2890	5570	8880	15 700	32 000
1400	229	480	903	1850	2780	5350	8530	15 100	30 800
1500	221	462	870	1790	2680	5160	8220	14 500	29 600
1600	213	446	840	1730	2590	4980	7940	14 000	28 600
1700	206	432	813	1670	2500	4820	7680	13 600	27 700
1800	200	419	789	1620	2430	4670	7450	13 200	26 900
1900	194	407	766	1570	2360	4540	7230	12 800	26 100
2000	189	395	745	1530	2290	4410	7030	12 400	25 400

For SI units: 1 inch = 25 mm, 1 foot = 304.8 mm, 1000 British thermal units per hour = 0.293 kW, 1 pound-force per square inch = 6.8947 kPa

* Table entries are rounded to 3 significant digits.

FUEL GAS PIPING

TABLE 1215.2(25)
SCHEDULE 40 METALLIC PIPE [NFPA 54: TABLE 6.3(b)]*

						GAS:	UNDILUTED PROPANE
						INLET PRESSURE:	10.0 psi
						PRESSURE DROP:	3.0 psi
						SPECIFIC GRAVITY:	1.50

INTENDED USE: PIPE SIZING BETWEEN FIRST STAGE (HIGH PRESSURE) REGULATOR AND SECOND STAGE (LOW PRESSURE) REGULATOR										
	PIPE SIZE (inch)										
NOMINAL INSIDE:	½	¾	1	1¼	1½	2	2½	3	4
ACTUAL:	0.622	0.824	1.049	1.380	1.610	2.067	2.469	3.068	4.026
LENGTH (feet)	CAPACITY IN THOUSANDS OF BTU PER HOUR								
10	5890	12 300	23 200	47 600	71 300	137 000	219 000	387 000	789 000
20	4050	8460	15 900	32 700	49 000	94 400	150 000	266 000	543 000
30	3250	6790	12 800	26 300	39 400	75 800	121 000	214 000	436 000
40	2780	5810	11 000	22 500	33 700	64 900	103 000	183 000	373 000
50	2460	5150	9710	19 900	29 900	57 500	91 600	162 000	330 000
60	2230	4670	8790	18 100	27 100	52 100	83 000	147 000	299 000
70	2050	4300	8090	16 600	24 900	47 900	76 400	135 000	275 000
80	1910	4000	7530	15 500	23 200	44 600	71 100	126 000	256 000
90	1790	3750	7060	14 500	21 700	41 800	66 700	118 000	240 000
100	1690	3540	6670	13 700	20 500	39 500	63 000	111 000	227 000
125	1500	3140	5910	12 100	18 200	35 000	55 800	98 700	201 000
150	1360	2840	5360	11 000	16 500	31 700	50 600	89 400	182 000
175	1250	2620	4930	10 100	15 200	29 200	46 500	82 300	167 800
200	1160	2430	4580	9410	14 100	27 200	43 300	76 500	156 100
250	1030	2160	4060	8340	12 500	24 100	38 400	67 800	138 400
300	935	1950	3680	7560	11 300	21 800	34 800	61 500	125 400
350	860	1800	3390	6950	10 400	20 100	32 000	56 500	115 300
400	800	1670	3150	6470	9690	18 700	29 800	52 600	107 300
450	751	1570	2960	6070	9090	17 500	27 900	49 400	100 700
500	709	1480	2790	5730	8590	16 500	26 400	46 600	95 100
550	673	1410	2650	5450	8160	15 700	25 000	44 300	90 300
600	642	1340	2530	5200	7780	15 000	23 900	42 200	86 200
650	615	1290	2420	4980	7450	14 400	22 900	40 500	82 500
700	591	1240	2330	4780	7160	13 800	22 000	38 900	79 300
750	569	1190	2240	4600	6900	13 300	21 200	37 400	76 400
800	550	1150	2170	4450	6660	12 800	20 500	36 200	73 700
850	532	1110	2100	4300	6450	12 400	19 800	35 000	71 400
900	516	1080	2030	4170	6250	12 000	19 200	33 900	69 200
950	501	1050	1970	4050	6070	11 700	18 600	32 900	67 200
1000	487	1020	1920	3940	5900	11 400	18 100	32 000	65 400
1100	463	968	1820	3740	5610	10 800	17 200	30 400	62 100
1200	442	923	1740	3570	5350	10 300	16 400	29 000	59 200
1300	423	884	1670	3420	5120	9870	15 700	27 800	56 700
1400	406	849	1600	3280	4920	9480	15 100	26 700	54 500
1500	391	818	1540	3160	4740	9130	14 600	25 700	52 500
1600	378	790	1490	3060	4580	8820	14 100	24 800	50 700
1700	366	765	1440	2960	4430	8530	13 600	24 000	49 000
1800	355	741	1400	2870	4300	8270	13 200	23 300	47 600
1900	344	720	1360	2780	4170	8040	12 800	22 600	46 200
2000	335	700	1320	2710	4060	7820	12 500	22 000	44 900

For SI units: 1 inch = 25 mm, 1 foot = 304.8 mm, 1000 British thermal units per hour = 0.293 kW, 1 pound-force per square inch = 6.8947 kPa

* Table entries are rounded to 3 significant digits.

FUEL GAS PIPING

TABLE 1215.2(26)
SCHEDULE 40 METALLIC PIPE [NFPA 54: TABLE 6.3(c)]*

	GAS:	UNDILUTED PROPANE
	INLET PRESSURE:	2.0 psi
	PRESSURE DROP:	1.0 psi
	SPECIFIC GRAVITY:	1.50

INTENDED USE: PIPE SIZING BETWEEN 2 PSI SERVICE AND LINE PRESSURE REGULATOR									
	PIPE SIZE (inch)								
NOMINAL:	½	¾	1	1¼	1½	2	2½	3	4
ACTUAL ID:	0.622	0.824	1.049	1.380	1.610	2.067	2.469	3.068	4.026
LENGTH (feet)	CAPACITY IN THOUSANDS OF BTU PER HOUR								
10	2680	5590	10 500	21 600	32 400	62 400	99 500	176 000	359 000
20	1840	3850	7240	14 900	22 300	42 900	68 400	121 000	247 000
30	1480	3090	5820	11 900	17 900	34 500	54 900	97 100	198 000
40	1260	2640	4980	10 200	15 300	29 500	47 000	83 100	170 000
50	1120	2340	4410	9060	13 600	26 100	41 700	73 700	150 000
60	1010	2120	4000	8210	12 300	23 700	37 700	66 700	136 000
70	934	1950	3680	7550	11 300	21 800	34 700	61 400	125 000
80	869	1820	3420	7020	10 500	20 300	32 300	57 100	116 000
90	815	1700	3210	6590	9880	19 000	30 300	53 600	109 000
100	770	1610	3030	6230	9330	18 000	28 600	50 600	103 000
125	682	1430	2690	5520	8270	15 900	25 400	44 900	91 500
150	618	1290	2440	5000	7490	14 400	23 000	40 700	82 900
175	569	1190	2240	4600	6890	13 300	21 200	37 400	76 300
200	529	1110	2080	4280	6410	12 300	19 700	34 800	71 000
250	469	981	1850	3790	5680	10 900	17 400	30 800	62 900
300	425	889	1670	3440	5150	9920	15 800	27 900	57 000
350	391	817	1540	3160	4740	9120	14 500	25 700	52 400
400	364	760	1430	2940	4410	8490	13 500	23 900	48 800
450	341	714	1340	2760	4130	7960	12 700	22 400	45 800
500	322	674	1270	2610	3910	7520	12 000	21 200	43 200
550	306	640	1210	2480	3710	7140	11 400	20 100	41 100
600	292	611	1150	2360	3540	6820	10 900	19 200	39 200
650	280	585	1100	2260	3390	6530	10 400	18 400	37 500
700	269	562	1060	2170	3260	6270	9990	17 700	36 000
750	259	541	1020	2090	3140	6040	9630	17 000	34 700
800	250	523	985	2020	3030	5830	9300	16 400	33 500
850	242	506	953	1960	2930	5640	9000	15 900	32 400
900	235	490	924	1900	2840	5470	8720	15 400	31 500
950	228	476	897	1840	2760	5310	8470	15 000	30 500
1000	222	463	873	1790	2680	5170	8240	14 600	29 700
1100	210	440	829	1700	2550	4910	7830	13 800	28 200
1200	201	420	791	1620	2430	4680	7470	13 200	26 900
1300	192	402	757	1550	2330	4490	7150	12 600	25 800
1400	185	386	727	1490	2240	4310	6870	12 100	24 800
1500	178	372	701	1440	2160	4150	6620	11 700	23 900
1600	172	359	677	1390	2080	4010	6390	11 300	23 000
1700	166	348	655	1340	2010	3880	6180	10 900	22 300
1800	161	337	635	1300	1950	3760	6000	10 600	21 600
1900	157	327	617	1270	1900	3650	5820	10 300	21 000
2000	152	318	600	1230	1840	3550	5660	10 000	20 400

For SI units: 1 inch = 25 mm, 1 foot = 304.8 mm, 1000 British thermal units per hour = 0.293 kW, 1 pound-force per square inch = 6.8947 kPa

* Table entries are rounded to 3 significant digits.

FUEL GAS PIPING

TABLE 1215.2(27)
SCHEDULE 40 METALLIC PIPE [NFPA 54: TABLE 6.3(d)]*

					GAS:	UNDILUTED PROPANE
					INLET PRESSURE:	11.0 in. w.c.
					PRESSURE DROP:	0.5 in. w.c.
					SPECIFIC GRAVITY:	1.50

INTENDED USE: PIPE SIZING BETWEEN SINGLE OR SECOND STAGE (LOW PRESSURE) REGULATOR AND APPLIANCE										
	PIPE SIZE (inch)									
NOMINAL INSIDE:	½	¾	1	1¼	1½	2	2½	3	4
ACTUAL ID:	0.622	0.824	1.049	1.380	1.610	2.067	2.469	3.068	4.026
LENGTH (feet)	CAPACITY IN THOUSANDS OF BTU PER HOUR								
10	291	608	1150	2350	3520	6790	10 800	19 100	39 000
20	200	418	787	1620	2420	4660	7430	13 100	26 800
30	160	336	632	1300	1940	3750	5970	10 600	21 500
40	137	287	541	1110	1660	3210	5110	9030	18 400
50	122	255	480	985	1480	2840	4530	8000	16 300
60	110	231	434	892	1340	2570	4100	7250	14 800
80	101	212	400	821	1230	2370	3770	6670	13 600
100	94	197	372	763	1140	2200	3510	6210	12 700
125	89	185	349	716	1070	2070	3290	5820	11 900
150	84	175	330	677	1010	1950	3110	5500	11 200
175	74	155	292	600	899	1730	2760	4880	9950
200	67	140	265	543	814	1570	2500	4420	9010
250	62	129	243	500	749	1440	2300	4060	8290
300	58	120	227	465	697	1340	2140	3780	7710
350	51	107	201	412	618	1190	1900	3350	6840
400	46	97	182	373	560	1080	1720	3040	6190
450	42	89	167	344	515	991	1580	2790	5700
500	40	83	156	320	479	922	1470	2600	5300
550	37	78	146	300	449	865	1380	2440	4970
600	35	73	138	283	424	817	1300	2300	4700
650	33	70	131	269	403	776	1240	2190	4460
700	32	66	125	257	385	741	1180	2090	4260
750	30	64	120	246	368	709	1130	2000	4080
800	29	61	115	236	354	681	1090	1920	3920
850	28	59	111	227	341	656	1050	1850	3770
900	27	57	107	220	329	634	1010	1790	3640
950	26	55	104	213	319	613	978	1730	3530
1000	25	53	100	206	309	595	948	1680	3420
1100	25	52	97	200	300	578	921	1630	3320
1200	24	50	95	195	292	562	895	1580	3230
1300	23	48	90	185	277	534	850	1500	3070
1400	22	46	86	176	264	509	811	1430	2930
1500	21	44	82	169	253	487	777	1370	2800
1600	20	42	79	162	243	468	746	1320	2690
1700	19	40	76	156	234	451	719	1270	2590
1800	19	39	74	151	226	436	694	1230	2500
1900	18	38	71	146	219	422	672	1190	2420
2000	18	37	69	142	212	409	652	1150	2350

For SI units: 1 inch = 25 mm, 1 foot = 304.8 mm, 1000 British thermal units per hour = 0.293 kW, 1 inch water column = 0.249 kPa

* Table entries are rounded to 3 significant digits.

TABLE 1215.2(28)
SEMI-RIGID COPPER TUBING [NFPA 54: TABLE 6.3(e)][2]

		GAS:	UNDILUTED PROPANE
		INLET PRESSURE:	10.0 psi
		PRESSURE DROP:	1.0 psi
		SPECIFIC GRAVITY:	1.50

INTENDED USE: TUBE SIZING BETWEEN FIRST STAGE (HIGH PRESSURE) REGULATOR AND SECOND STAGE (LOW PRESSURE) REGULATOR

NOMINAL:	K & L:	¼	⅜	½	⅝	¾	1	1¼	1½	2
	ACR:	⅜	½	⅝	¾	⅞	1⅛	1⅜	–	–
OUTSIDE:		0.375	0.500	0.625	0.750	0.875	1.125	1.375	1.625	2.125
INSIDE:[1]		0.305	0.402	0.527	0.652	0.745	0.995	1.245	1.481	1.959
LENGTH (feet)		CAPACITY IN THOUSANDS OF BTU PER HOUR								
10		513	1060	2150	3760	5330	11 400	20 500	32 300	67 400
20		352	727	1480	2580	3670	7830	14 100	22 200	46 300
30		283	584	1190	2080	2940	6290	11 300	17 900	37 200
40		242	500	1020	1780	2520	5380	9690	15 300	31 800
50		215	443	901	1570	2230	4770	8590	13 500	28 200
60		194	401	816	1430	2020	4320	7780	12 300	25 600
70		179	369	751	1310	1860	3980	7160	11 300	23 500
80		166	343	699	1220	1730	3700	6660	10 500	21 900
90		156	322	655	1150	1630	3470	6250	9850	20 500
100		147	304	619	1080	1540	3280	5900	9310	19 400
125		131	270	549	959	1360	2910	5230	8250	17 200
150		118	244	497	869	1230	2630	4740	7470	15 600
175		109	225	457	799	1130	2420	4360	6880	14 300
200		101	209	426	744	1060	2250	4060	6400	13 300
250		90	185	377	659	935	2000	3600	5670	11 800
300		81	168	342	597	847	1810	3260	5140	10 700
350		75	155	314	549	779	1660	3000	4730	9840
400		70	144	292	511	725	1550	2790	4400	9160
450		65	135	274	480	680	1450	2620	4130	8590
500		62	127	259	453	643	1370	2470	3900	8120
550		59	121	246	430	610	1300	2350	3700	7710
600		56	115	235	410	582	1240	2240	3530	7350
650		54	111	225	393	558	1190	2140	3380	7040
700		51	106	216	378	536	1140	2060	3250	6770
750		50	102	208	364	516	1100	1980	3130	6520
800		48	99	201	351	498	1060	1920	3020	6290
850		46	96	195	340	482	1030	1850	2920	6090
900		45	93	189	330	468	1000	1800	2840	5910
950		44	90	183	320	454	970	1750	2750	5730
1000		42	88	178	311	442	944	1700	2680	5580
1100		40	83	169	296	420	896	1610	2540	5300
1200		38	79	161	282	400	855	1540	2430	5050
1300		37	76	155	270	383	819	1470	2320	4840
1400		35	73	148	260	368	787	1420	2230	4650
1500		34	70	143	250	355	758	1360	2150	4480
1600		33	68	138	241	343	732	1320	2080	4330
1700		32	66	134	234	331	708	1270	2010	4190
1800		31	64	130	227	321	687	1240	1950	4060
1900		30	62	126	220	312	667	1200	1890	3940
2000		29	60	122	214	304	648	1170	1840	3830

For SI units: 1 inch = 25 mm, 1 foot = 304.8 mm, 1000 British thermal units per hour = 0.293 kW, 1 pound-force per square inch = 6.8947 kPa

Notes:
[1] Table capacities are based on Type K copper tubing inside diameter (shown), which has the smallest inside diameter of the copper tubing products.
[2] Table entries are rounded to 3 significant digits.

FUEL GAS PIPING

TABLE 1215.2(29)
SEMI-RIGID COPPER TUBING [NFPA 54: TABLE 6.3(f)][2, 3]

						GAS:	UNDILUTED PROPANE			
						INLET PRESSURE:	11.0 in. w.c.			
						PRESSURE DROP:	0.5 in. w.c.			
						SPECIFIC GRAVITY:	1.50			
INTENDED USE: TUBE SIZING BETWEEN SINGLE OR SECOND STAGE (LOW PRESSURE) REGULATOR AND APPLIANCE										
		TUBE SIZE (inch)								
NOMINAL:	K & L:	¼	⅜	½	⅝	¾	1	1¼	1½	2
	ACR:	⅜	½	⅝	¾	⅞	1⅛	1⅜	–	–
OUTSIDE:		0.375	0.500	0.625	0.750	0.875	1.125	1.375	1.625	2.125
INSIDE:[1]		0.305	0.402	0.527	0.652	0.745	0.995	1.245	1.481	1.959
LENGTH (feet)		**CAPACITY IN THOUSANDS OF BTU PER HOUR**								
10		45	93	188	329	467	997	1800	2830	5890
20		31	64	129	226	321	685	1230	1950	4050
30		25	51	104	182	258	550	991	1560	3250
40		21	44	89	155	220	471	848	1340	2780
50		19	39	79	138	195	417	752	1180	2470
60		17	35	71	125	177	378	681	1070	2240
70		16	32	66	115	163	348	626	988	2060
80		15	30	61	107	152	324	583	919	1910
90		14	28	57	100	142	304	547	862	1800
100		13	27	54	95	134	287	517	814	1700
125		11	24	48	84	119	254	458	722	1500
150		10	21	44	76	108	230	415	654	1360
175		NA	20	40	70	99	212	382	602	1250
200		NA	18	37	65	92	197	355	560	1170
250		NA	16	33	58	82	175	315	496	1030
300		NA	15	30	52	74	158	285	449	936
350		NA	14	28	48	68	146	262	414	861
400		NA	13	26	45	63	136	244	385	801
450		NA	12	24	42	60	127	229	361	752
500		NA	11	23	40	56	120	216	341	710
550		NA	11	22	38	53	114	205	324	674
600		NA	10	21	36	51	109	196	309	643
650		NA	NA	20	34	49	104	188	296	616
700		NA	NA	19	33	47	100	180	284	592
750		NA	NA	18	32	45	96	174	274	570
800		NA	NA	18	31	44	93	168	264	551
850		NA	NA	17	30	42	90	162	256	533
900		NA	NA	17	29	41	87	157	248	517
950		NA	NA	16	28	40	85	153	241	502
1000		NA	NA	16	27	39	83	149	234	488
1100		NA	NA	15	26	37	78	141	223	464
1200		NA	NA	14	25	35	75	135	212	442
1300		NA	NA	14	24	34	72	129	203	423
1400		NA	NA	13	23	32	69	124	195	407
1500		NA	NA	13	22	31	66	119	188	392
1600		NA	NA	12	21	30	64	115	182	378
1700		NA	NA	12	20	29	62	112	176	366
1800		NA	NA	11	20	28	60	108	170	355
1900		NA	NA	11	19	27	58	105	166	345
2000		NA	NA	11	19	27	57	102	161	335

For SI units: 1 inch = 25 mm, 1 foot = 304.8 mm, 1000 British thermal units per hour = 0.293 kW, 1 inch water column = 0.249 kPa

Notes:

[1] Table capacities are based on Type K copper tubing inside diameter (shown), which has the smallest inside diameter of the copper tubing products.

[2] Table entries are rounded to 3 significant digits.

[3] NA means a flow of less than 10 000 Btu/h (2.93 kW).

TABLE 1215.2(30)
SEMI-RIGID COPPER TUBING [NFPA 54: TABLE 6.3(g)][2]

							GAS:	UNDILUTED PROPANE		
							INLET PRESSURE:	2.0 psi		
							PRESSURE DROP:	1.0 psi		
							SPECIFIC GRAVITY:	1.50		
		INTENDED USE: TUBE SIZING BETWEEN 2 PSIG SERVICE AND LINE PRESSURE REGULATOR								
		TUBE SIZE (inch)								
NOMINAL:	**K & L:**	¼	⅜	½	⅝	¾	1	1¼	1½	2
	ACR:	⅜	½	⅝	¾	⅞	1⅛	1⅜	–	–
OUTSIDE:		0.375	0.500	0.625	0.750	0.875	1.125	1.375	1.625	2.125
INSIDE:[1]		0.305	0.402	0.527	0.652	0.745	0.995	1.245	1.481	1.959
LENGTH (feet)		**CAPACITY IN THOUSANDS OF BTU PER HOUR**								
10		413	852	1730	3030	4300	9170	16 500	26 000	54 200
20		284	585	1190	2080	2950	6310	11 400	17 900	37 300
30		228	470	956	1670	2370	5060	9120	14 400	29 900
40		195	402	818	1430	2030	4330	7800	12 300	25 600
50		173	356	725	1270	1800	3840	6920	10 900	22 700
60		157	323	657	1150	1630	3480	6270	9880	20 600
70		144	297	605	1060	1500	3200	5760	9090	18 900
80		134	276	562	983	1390	2980	5360	8450	17 600
90		126	259	528	922	1310	2790	5030	7930	16 500
100		119	245	498	871	1240	2640	4750	7490	15 600
125		105	217	442	772	1100	2340	4210	6640	13 800
150		95	197	400	700	992	2120	3820	6020	12 500
175		88	181	368	644	913	1950	3510	5540	11 500
200		82	168	343	599	849	1810	3270	5150	10 700
250		72	149	304	531	753	1610	2900	4560	9510
300		66	135	275	481	682	1460	2620	4140	8610
350		60	124	253	442	628	1340	2410	3800	7920
400		56	116	235	411	584	1250	2250	3540	7370
450		53	109	221	386	548	1170	2110	3320	6920
500		50	103	209	365	517	1110	1990	3140	6530
550		47	97	198	346	491	1050	1890	2980	6210
600		45	93	189	330	469	1000	1800	2840	5920
650		43	89	181	316	449	959	1730	2720	5670
700		41	86	174	304	431	921	1660	2620	5450
750		40	82	168	293	415	888	1600	2520	5250
800		39	80	162	283	401	857	1540	2430	5070
850		37	77	157	274	388	829	1490	2350	4900
900		36	75	152	265	376	804	1450	2280	4750
950		35	72	147	258	366	781	1410	2220	4620
1000		34	71	143	251	356	760	1370	2160	4490
1100		32	67	136	238	338	721	1300	2050	4270
1200		31	64	130	227	322	688	1240	1950	4070
1300		30	61	124	217	309	659	1190	1870	3900
1400		28	59	120	209	296	633	1140	1800	3740
1500		27	57	115	201	286	610	1100	1730	3610
1600		26	55	111	194	276	589	1060	1670	3480
1700		26	53	108	188	267	570	1030	1620	3370
1800		25	51	104	182	259	553	1000	1570	3270
1900		24	50	101	177	251	537	966	1520	3170
2000		23	48	99	172	244	522	940	1480	3090

For SI units: 1 inch = 25 mm, 1 foot = 304.8 mm, 1000 British thermal units per hour = 0.293 kW, 1 pound-force per square inch = 6.8947 kPa

Notes:

[1] Table capacities are based on Type K copper tubing inside diameter (shown), which has the smallest inside diameter of the copper tubing products.

[2] Table entries are rounded to 3 significant digits.

FUEL GAS PIPING

TABLE 1215.2(31)
CORRUGATED STAINLESS STEEL TUBING (CSST) [NFPA 54: TABLE 6.3(h)][1, 2]

						GAS:	UNDILUTED PROPANE
						INLET PRESSURE:	11.0 in. w.c.
						PRESSURE DROP:	0.5 in. w.c.
						SPECIFIC GRAVITY:	1.50

INTENDED USE: CSST SIZING BETWEEN SINGLE OR SECOND STAGE (LOW PRESSURE) REGULATOR AND APPLIANCE SHUTOFF VALVE

FLOW DESIGNATION:	\multicolumn{14}{c}{TUBE SIZE (EHD)[3]}

FLOW DESIGNATION:	13	15	18	19	23	25	30	31	37	39	46	48	60	62
LENGTH (feet)	\multicolumn{14}{c}{CAPACITY IN THOUSANDS OF BTU PER HOUR}
5	72	99	181	211	355	426	744	863	1420	1638	2830	3270	5780	6550
10	50	69	129	150	254	303	521	605	971	1179	1990	2320	4110	4640
15	39	55	104	121	208	248	422	490	775	972	1620	1900	3370	3790
20	34	49	91	106	183	216	365	425	661	847	1400	1650	2930	3290
25	30	42	82	94	164	192	325	379	583	762	1250	1480	2630	2940
30	28	39	74	87	151	177	297	344	528	698	1140	1350	2400	2680
40	23	33	64	74	131	153	256	297	449	610	988	1170	2090	2330
50	20	30	58	66	118	137	227	265	397	548	884	1050	1870	2080
60	19	26	53	60	107	126	207	241	359	502	805	961	1710	1900
70	17	25	49	57	99	117	191	222	330	466	745	890	1590	1760
80	15	23	45	52	94	109	178	208	307	438	696	833	1490	1650
90	15	22	44	50	90	102	169	197	286	414	656	787	1400	1550
100	14	20	41	47	85	98	159	186	270	393	621	746	1330	1480
150	11	15	31	36	66	75	123	143	217	324	506	611	1090	1210
200	9	14	28	33	60	69	112	129	183	283	438	531	948	1050
250	8	12	25	30	53	61	99	117	163	254	390	476	850	934
300	8	11	23	26	50	57	90	107	147	234	357	434	777	854

For SI units: 1 foot = 304.8 mm, 1000 British thermal units per hour = 0.293 kW, 1 inch water column = 0.249 kPa

Notes:

[1] Table includes losses for four 90 degree (1.57 rad) bends and two end fittings. Tubing runs with larger numbers of bends, fittings, or both shall be increased by an equivalent length of tubing to the following equation: $L = 1.3\,n$, where L is additional length (ft) of tubing and n is the number of additional fittings, bends, or both.

[2] Table entries are rounded to 3 significant digits.

[3] EHD = Equivalent Hydraulic Diameter, which is a measure of the relative hydraulic efficiency between different tubing sizes. The greater the value of EHD, the greater the gas capacity of the tubing.

TABLE 1215.2(32)
CORRUGATED STAINLESS STEEL TUBING (CSST) [NFPA 54: TABLE 6.3(i)][1, 2, 3, 4]

							GAS:	UNDILUTED PROPANE
							INLET PRESSURE:	2.0 psi
							PRESSURE DROP:	1.0 psi
							SPECIFIC GRAVITY:	1.50

INTENDED USE: CSST SIZING BETWEEN 2 PSI SERVICE AND LINE PRESSURE REGULATOR

	TUBE SIZE (EHD)[5]													
FLOW DESIGNATION:	13	15	18	19	23	25	30	31	37	39	46	48	60	62
LENGTH (feet)	CAPACITY IN THOUSANDS OF BTU PER HOUR													
10	426	558	927	1110	1740	2170	4100	4720	7130	7958	15 200	16 800	29 400	34 200
25	262	347	591	701	1120	1380	2560	2950	4560	5147	9550	10 700	18 800	21 700
30	238	316	540	640	1030	1270	2330	2690	4180	4719	8710	9790	17 200	19 800
40	203	271	469	554	896	1100	2010	2320	3630	4116	7530	8500	14 900	17 200
50	181	243	420	496	806	986	1790	2070	3260	3702	6730	7610	13 400	15 400
75	147	196	344	406	663	809	1460	1690	2680	3053	5480	6230	11 000	12 600
80	140	189	333	393	643	768	1410	1630	2590	2961	5300	6040	10 600	12 200
100	124	169	298	350	578	703	1260	1450	2330	2662	4740	5410	9530	10 900
150	101	137	245	287	477	575	1020	1180	1910	2195	3860	4430	7810	8890
200	86	118	213	248	415	501	880	1020	1660	1915	3340	3840	6780	7710
250	77	105	191	222	373	448	785	910	1490	1722	2980	3440	6080	6900
300	69	96	173	203	343	411	716	829	1360	1578	2720	3150	5560	6300
400	60	82	151	175	298	355	616	716	1160	1376	2350	2730	4830	5460
500	53	72	135	158	268	319	550	638	1030	1237	2100	2450	4330	4880

For SI units: 1 foot = 304.8 mm, 1000 British thermal units per hour = 0.293 kW, 1 pound-force per square inch = 6.8947 kPa

Notes:

[1] Table does not include effect of pressure drop across the line regulator. Where regulator loss exceeds 0.5 psi (3.4 kPa) [based on 13 inch water column (3.2 kPa) outlet pressure], DO NOT USE THIS TABLE. Consult with regulator manufacturer for pressure drops and capacity factors. Pressure drops across a regulator are capable of varying with flow rate.

[2] CAUTION: Capacities shown in table are capable of exceeding the maximum capacity for a selected regulator. Consult with regulator or tubing manufacturer for guidance.

[3] Table includes losses for four 90 degree (1.57 rad) bends and two end fittings. Tubing runs with larger numbers of bends, fittings, or both shall be increased by an equivalent length of tubing to the following equation: $L = 1.3\,n$, where L is additional length (ft) of tubing and n is the number of additional fittings, bends, or both.

[4] Table entries are rounded to 3 significant digits.

[5] EHD = Equivalent Hydraulic Diameter, which is a measure of the relative hydraulic efficiency between different tubing sizes. The greater the value of EHD, the greater the gas capacity of the tubing.

TABLE 1215.2(33)
CORRUGATED STAINLESS STEEL TUBING (CSST) [NFPA 54: TABLE 6.3(j)][1, 2, 3, 4]

	GAS:	UNDILUTED PROPANE
	INLET PRESSURE:	5.0 psi
	PRESSURE DROP:	3.5 psi
	SPECIFIC GRAVITY:	1.50

	TUBE SIZE (EHD)[5]													
FLOW DESIGNATION:	13	15	18	19	23	25	30	31	37	39	46	48	60	62
LENGTH (feet)	CAPACITY IN THOUSANDS OF BTU PER HOUR													
10	826	1070	1710	2060	3150	4000	7830	8950	13 100	14 441	28 600	31 200	54 400	63 800
25	509	664	1090	1310	2040	2550	4860	5600	8400	9339	18 000	19 900	34 700	40 400
30	461	603	999	1190	1870	2340	4430	5100	7680	8564	16 400	18 200	31 700	36 900
40	396	520	867	1030	1630	2030	3820	4400	6680	7469	14 200	15 800	27 600	32 000
50	352	463	777	926	1460	1820	3410	3930	5990	6717	12 700	14 100	24 700	28 600
75	284	376	637	757	1210	1490	2770	3190	4920	5539	10 300	11 600	20 300	23 400
80	275	363	618	731	1170	1450	2680	3090	4770	5372	9990	11 200	19 600	22 700
100	243	324	553	656	1050	1300	2390	2760	4280	4830	8930	10 000	17 600	20 300
150	196	262	453	535	866	1060	1940	2240	3510	3983	7270	8210	14 400	16 600
200	169	226	393	464	755	923	1680	1930	3050	3474	6290	7130	12 500	14 400
250	150	202	352	415	679	828	1490	1730	2740	3124	5620	6390	11 200	12 900
300	136	183	322	379	622	757	1360	1570	2510	2865	5120	5840	10 300	11 700
400	117	158	279	328	542	657	1170	1360	2180	2498	4430	5070	8920	10 200
500	104	140	251	294	488	589	1050	1210	1950	2247	3960	4540	8000	9110

For SI units: 1 foot = 304.8 mm, 1000 British thermal units per hour = 0.293 kW, 1 pound-force per square inch = 6.8947 kPa

Notes:
[1] Table does not include effect of pressure drop across the line regulator. Where regulator loss exceeds 0.5 psi (3.4 kPa) [based on 13 inch water column (3.2 kPa) outlet pressure], DO NOT USE THIS TABLE. Consult with regulator manufacturer for pressure drops and capacity factors. Pressure drops across a regulator are capable of varying with flow rate.
[2] CAUTION: Capacities shown in table are capable of exceeding the maximum capacity for a selected regulator. Consult with regulator or tubing manufacturer for guidance.
[3] Table includes losses for four 90 degree (1.57 rad) bends and two end fittings. Tubing runs with larger numbers of bends, fittings, or both shall be increased by an equivalent length of tubing to the following equation: $L = 1.3\,n$, where L is additional length (ft) of tubing and n is the number of additional fittings, bends, or both.
[4] Table entries are rounded to 3 significant digits.
[5] EHD = Equivalent Hydraulic Diameter, which is a measure of the relative hydraulic efficiency between different tubing sizes. The greater the value of EHD, the greater the gas capacity of the tubing.

TABLE 1215.2(34)
POLYETHYLENE PLASTIC PIPE [NFPA 54: TABLE 6.3(k)]*

					GAS:	UNDILUTED PROPANE
					INLET PRESSURE:	11.0 in. w.c.
					PRESSURE DROP:	0.5 in. w.c.
					SPECIFIC GRAVITY:	1.50

INTENDED USE: PE SIZING BETWEEN INTEGRAL SECOND-STAGE REGULATOR AT TANK OR SECOND-STAGE (LOW PRESSURE) REGULATOR AND BUILDING								
	PIPE SIZE (inch)							
NOMINAL OD:	½	¾	1	1¼	1½	2	3	4
DESIGNATION:	SDR 9.3	SDR 11	SDR 11	SDR 10	SDR 11	SDR 11	SDR 11	SDR 11
ACTUAL ID:	0.660	0.860	1.077	1.328	1.554	1.943	2.864	3.682
LENGTH (feet)	CAPACITY IN THOUSANDS OF BTU PER HOUR							
10	340	680	1230	2130	3210	5770	16 000	30 900
20	233	468	844	1460	2210	3970	11 000	21 200
30	187	375	677	1170	1770	3180	8810	17 000
40	160	321	580	1000	1520	2730	7540	14 600
50	142	285	514	890	1340	2420	6680	12 900
60	129	258	466	807	1220	2190	6050	11 700
70	119	237	428	742	1120	2010	5570	10 800
80	110	221	398	690	1040	1870	5180	10 000
90	103	207	374	648	978	1760	4860	9400
100	98	196	353	612	924	1660	4590	8900
125	87	173	313	542	819	1470	4070	7900
150	78	157	284	491	742	1330	3690	7130
175	72	145	261	452	683	1230	3390	6560
200	67	135	243	420	635	1140	3160	6100
250	60	119	215	373	563	1010	2800	5410
300	54	108	195	338	510	916	2530	4900
350	50	99	179	311	469	843	2330	4510
400	46	92	167	289	436	784	2170	4190
450	43	87	157	271	409	736	2040	3930
500	41	82	148	256	387	695	1920	3720

For SI units: 1 inch = 25 mm, 1 foot = 304.8 mm, 1000 British thermal units per hour = 0.293 kW, 1 inch water column = 0.249 kPa

* Table entries are rounded to 3 significant digits.

FUEL GAS PIPING

TABLE 1215.2(35)
POLYETHYLENE PLASTIC PIPE [NFPA 54: TABLE 6.3(I)]*

					GAS:	UNDILUTED PROPANE	
					INLET PRESSURE:	2.0 psi	
					PRESSURE DROP:	1.0 psi	
					SPECIFIC GRAVITY:	1.50	
INTENDED USE: PE PIPE SIZING BETWEEN 2 PSI SERVICE REGULATOR AND LINE PRESSURE REGULATOR							
PIPE SIZE (inch)							
NOMINAL OD:	½	¾	1	1¼	1½	2	3	4
DESIGNATION:	SDR 9.3	SDR 11	SDR 11	SDR 10	SDR 11	SDR 11	SDR 11	SDR 11
ACTUAL ID:	0.660	0.860	1.077	1.328	1.554	1.943	2.864	3.682
LENGTH (feet)	CAPACITY IN THOUSANDS OF BTU PER HOUR						
10	3130	6260	11 300	19 600	29 500	53 100	147 000	284 000
20	2150	4300	7760	13 400	20 300	36 500	101 000	195 000
30	1730	3450	6230	10 800	16 300	29 300	81 100	157 000
40	1480	2960	5330	9240	14 000	25 100	69 400	134 100
50	1310	2620	4730	8190	12 400	22 200	61 500	119 000
60	1190	2370	4280	7420	11 200	20 100	55 700	108 000
70	1090	2180	3940	6830	10 300	18 500	51 300	99 100
80	1010	2030	3670	6350	9590	17 200	47 700	92 200
90	952	1910	3440	5960	9000	16 200	44 700	86 500
100	899	1800	3250	5630	8500	15 300	42 300	81 700
125	797	1600	2880	4990	7530	13 500	37 500	72 400
150	722	1450	2610	4520	6830	12 300	33 900	65 600
175	664	1330	2400	4160	6280	11 300	31 200	60 300
200	618	1240	2230	3870	5840	10 500	29 000	56 100
250	548	1100	1980	3430	5180	9300	25 700	49 800
300	496	994	1790	3110	4690	8430	23 300	45 100
350	457	914	1650	2860	4320	7760	21 500	41 500
400	425	851	1530	2660	4020	7220	12 000	38 600
450	399	798	1440	2500	3770	6770	18 700	36 200
500	377	754	1360	2360	3560	6390	17 700	34 200
550	358	716	1290	2240	3380	6070	16 800	32 500
600	341	683	1230	2140	3220	5790	16 000	31 000
650	327	654	1180	2040	3090	5550	15 400	29 700
700	314	628	1130	1960	2970	5330	14 700	28 500
750	302	605	1090	1890	2860	5140	14 200	27 500
800	292	585	1050	1830	2760	4960	13 700	26 500
850	283	566	1020	1770	2670	4800	13 300	25 700
900	274	549	990	1710	2590	4650	12 900	24 900
950	266	533	961	1670	2520	4520	12 500	24 200
1000	259	518	935	1620	2450	4400	12 200	23 500
1100	246	492	888	1540	2320	4170	11 500	22 300
1200	234	470	847	1470	2220	3980	11 000	21 300
1300	225	450	811	1410	2120	3810	10 600	20 400
1400	216	432	779	1350	2040	3660	10 100	19 600
1500	208	416	751	1300	1960	3530	9760	18 900
1600	201	402	725	1260	1900	3410	9430	18 200
1700	194	389	702	1220	1840	3300	9130	17 600
1800	188	377	680	1180	1780	3200	8850	17 100
1900	183	366	661	1140	1730	3110	8590	16 600
2000	178	356	643	1110	1680	3020	8360	16 200

For SI units: 1 inch = 25 mm, 1 foot = 304.8 mm, 1000 British thermal units per hour = 0.293 kW, 1 pound-force per square inch = 6.8947 kPa
* Table entries are rounded to 3 significant digits.

TABLE 1215.2(36)
POLYETHYLENE PLASTIC TUBING [NFPA 54: TABLE 6.3(m)][2]

		GAS:	UNDILUTED PROPANE
		INLET PRESSURE:	11.0 in. w.c.
		PRESSURE DROP:	0.5 in. w.c.
		SPECIFIC GRAVITY:	1.50
INTENDED USE: PE TUBE SIZING BETWEEN INTEGRAL SECOND-STAGE REGULATOR AT TANK OR SECOND-STAGE (LOW PRESSURE) REGULATOR AND BUILDING			
	PLASTIC TUBING SIZE (CTS)[1] (inch)		
NOMINAL OD:	½	1	
DESIGNATION:	SDR 7	SDR 11	
ACTUAL ID:	0.445	0.927	
LENGTH (feet)	CAPACITY IN THOUSANDS OF BTU PER HOUR		
10	121	828	
20	83	569	
30	67	457	
40	57	391	
50	51	347	
60	46	314	
70	42	289	
80	39	269	
90	37	252	
100	35	238	
125	31	211	
150	28	191	
175	26	176	
200	24	164	
225	22	154	
250	21	145	
275	20	138	
300	19	132	
350	18	121	
400	16	113	
450	15	106	
500	15	100	

For SI units: 1 inch = 25 mm, 1 foot = 304.8 mm, 1000 British thermal units per hour = 0.293 kW, 1 inch water column = 0.249 kPa

Notes:

[1] CTS = Copper tube size.

[2] Table entries are rounded to 3 significant digits.

CHAPTER 13

HEALTH CARE FACILITIES AND MEDICAL GAS AND VACUUM SYSTEMS

Part I – General Requirements.

1301.0 General.

1301.1 Applicability. This chapter applies to the special fixtures and systems in health care facilities; the special plumbing requirements for such facilities; and the installation, testing, and verification of Categories 1, 2, and 3 medical gas and medical vacuum piping systems, except as otherwise indicated in this chapter, from the central supply system to the station outlets or inlets in hospitals, clinics, and other health care facilities. Other plumbing in such facilities shall comply with other applicable sections of this code. For Category 3 medical gas systems, only oxygen and nitrous oxide shall be used.

Special fixtures and systems within health care facilities require special attention. In order for these fixtures and systems to provide safe operation and safety for the patients, the following requirements in this chapter shall apply.

1301.2 Where Not Applicable. This chapter does not apply to the following except as otherwise addressed in this chapter:

(1) Cylinder and container management, storage, and reserve requirements
(2) Bulk supply systems
(3) Electrical connections and requirements
(4) Motor requirements and controls
(5) Systems having nonstandard operating pressures
(6) Waste anesthetic gas disposal (WAGD) systems
(7) Surface-mounted medical gas rail systems
(8) Breathing air replenishment (BAR) systems
(9) Portable compressed gas systems
(10) Medical support gas systems
(11) Gas-powered device supply systems
(12) Scavenging systems

Chapter 13 covers the installation of medical gas and vacuum piping from the point of supply to the user inlet/outlet. Although important to the operation of medical gas systems, requirements for portable systems, cylinder storage, bulk supply systems, electrical requirements, waste anesthetic gas disposal (WAGD), surface-mounted medical gas rail systems and systems with nonstandard pressures are not covered in depth, as those provisions exist in other codes and documents. This chapter addresses medical gas and vacuum piping systems installed in hospitals, outpatient surgery facilities, emergency rooms, medical clinics, trauma centers and other facilities that require life support.

1301.3 Conflict of Requirements. The requirements of this chapter shall not be interpreted to conflict with the requirements of NFPA 99. For requirements of portions of medical gas and vacuum systems not addressed in this chapter or medical gas and vacuum systems beyond the scope of this chapter refer to NFPA 99.

1301.4 Terms. Where the terms medical gas or medical support gas occur, the provisions shall apply to all piped systems for oxygen, nitrous oxide, medical air, carbon dioxide, helium, nitrogen, instrument air, and mixtures thereof. Wherever the name of a specific gas service occurs, the provision shall apply only to that gas. [NFPA 99:5.1.1.3]

1301.5 Where Required. Construction and equipment requirements shall be applied only to new construction and new equipment, except as modified in individual chapters. [NFPA 99:1.3.2]

1301.6 Existing Systems. Only the altered, renovated, or modernized portion of an existing system or individual component shall be required to meet the installation and equipment requirements stated in this code. If the alteration, renovation, or modernization adversely impact the existing performance requirements of a system or component, additional upgrading shall be required. An existing system that is not in strict compliance with the provisions of this code shall be permitted to be continued in use, unless the Authority Having Jurisdiction has determined that such use constitutes a distinct hazard to life. [NFPA 99:1.3.2.1 – 1.3.2.3]

Alterations to medical gas systems shall comply with the requirements of Chapter 13, while portions of the piping system unaffected by the alterations should comply with the standards in effect at the time of the installation. Exceptions to this practice would be where existing systems adversely impact the performance of the system.

1302.0 Design Requirements.

1302.1 Building System Risk Categories. Activities, systems, or equipment shall be designed to meet Category 1 through Category 4 requirements as detailed in this code. [NFPA 99:4.1]

1302.1.1 Risk Assessment. Categories shall be determined by following and documenting a defined risk assessment procedure. [NFPA 99:4.2.1]

1302.1.2 Documented Risk Assessment. A documented risk assessment shall not be required for Category 1. [NFPA 99:4.2.2]

1302.2 Patient Care Spaces. The governing body of the facility or its designee shall establish the following areas in accordance with the type of patient care anticipated (see definition of patient care spaces in Chapter 2):

(1) Category 1 spaces
(2) Category 2 spaces
(3) Category 3 spaces
(4) Category 4 spaces [NFPA 99:1.3.4.1]

HEALTH CARE FACILITIES AND MEDICAL GAS AND VACUUM SYSTEMS

1302.3 Anesthesia. It shall be the responsibility of the governing body of the health care organization to designate anesthetizing locations. [NFPA 99:1.3.4.2]

1302.4 Wet Procedure Locations. It shall be the responsibility of the governing body of the health care organization to designate wet procedure locations. [NFPA 99:1.3.4.3]

1303.0 Health Care Facilities.

1303.1 Drinking Fountain Control Valves. Drinking fountain control valves shall be flush-mounted or fully recessed where installed in corridors or other areas where patients are transported on a gurney, bed, or wheelchair.

1303.2 Psychiatric Patient Rooms. Piping and drain traps in psychiatric patient rooms shall be concealed. Fixtures and fittings shall be resistant to vandalism.

In order to protect the safety of psychiatric patients, all fixtures and fittings shall be of the vandal proof type; all piping and fixture traps shall be concealed. This may be accomplished by locking the cabinet/vanity doors that allow access to the piping and fixture traps.

1303.3 Locations for Ice Storage. Ice makers or ice storage containers shall be located in nursing stations or similarly supervised areas to minimize potential contamination.

In order to protect the making and storage of ice from possible contamination from any source, ice makers and ice storage chests shall be located in an area that can be constantly observed by staff members. Nursing stations are one such area.

1303.4 Sterilizers and Bedpan Steamers. Sterilizers and bedpan steamers shall be installed in accordance with the manufacturer's installation instructions and comply with Section 1303.4.1 and Section 1303.4.2.

1303.4.1 Drainage Connections. Sterilizers and bedpan steamers shall be connected to the sanitary drainage system through an air gap in accordance with Section 801.2. The size of indirect waste piping shall be not less than the size of the drain connection on the fixture. Each such indirect waste pipe shall not exceed 15 feet (4572 mm) in length and shall be separately piped to a receptor. Such receptors shall be located in the same room as the equipment served. Except for bedpan steamers, such indirect waste pipes shall not require traps. A trap having a seal of not less than 3 inches (76 mm) shall be provided in the indirect waste pipe for a bedpan steamer.

The drain connection requirements are consistent for an indirect connection according to Chapter 8. Note that traps are required only for bedpan steamers. The trap seal cannot be less than three inches. This does not mean that the trap has be three inches or greater. A typical PVC 1½-inch trap has a trap seal depth of 3 15/16-inches.

1303.4.2 Vapor Vents and Stacks. Where a sterilizer or bedpan steamer has provision for a vapor vent and such a vent is required by the manufacturer, the vent shall be extended to the outdoors above the roof. Sterilizer and bedpan steamer vapor vents shall be installed in accordance with the manufacturer's installation instructions and shall not be connected to a drainage system vent.

1303.5 Aspirators. Provisions for aspirators or other water-supplied suction devices shall be installed with the specific approval of the Authority Having Jurisdiction. Where aspirators are used for removing body fluids, they shall include a collection container to collect liquids and solid particles. Aspirators shall indirectly discharge to the sanitary drainage system through an air gap in accordance with Section 806.1. The potable water supply to an aspirator shall be protected by a vacuum breaker or equivalent backflow protection device in accordance with Section 603.5.9.

Prior approval is required before any aspirator can be installed. All aspirators or other water-supplied suction devices shall be installed in accordance with Chapters 6 and 8 of the UPC.

1303.6 Drains. Drains shall be installed on dryers, aftercoolers, separators, and receivers.

Drains must be installed on dryers, after-coolers, separators and receivers for the purpose of water removal and must be extended to approved locations.

1303.7 Clinical Sinks. Clinical sinks shall be installed in accordance with the manufacturer's installation instructions and shall comply with Section 1303.7.1.

1303.7.1 Drainage Connection. Clinical sinks shall be directly connected to the sanitary drainage system and shall be provided with approved flushing devices installed in accordance with Section 413.1.

1303.8 Water Supply for Hospitals. Hospitals shall be provided with not less than two approved potable water sources that are installed in such a manner as to prevent the interruption of water service.

Sources of potable water, such as a building supply line and emergency potable water tank(s), are to be provided for a hospital as approved by the Authority Having Jurisdiction (AHJ). If there is a failure with the water main or building supply, the hospital would still have one source of potable water for patient care. This is also consistent with provisions found in the Centers for Disease Control and Prevention's (CDC), "Emergency Water Supply Planning Guide for Hospitals and Health Care Facilities."

1304.0 Medical Gas and Medical Vacuum Piping Systems.

1304.1 General. The installation of medical gas and medical vacuum piping systems shall comply with the requirements of this chapter.

1304.2 Manufacturer's Instructions. The installation of individual components shall be made in accordance with the instructions of the manufacturer. Manufacturer's instructions shall include directions and information deemed by the manufacturer to be adequate for attaining proper operation, testing, and maintenance of the medical gas and vacuum systems. Copies of the manufacturer's instructions shall be left with the system owner. [NFPA 99:5.1.10.11.8.1 – 5.1.10.11.8.3]

1304.3 Category 2 Piped Medical Gas and Medical Vacuum. Category 2 piped gas or piped vacuum system requirements shall be permitted when all of the following criteria are met:

(1) Only moderate sedation; minimal sedation, as defined in Chapter 2; or no sedation is performed. Deep sedation and general anesthesia shall not be permitted.
(2) The loss of the piped gas or piped vacuum systems is likely to cause minor injury to patients, staff, or visitors.
(3) The facility piped gas or piped vacuum systems are intended for Category 2 patient care space as defined in Chapter 2 [NFPA 99:5.2.1.2]

1304.4 Category 3 Piped Medical Gas and Medical Vacuum. Category 3 piped gas and vacuum systems shall be permitted when all of the following criteria are met:

(1) Only moderate sedation; minimal sedation, as defined in Chapter 2; or no sedation is performed. Deep sedation and general anesthesia shall not be permitted.
(2) The loss of the piped gas and vacuum systems is not likely to cause injury to patients, staff, or visitors, but can cause discomfort.
(3) The facility piped gas and vacuum systems are intended for Category 3 or Category 4 patient care rooms per Chapter 2. [NFPA 99:5.3.1.2]

1304.5 Certification of Systems. Certification of medical gas and vacuum systems shall comply with the requirements of Section 1319.0.

1304.6 Construction Documents. Before a medical gas or medical vacuum system is installed or altered in a hospital, medical facility, or clinic, duplicate construction documents shall be filed with the Authority Having Jurisdiction. Approval of the plans shall be obtained before issuance of a permit by the Authority Having Jurisdiction.

All new and altered medical gas and vacuum systems must have specific site and piping plans drawn for review prior to installation. The installer needs the plans to guide the work. The AHJ must be able to refer to plan specifications to ensure that code requirements are met, and a copy of the approved plans must stay on the job site for the owner's future reference. There is very little room for deviation from the plans since proper installation is critical to life safety.

1304.6.1 Requirements. Construction documents shall show the following:

In the plot plans, it is important to show relationships of pipe lines and storage areas to other buildings, facilities, hazardous areas and surgical areas. It is always necessary to indicate the location of driveways, property lines and streets and their relationship to the storage of medical gas cylinders.

(1) Plot plan of the site, drawn to scale, indicating the location of existing or new cylinder storage areas, property lines, driveways, and existing or proposed buildings.
(2) Piping layout of the proposed piping system or alternation, including alarms, valves, the origin of gases, user outlets, and user inlets. The demand and loading of piping, existing or future, shall also be indicated.

Installing the piping and components correctly the first time eliminates the need for corrections and results in a safe system. Also, plans showing the exact location of pipelines, valves and inlets/outlets may aid in finding system problems in the case of an emergency and will save future installers time.

(3) Complete specification of materials.

Another component to thorough plan development is the incorporation of lists of all material specifications. Not only does the installer need to be able to reference the correct materials for installation, but so do clinic or hospital officials, health officials and fire safety officials. The more detailed a medical gas and vacuum piping plan is, the more assurance everyone has that the system will be complete and free from error.

1304.6.2 Extent of Work. Construction documents submitted to the Authority Having Jurisdiction shall clearly indicate the nature and extent of the work proposed and shall show in detail that such work will be in accordance with the provisions of this chapter.

1304.6.3 Record. A record of as-built plans and valve identification records shall remain on the site.

1305.0 System Performance.

1305.1 Required Operating Pressures. Medical gas and vacuum systems shall be capable of delivering service in the pressure ranges listed in Table 1305.1.

Medical gas systems are designed to operate at certain pressures. The user inlets/outlets require a certain amount of flow to provide enough oxygen, medical vacuum, nitrous oxide, compressed air, nitrogen, helium or carbon dioxide to ensure that the medical gas and medical vacuum needs of the patient and the medical equipment are met. See Table 1305.1 for the required operating pressures.

1305.2 Minimum Flow Rates. Medical gas and vacuum systems shall be capable of supplying the flow rates listed in Table 1305.2.

Table 1305.2 lists the minimum flow rates required for each medical gas and vacuum. Always be sure to read the footnotes in the tables. Footnotes may provide performance limitations or installation procedures.

1305.3 Minimum Station Outlets and Inlets. Station outlets and inlets for medical gas and vacuum systems shall be provided as listed in Table 1305.3.

Table 1305.3 lists the minimum number of station inlets/outlets needed for medical gas and vacuum systems. Some rooms are required to have multiple inlets/outlets because many medical procedures require the use of more than one inlet/outlet.

1306.0 Qualifications of Installers.

1306.1 General. The installation of medical gas and vacuum systems shall be made by qualified, competent technicians who are experienced in performing such installations, including all personnel who actually install the piping system. Installers of medical gas and vacuum piped distribution systems, all appurtenant piping supporting pump and compressor source systems, and appurtenant piping supporting source gas manifold systems, not including permanently installed bulk source systems shall be certified in accordance with ASSE 6010. [NFPA 99:5.1.10.11.10.1, 5.1.10.11.10.2]

It is required that anyone either installing or verifying a medical gas or vacuum system show evidence of competency, such as a certificate issued by a recognized agency acceptable to the AHJ.

The plumbers, installers or technicians must be certified

TABLE 1305.1
STANDARD DESIGNATION COLORS AND OPERATING PRESSURES FOR GAS AND VACUUM SYSTEMS
[NFPA 99: TABLE 5.1.11]

GAS SERVICE	ABBREVIATED NAME	COLORS (BACKGROUND/ TEXT)	STANDARD GAUGE PRESSURE
Medical air	Med Air	Yellow/black	50–55 psi
Carbon dioxide	CO_2	Gray/black or gray/white	50–55 psi
Helium	He	Brown/white	50–55 psi
Nitrogen	N_2	Black/white	160–185 psi
Nitrous oxide	N_2O	Blue/white	50–55 psi
Oxygen	O_2	Green/white or white/green	50–55 psi
Oxygen/carbon dioxide mixtures	O_2/CO_2 $n\%$ (n = % of CO_2)	Green/white	50–55 psi
Medical–surgical vacuum	Med Vac	White/black	15 inch to 30 inch HgV
Waste anesthetic gas disposal	WAGD	Violet/white	Varies with system type
Other mixtures	Gas A% / Gas B%	Colors as above; major gas for background/minor gas for text	None
Nonmedical air (Category 3 gas-powered device)	—	Yellow-and-white diagonal stripe/black	None
Nonmedical and Category 3 vacuum	—	White-and-black diagonal stripe/black boxed	None
Laboratory air	—	Yellow-and-white checker board/black	None
Laboratory vacuum	—	White-and-black checkerboard/black boxed	None
Instrument air	—	Red/white	160–185 psi

For SI units: 1 pound-force per square inch = 6.8947 kPa, 1 inch of mercury vacuum (HgV) = 3.386 kPa

TABLE 1305.2
MINIMUM FLOW RATES (cubic feet per minute)

MEDICAL SYSTEM	FLOW RATE
Oxygen	.71 CFM per outlet[1]
Nitrous Oxide	.71 CFM per outlet[1]
Medical Compressed Air	.71 CFM per outlet[1]
Nitrogen	15 CFM free air per outlet
Vacuum	1 SCFM per inlet[2]
Carbon Dioxide	.71 CFM per outlet[1]
Helium	.71 CFM per outlet

For SI units: 1 cubic foot per minute (CFM) = 0.47 L/s

Notes:
[1] A room designed for a permanently located respiratory ventilator or anesthesia machine shall have an outlet capable of a flow rate of 6.36 CFM (3.0 L/s) at the station outlet.
[2] For testing and certification purposes, individual station inlets shall be capable of a flow rate of 3 SCFM (1.4 L/s), while maintaining a system pressure of not less than 12 inches of mercury (41 kPa) at the nearest adjacent vacuum inlet.

to the requirements of ANSI/ASSE 6010, *Medical Gas Systems Installers of the Professional Qualifications Standard for Medical Gas Systems Personnel*. This standard currently requires not only classroom instruction in medical gas systems but also passing a practical brazing exam to demonstrate proficiency.

1306.2 Brazing. Brazing shall be performed by individuals who are qualified in accordance with Section 1307.0. [NFPA 99:5.1.10.11.10.4]

Certified brazers are further required to maintain certification through continuity, at least on a semiannual basis in accordance with ASME Section IX.

1306.2.1 Documentation. Prior to installation work, the installer of medical gas and vacuum piping shall provide and maintain documentation on the job site for the qualification of brazing procedures and individual brazers that is required under Section 1307.0. [NFPA 99:5.1.10.11.10.5]

1306.3 Health Care Organization Personnel. Health care organization personnel shall be permitted to install piping systems if all of the requirements of Section 1306.0 are met during the installation. [NFPA 99:5.1.10.11.10.6]

1307.0 Brazing Procedures.

1307.1 General. Brazing procedures and brazer performance for the installation of medical gas and vacuum piping shall be qualified in accordance with either Section IX, "Welding and Brazing Qualifications" of the ASME Boiler and Pressure Vessel Code or AWS B2.2, both as modified by Section 1307.2 through Section 1307.5 [NFPA 99:5.1.10.11.11.1]

1307.2 Examination. Brazers shall be qualified by visual examination of the test coupon followed by sectioning. [NFPA 99:5.1.10.11.11.2]

1307.3 Brazing Procedure Specification. The brazing procedure specification shall address cleaning, joint clearance, overlap, internal purge gas, purge gas flow rate, and filler metal. [NFPA 99:5.1.10.11.11.3]

TABLE 1305.3
MINIMUM OUTLETS AND INLETS PER STATION

LOCATION	OXYGEN	MEDICAL VACUUM	MEDICAL AIR	NITROUS OXIDE	NITROGEN	HELIUM	CARBON DIOXIDE
Patient rooms for medical/surgical, obstetrics, and pediatrics	1/bed	1/bed	1/bed	—	—	—	—
Examination/treatment for nursing units	1/bed	1/bed	—	—	—	—	—
Intensive care (all)	3/bed	3/bed	2/bed	—	—	—	—
Nursery[1]	2/bed	2/bed	1/bed	—	—	—	—
General operating rooms	2/room	3/room[4]	2/room	1/room	1/room	—	—
Cystoscopic and invasive special procedures	2/room	3/room[4]	2/room	—	—	—	—
Recovery delivery and labor/delivery/recovery rooms[2]	2/bed 2/room	2/bed 3/room[4]	1/bed 1/room	—	—	—	—
Labor rooms	1/bed	1/bed	1/bed	—	—	—	—
First aid and emergency treatment[3]	1/bed	1/bed[4]	1/bed	—	—	—	—
Autopsy	—	1/station	1/station	—	—	—	—
Anesthesia workroom	1/station	—	1/station	—	—	—	—

Notes:
[1] Includes pediatric nursery.
[2] Includes obstetric recovery.
[3] Emergency trauma rooms used for surgical procedures shall be classified as general operating rooms.
[4] Vacuum inlets required are in addition to inlets used as part of a scavenging system for removal of anesthetizing gases.

1307.4 Documentation. The brazing procedure qualification record and the record of brazer performance qualification shall document filler metal used, cleaning, joint clearance, overlap, internal purge gas and flow rate during brazing of coupon, and absence of internal oxidation in the completed coupon. [NFPA 99:5.1.10.11.11.4]

1307.5 Procedures. Brazing procedures qualified by a technically competent group or agency shall be permitted under the following conditions:

(1) The brazing procedure specification and the procedure qualification records meet the requirements of this code.
(2) The employer obtains a copy of both the brazing procedure specification and the supporting qualification records from the group or agency and signs and dates these records, thereby accepting responsibility for the qualifications that were performed by the group or agency.
(3) The employer qualifies at least one brazer following each brazing procedure specification used. [NFPA 99:5.1.10.11.11.5]

1307.6 Conditions of Acceptance. An employer shall be permitted to accept brazer qualification records of a previous employer under the following conditions:

(1) The brazer has been qualified following the same or an equivalent procedure that the new employer uses.
(2) The new employer obtains a copy of the record of brazer performance qualification tests from the previous employer and signs and dates these records, thereby accepting responsibility for the qualifications performed by the previous employer. [NFPA 99:5.1.10.11.11.6]

1307.7 Qualifications. Performance qualifications of brazers shall remain in effect indefinitely unless the brazer does not braze with the qualified procedure for a period exceeding 6 months or there is a specific reason to question the ability of the brazer. [NFPA 99:5.1.10.11.11.7]

Part II – Medical Gas and Vacuum System Piping.

1308.0 Pipe Materials.

1308.1 General. The provisions of this section shall apply to field-installed piping for the distribution of medical gases and vacuum systems.

1308.2 Cleaning. Tubes, valves, fittings, station outlets, and other piping components in medical gas systems shall have been cleaned for oxygen service by the manufacturer prior to installation in accordance with the mandatory requirements of CGA G-4.1, except that fittings shall be permitted to be cleaned by a supplier or agency other than the manufacturer. [NFPA 99:5.1.10.1.1]

Where tube ends, fittings or other components become contaminated before installation they shall be recleaned in accordance with Section 1311.0.

🔧 Cleanliness is especially important when installing medical gas systems. Care must be taken to ensure these systems are not contaminated by the installer or the surrounding environment.

It is recommended, but not required by code, that installers wear clean cotton gloves and change them when they are soiled. Disposable rubber gloves should be worn when washing fittings to protect your hands from irritating cleaners and to reduce the possibility of tube or fitting contamination.

Tubing and fittings used for medical gas systems are conventional Type K or L copper tube and fittings regularly used on plumbing jobs, except they have been cleaned and then sealed to keep them free from dirt and debris until ready to be installed.

1308.3 Delivery. Each length of tube shall be delivered plugged or capped by the manufacturer and kept sealed until prepared for installation. Fittings, valves, and other components shall be delivered sealed and labeled, and kept sealed until prepared for installation. [NFPA 99:5.1.10.1.2, 5.1.10.1.3]

The storage area for tubing, fittings, valves and appliances should be onsite in an area set aside, away from normal construction activity. Storage areas for tubing and fittings should be sealed off as much as practical from smoke, dust, dirt and moisture. A separate job trailer might be necessary for a large project that is expected to last for many weeks. For projects of short duration, an area tented with plastic sheathing may be adequate. Preplanning is essential to the successful installation of medical gas and vacuum systems.

1308.4 Tubes for Medical Gas Systems. Tubes shall be hard-drawn seamless copper ASTM B819 medical gas tube, Type L, except Type K shall be used where operating pressures are above a gauge pressure of 185 psi (1276 kPa) and the pipe sizes are larger than DN80 [(NPS 3) ($3\frac{1}{8}$ inches O.D.)].

ASTM B819 medical gas tube shall be identified by the manufacturer's markings "OXY," "MED," "OXY/MED," "OXY/ACR," or "ACR/MED" in blue (Type L) or green (Type K). [NFPA 99:5.1.10.1.4 – 5.1.10.1.5]

1308.5 Tubes for Medical Vacuum Systems. Piping for vacuum systems shall be constructed of any of the following:

(1) Hard-drawn seamless copper tube in accordance with the following:

(a) ASTM B88, copper tube (Type K, Type L, or Type M)

(b) ASTM B280, copper ACR tube

(c) ASTM B819, copper medical gas tubing (Type K or Type L)

(2) Stainless steel tube in accordance with the following:

(a) ASTM A269 TP304L or 316L.

(b) ASTM A312 TP304L or 316L.

(c) ASTM A312 TP 304L/316L, Sch 5S pipe, and ASTM A403 WP304L/316L, Sch 5S fittings [NFPA 99:5.1.10.2.1]

1308.6 Category 3 Systems. Category 3 systems shall comply with Section 1308.0 through Section 1309.0, except as follows:

(1) Dental air and dental vacuum shall comply with Section 1308.5, except the tubing shall be permitted to be annealed (soft temper).

(2) Dental vacuum tubing shall be permitted to be:

(a) PVC plastic pipe shall be Schedule 40 or Schedule 80, complying with ASTM D1785.

(b) PVC plastic fittings shall be Schedule 40 or Schedule 80 to match the pipe, complying with ASTM D2466 or ASTM D2467.

(c) Joints in PVC plastic piping shall be solvent-cemented in accordance with ASTM D2672.

(d) CPVC IPS plastic pipe shall be Schedule 40 or Schedule 80, complying with ASTM F441.

(e) CPVC IPS plastic fittings shall be Schedule 40 or Schedule 80 to match the pipe, complying with ASTM F438 or ASTM F439.

(f) CPVC CTS plastic pipe and fittings ½ of an inch (15 mm) through 2 inches (50 mm) in size shall be SDR 11, complying with ASTM D2846.

(g) Solvent cement for joints in CPVC plastic piping shall comply with ASTM F493.

(3) Dental air and dental vacuum fittings shall be permitted to be:

(a) Soldered complying with ASME B16.22.

(b) Flared fittings complying with ASME B16.26.

(c) Compression fittings (¾ of an inch (20 mm) maximum size)

(4) Soldered joints in Category 3 dental air supply piping shall be made in accordance with ASTM B828, using a "lead-free" solder filler metal containing not more than 0.2 percent lead by volume that complies with ASTM B32.

(5) Where required, gas and vacuum equipment and piping shall be seismically restrained against earthquakes in accordance with the applicable building code.

(6) Gas and vacuum piping systems shall be designed and sized to deliver the required flow rates at the utilized pressures. [NFPA 99:5.3.10]

1309.0 Joints and Connections.

1309.1 General. This section sets forth the requirements for pipe joint installations for a medical gas or vacuum system.

1309.2 Changes in Direction. Positive pressure patient gas systems, medical support gas systems, and vacuum systems shall have all turns, offsets, and other changes in direction made using fittings or techniques appropriate to any of the following acceptable joining methods:

(1) Brazed as described in Section 1309.4.

(2) Welding as described in Section 1309.16.

(3) Memory metal fittings as described in Section 1309.11.

(4) Axially swaged, elastic preload fittings as described in Section 1309.12.

(5) Threaded as described under Section 1309.13. [NFPA 99:5.1.10.3.1]

1309.3 Medical Vacuum Systems. Vacuum systems shall be permitted to have branch connections made using mechanically formed, drilled, and extruded tee-branch connections that are formed in accordance with the tool manufacturer's instructions. Such branch connections shall be joined by brazing, as described in Section 1309.4. [NFPA 99:5.1.10.3.2]

1309.4 Brazed Joints and Fittings. Fittings shall be wrought-copper capillary fittings complying with ASME B16.22, or brazed fittings complying with ASME B16.50. Cast copper alloy fittings shall not be permitted.

Brazed joints shall be made using a brazing alloy that exhibits a melting temperature in excess of 1000°F (538°C) to retain the integrity of the piping system in the event of fire exposure. [NFPA 99:5.1.10.4.1.1 – 5.1.10.4.1.3]

Proper brazing is important in medical gas piping. Improper brazing or rapid (shock) cooling of the joint can result in failure of the joint. The keys to brazing are the cleanliness of the metal, how much heat is applied to the joint and the manner in which the heat is applied.

Copper-to-copper joints must be brazed without flux by using a copper-phosphorus or copper-phosphorus-silver brazing filler metal.

During brazing, an oil-free, dry nitrogen purge must be continuously flowing to avoid oxidation and prevent other contaminates from entering the piping system. The purge gas must be analyzed prior to brazing with equipment that can establish that the oxygen content of the purge gas is less than 1 percent.

Dissimilar metals, such as copper and brass, must be brazed using an approved or listed brazing flux and a silver brazing filler metal. Flux-coated filler metals may be used to avoid flux penetration to the interior of the piping system.

1309.4.1 Tube Joints. Brazed tube joints shall be the socket type. [NFPA 99:5.1.10.4.1.4]

1309.4.2 Filler Metals. Filler metals shall bond with and be metallurgically compatible with the base metals being joined.

Filler metals shall comply with AWS A5.8. [NFPA 99:5.1.10.4.1.5, 5.1.10.4.1.6]

1309.4.3 Copper-to-Copper Joints. Copper-to-copper joints shall be brazed using a copper–phosphorus or copper-phosphorus-silver brazing filler metal (BCuP series) without flux. [NFPA 99:5.1.10.4.1.7]

1309.4.4 Accessible. Joints to be brazed in place shall be accessible for necessary preparation, assembly, heating, filler application, cooling, cleaning, and inspection. [NFPA 99:5.1.10.4.1.9]

1309.5 Tube Ends. Tube ends shall be cut square using a sharp tubing cutter to avoid deforming the tube. [NFPA 99:5.1.10.4.2.1]

1309.5.1 Cutting Wheels. The cutting wheels on tubing cutters shall be free from grease, oil, or other lubricant not suitable for oxygen service. [NFPA 99:5.1.10.4.2.2]

1309.5.2 Cut Ends. The cut ends of the tube shall be permitted to be rolled smooth or deburred with a sharp, clean deburring tool, taking care to prevent chips from entering the tube. [NFPA 99:5.1.10.4.2.3]

1309.6 Cleaning Procedures. The interior surfaces of tubes, fittings, and other components that are cleaned for oxygen service shall be stored and handled to avoid contamination prior to assembly and brazing. [NFPA 99:5.1.10.4.3.1]

1309.6.1 Exterior Surfaces. The exterior surfaces of tube ends shall be cleaned prior to brazing to remove any surface oxides. When cleaning the exterior surfaces of tube ends, no matter shall be allowed to enter the tube. [NFPA 99:5.1.10.4.3.2, 5.1.10.4.3.3]

1309.6.2 Interior Surfaces. If the interior surfaces of fitting sockets become contaminated prior to brazing, they shall be recleaned for oxygen in accordance with Section 1309.6.7 and be cleaned for brazing with a clean, oil-free wire brush. [NFPA 99:5.1.10.4.3.4]

1309.6.3 Abrasive Pads. Clean, nonshedding, abrasive pads shall be used to clean the exterior surfaces of the tube ends. [NFPA 99:5.1.10.4.3.5]

1309.6.4 Prohibited. The use of steel wool or sand cloth shall be prohibited. The cleaning process shall not result in grooving of the surfaces to be joined. [NFPA 99:5.1.10.4.3.6, 5.1.10.4.3.7]

1309.6.5 Wiped. After being abraded, the surfaces shall be wiped using a clean, lint-free white cloth. [NFPA 99:5.1.10.4.3.8]

1309.6.6 Examination. Tubes, fittings, valves, and other components shall be visually examined internally before being joined to verify that they have not become contaminated for oxygen service and that they are free of obstructions or debris. [NFPA 99:5.1.10.4.3.9]

1309.6.7 On-Site Recleaning. The interior surfaces of tube ends, fittings, and other components that were cleaned for oxygen service by the manufacturer, but that became contaminated prior to being installed, shall be permitted to be recleaned on-site by the installer by thoroughly scrubbing the interior surfaces with a clean, hot water–alkaline solution, such as sodium carbonate or trisodium phosphate, using a solution of 1 pound (0.5 kg) of sodium carbonate or trisodium phosphate to 3 gallons (11 L) of potable water and thoroughly rinsing them with clean, hot, potable water.

Other aqueous cleaning solutions shall be permitted to be used for on-site recleaning provided that they are as recommended in the mandatory requirements of CGA G-4.1. [NFPA 99:5.1.10.4.3.10, 5.1.10.4.3.11]

Cleaning is critical to preventing contamination of the system. All piping, valves, fittings and other components used in medical gas systems must be suitable for and compatible with oxygen service. This section requires that they be internally cleaned prior to delivery to the installation site to remove oil, grease and other materials. In medical vacuum systems, piping neither has to comply with cleanliness requirements, nor have to be compatible with oxygen.

HEALTH CARE FACILITIES AND MEDICAL GAS AND VACUUM SYSTEMS

1309.6.8 Contaminated Materials. Material that has become contaminated internally and is not clean for oxygen service shall not be installed. [NFPA 99:5.1.10.4.3.12]

Precleaned and sealed copper tubing is readily available from most suppliers with advance notice. Cleaned and bagged fittings are similarly available. Section 1311.1 limits on-site cleaning to recleaning joints that have become contaminated immediately prior to assembly.

Figure 1309.6.8 shows a typical cleaning station. Clean the fittings by immersing them in a bath of one of the chemical mixtures indicated in Section 1309.6.7. Rinse the fittings in hot water and blow dry immediately with dry nitrogen. After cleaning, place the fittings in sealable plastic bags, either individually or in small quantities. Fittings must remain bagged until use. Never clean beforehand more than one day's supply of fittings. Several commercial versions of presealable plastic bags are available. Zipper-type bags are also acceptable.

After cleaning, cap all piping, including short nipples. Fabricated piping can be protected from infiltration of dust and contaminants by capping all unbrazed joints. All such fabricated pipe should be brazed or taped within 24 hours of fabrication.

Clean the outside of all tubes, joints and fittings by washing with hot water after assembly and brazing. Threaded joints in piping systems must be made up with polytetrafluoroethylene (such as Teflon®) tape suitable for oxygen service or other thread sealants suitable for oxygen service. Sealants are applied to the male threads only.

1309.6.9 Timeframe for Brazing. Joints shall be brazed within 8 hours after the surfaces are cleaned for brazing. [NFPA 99:5.1.10.4.3.13]

1309.7 Brazing Dissimilar Metals. Flux shall only be used when brazing dissimilar metals, such as copper and bronze or brass, using a silver (BAg series) brazing filler metal. [NFPA 99:5.1.10.4.4.1]

1309.7.1 Surface Cleaning. Surfaces shall be cleaned for brazing in accordance with Section 1309.6. [NFPA 99:5.1.10.4.4.2]

1309.7.2 Flux. Flux shall be applied sparingly to minimize contamination of the inside of the tube with flux. The flux shall be applied and worked over the cleaned surfaces to be brazed using a stiff bristle brush to ensure complete coverage and wetting of the surfaces with flux. [NFPA 99:5.1.10.4.4.3, 5.1.10.4.4.4]

1309.7.3 Short Sections of Copper. Where possible, short sections of copper tube shall be brazed onto the non-copper component, and the interior of the subassembly shall be cleaned of flux prior to installation in the piping system. [NFPA 99:5.1.10.4.4.5]

1309.7.4 Flux-Coated Brazing Rods. On joints DN20 (NPS ¾) (⅞ inch O.D.) size and smaller, flux-coated brazing rods shall be permitted to be used in lieu of applying flux to the surfaces being joined. [NFPA 99:5.1.10.4.4.6]

1309.8 Nitrogen Purge. When brazing, joints shall be continuously purged with oil-free, dry nitrogen NF to

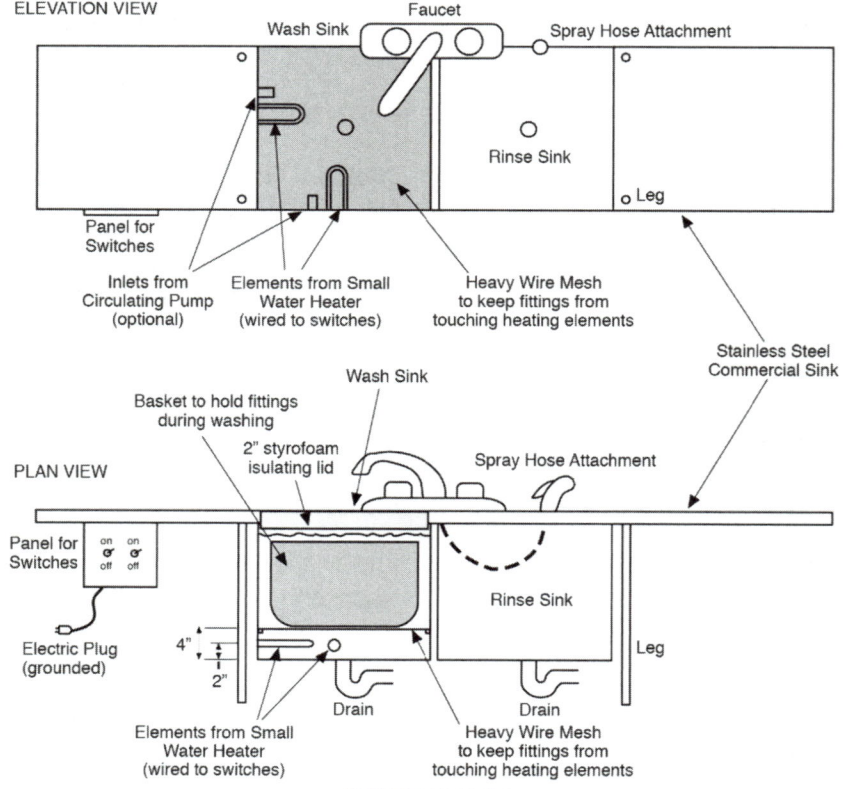

**FIGURE 1309.6.8
TYPICAL CLEANING STATION**

prevent the formation of copper oxide on the inside surfaces of the joint. [NFPA 99:5.1.10.4.5.1]

1309.8.1 Source. The source of the purge gas shall be monitored, and the installer shall be audibly alerted when the source content is low. [NFPA 99:5.1.10.4.5.2]

1309.8.2 Flow Rate Control. The purge gas flow rate shall be controlled by the use of a pressure regulator and flowmeter, or combination thereof.

Pressure regulators alone shall not be used to control purge gas flow rates. [NFPA 99:5.1.10.4.5.3, 5.1.10.4.5.4]

1309.8.3 Oxygen Analyzer. In order to ensure that all ambient air has been removed from the pipeline prior to brazing; an oxygen analyzer shall be used to verify the effectiveness of the purge. The oxygen analyzer shall read below 1 percent oxygen concentration before brazing begins. [NFPA 99:5.1.10.4.5.5]

1309.8.4 During Installation. During and after installation, openings in the piping system shall be kept sealed to maintain a nitrogen atmosphere within the piping to prevent debris or other contaminants from entering the system. [NFPA 99:5.1.10.4.5.6]

1309.8.5 Discharge Opening. While a joint is being brazed, a discharge opening shall be provided on the opposite side of the joint from where the purge gas is being introduced. [NFPA 99:5.1.10.4.5.7]

1309.8.6 Temperature of Joint. The flow of purge gas shall be maintained until the joint is cool to the touch. [NFPA 99:5.1.10.4.5.8]

1309.8.7 Opening to be Sealed. After the joint has cooled, the purge discharge opening shall be sealed to prevent contamination of the inside of the tube and maintain the nitrogen atmosphere within the piping system. [NFPA 99:5.1.10.4.5.9]

1309.8.8 Final Brazed Connection. The final brazed connection of new piping to an existing pipeline containing the system gas shall be permitted to be made without the use of a nitrogen purge. [NFPA 99:5.1.10.4.5.10]

1309.8.9 Final Tie-In Test. After a final brazed connection in a positive pressure medical gas pipeline is made without a nitrogen purge, an outlet in the immediate downstream zone of the affected portion(s) of both the new and existing piping shall be tested in accordance with the final tie-in test in Section 1309.8.9(1) through Section 1309.8.9(6). [NFPA 99:5.1.10.4.5.11]

(1) Each joint in the final connection between the new work and the existing system shall be leak-tested with the gas of system designation at the normal operating pressure by means of a leak detectant that is safe for use with oxygen and does not contain ammonia. [NFPA 99:5.1.12.3.9.2]

(2) Vacuum joints shall be tested using an ultrasonic leak detector or other means that will allow detection of leaks in an active vacuum system. [NFPA 99:5.1.12.3.9.3]

(3) For pressure gases, immediately after the final brazed connection is made and leak-tested, an outlet in the new piping and an outlet in the existing piping that are immediately downstream from the point or area of intrusion shall be purged in accordance with the applicable requirements of Section 1309.8.9(4). [NFPA 99:5.1.12.3.9.4]

(4) In order to remove any traces of particulate matter deposited in the pipelines as a result of construction, a heavy, intermittent purging of the pipeline shall be done. [NFPA 99:5.1.12.3.6]

(5) Before the new work is used for patient care, positive pressure gases shall be tested for operational pressure and gas concentration in accordance with Section 1319.13 and Section 1319.14. [NFPA 99:5.1.12.3.9.5]

(6) Permanent records of these tests shall be maintained in accordance with NFPA 99. [NFPA 99:5.12.3.9.6]

1309.8.10 Autogenous Orbital Welding Process. When using the autogenous orbital welding process, joints shall be continuously purged inside and outside with inert gas(es) in accordance with the qualified welding procedure. [NFPA 99:5.1.10.4.5.12]

1309.9 Assembling and Heating Brazed Joints. Tube ends shall be inserted into the socket, either fully or to a mechanically limited depth that is not less than the minimum cup depth (overlap) specified by ASME B16.50. [NFPA 99:5.1.10.4.6.1]

1309.9.1 Heating of Joint. Where flux is permitted, the joint shall be heated slowly until the flux has liquefied. After flux is liquefied, or where flux is not permitted to be used, the joint shall be heated quickly to the brazing temperature, taking care not to overheat the joint. [NFPA 99:5.1.10.4.6.2, 5.1.10.4.6.3]

1309.10 Inspection of Brazed Joints. After brazing, the outside of all joints shall be cleaned by washing with water and a wire brush to remove any residue and allow clear visual inspection of the joint. [NFPA 99:5.1.10.4.7.1]

1309.10.1 Where Flux is Used. Where flux has been used, the wash water shall be hot. [NFPA 99:5.1.10.4.7.2]

1309.10.2 Visually Inspected. Each brazed joint shall be visually inspected after cleaning the outside surfaces. [NFPA 99:5.1.10.4.7.3]

1309.10.3 Prohibited Brazed Joints. Joints exhibiting the following conditions shall not be permitted:

(1) Flux or flux residue (when flux or flux-coated BAg series rods are used with dissimilar metals).

(2) Base metal melting or erosion.

(3) Unmelted filler metal.

(4) Failure of the filler metal to be clearly visible all the way around the joint at the interface between the socket and the tube.

(5) Cracks in the tube or component.

(6) Cracks in the braze filler metal.

(7) Failure of the joint to hold the test pressure under the installer-performed initial pressure test (see Section 1318.5, Section 1318.6 or Section 1318.7) and standing pressure test (see Section 1318.9, Section 1318.10 or Section 1318.13). [NFPA 99:5.1.10.4.7.4]

1309.10.4 Defective Brazed Joints. Brazed joints that are identified as defective under the conditions of Section 1309.10.3(2) or Section 1309.10.3(5) shall be replaced.

Brazed joints that are identified as defective under the conditions of Section 1309.10.3(1), 1309.10.3(3), 1309.10.3(4), 1309.10.3(6) or 1309.10.3(7) shall be permitted to be repaired, except that no joint shall be reheated more than once before being replaced. [NFPA 99:5.1.10.4.7.5, 5.1.10.4.7.6]

1309.11 Memory Metal Fittings. Memory metal fittings having a temperature rating not less than 1000°F (538°C) and a pressure rating not less than 300 psi (2068 kPa) shall be permitted to be used to join copper or stainless steel tube. Memory metal fittings shall be installed by qualified technicians in accordance with the manufacturer's instructions. [NFPA 99:5.1.10.6.1, 5.1.10.6.2]

1309.12 Axially Swaged Fittings. Axially swaged, elastic strain preload fittings providing metal-to-metal seals, having a temperature rating not less than 1000°F (538°C) and a pressure rating not less than 300 psi (2068 kPa), and that, when complete, are permanent and nonseparable shall be permitted to be used to join copper or stainless steel tube. Axially swaged, elastic strain preload fittings shall be installed by qualified technicians in accordance with the manufacturer's instructions. [NFPA 99:5.1.10.7.1, 5.1.10.7.2]

1309.13 Threaded Fittings. Threaded fittings shall meet the following criteria:

(1) They shall be limited to connections for pressure and vacuum indicators, alarm devices, gas-specific demand check fittings, and source equipment on the source side of the source valve.

(2) They shall be tapered pipe threads complying with ASME B1.20.1.

(3) They shall be made up with polytetrafluoroethylene (PTFE) tape or other thread sealant recommended for oxygen service, with sealant applied to the male threads only and care taken to ensure sealant does not enter the pipe. [NFPA 99:5.1.10.8]

1309.14 Dielectric Fittings. Dielectric fittings that comply with the following shall be permitted only where required by the manufacturer of special medical equipment to electrically isolate the equipment from the system distribution piping:

(1) They shall be of brass or copper construction with an approved dielectric.

(2) They shall be permitted to be a union.

(3) They shall be clean for oxygen where used for medical gases and medical support gases. [NFPA 99:5.1.10.9.2]

1309.15 Other Types of Fittings. Listed or approved metallic gas tube fittings that, when made up, provide a permanent joint having the mechanical, thermal, and sealing integrity of a brazed joint shall be permitted to be used. [NFPA 99:5.1.10.9.1]

1309.16 Welded Joints. Welded joints for medical gas and medical-surgical vacuum systems shall be permitted to be made using a gas tungsten arc welding (GTAW) autogenous orbital procedure. [NFPA 99:5.1.10.5.1.1]

1309.16.1 Qualifications. Welders shall be qualified in accordance with Section IX of the ASME Boiler and Pressure Vessel Code. [NFPA 99:5.1.10.5.2.2]

1309.16.2 Welder Qualification Procedure. The GTAW autogenous orbital procedure and the welder qualification procedure shall be qualified in accordance with Section IX of the ASME Boiler and Pressure Vessel Code. Welder qualification procedures shall include a bend test and a tensile test in accordance with Section IX of the ASME Boiler and Pressure Vessel Code on each tube size diameter. [NFPA 99:5.1.10.5.1.2, 5.1.10.5.1.3]

1309.16.3 Purging of Joints. GTAW autogenous orbital welded joints shall be purged during welding with a commercially available mixture of 75 percent helium (+/- 5 percent) and 25 percent argon (+/- 5 percent). [NFPA 99:5.1.10.5.1.5]

1309.16.4 Test Coupons. Test coupons shall be welded and inspected, as a minimum, at the start of work and every 4 hours thereafter, or when the machine is idle for more than 30 minutes, and at the end of the work period. Test coupons shall be inspected on the I.D. and O.D. by a qualified quality control inspector. Test coupons shall also be welded at change of operator, weld head, welding power supply, or gas source. [NFPA 99:5.1.10.5.1.7 – 5.1.10.5.1.9]

1309.17 Welding for Stainless Tube. Stainless tube shall be welded using metal inert gas (MIG) welding, tungsten inert gas (TIG) welding, or other welding techniques suited to joining stainless tube. [NFPA 99:5.1.10.5.2.1]

1309.18 Prohibited Joints. The following joints shall be prohibited throughout medical gas and vacuum distribution pipeline systems:

(1) Flared and compression-type connections, including connections to station outlets and inlets, alarm devices, and other components.

(2) Other straight-threaded connections, including unions.

(3) Pipe-crimping tools used to permanently stop the flow of medical gas and vacuum piping.

(4) Removable and nonremovable push-fit fittings that employ a quick assembly push fit connector. [NFPA 99:5.1.10.10]

1310.0 Installation of Piping.

See **Figure 1310.0** for a drawing of a typical medical gas piping system.

See Section 1309.0 of this chapter on joints for brazing instructions.

HEALTH CARE FACILITIES AND MEDICAL GAS AND VACUUM SYSTEMS

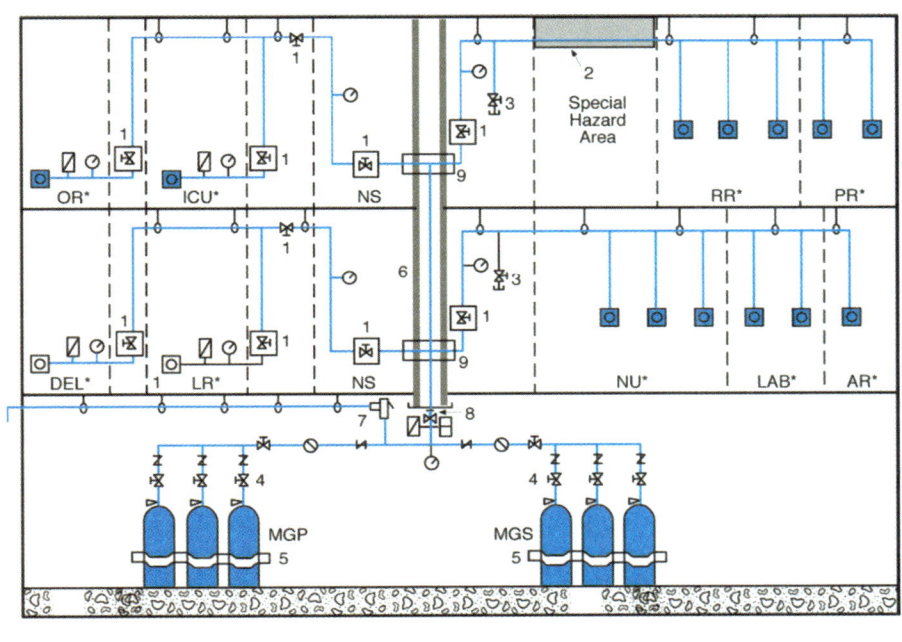

OR   Operating room
ICU   Intensive care unit
NS   Nursing station
RR   Recovery room
PR   Patient room
DEL   Delivery room
LR   Labor room
NU   Nursery
LAB   Laboratory
AR   Autopsy room
MGP   Medical gas, primary
MGS   Medical gas, secondary

1 Zone valve
2 Enclosure for piping in hazardous area
3 Valve and cap for future extensions
4 Source valves
5 Support for bottles
6 Fire-stop shaft
7 Pressure relief valve vents to outside
8 Supply/riser valve
9 Fire-stop caps

○ Pressure gauge
⋈ Check valve
▣ Station outlet
⌐ Safety relief
⋈ Fullway valve
⊘ Pressure regulator
▯ Hi/lo pressure switch
⊏⊐ Relief valve
◫ Changeover switch
◊ Hanger

Notes:
Required alarms are not shown.
Pipes are sized per Section 1310.2.5.
Pipes are labeled per Section 1310.11.
Hangers support piping per Section 1310.5.
The number of required gas outlets per area is listed in table 1305.3.

**FIGURE 1310.0
MEDICAL GAS SYSTEM**

1310.1 General. The installer shall furnish documentation certifying that all installed piping materials comply with the requirements of Section 1308.2. [NFPA 99:5.1.10.1.6]

1310.2 Required Pipe Sizing. Piping systems shall be designed and sized to deliver the required flow rates at the utilization pressures. [NFPA 99:5.1.10.11.1.1]

1310.2.1 Mains and Branches. Mains and branches in medical gas piping systems shall be not less than DN15 (NPS ½) (⅝ inch O.D.) size. Mains and branches in medical-surgical vacuum systems shall be not less than DN20 (NPS ¾) (⅞ inch O.D.) size. [NFPA 99:5.1.10.11.1.2, 5.1.10.11.1.3]

1310.2.2 Drops to Individual Stations. Drops to individual station outlets and inlets shall be not less than DN15 (NPS ½) (⅝ inch O.D.) size. [NFPA 99:5.1.10.11.1.4]

1310.2.3 Runouts and Connecting Tubing. Runouts to alarm panels and connecting tubing for gauges and alarm devices shall be permitted to be DN8 (NPS ¼) (⅜ inch O.D.) size. [NFPA 99:5.1.10.11.1.5]

1310.2.4 Maximum Demand. Where the maximum demand for each medical gas or vacuum system does not exceed the values in Table 1310.2.1(2) through Table 1310.2.1(7), the size of pipe of each section of the system shall be determined in accordance with Section 1310.2.5. The size for systems beyond the range of Table 1310.2.1(2) through Table 1310.2.1(7) shall be determined in accordance with Section 1310.2.6.

1310.2.5 Sizing Procedures. The size of each section of pipe in a system within the range of Table 1310.2.1(2) through Table 1310.2.1(7) shall be determined in accordance with the following:

(1) Determine the total flow rate and number of outlets or inlets for each section of pipe in accordance with Table 1305.2 and Table 1305.3.

(2) Measure the length of the section of pipe to each

station outlet or inlet on the system. Multiply the measured pipe length by 1.5 (150 percent), to account for the number of fittings in the system, to determine the pipe equivalent length.

(3) Beginning with the most remote outlet or inlet, multiply the total flow rate by the diversity factor specified in Table 1310.2.1(1) for each section of pipe to determine the sizing flow rate for the piping.

(4) Select Table 1310.2.1(2) through Table 1310.2.1(7) based on the medical gas or vacuum being transported through the piping.

(5) Select an estimated pipe size for determining the system pressure loss. Multiply the pipe equivalent length, for a given section of pipe, by the pressure loss for the sizing flow rate in the applicable table. Divide that number by 100 to determine the system pressure loss for the section of pipe.

(6) Add the pressure loss for each section of piping, from the source equipment location to the outlet or inlet, to determine the total system pressure loss to each outlet or inlet. The total system pressure loss in the piping to each outlet or inlet shall not exceed the values specified in Table 1310.2.2(1).

1310.2.6 Engineering Methods. For conditions other than those covered by Section 1310.2.4, such as longer runs of greater gas or vacuum demands, the size of each medical gas or vacuum piping system shall be determined by standard engineering methods acceptable to the Authority Having Jurisdiction, and each system shall be so designed that the total pressure drop or gain between the source equipment and an outlet or inlet shall not exceed the allowable pressures shown in Table 1305.1.

1310.3 Pipe Protection. Piping shall be protected against freezing, corrosion, and physical damage. [NFPA 99:5.1.10.11.2]

1310.3.1 Exposed Piping. Piping exposed in corridors and other areas where subject to physical damage from the movement of carts, stretchers, portable equipment, or vehicles shall be protected. [NFPA 99:5.1.10.11.2.1]

1310.3.2 Underground Piping. Piping underground within buildings or embedded in concrete floors or walls shall be installed in a continuous conduit. [NFPA 99:5.1.10.11.2.2]

1310.3.3 Frost Protection. Buried piping outside of buildings shall be installed below the local level of frost penetration. [NFPA 99:5.1.10.11.5.1]

1310.4 Location of Piping. Piping risers shall be permitted to be installed in pipe shafts if protected from physical damage, effects of excessive heat, corrosion, or contact with oil. [NFPA 99:5.1.10.11.3.1]

1310.4.1 Prohibited Locations. Piping shall not be installed in kitchens, elevator shafts, elevator machine rooms, areas with open flames, electrical service equipment over 600 volts, and areas prohibited under NFPA 70 except for the following locations:

(1) Room locations for medical air compressor supply systems and medical-surgical vacuum pump supply systems.

(2) Room locations for secondary distribution circuit panels and breakers having a voltage rating of 600 volts. [NFPA 99:5.1.10.11.3.2]

Medical gas piping shall not be installed in kitchens, elevator shafts, elevator machine rooms, areas with open flame, electrical service equipment rooms with voltage over 600 volts or any other areas prohibited by the National Electrical Code (NEC). Guidance is given in these code sections on installing medical gas piping with other unrelated piping in chases and in service tunnels.

1310.4.2 Prohibited Contact with Oil. Medical gas piping shall not be located where subject to contact with oil, including a possible flooding area in the case of a major oil leak. [NFPA 99:5.1.10.11.3.4]

1310.4.3 Approved Locations. Medical gas piping shall be permitted to be installed in the same service trench or tunnel with fuel gas lines, fuel oil lines, electrical lines, steam lines, and similar utilities, provided that the space is ventilated (naturally or mechanically) and the ambient temperature around the medical gas piping is limited to 130°F (54°C) maximum. [NFPA 99:5.1.10.11.3.3]

TABLE 1310.2.1(1)
SYSTEM SIZING – FLOW REQUIREMENTS FOR STATION OUTLETS AND INLETS[1]

NUMBER OF OUTLETS AND INLETS TERMINAL UNITS PER FACILITY	DIVERSITY PERCENTAGE OF AVERAGE FLOW PER OUTLETS AND INLETS TERMINAL UNITS	MINIMUM PERMISSIBLE SYSTEM FLOW OF ALL PRESSURIZED MEDICAL GAS SYSTEMS[2] (standard cubic feet per minute)
1–10	100%	Actual Demand
11–25	75%	7.0
26–50	50%	13.1
51–100	50%	17.5

Notes:
[1] Flow rates of station outlets and inlets in accordance with Table 1305.2.
[2] The minimum system flow is the average outlets and inlets flow times the number of station outlets and inlets times the diversity percentage.

TABLE 1310.2.2(1)
MAXIMUM PERMITTED PRESSURE LOSS IN MEDICAL GAS AND MEDICAL VACUUM SYSTEMS

TYPE OF SYSTEM	MAXIMUM ALLOWABLE SYSTEM PRESSURE LOSS (psi)
Medical Air	5
Nitrogen	15
Nitrous Oxide	5
Carbon Dioxide	5
Oxygen	5
Medical Vacuum	4 inches of mercury

For SI units: 1 pound-force per square inch = 6.8947 kPa, 1 inch of mercury = 3.386 kPa

TABLE 1310.2.1(2)
PRESSURE LOSS FOR MEDICAL AIR

FLOW RATE (SCFM)[1]	PRESSURE DROP (psi) PER 100 FEET[2]		
	½ INCH PIPE	¾ INCH PIPE	1 INCH PIPE
0.35	0.004	0.001	–
0.71	0.012	0.003	–
1.06	0.023	0.005	–
1.41	0.037	0.007	–
1.77	0.055	0.011	–
2.12	0.075	0.015	–
2.47	0.097	0.019	–
2.82	0.123	0.024	–
3.18	0.151	0.029	–
3.53	0.181	0.035	–
4.24	0.249	0.048	–
4.94	0.326	0.063	–
5.65	0.413	0.080	–
6.36	0.507	0.098	–
7.06	0.611	0.118	0.030
7.77	0.723	0.139	0.035
8.47	0.843	0.162	0.041
9.18	0.969	0.187	0.047
9.89	1.108	0.212	0.053
10.59	1.252	0.240	0.060
12.36	1.647	0.315	0.079
14.12	2.090	0.398	0.100
15.89	2.580	0.490	0.123
17.66	3.116	0.591	0.148
19.42	–	0.701	0.176
21.19	–	0.818	0.205
22.95	–	0.944	0.236
24.72	–	1.078	0.268
28.25	–	1.369	0.341
31.78	–	1.690	0.421
35.31	–	2.043	0.509
38.84	–	2.425	0.603
42.37	–	2.838	0.705
45.90	–	3.280	0.814
49.43	–	3.751	0.929
52.97	–	4.249	1.052
56.50	–	–	1.181
60.03	–	–	1.318
63.56	–	–	1.461
67.09	–	–	1.611
70.62	–	–	1.768
81.21	–	–	2.276
88.28	–	–	2.647
95.34	–	–	3.044

For SI units: 1 standard cubic foot per minute = 28.32 SLPM, 1 inch = 25 mm, 1 foot = 304.8 mm, 1 pound-force per square inch = 6.8947 kPa

Notes:
[1] Based on pressure of 14.7 psig (101 kPa) at 68°F (20°C).
[2] Based on pressure of 55 psig (379 kPa) at 68°F (20°C).

TABLE 1310.2.1(3)
PRESSURE LOSS FOR NITROGEN

FLOW RATE (SCFM)[1]	PRESSURE DROP (psi) PER 100 FEET[2]		
	½ INCH PIPE	¾ INCH PIPE	1 INCH PIPE
5.30	0.126	0.024	–
10.59	0.430	0.082	–
15.89	0.886	0.168	–
21.19	1.485	0.281	–
26.48	2.220	0.419	–
31.78	3.089	0.581	–
37.08	4.087	0.766	–
42.37	–	0.975	–
47.67	–	1.206	–
52.97	–	1.460	0.361
58.26	–	1.736	0.429
63.56	–	2.033	0.502
68.85	–	2.352	0.580
74.15	–	2.692	0.663
79.45	–	3.054	0.752
84.74	–	3.436	0.845
90.04	–	3.840	0.943
95.34	–	4.264	1.046
100.63	–	4.709	1.154
105.93	–	–	1.267
116.52	–	–	1.508
127.12	–	–	1.768
137.71	–	–	2.046
148.30	–	–	2.344
158.90	–	–	2.660
169.49	–	–	2.994
180.08	–	–	3.347
190.67	–	–	3.719
201.27	–	–	4.108
211.86	–	–	4.516
222.45	–	–	4.942
233.05	–	–	–
243.64	–	–	–
254.23	–	–	–
264.83	–	–	–
275.42	–	–	–
286.01	–	–	–
296.60	–	–	–
307.20	–	–	–
317.79	–	–	–

For SI units: 1 standard cubic foot per minute = 28.32 SLPM, 1 inch = 25 mm, 1 foot = 304.8 mm, 1 pound-force per square inch = 6.8947 kPa

Notes:
[1] Based on pressure of 14.7 psig (101 kPa) at 68°F (20°C).
[2] Based on pressure of 55 psig (379 kPa) at 68°F (20°C).

HEALTH CARE FACILITIES AND MEDICAL GAS AND VACUUM SYSTEMS

TABLE 1310.2.1(4)
PRESSURE LOSS FOR NITROUS OXIDE AND CARBON DIOXIDE

FLOW RATE (SCFM)[1]	PRESSURE DROP (psi) PER 100 FEET[2]		
	½ INCH PIPE	¾ INCH PIPE	1 INCH PIPE
0.35	0.004	–	–
0.71	0.014	–	–
1.06	0.029	–	–
1.41	0.047	–	–
1.77	0.070	–	–
2.12	0.096	–	–
2.47	0.125	–	–
2.82	0.159	–	–
3.18	0.195	–	–
3.53	0.235	0.045	–
4.24	0.324	0.062	–
4.94	0.425	0.081	–
5.65	0.539	0.103	–
6.36	0.664	0.127	–
7.06	0.802	0.153	0.038
7.77	0.950	0.181	0.045
8.47	1.110	0.211	0.053
9.18	1.281	0.243	0.061
9.89	1.463	0.278	0.070
10.59	1.656	0.314	0.079
12.36	2.186	0.413	0.103
14.12	2.752	0.525	0.131
15.89	3.442	0.648	0.162
17.66	4.166	0.783	0.195
19.42	–	0.929	0.231
21.19	–	0.744	0.270
22.95	–	0.858	0.312
24.72	–	0.980	0.356
28.25	–	1.244	0.453
31.78	–	1.537	0.560
35.31	–	1.858	0.677
38.84	–	2.205	0.804
42.37	–	2.581	0.941
45.90	–	2.982	1.088
49.43	–	3.411	1.245
52.97	–	4.249	1.411
56.50	–	–	1.587
60.03	–	–	1.772
63.56	–	–	1.967
67.09	–	–	2.174
70.62	–	–	2.385
79.45	–	–	2.959
88.28	–	–	3.589

For SI units: 1 standard cubic foot per minute = 28.32 SLPM, 1 inch = 25 mm, 1 foot = 304.8 mm, 1 pound-force per square inch = 6.8947 kPa

Notes:
[1] Based on pressure of 14.7 psig (101 kPa) at 68°F (20°C).
[2] Based on pressure of 55 psig (379 kPa) at 68°F (20°C).

TABLE 1310.2.1(5)
PRESSURE LOSS FOR OXYGEN

FLOW RATE (SCFM)[1]	PRESSURE DROP (psi) PER 100 FEET[2]		
	½ INCH PIPE	¾ INCH PIPE	1 INCH PIPE
0.35	0.004	–	–
0.71	0.013	0.003	–
1.06	0.025	0.005	–
1.41	0.041	0.008	–
1.77	0.060	0.012	–
2.12	0.082	0.016	–
2.47	0.107	0.021	–
2.82	0.135	0.026	–
3.18	0.166	0.032	–
3.53	0.199	0.038	–
4.24	0.274	0.053	–
4.94	0.359	0.069	–
5.65	0.454	0.087	–
6.36	0.558	0.107	–
7.06	0.672	0.129	0.033
7.77	0.795	0.153	0.039
8.47	0.927	0.179	0.045
9.18	1.066	0.205	0.052
9.89	1.218	0.233	0.059
10.59	1.377	0.263	0.066
12.36	1.811	0.346	0.087
14.12	2.298	0.438	0.110
15.89	2.837	0.539	0.135
17.66	3.456	0.650	0.163
19.42	–	0.771	0.193
21.19	–	0.900	0.225
22.95	–	1.038	0.260
24.72	–	1.185	0.295
28.25	–	1.505	0.375
31.78	–	1.859	0.463
35.31	–	2.247	0.559
38.84	–	2.667	0.663
42.37	–	3.121	0.775
45.90	–	3.607	0.895
49.43	–	4.125	1.022
52.97	–	–	1.157
56.50	–	–	1.299
60.03	–	–	1.449
63.56	–	–	1.607
67.09	–	–	1.772
70.62	–	–	1.944
81.21	–	–	2.503
91.81	–	–	3.127
102.40	–	–	3.813

For SI units: 1 standard cubic foot per minute = 28.32 SLPM, 1 inch = 25 mm, 1 foot = 304.8 mm, 1 pound-force per square inch = 6.8947 kPa

Notes:
[1] Based on pressure of 14.7 psig (101 kPa) at 68°F (20°C).
[2] Based on pressure of 55 psig (379 kPa) at 68°F (20°C).

TABLE 1310.2.1(6)
PRESSURE LOSS FOR VACUUM

FLOW RATE (SCFM)[1]	VACUUM LOSS (inch of mercury) PER 100 FEET FOR COPPER TUBE[2]				
	¾ INCH TUBE	1 INCH TUBE	1¼ INCH TUBE	1½ INCH TUBE	2 INCH TUBE
0.35	0.019	–	–	–	–
0.71	0.061	–	–	–	–
1.06	0.120	–	–	–	–
1.41	0.194	–	–	–	–
1.77	0.284	–	–	–	–
2.12	0.387	–	–	–	–
2.47	0.504	–	–	–	–
2.82	0.634	–	–	–	–
3.18	0.777	–	–	–	–
3.53	0.932	0.238	–	–	–
4.24	1.277	0.325	–	–	–
4.94	1.669	0.424	–	–	–
5.65	2.106	0.534	–	–	–
6.36	2.586	0.655	–	–	–
7.06	3.110	0.787	0.272	–	–
7.77	3.674	0.929	0.321	–	–
8.47	4.280	1.081	0.373	–	–
9.18	4.927	1.243	0.429	–	–
9.89	–	1.416	0.488	–	–
10.59	–	1.597	0.551	0.242	–
11.30	–	1.789	0.616	0.270	–
12.01	–	1.990	0.685	0.300	–
12.71	–	2.200	0.757	0.332	–
13.42	–	2.419	0.832	0.365	–
14.12	–	2.648	0.911	0.399	–
14.83	–	2.886	0.992	0.435	–
15.54	–	3.132	1.077	0.471	–
16.24	–	3.388	1.164	0.510	–
16.95	–	3.652	1.254	0.549	–
17.66	–	3.925	1.348	0.590	–
18.36	–	4.207	1.444	0.632	0.167
19.07	–	4.498	1.543	0.675	0.179
19.77	–	4.797	1.646	0.720	0.190
20.48	–	–	1.751	0.766	0.202
21.19	–	–	1.859	0.813	0.214
24.72	–	–	2.441	1.066	0.281
28.25	–	–	3.092	1.350	0.356
31.78	–	–	3.811	1.662	0.438
35.31	–	–	4.596	2.004	0.527
38.84	–	–	–	2.373	0.624
42.37	–	–	–	2.770	0.728
45.90	–	–	–	3.194	0.838
49.43	–	–	–	3.645	0.956
52.97	–	–	–	4.122	1.081
56.50	–	–	–	4.626	1.212
63.56	–	–	–	–	1.495
70.62	–	–	–	–	1.803
77.68	–	–	–	–	2.138
84.74	–	–	–	–	2.497
91.81	–	–	–	–	2.882
98.87	–	–	–	–	3.291
105.93	–	–	–	–	3.724
112.99	–	–	–	–	4.181

For SI units: 1 standard cubic foot per minute = 28.32 SLPM, 1 inch = 25 mm, 1 foot = 304.8 mm, 1 inch of mercury = 3.386 kPa

Notes:
[1] Based on the pressure of 14.7 psig (101 kPa) at 68°F (20°C).
[2] Based on the pressure of 19 inches of mercury gauge vacuum (64 kPa) at 68°F (20°C).

TABLE 1310.2.1(7)
PRESSURE LOSS FOR VACUUM (CATEGORY 3)

FLOW RATE (SCFM)[1]	VACUUM LOSS (inch of mercury) PER 100 FEET FOR PLASTIC TUBE[2]				
	¾ INCH TUBE	1 INCH TUBE	1¼ INCH TUBE	1½ INCH TUBE	2 INCH TUBE
0.35	0.005	–	–	–	–
0.71	0.010	–	–	–	–
1.06	0.015	–	–	–	–
1.41	0.021	–	–	–	–
1.77	0.026	–	–	–	–
2.12	0.060	0.010	–	–	–
2.47	0.077	0.020	–	–	–
2.82	0.096	0.025	–	–	–
3.18	0.118	0.031	0.011	–	–
3.53	0.141	0.036	0.013	–	–
4.24	0.192	0.050	0.017	–	–
4.94	0.249	0.064	0.023	0.010	–
5.65	0.313	0.081	0.028	0.012	–
6.36	0.383	0.099	0.035	0.015	–
7.06	0.459	0.118	0.041	0.018	–
7.77	0.541	0.139	0.049	0.021	–
8.47	0.628	0.161	0.056	0.024	–
9.18	0.722	0.185	0.065	0.027	–
9.89	0.821	0.210	0.073	0.031	–
10.59	0.925	0.237	0.083	0.035	–
11.30	1.035	0.265	0.092	0.039	0.010
12.01	1.151	0.294	0.102	0.043	0.011
12.71	1.270	0.324	0.113	0.048	0.012

1310.5 Pipe Support. Piping shall be supported from the building structure. [NFPA 99:5.1.10.11.4.1]

1310.5.1 Hangers and Supports. Hangers and supports shall comply with and be installed in accordance with MSS SP-58. [NFPA 99:5.1.10.11.4.2]

1310.5.2 Copper Tube. Supports for copper tube shall be sized for copper tube. [NFPA 99:5.1.10.11.4.3]

1310.5.3 Damp Locations. In potentially damp locations, copper tube hangers or supports that are in contact with the tube shall be plastic-coated or otherwise be electrically insulated from the tube by a material that will not absorb moisture. [NFPA 99:5.1.10.11.4.4]

1310.5.4 Maximum Spacing. Maximum support spacing shall be in accordance with Table 1310.5.4(1). [NFPA 99:5.1.10.11.4.5] Maximum support spacing for plastic pipe shall be in accordance with Table 1310.5.4(2). [NFPA 99:5.3.10.1.4]

1310.5.5 Seismic Provisions. Where required, medical gas and vacuum piping shall be seismically restrained against earthquakes in accordance with the applicable building code. [NFPA 99:5.1.10.11.4.6]

Piping support is important because failure of the tube could cause a dangerous situation. All piping, fittings and other components must be adequately supported and braced. Installations must conform to seismic requirements of the applicable building code.

1310.6 Backfilling and Trenching. The installation procedure for underground piping shall protect the piping from physical damage while being backfilled. [NFPA 99:5.1.10.11.5.2]

1310.6.1 Conduit, Cover, or Enclosure. If under-

TABLE 1310.2.1(7)
PRESSURE LOSS FOR VACUUM (CATEGORY 3) (continued)

FLOW RATE (SCFM)[1]	VACUUM LOSS (inch of mercury) PER 100 FEET FOR PLASTIC TUBE[2]				
	¾ INCH TUBE	1 INCH TUBE	1¼ INCH TUBE	1½ INCH TUBE	2 INCH TUBE
13.42	1.396	0.356	0.124	0.052	0.014
14.12	1.525	0.389	0.135	0.057	0.015
14.83	1.662	0.424	0.147	0.062	0.016
15.54	1.803	0.460	0.160	0.068	0.017
16.24	1.948	0.496	0.172	0.073	0.019
16.95	2.099	0.535	0.186	0.078	0.020
17.66	2.256	0.574	0.199	0.084	0.022
18.36	2.415	0.615	0.213	0.090	0.023
19.07	2.581	0.657	0.228	0.096	0.025
19.77	2.750	0.699	0.243	0.102	0.026
20.48	2.925	0.744	0.258	0.109	0.028
21.19	3.106	0.790	0.274	0.115	0.030
24.72	4.074	1.034	0.358	0.151	0.039
28.25	–	1.307	0.452	0.190	0.049
31.78	–	1.608	0.556	0.234	0.060
35.31	–	1.936	0.669	0.281	0.072
38.84	–	2.291	0.791	0.332	0.085
42.37	–	2.672	0.922	0.387	0.099
45.90	–	3.078	1.062	0.446	0.113
49.43	–	3.510	1.211	0.508	0.129
52.97	–	3.969	1.368	0.574	0.146
56.50	–	4.450	1.534	0.643	0.163
63.56	–	–	1.890	0.792	0.201
70.62	–	–	2.278	0.954	0.242
77.68	–	–	2.699	1.130	0.286
84.74	–	–	3.151	1.318	0.334
91.81	–	–	3.634	1.520	0.385
98.87	–	–	4.148	1.734	0.439
105.93	–	–	4.691	1.961	0.496
112.99	–	–	–	2.200	0.556

For SI units: 1 standard cubic foot per minute = 28.32 SLPM, 1 inch = 25 mm, 1 foot = 304.8 mm, 1 inch of mercury = 3.386 kPa

Notes:
[1] Based on the pressure of 14.7 psig (101 kPa) at 68°F (20°C).
[2] Based on the pressure of 19 inches of mercury gauge vacuum (64 kPa) at 68°F (20°C).

TABLE 1310.5.4(1)
MAXIMUM PIPE SUPPORT SPACING
[NFPA 99: TABLE 5.1.10.11.4.5, 5.3.10.1.3]

PIPE SIZE			HANGER SPACING (feet)
DN8	(NPS ¼)	(⅜ of an inch O.D.)	5
DN10	(NPS ⅜)	(½ of an inch O.D.)	6
DN15	(NPS ½)	(⅝ of an inch O.D.)	6
DN20	(NPS ¾)	(⅞ of an inch O.D.)	7
DN25	(NPS 1)	(1⅛ of an inch O.D.)	8
DN32	(NPS 1¼)	(1⅜ of an inch O.D.)	9
DN40 and larger	(NPS 1½)	(1⅝ of an inch O.D.)	10
Vertical risers, all sizes, every floor, but not to exceed:			15

For SI units: 1 inch = 25 mm, 1 foot = 304.8 mm

TABLE 1310.5.4(2)
MAXIMUM PLASTIC PIPE SUPPORT SPACING
[NFPA 99: TABLE 5.3.10.1.4]

PIPE SIZE			HANGER SPACING (feet)
DN15	(NPS ½)	(⅝ of an inch O.D.)	4
DN20	(NPS ¾)	(⅞ of an inch O.D.)	4
DN25	(NPS 1)	(1⅛ of an inch O.D.)	4.33
DN32	(NPS 1¼)	(1⅜ of an inch O.D.)	4.33
DN40	(NPS 1⅛)	(1⅝ of an inch O.D.)	4.66
DN50	(NPS 2)	(2⅜ of an inch O.D.)	4.66
DN65 and larger	(NPS 2½)	(2⅞ of an inch O.D.)	5
Vertical risers, all sizes, every floor, but not to exceed:			10

For SI units: 1 inch = 25 mm, 1 foot = 304.8 mm

ground piping is protected by a conduit, cover, or other enclosure, the following requirements shall be met:

(1) Access shall be provided at the joints for visual inspection and leak testing.

(2) The conduit, cover, or enclosure shall be self-draining and not retain groundwater in prolonged contact with the pipe. [NFPA 99:5.1.10.11.5.3]

1310.6.2 Excessive Stresses. Buried piping that will be subject to surface loads shall be buried at a depth that will protect the piping or its enclosure from excessive stresses. [NFPA 99:5.1.10.11.5.4]

1310.6.3 Minimum Backfill. The minimum backfilled cover above the top of the pipe or its enclosure for buried piping outside of buildings shall be 36 inches (914 mm), except that the minimum cover shall be permitted to be reduced to 18 inches (457 mm) where there is no potential for damage from surface loads or surface conditions. [NFPA 99:5.1.10.11.5.5]

1310.6.4 Trenches. Trenches shall be excavated so that the pipe or its enclosure has firm, substantially continuous bearing on the bottom of the trench. [NFPA 99:5.1.10.11.5.6]

1310.6.5 Composition of Backfill. Backfill shall be clean, free from material that can damage the pipe, and compacted. [NFPA 99:5.1.10.11.5.7]

1310.6.6 Marker. A continuous tape or marker placed immediately above the pipe, or its enclosure shall clearly identify the pipeline by specific name. [NFPA 99:5.1.10.11.5.8]

1310.6.7 Warning. A continuous warning means shall also be provided above the pipeline at approximately one-half the depth of burial. [NFPA 99:5.1.10.11.5.9]

1310.6.8 Wall Sleeve. Where underground piping is installed through a wall sleeve, the outdoor end of the sleeve shall be sealed to prevent the entrance of groundwater into the building. [NFPA 99:5.1.10.11.5.10]

1310.7 Connectors. Hose and flexible connectors, both metallic and nonmetallic, shall be no longer than necessary and shall not penetrate or be concealed in walls, floors, ceilings, or partitions. [NFPA 99:5.1.10.11.6.1]

1310.7.1 Flexible Connectors. Flexible connectors, metallic or nonmetallic, shall have a minimum burst pressure with a gauge pressure of 1000 psi (6895 kPa). [NFPA 99:5.1.10.11.6.2]

1310.7.2 Metallic Flexible Joints. Metallic flexible joints shall be permitted in the pipeline where required for expansion joints, seismic protection, thermal expansion, or vibration control and shall be as follows:

(1) For all wetted surfaces, made of bronze, copper, or stainless steel.

(2) Cleaned at the factory for oxygen service and received on the job site with certification of cleanliness.

(3) Suitable for service at 300 psig (2068 kPa) or above and able to withstand temperatures of 1000°F (538°C).

(4) Provided with brazing extensions to allow brazing into the pipeline per Section 1309.4.

(5) Supported with pipe hangers and supports as required for their additional weight. [NFPA 99:5.1.10.11.6.3]

1310.8 Prohibited System Interconnections. Two or more medical gas or vacuum piping systems shall not be interconnected for installation, testing, or any other reason except as permitted by Section 1310.8.1. [NFPA 99:5.1.10.11.7.1]

1310.8.1 Medical Gas and Medical Vacuum. Medical gas and vacuum systems with the same contents shall be permitted to be interconnected with an inline valve installed between the systems. [NFPA 99:5.1.10.11.7.2]

1310.8.2 Leak Testing. Leak testing shall be accomplished by separately charging and testing each individual piping system. [NFPA 99:5.1.10.11.7.3]

Individual piping systems cannot be tied and tested together, even if the testing manifold is to be cut out later. Each system must be piped and tested individually.

1310.9 Changes in System Use. Where a positive-pressure medical gas piping distribution system, originally used or constructed for use at one pressure and for one gas is converted for operation at another pressure or for another gas, all provisions of Section 1308.0 shall apply as if the system were new. [NFPA 99:5.1.10.11.9.1]

1310.9.1 Medical Vacuum System. A vacuum system shall not be permitted to be converted for use as a gas system. [NFPA 99:5.1.10.11.9.2]

1310.10 Breaching or Penetrating Medical Gas Piping. Positive pressure patient medical gas piping and medical support gas piping shall not be breached or penetrated by any means or process that will result in residual copper particles or other debris remaining in the piping or affect the oxygen-clean interior of the piping. The breaching or penetrating process shall ensure that any debris created by the process remains contained within the work area. [NFPA 99:5.1.10.11.12.1, 5.1.10.11.12.2]

1310.11 Labeling, Identification and Operating Pressure. Color and pressure requirements shall be in accordance with Table 1305.1. [NFPA 99:5.1.11] Medical gas piping shall not be painted. [NFPA 99:5.1.11.1.3]

There have been cases in which medical gas piping was incorrectly labeled, causing the deaths of patients who received the wrong gas. This section covers the code requirements on labeling pipe, valves and manifolds in medical piping systems.

All piping, valves and station inlets/outlets should be labeled, and ensure that the proper gas is flowing in the designated pipe. All enclosures housing valves must also be labeled and color coded to match the type of gas.

The following table shows some of the required labeling colors for medical gas and vacuum systems as required by Table 1305.1. These colors are used throughout the United States.

HEALTH CARE FACILITIES AND MEDICAL GAS AND VACUUM SYSTEMS

Marking Colors		
Gas	Background	Letters
Medical air	Yellow	Black
Nitrogen	Black	White
Nitrous oxide	Blue	White
Oxygen	Green	White
Vacuum	White	Black
Helium	Brown	White
Carbon dioxide	Gray	Black or white

1310.11.1 Pipe Labeling. Piping shall be labeled by stenciling or adhesive markers that identify the patient medical gas, the support gas or the vacuum system and include the following:

(1) Name of the gas or vacuum system or the chemical symbol per Table 1305.1.

(2) Gas or vacuum system color code per Table 1305.1.

(3) Where positive-pressure gas piping systems operate at pressures other than the standard gauge pressure in Table 1305.1, the operating pressure in addition to the name of the gas. [NFPA 99:5.1.11.1.1]

1310.11.2 Location of Pipe Labeling. Pipe labels shall be located as follows:

(1) At intervals of not more than 20 feet (6096 mm).

(2) At least once in or above every room.

(3) On both sides of walls or partitions penetrated by the piping.

(4) At least once in every story height traversed by risers. [NFPA 99:5.1.11.1.2]

1311.0 Valves.

🔧 See **Figures 1311.0a** and **1311.0b** for examples of valves. Station inlets/outlets are considered valves. However, gate valves are not allowed in medical gas systems.

1311.1 General. New or replacement valves shall be permitted to be of any type as long as they meet the following conditions:

(1) They have a maximum pressure drop at intended maximum flow of 0.2 psig (1.4 kPa) in pressure service and 0.15 Hg (3.8 mm) in vacuum service.

(2) They use a quarter turn to off.

(3) They are constructed of materials suitable for the service.

(4) They are provided with copper tube extensions by the manufacturer for brazing.

(5) They indicate to the operator if the valve is open or closed.

(6) They permit in-line serviceability.

(7) They are cleaned for oxygen service by the manufacturer if used for any positive pressure service. [NFPA 99:5.1.4.1.6]

🔧 Any type of valve may be used in medical gas and vacuum systems when it meets all the above referenced items. Valves must meet all the requirements found in Section 1308.2 for cleaning. All valves shall be labeled to identify the type of gas the valve controls according to Section 1311.10.2.

1311.1.1 Security. All valves, except valves in zone valve box assemblies, shall be secured by any of the following means:

(1) Located in secured areas.

(2) Locked or latched in their operating position.

(3) Located above ceilings, but remaining accessible and not obstructed. [NFPA 99:5.1.4.1.2]

1311.1.2 Accessibility. Zone valves shall be installed in valve boxes with removable covers large enough to allow manual operation of valves.

Zone valves for use in certain areas, such as psychiatric or pediatric areas, shall be permitted to be secured with the approval of the Authority Having Jurisdiction to prevent inappropriate access. [NFPA 99:5.1.4.1.4]

🔧 Zone valves shall be arranged that shutting off the supply of medical gas to one zone will not affect the supply of medical gas to the rest of the system. All station outlets and inlets shall be supplied through zone valves located on the same story. Zone valves cannot be located within the same room as outlets and inlets and zone valves should be placed so that a wall provides separation. Zone valves shall be installed where they are visible and accessible at all times and not installed in closed or locked rooms. Also, be careful not to install zone valves behind normally open or normally closed doors or otherwise hidden from plain view.

1311.1.3 Labeled. All valves shall be labeled as to gas supplied and the area(s) controlled, in accordance with Section 1311.10. [NFPA 99:5.1.4.1.3]

1311.2 Source Valves. A shutoff valve shall be placed at the immediate connection of each source system to the

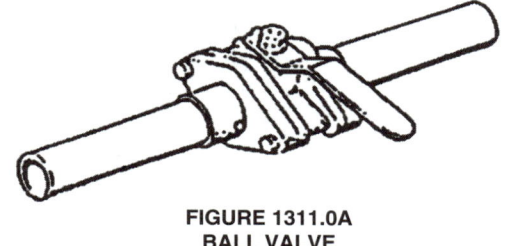

FIGURE 1311.0A
BALL VALVE

FIGURE 1311.0B
BUTTERFLY VALVE

piped distribution system to allow the entire source, including all accessory devices (e.g. air dryers, final line regulators), to be isolated from the facility. [NFPA 99:5.1.4.2.1]

1311.2.1 Location. The source valve shall be located in the immediate vicinity of the source equipment. [NFPA 99:5.1.4.2.2]

1311.3 Main Line Valve. A shutoff valve shall be provided in the main supply line inside of the buildings being served, except where one or more of the following conditions exist:

(1) The source and source valve are located inside the building served.

(2) The source system is physically mounted to the wall of the building served, and the pipeline enters the building in the immediate vicinity of the source valve. [NFPA 99:5.1.4.3.1]

1311.3.1 Location. The main line valve shall be located on the facility side of the source valve and outside of the source room, the enclosure, or where the main line first enters the building. [NFPA 99:5.1.4.3.2]

1311.4 Riser Valves. Each riser supplied from the main line shall be provided with a shutoff valve in the riser adjacent to the main line. [NFPA 99:5.1.4.4]

Shutoff valves are located at each floor of each horizontal branch between the riser and the first station inlet/outlet.

1311.5 Service Valves. Service valves shall be installed to allow servicing or modification of lateral branch piping from a main or riser without shutting down the entire main, riser, or facility. [NFPA 99:5.1.4.5.1]

1311.5.1 Branch Piping. Only one service valve shall be required for each branch off of a riser, regardless of how many zone valve boxes are installed on that lateral.

Service valves shall be placed in the branch piping prior to any zone valve box assembly on that branch. [NFPA 99:5.1.4.5.2-5.1.4.5.3]

1311.6 Zone Valves. All station outlets/inlets shall be supplied through a zone valve as follows:

(1) The zone valve shall be placed such that a wall intervenes between the valve and outlets/inlets that it controls.

(2) The zone valve shall serve only outlets/inlets located on that same story.

(3) The zone valve shall not be located in a room with the station outlets/inlets that it controls. [NFPA 99:5.1.4.6.1]

1311.6.1 Readily Operable. Zone valves shall be readily operable from a standing position in the corridor on the same floor they serve. [NFPA 99:5.1.4.6.2]

1311.6.2 Arrangement. Zone valves shall be so arranged that shutting off the supply of medical gas or vacuum to one zone will not affect the supply of medical gas or vacuum to another zone or the rest of the system. [NFPA 99:5.1.4.6.3]

Zone valves are important for isolating the gas supplying special-use areas. When a zone valve is closed, it must not affect any other zone.

1311.6.3 Indicators. A pressure/vacuum indicator shall be provided on the station outlet/inlet side of each zone valve. [NFPA 99:5.1.4.6.4]

1311.6.4 Location. A zone valve shall be located immediately outside each vital life-support area, critical care area, and anesthetizing location of moderate sedation, deep sedation, or general anesthesia, in each medical gas, or vacuum line, or both, and located so as to be readily accessible in an emergency. [NFPA 99:5.1.4.6.8]

1311.6.5 Special Installations. All gas-delivery columns, hose reels, ceiling tracks, control panels, pendants, booms, or other special installations shall be located downstream of the zone valve. [NFPA 99:5.1.4.6.9]

1311.7 In-Line Shutoff Valves. Optional in-line valves shall be permitted to be installed to isolate or shut off piping for servicing of individual rooms or areas. [NFPA 99:5.1.4.7]

1311.8 Emergency Shutoff Valves. Category 3 systems shall comply with Section 1311.0, except as follows:

(1) Where a central Category 3 medical gas supply is remote from a single treatment facility, the main supply line shall be provided with an emergency shutoff valve so located in the single treatment facility to be accessible from all use-point locations in an emergency.

(2) Where a central Category 3 medical gas supply system supplies two treatment facilities, each facility shall be provided with an emergency shutoff valve so located in the treatment facility to be accessible from all use-point locations in an emergency.

(3) Emergency shutoff valves shall be labeled to indicate the gas they control and shall shut off only the gas to the treatment facility that they serve.

(4) A remotely activated shutoff valve at a supply manifold shall not be used for emergency shutoff. For clinical purposes, such a remote valve actuator shall not fail-closed in the event of a loss of electric power. Where remote actuators are the type that fail-open, it shall be mandatory that cylinder shutoff valves be closed whenever the system is not in use. [NFPA 99:5.3.4.1]

1311.9 Valves for Future Connections. Future connection valves shall be labeled as to gas content. [NFPA 99:5.1.4.8.1]

1311.9.1 Downstream Piping. Downstream piping shall be closed with a brazed cap with tubing allowance for cutting and rebrazing. [NFPA 99:5.1.4.8.2]

1311.10 Identification of Shutoff Valves. Shutoff valves shall be identified with the following:

HEALTH CARE FACILITIES AND MEDICAL GAS AND VACUUM SYSTEMS

(1) Name or chemical symbol for the specific medical gas or vacuum system.

(2) Room or areas served.

(3) Caution to not close or open valve except in emergency. [NFPA 99:5.1.11.2.1]

1311.10.1 Nonstandard Operating Pressures. Where positive pressure gas piping systems operate at pressures other than the standard gauge pressure of 50 psi (345 kPa) to 55 psi (379 kPa), or a gauge pressure of 160 psi (1103 kPa) to 185 psi (1276 kPa) for nitrogen or instrument air, the valve identification shall also include the nonstandard operating pressure. [NFPA 99:5.1.11.2.2]

1311.10.2 Source Valves. Source valves shall be labeled in substance as follows:

SOURCE VALVE
FOR THE (SOURCE NAME)
[NFPA 99:5.1.11.2.3]

1311.10.3 Main Line Valves. Main line valves shall be labeled in substance as follows:

MAIN LINE VALVE FOR THE
(GAS/VACUUM NAME)
SERVING (NAME OF BUILDING)
[NFPA 99:5.1.11.2.4]

1311.10.4 Riser Valves. The riser valves shall be labeled in substance as follows:

RISER FOR THE (GAS/
VACUUM NAME) SERVING (NAME OF THE
AREA/BUILDING SERVED BY THE
PARTICULAR RISER)
[NFPA 99:5.1.11.2.5]

1311.10.5 Service Valves. The service valves shall be labeled in substance as follows:

SERVICE VALVE FOR THE
(GAS/VACUUM NAME) SERVING
(NAME OF THE AREA/BUILDING
SERVED BY THE PARTICULAR VALVE)
[NFPA 99:5.1.11.2.6]

1311.10.6 Zone Valve Box. Zone valve box assemblies shall be labeled outside of the valve box as to the areas that they control as follows:

ZONE VALVES FOR THE (GAS/VACUUM NAME)
SERVING (NAME OF AREA SERVED BY THE
PARTICULAR VALVE)
[NFPA 99:5.1.11.2.7]

Part III – Systems, Equipment, and Components.

1312.0 Central Supply Systems.

1312.1 Permitted Locations for Medical Gas. Central supply systems and medical gas outlets for oxygen, medical air, nitrous oxide, carbon dioxide, and all other patient medical gases shall be piped only into areas where the gases will be used under the direction of licensed medical professionals for purposes congruent with the following:

(1) Direct respiration by patients.

(2) Clinical application of the gas to a patient, such as the use of an insufflator to inject carbon dioxide into patient body cavities during laparoscopic surgery and carbon dioxide used to purge heart-lung machine blood flow ways.

(3) Medical device applications directly related to respiration.

(4) Power for medical devices used directly on patients.

(5) Calibration of medical devices intended for Section 1312.1(1) through Section 1312.1(4). [NFPA 99:5.1.3.5.2]

1312.2 Materials. Materials used in central supply systems shall meet the following requirements:

(1) In those portions of systems intended to handle oxygen at gauge pressures greater than 350 pounds-force per square inch (psi) (2413 kPa), interconnecting hose shall contain no polymeric materials.

(2) In those portions of systems intended to handle oxygen or nitrous oxide material, construction shall be compatible with oxygen under the temperatures and pressures to which the components can be exposed in the containment and use of oxygen, nitrous oxide, mixtures of these gases, or mixtures containing more than 23.5 percent oxygen.

(3) If potentially exposed to cryogenic temperatures, materials shall be designed for low temperature service.

(4) If intended for outdoor installation, materials shall be installed per the manufacturer's requirements. [NFPA 99:5.1.3.5.4]

1312.3 Pressure-Relief Valve Requirements. Central supply systems for positive pressure gases shall include one or more relief valves, all meeting the following requirements:

(1) They shall be located between each final line regulator and the source valve.

(2) They shall have a relief setting that is 50 percent above the normal system operating pressure, as indicated in Table 1305.1. [NFPA 99:5.1.3.5.6.3]

1313.0 Medical Air Supply Systems.

1313.1 Quality of Medical Air. Medical air shall be required to have the following characteristics:

(1) It shall be supplied from cylinders, bulk containers, or medical air compressor sources, or it shall be reconstituted from oxygen USP and oil-free, dry nitrogen NF.

(2) It shall meet the requirements of medical air USP.

(3) It shall have no detectable liquid hydrocarbons.

(4) It shall have less than 25 ppm gaseous hydrocarbons.

(5) It shall have equal to or less than $1 mg/m^3$ (6.85×10^{-07} lb/yd^3) of permanent particulates sized 1 micron or

larger in the air at normal atmospheric pressure. [NFPA 99:5.1.3.6.1]

1313.2 Medical Air Compressors. Medical air compressors shall be installed in a well-lit, ventilated, and clean location and shall be accessible. The location shall be provided with drainage facilities in accordance with this code. The medical air compressor area shall be located separately from medical gas cylinder system sources, and shall be readily accessible for maintenance.

Medical air is ordinary air that has been compressed by an air compressor. It must be filtered and dehumidified to remove the contaminants and water that are present in ordinary air. Air compressors must be installed in well-lit, ventilated, clean and accessible locations.

1313.2.1 Category 1 Medical Air Compressor. Medical air compressors shall be sufficient to serve the peak calculated demand with the largest single compressor out of service. In no case shall there be fewer than two compressors. [NFPA 99:5.1.3.6.3.9(B)]

Medical air systems need two or more air compressors, which work separately, to provide continuous service in the event that one unit should fail. Each system must have separate shutoff valves so that the system will not have to be completely shut down when it is being serviced.

1313.2.2 Required Components. Medical air compressor systems shall consist of the following:

(1) Components shall be arranged to allow service and a continuous supply of medical air in the event of a single fault failure.

Component arrangement shall be permitted to vary as required by the technology(ies) employed, provided that an equal level of operating redundancy and medical air quality is maintained. [NFPA 99:5.1.3.6.3.9(A)]

(2) Automatic means to prevent backflow from all on-cycle compressors through all off-cycle compressors.

(3) Manual shutoff valve to isolate each compressor from the centrally piped system and from other compressors for maintenance or repair without loss of pressure in the system.

(4) Intake filter-muffler(s) of the dry type.

(5) Pressure relief valve(s) set at 50 percent above line pressure.

(6) Piping and components between the compressor and the source shutoff valve that do not contribute to contaminant levels.

(7) Except as defined in Section 1313.2.2(1) through Section 1313.2.2(6), materials and devices used between the medical air intake and the medical air source valve that are of any design or construction appropriate for the service as determined by the manufacturer. [NFPA 99:5.1.3.6.3.2 (2-7)]

1313.3 Category 2 Medical Air Supply Systems. Category 2 systems shall comply with Section 1313.0, except as follows:

(1) Medical air compressors, dryers, aftercoolers, filters, and regulators shall be permitted to be simplex.

(2) The facility staff shall develop their emergency plan to deal with the loss of medical air. [NFPA 99:5.2.3.5]

1313.4 Category 3 Systems. Category 3 dental air compressor supply systems shall include the following:

(1) Disconnect switch(es).
(2) Motor starting device(s).
(3) Motor overload protection device(s).
(4) One or more compressors.
(5) For single, duplex, or multiple compressor systems, means for activation/deactivation of each individual compressor.
(6) When multiple compressors are used, manual or automatic means to alternate individual compressors.
(7) When multiple compressors are used, manual or automatic means to activate the additional unit(s) should the in-service unit(s) be incapable of maintaining adequate pressure.
(8) Intake filter-muffler(s) of the dry type.
(9) Receiver(s) with a manual or automatic drain.
(10) Shutoff valves.
(11) Compressor discharge check valve(s) (for multiple compressors).
(12) Air dryers that maintains a minimum of 40 percent relative humidity at operating pressure and temperature.
(13) In-line final particulate/coalescing filters rated at 0.01 micron (0.01 μm), with filter status indicator to ensure the delivery of dental air with a maximum allowable 0.05 ppm liquid oil.
(14) Pressure regulator(s).
(15) Pressure relief valve.
(16) Pressure indicator.
(17) Moisture indicator. [NFPA 99:5.3.3.6.1.1]

1313.5 Air Sources. Air sources for medical air compressors shall comply with Section 1313.5.1 or Section 1313.5.2.

1313.5.1 Medical Air Compressor Source. The medical air compressors shall draw their air from a source of clean air. [NFPA 99:5.1.3.6.3.11(A)]

If an air source equal to or better than outside air (e.g., air already filtered for use in operating room ventilating systems) is available, it shall be permitted to be used for the medical air compressors with the following provisions:

(1) This alternate source of supply air shall be available on a continuous 24 hours-per-day, 7 days-per-week basis.

(2) Ventilating systems having fans with motors or drive belts located in the airstream shall not be used as a source of medical air intake. [NFPA 99:5.1.3.6.3.11(E)]

1313.5.2 Source of Dental Air Compressor Intake.
Dental air sources for a compressor(s) shall meet the following requirements:

(1) If the intake is located inside the building, it shall be located within a space where no chemical-based materials are stored or used.

(2) If the intake is located inside the building, it shall be located in a space that is not used for patient medical treatment.

(3) If the intake is located inside the building, it shall not be taken from a room or space in which there is an open or semi-open discharge from a Category 3 vacuum system.

(4) If the intake is located outside the building, it shall be drawn from locations where no contamination from vacuum exhaust discharges or particulate matter is anticipated. [NFPA 99:5.3.3.6.1.5]

1313.6 Air Intakes.
Compressor intake piping shall be permitted to be made of materials and use a joining technique as permitted under Section 1308.5 and Section 1309.2. [NFPA 99:5.1.3.6.3.11(F)]

1313.6.1 Location.
Medical air intakes shall be located as follows:

(1) The medical air intake shall be located a minimum of 25 feet (7620 mm) from ventilating system exhausts, fuel storage vents, combustion vents, plumbing vents, and vacuum discharges, or areas that can collect vehicular exhausts or other noxious fumes.

(2) The medical air intake shall be located a minimum of 20 feet (6096 mm) above ground level.

(3) The medical air intake shall be located a minimum of 10 feet (3048 mm) from any door, window, or other opening in the building. [NFPA 99:5.1.3.6.3.11(B-D)]

1313.6.2 Separate Compressors.
Air intakes for separate compressors shall be permitted to be joined together to one common intake where the following conditions are met:

(1) The common intake is sized to minimize backpressure in accordance with the manufacturer's recommendations.

(2) Each compressor can be isolated by manual or check valve, blind flange, or tube cap to prevent open inlet piping when the compressor(s) is removed for service from the consequent backflow of room air into the other compressor(s). [NFPA 99:5.1.3.6.3.11(G)]

1313.6.3 Screening.
The end of the intake shall be turned down and screened or otherwise be protected against the entry of vermin, debris, or precipitation by screening fabricated or composed of a noncorroding material. [NFPA 99:5.1.3.6.3.11(H)]

Air intakes for medical air compressors must be located outside of the building above roof level in an area that is free from contamination and moisture. Air intakes must be turned down and screened as shown in **Figure 1313.6.3**.

1313.7 Medical Air Receivers.
Receivers for medical air shall meet the following requirements:

(1) They shall be made of corrosion-resistant materials or otherwise be made corrosion resistant.

(2) They shall comply with Section VIII, "Unfired Pressure Vessels" of the ASME Boiler and Pressure Vessel Code.

(3) They shall be equipped with a pressure relief valve, automatic drain, manual drain, sight glass, and pressure indicator.

(4) They shall be of a capacity sufficient to prevent the compressor from short-cycling. [NFPA 99:5.1.3.6.3.6]

1313.7.1 Category 3 Dental Air.
Receivers shall have the following:

(1) The capacity to prevent short-cycling of the compressor(s)

(2) Compliance with Section VIII, "Unfired Pressure Vessels" of the ASME Boiler and Pressure Vessel Code. [NFPA 99:5.3.3.6.1.2]

1313.7.2 Valves.
A medical air receiver(s) shall be provided with proper valves to allow the flow of compressed air to enter and exit out of separate receiver ports during normal operation and allow the receiver to be bypassed during service without shutting down the supply of medical air. [NFPA 99:5.1.3.6.3.9(D)]

1314.0 Medical Vacuum System.

1314.1 General.
The vacuum plant shall be installed in a well-lit, ventilated, and clean location with accessibility. The location shall be provided with drainage facilities in accordance with this code. The vacuum plant, where installed as a source, shall be located separately from other medical vacuum system sources and shall be readily accessible for maintenance.

Vacuum plants (**Figure 1314.1a** shows a duplex vacuum pump assembly) must be installed in well-lit, ventilated, clean and accessible locations. Adequate drainage facilities must be provided, located separately from other medical vacuum system sources, and readily accessible for maintenance.

Figure 1314.1b shows the piping layout of a typical medical vacuum system.

1314.2 Medical-Surgical Vacuum Sources.
Medical-surgical vacuum sources shall consist of the following:

(1) Two or more vacuum pumps sufficient to serve the peak calculated demand with the largest single vacuum pump out of service.

(2) Automatic means to prevent backflow from any on-cycle vacuum pump through any off-cycle vacuum pumps.

(3) Shutoff valve or other isolation means to isolate each vacuum pump from the centrally piped system, and other vacuum pumps for maintenance or repair without loss of vacuum in the system.

(4) Vacuum receiver.

(5) Piping between the vacuum pump(s), discharge(s),

HEALTH CARE FACILITIES AND MEDICAL GAS AND VACUUM SYSTEMS

receiver(s), and vacuum source shutoff valve in accordance with Section 1308.5, except brass, galvanized, or black steel pipe which is permitted to be used as recommended by the manufacturer.

(6) Except as defined in Section 1314.2(1) through Section 1314.2(5), materials and devices used between the medical vacuum exhaust and the medical vacuum source that are permitted to be of any design or construction appropriate for the service, as determined by the manufacturer. [NFPA 99:5.1.3.7.1.2]

Medical vacuum systems require two vacuum pumps to provide continuous service in case one unit fails. In addition, a receiver tank piped to the system must be installed. Medical vacuum systems must be equipped with isolation valves so that each pump and receiver can be serviced without disruption to any other area. Every vacuum pump shall have an isolation valve installed so that when it is closed during pump failure or repair it will not affect the operation of any other units.

1314.2.1 Category 2 Medical-Surgical Vacuum. Category 2 systems shall comply with Section 1314.2, except as follows:

(1) Medical-surgical vacuum systems shall be permitted to be simplex.
(2) The facility shall develop their emergency plan to deal with the loss of medical-surgical vacuum. [NFPA 99:5.2.3.6]

1314.2.2 Category 3 Medical-Surgical Vacuum. Category 3 medical-surgical vacuum systems if used, shall comply with Section 1314.2. [NFPA 99:5.3.3.9]

1314.3 Vacuum Pumps. Additional pumps shall automatically activate when the pump(s) in operation is inca-

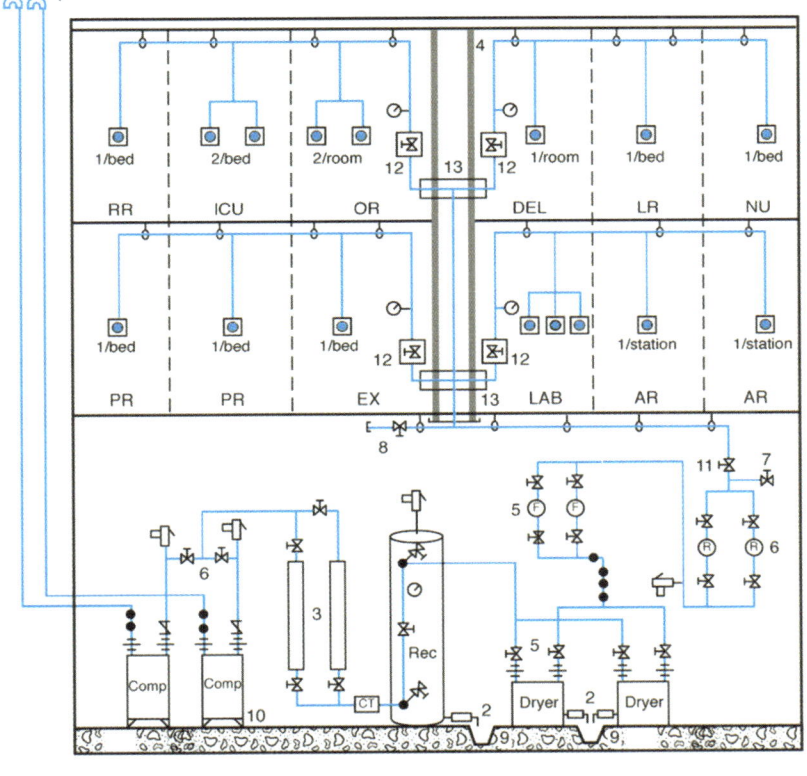

**FIGURE 1313.6.3
MEDICAL AIR COMPRESSOR SUPPLY SYSTEM**

HEALTH CARE FACILITIES AND MEDICAL GAS AND VACUUM SYSTEMS

FIGURE 1314.1A
VACUUM PUMP ASSEMBLY

pable of adequately maintaining the required vacuum.

Automatic or manual alternation of pumps shall allow division of operating time. If automatic alternation of pumps is not provided, the facility staff shall arrange a schedule for manual alternation. [NFPA 99:5.1.3.7.5.1, 5.1.3.7.5.2]

1314.4 Vacuum Receivers. Receivers for vacuum shall meet the following requirements:

(1) They shall be made of materials deemed suitable by the manufacturer.

(2) They shall comply with Section VIII, "Unfired Pressure Vessels," of the ASME Boiler and Pressure Vessel Code.

(3) They shall be capable of withstanding a gauge pressure of 60 psi (414 kPa) and 30 inch (762 mm) gauge HgV.

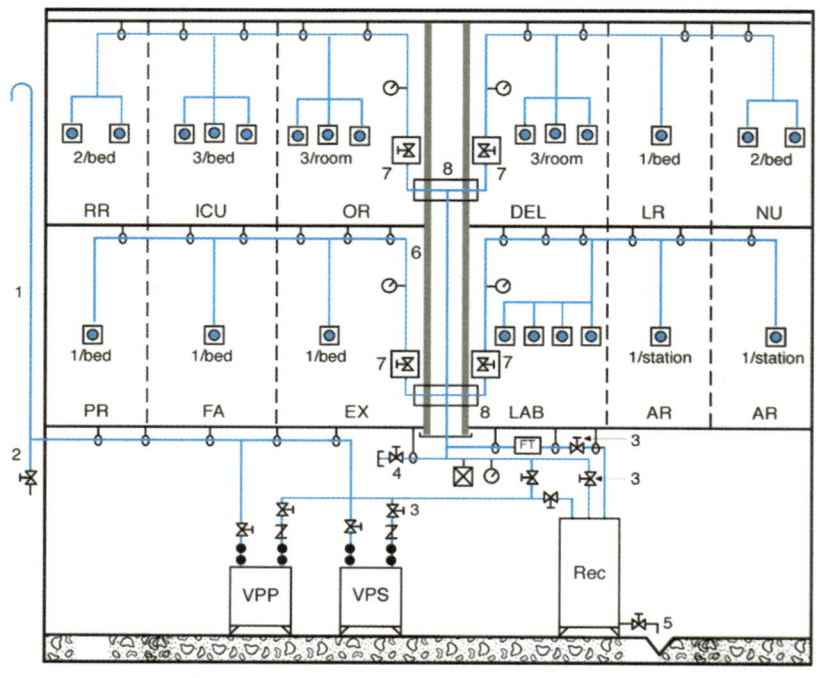

OR	Operating room
ICU	Intensive care unit
NS	Nursing station
RR	Recovery room
PR	Patient room
DEL	Delivery room
LR	Labor room
EX	Examination room
AR	Autopsy room
NU	Nursery
LAB	Laboratory
Rec	Receiver
FA	First aid
VPP	Vacuum pump, primary
VPS	Vacuum pump, secondary

- Pressure gauge
- N Check valve
- Station outlet
- Hanger
- Fullway valve
- Flexible connector
- FT Fluid trap
- Vacuum Switch
- Cap

1. Vacuum pump exhaust – must be 30 feet horizontally from door windows and 50 feet horizontally from any air intake
2. Drip leg – installed as close to pumps as possible
3. Isolation valves
4. Valve and cap for future extension
5. Floor drain required
6. Fire-stop shaft
7. Zone valves
8. Fire-stop caps

Notes:
Required alarms are not shown.
Pipes are sized per Section 1306.0.
Pipes are labeled per Section 1316.0.
Hangers support piping per Section 1311.9.

FIGURE 1314.1B
MEDICAL VACUUM SYSTEM

(4) They shall be equipped with a manual drain.

(5) They shall be of a capacity based on the technology of the pumps. [NFPA 99:5.1.3.7.3]

1314.5 Vacuum Source Exhausts. The medical-surgical vacuum pumps shall exhaust in a manner and location that minimizes the hazards of noise and contamination to the facility and its environment. [NFPA 99:5.1.3.7.6.1]

1314.5.1 Location. The exhaust shall be located as follows:

(1) Outdoors.

(2) At least 25 feet (7620 mm) from any door, window, air intake, or other openings in buildings or places of public assembly.

(3) At a level different from air intakes.

(4) Where prevailing winds, adjacent buildings, topography, or other influences will not divert the exhaust into occupied areas or prevent dispersion of the exhaust. [NFPA 99:5.1.3.7.6.2]

Termination of the vacuum exhaust must be outside the building, located at least 10 feet from any door, window, air intake or other openings into the building, provided further that the termination shall be at a different level from air intakes. This method should eliminate the possibility of contaminated waste entering any other systems. The exhaust pipe terminus shall be turned downward and screened to prevent the entry of anything that could cause a blockage.

1314.5.2 Screening. The end of the exhaust shall be turned down and screened or otherwise be protected against the entry of vermin, debris, or precipitation by screening fabricated or composed of a noncorroding material. [NFPA 99:5.1.3.7.6.3]

1314.5.3 Dips and Loops. The exhaust shall be free of dips and loops that might trap condensate or oil or provided with a drip leg and valved drain at the bottom of the low point. [NFPA 99:5.1.3.7.6.4]

1314.5.4 Multiple Pumps. Vacuum exhausts from multiple pumps shall be permitted to be joined together to one common exhaust where the following conditions are met:

(1) The common exhaust is sized to minimize backpressure in accordance with the pump manufacturer's recommendations.

(2) Each pump can be isolated by manual or check valve, blind flange, or tube cap to prevent open exhaust piping when the pump(s) is removed for service from consequent flow of exhaust air into the room. [NFPA 99:5.1.3.7.6.5]

1315.0 Pressure-Regulating Equipment.

1315.1 Where Required. Pressure-regulating equipment shall be installed in the supply main upstream of the final line-pressure valve. Where multiple piping systems for the same gas at different operating pressures are required, separate pressure-regulating equipment, relief valves, and source shutoff valves shall be provided for each pressure.

1315.2 Pressure Relief Valves. All pressure relief valves shall meet the following requirements:

(1) They shall be of brass, bronze, or stainless steel construction.

(2) They shall be designed for the specific gas service.

(3) They shall have a relief pressure setting not higher than the maximum allowable working pressure (MAWP) of the component with the lowest working pressure rating in the portion of the system being protected.

(4) They shall be vented to the outside of the building, except that relief valves for compressed air systems having less than 3000 cubic feet (84 950 L) at STP shall be permitted to be diffused locally by means that will not restrict the flow.

(5) They shall have a vent discharge line that is not smaller than the size of the relief valve outlet.

(6) Where two or more relief valves discharge into a common vent line, its internal cross-sectional area shall be not less than the aggregate cross-sectional area of all relief valve vent discharge lines served.

(7) They shall not discharge into locations creating potential hazards.

(8) They shall have the discharge terminal turned down and screened to prevent the entry of rain, snow, or vermin.

(9) They shall be designed in accordance with ASME B31.3. [NFPA 99:5.1.3.5.6.1]

1315.2.1 Category 3 Dental Air Pressure Relief Valve Discharge. Pressure relief valves for dental air systems having less than 3000 cubic feet (84 950 L) at STP shall be permitted to discharge locally indoors in a safe manner that will not restrict the flow. [NFPA 99:5.3.3.6.1.4]

1315.2.2 Isolation. A pressure-relief valve shall not be isolated from its intended use by a valve.

1315.3 Pressure and Vacuum Indicator Locations. Pressure/vacuum indicators shall be readable from a standing position. Pressure/vacuum indicators shall be provided at the following locations, as a minimum:

(1) Adjacent to the alarm-initiating device for source main line pressure and vacuum alarms in the master alarm system.

(2) At or in area alarm panels to indicate the pressure/vacuum at the alarm activating device for each system that is monitored by the panel.

(3) On the station outlet/inlet side of zone valves. [NFPA 99:5.1.8.2.1, 5.1.8.2.2]

Pressure-regulating equipment is necessary to monitor and control the pressure in medical gas and vacuum systems. Medical and maintenance personnel should be able to tell what the pressure is at a glance. The medical gas and vacuum pressures must be maintained within the range of pressures listed in Table 1305.1. This section details the requirements for pressure regulators and pressure gauges. See **Figures 1315.3a** and **1315.3b**.

1316.0 Station Outlets and Inlets.

1316.1 General. Each station outlet/inlet for medical gases or vacuums shall be gas-specific, whether the outlet/inlet is threaded or is a noninterchangeable quick coupler. [NFPA 99:5.1.5.1]

Because station inlets and outlets vary in size and design from manufacturer to manufacturer, it is important to follow the manufacturer's guidelines and installation procedures for installing inlets and outlets.

All systems must be purged prior to the installation of station inlets and outlets. Station inlets and outlets are considered valves.

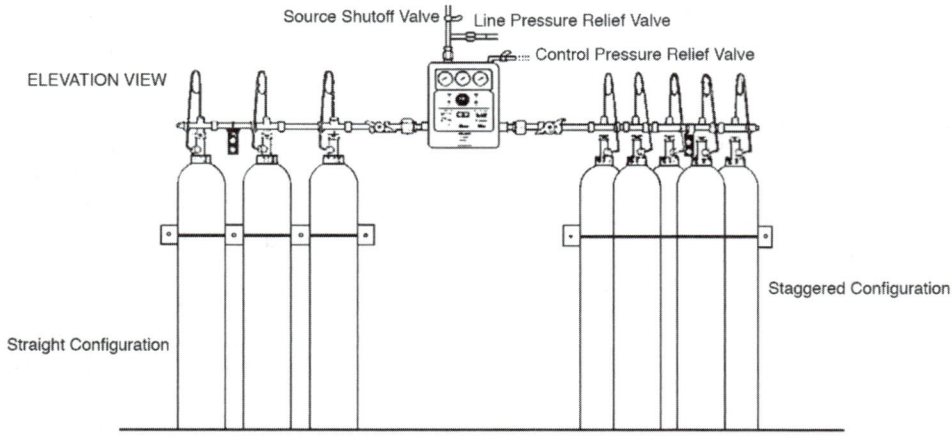

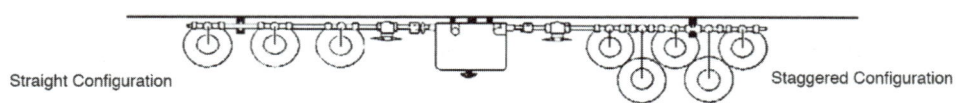

FIGURE 1315.3A
LOCATION OF PRESSURE-RELIEF VALVES ON AUTOMATIC MANIFOLDS

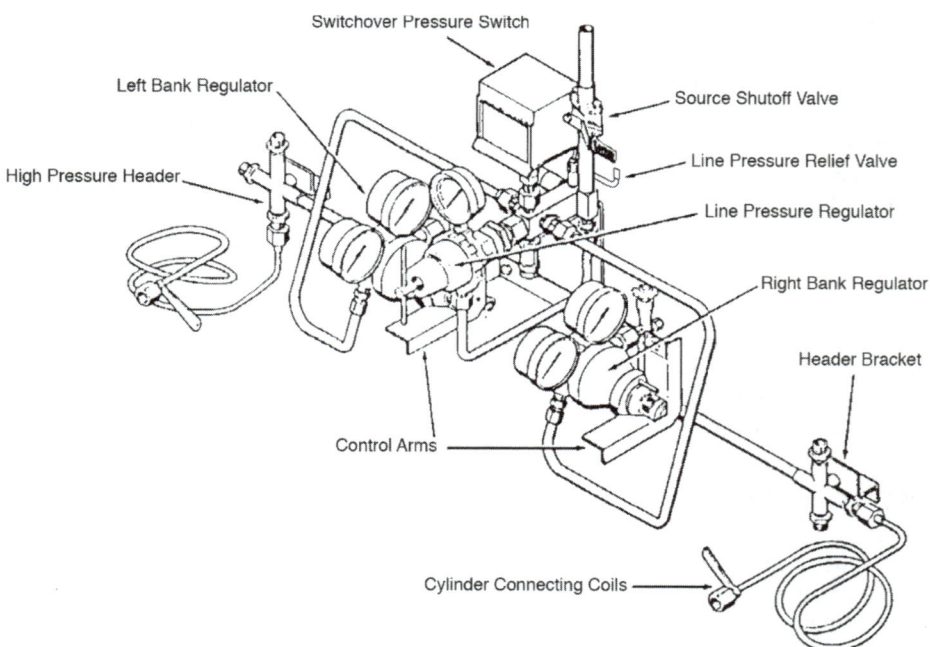

FIGURE 1315.3B
LOCATION OF PRESSURE-RELIEF VALVES ON A DUPLEX MANIFOLD

1316.2 Required Valves. Each station outlet shall consist of a primary and secondary valve (or assembly).

Each station inlet shall consist of a primary valve (or assembly) and shall be permitted to include a secondary valve (or assembly). [NFPA 99:5.1.5.2, 5.1.5.3]

1316.2.1 Secondary Valve. The secondary valve (or assembly) shall close automatically to stop the flow of gas (or vacuum, if provided) when the primary valve (or assembly) is removed. [NFPA 99:5.1.5.4]

1316.3 Post Installation. After installation of the piping, but before installation of the station outlets and inlets and other medical gas and medical gas system components (e.g., pressure-actuating switches for alarms, manifolds, pressure gauges, or pressure relief valves), the line shall be blown clear using oil-free, dry nitrogen NF.

1316.4 Identification. Station outlets and inlets shall be identified as to the name or chemical symbol for the specific medical gas or vacuum provided.

In sleep labs, where the outlet is downstream of a flow control device, the station outlet identification shall include a warning not to use the outlet for ventilating patients.

Where medical gas systems operate at pressures other than the standard gauge pressure of 50 psi to 55 psi (345 kPa to 380 kPa) or a gauge pressure of 160 psi to 185 psi (1103 kPa to 1275 kPa) for nitrogen, the station outlet identification shall include the nonstandard operating pressure in addition to the name of the gas. [NFPA 99:5.1.11.3.1-5.1.11.3.2]

1317.0 Warning Systems.

Piping systems are not complete until audible and visual alarms and alarm panels are installed and are operable. Alarms indicate any change in pressure above or below the normal operating pressure ranges listed in Table 1305.1 and also detect change in pressure from one supply source to another or any failure in the system of the emergency reserve.

Alarms for medical gas and vacuum systems must be approved prior to their use. Functioning of all alarm components must be verified in accordance with the testing and monitoring requirements of the manufacturer and the AHJ. **Figure 1317.0** shows an alarm panel.

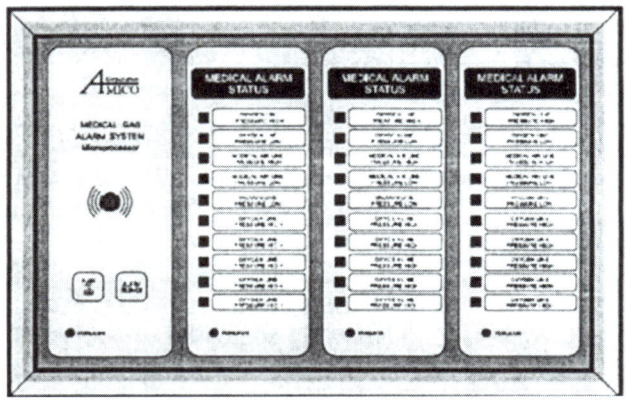

**FIGURE 1317.0
MASTER ALARM
(Courtesy of Amico Corporation)**

1317.1 Category 1. All master, area, and local alarm systems used for medical gas and vacuum systems shall include the following:

(1) Separate visual indicators for each condition monitored, except as permitted in Section 1317.1.1 for local alarms that are displayed on master alarm panels.

(2) Visual indicators that remain in alarm until the situation that has caused the alarm is resolved.

(3) Cancelable audible indication of each alarm condition that produces a sound with a minimum level of 80 decibels at 3 feet (914 mm).

(4) Means to indicate a lamp or LED failure and audible failure.

(5) Visual and audible indication that the communication with an alarm-initiating device is disconnected.

(6) Labeling of each indicator, indicating the condition monitored.

(7) Labeling of each alarm panel for its area of surveillance.

(8) Reinitiation of the audible signal if another alarm condition occurs while the audible alarm is silenced.

(9) Power for master, area alarms, sensors, and switches from the life safety branch of the essential electrical system as described in NFPA 99.

(10) Power for local alarms, dew point sensors, and carbon monoxide sensors permitted to be from the same essential electrical branch as is used to power the air compressor system.

(11) Where used for communications, wiring from switches or sensors that is supervised or protected as required by NFPA 70 for life safety and critical branch circuits in which protection is any of the following types:

(a) Conduit

(b) Free air

(c) Wire

(d) Cable tray

(e) Raceways

(12) Communication devices that do not use electrical wiring for signal transmission will be supervised such that failure of communication shall initiate an alarm.

(13) Assurance by the responsible authority of the facility that the labeling of alarms, where room numbers or designations are used, is accurate and up-to-date.

(14) Provisions for automatic restart after a power loss of 10 seconds (e.g., during generator startup) without giving false signals or requiring manual reset.

(15) Alarm switches/sensors installed so as to be removable. [NFPA 99:5.1.9.1]

1317.1.1 Master Alarm. The master alarm shall include at least one signal from the source equipment to indicate a problem with the source equipment at this location. This master alarm signal shall activate when any of the required local alarm signals for this source equipment activates. [NFPA 99:5.1.9.5.2]

1317.2 Category 2 Systems. Warning systems associated with Category 2 systems shall provide the master, area, and local alarm functions of a Category 1 system as required in Section 1317.1, except as follows:

(1) Warning systems shall be permitted to be a single alarm panel.

(2) The alarm panel shall be located in an area of continuous surveillance while the facility is in operation.

(3) Pressure and vacuum switches/sensors shall be mounted at the source equipment with a pressure indicator at the master alarm panel. [NFPA 99:5.2.9]

1317.3 Category 3 Systems. Category 3 warning systems shall comply with Section 1317.2 except as follows:

(1) Warning systems shall be permitted to be a single alarm panel.

(2) The alarm panel shall be located in an area of continuous surveillance while the facility is in operation.

(3) Pressure and vacuum switches/sensors shall be mounted at the source equipment with a pressure indicator at the master alarm panel.

(4) Warning systems for medical gas systems shall provide the following alarms:

(a) Oxygen main line pressure low.

(b) Oxygen main line pressure high.

(c) Oxygen changeover to secondary bank or about to changeover (if automatic).

(d) Nitrous oxide main line pressure low.

(e) Nitrous oxide main line pressure high.

(f) Nitrous oxide changeover to secondary bank or about to changeover (if automatic).

(5) Audible and noncancelable alarm visual signals shall indicate if the pressure in the main line increases or decreases 20 percent from the normal operating pressure.

(6) Visual indications shall remain until the situation that caused the alarm is resolved.

(7) Pressure switches/sensors shall be installed downstream of any emergency shutoff valves and any other shutoff valves in the system and shall cause an alarm for the medical gas if the pressure decreases or increases 20 percent from the normal operating pressure.

(8) A cancelable audible indication of each alarm condition that produces a sound at the alarm panel shall reinitiate the audible signal if another alarm condition occurs while the audible signal is silenced. [NFPA 99:5.3.9]

1317.4 Components. Functioning of alarm components shall be verified in accordance with the testing and monitoring requirements of the manufacturer and the Authority Having Jurisdiction.

Part IV – Testing, Inspection, and Certification.

1318.0 Testing and Inspection.

1318.1 Where Required. Inspection and testing shall be performed on components, or portions thereof, of new, piped medical gas or vacuum systems, additions, renovations, temporary installations, or repaired systems in accordance with Section 1318.2 through Section 1318.16, and certified in accordance with Section 1319.0.

All new and changed existing medical gas and vacuum systems are required to be inspected. Portions of existing systems may need to be inspected if they are affected by new or changed piping.

It is the duty of the AHJ to provide notice of corrections needed and violations found during the test. The correction notice may be posted at the job site or mailed or delivered to the permittee or an authorized representative. Refusal or failure to comply with a correction notice or order within 10 days is considered a violation of this code and is subject to the penalties for violations set forth elsewhere in this code.

Each AHJ has its own set of laws, ordinances or guidelines regarding inspection. Prior to doing any work within a jurisdiction, the installer needs to find out the necessary procedures and requirements to get work inspected and the requirements for final approval.

Medical gas or vacuum piping systems should never be covered until all piping has been tested, inspected and approved. If the system is covered or concealed before being tested and inspected, notice to uncover the work will be issued to the permittee by the AHJ. The work must then be uncovered for inspection. Remember, it is very costly and time-consuming to remove walls and floors to do an inspection that should have been done prior to concealing the piping.

1318.2 Breached Systems. All systems that are breached and components that are subject to additions, renovations, or replacement (e.g., new gas sources: bulk, manifolds, compressors, dryers, alarms) shall be inspected and tested. Systems shall be deemed breached at the point of pipeline intrusion by physical separation or by system component removal, replacement, or addition. Breached portions of the systems subject to inspection and testing shall be confined to only the specific altered zone and components in the immediate zone or area that is located upstream for vacuum systems and downstream for pressure gases at the point or area of intrusion. [NFPA 99:5.1.12.1.3 – 5.1.12.1.5]

1318.3 Reports. The inspection and testing reports shall be submitted directly to the party that contracted for the testing, who shall submit the report through channels to the responsible facility authority and others that are required. Reports shall contain detailed listings of all findings and results. [NFPA 99:5.1.12.1.6, 5.1.12.1.7]

1318.4 Initial Piping Blowdown. Piping in medical gas and vacuum distribution systems shall be blown clear by means of oil-free, dry nitrogen NF after installation of the distribution piping but before installation of station outlet/inlet rough-in assemblies and other system components (e.g., pressure/vacuum alarm devices, pressure/vacuum indicators, pressure relief valves, manifolds, source equipment). [NFPA 99:5.1.12.2.2]

1318.4.1 Test Gas. The test gas shall be oil-free, dry nitrogen NF. [NFPA 99:5.1.12.2.1.2]

1318.5 Initial Pressure Tests – Medical Gas and Vacuum Systems. Each section of the piping in medical gas and vacuum systems shall be pressure tested. Initial pressure tests shall be conducted as follows:

(1) After blowdown of the distribution piping.

(2) After installation of station outlet/inlet rough-in assemblies.

(3) Prior to the installation of components of the distribution piping system that would be damaged by the test pressure (e.g., pressure/vacuum alarm devices, pressure/vacuum indicators, line pressure relief valves). [NFPA 99:5.1.12.2.3.1, 5.1.12.2.3.2]

A rough test is required of all piping prior to the attachment of system components, but after installation of the station outlets and inlets.

In preparation for the rough test, test caps are installed. The source valve must be in the closed position. Testing pressure must be maintained at 150 psig or one and one-half times the working pressure with oil-free dry nitrogen until every joint is checked. Any leaks must be located by means of a leak detection that is safe for use. Dishwashing liquid or other noncaustic soap is commonly used. To do the "soapy water" test, mix soap with water in a bucket, then brush the soapy water over the fittings. When a hole or improper joint is present, the leaking air (nitrogen) will cause the soapy water to bubble in that location.

1318.5.1 Shutoff Valve. The source shutoff valve shall remain closed during tests specified in Section 1318.5. [NFPA 99:5.1.12.2.3.3]

1318.5.2 Required Test Pressure. The test pressure for pressure gases and vacuum systems shall be 1.5 times the system operating pressure, but not less than a gauge pressure of 150 psi (1034 kPa). The test pressure shall be maintained until each joint has been examined for leakage by means of a leak detectant that is safe for use with oxygen and does not contain ammonia. [NFPA 99:5.1.12.2.3.4, 5.1.12.2.3.5]

1318.5.3 Leaks. Leaks, if any, shall be located, repaired (if permitted), replaced (if required), and retested. [NFPA 99:5.1.12.2.3.6]

1318.6 Initial Pressure Test - Category 3 Copper Piping Systems. Initial pressure tests shall be conducted as follows:

(1) After blowdown of the distribution piping.

(2) After installation of outlet and inlet shutoff valves station outlets and inlets.

(3) Prior to the installation of components of the distribution piping system that would be damaged by the test pressure (e.g., pressure/vacuum indicators, line pressure relief valves).

(4) The source shutoff valves for the piping systems shall remain closed during the tests, unless being used for the pressure test gas.

(5) With test pressure 1.5 times the system operating pressure but not less than a gauge pressure of 150 psi (1034 kPa).

(6) With test pressure maintained until each joint is examined for leakage by means of a detectant that is safe for use with oxygen and that does not contain ammonia.

(7) With leaks, if any, located, repaired (if permitted), or replaced (if required) by the installer and retested. [NFPA 99:5.3.12.2.4]

1318.7 Initial Leak Test – Category 3 Plastic Vacuum Piping Systems. Initial leak tests shall be conducted as follows:

(1) Each section of the piping in Category 3 vacuum systems with plastic piping shall be leak tested using a test vacuum or the vacuum source equipment.

(2) If installed, the vacuum source shutoff valves for the piping systems shall remain closed during the tests, unless being used for the leak test vacuum source.

(3) The leak test vacuum shall be a minimum of 12 inch (305 mm) HgV.

(4) The test vacuum shall be maintained until each joint has been examined for leakage. An ultrasonic leak detector shall be permitted to be used.

(5) Leaks, if any, shall be located, repaired, or replaced (if required) by the installer and retested. [NFPA 99:5.3.12.2.5]

1318.8 Cross-Connection Tests – Medical Gas and Vacuum Systems. It shall determine that no cross-connections exist between the various medical gas and vacuum piping systems. [NFPA 99:5.1.12.2.4]

1318.8.1 Atmospheric Pressure. All piping systems shall be reduced to atmospheric pressure. [NFPA 99:5.1.12.2.4.1]

1318.8.2 Sources of Test Gas. Sources of test gas shall be disconnected from all piping systems except for the one system being tested. [NFPA 99:5.1.12.2.4.2]

1318.8.3 System to be Charged. The system under test shall be charged with oil-free, dry nitrogen NF to a gauge pressure of 50 psi (345 kPa). [NFPA 99:5.1.12.2.4.3]

1318.8.4 Check Outlets and Inlets. After the installation of the individual faceplates with appropriate adapters matching outlet/inlet labels, each individual outlet/inlet in each installed medical gas and vacuum piping system shall be checked to determine that the test gas is being dispensed only from the piping system being tested. [NFPA 99:5.1.12.2.4.4]

1318.8.5 Repeat Test. The cross-connection test referenced in Section 1318.8 shall be repeated for each installed medical gas and vacuum piping system. [NFPA 99:5.1.12.2.4.5]

1318.8.6 Identification of System. The proper labeling and identification of system outlets/inlets shall be confirmed during these tests. [NFPA 99:5.1.12.2.4.6]

1318.8.7 Initial Cross-Connection Test —Category 3 Copper Piping Systems. Initial cross-connection tests for copper piping systems shall be conducted as follows:

(1) Tests shall be conducted to determine that no cross-connections exist between the Category 3 copper piping systems and Category 3 copper vacuum piping systems.

(2) The piping systems shall be at atmospheric pressure.

(3) The test gas shall be oil-free, dry nitrogen NF or dental air.

(4) The source of test gas shall be connected only to the piping system being tested.

(5) The piping system being tested shall be pressurized to a gauge pressure of 50 psi (345 kPa).

(6) The individual system gas outlet and vacuum inlet in each installed gas-powered device and copper vacuum or copper piping system shall be checked to determine that the test gas pressure is present only at the piping system being tested.

(7) The cross-connection test shall be repeated for each installed Category 3 piping system for gas-powered devices and for vacuum with copper piping.

(8) The proper labeling and identification of system outlets/inlets shall be confirmed during the tests. [NFPA 99:5.3.12.2.6]

1318.8.8 Cross-Connection Test - Category 3 Plastic Vacuum Piping Systems. Initial cross-connection tests for plastic vacuum piping systems shall be conducted as follows:

(1) Tests shall be conducted to determine that no cross connections exist between any Category 3 plastic vacuum piping systems or Category 3 copper piping systems.

(2) The vacuum source shutoff valves for the vacuum piping systems shall remain closed during the tests, unless they are being used for the cross-connection test vacuum source.

(3) The cross-connection test vacuum shall be a minimum of 12 inch (305 mm) HgV.

(4) The source of test vacuum shall be connected only to the vacuum piping system being tested.

(5) The individual gas-powered device system gas outlets and vacuum system inlets shall be checked to determine that the test vacuum is only present at the vacuum piping system being tested.

(6) The cross-connection tests shall be repeated for each installed vacuum system with plastic piping.

(7) The proper labeling and identification of system outlets/inlets shall be confirmed during the tests. [NFPA 99:5.3.12.2.7]

1318.9 Standing Pressure Tests – for Positive Pressure Medical Gas Piping Systems. After successful completion of the initial pressure tests under Section 1318.5 through Section 1318.5.3, medical gas distribution piping shall be subjected to a standing pressure test. [NFPA 99:5.1.12.2.6]

The final testing of medical gas and vacuum systems is done after installation of all system components, including alarms, compressors and tanks. As in the rough testing, oil-free dry nitrogen is used. A 24-hour standing pressure test is performed at 20 percent above the normal operating line pressure. Pressure changes due to variations in the ambient temperature are allowed, usually up to 1 percent or 0.66 psi (4.6 kPa), or 10 percent of the allowable pressure. If any leaks are found, the system must be retested as required by the AHJ.

1318.9.1 Time Frame for Testing. Tests shall be conducted after the final installation of station outlet valve bodies, faceplates, and other distribution system components (e.g. pressure alarm devices, pressure indicators, line pressure relief valves, manufactured assemblies, hose). [NFPA 99:5.1.12.2.6.1]

1318.9.2 Source Valve. The source valve shall be closed during this test. [NFPA 99:5.1.12.2.6.2]

1318.9.3 Length of Testing. The piping systems shall be subjected to a 24 hour standing pressure test using oil-free, dry nitrogen NF. [NFPA 99:5.1.12.2.6.3]

1318.9.4 Test Pressure. Test pressures shall be 20 percent above the normal system operating line pressure. [NFPA 99:5.1.12.2.6.4]

1318.9.5 Conclusion of Test. At the conclusion of the tests, there shall be no change in the test pressure except that attributed to specific changes of ambient temperature. [NFPA 99:5.1.12.2.6.5]

1318.9.6 Leaks. Leaks, if any, shall be located, repaired (if permitted), or replaced (if required), and retested. [NFPA 99:5.1.12.2.6.6]

1318.9.7 Proof of Testing. The 24 hour standing pressure test of the positive pressure system shall be witnessed by the Authority Having Jurisdiction or its designee. A form indicating that this test has been performed and witnessed shall be provided to the verifier at the start of the tests required in Section 1318.16 [NFPA 99:5.1.12.2.6.7]

1318.10 Standing Pressure Tests – Medical Vacuum Piping Systems. After successful completion of the initial pressure tests under Section 1318.5 through Section 1318.5.3, vacuum distribution piping shall be subjected to a standing vacuum test. [NFPA 99:5.1.12.2.7]

1318.10.1 Timeframe for Testing. Tests shall be conducted after installation of all components of the vacuum system. [NFPA 99:5.1.12.2.7.1]

1318.10.2 Length of Testing. The piping systems shall be subjected to a 24 hour standing vacuum test. [NFPA 99:5.1.12.2.7.2]

1318.10.3 Test Pressure. Test pressure shall be between 12 inch HgV (41 kPa) and full vacuum. [NFPA 99:5.1.12.2.7.3]

1318.10.4 Disconnection of Testing Source. During the test, the source of test vacuum shall be disconnected from the piping system. [NFPA 99:5.1.12.2.7.4]

1318.10.5 Conclusion of Test. At the conclusion of the test, there shall be no change in the vacuum other than that attributed to changes in ambient temperature. [NFPA 99:5.1.12.2.7.5]

1318.10.6 Leaks. Leaks, if any, shall be located, repaired (if permitted), or replaced (if required), and retested. [NFPA 99:5.1.12.2.7.7]

1318.10.7 Proof of Testing. The 24 hour standing pressure test of the vacuum system shall be witnessed by the Authority Having Jurisdiction or its designee. A form indicating that this test has been performed and witnessed shall be provided to the verifier at the start of the tests required in Section 1318.16. [NFPA 99:5.1.12.2.7.6]

1318.11 Standing Pressure Tests – for Category 3 Gas Powered Device Distribution Piping. After successful completion of the initial pressure tests under Section 1318.8.7, Category 3 gas-powered device distribution piping shall be subjected to a standing pressure test, which includes the following:

(1) Tests shall be conducted after the installation of outlet valves and other distribution system components (e.g. pressure indicators and line pressure relief valves).

(2) The source valve shall be closed unless the source gas is being used for the test.

(3) The piping systems shall be subjected to a 24 hour standing pressure testing using oil-free, dry nitrogen NF or the system gas.

(4) Test pressures shall be 20 percent above the normal system operating line pressure.

(5) At the conclusion of the tests, there shall be no change in the test pressure greater than a gauge pressure of 5 psi (34 kPa).

(6) Leaks, if any, shall be located, repaired (if permitted), or replaced (if required), and retested. [NFPA 99:5.3.12.2.9]

1318.12 Medical Gas Piping System Purge Tests. The outlets in each medical gas piping system shall be purged to remove any particulate matter from the distribution piping. [NFPA 99:5.1.12.2.5]

1318.12.1 Procedure. Using appropriate adapters, each outlet shall be purged with an intermittent high-volume flow of test gas until the purge produces no discoloration in a clean white cloth. [NFPA 99:5.1.12.2.5.1]

1318.12.2 Location. The purging required in Section 1318.12.1 shall be started at the closest outlet/inlet to the zone valve and continue to the furthest outlet/inlet within the zone. [NFPA 99:5.1.12.2.5.2]

1318.13 Category 3 Dental Air and Nitrogen Supply Systems Purge Tests. The purge tests for dental air and nitrogen supply systems shall be conducted as follows:

(1) The outlets in each Category 3 dental air and nitrogen supply piping system shall be purged to remove any particulate matter from the distribution piping.

(2) The test gas shall be oil-free, dry nitrogen NF or the system gas.

(3) Each outlet shall be purged with an intermittent high-volume flow of test gas until the purge produces no discoloration in a clean white cloth.

(4) The purging shall be started at the furthest outlet in the system and proceed toward the source equipment. [NFPA 99:5.3.12.2.8]

1318.14 Operational Pressure Test. Operational pressure tests shall be performed at each station outlet/inlet or terminal where the user makes connections and disconnections. [NFPA 99:5.1.12.3.10]

1318.14.1 Test Gas. Tests shall be performed with the gas of system designation or the operating vacuum. [NFPA 99:5.1.12.3.10.1]

1318.14.2 Medical Gas Outlets. All gas outlets with a gauge pressure of 50 psi (345 kPa), including, but not limited to, oxygen, nitrous oxide, medical air, and carbon dioxide, shall deliver 3.5 standard cubic feet per minute (SCFM) (100 SLPM) with a pressure drop of not more than 5 psi (34 kPa) and static pressure of 50 psi (345 kPa) to 55 psi (379 kPa). [NFPA 99:5.1.12.3.10.2]

1318.14.3 Medical-Surgical Vacuum Inlets. Medical-surgical vacuum inlets shall draw 3 SCFM (85 Nl/min) without reducing the vacuum pressure below 12 inch mercury gauge (HgV) (41 kPa) at any adjacent station inlet. [NFPA 99:5.1.12.3.10.4]

1318.14.4 Oxygen and Medical Air Outlets. Oxygen and medical air outlets serving critical care areas shall allow a transient flow rate of 6 SCFM (170 SLPM) for 3 seconds. [NFPA 99:5.1.12.3.10.5]

1318.15 Medical Gas Concentration Test. After purging each system with the gas of system designation the following shall be performed:

(1) Each pressure gas source and outlet shall be analyzed for concentration of gas, by volume.

(2) Analysis shall be conducted with instruments designed to measure the specific gas dispensed.

(3) Allowable concentrations shall be as indicated in Table 1318.15. [NFPA 99:5.1.12.3.11]

TABLE 1318.15
GAS CONCENTRATIONS
[NFPA 99:5.1.12.3.11]

MEDICAL GAS	CONCENTRATION
Oxygen	≥99% oxygen
Nitrous oxide	≥99% nitrous oxide
Nitrogen	≤1% oxygen or ≥99% nitrogen
Medical air	19.5% - 23.5% oxygen
Other gases	As specified by +/-1%, unless otherwise specified

1318.16 System Verification. Verification tests shall be performed only after all tests required in Section 1318.4 through Section 1318.12, Installer Performed Tests have been completed. [NFPA 99:5.1.12.3.1.1]

1318.16.1 Test Gas. The test gas shall be oil-free, dry nitrogen NF or the system gas where permitted. [NFPA 99:5.1.12.3.1.2]

1318.16.2 Approved Tester. Testing shall be conducted by a party technically competent and experienced in the field of medical gas and vacuum pipeline testing and meeting the requirements of ASSE 6030. [NFPA 99:5.1.12.3.1.3]

Testing shall be performed by a party other than the installing contractor. [NFPA 99:5.1.12.3.1.4]

When systems have not been installed by in-house personnel, testing shall be permitted by personnel of that organization who meet the requirements of this section. [NFPA 99:5.1.12.3.1.5]

The certification must be done by a qualified third-party verification agency that is approved by the AHJ. The verifiers must be certified to the requirements of ANSI/ASSE 6030, Medical Gas Systems Verifiers. In addition to an extensive background and knowledge of the systems vital to successful health care, each verifier must pass a comprehensive exam, demonstrating these skills and abilities. The ASSE standard includes a detailed, thorough list of procedures to properly and safely verify medical gas systems. An independent third party, not the installing contractor, must perform system verification. The only exception is made for in-house personnel; however, verifiers must still be certified.

1319.0 System Certification.

1319.1 Certification. Prior to a medical gas or vacuum system being placed in service, such system shall be certified in accordance with Section 1319.2.

Because of the safety problems resulting from improperly installed medical gas and vacuum systems, these systems not only must be tested and inspected, but must also be certified.

1319.2 Certification Tests. Certification tests, verified and attested to by the certification agency, shall include the following:

(1) Verifying in accordance with the installation requirements.

(2) Testing and checking for leakage, correct zoning, and identification of control valves.

(3) Checking for identification and labeling of pipelines, station outlets, and control valves.

(4) Testing for cross-connection, flow rate, system pressure drop, and system performance.

(5) Functional testing of pressure relief valves and safety valves.

(6) Functional testing of sources of supply.

(7) Functional testing of alarm systems, including accuracy of system components.

(8) Purge flushing of system and filling with specific source gases.

(9) Testing for purity and cleanliness of source gases.

(10) Testing for specific gas identity at each station outlet.

1319.3 Report Items. A report that includes the specific items addressed in Section 1319.2, and other information required by this chapter, shall be delivered to the Authority Having Jurisdiction prior to acceptance of the system.

CHAPTER 14

FIRESTOP PROTECTION

1401.0 General.

1401.1 Applicability. Piping penetrations of required fire-resistance-rated walls, partitions, floors, floor/ceiling assemblies, roof/ceiling assemblies, or shaft enclosures shall be protected in accordance with the requirements of the building code, and this chapter.

As pipe is installed in buildings and homes it penetrates walls, ceilings and floors. Some of these vertical and horizontal partitions are constructed to withstand or at least delay the spread of fire; however, the penetrations of piping and other materials through these fire-resistance partitions compromise the integrity of the structure to resist the spread of fire. Firestopping of these penetrations is required to restore the fire-resistance rating of the wall, floor or ceiling, and in doing so, protect the safety of the occupants of the building or home.

Building codes address firestopping penetrations of rated walls, floors, and ceilings. They also address the fire-resistance ratings of those walls, floors, and ceilings and the construction type of a particular building. This chapter contains the requirements for firestopping piping penetrations.

1402.0 Construction Documents.

1402.1 Penetrations. Construction documents shall indicate with sufficient detail how penetrations of fire-resistance-rated assemblies shall be firestopped prior to obtaining design approval.

As with any other part of the construction of a building, firestopping must be documented on the submitted plans for approval by the Authority Having Jurisdiction (AHJ) (see **Figure 1403.1**). These plans must include exact documentation, with complete inventory, of all firestopped penetrations in the building, identification tags (if required) and a separate set of drawings showing each firestop with materials used and their certification listings and manufacturers' instructions. Unless this documentation is provided during the planning stages, or created during construction and then turned over in its entirety to the owner, the building maintenance staff will have no way of knowing which penetration is firestopped and with what firestop assembly. Without these documents in hand, proper repairs are but guesswork, which could violate the fire code and most importantly, denigrate the integrity of the fire-resistance ability of the structure.

1403.0 Installation.

1403.1 Materials. Firestop systems shall be installed in accordance with this chapter, the building code, and the manufacturer's installation instructions.

The information to determine when selecting an acceptable firestop is:

- The type of barrier, wall or floor and its material, thickness, and fire-resistance rating.
- The pipe material, size of pipe and schedule of pipe, which indicates wall thickness.
- The size of the hole and the annular space.
- The pipe's position in the hole (centered or off-center).
- It will then be necessary to match each opening in a fire-resistance-rated wall or floor in a building with a firestop assembly certification listing.

There are literally thousands of listings of various possible penetrations created by several certification and testing laboratories. Both the Canadian and U.S. Underwriters Laboratories each publish separate books containing their own listings of just those firestop manufacturers they have under contract. There is at any given time, just in North America, up to 60 different manufacturers, each with dozens of products, which in turn also have many listings. An architect or engineer will usually specify which manufacturer's products will be used on a particular project.

Once the decision is made as to which products for which application are to be used, the manufacturer's technical documents, which include installation instructions, should be cataloged. These must be followed exactly; otherwise the fire-resistance rating of the firestop assembly will not be ensured. An example of this documentation is illustrated in **Figure 1403.1**.

1404.0 Combustible Piping Installations.

1404.1 General Requirements. Combustible piping installations shall be protected in accordance with the appropriate fire resistance rating requirements in the building code that list the acceptable area, height, and type of construction for use in specific occupancies to assure compliance and integrity of the fire resistance rating prescribed.

Combustible piping installations, which consist of pipe that can flame at relatively low temperatures such as nonmetallic pipe, represent a special challenge to the installer when penetrating fire-rated assemblies. Special firestop systems have been designed to fill the void where combustible piping has been consumed at the penetration of the fire-rated assembly (see **Figure 1404.1**).

1404.2 Fire-Resistance Rating. Where penetrating a fire-resistance-rated wall, partition, floor, floor-ceiling assembly, roof-ceiling assembly, or shaft enclosure, the fire resistance rating of the assembly shall be restored to its original rating.

1404.3 Firestop Systems. Penetrations shall be protected by an approved penetration firestop system installed as tested in accordance with ASTM E119, ASTM E814, UL 263, or UL 1479 with a positive pressure differential of not less than 0.01 of an inch of water (0.002 kPa).

FIRESTOP PROTECTION

System No. F-A-1051
F Rating - 2 Hr
T Rating - 0 Hr

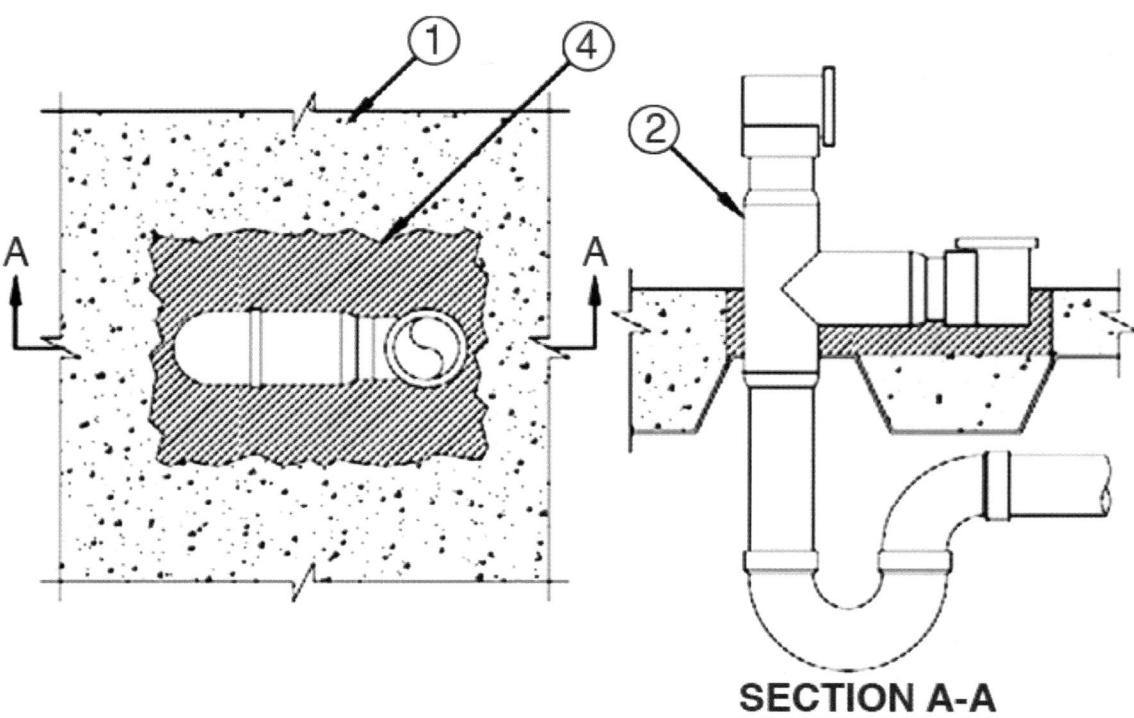

1. Floor Assembly - The fire rated unprotected concrete and steel floor assembly shall be constructed of the materials and in the manner specified in the individual D900 Series designs in the UL Fire Resistance Directory and is summarized below:

   A. Concrete - Min. 2-1/2 in. thick reinforced lightweight to normal weight (100-150pcf) concrete. Concrete to be recessed around opening to a depth of 2-1/2 in.. Size of recess in concrete to be 102 sq. in. with max dimension of 12 in.

   B. Steel Floor and Forms Units* – Composite or non-composite[posite max 3 in. deep galv steel fluted units as specified in the individual Floor-Ceiling Design. Mas diam of cored opening in steel deck is 100 mm.

2. Waste/Overflow/Drain Fitting _ Nom 2 in. diam copper, brass or cast iron waste/overflow/drain fitting installed concentrically or eccentrically through cored opening. Piping system to be rigidly supported on both sides of floor-ceiling assembly. The annular space between piping system and periphery of cored opening shall be a min. of 0 mm (point contact) to max of 1 7/8 in.

3. Forms _ (Not Shown) Used as a form to prevent leakage of fill material during installation. Forms to be rigid sheet metal, cut to fit the contour of the penetrating item and positioned beneath steel deck as required to accommodate the required thickness of fill material. Forms may be removed after fill material has cured.

4. Fill, Void or Cavity Materials* – Mortar – Min. 2-1/2 in. thickness of fill material applied within annulas, flush with top surface of floor. Fill material is mixed at a rate of 2-1/2 in. parts dry mix to one part water by volume in accordance with the fill manufacturer's installation instruction

*earing the UL Classification Mark

**FIGURE 1403.1
EXAMPLE OF MANUFACTURER'S INSTRUCTIONS**

**FIGURE 1404.1
TESTING OF FIRESTOPPING FOR COMBUSTIBLE MATERIALS**

Systems shall have an F rating of not less than 1 hour but not less than the required fire resistance rating of the assembly being penetrated. Systems protecting floor penetrations shall have a T rating of not less than 1 hour but not less than the required fire resistance rating of the floor being penetrated. Floor penetrations contained within the cavity of a wall at the location of the floor penetration do not require a T rating. No T rating shall be required for floor penetrations by piping that is not in direct contact with combustible material.

The F and T rating (see Chapter 2, Definitions) of a firestop are dependent upon various factors such as, the pipe material and whether it is insulated or not; the pipe size; the wall material (concrete block, wood or steel-framed gypsum board); and wall thickness or floor thickness (fire rating hours). Whatever the ratings are, the installer must be sure to have the correct firestop assembly for the application and, of course, adhere to the manufacturer's installation instructions.

1404.4 Connections. Where piping penetrates a rated assembly, combustible piping shall not connect to noncombustible piping unless it is capable of being demonstrated that the transition is in accordance with Section 1404.3.

1404.5 Insulation and Coverings. Insulation and coverings on or in the penetrating item shall not be permitted unless the specific insulating or covering material has been tested as part of the penetrating firestop system.

Pipe covering will neither be allowed to substitute nor be allowed to be part of the firestop unless it has been rated for the appropriate T and F ratings. Firestops contain materials that expand in the presence of heat, and in turn, seal the penetration if the heat softens the pipe. It will not expand enough to fill the void left by pipe covering unless it is designed and tested to do so (see **Figures 1404.5a and 1404.5b**).

1404.6 Sleeves. Where sleeves are used, the sleeves shall be securely fastened to the fire-resistance-rated assembly. The (inside) annular space between the sleeve and the penetrating item and the (outside) annular space between the

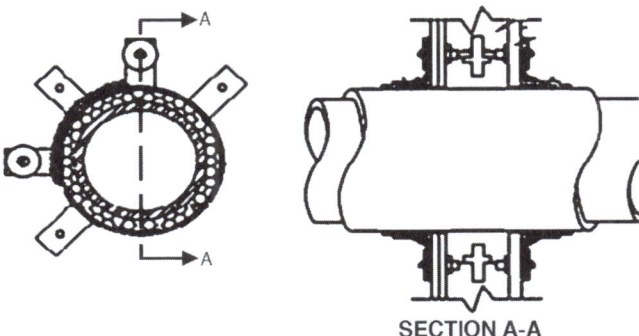

**FIGURE 1404.5A
EXAMPLE OF FIRESTOP FOR INSULATED PIPE**

sleeve and the fire-resistance-rated assembly shall be firestopped in accordance with this chapter.

Where sleeves are used to penetrate fire-resistant assemblies, the sleeve becomes part of the firestop assembly and must be included in the assembly listing and rating.

1405.0 Noncombustible Piping Installations.

1405.1 General Requirements. Noncombustible piping installations shall be protected in accordance with the appropriate fire resistance rating requirements in the building code that list the acceptable area, height, and type of construction for use in specific occupancies to ensure compliance and integrity of the fire-resistance rating prescribed.

Noncombustible piping is metallic piping such as steel, cast iron and copper pipe and tube, which will not flame at relatively low temperatures.

1405.2 Fire-Resistance Rating. Where penetrating a fire-resistance-rated wall, partition, floor, floor-ceiling assembly, roof-ceiling assembly, or shaft enclosure, the fire-resistance rating of the assembly shall be restored to its original rating.

Exceptions:

(1) Concrete, mortar, or grout shall be permitted to be used to fill the annular spaces around cast-iron, copper, copper

FIRESTOP PROTECTION

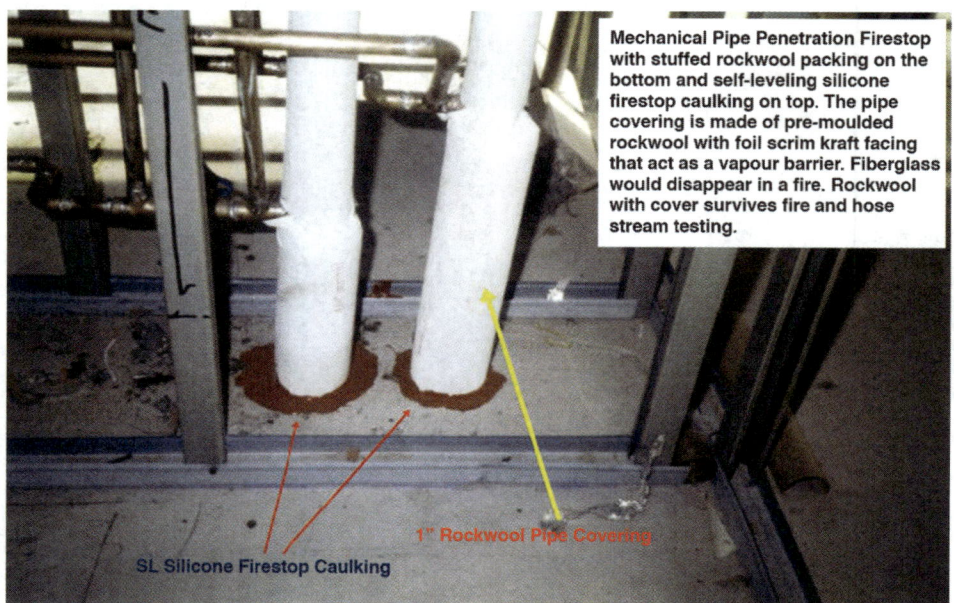

FIGURE 1404.5B
FIRESTOP OF FLOOR PENETRATION

alloy, or steel piping that penetrates concrete or masonry fire-resistant-rated assemblies. The nominal diameter of the penetrating item shall not exceed 6 inches (150 mm), and the opening size shall not exceed 144 square inches (0.093 m^2).

The thickness of concrete, mortar, or grout shall be the full thickness of the assembly or the thickness necessary to provide a fire-resistance rating not less than the required fire-resistance rating of the assembly penetrated.

(2) The material used to fill the annular space shall prevent the passage of flame and hot gases capable of igniting cotton waste for the time period equivalent to the fire-resistance rating of the assembly, where tested to standard(s) referenced in Section 1405.3.

See Figure 1405.2.

1405.3 Firestop Systems. Penetrations shall be protected by an approved penetration firestop system installed as tested in accordance with ASTM E119, ASTM E814, UL 263, or UL 1479 with a positive pressure differential of not less than 0.01 of an inch of water (0.002 kPa). Systems shall have an F rating of not less than 1 hour but not less than the required fire-resistance rating of the assembly being penetrated. Systems protecting floor penetrations shall have a T rating of not less than 1 hour but not less than the required fire-resistance rating of the floor being penetrated. Floor penetrations contained within the cavity of a wall at the location of the floor penetration do not require a T rating. No T rating shall be required for floor penetrations by piping that is not in direct contact with combustible material.

1405.4 Connections. Where piping penetrates a rated assembly, combustible piping shall not connect to noncombustible piping unless it is capable of being demonstrated that the transition is in accordance with the requirements of Section 1405.3.

1405.5 Unshielded Couplings. Unshielded couplings shall not be used to connect noncombustible piping unless it is capable of being demonstrated that the fire-resistive integrity of the penetration is maintained.

1405.6 Sleeves. Where sleeves are used, the sleeves shall be securely fastened to the fire-resistance-rated assembly. The (inside) annular space between the sleeve and the penetrating item and the (outside) annular space between the sleeve and the fire-resistance-rated assembly shall be firestopped in accordance with this chapter.

1405.7 Insulation and Coverings. Insulation and coverings on or in the penetrating item shall not be permitted unless the specific insulating or covering material has been tested as part of the penetrating firestop system.

See Figures 1404.5a and 1404.5b.

1406.0 Required Inspection.

1406.1 General. Prior to being concealed, piping penetrations shall be inspected by the Authority Having Jurisdiction to verify compliance with the fire-resistance rating prescribed in the building code.

1406.2 Examination. The Authority Having Jurisdiction shall conduct a thorough examination of sufficient representative installations, including destructive inspection, to provide verification of satisfactory compliance with this chapter, the appropriate manufacturer's installation instructions applied by the installer, construction documents, specifications, and applicable manufacturer's product information.

All firestop systems shall not be covered until the system has been inspected by the AHJ. The inspection may require destructive inspection, which is when a firestop

FIRESTOP PROTECTION

System No. C-AJ-1027
January 18, 2005
(Formerly System No. 202)
F Rating - 3 Hr
T Rating - 0 Hr

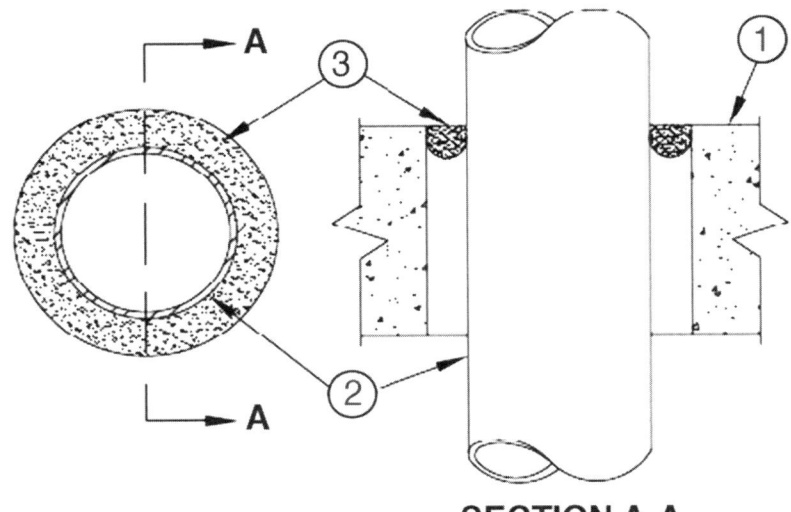

SECTION A-A

1. **Floor or Wall Assembly** – Min 1-1/2 in. thick lightweight or normal weight (100 150 pcf) concrete. Wall may also be constructed of any UL Classified Concrete Blocks*. Max diam of through openings is 12-1/4 in..

    See Concrete Blocks (CAZT) category in Fire Resistant Directory for names of manufacturers.

2. **Through Penetrants** – One metallic pipe, conduit or tubing to be installed either concentrically or eccentrically within the firestop system. Min annular space between pipe, conduit or tubing and edge of opening is 0 mm (point contact). Max annular space is dependent on pipe, conduit or tubing type and size as well as the F Rating of system, as shown in the table below. Pipe, conduit or tubing to be rigidly supported on both sides of wall assembly. The following types and sizes of metallic pipes, conduits or tubing may be used.

    A. **Steel Pipe** – No 10 in.. diam (or smaller) Schedule 10 (or heavier) steel pipe.
    B. **Conduit** – No 6 in. dam (or smaller) rigid steel conduit.
    C. **Conduit** – No 4 in. dam (or smaller) steel electrical metallic tubing or steel conduit.
    D. **Copper** – Tubing Nom 3 in. diam (or smaller) Type L (or heavier) copper tubing.
    E. **Copper** – Pipe Nom 3 in. diam (or smaller) Regular (or heavier) copper pipe.
    F. **Iron Pipe** – Nom 10 in. diam (or smaller) cast or ductile iron pipe.

Pipe Conduit or Tubing Type	Max Nom Pipe Conduit or Tubing Daim mm	F Rating Hr	Max Annular Space mm
A or F	10	3	3/4
B	6	3	3/4
C	4	3	1-1/2
D and E	3	3	3/4
D and E	3	2	7/8

3. **Fill, Void or Cavity Materials*** – **Putty** – Moldable putty material kneaded by hand and applied to full annular space to a min depth of 1 in., flush with top surface of floor. In wall assemblies, required putty thickness to be installed symmetrically on both sides of wall.

*Bearing the UL Classification Marking

This material was extracted and drawn by 3M Fire Protection Products form the 2006 edition of the UL Fire Resistance Directory

FIGURE 1405.2
FIRESTOP FOR METALLIC PIPE THROUGH CONCRETE FLOOR

FIRESTOP PROTECTION

penetration is completely exposed, to verify satisfactory compliance. The AHJ will compare the firestop systems used with the required fire-resistance rating of the construction being penetrated and ensure that the firestop systems are being installed per the manufacturer's installation instructions. Tagging or labeling the assembly may also be required (see **Figure 1406.2**).

1406.3 Penetrations. The Authority Having Jurisdiction shall determine the type, size, and quantity of penetrations to be inspected.

1406.4 Field Installations. The Authority Having Jurisdiction shall compare the field installations with the documentation supplied by the installer to determine the following:

(1) The required F ratings (1 hour, 2 hour, 3 hour, or 4 hour) and T ratings (0 hour, 1 hour, 2 hour, 3 hour, or 4 hour) of the penetration firestop systems are at least the same as the hourly rating of the assembly being penetrated.

(2) The penetrating firestop system includes the penetrating item as documented through testing of the systems conducted by an independent testing agency.

(3) The penetrating firestop system is installed as tested.

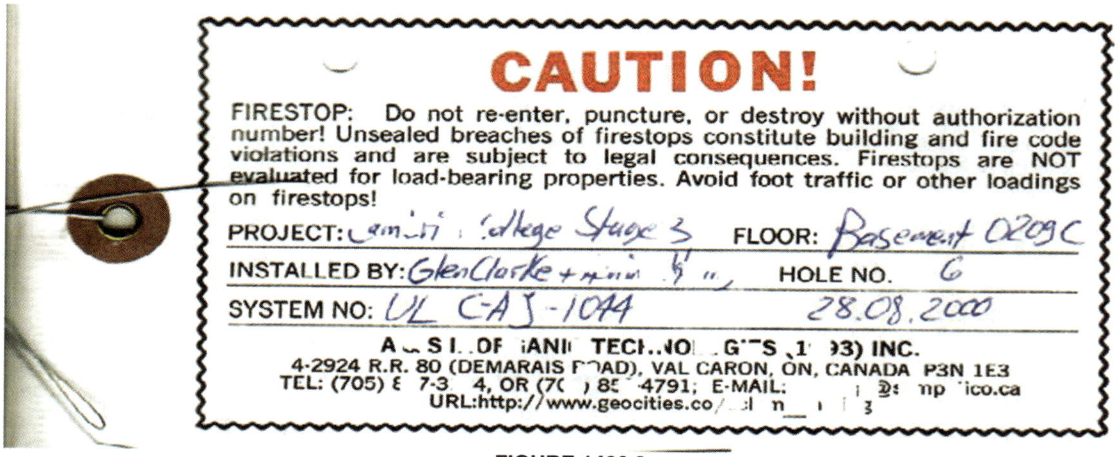

FIGURE 1406.2
FIRESTOP IDENTIFICATION TAG

CHAPTER 15
ALTERNATE WATER SOURCES FOR NONPOTABLE APPLICATIONS

1501.0 General.
🔧 This chapter focuses on the regulations for gray water, reclaimed (recycled) water, and on-site treated nonpotable water systems. Although the definition of alternate water source is not limited to these three systems, and that this code intends to allow room for other sources as technology develops (see **Figure 1501.0** for technology for a stormwater collection system), only these three systems are regulated by this chapter.

The demand for fresh water conservation around the world is growing while supplies are dwindling, and the cost of producing potable water is increasing. Alternative water sources are a viable means to lessen the demand for potable water in applications where drinking water standards are not required. It is just as important to develop alternate water sources as it is to develop alternate sources of energy. These provisions are a move toward this conservation effort and will ensure the safe use of alternate water sources.

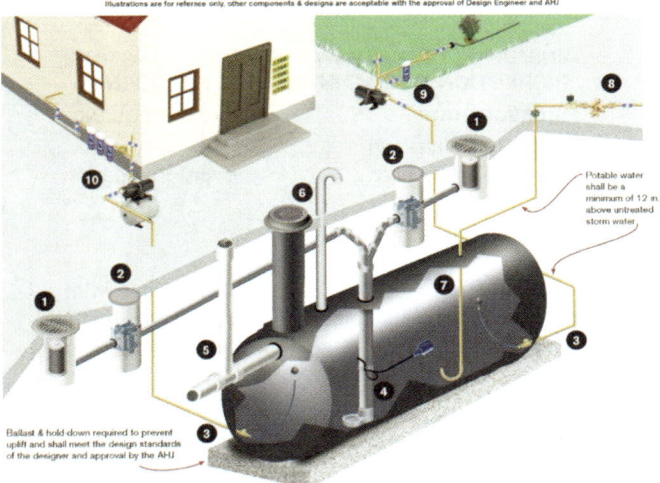

FIGURE 1501.0
STORMWATER COLLECTION SYSTEM
Courtesy of Building in California

1501.1 Applicability. The provisions of this chapter shall apply to the construction, alteration, and repair of alternate water source systems for nonpotable applications.

1501.1.1 Allowable Use of Alternate Water. Where approved or required by the Authority Having Jurisdiction, alternate water sources [reclaimed (recycled) water, gray water, and on-site treated nonpotable water] shall be permitted to be used instead of potable water for the applications identified in this chapter.

1501.2 System Design. Alternate water source systems shall be designed in accordance with this chapter by a registered design professional or licensed person who demonstrates competency to design the alternate water source system as required by the Authority Having Jurisdiction. Components, piping, and fittings used in an alternate water source system shall be listed.

🔧 The intent of this section is to ensure that the person designing alternate water source systems is qualified and competent to perform the work. For example, most jurisdictions will allow landscape architects to design the downstream side of the gray water system (typically beyond two feet outside the building after alterations to the existing plumbing system). They have a thorough understanding of soil types, root zones of specific plants, erosion control, and are one of the professional categories to become Storm Water Pollution Prevention Plan (SWPPP) developers and practitioners. This type of expertise is required to prevent the potential hazards associated with gray water ponding and runoff.

Exceptions:

(1) A registered design professional is not required to design gray water systems having a maximum discharge capacity of 250 gallons per day (gal/d) (0.011 L/s) for single family and multi-family dwellings.

🔧 Many gray water systems are very simple and are often based on gravity drainage with no tanks or pumps. Even the small pumped and filtered manufactured type gray water systems are very simple to install, usually taking only a few hours. If the system has larger volumes of water, or is from commercial buildings, a licensed person is required to design the system.

(2) A registered design professional is not required to design an on-site treated nonpotable water system for single-family dwellings having a maximum discharge capacity of 250 gal/d (0.011 L/s).

🔧 Like exception (1), small capacity residential systems are not required to be designed by licensed persons.

1501.3 Permit. It shall be unlawful for a person to construct, install, alter, or cause to be constructed, installed, or altered an alternate water source system in a building or on a premise without first obtaining a permit to do such work from the Authority Having Jurisdiction.

🔧 Regardless of whether a design professional is required or not, alternate water source systems still require a permit, inspection, and approval by the Authority Having Jurisdiction (AHJ). This provision assures that the system installed meets all standards required for the particular installation. **Figure 1501.3** is an example of a permit checklist for a graywater irrigation system.

ALTERNATE WATER SOURCES FOR NONPOTABLE APPLICATIONS

1501.4 Component Identification. System components shall be properly identified as to the manufacturer.

 System components comprise all the constituent parts of the system including tanks, pipes, fittings, valves, controls, etc. Proper identification is required (see **Figure 1501.4**).

1501.5 Maintenance and Inspection. Alternate water source systems and components shall be inspected and maintained in accordance with Section 1501.5.1 through Section 1501.5.3.

 These maintenance requirements will help ensure the systems continue to function as designed and do not pose any problems or hazards to the user.

1501.5.1 Frequency. Alternate water source systems and components shall be inspected and maintained in accordance with Table 1501.5 unless more frequent inspection and maintenance are required by the manufacturer.

1501.5.2 Maintenance Log. A maintenance log for gray water and on-site treated nonpotable water systems is required to have a permit in accordance with Section 1501.3 and shall be maintained by the property owner and be available for inspection. The property owner or designated appointee shall ensure that a record of testing, inspection, and maintenance in accordance with Table 1501.5 is maintained in the log. The log will indicate the frequency of inspection and maintenance for each system.

1501.5.3 Maintenance Responsibility. The required maintenance and inspection of alternate water source systems shall be the responsibility of the property owner unless otherwise required by the Authority Having Jurisdiction.

 It is ultimately the responsibility of the property owner to maintain their system. Some property owners will chose to have a maintenance contract with the installer. At some point in time it may be necessary and required by the AHJ to provide post construction maintenance agreements, yearly inspections, and compliance reporting to the local authority.

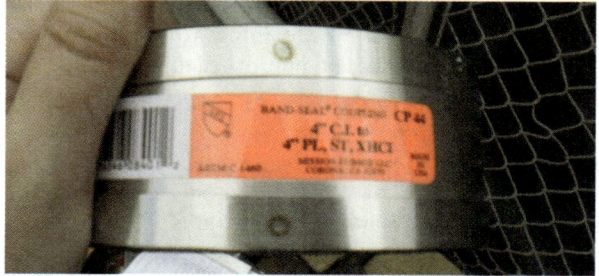

FIGURE 1501.3
SAMPLE OF PERMIT CHECKLIST
(Extracted From Community and Development Agency Building Services Dept-Oakland, CA)

FIGURE 1501.4
COMPONENT MARKING

TABLE 1501.5
MINIMUM ALTERNATE WATER SOURCE TESTING, INSPECTION, AND MAINTENANCE FREQUENCY

DESCRIPTION	MINIMUM FREQUENCY
Inspect and clean filters and screens, and replace (where necessary).	Every 3 months
Inspect and verify that disinfection, filters, and water quality treatment devices and systems are operational and maintaining minimum water quality requirements as determined by the Authority Having Jurisdiction.	In accordance with manufacturer's instructions, and the Authority Having Jurisdiction.
Inspect pumps and verify operation.	After initial installation and every 12 months thereafter
Inspect valves and verify operation.	After initial installation and every 12 months thereafter
Inspect pressure tanks and verify operation.	After initial installation and every 12 months thereafter
Clear debris from and inspect storage tanks, locking devices, and verify operation.	After initial installation and every 12 months thereafter
Inspect caution labels and marking.	After initial installation and every 12 months thereafter
Inspect and maintain mulch basins for gray water irrigation systems.	As needed to maintain mulch depth and prevent ponding and runoff.
Cross-connection inspection and test*	After initial installation and every 12 months thereafter

* The cross-connection test shall be performed in the presence of the Authority Having Jurisdiction in accordance with the requirements of this chapter.

1501.6 Operation and Maintenance Manual. An operation and maintenance manual for gray water and on-site treated water systems required to have a permit in accordance with Section 1501.3 shall be supplied to the building owner by the system designer. The operating and maintenance manual shall include the following:

(1) Detailed diagram of the entire system and the location of system components.

(2) Instructions for operating and maintaining the system.

(3) Details on maintaining the required water quality as determined by the Authority Having Jurisdiction.

(4) Details on deactivating the system for maintenance, repair, or other purposes.

(5) Applicable testing, inspection, and maintenance frequencies in accordance with Table 1501.5.

(6) A method of contacting the manufacturer(s).

The operations and maintenance (O&M) manual will help owners understand their system and the maintenance needs. If the dwelling changes ownership the O&M manual is critical to transfer key information to the new owner.

1501.7 Minimum Water Quality Requirements. The minimum water quality for alternate water source systems shall meet the applicable water quality requirements for the intended application as determined by the Authority Having Jurisdiction. In the absence of water quality requirements, the EPA/625/R-04/108 contains recommended water reuse guidelines to assist regulatory agencies to develop, revise, or expand alternate water source water quality standards.

Exception: Water treatment is not required for gray water used for subsurface irrigation.

Most States have guidelines for water quality, and the National Sanitation Foundation (NSF) has developed *Standard 350 Onsite Residential and Commercial Water Reuse Treatment Systems* for indoor nonpotable reuse. It's imperative that local jurisdictions follow their state guidelines to prevent liability concerns. California guidelines for tertiary water can be found in the California Code of Regulations. EPA guidelines can be found at www.waterreuseguidelines.org.

Water treatment is needed in systems where there is a potential for direct contact from the occupants of the dwellings. The exception for water treatment is for gray water subsurface irrigation where there is little to no risk of human contact. Subsurface irrigation distributes the nonpotable water below the surface of the soil, out of reach for human consumption. Reclaimed water is tertiary treated water that is delivered from a municipal treatment plant. It is not considered an on-site treated alternative water source.

1501.8 Material Compatibility. Alternate water source systems shall be constructed of materials that are compatible with the type of pipe and fitting materials, water treatment, and water conditions in the system.

1501.9 Commercial, Industrial, and Institutional Restroom Signs. A sign shall be installed in restrooms in commercial, industrial, and institutional occupancies using reclaimed (recycled) water and on-site treated water, for water closets, urinals, or both. Each sign shall contain ½ of an inch (12.7 mm) letters of a highly visible color on a contrasting background. The location of the sign(s) shall be such that the sign(s) are visible to users. The location of the sign(s) shall be approved by the Authority Having Jurisdiction and shall contain the following text:

TO CONSERVE WATER, THIS BUILDING USES *_____* TO FLUSH TOILETS AND URINALS.

Signage is required in restrooms and equipment rooms containing on-site treated water to prevent an accidental ingestion and cross-connection when the plumbing system is being altered (see **Figure 1501.9**). Colors and labeling requirements can be found Section 601.3 of the UPC.

FIGURE 1501.9
RESTROOM SIGNAGE
(Courtesy of Building in California)

1501.9.1 Equipment Room Signs. Each room containing reclaimed (recycled) water and on-site treated water equipment shall have a sign posted in a location that is visible to anyone working on or near nonpotable water equipment with the following wording in 1 inch (25.4 mm) letters:

CAUTION: NONPOTABLE *_____*, DO NOT DRINK. DO NOT CONNECT TO DRINKING WATER SYSTEM. NOTICE: CONTACT BUILDING MANAGEMENT BEFORE PERFORMING ANY WORK ON THIS WATER SYSTEM.

*_____*Shall indicate RECLAIMED (RECYCLED) WATER or ON-SITE TREATED WATER, accordingly.

ALTERNATE WATER SOURCES FOR NONPOTABLE APPLICATIONS

1501.10 System Controls. Controls for pumps, valves, and other devices that contain mercury that come in contact with alternate water source water supply shall not be permitted.

Many pre-manufactured sump and ejector systems contain float switch activated pumps. Switches containing mercury are prohibited to prevent accidental contamination of the collected gray water and irrigation zones.

1502.0. Inspection and Testing.

1502.1 General. Alternate water source systems shall be inspected and tested in accordance with Section 1502.2 through Section 1502.3.4.

1502.2 Supply System Inspection and Test. Alternate water source systems shall be inspected and tested in accordance with this code for testing of potable water piping.

1502.3 Annual Cross-Connection Inspection and Testing. An initial and subsequent annual inspection and test shall be performed on both the potable and alternate water source systems. The potable and alternate water source system shall be isolated from each other and independently inspected and tested to ensure there is no cross-connection in accordance with Section 1502.3.1 through Section 1502.3.4.

See Figure 1502.3

1502.3.1 Visual System Inspection. Before commencing the cross-connection testing, a dual system inspection shall be conducted by the Authority Having Jurisdiction and other authorities having jurisdiction as follows:

(1) Meter locations of the alternate water source and potable water lines shall be checked to verify that no modifications were made and that no cross-connections are visible.

(2) Pumps and equipment, equipment room signs and exposed piping in equipment room shall be checked.

(3) Valves shall be checked to ensure that the valve lock seals are still in place and intact. Valve control door signs shall be checked to verify that no signs have been removed.

See Figure 1502.3,1 for an example of a valve lock seal.

1502.3.2 Cross-Connection Test. The procedure for determining cross-connection shall be followed by the applicant in the presence of the Authority Having Jurisdiction and other authorities having jurisdiction to determine whether a cross-connection has occurred as follows:

(1) The potable water system shall be activated and pressurized. The alternate water source system shall be shut down, depressurized, and drained.

(2) The potable water system shall remain pressurized for a minimum period specified by the Authority Having Jurisdiction while the alternate water source system is empty. The minimum period the alternate water source system is to remain depressurized shall be determined

**FIGURE 1502.3
CROSS-CONNECTION TEST**

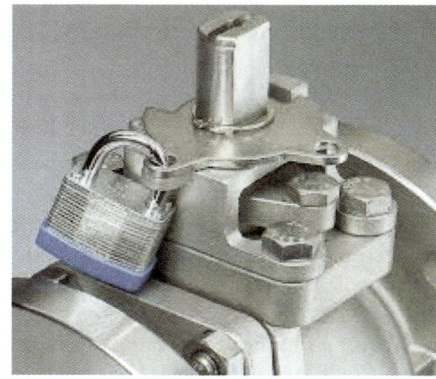

**FIGURE 1501.11.2.1
VALVE LOCK SEAL**

on a case-by-case basis, taking into account the size and complexity of the potable and the alternate water source distribution systems, but in no case shall that period be less than 1 hour.

(3) The drain on the alternate water source system shall be checked for flow during the test and fixtures, potable and alternate water source, shall be tested and inspected for flow. Flow from an alternate water source system outlet indicates a cross-connection. No flow from a potable water outlet shall indicate that it is connected to the alternate water source system.

(4) The potable water system shall then be depressurized and drained.

(5) The alternate water source system shall then be activated and pressurized.

(6) The alternate water source system shall remain pressurized for a minimum period specified by the Authority Having Jurisdiction while the potable water system is empty. The minimum period the potable water system is to remain depressurized shall be determined on a case-by-case basis, but in no case shall that period be less than 1 hour.

(7) Fixtures, potable, and alternate water source shall be tested and inspected for flow. Flow from a potable water system outlet indicates a cross-connection. No

flow from an alternate water source outlet will indicate that it is connected to the potable water system.

(8) The drain on the potable water system shall be checked for flow during the test and at the end of the test.

(9) Where there is no flow detected in the fixtures which would indicate a cross-connection, the potable water system shall be repressurized.

1502.3.3 Discovery of Cross-Connection. If a cross-connection is discovered, the following procedure, in the presence of the Authority Having Jurisdiction, shall be activated immediately:

(1) The alternate water source piping to the building shall be shutdown at the meter, and the alternate water source riser shall be drained.

(2) Potable water piping to the building shall be shutdown at the meter.

(3) The cross-connection shall be uncovered and disconnected.

(4) The building shall be retested in accordance with Section 1502.3.1 and Section 1502.3.2.

(5) The potable water system shall be chlorinated with 50 parts-per-million (ppm) chlorine for 24 hours.

(6) The potable water system shall be flushed after 24 hours, and a standard bacteriological test shall be performed. Where test results are acceptable, the potable water system shall be permitted to be recharged.

1502.3.4 Annual Inspection. An annual inspection of the alternate water source system, following the procedures listed in Section 1502.3.1 shall be required. Annual cross-connection testing, following the procedures listed in Section 1502.3.2 shall be required by the Authority Having Jurisdiction, unless site conditions do not require it. In no event shall the test occur less than once in 4 years. Alternate testing requirements shall be permitted by the Authority Having Jurisdiction.

Additional requirements for testing alternate water sources for cross-connection and frequency are specified in Table 1501.5.

1502.4 Separation Requirements. Underground alternate water source service piping other than gray water shall be separated from the building sewer in accordance with this code. Treated nonpotable water pipes shall be permitted to be run or laid in the same trench as potable water pipes with a 12 inch (305 mm) minimum vertical and horizontal separation where both pipe materials are approved for use within a building. Where horizontal piping materials do not comply with this requirement, the minimum separation shall be increased to 60 inches (1524 mm). The potable water piping shall be installed at an elevation above the treated nonpotable water piping.

1502.5 Abandonment. Alternate water source systems that are no longer in use or fail to be maintained in accordance with Section 1501.5 shall be abandoned. Abandonment shall comply with Section 1502.5.1 and Section 1502.5.2.

1502.5.1 General. An abandoned system or part thereof covered under the scope of this chapter shall be disconnected from remaining systems, drained, plugged, and capped in an approved manner.

1502.5.2 Underground Tank. An underground water storage tank that has been abandoned or otherwise discontinued from use in a system covered under the scope of this chapter shall be completely drained and filled with earth, sand, gravel, concrete, or other approved material or removed in a manner satisfactory to the Authority Having Jurisdiction.

1502.6 Sizing. Unless otherwise provided for in this chapter, alternate water source piping shall be sized in accordance with Chapter 6 for sizing potable water piping.

1503.0 Gray Water Systems.

1503.1 General. The provisions of this section shall apply to the construction, alteration, and repair of gray water systems.

This section identifies three types of gray water systems: 1) subsurface irrigation where drip emitters are buried 2 inches below finished grade; 2) subsoil irrigation where large diameter pipe is buried in deep trenches to permit deep root irrigation and rapid dispersal of gray water; and 3) mulch basin where gray water discharges into a mulch filled basin to irrigate deep rooted plants.

There have been many studies regarding the amount of coliform bacteria, nitrogen, biochemical oxygen demand (BOD), phosphorus, pharmaceuticals, and endocrine suppressors that are present in gray water. To limit these contaminants, gray water sources are restricted to only bathtubs, showers, lavatories, clothes washers, and laundry tubs (see definition for Gray Water in Section 209.0 and **Figure 1503.1**); and may only discharge to subsurface or subsoil irrigation systems, or a mulch basin. Gray water systems are not permitted for any above ground irrigation that may allow any potential for human aspiration, nor for many food crops. These restrictions reduce the risk for direct human contact with gray water, while promoting beneficial use from this source of water.

**FIGURE 1503.1
GRAY WATER VERSUS BLACK WATER**

ALTERNATE WATER SOURCES FOR NONPOTABLE APPLICATIONS

1503.2 System Requirements. Gray water shall be permitted to be diverted away from a sewer or private sewage disposal system, and discharge to a subsurface irrigation or subsoil irrigation system. The gray water shall be permitted to discharge to a mulch basin for single-family and multi-family dwellings. Gray water shall not be used to irrigate root crops or food crops intended for human consumption that comes in contact with soil.

Figure 1503.2a is an example of a clothes washer discharging to a subsurface irrigation system. In Multiunit/Multistory buildings where gray water requires suds relief, care should be taken to assure that the suds relief is accomplished before discharging to the irrigation system.

When California adopted emergency provisions for gray water during the 2009 code cycle, the laundry was allowed to be installed without a permit as long as the system did not alter the existing DWV system. **Figure 1503.2b** shows the most common system being installed and taught by gray water advocates. Typically, systems are installed without a trap and use a vacuum breaker to facilitate flow (depicted on the left hand side of the drawing). The trap prevents pests and insects from entering the building through the irrigation piping. Also, note that there is no air break for the washer when the valve diverts gray water to the irrigation field. The intent for the air break is to prevent sewage (black water) from entering into the washer. If the irrigation line backs up, the only material to enter the washer is material that was discharged from the washer to begin with. Many similar systems use a check valve to prevent backups through the irrigation piping. Since there is grit and dirt being discharged, check valves would need frequent maintenance and to be installed with unions. This system is non-compliant with Section 1503.2.2 that requires the discharge of gray water to occur downstream of the fixture trap and vent as shown in **Figure 1503.2a**.

It is not recommended to use a filter in this system since

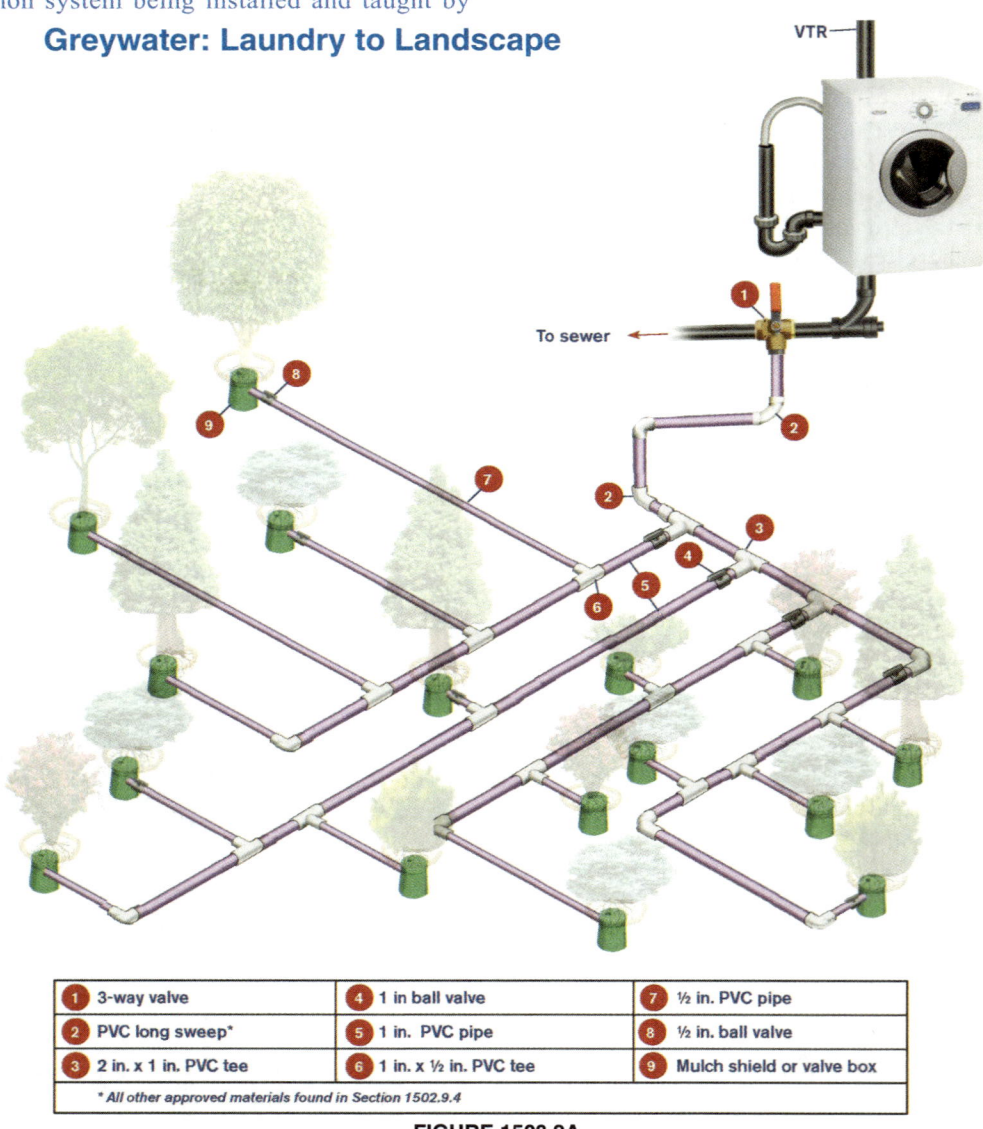

FIGURE 1503.2A
GRAY WATER DISCHARGE
(Courtesy of Building in California)

the pressure drop across the filter would strain the pump leading to premature failure. The system uses the pump from the laundry machine to discharge the gray water to the landscape area under pressure and allows for the use of smaller piping.

Another alternative is the dual standpipe system (see **Figure 1503.2c**). One drain is for the irrigation line and the other is connected to the sewer. In time of infrequent use (seasonal irrigation) a "jim cap" or wing nut test plug can be used to cap the unused drain to prevent the introduction of sewer gas into the building.

Since gray water is nonpotable, it is important that it does not contact the edible portion of any food plant so as to avoid accidental ingestion of gray water via a food source. This includes root crops (foods that grow underground) such as carrots, beets, radishes, and turnips; and any food crop that comes in direct contact with the soil such as pumpkins, melons, squash, cucumbers, and potatoes. Above ground edible portions of food plants that do not contact the soil, like fruit, corn, or tomatoes, are permitted to be irrigated with subsurface gray water and are safe to eat.

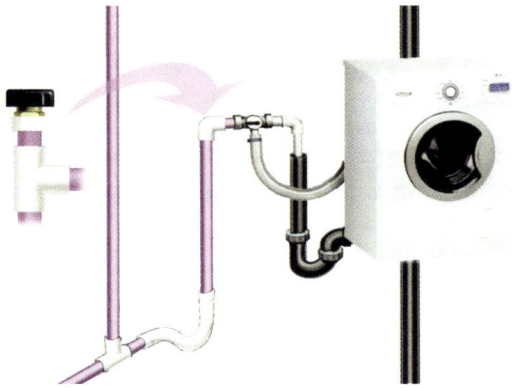

FIGURE 1503.2B
GRAY WATER DISCHARGE (NON-COMPLIANT)
(Courtesy of Building in California)

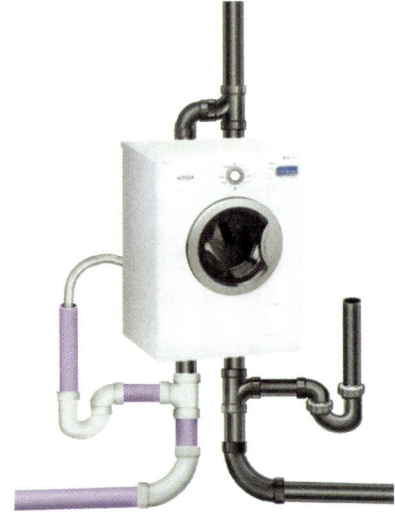

FIGURE 1503.2C
GRAY WATER DISCHARGE ALTERNATE
(Courtesy of Building in California)

1503.2.1 Surge Capacity. Gray water systems shall be designed to have the capacity to accommodate peak flow rates and distribute the total amount of estimated gray water on a daily basis to a subsurface irrigation field, subsoil irrigation field, or mulch basin without surfacing, ponding, or runoff. A surge tank is required for systems that are unable to accommodate peak flow rates and distribute the total amount of gray water by gravity drainage. The water discharge for gray water systems shall be determined in accordance with Section 1503.8.1 or Section 1503.8.2.

🔧 Due to the organic matter and nutrients typically found in untreated gray water, storage longer than one day is undesirable and unhealthy. The nutrients break down, using oxygen and causing malodorous emissions. Bacteria can also breed in the tank, decreasing the quality of the gray water and increasing the potential for unhealthy constituents. However, once gray water has been distributed into the ground, nutrients are taken up by plants and the soil bacteria break down other constituents, preventing any development of odor or health hazards.

Pooling and runoff should not occur in gray water systems as both will cause a potential for the gray water leaving the property and entering a storm drain. The pooling of gray water could also provide a mosquito breeding habitat.

1503.2.2 Diversion. The gray water system shall connect to the sanitary drainage system downstream of fixture traps and vent connections through an approved gray water diverter valve. The gray water diverter valve shall be installed in an accessible location and clearly indicate the direction of flow.

🔧 Gray water diverter valves are needed to divert the gray water to the sanitary drainage when either the gray water is not needed (e.g. plant dormancy, rainy periods or winter season), or when the gray water is hazardous, containing unhealthy contaminants (e.g. hair dye, pesticides, herbicides, cleaning chemicals, solvents, toxins, medications, etc). Where valves are not physically accessible, a simple electronic actuator can be included to allow for a switch inside the dwelling to control the valve. See **Figure 1503.2.2a** for a proper placement of a 3-way diverter valve.

Note: There are no known diverter valves specifically labeled for "gray water" in the US market. Typically, valves which are manufactured for the pool and spa industry are used (see **Figure 1503.2.2b**). They are not manufactured in ABS. They come in brass, CPVC, or PVC. CPVC valves are recommended for temperatures over 105°F. Both CPVC and PVC valves are manufactured in black color. Care must be taken to determine the material and correct joining and transition methods. ABS pipe shall not be directly glued to CPVC or PVC valve bodies. There are brass valves on the market, but they are not common on gray water systems.

1503.2.3 Backwater Valves. Gray water drains subject to backflow shall be provided with a backwater valve so located as to be accessible for inspection and maintenance.

ALTERNATE WATER SOURCES FOR NONPOTABLE APPLICATIONS

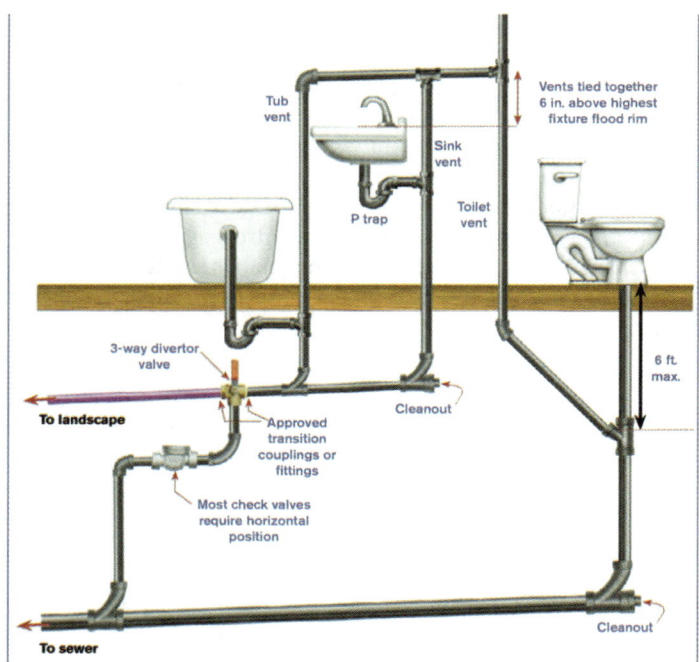

**FIGURE 1503.2.2A
THREE-WAY DIVERTER VALVE PLACEMENT
(Courtesy of Building in California)**

**FIGURE 1503.2.2B
THREE-WAY DIVERTER VALVE
(Courtesy of Building in California)**

🔧 Backwater valves prevent the contamination of the gray water system or surge tank from black water intrusion in the event of a sanitary sewer or building drain stoppages.

Backwater valves are designed to be installed in the horizontal position (see **Figure 1503.2.3**). The use of backwater valves in the vertical position must be installed per the terms of their listing and approved by the AHJ. Equally important is the direction of flow. Many ejector backwater valves are designed for vertical installation but only when the valve is closed when the pump is not active. Some gray water guidelines portray the valve in the vertical position and stay open unless there is a sewer back up. This installation is not recommended and the valve will not shut properly when required.

1503.3 Connections to Potable and Reclaimed (Recycled) Water Systems. Gray water systems shall

**FIGURE 1503.2.3
BACKWATER VALVE
(Courtesy of Rectorseal)**

have no direct connection to a potable water supply, on-site treated nonpotable water supply, or reclaimed (recycled) water systems. Potable, on-site treated nonpotable, or reclaimed (recycled) water is permitted to be used as makeup water for a non-pressurized storage tank provided the connection is protected by an air gap in accordance with this code.

1503.4 Location. No gray water system or part thereof shall be located on a lot other than the lot that is the site of the building or structure that discharges the gray water, nor shall a gray water system or part thereof be located at a point having less than the minimum distances indicated in Table 1503.4.

🔧 These setbacks are designed to prevent gray water from causing potential problems, either physical (e.g. water damage to buildings from being too close), or environmental (e.g. polluting a lake with the excessive nutrients in gray water).

1503.5 Plot Plan Submission. No permit for a gray water system shall be issued until a plot plan with data satisfactory to the Authority Having Jurisdiction has been submitted and approved.

1503.6 Prohibited Location. Where there is insufficient lot area or inappropriate soil conditions for adequate absorption to prevent the ponding, surfacing, or runoff of the gray water, as determined by the Authority Having Jurisdiction, no gray water system shall be permitted. A gray water system is not permitted on a property in a geologically sensitive area as determined by the Authority Having Jurisdiction.

🔧 Geologically sensitive sites may include steep hill side lots, lots with water courses, and sites subject to landslide. This is not all inclusive and each State may have its own guidelines or refer to federal standards. High water levels may prohibit proper absorption as seen in **Figure 1503.6**.

ALTERNATE WATER SOURCES FOR NONPOTABLE APPLICATIONS

TABLE 1503.4
LOCATION OF GRAY WATER SYSTEM[7]

MINIMUM HORIZONTAL DISTANCE IN CLEAR REQUIRED FROM	SURGE TANK (feet)	SUBSURFACE AND SUBSOIL IRRIGATION FIELD AND MULCH BED (feet)
Building structures[1]	5[2, 9]	2[3, 8]
Property line adjoining private property	5	5[8]
Water supply wells[4]	50	100
Streams and lakes[4]	50	50[5]
Sewage pits or cesspools	5	5
Sewage disposal field[10]	5	4[6]
Septic tank	0	5
On-site domestic water service line	5	5
Pressurized public water main	10	10[7]

For SI units: 1 foot = 304.8 mm

Notes:
[1] Including porches and steps, whether covered or uncovered, breezeways, roofed carports, roofed patios, carports, covered walks, covered driveways, and similar structures or appurtenances.
[2] The distance shall be permitted to be reduced to 0 feet for aboveground tanks where first approved by the Authority Having Jurisdiction.
[3] Reference to a 45 degree (0.79 rad) angle from the foundation.
[4] Where special hazards are involved, the distance required shall be increased as directed by the Authority Having Jurisdiction.
[5] These minimum clear horizontal distances shall apply between the irrigation or disposal field and the ocean mean higher high tide line.
[6] Add 2 feet (610 mm) for each additional foot of depth more than 1 foot (305 mm) below the bottom of the drain line.
[7] For parallel construction or crossings, approval by the Authority Having Jurisdiction shall be required.
[8] The distance shall be permitted to be reduced to 1½ feet (457 mm) for drip and mulch basin irrigation systems.
[9] The distance shall be permitted to be reduced to 0 feet for surge tanks of 75 gallons (284 L) or less.
[10] Where irrigation or disposal fields are installed in the sloping ground, the minimum horizontal distance between a part of the distribution system and the ground surface shall be 15 feet (4572 mm).

FIGURE 1503.6
PROHIBITED LOCATION - INADEQUATE ABSORPTION

1503.7 Drawings and Specifications. The Authority Having Jurisdiction shall require the following information to be included with or in the plot plan before a permit is issued for a gray water system, or at a time during the construction thereof:

(1) Plot plan drawn to scale and completely dimensioned, showing lot lines and structures, direction and approximate slope of surface, location of present or proposed retaining walls, drainage channels, water supply lines, wells, paved areas and structures on the plot, number of bedrooms and plumbing fixtures in each structure, location of private sewage disposal system and expansion area or building sewer connecting to the public sewer, and location of the proposed gray water system.

Figure 1503.7a shows a sample plot plan with setbacks for a simple residential system.

(2) Details of construction necessary to ensure compliance with the requirements of this chapter, together with a full description of the complete installation, including installation methods, construction, and materials in accordance with the Authority Having Jurisdiction.

(3) Details for holding tanks shall include dimensions, structural calculations, bracings, and such other pertinent data as required.

(4) A log of soil formations and groundwater level as determined by test holes dug in proximity to proposed irrigation area, together with a statement of water absorption characteristics of the soil at the proposed site as determined by approved percolation tests.

The gray water system depends on the absorption of gray water by the native soil. Different compositions of soil have different absorption rates, or the time it takes the soil to absorb and disperse a certain amount of gray water. To determine the type of soil and the ground water level, test holes have to be dug into the location where the gray water irrigation/disposal trenches will be dug. Ground water levels, as illustrated in Figure 1503.7b, must be a minimum of 3 feet vertical distance below the irrigation pipe (see Section 1504.4).

Percolation tests that determine the exact absorption rate of the soil can be conducted in order to determine the required length and absorption area needed per Section 1504.2. This percolation test can be quite costly so in lieu of the percolation test, if the soil type is known, Table 1504.2, Design of Six Typical Soils, can be used.

Figure 1503.7c is an example of soil specifications for permit.

Exception: The Authority Having Jurisdiction shall permit the use of Table 1504.2 instead of percolation tests.

(5) Distance between the plot and surface waters such as lakes, ponds, rivers or streams, and the slope of the plot and the surface water, wherein close proximity.

1503.8 Procedure for Estimating Gray Water Discharge. Gray water systems shall be designed to distribute the total amount of estimated gray water on a daily basis. The water discharge for gray water systems

ALTERNATE WATER SOURCES FOR NONPOTABLE APPLICATIONS

shall be determined in accordance with Section 1503.8.1 or Section 1503.8.2.

1503.8.1 Single Family Dwellings and Multi-Family Dwellings. The gray water discharge for single family and multi-family dwellings shall be calculated by water use records, calculations of local daily per person interior water use, or the following procedure:

(1) The number of occupants of each dwelling unit shall be calculated as follows:

First Bedroom   2 occupants

Each additional bedroom   1 occupant

(2) The estimated gray water flows of each occupant shall be calculated as follows:

Showers, bathtubs, and lavatories   25 gallons (95 L) per day/occupant

Laundry 15 gallons (57 L) per day/occupant

(3) The total number of occupants shall be multiplied by the applicable estimated gray water discharge as provided above and the type of fixtures connected to the gray water system.

 The estimated flows in this section reflect long standing procedures for conservatively calculating the gray water discharge of plumbing fixtures and fixture fittings permitted by this Code. Other means for calculating gray water discharge for reduced volumes are permitted such as actual water use records and calculating the interior water use per person per day on a local level, which could take into account the use of high-efficiency fixtures. Check with

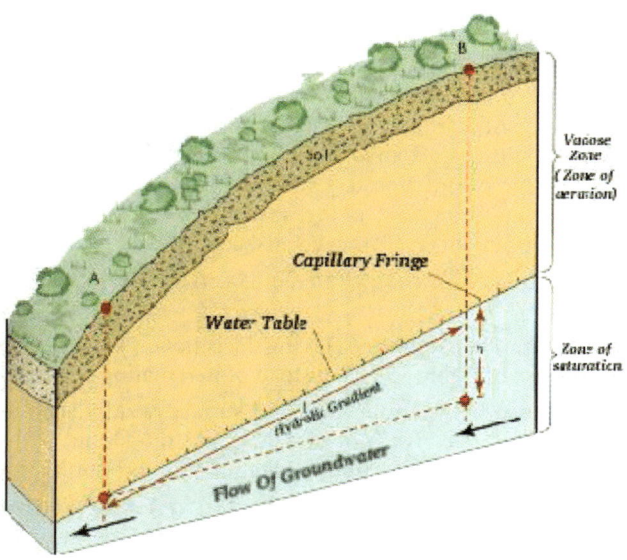

FIGURE 1502.7A
SAMPLE PLOT PLAN

FIGURE 1502.7B
ILLUSTRATION OF GROUND WATER LEVEL

FIGURE 1503.7C
SOIL TEST SPECIFICATION

the AHJ for any other permitted means for estimating gray water discharge especially where there is adoption of Green code requirements for water saving fixtures.

1503.8.2 Commercial, Industrial, and Institutional Occupancies. The gray water discharge for commercial, industrial, and institutional occupancies shall be calculated by utilizing the procedure in Section 1503.8.1, water use records or other documentation to estimate gray water discharge.

1503.9 Gray Water System Components. Gray water system components shall comply with Section 1503.9.1 through Section 1503.9.7.

1503.9.1 Surge Tanks. Where installed, surge tanks shall be in accordance with the following:

The vent on the surge tank is required and will prevent the tank from getting air bound (see **Figure 1503.9.1a**). Vents should be screened to prevent mosquito habitation. Mosquitoes would be able to travel to the exterior open air through the gray water discharge line.

It is highly unlikely to require a running trap and vent as shown in **Figures 1503.9.1a** and **1503.9.1b**. As long as the upstream fixtures discharging into the surge tank are trapped and vented to code, the running trap would not be needed. The vent on the surge tank will prevent the tank from getting air bound. Consult the AHJ to verify if this is required.

(1) Surge tanks shall be constructed of solid, durable materials not subject to excessive corrosion or decay and shall be watertight. Surge tanks constructed of steel shall be approved by the Authority Having Jurisdiction, provided such tanks are in accordance with approved applicable standards.

(2) Each surge tank shall be vented in accordance with this code. The vent size shall be determined based on the total gray water fixture units as outlined in this code.

(3) Each surge tank shall have an access opening with lockable gasketed covers or approved equivalent to allow for inspection and cleaning.

(4) Each surge tank shall have its rated capacity permanently marked on the unit. Also, a sign stating GRAY WATER, DANGER — UNSAFE WATER shall be permanently marked on the holding tank.

(5) Each surge tank shall have an overflow drain. The overflow drains shall have permanent connections to the building drain or building sewer, upstream of septic tanks. The overflow drain shall not be equipped with a shutoff valve.

(6) The overflow drain pipes shall not be less in size than the inlet pipe. Unions or equally effective fittings shall be provided for piping connected to the surge tank.

(7) Surge tank shall be structurally designed to withstand anticipated earth or other loads. Surge tank covers shall be capable of supporting an earth load of not less than 300 pounds per square foot (lb/ft^2) (1465 kg/m^2) where the tank is designed for underground installation.

(8) Where a surge tank is installed underground, the system shall be designed so that the tank overflow will gravity drain to the existing sewer line or septic tank. The tank shall be protected against sewer line backflow by a backwater valve installed in accordance with this code.

(9) Surge tanks shall be installed on dry, level, well-compacted soil where underground or on a level 3 inch (76 mm) thick concrete slab where aboveground.

(10) Surge tanks shall be anchored to prevent against overturning where installed aboveground. Underground tanks shall be ballasted, anchored, or otherwise secured, to prevent the tank from floating out of the ground where empty. The combined weight of the tank and hold down system shall meet or exceed the buoyancy forces of the tank.

See **Figures 1503.9.1a** and **1503.9.1b** for different configurations for surge tanks.

1503.9.2 Gray Water Pipe and Fitting Materials. Aboveground and underground building drainage and vent pipe and fittings for gray water systems shall comply with the requirements for aboveground and underground sanitary building drainage and vent pipe and fittings in this code. These materials shall extend not less than 2 feet (610 mm) outside the building.

1503.9.3 Subsoil Irrigation Field Materials. Subsoil irrigation field piping shall be constructed of perforated high-density polyethylene pipe, perforated ABS pipe, perforated PVC pipe, or other approved materials, provided that sufficient openings are available for distribution of the gray water into the trench area. Material, construction, and perforation of the pipe shall be in accordance with the appropriate absorption field drainage piping standards and shall be approved by the Authority Having Jurisdiction.

1503.9.4 Subsurface Irrigation Field and Mulch Basin Supply Line Materials. Materials for gray water piping outside the building shall be polyethylene or PVC. Drip feeder lines shall be PVC or polyethylene tubing.

High-density polyethylene (HDPE) pipe is becoming more popular and has excellent characteristics for gray water piping and tubing. The AHJ will have to approve this as an alternate material since it is not yet listed as an approved material in this section.

1503.9.5 Valves. Valves shall be accessible.

1503.9.6 Trap. Gray water piping discharging into the surge tank or having a direct connection to the sanitary drain or sewer piping shall be downstream of an approved water seal type trap(s). Where no such trap(s) exists, an approved vented running trap shall be installed upstream of the connection to protect the building from possible waste or sewer gases.

1503.9.7 Backwater Valve. A backwater valve shall be installed on gray water drain connections to the sanitary drain or sewer.

Backwater valves are designed to be installed in the horizontal position. The use of backwater valves in the

ALTERNATE WATER SOURCES FOR NONPOTABLE APPLICATIONS

vertical position is not recommended even though some gray water guidelines portray the valve in the vertical position. Backwater valves in the vertical position used for ejector pumps may not be applicable for gray water systems since they are designed to be normally closed when the pump is not active. Check with the manufacturer and the AHJ for approved uses for vertical backwater valves.

1504.0 Subsurface Irrigation System Zones.

See **Figures 1504.0a** and **1504.0b**.

1504.1 General. Irrigation or disposal fields shall be permitted to have one or more valved zones. Each zone shall be of a size to receive the gray water anticipated in that zone.

1504.2 Required Area of Subsurface Irrigation Fields, Subsoil Irrigation Fields, and Mulch Basins. The minimum effective irrigation area of subsurface irrigation fields, subsoil irrigation fields, and mulch basins shall be determined by Table 1504.2 for the type of soil found in the excavation, based upon a calculation of estimated gray water discharge under Section 1503.8. For a subsoil irrigation field, the area shall be equal to the aggregate length of the perforated pipe sections within the valved zone multiplied by the width of the proposed subsoil irrigation field.

1504.3 Determination of Maximum Absorption Capacity. The irrigation field and mulch basin size shall be based on the maximum absorption capacity of the soil and determined using Table 1504.2. For soils not listed in

TABLE 1504.2
DESIGN OF SIX TYPICAL SOILS

TYPE OF SOIL	MINIMUM SQUARE FEET OF IRRIGATION AREA PER 100 GALLONS OF ESTIMATED GRAY WATER DISCHARGE PER DAY	MAXIMUM ABSORPTION CAPACITY IN GALLONS PER SQUARE FOOT OF IRRIGATION/ LEACHING AREA FOR A 24-HOUR PERIOD
Coarse sand or gravel	20	5.0
Fine sand	25	4.0
Sandy loam	40	2.5
Sandy clay	60	1.7
Clay with considerable sand or gravel	90	1.1
Clay with small amounts of sand or gravel	120	0.8

For SI units: 1 square foot = 0.0929 m^2, 1 gallon per day = 0.000043 L/s

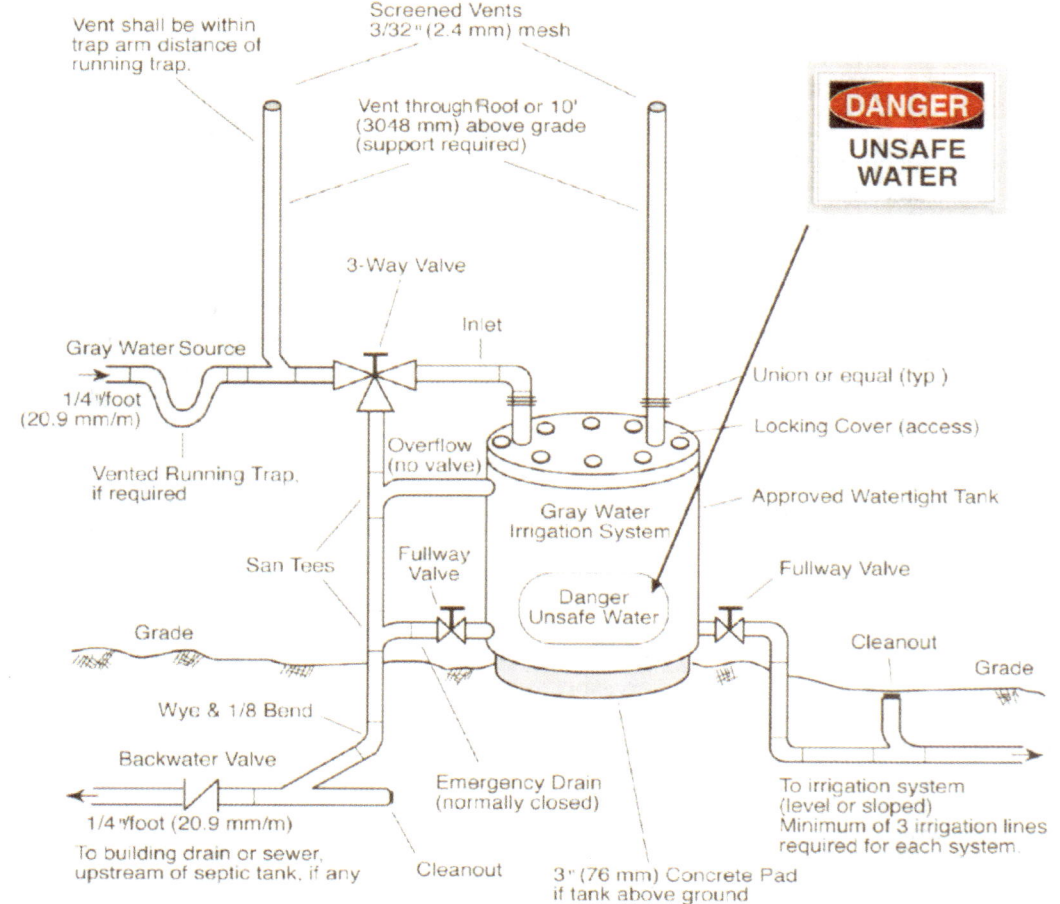

FIGURE 1503.9.1A
GRAY WATER SURGE TANK - GRAVITY

ALTERNATE WATER SOURCES FOR NONPOTABLE APPLICATIONS

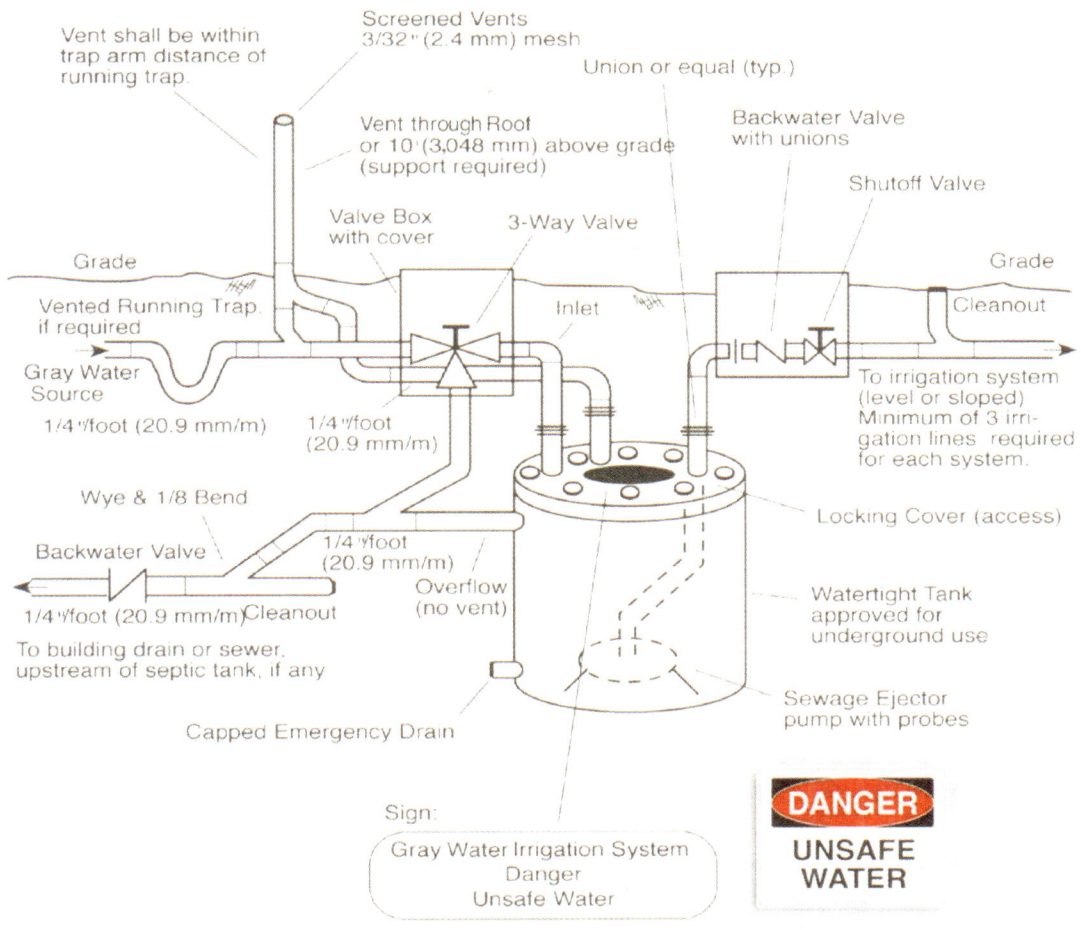

FIGURE 1503.9.1B
GRAY WATER SURGE TANK WITH PUMP

Table 1504.2, the maximum absorption capacity for the proposed site shall be determined by percolation tests or another method acceptable to the Authority Having Jurisdiction. A gray water system shall not be permitted, where the percolation test shows the absorption capacity of the soil is unable to accommodate the maximum discharge of the proposed gray water irrigation system.

1504.4 Groundwater Level. No excavation for an irrigation field, disposal field, or mulch basin shall extend within 3 feet (914 mm) vertical of the highest known seasonal groundwater level, nor to a depth where gray water contaminates the groundwater or surface water. The applicant shall supply evidence of groundwater depth to the satisfaction of the Authority Having Jurisdiction.

1504.5 Subsurface and Subsoil Irrigation Field Design and Construction. Subsurface and subsoil irrigation field design and construction shall be in accordance with Section 1504.5.1 through Section 1504.7.3. Where a gray water irrigation system design is predicated on soil tests, the subsurface or subsoil irrigation field or mulch basin shall be installed at the same location and depth as the tested area.

1504.5.1 Subsurface Irrigation Field. A subsurface irrigation field shall comply with Section 1504.5.2 through Section 1504.5.7.

1504.5.2 Minimum Depth. Supply piping, including drip feeders, shall be not less than 2 inches (51 mm) below finished grade and covered with mulch or soil.

🔧 This provision ensures that gray water discharges below the finished grade.

1504.5.3 Filter. Not less than 140 mesh (105 microns) filter with a capacity of 25 gallons per minute (gpm) (1.58 L/s), or equivalent shall be installed. Where a filter backwash is installed, the backwash and flush discharge shall discharge into the building sewer or private sewage disposal system. Filter backwash and flush water shall not be used.

🔧 These filters are easily obtained and are needed to prevent the emitters from either getting clogged or from discharging unwanted solids into the irrigation zone (see **Figure 1504.5.3**).

1504.5.4 Emitter Size. Emitters shall be installed in accordance with the manufacturer's installation instructions. Emitters shall have a flow path of not less than 1200 microns (μ) (1200 μm) and shall not have a coefficient of manufacturing variation (Cv) exceeding 7 percent. Irriga-

ALTERNATE WATER SOURCES FOR NONPOTABLE APPLICATIONS

tion system design shall be such that emitter flow variation shall not exceed 10 percent.

🔧 Emitter flow data and specifications can be obtained through the manufacturer or the International Center for Water Technology (ICWT). For lab inquiries see the ICWT site at www.icwt.net/TestLab.htm.

1504.5.5 Number of Emitters. The minimum number of emitters and the maximum discharge of each emitter in an irrigation field shall be in accordance with Table 1504.5.5.

1504.5.6 Controls. The system design shall provide user controls, such as valves, switches, timers, and other controllers, to rotate the distribution of gray water between irrigation zones.

1504.5.7 Maximum Pressure. Where pressure at the discharge side of the pump exceeds 20 pounds-force per square inch (psi) (138 kPa), a pressure-reducing valve able to maintain downstream pressure not exceeding 20 psi (138 kPa) shall be installed downstream from the pump and before an emission device.

1504.6 Mulch Basin Design and Construction. A mulch basin shall comply with Section 1504.6.1 through Section 1504.6.4.

1504.6.1 Single Family and Multi-Family Dwellings. The gray water discharge to a mulch basin is limited to single family and multi-family dwellings.

1504.6.2 Size. Mulch basins shall be of sufficient size to accommodate peak flow rates and distribute the total amount of estimated gray water on a daily basis without surfacing, ponding or runoff. Mulch basins shall have a depth of not less than 10 inches (254 mm) below finished grade. The mulch basin size shall be based on the maximum absorption capacity of the soil and determined using Table 1504.2.

1504.6.3 Minimum Depth. Gray water supply piping, including drip feeders, shall be not less than 2 inches (51 mm) below finished grade and covered with mulch.

1504.6.4 Maintenance. The mulch basin shall be maintained periodically to retain the required depth and area, and to replenish the required mulch cover.

🔧 Below grade gray water outlets must be protected from root intrusion of plants. The most common way to do this is to create an air gap around the outlet by using some sort of solid shield, like an adapted irrigation valve box. If gray water is required to be covered in mulch it will encourage systems that are prone to clogging, or encourage people to bury their irrigation valve boxes and have a harder time doing maintenance. A simple fix would be to include a "solid shield" to the allowable coverings (see **Figure 1504.6.4**).

1504.7 Subsoil Irrigation Field. Subsoil irrigation fields shall comply with Section 1504.7.1 through Section 1504.7.3.

🔧 Subsoil irrigation fields are not as efficient as a mulch or emitter system due to the required depth of the discharge. This system acts more as a disposal leech field than an irrigation system.

1504.7.1 Minimum Pipe Size. Subsoil irrigation field distribution piping shall be not less than 3 inches (80 mm) diameter.

🔧 See **Figure 1504.7.2** for an example of a perforated subsoil irrigation distribution pipe.

1504.7.2 Filter Material and Backfill. Filter material, clean stone, gravel, slag, or similar material acceptable to the Authority Having Jurisdiction, varying in size from ¾ of an inch (19.1 mm) to 2½ inches (64 mm) shall be placed in

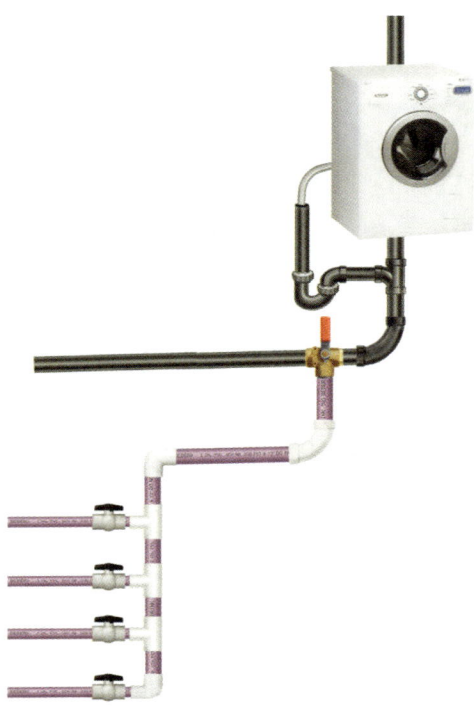

**FIGURE 1504.0A
VALVED ZONES
(Courtesy of Building in California)**

**TABLE 1504.5.5
SUBSURFACE IRRIGATION DESIGN
CRITERIA FOR SIX TYPICAL SOILS**

TYPE OF SOIL	MAXIMUM EMITTER DISCHARGE (gallons per day)	MINIMUM NUMBER OF EMITTERS PER GALLON OF ESTIMATED GRAY WATER DISCHARGE PER DAY* (gallons per day)
Sand	1.8	0.6
Sandy loam	1.4	0.7
Loam	1.2	0.9
Clay loam	0.9	1.1
Silty clay	0.6	1.6
Clay	0.5	2.0

For SI units: 1 gallon per day = 0.000043 L/s

* The estimated gray water discharge per day shall be determined in accordance with Section 1503.8 of this code.

ALTERNATE WATER SOURCES FOR NONPOTABLE APPLICATIONS

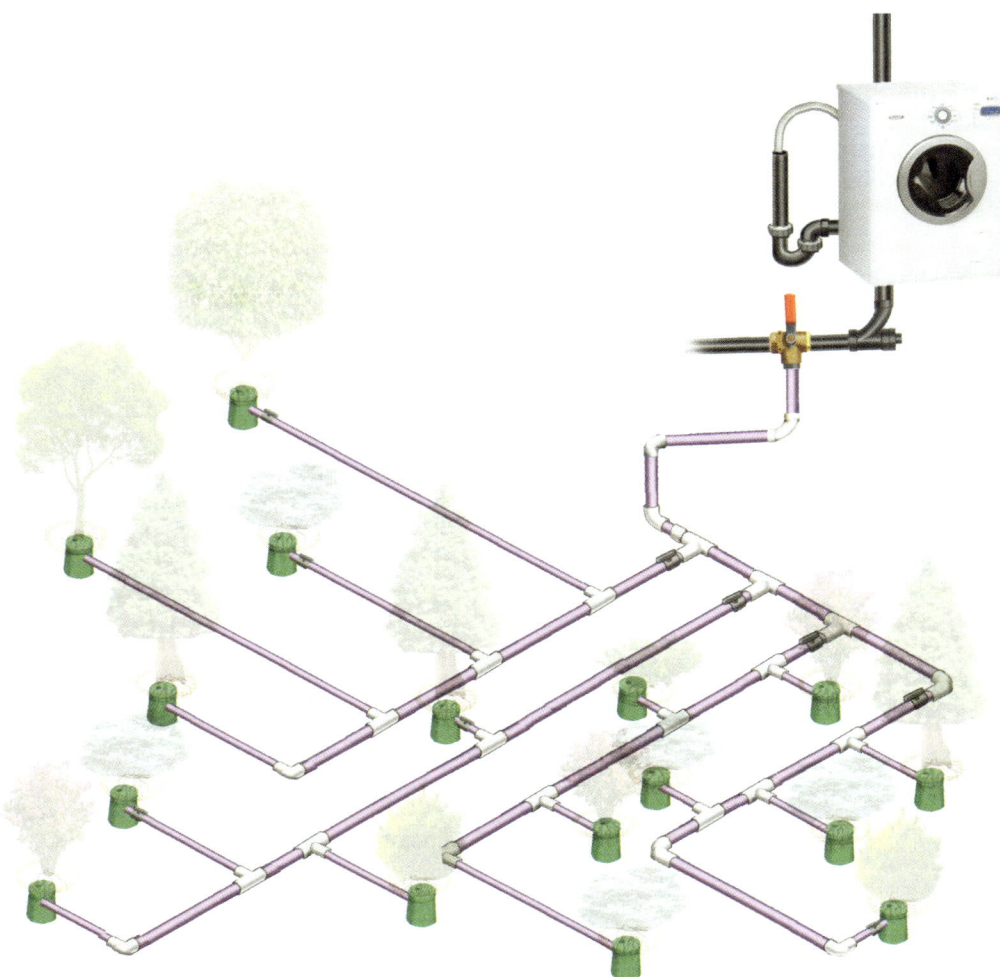

FIGURE 1504.0B
EXAMPLE OF VALVED ZONES
(Courtesy of Building in California)

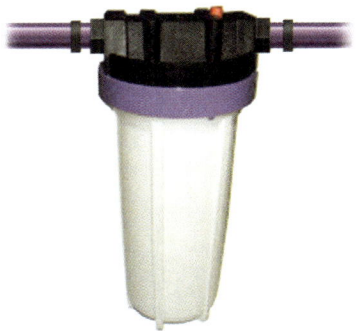

FIGURE 1504.5.3
FILTER
(Courtesy of Building in California)

the trench to the depth and grade in accordance with Table 1504.7.3. The perforated section of subsoil irrigation field distribution piping shall be laid on the filter material in an approved manner. The perforated section shall then be covered with filter material to the minimum depth in accordance with Table 1504.7.3. The filter material shall then be covered with porous material to prevent the closure of voids with earth backfill. No earth backfill shall be placed over the filter material cover until after inspection and acceptance.

🔧 See **Figure 1504.7.2**.

1504.7.3 Subsoil Irrigation Field Construction.
Subsoil irrigation fields shall be constructed in accordance with Table 1504.7.3. Where necessary on sloping ground to prevent excessive line slopes, irrigation lines shall be stepped. The lines between each horizontal leaching section shall be made with approved watertight joints and installed on the natural or unfilled ground.

🔧 This provision is extremely important to prevent gray water from ponding and preventing runoff (see **Figure 1504.7.3**).

1504.8 Gray Water System Color and Marking Information.
Pressurized gray water distribution systems shall be identified as containing nonpotable water in accordance with Section 601.3 of this code.

1504.9 Other Collection and Distribution Systems.
Other collection and distribution systems shall be approved

ALTERNATE WATER SOURCES FOR NONPOTABLE APPLICATIONS

by the local Authority Having Jurisdiction, as allowed by Section 301.3 of this code.

1504.9.1 Higher Requirements. Nothing contained in this chapter shall be construed to prevent the Authority Having Jurisdiction from requiring compliance with higher requirements than those contained herein, where such higher requirements are essential to maintaining a safe and sanitary condition.

1504.10 Testing. Building drains and vents for gray water systems shall be tested in accordance with this code. Surge tanks shall be filled with water to the overflow line prior to and during the inspection. Seams and joints shall be left exposed, and the tank shall remain watertight. A flow test shall be performed through the system to the point of gray water discharge. Lines and components shall be watertight up to the point of the irrigation perforated and drip lines.

1504.11 Maintenance. Gray water systems and components shall be maintained in accordance with Table 1501.5.

1505.0 Reclaimed (Recycled) Water Systems.

1505.1 General. The provisions of this section shall apply to the installation, construction, alteration, and repair of reclaimed (recycled) water systems intended to supply uses such as water closets, urinals, trap primers for floor drains and floor sinks, aboveground and subsurface irrigation, industrial or commercial cooling or air conditioning and other uses approved by the Authority Having Jurisdiction.

🔧 In most cases, allowable uses of reclaimed water will be defined by the AHJ such as the State and Water Purveyor.

1505.2 Permit. It shall be unlawful for a person to construct, install, alter, or cause to be constructed, installed, or altered a reclaimed (recycled) water system within a building or on premises without first obtaining a permit to do such work from the Authority Having Jurisdiction.

1505.2.1 Plumbing Plan Submission. No permit for a reclaimed (recycled) water system shall be issued until complete plumbing plans, with data satisfactory to the Authority Having Jurisdiction, have been submitted and approved.

1505.3 System Changes. No changes or connections shall be made to either the reclaimed (recycled) water system or the potable water system within site containing a reclaimed (recycled) water system without approval by the Authority Having Jurisdiction.

1505.4 Connections to Potable or Reclaimed (Recycled) Water Systems. Reclaimed (recycled) water systems shall have no connection to a potable water supply or alternate water source system. Potable water is permitted to be used as makeup water for a reclaimed

TABLE 1504.7.3
SUBSOIL IRRIGATION FIELD CONSTRUCTION

DESCRIPTION	MINIMUM	MAXIMUM
Number of drain lines per valved zone	1	–
Length of each perforated line	–	100 feet
Bottom width of trench	12 inches	18 inches
Spacing of lines, center to center	4 feet	–
Depth of earth covers of lines	10 inches	–
Depth of filter material cover of lines	2 inches	–
Depth of filter material beneath lines	3 inches	–
Grade of perforated lines level	level	3 inches per 100 feet

For SI units: 1 inch = 25.4 mm, 1 foot = 304.8 mm, 1 inch per foot = 83.3 mm/m

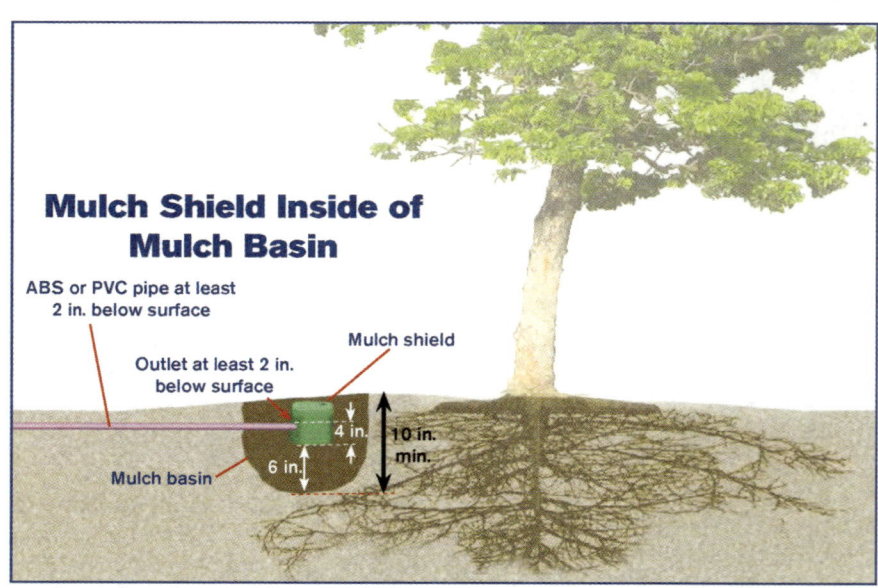

Roots of a real tree would extend under basin and outside of drip line by many feet.

FIGURE 1504.6.4
MULCH BASIN
(Courtesy of Building in California)

ALTERNATE WATER SOURCES FOR NONPOTABLE APPLICATIONS

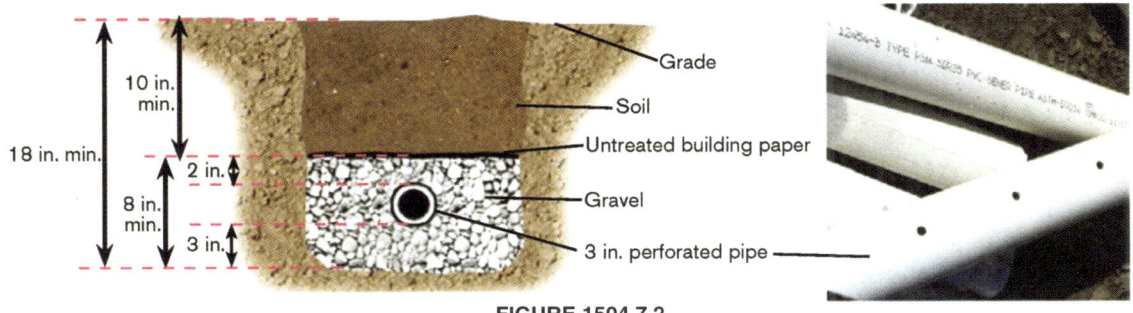

**FIGURE 1504.7.2
SUBSOIL IRRIGATION**
(Courtesy of Building in California)

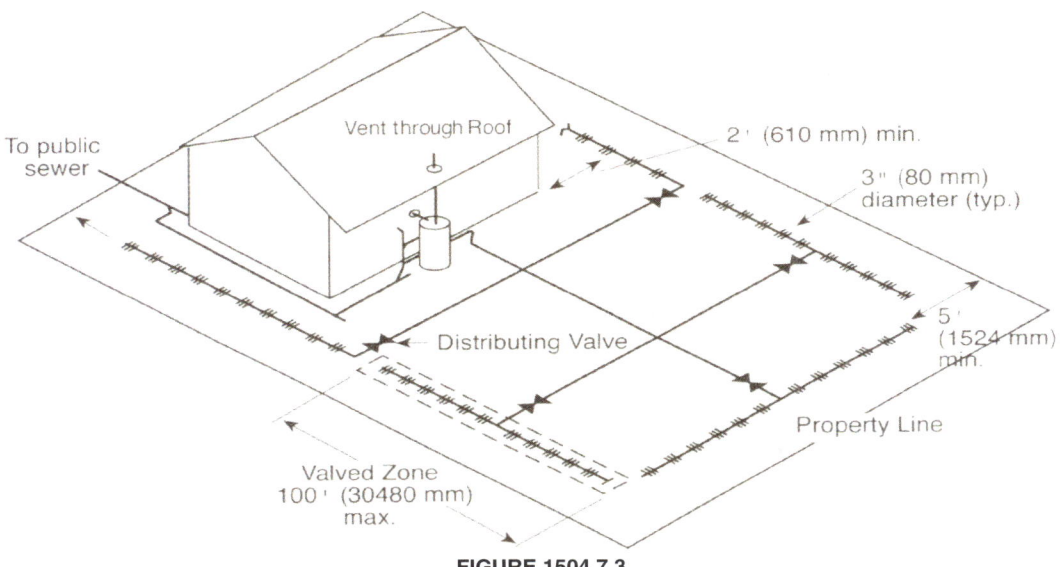

**FIGURE 1504.7.3
TYPICAL IRRIGATION LAYOUT**

(recycled) water storage tank provided the water supply inlet is protected by an air gap or reduced-pressure principle backflow preventer in accordance with this code.

1505.5 Initial Cross-Connection Test. A cross-connection test is required in accordance with Section 1502.3. Before the building is occupied or the system is activated, the installer shall perform the initial cross-connection test in the presence of the Authority Having Jurisdiction and other authorities having jurisdiction. The test shall be ruled successful by the Authority Having Jurisdiction before final approval is granted.

1505.6 Reclaimed (Recycled) Water System Materials. Reclaimed (recycled) water supply and distribution system materials shall comply with the requirements of this code for potable water supply and distribution systems unless otherwise provided for in this section.

1505.7 Reclaimed (Recycled) Water System Color and Marking Information. Reclaimed (recycled) water systems shall have a colored background and marking information in accordance with Section 601.3 of this code.

According to Section 601.3, when dual (potable and nonpotable) systems are installed in the same building, the potable water piping shall be labeled a green background with white lettering. The background colors and required information shall be indicated every 20 feet and shall be labeled in each room and visible from the floor. (See the Reclaimed (Recycled) Water System Color and Marking Information Quick Reference Table following this section.) Many manufactures are already supplying pre-lettered and colored piping for most applications.

ALL ALTERNATE WATER SYSTEMS SHALL HAVE A PURPLE (Pantone color No. 512, 522C, or equivalent) background with upper case lettering and shall be field or factory marked as follows in the Quick Reference Table (see **Figure 1505.7a** and **1505.7b**).

Figure 1503.2a shows the color requirements for a gray water subsoil irrigation system. Although it was not the intention of this code section to require the applicable color and markings for existing landscape irrigation piping when it converts to an alternate water source, the language does not include an exception for pre-existing uncolored and unmarked piping. Until this is further clarified in the code, check with the AHJ for existing piping requirements when converting to an alternate water source.

ALTERNATE WATER SOURCES FOR NONPOTABLE APPLICATIONS

Reclaimed (Recycled) Water System Color and Marking Information Quick Reference Table		
Gray water	"CAUTION: NON-POTABLE GRAY WATER, DO NOT DRINK"	Yellow letters (pantone 108 or equivalent)
Reclaimed	"CAUTION: NON-POTABLE RECLAIMED WATER, DO NOT DRINK"	Black lettering
On-site treated water	"CAUTION: ON-SITE TREATED NON-POTABLE WATER, DO NOT DRINK"	Yellow letters (pantone 108 or equivalent
Rainwater systems	"CAUTION: NONPOTABLE RAIN-WATER, DO NOT DRINK"	Yellow letters (pantone 108 or equivalent

1505.8 Valves. Valves, except fixture supply control valves, shall be equipped with a locking feature.

1505.9 Hose Bibbs. Hose bibbs shall not be allowed on reclaimed (recycled) water piping systems located in areas accessible to the public. Access to reclaimed (recycled) water at points in the system accessible to the public shall be through a quick-disconnect device that differs from those installed on the potable water system. Hose bibbs supplying reclaimed (recycled) water shall be marked with the words: "CAUTION: NONPOTABLE RECLAIMED WATER, DO NOT DRINK," and the symbol in Figure 1505.9.

1505.10 Required Appurtenances. The reclaimed (recycled) water system and the potable water system within the building shall be provided with the required appurtenances (e.g., valves, air/vacuum relief valves, etc.) to allow for deactivation or drainage as required for a cross-connection test in accordance with Section 1502.3.

1505.11 Same Trench as Potable Water Pipes. Reclaimed (recycled) water pipes shall be permitted to be run or laid in the same trench as potable water pipes with 12 inches (305 mm) minimum vertical and horizontal separation where both pipe materials are approved for use within a building. Where piping materials do not meet this requirement, the minimum horizontal separation shall be increased to 60 inches (1524 mm). The potable water piping shall be installed at an elevation above the reclaimed (recycled) water piping. Reclaimed (recycled) water pipes laid in the same trench or crossing building sewer or drainage piping shall be installed in accordance with this code for potable water piping.

🔧 This provision is required to prevent possible contamination of the potable water system.

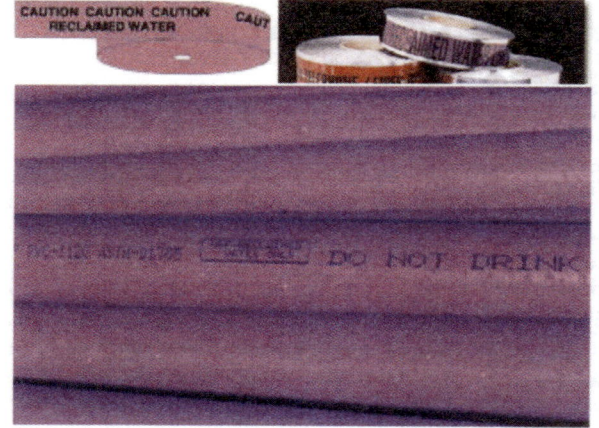

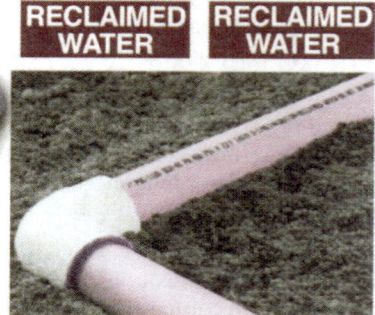

**FIGURE 1505.7A
COLOR AND MARKING FOR RECLAIMED WATER PIPE**

ALTERNATE WATER SOURCES FOR NONPOTABLE APPLICATIONS

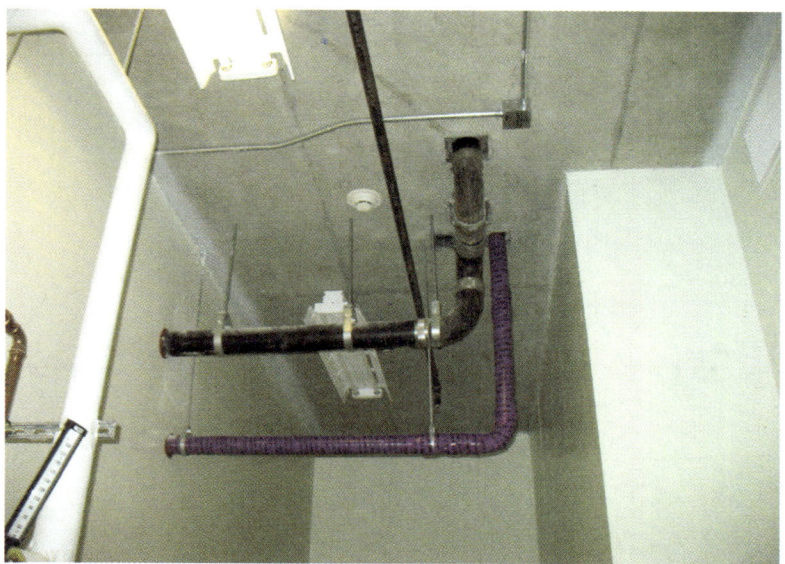

**FIGURE 1505.7B
ALTERNATE WATER MARKINGS**

1505.12 Signs. Signs in rooms and water closet tanks in buildings using reclaimed (recycled) water shall be in accordance with Section 1501.9 and Section 1501.9.1.

1505.13 Inspection and Testing. Reclaimed (recycled) water systems shall be inspected and tested in accordance with Section 1502.1.

1506.0 On-Site Treated Nonpotable Water Systems.

1506.1 General. The provisions of this section shall apply to the installation, construction, alteration, and repair of on-site treated nonpotable water systems intended to supply uses such as water closets, urinals, trap primers for floor drains and floor sinks, above and belowground irrigation, and other uses approved by the Authority Having Jurisdiction.

The sources for on-site treated nonpotable water include but are not limited to gray water; black water; rainwater; stormwater; reclaimed (recycled) water; swimming pool backwash; condensate; cooling tower blow-down water; foundation drainage; fluid cooler discharge water; food steamer discharge water; combination oven discharge water; industrial process water; fire pump test water; and dry weather runoff. This is intended to cover all potential on-site water reuse possibilities.

1506.2 Plumbing Plan Submission. No permit for an on-site treated nonpotable water system shall be issued until complete plumbing plans, with data satisfactory to the Authority Having Jurisdiction, have been submitted and approved.

The system shown in **Figures 1506.2a** and **1506.2b** is for an on-site treated water system for trap primers, toilets, and urinals. Ozone sterilization is rarely used because of the expense and the corrosive properties of ozone toward pipe materials. The preferred method of sterilization is ultraviolet radiation or chlorine.

1506.3 System Changes. No changes or connections shall be made to either the on-site treated nonpotable water system or the potable water system within a site containing an on-site treated nonpotable water system without approval by the Authority Having Jurisdiction.

1506.4 Connections to Potable or Reclaimed (Recycled) Water Systems. On-site treated nonpotable water systems shall have no connection to a potable water supply or reclaimed (recycled) water source system. Potable or reclaimed (recycled) water is permitted to be used as makeup water for a non-pressurized storage tank provided the makeup water supply is protected by an air gap in accordance with this code.

The level of hazard for potential cross connection makes it imperative that an air gap be provided. A Reduced Pressure Backflow device cannot be substituted for this non-connection. It isn't as fool-proof as an airgap and eliminates the need for annual testing of the device.

1506.5 Initial Cross-Connection Test. A cross-connection test is required in accordance with Section 1502.3. Before the building is occupied or the system is activated, the installer shall perform the initial cross-connection test in the presence of the Authority Having Jurisdiction and other authorities having jurisdiction. The test shall be ruled successful by the Authority Having Jurisdiction before final approval is granted.

1506.6 On-Site Treated Nonpotable Water System Materials. On-site treated nonpotable water supply, and distribution system materials shall comply with the requirements of this code for potable water supply and distribution systems unless otherwise provided for in this section.

1506.7 On-Site Treated Nonpotable Water Devices and Systems. Devices or equipment used to treat on-site treated nonpotable water to maintain the minimum water

ALTERNATE WATER SOURCES FOR NONPOTABLE APPLICATIONS

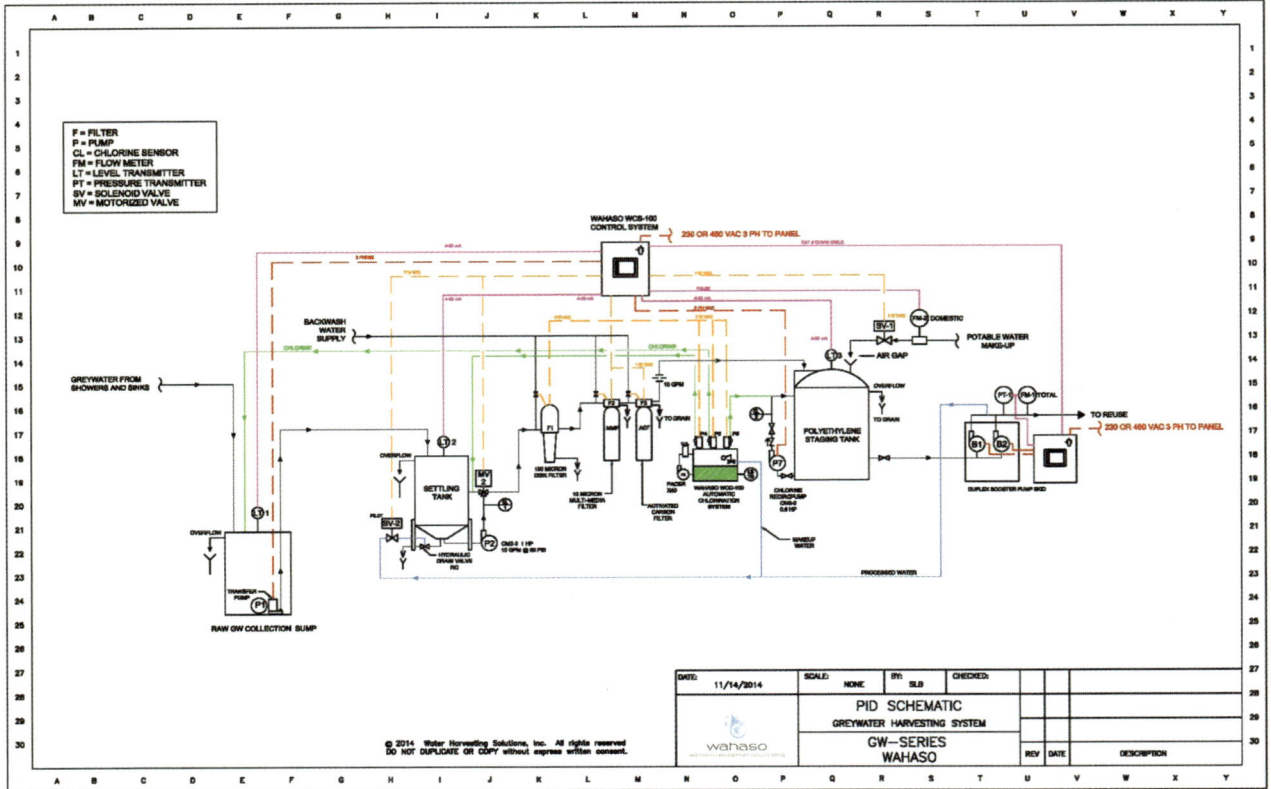

FIGURE 1506.2A
ON-SITE TREATED WATER SYSTEM SCHEMATIC
(Courtesy of Wahaso - Water Harvesting Solutions)

quality requirements determined by the Authority Having Jurisdiction shall be listed or labeled (third-party certified) by a listing agency (accredited conformity assessment body) or approved for the intended application. Devices or equipment used to treat on-site treated nonpotable water for use in the water closet and urinal flushing, surface irrigation, and similar applications shall comply with NSF 350 or approved by the Authority Having Jurisdiction.

1506.8 On-Site Treated Nonpotable Water System Color and Marking Information. On-site treated water systems shall have a colored background and marking information in accordance with Section 601.3 of this code.

1506.9 Design and Installation. The design and installation of on-site treated nonpotable systems shall be in accordance with Section 1506.9.1 through Section 1506.9.5.

1506.9.1 Listing Terms and Installation Instructions. On-site treated nonpotable water systems shall be installed in accordance with the terms of its listing and the manufacturer's installation instructions.

1506.9.2 Minimum Water Quality. On-site treated nonpotable water supplied to toilets or urinals or for other uses in which it is sprayed or exposed shall be disinfected. Acceptable disinfection methods shall include chlorination, ultraviolet sterilization, ozone, or other methods as approved by the Authority Having Jurisdiction. The minimum water quality for on-site treated nonpotable water systems shall meet the applicable water quality requirements for the intended applications as determined by the public health Authority Having Jurisdiction.

1506.9.3 Deactivation and Drainage. The on-site treated nonpotable water system and the potable water system within the building shall be provided with the required appurtenances (e.g., valves, air/vacuum relief valves, etc.) to allow for deactivation or drainage as required for a cross-connection test in accordance with Section 1502.3.

1506.9.4 Near Underground Potable Water Pipe. On-site treated nonpotable water pipes shall be permitted to be run or laid in the same trench as potable water pipes with a 12 inch (305 mm) minimum vertical and horizontal separation where both pipe materials are approved for use within a building. Where piping materials do not meet this requirement the minimum separation shall be increased to 60 inches (1524 mm). The potable water piping shall be installed at an elevation above the on-site treated nonpotable water piping.

It is important to note that BOTH materials must be approved for use within the building AND separation is required. This differs from the sewer and water in the same trench requirements in section 720.0.

**FIGURE 1506.2B
ON-SITE TREATED WATER SYSTEM
(Courtesy of Wahaso - Water Harvesting Solutions)**

1506.9.5 Required Filters. A filter permitting the passage of particulates no larger than 100 microns (100 µm) shall be provided for on-site treated nonpotable water supplied to water closets, urinals, trap primers, and drip irrigation system.

There are numerous filters and screens readily manufactured that can handle large flow rates. The simplest and least expensive are cartridge filters that fit into a plastic housing (see **Figure 1504.5.3**). There are also inexpensive newer cartridges that are made of polypropylene string.

1506.10 Valves. Valves, except fixture supply control valves, shall be equipped with a locking feature.

1506.11 Signs. Signs in buildings using on-site treated nonpotable water shall comply with Section 1501.9 and Section 1501.9.1.

1506.12 Inspection and Testing. On-site treated nonpotable water systems shall be inspected and tested in accordance with Section 1502.1.

CHAPTER 16
NONPOTABLE RAINWATER CATCHMENT SYSTEMS

1601.0 General.

1601.1 Applicability. The provisions of this chapter shall apply to the installation, construction, alteration, and repair of nonpotable rainwater catchment systems.

Rainwater catchment is a viable alternate water source used around the world to supplement the existing water supply. Nonpotable rainwater catchment systems are included in the Uniform Plumbing Code (UPC) since water supplied from these systems are used for plumbing fixtures, appliances, and other uses typically supplied by the plumbing system. Being regulated under the standards of the plumbing code, rainwater catchment systems will safely provide nonpotable water to end uses not intended for potable drinking water, sanitation, or bathing (see Section 1602.1).

1601.1.1 Allowable Use of Alternate Water. Where approved or required by the Authority Having Jurisdiction, rainwater shall be permitted to be used instead of potable water for the applications identified in this chapter.

1601.2 System Design. Rainwater catchment systems shall be designed in accordance with this chapter by a person registered or licensed to perform plumbing design work or who demonstrates competency to design the rainwater catchment system as required by the Authority Having Jurisdiction. Components, piping, and fittings used in a rainwater catchment system shall be listed.

Exceptions:

(1) A person registered or licensed to perform plumbing design work is not required to design rainwater catchment systems used for irrigation with a maximum storage capacity of 360 gallons (1363 L).

(2) A person registered or licensed to perform plumbing design work is not required to design rainwater catchment systems for single family dwellings where outlets, piping, and system components are located on the exterior of the building.

1601.3 Permit. It shall be unlawful for a person to construct, install, alter, or cause to be constructed, installed, or altered a rainwater catchment system in a building or on a premise without first obtaining a permit to do such work from the Authority Having Jurisdiction.

Exceptions:

(1) A permit is not required for exterior rainwater catchment systems used for outdoor drip and subsurface irrigation with a maximum storage capacity of 360 gallons (1363 L).

(2) A plumbing permit is not required for rainwater catchment systems for single family dwellings where outlets, piping, and system components are located on the exterior of the building. This does not exempt the need for permits where required for electrical connections, tank supports, or enclosures.

A permit is neither required for exterior rainwater catchment systems used for outdoor drip and subsurface irrigation with a maximum storage capacity of 360 gallons, nor for single family dwellings where all outlets, piping, and components are located exterior of the building. The exceptions intend not to regulate rain barrels and small systems that do not interface with building plumbing systems. The 360 gallons was determined by the consideration of a residential dwelling having a 90 gallon rain barrel at each corner of the house (4 x 90 gallons).

1601.4 Component Identification. System components shall be properly identified as to the manufacturer.

1601.5 Maintenance and Inspection. Rainwater catchment systems and components shall be inspected and maintained in accordance with Section 1601.5.1 through Section 1601.5.3.

1601.5.1 Frequency. Rainwater catchment systems and components shall be inspected and maintained in accordance with Table 1601.5 unless more frequent inspection and maintenance are required by the manufacturer.

1601.5.2 Maintenance Log. A maintenance log for rainwater catchment systems is required to have a permit in accordance with Section 1601.3 and shall be maintained by the property owner and be available for inspection. The property owner or designated appointee shall ensure that a record of testing, inspection, and maintenance in accordance with Table 1601.5 is maintained in the log. The log will indicate the frequency of inspection and maintenance for each system.

1601.5.3 Maintenance Responsibility. The required maintenance and inspection of rainwater catchment systems shall be the responsibility of the property owner unless otherwise required by the Authority Having Jurisdiction.

1601.6 Operation and Maintenance Manual. An operation and maintenance manual for rainwater catchment systems required to have a permit in accordance with Section 1601.3, shall be supplied to the building owner by the system designer. The operating and maintenance manual shall include the following:

(1) Detailed diagram of the entire system and the location of system components.

(2) Instructions for operating and maintaining the system.

(3) Details on maintaining the required water quality as determined by the Authority Having Jurisdiction.

(4) Details on deactivating the system for maintenance, repair, or other purposes.

(5) Applicable testing, inspection, and maintenance frequencies in accordance with Table 1601.5.

(6) A method of contacting the manufacturer(s).

1601.7 Minimum Water Quality Requirements. The minimum water quality for rainwater catchment systems shall

TABLE 1601.5
MINIMUM ALTERNATE WATER SOURCE TESTING, INSPECTION, AND MAINTENANCE FREQUENCY

DESCRIPTION	MINIMUM FREQUENCY
Inspect and clean filters and screens, and replace (where necessary).	Every 3 months
Inspect and verify that disinfection, filters, and water quality treatment devices and systems are operational and maintaining minimum water quality requirements as determined by the Authority Having Jurisdiction.	In accordance with manufacturer's instructions and the Authority Having Jurisdiction.
Inspect and clear debris from rainwater gutters, downspouts, and roof washers.	Every 6 months
Inspect and clear debris from the roof or another aboveground rainwater collection surfaces.	Every 6 months
Remove tree branches and vegetation overhanging a roof or other aboveground rainwater collection surfaces.	As needed
Inspect pumps and verify operation.	After initial installation and every 12 months thereafter
Inspect valves and verify operation.	After initial installation and every 12 months thereafter
Inspect pressure tanks and verify operation.	After initial installation and every 12 months thereafter
Clear debris from and inspect storage tanks, locking devices, and verify operation.	After initial installation and every 12 months thereafter
Inspect caution labels and marking.	After initial installation and every 12 months thereafter
Cross-connection inspection and test*	After initial installation and every 12 months thereafter
Test water quality of rainwater catchment systems required by Section 1602.9.6 to maintain a minimum water quality	Every 12 months. After system renovation or repair.

* The cross-connection test shall be performed in the presence of the Authority Having Jurisdiction in accordance with the requirements of this chapter.

comply with the applicable water quality requirements for the intended application as determined by the Authority Having Jurisdiction. Water quality for nonpotable rainwater catchment systems shall comply with Section 1602.9.6.

Exceptions:
(1) Water treatment is not required for rainwater catchment systems used for aboveground irrigation with a maximum storage capacity of 360 gallons (1363 L).
(2) Water treatment is not required for rainwater catchment systems used for subsurface or drip irrigation.

1601.8 Material Compatibility. Rainwater catchment systems shall be constructed of materials that are compatible with the type of pipe and fitting materials, water treatment, and water conditions in the system.

1601.9 System Controls. Controls for pumps, valves, and other devices that contain mercury that come in contact with rainwater supply shall not be permitted.

1601.10 Separation Requirements. Underground rainwater catchment service piping shall be separated from the building sewer in accordance with Section 609.2. Treated nonpotable water pipes shall be permitted to be run or laid in the same trench as potable water pipes with a 12 inch (305 mm) minimum vertical and horizontal separation where both pipe materials are approved for use within a building. Where horizontal piping materials do not meet this requirement, the minimum separation shall be increased to 60 inches (1524 mm). The potable water piping shall be installed at an elevation above the treated nonpotable water piping.

1601.11 Abandonment. Rainwater catchment systems that are no longer in use, or fail to be maintained in accordance with Section 1601.5, shall be abandoned. Abandonment shall comply with Section 1601.11.1 and Section 1601.11.2.

1601.11.1 General. An abandoned system or part thereof covered under the scope of this chapter shall be disconnected from remaining systems, drained, plugged, and capped in an approved manner.

1601.11.2 Underground Tank. An underground water storage tank that has been abandoned or otherwise discontinued from use in a system covered under the scope of this chapter shall be completely drained and filled with earth, sand, gravel, concrete, or other approved material or removed in a manner satisfactory to the Authority Having Jurisdiction.

1601.12 Sizing. Unless otherwise provided for in this chapter, rainwater catchment piping shall be sized in accordance with Chapter 6 for sizing potable water piping.

1602.0 Nonpotable Rainwater Catchment Systems.

1602.1 General. The installation, construction, alteration, and repair of rainwater catchments systems intended to supply uses such as water closets, urinals, trap primers for floor drains and floor sinks, irrigation, industrial processes, water features, cooling tower makeup and other uses shall be approved by the Authority Having Jurisdiction.

1602.2 Plumbing Plan Submission. No permit for a rainwater catchment system shall be issued until complete plumbing plans, with data satisfactory to the Authority Having Jurisdiction, have been submitted and approved.

1602.3 System Changes. No changes or connections shall be made to either the rainwater catchment system or the potable water system within a site containing a rainwater catchment system requiring a permit without approval by the Authority Having Jurisdiction.

1602.4 Connections to Potable or Reclaimed (Recycled) Water Systems. Rainwater catchment systems shall have no direct connection to a potable water supply or alternate water source system. Potable or reclaimed (recycled) water is permitted to be used as makeup water for a rainwater catchment system provided the potable or reclaimed (recycled) water supply connection is protected by an air gap or reduced-pressure principle backflow preventer in accordance with this code.

Air gaps and backflow devices serve to protect the potable water supply from nonpotable sources that may contaminate the drinking water supply. It is important to note that the rainwater system shall have no direct connection to the potable water supply, i.e. it cannot supplement the potable water supply in any way. However, the potable water supply may be used to supplement the rainwater system under the most stringent backflow prevention method of an air gap or a reduced-pressure principle backflow prevention assembly (RPZ).

1602.5 Initial Cross-Connection Test. Where a portion of a rainwater catchment system is installed within a building, a cross-connection test is required in accordance with Section 1605.3. Before the building is occupied or the system is activated, the installer shall perform the initial cross-connection test in the presence of the Authority Having Jurisdiction and other authorities having jurisdiction. The test shall be ruled successful by the Authority Having Jurisdiction before final approval is granted.

1602.6 Sizing. The design and size of rainwater drains, gutters, conductors, and leaders shall comply with Chapter 11 of this code.

The rainwater collection system is an extension of the basic building rainwater collection system and not the sanitary system. Therefore, pipe sizing is regulated by Chapter 11 Storm Drainage, and not Chapter 7 Sanitary Drainage. All piping needs to be sized to accommodate the proper drainage of the rainwater from the building. The rainwater distribution pipe sizing for indoor plumbing fixtures and appliances is regulated by Chapter 6 Water Supply and Distribution.

1602.7 Rainwater Catchment System Materials. Rainwater catchment system materials shall comply with Section 1602.7.1 through Section 1602.7.4.

1602.7.1 Water Supply and Distribution Materials. Rainwater catchment water supply and distribution materials shall comply with the requirements of this code for potable water supply and distribution systems unless otherwise provided for in this section.

The specification of potable water quality material for nonpotable use allows for,

(1) An emergency situation where nonpotable water can be safely made potable by boiling or through chemical additives,

(2) The prevention of toxic elements from being inadvertently applied to edible garden products, or

(3) Future conversion of a nonpotable system to a potable water application.

1602.7.2 Rainwater Catchment System Drainage Materials. Materials used in rainwater catchment drainage systems, including gutters, downspouts, conductors, and leaders shall be in accordance with the requirements of this code for storm drainage.

1602.7.3 Storage Tanks. Rainwater storage tanks shall comply with Section 1603.1.

1602.7.4 Collections Surfaces. The collection surface shall be constructed of a hard, impervious material.

1602.8 Rainwater Catchment System Color and Marking Information. Rainwater catchment systems shall have a colored background in accordance with Section 601.3. Rainwater catchment systems shall be marked, in lettering in accordance with Section 601.3.3, with the words: "CAUTION: NONPOTABLE RAINWATER WATER, DO NOT DRINK."

1602.9 Design and Installation. The design and installation of nonpotable rainwater catchment systems shall be in accordance with Section 1602.9.1 through Section 1603.16.

1602.9.1 Outside Hose Bibbs. Outside hose bibbs shall be allowed on rainwater piping systems. Hose bibbs supplying rainwater shall be marked with the words: "CAUTION: NONPOTABLE WATER, DO NOT DRINK" and **Figure 1602.9.1**.

FIGURE 1602.9.1

1602.9.2 Deactivation and Drainage for Cross-Connection Test. The rainwater catchment system and the potable water system within the building shall be provided with the required appurtenances (e.g., valves, air or vacuum relief valves, etc.) to allow for deactivation or drainage as required for a cross-connection test in accordance with Section 1605.3.

1602.9.3 Rainwater Catchment System Surfaces. Rainwater shall be collected from roof surfaces or other manmade, aboveground collection surfaces.

Care must be given that rainwater is not gathered from ground level surfaces to assure that toxic elements, bacteria, or oil do not endanger local occupants or result in environmental contamination as a result of the water use.

1602.9.4 Other Surfaces. Natural precipitation collected from surface water runoff, vehicular parking surfaces, or manmade surfaces at or below grade shall be in accordance with the stormwater requirements for on-site treated nonpotable water systems in Section 1506.0.

NONPOTABLE RAINWATER CATCHMENT SYSTEMS

1602.9.5 Prohibited Discharges. Overflows and bleed-off pipes from roof-mounted equipment and appliances shall not discharge onto roof surfaces that are intended to collect rainwater.

1602.9.6 Minimum Water Quality. The minimum water quality for harvested rainwater shall meet the applicable water quality requirements for the intended applications as determined by the Authority Having Jurisdiction. In the absence of water quality requirements determined by the Authority Having Jurisdiction, the minimum treatment and water quality shall be in accordance with Table 1602.9.6.

🔧 Table 1602.9.6 is intended as a guideline for minimum water quality for the intended application. If the quality of the tested water cannot consistently be maintained at the minimum levels specified in Table 1602.9.6, then the system should be equipped with an appropriate treatment device meeting the applicable NSF/ANSI Standard referenced in Table 1701.1.

1603.0 Rainwater Storage Tanks.

1603.1 General. Rainwater storage tanks shall be constructed and installed in accordance with Section 1603.2 through Section 1603.9.

1603.2 Construction. Rainwater storage tanks shall be constructed of solid, durable materials not subject to excessive corrosion or decay and shall be watertight.

1603.3 Location. Rainwater storage tanks shall be permitted to be installed above or below grade.

1603.4 Above Grade. Above grade, storage tanks shall be of an opaque material, approved for aboveground use in direct sunlight or shall be shielded from direct sunlight. Tanks shall be installed in an accessible location to allow for inspection and cleaning. The tank shall be installed on a foundation or platform that is constructed to accommodate loads in accordance with the building code.

🔧 It is important to avoid sunlight, either passing through the tank walls or through tank vents and fittings, in order to reduce the incidence of algae growing in the stored water. Additionally, the installation of tanks in the sun is to be avoided to reduce the possibility of Legionella and other detrimental organisms growing in the stored water. For this reason, where water quality is not periodically tested, applications of above-ground irrigation are to be avoided. **Figure 1603.4** shows an above-ground application in a non-freeze environment. In an environment where freezing is possible, tanks should be moved to a heated environment or buried below the frost line.

1603.5 Below Grade. Rainwater storage tanks installed below grade shall be structurally designed to withstand anticipated earth or other loads. Holding tank covers shall be capable of supporting an earth load of not less than 300 pounds per square foot (lb/ft²) (1465 kg/m²) where the tank is designed for underground installation. Below grade rainwater tanks installed underground shall be provided with manholes. The manhole opening shall be not less than 20 inches (508 mm) in diameter and located not less than 4 inches (102 mm) above the surrounding grade. The surrounding grade shall be sloped away from the manhole. Underground tanks shall be ballasted, anchored, or otherwise secured, to prevent the tank from floating out of the ground where empty. The combined weight of the tank and hold down system shall meet or exceed the buoyancy force of the tank.

🔧 The objective for the manhole to be above grade is to avoid rainwater from flowing into the cistern and contaminating the stored rainwater. Where this is not possible or advisable (e.g. cistern installed under a parking lot), waterproof

TABLE 1602.9.6
MINIMUM WATER QUALITY

APPLICATION	MINIMUM TREATMENT	MINIMUM WATER QUALITY
Car washing	Debris excluder or other approved means in accordance with Section 1603.14, and 100 microns in accordance with Section 1603.15 for drip irrigation.	N/A
Subsurface and drip irrigation	Debris excluder or other approved means in accordance with Section 1603.14, and 100 microns in accordance with Section 1603.15 for drip irrigation.	N/A
Spray irrigation where the maximum storage volume is less than 360 gallons	Debris excluder or other approved means in accordance with Section 1603.14, and disinfection in accordance with Section 1603.12.	N/A
Spray irrigation where the maximum storage volume is equal to or more than 360 gallons	Debris excluder or other approved means in accordance with Section 1603.14.	Escherichia coli: < 100 CFU/100 mL, and Turbidity: < 10 NTU
Urinal and water closet flushing, clothes washing, and trap priming	Debris excluder or other approved means in accordance with Section 1603.14, and 100 microns in accordance with Section 1603.15.	Escherichia coli: < 100 CFU/100 mL, and Turbidity: < 10 NTU
Ornamental fountains and other water features	Debris excluder or other approved means in accordance with Section 1603.14.	Escherichia coli: < 100 CFU/100 mL, and Turbidity: < 10 NTU
Cooling tower make-up water	Debris excluder or other approved means in accordance with Section 1603.14, and 100 microns in accordance with Section 1603.15.	Escherichia coli: < 100 CFU/100 mL, and Turbidity: < 10 NTU

For SI units: 1 micron = 1 µm, 1 gallon = 3.785 L

NONPOTABLE RAINWATER CATCHMENT SYSTEMS

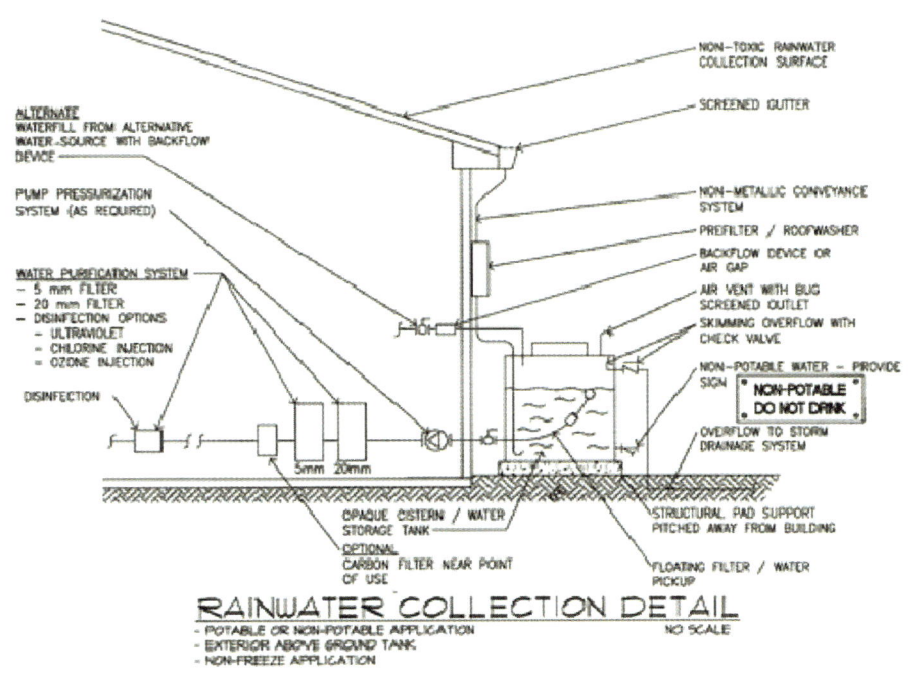

FIGURE 1603.4
ABOVE-GROUND APPLICATION IN A NON-FREEZE ENVIRONMENT
(Courtesy of the American Rainwater Catchment Systems Association)

gasketed seals to tank access shall be provided as a viable alternative to extending tank access above surrounding grade. Where soil saturation is a possibility, it is recommended that the combined weight of the tank and ballast must meet or exceed the buoyancy upward force of an empty cistern (see **Figure 1603.5**). This buoyant force (lb-force) is equal to the volume of the tank (in cubic feet) multiplied by 62.4 (lbs/ft^3), or tank volume (in gallons) multiplied by 8.34 (lbs/gallon).

1603.6 Drainage and Overflow. Rainwater storage tanks shall be provided with a means of draining and cleaning. The overflow drain shall not be equipped with a shutoff valve. The overflow outlet shall discharge in accordance with this code for storm drainage systems. Where discharging to the storm drainage system, the overflow drain shall be protected from backflow of the storm drainage system by a backwater valve or other approved method.

For tanks installed below grade, provision for using a sump pump to drain the tank is a suitable alternative if gravity draining of the tank is not possible.

1603.6.1 Overflow Outlet Size. The overflow outlet shall be sized to accommodate the flow of the rainwater entering the tank and not less than the aggregate cross-sectional area of inflow pipes.

1603.7 Opening and Access Protection. Rainwater tank openings shall be protected to prevent the entrance of insects, birds, or rodents into the tank.

Rainwater tank access openings exceeding 12 inches (305 mm) in diameter shall be secured to prevent tampering and unintended entry by either a lockable device or other approved method.

The possibility of West Nile Virus and other mosquito borne illness makes protection of tank inlets and outlets critical to avoid the stored water becoming a health hazard.

Access openings greater than 12 inches in diameter may allow unintended entry if tampered with and left open. The possibility for a child to enter is a safety risk to be avoided by requiring the access opening to be locked. A weighted cover with a lifting weight of 40 lbs (18 kg) may also serve as an alternative to a lockable device if approved by the Authority Having Jurisdiction (AHJ).

Other OSHA requirements may also apply such as tie-off points and ladder requirements.

1603.8 Marking. Rainwater tanks shall be permanently marked with the capacity and the language: "NONPOTABLE RAINWATER." Where openings are provided to allow a person to enter the tank, the opening shall be marked with the following language: "DANGER-CONFINED SPACE."

1603.9 Storage Tank Venting. Where venting using drainage or overflow piping is not provided or is considered insufficient, a vent shall be installed on each tank. The vent shall extend from the top of the tank and terminate not less than 6 inches (152 mm) above grade and shall be not less than 1½ inches (40 mm) in diameter. The vent terminal shall be directed downward and covered with a 3/32 of an inch (2.4 mm) mesh screen to prevent the entry of vermin and insects.

1603.10 Pumps. Pumps serving rainwater catchment systems shall be listed. Pumps supplying water to water closets, urinals, and trap primers shall be capable of delivering not less than 15 pounds-force per square inch (psi) (103 kPa)

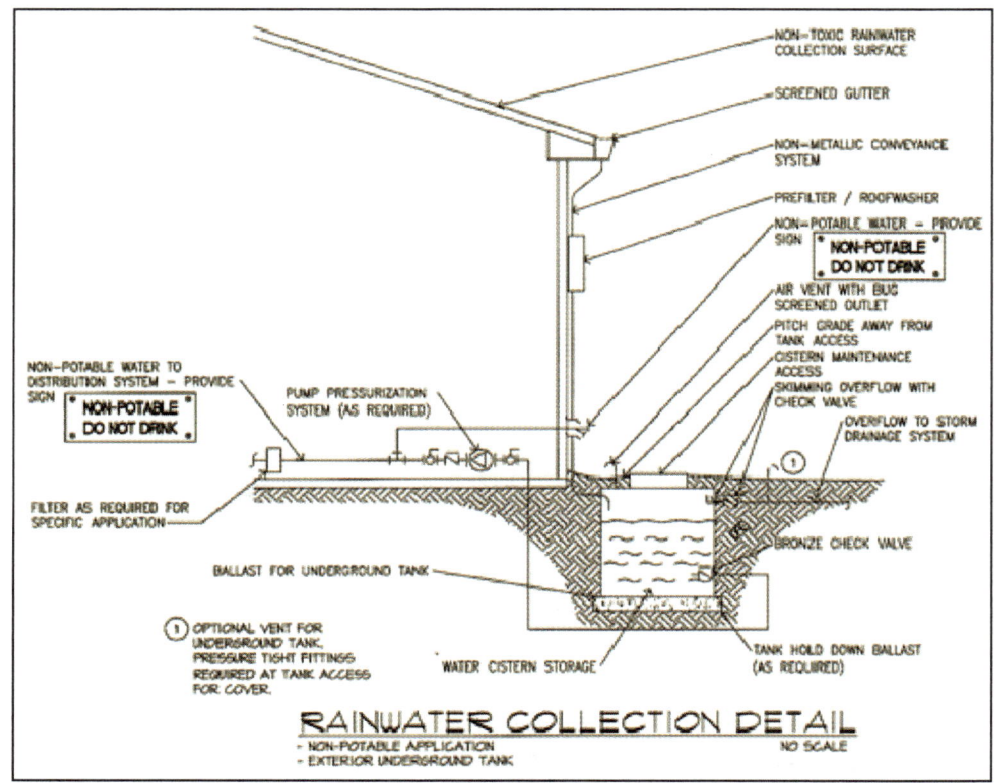

**FIGURE 1603.5
BELOW-GROUND APPLICATION**
(Courtesy of the American Rainwater Catchment Systems Association)

residual pressure at the highest and most remote outlet served. Where the water pressure in the rainwater supply system within the building exceeds 80 psi (552 kPa), a pressure reducing valve reducing the pressure to 80 psi (552 kPa) or less to water outlets in the building shall be installed in accordance with this code.

1603.11 Roof Drains. Primary and secondary roof drains, conductors, leaders, and gutters shall be designed and installed in accordance with this code.

1603.12 Water Quality Devices and Equipment. Devices and equipment used to treat rainwater to maintain the minimum water quality requirements determined by the Authority Having Jurisdiction shall be listed or labeled (third-party certified) by a listing agency (accredited conformity assessment body) and approved for the intended application.

Where carbon filters are used, they may be put down stream of chlorine and ozone disinfection systems, but are recommended to be upstream of Ultraviolet disinfection systems (see **Figure 1603.12**). Filters to remove particulate may be added to improve water quality in order to avoid problems with sprinkler or process devices (see also **Figure 1603.4**).

1603.13 Freeze Protection. Tanks and piping installed in locations subject to freezing shall be provided with an approved means of freeze protection.

Installing a water storage tank in a heated environment is preferred for an installation subject to freezing (see **Figure 1603.13**). Appropriate signage is necessary to label nonpotable water outlets.

1603.14 Debris Removal. The rainwater catchment conveyance system shall be equipped with a debris excluder or other approved means to prevent the accumulation of leaves, needles, other debris and sediment from entering the storage tank. Devices or methods used to remove debris or sediment shall be accessible and sized and installed in accordance with manufacturer's installation instructions.

1603.15 Required Filters. A filter permitting the passage of particulates not larger than 100 microns (100 μm) shall be provided for rainwater supplied to water closets, urinals, trap primers, and drip irrigation system.

Outlet water quality is dependent upon the degree of filtration provided to remove organic material prior the water entering the water storage tank if objectionable odor and water color are to be avoided.

1603.16 Roof Gutters. Gutters shall maintain a minimum slope and be sized in accordance with Section 1103.3.

It is important that dips and sags be avoided so that gutters drain completely to eliminate breeding grounds for mosquitoes and other airborne vectors.

1604.0 Signs.

1604.1 General. Signs in buildings using rainwater shall be in accordance with Section 1604.2 and Section 1604.3.

1604.2 Commercial, Industrial, and Institutional Restroom Signs. A sign shall be installed in restrooms in

NONPOTABLE RAINWATER CATCHMENT SYSTEMS

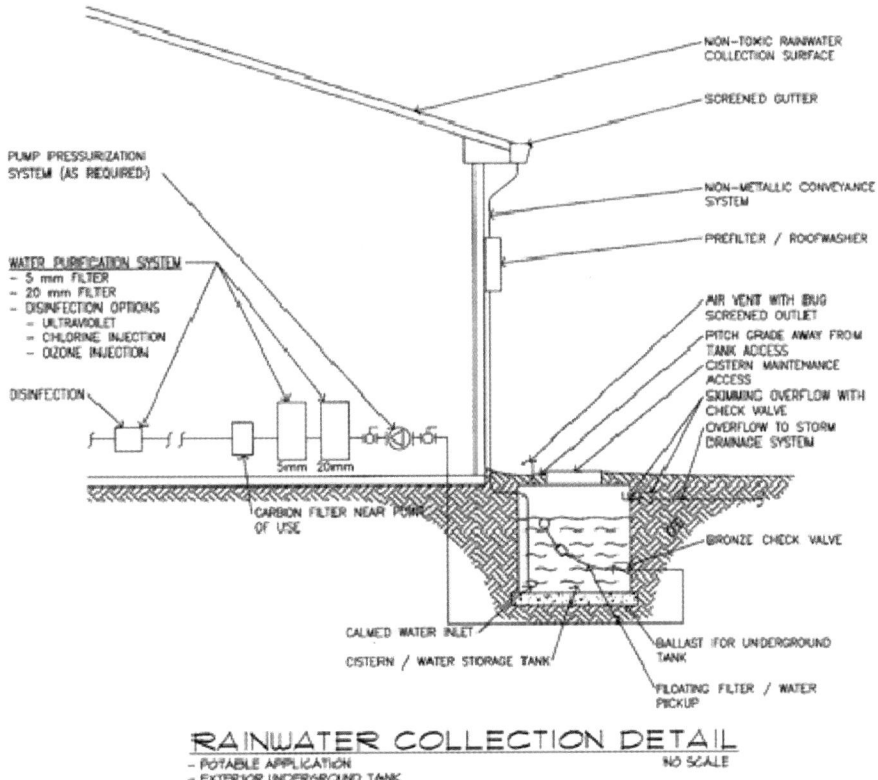

FIGURE 1603.12
WATER TREATMENT DEVICES FOR RAINWATER SYSTEMS
(Courtesy of the American Rainwater Catchment Systems Association)

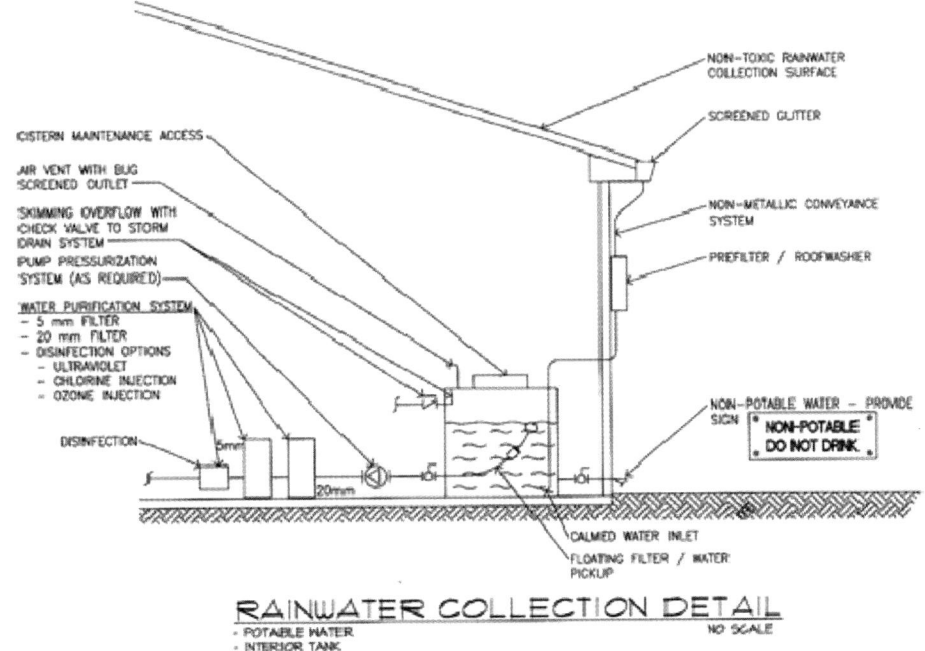

FIGURE 1603.13
INDOOR HEATED ENVIRONMENT APPLICATION FOR FREEZE PROTECTION
(Courtesy of the American Rainwater Catchment Systems Association)

commercial, industrial, and institutional occupancies using nonpotable rainwater for water closets, urinals, or both. Each sign shall contain ½ of an inch (12.7 mm) letters of a highly visible color on a contrasting background. The location of the sign(s) shall be such that the sign(s) shall be visible to users. The number and location of the signs shall be approved by the Authority Having Jurisdiction and shall contain the following text:

TO CONSERVE WATER, THIS BUILDING USES RAINWATER TO FLUSH TOILETS AND URINALS.

🔧 This section requires public notice when a rainwater system supplies water to a restroom. The user is informed that only the toilets and urinals are supplied with nonpotable rainwater and not the lavatories. Lavatories are used for hygiene and therefore are required to be supplied with treated potable water.

1604.3 Equipment Room Signs. Each equipment room containing nonpotable rainwater equipment shall have a sign posted with the following wording in 1 inch (25.4 mm) letters:

CAUTION NONPOTABLE RAINWATER, DO NOT DRINK. DO NOT CONNECT TO DRINKING WATER SYSTEM. NOTICE: CONTACT BUILDING MANAGEMENT BEFORE PERFORMING ANY WORK ON THIS WATER SYSTEM.

This sign shall be posted in a location that is visible to anyone working on or near rainwater water equipment.

🔧 This is to give notice to service mechanics to alert them to confirm the piping they are working with to avoid cross connection of water piping.

1605.0 Inspection and Testing.

1605.1 General. Rainwater catchment systems shall be inspected and tested in accordance with Section 1605.2 and Section 1605.3

1605.2 Supply System Inspection and Test. Rainwater catchment systems shall be inspected and tested in accordance with the applicable provisions of this code for testing of potable water and storm drainage systems. Storage tanks shall be filled with water to the overflow opening for a period of 24 hours, and during the inspection, or by other means as approved by the Authority Having Jurisdiction. Seams and joints shall be exposed during the inspection and checked for watertightness.

🔧 This is to test the cross connection protection between the potable water and alternative water systems.

1605.3 Annual Cross-Connection Inspection and Testing. An initial and subsequent annual inspection and test in accordance with Section 1602.5 shall be performed on both the potable and rainwater catchment water systems. The potable and rainwater catchment water systems shall be isolated from each other and independently inspected and tested to ensure there is no cross-connection in accordance with Section 1605.3.1 through Section 1605.3.4.

1605.3.1 Visual System Inspection. Prior to commencing the cross-connection testing, a dual system inspection shall be conducted by the Authority Having Jurisdiction and other authorities having jurisdiction as follows:

(1) Pumps, equipment, equipment room signs, and exposed piping in an equipment room shall be checked.

1605.3.2 Cross-Connection Test. The procedure for determining cross-connection shall be followed by the applicant in the presence of the Authority Having Jurisdiction and other authorities having jurisdiction to determine whether a cross-connection has occurred as follows:

(1) The potable water system shall be activated and pressurized. The rainwater catchment water system shall be shut down and completely drained.

(2) The potable water system shall remain pressurized for a minimum period of time specified by the Authority Having Jurisdiction while the rainwater catchment water system is empty. The minimum period the rainwater catchment water system is to remain depressurized shall be determined on a case-by-case basis, taking into account the size and complexity of the potable and rainwater catchment water distribution systems, but in no case shall that period be less than 1 hour.

(3) Fixtures, potable, and rainwater shall be tested and inspected for flow. Flow from a rainwater catchment water system outlet shall indicate a cross-connection. No flow from a potable water outlet shall indicate that it is connected to the rainwater water system.

(4) The drain on the rainwater catchment water system shall be checked for flow during the test and at the end of the period.

(5) The potable water system shall then be completely drained.

(6) The rainwater catchment water system shall then be activated and pressurized.

(7) The rainwater catchment water system shall remain pressurized for a minimum period of time specified by the Authority Having Jurisdiction while the potable water system is empty. The minimum period the potable water system is to remain depressurized shall be determined on a case-by-case basis, but in no case shall that period be less than 1 hour.

(8) Fixtures, potable and rainwater catchment, shall be tested and inspected for flow. Flow from a potable water system outlet shall indicate a cross-connection. No flow from a rainwater catchment water outlet shall indicate that it is connected to the potable water system.

(9) The drain on the potable water system shall be checked for flow during the test and at the end of the period.

(10) Where there is no flow detected in the fixtures which would indicate a cross-connection, the potable water system shall be repressurized.

1605.3.3 Discovery of Cross-Connection. In the event that a cross-connection is discovered, the following procedure, in the presence of the Authority Having Jurisdiction, shall be activated immediately:

(1) Rainwater catchment water piping to the building shall be shutdown at the meter, and the rainwater water riser shall be drained.

(2) Potable water piping to the building shall be shutdown at the meter.

(3) The cross-connection shall be uncovered and disconnected.

(4) The building shall be retested following procedures listed in Section 1605.3.1 and Section 1605.3.2.

(5) The potable water system shall be chlorinated with 50 ppm chlorine for 24 hours.

(6) The potable water system shall be flushed after 24 hours, and a standard bacteriological test shall be performed. Where test results are acceptable, the potable water system shall be permitted to be recharged.

1605.3.4 Annual Inspection. An annual inspection of the rainwater catchment water system, following the procedures listed in Section 1605.3.1 shall be required. Annual cross-connection testing, following the procedures listed in Section 1605.3.2 shall be required by the Authority Having Jurisdiction, unless site conditions do not require it. In no event shall the test occur less than once in 4 years.

Alternate testing requirements shall be permitted by the Authority Having Jurisdiction.

CHAPTER 17
REFERENCED STANDARDS

1701.0 General.

Code Requirements and Mandatory Referenced Standards: Section 301.2 provides for the minimum standards for all material related to pipe, fittings, fixtures and devices used in a plumbing system. Each device must be listed (third-party certified) by a listing agency (accredited conformity assessment body) and must conform to approved applicable recognized standards referenced in this code. In order to conform to minimum standards, all devices must be free from defects and meet the performance requirements contained in the standards. Products are referred to as "listed" indicating that the minimum quality required in the referenced standard is upheld through the rigors of the third-party certification process.

A standard is a set of technical definitions, requirements and guidelines for the manufacture of devices and products that establishes test methods, specifications, classifications, practices and scoping requirements. Standards are documents that provide specific guidelines for manufacturing all sorts of products. Their purpose is to ensure that products are made to be safe and efficient for the consumer. They also provide dimensional requirements that ensure that the product can be installed and interchanged as required.

Although each standard is different, there are common requirements. All standards must:

- Define the scope of the product, system or process that is covered.
- Be able to be used repeatedly and not so specific that it does not apply to many applications.
- Include a test methods and performance criteria, with a clear and concise description for measurement and evaluation of one or more properties, qualities or characteristics. A test method is a kind of standard that produces a test result, and it is recommended practice to reference a test method along with the applicable performance requirements. (e.g. A test method may describe how to measure thickness. A performance standard will specify what the thickness must be when measured in accordance with the test method).
- Specify the requirements in simple, precise understandable language free from ambiguous terminology.
- Contain terms and explanations of symbols, abbreviations or acronyms that are relevant to the standard and its application.
- Be enforceable to the extent of its use and application, which includes being written in mandatory language and does not mandate the use of proprietary materials or agencies.

Standard requirements are typically based on unambiguous specifications, which include physical, mechanical and chemical properties. Additionally, standards identify the applicable test methods and performance criteria to determine that each requirement is met or satisfied.

Codes and standards work together to protect public health and safety. A standard is considered a basis of comparison or an approved model. Simply stating, codes tell the user what to do and when and under what circumstances to do it. Codes are often legal requirements that are enforced by local jurisdictions that enforce their provisions. Standards provide the user with approved materials and tell the user that such products have been tested to performance requirements and are applicable only to the extent the code references the standard. For example, a standard may have performance specifications for materials and use of application or installation requirements. The standard is only applicable to the extent suggested in the text of the code. The code text takes precedence when the requirements of the standard conflict with the requirements of the code.

As an example, Section 604.1 requires all water piping material must conform to Table 604.1 where the applicable standards are referenced. ASTM F 1281 is one of the standards for PEX-AL-PEX pressure pipe for water distribution systems. This standard regulates the performance requirements for materials that cover PEX-AL-PEX composite pressure pipe. In addition, the performance requirements include dimensions, burst and sustained pressure performance. Section 604.13 requires that PEX-AL-PEX must not be installed within the first 18 inches (457 mm) of piping connected to the water heater. The mandatory information for connectors in this standard suggests if water-heating equipment malfunctions, the assemblies (connectors) shall have satisfactory strength to allow short-term overheating conditions. However, this standard does not restrict installing PEX-AL-PEX within the first 18 inches (457 mm) of piping connected to the water heater; therefore, Section 604.13 would take precedence over the standard. Additionally, Section 604.1 requires that pipe, tube and fittings carrying water used in potable water systems intended to supply drinking water must meet the requirements of NSF61. However, assuming ASTM F1281 did not contain mandatory language that this material must conform to NSF 61, Section 604.1 would take precedence over the standard and requires conformance to NSF 61 for potable water piping used to supply drinking water. Remember, the referenced standard is a guide to aid the user in deciding whether a product, device, joining methods or installation complies with the code.

Section 301.2.1 requires that each pipe and each pipe fitting, trap, fixture, material and device used in the plumbing system must have the manufacturer's mark or name cast, stamped or permanently marked on it, which must readily identify the manufacturer to the end user of the product when such marking is required by the approved standard that applies. The code does not specifically state the marking requirements, except for the manufacturer's identification, but the referenced standard does. The identification requirements vary by standard but typically include any of the following: the name of the manufacturer or trademark; type or model number; maximum rated pressure and temperature; serial number; nominal size and standard designation.

Section 301.2.2 requires that standards listed or referred to in this chapter or other chapters cover materials that will conform to the requirements of this code, when used in accordance with limits imposed in this or other chapters and their listing. This section reinforces that only a portion of the listed standard is applicable to the extent of the code reference and application.

The terms "Listed," "Listing Agency," "Approved," and "Approved Testing Agency" are defined in Chapter 2 of the Uniform Plumbing Code (UPC). A pipe or plumbing product must be listed and presented to the Authority Having Jurisdiction (AHJ) for approval. Listing provides independent confirmation that a product conforms to standards and that it is verified to comply with those standards. For a product to be "listed," it must be tested to the applicable recognized standards for that product, be found safe for use in a specific manner, and then be put on a list of such products by a "listing agency." These lists (usually called "directories" of listed plumbing products) are used by juris- dictions when inspecting installations for code compliance. The listing agency must do periodic inspections to ensure the product continues to be manufactured to the correct standards. A listed product is labeled with the mark of the listing agency so inspectors in the field can identify it as listed.

In order to be listed, plumbing products are required to comply with standards that are designed to prove the products ability to provide safe operation over long periods. A product typically undergoes many different tests that simulate use and extended wear and tear. Testing criteria are used to develop maximum tolerances for products and materials. All testing procedures and results must be verifiable.

Types of Standards

Nationally Recognized Consensus Standards: The following is a partial list in Table 1701.1 of different standards developing organizations (SDO) that create Nationally Recognized Consensus Standards with the typical designation of the American National Standards Institute (ANSI). ANSI is not the developer of the standards, but is the accreditation body of SDOs that use the consensus process for development of national standards. It promotes developing American National Standards by accrediting the procedures of standards developing organizations. Accreditation by ANSI represents the procedures used by the standards body with developing American National Standards that meet the essential requirements for openness, balance, consensus and due process.

- American Society of Mechanical Engineers (ASME)
- American Society of Sanitary Engineers (ASSE)
- American Welding Society (AWS)
- American Water Works Association (AWWA)
- Canadian Standards Association (CSA)
- ASTM International
- International Association of Plumbing and Mechanical Officials (IAPMO)
- Industry Safety Equipment Association (ISEA)
- National Fire Protection Association (NFPA)
- NSF International (NSF)
- Underwriters Laboratories (UL)

The above are examples of standard developing organizations (SDOs) that provide a consensus by a group or body that is open to representatives from all affected and interested parties. Standards are developed by presenting a rationale for the need of a particular standard. If the proposed rationale for the standard is approved, the next process is development of a draft text by a committee of subject matter experts, which constitute a wide spectrum of stakeholders, including manufacturers, testing agencies, environmental experts and end users. When the developing committee reaches a consensus on the content of the draft, it is sent out to all interested parties (public review period) and the standards committee for review and comments (ballot form). All comments are reviewed and considered during this period and are responded to through ballot form back to the standards committee for review. All changes that are found to be persuasive are incorporated into the draft and further public review and ballots are sent out to the standards committee. A vote is taken and any comments received are resolved. Once a level of consensus is gained among the committee the standard is then presented to the Board of Standards Review (BSR) for approval as an American National Standards from ANSI. Once the standard is published, it will then be reviewed every five years for possible revision, reaffirmation or withdrawal.

Other Standards: Standards written by other organizations are listed in Table 1701.1. IAPMO has developed more than 300 quasi-consensus standards which:

- cover innovative new plumbing products not covered by existing standards;
- are widely accepted by industry and regulators (jurisdictions);
- are jointly developed by IAPMO staff, stakeholders, and public comments;
- are approved by the IAPMO Standards Review Committee (SRC); and
- can evolve into ANSI standards.

The development process for IAPMO quasi-consensus standards parallels the ANSI process for consensus standards, but it is much shorter (usually 1 to 3 months) thus allowing manufacturers timely and efficient access to the marketplace.

IAPMO Guide Criteria

The IAPMO Standards Department works with stakeholders to provide an opportunity for the development of a new performance test standard known as an IAPMO Guide Criteria (IGC) when no applicable standard exists for an innovatively new product. Often, new products or new technologies surge ahead far faster than most standards can keep pace. Through IAPMO Guide Criteria (IGC), IAPMO provides manufacturers and product developers an opportunity to use IGC standards as a vehicle for testing, listing, and introducing new products to the marketplace. Once an IGC is accepted, IAPMO R&T can list products manufactured and tested in compliance with the new requirements.

The IAPMO Standards Review Committee (SRC) meets every month to review proposals for IGCs and various types of standards. SRC meetings are open to the public whereby all proponents and participants are given the opportunity to present their views and supporting information on the proposals discussed at these meetings. IAPMO standards, including IGCs, approved by the SRC are posted on the IAPMO Standards public review webpage for public comment over a 20-day period.

IAPMO Material and Property Standards: If after three years an IGC continued in use in active listings, then the IGC evolved to become an IAPMO material and property standards (PS) to continue acceptable quality and performance level for plumbing products in the absence of approved nationally recognized standards. To date, PS type standards are no longer developed but those in existence continue to be used.

IAPMO Installation Standards: Appendix I is the location for installation standards that provide guidelines and best practices.

As with all the appendices, the installation standards are only mandatory if a local jurisdiction adopts them as part of the code.

Form and Style for Standards: Each standard includes a title; scope; purpose; referenced documents; terminology or definitions; materials or classification; test methods and performance requirements; mandatory installation practices; product marking and quality assurance. In addition, mandatory and non-mandatory information with recommended installation practices are included.

Scope: The scope defines without ambiguity the subject of the document and the aspects covered, thereby indicating the limits of applicability of the standard. It provides a common understanding of the standard with defining the standard's overall boundaries by suggesting what it will and will not perform (objectives) on the subject product. In addition, scoping sections include physical and performance testing methods with material component references.

Reference Standards: Many standards make reference to other standards or documents as a part of the entire standard for compliance with physical or performance requirements for various material or device components listed in the standard. For example, the standard may reference another standard for a test method for dimensions. Therefore, the test method does not need to be repeated in the standard; only the referenced standard test method is mentioned and collaborated. In the plumbing industry, the code references various product standards, which define the product's performance and design.

Terminology or Definitions: Standards have definitions that are specific to the applicability of the standard and apply only to the requirements of the standard. These definitions aid the user in understanding the application and extent of requirements within the standard. It is important to stress that these definitions are applicable to this specific standard. There may be instances where the standard promulgator uses the same definitions in many referenced standards; however, the user should be aware that only those definitions indicated within the standards are applicable to the performance and testing requirements.

Material or Physical Requirements: Material or physical requirements specifically define the materials used, such as its origin, composition or properties. Within this, such specific requirements may include physical properties, mechanical or chemical properties. In addition, many physical and chemical properties reference various standards or cell classifications and that such materials must be in accordance with the applicable standard. The material or physical characteristics must be precise and specific to satisfy the performance criteria.

Test Methods and Performance Requirements

Test methods and performance requirements identify the methods for deciding whether each of the requirements is satisfied. Many standards include provisions for sampling or test specimens; test conditions and procedures; calculations and acceptance criteria. Each standard must have standardized methods for testing and evaluating each product or device. Test methods are the key to ensuring product performance is measured accurately and consistently. As with all standards, test requirements include mandatory and optional elements that vary according to the test methods.

Installation Practices: Many standards include some installation instructions that give specific application, cleaning, installation, preparation, and training methods. Typically, these are in the form of a set of instructions for performing specific methods to ensure installing the product or device complies with recommended practices. These practices are important for people who install such products or devices, as they contain significant input from many subject experts in the field and ensure the health and safety of the consumer. Codes contain many of these installation practices as specific code requirements based on the specific standard.

Product Marking and Quality Assurance: Standards specify the required product markings and detail, especially if the markings are required to be permanently applied. Methods such as stamping or embossing with specific lettering and coloring are often required. Each standard has specific requirements, which may include the manufacturer's name or trademark, standard designation, model number, sizing and spacing intervals.

Nonmandatory Information: Nonmandatory information is located in the appendix and provides a guide without recommending a specific action. The purpose of this is to provide the user with help based on a consensus of subject matter experts. This information is intended to provide added techniques in supplementing the standard. For example, the installation guidelines and joining methods for a particular piping material may be covered with information about other uses and practices.

REFERENCED STANDARDS

In summary, standards are important and work with the plumbing code to promote health and safety. They help to produce consistent performance requirements and lead to better specification of materials so the end user knows the properties of that material. This means designers and installers can specify the material that best suits their needs. Standards lead to harmonization of performance requirements, which provides for the greatest opportunities for consumers, businesses and the industry.

IAPMO Product Markings: The Research and Testing (R&T) division of IAPMO tests and evaluates products to determine whether they comply with select standards and applicable codes. Each tested product is then assessed as meeting or not meeting the standards. Compliant products may qualify for application of one of seven possible IAPMO marking labels (see **Figure 1701.1a**).

By its shape or markings, an IAPMO R&T certification mark may denote that a product is either "listed" or "classified." Each of these labels carries considerably different implications. A product that bears the "listed" label has demonstrated compliance with the specific standard or standards to which it was tested. Additionally, a listed product is defined as compliant with code stipulations, engineering concepts or other fundamental principles found in the UPC or Uniform Mechanical Code (UMC).

Conversely, a "classified" label represents compliance with all standards to which the product was tested but does not ensure compliance with UPC or UMC code stipulations or requirements. Classified products are published in a separate product directory, the objective being to avoid user confusion regarding the propriety of their application. Section 301.2.2 of the UPC allows the use of these products, but only with specific approval from the local AHJ. Listed products require no special approval or acceptance by local authorities when installed within the limits of their listing.

The remaining five IAPMO R&T labels shown in Figure 1701.1a have specialty applications, e.g., solar, recreational vehicles, etc. These five labels represent specialty listings that have limited applications but that are code compliant when applied within the limitations of their listings and each applicable code.

There are no approvals issued by IAPMO R&T. In all cases, R&T evaluations are solely for the purpose of verifying that the tested product has demonstrated compliance with a specified test standard. Satisfying a given standard does not ensure compliance with applicable codes. Products labeled as "listed" products will not conflict with either the language or the intent of the UPC or the UMC. Other products labeled as "classified" products meet the standard to which they were tested, but cannot comply with certain requirements found in the UPC or UMC.

The use of specialized labeling is intended to provide the consumer with guidance regarding the limits of product application.

Compliance with an evaluation standard neither suggests that a product is suitable for any specific application nor that a product is code compliant nor approved for use in a particular application. IAPMO R&T tests to standards specified by the person who submits the product for evaluation (see **Figure 1701.1b**). Because all test standards are not necessarily related to code requirements or standard engineering practices, test compliance is essentially unrelated to the appropriate utilization of a particular product.

1701.1 Standards. The standards listed in Table 1701.1 are referenced in various sections of this code and shall be considered part of the requirements of this document. The standards are listed herein by the standard number and effective date, the title, application and the section(s) of this code that references the standard. The application of the referenced standard(s) shall be as specified in Section 301.2.2. The promulgating agency acronyms referred to in Table 1701.1 are defined in a list found at the end of the tables.

The table, referencing installation and testing standards applicable to plumbing systems, found in Chapter 17 prior to the publication of the 2018 UPC has been divided into two separate tables. Table 1701.1 lists standards referenced within the body of Code and are considered as part of the requirements of the Code. Table 1701.2 consists of installation and testing standards that are not referenced in other sections of this code but may aid in the user's ability to quickly find an applicable standard. The application of both sets of referenced standards are specified in Section 301.2.2 of this Code.

REFERENCED STANDARDS

NSF 61-9
The product complies with Section 9 of the NSF 61 standard.

 **IAPMO**

Classified Marking Certification Marks
The product complies with only the product's performance standard BUT not acceptable or recognized by the UPC.

 IAPMO - UMC™

Uniform Mechanical Code Certification Marks
The product complies with BOTH the product's performance AND the UMC

 UPC®

Uniform Plumbing Code Certification Marks
The product complies with BOTH the product's performance AND the UPC

 USEC®

Uniform Solar Energy Code Certification Marks
The product complies with BOTH the product's performance AND the Uniform Solar Energy Code

 USPC®

Uniform Pool, Spa & Hot Tub Code Certification Marks
The product complies with BOTH the product's performance AND the Pool, Spa and Hot Tub Code.

 IAPMO-T®

Manufactured Housing/Recreational Vehicle Certification Marks
The product complies with IAMPO's MHRV product's performance standard.

Materials/Components Certification Marks
The product material and/or component complies with applicable sections of the product's performance standard.

**FIGURE 1701.1A
IAPMO MARKS**

**FIGURE 1701.1B
IAPMO R&T TESTING**

REFERENCED STANDARDS

TABLE 1701.1
REFERENCED STANDARDS

STANDARD NUMBER	STANDARD TITLE	APPLICATION	REFERENCED SECTIONS
ASME A112.1.2-2012	Air Gaps in Plumbing Systems (For Plumbing Fixtures and Water-Connected Receptors)	Fittings	Table 603.2
ASME A112.1.3-2000 (R2015)	Air Gap Fittings for Use with Plumbing Fixtures, Appliances, and Appurtenances	Fittings	Table 603.2
ASME A112.3.1-2007 (R2012)	Stainless Steel Drainage Systems for Sanitary DWV, Storm, and Vacuum Applications, Above- and Below-Ground	Piping	418.1, Table 701.2, 705.7.2, 1102.1
ASME A112.3.4-2013/CSA B45.9-2013	Plumbing Fixtures with Pumped Waste and Macerating Toilet Systems	Fixtures	710.13
ASME A112.4.1-2009 (R2014)	Water Heater Relief Valve Drain Tubes	Appliances	608.5
ASME A112.4.2-2015/CSA B45.16-2015	Personal Hygiene Devices for Water Closets	Fixtures	411.4
ASME A112.4.14-2004 (R2010)	Manually Operated, Quarter-Turn Shutoff Valves for Use in Plumbing Systems	Valves	606.1
ASME A112.6.1M-1997 (R2012)	Floor-Affixed Supports for Off-the-Floor Plumbing Fixtures for Public Use	Fixtures	402.4
ASME A112.6.2-2000 (R2010)	Framing-Affixed Supports for Off-the-Floor Water Closets with Concealed Tanks	Fixtures	402.4
ASME A112.6.3-2001 (R2007)	Floor and Trench Drains	Fixtures	418.1
ASME A112.6.4-2003 (R2012)	Roof, Deck, and Balcony Drains	Fixtures	1102.1
ASME A112.6.7-2010 (R2015)	Sanitary Floor Sinks	Fixtures	421.1
ASME A112.6.9-2005 (R2015)	Siphonic Roof Drains	DWV Components	1106.3
ASME A112.14.1-2003 (R2012)	Backwater Valves	Valves	710.6
ASME A112.14.3-2000 (R2014)	Grease Interceptors	Fixtures	1014.1
ASME A112.14.4-2001 (R2012)	Grease Removal Devices	Fixtures	1014.1
ASME A112.14.6-2010 (R2015)	FOG (Fats, Oils, and Greases) Disposal Systems	Fixtures	1015.2
ASME A112.18.1-2012/CSA B125.1-2012	Plumbing Supply Fittings	Fittings	408.3, 417.1, 417.2, 417.3, 417.4, 603.5.19
ASME A112.18.2-2015/CSA B125.2-2015	Plumbing Waste Fittings	Fittings	404.1
ASME A112.18.3-2002 (R2012)	Performance Requirements for Backflow Protection Devices and Systems in Plumbing Fixture Fittings	Backflow Protection	417.3, 417.4
ASME A112.18.6-2009/CSA B125.6-2009 (R2014)	Flexible Water Connectors	Piping	604.5, 604.13
ASME A112.18.9-2011	Protectors/Insulators for Exposed Waste and Supplies on Accessible Fixtures	Miscellaneous	403.3
ASME A112.19.1-2013/CSA B45.2-2013	Enameled Cast Iron and Enameled Steel Plumbing Fixtures	Fixtures	407.1, 408.1, 409.1, 415.1, 420.1
ASME A112.19.2-2013/CSA B45.1-2013	Ceramic Plumbing Fixtures	Fixtures	407.1, 408.1, 409.1, 410.1, 411.1, 412.1, 415.1, 420.1

REFERENCED STANDARDS

TABLE 1701.1 (continued)
REFERENCED STANDARDS

Standard	Title	Application	References
ASME A112.19.3-2008/CSA B45.4-2008 (R2013)	Stainless Steel Plumbing Fixtures	Fixtures	407.1, 408.1, 409.1, 410.1, 411.1, 415.1, 420.1
ASME A112.19.5-2011/CSA B45.15-2011 (R2016)	Flush Valves and Spuds for Water Closets, Urinals, and Tanks	Fixtures	413.3
ASME A112.19.7-2012/CSA B45.10-2012	Hydromassage Bathtub Systems	Fixtures	409.1, 409.6
ASME A112.19.12-2014	Wall Mounted, Pedestal Mounted, Adjustable, Elevating, Tilting, and Pivoting Lavatory, Sink, and Shampoo Bowl Carrier Systems and Drain Waste Systems	Fixtures	407.1, 420.1
ASME A112.19.14-2013	Six-Liter Water Closets Equipped with a Dual Flushing Device	Fixtures	411.2.1
ASME A112.19.15-2012	Bathtubs/Whirlpool Bathtubs with Pressure Sealed Doors	Fixtures	409.1
ASME A112.19.19-2016	Vitreous China Nonwater Urinals	Fixtures	412.1
ASME B1.20.1-2013	Pipe Threads, General Purpose, Inch	Joints	605.1.5, 605.2.3, 605.5.2, 605.12.3, 705.1.3, 705.3.4, 705.4.2, 705.6.3, 1208.6.10, 1309.13(2)
ASME B16.1-2015	Gray Iron Pipe Flanges and Flanged Fittings: Classes 25, 125, and 250	Fittings	1208.6.13.1
ASME B16.3-2011	Malleable Iron Threaded Fittings: Classes 150 and 300	Fittings	Table 604.1, Table 701.2
ASME B16.4-2011	Gray Iron Threaded Fittings: Classes 125 and 250	Fittings	Table 604.1
ASME B16.5-2013	Pipe Flanges and Flanged Fittings: NPS 1/2 through NPS 24 Metric/Inch	Fittings	1208.6.13.2(1)
ASME B16.12-2009 (R2014)	Cast Iron Threaded Drainage Fittings	Fittings	Table 701.2
ASME B16.15-2013	Cast Copper Alloy Threaded Fittings: Classes 125 and 250	Fittings	Table 604.1
ASME B16.18-2012	Cast Copper Alloy Solder Joint Pressure Fittings	Fittings	Table 604.1
ASME B16.20-2012	Metallic Gaskets For Pipe Flanges: Ring-Joint, Spiral-Wound, and Jacketed	Joints	1208.6.14.2
ASME B16.21-2011	Nonmetallic Flat Gaskets for Pipe Flanges	Joints	1208.6.14.3
ASME B16.22-2013	Wrought Copper and Copper Alloy Solder-Joint Pressure Fittings	Fittings	1308.6(3)(a), 1309.4, Table 604.1
ASME B16.23-2011	Cast Copper Alloy Solder Joint Drainage Fittings: DWV	Fittings	Table 701.2
ASME B16.24-2011	Cast Copper Alloy Pipe Flanges and Flanged Fittings: Classes 150, 300, 600, 900, 1500, and 2500	Fittings	1208.6.13.3
ASME B16.26-2013	Cast Copper Alloy Fittings for Flared Copper Tubes	Fittings	Table 604.1, 1308.6(3)(b)
ASME B16.29-2012	Wrought Copper and Wrought Copper Alloy Solder-Joint Drainage Fittings – DWV	Fittings	Table 701.2
ASME B16.34-2013	Valves-Flanged, Threaded, and Welding End	Valves	606.1
ASME B16.42-2011	Ductile Iron Pipe and Flanged Fittings	Fuel Gas Piping	1208.6.13.4
ASME B16.47-2011	Large Diameter Steel Flanges: NPS 26 through NPS 60 Metric/Inch	Fittings	1208.6.13.2(2)
ASME B16.50-2013	Wrought Copper and Copper Alloy Braze-Joint Pressure Fittings	Fittings	Table 604.1, 1309.4, 1309.9
ASME B16.51-2013	Copper and Copper Alloy Press-Connect Pressure Fittings	Fittings	Table 604.1
ASME B31.3-2014	Process Piping	Piping	1315.2(9)

REFERENCED STANDARDS

TABLE 1701.1 (continued)
REFERENCED STANDARDS

STANDARD NUMBER	STANDARD TITLE	APPLICATION	REFERENCED SECTIONS
ASME B36.10M-2015	Welded and Seamless Wrought Steel Pipe	Fuel Gas, Piping	1208.6.3.1(1)
ASME BPVC Section VIII-2015	Rules for Construction of Pressure Vessels Division 1	Miscellaneous	505.4, 1313.7(2), 1314.4(2), 1313.7.1(2)
ASME BPVC Section IX-2015	Welding, Brazing, and Fusing Qualifications Qualification Standard for Welding, Brazing, and Fusing Procedures; Welders; Brazers; and Welding, Brazing, and Fusing Operators	Certification	225.0, 1307.1, 1309.16.1, 1309.16.2
ASPE 45-2013	Siphonic Roof Drainage	Storm Drainage	1106.2
ASSE 1001-2008	Atmospheric Type Vacuum Breakers	Backflow Protection	Table 603.2
ASSE 1002/ASME A112.1002/CSA B125.12-2015	Anti-Siphon Fill Valves for Water Closet Tanks	Backflow Protection	413.3, Table 603.2
ASSE 1004-2008	Backflow Prevention Requirements for Commercial Dishwashing Machines	Backflow Protection	414.2
ASSE 1008-2006	Plumbing Aspects of Residential Food Waste Disposer Units	Appliances	419.1
ASSE 1010-2004	Water Hammer Arresters	Appliances	609.10
ASSE 1011-2004	Hose Connection Vacuum Breakers	Backflow Protection	Table 603.2
ASSE 1013-2011	Reduced Pressure Principle Backflow Preventers and Reduced Pressure Principle Fire Protection Backflow Preventers	Backflow Protection	Table 603.2
ASSE 1014-2005	Backflow Prevention Devices for Hand-Held Shower	Backflow Protection	417.3
ASSE 1015-2011	Double Check Backflow Prevention Assemblies and Double Check Fire Protection Backflow Prevention Assemblies	Backflow Protection	Table 603.2
ASSE 1016-2017/ASME A112.1016-2017/CSA B125.16-2017	Automatic Compensating Valves for Individual Showers and Tub/Shower Combinations	Valves	408.3
ASSE 1018-2001	Trap Seal Primer Valves - Potable Water Supplied	Valves	1007.2
ASSE 1019-2011 (R2016)	Wall Hydrant with Backflow Protection and Freeze Resistance	Backflow Protection	Table 603.2
ASSE 1020-2004	Pressure Vacuum Breaker Assembly	Backflow Protection	Table 603.2
ASSE 1022-2003	Backflow Preventer for Beverage Dispensing Equipment	Backflow Protection	Table 603.2, 603.5.12
ASSE 1037-2015/ASME A112.1037-2015/CSA B125.37-2015	Pressurized Flushing Devices for Plumbing Fixtures	Backflow Protection	413.2
ASSE 1044-2015	Trap Seal Primer - Drainage Types and Electric Design Types	DWV Components	1007.2
ASSE 1047-2011	Reduced Pressure Detector Fire Protection Backflow Prevention Assemblies	Backflow Protection	Table 603.2
ASSE 1048-2011	Double Check Detector Fire Protection Backflow Prevention Assemblies	Backflow Protection	Table 603.2
ASSE 1052-2016	Hose Connection Backflow Preventers	Backflow Protection	Table 603.2
ASSE 1053-2004	Dual Check Backflow Preventer Wall Hydrants – Freeze Resistant Type	Backflow Protection	Table 603.2
ASSE 1055-2016	Chemical Dispensing Systems	Backflow Protection	603.5.21
ASSE 1056-2013	Spill Resistant Vacuum Breaker Assemblies	Backflow Protection	Table 603.2
ASSE 1057-2012	Freeze Resistant Sanitary Yard Hydrants with Backflow Protection	Backflow Protection	Table 603.2
ASSE 1060-2017	Outdoor Enclosures for Fluid Conveying Components	Miscellaneous	603.4.7

TABLE 1701.1 (continued)
REFERENCED STANDARDS

STANDARD NUMBER	STANDARD TITLE	APPLICATION	REFERENCED SECTIONS
ASSE 1061-2015	Push-Fit Fittings	Fittings	Table 604.1, 605.1.3.3, 605.2.1.1, 605.3.2.1, 605.9.3
ASSE 1069-2005	Automatic Temperature Control Mixing Valves	Valves	408.3
ASSE 1070-2015/ASME A112.1070-2015/CSA B125.70-2015	Water Temperature Limiting Devices	Valves	407.3, 409.4, 410.3
ASSE 1071-2012	Temperature Actuated Mixing Valves for Plumbed Emergency Equipment	Valves	416.2
ASSE 1079-2012	Dielectric Pipe Unions	Fittings	605.15, 605.16.1, 605.16.3
ASSE Series 5000-2015	Cross-Connection Control Professional Qualifications Standard	Certification	603.2, 603.4.2
ASSE Series 6000-2015	Professional Qualifications Standard for Medical Gas Systems Personnel	Certification	1306.1, 1318.16.2
ASSE Series 7000-2013	Residential Potable Water Fire Protection System Installers & Inspectors for One and Two Family Dwellings	Miscellaneous	612.1
ASTM A53/A53M-2012	Pipe, Steel, Black and Hot-Dipped, Zinc-Coated, Welded and Seamless	Piping	Table 604.1, Table 701.2, 1208.6.3.1(2)
ASTM A74-2016	Cast Iron Soil Pipe and Fittings	Piping	301.2.4, Table 701.2
ASTM A106/A106M-2015	Seamless Carbon Steel Pipe for High-Temperature Service	Piping	1208.6.3.1(3)
ASTM A254/A254M-2012	Copper-Brazed Steel Tubing	Piping	1208.6.4.1
ASTM A269/A269M-2015a	Seamless and Welded Austenitic Stainless Steel Tubing for General Service	Piping	1308.5(2)(a), Table 604.1
ASTM A312/A312M-2016a	Seamless, Welded, and Heavily Cold Worked Austenitic Stainless Steel Pipes	Piping	Table 604.1, 1308.5(2)(b)
ASTM A403/A403M-2011	Wrought Austenitic Stainless Steel Pipe Fittings	Fittings	1308.5(2)(c)
ASTM A861-2004 (R2013)	High-Silicon Iron Pipe and Fittings	Piping	811.2
ASTM A888-2015	Hubless Cast Iron Soil Pipe and Fittings for Sanitary and Storm Drain, Waste, and Vent Piping Applications	Piping	301.2.4, Table 701.2
ASTM A1056-2012	Cast Iron Couplings used for Joining Hubless Cast Iron Soil Pipe and Fittings	Fittings	705.2.2
ASTM B32-2008 (R2014)	Solder Metal	Joints	605.1.4, 705.3.3, 1308.6(4)
ASTM B42-2015a	Seamless Copper Pipe, Standard Sizes	Piping	Table 604.1
ASTM B43-2015	Seamless Red Brass Pipe, Standard Sizes	Piping	Table 604.1, Table 701.2
ASTM B75/B75M-2011	Seamless Copper Tube	Piping	Table 604.1, Table 701.2
ASTM B88-2016	Seamless Copper Water Tube	Piping	Table 604.1, 604.4, 903.2.3, 1208.6.4.2, 1308.5(1)(a)
ASTM B135-2010	Seamless Brass Tube	Piping	Table 604.1
ASTM B152/B152M-2013	Copper Sheet, Strip, Plate, and Rolled Bar	Miscellaneous	408.7.4
ASTM B210-2012	Aluminum and Aluminum-Alloy Drawn Seamless Tubes	Piping	1208.6.4.3
ASTM B241/B241M-2016	Aluminum and Aluminum-Alloy Seamless Pipe and Seamless Extruded Tube	Piping	1208.6.3.3, 1208.6.4.3
ASTM B251-2010	General Requirements for Wrought Seamless Copper and Copper-Alloy Tube	Piping	Table 604.1, Table 701.2
ASTM B280-2016	Seamless Copper Tube for Air Conditioning and Refrigeration Field Service	Piping	1208.6.4.2, 1308.5(1)(b)

REFERENCED STANDARDS

TABLE 1701.1 (continued)
REFERENCED STANDARDS

STANDARD NUMBER	STANDARD TITLE	APPLICATION	REFERENCED SECTIONS
ASTM B302-2012	Threadless Copper Pipe, Standard Sizes	Piping	Table 604.1, Table 701.2
ASTM B306-2013	Copper Drainage Tube (DWV)	Piping	Table 701.2, 903.2.3
ASTM B447-2012a	Welded Copper Tube	Piping	Table 604.1
ASTM B813-2016	Liquid and Paste Fluxes for Soldering of Copper and Copper Alloy Tube	Joints	605.1.4, 705.3.3
ASTM B819-2000 (R2011)	Seamless Copper Tube for Medical Gas Systems	Piping	1308.4, 1308.5(1)(c)
ASTM B828-2016	Making Capillary Joints by Soldering of Copper and Copper Alloy Tube and Fittings	Joints	605.1.4, 705.3.3, 1308.6(4)
ASTM C4-2004 (R2014)	Clay Drain Tile and Perforated Clay Drain Tile	Piping	Table 1101.4.6
ASTM C425-2004 (R2013)	Compression Joints for Vitrified Clay Pipe and Fittings	Joints	705.8.1
ASTM C564-2014	Rubber Gaskets for Cast Iron Soil Pipe and Fittings	Joints	705.2.2
ASTM C700-2013	Vitrified Clay Pipe, Extra Strength, Standard Strength, and Perforated	Piping	Table 701.2, Table 1101.4.6
ASTM C1053-2000 (R2015)	Borosilicate Glass Pipe and Fittings for Drain, Waste, and Vent (DWV) Applications	Piping	811.2
ASTM C1173-2010(R2014)	Flexible Transition Couplings for Underground Piping Systems	Fittings	705.10
ASTM C1277-2015	Shielded Couplings Joining Hubless Cast Iron Soil Pipe and Fittings	Fixtures	301.2.4, 705.2.2
ASTM C1460-2012	Shielded Transition Couplings for Use With Dissimilar DWV Pipe and Fittings Above Ground	Joints	705.10
ASTM C1461-2008 (R2013)	Mechanical Couplings Using Thermoplastic Elastomeric (TPE) Gaskets for Joining Drain, Waste, and Vent (DWV), Sewer, Sanitary, and Storm Plumbing Systems for Above and Below Ground Use	Joints	705.10
ASTM C1540-2015	Heavy Duty Shielded Couplings Joining Hubless Cast Iron Soil Pipe and Fittings	Joints	705.2.2
ASTM C1563-2008 (R2013)	Gaskets for Use in Connection with Hub and Spigot Cast Iron Soil Pipe and Fittings for Sanitary Drain, Waste, Vent, and Storm Piping Applications	Joints	705.2.2
ASTM C1822-2015	Insulating Covers on Accessible Lavatory Piping	Miscellaneous	403.3
ASTM D1785-2015	Poly (Vinyl Chloride) (PVC) Plastic Pipe, Schedules 40, 80, and 120	Piping	Table 604.1, Table 701.2, 1308.6(2)(a)
ASTM D2235-2004(R2016)	Solvent Cement for Acrylonitrile-Butadiene-Styrene (ABS) Plastic Pipe and Fittings	Joints	705.1.2
ASTM D2239-2012a	Polyethylene (PE) Plastic Pipe (SIDR-PR) Based on Controlled Inside Diameter	Piping	Table 604.1
ASTM D2241-2015	Poly (Vinyl Chloride) (PVC) Pressure-Rated Pipe (SDR Series)	Piping	Table 604.1
ASTM D2464-2015	Threaded Poly (Vinyl Chloride) (PVC) Plastic Pipe Fittings, Schedule 80	Fittings	Table 604.1
ASTM D2466-2015	Poly (Vinyl Chloride) (PVC) Plastic Pipe Fittings, Schedule 40	Fittings	Table 604.1, 1308.6(2)(b)
ASTM D2467-2015	Poly (Vinyl Chloride) (PVC) Plastic Pipe Fittings, Schedule 80	Fittings	Table 604.1, 1308.6(2)(b)
ASTM D2513-2014e1	Polyethylene (PE) Gas Pressure Pipe, Tubing, and Fittings	Piping	1208.6.5, 1208.6.7(2), 1208.6.12.2, 1210.1.7.1(1)
ASTM D2564-2012	Solvent Cements for Poly (Vinyl Chloride) (PVC) Plastic Piping Systems	Joints	605.12.2, 705.6.2

TABLE 1701.1 (continued)
REFERENCED STANDARDS

STANDARD NUMBER	STANDARD TITLE	APPLICATION	REFERENCED SECTIONS
ASTM D2609-2015	Plastic Insert Fittings for Polyethylene (PE) Plastic Pipe	Fittings	Table 604.1
ASTM D2661-2014	Acrylonitrile-Butadiene-Styrene (ABS) Schedule 40 Plastic Drain, Waste, and Vent Pipe and Fittings	Piping	Table 701.2
ASTM D2665-2014	Poly (Vinyl Chloride) (PVC) Plastic Drain, Waste, and Vent Pipe and Fittings	Piping	Table 701.2
ASTM D2672-2014	Joints for IPS PVC Pipe Using Solvent Cement	Joints	1308.6(2)(c)
ASTM D2680-2001 (R2014)	Acrylonitrile-Butadiene-Styrene (ABS) and Poly (Vinyl Chloride) (PVC) Composite Sewer Piping	Piping	Table 701.2
ASTM D2683- 2014	Socket-Type Polyethylene Fittings for Outside Diameter-Controlled Polyethylene Pipe and Tubing	Fittings	Table 604.1
ASTM D2729-2011	Poly (Vinyl Chloride) (PVC) Sewer Pipe and Fittings	Piping	Table 701.2, Table 1101.4.6
ASTM D2737-2012a	Polyethylene (PE) Plastic Tubing	Piping, Plastic	Table 604.1
ASTM D2846/D2846M-2014	Chlorinated Poly (Vinyl Chloride) (CPVC) Plastic Hot- and Cold-Water Distribution Systems	Piping	Table 604.1, 605.2.2, 605.3.1, 1308.6(2)
ASTM D3034-2014a	Type PSM Poly (Vinyl Chloride) (PVC) Sewer Pipe and Fittings	Piping, Plastic	Table 701.2
ASTM D3035-2015	Polyethylene (PE) Plastic Pipe (DR-PR) Based on Controlled Outside Diameter	Piping	Table 604.1
ASTM D3138-2004 (R2016)	Solvent Cement for Transition Joints Between Acrylonitrile-Butadiene-Styrene (ABS) and Poly (Vinyl Chloride) (PVC) Non-Pressure Piping Components	Joints	705.9.4
ASTM D3139-1998 (R2011)	Joints for Plastic Pressure Pipes Using Flexible Elastomeric Seals	Joints	605.12.1
ASTM D3212-2007 (R2013)	Joints for Drain and Sewer Plastic Pipes Using Flexible Elastomeric Seals	Joints	705.1.1, 705.6.1
ASTM D3261-2016	Butt Heat Fusion Polyethylene (PE) Plastic Fittings for Polyethylene (PE) Plastic Pipe and Tubing	Fittings	Table 604.1
ASTM D4068-2015	Chlorinated Polyethylene (CPE) Sheeting for Concealed Water-Containment Membrane	Miscellaneous	408.7.2
ASTM D4551-2012	Poly (Vinyl Chloride) (PVC) Plastic Flexible Concealed Water-Containment Membrane	Miscellaneous	408.7.1
ASTM E84-2016	Surface Burning Characteristics of Building Materials	Miscellaneous	701.2(2), 903.1(2), 1101.4
ASTM E119- 2016a	Fire Tests of Building Construction and Materials	Miscellaneous	1404.3, 1405.3
ASTM E814-2013a	Fire Tests of Penetration Firestop Systems	Miscellaneous	208.0, 222.0, 1404.3, 1405.3
ASTM F409-2012	Thermoplastic Accessible and Replaceable Plastic Tube and Tubular Fittings	Piping, Plastic	404.1
ASTM F437-2015	Threaded Chlorinated Poly (Vinyl Chloride) (CPVC) Plastic Pipe Fittings, Schedule 80	Fittings	Table 604.1
ASTM F438-2015	Socket-Type Chlorinated Poly (Vinyl Chloride) (CPVC) Plastic Pipe Fittings, Schedule 40	Fittings	Table 604.1, 1308.6(2)(e),
ASTM F439-2013	Chlorinated Poly (Vinyl Chloride) (CPVC) Plastic Pipe Fittings, Schedule 80	Fittings	Table 604.1, 1308.6(2)(e)
ASTM F441/F441M-2015	Chlorinated Poly (Vinyl Chloride) (CPVC) Plastic Pipe, Schedules 40 and 80	Piping	Table 604.1, 1308.6(2)(a)
ASTM F442/F442M-2013^{e1}	Chlorinated Poly (Vinyl Chloride) (CPVC) Plastic Pipe (SDR-PR)	Piping	Table 604.1, 605.2.2
ASTM F493-2014	Solvent Cements for Chlorinated Poly (Vinyl Chloride) (CPVC) Plastic Pipe and Fittings	Joints	605.2.2, 605.3.1, 1308.6(2)(g)

REFERENCED STANDARDS

TABLE 1701.1 (continued)
REFERENCED STANDARDS

STANDARD NUMBER	STANDARD TITLE	APPLICATION	REFERENCED SECTIONS
ASTM F628-2012e1	Acrylonitrile-Butadiene-Styrene (ABS) Schedule 40 Plastic Drain, Waste, and Vent Pipe with a Cellular Core	Piping	Table 701.2
ASTM F656-2015	Primers for Use in Solvent Cement Joints of Poly (Vinyl Chloride) (PVC) Plastic Pipe and Fittings	Joints	605.2.2, 605.3.1, 605.12.2, 705.6.2
ASTM F667/F667M-2016	3 through 24 in. Corrugated Polyethylene Pipe and Fittings	Piping, Plastic	Table 1101.4.6
ASTM F714-2013	Polyethylene (PE) Plastic Pipe (DR-PR) Based on Outside Diameter	Piping	Table 701.2
ASTM F794-2003 (R2014)	Poly (Vinyl Chloride) (PVC) Profile Gravity Sewer Pipe and Fittings Based on Controlled Inside Diameter	Piping	Table 701.2
ASTM F876-2015a	Crosslinked Polyethylene (PEX) Tubing	Piping	Table 604.1, 605.9.1
ASTM F877-2011a	Crosslinked Polyethylene (PEX) Hot- and Cold-Water Distribution Systems	Piping	Table 604.1
ASTM F891-2010	Coextruded Poly (Vinyl Chloride) (PVC) Plastic Pipe with a Cellular Core	Piping	Table 701.2
ASTM F894-2013	Polyethylene (PE) Large Diameter Profile Wall Sewer and Drain Pipe	Piping, Plastic	Table 701.2
ASTM F1055-2016a	Electrofusion Type Polyethylene Fittings for Outside Diameter Controlled Polyethylene and Crosslinked Polyethylene (PEX) Pipe and Tubing	Fittings	Table 604.1, 705.5.1.2
ASTM F1216-2016	Rehabilitation of Existing Pipelines and Conduits by the Inversion and Curing of a Resin-Impregnated Tube	Piping	715.3
ASTM F1281-2011	Crosslinked Polyethylene/Aluminum/Crosslinked Polyethylene (PEX-AL-PEX) Pressure Pipe	Piping	Table 604.1
ASTM F1282-2010	Polyethylene/Aluminum/Polyethylene (PE-AL-PE) Composite Pressure Pipe	Piping	Table 604.1
ASTM F1336-2015	Poly (Vinyl Chloride) (PVC) Gasketed Sewer Fittings	Fittings	Table 701.2
ASTM F1412-2016	Polyolefin Pipe and Fittings for Corrosive Waste Drainage Systems	Piping	811.2
ASTM F1488-2014	Coextruded Composite Pipe	Piping	Table 701.2
ASTM F1673-2010 (R2016)	Polyvinylidene Fluoride (PVDF) Corrosive Waste Drainage Systems	Piping	811.2
ASTM F1760-2001 (R2011)	Coextruded Poly(Vinyl Chloride) PVC Non-Pressure Plastic Pipe Having Reprocessed Recycled Content	Piping	Table 701.2
ASTM F1807-2015	Metal Insert Fittings Utilizing a Copper Crimp Ring for SDR9 Cross-linked Polyethylene (PEX) Tubing and SDR9 Polyethylene of Raised Temperature (PE-RT) Tubing	Fittings	Table 604.1
ASTM F1866-2013	Poly (Vinyl Chloride) (PVC) Plastic Schedule 40 Drainage and DWV Fabricated Fittings	Fittings	Table 701.2
ASTM F1960-2015	Cold Expansion Fittings with PEX Reinforcing Rings for Use with Cross-linked Polyethylene (PEX) Tubing	Fittings	Table 604.1
ASTM F1961-2009	Metal Mechanical Cold Flare Compression Fittings with Disc Spring for Crosslinked Polyethylene (PEX) Tubing	Fittings	Table 604.1
ASTM F1970-2012e1	Special Engineered Fittings, Appurtenances or Valves for Use in Poly (Vinyl Chloride) (PVC) or Chlorinated Poly (Vinyl Chloride) (CPVC) Systems	Piping	Table 604.1, 606.1
ASTM F1973-2013e1	Factory Assembled Anodeless Risers and Transition Fittings in Polyethylene (PE) and Polyamide 11 (PA11) and Polyamide 12 (PA12) Fuel Gas Distribution Systems	Fuel Gas	1210.1.7.1(2)
ASTM F1974-2009 (R2015)	Metal Insert Fittings for Polyethylene/ Aluminum/Polyethylene and Crosslinked Polyethylene/Aluminum/Crosslinked Polyethylene Composite Pressure Pipe	Fittings	Table 604.1, 605.7.1, 605.10.1

REFERENCED STANDARDS

TABLE 1701.1 (continued)
REFERENCED STANDARDS

STANDARD NUMBER	STANDARD TITLE	APPLICATION	REFERENCED SECTIONS
ASTM F1986-2001 (R2011)	Multilayer Pipe Type 2, Compression Fittings, and Compression Joints for Hot and Cold Drinking-Water Systems	Fittings	Table 604.1
ASTM F2080-2016	Cold-Expansion Fittings with Metal Compression-Sleeves for Crosslinked Polyethylene (PEX) Pipe and SDR9 Polyethylene of Raised Temperature (PE-RT) Pipe	Fittings	Table 604.1
ASTM F2098-2015	Stainless Steel Clamps for Securing SDR9 Cross-linked Polyethylene (PEX) Tubing to Metal Insert and Plastic Insert Fittings	Fittings	Table 604.1
ASTM F2159-2014	Plastic Insert Fittings Utilizing a Copper Crimp Ring for SDR9 Cross-linked Polyethylene (PEX) Tubing and SDR9 Polyethylene of Raised Temperature (PE-RT) Tubing	Fittings	Table 604.1
ASTM F2262-2009	Crosslinked Polyethylene/Aluminum/ Crosslinked Polyethylene Tubing OD Controlled SDR9	Piping, Plastic	Table 604.1
ASTM F2389-2015	Pressure-Rated Polypropylene (PP) Piping Systems	Piping	Table 604.1, 605.11.1, 606.1,
ASTM F2434-2014	Metal Insert Fittings Utilizing a Copper Crimp Ring for SDR9 Cross-linked Polyethylene (PEX) Tubing and SDR9 Cross-linked Polyethylene/Aluminum/Cross-linked Polyethylene (PEX-AL-PEX) Tubing	Fittings	Table 604.1, 605.10.1
ASTM F2509-2015	Field-Assembled Anodeless Riser Kits for Use on Outside Diameter Controlled Polyethylene and Polyamide-11 (PA11) Gas Distribution Pipe and Tubing	Fuel Gas	1210.1.7.1(3)
ASTM F2618-2015	Chlorinated Poly(Vinyl Chloride) CPVC Pipe and Fittings for Chemical Waste Drainage Systems	Piping	811.2
ASTM F2620-2013	Heat Fusion Joining of Polyethylene Pipe and Fittings	Joints	605.6.1.1, 605.6.1.3, 705.5.1.1, 705.5.1.3
ASTM F2735-2009 (R2016)	Plastic Insert Fittings for SDR9 Cross-linked Polyethylene (PEX) and Polyethylene of Raised Temperature (PE-RT) Tubing	Fittings	Table 604.1
ASTM F2769-2016	Polyethylene of Raised Temperature (PE-RT) Plastic Hot and Cold-Water Tubing and Distribution Systems	Piping, Fittings	Table 604.1
ASTM F2831-2012	Internal Non-Structural Epoxy Barrier Coating Material Used in Rehabilitation of Metallic Pressurized Piping Systems	Miscellaneous	320.1
ASTM F2855-2012	Chlorinated Poly (Vinyl Chloride)/Aluminum/Chlorinated Poly (Vinyl Chloride) (CPVC/AL/CPVC) Composite Pressure Tubing	Piping	Table 604.1, 605.3.1
AWS A5.8-2011	Filler Metals for Brazing and Braze Welding	Joints	605.1.1, 705.3.1, 1309.4.2
AWS A5.9-2012	Bare Stainless Steel Welding Electrodes and Rods	Joints	605.13.2
AWS B2.2-2010	Brazing Procedure and Performance Qualification	Certification	1307.1
AWWA C110-2012	Ductile-Iron and Gray-Iron Fittings	Fittings	Table 604.1
AWWA C111-2012	Rubber-Gasket Joints for Ductile-Iron Pressure Pipe and Fittings	Joints	605.4.1, 605.4.2
AWWA C151-2009	Ductile-Iron Pipe, Centrifugally Cast	Piping	Table 604.1
AWWA C153-2011	Ductile-Iron Compact Fittings	Fittings	Table 604.1
AWWA C210-2015	Liquid-Epoxy Coatings and Linings for Steel Water Pipe and Fittings	Miscellaneous	604.9
AWWA C500-2009	Metal-Seated Gate Valves for Water Supply Service	Valves	606.1
AWWA C504-2015	Rubber-Seated Butterfly Valves	Valves	606.1
AWWA C507- 2015	Ball Valves, 6 in. through 60 in. (150 mm through 1,500 mm)	Valves	606.1
AWWA C510-2007	Double Check Valve Backflow Prevention Assembly	Backflow Protection	Table 603.2
AWWA C511-2007	Reduced-Pressure Principle Backflow Prevention Assembly	Backflow Protection	Table 603.2
AWWA C900-2016	Polyvinyl Chloride (PVC) Pressure Pipe and Fabricated Fittings, 4 in. through 12 in. (100 mm through 300 mm)	Piping	Table 604.1

REFERENCED STANDARDS

TABLE 1701.1 (continued)
REFERENCED STANDARDS

STANDARD NUMBER	STANDARD TITLE	APPLICATION	REFERENCED SECTIONS
AWWA C901-2008	Polyethylene (PE) Pressure Pipe and Tubing, 1/2 in. (13 mm) through 3 in. (76 mm), for Water Service	Piping	Table 604.1
AWWA C904-2016	Crosslinked Polyethylene (PEX) Pressure Tubing, 1/2 in. through 3 in. (13 mm through 76 mm), for Water Service	Piping	Table 604.1
AWWA C907-2012	Injection-Molded Polyvinyl Chloride (PVC) Pressure Fittings, 4 in. through 12 in. (100 mm through 300 mm), for Water, Wastewater, and Reclaimed Water Service	Fittings	Table 604.1
CGA G-4.1-2009	Cleaning Equipment for Oxygen Service	Miscellaneous	1308.2, 1309.6
CISPI 301-2012	Hubless Cast Iron Soil Pipe and Fittings for Sanitary and Storm Drain, Waste, and Vent Piping Applications	Piping, Ferrous	301.2.4, Table 701.2
CISPI 310-2012	Couplings for Use in Connection with Hubless Cast Iron Soil Pipe and Fittings for Sanitary and Storm Drain, Waste, and Vent Piping Applications	Joints	301.2.4, 705.2.2
CSA B45.5-2011/IAPMO Z124-2011 (R2016)	Plastic Plumbing Fixtures	Fixtures	407.1, 408.1, 409.1, 411.1, 412.1, 420.1
CSA B45.8-2013/IAPMO Z403-2013	Terrazzo, Concrete, and Natural Stone Plumbing Fixtures	Fixtures	407.1, 420.1
CSA B45.11-2011/IAPMO Z401-2011	Glass Plumbing Fixtures	Fixtures	407.1
CSA B45.12-2013/IAPMO Z402-2013	Aluminum and Copper Plumbing Fixtures	Fixtures	407.1, 408.1, 409.1, 420.1
CSA B64.1.1-2011 (R2016)	Atmospheric Vacuum Breakers (AVB)	Backflow Protection	Table 603.2
CSA B64.1.2-2011 (R2016)	Pressure Vacuum Breakers (PVB)	Backflow Protection	Table 603.2
CSA B64.2.1.1-2011 (R2016)	Hose Connection Dual Check Vacuum Breakers (HCDVB)	Backflow Protection	Table 603.2
CSA B64.4-2011 (R2016)	Reduced Pressure Principle (RP) Backflow Preventers	Backflow Protection	Table 603.2
CSA B64.4.1-2011 (R2016)	Reduced Pressure Principle Backflow Preventers for Fire Protection Systems (RPF)	Backflow Protection	Table 603.2
CSA B64.5-2011 (R2016)	Double Check Valve (DCVA) Backflow Preventers	Backflow Protection	Table 603.2
CSA B64.5.1-2011 (R2016)	Double Check Valve Backflow Preventers for Fire Protection Systems (DCVAF)	Backflow Protection	Table 603.2
CSA B79-2008 (R2013)	Commercial and Residential Drains and Cleanouts	Fixtures	418.1
CSA B125.5/IAPMO Z2600-2011	Flexible Water Connectors with Excess Flow Shutoff Device	Miscellaneous	604.5
CSA B137.1-2013	Polyethylene (PE) Pipe, Tubing, and Fittings for Cold-Water Pressure Services	Piping	Table 604.1
CSA B137.5-2013	Crosslinked Polyethylene (PEX) Tubing Systems for Pressure Applications	Piping	Table 604.1
CSA B137.6-2013	Chlorinated Polyvinylchloride (CPVC) Pipe, Tubing, and Fittings for Hot- and Cold-Water Distribution Systems	Piping, Fittings	Table 604.1
CSA B137.9-2013	Polyethylene/Aluminum/Polyethylene (PE-AL-PE) Composite Pressure-Pipe Systems	Piping	Table 604.1
CSA B137.10-2013	Crosslinked Polyethylene/Aluminum/Crosslinked Polyethylene (PEX-AL-PEX) Composite Pressure-Pipe Systems	Piping	Table 604.1
CSA B137.11-2013	Polypropylene (PP-R) Pipe and Fittings for Pressure Applications	Piping	Table 604.1, 605.11.1
CSA B137.18-2013	Polyethylene of Raised Temperature (PE-RT) Tubing Systems for Pressure Applications	Piping, Fittings	Table 604.1
CSA B181.3-2015	Polyolefin and Polyvinylidene Fluoride (PVDF) Laboratory Drainage Systems	Piping	811.2
CSA B481-2012	Grease Interceptors	Fixtures	1014.1
CSA LC 1-2016	Fuel Gas Piping Systems Using Corrugated Stainless Steel Tubing (same as CSA 6.26)	Fuel Gas	1208.6.4.4

REFERENCED STANDARDS

TABLE 1701.1 (continued)
REFERENCED STANDARDS

STANDARD NUMBER	STANDARD TITLE	APPLICATION	REFERENCED SECTIONS
CSA LC 4a-2013	Press-Connect Metallic Fittings for Use in Fuel Gas Distribution Systems (same as CSA 6.32a)	Fuel Gas	1208.6.11.1, 1208.6.11.2
CSA Z21.10.1-2014	Gas Water Heaters, Volume I, Storage Water Heaters with Input Ratings of 75,000 Btu Per Hour or Less (same as CSA 4.1)	Fuel Gas, Appliances	Table 501.1(1)
CSA Z21.10.3-2015	Gas-Fired Water Heaters, Volume III, Storage Water Heaters with Input Ratings Above 75,000 Btu Per Hour, Circulating and Instantaneous (same as CSA 4.3)	Fuel Gas, Appliances	Table 501.1(1)
CSA Z21.22-2015	Relief Valves for Hot Water Supply Systems (same as CSA 4.4)	Valves	607.5, 608.7
CSA Z21.24-2015	Connectors for Gas Appliances (same as CSA 6.10)	Fuel Gas	1212.1(3), 1212.2
CSA Z21.41-2014	Quick-Disconnect Devices for Use with Gas Fuel Appliances (same as CSA 6.9)	Fuel Gas	1212.7
CSA Z21.54-2014	Gas Hose Connectors for Portable Outdoor Gas-Fired Appliances (same as CSA 8.4)	Fuel Gas	1212.3.2
CSA Z21.69-2015	Connectors for Moveable Gas Appliances (same as CSA 6.16)	Fuel Gas	1212.1.1
CSA Z21.75-2016	Connectors for Outdoor Gas Appliances and Manufactured Homes (same as CSA 6.27)	Fuel Gas	1212.1(4)
CSA Z21.80a-2012	Line Pressure Regulators (same as CSA 6.22a)	Fuel Gas	1208.8.1, 1208.8.4(1)
CSA Z21.90-2015	Gas Convenience Outlets and Optional Enclosures (same as CSA 6.24)	Fuel Gas	1212.8
CSA Z21.93-2013	Excess Flow Valves for Natural and LP Gas with Pressures up to 5 psig (same as CSA 6.30)	Fuel Gas	1209.1
IAPMO PS 65-2002	Airgap Units for Water Conditioning Equipment Installation	Backflow Protection	611.2
IAPMO Z124.5-2013e1	Plastic Toilet Seats	Appurtenance	411.3
IAPMO Z1001-2016	Prefabricated Gravity Grease Interceptors	Fixtures	1014.3.4
IAPMO Z1033-2015	Flexible PVC Hoses and Tubing for Pools, Hot Tubs, Spas, and Jetted Bathtubs	Tubing	409.6.1
IAPMO Z1157-2014e1	Ball Valves	Valves	606.1
ICC A117.1-2009	Accessible and Usable Buildings and Facilities	Miscellaneous	403.2, 408.6
ISEA Z358.1-2014	Emergency Eyewash and Shower Equipment	Miscellaneous	416.1, 416.2
MSS SP-58-2009	Pipe Hangers and Supports – Materials, Design, Manufacture, Selection, Application, and Installation	Miscellaneous	1210.2.4, 1310.5.1
MSS SP-67-2016	Butterfly Valves	Valves	606.1
MSS SP-70-2011	Gray Iron Gate Valves, Flanged and Threaded Ends	Valves	606.1
MSS SP-71-2011	Gray Iron Swing Check Valves, Flanged and Threaded Ends	Valves	606.1
MSS SP-72-2010a	Ball Valves with Flanged or Butt-Welding Ends for General Service	Valves	606.1
MSS SP-78-2011	Gray Iron Plug Valves, Flanged and Threaded Ends	Valves	606.1
MSS SP-80-2013	Bronze Gate, Globe, Angle, and Check Valves	Valves	606.1
MSS SP-110-2010	Ball Valves Threaded, Socket-Welding, Solder Joint, Grooved and Flared Ends	Valves	606.1
MSS SP-122-2012	Plastic Industrial Ball Valves	Valves	606.1
NFPA 13D-2016	Installation of Sprinkler Systems in One- and Two-Family Dwellings and Manufactured Homes	Miscellaneous	612.1, 612.5.3.1
NFPA 30A-2015	Motor Fuel Dispensing Facilities and Repair Garages	Miscellaneous	507.14.2
NFPA 31-2016	Installation of Oil-Burning Equipment	Fuel Gas, Appliances	505.3, 1201.1
NFPA 51-2018	Design and Installation of Oxygen-Fuel Gas Systems for Welding, Cutting, and Allied Processes	Fuel Gas	507.9

REFERENCED STANDARDS

TABLE 1701.1 (continued)
REFERENCED STANDARDS

STANDARD NUMBER	STANDARD TITLE	APPLICATION	REFERENCED SECTIONS
NFPA 54/Z223.1-2015	National Fuel Gas Code	Fuel Gas	Chapter 5, Chapter 12
NFPA 58-2017	Liquefied Petroleum Gas Code	Fuel Gas	1208.5(6), 1208.6.7(3), 1208.6.12.4, 1212.11
NFPA 70-2017	National Electrical Code	Miscellaneous	1210.12.5(2), 1211.2.4, 1211.6, 1310.4.1, 1317.1(11)
NFPA 88A-2015	Parking Structures	Miscellaneous	507.14.1
NFPA 99-2015	Health Care Facilities Code	Miscellaneous	1301.3, 1309.8.9(6), 1317.1(9)
NFPA 211-2016	Chimneys, Fireplaces, Vents, and Solid Fuel-Burning Appliances	Fuel Gas, Appliances	509.5.2, 509.5.3, 509.5.6.1, 509.5.6.3
NFPA 409-2016	Aircraft Hangars	Miscellaneous	507.15
NFPA 780-2017	Installation of Lightning Protection Systems	Fuel Gas	1211.4
NFPA 1192-2015	Recreational Vehicles	Fuel Gas	1202.3
NSF 3-2012	Commercial Warewashing Equipment	Appliances	414.1
NSF 14-2016	Plastics Piping System Components and Related Materials	Miscellaneous	301.2.3, 604.1
NSF 42-2015	Drinking Water Treatment Units – Aesthetic Effects	Appliances	611.1
NSF 44-2015	Residential Cation Exchange Water Softeners	Appliances	611.1
NSF 53-2015	Drinking Water Treatment Units-Health Effects	Appliances	611.1
NSF 55-2016	Ultraviolet Microbiological Water Treatment Systems	Appliances	611.1
NSF 58-2015	Reverse Osmosis Drinking Water Treatment Systems	Appliances	611.1, 611.2
NSF 61-2016	Drinking Water System Components – Health Effects	Miscellaneous	415.1, 417.1, 604.1, 604.9, 606.1, 607.2, 608.2
NSF 62-2015	Drinking Water Distillation Systems	Appliances	611.1
NSF 350-2014	Onsite Residential and Commercial Water Reuse Treatment Systems	Miscellaneous	1506.7
NSF 359-2016	Valves for Crosslinked Polyethylene (PEX) Water Distribution Tubing Systems	Valves	606.1
PDI G-101-2015	Testing and Rating Procedure for Hydro Mechanical Grease Interceptors with Appendix of Installation and Maintenance	DWV Components	1014.1
PDI G-102-2010	Testing and Certification for Grease Interceptors with FOG Sensing and Alarm Devices	Certification	1014.1
PDI-WH 201-2010	Water Hammer Arresters	Water Supply Components	609.10
UL 17-2008	Vent or Chimney Connector Dampers for Oil-Fired Appliances (with revisions through September 25, 2013)	Fuel Gas, Vent Dampers	509.14.1
UL 103-2010	Factory-Built Chimneys for Residential Type and Building Heating Appliances (with revisions through July 27, 2012)	Fuel Gas, Appliances	509.5.1.1, 509.5.1.2
UL 174-2004	Household Electric Storage Tank Water Heaters (with revisions through April 10, 2015)	Appliances	Table 501.1(1)
UL 263-2011	Fire Tests of Building Construction and Materials (with revisions through June 2, 2015)	Miscellaneous	1404.3, 1405.3
UL 378-2006	Draft Equipment (with revisions through September 17, 2013)	Fuel Gas, Appliances	509.14.1
UL 399-2008	Drinking Water Coolers (with revisions through October 18, 2013)	Fixtures	415.1
UL 430-2015	Waste Disposers	Appliances	419.1
UL 441-2016	Gas Vents (with revisions through July 27, 2016)	Fuel Gas, Vents	509.1
UL 467-2013	Grounding and Bonding Equipment	Miscellaneous	1211.2.5

REFERENCED STANDARDS

TABLE 1701.1 (continued)
REFERENCED STANDARDS

STANDARD NUMBER	STANDARD TITLE	APPLICATION	REFERENCED SECTIONS
UL 641-2010	Type L Low-Temperature Venting Systems (with revisions through June 12, 2013)	Fuel Gas	509.1
UL 651-2011	Schedule 40, 80, Type EB and A Rigid PVC Conduit and Fittings (with revisions through June 15, 2016)	Piping	1208.6.6
UL 723-2008	Test for Surface Burning Characteristics of Building Materials (with revisions through August 12, 2013)	Miscellaneous	701.2(2), 903.1(2), 1101.4
UL 732-1995	Oil-Fired Storage Tank Water Heaters (with revisions through October 9, 2013)	Fuel Gas, Appliances	Table 501.1(1)
UL 749-2013	Household Dishwashers (with revisions through May 24, 2013)	Appliances	414.1
UL 778-2016	Motor-Operated Water Pumps (with revisions through November 14, 2016)	Appliances	1101.14
UL 921-2016	Commercial Dishwashers	Appliances	414.1
UL 959-2010	Medium Heat Appliance Factory-Built Chimneys (with revisions through June 12, 2014)	Fuel Gas, Appliances	509.5.1.2
UL 1453-2016	Electric Booster and Commercial Storage Tank Water Heaters	Appliances	Table 501.1(1)
UL 1479- 2015	Fire Tests of Penetration Firestops	Miscellaneous	208.0, 222.0, 1404.3, 1405.3
UL 2523-2009	Solid Fuel-Fired Hydronic Heating Appliances, Water Heaters, and Boilers (with revisions through February 8, 2013)	Fuel Gas, Appliances	Table 501.1(1)

1701.2 Standards, Publications, Practices, and Guides. The standards, publications, practices, and guides listed in Table 1701.2 are not referenced in other sections of this code. The application of the referenced standards, publications, practices, and guides shall be as specified in Section 301.2.2. The promulgating agency acronyms are found at the end of the tables.

A list of additional standards, publications, and guides that are not referenced in specific sections of this code appear in Table 1701.2. The standards from Table 1701.2 shall be permitted only after they have been approved by the Authority Having Jurisdiction. See also Section 301.2.2.

TABLE 1701.2
STANDARDS, PUBLICATIONS, PRACTICES, AND GUIDES

STANDARD NUMBER	STANDARD TITLE	APPLICATION
AHAM FWD-1-2009	Food Waste Disposers	Appliances
ASCE 25-2006	Earthquake-Actuated Automatic Gas Shutoff Devices	Fuel Gas
ASHRAE 90.1-2016	Energy Standard for Buildings Except Low-Rise Residential Buildings	Miscellaneous
ASHRAE 90.2-2007	Energy-Efficient Design of Low-Rise Residential Buildings	Miscellaneous
ASME A13.1-2007 (R2013)	Scheme for the Identification of Piping Systems	Piping
ASME A112.4.3-1999 (R2015)	Plastic Fittings for Connecting Water Closets to the Sanitary Drainage System	Fittings
ASME A112.19.10-2003 (R2008)	Dual Flush Devices for Water Closets	Fixtures
ASME A112.21.3M-1985 (R2007)	Hydrants for Utility and Maintenance Use	Valves
ASME A112.36.2M-1991 (R2012)	Cleanouts	DWV Components
ASME B1.20.3-1976 (R2013)	Dryseal Pipe Threads, (Inch)	Joints
ASME B16.33-2012	Manually Operated Metallic Gas Valves for Use in Gas Piping Systems Up to 175 psi (Sizes NPS ½ through NPS 2)	Valves
ASME B16.39-2014	Malleable Iron Threaded Pipe Unions: Classes 150, 250 and 300	Fittings
ASME B16.40-2013	Manually Operated Thermoplastic Gas Shutoffs and Valves in Gas Distribution Systems	Valves
ASME B31.1-2016	Power Piping	Piping
ASME B36.19M-2004 (R2015)	Stainless Steel Pipe	Piping, Ferrous

REFERENCED STANDARDS

TABLE 1701.2 (continued)
STANDARDS, PUBLICATIONS, PRACTICES, AND GUIDES

STANDARD NUMBER	STANDARD TITLE	APPLICATION
ASME BPVC Section IV-2015	Rules for Construction of Heating Boilers	Miscellaneous
ASSE 1003-2009	Water Pressure Reducing Valves for Domestic Water Distribution Systems	Valves
ASSE 1012-2009	Backflow Preventers with an Intermediate Atmospheric Vent	Backflow Protection
ASSE 1017-2009	Temperature Actuated Mixing Valves for Hot Water Distribution Systems	Valves
ASSE 1023-1979	Hot Water Dispensers Household Storage Type - Electrical	Appliances
ASSE 1024-2004	Dual Check Backflow Preventers	Backflow Protection
ASSE 1032-2004 (R2011)	Dual Check Valve Type Backflow Preventers for Carbonated Beverage Dispensers, Post Mix Type	Backflow Protection
ASSE 1035-2008	Laboratory Faucet Backflow Preventers	Backflow Protection
ASSE 1062-2006	Temperature Actuated, Flow, Reduction (TAFR) Valves for Individual Supply Fittings	Valves
ASSE 1066-1997	Individual Pressure Balancing In-Line Valves for Individual Fixture Fittings	Valves
ASTM A48/A48M-2003 (R2016)	Gray Iron Castings	Piping, Ferrous
ASTM A126-2004 (R2014)	Gray Iron Castings for Valves, Flanges, and Pipe Fittings	Piping, Ferrous
ASTM A377-2003 (R2014)	Ductile-Iron Pressure Pipe	Piping, Ferrous
ASTM A479/A479M-2016a	Stainless Steel Bars and Shapes for Use in Boilers and Other Pressure Vessels	Piping, Ferrous
ASTM A536-1984 (R2014)	Ductile Iron Castings	Piping, Ferrous
ASTM A733-2015	Welded and Seamless Carbon Steel and Austenitic Stainless Steel Pipe Nipples	Piping, Ferrous
ASTM A1045-2010 (R2014)	Flexible Poly (Vinyl Chloride) (PVC) Gaskets used in Connection of Vitreous China Plumbing Fixtures to Sanitary Drainage Systems	Piping, Plastic
ASTM B29-2014	Refined Lead	Joints
ASTM B370-2012	Standard Specification for Copper Sheet and Strip for Building Construction	Miscellaneous
ASTM B687-1999 (R2016)	Brass, Copper, and Chromium-Plated Pipe Nipples	Piping, Copper Alloy
ASTM C14-2015a	Nonreinforced Concrete Sewer, Storm Drain, and Culvert Pipe	Piping, Non-Metallic
ASTM C412-2015	Concrete Drain Tile	Piping, Non-Metallic
ASTM C443-2012	Joints for Concrete Pipe and Manholes, Using Rubber Gaskets	Joints
ASTM C444-2003 (R2009)	Perforated Concrete Pipe	Piping, Non-Metallic
ASTM C478-2015a	Circular Precast Reinforced Concrete Manhole Sections	Miscellaneous
ASTM C1440-2008 (R2013)[e1]	Thermoplastic Elastomeric (TPE) Gasket Materials for Drain, Waste, and Vent (DWV), Sewer, Sanitary and Storm Plumbing Systems	Joints
ASTM D1784-2011	Rigid Poly (Vinyl Chloride) (PVC) Compounds and Chlorinated Poly (Vinyl Chloride) (CPVC) Compounds	Piping, Plastic
ASTM D2321-2014[e1]	Underground Installation of Thermoplastic Pipe for Sewers and Other Gravity-Flow Applications	Piping, Plastic
ASTM D2517-2006 (R2011)	Reinforced Epoxy Resin Gas Pressure Pipe and Fittings	Piping, Plastic
ASTM D2657-2007 (R2015)	Heat Fusion Joining of Polyolefin Pipe and Fittings	Joints
ASTM D2774-2012	Underground Installation of Thermoplastic Pressure Piping	Piping, Plastic
ASTM D2855-2015	Two-Step (Primer and Solvent Cement) Method of Joining Poly (Vinyl Chloride) (PVC) or Chlorinated Poly (Vinyl Chloride) (CPVC) Pipe and Piping Components with Tapered Sockets	Joints
ASTM D3122-2015	Solvent Cement for Styrene-Rubber (SR) Plastic Pipe and Fittings	Joints
ASTM D3311-2011(R2016)	Drain, Waste, and Vent (DWV) Plastic Fittings Patterns	Fittings
ASTM F402-2005 (R2012)	Safe Handling of Solvent Cements, Primers, and Cleaners Used for Joining Thermoplastic Pipe and Fittings	Joints
ASTM F446-1985 (R2009)	Grab Bars and Accessories Installed in the Bathing Area	Miscellaneous
ASTM F480-2014	Thermoplastic Well Casing Pipe and Couplings Made in Standard Dimension Ratios (SDR), SCH 40 and SCH 80	Piping, Plastic

REFERENCED STANDARDS

TABLE 1701.2 (continued)
STANDARDS, PUBLICATIONS, PRACTICES, AND GUIDES

STANDARD NUMBER	STANDARD TITLE	APPLICATION
ASTM F810-2012	Smoothwall Polyethylene (PE) Pipe for Use in Drainage and Waste Disposal Absorption Fields	Piping, Plastic
ASTM F949-2015	Poly (Vinyl Chloride) (PVC) Corrugated Sewer Pipe with a Smooth Interior and Fittings	Piping, Plastic
ASTM F1476-2007 (R2013)	Performance of Gasketed Mechanical Couplings for Use in Piping Applications	Joints
ASTM F1499-2012	Coextruded Composite Drain, Waste, and Vent Pipe (DWV)	Piping, Plastic
ASTM F1743-2008 (R2016)	Rehabilitation of Existing Pipelines and Conduits by Pulled-in-Place Installation of Cured-in-Place Thermosetting Resin Pipe (CIPP)	Piping, Plastic
ASTM F1924-2012	Plastic Mechanical Fittings for Use on Outside Diameter Controlled Polyethylene Gas Distribution Pipe and Tubing	Fittings
ASTM F1948-2015	Metallic Mechanical Fittings for Use on Outside Diameter Controlled Thermoplastic Gas Distribution Pipe and Tubing	Fittings
ASTM F2165-2013	Flexible Pre-Insulated Piping	Piping, Plastic
ASTM F2206-2014	Fabricated Fittings of Butt-Fused Polyethylene (PE)	DWV Components
ASTM F2306/F2306M-2014^{e1}	12 to 60 in. [300 to 1500 mm] Annular Corrugated Profile-Wall Polyethylene (PE) Pipe and Fittings for Gravity-Flow Storm Sewer and Subsurface Drainage Applications	Piping, Plastic
AWS B2.4-2012	Welding Procedure and Performance Qualification for Thermoplastics	Joints, Certification
AWWA C203-2015	Coal-Tar Protective Coatings and Linings for Steel Water Pipe	Miscellaneous
AWWA C213-2015	Fusion-Bonded Epoxy Coatings and Linings for Steel Water Pipe and Fittings	Miscellaneous
AWWA C215-2016	Extruded Polyolefin Coatings for Steel Water Pipe	Miscellaneous
AWWA C606-2015	Grooved and Shouldered Joints	Joints
CGA C-9-2013	Standard Color Marking of Compressed Gas Containers for Medical Use	Miscellaneous
CGA S-1.3-2008	Pressure Relief Device Standards-Part 3-Stationary Storage Containers for Compressed Gases	Fuel Gas
CGA V-1-2013	Compressed Gas Cylinder Valve Outlet and Inlet Connections	Valves
CSA A257-2014	Concrete Pipe and Manhole Sections	Piping
CSA B64.7-2011 (R2016)	Laboratory Faucet Vacuum Breakers (LFVB)	Backflow Protection
CSA B66-2010 (R2015)	Design, Material, and Manufacturing Requirements for Prefabricated Septic Tanks and Sewage Holding Tanks	DWV Components
CSA B128.1-2006/B128.2-2006 (R2016)	Design and Installation of Non-Potable Water Systems/Maintenance and Field Testing of Non-Potable Water Systems	Miscellaneous
CSA B242-2005 (R2016)	Groove- and Shoulder-Type Mechanical Pipe Couplings	Fittings
CSA B356-2010 (R2015)	Water Pressure Reducing Valves for Domestic Water Supply Systems	Valves
CSA G401-2014	Corrugated Steel Pipe Products	Piping, Ferrous
CSA Z21.12b-1994 (R2010)	Draft Hoods	Fuel Gas, Appliances
CSA Z21.13-2014	Gas-Fired Low-Pressure Steam and Hot Water Boilers (same as CSA 4.9)	Fuel Gas, Appliances
CSA Z21.15b-2013 (R2014)	Manually Operated Gas Valves for Appliances, Appliance Connector Valves, and Hose End Valves (same as CSA 9.1b)	Fuel Gas
CSA Z21.81a-2007 (R2015)	Cylinder Connection Devices (same as CSA 6.25a)	Fuel Gas
CSA Z21.86-2008 (R2014)	Vented Gas-Fired Space Heating Appliances (same as CSA 2.32)	Fuel Gas, Appliances
CSA Z83.11-2016	Gas Food Service Equipment (same as CSA 1.8)	Fuel Gas, Appliances
CSA Z317.1-2016	Special Requirements for Plumbing Installations in Health Care Facilities	Miscellaneous
Energy Star-2007 (version 2.0)	Program Requirements for Commercial Dishwashers	Appliances
Energy Star-2015 (version 7.1)	Product specification for Clothes Washers	Appliances
Energy Star-2016 (version 6.0)	Program Requirements for Residential Dishwashers	Appliances
EPA/625/R-04/108-2004	Guidelines for Water Reuse	Miscellaneous

REFERENCED STANDARDS

TABLE 1701.2 (continued)
STANDARDS, PUBLICATIONS, PRACTICES, AND GUIDES

STANDARD NUMBER	STANDARD TITLE	APPLICATION
EPA WaterSense-2007	High-Efficiency Lavatory Faucet Specification	Fixtures
EPA WaterSense-2009	Specification for Flushing Urinals	Fixtures
EPA WaterSense-2014	Specification for Tank-Type Toilets	Fixtures
IAPMO IGC 154-2016	Shower and Tub/Shower Enclosures, Bathtubs with Glass Pressure-Sealed Doors, and Shower/Steam Panels	Fixtures
IAPMO Z1157- 2014[e1]	Ball Valves	Valves
IAPMO IGC 193-2010	Safety Plates, Plate Straps, Notched Plates and Safety Collars	Miscellaneous
IAPMO IGC 226-2006a	Drinking Water Fountains With or Without Chiller or Heater	Fixtures
IAPMO IGC 276-2011	Bundled Expanded Polystyrene Synthetic Aggregate Units	DWV Components
IAPMO PS 23-2006a	Dishwasher Drain Airgaps	Backflow Protection
IAPMO PS 25-2002	Metallic Fittings for Joining Polyethylene Pipe for Water Service and Yard Piping	Joints
IAPMO PS 34-2003	Encasement Sleeve for Potable Water Pipe and Tubing	Piping
IAPMO PS 36-2014[e1]	Lead-Free Sealing Compounds for Threaded Joints	Joints
IAPMO PS 37-1990	Black Plastic PVC or PE Pressure-Sensitive Corrosion Preventive Tape	Miscellaneous
IAPMO PS 42-2013[e1]	Pipe Alignment and Secondary Support Systems	Miscellaneous
IAPMO PS 50-2010	Flush Valves with Dual Flush Device For Water Closets or Water Closet Tank with an Integral Flush Valves with a Dual Flush Device	Fixtures
IAPMO PS 51-2016	Expansion Joints and Flexible Expansion Joints for DWV Piping Systems	Joints
IAPMO PS 52-2009	Pump/Dose, Sumps and Sewage Ejector Tanks with or without a Pump	DWV Components
IAPMO PS 53-2016	Grooved Mechanical Pipe Couplings and Grooved Fittings	Joints
IAPMO PS 54-2014a	Metallic and Plastic Utility Boxes	Miscellaneous
IAPMO PS 57-2002	PVC Hydraulically Actuated Diaphragm Type Water Control Valves	Valves
IAPMO PS 59-2016	Wastewater Diverter Valves and Diversion Systems	DWV Components
IAPMO PS 60-1996	Sewage Holding Tank Containing Sewage Ejector Pump for Direct Mounted Water Closet	DWV Components
IAPMO PS 63-2014	Plastic Leaching Chambers	DWV Components
IAPMO PS 64-2012a[e1]	Roof Pipe Flashings	Miscellaneous
IAPMO PS 66-2015	Dielectric Fittings	Fittings
IAPMO PS 67-2010	Early-Closure Replacement Flappers or Early-Closure Replacement Flapper with Mechanical Assemblies	Fixtures
IAPMO PS 69-2006	Bathwaste and Overflow Assemblies with Tub Filler Spout	DWV Components
IAPMO PS 72-2007[e1]	Valves with Atmospheric Vacuum Breakers	Valves
IAPMO PS 73-2015	Dental Liquid-Ring Vacuum Pumps	Miscellaneous
IAPMO PS 76-2012a	Trap Primers for Fill Valves and Flushometer Valves	DWV Components
IAPMO PS 79-2005	Multiport Electronic Trap Primer	DWV Components
IAPMO PS 80-2008	Clarifiers	DWV Components
IAPMO PS 81-2006	Precast Concrete Seepage Pit Liners and Covers	DWV Components
IAPMO PS 82-1995	Fiberglass (Glass Fiber Reinforced Thermosetting Resin) Fittings	Fittings
IAPMO PS 85-1995	Tools for Mechanically Formed Tee Connections in Copper Tubing	Miscellaneous
IAPMO PS 86-1995	Rainwater Diverter Valve for Non-Roofed Area Slabs	DWV Components
IAPMO PS 89-1995	Soaking and Hydrotherapy (Whirlpool) Bathtubs with Hydraulic Seatlift	Fixtures
IAPMO PS 90-2014	Elastomeric Test Caps, Cleanout Caps, and Combination Test Caps/Shielded Couplings	DWV Components
IAPMO PS 91-2005a	Plastic Stabilizers for Use with Plastic Closet Bends	DWV Components
IAPMO PS 92-2013	Heat Exchangers and Indirect Water Heaters	Miscellaneous
IAPMO PS 94-2012	Insulated Protectors for P-Traps, Supply Stops and Risers	Miscellaneous
IAPMO PS 95-2001	Drain, Waste, and Vent Hangers and Plastic Pipe Support Hooks	DWV Components
IAPMO PS 98-1996	Prefabricated Fiberglass Church Baptisteries	Fixtures
IAPMO PS 100-1996	Porous Filter Protector for Sub-Drain Weep Holes	DWV Components

REFERENCED STANDARDS

TABLE 1701.2 (continued)
STANDARDS, PUBLICATIONS, PRACTICES, AND GUIDES

STANDARD NUMBER	STANDARD TITLE	APPLICATION
IAPMO PS 101-1997	Suction Relief Valves	Valves
IAPMO PS 104-1997	Pressure Relief Connection for Dispensing Equipment	Valves
IAPMO PS 105-1997	Polyethylene Distribution Boxes	DWV Components
IAPMO PS 106-2015[e1]	Tileable Shower Receptors and Shower Kits	Fixtures
IAPMO PS 110-2006a	PVC Cold Water Compression Fittings	Fittings
IAPMO PS 111-1999	PVC Cold Water Gripper Fittings	Fittings
IAPMO PS 112-1999	PVC Plastic Valves for Cold Water Distribution Systems Outside a Building and CPVC Plastic Valves for Hot and Cold Water Distribution Systems	Valves
IAPMO PS 113-2010	Hydraulically Powered Household Food Waste Disposers	Appliances
IAPMO PS 114-1999[e1]	Remote Floor Box Industrial Water Supply, Air Supply, Drainage	Miscellaneous
IAPMO PS 115-2007	Hot Water On-Demand or Automatic Activated Hot Water Pumping Systems	Miscellaneous
IAPMO PS 116-1999	Hot Water Circulating Devices Which Do Not Use a Pump	Miscellaneous
IAPMO PS 117-2016	Press and Nail Connections	Fittings
IAPMO PS 119-2012a[e1]	Water-Powered Sump Pumps	Miscellaneous
IAPMO Z124.7-2013	Prefabricated Plastic Spa Shells	Fixtures, Swimming Pools, Spas, and Hot Tubs
IAPMO Z124.8-2013[e2]	Plastic Liners for Bathtubs and Shower Receptors	Fixtures
IAPMO Z1000-2013	Prefabricated Septic Tanks	DWV Components
IAPMO Z1088-2013	Pre-Pressurized Water Expansion Tanks	Miscellaneous
MSS SP-25-2013	Marking System for Valves, Fittings, Flanges, and Unions	Miscellaneous
MSS SP-42-2013	Corrosion-Resistant Gate, Globe, Angle, and Check Valves with Flanged and Butt Weld Ends (Classes 150, 300, & 600)	Piping, Ferrous
MSS SP-44-2016	Steel Pipeline Flanges	Fittings
MSS SP-83-2014	Class 3000 and 6000 Pipe Unions, Socket Welding and Threaded (Carbon Steel, Alloy Steel, Stainless Steels, and Nickel Alloys)	Joints
MSS SP-104-2012	Wrought Copper Solder-Joint Pressure Fittings	Fittings
MSS SP-106-2012	Cast Copper Alloy Flanges and Flanged Fittings: Class 125, 150, and 300	Fittings
MSS SP-109-2012	Weld-Fabricated Copper Solder-Joint Pressure Fittings	Fittings
MSS SP-123-2013	Non-Ferrous Threaded and Solder-Joint Unions for Use with Copper Water Tube	Joints
NFPA 13R-2016	Installation of Sprinkler Systems in Low-Rise Residential Occupancies	Miscellaneous
NFPA 80-2016	Fire Doors and Other Opening Protectives	Miscellaneous
NFPA 501A-2017	Fire Safety Criteria for Manufactured Home Installations, Sites, and Communities	Miscellaneous
NFPA 1981-2013	Open-Circuit Self-Contained Breathing Apparatus (SCBA) for Emergency Services	Miscellaneous
NFPA 1989-2013	Breathing Air Quality for Emergency Services Respiratory Protection	Miscellaneous
NFPA 5000-2015	Building Construction and Safety Code	Miscellaneous
NSF 2-2015	Food Equipment	Appliances
NSF 4-2014	Commercial Cooking, Rethermalization, and Powered Hot Food Holding and Transport Equipment	Appliances
NSF 5-2016	Water Heaters, Hot Water Supply Boilers, and Heat Recovery Equipment	Appliances
NSF 12-2012	Automatic Ice Making Equipment	Appliances
NSF 18-2016	Manual Food and Beverage Dispensing Equipment	Appliances
NSF 29-2012	Detergent and Chemical Feeders for Commercial Spray-Type Dishwashing Machines	Appliances
NSF 40-2013	Residential Wastewater Treatment Systems	DWV Components
NSF 41-2016	Non-Liquid Saturated Treatment Systems	DWV Components
NSF 46-2016	Evaluation of Components and Devices Used in Wastewater Treatment Systems	DWV Components
NSF 169-2012	Special Purpose Food Equipment and Devices	Appliances
PSAI Z4.1-2005	For Sanitation – In Places of Employment – Minimum Requirements	Miscellaneous
SAE J512-1997	Automotive Tube Fittings	Fittings
SAE J1670-2008	Type "F" Clamps for Plumbing Applications	Joints

REFERENCED STANDARDS

TABLE 1701.2 (continued)
STANDARDS, PUBLICATIONS, PRACTICES, AND GUIDES

STANDARD NUMBER	STANDARD TITLE	APPLICATION
TCNA A118.10-2014	Load Bearing, Bonded, Waterproof Membranes for Thin-Set Ceramic Tile and Dimension Stone Installation	Miscellaneous
Title 49, Code of Federal Regulations, Part 192	Transportation of Natural and Other Gas by Pipeline: Minimum Federal Standards	Miscellaneous
UL 70-2001	Septic Tanks, Bituminous-Coated Metal	DWV Components
UL 80-2007	Steel Tanks for Oil-Burner Fuels and Other Combustible Liquids (with revisions through January 16, 2014)	Fuel Gas
UL 144-2012	LP-Gas Regulators (with revisions through November 5, 2014)	Fuel Gas
UL 252-2010	Compressed Gas Regulators (with revisions through January 28, 2015)	Fuel Gas
UL 296-2003	Oil Burners (with revisions through June 11, 2015)	Fuel Gas, Appliances
UL 404-2010	Gauges, Indicating Pressure, for Compressed Gas Service (with revisions through February 11, 2015)	Fuel Gas
UL 429-2013	Electrically Operated Valves	Valves
UL 536-2014	Flexible Metallic Hose	Fuel Gas
UL 563-2009	Ice Makers (with revisions through November 29, 2013)	Appliances
UL 569-2013	Pigtails and Flexible Hose Connectors for LP-Gas	Fuel Gas
UL 726-1995	Oil-Fired Boiler Assemblies (with revisions through October 9, 2013)	Fuel Gas, Appliances
UL 1206-2003	Electric Commercial Clothes-Washing Equipment (with revisions through November 30, 2012)	Appliances
UL 1331-2005	Station Inlets and Outlets (with revisions through August 25, 2014)	Medical Gas
UL 1795-2016	Hydromassage Bathtubs (with revisions through October 7, 2016)	Fixtures
UL 1951-2011	Electric Plumbing Accessories (with revisions through October 7, 2016)	Miscellaneous
UL 2157-2015	Electric Clothes Washing Machines and Extractors	Appliances
WQA S-300-2000	Point-of-Use Low-Pressure Reverse Osmosis Drinking Water Systems	Appliances

REFERENCED STANDARDS

ABBREVIATIONS IN TABLE 1701.1 AND TABLE 1701.2

AHAM	Association of Home Appliance Manufacturers, 1111 19th Street, NW, Suite 402, Washington, DC 20036.
ANSI	American National Standards Institute, Inc., 25 W. 43rd Street, 4th Floor, New York, NY 10036.
ASCE	American Society of Civil Engineers, 1801 Alexander Bell Drive, Reston, VA 20191-4400.
ASHRAE	American Society of Heating, Refrigerating, and Air Conditioning Engineers, Inc., 1791 Tullie Circle, NE, Atlanta, GA 30329-2305.
ASME	American Society of Mechanical Engineering, Two Park Avenue, New York, NY 10016-5990.
ASPE	American Society of Plumbing Engineers, 6400 Shafer Court, Suite 350, Rosemont, IL 60018.
ASSE	American Society of Sanitary Engineering, 18927 Hickory Creek Drive, Suite 220, Mokena, IL 60448.
ASTM	ASTM International, 100 Barr Harbor Drive, West Conshohocken, PA 19428-2959.
AWS	American Welding Society, 8669 NW 36 Street, #130 Miami, FL 33166-6672.
AWWA	American Water Works Association, 6666 W. Quincy Avenue, Denver, CO 80235.
CFR	U.S Government Publishing Office, 723 North Capitol Street, N.W. Washington, DC 20401-001
CGA	Compressed Gas Association, 14501 George Carter Way, Suite 103, Chantilly, VA 20151.
CISPI	Cast-Iron Soil Pipe Institute, 2401 Fieldcrest Drive, Mundelein, IL 60060.
CSA	Canadian Standards Association, 5060 Spectrum Way, Suite 100, Mississauga, Ontario, Canada, L4W 5N6.
e1	An editorial change since the last revision or reapproval.
ENERGY STAR	1200 Pennsylvania Avenue, N.W., Washington, D.C. 20460.
EPA WATERSENSE	U.S. Environmental Protection Agency, Office of Wastewater Management (4204M), 1200 Pennsylvania Avenue, N.W., Washington, D.C. 20460.
IAPMO	International Association of Plumbing and Mechanical Officials, 4755 E. Philadelphia Street, Ontario, CA 91761.
ICC	International Code Council, 500 New Jersey Avenue, NW, 6th Floor, Washington, DC 20001.
ISEA	International Safety Equipment Association, 1901 N. Moore Street, Suite 808, Arlington, VA 22209-1762.
ISO	International Organization for Standardization, 1 ch. de la Voie-Creuse, Casa Postale 56, CH-1211 Geneva 20, Switzerland.
MSS	Manufacturers Standardization Society of the Valve and Fittings Industry, 127 Park Street, NE, Vienna, VA 22180.
NFPA	National Fire Protection Association, 1 Batterymarch Park, Quincy, MA 02169-7471.
NSF	NSF International, 789 N. Dixboro Road, Ann Arbor, MI 48105.
PDI	Plumbing and Drainage Institute, 800 Turnpike Street, Suite 300, North Andover, MA 01845.
PSAI	Portable Sanitation Association International, 2626 E 82nd Street, Suite 175, Bloomington, MN 55425.
SAE	Society of Automotive Engineers, 400 Commonwealth Drive, Warrendale, PA 15096.
TCNA	Tile Council of North America, Inc. 100 Clemson Research Blvd., Anderson, SC 29625.
UL	Underwriters Laboratories, Inc., 333 Pfingsten Road, Northbrook, IL 60062.
WQA	Water Quality Association, 4151 Naperville Road, Lisle, IL 60532-3696.

APPENDIX A
RECOMMENDED RULES FOR SIZING THE WATER SUPPLY SYSTEM

A 101.0 General.

A 101.1 Applicability. This appendix provides a general procedure for sizing a water supply system. Because of the variable conditions encountered, it is impractical to lay down definite detailed rules of procedure for determining the sizes of water supply pipes in an appendix, which shall necessarily be limited in length. For a more adequate understanding of the problems involved, refer to Water-Distributing Systems for Buildings, Report BMS 79 of the National Bureau of Standards; and Plumbing Manual, Report BMS 66, also published by the National Bureau of Standards.

Section 610.0, Size of Potable Water Piping, of Chapter 6 provides the plumber or design professional with three methods to size the water distribution system. The most commonly used method for small to moderately sized installations is by use of Table 610.4, Fixture Unit Table for Determining Water Pipe and Meter Sizes. The use of this table and Table 610.3, Water Supply Fixture Units and Minimum Fixture Branch Pipe Sizes, is described in Section 610.0.

Another method of sizing the water supply system is by use of Appendix C, Alternate Plumbing Systems. Section 610.5 directs the user to Appendix C for an alternative method of sizing. In Appendix C, Section C 303.2 allows the use of Table C 303.2, Water Supply Fixture Units for Bathroom Groups, in conjunction with Tables 610.3 and 610.4. This method is used mostly for multiple dwellings (apartment and condominium) where smaller sizes of pipe can be obtained by grouping fixtures together rather than by using individual fixture unit values. This method of sizing is allowed only with permission of the Authority Having Jurisdiction (AHJ).

The third method of sizing the water supply system is by the use of this appendix and its various tables and charts. The plumber or design professional may use this method of sizing for large, intricate or complicated systems that will require precise calculations to ensure the proper pressures and flow to the system's fixtures. The appendix provides the proper steps to size the water supply system and gives an example of sizing such systems.

A 102.0 Preliminary Information.

A 102.1 Daily Service Pressure. Obtain the necessary information regarding the minimum daily service pressure in the area where the building is to be located.

The minimum daily service pressure can be obtained from water districts, fire department hydrant checks, and water departments.

A 102.2 Water Meter. Where the building supply is to be metered, obtain information regarding friction loss relative to the rate of flow for meters in the range of sizes likely to be used. Friction-loss data is capable of being obtained from most manufacturers of water meters. Friction losses for disk-type meters shall be permitted to be obtained from Chart A 102.2.

Friction loss through the water meter, water filters, water conditioners and backflow prevention devices must also be included. Listed devices have the pressure loss on a permanently attached label or it is contained on the technical specification document (see **Figure A 102.2**).

A 102.3 Local Information. Obtain available local information regarding the use of different kinds of pipe with respect both to durability and to decrease in capacity with length of service in the particular water supply.

CHART A 102.2
FRICTION LOSSES FOR DISK-TYPE WATER METERS

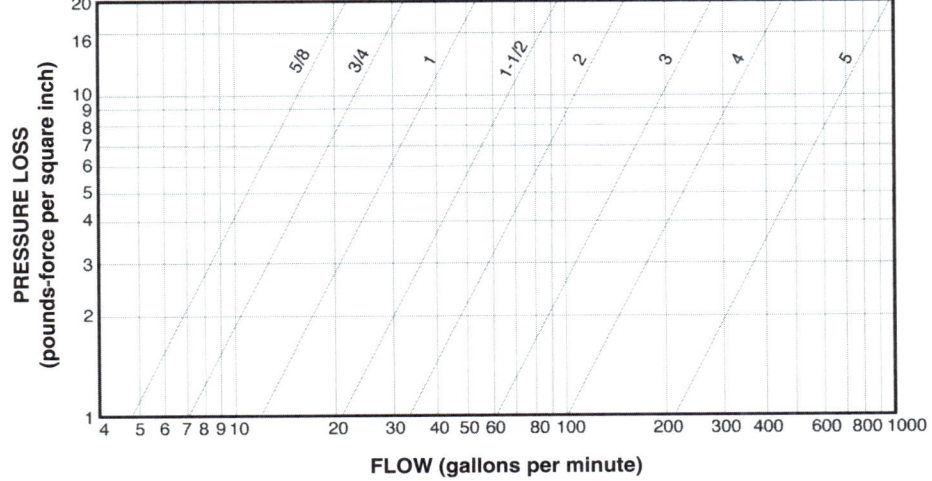

For SI units: 1 inch = 25 mm, 1 pound-force per square inch = 6.8947 kPa, 1 gallon per minute = 0.06 L/s

APPENDIX A

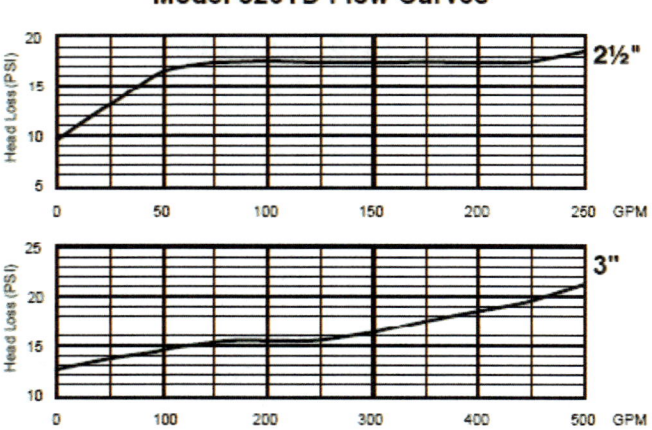

FIGURE A 102.2
EXAMPLE OF PRESSURE LOSS TABLE FROM SPECIFICATION SHEET FOR AN RPZ

🔧 Local water and soil conditions may dictate the use of some materials or preclude the use of others. Some soils may be corrosive to certain metals such as copper, and water conditions may be hard or softened that may cause a deleterious effect on certain pipes. More serious conditions require additional allowances for hot water caking. This is a case where knowledge of local conditions is necessary to properly plan the work.

A 103.0 Demand Load.

A 103.1 Supply Demand. Estimate the supply demand for the building main, the principal branches and risers of the system by totaling the fixture units on each, Table A 103.1, and then by reading the corresponding ordinate from Chart A 103.1(1) or Chart A 103.1(2), whichever is applicable.

🔧 Table A 103.1 is used to find the water supply fixture units for all fixtures on branches and risers. Note the following columns that may have different fixture unit values. These differing classifications indicate use patterns that will affect fixture unit values when exceeding typical private use patterns.

Private use, as defined in Chapter 2, "applies to plumbing fixtures in residences and apartments, to private bathrooms in hotels and hospitals, and to restrooms in commercial establishments where the fixtures are intended for the use of a family or an individual."

Public use, as defined in Chapter 2, "applies to plumbing fixtures that are not defined as private or private use".

Assembly applies to plumbing fixtures in types of occupancies indicated as A-1 through A-5 in Chapter 4, Table 422.1.

Chart A 103.1(1) and the enlarged scale Chart A 103.1(2) are used to convert fixture units to water supply demand in gallons per minute (gpm) for either a flush tank water closet system or flushometer valve system.

Example 1
200 fixture units are equivalent to a flow rate of 64 gpm (4 L/s) in a flush tank water closet system, or 90 gpm (5.7 L/s) in a flushometer valve system.

Example 2
1,500 fixture units are equal to a demand of 270 gpm (17.04 L/s).

A 103.2 Continuous Supply Demand. Estimate continuous supply demands in gallons per minute (gpm) (L/s) for lawn sprinklers, air conditioners, etc., and add the sum to the total demand for fixtures. The result is the estimated supply demand of the building supply.

🔧 The continuous demand load, i.e., lawn sprinklers, water towers, water-cooled equipment, etc., shall be estimated in gallons per minute and added to the fixture demand load previously determined. This gives the total demand load on the building supply.

A 104.0 Permissible Friction Loss.

🔧 Friction loss occurs as fluid flows through a pipe. Factors affecting friction loss due to fluid flow are velocity and roughness of pipe walls. With respect to pipe roughness, the National Bureau of Standards published a report, *Water-Distributing Systems for Buildings, BMS79*, recommending estimates of pipe capacity due to varying degrees of roughness and aging. The investigation resulted in a set of nomographs showing friction loss relative to flow rate, velocity, and pipe diameter for varying degrees of pipe roughness. These nomographs have been reproduced in Charts A 105.1(1) through A 105.1(7). Furthermore, the report offered allowances for decreasing pipe capacity due to various levels of caking and corrosion that result from aging. Lime, calcium and other products suspended in potable water will eventually attach themselves and, in time, will line the interior of the water pipe system, thus resulting in an unintended pipe size reduction and an equally unintended increase in flow velocity as the system attempts to supply the desired quantity of water at each point of supply (see **Figure A 104.0**). Aging pipe and higher flow velocities are not a desirable combination. There is virtually nothing in a water piping system that improves with age; consequently, the further the system moves from optimal design parameters, the greater the risk of failure or impaired performance. Premature "aging" of the piping system will commonly result in premature pipe and component failure.

With respect to friction loss due to fluid flow velocity, it is a well known hydraulic principle that frictional resistance varies as the square of the velocity and is mathematically expressed as $F=KV^2$, where F is the frictional resistance, K is a constant and V is velocity. For example, assume flow velocity is 5 feet per second (fps) and the frictional resistance at this velocity is 3 pounds per square inch (psi). Expressing this mathematically using the equation $F=KV^2$ and substituting the known values, would look like this: $3 \text{ psi} = K(5^2)$. If flow velocity is doubled to 10 fps, the frictional resistance will then be calculated by the square of this change in velocity. The equation would then be modified by doubling the velocity: $3 \text{ psi} = K(5 \times 2)^2$. Since we are increasing one side of the equation by 2^2, we need to increase the other side of the equation by the same value to maintain equality. Hence, $(3 \times 2^2) \text{ psi} = K(5 \times 2)^2$, or 12 psi = K100. The resultant change in resistance is equal to 12 psi.

High flow velocities should be avoided not only

because of the exponential increase of frictional resistance that will affect pressure, but also for the reasons that they create undesirable noise in the piping system, reduce residual head pressure and cause excessive shock in reaction to quick-closing valves, resulting in damage to the piping materials. Velocities should not exceed 10 feet per second or the maximum values given by the manufacturers. Various plastics and copper have a specific prohibition against velocities in excess of 8 fps.

Flow restrictions in water pipe cause an increase in flow velocity and a reduction in pressure. The pressure may drop below 0 psig causing a vacuum in the piping system.

FIGURE A 104.0
EXAMPLE OF FRICTION LOSS

A 104.1 Residual Pressure. Decide what is the desirable minimum residual pressure that shall be maintained at the highest fixture in the supply system. Where the highest group of fixtures contains flushometer valves, the residual pressure for the group shall be not less than 15 pounds-force per square inch (psi) (103 kPa). For flush tank supplies, the available residual pressure shall be not less than 8 psi (55 kPa).

Residual pressure differs from static, flow, and normal operating pressure. Static pressure is pressure exerted on water when the water is not flowing. Factors effecting static pressure in a municipal water distribution system are atmospheric pressure, the height of water towers, and pumping stations. Static pressure when applied to a plumbing system within a building is the pressure exerted on the water supply when no water is flowing from fixtures. A static pressure reading within a building may be conducted by placing a pressure gauge on a faucet spout or hose bibb. With the faucet open, the gauge will read the static pressure. Static pressure in a plumbing system will vary relative to the height above or below the water service (see Section A 104.2). Municipal water systems will rarely read a true static pressure when conducting a hydrant flow test since there is always some flow of water in the pipe from local usage. The minimum daily service pressure in Section A 102.1 is actually the residual pressure of the municipal water main when taking into account pressure losses during peak periods of customer usage.

Residual pressure, therefore, is the remaining pressure available during peak demand periods. When applied to a plumbing system within a building, residual pressure is the pressure available at a fixture when the system is flowing at its estimated peak demand (see Section A 103.0). Factors affecting residual pressure are pressure drops due to friction and head loss, and losses due to meters and other appurtenances.

Flow pressure is the forward velocity pressure at a discharge opening when water is flowing. This measurement is done with the use of a Pitot tube. Faucets and showerheads are provided with flow restrictors and are labeled with maximum flow rates at a given pressure, typically at 80 psi. The flow restrictors guarantee that the flow rate will not increase relative to increasing flow pressure above the maximum listing.

Normal operating pressure is the pressure that flowing water exerts against the wall of the pipe. Normal operating pressure differs from static pressure by the friction loss caused by flowing water. A piezometer gauge is used to determine this kind of pressure.

Caution needs to be applied when determining the minimum residual pressure that should be maintained under the estimated maximum demand. The assumption is that the highest group of fixtures will contain a water closet, which will require either a minimum 15 psi for flushometer valve, or a minimum of 8 psi for a flush tank. Although this is typically true for residential designs, this may not be applicable for commercial designs. For example, in a commercial kitchen, a water supply branch for a dishwasher having an operating pressure of 25 psi may be preceded by a water heater and a water softener provided for the dishwasher with a pressure loss of 5 psi and 10psi respectively. The branch line now requires a minimum residual pressure of 40 psi. This exceeds the minimum pressure required for a flushometer. Plumbing fixtures and appliances must be evaluated for the minimum operating pressure. Any appurtenances or devices that precede a fixture or appliance that may affect its operating pressure needs to be considered when determining residual pressure.

A 104.2 Elevation. Determine the elevation of the highest fixture or group of fixtures above the water (street) main. Multiply this difference in elevation by 0.43. The result is the loss in static pressure in psi (kPa).

The "head loss" is calculated at 0.43 times the difference in elevation between the street main and the highest outlet in feet. One foot water column is equivalent to 0.43 foot pounds per square inch (psi). The calculated result will give the static pressure loss (or gain) in pounds per square inch. For SI units, multiply the difference of elevation in meters by the factor 9.8 for kilopascal units and 0.1 for bar units.

In **Figure A 104.2**, the difference in elevation between the street main A and the highest fixture, X is denoted by A, A'. Multiplying the difference by 0.43 will give the static pressure gain (in psi) since the street main is elevated above the highest fixture. The difference in elevation between the street main B and the highest fixture X' is denoted by B, B'. Multiplying the difference by 0.43 psi will give the static pressure loss since the highest fixture is elevated above the street main.

A 104.3 Available Pressure. Subtract the sum of loss in static pressure and the residual pressure to be maintained

APPENDIX A

at the highest fixture from the average minimum daily service pressure. The result will be the pressure available for friction loss in the supply pipes, where no water meter is used. Where a meter is to be installed, the friction loss in the meter for the estimated maximum demand should also be subtracted from the service pressure to determine the pressure loss available for friction loss in the supply pipes.

A 104.4 Developed Length. Determine the developed length of pipe from the water (street) main to the highest fixture. Where close estimates are desired, compute with the aid of Table A 104.4(1), Table A 104.4(2), or Table A104.4(3), whichever is applicable, the equivalent length of pipe for fittings in the line from the water (street) main to the highest fixture and add the sum to the developed length. The pressure available for friction loss in psi (kPa), divided by the developed lengths of pipe from the water (street) main to the highest fixture, times 100, will be the average permissible friction loss per 100 feet (30 480 mm) length of pipe.

The purpose for determining the developed length of pipe from the street main to the most distant fixture is to derive the average permissible friction loss per 100 feet of pipe length for ferrous metal, non-ferrous metal, CPVC or PVC piping materials. Begin by laying out a sketch of the building water supply system from the street main and determine the following:

1. The developed length of the building main (indicated by W in **Figure A 104.4**);

2. The developed length from the street main to the foot of each riser branching from the building main (the three separate sections of W in **Figure A 104.4**);

3. The greatest developed length of pipe from the street main to the top of any riser. In **Figure A 104.4**, there are three risers from the building main. Determine the developed length for each riser.

4. Table 104.4 provides equivalent lengths of pipe for the various fittings and valves. When a close estimate is required, compute the equivalent length of pipe for all fittings in each developed length of pipe and add the sum to the developed length.

Calculating the Permissible Friction Loss in Pounds per Square Inch per 100 Feet.

The following summarizes the steps needed derive the permissible friction loss in order to determine the diameter of the building supply, branches and risers.

1. Start with the minimum daily service pressure.
2. Subtract from the minimum daily service pressure:
a. Friction loss due to the water meter, water treatment devices, and backflow devices.
b. The residual pressure required at the highest fixture, or the fixture or appliance that has the highest total pressure drop.
c. The static pressure when it is a loss (add to the service pressure when it is a gain).

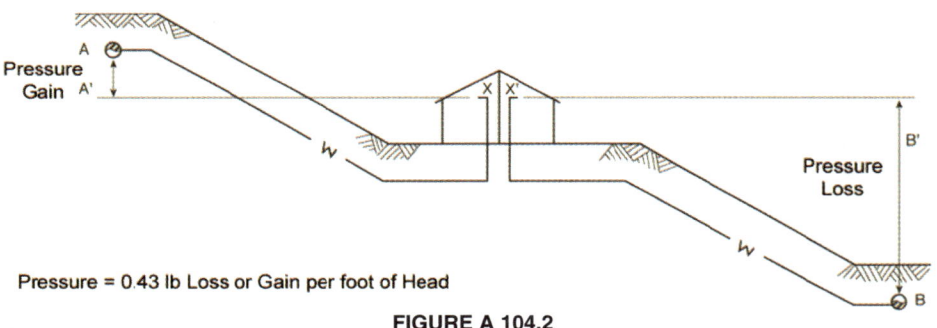

Pressure = 0.43 lb Loss or Gain per foot of Head

**FIGURE A 104.2
HEAD LOSS OR GAIN DUE TO ELEVATION**

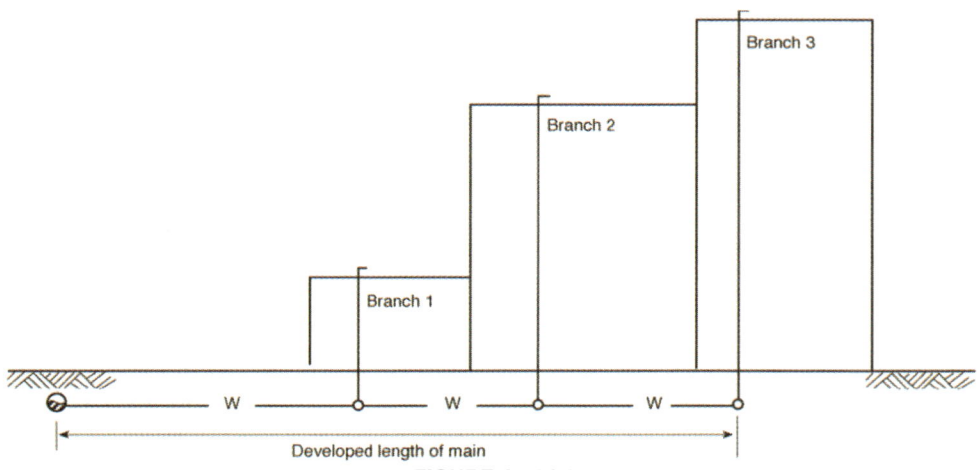

**FIGURE A 104.4
SKETCH OF BUILDING WATER SUPPLY SYSTEM**

3. Divide the resulting pressure available for friction loss by the greatest developed length in feet. This will yield the average permissible friction loss per foot of pipe. Multiply the result by 100. This will yield the average permissible friction loss per 100 feet of pipe and will be useful when applied to the friction loss charts to calculate velocity and pipe diameter. The use of the friction loss charts is explained in Section A 105.1.

Knowing the supply demand load (gpm) and the allowable friction loss per 100 feet (in psi), the building supply, risers, and the various branches can be determined.

A 105.0 Size of Building Supply.

A 105.1 Diameter. Knowing the permissible friction loss per 100 feet (30 480 mm) of pipe and the total demand, the diameter of the building supply pipe shall be permitted to be obtained from Chart A 105.1(1), Chart A 105.1(2), Chart A 105.1(3), Chart A 105.1(4) Chart A 105.1(5), Chart A 105.1(6), or Chart A 105.1(7), whichever is applicable. The diameter of pipe on or next above the coordinate point corresponding to the estimated total demand and the permissible friction loss will be the size needed up to the first branch from the building supply pipe.

The use of this appendix for sizing piping systems designed to consist of ferrous metal, non-ferrous metal CPVC or PVC piping materials, does not exclude the prescriptive requirements of Chapter 6 in the UPC or the associated referenced standards. Solvents contained in CPVC plastics are classified as airborne contaminants along with being flammable and combustible liquids. Worker safety training is essential to train personnel about the hazards associated with CPVC plumbing pipe installations. The manufacturer's website and installation instructions provide the information on the health and safety hazards associated with CPVC plumbing installations, and the solvents used for such installations.

In order to determine the correct diameter of piping for the building supply (the piping after the water meter other than branches and risers), the permissible friction loss per 100 feet as determined by Section A 104.4 is to be applied to Charts A 105.1(1) through A 105.1(7). Select the chart applicable to the type of pipe material and condition of service. Chart A 105.1(1) applies to smooth new copper tubing (Type M, L and K) and may be applied to any corresponding smooth pipe with recessed fittings such as brass and plastic. Chart A 105.1(2) applies to new wrought iron or steel pipe and is considered fairly smooth. Chart A 105.1(3) applies to fairly rough pipe. This chart is applicable when needing to estimate the probable small change in diameter of the pipe in service due to corrosion or caking that will present a roughness intermediate between fairly smooth and rough pipe. Chart A 105.1(4) applies to rough pipe where the character of the water will produce a significant change in the diameter of the pipe due to corrosion and caking. Charts A 105.1(5) through A 105.1(7) apply to varying plastic pipe schedules under very smooth conditions. The C-factor for the plastics charts is a Hazen-Williams roughness co-efficient. The C-factor of 150 is the industry standard for smoothness of newly manufactured pipe.

Having selected the appropriate chart, find the permissible friction loss per 100 feet of pipe at the bottom of the chart. **Figure A 105.1** shows the smooth pipe chart and that the permissible friction loss per 100 feet of pipe is 23.53 psi. If the estimated demand according to Section A 103.0 is 88 gpm, then a vertical line is to be drawn from the bottom of the chart at 23.53 psi to 88 gpm (designated on the right-hand side of the chart). Pipe diameters are indicated by the diagonal lines having a positive slope (left to right) and velocity is indicated by diagonal lines having negative slope (top to bottom). Notice that the vertical line for friction loss and the horizontal line for demand intersect at the 1½" pipe diameter diagonal line. To determine if 1½-inch diameter is the correct pipe size, check the diagonal velocity line. In **Figure A 105.1**, the velocity at the intersection point is 16 feet per second. This exceeds the limitation for smooth copper pipe. To correct the velocity at the demand rate of 88gpm, continue the horizontal line to the left until it intersects the diagonal velocity line at 8 feet per second. This new intersection point falls between the 2-inch diameter diagonal line and the 2½-inch diameter diagonal line, therefore the greater pipe size determines the diameter for the building supply.

A 105.2 Copper and Copper Alloy Piping. Where copper tubing or copper alloy pipe is to be used for the supply piping and where the character of the water is such that slight changes in the hydraulic characteristics are expected, Chart A 105.1(1) shall be permitted to be used.

A 105.3 Hard Water. Chart A 105.1(2) shall be used for ferrous pipe with the most favorable water supply in regards to corrosion and caking. Where the water is hard or corrosive, Chart A 105.1(3) or Chart A 105.1(4) will be applicable. For extremely hard water, it will be advisable to make additional allowances for the reduction of capacity of hot-water lines in service.

A 106.0 Size of Principal Branches and Risers.

Figure A 105.1 also shows the maximum allowed demand for all smaller pipe diameters to size branches and risers. Notice the darkened velocity line at 8 feet per second below the intersection of 120 gpm and the 2½-inch diameter diagonal line. Where the velocity line intersects the pipe diameter line a horizontal line is drawn to the right-hand side of the graph to determine the demand for each pipe size. The velocity line cannot exceed the vertical permissible friction loss line showing that 3/8 inch diameter pipe cannot exceed 3gpm in this application.

A 106.1 Size. The required size of branches and risers shall be permitted to be obtained in the same manner as the building supply, by obtaining the demand load on each branch or riser and using the permissible friction loss computed in Section A 104.0.

A 106.2 Branches. Where fixture branches to the building supply are sized for the same permissible friction loss per 100 feet (30 480 mm) of pipe as the branches and risers to the highest level in the building, and lead to inadequate water supply to the upper floor of a building, one of the following shall be provided:

(1) Selecting the sizes of pipe for the different branches so that the total friction loss in each lower branch is approximately equal to the total loss in the riser, including both friction loss and loss in static pressure.

APPENDIX A

(2) Throttling each such branch by means of a valve until the preceding balance is obtained.

(3) Increasing the size of the building supply and risers above the minimum required to meet the maximum permissible friction loss.

Sizing the distribution system according to the method outlined in the prior sections should result in pipe sizes that are capable of adequate water supply throughout the whole system. When the sum of pressure loss subtracted from the average minimum daily service pressure results in a negative value, a booster pump is the typical method to provide adequate water supply.

A 106.3 Water Closets. The size of branches and mains serving flushometer tanks shall be consistent with sizing procedures for flush tank water closets.

A 107.0 General.

A 107.1 Velocities. Velocities shall not exceed 10 feet per second (ft/s) (3 m/s), except as otherwise approved by the Authority Having Jurisdiction.

As explained in Section A 105.1, the diameter of pipe selected in the friction loss charts must be checked by the diagonal velocity lines. This section limits velocity in any pipe material to 10 feet per second. However, the manufacturer may stipulate a lesser velocity. The lesser velocity shall be required for pipe sizing, or otherwise approved by the AHJ.

A 107.2 Pressure-Reducing Valves. Where a pressure-reducing valve is used in the building supply, the developed length of supply piping and the permissible friction loss shall be computed from the building side of the valve.

A 107.3 Fittings. The allowances in Table A 104.4 for fittings are based on non-recessed threaded fittings. For recessed threaded fittings and streamlined soldered fittings, one-half of the allowances given in the table will be ample.

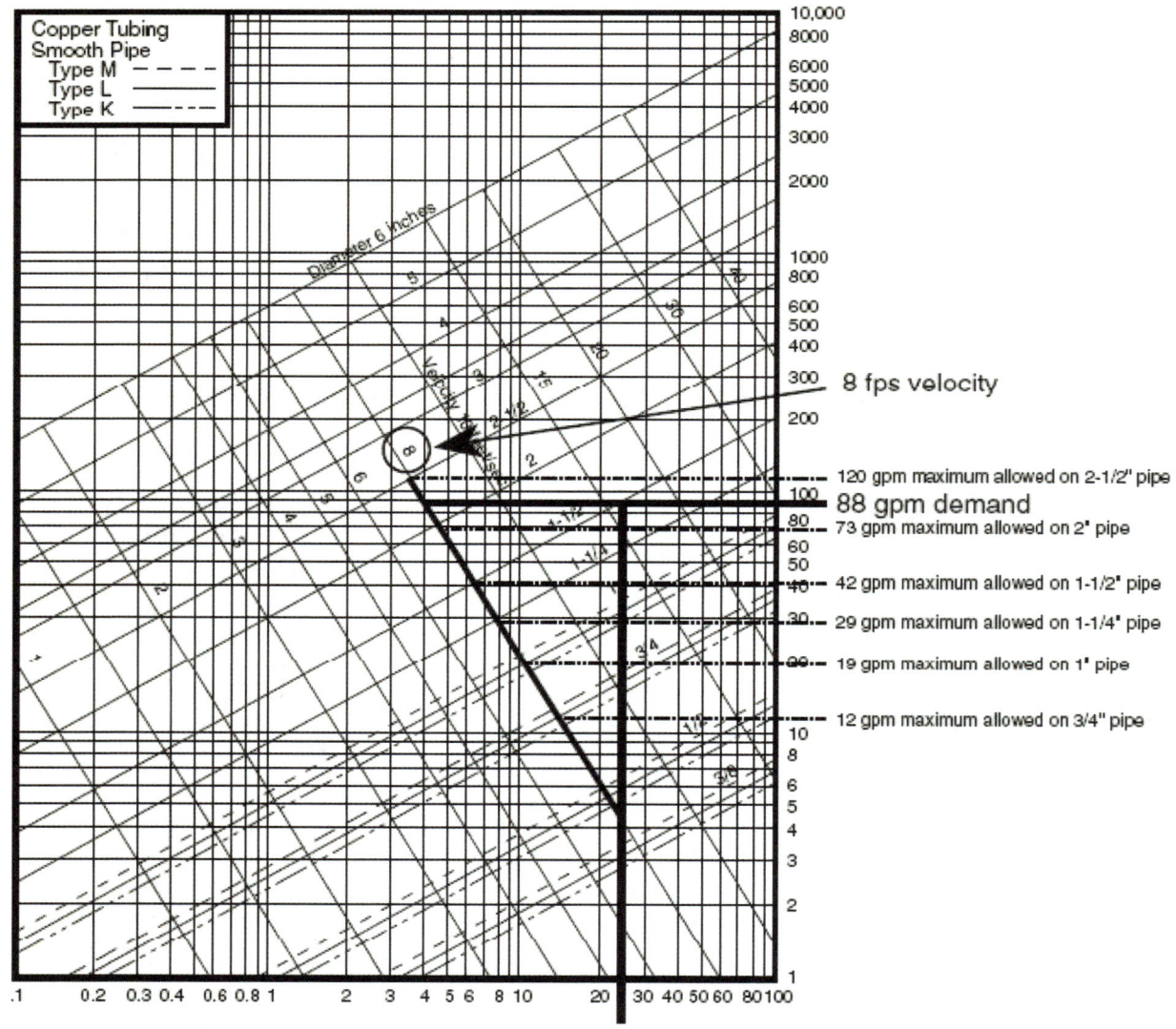

FIGURE A 105.1
EXAMPLE OF USE OF CHART A 105.1(1) FOR SOLUTION TO FIGURE A 108.1A

APPENDIX A

TABLE A 103.1
WATER SUPPLY FIXTURE UNITS (WSFU) AND MINIMUM FIXTURE BRANCH PIPE SIZES[3]

APPLIANCES, APPURTENANCES, OR FIXTURES[2]	MINIMUM FIXTURE BRANCH PIPE SIZE[1,4] (inches)	PRIVATE	PUBLIC	ASSEMBLY[6]
Bathtub or Combination Bath/Shower (fill)	½	4.0	4.0	–
¾ inch Bathtub Fill Valve	¾	10.0	10.0	–
Bidet	½	1.0	–	–
Clothes Washer	½	4.0	4.0	–
Dental Unit, cuspidor	½	–	1.0	–
Dishwasher, domestic	½	1.5	1.5	–
Drinking Fountain or Water Cooler	½	0.5	0.5	0.75
Hose Bibb	½	2.5	2.5	–
Hose Bibb, each additional[7]	½	1.0	1.0	–
Lavatory	½	1.0	1.0	1.0
Lawn Sprinkler, each head[5]	–	1.0	1.0	–
Mobile Home, each (minimum)	–	12.0	–	–
Sinks	–	–	–	–
Bar	½	1.0	2.0	–
Clinical Faucet	½	–	3.0	–
Clinical Flushometer Valve with or without faucet	1	–	8.0	–
Kitchen, domestic	½	1.5	1.5	–
Laundry	½	1.5	1.5	–
Service or Mop Basin	½	1.5	3.0	–
Washup, each set of faucets	½	–	2.0	–
Shower per head	½	2.0	2.0	–
Urinal, 1.0 GPF Flushometer Valve	¾	3.0	4.0	5.0
Urinal, greater than 1.0 GPF Flushometer Valve	¾	4.0	5.0	6.0
Urinal, flush tank	½	2.0	2.0	3.0
Wash Fountain, circular spray	¾	–	4.0	–
Water Closet, 1.6 GPF Gravity Tank	½	2.5	2.5	3.5
Water Closet, 1.6 GPF Flushometer Tank	½	2.5	2.5	3.5
Water Closet, 1.6 GPF Flushometer Valve	1	5.0	5.0	8.0
Water Closet, greater than 1.6 GPF Gravity Tank	½	3.0	5.5	7.0
Water Closet, greater than 1.6 GPF Flushometer Valve	1	7.0	8.0	10.0

For SI units: 1 inch = 25 mm

Notes:
[1] Size of the cold branch pipe, or both the hot and cold branch pipes.
[2] Appliances, appurtenances, or fixtures not included in this table shall be permitted to be sized by reference to fixtures having a similar flow rate and frequency of use.
[3] The listed fixture unit values represent their total load on the cold water building supply. The separate cold water and hot water fixture unit value for fixtures having both cold and hot water connections shall be permitted to each be taken as three-quarters of the listed total value of the fixture.
[4] The listed minimum supply branch pipe sizes for individual fixtures are the nominal (I.D.) pipe size.
[5] For fixtures or supply connections likely to impose continuous flow demands, determine the required flow in gallons per minute (gpm) (L/s) and add it separately to the demand in gpm (L/s) for the distribution system or portions thereof.
[6] Assembly [Public Use (see Table 422.1)].
[7] Reduced fixture unit loading for additional hose bibbs is to be used where sizing total building demand and for pipe sizing where more than one hose bibb is supplied by a segment of water distribution pipe. The fixture branch to each hose bibb shall be sized on the basis of 2.5 fixture units.

APPENDIX A

**CHART A 103.1(1)
ESTIMATE CURVES FOR DEMAND LOAD**

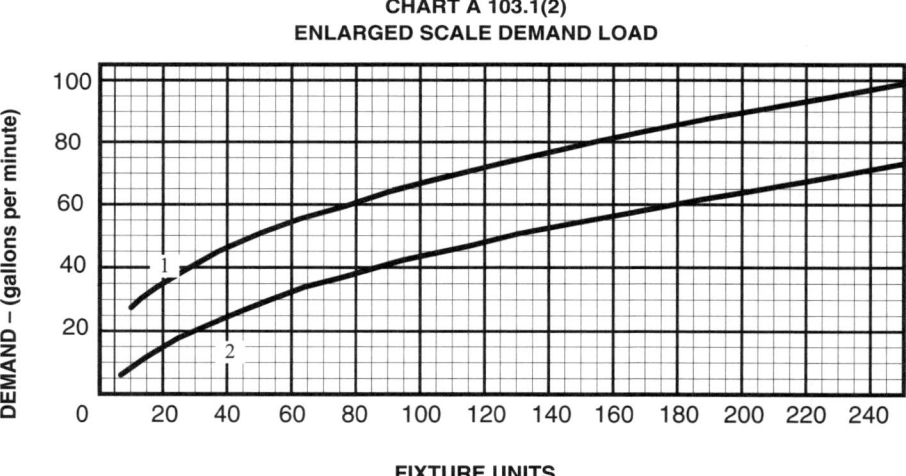

For SI units: 1 gallon per minute = 0.06 L/s

**CHART A 103.1(2)
ENLARGED SCALE DEMAND LOAD**

For SI units: 1 gallon per minute = 0.06 L/s

TABLE A 104.4(1)
ALLOWANCE IN EQUIVALENT LENGTH OF PIPE FOR FRICTION LOSS IN VALVES AND THREADED FITTINGS*
EQUIVALENT LENGTH OF PIPE FOR VARIOUS FITTINGS

DIAMETER OF FITTING (inches)	90° STANDARD ELBOW (feet)	45° STANDARD ELBOW (feet)	90° STANDARD TEE (feet)	COUPLING OR STRAIGHT RUN OF TEE (feet)	GATE VALVE (feet)	GLOBE VALVE (feet)	ANGLE VALVE (feet)
3/8	1.0	0.6	1.5	0.3	0.2	8	4
1/2	2.0	1.2	3.0	0.6	0.4	15	8
3/4	2.5	1.5	4.0	0.8	0.5	20	12
1	3.0	1.8	5.0	0.9	0.6	25	15
1 1/4	4.0	2.4	6.0	1.2	0.8	35	18
1 1/2	5.0	3.0	7.0	1.5	1.0	45	22
2	7.0	4.0	10.0	2.0	1.3	55	28
2 1/2	8.0	5.0	12.0	2.5	1.6	65	34
3	10.0	6.0	15.0	3.0	2.0	80	40
4	14.0	8.0	21.0	4.0	2.7	125	55
5	17.0	10.0	25.0	5.0	3.3	140	70
6	20.0	12.0	30.0	6.0	4.0	165	80

For SI units: 1 inch = 25 mm, 1 foot = 304.8 mm, 1 degree = 0.017 rad

* Allowances are based on nonrecessed threaded fittings. Use one-half the allowances for recessed threaded fittings or streamlined solder fittings.

TABLE A 104.4(2)
EQUIVALENT LENGTH OF COPPER TUBE SIZE CPVC PIPE FOR VARIOUS FITTINGS

DIAMETER OF FITTING (inches)	90 DEGREE ELBOW (feet)	45 DEGREE ELBOW (feet)	COUPLING OR STRAIGHT RUN OF TEE (feet)	90 DEGREE STANDARD TEE (feet)
1/2	1.6	0.8	1.0	3.1
3/4	2.1	1.1	1.4	4.1
1	2.6	1.4	1.7	5.3
1 1/4	3.5	1.8	2.3	6.9
1 1/2	4.0	2.1	2.7	8.1
2	5.2	2.8	3.5	10.3

For SI units: 1 inch = 25 mm, 1 foot = 304.8 mm

TABLE A 104.4(3)
EQUIVALENT LENGTH OF SCHEDULE 40 AND 80 CPVC PIPE FOR VARIOUS FITTINGS

DIAMETER OF FITTING (inches)	90 DEGREE ELBOW (feet)	45 DEGREE ELBOW (feet)	COUPLING OR STRAIGHT RUN OF TEE (feet)	90 DEGREE STANDARD TEE (feet)
1/2	1.5	0.8	1.0	4.0
3/4	2.0	1.1	1.4	5.0
1	2.5	1.4	1.7	6.0
1 1/4	3.8	1.8	2.3	7.0
1 1/2	4.0	2.1	2.7	8.0
2	5.7	2.6	4.3	12.0

For SI units: 1 inch = 25 mm, 1 foot = 304.8 mm

APPENDIX A

CHART A 105.1(1)

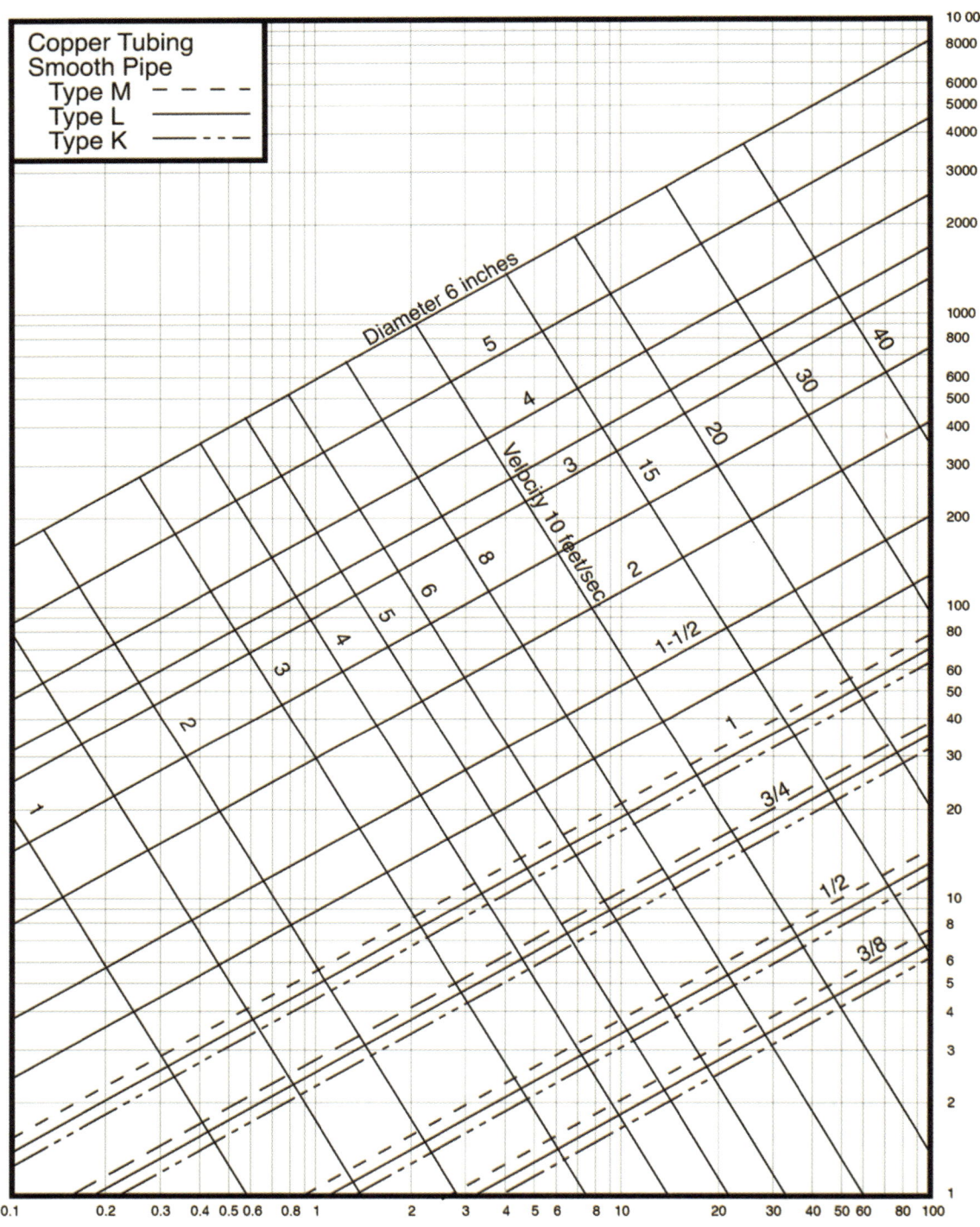

FRICTION LOSS IN HEAD (pounds-force per square inch) PER 100-FOOT LENGTH

For SI units: 1 inch = 25 mm, 1 gallon per minute = 0.06 L/s, 1 pound-force per square inch = 6.8947 kPa, 1 foot = 304.8 mm, 1 foot per second = 0.3048 m/s

APPENDIX A

CHART A 105.1(2)

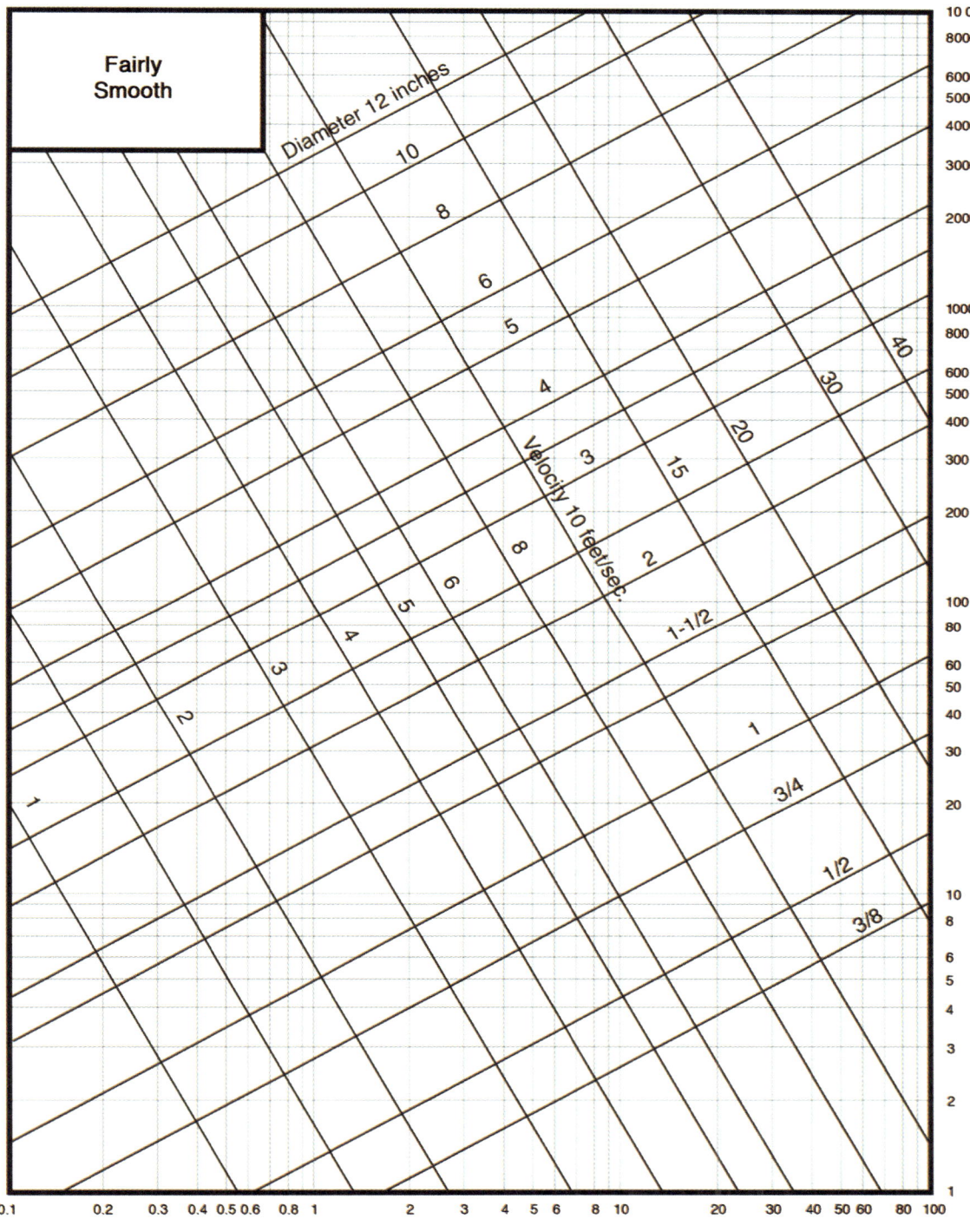

FRICTION LOSS IN HEAD (pounds-force per square inch) PER 100-FOOT LENGTH

For SI units: 1 inch = 25 mm, 1 gallon per minute = 0.06 L/s, 1 pound-force per square inch = 6.8947 kPa, 1 foot = 304.8 mm, 1 foot per second = 0.3048 m/s

APPENDIX A

CHART A 105.1(3)

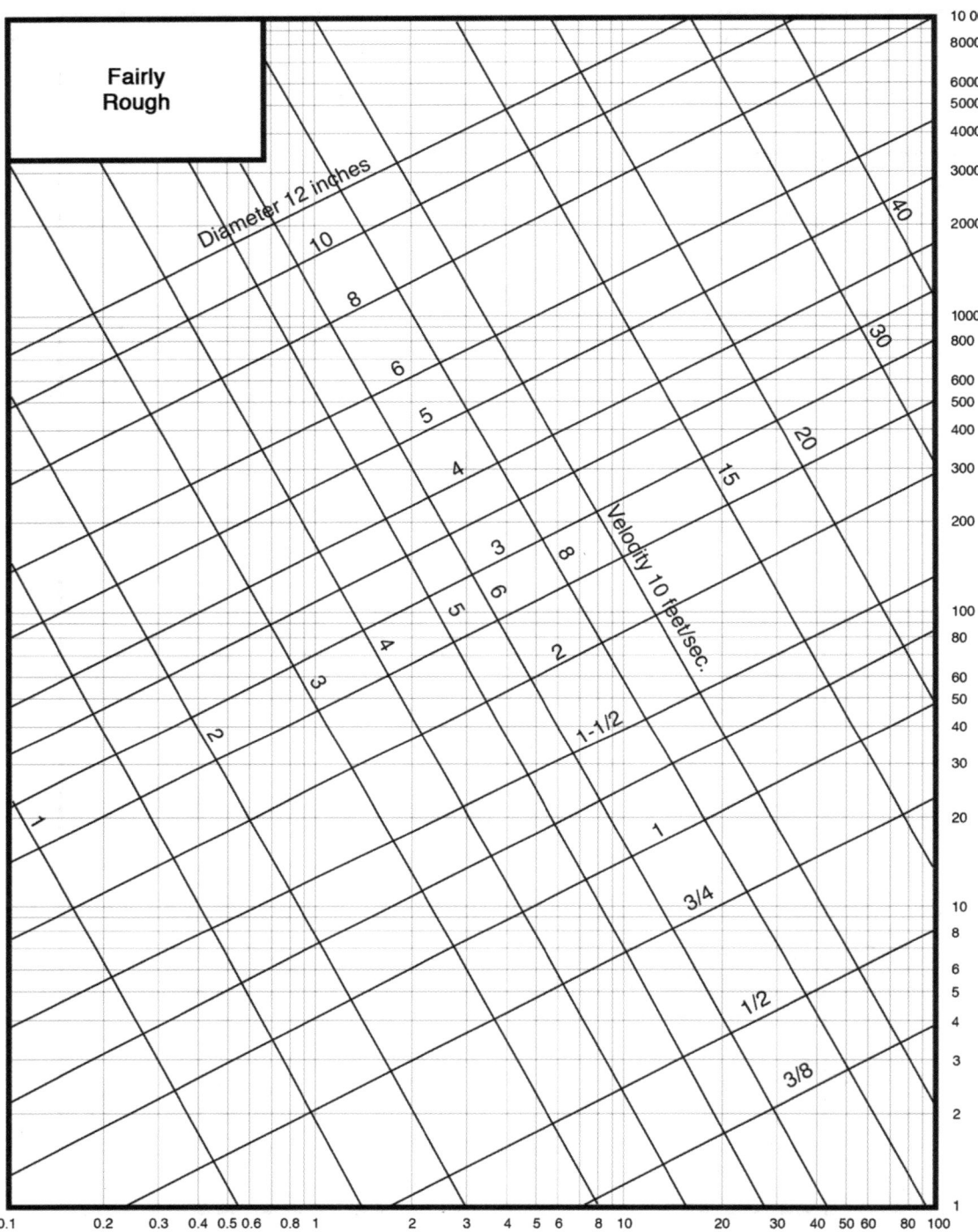

For SI units: 1 inch = 25 mm, 1 gallon per minute = 0.06 L/s, 1 pound-force per square inch = 6.8947 kPa, 1 foot = 304.8 mm, 1 foot per second = 0.3048 m/s

APPENDIX A

CHART A 105.1(4)

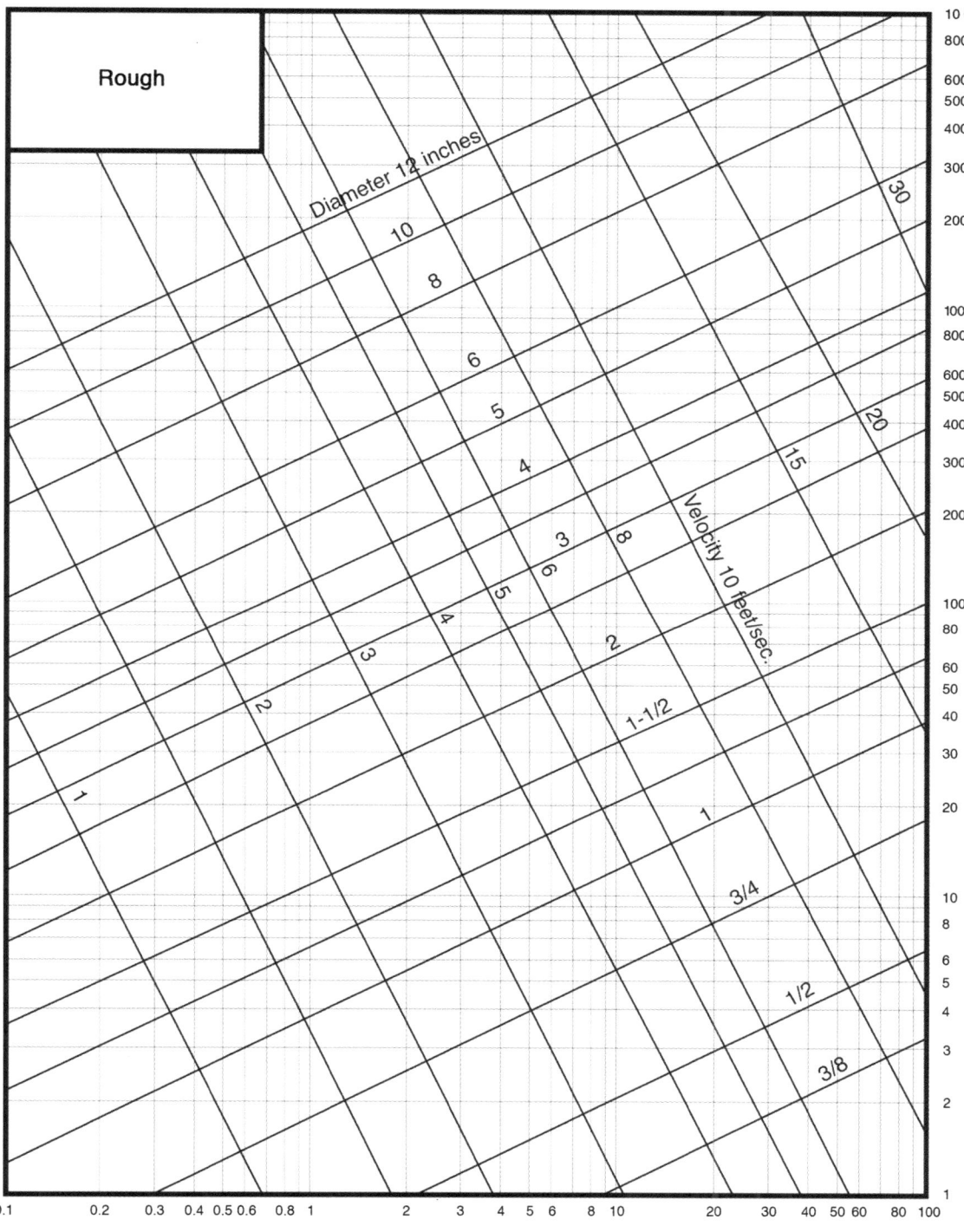

FRICTION LOSS IN HEAD (pounds-force per square inch) PER 100-FOOT LENGTH

For SI units: 1 inch = 25 mm, 1 gallon per minute = 0.06 L/s, 1 pound-force per square inch = 6.8947 kPa, 1 foot = 304.8 mm, 1 foot per second = 0.3048 m/s

APPENDIX A

CHART A 105.1(5)

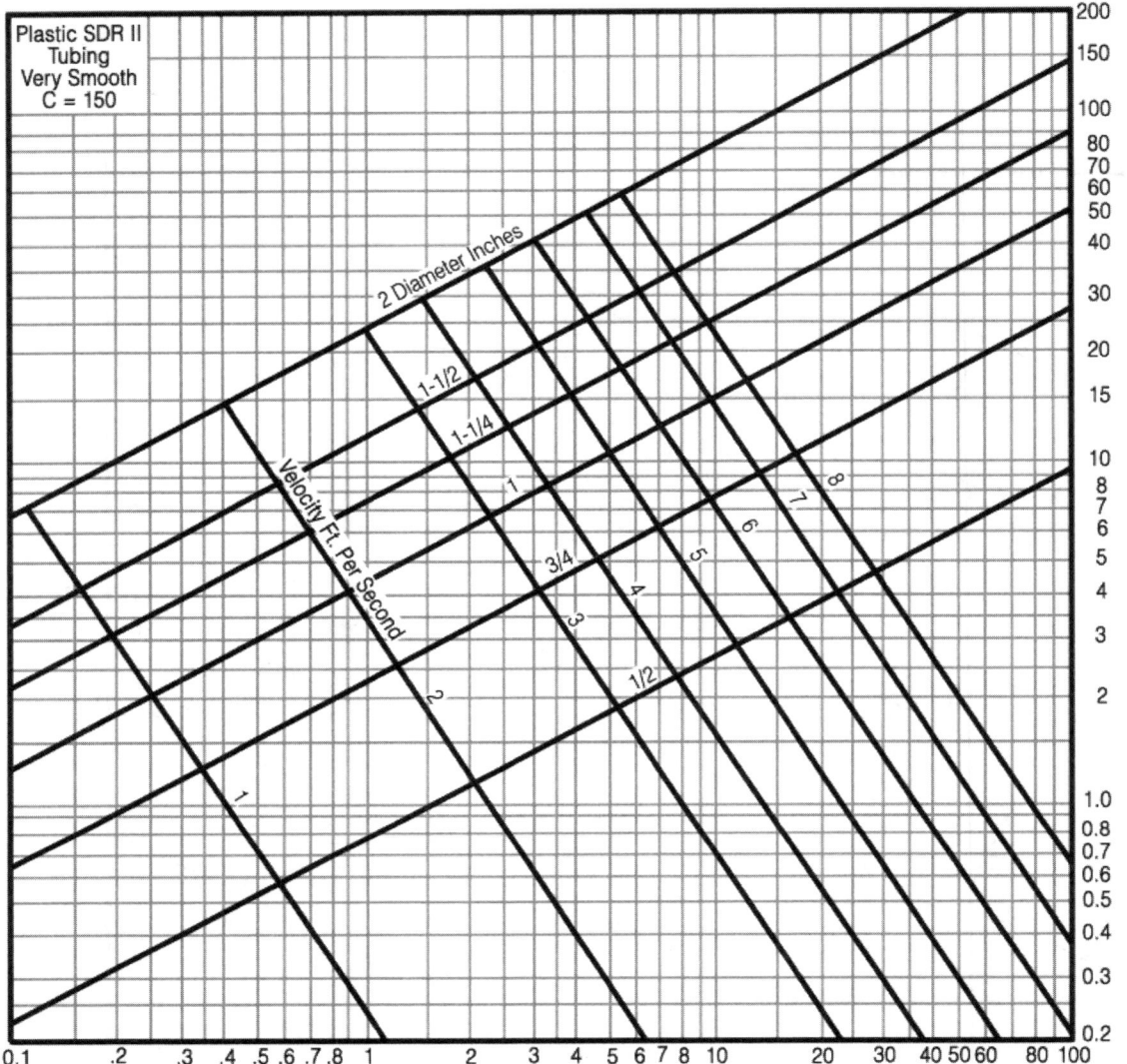

FRICTION LOSS IN HEAD (pounds-force per square inch) PER 100-FOOT LENGTH

For SI units: 1 inch = 25 mm, 1 gallon per minute = 0.06 L/s, 1 pound-force per square inch = 6.8947 kPa, 1 foot = 304.8 mm, 1 foot per second = 0.3048 m/s

APPENDIX A

CHART A 105.1(6)

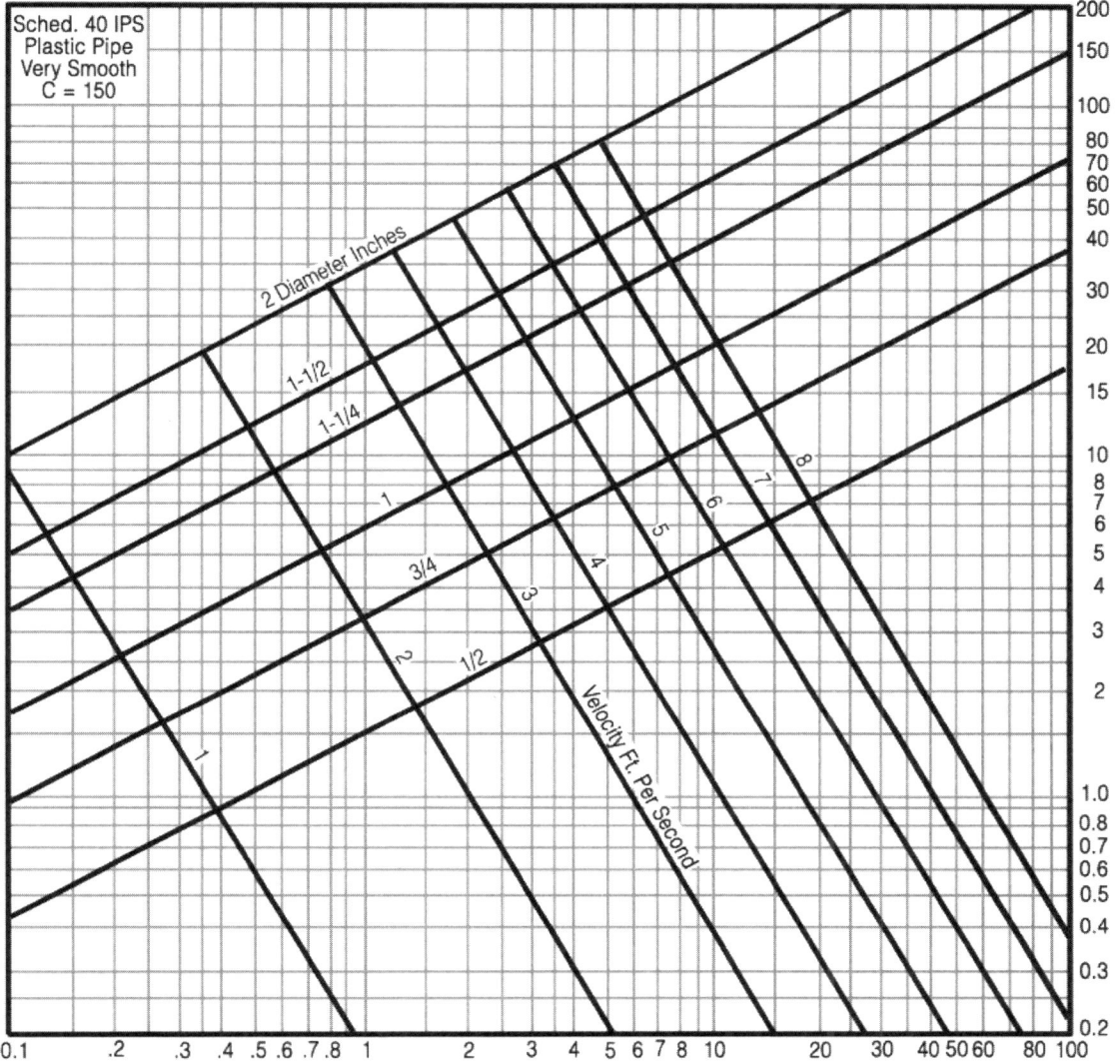

FRICTION LOSS IN HEAD (pounds-force per square inch) PER 100-FOOT LENGTH

For SI units: 1 inch = 25 mm, 1 gallon per minute = 0.06 L/s, 1 pound-force per square inch = 6.8947 kPa, 1 foot = 304.8 mm, 1 foot per second = 0.3048 m/s

CHART A 105.1(7)

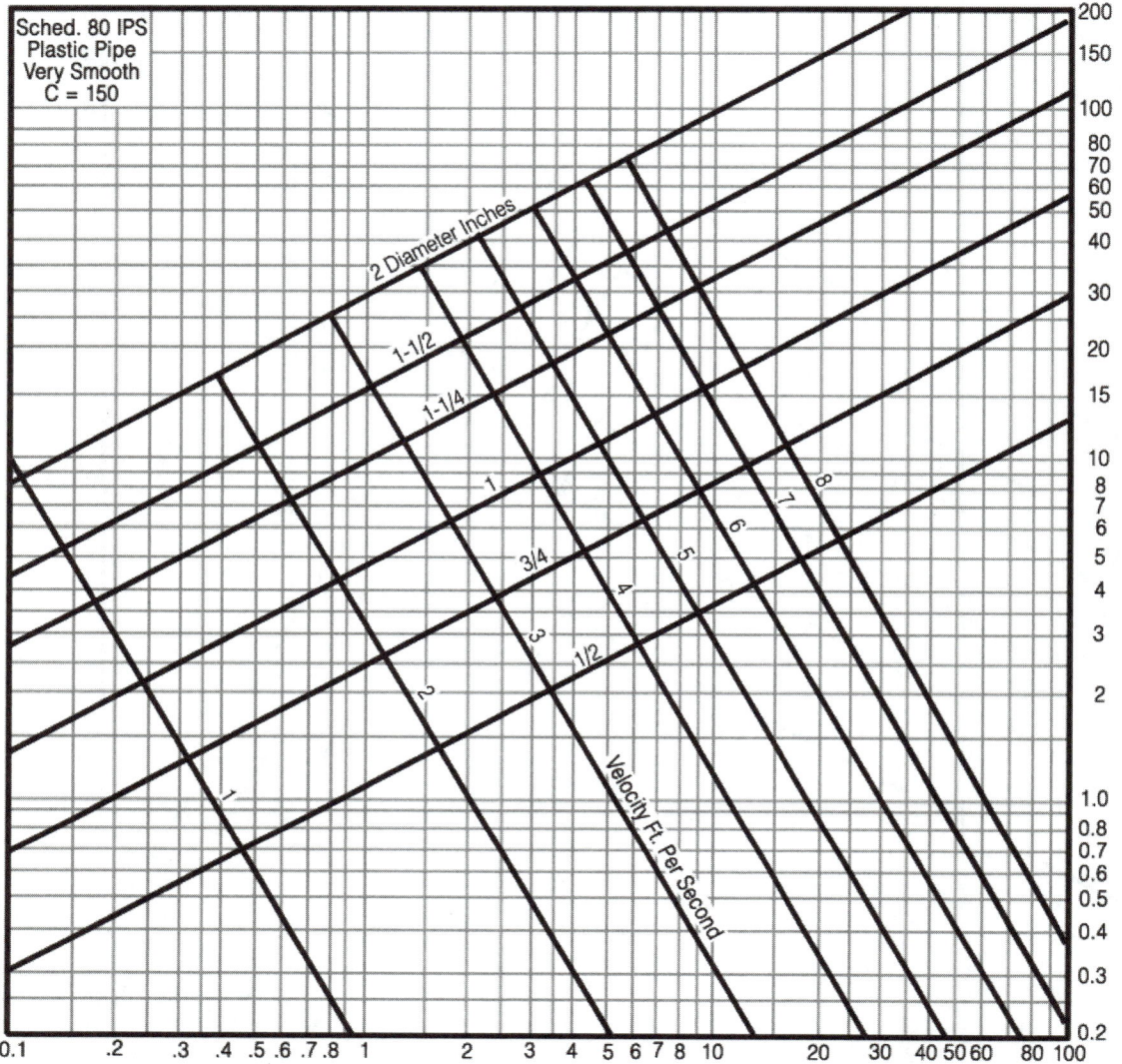

For SI units: 1 inch = 25 mm, 1 gallon per minute = 0.06 L/s, 1 pound-force per square inch = 6.8947 kPa, 1 foot = 304.8 mm, 1 foot per second = 0.3048 m/s

A 108.0 Sizing.

A 108.1 Example. Assume an office building of four stories and basement; pressure on the building side of the pressure-reducing valve of 55 psi (379 kPa) (after an allowance for reduced pressure falloff at peak demand); an elevation of highest fixture above the pressure-reducing valve of 45 feet (13 716 mm); a developed length of pipe from the pressure-reducing valve to the most distant fixture of 200 feet (60 960 mm); and fixtures to be installed with flush valves for water closets and stall urinals as follows:

Where the pipe material and water supply are such that Chart A 105.1(2) applies, the required diameter of the building supply is 3½ inches (90 mm) and the required diameter of the branch to the hot-water heater is 1½ inches (40 mm).

The sizes of the various branches and risers shall be permitted to be determined in the same manner as the size of the building supply or the branch to the hot-water system, by estimating the demand for the riser or branch from Chart A 103.1(1) or Chart A 103.1(2) and applying the total demand estimate from the branch, riser, or section thereof to the appropriate flowchart.

Example of Pipe Sizing

The following example demonstrates the procedure outlined in Appendix A using **Figure A 108.1a**.

APPENDIX A

1. Demand – The demand for this system is computed from the fixture unit values given in Table A 103.1 under the "Assembly" column. Each 1.6 GPF water closet flushometer valve is rated at 8 fixture units. Each 1.0 GPF urinal flushometer valve is rated at 5 fixture units. Therefore,

15 x 8 fu = 120 fixture units

13 x 5 fu = 65 fixture units

Total Demand = 185 fixture units

2. Distance from meter to most remote fixture is given as **150 feet**. If close estimates are desired, use A 104.4 to calculate the equivalent pressure losses for valves and fitting.

3. Elevation of highest outlet above meter is given as **40 feet**.

4. Minimum pressure at the meter is given as **74 psi**.

5. Using Chart A 103.1(1) or A 103.1(2) convert fixture units to gallons per minute. Use Curve 1 for flushometer valve fixtures. See also **Figure A 108.1b** for a conversion table of fixture units to gallons per minute. From Chart A 103.1(2), **185 FU is equivalent to 88 gallons per minute**. Use Chart A 103.1(2) or **Figure 108.1b** to convert all pipe sections from fixture units to gallons per minute.

6. Determine the total pressure available for friction loss in pounds per square inch (psi) per 100 feet of pipe. There are several steps involved in this calculation. We must first determine the total pressure losses from factors other than friction loss.

a. Determine the minimum pressure to be maintained at the highest outlet. Flushometer valves require a minimum available pressure of 15 psi to operate properly. Therefore, the **minimum residual pressure is 15.0 psi**.

b. Total static pressure loss (head loss) – This is determined by multiplying the elevation difference by 0.43. The elevation from the meter to the highest fixture outlet is 40 feet. Therefore, **40 ft x 0.43 psi/ft = 17.2 psi**.

c. Calculate pressure losses through the water meter, filters, reducing valves, backflow preventers, etc. In this illustration, we have chosen a 2 inch disc meter. From Chart A 102.2, this meter has a pressure loss of about 6.8 psi at 88 gpm. Therefore:

Meter loss = 6.8 psi

Total pressure loss is 15.0 + 17.2 + 6.8 = 39 psi

Subtracting pressure loss as calculated above from the minimum static pressure at the meter yields **35 psi (74 psi - 39 psi)**.

d. This remaining pressure of 35 psi is the maximum amount of pressure we have available for pressure loss due to friction in our system while maintaining a minimum pressure of 15 psi at the highest outlet. However, this 35 psi of pressure must be distributed throughout the total length of the system, which, in this case, has a total length of 150 feet. Before we are able to use the appropriate chart for determining pipe sizes, we must determine the pressure loss per 100 feet of pipe. Since our total loss is 35 psi for 150 feet of pipe,

A 108.1 EXAMPLE

FIXTURE UNITS AND ESTIMATED DEMANDS							
BUILDING SUPPLY DEMAND					BRANCH TO HOT WATER SYSTEM		
KIND OF FIXTURES	NUMBER OF FIXTURES	FIXTURE UNIT DEMAND	TOTAL UNITS	BUILDING SUPPLY DEMAND (gallons per minute)	NUMBER OF FIXTURES	FIXTURE UNIT DEMAND CALCULATION	DEMAND (gallons per minute)
Water Closets	130	8.0	1040	–	–	–	–
Urinals	30	4.0	120	–	–	–	–
Showerheads	12	2.0	24	–	12	12 x 2 x ¾ = 18	–
Lavatories	100	1.0	100	–	100	100 x 1 x ¾ = 75	–
Service Sinks	27	3.0	81	–	27	27 x 3 x ¾ = 61	–
Total	–	–	1365	252	–	154	55

For SI units: 1 gallon per minute = 0.06 L/s, 1 pound-force per square foot = 6.8947 kPa

Allowing for 15 psi (103 kPa) at the highest fixture under the maximum demand of 252 gallons per minute (15.90 L/s), the pressure available for friction loss is found by the following:

$$55 - [15 + (45 \times 0.43)] = 20.65 \text{ psi } (142.38 \text{ kPa})$$

The allowable friction loss per 100 feet (30 480 mm) of the pipe is, therefore:

$$100 \times 20.65 \div 200 = 10.32 \text{ psi } (71.15 \text{ kPa})$$

we would calculate our loss per 100 feet as follows:

(35 ÷ 150) x 100 = 23.3 psi loss per 100 ft of pipe.

7. Since our system is constructed of Type L copper pipe, we will select Chart A 105.1(1) for determining our pipe sizes (see **Figure A 105.1**).

   a. Our first step is to follow from left to right along the bottom of the chart until we come to our calculated allowable loss per 100 feet of pipe – **23.3 psi**. We then follow straight up the chart until we intersect the line for demand – **88 gpm**. Note that the intersection of these two lines is right above a line that runs from the upper left to the lower right of the chart with the number 15 above it. This diagonal line represents velocity. If we use this intersection as a starting point for our sizing, we will be designing our mains and branches to deliver their rated flow at a velocity of about 16 feet per second. Chapter 6, Section 610.12.1 limit the velocity of cold water in copper pipe to a maximum of 8 feet per second (fps). Good engineering practices would further reduce the velocity to 4-6 fps. Section 610.12.1 requires that copper tube systems conveying hot water be limited to 5 fps velocity. Since the system will be designed for 8 fps, we must move our starting point to the left along our 88 gpm horizontal line until we intersect the 8 fps velocity diagonal line.

   b. Note that the intersection of the new starting point falls above another diagonal line running from the lower left to the upper right. These lines represent our pipe sizes. The new starting point falls above the 2 inch pipe line, but below the 2½ inch pipe line. When the point falls between two pipe sizes, choose the larger pipe size. This is the size for the building supply. If we were to follow up and to the left along our 8fps velocity line until we intersect the 2½ inch pipe size line, and from that point draw a horizontal line to the right edge of Chart A 105.1(1), then we would find the upper limit for maximum loading allowed on a 2½ inch pipe at 8 fps. The upper limit would be approximately 120 gpm.

   Therefore, the maximum demand for **2½ inch pipe** at 8 fps is **120 gpm**.

   c. Going back to our 8 fps diagonal line, we follow it down to the lower right until we intersect the 2 inch pipe diagonal line. From this intersection, we draw another horizontal line to the right edge of the chart. This gives us the maximum demand allowed on a 2 inch pipe at 8 feet per second.

   Therefore, maximum demand for **2 inch pipe** at 8 fps is **73 gpm**.

   d. Further down the 8 fps diagonal line, at the intersection of the 1½ inch pipe, and reading horizontally to the right edge of Chart A 105.1(1) yields a maximum load of 42 gpm.

   Therefore, maximum demand for **1½ inch pipe** at 8 fps is **42 gpm**.

   e. Following down the 8 fps line until we intersect the 1¼ inch line and reading over to the right edge of the chart gives us a maximum load of 29 gpm.

   Therefore, maximum demand for **1¼ inch pipe** at 8 fps is **29 gpm**.

   f. Likewise, the maximum demand for **1 inch pipe** at 8 fps is **19 gpm**.

   g. The maximum demand for **¾ inch pipe** at 8 fps is **12 gpm**.

   h. As you follow the 8 fps diagonal line from the upper left to the lower right, you will intersect the original vertical line established as the maximum allowable pressure loss per 100 feet of pipe due to friction. We are not allowed to go past this line since there is no more pressure to lose. At this juncture, we must follow down vertically along this original line and read all subsequent pipe sizes at the point where they intersect the original friction loss line. An example of this happens if we want to calculate the maximum demand allowed on 3/8 inch L-copper pipe at 8 fps. Following from upper left to lower right along the 8 fps line, we intersect our originally established friction loss line before we reach the 3/8 inch diagonal line. At this point we must follow vertically down the friction loss line until we intersect the 3/8 inch L-copper line. Reading over to the right side of the chart would give us a maximum load of approximately 3 gallons per minute.

   i. Note that as you follow the 8 fps line from the upper left to the lower right and reach the 1 inch and below pipe sizes, you have three lines to choose from. Each line represents a different wall thickness (hence a different inside diameter). Be careful to follow the proper line as indicated in the legend in the upper left-hand corner of the chart.

8. A table at the bottom of **Figure A 108.1a** shows the pipe section to be sized, the fixture unit total for that pipe section, and the conversions from fixture units to gallons per minute using Chart A 103.1(2) (see step 5 above). The last column lists the appropriate pipe size for each pipe section.

9. A smaller chart listing the various pipe sizes and their maximum gallons per minute loading from Chart A 105.1(1) is shown in **Figure A 108.1a**. This smaller chart allows sizing by demand for each pipe section in **Figure A 108.1a**.

APPENDIX A

Flushometer Valve Sizing
Sized in accordance with Appendix A
Occupancy - Heavy Use Assembly

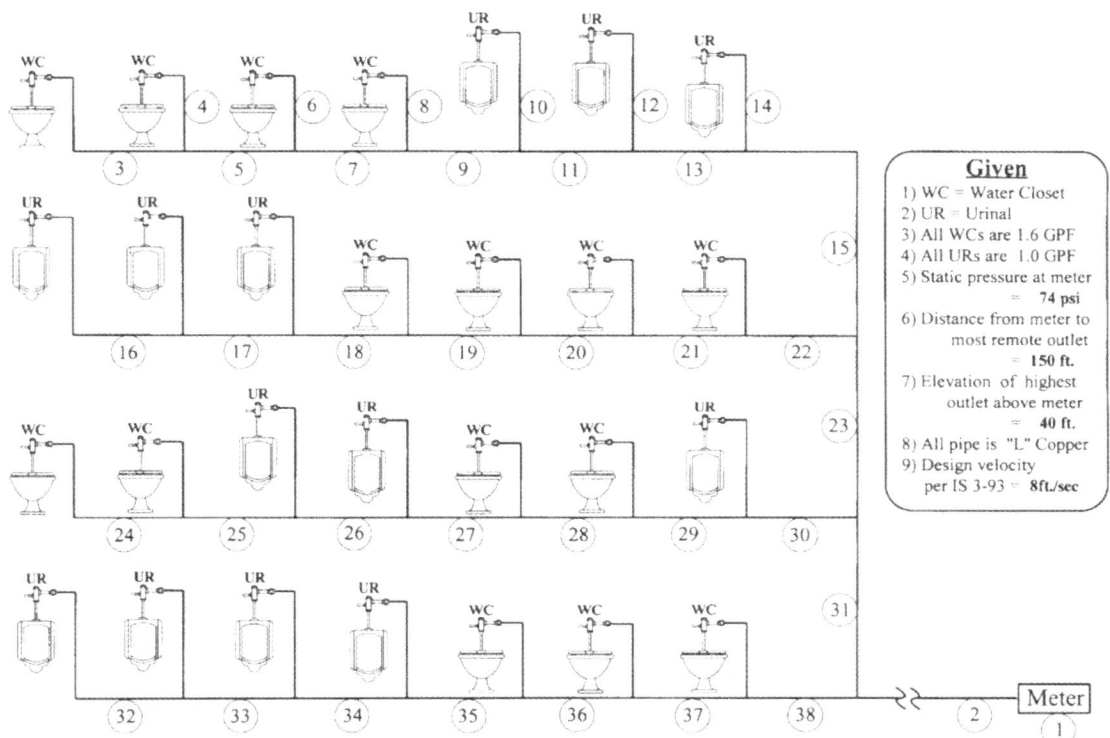

Given
1) WC = Water Closet
2) UR = Urinal
3) All WCs are 1.6 GPF
4) All URs are 1.0 GPF
5) Static pressure at meter = **74 psi**
6) Distance from meter to most remote outlet = **150 ft.**
7) Elevation of highest outlet above meter = **40 ft.**
8) All pipe is "L" Copper
9) Design velocity per IS 3-93 = 8ft./sec

...from Chart A-4

Pipe Size	Maximum GPM
2-1/2"	120
2"	73
1-1/2"	42
1-1/4"	29
1"	19
3/4"	12

Pipe Section	Fixture Units	Demand in gpm	Pipe Size	Pipe Section	Fixture Units	Demand in gpm	Pipe Size	Pipe Section	Fixture Units	Demand in gpm	Pipe Size
1 meter	185	88	2"	14	5	22	1-1/4"	27	26	39	1-1/2"
2	185	88	2-1/2"	15	47	50	2"	28	34	44	2"
3	8	25	1-1/4"	16	5	22	1-1/4"	29	42	47	2"
4	8	25	1-1/4"	17	10	27	1-1/4"	30	47	50	2"
5	16	33	1-1/2"	18	15	32	1-1/2"	31	141	78	2-1/2"
6	8	25	1-1/4"	19	23	37	1-1/2"	32	5	22	1-1/4"
7	24	38	1-1/2"	20	31	42	1-1/2"	33	10	27	1-1/4"
8	8	25	1-1/4"	21	39	46	2"	34	15	32	1-1/2"
9	32	43	2"	22	47	50	2"	35	20	35	1-1/2"
10	5	22	1-1/4"	23	94	66	2"	36	28	40	1-1/2"
11	37	45	2"	24	8	25	1-1/4"	37	36	45	2"
12	5	22	1-1/4"	25	16	33	1-1/2"	38	44	48	2"
13	42	47	2"	26	21	36	1-1/2"				

FIGURE A 108.1A
FLUSHOMETER SIZING EXAMPLE USING APPENDIX A

GPM TO FIXTURE UNIT CONVERSION TABLE

These tables may be interpolated

FLOW GPM	FIXTURE UNITS FLUSH VALVE TANKS	FIXTURE UNITS FLUSHOMETER VALVES	FLOW GPM	FIXTURE UNITS FLUSH VALVE TANKS	FIXTURE UNITS FLUSHOMETER VALVES	FLOW GPM	FIXTURE UNITS FLUSH VALVE TANKS	FIXTURE UNITS FLUSHOMETER VALVES
1	0	—	43	99	33	125	506	396
2	1	—	44	103	35	130	533	430
3	3	—	45	107	37	135	559	460
4	4	—	46	111	39	140	585	490
5	6	—	47	115	42	145	611	521
6	7	—	48	119	44	150	638	559
7	8	—	49	123	46	155	665	596
8	10	—	50	127	48	160	692	631
9	12	—	51	130	50	165	719	666
10	13	—	52	135	52	170	748	700
11	15	—	53	141	54	175	778	739
12	16	—	54	146	57	180	809	775
13	18	—	55	151	60	185	840	811
14	20	—	56	155	63	190	874	850
15	21	—	57	160	66	200	945	931
16	23	—	58	165	69	210	1018	1009
17	24	—	59	170	73	220	1091	1091
18	26	—	60	175	76	230	1173	1173
19	28	—	62	185	82	240	1254	1254
20	30	—	64	195	88	250	1335	1335
21	32	—	66	205	95	260	1418	1418
22	34	5	68	215	102	270	1500	1500
23	36	6	70	225	108	280	2583	2583
24	39	7	72	236	116	290	1668	1668
25	42	8	74	245	124	300	1755	1755
26	44	9	76	254	132	310	1845	1845
27	46	10	78	264	140	320	1926	1926
28	49	11	80	275	148	330	2018	2018
29	51	12	82	284	158	340	2110	2110
30	54	13	84	294	168	350	2204	2204
31	56	14	86	305	176	360	2298	2298
32	58	15	88	315	186	370	2388	2388
33	60	16	90	325	195	380	2480	2480
34	63	18	92	337	205	390	2575	2575
35	66	20	94	348	214	400	2670	2670
36	69	21	96	359	223	410	2765	2765
37	74	23	98	370	234	420	2862	2862
38	78	25	100	380	245	430	2960	2960
39	83	26	105	405	270	440	3060	3060
40	86	28	110	431	295	450	3150	3150
41	90	30	115	455	329	500	3620	3620
42	95	31	120	479	365			

FIGURE A 108.1B
INTERPOLATION OF GPM TO FIXTURE UNITS FOR CHART A 103.1(1)

APPENDIX B

EXPLANATORY NOTES ON COMBINATION WASTE AND VENT SYSTEMS

(See Section 910.0 for specific limitations)

B 101.0 General.

B 101.1 Applicability. This appendix provides general guidelines for the design and installation of a combination waste and vent system.

B 101.2 General Requirements. Combination waste and vent systems, as outlined in Section 910.0 of this code, cover the horizontal wet venting of a series of traps by means of a common waste and vent pipe. Pipe sizes not less than two pipe sizes larger than those required for a conventional system are designed to maintain a wetted perimeter or flow line low enough in the waste pipe to allow adequate air movement in the upper portion, thus balancing the system. Sinks, lavatories, and other fixtures that rough in above the floor, shall not be permitted on a combination waste and vent system, which, at best, is merely an expedient designed to be used in locations where it would be structurally impractical to provide venting in the conventional manner.

Combination waste and vent systems are intended primarily for extensive floor or shower drain installations where separate venting is not practical, for floor sinks in markets, demonstration or work tables in school buildings, or for similar applications where the fixtures are not adjacent to walls or partitions. Due to its oversize characteristics, such a waste system is not self-scouring and, consequently, care shall be exercised as to the type of fixtures connected thereto and to the location of cleanouts. In view of its grease-producing potential, restaurant kitchen equipment shall not be connected to a combination waste and vent system.

B 101.3 Caution. Caution shall be exercised to exclude appurtenances delivering large quantities or surges of water (such as pumps, sand interceptors, etc.) from combination waste and vent systems in order that adequate venting will be maintained. Small fixtures with a waste-producing potential of less than 7½ gallons per minute (gpm) (0.47 L/s) shall be permitted to be safely assigned a loading value of one unit. Long runs shall be laid at the minimum permissible slope in order to keep tailpieces as short as possible. Tailpieces shall not exceed 2 feet (610 mm) in length, which shall necessitate slopes up to 45 degrees (0.79 rad) (see definition of horizontal pipe) on some branches.

B 101.4 Pneumatics. It is essential that the pneumatics of such a system be properly engineered, as the air pressure within the line shall at all times balance that of outside atmosphere in order to prevent either trap seal loss or air locking between traps. Long mains shall be provided with additional relief vents located at intervals not exceeding 100 feet (30 480 mm). Each such relief vent shall equal not less than one-half of the inside cross-sectional area of the drain pipe served.

B 101.5 Trap Sizes. Trap sizes are required to be equivalent to the branches they serve (two pipe sizes larger than normal), and tailpieces between fixtures or floor drains and such traps shall be reduced to normal size.

B 101.6 Layout Drawings. Duplicate layout drawings of each such proposed piping system shall be presented to the Authority Having Jurisdiction and approval obtained before an installation is made. Complicated layouts shall be checked by qualified personnel.

B 101.6.1 Example of Sizing. A floor drain normally requires a 2 inch (50 mm) trap and waste. On a combination waste and vent system, both trap and waste shall be increased two pipe sizes (through 2½ inches and 3 inches) (65 mm and 80 mm), which would make the trap 3 inches (80 mm). Pipe sizes recognized for this purpose are 2 inches, 2½ inches, 3 inches, 3½ inches, 4 inches, 4½ inches, 5 inches, 6 inches, etc. (50 mm, 65 mm, 80 mm, 90 mm, 100 mm, 115 mm, 125 mm, 150 mm, etc.). The tailpiece between the floor drain and its trap shall be 2 inches (50 mm) (or normal size) to ensure that the amount of wastewater entering the trap partially fills the waste branch. A 3 inch (80 mm) floor drain would thus require a 4 inch (100 mm) trap, and a 4 inch (100 mm) floor drain would require a 5 inch (125 mm) trap for the reasons previously stated.

WHERE IN DOUBT, CHECK WITH YOUR LOCAL Authority Having Jurisdiction.

APPENDIX C
ALTERNATE PLUMBING SYSTEMS

🔧 This appendix was added to the Uniform Plumbing Code (UPC) to allow the use of alternate plumbing systems and specifically, engineered systems. Section 301.3 already permits the use of alternate methods and materials including engineered systems when approved by the Authority Having Jurisdiction (AHJ); however, this appendix provides extra justification and guidelines for using these systems that are not covered in the body of the code.

The methods described in this appendix can be classified as "green technology" in that the use of some of these special systems can lead to water and resource reductions. Use of these systems could lead to points in the LEED Rating System.

As with all the appendices to the code, the systems in Appendix C are permitted only if it is adopted as part of the code and if the individual section is also adopted. If adopted, these systems may be used, but are not required.

C 101.0 General.

C 101.1 Applicability. The intent of this appendix is to provide clarification of procedures for the design and approval of engineered plumbing systems, alternate materials, and equipment not specifically covered in other parts of the code.

C 101.2 Provisions. The provisions of this appendix apply to the design, installation, and inspection of an engineered plumbing system, alternate material, and equipment.

C 101.3 Authority Having Jurisdiction. The Authority Having Jurisdiction has the right to require descriptive details of an engineered plumbing system, alternate material, or equipment including pertinent technical data to be filed.

C 101.4 Standards and Specifications. Components, materials, and equipment shall comply with standards and specifications listed in Table 1701.1 of this code and other national consensus standards applicable to plumbing systems and materials.

C 101.5 Alternate Materials and Equipment. Where such standards and specifications are not available, alternate materials and equipment shall be approved in accordance with the provisions of Section 301.3 of this code.

C 201.0 Definitions.

C 201.1 General. For the purposes of this code, these definitions shall apply to this appendix:

Branch Interval. A length of soil or waste stack corresponding in general to a story height, but in no case less than 8 feet (2438 mm), within which the horizontal branches from one floor or story of the building are connected to the stack.

Engineered Plumbing System. A system designed for a specific building project with drawings and specifications indicating plumbing materials to be installed, all as prepared by a registered design professional.

🔧 The term "registered design professional" is the accepted term for an architect, engineer, or other professional who is licensed and certified by a state to practice their respective design profession. This is also the term as used and defined in the construction and safety codes adopted by almost all state and local jurisdictions throughout the U.S.

The plumbing system designed per the parameters of Appendix C must be documented by plans and specifications and possibly by engineering calculations or other documentation certifying that the system meets appropriate national plumbing standards. If this is not possible, then the system must be approved by the process outlined in Chapter 3 for alternate materials and methods.

C 301.0 Engineered Plumbing Systems.

C 301.1 Inspection and Installation. In other than one- and two-family dwellings, the designer of the system is to provide periodic inspection of the installation on a schedule approved by the Authority Having Jurisdiction. Prior to the final approval, the designer shall verify to the Authority Having Jurisdiction that the installation is in accordance with the approved plans, specifications, and data and such amendments thereto. The designer shall certify to the Authority Having Jurisdiction that the installation is in accordance with the applicable engineered design criteria.

🔧 The requirements for designer inspection and certification of the plumbing system designed per Appendix C ensure that the professional who designs the system is taking responsibility not only for the system design but also for its proper installation. Because these types of plumbing systems may not be familiar to maintenance personnel, the designer must also provide information on how the system works and requirements for additions to or maintenance of the system.

C 301.2 Owner Information. The designer of the system shall provide the building owner with information concerning the system, considerations applicable for subsequent modifications to the system, and maintenance requirements as applicable.

C 302.0 Water Heat Exchangers.

C 302.1 Protection from Contamination. Heat exchangers used for heat transfer, heat recovery, or solar heating shall protect the potable water system from being contaminated by the heat-transfer medium.

C 302.2 Single-Wall Heat Exchangers. Single-wall heat exchangers shall comply with the following requirements:

(1) The heat-transfer medium is either potable water or contains essentially nontoxic transfer fluids having a toxicity rating or class of 1 (see Section 207.0).

(2) The pressure of the heat-transfer medium is maintained at less than the normal minimum operating pressure of the potable water system.

Exception: Steam in accordance with Section C 302.2(1) above.

(3) The equipment is permanently labeled to indicate that only additives recognized as safe by the FDA shall be used in the heat-transfer medium.

🔧 Double-wall heat exchangers are allowed per Section 603.5.4. Single-wall water heat exchangers are also allowed under that section but must meet the parameters set forth under Section 505.4.1 for indirect-fired water heaters that incorporate a single-wall heat exchanger. Section 505.4.1 mirrors most of the requirements contained in this section of Appendix C. It must be remembered that single-wall heat exchangers, as opposed to the double-wall heat exchanger shown in **Figure C 302.2**, pose more of a potential safety problem. This is due to the possibility of the single wall corroding or otherwise failing to prevent contamination of the potable water with the treated water flowing through the heat exchanger. The double-wall exchanger is designed so that a leak in one of the walls will escape between the double-wall without contaminating the potable water.

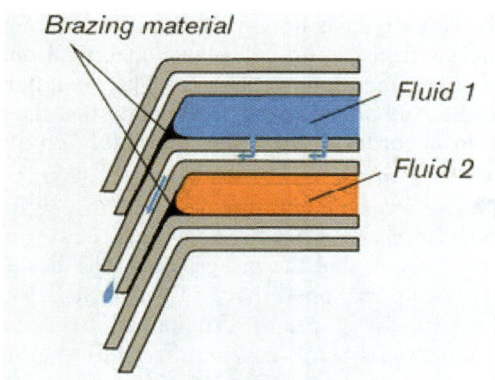

FIGURE C 302.2
CUTAWAY OF DOUBLE WALL HEAT EXCHANGER

C 302.3 Alternate Designs. Other heat exchanger designs shall be permitted where approved by the Authority Having Jurisdiction.

C 303.0 Fixture Unit Values for Private or Private Use Bathroom Groups.

C 303.1 Fixtures. Table C 303.2 and Table C 303.3 reflect the fixture unit loads for the fixtures in bathrooms as groups, rather than as individual fixtures. Such fixtures include water closets, lavatories, and bathtubs or showers. The tables reflect diversity in the use of fixtures within a bathroom and between multiple bathrooms.

🔧 This alternate method of pipe sizing was developed as a result of research by the Stevens Institute of Technology, Hoboken, New Jersey. Instead of counting fixture units individually, as in Tables 610.3 and 702.1, bathroom fixtures are grouped and assigned fixture unit values as a group. The underlying assumption is that fixtures located in the same bathroom are most likely to be used by one person, one fixture at a time, rather than simultaneously. This decreased probability of simultaneous use is calculated into Tables C 303.2 and C 303.3, allowing for decreased values for bathroom fixture groups as compared to sizing the system by individual fixtures and their fixture unit values. This method of sizing may result in smaller sizing of the branch and mains serving these groups, compared to sizing using the traditional methods in Chapters 6 and 7. Note that the use of these tables is for private use installations only. See the definitions for "Private or Private use" in Chapter 2.

Notes 1 and 2 in Tables C 303.2 and C 303.3 define "bathroom group" differently than in Chapter 2. The definition for bathroom group in Tables C 303.2 and C 303.3 are for use only with these tables. Therefore, a bathroom group consists of not more than one water closet, up to two lavatories and either one bathtub or one shower or a combination bath/shower.

C 303.2 Water Supply Fixture Unit Values. The listed water supply fixture unit values in Table C 303.2 reflect the load of entire bathroom groups on the cold water building supply. Individual hot and cold water branch piping to the fixtures shall be permitted to be sized in accordance with Chapter 6 and Appendix A.

🔧 Table C 303.2 is a modification of Table B.5.3 found in the National Standard Plumbing Code that was designed by Thomas Konen from the Steven's Institute of Technology. The fixture unit values listed under the Cold columns in Table C 303.2 indicates the total fixture unit load of the entire group when sizing the building supply and not for the separate cold water supply branches. The Hot column is ¾ the fixture unit value from Table 610.3 or Table A 103.1 for individual fixtures with a further diversity factor reduction. The diversity factor is a ratio reduction between the fixture unit values in the Cold column and the fixture units as determined by Table 610.3 or Table A 103.1. For example, Table 610.3 determines a total 7.5 fixture units for one bathroom containing a combination bath/shower, a 1.6 gpf gravity toilet and a lavatory. Table C 303.1(1) determines 5 fixture units for one bathroom group, which is a ratio deduction of .67 (5/7.5). Multiplying .67 to 7.5 yields the reduced fixture unit value of 5 as listed in Table C 303.1(1). The total hot supply fixture unit value for one bathroom is ¾ the fixture unit value as determined by Table 610.3 (.75 x (4.0+1.0) = 3.75). The fixture unit value of 3.75 is further reduced by the diversity factor of .67, which yields 2.5 and is the hot fixture unit value in Table C 303.2. **Figure C 303.2** shows the diversity factors used in Table C 303.2 to determine the Hot column.

C 303.3 Drainage Fixture Unit Values. The listed drainage fixture unit values in Table C 303.3 reflect the load of entire bathroom groups on the sanitary drainage system. Where fixtures within bathrooms connect to differ-

APPENDIX C

TABLE C 303.2
WATER SUPPLY FIXTURE UNITS (WSFU) FOR BATHROOM GROUPS[1, 2]

	PRIVATE USE BATHROOM GROUP		SERVING 3 OR MORE PRIVATE USE BATHROOM GROUPS	
	COLD	HOT[3]	COLD	HOT
Bathroom Groups Having up to 1.6 GPF Gravity-Tank Water Closets				
Half-Bath or Powder Room	3.5	0.8	2.5	0.5
1 Bathroom Group	5.0	2.5	3.5	1.8
1½ Bathrooms	6.0	2.5	–	–
2 Bathrooms	7.0	3.5	–	–
2½ Bathrooms	8.0	3.6	–	–
3 Bathrooms	9.0	4.5	–	–
Each Additional ½ Bath	0.5	0.1	–	–
Each Additional Bathroom Group	1.0	0.5	–	–
Bathroom Groups Having up to 1.6 GPF Pressure-Tank Water Closets				
Half-Bath or Powder Room	3.5	0.8	2.5	0.5
1 Bathroom Group	5.0	2.5	3.5	1.8
1½ Bathrooms	6.0	2.5	–	–
2 Bathrooms	7.0	3.5	–	–
2½ Bathrooms	8.0	3.6	–	–
3 Bathrooms	9.0	4.5	–	–
Each Additional ½ Bath	0.5	0.1	–	–
Each Additional Bathroom Group	1.0	0.5	–	–
Bathroom Group (1.6 GPF Flushometer Value)	6.0	2.5	4.0	1.7
Kitchen Group (Sink and Dishwasher)	2.0	2.0	1.5	1.5
Laundry Group (Sink and Clothes Washer)	5.0	5.0	3.0	3.0

Notes:
[1] A bathroom group, for the purposes of this table, consists of one water closet, up to two lavatories, and either one bathtub or one shower.
[2] A half-bath or powder room, for the purposes of this table, consists of one water closet and one lavatory.
[3] Multi-unit dwellings with individual water heaters use the same WSFU as for individual dwellings.

ent branches of the drainage system, the fixture unit values for the individual fixtures shall be used, as listed in Table 702.1 of this code.

Table C 303.3 reduces the fixture unit load on the drainage system and branches by bathroom groupings. The system total load can be calculated using the bathroom group values from the table. When the individual fixture drains connect to a branch serving the bathroom groups, they then may be sized using the fixture unit values in Table C 303.3. However, drainage from individual fixtures outside the bathroom groups should be sized by their individual drainage fixture unit values from Table 702.1. If the fixture drains from the bathroom group connect to different branches of the drainage system that are not serving the bathroom group, they should be sized using the fixture unit values of Table 702.1.

C 304.0 Drainage System Sizing.

The method of sizing the drainage and vent system in Sections C 304.0 and C 401.0 is based on that used in the American Society of Plumbing Engineers (ASPE) *Design Handbook Volume 2*, 2006 edition. In fact, the tables contained here, Table C 304.2, C 304.3 and C 401.1, are exact duplications of the corresponding tables in the *ASPE Design Handbook*, provided with the permission of ASPE.

The rationality for including this method of sizing in Appendix C is to allow design professionals access to it since it is used in many areas, primarily for sizing the drainage and vent systems in high-rise buildings.

C 304.1 Drainage Fixture Units. Drainage fixture unit values shall be sized in accordance with Section 702.0 and Table 702.1.

C 304.2 Size of Building Drain and Building Sewer. The maximum number of drainage fixture units allowed on the building drain or building sewer of a given size shall be in accordance with Table C 304.2. The size of a building drain or building sewer serving a water closet shall be not less than 3 inches (80 mm).

Table 304.2 is used for sizing horizontal building drains and sewers only. Horizontal branches connecting to

APPENDIX C

TABLE C 303.3
DRAINAGE FIXTURE UNIT VALUES (DFU) FOR BATHROOM GROUPS[1, 2]

	PRIVATE USE BATHROOM GROUP	SERVING 3 OR MORE PRIVATE USE BATHROOM GROUP
Bathroom Groups having 1.6 GPF Gravity-Tank Water Closets		
Half-Bath or Powder Room	3.0	2.0
1 Bathroom Group	5.0	3.0
1½ Bathrooms	6.0	—
2 Bathrooms	7.0	—
2½ Bathrooms	8.0	—
3 Bathrooms	9.0	—
Each Additional ½ Bath	0.5	—
Each Additional Bathroom Group	1.0	—
Bathroom Groups having 1.6 GPF Pressure-Tank Water Closets		
Half-Bath or Powder Room	3.5	2.5
1 Bathroom Group	5.5	3.5
1½ Bathrooms	6.5	—
2 Bathrooms	7.5	—
2½ Bathrooms	8.5	—
3 Bathrooms	9.5	—
Each Additional ½ Bath	0.5	—
Each Additional Bathroom Group	1.0	—
Bathroom Groups having 3.5 GPF Gravity-Tank Water Closets		
Half-Bath or Powder Room	3.0	2.0
1 Bathroom Group	6.0	4.0
1½ Bathrooms	8.0	—
2 Bathrooms	10.0	—
2½ Bathrooms	11.0	—
3 Bathrooms	12.0	—
Each Additional ½ Bath	0.5	—
Each Additional Bathroom	1.0	—
Bathroom Group (1.6 GPF Flushometer Valve)	3.0	—
Bathroom Group (3.5 GPF Flushometer Valve)	4.0	—

Notes:
[1] A bathroom group, for the purposes of this table, consists of not more than one water closet, up to two lavatories, and either one bathtub or one shower.
[2] A half-bath or powder room, for the purposes of this table, consists of one water closet and one lavatory.

the building drain may also be sized using this table. Horizontal branches connecting to a stack above the building drain must be sized using either column 2 or column 5 of Table C 304.3.

Use of this table is fairly straightforward. For example, 4-inch (100 mm) pipe may serve 180 DFUs at ⅛ inch slope, 216 DFUs at ¼ inch slope and 250 DFUs at ½ inch slope.

Note 2 limits public use installations of 3-inch (80 mm) pipe to only two water closets (WCs) or two bathroom groups. Single-family use may have three WCs or three bathroom groups served by 3-inch (80 mm) pipe.

C 304.3 Size of Horizontal Branch or Vertical Stack. The maximum number of drainage fixture units allowed on a horizontal branch or vertical soil or waste stack of a given size shall be in accordance with Table C 304.3. Stacks shall be sized based on the total accumulated connected load at each story or branch interval.

One of the features of this method of sizing the drainage system is the inclusion of the branch interval (see **Figure C 304.3**) as a sizing parameter in Table C 304.3. Notice that in the table there are varying drainage fixture unit (DFU) values for branch piping, depending on how many branch intervals are contained in the drainage stack. This is due to the fact that the stack is allowed to serve more fixture units depending on how many branch intervals or how tall the building is. Larger stacks serving more than three branch intervals allow a greater load than smaller stacks. However, the discharge load from branches into the larger stack is more restricted than from branches into smaller stacks. Looking at Table C 304.3, any 4-inch (100 mm) horizontal branch may serve 160 DFUs. If the same

APPENDIX C

	Total Fixture Units from Table 610.3	Total Fixture Units from Table C 303.1(1)	Diversity Factor	Cold	Hot
1/2 bath	3.5	3.5	1	3.3	0.8
1 Bathroom Group	7.5	5	0.67	4.2	2.5
1 1/2 Bathroom Group	11	6	0.55	5.2	2.5
2 Bathroom Groups	15	7	0.47	5.8	3.5
2 1/2 Bathroom Groups	18.5	8	0.43	6.8	3.6
3 Bathroom Groups	22.5	9	0.40	7.5	4.5
Additional 1/2 Bath	3.5	0.5	0.14	0.5	0.1
Additional Bathroom Group	7.5	1	0.13	0.8	0.5

FIGURE C 303.2
DIVERSITY FACTORS USED IN TABLE C 303.1(1)

TABLE C 304.2
BUILDING DRAINS AND BUILDING SEWERS[1]

DIAMETER OF PIPE (inches)	MAXIMUM NUMBER OF DRAINAGE FIXTURE UNITS FOR SANITARY BUILDING DRAINS AND RUNOUTS FROM STACKS			
	SLOPE (inches per foot)			
	1/16	1/8	1/4	1/2
2	–	–	21	26
2½	–	–	24	31
3	–	20	42[2]	50[2]
4	–	180	216	250
5	–	390	480	575
6	–	700	840	1000
8	1400	1600	1920	2300
10	2500	2900	3500	4200
12	3900	4600	5600	6700
15	7000	8300	10 000	12 000

For SI units: 1 inch = 25 mm, 1 inch per foot = 83.3 mm/m

Notes:
[1] On-site sewers that serve more than one building shall be permitted to be sized according to the current standards and specifications of the administrative authority for public sewers.
[2] A maximum of two water closets or two bathroom groups, except in single-family dwellings, where a maximum of three water closets or three bathroom groups shall be permitted to be installed.

horizontal branch is connected to a stack of more than three branch intervals, then the horizontal branch may only serve 90 DFUs.

The *ASPE Design Handbook* explains why this is so as follows:

It should be noted that there is a restriction of the amount of flow permitted to enter a stack from any branch when the stack is more than three branch intervals. If an attempt is made to introduce too large a flow into the stack at any one level, the inflow will fill the stack at that level and will even back up the water above the elevation of inflow, which will cause violent pressure fluctuations in the stack – resulting in the siphoning of trap seals – and may also cause sluggish flow in the horizontal branch.

According to Footnote 1, Table C 304.3 does not include horizontal branches connecting to the building drain, which may be sized using Table C 304.2 (see Section C 304.2). Footnote 2, as in Table C 304.2, limits 3-inch diameter pipes. Three-inch stacks are limited to a maximum of six water closets or bathroom groups, and 3-inch branches are limited to a maximum of two water closets or bathroom groups.

In Table C 304.3 there is one column for sizing horizontal branches (column 2), one for sizing a single branch interval (column 5) and two for sizing drainage stacks (columns 3 and 4). Using the 4-inch row, the table reads as follows:

- Any horizontal branch other than a branch connecting to the building drain is limited to 160 DFUs.

APPENDIX C

TABLE C 304.3
HORIZONTAL FIXTURE BRANCHES AND STACKS

DIAMETER OF PIPE (inches)	MAXIMUM NUMBER OF DRAINAGE FIXTURE UNITS			
	HORIZONTAL FIXTURE BRANCH[1]	ONE STACK OF THREE OR FEWER BRANCH INTERVALS	STACKS WITH MORE THAN THREE BRANCH INTERVALS	
			TOTAL FOR STACK	TOTAL AT ONE BRANCH INTERVAL
1½	3	4	8	2
2	6	10	24	6
2½	12	20	42	9
3	20[2]	48[2]	72[2]	20[2]
4	160	240	500	90
5	360	540	1100	200
6	620	960	1900	350
8	1400	2200	3600	600
10	2500	3800	5600	1000
12	3900	6000	8400	1500

For SI units: 1 inch = 25 mm

Notes:
[1] Does not include branches of the building drain.
[2] A maximum of two water closets or bathroom groups within each branch interval or more than six water closets or bathroom groups on the stack.

- A 4-inch drainage stack of three or less branch intervals is limited to a maximum of 240 DFUs.
- The maximum amount of DFUs permitted on a 4-inch drainage stack having more than three branch intervals is 500 DFUs.
- A branch interval connecting to a 4-inch drainage stack having more than three branch intervals connecting to it is limited to a maximum of 90 DFUs.

C 304.3.1 Horizontal Stack Offsets. Horizontal stack offsets shall be sized in accordance with Table C 304.2 as required for building drains.

🔧 If the stack is offset 45 degrees or more from the vertical, then the offset is considered a horizontal pipe. In this type of an offset the vertical flow of waste and water down the stack is interrupted, creating a hydraulic jump exactly like that created at the base of the stack (see commentary in Sections 706.4 and 901.2). Indeed, the horizontal offset creates two stacks — one above the offset and one below the offset. The hydraulic jump creates pressure fluctuations as discussed in Chapter 9. Because of the increased amount of flow allowed in this system of drainage piping, the pressure fluctuations can be even greater.

In order to eliminate these pressure fluctuations in the horizontal offset, the offset must be sized as if it was a part of the building drain. Thus, the offset must be sized using the appropriate row and column in Table C 304.2. Note that if an increase in size of the offset is required by using Table C 304.2, the lower stack connected to the offset may not remain as originally sized. Because of the requirements in Chapter 3 for elimination of flow restriction, the lower stack must not be less is size than the offset.

C 304.3.2 Vertical Stack Offsets. Vertical stack offsets shall be sized in accordance with Table C 304.3 as required for stacks.

🔧 A vertical stack offset is an offset of less than 45 degrees from the vertical. This offset is still considered and sized as a stack in Table C 304.3 and does not need to be vented.

C 304.4 Horizontal Stack Offset and Horizontal Branch Connections. Horizontal branch connections shall not connect to a horizontal stack offset or within 2 feet (610 mm) above or below the offset where such horizontal offset is located more than four branch intervals below the top of the stack.

🔧 Connections of horizontal branches to the offset or within 2 feet above or below the offset are eliminated in stacks below the fourth branch interval. This is to eliminate increased volumes of flow at the point of the hydraulic jump in the offset.

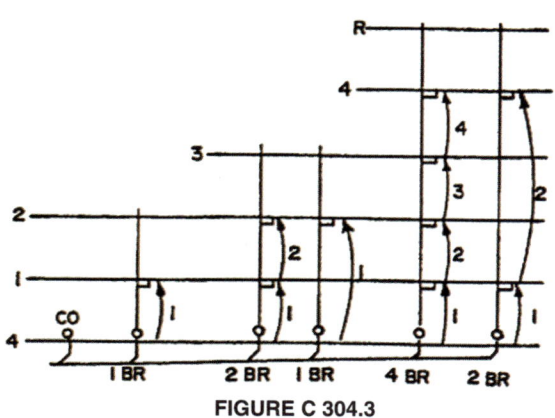

FIGURE C 304.3
ILLUSTRATION OF BRANCH INTERVAL

C 401.0 Vent System Sizing.

🔧 The venting systems contained in Sections C 401.0, C 501.0 and C 601.0 deviate from the conventional method of venting contained in Chapter 9. There is nothing new or radical about these alternate venting systems for they have been used in many areas of the country for decades. For example, the Single-stack Vent System in Section C 601.0 is quite similar to the Philadelphia Single-stack System, which has been widely used in that city.

These venting systems have been extensively researched and found to be sound venting systems if designed and installed properly. They are placed in this appendix to provide the design professional with alternatives to the venting systems in Chapter 9. Sections C 501.0 and C 601.0 specifically require a registered professional engineer to design the system.

The requirements of these systems are outlined below. Little explanation is needed for most of these sections and subsections. In all cases, the distance of the fixture trap to the horizontal waste or wet vent branch must be per Table 1002.2, Horizontal Lengths of Trap Arms. It is of utmost importance to adhere to these distances. Much of the research done on these systems has found that deviating from these distances and severe grading of the trap arms has led to trap seal loss and failure of the system.

C 401.1 Size of Vents. The size of vent piping shall be determined from the developed length and the total number of drainage fixture units connected in accordance with Table C 401.1. Vents shall be not less than one-half the required size of the drainage pipe size served as determined by Table C 304.3 for horizontal fixture branches and stacks nor less than 1¼ inches (32 mm) in diameter. The drainage system shall be vented by not less than one vent pipe which shall be not less than one-half the size of the required building drain and which shall extend from the building drain or extension of building drain to the outdoors. Vents shall be installed in accordance with Chapter 9.

🔧 To determine the size of vent needed for any drainage piping, the developed length of the vent and the total number of fixture units served by the drainage pipe must be known. The developed length of the vent is determined from the point of connection to the waste pipe to the point being sized. For an individual fixture, this would be from the inner edge of the vent to the connection with the branch vent or vent stack. For a vent stack, this would be from the point of connection at the base of the drainage stack to the top of the vent stack through the roof.

TABLE C 401.1
SIZE AND LENGTH OF VENTS

SIZE OF SOIL OR WASTE STACK (inches)	FIXTURE UNITS CONNECTED	DIAMETER OF VENT REQUIRED (inches)								
		1¼	1½	2	2½	3	4	5	6	8
		MAXIMUM LENGTH OF VENT (feet)								
1½	8	50	150	—	—	—	—	—	—	—
2	12	30	75	200	—	—	—	—	—	—
2	20	26	50	150	—	—	—	—	—	—
2½	42	—	30	100	300	—	—	—	—	—
3	10	—	30	100	100	600	—	—	—	—
3	30	—	—	60	200	500	—	—	—	—
3	60	—	—	50	80	400	—	—	—	—
4	100	—	—	35	100	260	1000	—	—	—
4	200	—	—	30	90	250	900	—	—	—
4	500	—	—	20	70	180	700	—	—	—
5	200	—	—	—	35	80	350	1000	—	—
5	500	—	—	—	30	70	300	900	—	—
5	1100	—	—	—	20	50	200	700	—	—
6	350	—	—	—	25	50	200	400	1300	—
6	620	—	—	—	15	30	125	300	1100	—
6	960	—	—	—	—	24	100	250	1000	—
6	1900	—	—	—	—	20	70	200	700	—
8	600	—	—	—	—	—	50	150	500	1300
8	1400	—	—	—	—	—	40	100	400	1200
8	2200	—	—	—	—	—	30	80	350	1100
8	3600	—	—	—	—	—	25	60	250	800
10	1000	—	—	—	—	—	—	75	125	1000
10	2500	—	—	—	—	—	—	50	100	500
10	3800	—	—	—	—	—	—	30	80	350
10	5600	—	—	—	—	—	—	25	60	250

For SI units: 1 inch = 25 mm, 1 foot = 304.8 mm

Table C 401.1 contains two columns at the left that represent the size of the soil or waste stack and the total number of fixture units served by the drainage piping at the point of sizing. The columns to the right represent the maximum length of vent in feet allowed for the diameter of vent required.

For example, to find the required diameter for a vent pipe with a developed length of 57 feet serving 75 fixture units, follow the second column down until the fixture unit value is equal to or greater than 75. Select the row showing 100 fixture units. Read horizontally to the right until finding the proper developed length equal to or greater than 57 feet. This will be 100 feet. Then read the pipe size at the top of the column for the diameter of vent required. This would be 2½ inches.

C 401.2 Vent Stack. A vent stack shall be required for a drainage stack that extends five or more branch intervals above the building drain or horizontal branch. The developed length of the vent stack shall be measured from the lowest connection of a branch vent to the termination outdoors.

A separate parallel vent stack shall be provided for a drainage stack that extends five or more branch intervals, which is similar to what is required in Chapter 9, Section 907.1. The vent stack should remain undiminished in size for its entire length. The size of the vent stack is based on the full developed length of the vent and the total fixture units for the drainage stack served by the vent. See the commentary in Section 907.1 for further details.

C 401.3 Branch Vents. Where branch vents exceed 40 feet (12 192 mm) in developed length, such vent shall be increased by one pipe size for the entire developed length of the vent pipe.

Branch vents connect individual or common vents to a vent stack. They are sized as described in Section C 401.1; however, when they exceed 40 feet (12 m) in developed length, they are required to be increased by one pipe size.

C 401.4 Venting Horizontal Offsets. Drainage stacks with horizontal offsets shall be vented where five or more branch intervals are located above the offset. The upper and lower section of the horizontal offset shall be vented in accordance with Section C 401.4.1 and Section C 401.4.2.

The drainage stack offset, as stated in C 304.4, creates a hydraulic jump at the offset. The offset must be vented to eliminate the increased pressure ahead of the waste flow and to allow air into the lower stack (see **Figure C 401.4**).

C 401.4.1 Venting Upper Section. The vent for the upper section of the stack shall be vented as a separate stack with a vent stack connection installed at the base of the drainage stack. Such vent stack shall connect below the lowest horizontal branch or building drain. Where vent stack connects to the building drain, the connection shall be located downstream of the drainage stack and within a distance of 10 times the diameter of the drainage stack.

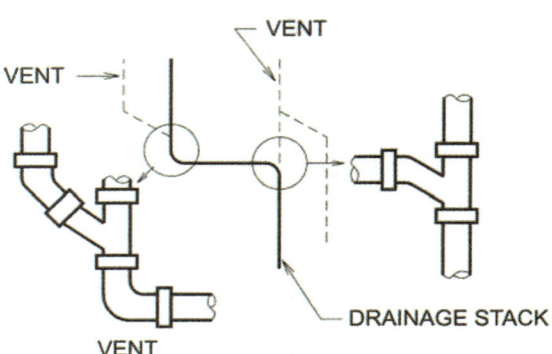

FIGURE C 401.4
VENTING THE DRAINAGE STACK OFFSET

C 401.4.2 Venting Lower Section. The vent for the lower section of the stack shall be vented by a yoke vent connecting between the offset and the next lower horizontal branch by means of a wye-branch fitting. The size of the yoke vent and connection shall be not less in diameter than the required size for the vent serving the drainage stack. The yoke vent connection shall be permitted to be a vertical extension of the drainage stack.

C 501.0 Vacuum Drainage Systems.

Vacuum drainage systems are specially designed engineered systems utilizing vacuum and air to transport waste from fixtures through piping systems to waste collection tanks. The waste is then conveyed to the public or private sewage system by gravity. Fixtures, especially water closets, are specifically designed to be used with the vacuum drainage system.

These systems can be described as "green systems" because they utilize fixtures with reduced water usage and output. One such system uses a ½-gallon per flush water closet, which can lead to significant water savings.

The system can also lead to space and material cost savings by using smaller pipe sizes and eliminating underground pipe installations. These characteristics could provide significant "points" in the LEED Rating System.

The manufacturer's specifications and installation requirements must always be followed. Only the piping materials and fixtures specified can be used in the system. They must meet standards developed for the system or comply with Section 301.3. See **Figure C 501.0** for an example of one such system.

C 501.1 General. This section regulates the design and installation provisions for vacuum waste drainage systems. Plans for vacuum waste drainage systems shall be submitted to the Authority Having Jurisdiction for approval and shall be considered an engineered designed system. Such plans shall be prepared by a registered design professional to perform plumbing design work. Details are necessary to ensure compliance with the requirements of this section, together with a full description of the complete installation including quality, grade of materials, equipment, construction, and methods of assembly and installation.

APPENDIX C

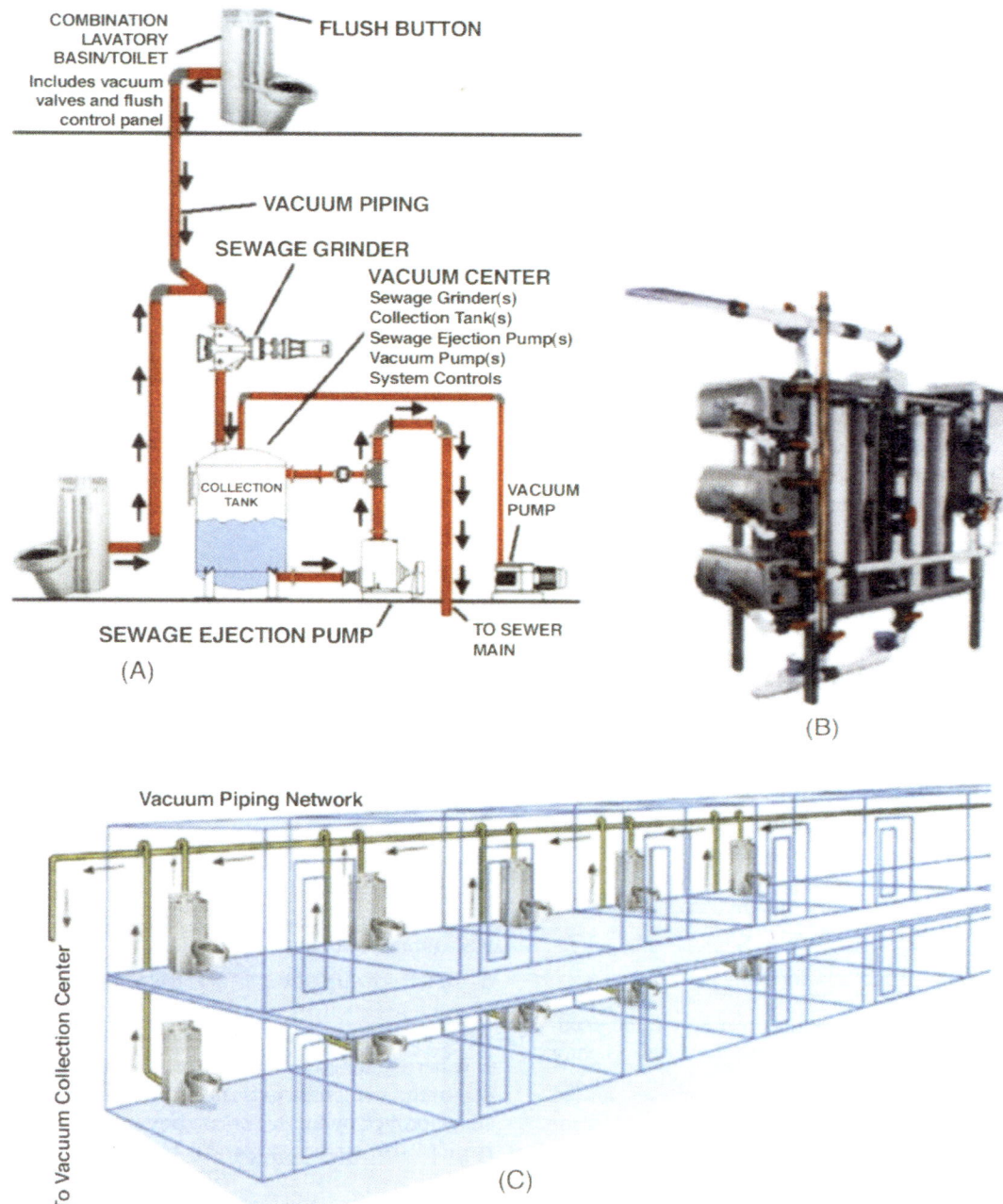

FIGURE C 501.0
ACORN ENGINEERING VACUUM PLUMBING AND DRAINAGE SYSTEM
(A) PENAL APPLICATION IS A VISUALIZATION OF HOW THE SYSTEM WORKS. (B) THE VACUUM CENTER.
(C) ILLUSTRATES JAIL CELL PIPING SYSTEM.

Components, materials, and equipment shall comply with standards and specifications listed in Table 1701.1 of this code or approved by the Authority Having Jurisdiction and other national consensus standards applicable to plumbing systems and materials. Where such standards and specifications are not available, alternate materials and equipment shall be approved in accordance with Section 301.3.

C 501.2 System Design. Vacuum waste drainage systems shall be designed and installed in accordance with the manufacturer's installation instructions. A vacuum waste drainage system shall include a vacuum generating system, waste collection center, piping network, vacuum valve, and control components used to isolate the vacuum piping network from atmospheric pressure and to collect waste at its point of origin. Where a vacuum system provides the only means of sanitation, duplicate vacuum generating equipment set to operate automatically shall be installed to allow the system to continue in operation during periods of maintenance.

C 501.2.1 Vacuum Generating System. The vacuum generating station shall include vacuum pumps to create a constant vacuum pressure within the piping network and storage tanks. Operation of pumps, collection tanks, and alarms shall be automated by controls. The vacuum pumps shall be activated on demand and accessible for repair or replacement. The vent from the vacuum pump shall be provided for vacuum pump air exhaust, and shall be of a size capable of handling the total air volume of the vacuum pump.

C 501.2.2 Waste Collection Center or Storage Tanks. Vacuum collection center or storage tanks shall be of such capacity as to provide storage of waste to prevent fouling of the system. Such collection or storage tank shall be capable of withstanding 150 percent of the rated vacuum (negative pressure) created by the vacuum source without leakage or collapse. Waste collection center or storage tanks shall be accessible for adjustment, repair, or replacement.

C 501.2.3 Piping Network. The piping network shall be under a continuous vacuum and shall be designed to withstand 150 percent of the vacuum (negative pressure) created by the vacuum source within the system without leakage or collapse. Sizing the piping network shall be in accordance with the manufacturer's instructions. The water closet outlet fitting shall connect with a piping network having not less than a 1½ inch (40 mm) nominal inside diameter.

C 501.2.4 Vacuum Interface Valve. A closed vacuum interface valve shall be installed to separate the piping network vacuum from atmospheric pressure. A control device shall open the vacuum interface valve where a signal is generated to remove waste from the plumbing fixture.

C 501.2.5 Control Components. Where a pneumatic signal is generated at the controller, a vacuum from the system to open the extraction valve shall be designed to operate where vacuum pressure exists to remove the accumulated waste. Each tank shall incorporate a level indicator switch that automatically controls the discharge pump and warns of malfunction or blockage as follows:

(1) Start discharge.

(2) Stop discharge.

(3) Activate an audible alarm where the level of effluent is usually high.

(4) Warning of system shutdown where tank is full.

C 501.3 Fixtures. Fixtures utilized in a vacuum waste drainage system shall be in accordance with referenced standards listed in Table 1701.1. Components shall be of corrosion resistant materials. The water closet outlet shall be able to pass a 1 inch (25.4 mm) diameter ball and shall have a smooth, impervious surface. The waste outlet and passages shall be free of obstructions, recesses, or chambers that are capable of permitting fouling. The mechanical valve and its seat shall be of such materials and design to provide a leak-free connection where at atmospheric pressure or under vacuum. The flushing mechanism shall be so designed as to ensure proper cleansing of the interior surfaces during the flushing cycle at a minimum operating flow rate. Mechanical seal mechanisms shall withdraw completely from the path of the waste discharge during flushing operation. Each mechanical seal vacuum water closet shall be equipped with a listed vacuum breaker. The vacuum breaker shall be mounted with the critical level or marking not less than 1 inch (25.4 mm) above the flood-level rim of the fixture. Vacuum breakers shall be installed on the discharge side of the last control valve in the potable water supply line and shall be located so as to be protected from physical damage and contamination.

C 501.4 Drainage Fixture Units. Drainage fixture units shall be determined by the manufacturer's instructions. The pump discharge load from the collector tanks shall be in accordance with this appendix.

C 501.5 Water Supply Fixture Units. Water supply fixture units shall be determined by the manufacturer's instructions.

C 501.6 Materials. Materials used for water distribution pipe and fittings shall be in accordance with Table 604.1. Materials used for aboveground drainage shall be in accordance with Table 701.2 and shall have a smooth bore, and be constructed of non-porous material.

C 501.7 Traps and Cleanouts. Traps and cleanouts shall be installed in accordance with Chapter 7 and Chapter 10.

C 501.8 Testing. The entire vacuum waste system shall be subjected to a vacuum test of 29 inches of mercury (98 kPa) or not less than the working pressure of the system for 30 minutes. The system shall be gastight and watertight at all points. Verification of test results shall be submitted to the Authority Having Jurisdiction.

C 501.9 Manufacturer's Instructions. Manufacturer's instructions shall be provided for the purpose of providing information regarding safe and proper operating instructions whether or not as part of the condition of listing in order to determine compliance. Such instructions shall be submitted and approved by the Authority Having Jurisdiction.

C 601.0 Single-Stack Vent System.

The single-stack vent system was originally designed for residential one- and two-story or apartment building plumbing systems. It allowed the fixtures for a bathroom group and kitchen to be installed without vents and to connect with a single-stack, saving material and manpower cost. The system for one- and two-story housing was the original drainage system researched and recommended by Dr. Roy Hunter in the 1920s.

The provisions in this section are more applicable to what is commonly referred to as the Philadelphia Single Stack System, which is derived from the Philadelphia Plumbing Code. This system shall be designed only by a registered professional engineer.

As with the previous systems, care must be taken to comply with sizing and branch and trap arm distances.

Siphonage or blowout of fixture traps occurs when these limits are exceeded.

C 601.1 Where Permitted. Single-stack venting shall be designed by a registered design professional as an engineered design. A drainage stack shall be permitted to serve as a single-stack vent system where sized and installed in accordance with Section C 601.2 through Section C 601.9. The drainage stack and branch piping in a single-stack vent system shall provide for the flow of liquids, solids, and air without the loss of fixture trap seals.

C 601.2 Stack Size. Drainage stacks shall be sized in accordance with Table C 601.2. Not more than two water closets shall be permitted to discharge to a 3 inch (80 mm) stack. Stacks shall be uniformly sized based on the total connected drainage fixture unit load, with no reductions in size.

C 601.2.1 Stack Vent. The drainage stack vent shall have a stack vent of the same size terminating to the outdoors.

The single-stack vent system is basically a vertical wet vent system (see **Figure C 601.2.1**). The stack is oversized to allow constant airflow throughout the stack. An important aspect, therefore, is to ensure that the stack size is maintained from the bottom of the stack, where it is sized for the total load on the stack, to the stack vent outlet through the roof. If the stack is sized at 5 inches (125 mm) at the lowest portion of the stack, the size of the stack vent should be 5 inches (125 mm). In this manner, the flow of air is ensured.

C 601.3 Branch Size. Horizontal branches connecting to a single-stack vent system shall be sized in accordance with Table 703.2.

Exceptions:

(1) Not more than one water closet within 18 inches (457 mm) of the stack horizontally shall be permitted on a 3 inch (80 mm) horizontal branch.

(2) A water closet within 18 inches (457 mm) of a stack

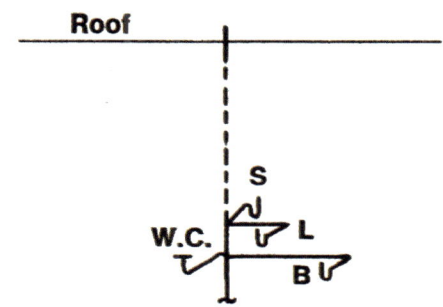

FIGURE C 601.2.1
TOP FLOOR OF A SINGLE-STACK VENT SYSTEM

horizontally and one other fixture with up to 1½ inch (40 mm) fixture drain size shall be permitted on a 3 inch (80 mm) horizontal branch where connected to the stack through a sanitary tee.

C 601.4 Length of Horizontal Branches. Water closets shall be not more than 4 feet (1219 mm) horizontally from the stack.

Exception: Water closets shall be permitted to be up to 8 feet (2438 mm) horizontally from the stack where connected to the stack through a sanitary tee.

C 601.4.1 Other Fixtures. Fixtures other than water closets shall be not more than 12 feet (3658 mm) horizontally from the stack.

C 601.4.2 Length of Vertical Piping. The length of a vertical piping from a fixture trap to a horizontal branch shall not be considered in computing the fixture's horizontal distance from the stack.

C 601.5 Maximum Vertical Drops from Fixtures. Vertical drops from fixture traps to horizontal branch piping shall be one size larger than the trap size, but not less than 2 inch (50 mm) in diameter. Vertical drops shall be 4 feet (1219 mm) maximum length. Fixture drains that are not increased in size, or have a vertical drop exceeding 4 feet (1219 mm) shall be individually vented.

TABLE C 601.2
SINGLE STACK SIZE*

STACK SIZE (inches)	MAXIMUM CONNECTED DRAINAGE FIXTURE UNITS		
	STACKS LESS THAN 75 FEET IN HEIGHT	STACK 75 FEET TO LESS THAN 160 FEET IN HEIGHT	STACK 160 FEET OR GREATER IN HEIGHT
3	24	NP	NP
4	225	24	NP
5	480	225	24
6	1015	480	225
8	2320	1015	480
10	4500	2320	1015
12	8100	4500	2320
15	13 600	8100	4500

For SI units: 1 inch = 25 mm, 1 foot = 304.8 mm
* NP = Not permitted

APPENDIX C

C 601.6 Additional Venting Required. Additional venting shall be provided where more than one water closet is on a horizontal branch and where the distance from a fixture trap to the stack exceeds the limits in Section C 601.4. Where additional venting is required, the fixture(s) shall be vented by individual vents, common vents, wet vents, circuit vents, or a combination waste and vent pipe. The dry vent extensions for the additional venting shall connect to a branch vent, vent stack, stack vent, or be extended outdoors and terminate to the open air.

C 601.7 Stack Offsets. Where there are no fixture drain connections below a horizontal offset in a stack, the offset does not need to be vented. Where there are fixture drain connections below a horizontal offset in a stack, the offset shall be vented. There shall be no fixture connections to a stack within 2 feet (610 mm) above and below a horizontal offset.

Because the stack provides the only venting for this system, when a stack offsets, the offset needs to be vented. When the stack turns horizontally, the waste flow effectively closes the airflow through the system. Venting the offset continues the system venting.

C 601.8 Separate Stack Required. Where stacks are more than two stories high, a separate stack shall be provided for the fixtures on the lower two stories. The stack for the lower two stories shall be permitted to be connected to the branch of the building drain that serves the stack for the upper stories at a point that is not less than 10 pipe diameters downstream from the base of the upper stack.

The lower two floors of fixtures are separated from the system in this subsection. Most of the positive pressures in the system occur at the bottom of the stack where the hydraulic jump is created by the effluent flow going from vertical to horizontal. This separation eliminates the effect of the hydraulic jump from the floors that are mostly affected by this phenomenon.

C 601.9 Sizing Building Drains and Sewers. In a single-stack vent system, the building drain and branches thereof shall be sized in accordance with Table 703.2, and the building sewer shall be sized in accordance with Table 717.1.

The horizontal building drain, collecting the waste from the drainage stack or stacks, and its horizontal branches shall be sized per Table 703.2. There is no need for any special sizing once the system is connected to the "normal" drainage system.

APPENDIX D
SIZING STORM WATER DRAINAGE SYSTEMS

D 101.0 General.

D 101.1 Applicability. This appendix provides general guidelines for the sizing of storm water drainage systems based on maximum rates of rainfall for various cities. The rainfall rates in Table D 101.1 shall be permitted to be used for design unless higher values are established locally.

Table D 101.1 in the UPC contains rainfall rates for various U.S. locations. According to Table D 101.1, these values should be used in designing storm water drainage systems unless higher values are established locally. In some localities, extremely heavy showers may occur for short durations. In areas where this occurs, maximum rainfall rates may be used instead of the averages shown in the table to prevent roof failure due to ponding.

For complete installation and sizing requirements for storm water drainage systems, refer to Chapter 11, Storm Drainage.

D 102.0 Sizing by Flow Rate.

D 102.1 General. Storm drainage systems shall be permitted to be sized by storm water flow rates, using the gallons per minute per square foot [(L/s)/m^2] of rainfall listed in Table D 101.1 for the local area. Multiplying the listed gallons per minute per square foot [(L/s)/m^2] by the roof area being drained (in square feet) (m^2) by each inlet produces the gallons per minute (gpm) (L/s) of required flow for sizing each drain inlet. The flow rates shall be permitted to be added to determine the flows in each of the drainage systems. Required pipe sizes for various flow rates are listed in Table 1101.8 and Table 1101.12.

D 103.0 Sizing by Roof Area.

D 103.1 General. Storm drainage systems shall be permitted to be sized using the roof area served by each of the drainage system. Maximum allowable roof areas with various rainfall rates are listed in Table 1101.8 and Table 1101.12, along with the required pipe sizes. Using this method, it shall be permitted to interpolate between two listed rainfall rate columns (inches per hour) (mm/h). To determine the allowable roof area for a listed pipe size at a listed slope, divide the allowable square feet (m^2) of roof for a 1 inch per hour (in/h) (25.4 mm/h) rainfall rate by the listed rainfall rate for the local area. For example, the allowable roof area for a 6 inch (150 mm) drain at ⅛ inch per foot (10.4 mm/m) slope with a rainfall rate of 3.2 in/h (81 mm/h) is 21 400/3.2 = 6688 square feet (621.3 m^2).

D 104.0 Capacity of Rectangular Scuppers.

D 104.1 General. Table D 104.1 lists the discharge capacity of rectangular roof scuppers of various widths with various heads of water. The maximum allowable level of water on the roof shall be obtained from the registered design professional, based on the design of the roof.

TABLE D 101.1
MAXIMUM RATES OF RAINFALL FOR VARIOUS CITIES*

STATES AND CITIES	STORM DRAINAGE 60-MINUTE DURATION, 100-YEAR RETURN	
	inches per hour	gallons per minute per square foot
ALABAMA	–	–
Birmingham	3.7	0.038
Huntsville	3.3	0.034
Mobile	4.5	0.047
Montgomery	3.8	0.039
ALASKA	–	–
Aleutian Islands	1.0	0.010
Anchorage	0.6	0.006
Bethel	0.8	0.008
Fairbanks	1.0	0.010
Juneau	0.6	0.006
ARIZONA	–	–
Flagstaff	2.3	0.024
Phoenix	2.2	0.023
Tucson	3.0	0.031
ARKANSAS	–	–
Eudora	3.8	0.039
Ft. Smith	3.9	0.041
Jonesboro	3.5	0.036
Little Rock	3.7	0.038

APPENDIX D

TABLE D 101.1
MAXIMUM RATES OF RAINFALL FOR VARIOUS CITIES* (continued)

STATES AND CITIES	STORM DRAINAGE 60-MINUTE DURATION, 100-YEAR RETURN	
	inches per hour	gallons per minute per square foot
CALIFORNIA	–	–
Eureka	1.5	0.016
Lake Tahoe	1.3	0.014
Los Angeles	2.0	0.021
Lucerne Valley	2.5	0.026
Needles	1.5	0.016
Palmdale	3.0	0.031
Redding	1.5	0.016
San Diego	1.5	0.016
San Francisco	1.5	0.016
San Luis Obispo	1.5	0.016
COLORADO	–	–
Craig	1.5	0.016
Denver	2.2	0.023
Durango	1.8	0.019
Stratton	3.0	0.031
CONNECTICUT	–	–
Hartford	2.8	0.029
New Haven	3.0	0.031
DELAWARE	–	–
Dover	3.5	0.036
Rehobeth Beach	3.6	0.037
DISTRICT OF COLUMBIA	–	–
Washington	4.0	0.042
FLORIDA	–	–
Daytona Beach	4.0	0.042
Ft. Myers	4.0	0.042
Jacksonville	4.3	0.045
Melbourne	4.0	0.042
Miami	4.5	0.047
Palm Beach	5.0	0.052
Tampa	4.2	0.044
Tallahassee	4.1	0.043
GEORGIA	–	–
Atlanta	3.5	0.036
Brunswick	4.0	0.042
Macon	3.7	0.038
Savannah	4.0	0.042
Thomasville	4.0	0.042
HAWAII	–	–
Rainfall rates in the Hawaiian Islands vary from 1½ inches per hour to 8 inches per hour, depending on location and elevation. Consult local data.		
IDAHO	–	–
Boise	1.0	0.010
Idaho Falls	1.2	0.012
Lewiston	1.0	0.010
Twin Falls	1.1	0.011

APPENDIX D

TABLE D 101.1
MAXIMUM RATES OF RAINFALL FOR VARIOUS CITIES* (continued)

STATES AND CITIES	STORM DRAINAGE 60-MINUTE DURATION, 100-YEAR RETURN	
	inches per hour	gallons per minute per square foot
ILLINOIS	–	–
Chicago	2.7	0.028
Harrisburg	3.1	0.032
Peoria	2.9	0.030
Springfield	3.0	0.031
INDIANA	–	–
Evansville	3.0	0.031
Indianapolis	2.8	0.029
Richmond	2.7	0.028
South Bend	2.7	0.028
IOWA	–	–
Council Bluffs	3.7	0.038
Davenport	3.0	0.031
Des Moines	3.4	0.035
Sioux City	3.6	0.037
KANSAS	–	–
Goodland	3.5	0.036
Salina	3.8	0.039
Topeka	3.8	0.039
Wichita	3.9	0.041
KENTUCKY	–	–
Bowling Green	2.9	0.030
Lexington	2.9	0.030
Louisville	2.8	0.029
Paducah	3.0	0.031
LOUISIANA	–	–
Monroe	3.8	0.039
New Orleans	4.5	0.047
Shreveport	4.0	0.042
MAINE	–	–
Bangor	2.2	0.023
Kittery	2.4	0.025
Millinocket	2.0	0.021
MARYLAND	–	–
Baltimore	3.6	0.037
Frostburg	2.9	0.030
Ocean City	3.7	0.038
MASSACHUSETTS	–	–
Adams	2.6	0.027
Boston	2.7	0.028
Springfield	2.7	0.028
MICHIGAN	–	–
Detroit	2.5	0.026
Grand Rapids	2.6	0.027
Kalamazoo	2.7	0.028

2018 UNIFORM PLUMBING CODE ILLUSTRATED TRAINING MANUAL

APPENDIX D

TABLE D 101.1
MAXIMUM RATES OF RAINFALL FOR VARIOUS CITIES* (continued)

STATES AND CITIES	STORM DRAINAGE 60-MINUTE DURATION, 100-YEAR RETURN	
	inches per hour	gallons per minute per square foot
Sheboygan	2.1	0.022
Traverse City	2.2	0.023
MINNESOTA	–	–
Duluth	2.6	0.027
Grand Forks	2.5	0.026
Minneapolis	3.0	0.031
Worthington	3.4	0.035
MISSISSIPPI	–	–
Biloxi	4.5	0.047
Columbus	3.5	0.036
Jackson	3.8	0.039
MISSOURI	–	–
Independence	3.7	0.038
Jefferson City	3.4	0.035
St. Louis	3.2	0.033
Springfield	3.7	0.038
MONTANA	–	–
Billings	1.8	0.019
Glendive	2.5	0.026
Great Falls	1.8	0.019
Missoula	1.3	0.014
NEBRASKA	–	–
Omaha	3.6	0.037
North Platte	3.5	0.036
Scotts Bluff	2.8	0.029
NEVADA	–	–
Las Vegas	1.5	0.016
Reno	1.2	0.012
Winnemucca	1.0	0.010
NEW HAMPSHIRE	–	–
Berlin	2.2	0.023
Manchester	2.5	0.026
NEW JERSEY	–	–
Atlantic City	3.4	0.035
Paterson	3.0	0.031
Trenton	3.2	0.033
NEW MEXICO	–	–
Albuquerque	2.0	0.021
Carlsbad	2.6	0.027
Gallup	2.1	0.022
NEW YORK	–	–
Binghamton	2.4	0.025
Buffalo	2.3	0.024

TABLE D 101.1
MAXIMUM RATES OF RAINFALL FOR VARIOUS CITIES* (continued)

STATES AND CITIES	STORM DRAINAGE 60-MINUTE DURATION, 100-YEAR RETURN	
	inches per hour	gallons per minute per square foot
New York City	3.1	0.032
Schenectady	2.5	0.026
Syracuse	2.4	0.025
NORTH CAROLINA	–	–
Asheville	3.2	0.033
Charlotte	3.4	0.035
Raleigh	4.0	0.042
Wilmington	4.4	0.046
NORTH DAKOTA	–	–
Bismarck	2.7	0.028
Fargo	2.9	0.030
Minot	2.6	0.027
OHIO	–	–
Cincinnati	2.8	0.029
Cleveland	2.4	0.025
Columbus	2.7	0.028
Toledo	2.6	0.027
Youngstown	2.4	0.025
OKLAHOMA	–	–
Boise City	3.4	0.035
Muskogee	4.0	0.042
Oklahoma City	4.1	0.043
OREGON	–	–
Medford	1.3	0.014
Ontario	1.0	0.010
Portland	1.3	0.014
PENNSYLVANIA	–	–
Erie	2.4	0.025
Harrisburg	2.9	0.030
Philadelphia	3.2	0.033
Pittsburgh	2.5	0.026
Scranton	2.8	0.029
RHODE ISLAND	–	–
Newport	3.0	0.031
Providence	2.9	0.030
SOUTH CAROLINA	–	–
Charleston	4.1	0.043
Columbia	3.5	0.036
Greenville	3.3	0.034
SOUTH DAKOTA	–	–
Lemmon	2.7	0.028
Rapid City	2.7	0.028
Sioux Falls	3.4	0.035

TABLE D 101.1
MAXIMUM RATES OF RAINFALL FOR VARIOUS CITIES* (continued)

STATES AND CITIES	STORM DRAINAGE 60-MINUTE DURATION, 100-YEAR RETURN	
	inches per hour	gallons per minute per square foot
TENNESSEE	–	–
Knoxville	3.1	0.032
Memphis	3.5	0.036
Nashville	3.0	0.031
TEXAS	–	–
Corpus Christi	4.6	0.048
Dallas	4.2	0.044
El Paso	2.0	0.021
Houston	4.6	0.048
Lubbock	3.3	0.034
San Antonio	4.4	0.046
UTAH	–	–
Bluff	2.0	0.021
Cedar City	1.5	0.016
Salt Lake City	1.3	0.014
VERMONT	–	–
Bennington	2.5	0.026
Burlington	2.3	0.024
Rutland	2.4	0.025
VIRGINIA	–	–
Charlottesville	3.4	0.035
Norfolk	4.0	0.042
Richmond	4.0	0.042
Roanoke	3.3	0.034
WASHINGTON	–	–
Seattle	1.0	0.010
Spokane	1.0	0.010
Walla Walla	1.0	0.010
WEST VIRGINIA	–	–
Charleston	2.9	0.030
Martinsburg	3.0	0.031
Morgantown	2.7	0.028
WISCONSIN	–	–
Green Bay	2.5	0.026
Lacrosse	2.9	0.030
Milwaukee	2.7	0.028
Wausau	2.5	0.026
WYOMING	–	–
Casper	1.9	0.020
Cheyenne	2.5	0.026
Evanston	1.3	0.014
Rock Springs	1.4	0.015

For SI units: 1 inch per hour = 25.4 mm/h, 1 gallon per minute per square foot = 0.618 [(L/s)m^2]

*The rainfall rates in this table are based on U.S. Weather Bureau Technical Paper No. 40, Chart 14: 100-Year 60-Minute Rainfall (inches).

TABLE D 104.1
DISCHARGE FROM RECTANGULAR SCUPPERS (gallons per minute)[1, 2, 3, 4]

WATER HEAD (inches)	WIDTH OF SCUPPER (inches)					
	6	12	18	24	30	36
½	6	13	19	25	32	38
1	17	35	53	71	89	107
1½	31	64	97	130	163	196
2	–	98	149	200	251	302
2½	–	136	207	278	349	420
3	–	177	271	364	458	551
3½	–	–	339	457	575	693
4	–	–	412	556	700	844

For SI units: 1 inch = 25.4 mm, 1 gallon per minute = 0.06 L/s

Notes:
[1] Table D 104.1 is based on discharge over a rectangular weir with end contractions.
[2] Head is the depth of water above bottom of the scupper opening.
[3] The height of the scupper opening shall be not less than two times the design head.
[4] Coordinate the allowable head of water with the structural design of the roof.

APPENDIX E

MANUFACTURED/MOBILE HOME PARKS AND RECREATIONAL VEHICLE PARKS

E 101.0 Manufactured/Mobile Home Park. Construction of mobile home and recreational vehicle parks requires prior approval through review and evaluation of land-use policies. When approved, agencies will impose density restrictions that are similar to those for planned unit developments for other uses (see **Figure E 101.0**).

National Mobile Home Construction and Safety Standards Act of 1974, Title VI

The Federal Department of Housing and Urban Development (HUD) has established federal mobile home construction and safety standards. The standards apply to all mobile homes and recreational vehicles intended for dwelling units manufactured for sale on or after June 15, 1976. The "Federal Procedural and Enforcement Regulation" part of the document provides for design review, in-plant inspections and labeling of the units. An appeal process is also included.

Data Plate

Mobile homes that are in compliance with HUD regulations must bear a permanent data plate located in a readily accessible location, which includes all of the following information:

1. Name and address of the manufacturer of the unit;
2. Serial number and model number of the unit;
3. Date of manufacture of the unit;
4. The statement, "This mobile home is designed to comply with the federal mobile home construction and safety standard in force at the time of manufacture";
5. The name of the manufacturer of factory-installed equipment in the unit;
6. The model designation of major factory-installed appliances in the unit; and
7. Reference to the structural and wind zone for which the home is designed, including a duplicate of the wind zone map. This information may be combined with the heating/cooling certificate and the insulation zone map.

E 101.1 Applicability. The manufactured home park plumbing and drainage systems shall be designed and installed in accordance with the requirements of this appendix and the requirements of this code.

E 101.2 Construction Documents. Before plumbing or sewage disposal facilities are installed or altered in a manufactured home park, duplicate construction documents shall be filed and proper permits obtained from the depart-

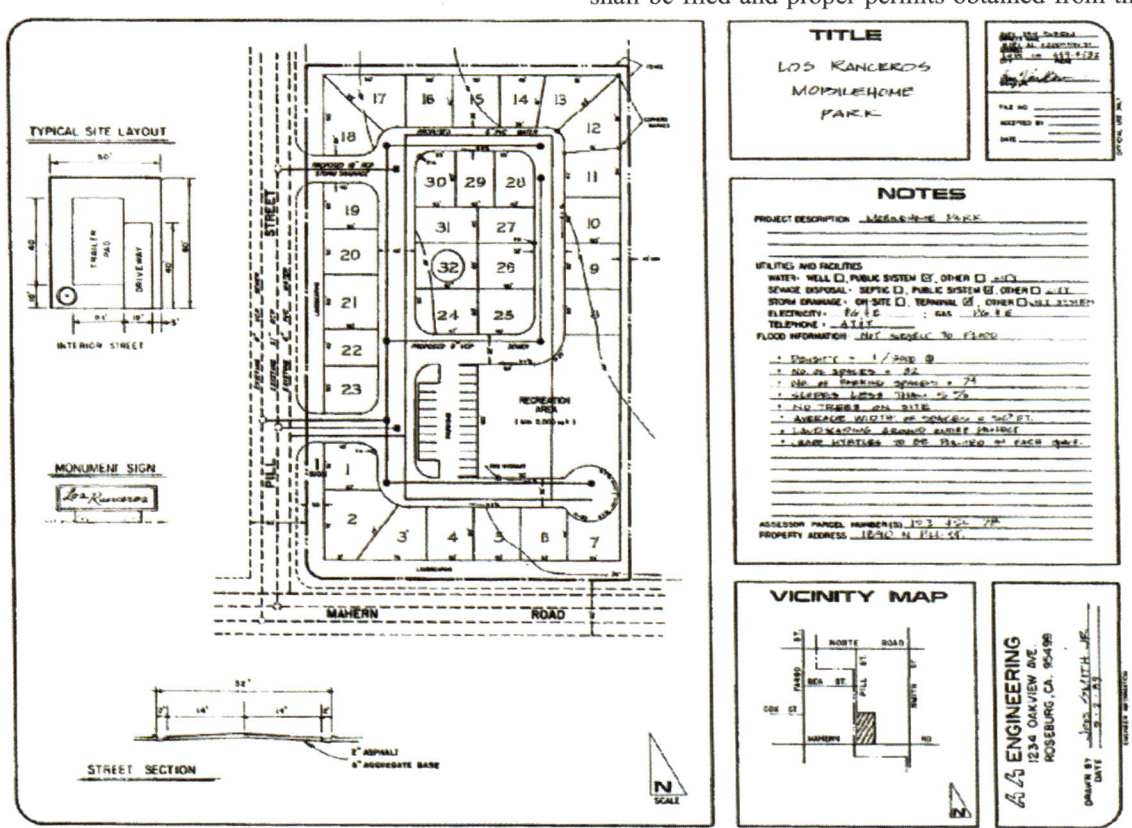

FIGURE E 101.0
EXAMPLE OF PLOT PLAN FOR PROPOSED MOBILE HOME PARK

ment or departments having jurisdiction. Plans shall show in detail:

(1) Plot plan of the park drawn to scale, indicating elevations, property lines, driveways, existing or proposed buildings, and the sizes of manufactured home lots.
(2) Complete specification and piping layout of proposed plumbing systems or alteration.
(3) Complete specification and layout of proposed sewage disposal system or alteration.
(4) The nature and extent of the work proposed, showing clearly that such work will conform to the provisions of this code.

🔧 Plans for the park utility system are required on scaled drawings indicating elevations, property lines, driveways, all buildings and individual lots. Proposed installations of drainage, potable water, fuel gas, electricity and fire protection systems should also be shown. Plans are also required for park alterations.

The individual units intended for installation at a mobile home lot differ from job-site-constructed dwellings in that they are manufactured and delivered to the lot where, depending if they are single-wide or double-wide, they require assembling. The units may contain various amenities, including kitchens, one or more bathrooms, central heat, air conditioning and major appliances. All of the units are plumbed and wired at the manufacturing location. Once installed on an approved support system, connections to the park utility system activate the various systems in the unit.

The design of park plumbing systems requires knowledge of mobile home construction standards. The park drainage system, for example, will only function properly if it is known whether the plumbing fixtures in the units are effectively trapped and vented.

E 201.0 Definitions.

E 201.1 General. For purposes of this chapter, the following definitions shall apply:

Manufactured/Mobile Home. A structure transportable in one or more sections, which in the traveling mode is 8 feet (2438 mm) or more in width and 40 feet (12 192 mm) or more in length or, where erected on site, is 320 square feet (29.73 m²) or more, and which is built on a permanent chassis, and designed to be used as a dwelling with or without a permanent foundation where connected to the required utilities. It includes the plumbing, heating, air-conditioning, and electrical systems contained therein. For further clarification of definition, see Federal Regulation 24 CFR.

Manufactured/Mobile Home Accessory Building or Structure. A building or structure that is an addition to or supplements the facilities provided to a manufactured home. It is not a self-contained, separate, habitable building or structure. Examples are awnings, cabanas, ramadas, storage structures, carports, fences, windbreaks, or porches.

Manufactured/Mobile Home Lot. A portion of a manufactured home park designed for the accommodation of one manufactured home and its accessory buildings or structures for the exclusive use of the occupants.

Manufactured/Mobile Home Park. A parcel (or contiguous parcels) of land that has been so designated and improved that it contains two or more manufactured home lots available to the general public for the placement thereon of a manufactured home for occupancy.

Recreational Vehicle (RV). A vehicular-type unit primarily designed as temporary living quarters for recreational, camping, travel, or seasonal use that either has its own motive power or is mounted on or towed by another vehicle. The basic entities are camping trailer, fifth-wheel trailer, motor home, park trailer, travel trailer, and truck camper.

🔧 Recreational vehicles (RV) include travel trailers, camping trailers, truck campers and motor homes. They may be fully equipped with appliances and plumbing fixtures, be self-contained or have connections for utility service. In RV parks, a given number of sites are established as permitted through approved plot plans. Sites are used temporarily by campers in vehicles or tents and sites may or may not have utilities.

Recreational Vehicle Park. A plot of land upon which two or more recreational vehicle sites are located, established or maintained for occupancy by recreational vehicles of the general public as temporary living quarters for recreation or vacation purpose.

Recreational Vehicle Site. Within a recreational vehicle park, a plot of ground intended for the accommodation of a recreational vehicle, a tent, or another individual camping unit on a temporary basis.

E 301.0 Manufactured/Mobile Home Park Drainage System Construction.

E 301.1 General. A drainage system shall be provided in manufactured home parks for conveying and disposing of sewage. Where feasible, the connection shall be made to a public system. New improvements shall be designed, constructed, and maintained in accordance with applicable laws and regulations. Where the drainage lines of the manufactured home park are not connected to a public sewer, the Authority Having Jurisdiction shall approve sewage disposal facilities prior to construction.

E 301.2 Underground Drainage System Location, Size, and Slope. Drainage (sewage) collection lines shall be located in trenches at an approved depth to be free of breakage from traffic or other movements and shall be separated from the park water supply system as specified in this code. Drainage (sewage) lines shall have a minimum size and slope as specified in Table E 301.2(1) and Table E 301.2(2).

E 301.2.1 Inlet, System, and Lateral Sizing. Each manufactured home lot drainage inlet shall be assigned a waste loading value of 12 drainage fixture units, and each park drainage system shall be sized in accordance with Table E 301.2(1) or as provided herein. Drainage laterals shall be not less than 3 inches (80 mm) in diameter.

🔧 Each mobile home lot shall be provided with a drain inlet not less than 3 inches in diameter that will receive the waste of units. Such a waste loading inlet shall be assigned a value of 12 fixture units. This is consistent with Table 703.2 for an individual mobile home: a 3-inch trap arm and trap for each unit. Use Table E 301.2(1) to size the sewage collection lines for multiple units.

E 301.2.2 Engineered Design. A park drainage system that exceeds the fixture unit loading of Table E 301.2(1) or in which the grade and slope of drainage pipe do not meet the minimum specified in Table E 301.2(2) shall be designed by a registered design professional.

E 301.2.3 Materials. Pipe and fittings installed underground in manufactured home park drainage systems shall be of a material approved for the purpose. Manufactured home lot drainage inlets and extensions to grade shall be of a material approved for underground use within a building.

E 301.3 Lot Drainage Inlet. Provision shall be made for plugging or capping the sewage drain inlet where a manufactured home does not occupy the lot. Surface drainage shall be diverted away from the inlet. The rim of the inlet shall extend to a maximum of 4 inches (102 mm) aboveground elevation.

E 301.3.1 Location. Each lot drainage inlet shall be located in the third rear section and within 4 feet (1219 mm) of the proposed location of the manufactured home.

🔧 Each mobile home shall have only one drain outlet, which shall terminate in the rear third section and within 4 feet of the proposed location (see Figures E 301.3.1a and E 301.3.1b).

E 301.3.2 Materials. Materials used for drainage connections between a manufactured home, and the lot drainage inlet shall be semi-rigid, corrosion resistant, nonabsorbent, and durable. The inner surface shall be smooth.

E 301.4 Drain Connector. A manufactured home shall be connected to the lot drainage inlet using a drain connector consisting of approved pipe not less than Schedule 40, approved fittings and connectors, and not less in size than the manufactured home drainage outlet. An approved cleanout shall be provided between the manufactured home and the lot drainage inlet. The fitting connected to the lot drainage inlet shall be a directional fitting to discharge the flow into the drainage inlet.

E 301.4.1 Grade and Gastightness. A drain connector shall be installed or maintained with a grade not less than ¼ inch per foot (20.8 mm/m). A drain connector shall be gastight and no longer than necessary to make the connection between the manufactured home outlet and the drain inlet on the lot. A flexible connector shall be permitted to be used at the lot drainage inlet area. Each lot drainage inlet shall be capped gastight where not in use.

E 302.0 Manufactured/Mobile Home Park Water Supply.

E 302.1 Potable Water Supply. An accessible and approved supply of potable water shall be provided in each

TABLE E 301.2(1)
DRAINAGE PIPE DIAMETER AND NUMBER OF FIXTURE UNITS ON DRAINAGE SYSTEM

SIZE OF DRAINAGE (inches)	MAXIMUM NUMBER OF FIXTURE UNITS
2*	8
3	35
4	256
5	428
6	720
8	2640
10	4680
12	8200

For SI units: 1 inch = 25 mm
* Except for six unit fixtures

TABLE E 301.2(2)
MINIMUM GRADE AND SLOPE OF DRAINAGE PIPE

PIPE SIZE	SLOPE (per 100 feet)	PIPE SIZE	SLOPE (per 100 feet)
inches	inches	inches	inches
2	25	6	8
3	25	8	4
4	15	10	3½
5	11	12	3

For SI units: 1 inch = 25 mm, 1 inch per foot = 83.3 mm/m

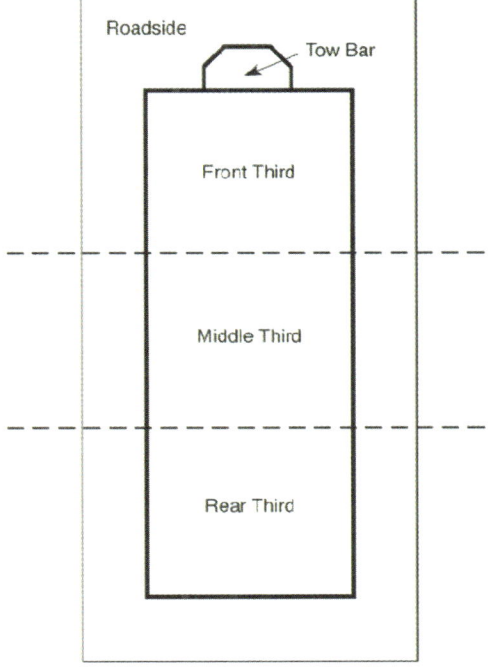

Tow Bar or Tongue – The attachment on the front of a mobile home unit used to pull the unit.

Roadside – The section of the mobile home unit on the left side of the tow bar as it is towed.

**FIGURE E 301.3.1A
DIAGRAM OF MOBILE HOME UNIT**

APPENDIX E

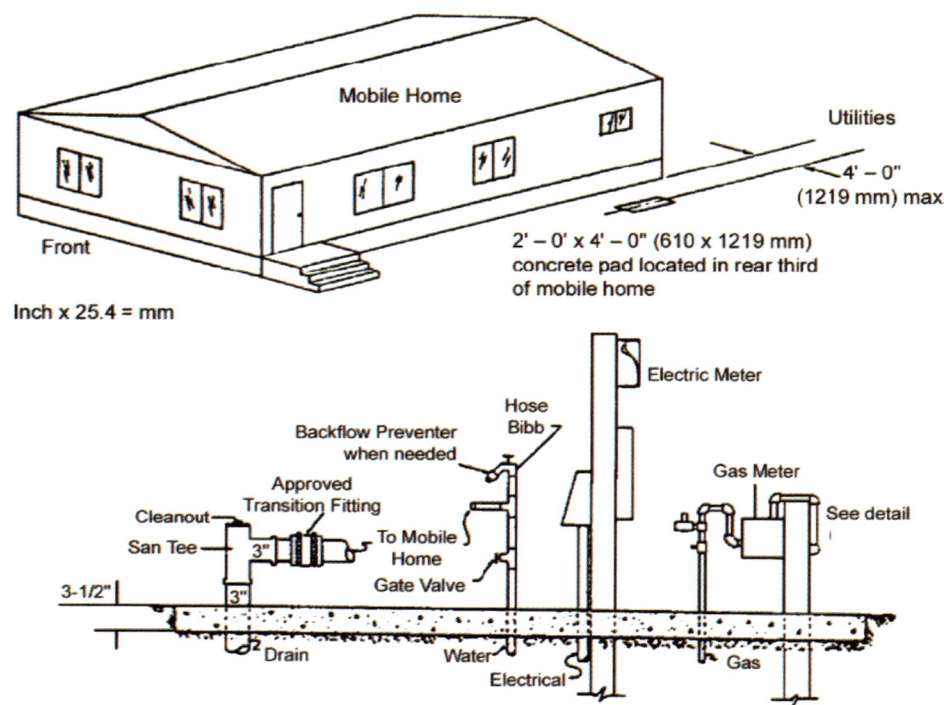

FIGURE E 301.3.1B
LOCATION OF UTILITY PAD (TOP) AND COMMON UTILITY PAD (BOTTOM)

manufactured home park. Where a public supply of water of approved quantity, quality, and pressure is available at or within the boundary of the park site, the connection shall be made to it and its supply used exclusively. Where an approved public water supply is not available, a private water supply system shall be developed and used as approved by the Authority Having Jurisdiction.

E 302.2 Water Service Outlet. Each manufactured home lot shall be provided with a water service outlet delivering potable water. The water service outlet riser shall be not less than ¾ of an inch (20 mm) nominal pipe size and capable of delivering 12 water supply fixture units.

E 302.2.1 Connection. A manufactured home shall be connected to the park water service outlet by a flexible connector, such as copper or copper alloy or other approved material not less than ¾ of an inch nominal (20 mm) interior diameter.

E 302.2.2 Water Supply Fixture Units. Park water distribution systems shall be designed to deliver a minimum of 12 water supply fixture units to each lot and installed with materials in accordance with Chapter 6, Appendix A, or both of this code.

E 302.2.3 Pressure. Each manufactured home park water distribution system shall be so designed and maintained as to provide a pressure of not less than 20 pounds-force per square inch (psi) (138 kPa) at each manufactured home lot at maximum operating conditions.

E 302.2.4 Location. Each lot water service outlet shall be located in the third rear section and within 4 feet (1219 mm) of the proposed location of the proposed home.

See **Figures E 301.3.1a** and **E 301.3.1b**.

E 302.3 Shutoff Valve. A separate water shutoff valve shall be installed in each water service outlet at each manufactured home lot. Where a listed backflow protective device is installed, the service shutoff shall be located on the supply side of such device.

E 302.4 Backflow Preventer. Where a condition exists in the plumbing of a manufactured home that creates a cross-connection, a listed backflow preventer shall be installed in the water service line to the manufactured home at or near the water service outlet. Where a hose bibb or outlet is installed on the supply outlet riser in addition to the service connector, a listed backflow preventer shall be installed on each additional outlet.

E 302.5 Pressure-Relief Valve. Where it is required to install a backflow preventer at the manufactured home lot service outlet, a listed pressure-relief valve shall be installed in the water service line on the discharge side of the backflow preventer. Pressure-relief valves shall be set to release at a pressure at a maximum of 150 psi (1034 kPa). Pressure-relief valves shall discharge toward the ground. Backflow preventers and pressure-relief valves shall be not less than 12 inches (305 mm) above the ground.

E 302.6 Mechanical Protection. Park water service outlets, backflow preventers, and pressure-relief valves shall be protected from damage by vehicles or other causes. Such protection shall be permitted to consist of posts, fencing, or other permanent barriers.

E 302.7 Water-Conditioning Equipment. A permit shall be obtained from the Authority Having Jurisdiction before installing water-conditioning equipment on a manufactured home lot. Approval of the park operator is required on applications for a permit to install such equipment. Where the water-conditioning equipment is of the regenerating type, and the park drainage system discharges into a public sewer, approval of the sanitary district or agency having jurisdiction over the public sewer is required.

E 302.7.1 Approval. Regenerating water-conditioning equipment shall be listed and labeled by an approved listing agency.

E 302.7.2 Installation. Regenerating units shall discharge the effluent of regeneration into a trap not less than 1½ inches (40 mm) in diameter connected to the manufactured home park drainage system. An approved air gap shall be installed on the discharge line a minimum of 12 inches (305 mm) above the ground.

E 302.8 Testing. Installations shall be tested and inspected in accordance with Chapter 3 of this code.

E 401.0 Fuel Supply.

E 401.1 Fuel Gas Piping Systems. All fuel gas piping systems serving manufactured homes, accessory buildings, or structures and communities shall be designed and constructed in accordance with any applicable provisions of NFPA 54 and NFPA 58. NFPA 31 shall apply to oil fuel-burning systems and shall conform to the criteria of the Authority Having Jurisdiction. [NFPA 501A:4.1.1.1 – 4.1.1.2]

E 401.2 Gas Supply Connections. Gas supply connections at sites, where provided from an underground gas supply piping system, shall be located and arranged to permit attachment to a manufactured home occupying the site.

For the installation of liquefied petroleum gas (LP-Gas) storage systems, the provisions of NFPA 58 shall be followed. [NFPA 501A:4.1.2.1 – 4.1.2.2]

E 401.3 Location of Gas Supply Connection. The gas supply to the manufactured home shall be located within 4 feet (1219 mm) of the manufactured home stand.

Exception: The requirements of E 401.3 shall not apply to gas supply connections for manufactured homes located on all-weather wood, concrete, concrete block foundation systems or on foundations constructed in accordance with the local building code or, in the absence of a local code, with a recognized model building code. [NFPA 501A:4.1.3]

E 401.4 Recreational Vehicle Park Fuel Gas Equipment and Installations. Fuel gas equipment and installations shall comply with this appendix, except as otherwise permitted or required by this code.

E 402.0 Single and Multiple Manufactured Home Site Fuel Supply Systems.

E 402.1 Underground Installations. Underground gas piping system installations shall comply with any applicable building code, Section E 402.2 and Section E 402.2.1. [NFPA 501A:4.2.1]

E 402.2 Open-Ended Gastight Conduit. Underground gas piping shall not be installed beneath that portion of a manufactured home site reserved for the location of a manufactured home or manufactured home accessory building or structure unless installed in the open-ended gastight conduit of Section E 402.2.1. [NFPA 501A:4.2.1.1]

E 402.2.1 Requirements. The open-ended gastight conduit shall conform to the requirements in the following:

(1) The conduit shall be not less than Schedule 40 pipe that is approved for underground installation beneath buildings.

(2) The interior diameter of the conduit shall be not less than ½ of an inch (15 mm) larger than the outside diameter of the gas piping.

(3) The conduit shall extend to a point not less than 4 inches (102 mm) beyond the outside wall of the manufactured home or accessory building or structure, and the outer ends shall not be sealed.

(4) Where the conduit terminates within a manufactured home or accessory building or structure, it shall be accessible, and the space between the conduit and the gas piping shall be sealed to prevent leakage of gas into the building. [NFPA 501A:4.2.1.2 – 4.2.1.2.4]

E 402.3 Shutoff Valve. Each manufactured home site shall have a listed gas shutoff valve installed upstream of the manufactured home site gas outlet. The gas shutoff valve shall be located on the outlet riser at the height of not less than 6 inches (152 mm) above grade. A gas shutoff valve shall not be located under any manufactured home. The outlet shall be equipped with a cap or plug to prevent discharge of gas whenever the manufactured home site outlet is not connected to a manufactured home. [NFPA 501A:4.2.2.1 – 4.2.2.4]

Exception: Gas shutoff valves shall conform to Section E 402.3, except for manufactured homes located on foundations constructed in accordance with the local building code or, in the absence of a local code, with a recognized model building code. [NFPA 501A:4.2.2]

E 402.4 Gas Meters. Where installed, gas meters shall be supported by a post or bracket placed on a firm footing or other means providing equivalent support and shall not depend on the gas outlet riser for support. [NFPA 501A:4.2.3.1]

E 402.4.1 Location of Meters. Each gas meter shall be installed in an accessible location and shall be provided with unions or other fittings so that the meter can be removed easily and placed in an upright position. Meters shall not be installed in unventilated or inaccessible locations or closer than 3 feet (914 mm) to sources of ignition. [NFPA 501A:4.2.3.2 – 4.2.3.2.2]

E 402.4.2 Meter Shutoff Valve or Cock. All gas meter installations shall be provided with shutoff valves or cocks located adjacent to and on the inlet side of the meters. In the case of a single meter installation utilizing an LP-Gas container, the container service valve shall be permitted to be used in lieu of the shutoff valve or cock. All gas meter installations shall be provided with test tees located adjacent to and on the outlet side of the meters. [NFPA 501A:4.2.4.1 – 4.2.4.3]

E 403.0 Multiple Manufactured Home Site Fuel Distribution and Supply Systems.

E 403.1 Manufactured Home Community LP-Gas Supply Systems. Where 10 or more customers are served by one LP-Gas supply system, the installation of the gas supply system shall be in accordance with 49 CFR 192. Other types of liquefied petroleum gas supply systems and the storage and handling of LP-Gas shall be in accordance with NFPA 58 (see Section E 403.10). [NFPA 501A:4.3.2.1 – 4.3.2.2]

E 403.2 Required Gas Supply. The minimum hourly volume of gas required at each manufactured home site outlet or any section of the manufactured home community gas piping system shall be calculated as shown in Table E 403.2.

In extreme climate areas, additional capacities other than those in Table E 403.2 shall be considered. [NFPA 501A:4.3.4.1 – 4.3.4.2]

TABLE E 403.2
DEMAND FACTORS FOR USE IN CALCULATING GAS PIPING SYSTEMS IN M/H COMMUNITIES
[NFPA 501A: TABLE 4.3.4.1]

NUMBER OF M/H SITES	BRITISH THERMAL UNITS PER HOUR PER M/H SITE
1	125 000
2	117 000
3	104 000
4	96 000
5	92 000
6	87 000
7	83 000
8	81 000
9	79 000
10	77 000
11–20	66 000
21–30	62 000
31–40	58 000
41–60	55 000
Over 60	50 000

For SI units: 1000 British thermal units per hour = 0.293 kW

E 403.3 Size. The size of each section of a gas piping system shall be determined in accordance with NFPA 54 or by other standard engineering methods acceptable to the Authority Having Jurisdiction. [NFPA 501A:4.3.5.1]

E 403.4 Pressure. Where all connected appliances are operated at their rated capacity, the gas supply pressure shall be not less than 7 inches water column (1.7 kPa). The gas supply pressure shall not exceed 14 inches water column (3.5 kPa). [NFPA 501A:4.3.5.2]

E 403.5 Metal Gas Piping. Metal gas pipe shall be standard-weight wrought iron or steel (galvanized or black), yellow brass containing not more than 75 percent copper, or internally tinned or treated copper of iron pipe size. Galvanizing shall not be considered protection against corrosion.

Seamless copper or steel tubing shall be permitted to be used with gases not corrosive to such material. Steel tubing shall comply with ASTM A254. Copper tubing shall comply with ASTM B88 (Type K or Type L) or ASTM B280. Copper tubing (unless tin-lined) shall not be used if the gas contains more than an average of 0.3 grains of hydrogen sulfide per 100 standard cubic feet (scf) of gas (0.7 mg/100 L). [NFPA 501A:4.3.6.1.1 – 4.3.6.1.6]

E 403.6 Protection Coatings for Metal Gas Piping. All buried or submerged metallic gas piping shall be protected from corrosion by approved coatings or wrapping materials. All gas pipe protective coatings shall be approved types, shall be machine applied, and shall conform to recognized standards. Field wrapping shall provide equivalent protection and is restricted to those short sections and fittings that are necessarily stripped for threading or welding. Risers shall be coated or wrapped to a point at least 6 inches (152 mm) above ground. [NFPA 501A:4.3.6.2.1 – 4.3.6.2.4]

E 403.7 Plastic Piping. Plastic piping shall only be used underground and shall meet the requirements of ASTM D2513 or ASTM D2517, as well as the design pressure and design limitations of 49 CFR (Section 192.123), and shall otherwise conform to the installation requirements thereof. [NFPA 501A:4.3.6.3]

E 403.8 Gas Piping Installation. All gas piping installed below ground level shall have a minimum earth cover of 18 inches (457 mm) and shall be installed with at least 12 inches (305 mm) of clearance in any direction from any other underground utility systems. [NFPA 501A:4.3.7.1]

E 403.8.1 Metallic Gas Piping. All metallic gas piping systems shall be installed in accordance with approved plans and specifications, including provisions for cathodic protection. Each cathodic protection system shall be designed and installed to conform to the provisions of 49 CFR 192. [NFPA 501A:4.3.7.2.1, 4.3.7.2.2]

E 403.8.2 Cathodic Protection. Where the cathodic protection system is designed to protect only the gas piping system, the gas piping system shall be electrically isolated from all other underground metallic systems or installations. Where only the gas piping system is cathodically protected against corrosion, a dielectric fitting shall be used in the manufactured home gas connection to insulate the manufactured home from the underground gas piping system. [NFPA 501A:4.3.7.2.3, 4.3.7.2.4]

E 403.8.3 Underground Metallic Systems. Where a cathodic protection system is designed to provide all underground metallic systems and installations with protection against corrosion, all such systems and installations shall be electrically bonded together and protected as a whole. [NFPA 501A:4.3.7.2.5]

E 403.8.4 Plastic Gas Piping. Plastic gas piping shall be used only underground and shall be installed with an electrically conductive wire for locating the pipe. The wire used to locate the plastic pipe shall be copper, not smaller in size than No. 18 AWG, with insulation approved for direct burial. Every portion of a plastic gas piping system consisting of metallic pipe shall be cathodically protected against corrosion. [NFPA 501A:4.3.7.3.1 – 4.3.7.3.3]

E 403.9 Gas Piping System Shutoff Valve. An accessible and identifiable shutoff valve controlling the flow of gas to the entire manufactured home community gas piping system shall be installed in a location acceptable to the Authority Having Jurisdiction and near the point of connection to the service piping or the supply connection of an LP-Gas container. [NFPA 501A:4.3.7.4]

E 403.10 Liquefied Petroleum Gas Equipment. LP-Gas equipment shall be installed in accordance with the applicable provisions of NFPA 58. [NFPA 501A:4.3.8]

E 403.11 Oil Supply. The following three methods of supplying oil to an individual manufactured home site shall be permitted:

(1) Supply from an outside underground tank (see Section E 404.6).

(2) Supply from a centralized oil distribution system designed and installed in accordance with accepted engineering practices and in compliance with NFPA 31.

(3) Supply from an outside aboveground tank (see Section E 404.6). [NFPA 501A:4.3.9]

E 403.12 Minimum Oil Supply Tank Size. Oil supply tanks shall have a minimum capacity equal to 20 percent of the average annual oil consumption. [NFPA 501A:4.3.10]

E 403.13 Oil Supply Connections. Oil supply connections at manufactured home sites, where provided from a centralized oil distribution system, shall be located and arranged to permit attachment to a manufactured home utilizing the stand. [NFPA 501A:4.3.11.1] The installation of such facilities shall comply with the following requirements:

(1) The main distribution pipeline shall be permitted to be connected to a tank or tanks having an aggregate capacity not exceeding 20 000 gallons (75 708 L) at a point below the liquid level.

(2) Where this piping is so connected, a readily accessible internal or external shutoff valve shall be installed in the piping as close as practicable to the tank.

(3) If external and aboveground, the shutoff valve and its tank connections shall be made of steel.

(4) Connections between the tank(s) and the main pipeline shall be made with double swing joints or flexible connectors, or shall otherwise be arranged to permit the tank(s) to settle without damaging the system.

(5) If located aboveground, the connections specified in Section E 403.13(4) shall be located within the diked area.

(6) A readily accessible and identified manual shutoff valve shall be installed either inside or outside of the structure in each branch supply pipeline that enters a building, mobile home, travel trailer, or other structure. If outside, the valve shall be protected from weather and damage. If inside, the valve shall be located directly adjacent to the point at which the supply line enters the structure.

(7) A device shall be provided in the supply line at or ahead of the point where it enters the interior of the structure that will automatically shut off the oil supply, if the supply line between this device and the appliance is broken. This device shall be located on the appliance side of the manual shutoff valve required in Section E 403.13(6) and shall be solidly supported and protected from damage.

(8) Means shall be provided to limit the oil pressure at the appliance inlet to a maximum gauge pressure of 3 pounds-force per square inch gauge (psig) (21 kPa). If a pressure-reducing valve is used, it shall be a type approved for the service.

(9) A device shall be provided that will automatically shut off the oil supply to the appliance if the oil pressure at the appliance inlet exceeds a gauge pressure of 8 psig (55 kPa). This device shall not be required under either of the following conditions:

(a) Where the distribution system is supplied from a gravity tank and the maximum hydrostatic head of oil in the tank is such that the oil pressure at the appliance inlet will not exceed a gauge pressure of 8 psig (55 kPa).

(b) Where a means is provided to automatically shut off the oil supply if the pressure-regulating device provided in accordance with Section E 403.13(8) fails to regulate the pressure as required.

(10) Only appliances equipped with primary safety controls specifically listed for the appliance shall be connected to a centralized oil distribution system. [NFPA 31:9.2.10 – 9.2.15]

E 404.0 Fuel Supply Systems Installation.

E 404.1 Flexible Gas Connector. Except for manufactured homes located on an all-weather wood, concrete, concrete block foundation system or on a foundation constructed in accordance with the local building code or, in the absence of a local code, with a recognized model building code, each gas supply connector shall be listed for outside manufactured home use, shall be not more than 6 feet (1829 mm) in length, and shall have a capacity rating to supply the connected load. [NFPA 501A:4.4.1]

E 404.2 Use of Approved Pipe and Fittings of Extension. Where it is necessary to extend a manufactured home inlet to permit connection of the 6 foot (1829 mm) listed connector to the site gas outlet, the extension shall be of approved materials of the same size as the manufactured home inlet and shall be adequately supported at no more than 4 foot (1219 mm) intervals to the manufactured home. [NFPA 501A:4.4.2]

E 404.3 Mechanical Protection. All gas outlet risers, regulators, meters, valves, or other exposed equipment shall be protected against accidental damage. [NFPA 501A:4.4.3]

E 404.4 Special Rules on Atmospherically Controlled Regulators. Atmospherically controlled regulators shall be installed in such a manner that moisture cannot enter the regulator vent and accumulate above the diaphragm. Where the regulator vent is obstructed due to snow and icing conditions, shields, hoods, or other suitable devices shall be provided to guard against closing of the vent opening. [NFPA 501A:4.4.4.1 – 4.4.4.2]

E 404.5 Fuel Gas Piping Test. The manufactured home fuel gas piping system shall be tested only with air before it is connected to the gas supply. The manufactured home gas piping system shall be subjected to a pressure test with all appliance shutoff valves in their closed positions. [NFPA 501A:4.4.5]

E 404.5.1 Procedures. The fuel gas piping test shall consist of air pressure at not less than 10 inches water column or more than 14 inches water column (2.5 kPa to 3.5 kPa). The fuel gas piping system shall be isolated from the air pressure source and shall maintain this pressure for not less than 10 minutes without perceptible leakage. Upon satisfactory completion of the fuel gas piping test, the appliance valves shall be opened, and the gas appliance connectors shall be tested with soapy water or bubble solution while under the pressure remaining in the piping system. Solutions used for testing for leakage shall not contain corrosive chemicals. Pressure shall be measured with a manometer, slope gauge, or gauge that is calibrated in either water inch (mm) or psi (kPa) with increments of either $1/10$ inch (2.5 mm) or $1/10$ psi (0.7 kPa) gauge, as applicable. Upon satisfactory completion of the fuel gas piping test, the manufactured home gas supply connector shall be installed, and the connections shall be tested with soapy water or bubble solution. [NFPA 501A:4.4.5.1.1 – 4.4.5.1.6]

E 404.5.2 Warning. The following warning shall be supplied to the installer:

WARNING: Do not overpressurize the fuel gas piping system. Damage to valves, regulators, and appliances can occur due to pressurization beyond the maximums specified. [NFPA 501A:4.4.5.2]

E 404.5.3 Vents. Gas appliance vents shall be visually inspected to ensure that they have not been dislodged in transit and are connected securely to the appliance. [NFPA 501A:4.4.5.3]

E 404.6 Oil Tanks. Oil tanks shall comply with the following:

(1) No more than one 660 gallon (2498 L) tank or two tanks with an aggregate capacity of 660 gallons (2498 L) or less shall be connected to one oil-burning appliance.

(2) Two supply tanks, where used, shall be cross-connected and provided with a single fill and single vent, as described in NFPA 31 and shall be on a common slab and rigidly secured one to the other.

(3) Tanks having a capacity of 660 gallons (2498 L) or less shall be securely supported by rigid, noncombustible supports to prevent settling, sliding, or lifting. [NFPA 501A:4.4.6]

E 404.6.1 Installation. Oil supply tanks shall be installed in accordance with the applicable provisions of NFPA 31. [NFPA 501A:4.4.6.1]

E 404.6.2 Capacity. A tank with a capacity no larger than 60 gallons (227 L) shall be permitted to be a DOT-5 shipping container (drum), and so marked, or a tank meeting the provisions of UL 80. Tanks other than DOT-5 shipping containers having a capacity of not more than 660 gallons (2498 L) shall meet the provisions of UL 80. Pressure tanks shall be built in accordance with Section VIII, Pressure Vessels of the ASME Boiler and Pressure Vessel Code. [NFPA 501A:4.4.6.2.1 – 4.4.6.2.2]

E 404.6.3 Location. Tanks, as described in Section E 404.6 and Section E 404.6.2, that are adjacent to buildings shall be located not less than 10 feet (3048 mm) from a property line that is permitted to be built upon. [NFPA 501A:4.4.6.3]

E 404.6.4 Vent. Tanks with a capacity no larger than 660 gallons (2498 L) shall be equipped with an open vent no smaller than 1½ inch (40 mm) iron pipe size; tanks with a 500 gallon (1892 L) or less capacity shall have a vent of 1¼ inch (32 mm) iron pipe size. [NFPA 501A:4.4.6.4]

E 404.6.5 Liquid Level. Tanks shall be provided with a means of determining the liquid level. [NFPA 501A:4.4.6.5]

E 404.6.6 Fill Opening. The fill opening shall be a size and in a location that permits filling without spillage. [NFPA 501A:4.4.6.6]

E 405.0 Manufactured Home Accessory Building Fuel Supply Systems.

E 405.1 General. Fuel gas supply systems installed in a manufactured home accessory building or structure shall comply with the applicable provisions of NFPA 54 and NFPA 58. Fuel oil supply systems shall comply with the applicable provisions of NFPA 31. [NFPA 501A:4.5.1 – 4.5.2]

E 406.0 Community Building Fuel Supply Systems in Manufactured Home Communities.

E 406.1 Fuel Gas Piping and Equipment Installations. Fuel gas piping and equipment installed within a permanent building in a manufactured home community shall comply with nationally recognized appliance and fuel gas piping codes and standards adopted by the Authority Having Jurisdiction. Where the state or other political subdivision does not assume jurisdiction, such fuel gas piping and equipment installations shall be designed and installed in accordance with the applicable provisions of NFPA 54 or NFPA 58. [NFPA 501A:4.6.1.1 – 4.6.1.2]

E 406.2 Oil Supply Systems. Oil burning equipment and installation within a manufactured home community shall be designed and constructed in accordance with the applicable codes and standards adopted by the Authority Having Jurisdiction. Where the state or other political subdivision does not assume jurisdiction, such installations shall be designed and constructed in accordance with the applicable provisions of NFPA 31. [NFPA 501A:4.6.2.1 – 4.6.2.2]

E 406.3 Oil-Burning Equipment and Installation. Oil burning equipment and installation within a building constructed in a manufactured home community in accordance with the local building code or a nationally recognized building code shall comply with nationally recognized codes and standards adopted by the Authority Having Jurisdiction. Where the state or other political subdivision does not assume jurisdiction, such oil-burning equipment and installations shall be designed and installed in accordance with the applicable provisions of NFPA 31. [NFPA 501A:4.6.3.1 – 4.6.3.2]

E 406.4 Inspections and Tests. Inspections and tests for fuel gas piping shall be made in accordance with Chapter 1 and Chapter 12 of this code.

E 501.0 Recreational Vehicle Parks.

E 501.1 Plumbing Systems. Plumbing systems shall be installed in accordance with the plumbing codes of the Authority Having Jurisdiction and with this appendix.

E 501.2 Toilet Facilities. Water closets and urinals shall be provided at one or more locations in a recreational vehicle park. They shall be of convenient access and shall be located within a 500 foot (152 m) radius from a recreational vehicle site not provided with an individual sewer connection.

E 501.2.1 Signage. Facilities for males and females shall be appropriately marked.

E 501.2.2 Interior Finish. The interior finish of walls shall be moisture resistant to a height of not less than 4 feet (1219 mm) to facilitate washing and cleaning.

E 501.2.3 Receptacle. Each toilet room for women shall be provided with a receptacle for sanitary napkins. The receptacle shall be of durable, impervious, readily cleanable material, and shall be provided with a lid.

E 501.2.4 Ceiling Height and Doors. A toilet facility shall have a ceiling height of not less than 7 feet (2134 mm) and, unless the artificial light is provided, a window or skylight area equal to not less than 10 percent of the floor area shall be provided.

Doors to the exterior shall open outward, be self-closing, and shall be visually screened using a vestibule or wall to prevent a direct view of the interior where the exterior doors are open. Such screening shall not be required on single toilet units.

E 501.2.5 Ventilation. A toilet facility shall have permanent, non-closable, screened opening(s), having a total area not less than 5 percent of the floor area and opening directly to the exterior to provide proper ventilation. A listed exhaust fan(s), vented to the exterior, the rating of which in cubic feet per minute (L/s) is not less than 25 percent of the total volume of the room(s) served, shall be considered as meeting the requirements of this section. Openable windows and vents to the outside shall be provided with fly-proof screens of not less than number 16 mesh.

E 501.3 Water Closets. Not less than one water closet shall be provided for each sex up to the first 25 sites. For each additional 25 sites not provided with sewer connections, an additional water closet shall be provided.

E 501.3.1 Application. Water closets shall be of an approved, elongated bowl type and shall be provided with seats with open fronts.

E 501.3.2 Compartment. Each water closet shall be in a separate compartment and shall be provided with a latched door for privacy. A holder or dispenser for toilet paper shall be provided. Dividing walls or partitions shall be not less than 5 feet (1524 mm) high and shall be separated from the floor by a space not exceeding 12 inches (305 mm).

E 501.3.3 Size. Water closet compartments shall be not less than 30 inches (762 mm) in width [no water closet shall be set closer than 15 inches (381 mm) from its center to a side wall] and shall be not less than 30 inches (762 mm) of clear space in front of each water closet.

Walls in the toilet room shall have a finish resistant to moisture for easy cleaning.

E 501.4 Lavatories. Where water-supplied water closets are provided, an equal number of lavatories shall be provided for up to six water closets. One additional lavatory shall be provided for each two water closets where more than six water closets are required. Each lavatory basin shall have a piped supply of potable water and shall drain into the drainage system.

E 501.5 Urinals. Where separate facilities are provided for men and women, urinals shall be acceptable for not more than one-third of the water closets required in the men's facilities, except that one urinal shall be permitted to be used to replace a water closet in a minimum park. Individual stall or wall-hung types of urinals shall be installed. Floor-type trough units shall be prohibited.

E 501.6 Floors and Drains. The floors shall be constructed of material impervious to water and shall be easily cleanable. A building having water-supplied water closets shall be provided with a floor drain in the toilet room. This drain shall be provided with means to protect the trap seal in accordance with this code.

E 501.7 Shower Size. Each shower, where provided, shall have a floor area of 36 inches by 36 inches (914 mm by 914 mm), shall be capable of encompassing a 30 inch (762 mm) diameter circle and shall be of the individual type. The shower area shall be visually screened from view with a minimum floor area of 36 by 36 inches (914 mm by 914 mm) per shower. Each shall be provided with individual dressing areas screened from view and shall contain a minimum of one clothing hook and stool (or bench area).

E 501.7.1 Drainage Connection. Each shower area shall be designed to minimize the flow of water into the dressing area and shall be connected to the drainage system using a properly trapped and vented inlet. Each such area shall have an impervious, skid-resistant surface; wooden racks (duck boards) over shower floors shall be prohibited.

See Section 408.3 for information on shower valves with hot water.

E 501.8 Drinking Fountains. Where provided, drinking fountains shall be in accordance with the requirements of this code.

E 502.0 Recreational Vehicle Park Potable Water Supply and Distribution.

E 502.1 Quality. The supply or supplies of water shall comply with the potable water standards of the state, local health authority or, in the absence thereof, with the Drinking Water Standard of the Federal Environmental Protection Agency.

Systems used on a seasonal basis shall be provided with a means of draining the potable water supply to prevent freezing and damage of piping and fixtures.

E 502.2 Sources. Water approved by a regulating agency shall be acceptable. Where an approved public water supply system is available, it shall be used. Where the park has its own water supply system, the components of the system shall be approved. A water supply system that is used on a seasonal basis shall be provided with means for draining.

E 502.3 Prohibited Connections. The potable water supply shall not be connected to a nonpotable or unapproved water supply, nor be subjected to backflow or backsiphonage.

E 502.4 Supply. The water supply system shall be designed and constructed in accordance with the following:

(1) A minimum of 25 gallons (95 L) per day per site for sites without individual water connections.

(2) A minimum of 50 gallons (189 L) per day per site for sites with individual water connections.

(3) A minimum of 50 gallons (189 L) per day per site where water-supplied water closets are provided in restrooms.

E 502.5 Pressure and Volume. Where water is distributed under pressure to an individual site, the water supply system shall be designed to provide a minimum flow pressure of not less than 20 psi (138 kPa) with a minimum flow of 2 gallons per minute (gpm) (0.1 L/s) at an outlet. The pressure shall not exceed 80 psi (552 kPa).

E 502.6 Outlets. Water outlets shall be convenient to access and, where not piped to individual recreational vehicle sites, shall not exceed 300 feet (91 m) from a site. Provisions shall be made to prevent accumulation of standing water or the creation of muddy conditions at each water outlet.

E 502.7 Storage Tanks. Water storage tanks shall be constructed of impervious materials, protected against contamination, and provided with locked, watertight covers. Overflow or ventilation openings shall be down-facing and provided with a corrosion-resistant screening of not less than number 24 mesh to prevent the entrance of insects and vermin. Water storage tanks shall not have direct connections to sewers.

503.0 Recreational Vehicle Park Water Connections for Individual Recreational Vehicles.

E 503.1 Location. Where provided, the water connections for potable water to individual recreational vehicle sites shall be located on the left rear half of the site (left side of recreational vehicle) within 4 feet (1219 mm) of the stand.

E 503.2 Water Riser Pipe. Each potable water connection shall consist of a water riser pipe that is equipped with a threaded male spigot located not less than 12 inches (305 mm) but not more than 24 inches (610 mm) above grade level for the attachment of a standard water hose. The water riser pipe shall be protected from physical damage in accordance with this code. This connection shall be equipped with a listed antisiphon backflow prevention device.

E 504.0 Recreational Vehicle Park Drainage System.

E 504.1 Where Required. An approved drainage system shall be provided in recreational vehicle parks for conveying and disposing of sewage. Where available, parks shall be connected to a public sewer system.

E 504.2 Location. Sewer lines shall be located to prevent damage from vehicular traffic.

E 504.3 Materials. Pipe and fittings installed in the drainage system shall be of material listed, approved, and installed in accordance with this code.

E 504.4 Pipe Sizes. The minimum diameters of drainage laterals, branches, and mains serving recreational vehicle sites shall be in accordance with Table E 504.4.

E 504.5 Cleanouts. Cleanouts shall be provided in accordance with Chapter 7 of this code.

Cleanouts shall be provided at the upper terminal of each sewer main or branch and at intervals not exceeding

TABLE E 504.4
PIPE SIZES

MAXIMUM NUMBER OF RECREATIONAL VEHICLE STANDS SERVED	MINIMUM PIPE SIZES (inches)
5	3
36	4
71	5
120	6
440	8

For SI units: 1 inch = 25 mm

100 feet (30.5 m) along any straight run or portion thereof. Manholes may be used in lieu of cleanouts and shall not be spaced more than 300 feet (91.4 m) apart, per Chapter 7 (see Section 719.0, Cleanouts).

The inlet and outlet connections shall be made by the use of a flexible compression joint no closer than 12 inches (305 mm) to and not farther than 3 feet (914 mm) from the manhole. Flexible connections shall not be installed in the manhole base.

E 504.6 Drainage Inlet. Where provided, the site drainage system inlet connections for individual recreational vehicles shall be located to prevent damage by the parking of recreational vehicles or automobiles and shall consist of a sewer riser extending vertically to grade. The minimum diameter of the sewer riser pipe shall be not less than 3 inches (80 mm) in diameter, and shall be provided with a 4 inch (100 mm) inlet or not less than a 3 inch (80 mm) female fitting.

504.6.1 Location. Where provided, the sewer inlet to individual recreational vehicle sites shall be located on the left rear half of the site (left side of the recreational vehicle) within 4 feet (1219 mm) of the stand.

504.6.2 Protection. The sewer riser pipe shall be firmly imbedded in the ground and protected against damage from movement. It shall be provided with a tight-fitting plug or cap, which shall be secured by a durable chain (or equivalent) to prevent loss.

E 505.0 Recreational Vehicle Park Sanitary Disposal Stations.

E 505.1 Where Required. One recreational vehicle sanitary disposal station shall be provided for each 100 recreational vehicle sites, or part thereof, which are not equipped with individual drainage system connections.

E 505.2 Access. Each station shall be level and convenient of access from the service road and shall provide easy ingress and egress for recreational vehicles.

E 505.3 Construction. Unless other approved means are used, each station shall have a concrete slab with the drainage system inlet located to be on the road (left) side of the recreational vehicle. The slab shall be not less than 3 feet by 3 feet (914 mm by 914 mm), not less than 3½ inches (89 mm) thick and properly reinforced. The slab surface is to be troweled to a smooth finish and sloped from each side inward to a drainage system inlet.

The drainage system inlet shall consist of a 4 inch (102 mm), self-closing, foot-operated hatch of approved material with the cover milled to fit tight. The hatch body shall be set in the concrete of the slab with the lip of the opening flush with its surface to facilitate the cleansing of the slab with water. The hatch shall be properly connected to a drainage system inlet, which shall discharge to an approved sanitary sewage disposal facility.

E 505.4 Flushing Device. Where the recreational vehicle park is provided with a piped water supply system, means for flushing the recreational vehicle holding tank and the sanitary disposal station slab shall be provided that consists of a piped supply of water under pressure, terminating in an outlet located and installed to prevent damage by automobiles or recreational vehicles. The flushing device shall consist of a properly supported riser terminating not less than 2 feet (610 mm) above the ground surface, with a ¾ of an inch (20 mm) valved outlet adaptable for a flexible hose.

The water supply to the flushing device shall be protected from backflow using a listed vacuum breaker or backflow prevention device located downstream from the last shutoff valve.

Adjacent to the flushing arrangement shall be posted a sign of durable material not less than 2 feet by 2 feet (610 mm by 610 mm) in size. Inscribed thereon in clearly legible letters shall be the following:

"DANGER – NOT TO BE USED FOR DRINKING OR DOMESTIC PURPOSES."

E 506.0 Recreational Vehicle Park Water Supply Stations.

E 506.1 Potable Watering Stations. A potable watering station, where provided for filling recreational vehicle potable water tanks, shall be located not less than 50 feet (15 240 mm) from a sanitary disposal station. Where such is provided, adjacent to the potable water outlet there shall be a posted sign of durable material not less than 2 feet by 2 feet (610 mm by 610 mm) in size. Inscribed thereon in clear, legible letters on a contrasting background shall be:

"POTABLE WATER. NOT TO BE USED FOR FLUSHING WASTE TANKS."

The potable water shall be protected from backflow using a listed vacuum breaker located downstream from the last shutoff valve.

APPENDIX F
FIREFIGHTER BREATHING AIR REPLENISHMENT SYSTEMS

F 101.0 General.

F 101.1 Applicability. This chapter covers minimum requirements for the installation of firefighter breathing air replenishment systems.

F 201.0 Definitions.

F 201.1 General. For purposes of this chapter, the following definitions shall apply:

High-Rise Building. A building where the floor of an occupiable story exceeds 75 feet (22 860 mm) above the lowest level of fire department vehicle access. [NFPA 5000:3.3.69.10]

Interior Cylinder Fill Panels. Lockable interior panels that provide firefighters the ability to regulate breathing air pressure and refill self-contained breathing apparatus (SCBA) cylinders.

Interior Cylinder Fill Stations and Enclosures. Free-standing fill containment stations that provide firefighters the ability to regulate breathing air pressure and refill SCBA cylinders.

Open-Circuit Self-Contained Breathing Apparatus. A SCBA in which exhalation is vented to the atmosphere and not rebreathed. [NFPA 1981:3.3.34]

Self-Contained Breathing Apparatus (SCBA). An atmosphere-supplying respirator that supplies a respirable air atmosphere to the user from a breathing air source that is independent of the ambient environment and designed to be carried by the user. [NFPA 1981:3.3.46]

For the purposes of this appendix, where this term is used without a qualifier, it indicates an open-circuit self-contained breathing apparatus or combination SCBA/SARs. For the purposes of this appendix, combination SCBA/SARs are encompassed by the terms self-contained breathing apparatus or SCBA.

Welding Procedure Specification (WPS). A written qualified welding procedure prepared to provide direction for making production welds to code requirements. [ASME B31.1:100.2]

F 301.0 System Components.

F 301.1 General. Firefighter breathing air replenishment systems shall contain, as a minimum, the following components:

(1) Exterior fire department connection panel
(2) Interior fire department air fill panel or station
(3) Interconnected piping distribution system
(4) Pressure monitoring switch

F 401.0 Required Installations.

F 401.1 General. A firefighter air system shall be installed in the following buildings:

(1) High-rise buildings.
(2) Underground structures that are three or more floors below grade with an area greater than 20 000 square feet (1858 m^2).
(3) Large area structures with an area greater than 200 000 square feet (18 580 m^2) and where the travel distance from the building centerline to the closest exit is greater than 500 feet (152 m), such as warehouses, manufacturing complexes, malls, or convention centers.
(4) Underground transportation or pedestrian tunnels exceeding 500 feet (152 m) in length.

F 501.0 Exterior Fire Department Connection Panel and Enclosure.

F 501.1 Purpose. The exterior fire department connection panel shall provide the fire department's mobile air operator access to the system and shall be compatible with the fire department's mobile air unit.

F 501.2 Number of Panels. Each building or structure shall have a minimum of two panels.

F 501.3 Location. Each panel shall be attached to the building or on a remote monument at the exterior of the building with a minimum 6 foot (1829 mm) radius and 180-degree (3.14 rad) clear unobstructed access to the front of the panel. The panel shall be weather-resistant or secured inside of a weather-resistant enclosure. The panel shall be located on opposite sides of the building within 50 feet (15 240 mm) of an approved roadway or driveway, or other locations approved by the Authority Having Jurisdiction.

F 501.4 Construction. The fire department connection panel shall be installed in a metal cabinet constructed of not less than 18-gauge carbon steel or equivalent. The cabinet shall be provided with a coating or other means to protect the cabinet from corrosion.

F 501.5 Vehicle Protection. Where the panel is located in an area subject to vehicle traffic, impact protection shall be provided.

F 501.6 Enclosure Marking. The front of the enclosure shall be marked, "FIREFIGHTER AIR SYSTEM". The lettering shall be in a color that contrasts with the enclosure front and in letters that are not less than 2 inches (51 mm) high with a ⅜ of an inch (9.5 mm) brush stroke.

F 501.7 Enclosure Components. The exterior fire department connection panel shall contain the necessary gauges, isolation valves, pressure-relief valves, pressure-regulating valves, check valves, tubing, fittings, supports, connectors, adapters, and other necessary components as required to allow the fire department's mobile air unit to connect and augment the system with a constant source of breathing air. Each fire department connection panel shall contain not less than two inlet air connections.

APPENDIX F

F 501.8 Pressure-Relief Valve. Pressure-relief valves shall be installed downstream of the pressure regulator inlet. The relief valve shall meet the requirements of CGA S-1.3 and shall not be field adjustable. The relief valve shall have a set-to-open pressure not exceeding 1.1 times the design pressure of the system. Pressure-relief valve discharge shall terminate so that the exhaust air stream cannot impinge upon personnel in the area. Valves, plugs, or caps shall not be installed in the discharge of a pressure-relief valve. Where discharge piping is used, the end shall not be threaded.

F 501.9 Security. The fire department connection panel enclosure shall be locked by an approved means.

F 601.0 Interior Cylinder Fill Panels.

F 601.1 Cabinet Requirements. Each cylinder fill panel shall be installed in a metal cabinet constructed of not less than 18-gauge carbon steel or equivalent. The depth of the cabinet shall not create an exit obstruction where installed in building stairwells. With the exception of the shutoff valve, pressure gauges, fill hoses, and ancillary components; no system components shall be visible and shall be contained behind a not less than 18-gauge interior panel.

F 601.2 Clearance and Access. The panel shall be located not less than 36 inches (914 mm) but not more than 60 inches (1524 mm) above the finished floor or a stairway landing. Clear unobstructed access shall be provided to each panel.

F 601.3 Door. The door shall be arranged such that where the door is open, it does not reduce the required exit width or create an obstruction in the path of egress.

F 601.4 Cabinet Marking. The front of each cylinder fill panel shall be marked, "FIREFIGHTER AIR SYSTEM". The lettering shall be in a color that contrasts with the cabinet front and in letters that are not less than 2 inches (51 mm) high with a ⅜ of an inch (9.5 mm) brush stroke.

F 601.5 Cabinet Components. The cabinet shall be of a size to allow for the installation of the components in Section F 601.5.1.

F 601.5.1 Cylinder Fill Panel. The cylinder fill panel shall contain the gauges, isolation valves, pressure-relief valves, pressure-regulating valves, check valves, tubing, fittings, supports, connectors, hoses, adapters, and other components to refill SCBA cylinders.

F 601.6 Cylinder Filling Hose. The design of the cabinet shall provide a means for storing the hose to prevent kinking. Where the hose is coiled, the brackets shall be installed so that the hose bend radius is maintained at 4 inches (102 mm) or greater. Fill hose connectors for connection to SCBA cylinders shall comply with the requirements of CGA V-1, number 346 or 347. No other SCBA cylinder fill connections shall be permitted.

F 601.7 Security. Each panel cover shall be maintained and locked by an approved means.

F 701.0 Interior Cylinder Fill Stations and Enclosures.

F 701.1 Location. The location of the closet or room for each air fill station shall be approved by the Authority Having Jurisdiction. Where approved by the Authority Having Jurisdiction, the space shall be permitted to be utilized for other firefighting purposes. The door to each room enclosing the air filling station enclosure shall be readily accessible at all times. Not less than a 6 foot (1829 mm) radius and 180-degree (3.14 rad) clear unobstructed access to the front of the air filling station shall be provided. The enclosure shall have emergency lighting installed in accordance with NFPA 70.

F 701.2 Security. Each air fill station shall be installed within a lockable enclosure, closet, or room by an approved means. Access to fill equipment and controls shall be restricted to authorized personnel by key or other means.

F 701.3 Components. The air fill station shall contain the gauges, isolation valves, pressure-relief valves, pressure-regulating valves, check valves, tubing, fittings, supports, connectors, hoses, adapters, and other components to refill SCBA cylinders.

F 701.4 Cylinder Filling Hose. Where hoses are used, the design of the cabinet shall provide a means for storing the hose to prevent kinking. Where the hose is coiled, the brackets shall be installed so that the hose bend radius is maintained at 4 inches (102 mm) or greater. Fill hose connectors for connection to SCBA cylinders shall comply with the requirements of CGA V-1, no. 346 or 347. For high-pressure SCBA cylinders of 4500 pounds-force per square inch (psi) (31 026 kPa), no. 347 connectors shall be used. For low-pressure SCBA cylinders of 3000 psi (20 684 kPa) and 2200 psi (15 168 kPa), no. 346 connectors shall be used. No other SCBA cylinder fill connections shall be permitted.

F 701.5 Enclosure and Air Filling Station Marking. Each enclosure, closet, or room shall be marked, "FIREFIGHTERS AIR SYSTEM." The lettering shall be in a color that contrasts with the cabinet front and in letters that are not less than 2 inches (51 mm) high with a ⅜ of an inch (9.5 mm) brush stroke.

F 801.0 Materials.

F 801.1 General. Pressurized components shall be compatible for use with high-pressure breathing air equipment and self-contained breathing air apparatus. Pressurized breathing air components shall be rated for not less than a working pressure of 5000 psi (34 474 kPa).

F 801.2 Tubing. Tubing shall be stainless steel in accordance with ASTM A269, or other approved materials that are compatible with breathing air at the system pressure. Routing of tubing and bends shall be such as to protect the tubing from mechanical damage.

F 801.3 Fittings. Fittings shall be constructed of stainless steel in accordance with ASTM A479, or other approved materials that are compatible with breathing air at the system pressure.

F 801.4 Prohibited Materials. The use of nonmetallic materials, carbon steel, iron pipe, malleable iron, high-strength gray iron, or alloy steel shall be prohibited for breathing air pipe and tubing materials.

F 801.5 Pressure Monitoring Switch. An electric low-pressure monitoring switch shall be installed in the piping system to monitor the air pressure. The pressure switch shall transmit a supervisory signal to the central alarm monitoring station where the pressure of the breathing air system is less than 80 percent of the system operating pressure. Activation of the pressure switch shall also activate an audible alarm and visual strobe located at the building annunciator panel. A weather-resistant sign shall be provided in conjunction with the audible alarm stating, "FIREFIGHTER AIR SYSTEM – LOW AIR PRESSURE ALARM." Where not part of a building annunciator panel, the lettering shall be in a contrasting color, and the letters shall be not less than 2 inches (51 mm) high with a ⅜ of an inch (9.5 mm) brush stroke.

F 801.6 Isolation Valve. A system isolation valve shall be installed downstream of each air fill station and shall be located in the panel or within 3 feet (914 mm) of the station. The isolation valve shall be marked with its function in letters that are not less than ³⁄₁₆ of an inch (4.8 mm) high with a ¹⁄₁₆ of an inch (1.6 mm) brush stroke.

F 901.0 System Requirements.

F 901.1 Protection. Components of the firefighter breathing air replenishment system installed in a building or structure shall be protected by not less than a 2 hour fire-resistive construction. Components shall be protected from physical damage.

F 901.2 Markings. Components shall be clearly identified by means of stainless steel or plastic labels or tags indicating their function. This shall include not less than all fire department connection panels, air fill stations, air storage system, gauges, valves, air connections, air outlets, enclosures, and doors.

F 901.3 Tubing Markings. Tubing shall be clearly marked, "FIREFIGHTERS AIR SYSTEM" and "HIGH PRESSURE BREATHING AIR" by means of signs or self-adhesive labels. Signs shall be 1 inch (25.4 mm) high and shall be secured to the tubing. Signs shall be made of copper alloy, stainless steel, or plastic and engraved with ⅜ of an inch (9.5 mm) letters with a ¹⁄₁₆ of an inch (1.6 mm) stroke lettering. Signs or labels shall be placed at not less than 20 foot (6096 mm) intervals and at each fitting, whether the tubing is concealed or in plain view. Tubing shall have a sign or label at an accessible point.

F 901.4 Support. Pipe and tubing shall be supported at intervals not less than that shown in Table 313.3 of this code. Pipe and tubing shall be supported in accordance with Section 313.0 of this code.

F 1001.0 Design Criteria.

F 1001.1 Fill Time. The system shall be designed to fill, at the most remote fill station or panel, not less than two 66 cubic foot (ft³) (1.87 m³) compressed breathing air cylinders to a pressure not to exceed 4500 psi (31 026 kPa) simultaneously in 3 minutes or less. Where greater capacity is required, the Authority Having Jurisdiction shall specify the required system capacity.

F 1001.2 Fill Panels or Stations Location. Cylinder fill panels or stations shall be installed in the interior of buildings in accordance with Section F 1001.2.1 through Section F 1001.2.3.

F 1001.2.1 High-Rise Buildings. An interior cylinder fill panel or station shall be installed commencing on the third floor and every third floor thereafter above grade. For underground floors in buildings with more than five underground floors, an interior cylinder fill panel or station shall be installed commencing on the third floor below grade and every three floors below grade thereafter, except for the bottom-most floor.

F 1001.2.2 Underground Structures. For underground floors in buildings with more than five underground floors, an interior cylinder fill panel or station shall be installed commencing on the third floor below grade and every three floors below grade thereafter, except for the bottom-most floor.

F 1001.2.3 Installation Locations. The specific location or locations on each floor shall be approved by the Authority Having Jurisdiction.

F 1101.0 System Assembly Requirements.

F 1101.1 General. The system shall be an all-welded system except where the tubing joints are readily accessible and at the individual air fill panels or stations. Where mechanical high-pressure tube fittings are used, they shall be approved for the type of materials to be joined and rated for the maximum pressure of the system.

F 1101.2 Welding Requirements. Prior to and during the welding of sections of tubing, a continuous, regulated dry nitrogen or argon purge at 3 psi (21 kPa) shall be maintained to eliminate contamination with products of the oxidation or welding flux. The purge shall commence not less than 2 minutes prior to welding operations and continue until the welded joint is at ambient temperature. Welding procedures shall comply with the following requirements:

(1) Qualification of the WPS to be used, and of the performance of welders and operators, is required.

(2) No welding shall be done where there is impingement of rain, snow, sleet, or high wind on the weld area.

(3) Tack welds permitted to remain in the finished weld shall be made by a qualified welder. Tack welds made by an unqualified welder shall be removed. Tack welds which remain shall be made with an electrode and WPS which is the same as or equivalent to the electrode and WPS to be used for the first pass. The stopping and starting ends shall be prepared by grinding or other means so that they are capable of being satisfactorily incorporated into the final weld. Tack welds which have cracked shall be removed.

(4) Arc strikes outside the area of the intended weld shall be avoided on a base metal. [ASME B31.1:127.4.1]

F 1101.3 Prevention of Contamination. The system components shall not be exposed to contaminants, including but not limited to, oils, solvents, dirt, and construction materials. Where contamination of system components has occurred, the affected component shall not be installed in the system.

F 1201.0 System Acceptance and Certification.

F 1201.1 Static Pressure Testing. Following fabrication, assembly, and installation of the piping distribution system, exterior connection panel, and interior cylinder fill panels, the Authority Having Jurisdiction shall witness the pneumatic testing of the complete system at a test pressure of not less than 7500 psi (51 711 kPa) using oil-free dry air, nitrogen, or argon. A pneumatic test of not less than 24 hours shall be performed. During this test, all fittings, joints, and system components shall be inspected for leaks. A solution compatible with the system component materials shall be used on each joint and fitting. Defects in the system or leaks detected shall be documented on an inspection report, repaired or replaced. As an alternate, a pressure-decay test in accordance with ASME B31.3 shall be permitted.

F 1201.2 Low Pressure Switch Test. Upon successful completion of the 24 hour static pressure test, the system's low-pressure monitoring switch shall be calibrated to not less than 3000 psi (20 684 kPa) descending, and tested to verify that the signal is annunciated at the building's main fire alarm panel and by means of an audible alarm and visual strobe located in a visible location.

F 1201.3 Compatibility Check. Each air fill panel and station, and each exterior fire department connection panel, shall be tested for compatibility with the fire department's SCBA fill fittings.

F 1201.4 Material Certifications. The pipe or tubing material certifications shall be provided to the Authority Having Jurisdiction.

F 1201.5 Air Sampling. Before the system is placed into service, a minimum of two samples shall be taken from separate air fill panels and submitted to an independent certified gas analysis laboratory to verify the system's cleanliness and that the air is in accordance with the following requirements for breathing air:

(1) Breathing air shall have oxygen content not less than 19.5 percent and not greater than 23.5 percent by volume.

(2) Breathing air shall not have a concentration of carbon monoxide exceeding 5.0 parts per million (ppm) by volume.

(3) Breathing air shall not have a concentration of carbon dioxide exceeding 1000 ppm by volume.

(4) Breathing air shall not have a concentration of condensed oil and particulate exceeding 7.2 E-11 pounds per cubic inch (lb/in^3) (2.0 mg/m^3) at 72°F (22°C) and 30 inches of Hg (102 kPa).

(5) Where breathing air supply for respirators is stored at pressures exceeding 15 bar (1500 kPa), the breathing air shall not have a concentration of water exceeding 24 ppm by volume.

(6) Breathing air shall not have a nonmethane volatile organic compound (VOC) content exceeding 25 ppm as methane equivalents.

(7) Breathing air shall not have a pronounced or unusual odor.

(8) Breathing air shall have a concentration of nitrogen not less than 75 percent and not more than 81 percent. [NFPA 1989:5.6]

The written report of the analysis shall be submitted to the Authority Having Jurisdiction, documenting that the breathing air is in accordance with this section.

F 1201.5.1 Air Quality Analysis. During the period of air quality analysis, the air fill panel inlet shall be secured so that no air is capable of being introduced into the system and each air fill panel shall be provided with a sign stating, "AIR QUALITY ANALYSIS IN PROGRESS, DO NOT FILL OR USE ANY AIR FROM THIS SYSTEM." This sign shall be not less than 8½ inches (216 mm) by 11 inches (279 mm) with not less than 1 inch (25.4 mm) lettering.

F 1201.6 Annual Air Sampling. The breathing air within the system shall be sampled and certified annually and inspected in accordance with the procedure in Section F 1201.5.

F 1201.7 Final Proof Test. The Authority Having Jurisdiction shall witness the filling of two empty 66 cubic foot (ft^3) (1.87 m^3) capacity SCBA cylinders in 3 minutes or less, using compressed air supplied by fire department equipment connected to the exterior fire department connection panel. The SCBA cylinders shall be filled at the air fill panel or station farthest from the exterior fire department connection panel. Following this, not less than two air samples shall then be taken from separate air filling stations and submitted to an independent certified gas analyst laboratory to verify the system's cleanliness and that the air is in accordance with the requirements of NFPA 1989. The written report shall be provided to the Authority Having Jurisdiction certifying that the air analysis is in accordance with the above requirements.

APPENDIX G

SIZING OF VENTING SYSTEMS

(The content of this Appendix is based on Annex F of NFPA 54)

G 101.0 General.

G 101.1 Applicability. This appendix provides general guidelines for sizing venting systems serving appliances equipped with draft hoods, Category I appliances, and appliances listed for use with Type B vents.

G 101.2 Examples Using Single Appliance Venting Tables. See Figure G 101.2(1) through Figure G 101.2(14).

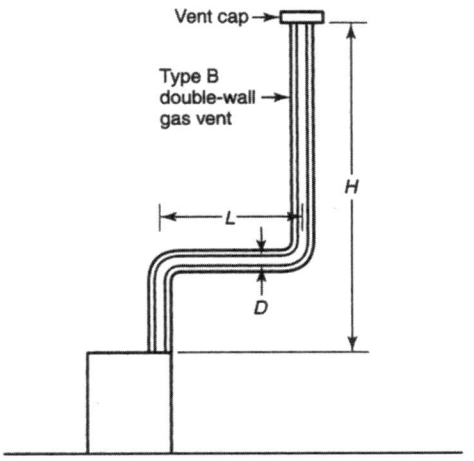

Table 510.1.2(1) is used where sizing a Type B double-wall gas vent connected directly to the appliance.

Note: The appliance is permitted to be either Category I draft hood-equipped or fan-assisted type.

FIGURE G 101.2(1)
TYPE B DOUBLE-WALL VENT SYSTEM SERVING A SINGLE APPLIANCE WITH A TYPE B DOUBLE-WALL VENT

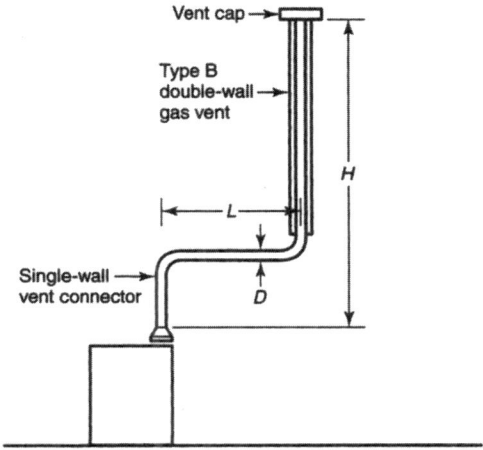

Table 510.1.2(2) is used where sizing a single-wall metal vent connector attached to a Type B double-wall gas vent.

Note: The appliance is permitted to be either Category I draft hood-equipped or fan-assisted type.

FIGURE G 101.2(2)
TYPE B DOUBLE-WALL VENT SYSTEM SERVING A SINGLE APPLIANCE WITH A SINGLE-WALL METAL VENT CONNECTOR

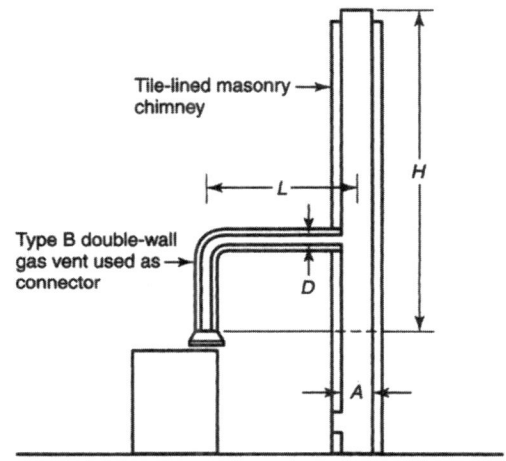

Table 510.1.2(3) is used where sizing a Type B double-wall gas vent connector attached to a tile-lined masonry chimney.

Notes:

1. *A* is the equivalent cross-sectional area of the tile liner.
2. The appliance is permitted to be either Category I draft-hood-equipped or fan-assisted type.

FIGURE G 101.2(3)
VENT SYSTEM SERVING A SINGLE APPLIANCE WITH A MASONRY CHIMNEY AND A TYPE B DOUBLE-WALL VENT CONNECTOR

APPENDIX G

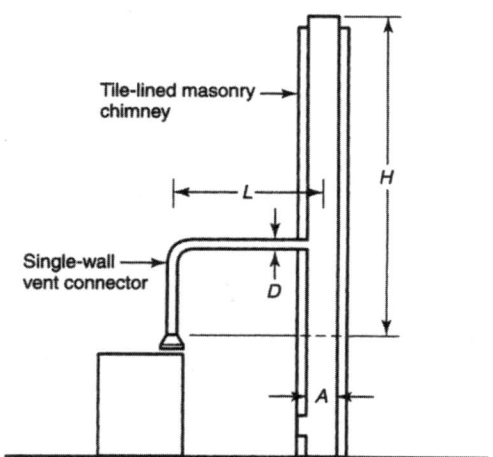

Table 510.1.2(4) is used where sizing a single-wall vent connector attached to a tile-lined masonry chimney.

Notes:
1. A is the equivalent cross-sectional area of the tile liner.
2. The appliance is permitted to be either Category I draft hood-equipped or fan-assisted type.

FIGURE G 101.2(4)
VENT SYSTEM SERVING A SINGLE APPLIANCE USING A MASONRY CHIMNEY AND A SINGLE-WALL METAL VENT CONNECTOR

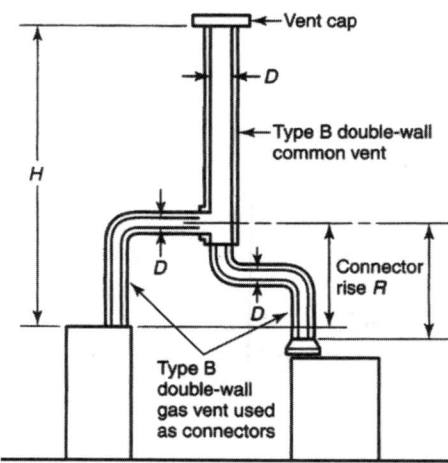

Table 510.2(1) is used where sizing Type B double-wall gas vent connectors attached to a Type B double-wall common vent.

Note: Each appliance is permitted to be either Category I draft hood-equipped or fan-assisted type.

FIGURE G 101.2(6)
VENT SYSTEM SERVING TWO OR MORE APPLIANCES WITH TYPE B DOUBLE-WALL VENT AND TYPE B DOUBLE-WALL VENT CONNECTORS

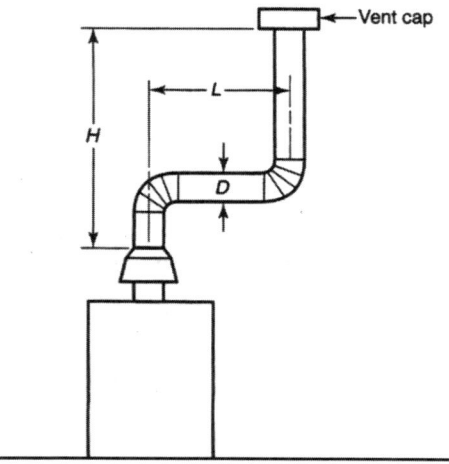

Asbestos cement Type B or single-wall metal vent serving a single draft-hood-equipped appliance. [See Table 510.1.2(5)]

FIGURE G 101.2(5)
ASBESTOS CEMENT TYPE B OR SINGLE-WALL METAL VENT SYSTEM SERVING A SINGLE DRAFT HOOD-EQUIPPED APPLIANCE

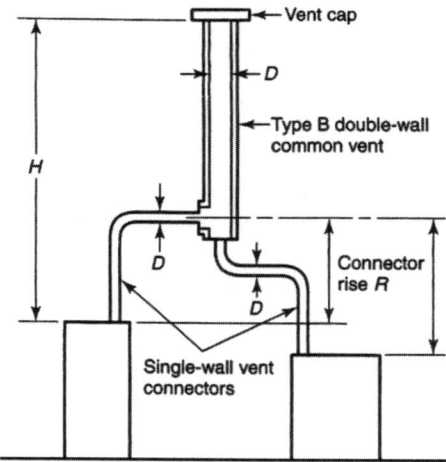

Table 510.2(2) is used where sizing single-wall vent connectors attached to a Type B double-wall common vent.

Note: Each appliance is permitted to be either Category I draft hood-equipped or fan-assisted type.

FIGURE G 101.2(7)
VENT SYSTEM SERVING TWO OR MORE APPLIANCES WITH TYPE B DOUBLE-WALL VENT AND SINGLE-WALL METAL VENT CONNECTORS

APPENDIX G

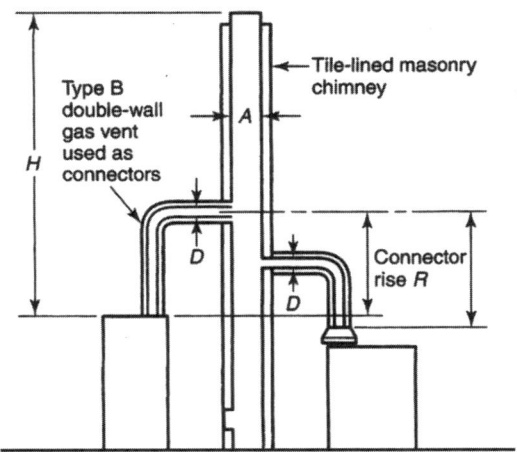

Table 510.2(3) is used where sizing Type B double-wall vent connectors attached to a tile-lined masonry chimney.

Notes:
1. *A* is the equivalent cross-sectional area of the tile liner.
2. The appliance is permitted to be either Category I draft hood- equipped or fan-assisted type.

FIGURE G 101.2(8)
MASONRY CHIMNEY SERVING TWO OR MORE APPLIANCES WITH TYPE B DOUBLE-WALL VENT CONNECTORS

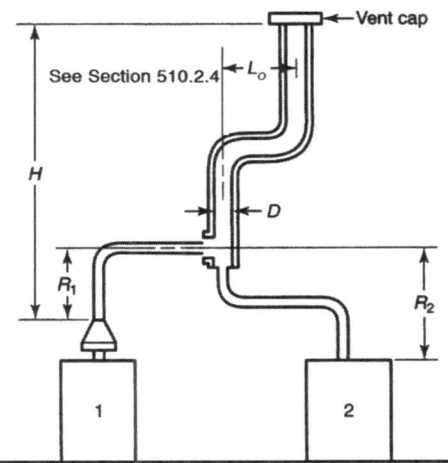

Asbestos cement Type B or single-wall metal pipe vent serving two or more draft-hood-equipped appliances. [See Table 510.2(5)]

FIGURE G 101.2(10)
ASBESTOS CEMENT TYPE B OR SINGLE-WALL METAL VENT SYSTEMS SERVING TWO OR MORE DRAFT HOOD-EQUIPPED APPLIANCES

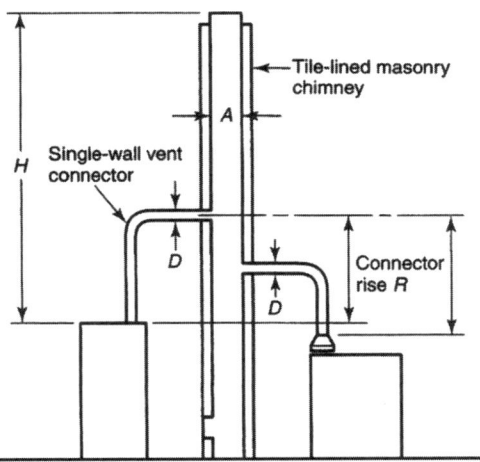

Table 510.2(4) is used where sizing single-wall metal vent connectors attached to a tile-lined masonry chimney.

Notes:
1. *A* is the equivalent cross-sectional area of the tile liner.
2. Each appliance is permitted to be either Category I draft-hood-equipped or fan-assisted type.

FIGURE G 101.2(9)
MASONRY CHIMNEY SERVING TWO OR MORE APPLIANCES WITH SINGLE-WALL METAL VENT CONNECTORS

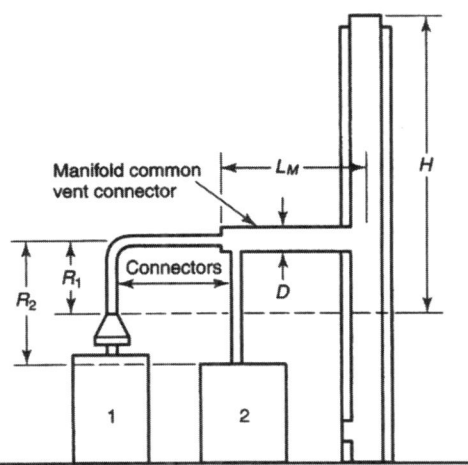

Example: Manifolded common vent connector LM can be no greater than 18 times the common vent connector manifold inside diameter; that is, a 4 inch (102 mm) inside diameter common vent connector manifold shall not exceed 72 inches (1829 mm) in length. [See Section 510.2.3]

Note: This is an illustration of a typical manifolded vent connector. Different appliance, vent connector, or common vent types are possible. [See Section 510.2]

FIGURE G 101.2(11)
USE OF MANIFOLDED COMMON VENT CONNECTORS

APPENDIX G

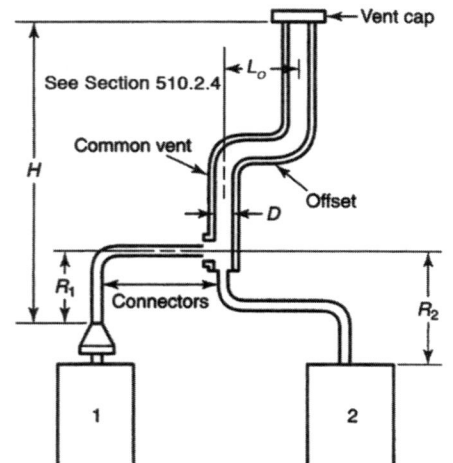

Example: Offset common vent

Note: This is an illustration of a typical offset vent. Different appliance, vent connector, or vent types are possible. [See Section 510.2.4 and Section 510.2.6]

**FIGURE G 101.2(12)
USE OF OFFSET COMMON VENT**

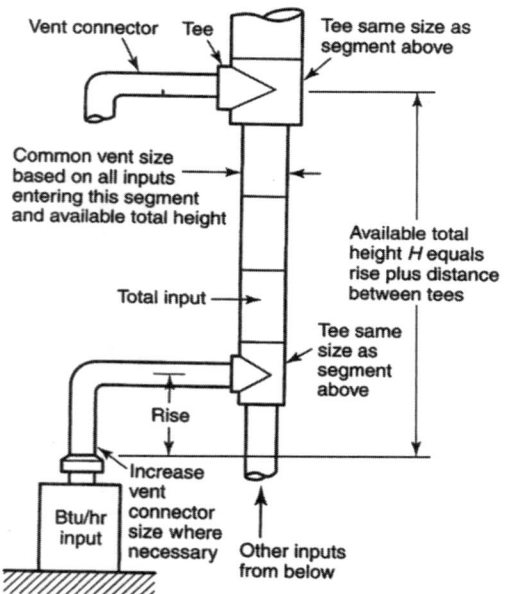

**FIGURE G 101.2(13)
MULTISTORY GAS VENT DESIGN PROCEDURE
FOR EACH SEGMENT OF SYSTEM**

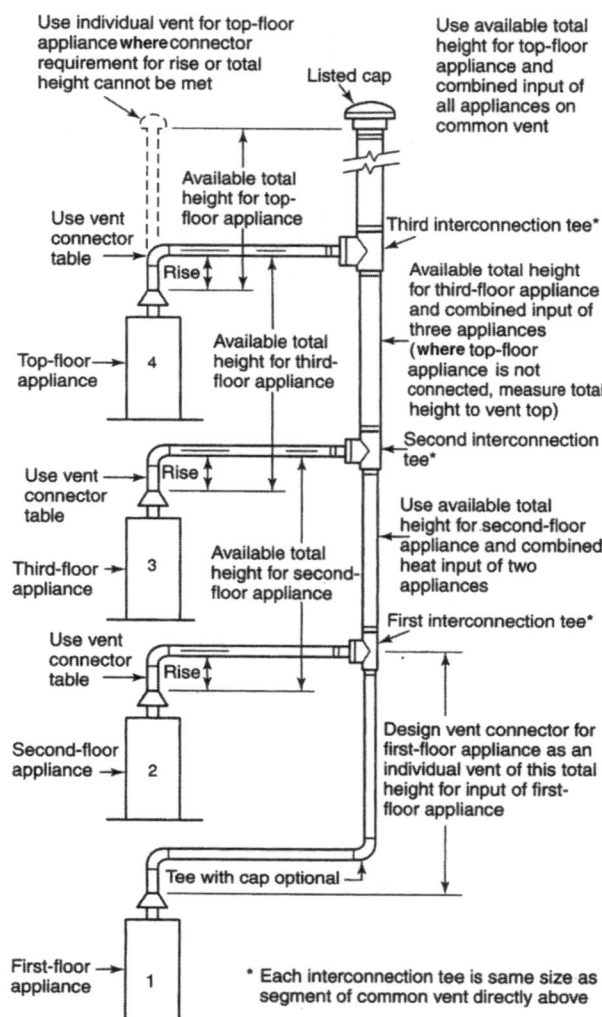

**FIGURE G 101.2(14)
PRINCIPLES OF DESIGN OF MULTISTORY VENTS USING
VENT CONNECTOR AND COMMON VENT DESIGN TABLES**

G 101.3 Example 1: Single Draft Hood-Equipped Appliance. An installer has a 120 000 British thermal units per hour (Btu/h) (35 kW) input appliance with a 5 inch (127 mm) diameter draft hood outlet that needs to be vented into a 10 foot (3048 mm) high Type B vent system. What size vent shall be used assuming: (1) a 5 foot (1524 mm) lateral single-wall metal vent connector is used with two 90 degree (1.57 rad) elbows or (2) a 5 foot (1524 mm) lateral single-wall metal vent connector is used with three 90 degree (1.57 rad) elbows in the vent system? (See **Figure G 101.3**)

Solution:

Table 510.1.2(2) shall be used to solve this problem because single-wall metal vent connectors are being used with a Type B vent, as follows:

(1) Read down the first column in Table 510.1.2(2) until the row associated with a 10 foot (3048 mm) height and 5 foot (1524 mm) lateral is found. Read across this row until a vent capacity exceeding 120 000 Btu/h (35 kW) is

located in the shaded columns labeled NAT Max for draft-hood-equipped appliances. In this case, a 5 inch (127 mm) diameter vent has a capacity of 122 000 Btu/h (35.7 kW) and shall be permitted to be used for this application.

(2) Where three 90 degree (1.57 rad) elbows are used in the vent system, the maximum vent capacity listed in the tables shall be reduced by 10 percent. This implies that the 5 inch (127 mm) diameter vent has an adjusted capacity of 110 000 Btu/h (32 kW). In this case, the vent system shall be increased to 6 inches (152 mm) in diameter. See the following calculations:

122 000 Btu/h (35.7 kW) x 0.90 = 110 000 Btu/h (32 kW) for 5 inch (127 mm) vent

From Table 510.1.2(2), select 6 inch (152 mm) vent.

186 000 Btu/h (54.5 kW) x 0.90 = 167 000 Btu/h (49 kW)

This figure is exceeding the required 120 000 Btu/h (35 kW). Therefore, use a 6 inch (152 mm) vent and connector where three elbows are used.

single-wall metal connector has a recommended maximum vent capacity of 144 000 Btu/h (42 kW). The 80 000 Btu/h (23.4 kW) fan-assisted appliance is outside this range, so the conclusion is that a single-wall metal connector shall not be used to vent the appliance using a 10 foot (3048 mm) of lateral for the connector. However, if the 80,000 Btu/hr (23.4 kW) input appliance is moved within 5 feet (1524 mm) of the vertical vent, a 4 inch (102 mm) single-wall metal connector shall be used to vent the appliance. Table 510.1.2(2) shows the acceptable range of vent capacities for a 4 inch (102 mm) vent with 5 feet (1524 mm) of lateral to be between 72 000 Btu/h (21.1 kW) and 157 000 Btu/h (46 kW).

Where the appliance cannot be moved closer to the vertical vent, then a Type B vent shall be used as the connector material. In this case, Table 510.1.2(1) shows that, for a 30 foot (9144 mm) high vent with 10 feet (3048 mm) of lateral, the acceptable range of vent capacities for a 4 inch (102 mm) diameter vent attached to a fan-assisted appliance is between 37 000 Btu/h (10.8 kW) and 150 000 Btu/h (44 kW).

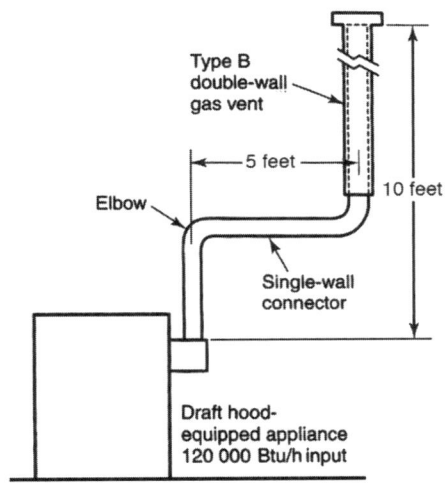

For SI units: 1 foot = 304.8 mm, 1000 British thermal units per hour = 0.293 kW

FIGURE G 101.3
SINGLE DRAFT HOOD-EQUIPPED APPLIANCE – EXAMPLE 1

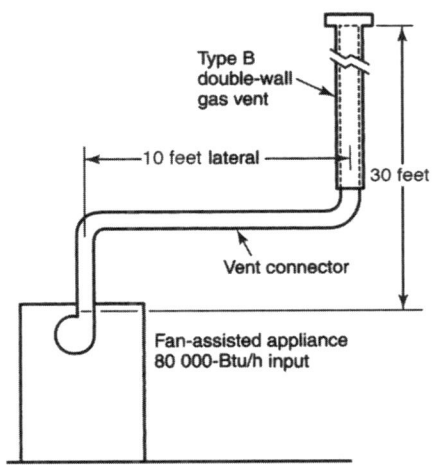

For SI units: 1 foot = 304.8 mm, 1000 British thermal units per hour = 0.293 kW

FIGURE G 101.4
SINGLE FAN-ASSISTED APPLIANCE – EXAMPLE 2

G 101.4 Example 2: Single Fan-Assisted Appliance.
An installer has an 80 000 Btu/h (23.4 kW) input fan-assisted appliance that shall be installed using 10 feet (3048 mm) of the lateral connector attached to a 30 foot high (9144 mm) Type B vent. Two 90 degree (1.57 rad) elbows are needed for the installation. Is a single-wall metal vent connector permitted to be used for this application? **(See Figure G 101.4)**

Solution:

Table 510.1.2(2) refers to the use of single-wall metal vent connectors with Type B vent. In the first column find the row associated with a 30 foot (9144 mm) height and a 10 foot (3048 mm) lateral. Read across this row, looking at the FAN Min and FAN Max columns, to find that a 3 inch (76 mm) diameter single-wall metal vent connector is not recommended. Moving to the next larger size single-wall connector [4 inch (102 mm)] we find that a 4 inch (102 mm) diameter

G 101.5 Example 3: Interpolating Between Table Values.
An installer has an 80 000 Btu/h (23.4 kW) input appliance with a 4 inch (102 mm) diameter draft hood outlet that needs to be vented into a 12 foot (3658 mm) high Type B vent. The vent connector has a 5 foot (1524 mm) lateral length and is also Type B. Can this appliance be vented using a 4 inch (102 mm) diameter vent?

Solution:

Table 510.1.2(1) is used in the case of an all Type B Vent system. However, Table 510.1.2(1) does not have an entry for a height of 12 feet (3658 mm), and interpolation must be used. Read down the 4 inch (102 mm) diameter NAT Max column to the row associated with a 10 foot (3048 mm) height and 5 foot (1524 mm) lateral to find the capacity value of 77 000 Btu/h (22.6 kW). Read further down to the 15 foot (4572 mm) height, 5 foot (1524 mm) lateral row to find the capacity value of 87 000 Btu/h (25.5 kW). The difference between the 15 foot (4572 mm) height capacity value and the 10 foot (3048 mm) height capacity value is 10 000 Btu/h (3 kW). The capacity for a vent system with a 12 foot (3658 mm) height is equal to the capacity for a 10 foot (3048 mm) height plus two-fifths of the difference between the 10 foot (3048 mm) and 15 foot (4572 mm) height values, or 77 000 Btu/h (22.6 kW) + ⅖ x 10 000 Btu/h (3 kW) = 81 000 Btu/h (23.7 kW). Therefore, a 4 inch (102 mm) diameter vent can be used in the installation.

G 101.6 Example 4: Common Venting Two Draft Hood-Equipped Appliances.
A 35 000 Btu/h (10.3 kW) water heater is to be common vented with a 150 000 Btu/h (44 kW) furnace, using a common vent with a total height of 30 feet (9144 mm). The connector rise is 2 feet (610 mm) for the water heater with a horizontal length of 4 feet (1219 mm). The connector rise for the furnace is 3 feet (914 mm) with a horizontal length of 8 feet (2438 mm). Assume single-wall metal connectors will be used with Type B vent. What size connectors and combined vent should be used in this installation? (See **Figure G 101.6**)

Solution:

Table 510.2(2) shall be used to size single-wall metal vent connectors attached to Type B vertical vents. In the vent connector capacity portion of Table 510.2(2), find the row associated with a 30 foot (9144 mm) vent height. For a 2 foot (610 mm) rise on the vent connector for the water heater, read the shaded columns for draft-hood-equipped appliances to find that a 3 inch (76 mm) diameter vent connector has a capacity of 37 000 Btu/h (10.8 kW). Therefore, a 3 inch (76 mm) single-wall metal vent connector shall be used for the water heater. For a draft-hood-equipped furnace with a 3 foot (914 mm) rise, read across the row to find that a 5 inch (127 mm) diameter vent connector has a maximum capacity of 120 000 Btu/h (35 kW) (which is too small for the furnace), and a 6 inch (152 mm) diameter vent connector has a maximum vent capacity of 172 000 Btu/h (50 kW). Therefore, a 6 inch (152 mm) diameter vent connector shall be used with the 150 000 Btu/h (44 kW) furnace. Since both vent connector, horizontal lengths are less than the maximum lengths listed in Section 510.2.1; the table values shall be used without adjustments.

In the common vent capacity portion of Table 510.2(2), find the row associated with a 30 foot (9144 mm) vent height and read over to the NAT + NAT portion of the 6 inch (152 mm) diameter column to find a maximum combined capacity of 257 000 Btu/h (75 kW). Since the two appliances total 185 000 Btu/h (54 kW), a 6 inch (152 mm) common vent shall be used.

G 101.7 Example 5(a): Common Venting a Draft Hood-Equipped Water Heater with a Fan-Assisted Furnace into a Type B Vent.
In this case, a 35 000 Btu/h (10.3 kW) input draft-hood-equipped water heater with a 4 inch (102 mm) diameter draft hood outlet, 2 feet (610 mm) of connector rise, and 4 feet (1219 mm) of horizontal length is to be common vented with a 100 000 Btu/h (29 kW) fan-assisted furnace with a 4 inch (102 mm) diameter flue collar, 3 feet (914 mm) of connector rise, and 6 feet (1829 mm) of horizontal length. The common vent consists of a 30 foot (9144 mm) height of Type B vent. What are the recommended vent diameters for each connector and the common vent? The installer would like to use a single-wall metal vent connector. (See **Figure G 101.7**)

Solution:

Water Heater Vent Connector Diameter. Since the water heater vent connector, horizontal length of 4 feet (1219 mm) is less than the maximum value listed in Table 510.2(2), the venting table values can be used without adjustment. Using the Vent Connector Capacity portion of Table 510.2(2), read

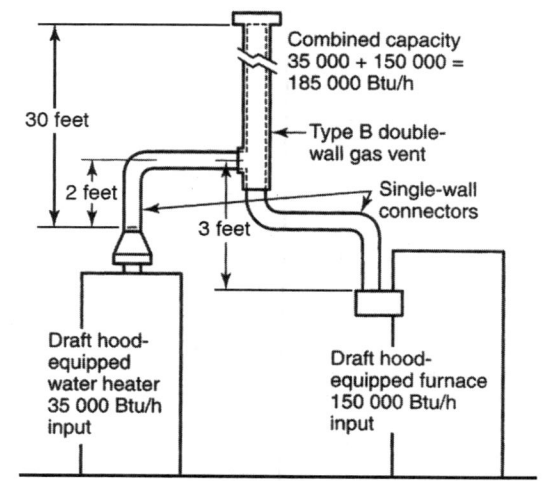

For SI units: 1 foot = 304.8 mm, 1000 British thermal units per hour = 0.293 kW

FIGURE G 101.6
COMMON VENTING TWO DRAFT HOOD-EQUIPPED APPLIANCES – EXAMPLE 4

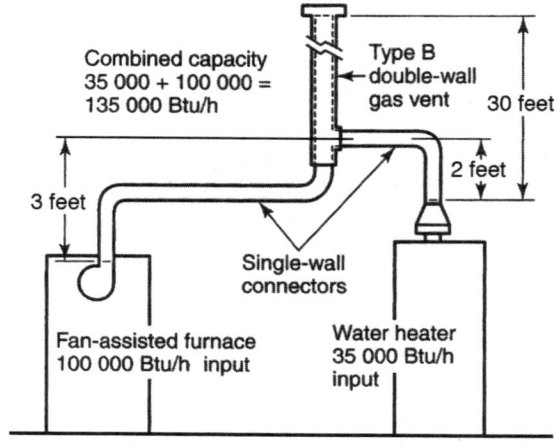

For SI units: 1 foot = 304.8 mm, 1000 British thermal units per hour = 0.293 kW

FIGURE G 101.7
COMMON VENTING A DRAFT HOOD-EQUIPPED WATER HEATER WITH A FAN-ASSISTED FURNACE INTO A TYPE B DOUBLE-WALL COMMON VENT – EXAMPLE 5(a)

down the Total Vent Height (H) column to 30 feet (9144 mm) and read across the 2 feet (610 mm) Connector Rise (R) row to the first Btu/h rating in the NAT Max column that is equal to or greater than the water heater input rating. The table shows that a 3 inch (76 mm) vent connector has a maximum input rating of 37 000 Btu/h (10.8 kW). Although this rating is greater than the water heater input rating, a 3 inch (76 mm) vent connector is prohibited by Section 510.2.18. A 4 inch (102 mm) vent connector has a maximum input rating of 67 000 Btu/h (19.6 kW) and is equal to the draft hood outlet diameter. A 4 inch (102 mm) vent connector is selected. Since the water heater is equipped with a draft hood, there are no minimum input rating restrictions.

Furnace Vent Connector Diameter. Using the Vent Connector Capacity portion of Table 510.2(2), read down the Total Vent Height (H) column to 30 feet (9144 mm) and across the 3 feet (914 mm) Connector Rise (R) row. Because the furnace has a fan-assisted combustion system, find the first FAN Max column with a Btu/h rating greater than the furnace input rating. The 4 inch (102 mm) vent connector has a maximum input rating of 119 000 Btu/h (34.9 kW) and a minimum input rating of 85 000 Btu/h (24.9 kW).

The 100 000 Btu/h (29 kW) furnace in this example falls within this range, so a 4 inch (102 mm) connector is adequate. Because of the furnace vent connector, the horizontal length of 6 feet (1829 mm) is less than the maximum value listed in Section 510.2.1; the venting table values can be used without adjustment. If the furnace had an input rating of 80 000 Btu/h (23.4 kW), a Type B vent connector would be needed to meet the minimum capacity limit.

Common Vent Diameter. The total input to the common vent is 135 000 Btu/h (40 kW). Using the Common Vent Capacity portion of Table 510.2(2), read down the Total Vent Height (H) column to 30 feet (9144 mm) and across this row to find the smallest vent diameter in the FAN + NAT column that has a Btu/h rating equal to or greater than 135 000 Btu/h (40 kW). The 4 inch (102 mm) common vent has a capacity of 132 000 Btu/h (39 kW) and the 5 inch (127 mm) common vent has a capacity of 202 000 Btu/h (59 kW). Therefore, the 5 inch (127 mm) common vent should be used in this example.

Summary: In this example, the installer can use a 4 inch (102 mm) diameter, single-wall metal vent connector for the water heater and a 4 inch (102 mm) diameter, single-wall metal vent connector for the furnace. The common vent should be a 5 inch (127 mm) diameter Type B vent.

G 101.8 Example 5(b): Common Venting into an Interior Masonry Chimney. In this case, the water heater and fan-assisted furnace of G 101.7 Example 5(a) are to be common-vented into a clay-tile-lined masonry chimney with a 30 foot (9144 mm) height. The chimney is not exposed to the outdoors below the roof line. The internal dimensions of the clay tile liner are nominally 8 inches (203 mm) by 12 inches (305 mm). Assuming the same vent connector heights, laterals, and materials found in Example 5(a), what are the recommended vent connector diameters, and is this an acceptable installation?

Solution:

Table 510.2(4) is used to size common venting installations involving single-wall connectors into masonry chimneys.

Water Heater Vent Connector Diameter. Using Table 510.2(4), Vent Connector Capacity, read down the Total Vent Height (H) column to 30 feet (9144 mm), and read across the 2 feet (610 mm) Connector Rise (R) row to the first Btu/h rating in the NAT Max column that is equal to or greater than the water heater input rating. The table shows that a 3 inch (76 mm) vent connector has a maximum input of only 31 000 Btu/h (9 kW), while a 4 inch (102 mm) vent connector has a maximum input of 57 000 Btu/h (16.7 kW). A 4 inch (102 mm) vent connector must therefore be used.

Furnace Vent Connector Diameter. Using the Vent Connector Capacity portion of Table 510.2(4), read down the Total Vent Height (H) column to 30 feet (9144 mm) and across the 3 feet (914 mm) Connector Rise (R) row. Because the furnace has a fan-assisted combustion system, find the first FAN Max column with a Btu/h rating greater than the furnace input rating. The 4 inch (102 mm) vent connector has a maximum input rating of 127 000 Btu/h (37 kW) and a minimum input rating of 95 000 Btu/h (27.8 kW). The 100 000 Btu/h (29 kW) furnace in this example falls within this range, so a 4 inch (102 mm) connector is adequate.

Masonry Chimney. From Table G 101.8, the equivalent area for a nominal liner size of 8 inches (203 mm) by 12 inches (305 mm) is 63.6 square inches (0.041 m^2). Using Table 510.2(4), Common Vent Capacity, read down the FAN + NAT column under the Minimum Internal Area of Chimney value of 63 to the row for 30 foot (9144 mm) height to find a capacity value of 739 000 Btu/h (217 kW). The combined input rating of the furnace and water heater, 135 000 Btu/h (40 kW), is less than the table value, so this is an acceptable installation.

Subsection 510.2.16 requires the common vent area to be no greater than seven times the smallest listed appliance categorized vent area, flue collar area, or draft hood outlet area. Both appliances in this installation have 4 inch (102 mm) diameter outlets. From Table G 101.8, the equivalent area for an inside diameter of 4 inches (102 mm) is 12.2 square inches (0.008 m^2). Seven times 12.2 equals 85.4, which is greater than 63.6, so this configuration is acceptable.

G 101.9 Example 5(c): Common Venting into an Exterior Masonry Chimney. In this case, the water heater and fan-assisted furnace of G 101.7 Example 5(a) and G 101.8 Example 5(b) are to be common-vented into an exterior masonry chimney. The chimney height, clay-tile-liner dimensions, and vent connector heights and laterals are the same as in G 101.8 Example 5(b). This system is being installed in Charlotte, North Carolina. Does this exterior masonry chimney need to be relined? If so, what corrugated metallic liner size is recommended? What vent connector diameters are recommended? [See Table G 101.8 and **Figure 510.1.10**]

Solution:

According to Section 510.2.20, Type B vent connectors are required to be used with exterior masonry chimneys. Use Table 510.2(8) and Table 510.2(9) to size FAN+NAT com-

APPENDIX G

TABLE G 101.8
MASONRY CHIMNEY LINER DIMENSIONS WITH CIRCULAR EQUIVALENTS*

NOMINAL LINER SIZE (inches)	INSIDE DIMENSIONS OF LINER (inches)	INSIDE DIAMETER OR EQUIVALENT DIAMETER (inches)	EQUIVALENT AREA (square inches)
4 x 8	2½ x 6½	4.0	12.2
		5.0	19.6
		6.0	28.3
		7.0	38.3
8 x 8	6¾ x 6¾	7.4	42.7
		8.0	50.3
8 x 12	6½ x 10½	9.0	63.6
		10.0	78.5
12 x 12	9¾ x 9¾	10.4	83.3
		11.0	95.0
12 x 16	9½ x 13½	11.8	107.5
		12.0	113.0
		14.0	153.9
16 x 16	13¼ x 13¼	14.5	162.9
		15.0	176.7
16 x 20	13 x 17	16.2	206.1
		18.0	254.4
20 x 20	16½ x 16¾	18.2	260.2
		20.0	314.1
20 x 24	16½ x 20½	20.1	314.2
		22.0	380.1
24 x 24	20¼ x 20¼	22.1	380.1
		24.0	452.3
24 x 28	20¼ x 24¼	24.1	456.2
28 x 28	24¼ x 24¼	26.4	543.3
		27.0	572.5
30 x 30	25½ x 25½	27.9	607.0
		30.0	706.8
30 x 36	25½ x 31½	30.9	749.9
		33.0	855.3
36 x 36	31½ x 31½	34.4	929.4
		36.0	1017.9

For SI units: 1 inch = 25.4 mm, 1 square inch = 0.000645 m^2

* Where liner sizes differ dimensionally from those shown in this table, equivalent diameters can be determined from published tables for square and rectangular ducts of equivalent carrying capacity or by other engineering methods.

mon venting installations involving Type-B double-wall connectors into exterior masonry chimneys.

The local 99 percent winter design temperature needed to use Table 510.2(8) and Table 510.2(9) can be found in ASHRAE Handbook – Fundamentals. For Charlotte, North Carolina, this design temperature is 19°F (-7.2°C).

Chimney Liner Requirement. As in Example 5(b), use the 63 square inches (0.04 m^2) Internal Area column for this size clay tile liner. Read down the 63 square inches (0.04 m^2) column of Table 510.2(8) to the 30 foot (9144 mm) height row to find that the combined appliance maximum input is 747 000 Btu/h (218.9 kW). The combined input rating of the appliances in this installation, 135 000 Btu/h (40 kW), is less than the maximum value, so this criterion is satisfied. Table 510.2(9), at a 19°F (-7.2°C) design temperature, and at the same vent height and the internal area used earlier, shows that the minimum allowable input rating of a space-heating appliance is 470 000 Btu/h (137.7 kW). The furnace input rating of 100 000 Btu/h (29 kW) is less than this minimum value. So this criterion is not satisfied, and an alternative venting design needs to be used, such as a Type B vent shown in Example 5(a) or a listed chimney liner system is shown in the remainder of the example.

According to Section 510.2.19, Table 510.2(1) or Table 510.2(2) is used for sizing corrugated metallic liners in masonry chimneys, with the maximum common vent capacities reduced by 20 percent. This example will be continued assuming Type B vent connectors.

Water Heater Vent Connector Diameter. Using Table 510.2(1) Vent Connector Capacity, read down the total Vent Height (H) column to 30 feet (9144 mm), and read across the 2 feet (610 mm) Connector Rise (R) row to the first Btu/hour greater than the water heater input rating. The table shows that a 3 inch (76 mm) vent connector has a maximum capacity of 39 000 Btu/h (11.4 kW). Although this rating is greater than the water heater input rating, a 3 inch (76 mm) vent connector is prohibited by Section 510.2.20. A 4 inch (102 mm) vent connector has a maximum input rating of 70 000 Btu/h (20.5 kW) and is equal to the draft hood outlet diameter. A 4 inch (102 mm) vent connector is selected.

Furnace Vent Connector Diameter. Using Table 510.2(1), Vent Connector Capacity, read down the total Vent Height (H) column to 30 feet (9144 mm), and read across the 3 feet (914 mm) Connector Rise (R) row to the first Btu/h rating in the FAN MAX column that is equal to or greater than the furnace input rating. The 100 000 Btu/h (29 kW) furnace in this example falls within this range, so a 4 inch (102 mm) connector is adequate.

Chimney Liner Diameter. The total input to the common vent is 135 000 Btu/h (40 kW). Using the Common Vent Capacity portion of Table 510.2(1), read down the total Vent Height (H) column to 30 feet (9144 mm) and across this row to find the smallest vent diameter in the FAN + NAT column that has a Btu/h rating greater than 135 000 Btu/h (40 kW). The 4 inch (102 mm) common vent has a capacity of 138 000 Btu/h (40.4 kW). Reducing the maximum capacity by 20 percent results in a maximum capacity for a 4 inch (102 mm) corrugated liner of 110 000 Btu/h (32 kW), less than the total input of 135 000 Btu/h (40 kW). So a larger liner is needed. The 5 inch (127 mm) common vent capacity listed in Table 510.2(1) is 210 000 Btu/h (62 kW), and after reducing by 20 percent is 168 000 Btu/h (49.2 kW). Therefore, a 5 inch (127 mm) corrugated metal liner should be used in this example.

Single Wall Connectors. Once it has been established that relining the chimney is necessary, Type B double-wall vent connectors are not specifically required. This example could be redone using Table 510.2(2) for single-wall vent connectors. For this case, the vent connector and liner diameters would be the same as found for Type B double-wall connectors.

🔧 **The following are sizing examples from the Uniform Mechanical Code Illustrated Training Manual.**

Examples Using Single-Appliance Venting Tables.

Procedure for Sizing Multistory Vents

Following is an example of the use of the tables in Section 510.0 to size a multistory vent. Assume that **Figure G 101.2(14)** represents a four-story apartment building that has a listed fan-assisted combustion furnace installed on each floor. Each furnace has a 4-inch (100-mm) flue collar and an input of 80,000 Btu/hr. (23 KW). All the furnaces are installed outside the conditioned space (i.e., isolated combustion) and are to be vented into the common vent, which is located 5 feet (1.5 m) from the furnaces. For the purpose of this calculation, the overall system is divided into smaller, simple vent systems for each level, as shown in **Figure G 101.2(13)**.

The structure is such that the vent connector rise is to be 2 feet (0.6 m) and the available total height for each level is 10 feet (3.0 m), except for the top floor, which is 6 feet (1.8 m). The vent connector for the first floor, or lowest furnace to the common vent, is considered to be an individual vent terminating at the first tee or interconnection, and this vent is sized in accordance with Table 510.1.2(2). Every other vent connector is sized in accordance with the "Vent Connector Capacity" section of Table 510.2(1). Each section of the common vent is sized in accordance with the "Common Vent Capacity" section of Table 510.2(1) to accommodate the accumulated total input of all appliances discharging into it, but sections are never smaller than the largest section below them.

Procedure for Sizing Connectors

First-Floor Connector (Furnace 1). The input is 80,000 Btu/hr., and the total height is 10 feet. This section of the vent is treated as a single-appliance vent with 5 feet of vent connector. Using Table 510.1.2(1), read across the line for a 10-foot height (H) and lateral (L) of 5 feet. Look under the FAN columns for a range that includes 80,000. The 4-inch column has a range of 32,000 Btu/hr. to 113,000 Btu/hr., and the 5-inch column has a range of 41,000 Btu/hr. to 187,000 Btu/hr. Both sizes are acceptable, and a 4-inch size is selected because it is smaller and less expensive.

Section 510.2.1 limits a 4-inch vent connector to a maximum length of 6 feet (1.5-foot length for each inch of diameter), which is not exceeded in this example.

Second- and Third-Floor Connectors (Furnaces 2 and 3). The input is 80,000 Btu/hr., the connector rise is 2 feet and the vent height is 10 feet. Using the "Vent Connector Capacity" section of Table 510.2(1), read across the 10-foot height, 2-foot rise row in the FAN columns. It shows that a 4-inch connector has a range of 36,000 Btu/hr. to 86,000 Btu/hr., a 5-inch connector has a range of 51,000 Btu/hr. to 136,000 Btu/hr., and a 6-inch connector has a range of 67,000 Btu/hr. to 206,000 Btu/hr. A connector with a diameter of 4 inches, 5 inches or 6 inches will work, but a 4-inch connector is selected.

Fourth-Floor Connector (Furnace 4). The input is 80,000 Btu/hr. the connector rise is 2 feet and the vent height is 6 feet. Using the "Vent Connector Capacity" section of Table 510.2(1), read across the 6-foot height, 2-foot rise row. It shows that a 5-inch connector has a range of 48,000 Btu/hr. to 121,000 Btu/hr. and a 6-inch connector has a range of 60,000 Btu/hr. to 183,000 Btu/hr. Either a 5-inch or a 6-inch connector will work, but a 5-inch connector is selected because it is smaller and less expensive.

Section 510.2.1 limits a 5-inch vent connector to a maximum length of 7½ feet (1.5 foot length for each inch of diameter), which is not exceeded by the connector size selected.

Procedure for Sizing Common Vent

Common Vent for Furnaces 1 and 2. The input is 160,000 Btu/hr. and the vent height is 10 feet. Using the "Common Vent Capacity" section of Table 510.2(1), read across the 10-foot line in the FAN + FAN columns only. It shows that a 5-inch vent has a capacity of 169,000 Btu/hr, which exceeds the 160,000 Btu/hr. input. A 5-inch vent is used.

Common Vent for Third-Floor Furnace (Furnace 3). The input is 240,000 Btu/hr. and the height is 10 feet. Using the "Common Vent Capacity" section of Table 510.2(1), read across the 10-foot line in the FAN + FAN columns. It shows that a 6-inch common vent has a capacity of 243,000 Btu/hr. A 6-inch vent is used.

Common Vent for Fourth-Floor Furnace (Furnace 4). The input is 320,000 Btu/hr. and the height is 6 feet. Using the FAN + FAN columns in the "Common Vent Capacity" section of Table 510.2(1), read across the 6-foot line. It shows that an 8-inch vent has a capacity of 404,000 Btu/hr. An 8-inch vent is selected.

The foregoing sizing procedures are summarized in **Commentary Table**, Common Vent Serving Appliances on Four Floors and Common Vent Serving Appliances on Three Floors: Independent Vent Serving Appliance on the Fourth Floor.

Check for Excessive Vent Area

The vent connector and common vent have been sized using Table 510.2(1) and Section 510.2. Refer to Section 510.2.16, which states:

Where two or more appliances are connected to a vertical vent or chimney, the flow area of the largest section of vertical vent or chimney shall not exceed seven times the smallest listed appliance categorized vent areas, flue collar area, or draft hood outlet area unless designed in accordance with approved engineering methods.

In the following example, the smallest vent connector diameter is 4 inches, and the largest common vent diameter is 8 inches.

For a 4-inch diameter connector:

Area = πr^2 = 3.14(2)^2 = 12.56 in^2

Where: π = 3.14

r = 2 (½ the 4-inch diameter)

For an 8-inch diameter common vent:

Area = πr^2 = 3.14(4)^2 = 50.25 in^2

Where: π = 3.14

r = 4 (½ the 8-inch diameter)

Section 510.2.16 limits the common vent to seven times the smallest connector area. The ratio of the area of the smallest connector to the common vent area in this example must be checked to be sure it is less than 7:

Ratio = area (largest common vent section)/area (smallest connector)

Ratio = 50.25/12.56 = 4

Because 4 is less than 7, the sizing of the common vent is acceptable.

COMMENTARY TABLE

COMMENTARY TABLE, Common Vent Serving Appliances on Four Floors

Furnace	Total Input to Common Vent (Btu/hr)	Available Total Height (ft)	Vent Connector Size (in.)	Common Vent Size (in.)
1	80,000	10	4	4
2	160,000	10	4	5
3	240,000	10	4	6
4	320,000	6	5	8

COMMENTARY TABLE, Common Vent Serving Appliances on Three Floors; Independent Vent Serving Appliance on the Fourth Floor

Furnace	Total Input to Common Vent (Btu/hr)	Available Total Height (ft)	Vent Connector Size (in.)	Common Vent Size (in.)
1	80,000	10	4	4
2	160,000	10	4	5
3	240,000	10	4	6
4	80,000	6	4	None

APPENDIX H
PRIVATE SEWAGE DISPOSAL SYSTEMS

H 101.0 General.

H 101.1 Applicability. This appendix provides general guidelines for the materials, design, and installation of private sewage disposal systems.

H 101.2 General Requirements. Where permitted by Section 713.0, the building sewer shall be permitted to be connected to a private sewage disposal system in accordance with the provisions of this appendix. The type of system shall be determined on the basis of location, soil porosity, and groundwater level, and shall be designed to receive all sewage from the property. The system, except as otherwise approved, shall consist of a septic tank with effluent discharging into a subsurface disposal field, into one or more seepage pits, or into a combination of subsurface disposal field and seepage pits. The Authority Having Jurisdiction shall be permitted to grant exceptions to the provisions of this appendix for permitted structures that have been destroyed due to fire or natural disaster and that cannot be reconstructed in compliance with these provisions provided that such exceptions are the minimum necessary.

Private sewage disposal systems must also comply with local conditions. The local Authority Having Jurisdiction (AHJ) may require more restrictive requirements than specified in this appendix. The type of system permitted will depend upon the topography of the site, geographical location, soil porosity, groundwater elevations and the system location with relation to streams, wells, property lines, water lines and other factors. Some jurisdictions also specify the minimum lot size required to install a private sewage disposal system.

Each system shall be designed to receive all the domestic sewage from the property. Rainwater must be excluded from any private sewage disposal system. More than one system may be used or be required when there is an abnormal amount of sewage, buildings with large floor area, separate buildings or poor soil conditions.

Generally, the system will consist of a septic tank with the effluent discharging into a subsurface disposal field consisting of either trenches or beds, one or more seepage pits or a combination of a subsurface disposal field and seepage pits. See **Figures H 101.2a, H 101.2b, H 101.2c** and **H 101.2d** for illustrations of systems covered in this appendix.

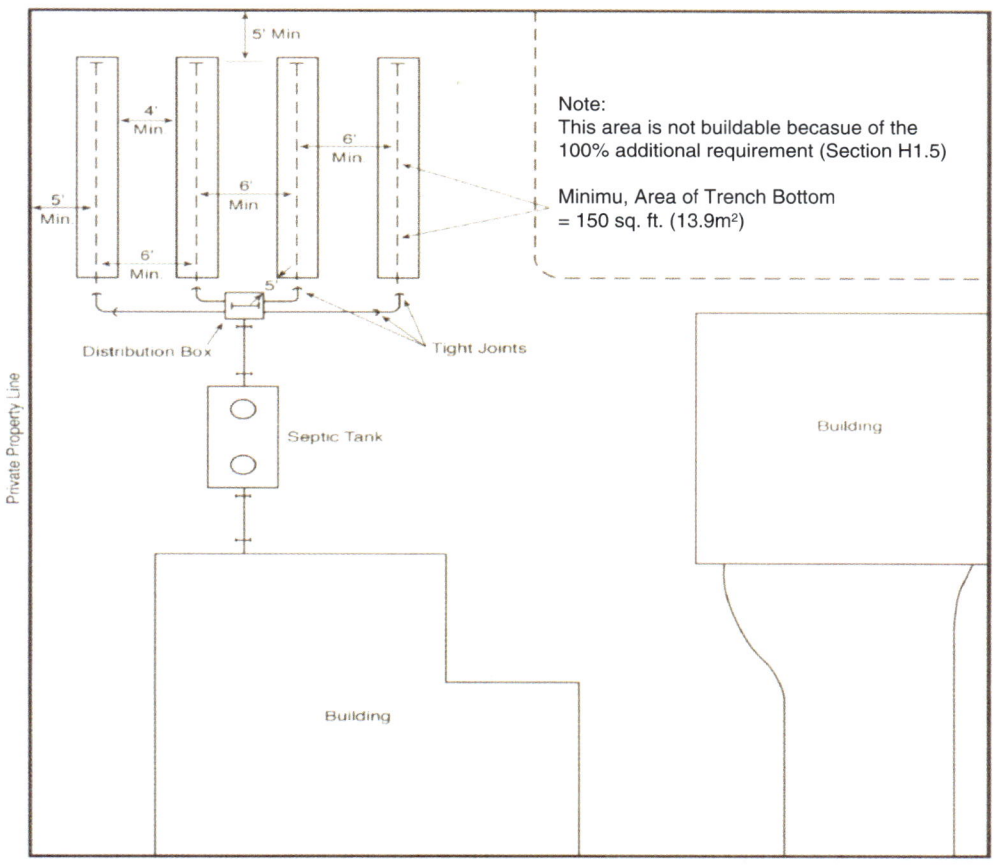

FIGURE H 101.2A
PRIVATE SEWAGE DISPOSAL SYSTEM UTILIZING LEACHING LINES FOR DISPOSAL

APPENDIX H

H 101.3 Quantity and Quality. Where the quantity or quality of the sewage is such that the above system cannot be expected to function satisfactorily for commercial, agricultural, and industrial plumbing systems; for installations where appreciable amounts of industrial or indigestible wastes are produced; for occupancies producing abnormal quantities of sewage or liquid waste; or where grease interceptors are required by other parts of this code, the method of sewage treatment and disposal shall be first approved by the Authority Having Jurisdiction. Special sewage disposal systems for minor, limited, or temporary uses shall be first approved by the Authority Having Jurisdiction.

There are other methods of private sewage disposal such as mound, evapotranspiration, electro-osmosis and sewage lagoons. These systems may better serve commercial or industrial installations and are designed by civil or sanitary engineers; therefore, these systems will not be discussed in this appendix.

H 101.4 Septic Tank and Disposal Field Systems. Disposal systems shall be designed to utilize the most porous or absorptive portions of the soil formation. Where the groundwater level extends to within 12 feet (3658 mm) or less of the ground surface or where the upper soil is porous and the underlying stratum is rock or impervious soil, a septic tank and disposal field system shall be installed.

The limitation contained in this subsection allowing only disposal field systems in shallow groundwater areas is due to the fact that the seepage pit is a deep underground installation by nature and not suited to a shallow groundwater area where it could lead to contamination of the groundwater.

H 101.5 Flood Hazard Areas. Disposal systems shall be located outside of flood hazard areas.

Exception: Where suitable sites outside of flood hazard areas are not available, disposal systems shall be permitted to be located in flood hazard areas on sites where the

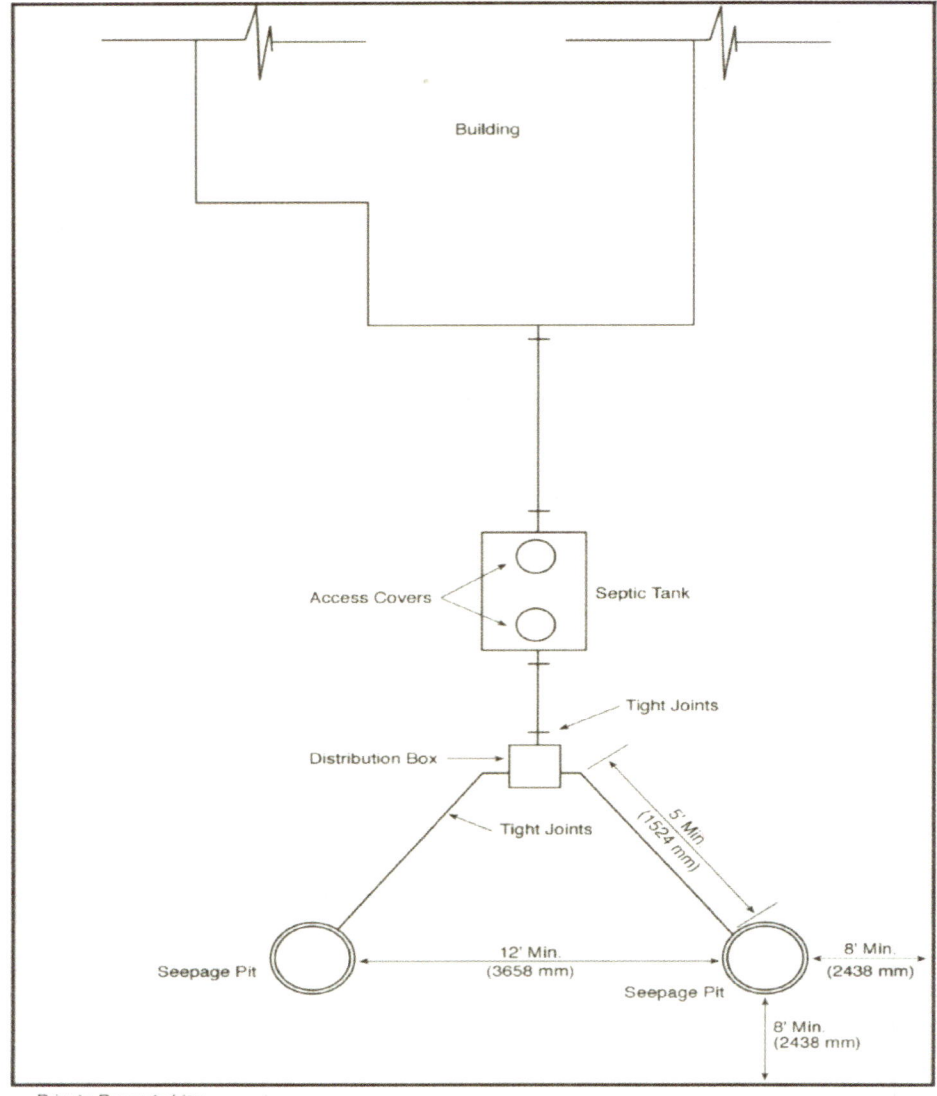

FIGURE H 101.2B
PRIVATE SEWAGE DISPOSAL SYSTEM UTILIZING SEEPAGE PITS FOR DISPOSAL

APPENDIX H

effects of inundation under conditions of the design flood are minimized.

🔧 See Section 301.4, Flood Hazard Areas.

H 101.6 Design. Private sewage disposal systems shall be so designed that additional seepage pits or subsurface drain fields, equivalent to not less than 100 percent of the required original system, shall be permitted to be installed where the original system cannot absorb all the sewage. No division of the lot or erection of structures on the lot shall be made where such division or structure impairs the usefulness of the 100 percent expansion area.

🔧 The requirement for 100 percent of the area of the system to be set aside is included in the event that the original system cannot absorb all the sewage or if the system fails and the site cannot be reused. A new system can then be installed on the property without disturbing any building on the premises (see **Figure H 101.2a**).

H 101.7 Capacity. No property shall be improved in excess of its capacity to properly absorb sewage effluent by the means provided in this code.

Exception: The Authority Having Jurisdiction shall be permitted to, at its discretion, approve an alternate system.

H 101.8 Location. No private sewage disposal system, or part thereof, shall be located in any lot other than the lot that is the site of the building or structure served by such private sewage disposal system, nor shall any private sewage disposal system or part thereof be located at any point having less than the minimum distances indicated in Table H 101.8.

Nothing contained in this code shall be construed to prohibit the use of all or part of an abutting lot to provide additional space for a private sewage disposal system or part thereof where proper cause, transfer of ownership, or change of boundary not in violation of other requirements has been first established to the satisfaction of the Authority Having Jurisdiction. The instrument recording such action shall constitute an agreement with the Authority Having Jurisdiction, which shall clearly state and show that the areas so joined or used shall be maintained as a unit during the time they are so used. Such agreement shall be recorded in the office of the County Recorder as part of the condi-

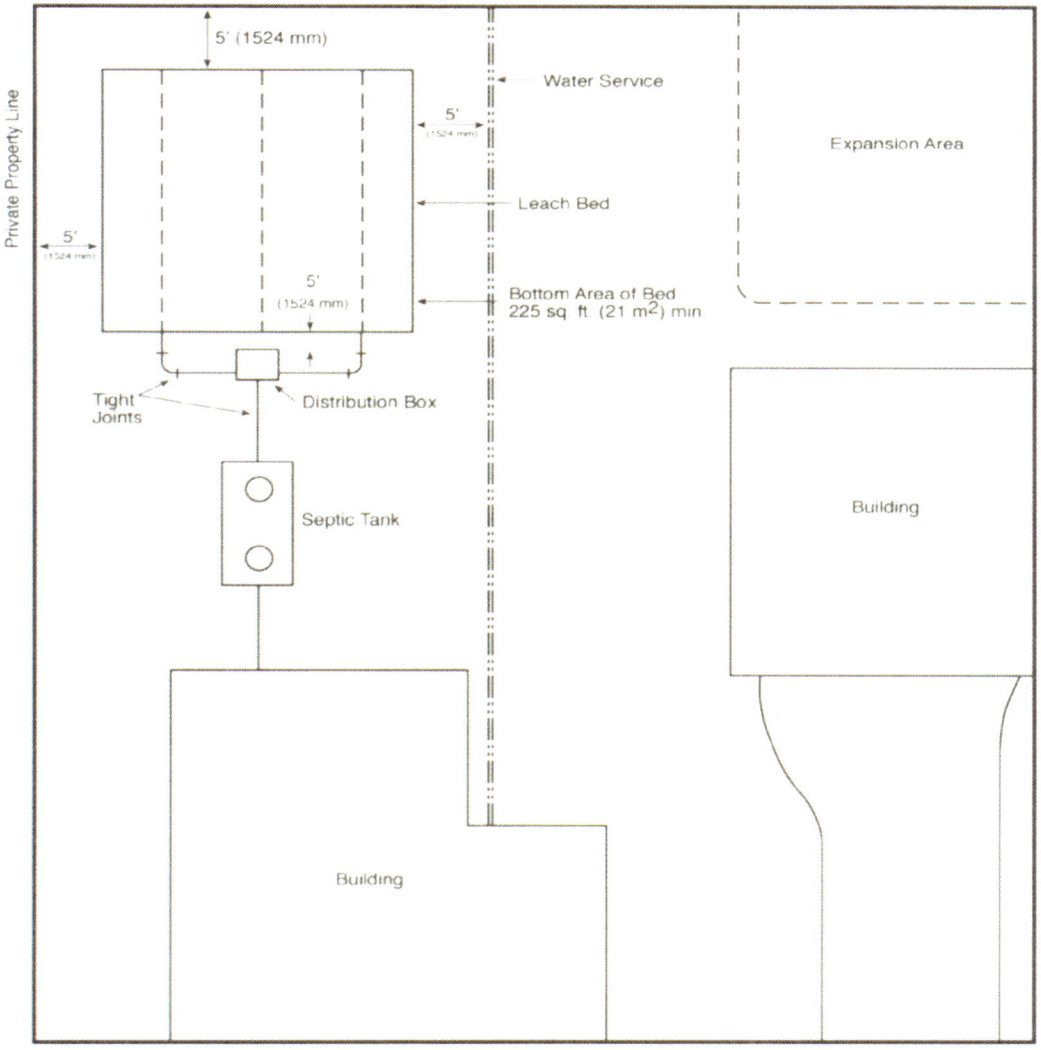

FIGURE H 101.2C
PRIVATE SEWAGE DISPOSAL SYSTEM UTILIZING LEACHING BEDS FOR DISPOSAL

APPENDIX H

tions of ownership of said properties and shall be binding on heirs, successors, and assigns to such properties. A copy of the instrument recording such proceedings shall be filed with the Authority Having Jurisdiction.

H 101.9 Building Permit. Where there is insufficient lot area or improper soil conditions for sewage disposal for the building or land use proposed, and the Authority Having Jurisdiction so finds, no building permit shall be issued and no private sewage disposal shall be permitted. Where space or soil conditions are critical, no building permit shall be issued until engineering data and test reports satisfactory to the Authority Having Jurisdiction have been submitted and approved.

H 101.10 Additional Requirements. Nothing contained in this appendix shall be construed to prevent the Authority Having Jurisdiction from requiring compliance with additional requirements than those contained herein, where such additional requirements are essential to maintain a safe and sanitary condition.

There are currently 20 different types or brands of aerobic treatment unit systems available on the market today. They are similar in function to the septic tank in that they break down organic matter, achieve decomposition of organic solids and reduce the concentration of pathogens in the wastewater and then distribute the effluent by ground distribution. Some may accomplish these functions more efficiently and in less area than the septic tank; however, they are more complicated than the septic tank system alone and require an air bubbler to facilitate the aerobic action. This could lead to higher maintenance costs than a properly installed septic tank system. In order for the AHJ to accept such a system, it must meet the septic tank effluent characteristics to be approved. See **Figure H 101.10a** for an illustration of one such system.

Septic Tanks: The septic tank is the centerpiece of the private sewage disposal system. Understanding the functions of the septic tank are of utmost importance when designing and installing a proper private sewage disposal system. If the system is not designed and installed properly, it could lead to system failure and a very serious health hazard as quickly than an incorrectly installed sewer system.

The domestic sewage received by the building sewer is

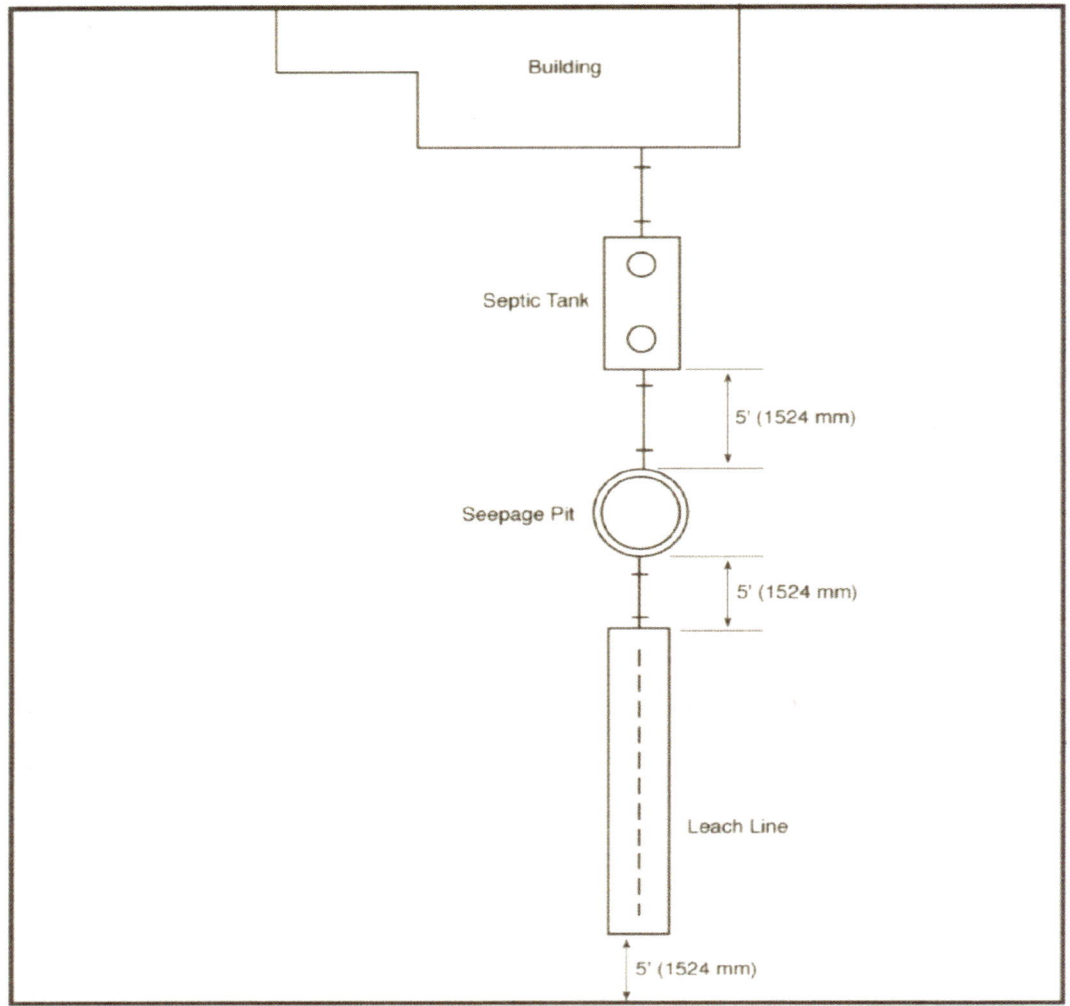

FIGURE H 101.2D
PRIVATE SEWAGE DISPOSAL SYSTEM UTILIZING A COMBINED SYSTEM FOR DISPOSAL

composed primarily of water and waste matter. There are two types of waste matter: suspended solids like coffee grounds and dissolved solids like sugar in coffee. Domestic sewage will quickly clog all but the most porous gravel soil formations unless treated properly. That treatment in a private sewage disposal system occurs in the septic tank. The septic tank has three functions and consequently three levels of effluent in the tank. The three functions of the septic tank are:

1. *To remove as many solids as possible from the sewage.* Once the solids are removed, the effluent from the tank will ultimately be distributed over an adequate area of land where it can be absorbed by the soil without plugging the piping or the soil itself.

2. *To treat the remaining solids in the tank with anaerobic bacteria (only active in the absence of oxygen).* Given enough time, these bacteria decompose the solids and, eventually, make them stable. This decomposition or treatment of sewage under anaerobic conditions is called "purification" or "septic," hence the name "septic tank."

3. *To store the remaining solids until decomposition.* A rock is chemically stable and cannot be dissolved by decomposition (and, by the way, should not be in the septic tank). An orange, for example, is unstable because it can be decomposed by bacteria, fungus, etc., until it is stable (dissolved) and is no longer subject to bacterial action. This process of bacterial breakdown is called "digestion."

There are three layers in the tank corresponding to the actions described above (see **Figure H 101.10b**):

1. The sludge at the bottom of the tank (heavier solids that have settled);
2. The scum at the top of the tank (fats, greases and light solids that have risen); and
3. In the middle of the tank, relatively clear sewage that still contains sugars, detergents and other dissolved solids.

Anaerobic bacteria work in all three layers, reducing the size and weight of the solids by turning a large part of them into liquids and gases that rise to the top or are suspended in the middle layer. Thus, a pound of solids entering the tank may be only a fraction of that weight three months later.

Every time raw sewage comes into the tank, it forces an equal amount of treated sewage out of the tank. Septic tank tees or baffles prevent the sludge and top scum layer from moving out with the treated sewage. The fact that the tees do not reach to the bottom of the tank also prevents any solids from leaving the tank so that only the "relatively clear sewage" in the middle layer is the only effluent that leaves the tank.

TABLE H 101.8
LOCATION OF SEWAGE DISPOSAL SYSTEM

MINIMUM HORIZONTAL DISTANCE	BUILDING SEWER	SEPTIC TANK	DISPOSAL FIELD	SEEPAGE PIT OR CESSPOOL
Building or structures[1]	2 feet	5 feet	8 feet	8 feet
Property line adjoining private property	Clear[2]	5 feet	5 feet	8 feet
Water supply wells	50 feet[3]	50 feet	100 feet	150 feet
Streams and other bodies of water	50 feet	50 feet	100 feet[7]	150 feet[7]
Trees	–	10 feet	–	10 feet
Seepage pits or cesspools[8]	–	5 feet	5 feet	12 feet
Disposal field[8]	–	5 feet	4 feet[4]	5 feet
On-site domestic water service line	1 foot[5]	5 feet	5 feet	5 feet
Distribution box	–	–	5 feet	5 feet
Pressure public water main	10 feet[6]	10 feet	10 feet	10 feet

For SI units: 1 foot = 304.8 mm

Notes:
1. Including porches and steps, whether covered or uncovered, breezeways, roofed porte cocheres, roofed patios, carports, covered walks, covered driveways, and similar structures or appurtenances.
2. See Section 312.3.
3. Drainage piping shall clear domestic water supply wells by not less than 50 feet (15 240 mm). This distance shall be permitted to be reduced to not less than 25 feet (7620 mm) where the drainage piping is constructed of materials approved for use within a building.
4. Plus 2 feet (610 mm) for each additional 1 foot (305 mm) of depth more than 1 foot (305 mm) below the bottom of the drain line. (See Section H 601.0)
5. See Section 720.0.
6. For parallel construction – For crossings, approval by the Health Department shall be required.
7. These minimum clear horizontal distances shall also apply to disposal fields, seepage pits, and the mean high-tide line.
8. Where disposal fields, seepage pits, or both are installed in sloping ground, the minimum horizontal distance between any part of the leaching system and ground surface shall be 15 feet (4572 mm).

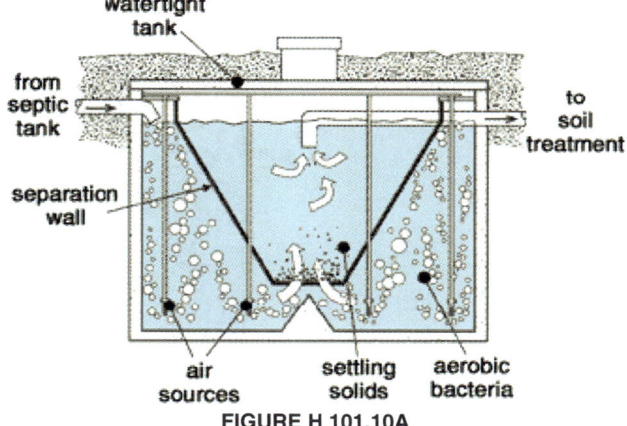

FIGURE H 101.10A
SUSPENDED GROWTH AEROBIC TREATMENT UNIT

APPENDIX H

The sewage leaving the tank (effluent) may still contain disease organisms. There is always solid matter that has not been completely stabilized or digested (both suspended and dissolved) in the effluent flowing out of the tank to the disposal, field or seepage pits. Further digestion of these organisms is carried on by bacteria present in the soil. The final treatment of effluent in the private sewage disposal system thus occurs in the native soils.

H 101.11 Alternate Systems. Alternate systems shall be permitted to be used by special permission of the Authority Having Jurisdiction after being satisfied of their adequacy. This authorization is based on extensive field and test data from conditions similar to those at the proposed site, or require such additional data as necessary to provide assurance that the alternate system will produce continuous and long-range results at the proposed site, not less than equivalent to systems which are specifically authorized.

Where demonstration systems are to be considered for installation, conditions for installation, maintenance, and monitoring at each such site shall first be established by the Authority Having Jurisdiction.

Approved aerobic systems shall be permitted to be substituted for conventional septic tanks provided the Authority Having Jurisdiction is satisfied that such systems will produce results not less than equivalent to septic tanks, whether their aeration systems are operating or not.

H 201.0 Capacity of Septic Tanks.

H 201.1 General. The liquid capacity of septic tanks shall comply with Table H 201.1(1) and Table H 201.1(4) as determined by the number of bedrooms or apartment units in dwelling occupancies and the estimated waste/sewage design flow rate or the number of plumbing fixture units as determined from Table 702.1 of this code, whichever is greater in other building occupancies. The capacity of any one septic tank and its drainage system shall be limited by the soil structure classification in Table H 201.1(2), and as specified in Table H 201.1(3).

Capacity is one of the most important considerations in septic tank design. Studies have proven that liberal tank capacity is not only important from a functional stand-

TABLE H 201.1(1)
CAPACITY OF SEPTIC TANKS[1, 2, 3, 4]

SINGLE-FAMILY DWELLINGS - NUMBER OF BEDROOMS	MULTIPLE DWELLING UNITS OR APARTMENTS - ONE BEDROOM EACH	OTHER USES: MAXIMUM FIXTURE UNITS SERVED PER TABLE 702.1	MINIMUM SEPTIC TANK CAPACITY (gallons)
1 or 2	–	15	750
3	–	20	1000
4	2 units	25	1200
5 or 6	3	33	1500
–	4	45	2000
–	5	55	2250
–	6	60	2500
–	7	70	2750
–	8	80	3000

For SI units: 1 gallon = 3.785 L

Notes:
[1] Extra bedroom, 150 gallons (568 L) each.
[2] Extra dwelling units over 10: 250 gallons (946 L) each.
[3] Extra fixture units over 100: 25 gallons (94.6 L) per fixture unit.
[4] Septic tank sizes in this table include sludge storage capacity and the connection of domestic food waste disposers without further volume increase.

TABLE H 201.1(2)
DESIGN CRITERIA OF FIVE TYPICAL SOILS

TYPE OF SOIL	REQUIRED SQUARE FEET OF LEACHING AREA PER 100 GALLONS	MAXIMUM ABSORPTION CAPACITY IN GALLONS PER SQUARE FEET OF LEACHING AREA FOR A 24 HOUR PERIOD
Coarse sand or gravel	20	5.0
Fine sand	25	4.0
Sandy loam or sandy clay	40	2.5
Clay with considerable sand or gravel	90	1.1
Clay with small amount of sand or gravel	120	0.8

For SI units: 1 square foot = 0.0929 m², 1 gallon = 3.785 L, 1 gallon per square foot = 40.7 L/m²

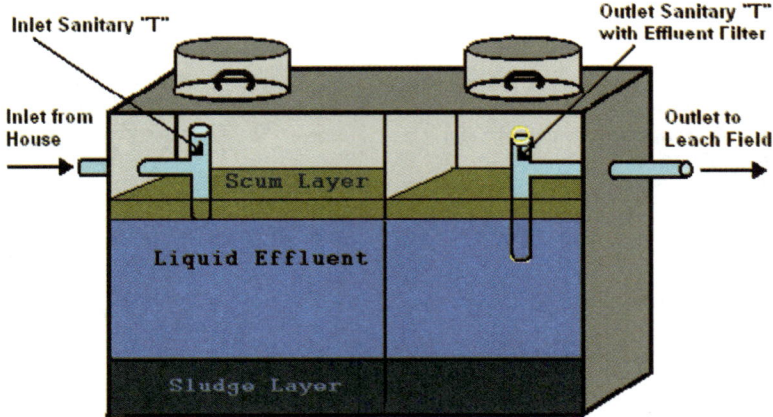

FIGURE H 101.10B
LAYERS OF EFFLUENT IN SEPTIC TANK

TABLE H 201.4(3)
LEACHING AREA SIZE BASED ON SEPTIC TANK CAPACITY

REQUIRED SQUARE FEET OF LEACHING AREA PER 100 GALLONS SEPTIC TANK CAPACITY (square feet per 100 gallons)	MAXIMUM SEPTIC TANK SIZE ALLOWABLE (gallons)
20–25	7500
40	5000
90	3500
120	3000

For SI units: 1 square foot per 100 gallons = 0.000245 m^2/L, 1 gallon = 3.785 L

TABLE H 201.1(4)
ESTIMATED WASTE /SEWAGE FLOW RATES[1, 2, 3]

TYPE OF OCCUPANCY	GALLONS PER DAY
Airports (per employee)	15
Airports (per passenger)	5
Auto washers – check with equipment manufacturer	-
Bowling alleys – with snack bar only (per lane)	75
Campground – with central comfort station (per person)	35
Campground – with flush toilets - no showers (per person)	25
Camps (day) – no meals served (per person)	15
Camps (summer and seasonal camps) – (per person)	50
Churches – sanctuary (per seat)	5
Churches – with kitchen waste (per seat)	7
Dance halls – (per person)	5
Factories – no showers (per employee)	25
Factories – with showers (per employee)	35
Factories – with cafeteria (per employee)	5
Hospitals – (per bed)	250
Hospitals – kitchen waste only (per bed)	25
Hospitals – laundry waste only (per bed)	40
Hotels – no kitchen waste (per bed)	60
Institutions – resident (per person)	75
Nursing home – (per person)	125
Rest home – (per person)	125
Laundries – self-service with minimum 10 hours per day (per wash cycle)	50
Laundries – commercial check with manufacturer's specification	-
Motel (per bed space)	50
Motel – with kitchen (per bed space)	60
Offices – (per employee)	20
Parks – mobile homes (per space)	250
Parks (picnic) – with toilets only (per parking space)	20
Parks (recreational vehicles) – without water hook-up (per space)	75
Parks (recreational vehicles) – with water and sewer hook-up (per space)	100
Restaurants – cafeteria (per employee)	20
Restaurants – with toilet waste (per customer)	7
Restaurants – with kitchen waste (per meal)	6
Restaurants – with kitchen waste disposable service (per meal)	2
Restaurants – with garbage disposal (per meal)	1
Restaurants – with cocktail lounge (per customer)	2
Schools staff and office (per person)	20
Schools – elementary (per student)	15
Schools – intermediate and high (per student)	20
Schools – with gym and showers (per student)	5

APPENDIX H

TABLE H 201.1(4) (continued)
ESTIMATED WASTE /SEWAGE FLOW RATES[1, 2, 3]

TYPE OF OCCUPANCY	GALLONS PER DAY
Schools – with cafeteria (per student)	3
Schools (boarding) – total waste (per person)	100
Service station – with toilets for 1st bay	1000
Service station – with toilets for each additional bay	500
Stores – (per employee)	20
Stores – with public restrooms (per 10 square feet of floor space)	1
Swimming pools – public (per person)	10
Theaters – auditoriums (per seat)	5
Theaters – with drive-in (per space)	10

For SI units: 1 square foot = 0.0929 m2, 1 gallon per day = 3.785 L/day

Notes:

[1] Sewage disposal systems sized using the estimated waste/sewage flow rates shall be calculated as follows:
  (a) Waste/sewage flow, up to 1500 gallons per day (5678 L/day)
    Flow x 1.5 = septic tank size
  (b) Waste/sewage flow, over 1500 gallons per day (5678 L/day)
    Flow x 0.75 + 1125 = septic tank size
  (c) Secondary system shall be sized for total flow per 24 hours.

[2] See Section H 201.1.

[3] Because of the many variables encountered, it is not possible to set absolute values for waste/sewage flow rates for all situations. The designer should evaluate each situation and, where figures in this table need modification; they should be made with the concurrence of the Authority Having Jurisdiction.

point, but is also good economy. In other words, having a bit more capacity is better than not having enough capacity. If the septic tank is sized too small, solids or fats and oils will flow from the tank rather than just the liquid effluent. The liquid capacity of all septic tanks is required to conform to Tables H 201.1(1) and H 201.1(3). For residences, Table H 201.1(1) will be used basing capacity by the number of bedrooms or apartments in the occupancy (see Examples 1 and 2 below). In other building occupancies, such as in Example 3 below (a movie theater), the capacity of the septic tank is determined by the estimated waste/sewage design flow rate from Table H 201.1(3) or the number of plumbing fixture units, as determined from Table 702.1, Drainage Fixture Unit Values. The largest tank required from a comparison of the results from the two methods will be the size required for that installation.

All of the following examples in sizing the septic tank will be to the minimum standards required by those tables. These examples used for septic tank sizing will also be used for sizing other parts of the system later in this chapter.

Examples:

1. What is the size of a septic tank for a four-bedroom single-family dwelling? Refer to "Single Family Dwellings" column in Table H 201.1(1) to determine that a **1,200-gallon** (4,542.5 L) tank is required. Occasionally, where space on a lot is critical, a builder will try to disguise a bedroom on the building plans by calling the room a den, office, sewing room, etc., in an attempt to reduce the system's size. It is best to resolve this kind of problem before plans are approved or the system has been installed. To undersize the septic tank would eventually be a very costly mistake that would require replacement of the system with a larger system or cause continual problems.

2. What is the size of a septic tank for a 12-unit apartment house with six one-bedroom units and six two-bedroom units?

   Refer to the "Multiple Dwelling Units" column in Table H 201.1(1), which shows that 10 units require a 3,500-gallon (13,249 L) tank. The footnote at the bottom of the table referring to extra dwelling units states that the extra two units will add 250 gallons (946.4 L) each. The capacities shown in the table are for one-bedroom units. The footnote for extra bedrooms states that an extra bedroom requires an additional 150 gallons (567.8 L) to be added to the tank size. Therefore, six times 150 gallons (567.8 L) must be added for the extra bedrooms in this example. Septic tank size would be:

   3,500 + (2 x 250) + (6 x 150) = **4,900 gal.**

   [13,249 L + (2 x 946.4 L) + (6 x 567.8 L) = 18,548.6 L]

   Refer to Tables H 201.1(3) and H 201.1(4), which show that this size system can only be installed where the soil is classified as Type 3 (sandy loam or sandy clay) or better. The maximum size septic tank permitted in Type 3 soil is 5,000 gallons (18,927 L). If soils are less porous than Type 3, plumbing in the building would have to be divided so that two smaller systems can be installed at different locations on the property.

3. What is the size of a septic tank for a movie theater with 700 seats?

   First, determine the total fixture unit loading on the building sewer from Table 702.1 and the estimated sewage waste per 24 hours from Table H 201(1).

   The number and type of plumbing fixtures and their fixture unit values from Table 702.1 are the following:
   Number of Fixture Unit Values

   Fixtures (from Table 702.1)

Totals
11 Water Closets x 6 = 66
4 Urinals x 2 = 8
6 Lavatories x 1 = 6
3 Drinking Fountains x 1 = 3
1 Service Sink x 3 = 3
2 Floor Sinks x 1 = 2
88 Fixture Units

Note that six fixture units also are used for private use toilets when sizing septic tanks. See the footnote at the end of Table 702.1.

To start, in Table H 201.1(1), 88 fixture units require a 3,250 gallon (12,302 L) septic tank.

From Table H 201.1(3) #20, a theater has a rate of flow of 5 gallons (18.93 L) per seat per day or 700 seats x 5 = 3,500 gallons sewage/day. (700 x 18.93 = 13,251 L)

From Table H 201.1(2) "(a) Recommended Design criteria" for a septic tank system over 1,500 gallons (5,678 L)[(2)] per day:

Flow x 0.75 + 1,125 = septic tank size or 3,500 x 0.75 + 1,125 = 3,750 gal. tank (13,248.9 L x 0.75 + 4,258.6 L = 14,192.3 L)

Comparing the two results, the second design method (from Table H 201.1(3) produces the largest tank — 3,750 gallons (14,192.3 L). This size would be the minimum required. Also, keep in mind the limitation on the maximum size tank from Table H 201.1(2).

Prefabricated concrete battery sectioned tanks usually do not come in the exact size needed. It is necessary to pick a combination that will produce a size equal to the requirements or the next available size above. In this case, it may be necessary to install a tank as large as 4,000 gallons (15,141.6 L) or job-construct a tank from approved engineered plans.

The designer of septic tank systems for commercial and industrial projects sometimes has difficulty establishing the actual flow rate before a building is constructed. The first paragraph in Table H 201.1(4) clearly notes that it is not possible to set absolute values for waste/sewage flow ratio for all situations. The designer should evaluate each situation, and if figures in this table need modification, they should be made with the concurrence of the local AHJ. The designer must also consider business expansion, peak loading and overloading where applicable. Also, any plumbing fixtures roughed-in, but not installed, should be included in the septic tank design. In one case of a premature system failure, it was discovered that a restaurant was being used by a cross-country bus company as a rest stop. This was in addition to normal patronage from a freeway nearby. The system was not designed or installed to accommodate this overloading and failed.

Disposal Fields and Seepage Pits: There are two common soil absorption systems: disposal fields, which consist of leaching lines, leaching beds or leaching chambers; and seepage pits. These are the only systems that the code has specific requirements for in Appendix H (see **Figures H 101.2a, H 101.2b, H 101.2c, H 101.2d** and **H 201.1**).

The selection of the type of absorption system will depend to some extent on the geographical location of the site under consideration and on local regulations. The system must also be designed to utilize the most porous or absorptive portions of the soil formation. For example, if there is soil that consists of clay with considerable gravel, the seepage pit might be a better disposal method than disposal fields in order to utilize deeper soils. However, if the groundwater is within 12 feet (3,658 mm) a seepage pit is not allowed and a disposal field must be used. These are the types of considerations that must be taken into account to properly design a private sewage disposal system.

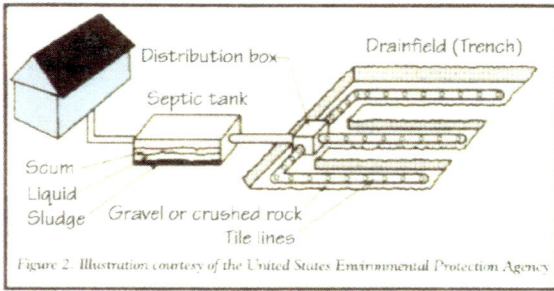

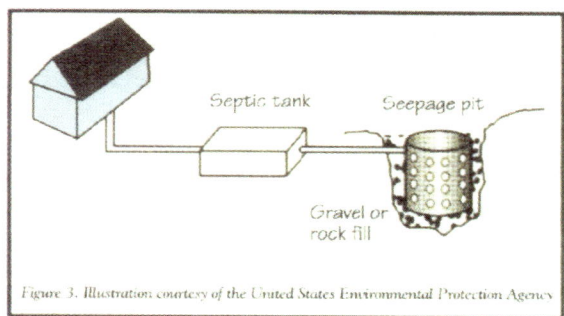

**FIGURE H 201.1
DISPOSAL FIELDS AND SEEPAGE PITS**

H 301.0 Area of Disposal Fields and Seepage Pits.

H 301.1 General. The minimum effective absorption area in disposal fields in square feet (m^2), and in seepage pits in square feet (m^2) of sidewall, shall be predicated on the required septic tank capacity in gallons (liters), estimated waste/sewage flow rate, or whichever is greater, and shall be in accordance with Table H 201.1(3) as determined for the type of soil found in the excavation, and shall be as follows:

(1) Where disposal fields are installed, not less than 150 square feet (13.9 m^2) of trench bottom shall be provided for each system exclusive of any hard pan, rock, clay, or other impervious formations. Sidewall area in excess of the required 12 inches (305 mm) and not exceeding 36 inches (914 mm) below the leach line shall be permitted to be added to the trench bottom area where computing absorption areas.

APPENDIX H

🔧 See **Figure H 301.1a**.

The minimum area for a disposal field consisting of trenches is 150 ft². If there is any clay or other impervious soils, they must be excluded from the trench area. For example, if within a trench that is 1 foot by 100 feet long there is section of that trench 20 feet long consisting of clay in the bottom of the trench, the 20 feet of clay bottom cannot be counted in the required amount of trench area.

Trenches deeper than 12 inches can use the added vertical side wall area to meet the required minimum trench area needed. For example, if there is a trench 12 inches wide and 24 inches deep, the square foot area per linear foot of that trench is 3 ft², not 1 ft². The 12 inches of side wall exceeding the 12 inch minimum on both sides of the trench add to the disposal area of the trench (see **Figure H 301.1a**).

When conditions permit, a narrow trench with 36 inches of filter material under the leach line has been found to be one of the best designs. This arrangement allows for a more compact but efficient installation.

(2) Where leaching beds are permitted in lieu of trenches, the area of each such bed shall be not less than 50 percent greater than the tabular requirements for trenches. Perimeter sidewall area in excess of the required 12 inches (305 mm) and not exceeding 36 inches (914 mm) below the leach line shall be permitted to be added to the trench bottom area where computing absorption areas.

🔧 The minimum bottom area for a leach bed would be 150 feet x 1.50 = 225 ft². The perimeter side wall area in excess of the required 12 inch depth and not to exceed 36 inches below the leach line may be added to the trench bottom area when computing absorption areas. For example, if there is a leach bed 100 feet by 100 feet the perimeter linear foot area would be 400 ft². If the leaching

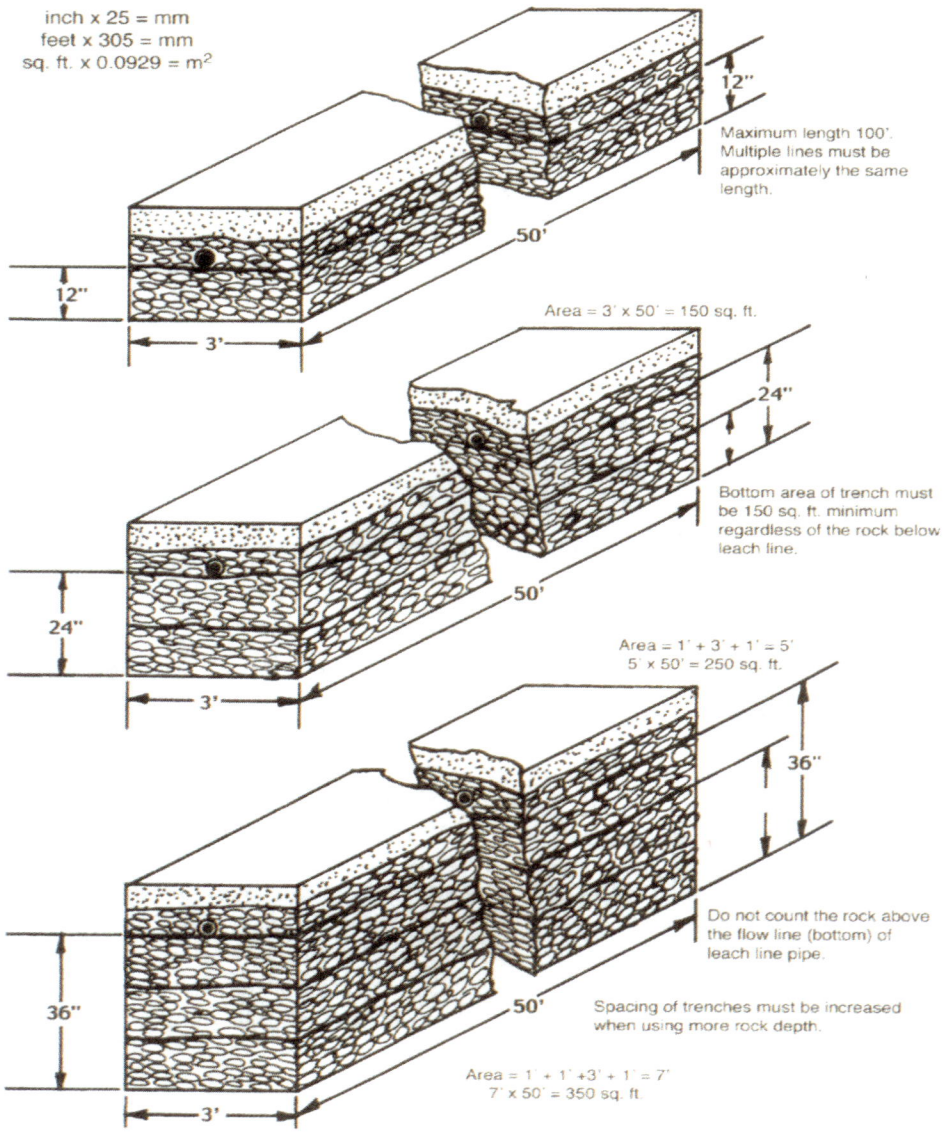

FIGURE H 301.1A
DESIGN AND COMPUTATION OF LEACH LINE AREAS

bed was 24 inches, deep only 400 ft² could be added to the bed bottom area to meet the required minimum of area needed (see **Figure H 301.1b**).

(3) No excavation for a leach line or leach bed shall be located within 5 feet (1524 mm) of the water table nor to a depth where sewage is capable of contaminating the underground water stratum that is usable for domestic purposes.

Exception: In areas where the records or data indicate that the groundwaters are grossly degraded, the 5 foot (1524 mm) separation requirement shall be permitted to be reduced by the Authority Having Jurisdiction. The applicant shall supply evidence of groundwater depth to the satisfaction of the Authority Having Jurisdiction.

(4) The minimum effective absorption area in any seepage pit shall be calculated as the excavated sidewall area below the inlet exclusive of any hardpan, rock, clay, or other impervious formations. The minimum required area of porous formation shall be provided in one or more seepage pits. No excavation shall extend within 10 feet (3048 mm) of the water table nor to a depth where sewage is capable of contaminating underground water stratum that is usable for domestic purposes.

The greater separation from groundwater is required for seepage pits because of the greater hydraulic pressure developed by the vertical columns of effluent in a seepage pit. Groundwater level should be that level known to be the highest height recorded, rather than the level taken from a percolation test made at the time of system installation.

Exception: In areas where the records or data indicate that the groundwaters are grossly degraded, the 10 foot (3048 mm) separation requirement shall be permitted to be reduced by the Authority Having Jurisdiction.

The applicant shall supply evidence of groundwater depth to the satisfaction of the Authority Having Jurisdiction.

(5) Leaching chambers that comply with IAPMO PS 63 and bundled expanded polystyrene synthetic aggregate units that comply with IAPMO IGC 276 shall be sized using the required area calculated using Table H 201.1(3) with a 0.70 multiplier.

Leaching chambers are specially designed products made to replace the traditional leach field discussed here. Because the disposal area required for the leaching chamber depends on the chamber itself and is not known, the absorption rates for soils in Table H 201.1(2) will be reduced by 70 percent. For the use of this appendix, the leaching chamber will not be discussed here other than to illustrate its use (see **Figure H 301.1c**).

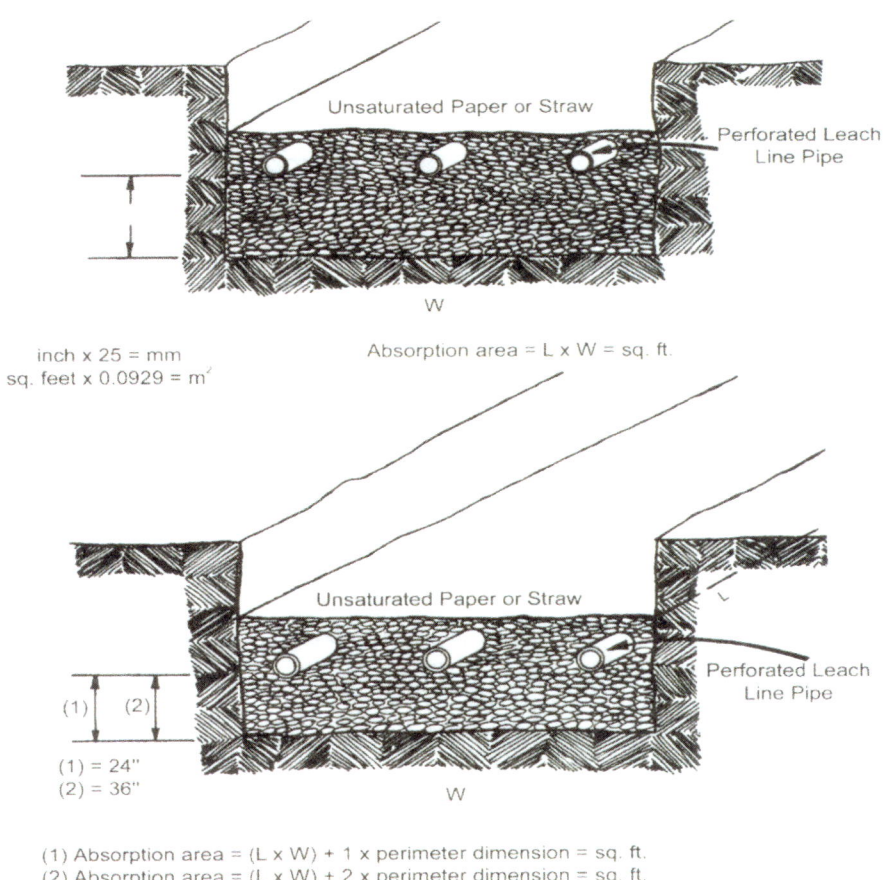

inch x 25 = mm
sq. feet x 0.0929 = m²

Absorption area = L x W = sq. ft.

(1) = 24"
(2) = 36"

(1) Absorption area = (L x W) + 1 x perimeter dimension = sq. ft.
(2) Absorption area = (L x W) + 2 x perimeter dimension = sq. ft.

**FIGURE H 301.1B
DESIGN AND COMPUTATION OF LEACH BED AREAS**

APPENDIX H

FIGURE H 301.1C
EXAMPLE OF LEACHING CHAMBER SYSTEM

Computing Absorption Areas

Examples: See **Figures H 301.1d through H 301.1g**.

Problem 1

A four-bedroom dwelling requires a 1,200-gallon septic tank. Assume the soil has been classified as Type 3 per Table H 201.1(2) (sandy loam or sandy clay). The required disposal field would be 40 ft² per each 100 gallons of the required septic tank.

Solution:

40 x 12 = 480 ft² of absorption area for a disposal field or

480 ft² x 1.5 = 720 ft² for a leaching bed.

1. Assume the designer wishes to use a leach line trench 2 feet wide with 12 inches of filter material below the leach line. With only 12 inches below the leaching line, the absorption area would equal 2 ft² per running foot of trench or bottom area of the trench.

480 ÷ 2 = 240 lineal feet of leach line required.

When the line is over 100 feet in length, multiple lines are required (see Table H 601.9). The lines should be of equal length so three lines 80 feet long must be connected to the septic tank by a distribution box (see **Figure H 301.1d**).

2. Using the same parameters, assume the designer wishes to use a different trench. Assume the trench to be 3 feet wide with 36 inches of filter material below the leach line pipe.

The absorption area would equal 7 ft²/running foot of the trench (see **Figure H 301.1b**).

480 ÷ 7 = 68.5 lineal feet of leach line.

Also evaluate the trench bottom area to determine that it is at least 150 ft². 3 x 68.5 = 205 ft², which is greater than 150 ft². This calculation is satisfactory (see **Figure H 301.1e**).

As one can see, the second design reduces the need for a distribution box, and the necessary land needed to be reserved for expansion has been greatly reduced.

3. Again using the same parameters, assume the designer wishes to install a leach bed with 12 inches of filter material covering the whole bed below the lines.

Seven-hundred-twenty square feet of bottom area is required. Section H 601.9 requires that lines be spaced 6 feet maximum and no more than 3 feet from the excavation perimeter.

A bed that is 18 feet wide by 40 feet long with three lines would meet these conditions. The bottom area also exceeds the minimum of 225 ft² (see **Figure H 301.1f**).

If the designer wishes to use 24 inches of filter material below the lines in the bed, the extra square foot of depth around the exterior perimeter may be added to the bottom area. This becomes a trial-and-error method to obtain the required precise amount. The 18 foot by 40 foot bed above with 24 inches would have an absorption area of:

(18 x 40) + 18 + 18 + 40 + 40 = 836 ft²

Problem 2

Compute the disposal field for **Figure H 301.1g** — the movie theater. The flow rate governing the design of the septic tank must be used in the design of the disposal field.

Solution:

Flow rate = 700 seats x 5 gal. per seat = 3,500 gal. per day discharge.

Assume the soil has been classified as Type 2 (fine sand) from Table H 201.1(2). This table has a column that is headed "Maximum absorption capacity gallons/ sq. ft. of leaching area for a 24 hr. period." Fine sand has an application rate of 4 gallons per ft² per day.

3,500 ÷ 4 = 875 ft² of absorption area in trenches.

Assume the designer wishes to use trenches 30 inches wide with 30 inches of filter material below the leach lines.

Absorption area of the trench would be:

2.5 ft. + (1.5 x 2) = 2.5 + 3 = 5.5 ft²/running foot of trench.

875 ÷ 5.5 = 159.1 feet or two leach lines 30 inches wide by 79.5 feet long with 30 inches of filter material under the

leach line. Both lines would be connected to the septic tank through a distribution box (see **Figure H 301.1g**).

What is the required spacing between the trenches? Per Section H 601.9, the spacing must be increased beyond the 4 foot minimum, because additional filter material was used beyond the 12 inches minimum. Increase spacing 2 feet for additional 1 foot of filter material.

Spacing = 4 + (2 x 1.5) = 7 feet

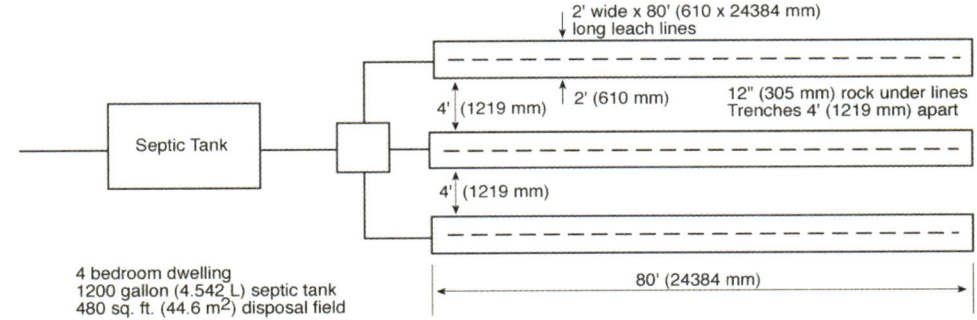

FIGURE H 301.1D
SOLUTION TO PROBLEM 1 - EXAMPLE 1

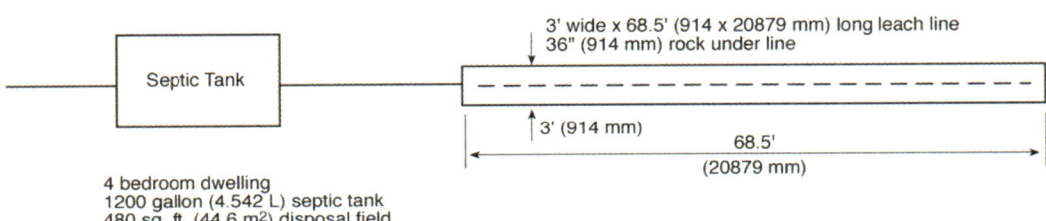

FIGURE H 301.1E
SOLUTION TO PROBLEM 1 - EXAMPLE 2

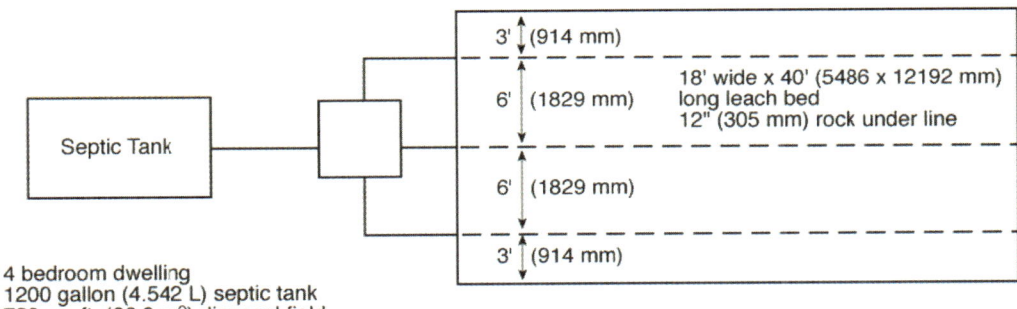

FIGURE H 301.1F
SOLUTION TO PROBLEM 1 - EXAMPLE 3

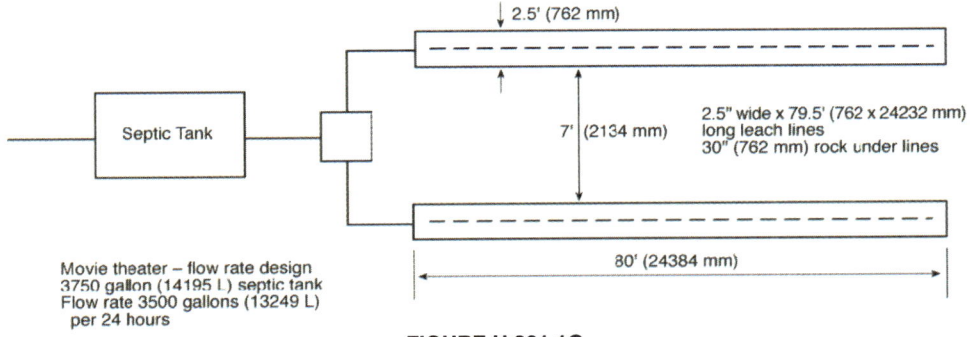

FIGURE H 301.1G
SOLUTION TO PROBLEM 2

APPENDIX H

H 401.0 Percolation Test.

H 401.1 Pit Sizes. Where practicable, disposal field and seepage pit sizes shall be computed from Table H 201.1(2). Seepage pit sizes shall be computed by percolation tests, unless use of Table H 201.1(2) is approved by the Authority Having Jurisdiction.

H 401.2 Absorption Qualities. In order to determine the absorption qualities of seepage pits and of questionable soils other than those listed in Table H 201.1(2), the proposed site shall be subjected to percolation tests acceptable to the Authority Having Jurisdiction.

H 401.3 Absorption Rates. Where a percolation test is required, no private disposal system shall be permitted to serve a building where that test shows the absorption capacity of the soil is less than 0.83 gallons per square foot (gal/ft^2) (33.8 L/m^2) or more than 5.12 gal/ft^2 (208.6 L/m^2) of leaching area per 24 hours. Where the percolation test shows an absorption rate greater than 5.12 gal/ft^2 (208.6 L/m^2) per 24 hours, a private disposal system shall be permitted where the site does not overlie groundwaters protected for drinking water supplies, a minimum thickness of 2 feet (610 mm) of the native soil below the entire proposed system is replaced by loamy sand, and the system design is based on percolation tests made in the loamy sand.

The percolation test is designed to determine the suitability of a site for a subsurface private sewage disposal system. More specifically, a percolation test measures the ability of the soil to absorb liquid. The percolation test is designed to simulate conditions in a septic system. The percolation test consists of a hole 6 to 12 inches in diameter dug in the area of the proposed septic system. The hole is initially filled with water (presoak) in an attempt to saturate the soil, allowed to drain away and then refilled. The rate at which the water drops in the hole is measured at intervals over a period of time ranging from 30-60 minutes. The uniform slowest rate of drop in the water level over a measured time interval is converted to minutes per inch and used as a basis of design in determining the septic system size. For example, if the water dropped uniformly ¼ inch every 5 minutes, the rate would be 20 minutes per inch. This test is usually performed by a soils engineering firm.

Some building, safety and health departments have established design rates for many areas within their jurisdiction. It may not be necessary to evaluate the soils, although an on-site inspection should be conducted to verify the physical conditions. When it is necessary to evaluate soils using Table H 201.1(4), some excavation work must be done at the site to determine the soils at the leaching depth. The excavation should be made to a minimum of 5 feet below the bottom of a proposed leach line or bed trench to check for groundwater and impervious strata. It may be necessary to choose a new location or change to a different type of absorption system when adverse conditions are found

Table H 201.1(4) shows the design rate for five basic soil types. There are many other soil classifications that are not listed. The designer must evaluate the soil and classify to the nearest material according to texture and permeability of soils in Table H 201.1(4). With proper excavation work, a qualified person can usually make a good judgment as to which design criteria to use. He or she also should be able to identify soils that are questionable or where a system should not be installed. A soil percolation test is required when the soil is questionable, on large projects or when required by the local AHJ.

Most local jurisdictions have adopted a standard testing procedure such as is found in the U.S. Public Health publication, *Manual of Septic Tank Practice*. Some administrative authorities have developed their own testing procedures. Test reports should be submitted to the local jurisdiction for approval prior to doing design work. The conclusions in the report may be adjusted by the jurisdiction, based upon area failure rates or other data.

H 501.0 Septic Tank Construction.

H 501.1 Plans. Plans for septic tanks shall be submitted to the Authority Having Jurisdiction for approval. Such plans shall show dimensions, reinforcing, structural calculations, and such other pertinent data as required.

H 501.2 Design. Septic tank design shall be such as to produce a clarified effluent consistent with accepted standards and shall provide adequate space for sludge and scum accumulations.

H 501.3 Construction. Septic tanks shall be constructed of solid durable materials not subject to excessive corrosion or decay and shall be watertight.

H 501.4 Compartments. Septic tanks shall have not less than two compartments unless otherwise approved by the Authority Having Jurisdiction. The inlet compartment of any septic tank shall be not less than two-thirds of the total capacity of the tank, nor less than 500 gallons (1892 L) liquid capacity, and shall be not less than 3 feet (914 mm) in width and 5 feet (1524 mm) in length. Liquid depth shall be not less than 2½ feet (762 mm) nor more than 6 feet (1829 mm). The secondary compartment of a septic tank shall have a capacity of not less than 250 gallons (946 L) and a capacity not exceeding one-third of the total capacity of such tank. In septic tanks having a 1500 gallon (5678 L) capacity, the secondary compartment shall be not less than 5 feet (1524 mm) in length.

See **Figure H 501.4a**.

Septic tanks must be watertight and constructed of materials not subject to excessive corrosion or decay, such as concrete, coated metal, concrete block plastered smooth inside, fiberglass or other suitable materials. Wooden septic tanks are not acceptable (see Table 1701.1 for approved materials).

Prefabricated concrete septic tanks over 1,500 gallons are usually installed in battery sections. These are standard dimensioned tanks with ends removed so that the tanks' shells may be fastened together end-to-end, making the required capacity. The completed tank still should be of two compartments. For example, two 1,200-gallon tanks may be fastened together to make a 2,400-gallon tank. It is important that the compartment location meet the requirements of the preceding paragraphs. The use of two 1,200-gallon septic tanks installed parallel to each other and discharging into a diversion box should not be used. There is no assurance that raw sewage will be equally divided into the two tanks (see **Figure H 501.4b**).

H 501.5 Access. Access to each septic tank shall be provided by not less than two manholes 20 inches (508 mm) in minimum dimension or by an equivalent removable cover slab. One access manhole shall be located over the inlet and one access manhole shall be located over the outlet. Where a first compartment exceeds 12 feet (3658 mm) in length, an additional manhole shall be provided over the baffle wall.

H 501.6 Pipe Opening Sizes. The inlet and outlet pipe openings shall not be larger in size than the connecting sewer pipe. The vertical leg of round inlet and outlet fittings shall not be less in size than the connecting sewer pipe nor less than 4 inches (102 mm). A baffle-type fitting shall have the equivalent cross-sectional area of the connecting sewer pipe and not less than a 4 inch (102 mm) horizontal dimension where measured at the inlet and outlet pipe inverts.

H 501.7 Pipe Extension. The inlet and outlet pipe or baffle shall extend 4 inches (102 mm) above and not less than 12 inches (305 mm) below the water surface. The invert of the inlet pipe shall be at a level not less than 2 inches (51 mm) above the invert of the outlet pipe.

The invert of the inlet fitting must be 2 inches above the invert of the outlet; this will be the flow line of the tank. This is to allow for a momentary rise in liquid level during the discharge of the building sewer into the tank. This freefall drop prevents sewage from standing in the building sewer at the tank connection.

A vented inlet tee or baffle must be provided to divert the incoming sewage downward into the tank. To do otherwise would disturb the floating mat. The size of the vertical leg of the tee must be not less than the size of the building sewer or 4 inches minimum. The outlet fitting is to be of the same configuration, except the invert is 2 inches lower than the inlet.

H 501.8 Free Vent Area. Inlet and outlet pipe fittings or baffles and compartment partitions shall have a free vent area equal to the required cross-sectional area of the house sewer or private sewer discharging therein to provide free ventilation above the water surface from the disposal field or seepage pit through the septic tank, house sewer, and stack to the outer air.

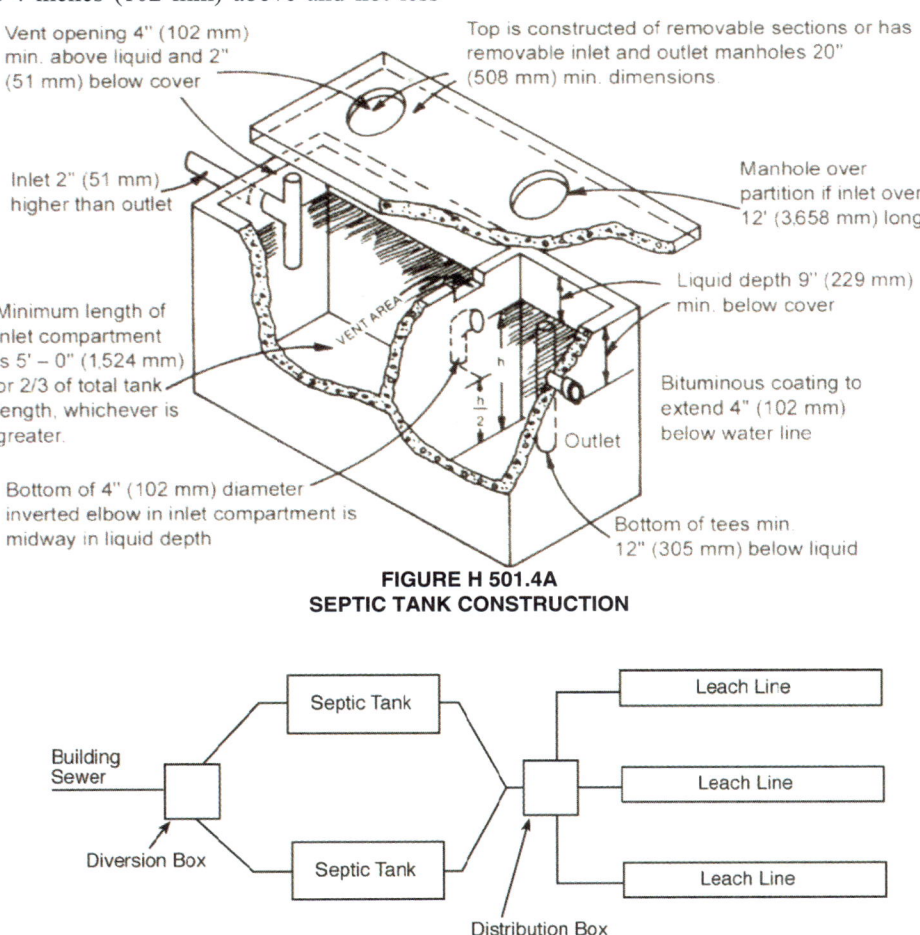

FIGURE H 501.4A
SEPTIC TANK CONSTRUCTION

FIGURE H 501.4B
PARALLEL SEPTIC TANK SYSTEM SHOULD NOT BE USED

H 501.9 Sidewalls. The sidewalls shall extend not less than 9 inches (229 mm) above the liquid depth. The cover of the septic tank shall be not less than 2 inches (51 mm) above the back vent openings.

H 501.10 Partitions and Baffles. Partitions or baffles between compartments shall be of solid, durable material and shall extend not less than 4 inches (102 mm) above the liquid level. The transfer port between compartments shall be a minimum size equivalent to the tank inlet, but in no case less than 4 inches (102 mm) in size, shall be installed in the inlet compartment side of the baffle so that the entry into the port is placed 65 percent to 75 percent in the depth of the liquid. Wooden baffles are prohibited.

H 501.11 Structural Design. The structural design of septic tanks shall comply with the following requirements:

(1) Each such tank shall be structurally designed to withstand all anticipated earth or other loads. Septic tank covers shall be capable of supporting an earth load of not less than 500 pounds per square foot (lb/ft^2) (2441 kg/m^2) where the maximum coverage does not exceed 3 feet (914 mm).

(2) In flood hazard areas, tanks shall be anchored to counter buoyant forces during conditions of the design flood. The vent termination and service manhole of the tank shall be not less than 2 feet (610 mm) above the design flood elevation or fitted with covers designed to prevent the inflow of floodwater or the outflow of the contents of the tanks during conditions of the design flood.

H 501.12 Manholes. Septic tanks installed under concrete or blacktop paving shall have the required manholes accessible by extending the manhole openings to grade in a manner acceptable to the Authority Having Jurisdiction.

H 501.13 Materials. The materials used for constructing a septic tank shall be in accordance with the following:

(1) Materials used in constructing a concrete septic tank shall be in accordance with applicable standards in Table 1701.1.

(2) The minimum wall thickness of a steel septic tank shall be number 12 U.S. gauge (0.109 of an inch) (2.77 mm), and each such tank shall be protected from corrosion both externally and internally by an approved bituminous coating or by other acceptable means.

(3) Septic tanks constructed of alternate materials shall be permitted to be approved by the Authority Having Jurisdiction where in accordance with approved applicable standards. Wooden septic tanks shall be prohibited.

Additional information on the manufacture of prefabricated concrete tanks and reinforced glass fiber tanks can be found in IAPMO Z1000-2013, *Prefabricated Septic Tanks*. Because chemically aggressive gases are generated in a septic tank, concrete must be protected by coating the wall to a point 4 inches below the liquid level and by covering all the internal area above that point. Steel tanks must be coated both externally and internally to protect against corrosion by an approved bituminous coating.

H 501.14 Prefabricated Septic Tanks. Prefabricated septic tanks shall comply with the following requirements:

(1) Manufactured or prefabricated septic tanks shall comply with approved applicable standards and be approved by the Authority Having Jurisdiction. Prefabricated bituminous coated septic tanks shall comply with UL 70.

(2) Independent laboratory tests and engineering calculations certifying the tank capacity and structural stability shall be provided as required by the Authority Having Jurisdiction.

H 601.0 Disposal Fields.

Currently, the most common piping in use for distribution is plastic: high-density polyethylene, ABS or PVC. Manufactured pipe for this purpose has two rows of ½ inch or 5/8 inch holes 120 degrees apart and spaced approximately 5 inches on center. The pipe is installed with holes downward and centered over the rock filter material. Pipe joints must be coupled together and fittings be used for changes in direction (see **Figures H 601.0a** and **H 601.0b**).

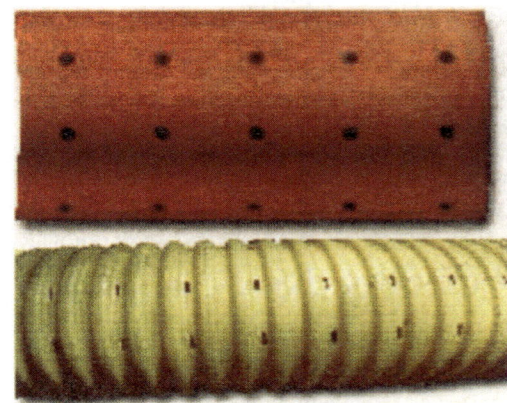

**FIGURE H 601.0A
PERFORATED TILE AND PIPE**

H 601.1 Distribution Lines. Distribution lines shall be constructed of clay tile laid with open joints, perforated clay pipe, perforated bituminous fiber pipe, perforated high-density polyethylene pipe, perforated ABS pipe, perforated PVC pipe, or other approved materials, provided that approved openings are available for distribution of the effluent into the trench area.

H 601.2 Filter Material. Before placing filter material or drain lines in a prepared excavation, smeared or compacted surfaces shall be removed from trenches by raking to a depth of 1 inch (25.4 mm) and the loose material removed. Clean stone, gravel, slag, or similar filter material acceptable to the Authority Having Jurisdiction, varying in size from ¾ of an inch to 2½ inches (19.1 mm to 64 mm), shall be placed in the trench to the depth and grade required by this section. Drain pipe shall be placed on filter material in an approved manner. The drain lines shall then be covered with filter material to the minimum depth required by this section, and this material covered with untreated building paper, straw, or

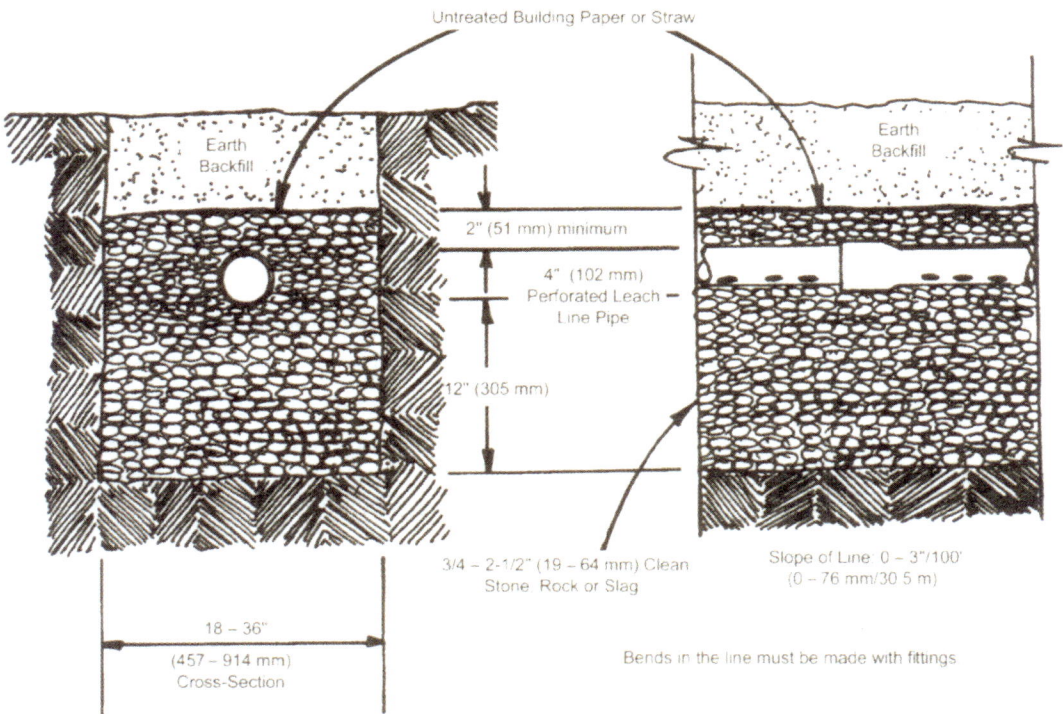

**FIGURE H 601.0B
LEACHING TRENCH CONSTRUCTION**

similar porous material to prevent closure of voids with earth backfill. No earth backfill shall be placed over the filter material cover until after inspection and acceptance.

Exception: Listed or approved plastic leaching chambers and bundled expanded polystyrene synthetic aggregate units shall be permitted to be used in lieu of pipe and filter material. Chamber and bundled expanded polystyrene synthetic aggregate unit installations shall follow the rules for disposal fields, where applicable, and shall be in accordance with the manufacturer's instructions.

H 601.3 Grade Board. A grade board staked in the trench to the depth of filter material shall be utilized where the distribution line is constructed with drain tile or a flexible pipe material that will not maintain alignment without continuous support.

H 601.4 Seepage Pits. Where seepage pits are used in combination with disposal fields, the filter material in the trenches shall terminate not less than 5 feet (1524 mm) from the pit excavation, and the line extending from such points to the seepage pit shall be approved pipe with watertight joints.

H 601.5 Distribution Boxes. Where two or more drain lines are installed, an approved distribution box of sufficient size to receive lateral lines shall be installed at the head of each disposal field. The inverts of outlets shall be level, and the invert of the inlet shall be not less than 1 inch (25.4 mm) above the outlets. Distribution boxes shall be designed to ensure equal flow and shall be installed on a level concrete slab in natural or compacted soil.

See **Figure H601.5**.

H 601.6 Laterals. Laterals from a distribution box to the disposal field shall be approved pipe with watertight joints. Multiple disposal field laterals, where practicable, shall be of uniform length.

H 601.7 Connections. Connections between a septic tank and a distribution box shall be laid with approved pipe with watertight joints on natural ground or compacted fill.

H 601.8 Dosing Tanks. Where the quantity of sewage exceeds the amount that is permitted to be disposed in 500 lineal feet (152.4 m) of leach line, a dosing tank shall be used. Dosing tanks shall be equipped with an automatic siphon or pump that discharges the tank once every 3 or 4 hours. The tank shall have a capacity equal to 60 to 75 percent of the interior capacity of the pipe to be dosed at one time. Where the total length of pipe exceeds 1000 lineal feet (304.8 m), the dosing tank shall be provided with two siphons or pumps dosing alternately and each serving one-half of the leach field.

This section establishes the requirements and design criteria for dosing tanks. A dosing tank is required whenever the lineal length of pipe in a leaching system exceeds 500 feet. These are seldom necessary, except for the very largest specially designed private sewage disposal systems because the size of a septic tank is limited by Table H 201.1(3). Most installations can be kept to the 500 foot maximum length.

The purpose of a dosing tank is to charge the leach line with effluent from 60 to 75 percent of the interior pipe

APPENDIX H

capacity and then to allow the system to rest. This provides equal distribution throughout the system, and the rest period is beneficial in extending the life of the system. This is accomplished with an automatic siphon or by a pump. On systems where line lengths exceed 1,000 linear feet, dual siphons or pumps are required to alternately dose each half of the system (see **Figure H 601.8**).

One comment about alternating leach line systems is needed. A septic tank effluent diverter valve is available commercially. This is a manually operated device. Its purpose is to divert the effluent into half the leaching system for a period of up to a year and then divert to the other half for another year. This has a very beneficial effect in allowing the resting system to rejuvenate. Where layout permits, when an existing system fails due to clogging and a new system must be installed, a diverter valve can be placed in the line, leaving the septic tank and diverting the sewage to the replaced system. After an extended rest period, the old system can again become effective. When installing a new dual system, each side of the system should be designed to 75 percent or greater of a single system. A major problem is remembering to change the valve at the predetermined time.

H 601.9 Construction. Disposal fields shall be constructed in accordance with Table H 601.9.

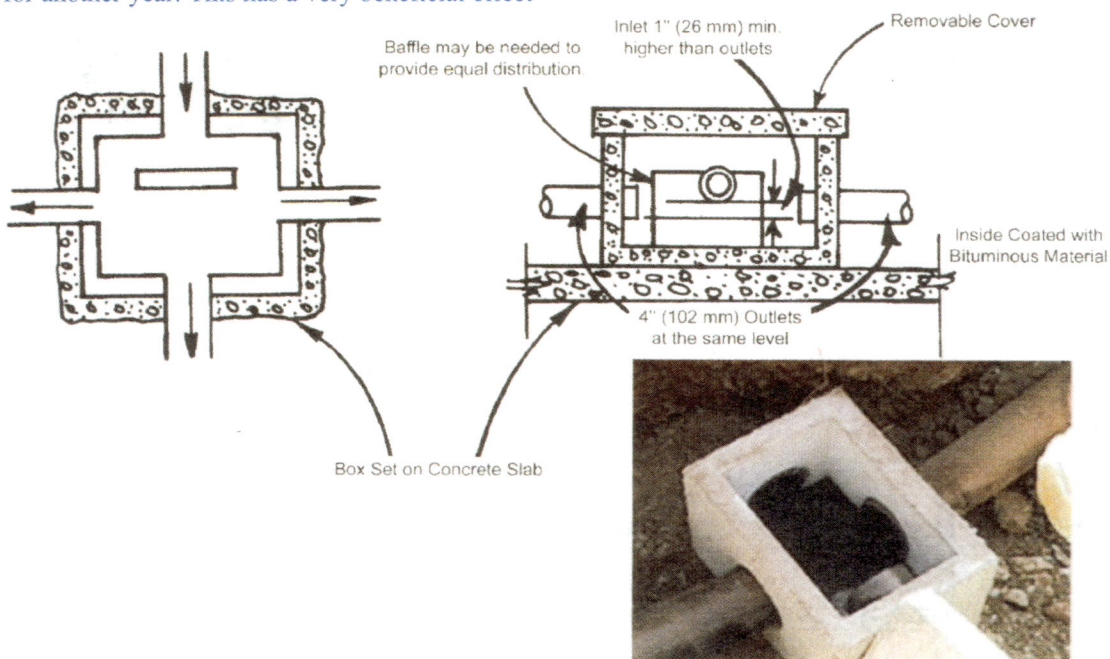

FIGURE H 601.5
DISTRIBUTION BOX CONSTRUCTION AND INSTALLATION

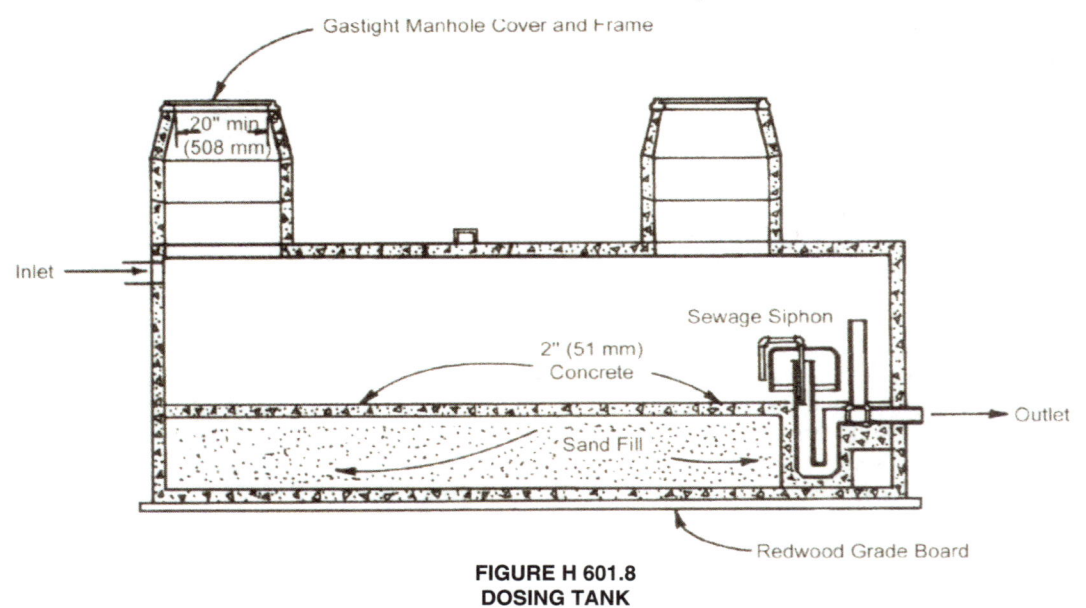

FIGURE H 601.8
DOSING TANK

Minimum spacing between trenches or leaching beds shall be not less than 4 feet (1219 mm) plus 2 feet (610 mm) for each additional foot (305 mm) of depth in excess of 1 foot (305 mm) below the bottom of the drain line. Distribution drain lines in leaching beds shall be not more than 6 feet (1829 mm) apart on centers, and no part of the perimeter of the leaching bed shall exceed 3 feet (914 mm) from a distribution drain line. Disposal fields, trenches, and leaching beds shall not be paved over or covered by concrete or a material that is capable of reducing or inhibiting a possible evaporation of sewer effluent.

TABLE H 601.9
GENERAL DISPOSAL FIELD REQUIREMENTS

ELEMENT	MINIMUM	MAXIMUM
Number of drain lines per field	1	-
Length of each line	-	100 feet
Bottom width of trench	18 inches	36 inches
Spacing of lines, center-to-center	6 feet	-
Depth of earth cover of lines (preferred 18 inches)	12 inches	-
Grade of lines	level	3 inches per 100 feet
Filter material under drain lines	12 inches	-
Filter material over drain lines	2 inches	-

For SI units: 1 inch = 25.4 mm, 1 foot = 304.8 mm, 1 inch per foot = 83.3 mm/m

🔧 The "General Disposal Field Requirements Table H 601.9" gives minimum and maximum tolerances for disposal field drain lines per this section of the appendix. The grade required for leach lines is contained here and must be laid as near level as possible, but not sloped more than 3 inches uniformly per 100 feet. This is to provide as much equal distribution of effluent as possible.

H 601.10 Joints. Where necessary on sloping ground to prevent excessive line slope, leach lines or leach beds shall be stepped. The lines between each horizontal section shall be made with watertight joints and shall be designed so each horizontal leaching trench or bed shall be utilized to the maximum capacity before the effluent shall pass to the next lower leach line or bed. The lines between each horizontal leaching section shall be made with approved watertight joints and installed on natural or unfilled ground.

🔧 When the site of a leach field is sloped, lines must follow the site contours to maintain the required minimum slope. When there is insufficient land width to accommodate the system design absorption area on one contour, the lines must be stepped. It becomes necessary to install one or more lines at a lower level on the site. This is known as a Serial Distribution System. The line between each horizontal section must be made watertight. It must be designed and fitted so each horizontal line will be utilized to its maximum capacity before the effluent passes to the next lower leach line. It is important that the connecting lines be installed on natural and unfilled ground. If the connection trench is over-dug, effluent will follow the trench down around the pipe, never allowing the upper trenches to completely fill. Therefore, a poured concrete dam should be used to correct an over-dug trench. Failure to follow this procedure will allow the effluent to immediately drain down to the lowest trench and overload it. This will lead to eventual early failure without the upper lines being used to any great degree (see **Figures H 601.10a** and **H 601.10b**).

Disposal Field Failure

As we conclude the topic on disposal fields, it is important to understand the major causes of disposal field failure. Failures may be identified by the submersion of the system; surfacing of effluent, which is evident by a run-off from the site; or by a sluggish or slow flow in the building drain (see **Figure H 601.10c**). Most of these failures can be prevented or delayed. Failures may be attributed to:

1. *Faulty construction methods, incorrect materials or improper physical condition of the site.*

2. *Overloading caused by undersizing the system or the improper evaluation of the soil's permeability.* Overloading can be caused by a leaking toilet or other defective valve. In public-use facilities, self-closing valves help to reduce the loading. Urinals should be equipped with flush valves to avoid the automatic cyclic flow from the flush tanks.

3. *High groundwater levels.* The groundwater level used in the design should be the highest ever recorded. Tests made during dry years or the dry time of the year may provide false data.

4. *The buildup of groundwater under the system.* This is caused by the system being situated just above bedrock or a hard pan layer of soil and by it being located in soils with a low permeability that will not provide for sufficient lateral drainage away from the area.

5. *The loss of infiltration capacity by alteration of the rocks in the soils to clay by wetting and drying or by chemical action.* A percolation test may show that soil has a good percolation rate due to cracks and fissures in the soil. Over a period of time, sewage effluent breaks down the soil and plugs the cracks and fissures. In these cases, the soil needs further evaluation.

6. *Plugging of the filter material by lack of septic tank maintenance.* A carryover of scum or sludge into a disposal field can soon effect the life of the system.

H 701.0 Seepage Pits.

H 701.1 Capacity. The capacity of seepage pits shall be based on the quantity of liquid waste discharging thereinto and on the character and porosity of the surrounding soil, and shall be in accordance with Section H 301.0 of this appendix.

🔧 Most of the requirements for seepage pit design are the same as those for disposal fields and beds. In choosing a site for a seepage pit system, Table H 101.8 requires greater horizontal distances in nearly all cases. The slope to the site is not as critical, although it is a factor when

APPENDIX H

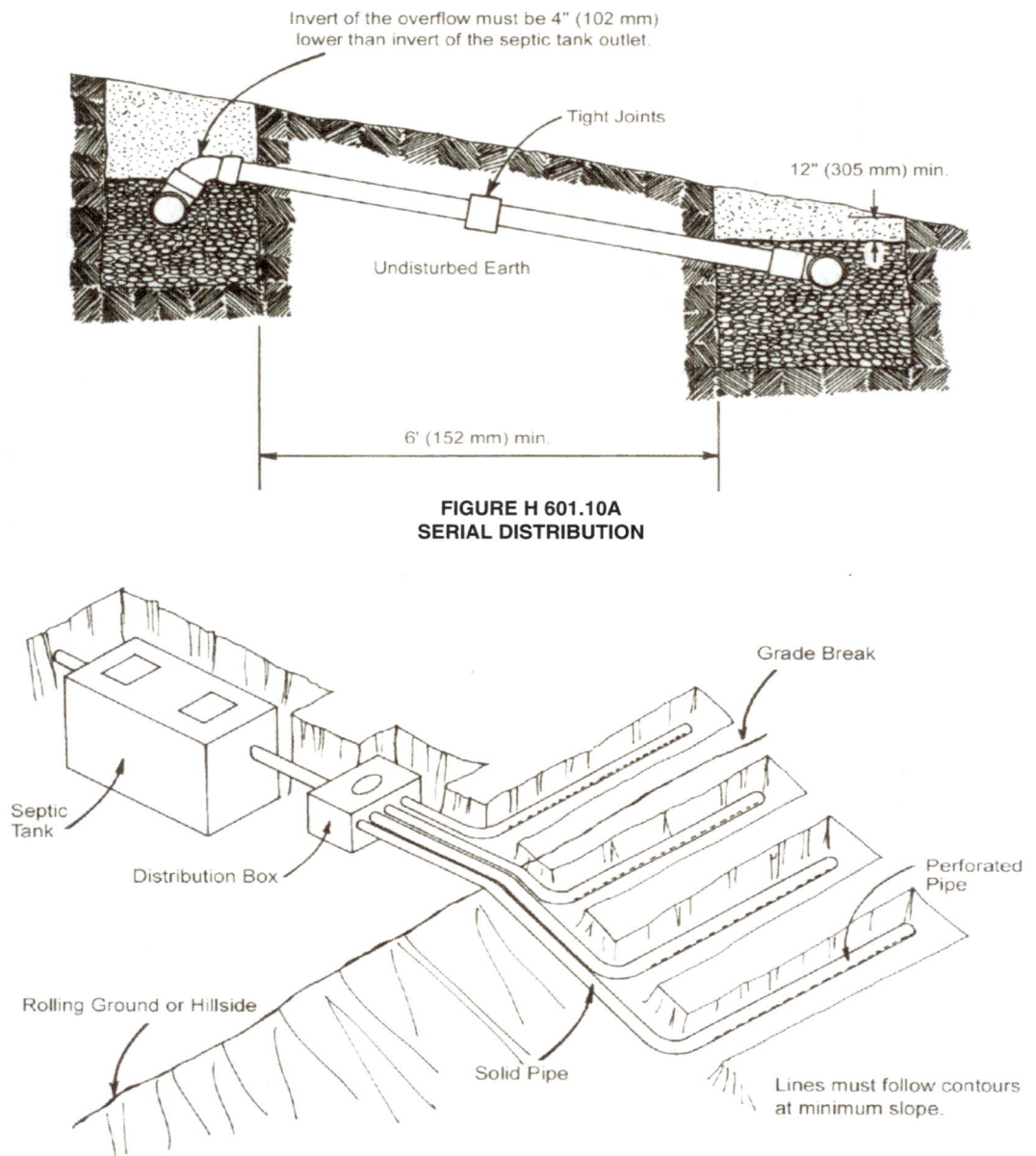

FIGURE H 601.10A
SERIAL DISTRIBUTION

FIGURE H 601.10B
SERIAL DISTRIBUTION ALTERNATIVE

moving, drilling or excavating equipment onto a site. In mountainous areas, many building or health departments restrict the secondary part of the disposal system to disposal fields. The absorption area of the seepage pit is based upon side wall area in square feet. The side wall area is predicated on the required septic tank capacity in gallons or estimated waste/sewage flow rate, whichever is greater. The seepage pit is then designed based on the circumference of the circle, which will provide the appropriate amount of area. See **Figure H 701.1** and the examples below.

H 701.2 Multiple Installations. Multiple seepage pit installations shall be served through an approved distribution box or be connected in series by means of a watertight connection laid on undisturbed or compacted soil. The outlet from the pit shall have an approved vented leg fitting extending not less than 12 inches (305 mm) below the inlet fitting.

H 701.3 Construction. A seepage pit shall be circular in shape and shall have an excavated diameter of not less than 4 feet (1219 mm). Each such pit shall be lined with approved-type whole new hard-burned clay brick, concrete brick, concrete circular-type cesspool blocks, or other approved materials. Approval shall be obtained prior to construction for any pit having an excavated diameter greater than 6 feet (1829 mm).

Larger diameter circles present a safety hazard due to the loading on the lid, as well as the unmortared bricks lining the wall.

H 701.4 Lining. The lining in a seepage pit shall be laid

APPENDIX H

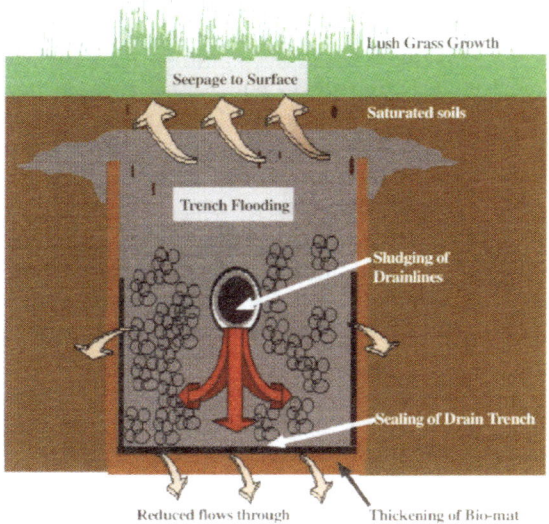

FIGURE H 601.10C
DISPOSAL FIELD FAILURE

on a firm foundation. Lining materials shall be placed tight together and laid with joints staggered. Except in the case of approved-type precast concrete circular sections, no brick or block shall be greater in height than its width, and shall be laid flat to form not less than a 4 inch (102 mm) wall. Brick or block greater than 12 inches (305 mm) in length shall have chamfered matching ends and be scored to provide for seepage. Excavation voids behind the brick, block, or concrete liner shall have not less than 6 inches (152 mm) of clean ¾ of an inch (19.1 mm) gravel or rock.

See **Figure H 701.1**.

The gravel or rock behind the pit liner has three functions:

1. It provides secondary treatment of the effluent to help prolong soil clogging.
2. It fills the voids behind the liner, avoiding ground settlement that sometimes occurs. Voids are created during excavation work by caving or slough-off due to rocks or loose strata.
3. It reduces silting of the pit. Soil with a considerable amount of fines or silt has a tendency to work into the pit through the pit wall by pumping action. As effluent rises and lowers due to changes in loading, some of the effluent, which has moved away from the pit laterally, will flow back into the pit as the level in the pit lowers, bringing with it fine silt from the surrounding soils. Over a period of time, this can significantly change the pit depth, as well as create voids behind the liner. Rock behind the liner helps to prevent this problem.

H 701.5 Brick and Block. Brick or block used in seepage pit construction shall have a compressive strength of not less than 2500 pounds per square inch (lb/in^2) (1 757 674 kg/m^2).

H 701.6 Sidewall. A seepage pit shall have a minimum sidewall (not including the arch) of 10 feet (3048 mm) below the inlet.

H 701.7 Arch and Dome. The arch or dome of a seepage pit shall be permitted to be constructed in one of three ways:

(1) Approved-type hard-burned clay brick or solid concrete brick or block laid in cement mortar.

(2) Approved brick or block laid dry. In both of the above methods, an approved cement mortar covering of not less than 2 inches (51 mm) in thickness shall be applied, said covering to extend not less than 6 inches (152 mm) beyond the sidewalls of the pit.

(3) Approved-type one or two-piece reinforced concrete slabs of not less than 2500 lb/in^2 (1 757 674 kg/m^2) minimum compressive strength, not less than 5 inches (127 mm) thick, and designed to support an earth load of not less than 400 pounds per square foot (lb/ft^2) (1953 kg/m^2). Each such cover shall be provided with a 9 inch (229 mm) minimum inspection hole with plug or cover and shall be coated on the underside with an approved bituminous or other nonpermeable protective compound.

H 701.8 Location. The top of the arch or cover shall be not less than 18 inches (457 mm) but not exceed 4 feet (1219 mm) below the surface of the ground.

H 701.9 Inlet Fitting. An approved vented inlet fitting shall be provided in the seepage pit so arranged as to prevent the inflow from damaging the sidewall.

Exception: Where using a one- or two-piece concrete slab cover inlet, fitting shall be permitted to be a one-fourth bend fitting discharging through an opening in the top of the slab cover. On multiple seepage pit installations, the outlet fittings shall comply with Section H 701.2 of this appendix.

See **Figure H 701.9a**.

Seepage Pit Failure

Seepage pit failure is usually evident by a sluggish or stopped-up building drain or backup in the lower fixtures in the building. Another indication is the surfacing of effluent, especially if the pit is located on a considerable downhill slope from the building. Caving or subsidence at the pit may also indicate a failure. The causes of failure are basically the same as for disposal fields:

1. Faulty construction.
2. Overloading caused by under sizing or improper evaluation of the soil's permeability.
3. High groundwater.
4. The loss of infiltration capacity by altering the rocks in the soil to clay by wetting and drying or by chemical action.
5. Silting of the pit, as mentioned earlier.
6. Lack of septic tank maintenance. Each seepage pit shall have a minimum side wall depth of 10 feet below inlet, not including the arch.
7. In old seepage pits structural failure due to the erosion of concrete by the attack of gases generated by the sewage.

APPENDIX H

Reasonably new systems can fail sometimes, due to a breakage of the building sewer connection to or from the septic tank. This is due to settlement of the septic tank or seepage pit. Replacement seepage pits have been dug only to find that the trouble is not with the existing seepage pit, but with building sewer piping.

Examples of Seepage Pit Sizing (see **Figure H 701.9b**):

1. A four-bedroom dwelling requires a 1,200-gallon septic tank. Assume the soil has been classified as Type 3 (sandy loam or sandy clay) 40 ft^2/100 gallons of septic tank.

Solution:

40 x 12 = 480 ft^2 side wall area.

Assume a 5-foot excavated diameter seepage pit is to be used. The full excavated diameter is used, not the diameter of the brick liner. The circumference of a 5-foot hole is 15.71 feet.

480 ÷ 15.71 = 30.55 feet effective depth below the inlet.

The seepage pit will be a 5-foot diameter by 30.55-feet deep excavation (see **Figure H 701.9b**).

2. An apartment house requires a 4,900-gallon septic

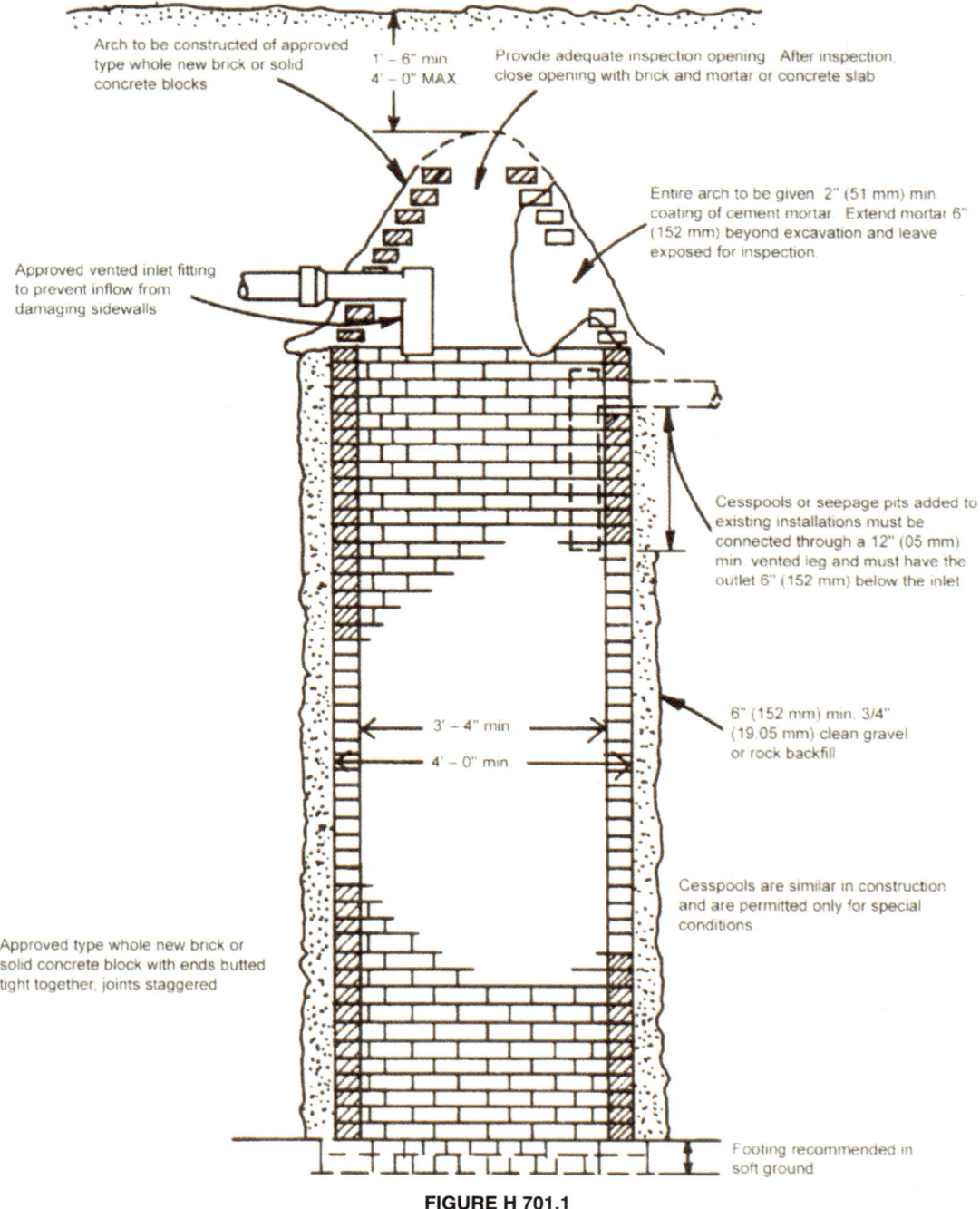

FIGURE H 701.1
SEEPAGE PIT CONSTRUCTION

tank. Assume the soil has been classified as Type 2 (fine sand) 25 ft²/100 gallons of septic tank.

Solution:

25 × 49 = 1,225 ft² side wall area. Assume that multiple 6-feet diameter seepage pits will be used. The circumference of a 6-foot diameter excavation is 18.85 feet. 1,225 ÷ 18.85 = 64.99 feet effective depth below the inlet needed.

Because the depth is excessive, it is advisable to use two pits, each 32.5 feet below inlet. These may be connected to the septic tank through a distribution box or be connected in series (see **Figure H 701.9b**).

3. In the movie theater problem earlier in this chapter, the flow rate for the septic tank design was required to be used so that method must also be used in sizing the seepage pit. The flow rate was 3,500 gallons/day. Assume the soil has been classified as Type 3 (sandy loam or sandy clay) 2.5 gallons/ft² per 24 hours.

Solution:

3,500 ÷ 2.5 = 1,400 ft² side wall area.

Assume that 5 feet diameter seepage pits will be used.

1,400 ÷ 15.71 = 89.1 ft. effective depth below inlet.

Because of excessive depth, it would be best to use three pits, 5 feet in diameter and 30 feet below inlet (see **Figure H 701.9b**).

H 801.0 Cesspools.

H 801.1 Limitations. A cesspool shall be considered as a temporary expedient pending the construction of a public sewer; as an overflow facility where installed in conjunction with an existing cesspool; or as a means of sewage disposal for limited, minor, or temporary uses, where first approved by the Authority Having Jurisdiction.

H 801.2 Septic Tanks. Where it is established that a public sewer system will be available in less than 2 years, and soil and groundwater conditions are favorable to cesspool disposal, cesspools without septic tanks shall be permitted to be installed for single-family dwellings or for other limited uses where first approved by the Authority Having Jurisdiction.

H 801.3 Construction. Each cesspool, where permitted, shall be in accordance with the construction requirements set forth in Section H 701.0 of this appendix for seepage pits and shall have a sidewall (not including arch) of not less than 20 feet (6096 mm) below the inlet, provided, however, that where a strata of gravel or equally pervious material of 4 feet (1219 mm) in thickness is found, the depth of such sidewall shall not exceed 10 feet (3048 mm) below the inlet.

H 801.4 Existing Installations. Where overflow cesspools or seepage pits are added to existing installations, the effluent shall leave the existing pit through an approved vented leg extending not less than 12 inches (305 mm) downward into such existing pit and having its outlet flow line not less than 6 inches (152 mm) below the inlet. Pipe between pits shall be laid with approved watertight joints.

The cesspool is simply a seepage pit installation without the septic tank. Without the septic tank, all of the solids deposited into the drainage system enter the cesspool. This is not the best system for solids treatment; indeed, this was the original treatment system for waste

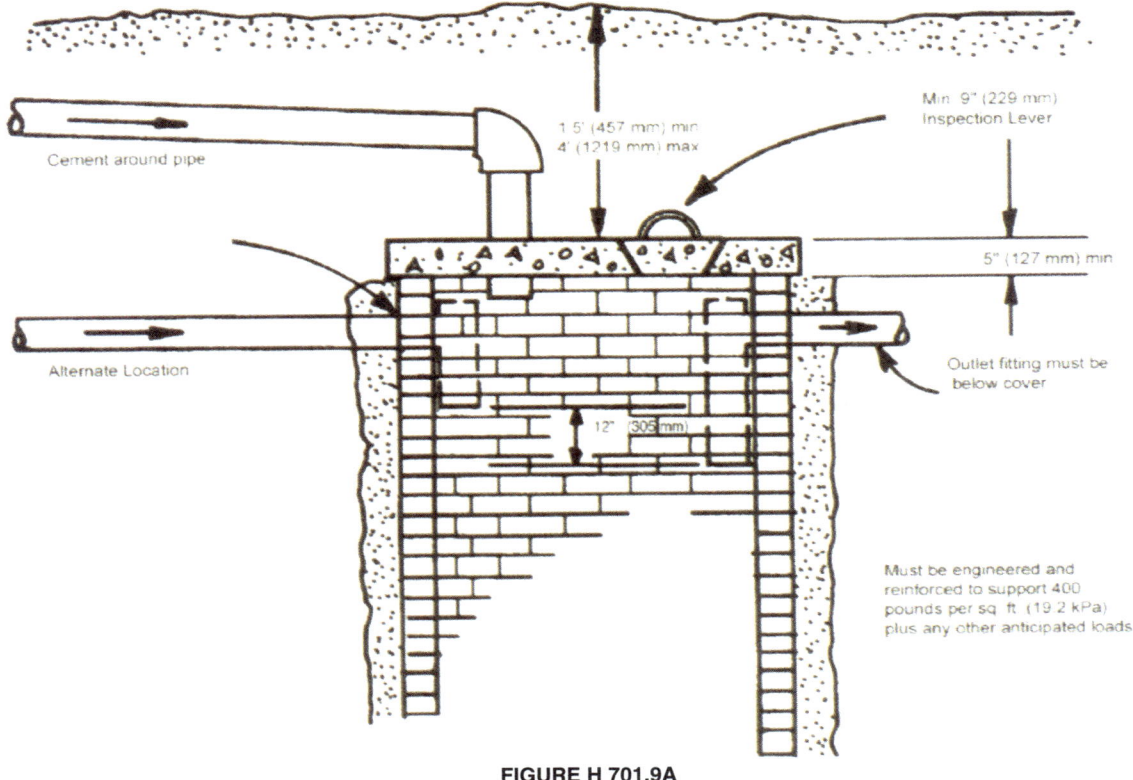

**FIGURE H 701.9A
SEEPAGE PIT WITH CONCRETE SLAB TOP**

centuries ago – dig a hole and dump the waste in it. The maturing plumbing industry prohibited this method long ago except for temporary or emergency situations. It is, however, an accepted method of disposal for clear liquid waste where a drainage system is not available. Examples of minor use of a cesspool are for condensate waste disposal and swimming pool filter backwash disposal.

H 901.0 Commercial or Industrial Special Liquid-Waste Disposal.

H 901.1 Interceptor. Where liquid wastes contain excessive amounts of grease, garbage, flammable wastes, sand, or other ingredients that affect the operation of a private sewage disposal system, an interceptor for such wastes shall be installed.

H 901.2 Installation. Installation of such interceptors shall comply with Section 1009.0 of this code, and their location shall comply with Table H 101.8 of this appendix.

H 901.3 Sampling Box. A sampling box shall be installed where required by the Authority Having Jurisdiction.

H 901.4 Design and Structural Requirement. Interceptors shall be of approved design and be not less than two compartments. Structural requirements shall comply with Section H 501.0 of this appendix.

H 901.5 Location. Interceptors shall be located as close to the source as possible and be accessible for servicing. Necessary manholes for servicing shall be at grade level and be gastight.

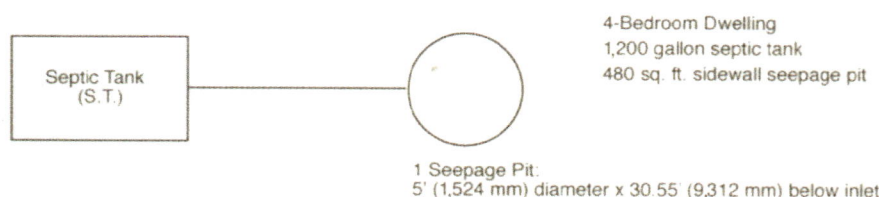

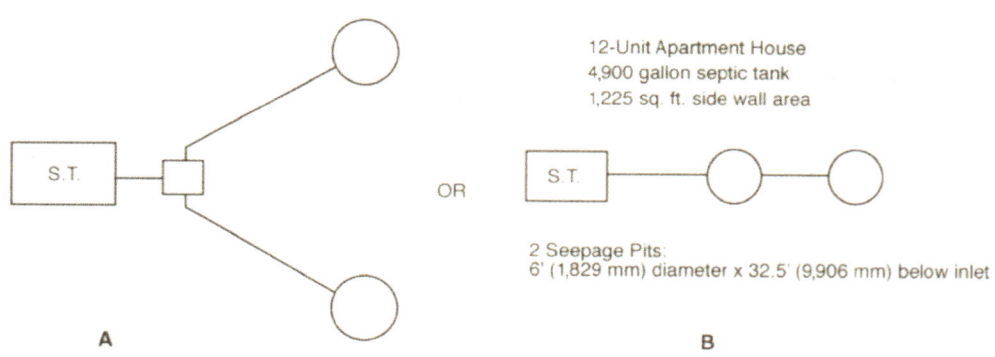

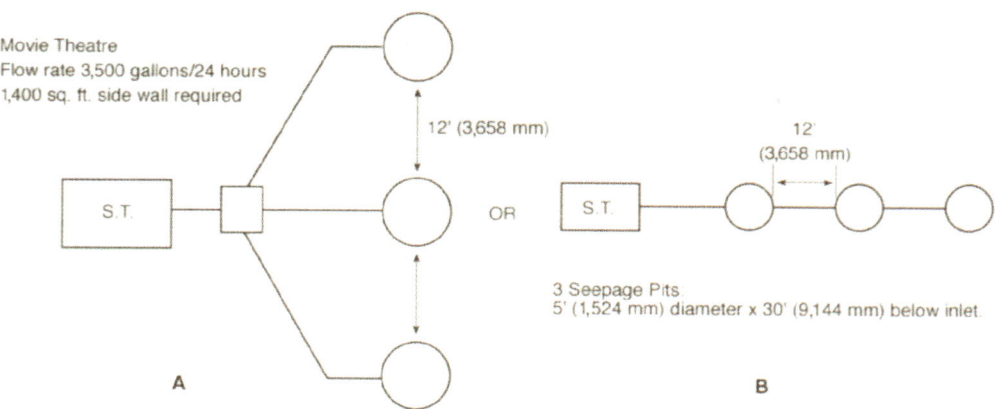

FIGURE H 701.9B
SOLUTIONS TO SEEPAGE PIT SIZING EXAMPLES

H 901.6 Waste Discharge. Waste discharge from interceptors shall be permitted to be connected to a septic tank or other primary system or be disposed into a separate disposal system.

H 901.7 Design Criteria. A formula shall be permitted to be adapted to other types of occupancies with similar wastes. (See Chart H 901.7)

The design of the above commercial interceptor installations are prepared by design professionals and the design parameters included here are written to assist them in that design. The journeyman plumber would not normally be involved in this design phase, although he or she would of course install such a system. However, except for sizing, these systems will be roughly the same as those discussed in Section 1009.0, Interceptors (Clarifiers) and Separators, of Chapter 10.

H 1001.0 Inspection and Testing.

H 1001.1 Inspection. Inspection requirements shall comply with the following:

(1) Applicable provisions of Section 105.0 of this code and this appendix shall be required. Plans shall be required in accordance with Section 103.3 of this code.

(2) System components shall be properly identified as to manufacturer. Septic tanks or other primary systems shall have the rated capacity permanently marked on the unit.

(3) Septic tanks or other primary systems shall be installed on dry, level, well-compacted soil.

(4) Where design is predicated on soil tests, the system shall be installed at the same location and depth as the tested area.

H 1001.2 Testing. Testing requirements shall comply with the following:

(1) Septic tanks or other primary components shall be filled with water to flow line prior to requesting inspection. Seams or joints shall be left exposed (except the bottom), and the tank shall remain watertight.

(2) A flow test shall be performed through the system to the point of effluent disposal. All lines and components shall be watertight. Capacities, required air space, and fittings shall comply with the provisions set forth in this appendix.

H 1101.0 Abandoned Sewers and Sewage Disposal Facilities.

H 1101.1 Plugged and Capped. An abandoned building (house) sewer, or part thereof, shall be plugged or capped in an approved manner within 5 feet (1524 mm) of the property line.

H 1101.2 Fill Material. A cesspool, a septic tank, or a seepage pit that has been abandoned or has been discontinued otherwise from further use, or to which no waste or soil pipe from a plumbing fixture is connected, shall have the sewage removed therefrom and be completely filled with the earth, sand, gravel, concrete, or other approved material.

H 1101.3 Filling Requirements. The top cover or arch over the cesspool, septic tank, or seepage pit shall be removed before filling, and the filling shall not extend above the top of the vertical portions of the sidewalls or above the level of any outlet pipe until inspection has been called and the cesspool, septic tank, or seepage pit has been inspected. After such inspection, the cesspool, septic tank, or seepage pit shall be filled to the level of the top of the ground.

H 1101.4 Owner. No person owning or controlling a cesspool, septic tank, or seepage pit on the premises of such person or in that portion of any public street, alley, or other public property abutting such premises shall fail, refuse, or neglect to be in accordance with the provisions of this section or upon receipt of notice so to be in accordance with the Authority Having Jurisdiction.

H 1101.5 Permittee. Where disposal facilities are abandoned consequent to connecting any premises with the public sewer, the permittee making the connection shall fill all abandoned facilities in accordance with the Authority Having Jurisdiction within 30 days from the time of connecting to the public sewer.

H 1201.0 Drawings and Specifications.

H 1201.1 General. The Authority Having Jurisdiction, Health Officer, or other department having jurisdiction shall be permitted to require the following information before a permit is issued for a private sewage disposal system or at a time during the construction thereof:

(1) Plot plan drawn to scale, completely dimensioned, showing direction and approximate slope of surface, location of present or proposed retaining walls, drainage channels, water supply lines or wells, paved areas and structures on the plot, number of bedrooms or plumbing fixtures in each structure, and location of the private sewage disposal system with relation to lot lines and structures.

(2) Details of construction necessary to ensure compliance with the requirements of this appendix together with a full description of the complete installation including quality, kind, and grade of materials, equipment, construction, workmanship, and methods of assembly and installation.

(3) A log of soil formations and groundwater levels as determined by test holes dug in close proximity to a proposed seepage pit or disposal field, together with a statement of water absorption characteristics of the soil at the proposed site, as determined by approved percolation tests.

APPENDIX H

CHART H 901.7
RECOMMENDED DESIGN CRITERIA

GREASE AND GARBAGE, COMMERCIAL KITCHENS									
Number of meals per peak hour	x	Waste flow rate[1]	x	Retention time[2]	x	Storage factor[3]	=	Interceptor size (liquid capacity)	
SAND-SILT OIL, AUTO WASHERS									
Number of vehicles per hour	x	Waste flow rate[1]	x	Retention time[2]	x	Storage factor[3]	=	Interceptor size (liquid capacity)	
SILT-LINT GREASE, LAUNDRIES, LAUNDROMATS									
Number of machines	x	2 cycles per hour	x	Waste flow rate[1]	x	Retention time[2]	=	Storage factor[3]	=	Interceptor size (liquid capacity)

Notes:
[1] For waste flow rate see Table H 201.1(4).
[2] Retention Times:
 (a) Kitchen (commercial) – with dishwasher, garbage disposal, or both = 2.5 hours
 (b) Kitchen (single service) – with garbage disposal = 1.5 hours
 (c) Auto Washers (sand-silt oil) = 2.0 hours
 (d) Laundries/Laundromats = 2.0 hours
[3] Storage Factors:
 (a) Kitchen (commercial) – with 8 hours operation = 1
 (b) Kitchen (commercial) – with 16 hours operation = 2
 (c) Kitchen (commercial) – with 24 hours operation = 3
 (d) Kitchen (single service) = 1.5
 (e) Auto Washers (sand-silt oil) – with self service = 1.5
 (e) Auto Washers (sand-silt oil) – with employee operated = 2
 (d) Laundries/Laundromats – with rock filter = 1.5 hours

"The information contained in this appendix is not part of this American National Standard (ANS) and has not been processed in accordance with ANSI's requirements for an ANS. As such, this appendix may contain material that has not been subjected to public review or a consensus process. In addition, it does not contain requirements necessary for conformance to the standard."

The following IAPMO Installation Standard is included here for the convenience of the users of the Uniform Plumbing Code. It is not considered as a part of the Uniform Plumbing Code unless formally adopted as such. This Installation Standard is an independent, stand-alone document published by the International Association of Plumbing and Mechanical Officials and is printed herein by the expressed written permission of IAPMO.

APPENDIX I

INSTALLATION STANDARD FOR PEX TUBING SYSTEMS FOR HOT- AND COLD-WATER DISTRIBUTION

IAPMO IS 31-2014

1.0 Scope.

1.1 General.

1.1.1 This Standard specifies requirements for the installation of SDR 9 CTS crosslinked polyethylene (PEX) tubing and fittings, including cold-expansion, crimp, press, and mechanical compression fittings, intended for hot- and cold-water distribution systems within buildings.

1.1.2 This Standards applies to

(a) SDR 9 CTS PEX tubing complying with ASTM F876 and pressure-rated in accordance with PPI TR-3; and

(b) PEX fitting systems complying with

(i) ASTM F877, for mechanical compression fittings and metal or plastic insert fittings with stainless steel press sleeves;

(ii) ASTM F1807 or ASTM F2159, for metal or plastic insert fittings with copper crimp rings;

(iii) ASTM F1960, for cold expansion fittings with PEX reinforced rings; or

(iv) ASTM F2080, for cold expansion fittings with metal compression sleeves.0

1.2 Terminology.

In this Standard,

(a) "shall" is used to express a requirement, i.e., a provision that the user is obliged to satisfy to comply with the Standard;

(b) "should" is used to express a recommendation, but not a requirement;

(c) "may" is used to express an option or something permissible within the scope of the Standard; and

(d) "can" is used to express a possibility or a capability.

Notes accompanying sections of the Standard do not specify requirements or alternative requirements; their purpose is to separate explanatory or informative material from the text. Notes to tables and figures are considered part of the table or figure and can be written as requirements.

1.3 Amendments.

Proposals for amendments to this Standard will be processed in accordance with the standards-writing procedures of IAPMO.

2.0 Reference Publications.

This Standard refers to the following publications, and where such reference is made, it shall be to the current edition of those publications, including all amendments published thereto.

ASTM F876 — Standard Specification for Crosslinked Polyethylene (PEX) Tubing

ASTM F877 — Standard Specification for Crosslinked Polyethylene (PEX) Hot- and Cold-Water Distribution Systems

ASTM F1807 — Standard Specification for Metal Insert Fittings Utilizing a Copper Crimp Ring for SDR9 Cross-linked Polyethylene (PEX) Tubing and SDR9 Polyethylene of Raised Temperature (PE-RT) Tubing

ASTM F1960 — Standard Specification for Cold Expansion Fittings with PEX Reinforcing Rings for Use with Cross-linked Polyethylene (PEX) Tubing

ASTM F2080 — Standard Specification for Cold-Expansion Fittings With Metal Compression-Sleeves for Cross-Linked Polyethylene (PEX) Pipe

ASTM F2159 — Standard Specification for Plastic Insert Fittings Utilizing a Copper Crimp Ring for SDR9 Cross-linked Polyethylene (PEX) Tubing and SDR9 Polyethylene of Raised Temperature (PE-RT) Tubing

ASTM F2657 — Standard Test Method for Outdoor Weathering Exposure of Crosslinked Polyethylene (PEX) Tubing

AWWA C904 — Cross-Linked Polyethylene (PEX) Pressure Tubing, ½ In. (12 mm) Through 3 In. (76 mm) for Water Service

IAPMO/ANSI UPC-1 — Uniform Plumbing Code

PPI TR-3   Policies and Procedures for Developing Hydrostatic Design Basis (HDB), Pressure Design Basis (PDB), Strength Design Basis (SDB), and Minimum Required Strength (MRS) Ratings for Thermoplastic Tubing Materials or Tubing

3.0 Abbreviations.

The following abbreviations apply in this Standard:

CTS — copper tube size

HDPE — high density polyethylene

IC — insulation contact

NTS — nominal tubing size

PEX — crosslinked polyethylene

SDR — standard dimension ratio

UV — ultraviolet light

4.0 General.

4.1 Tubing.

4.1.1 PEX tubing can be

(a) pigmented throughout (i.e., with color);

(b) non-pigmented (e.g., translucent or natural); or

(c) coated with a pigmented layer.

4.1.2 PEX tubing is typically available in NTS-$\frac{1}{4}$ to NTS-3.

4.1.3 Before installation, the installer shall review the tubing markings and verify that

(a) the standard designation(s) of the fittings to which the tube can be joined to is included in the markings;

(b) it bears a certification mark from an accredited certification organization; and

(c) pressure and temperature ratings meet or exceed that of the intended end-use.

4.2 Fittings.

4.2.1 Cold-Expansion Fittings.

Cold-expansion fittings typically

(a) are made of brass, stainless steel, or sulfone;

(b) consist of an insert and a PEX reinforcing ring; and

(c) are available in NTS-$\frac{3}{8}$ to NTS-3.

4.2.2 Crimp or Press Insert Fittings.

Crimp or press insert fittings typically

(a) are made of brass, stainless steel, or sulfone;

(b) consist of an insert and a copper crimp ring or a stainless steel press ring

(c) are available in NTS-$\frac{3}{8}$ to NTS-2.

4.2.3 Compression Fittings.

Compression (i.e., transition) fittings typically

(a) are made of brass; and

(b) consist of

(i) a nut, a compression ring, and an insert; or

(ii) an O-ring brass insert with a compression sleeve

(c) are available in NTS-$\frac{1}{4}$ to NTS-3.

4.3 Installation.

Only fittings systems marked on the tubing shall be used for installation with that particular tubing.

4.4 Tools.

Tools and tool accessories (e.g., tool heads) used for the installation of PEX tubing systems shall be in accordance with the manufacturer's specifications and written instructions.

4.5 Tubing Protection.

4.5.1 Abrasion.

PEX tubing passing through drilled or notched metal studs or metal joists, or hollow-shell masonry walls shall be protected from abrasion by elastomeric or plastic sleeves or grommets.

4.5.2 Puncture.

Steel-plate protection shall be installed in accordance with the local plumbing code.

5.0 Handling.

5.1 Receiving.

When receiving PEX tubing shipments, the receiver shall inspect and inventory each shipment, ensuring that there has been no loss or damage. In addition:

(a) At the time of unloading, the markings of all tubing, fittings, and accessories shall be verified to ensure that all items have been manufactured in accordance with the applicable product Standard and appropriately certified.

(b) An overall examination of the shipment shall be made. If the shipment is intact, ordinary inspection while unloading shall be sufficient to ensure that the items have arrived in good condition.

(c) If the load has shifted, has broken packaging, or shows evidence of rough treatment, each item shall be carefully inspected for damage.

(d) The total quantities of each shipment (e.g., tubing, gaskets, fittings, and accessories) shall be checked against shipping records.

(e) Any damaged or missing items shall be noted on the delivery slip. The carrier shall be notified immediately and a claim made in accordance with its instructions.

(f) No damaged material shall be disposed of. The carrier shall recommend the procedure to follow.

(g) Shortages and damaged materials are normally not reshipped without request. If replacement material is needed, it shall be reordered from the manufacturer, the distributor, or a manufacturer's representative.

5.2 Storage and UV Exposure.

5.2.1 PEX tubing and fittings shall be stored indoors and in its original packaging until the time of installation. Appropriate precautions to protect the tubing from damage, impact, and punctures shall be taken.

5.2.2 Accumulative exposure time to UV radiation during storage and installation shall not exceed the UV exposure limits recommended by the manufacturer or specified in ASTM F876.

Note: *ASTM F876 has four categories for UV-resistance, ranging from untested to 6 months of continuous exposure, as listed in the material designation code.*

5.3 Exposure to Heat.

5.3.1 PEX tubing and fittings shall not be exposed to open flames.

5.3.2 PEX tubing shall not be exposed to temperatures exceeding 93°C (200°F).

5.4 Exposure to Chemicals.

5.4.1 Chemical compatibility (e.g., with common construction materials) shall be verified with the manufacturer prior to direct contact.

5.4.2 In general, petroleum- or solvent-based chemicals (e.g., paints, greases, pesticides, or sealants) shall not be allowed to come in direct contact with PEX tubing or fittings.

6.0 Thermal Expansion and Contraction.

6.1 Horizontal Tubing Runs.

Thermal expansion and contraction forces on suspended horizontal runs of PEX tubing that can experience a 22°C (40°F) or greater change in temperature (operating temperature compared to ambient temperature) shall be controlled by a means of mitigating temperature-induced stresses to other parts of the water distribution system. Means for controlling thermal expansion and contraction include

(a) loops;

(b) offsets;

(c) arms with rigid anchor points; and

(d) supporting the tubing with continuous runs of CTS support channels with

(i) rigid anchor points installed every 20 m (65 ft); and

(ii) proper strapping (e.g., 27 kg (60 lb) straps or equivalent) spaced 1 m (3 ft) and rated for the maximum temperature and UV exposure of the PEX tubing application.

6.2 Vertical Tubing Runs.

Thermal expansion and contraction forces on vertical runs of PEX tubing that pass through more than one floor and can experience a 22°C (40°F) or greater change in temperature (operating temperature compared to ambient temperature) shall be controlled by installing

(a) a riser clamp at the top of every other floor; and

(b) mid-story guides to maintain the alignment of the vertical tubing.

Note: *Installing riser clamps isolates expansion and contraction to two-floor intervals allowing the PEX tubing to naturally compensate for the expansion and contraction.*

6.3 Clearance.

Adequate clearance shall be provided between PEX tubing and the building structure (e.g., using bored holes and sleeves) to allow for free longitudinal movement of the tubing.

6.4 Expansion Arms and Expansion Loops.

6.4.1 Expansion Arms (See Figure 1).

6.4.1.1 Expansion arms shall be installed as illustrated in Figure 1.

6.4.1.2 The minimum length of expansion arms shall be calculated using the following equation:

$$LB = C \times \sqrt{(D \times \Delta L)}$$

where

LB = length of flexible arm

C = material constant (12 for PEX)

D = nominal outside diameter of tubing

ΔL = thermal expansion length

6.4.2 Expansion Loops (See Figure 2).

6.4.2.1 Expansion loops shall be installed at the mid-point between anchors, as illustrated in Figure 2.

6.4.2. The minimum length of expansion loops shall be calculated using the equation in Section 6.4.1.2; however, the distance LB shall be divided into three sections, as illustrated in Figure 2, where

$L1 = LB \div 5$; and

$L2 = L1 \times 2$

7.0 Hangers and Supports.

7.1 Vertical Tubing.

Vertical PEX tubing shall

(a) be supported at each floor or as specified by the water-distribution system designer to allow for expansion and contraction; and

(b) have mid-story guides.

7.2 Horizontal Tubing.

Unless otherwise authorized by the authority having jurisdiction, suspended horizontal runs of PEX tubing

(a) NTS-1 and smaller shall be supported every 0.8 m (32 in), unless continuously supported by metallic CTS or V channels that

(i) are supported at intervals not exceeding 1.8 m (6 ft);

(ii) have a maximum cantilever, measured from the support to the end of the CTS support channel, of 0.5 m (1.5 ft); and

(b) NTS-1¼ and larger shall be supported every 1.2 m (4 ft), unless continuously supported by metallic CTS or V channels that

(i) are supported at intervals not exceeding 2.4 m (8 ft); and

(ii) have a maximum cantilever, measured from the support to the end of the CTS support channel, of 0.5 m (1.5 ft).

7.3 Anchors.

Anchors shall be

(a) used to restrict PEX tubing movement;

(b) made of materials that provide rigidity to the support system and utilize pipe clamps designed for plastic tubing capable of restraining the tubing; and

(c) installed in accordance with Figures 1 or 2, as applicable (i.e., anchor distances and size of arms and offsets).

Note: *Anchors are typically installed every 20 m (65 ft). See Section 6.*

8.0 Joints and Connections.

8.1 Assembly Procedure.

The procedure for making joints shall be as specified by the manufacturer.

8.2 Concealed Joints.

PEX tubing systems manufactured in accordance with the applicable standards referenced in Section 2 are deemed manufactured joints and may be installed in concealed spaces without the need for access panels.

9.0 Clearances.

9.1 Gas Vents.

Except for double-wall B-vents, which require a 25 mm (1 in) clearance, the clearance between gas appliance vents and PEX tubing shall be at least 150 mm (6 in).

9.2 Recessed Light Fixtures.

Except when the PEX tubing is protected with fiberglass or closed-cell insulation or the recessed light is IC-rated, the clearance between recessed light fixtures and PEX tubing shall be at least 300 mm (12 in).

9.3 Fluorescent Lighting.

When in direct view of the light source, the clearance between fluorescent lighting and PEX tubing shall be at least 1.5 m (5 ft). If the minimum clearance cannot be achieved, the PEX tubing shall be protected with a UV-blocking sleeve.

10.0 Other Considerations.

10.1 Hot-Work Joints.

Hot-work joints (e.g., soldering, brazing, welding, and fusion-welding) shall be

(a) made at least 500 mm (18 in) from PEX tubing in the same water line; and

(b) performed prior to completing the PEX joints.

10.2 Bending Radius.

10.2.1 The free (unsupported) bending radius for PEX tubing, measured at the outside of the bend, shall be not less than six times the actual outside diameter of the tubing, unless otherwise specified by the PEX manufacturer. Supports should be used to facilitate rigid bends and to alleviate stress on PEX joints when bends are needed in close proximity to such joints.

10.2.2 Tighter bends may be used when the PEX tubing is uniformly bent (supported) around a curved bracket or other rigid fixture. In this case, the minimum outside radius of the supported bend shall be as specified by the PEX manufacturer.

10.3 Directional Fittings.

Directional fittings (e.g., 90° and 45° elbows) should only be installed where necessary.

Note: *The flexible nature of PEX tubing allows for sweeping bends resulting in less fittings and joints.*

10.4 Direct Burial.

PEX tubing and fittings may be used in direct burial applications when allowed in the manufacturer's written installation instructions.

Note: *AWWA C904 should be consulted for water service applications.*

10.5 Fire-Resistive Construction.

Manufacturer's installation instructions shall be consulted prior to installation of PEX tubing in fire resistive constructions. PEX tubing penetrating a wall or floor-and-ceiling fire-rated assembly shall include a means of passive fire protection in accordance with the local codes.

10.6 Sizing and Flow Velocities.

10.6.1 PEX tubing shall be sized in accordance with IAPMO/ANSI UPC 1.

Note: *Potable water piping sizing is addressed in Section 610.0 and Appendix A of IAPMO/ANSI UPC 1-2012.*

10.6.2 The tubing manufacturer's pressure-loss data should be referenced when using Appendix A of IAPMO/ANSI UPC 1. In absence of such data, Figures 3 and 4 shall be used.

10.6.3 Flow velocities through the water distribution system, used for calculating flush tank and flush valve fixture units depending on the tubing sizes (see Table 1), shall not exceed

(a) 3.0 m/s (10 ft/s) for cold-water distribution systems; and

(b) 2.4 m/s (8 ft/s) for hot-water distribution systems.

Note: *The flow velocities in Items (a) and (b) account for the increased velocities through the fittings.*

10.6.4 Hot-water recirculation systems shall

(a) be balanced to maintain adequate system temperatures; and

(b) have flow velocities that do not exceed 0.6 m/s (2 ft/s) (see Table 2); and

(c) use only PEX tubing designated for hot, chlorinated water recirculation systems and rated for the maximum percentage of time during which the system is intended to be operated at elevated temperatures, in accordance with ASTM F876.

10.7 Installation Testing.

Installation of PEX water distribution systems may be tested with air when

(a) expressly allowed in the written instructions of the manufacturers of all plastic pipe and fittings installed at the time the PEX piping system is being tested; and

(b) compressed air or other gas testing is not prohibited by the authority having jurisdiction.

APPENDIX I

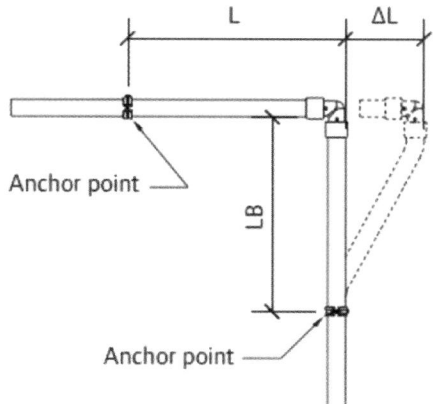

Note:

$LB = C \times \sqrt{(D \times \Delta L)}$

where

LB = length of flexible arm
C = material constant (12 for PEX)
D = nominal outside diameter of tubing
ΔL = thermal expansion length

**FIGURE 1
EXPANSION ARMS**
(See Sections 6.4.1 and 7.3)

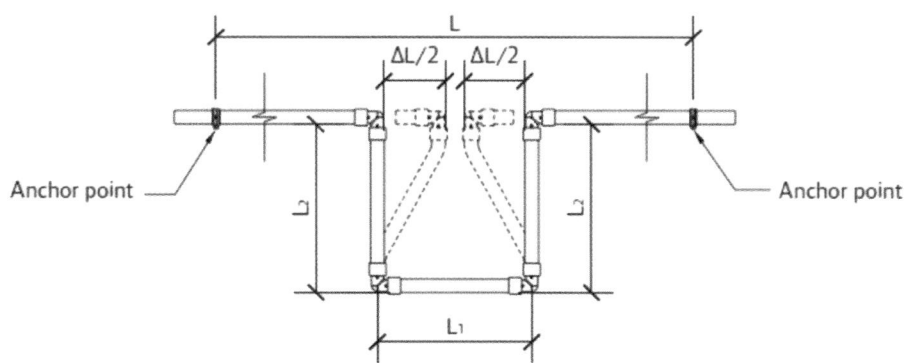

Note:
LB shall be calculated as specified in Figure 2 and divided into three sections, as follows:

$LB = L1 + (2 \times L2)$

where

$L1 = LB \div 5$; and
$L2 = L1 \times 2$.

**FIGURE 2
EXPANSION LOOPS**
(See Sections 6.4.2 and 7.3)

APPENDIX I

TABLE 1
CALCULATION OF FLUSH TANK AND FLUSH VALVE FIXTURE UNITS
(See Section 10.6.3)

Nominal Tubing Size	Flow Velocity: 3.0 m/s (10 ft/s)			Flow Velocity: 2.4 m/s (8 ft/s)		
	Flow Volume, L/min (gpm)	Flush Tank Fixture Units	Flush Valve Fixture Units	Flow, L/min (gpm)	Flush Tank Fixture Units	Flush Valve Fixture Units
1/2	20.8 (5.5)	6	—	16.7 (4.4)	4	—
3/4	41.6 (11.0)	15	—	33.3 (8.8)	11	—
1	68.9 (18.2)	26	—	55.3 (14.6)	20	—
1-1/4	103.0 (27.2)	46	10	82.5 (21.8)	33	5
1-1/2	143.5 (37.9)	77	24	114.7 (30.3)	54	13
2	246.1 (65.0)	200	91	196.8 (52.0)	135	52
3	533.0 (140.8)	590	495	426.2 (112.6)	443	310

TABLE 2
TUBING SIZES, FLOWS, AND FRICTION LOSSES FOR
HOT-WATER RECIRCULATION SYSTEMS
(See Section 10.6.4)

Nominal Tubing Size	Flow Velocity m/s (ft/s)	Flow Volume L/min (gpm)	Friction Losses at 49 °C (120°F) kPa/m (psi/ft)
½	0.6 (2)	4.2 (1.1)	0.4411 (0.0195)
¾	0.6 (2)	8.3 (2.2)	0.2850 (0.0126)
1	0.6 (2)	13.6 (3.6)	0.2081 (0.0092)
1¼	0.6 (2)	20.4 (5.4)	0.1629 (0.0072)
1½	0.6 (2)	28.4 (7.5)	0.1335 (0.0059)
2	0.6 (2)	48.8 (12.9)	0.0950 (0.0042)

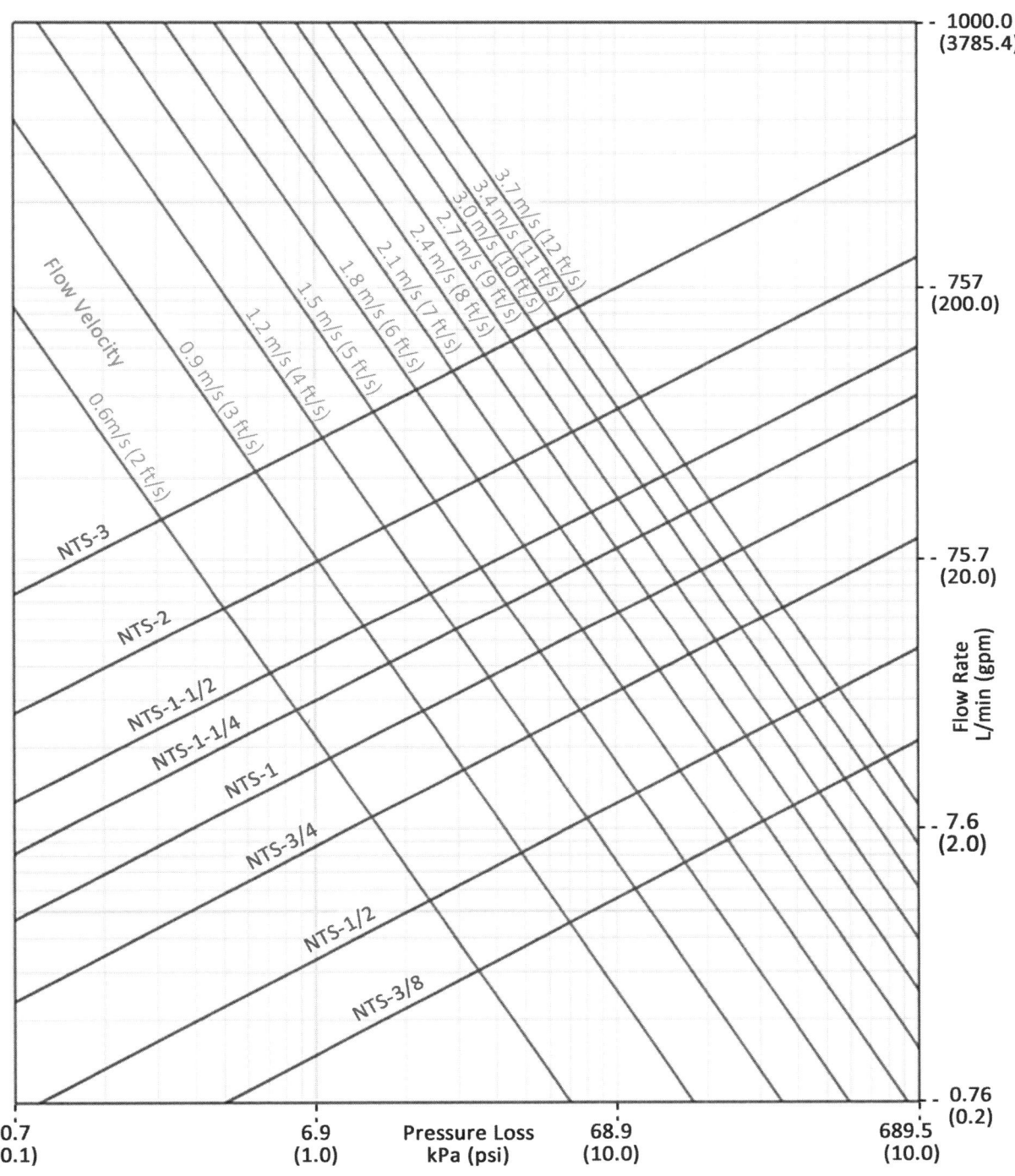

FIGURE 3
PRESSURE LOSS OF PEX TUBING AT 16 °C (60°F)
(See Section 10.6.2)

APPENDIX I

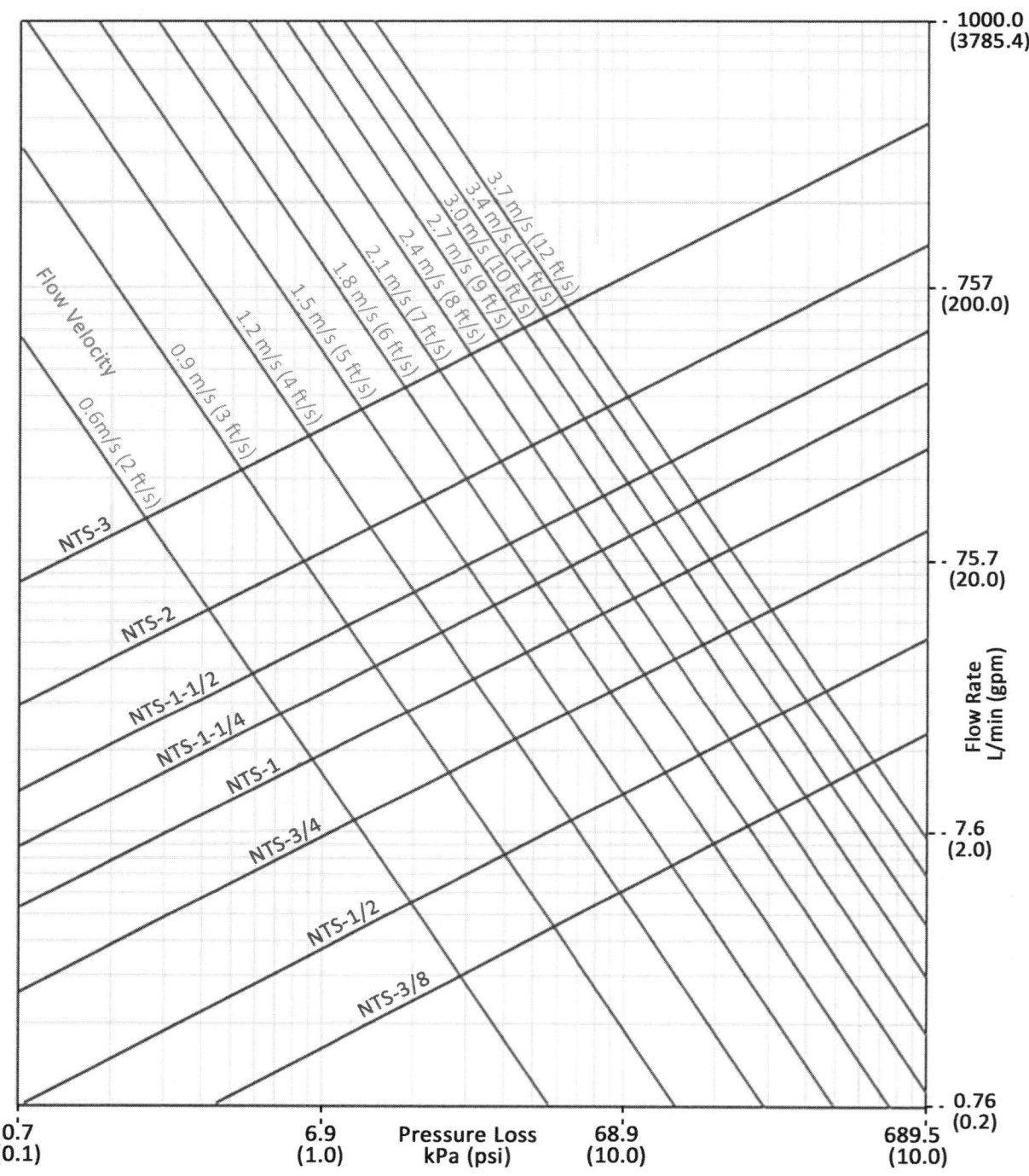

FIGURE 4
PRESSURE LOSS OF PEX TUBING AT 49 °C (120°F)
(See Section 10.6.2)

APPENDIX J

COMBINATION OF INDOOR AND OUTDOOR COMBUSTION AND VENTILATION OPENING DESIGN

(The content of this Appendix is based on Annex I of NFPA 54)

J 101.0 General.

J 101.1 Applicability. This appendix provides general guidelines for the sizing of combination indoor and outdoor combustion and ventilation air openings.

J 101.2 Example of Combination Indoor and Outdoor Combustion Air Opening. Determine the required combination of indoor and outdoor combustion air opening sizes for the following appliance installation example.

Example Installation: A fan-assisted furnace and a draft hood-equipped water heater with the following inputs are located in a 15 foot by 30 foot (4572 mm by 9144 mm) basement with an 8 foot (2438 mm) ceiling. No additional indoor spaces shall be used to help meet the appliance combustion air needs.

Fan-Assisted Furnace Input: 100 000 British thermal units per hour (Btu/h) (29 kW)

Draft Hood-Equipped Water Heater Input: 40 000 Btu/h (11.7 kW)

Solution:

(1) Determine the total available room volume. Appliance room volume:

15 feet by 30 feet (4572 mm by 9144 mm) with an 8 foot (2438 mm) ceiling = 3600 cubic feet (101.94 m^3)

(2) Determine the total required volume. The standard method to determine combustion air is used to calculate the required volume. The combined input for the appliances located in the basement is calculated as follows:

100 000 Btu/h (29 kW) + 40 000 Btu/h (11.7 kW) = 140 000 Btu/h (41 kW)

The standard method requires that the required volume be determined based on 50 cubic feet per 1000 Btu/h (4.83 m^3/kW). Using Table J 101.2, the required volume for a 140 000 Btu/h (41 kW) water heater is 7000 cubic feet (198.22 m^3).

Conclusion: Indoor volume is insufficient to supply combustion air since the total of 3600 cubic feet (101.94 m^3) does not meet the required volume of 7000 cubic feet (198.22 m^3). Therefore, additional combustion air shall be provided from the outdoors.

(3) Determine ratio of the available volume to the required volume.

$$\frac{3600 \text{ cubic feet}}{7000 \text{ cubic feet}} = 0.51$$

(4) Determine the reduction factor to be used to reduce the full outdoor air opening size to the minimum required based on ratio of indoor spaces.

1.00 – 0.51 (from Step 3) = 0.49

(5) Determine the single outdoor combustion air opening size as where combustion air is to come from outdoors. In this example, the combustion air opening directly communicates with the outdoors.

$$\frac{140\,000 \text{ Btu/h}}{3000 \text{ British thermal units per square inch (Btu/in}^2\text{)}} = 47 \text{ square inches } (0.03 \text{ m}^2)$$

(6) Determine the minimum outdoor combustion air opening area.

Outdoor opening area = 0.49 (from Step 4) x 47 square inches (0.03 m^2) = 23 square inches (0.01 m^2)

Section 506.5.3(3) requires the minimum dimension of the air opening shall be not less than 3 inches (76 mm).

APPENDIX J

TABLE J 101.2
STANDARD METHOD: REQUIRED VOLUME, ALL APPLIANCES*
[NFPA 54: TABLE A.9.3.2.1]

APPLIANCE INPUT (Btu/h)	REQUIRED VOLUME (cubic feet)
5000	250
10 000	500
15 000	750
20 000	1000
25 000	1250
30 000	1500
35 000	1750
40 000	2000
45 000	2250
50 000	2500
55 000	2750
60 000	3000
65 000	3250
70 000	3500
75 000	3750
80 000	4000
85 000	4250
90 000	4500
95 000	4750
100 000	5000
105 000	5250
110 000	5500
115 000	5750
120 000	6000
125 000	6250
130 000	6500
135 000	6750
140 000	7000
145 000	7250
150 000	7500
160 000	8000
170 000	8500
180 000	9000
190 000	9500
200 000	10 000
210 000	10 500
220 000	11 000
230 000	11 500
240 000	12 000
250 000	12 500
260 000	13 000
270 000	13 500
280 000	14 000
290 000	14 500
300 000	15 000

For SI units: 1000 British thermal units per hour = 0.293 kW, 1 cubic foot = 0.0283 m^3

* See Section 506.4.1, Section 506.4.1(1), Figure 506.4(1), Figure 506.4(2), Section 506.4.1(2), Figure 506.4(3), Section 506.4.2, and Figure 506.4.2.

APPENDIX K
POTABLE RAINWATER CATCHMENT SYSTEMS

K 101.0 General.

K 101.1 Applicability. The provisions of this appendix shall apply to the installation, construction, alteration, and repair of potable rainwater catchment systems.

K 101.2 System Design. Potable rainwater catchment systems in accordance with this appendix shall be designed by a registered design professional or person deemed competent by the Authority Having Jurisdiction to perform potable rainwater catchment system design work.

K 101.3 Permit. It shall be unlawful for a person to construct, install, or alter, or cause to be constructed, installed, or altered a potable rainwater catchment systems in a building or on a premise without first obtaining a permit to do such work from the Authority Having Jurisdiction.

K 101.3.1 Plumbing Plan Submission. No permit for a rainwater catchment system requiring a permit shall be issued until complete plumbing plans, with data satisfactory to the Authority Having Jurisdiction, have been submitted and approved. No changes or connections shall be made to either the rainfall catchment or the potable water system within a site containing a rainwater catchment water system without approval by the Authority Having Jurisdiction.

K 101.3.2 System Changes. No changes or connections shall be made to either the rainwater catchment system or the potable water system within a site containing a rainwater catchment system requiring a permit without approval by the Authority Having Jurisdiction.

K 101.4 Product and Material Approval. System components shall be properly identified as to the manufacturer.

K 101.4.1 Plumbing Materials and Systems. Pipe, pipe fittings, traps, fixtures, material, and devices used in a potable rainwater system shall be listed or labeled (third-party certified) by a listing agency (accredited conformity assessment body) and shall be in accordance with approved applicable recognized standards referenced within this code, and shall be free from defects. Unless otherwise provided for in this appendix, materials, fixtures, or devices used or entering into the construction of plumbing systems, or parts thereof shall be submitted to the Authority Having Jurisdiction for approval.

K 101.5 Maintenance and Inspection. Potable rainwater catchment systems and components shall be inspected and maintained in accordance with Section K 101.5.1 through Section K 101.5.3.

K 101.5.1 Frequency. Potable rainwater catchment systems and components shall be inspected and maintained in accordance with Table K 101.5.1 unless more frequent inspection and maintenance are required by the manufacturer.

TABLE K 101.5.1
MINIMUM POTABLE RAINWATER CATCHMENT SYSTEM TESTING, INSPECTION, AND MAINTENANCE FREQUENCY

DESCRIPTION	MINIMUM FREQUENCY
Inspect and clean filters and screens, and replace (where necessary).	Every 3 months
Inspect and verify that disinfection, filters and water quality treatment devices and systems are operational. Perform water quality tests in accordance with the Authority Having Jurisdiction.	In accordance with the manufacturer's instructions, and the Authority Having Jurisdiction.
Perform applicable water quality tests to verify compliance with Section K 104.3	Every 3 months
Perform a water quality test for E. Coli, Total Coliform, and Heterotrophic bacteria. For a system where 25 different people consume water from the system over a 60 day period, a water quality test for cryptosporidium shall be performed.	After initial installation and every 12 months thereafter, or as directed by the Authority Having Jurisdiction.
Inspect and clear debris from rainwater gutters, downspouts, and roof washers.	Every 6 months
Inspect and clear debris from the roof or other aboveground rainwater collection surface.	Every 6 months
Remove tree branches and vegetation overhanging roof or other aboveground rainwater collection surface.	As needed
Inspect pumps and verify operation.	After initial installation and every 12 months thereafter
Inspect valves and verify operation.	After initial installation and every 12 months thereafter
Inspect pressure tanks and verify operation.	After initial installation and every 12 months thereafter
Clear debris and inspect storage tanks, locking devices, and verify operation.	After initial installation and every 12 months thereafter
Inspect caution labels and marking.	After initial installation and every 12 months thereafter

APPENDIX K

K 101.5.2 Maintenance Log. A maintenance log for potable rainwater catchment systems shall be maintained by the property owner and be available for inspection. The property owner or designated appointee shall ensure that a record of testing, inspection, and maintenance in accordance with Table K 101.5.1 is maintained in the log. The log will indicate the frequency of inspection, and maintenance of each system. A record of the required water quality tests shall be retained for not less than 2 years.

K 101.5.3 Maintenance Responsibility. The required maintenance and inspection of potable rainwater catchment systems shall be the responsibility of the property owner unless otherwise required by the Authority Having Jurisdiction.

K 101.6 Operation and Maintenance Manual. An operation and maintenance manual for potable rainwater catchment systems shall be supplied to the building owner by the system designer. The operating and maintenance manual shall include the following:

(1) Detailed diagram of the entire system and the location of system components.

(2) Instructions for operating and maintaining the system.

(3) Details on maintaining the required water quality as determined by the Authority Having Jurisdiction.

(4) Details on deactivating the system for maintenance, repair, or other purposes.

(5) Applicable testing, inspection, and maintenance frequencies in accordance with Table K 101.5.1.

(6) A method of contacting the manufacturer(s).

K 101.7 Minimum Water Quality Requirements. The minimum water quality for potable rainwater catchment systems shall comply with the applicable water quality requirements as determined by the public health Authority Having Jurisdiction.

K 101.8 Material Compatibility. In addition to the requirements of this appendix, potable rainwater catchment systems shall be constructed of materials that are compatible with the type of pipe and fitting materials and water conditions in the system.

K 101.9 System Controls. Controls for pumps, valves, and other devices that contain mercury that come in contact with the water supply shall not be permitted.

K 102.0 Connection.

K 102.1 General. No water piping supplied by a potable rainwater catchment system shall be connected to a source of supply without the approval of the Authority Having Jurisdiction, Health Department, or other department having jurisdiction.

K 102.2 Connections to Public or Private Potable Water Systems. Potable rainwater catchment systems shall have no direct connection to a public or private potable water supply or alternate water source system. Potable water from a public or private potable water system is permitted to be used as makeup water to the rainwater storage tank provided the public, or private potable water supply connection is protected by an air gap or reduced-pressure principle backflow preventer in accordance with this code.

K 102.3 Backflow Prevention. The potable rainwater catchment system shall be protected against backflow in accordance with this code.

K 103.0 Potable Rainfall Catchment System Materials.

K 103.1 Collections Surfaces. The collection surface for potable applications shall be constructed of a hard, impervious material and shall be approved for potable water use. Roof coatings, paints, and liners shall comply with NSF Protocol P151.

K 103.1.1 Prohibited. Roof paints and coatings with lead, chromium, or zinc shall not be permitted. Wood roofing material and lead flashing shall not be permitted.

K 103.2 Rainwater Catchment System Drainage Materials. Materials used in rainwater catchment drainage systems, including gutters, downspouts, conductors, and leaders shall be in accordance with the requirements of this code for storm drainage.

K 103.3 Storage Tanks. Rainwater storage shall comply with Section K 105.0.

K 103.4 Water Supply and Distribution Materials. Potable rainwater supply and distribution materials shall comply with the requirements of this code for potable water supply and distribution systems.

K 104.0 Design and Installation.

K 104.1 Collection Surfaces. Rainwater shall be collected from a roof or other cleanable aboveground surfaces specifically designed for rainwater catchment. A rainwater catchment system shall not collect rainwater from:

(1) Vehicular parking surfaces

(2) Surface water runoff

(3) Bodies of standing water

K 104.2 Prohibited Discharges. Overflows, condensate, and bleed-off pipes from roof-mounted equipment and appliances shall not discharge onto roof surfaces that are intended to collect rainwater.

K 104.3 Minimum Water Quality. Upon initial system startup, the quality of the water for the intended application shall be verified at the point(s) of use as determined by the Authority Having Jurisdiction. In the absence of water quality requirements determined by the Authority Having Jurisdiction, the minimum water quality shall be in accordance with Table K 104.3(1).

Normal system maintenance will require system testing every 3 months. Systems shall comply with Table K 104.3(2).

K 104.4 Filtration Devices. Potable water filters shall comply with NSF 53 and shall be installed in accordance with the manufacturer's installation instructions.

K 104.5 Disinfection Devices. Chlorination, ozone, ultraviolet, or other disinfection methods approved by the Authority Having Jurisdiction, or the product is listed and certified according to a microbiological reduction performance standard for drinking water, shall be used to treat harvested rainwater to meet the required water quality permitted. The disinfection devices and systems shall be installed in accordance with the manufacturer's installation instructions and

TABLE K 104.3(1)
MINIMUM WATER QUALITY[1,2]

Escherichia coli (fecal coliform)	99.9 % reduction
Protozoan Cysts	99.99 % reduction
Viruses	99.99 % reduction
Turbidity	<0.3 NTU

Notes:
[1] Upon failure of the fecal coliform test, the system shall be re-commissioned involving cleaning and retesting in accordance with Section K 104.3.
[2] One sample shall be analyzed for applications serving up to 1000 persons. Where the treated water shall serve 1000 - 2500 persons two samples shall be analyzed, and for 2501 - 3300 persons three samples shall be analyzed.

TABLE K 104.3(2)
MINIMUM SYSTEM MAINTENANCE REQUIREMENTS[1,2]

Escherichia coli (fecal coliform)	99.9 % reduction
Turbidity	<0.3 NTU

Notes:
[1] Upon failure of the fecal coliform test, the system shall be re-commissioned involving cleaning and retesting in accordance with Section K 104.3.
[2] One sample shall be analyzed for applications serving up to 1000 persons. Where the treated water shall serve 1000 - 2500 persons two samples shall be analyzed, and for 2501 - 3300 persons three samples shall be analyzed.

the conditions of listing. Disinfection devices and systems shall be located downstream of the storage tank.

K 104.6 Overhanging Tree Branches and Vegetation. Tree branches and vegetation shall not be located over the roof or other aboveground rainwater collection surface. Where existing tree branch and vegetation growth extends over the rainwater collection surface, it shall be removed in accordance with Section K 101.5.

K 105.0 Rainwater Storage Tanks.

K 105.1 General. Rainwater storage tanks shall be installed in accordance with Section K 105.2 through Section K 105.10.

K 105.2 Construction. Rainwater storage tanks shall be constructed of solid, durable materials not subject to excessive corrosion or decay and shall be watertight. Storage tanks shall be approved by the Authority Having Jurisdiction for potable water applications, provided such tanks are in accordance with approved applicable standards.

K 105.3 Location. Rainwater storage tanks shall be permitted to be installed above or below grade.

K 105.3.1 Above Grade. Above grade, storage tanks shall be of an opaque material, approved for aboveground use in direct sunlight, or shall be shielded from direct sunlight. Tanks shall be installed in an accessible location to allow for inspection and cleaning. The tank shall be installed on a foundation or platform that is constructed to accommodate loads in accordance with the building code.

K 105.3.2 Below Grade. Rainwater storage tanks installed below grade shall be structurally designed to withstand anticipated earth or other loads. Holding tank covers shall be capable of supporting an earth load of not less than 300 pounds per square foot (lb/ft^2) (1465 kg/m^2) where the tank is designed for underground installation. Below grade rainwater tanks installed underground shall be provided with manholes. The manhole opening shall be not less than 20 inches (508 mm) in diameter and located not less than 4 inches (102 mm) above the surrounding grade. The surrounding grade shall be sloped away from the manhole. Underground tanks shall be ballasted, anchored, or otherwise secured, to prevent the tank from floating out of the ground where empty. The combined weight of the tank and hold down system shall meet or exceed the buoyancy force of the tank.

K 105.4 Drainage and Overflow. Rainwater storage tanks shall be provided with a means of draining and cleaning. The overflow drain shall not be equipped with a shutoff valve. The overflow outlet shall discharge in accordance with this code for storm drainage systems. Where discharging to the storm drainage system, the overflow drain shall be protected from backflow of the storm drainage system by a backwater valve or other approved method.

K 105.4.1 Overflow Outlet Size. The overflow outlet shall be sized to accommodate the flow of the rainwater entering the tank and not less than the aggregate cross-sectional area of the inflow pipes.

K 105.5 Animals and Insects. Rainwater tank openings to the atmosphere shall be protected to prevent the entrance of insects, birds, or rodents into the tank.

K 105.6 Human Access. Rainwater tank access openings exceeding 12 inches (305 mm) in diameter shall be secured to prevent tampering and unintended entry by either a lockable device or other approved method.

K 105.7 Exposure to Sunlight. Rainwater tank openings shall not be exposed to direct sunlight.

K 105.8 Inlets. A device or arrangement of fittings shall be installed at the inlet of the tank to prevent rainwater from disturbing sediment as it enters the tank.

K 105.9 Primary Tank Outlets. The primary tank outlet shall be located not less than 4 inches (102 mm) above the bottom of the tank, or shall be provided with a floating inlet to draw water from the cistern just below the water surface.

K 105.10 Storage Tank Venting. Where venting using drainage or overflow piping is not provided or is considered insufficient, a vent shall be installed on each tank. The vent shall extend from the top of the tank and terminate not less than 6 inches (152 mm) above grade and shall be not less than 1½ inches (40 mm) in diameter. The vent terminal shall be directed downward and covered with a 3/32 of an inch (2.4 mm) mesh screen to prevent the entry of vermin and insect.

K 105.11 Pumps. Pumps serving rainwater catchment systems shall be listed for potable water use. Pumps supplying water to water closets, urinals, and trap primers shall be capable of delivering not less than 15 pounds-force per square inch (psi) (103 kPa) residual pressure at the highest and most remote outlet served. Where the water pressure in the rainwater supply system within the building exceeds 80 psi (552 kPa), a pressure reducing valve reducing the pressure to 80 psi (552 kPa) or less to water outlets in the building shall be installed in accordance with this code.

K 105.12 Roof Drains. Primary and secondary roof drains, conductors, leaders, overflows, and gutters shall be designed and installed in accordance with this code.

K 106.0 Water Quality Devices and Equipment.

K 106.1 General. Devices and equipment used to treat rainwater to maintain the minimum water quality requirements determined by the Authority Having Jurisdiction shall be listed or labeled (third-party certified) by a listing agency (accredited conformity assessment body) and approved for the intended application.

K 106.2 Filtration and Disinfection Systems. Filtration and disinfection systems shall be located after the water storage tank. Where a chlorination system is installed, it shall be installed upstream of filtration systems. Where an ultraviolet disinfection system is installed, a filter not more than 5 microns (5 μm) shall be installed upstream of the disinfection system.

K 106.3 Freeze Protection. Tanks and piping installed in locations subject to freezing shall be provided with an approved means of freeze protection.

K 106.4 Roof Washer or Pre-Filtration System. Collected rainwater shall pass through a roof washer or pre-filtration system before the water enters the rainwater storage tank. Roof washer systems shall comply with Section K 106.4.1 through Section K 106.4.4.

K 106.4.1 Size. The roof washer shall be sized to direct rainwater containing debris that has accumulated on the collection surface away from the storage tank. ARCSA/ASPE 63 contains additional guidance on acceptable methods of sizing roof washers.

K 106.4.2 Debris Screen. The inlet to the roof washer shall be provided with a debris screen or other approved means that protects the roof washer from the intrusion of debris and vermin. Where the debris screen is installed, the debris screen shall be corrosion resistant and shall have openings not larger than ½ of an inch (12.7 mm).

K 106.4.3 Drain Discharge. Water drained from the roof washer, or pre-filter shall be diverted away from the storage tank and discharged to a disposal area that does not cause property damage or erosion. Roof washer drainage shall not drain over a public way.

K 106.4.4 Automatic Drain. Roof washing systems shall be provided with an automatic means of self-draining between rain events.

K 106.5 Roof Gutters. Gutters shall maintain a minimum slope and be sized in accordance with this code.

K 106.6 Drains, Conductors, and Leaders. The design and size of rainwater drains, conductors, and leaders shall comply with this code.

K 106.7 Size of Potable Water Piping. Potable rainwater system distribution piping shall be sized in accordance with this code for sizing potable water piping.

K 107.0 Cleaning.

K 107.1 General. The interior surfaces of tanks and equipment shall be clean before they are put into service.

K 108.0 Supply System Inspection and Test.

K 108.1 General. Rainwater catchment systems shall be inspected and tested in accordance with the applicable provisions of this code for testing of potable water and storm drainage systems. Storage tanks shall be filled with water to the overflow opening for 24 hours, and during the inspection, or by other means as approved by the Authority Having Jurisdiction. Seams and joints shall be exposed during the inspection and checked for water tightness.

APPENDIX L
SUSTAINABLE PRACTICES

L 101.0 General.

L 101.1 Applicability. The purpose of this appendix is to provide a comprehensive set of technically sound provisions that encourage sustainable practices and works towards enhancing the design and construction of plumbing systems that result in a positive long-term environmental impact. This appendix is not intended to circumvent the health, safety, and general welfare requirements of this code.

L 101.2 Definition of Terms. For the purposes of this code, the definitions in Section L 201.0 shall apply to this appendix.

No attempt is made to define ordinary words, which are used in accordance with their established dictionary meanings, except where a word has been used loosely, and it is necessary to define its meaning as used in this appendix to avoid misunderstanding.

The definitions of terms are arranged alphabetically according to the first word of the term.

L 201.0 Definitions.

Catch Can Test. Method to measure the precipitation rate of an irrigation system by placing catchment containers at various random positions in the irrigation zone for a prescribed amount of time during irrigation application. The volumes of water in the containers are measured, averaged, and calculated to determine precipitation rate. Tests are conducted using irrigation industry accepted practices.

Combination Ovens. A device that combines the function of hot air convection (oven mode) and saturated and superheated steam heating (steam mode), or both, to perform steaming, baking, roasting, rethermalizing, and proofing of various food products. In general, the term combination oven is used to describe this type of equipment, which is self-contained. The combination oven is also referred to as a combination oven/steamer, combi or combo.

Energy Star. A joint program of the U.S. Environmental Protection Agency and the U.S. Department of Energy. Energy Star is a voluntary program designed to identify and promote energy-efficient products and practices.

Evapotranspiration (ET). The combination of water transpired from vegetation and evaporated from the soil, water, and plant surfaces. Evapotranspiration rates are expressed in inches (mm) per day, week, month, or year. Evapotranspiration varies by climate and time of year. Common usage includes evapotranspiration as the base rate [water demand of 4 - 6 inch (102 mm - 152 mm) tall cool season grass], with coefficients for specific plant types. Evapotranspiration rates are used as a factor in estimating the irrigation water needs of landscapes. Local agriculture extension, state departments of agriculture, water agencies, irrigation professionals, and internet websites are common sources for obtaining local evapotranspiration rates.

Food Steamers (Steam Cookers). A cooking appliance wherein heat is imparted to food in a closed compartment by direct contact with steam. The compartment can be at or above atmospheric pressure. The steam can be static or circulated.

Gang Showers (Non-Residential). Shower compartments designed and intended for use by multiple persons simultaneously in non-residential occupancies.

Hydrozone. A grouping of plants with similar water requirements that are irrigated by the same irrigation zone.

Irrigation Emission Device. The various landscape irrigation equipment terminal fittings or outlets that emit water for irrigating vegetation in a landscape.

Irrigation Zone. The landscape area that is irrigated by a set of landscape irrigation emission devices installed on the same water supply line downstream of a single valve.

Kitchen and Bar Sink Faucets. A faucet that discharges into a kitchen or bar sinks in domestic or commercial installations. Supply fittings that discharge into other type sinks, including clinical sinks, floor sinks, service sinks and laundry trays are not included.

Lavatory. (1) A basin or vessel for washing. (2) A plumbing fixture, as defined in (1), especially placed for use in personal hygiene. Principally not used for laundry purposes and never used for food preparation, or utensils, in food services. (3) A fixture designed for the washing of the hands and face. Sometimes called a wash basin.

Lavatory Faucet. A faucet that discharges into a lavatory basin in a domestic or commercial installation.

Low Application Rate Irrigation. A means of irrigation using low precipitation rate sprinkler heads or low flow emitters in conjunction with cycling irrigation schedules to apply water at a rate less than the soil absorption rate.

Low Flow Emitter. Low-flow irrigation emission device designed to dissipate water pressure and discharge a small uniform flow or trickle of water at a constant flow rate. To be classified as a low flow emitter: drip emitters shall discharge water at less than 4 gallons (15 L) per hour per emitter; micro-spray, micro-jet, and misters shall discharge water at a maximum of 30 gallons (114 L) per hour per nozzle.

Low Precipitation Rate Sprinkler Heads. Landscape irrigation emission devices or sprinkler heads with a maximum precipitation rate of 1 inch per hour (25.4 mm/h) over the applied irrigation area.

Maintenance. The upkeep of property or equipment by the owner of the property in compliance with the requirements of this appendix.

Metering Faucet. A self-closing faucet that dispenses a specific volume of water for each actuation cycle. The volume or cycle duration can be fixed or adjustable.

Multi-Occupant Spaces. Indoor spaces used for presentations and training, including classrooms and conference rooms.

Precipitation Rate. The sprinkler head application rate of water applied to landscape irrigation zone, measured as inches per hour (mm/h). Precipitation rates of sprinkler heads are calculated according to the flow rate, pattern, and spacing of the sprinkler heads.

Pre-Rinse Spray Valve. A handheld device for use with commercial dishwashing and ware washing equipment that sprays water on dishes, flatware, and other food service items for the purpose of removing food residue before cleaning and sanitizing the items.

Recirculation System. A system of hot water supply and return piping with shutoff valves, balancing valves, circulating pumps, and a method of controlling the circulating system.

Reverse Osmosis Reject Water. Water that does not pass through a membrane of a reverse osmosis system.

Run Out. The developed length of pipe that extends away from the circulating loop system to a fixture(s).

Self Closing Faucet. A faucet that closes itself after the actuation or control mechanism is deactivated. The actuation or control mechanism can be mechanical or electronic.

Single Occupant Spaces. Private offices, workstations in open offices, reception workstations, and ticket booths.

Soil Absorption Rate. The rate of the soil's ability to allow water to percolate or infiltrate the soil and be retained in the root zone of the soil expressed as inches per hour (mm/h).

Sprinkler Head. Landscape irrigation emission device discharging water in the form of sprays or rotating streams, not including low flow emitters.

Storage Tank. The central component of the rainwater, stormwater, or dry weather runoff catchment system. Also known as a cistern or rain barrel.

Stormwater. Natural precipitation that has contacted a surface at grade or below grade and has not been put to beneficial use.

Stormwater Catchment System. A system that collects and stores stormwater for beneficial use.

Submeter. A meter installed subordinate to a site meter. Also known as a dedicated meter.

WaterSense. A voluntary program of the U.S., Environmental Protection Agency, designed to identify and promote water-efficient products and practices.

Water Closet. A fixture with a water-containing receptor that receives liquid and solid body waste and on actuation conveys the waste through an exposed integral trap into a drainage system. Also referred to as a toilet.

Water Factor (WF). A measurement and rating of appliance water efficiency, most often used for residential and light commercial clothes washers, as follows:

Water Factor (WF), Clothes Washer. The quantity of water in gallons used to complete a full wash and rinse cycle per measured cubic foot capacity of the clothes container.

L 301.0 General Regulations.

L 301.1 Installation. Plumbing systems covered by this appendix shall be installed in accordance with this code, other applicable codes, and the manufacturer's installation instructions.

L 301.2 Qualifications. Where permits are required, the Authority Having Jurisdiction shall have the authority to require contractors, installers, or service technicians to demonstrate competency. Where determined by the Authority Having Jurisdiction, the contractor, installer or service technician shall be licensed to perform such work.

L 302.0 Disposal of Liquid Waste.

L 302.1 Disposal. It shall be unlawful for a person to cause, suffer, or permit the disposal of sewage, human excrement, or other liquid wastes, in a place or manner, except through and by means of an approved drainage system, installed and maintained in accordance with the provisions of this code.

L 302.2 Connections to Plumbing System Required. Equipment and appliances, used to receive or discharge liquid wastes or sewage, shall be connected properly to the drainage system of the building or premises, in accordance with the requirements of this code.

L 303.0 Abandonment.

L 303.1 General. An abandoned system or part thereof covered under the scope of this appendix shall be disconnected from remaining systems, drained, plugged, and capped in an approved manner.

L 401.0 Water Conservation and Efficiency.

L 401.1 Scope. The provisions of this section establish the means of conserving potable and nonpotable water used in and around a building.

L 402.0 Water-Conserving Plumbing Fixtures and Fittings.

L 402.1 General. The maximum water consumption of fixtures and fixture fittings shall comply with the flow rates specified in Table L 402.1, and Section L 402.2 through Section L 402.9.

L 402.2 Water Closets. No water closet shall have a flush volume exceeding 1.6 gallons per flush (gpf) (6.0 Lpf).

L 402.2.1 Gravity, Pressure Assisted, and Electro-Hydraulic Tank Type Water Closets. Gravity, pressure assisted, and electro-hydraulic tank-type water closets shall have a maximum effective flush volume of not more than 1.28 gallons (4.8 Lpf) of water per flush in accordance with ASME A112.19.2/CSA B45.1 or ASME A112.19.14 and shall be listed to the EPA WaterSense Specification for Tank-Type Toilets. The effective flush volume for dual flush toilets is defined as the composite, average flush volume of two reduced flushes and one full flush.

L 402.2.2 Flushometer-Valve Activated Water Closets. Flushometer-valve activated water closets shall have a maximum flush volume of not more than 1.6 gallons (6.0 Lpf) of water per flush in accordance with ASME A112.112.19.2/CSA B45.1.

L 402.3 Urinals. Urinals shall have a maximum flush volume of not more than 0.5 gallon (1.9 Lpf) of water per flush in accordance with ASME A112.19.2/CSA B45.1 or CSA B45.5/IAPMO Z124. Flushing urinals shall be listed to the EPA WaterSense Flushing Urinal Specification.

APPENDIX L

TABLE L 402.1
MAXIMUM FIXTURE AND FIXTURE FITTINGS FLOW RATES

FIXTURE TYPE	FLOW RATE
Showerheads	2.0 gpm at 80 psi[1]
Kitchen faucets residential[5]	1.8 gpm at 60 psi
Lavatory faucets residential	1.5 gpm at 60 psi
Lavatory faucets other than residential	0.5 gpm at 60 psi
Metering faucets	0.25 gallons/cycle
Metering faucets for wash fountains	One 0.25 gallons/cycle fixture fitting for each 20 inches rim space
Wash fountains	One 2.2 gpm at 60 psi fixture fitting for each 20 inches rim space
Water Closets - other than remote locations[4]	1.28 gallons/flush[2]
Water Closets - remote locations[4]	1.6 gallons/flush
Urinals	0.5 gallons/flush[3]
Commercial Pre-Rinse Spray Valves	1.3 gpm at 60 psi

For SI units: 1 gallon per minute = 0.06 L/s, 1 pound-force per square inch = 6.8947 kPa, 1 inch = 25.4 mm, 1 gallon = 3.785 L

Notes:
[1] For multiple showerheads serving one shower compartment see Section L 402.6.1.
[2] Shall be listed to EPA WaterSense Tank-Type Toilet Specification.
[3] Shall be listed to EPA WaterSense Flushing Urinal Specification. Nonwater urinals shall comply with specifications listed in Section L 402.3.1.
[4] Remote location is where a water closet is located not less than 30 feet (9144 mm) upstream of the nearest drain line connections or fixtures and is located where less than 1.5 drainage fixture units are upstream of the water closet drain line connection.
[5] See Section L 402.4.

L 402.3.1 Nonwater Urinals. Nonwater urinals shall comply with ASME A112.19.3/CSA B45.4, ASME A112.19.19, or CSA B45.5/IAPMO Z124. Nonwater urinals shall be cleaned and maintained in accordance with the manufacturer's instructions after installation. Where nonwater urinals are installed, they shall have a water distribution line roughed-in to the urinal location at a height not less than 56 inches (1422 mm) to allow for the installation of an approved backflow prevention device in the event of a retrofit. Such water distribution lines shall be installed with shutoff valves located as close as possible to the distributing main to prevent the creation of dead ends. Where nonwater urinals are installed, not less than one water supplied fixture rated at not less than 1 drainage fixture unit (DFU) shall be installed upstream on the same drain line to facilitate drain line flow and rinsing.

L 402.4 Residential Kitchen Faucets. The maximum flow rate of residential kitchen faucets shall not exceed 1.8 gallons per minute (gpm) (6.8 L/m) at 60 pounds-force per square inch (psi) (414 kPa). Kitchen faucets are permitted to temporarily increase the flow above the maximum rate, but not to exceed 2.2 gpm (8.3 L/m) at 60 psi (414 kPa), and shall revert to a maximum flow rate of 1.8 gpm (6.8 L/m) at 60 psi (414 kPa) upon valve closure.

L 402.5 Lavatory Faucets. The maximum water flow rate of faucets shall comply with Section L 402.5.1 and Section L 402.5.2.

L 402.5.1 Lavatory Faucets in Residences, Apartments, and Private Bathrooms in Lodging Facilities, Hospitals, and Patient Care Facilities. The flow rate for lavatory faucets installed in residences, apartments, and private bathrooms in lodging, hospitals, and patient care facilities (including skilled nursing and long-term care facilities) shall not exceed 1.5 gpm (5.7 L/m) at 60 psi (414 kPa) in accordance with ASME A112.18.1/CSA B125.1 and shall be listed to the EPA WaterSense High Efficiency Lavatory Faucet Specification.

L 402.5.2 Lavatory Faucets in Other Than Residences, Apartments, and Private Bathrooms in Lodging Facilities. Lavatory faucets installed in bathrooms of buildings or occupancies other than those specified in Section L 402.5.1 shall be in accordance with Section L 402.5.2(1) or Section L 402.5.2(2).

(1) The flow rate shall not exceed 0.5 gpm (1.9 L/m) at 60 psi (414 kPa) in accordance with ASME A112.18.1/CSA B125.1.

(2) Metering faucets shall deliver not more than 0.25 gallons (1.0 L) of water per cycle.

L 402.6 Showerheads. Showerheads shall comply with the requirements of the Energy Policy Act of 1992, except that the flow rate shall not exceed 2.0 gpm (7.6 L/m) at 80 psi (552 kPa), where listed to ASME A112.18.1/CSA B125.1.

L 402.6.1 Multiple Showerheads Serving One Shower Compartment. The total allowable flow rate of water from multiple showerheads flowing at a given time, with or without a diverter, including rain systems, waterfalls, bodysprays, and jets, shall not exceed 2.0 gpm (7.6 L/m) per shower compartment, where the floor area of the shower compartment is less than 1800 square inches (1.161 m^2). For each increment of 1800 square inches (1.161 m^2) of floor area after that or part thereof, additional showerheads are allowed, provided the total flow rate of water from flowing devices shall not exceed 2.0 gpm (7.6 L/m) for each such increment.

Exceptions:

(1) Gang showers in non-residential occupancies. Singular showerheads or multiple shower outlets serving one showering position in gang showers shall not have more than 2.0 gpm (7.6 L/m) total flow.

(2) Where provided, accessible shower compartments shall not be permitted to have more than 4.0 gpm (15 L/m) total flow, where one outlet is the hand shower. The hand shower shall have control with a nonpositive shutoff feature.

L 402.6.2 Bath and Shower Diverters. The rate of leakage out of the tub spout of bath and shower diverters while operating in the shower mode shall not exceed 0.1 gpm (0.4 L/m) in accordance with ASME A112.18.1/CSA B125.1.

L 402.6.3 Shower Valves. Shower valves shall comply

APPENDIX L

with the temperature control performance requirements of ASSE 1016 or ASME A112.18.1/CSA B125.1 where tested at 2.0 gpm (7.6 L/m).

L 402.7 Commercial Pre-Rinse Spray Valves. The flow rate for a pre-rinse spray valve installed in a commercial kitchen to remove food waste from cookware and dishes before cleaning shall not be more than 1.3 gpm (4.9 L/m) at 60 psi (414 kPa). Where pre-rinse spray valves with maximum flow rates of 1.0 gpm (3.8 L/m) or less are installed, the static pressure shall be not less than 30 psi (207 kPa). Commercial kitchen pre-rinse spray valves shall be equipped with an integral automatic shutoff.

L 402.8 Emergency Safety Showers and Eye Wash Stations. Emergency safety showers and emergency eyewash stations shall not be limited to their water supply flow rates.

L 402.9 Drinking Fountains. Drinking fountains shall be self-closing.

L 403.0 Appliances.

L 403.1 Dishwashers. Residential and commercial dishwashers shall comply with the Energy Star program requirements.

L 403.2 Clothes Washers. Residential clothes washers shall comply with the Energy Star program requirements. Commercial clothes washers shall comply with Energy Star program requirements, where such requirements exist.

L 404.0 Occupancy Specific Water Efficiency Requirements.

L 404.1 Commercial Food Service. Commercial food service facilities shall comply with the water efficiency requirements in Section L 404.2 through Section L 404.6.

L 404.2 Ice Makers. Ice makers shall be air cooled and shall be in accordance with Energy Star for commercial ice machines.

L 404.3 Food Steamers. Steamers shall consume not more than 5.0 gallons (19 L) per hour per steamer pan in the fully operational mode.

L 404.4 Combination Ovens. Combination ovens shall not consume more than 3.5 gallons per hour (gph) (13.2 L/h) per pan in the fully operational mode.

L 404.5 Grease Interceptors. Grease interceptor maintenance procedures shall not include post-pumping/cleaning refill using potable water. Refill shall be by connected appliance accumulated discharge only.

L 404.6 Dipper Well Faucets. Where dipper wells are installed, the water supply to a dipper well shall have a shutoff valve and flow control. The flow of water into a dipper well shall be limited by not less than one of the following methods:

(1) Water flow shall not exceed the water capacity of the dipper well in one minute at a supply pressure of 60 psi (414 kPa), and the maximum flow shall not exceed 2.2 gpm (8.3 L/m) at a supply pressure of 60 psi (414 kPa). The water capacity of a dipper well shall be the maximum amount of water that the fixture can hold before water flows into the drain.

(2) The volume of water dispensed into a dipper well in each activation cycle of a self-closing fixture fitting shall not exceed the water capacity of the dipper well, and the maximum flow shall not exceed 2.2 gpm (8.3 L/m) at a supply pressure of 60 psi (414 kPa).

L 404.7 Medical and Laboratory Facilities. Medical and laboratory facilities shall comply with the water efficiency requirements in Section L 404.8 through Section L 404.10.

L 404.8 Steam Sterilizers. Controls shall be installed to limit the discharge temperature of condensate or water from steam sterilizers to 140°F (60°C) or less. A venturi-type vacuum system shall not be utilized with vacuum sterilizers.

L 404.9 X-Ray Film Processing Units. Processors for X-ray film exceeding 6 inches (152 mm) in any dimension shall be equipped with water recycling units.

L 404.10 Exhaust Hood Liquid Scrubber Systems. Liquid scrubber systems for exhaust hoods and ducts shall be of the recirculation type. Liquid scrubber systems for perchloric acid exhaust hoods and ducts shall be equipped with a timer-controlled water recirculation system. The collection sump for perchloric acid exhaust systems shall be designed to drain automatically after the wash down process has completed.

L 405.0 Leak Detection and Control.

L 405.1 General. Where installed, leak detection and control devices shall be approved by the Authority Having Jurisdiction. Leak detection and control devices help protect property from water damage and also conserve water by shutting off the flow when leaks are detected.

L 406.0 Fountains and Other Water Features.

L 406.1 Use of Alternate Water Source for Special Water Features. Special water features such as ponds and water fountains shall be provided with reclaimed (recycled) water, rainwater, or on-site treated nonpotable water where the source and capacity are available on the premises and approved by the Authority Having Jurisdiction.

L 407.0 Meters.

L 407.1 Required. A water meter shall be required for buildings connected to a public water system, including municipally supplied reclaimed (recycled) water. In other than single-family houses, multifamily structures of three stories or fewer above grade, and modular houses, a separate meter or submeter shall be installed in the following locations:

(1) The water supply for irrigated landscape with an accumulative area exceeding 2500 square feet (232.3 m^2).

(2) The water supply to a water-using process where the consumption exceeds 1000 gallons per day (gal/d) (0.0438 L/s), except for manufacturing processes.

(3) The water supply to each building on a property with multiple buildings where the water consumption exceeds 500 gals/d (0.021 L/s).

(4) The water supply to an individual tenant space on a property where one or more of the following applies:

(a) Water consumption exceeds 500 gals/d (0.021 L/s) for that tenant.

(b) Tenant space is occupied by a commercial laundry,

cleaning operation, restaurant, food service, medical office, dental office, laboratory, beauty salon, or barbershop.

(c) Total building area exceeds 50 000 square feet (4645 m^2).

(5) The makeup water supplies to a swimming pool.

L 407.2 Approval. Dedicated meters, other than water utility meters shall be approved by the Authority Having Jurisdiction for the intended use.

L 407.3 Consumption Data. A means of communicating water consumption data from submeters to the water consumer shall be provided.

L 407.4 Access. Meters and submeters shall be accessible.

L 408.0 Condensate Recovery.

L 408.1 General. Condensate is permitted to be used as on-site treated nonpotable water when collected, stored, and treated in accordance with Section 1506.0.

L 408.1.1 Condensate Drainage Recovery. Condensate from air-conditioning, boiler and steam systems used to supply water for non-potable water systems shall be in accordance with Section 1506.0.

L 409.0 Water-Powered Sump Pumps.

L 409.1 General. Sump pumps powered by potable or reclaimed (recycled) water pressure shall be used as an emergency backup pump. The water-powered pump shall be equipped with a battery powered alarm having a minimum rating of 85 dBa at 10 feet (3048 mm). Water-powered pumps shall have a water efficiency factor of pumping at least 1.4 gallons (5.3 L) of water to a height of 10 feet (3048 mm) for every gallon of water used to operate the pump, measured at a water pressure of 60 psi (414 kPa). Pumps shall be labeled as to the gallons of water pumped per gallon of potable water consumed.

Water-powered stormwater sump pumps shall be equipped with a reduced pressure principle backflow prevention assembly.

L 410.0 Water Softeners and Treatment Devices.

L 410.1 Water Softeners. Actuation of regeneration of water softeners shall be by demand initiation. Water softeners shall be listed to NSF 44. Water softeners shall have a rated salt efficiency exceeding 3400 grains (gr) (0.222 kg) of total hardness exchange per pound (0.5 kg) of salt, based on sodium chloride (NaCl) equivalency, and shall not generate more than 5 gallons (19 L) of water per 1000 grains (0.0647 kg) of hardness removed during the service cycle.

L 410.2 Water Softener Limitations. In residential buildings, where the supplied potable water hardness is equal to or less than 8 grains per gallon (gr/gal) (137 mg/L) measured as total calcium carbonate equivalents, water softening equipment that discharges water into the wastewater system during the service cycle shall not be allowed, except as required for medical purposes.

L 410.3 Point-of-Use Reverse Osmosis Water Treatment Systems. Reverse osmosis water treatment systems installed in residential occupancies shall be equipped with automatic shutoff valves to prevent discharge when there is no call for producing treated water. Reverse osmosis water treatment systems shall be listed in accordance with NSF 58.

L 411.0 Landscape Irrigation Systems.

L 411.1 General. Where landscape irrigation systems are installed, they shall use low application irrigation methods and shall be in accordance with Section L 411.2 through Section L 411.12. Requirements limiting the amount or type of plant material used in landscapes shall be established by the Authority Having Jurisdiction.

Exception: Plants grown for food production.

L 411.2 Backflow Protection. Potable water and supplies to landscape irrigation systems shall be protected from backflow in accordance with this code and the Authority Having Jurisdiction.

L 411.3 Use of Alternate Water Sources for Landscape Irrigation. Where available by pre-existing treatment, storage, or distribution network, and where approved by the Authority Having Jurisdiction, alternative water source(s) shall be utilized for landscape irrigation. Where adequate capacity and volumes of pre-existing alternative water sources are available, the irrigation system shall be designed to use a minimum of 75 percent of alternative water for the annual irrigation demand before supplemental potable water is used.

L 411.4 Irrigation Control Systems. Where installed as part of a landscape irrigation system, irrigation control systems shall:

(1) Automatically adjust the irrigation schedule to respond to plant water needs determined by weather or soil moisture conditions.

(2) Utilize sensors to suspend irrigation during a rainfall.

(3) Utilize sensors to suspend irrigation where adequate soil moisture is present for plant growth.

(4) Have the capability to program multiple and different run times for each irrigation zone to enable cycling of water applications and durations to mitigate water flowing off of the intended irrigation zone.

(5) The site-specific settings of the irrigation control system affecting the irrigation and shall be posted at the control system location. The posted data, where applicable to the settings of the controller, shall include:

(a) Precipitation rate for each zone.

(b) Plant evapotranspiration coefficients for each zone.

(c) Soil absorption rate for each zone.

(d) Rain sensor settings.

(e) Soil moisture setting.

(f) Peak demand schedule including run times for each zone and the number of cycles to mitigate runoff and monthly adjustments or percentage.

L 411.5 Low Flow Irrigation. Irrigation zones using low flow irrigation shall be equipped with filters sized for the irrigation emission devices and with a pressure regulator installed upstream of the irrigation emission devices as necessary to reduce the operating water pressure in accordance with the manufacturers' equipment requirements.

L 411.6 Mulched Planting Areas. Only low volume emitters are allowed to be installed in mulched planting areas with vegetation taller than 12 inches (305 mm).

L 411.7 System Performance Requirements. The landscape irrigation system shall be designed and installed to:

(1) Prevent irrigation water from runoff out of the irrigation zone.

(2) Prevent water in the supply line drainage from draining out between irrigation events.

(3) Not allow irrigation water to be applied onto or enter non-targeted areas including adjacent property and vegetation areas, adjacent hydrozones not requiring the irrigation water to meet its irrigation demand, non-vegetative areas, impermeable surfaces, roadways, and structures.

L 411.8 Narrow or Irregularly Shaped Landscape Areas. Narrow or irregularly shaped landscape areas, less than 4 feet (1219 mm) in any direction across opposing boundaries, shall not be irrigated by an irrigation emission device except low flow emitters.

L 411.9 Sloped Areas. Where soil surface rises more than 1 foot (305 mm) per 4 feet (1219 mm) of length, the irrigation zone system average precipitation rate shall not exceed 0.75 inches (19 mm) per hour as verified through either of the following methods:

(1) Manufacturer documentation that the precipitation rate for the installed sprinkler head does not exceed 0.75 inches (19 mm) per hour where the sprinkler heads are installed not closer than the specified radius and where the water pressure of the irrigation system is not more than the manufacturer's recommendations.

(2) Catch can test in accordance with the requirements of the Authority Having Jurisdiction and where emitted water volume is measured with a minimum of six catchment containers at random places within the irrigation zone for a minimum of 15 minutes to determine the average precipitation rate, expressed as inches per hour (mm/h).

L 411.10 Sprinkler Head Installations. Installed sprinkler heads shall be low precipitation rate sprinkler heads.

L 411.10.1 Sprinkler Heads in Common Irrigation Zones. Sprinkler heads installed in irrigation zones served by a common valve shall be limited to applying water to plants with similar irrigation needs, and shall have matched precipitation rates (identical inches of water application per hour as rated or tested, plus or minus 5 percent).

L 411.10.2 Sprinkler Head Pressure Regulation. Sprinkler heads shall utilize pressure regulating devices (as part of an irrigation system or integral to the sprinkler head) to maintain manufacturer's recommended operating pressure for each sprinkler and nozzle type.

L 411.10.3 Pop-up Type Sprinkler Heads. Where pop-up type sprinkler heads are installed, the sprinkler heads shall rise to a height of not less than 4 inches (102 mm) above the soil level where emitting water.

L 411.11 Irrigation Zone Performance Criteria. Irrigation zones shall be designed and installed to ensure the average precipitation rate of the sprinkler heads over the irrigated area does not exceed 1 inch per hour (25.4 mm/h) as verified through either of the following methods:

(1) Manufacturer's documentation that the precipitation rate for the installed sprinkler head does not exceed 1 inch per hour (25.4 mm/h) where the sprinkler heads are installed not closer that the specified radius and where the water pressure of the irrigation system is not more than the manufacturer's recommendations.

(2) Catch can test in accordance with the requirements of the Authority Having Jurisdiction and where emitted water volume is measured with a minimum of six catchment containers at random places within the irrigation zone for a minimum of 15 minutes to determine the average precipitation rate, expressed as inches per hour (mm/h).

L 411.12 Qualifications. The Authority Having Jurisdiction shall have the authority to require landscape irrigation contractors, installers, or designers to demonstrate competency. Where required by the Authority Having Jurisdiction, the contractor, installer, or designer shall be certified to perform such work.

L 412.0 Trap Seal Protection.

L 412.1 Water Supplied Trap Primers. Water supplied trap primers shall be electronic or pressure activated and shall use not more than 30 gallons (114 L) per year per drain. Where an alternate water source, as defined by this code, is used for fixture flushing or other uses in the same room, the alternate water source shall be used for the trap primer water supply.

Exception: Flushometer tailpiece trap primers in accordance with IAPMO PS 76.

L 412.2 Drainage Type Trap Seal Primer Devices. Drainage type trap seal primer devices shall not be limited in the amount of water they discharge.

L 413.0 Vehicle Wash Facilities.

L 413.1 Automatic. The maximum make-up water use for automobile washing shall not exceed 40 gallons (151 L) per vehicle for in-bay automatic car washes and 35 gallons (132 L) for conveyor and express type car washes.

L 413.2 Self-Service. Spray wands and foamy brushes shall use not more than 3.0 gpm (0.19 L/s).

L 413.3 Reverse Osmosis. Spot-free reverse osmosis discharge (reject) water shall be recycled.

L 413.4 Towel Ringers. Towel ringers shall have a positive shutoff valve. Spray nozzles shall be replaced annually.

Exception: Bus and large commercial vehicle washes are exempt from the requirements of this section.

L 501.0 Water Heating Design, Equipment, and Installation.

L 501.1 Scope. The provisions of this section shall establish the means of conserving potable and nonpotable water and energy associated with the generation and use of hot water in a building. This includes provisions for the hot water distribution system, which is the portion of the potable water distribution system between a water heating device and the plumbing fixtures, including dedicated return piping and appurtenances to the water heating device in a recirculation system.

L 501.2 Insulation. Hot water supply and return piping shall be thermally insulated. The wall thickness of the insulation shall be equal to the nominal diameter of the pipe up to 2 inches (50 mm). The wall thickness shall be not less than 2 inches (51 mm) for nominal pipe diameters exceeding 2 inches (50 mm). The conductivity of the insulation [k-factor (Btu·in/(h·ft^2·°F))], measured radially, shall not be more than 0.28 [Btu·in/(h·ft^2·°F)] [0.04 W/(m·k)]. Hot water piping to be insulated shall be installed such that insulation is continuous. Pipe insulation shall be installed to within ¼ of an inch (6.4 mm) of appliances, appurtenances, fixtures, structural members, or a wall where the pipe passes through to connect to a fixture within 24 inches (610 mm). Building cavities shall be large enough to accommodate the combined diameter of the pipe, the insulation, and other objects in the cavity that the piping shall cross. Pipe supports shall be installed on the outside of the pipe insulation.

Exceptions:
(1) Where the hot water pipe is installed in a wall that is not of a width to accommodate the pipe and insulation, the insulation thickness shall be permitted to have the maximum thickness that the wall is capable of accommodating and not less than ½ of an inch (12.7 mm) thick.
(2) Hot water supply piping exposed under sinks, lavatories, and similar fixtures.
(3) Where hot water distribution piping is installed within an attic, crawlspace, or wall insulation.
 (a) In attics and crawlspaces, the insulation shall cover the pipe not less than 5½ inches (140 mm) further away from the conditioned space.
 (b) In walls, the insulation shall completely surround the pipe with not less than 1 inch (25.4 mm) of insulation.
 (c) Where burial within the insulation will not completely or continuously surround the pipe, then these exceptions do not apply.

L 501.3 Recirculation Systems. Recirculation systems shall comply with Section L 501.3.1 and Section L 501.3.2.

L 501.3.1 For Low-Rise Residential Buildings. Circulating hot water systems shall be arranged so that the circulating pump(s) are capable of being turned off (automatically or manually) where the hot water system is not in operation. [ASHRAE 90.2:7.2]

L 501.3.2 For Pumps Between Boilers and Storage Tanks. Where used to maintain storage tank water temperature, recirculating pumps shall be equipped with controls limiting operation to a period from the start of the heating cycle to a maximum of 5 minutes after the end of the heating cycle. [ASHRAE 90.1:7.4.4.3]

L 501.4 Recirculation Pump Controls. Pump controls shall include on-demand activation or time clocks combined with temperature sensing. Time clock controls for pumps shall not let the pump operate more than 15 minutes every hour. Temperature sensors shall stop circulation where the temperature set point is reached and shall be located on the circulation loop at or near the last fixture. The pump, pump controls, and temperature sensors shall be accessible. Pump operation shall be limited to the building's hours of operation.

L 501.5 Temperature Maintenance Controls. Systems designed to maintain usage temperatures in hot-water pipes, such as recirculating hot water systems or heat trace, shall be equipped with automatic time switches or other controls that are capable of being set to switch off the usage temperature maintenance system during extended periods where hot water is not required. [ASHRAE 90.1:7.4.4.2]

L 501.6 System Balancing. Systems with multiple recirculation zones shall be balanced to distribute hot water uniformly, or they shall be operated with a pump for each zone. The circulation pump controls shall comply with the provisions of Section L 501.4.

L 501.7 Flow Balancing Valves. Flow balancing valves shall be a factory preset automatic flow control valve, a flow regulating valve, or a balancing valve with memory stop.

L 501.8 Air Elimination. Provision shall be made for the elimination of air from the return system.

L 501.9 Gravity or Thermosyphon Systems. Gravity or thermosyphon systems are prohibited.

L 502.0 Service Hot Water – Low-Rise Residential Buildings.

L 502.1 General. The service water heating system for single-family houses, multi-family structures of three stories or fewer above grade, and modular houses shall comply with Section L 502.2 through Section L 502.7.3. The service water heating system of all other buildings shall comply with Section L 503.0.

L 502.2 Water Heaters and Storage Tanks. Residential-type water heaters, pool heaters, and unfired water heater storage tanks shall comply with the minimum performance requirements specified by federal law.

Unfired storage water heating equipment shall have a heat loss through the tank surface area of less than 6.5 British thermal units per square foot hour [Btu/(ft^2·h)] (20.5 W/m^2). [ASHRAE 90.2:7.1]

L 502.3 Recirculation Systems. Recirculation systems shall comply with the provisions of Section L 501.3.

L 502.4 Central Water Heating Equipment. Service water heating equipment (central systems) that do not fall under the requirements for residential-type service water heating equipment addressed in Section L 502.0 shall comply with the applicable requirements for service water-heating equipment found in Section L 503.0. [ASHRAE 90.2:7.3]

L 502.5 Insulation. Insulation for hot water and return piping shall comply with the provisions of Section L 501.2.

L 502.6 Hard Water. Where water has hardness equal to or exceeding 9 grains per gallon (gr/gal) (154 mg/L) measured as total calcium carbonate equivalents, the water supply line to water heating equipment in new one- and two-family dwellings shall be roughed-in to allow for the installation of water treatment equipment.

L 502.7 Maximum Volume of Hot Water. The maximum volume of water contained in hot water distribution pipes shall be in accordance with Section L 502.7.1 or Section L 502.7.2. The water volume shall be calculated using Table L 502.7.

L 502.7.1 Maximum Volume of Hot Water Without Recirculation or Heat Trace. The maximum volume of water contained in hot water distribution pipe between the water heater and any fixture fitting shall not exceed 32 ounces (oz) (946 mL). Where a fixture fitting shutoff valve (supply stop) is installed ahead of the fixture fitting, the maximum volume of water is permitted to be calculated between the water heater and the fitting shutoff valve (supply stop).

L 502.7.2 Maximum Volume of Hot Water with Recirculation or Heat Trace. The maximum volume of water contained in the branches between the recirculation loop or electrically heat traced pipe, and the fixture fitting shall not exceed 16 oz (473 mL). Where a fixture fitting shutoff valve (supply stop) is installed ahead of the fixture fitting, the maximum volume of water is permitted to be calculated between the recirculation loop or electrically, heat traced pipe and the fixture fitting shutoff valve (supply stop).

Exception: Whirlpool bathtubs or bathtubs that are not equipped with a shower are exempted from the requirements of Section L 502.7.

L 502.7.3 Hot Water System Submeters. Where a hot water pipe from a circulation loop or electric heat trace line is equipped with a submeter, the hot water distribution system downstream of the submeter shall have either an end-of-line hot water circulation pump or shall be electrically heat traced. The maximum volume of water in a branch from the circulation loop or electric heat trace line downstream of the submeter shall not exceed 16 oz (473 mL).

Where there is no circulation loop or electric heat traced line downstream of the submeter, the submeter shall be located within 2 feet (610 mm) of the central hot water system; or the branch line to the submeter shall be circulated or heat traced to within 2 feet (610 mm) of the submeter. The maximum volume from the submeter to each fixture shall not exceed 32 oz (946 mL).

The circulation pump controls shall comply with the provisions of Section L 501.4.

L 503.0 Service Hot Water – Other Than Low-Rise Residential Buildings.

L 503.1 General. The service hot water, other than single-family houses, multifamily structures of three stories or fewer above grade, and modular houses shall comply with this section.

L 503.1.1 New Buildings. Service water-heating systems and equipment shall comply with the requirements of this section as described in Section L 503.2. [ASHRAE 90.1:7.1.1.1]

L 503.1.2 Additions to Existing Buildings. Service water heating systems and equipment shall comply with the requirements of this section.

Exception: Where the service water-heating to an addition is provided by existing service water-heating systems and equipment, such systems and equipment shall not be required to be in accordance with this appendix. However, new systems or equipment installed shall be in accordance with specific requirements applicable to those systems and equipment. [ASHRAE 90.1:7.1.1.2]

L 503.1.3 Alterations to Existing Buildings. Building service water-heating equipment installed as a direct replacement for existing building service water-heating equipment shall be in accordance with the requirements of Section L 503.0 applicable to the equipment being replaced. New and replacement piping shall comply with Section L 503.3.3.

Exception: Compliance shall not be required where there is insufficient space or access to meet these requirements. [ASHRAE 90.1:7.1.1.3]

L 503.2 Compliance Path(s). Compliance shall be achieved in accordance with the requirements of Section L 503.1, Section L 503.3, Section L 503.4, and Section L 503.5. [ASHRAE 90.1:7.2.1]

L 503.2.1 Energy Cost Budget Method. Projects using the energy cost budget method of ASHRAE 90.1 for demonstrating compliance with the standard shall be in accordance with the requirements of Section L 503.3 in conjunction with the energy cost budget method of ASHRAE 90.1. [ASHRAE 90.1:7.2.2]

L 503.3 Mandatory Provisions. The mandatory provisions of Section L 503.3.1 through Section L 503.3.7 shall be followed.

TABLE L 502.7
WATER VOLUME FOR DISTRIBUTION PIPING MATERIALS*

NOMINAL SIZE (inch)	COPPER M	COPPER L	COPPER K	CPVC CTS SDR 11	CPVC SCH 40	PEX-AL-PEX	PE-AL-PE	CPVC SCH 80	PEX CTS SDR 9	PE-RT SDR 9	PP SDR 6	PP SDR 7.3	PP SDR 11
3/8	1.06	0.97	0.84	NA	1.17	0.63	0.63	NA	0.64	0.64	0.91	1.09	1.24
1/2	1.69	1.55	1.45	1.25	1.89	1.31	1.31	1.46	1.18	1.18	1.41	1.68	2.12
3/4	3.43	3.22	2.90	2.67	3.38	3.39	3.39	2.74	2.35	2.35	2.23	2.62	3.37
1	5.81	5.49	5.17	4.43	5.53	5.56	5.56	4.57	3.91	3.91	3.64	4.36	5.56
1 1/4	8.70	8.36	8.09	6.61	9.66	8.49	8.49	8.24	5.81	5.81	5.73	6.81	8.60
1 1/2	12.18	11.83	11.45	9.22	13.20	13.88	13.88	11.38	8.09	8.09	9.03	10.61	13.47
2	21.08	20.58	20.04	15.79	21.88	21.48	21.48	19.11	13.86	13.86	14.28	16.98	21.39

For SI units: 1 ounce = 29.573 mL
* NA: Not Applicable

L 503.3.1 Load Calculations. Service water-heating system design loads for the purpose of sizing systems and equipment shall be determined in accordance with manufacturer's published sizing guidelines or accepted engineering standards and handbooks acceptable to the adopting authority (e.g., ASHRAE Handbook – HVAC Applications). [ASHRAE 90.1:7.4.1]

L 503.3.2 Equipment Efficiency. Water-heating equipment, hot-water supply boilers used solely for heating potable water, pool heaters, and hot water storage tanks shall comply with the criteria listed in Table L 503.3.2. Where multiple criteria are listed, all criteria shall be met. The omission of minimum performance requirements for certain classes of equipment does not preclude the use of such equipment where appropriate. Equipment not listed in Table L 503.3.2 has no minimum performance requirements.

Exceptions: Water heaters and hot-water supply boilers having more than 140 gallons (530 L) of storage capacity are not required to meet the standby loss (SL) requirements of Table L 503.3.2 where:

(1) The tank surface is thermally insulated to R-12.5.

(2) A standing pilot light is not installed.

(3) Gas- or oil-fired storage water heaters have a flue damper or fan-assisted combustion. [ASHRAE 90.1:7.4.2]

L 503.3.3 Insulation. The following piping shall be insulated in accordance with Table L 503.3.3:

(1) Recirculating system piping, including the supply and return piping of a circulating tank type water heater.

(2) The first 8 feet (2438 mm) of outlet piping for a constant temperature nonrecirculating storage system.

(3) The first 8 feet (2438 mm) of branch piping connecting to recirculated, heat-traced, or impedance heated piping.

(4) The inlet piping between the storage tank and a heat trap in a nonrecirculating storage system.

(5) Piping that is externally heated (such as heat trace or impedance heating). [ASHRAE 90.1:7.4.3]

L 503.3.4 Hot Water System Design. Hot water systems shall comply with Section L 503.3.4(1) and Section L 503.3.4(2).

(1) Recirculation systems shall comply with the provisions of Section L 501.3.

(2) The maximum volume of water contained in hot water distribution lines between the water heater and the fixture stop or connection to showers, kitchen faucets, and lavatories shall be determined in accordance with Section L 502.7.

L 503.3.5 Service Water Heating System Controls. Service water heating system controls shall comply with Section L 503.3.5(1) and Section L 503.3.5(2).

(1) Temperature controls shall be provided that allows for storage temperature adjustment from 120°F (49°C) or lower to a maximum temperature compatible with the intended use.

Exception: Where the manufacturer's installation instructions specify a higher minimum thermostat setting to minimize condensation and resulting corrosion. [ASHRAE 90.1:7.4.4.1]

(2) Temperature controlling means shall be provided to limit the maximum temperature of water delivered from lavatory faucets in public facility restrooms to 110°F (43°C). [ASHRAE 90.1:7.4.4.3]

L 503.3.6 Pools. Pool heating systems shall comply with Section L 503.3.6(1) through Section L 503.3.6(3).

(1) Pool heaters shall be equipped with a readily accessible ON/OFF switch to allow shutting off the heater without adjusting the thermostat setting. Pool heaters fired by natural gas shall not have continuously burning pilot lights. [ASHRAE 90.1:7.4.5.1]

(2) Heated pools shall be equipped with a vapor retardant pool cover on or at the water surface. Pools heated to more than 90°F (32°C) shall have a pool cover with a minimum insulation value of R-12.

Exception: Pools that are deriving over 60 percent of the energy for heating from site-recovered energy or solar energy. [ASHRAE 90.1:7.4.5.2]

(3) Time switches shall be installed on swimming pool heaters and pumps.

Exceptions:

(1) Where public health standards require 24-hour pump operation.

(2) Where pumps are required to operate solar and waste heat recovery pool heating systems. [ASHRAE 90.1:7.4.5.3]

L 503.3.7 Heat Traps. Vertical pipe risers serving storage water heaters and storage tanks not having integral heat traps and serving a nonrecirculating system shall have heat traps on both the inlet and outlet piping as close as practical to the storage tank. A heat trap is a means to counteract the natural convection of heated water in a vertical pipe run. The means is either of the following:

(1) A device specifically designed for the purpose or an arrangement of tubing that forms a loop of 360 degrees (6.28 rad).

(2) Piping that, from the point of connection to the water heater (inlet or outlet) includes a length of piping directed downward before connection to the vertical piping of the supply water or hot-water distribution system, as applicable. [ASHRAE 90.1:7.4.6]

L 503.4 Prescriptive Path. The prescriptive path for space or water heating efficiency shall comply with Section L 503.4.1 through Section L 503.4.5.

L 503.4.1 Space Heating and Water Heating. The use of a gas-fired or oil-fired space heating boiler system, otherwise in accordance with Section L 503.0, to provide the total space heating and service water heating for a building is allowed where one of the following conditions is met:

(1) The single space-heating boiler, or the component of a modular or multiple boiler system that is heating the service water, has a standby loss in Btu/h (kW) not

APPENDIX L

TABLE L 503.3.2
PERFORMANCE REQUIREMENTS FOR WATER-HEATING EQUIPMENT
MINIMUM EFFICIENCY REQUIREMENTS
[ASHRAE 90.1: TABLE 7.8]

EQUIPMENT TYPE	SIZE CATEGORY (INPUT)	SUBCATEGORY OR RATING CONDITION	PERFORMANCE REQUIRED[1]	TEST PROCEDURE[2,3]
Electric table top water heaters	≤12 kW	Resistance ≥20 gal	See footnote 7	—
Electric water heaters	≤12 kW[5]	Resistance ≥20 gal	See footnote 7	—
	>12 kW[5]	Resistance ≥20 gal	$0.3 + 27\sqrt{V_m}$ %/h	Section G.2 of CSA Z21.10.3
	≤24 Amps and ≤250 Volts	Heat Pump	See footnote 7	—
Gas storage water heaters	≤75 000 Btu/h	≥20 gal	See footnote 7	—
	>75 000 Btu/h[6]	<4000 (Btu/h)/gal	80% E_t (Q/800 + 110$\sqrt{V}$) SL, Btu/h	Sections G.1 and G.2 of CSA Z21.10.3
Gas instantaneous water heaters	>50 000 Btu/h and <200 000 Btu/h	≥4000 (Btu/h)/gal and <2 gal	See footnote 7	—
	≥200 000 Btu/h[4,6]	≥4000 (Btu/h)/gal and <10 gal	80% E_t	Sections G.1 and G.2 of CSA Z21.10.3
	≥200 000 Btu/h[6]	≥4000 (Btu/h)/gal and ≥10 gal	80% E_t (Q/799 + 16.6$\sqrt{V}$) SL, Btu/h	
Oil storage water heaters	≤105 000 Btu/h	≥20 gal	0.59-0.0005V EF	—
	>105 000 Btu/h	<4000 (Btu/h)/gal	80% E_t (Q/800 + 110$\sqrt{V}$) SL, Btu/h	Sections G.1 and G.2 of CSA Z21.10.3
Oil instantaneous water heaters	≤210 000 Btu/h	≥4000 (Btu/h)/gal and <2 gal	See footnote 7	—
	>210 000 Btu/h	≥4000 (Btu/h)/gal and <10 gal	80% E_t	Sections G.1 and G.2 of CSA Z21.10.3
	>210 000 Btu/h	≥4000 (Btu/h)/gal and ≥10 gal	78% E_t (Q/800 + 110$\sqrt{V}$) SL, Btu/h	
Hot-water supply boilers, gas and oil[6]	≥300 000 Btu/h and <12 500 000 Btu/h	≥4000 (Btu/h)/gal and <10 gal	80% E_t	
Hot-water supply boilers, gas[6]	—	≥4000 (Btu/h)/gal and ≥10 gal	80% E_t (Q/800 + 110$\sqrt{V}$) SL, Btu/h	Sections G.1 and G.2 of CSA Z21.10.3
Hot-water supply boilers, oil	—	≥4000 (Btu/h)/gal and ≥10 gal	78% E_t (Q/800 + 110$\sqrt{V}$) SL, Btu/h	
Pool heaters, oil and gas	All	—	See footnote 7	ASHRAE 146
Heat pump pool heaters	All	50°F db 44.2° wb Outdoor air 80.0°F entering water	4.0 COP	AHRI 1160
Unfired storage tanks	All	—	R-12.5	(none)

For SI units: 1 gallon = 3.785 L, 1000 British thermal units per hour = 0.293 kW, °C = (°F-32)/1.8

Notes:
[1] Thermal efficiency (E_t) is a minimum requirement, while standby loss (SL) is maximum Btu/h (kW) based on a 70°F (21°C) temperature difference between stored water and ambient requirements. In the SL equation, V is the rated volume in gallons and Q is the nameplate input rate in Btu/h (kW). V_m is the measured volume in the tank in gallons.
[2] ASHRAE 90.1 contains a complete specification, including the year version, of the referenced test procedure.
[3] Section G.1 is titled "Test Method for Measuring Thermal Efficiency" and Section G.2 is titled "Test Method for Measuring Standby Loss."
[4] Instantaneous water heaters with input rates below 200 000 Btu/h (58.6 kW) shall be in accordance with these requirements where the water heater is designed to heat water to temperatures of 180°F (82°C) or higher.
[5] Electric water heaters with input rates less than 40 946 Btu/h (12 kW) shall be in accordance with these requirements where the water heater is designed to heat water to temperatures of 180°F (82°C) or higher.
[6] Refer to Section L 503.4.3 for additional requirements for gas storage and instantaneous water heaters and gas hot water supply boilers.
[7] In the U.S., the efficiency requirements for water heaters or gas pool heaters in this category or subcategory are specified by the U.S. Department of Energy. Those requirements and applicable test procedures are found in the Code of Federal Regulations 10 CFR Part 430.

APPENDIX L

exceeding $(13.3 \times pmd + 400)/n$, where (pmd) is the probable maximum demand in gallons per hour, determined in accordance with the procedures described in generally accepted engineering standards and handbooks, and (n) is the fraction of the year where the outdoor daily mean temperature exceeds 64.9°F (18.28°C).

The standby loss is to be determined for a test period of 24 hours duration while maintaining a boiler water temperature of not less than 90°F (32°C) above ambient, with an ambient temperature between 60°F (16°C) and 90°F (32°C). For a boiler with a modulating burner, this test shall be conducted at the lowest input.

(2) It is demonstrated to the satisfaction of the Authority Having Jurisdiction that the use of a single heat source will consume less energy than separate units.

(3) The energy input of the combined boiler and water heater system is less than 150 000 British thermal units per hour (Btu/h) (44 kW). [ASHRAE 90.1:7.5.1]

L 503.4.2 Service Water Heating Equipment. Service water-heating equipment used to provide the additional function of space heating as part of a combination (integrated) system shall satisfy stated requirements for the service water-heating equipment. [ASHRAE 90.1:7.5.2]

L 503.4.3 Buildings with High-Capacity Service Water Heating Systems. New buildings with gas service hot-water systems with a total installed gas water-heating input capacity of 1 000 000 Btu/h (293 kW) or more, shall have gas service water-heating equipment with a thermal efficiency (E_t) of not less than 90 percent. Multiple units of gas water-heating equipment shall be permitted to comply with this requirement where the water-heating input provided by the equipment, with thermal efficiency (E_t) of more or less than 90 percent, provides an input capacity-weighted average thermal efficiency of not less than 90 percent.

Exceptions:

(1) Where 25 percent of the annual service water-heating requirement is provided by site-solar or site-recovered energy.

(2) Water heaters installed in individual dwelling units.

(3) Individual gas water heaters with input capacity, not more than 100 000 Btu/h (29.3 kW). [ASHRAE 90.1:7.5.3]

L 503.4.4 Heat Recovery for Service Water Heating. Condenser heat recovery systems shall be installed for heating or preheating of service hot water provided the following are true:

(1) The facility operates 24 hours a day.

(2) The total installed heat rejection capacity of the water-cooled systems exceeds 6 000 000 Btu/h (1758 kW) of heat rejection.

(3) The design service water-heating load exceeds 1 000 000 Btu/h (293 kW). [ASHRAE 90.1:6.5.6.2.1]

TABLE L 503.3.3
MINIMUM PIPING INSULATION THICKNESS FOR HEATING AND HOT-WATER SYSTEMS
(STEAM, STEAM CONDENSATE, HOT-WATER HEATING, AND DOMESTIC WATER SYSTEMS)
[ASHRAE 90.1: TABLE 6.8.3-1]

FLUID OPERATING TEMPERATURE RANGE AND USAGE (°F)	INSULATION CONDUCTIVITY		≥NOMINAL PIPE SIZE OR TUBE SIZE (inches)				
	CONDUCTIVITY Btu•inch/(h•ft²•°F)	MEAN RATING TEMPERATURE (°F)	<1	1 to <1½	1½ to <4	4 to <8	≥8
			INSULATION THICKNESS (inches)				
>350	0.32 to 0.34	250	4.5	5.0	5.0	5.0	5.0
251 to 350	0.29 to 0.32	200	3.0	4.0	4.5	4.5	4.5
201 to 250	0.27 to 0.30	150	2.5	2.5	2.5	3.0	3.0
141 to 200	0.25 to 0.29	125	1.5	1.5	2.0	2.0	2.0
105 to 140	0.22 to 0.28	100	1.0	1.0	1.5	1.5	1.5

For SI units: °C=(°F-32)/1.8, 1 British thermal unit inch per hour square foot degree Fahrenheit = [0.1 W/(m•K)], 1 inch = 25 mm

Notes:
1. For insulation outside the stated conductivity range, the minimum thickness (T) shall be determined as follows:
   $T = r\{(1 + t/r)^{K/k} - 1\}$ Where:
   T = minimum insulation thickness (inches) (mm).
   r = actual outside radius of pipe (inches) (mm).
   t = insulation thickness listed in this table for applicable fluid temperature and pipe size.
   K = conductivity of alternate material at mean rating temperature indicated for the applicable fluid temperature [Btu•in/(h•ft²•°F)] [W/(m•K)].
   k = the upper value of the conductivity range listed in this table for the applicable fluid temperature.
2. These thicknesses are based on energy efficiency considerations only. Additional insulation is sometimes required relative to safety issues or surface temperature.
3. For piping 1½ inches (40 mm) or less, and located in partitions within conditioned spaces, reduction of insulation thickness by 1 inch (25.4 mm) shall be permitted before thickness adjustment required in Footnote 1, but not a thickness less than 1 inch (25.4 mm).
4. For direct-buried heating and hot water system piping, reduction of insulation thickness by 1½ inch (38 mm) shall be permitted before thickness adjustment required in Footnote 1, but not a thickness less than 1 inch (25.4 mm).
5. Table L 503.3.3 is based on steel pipe. Non-metallic pipes, Schedule 80 thickness or less shall use the table values. For other non-metallic pipes having a thermal resistance more than that of steel pipe, reduced insulation thicknesses shall be permitted where documentation is provided showing that the pipe with the proposed insulation has no more heat transfer per foot (mm) than a steel pipe of the same size with the insulation thickness shown in Table L 503.3.3.

APPENDIX L

L 503.4.5 Capacity. The required heat recovery system shall have the capacity to provide the smaller of:

(1) Sixty percent of the peak heat-rejection load at design conditions.

(2) Preheat of the peak service hot-water draw to 85°F (29°C).

Exceptions:

(1) Facilities that employ condenser heat recovery for space heating with a heat recovery design exceeding 30 percent of the peak water-cooled condenser load at design conditions.

(2) Facilities that provide 60 percent of their service water heating from site-solar or site-recovered energy or other sources. [ASHRAE 90.1:6.5.6.2.2]

L 503.5 Submittals. The Authority Having Jurisdiction shall require submittal of compliance documentation and supplemental information in accordance with Section 104.3.1 of this code.

L 504.0 Solar Water Heating Systems.

L 504.1 General. The erection, installation, alteration, addition to, use or maintenance of solar water heating systems shall be in accordance with this section and the Uniform Solar Energy and Hydronics Code.

L 504.2 Annual Inspection and Maintenance. Solar energy systems that utilize a heat transfer fluid shall annually be inspected unless inspections are required on a more frequent basis by the solar energy system manufacturer.

L 505.0 Hard Water.

L 505.1 Softening and Treatment. Where water has a hardness equal to or exceeding 10 gr/gal (171 mg/L) measured as total calcium carbonate equivalents, the water supply line to water heating equipment and the circuit of boilers shall be softened or treated to prevent accumulation of limescale and consequent reduction in energy efficiency.

L 506.0 Drain Water Heat Exchangers.

L 506.1 General. Drain water heat exchangers shall comply with IAPMO PS 92. The heat exchanger shall be accessible.

L 601.0 Installer Qualifications.

L 601.1 Scope. The provisions of this section address minimum qualifications of installers of plumbing and mechanical systems covered within the scope of this appendix.

L 602.0 Qualifications.

L 602.1 General. Where permits are required, the Authority Having Jurisdiction shall have the authority to require contractors, installers, or service technicians to demonstrate competency. Where determined by the Authority Having Jurisdiction, the contractor, installer, or service technician shall be licensed to perform such work.

L 701.0 Method of Calculating Water Savings.

L 701.1 Purpose. The purpose of this section is to provide

TABLE L 701.2(1)
WATER USE BASELINE[5]

FIXTURE TYPE	MAXIMUM FLOW-RATE CONSUMPTION[2]	DURATION	ESTIMATED DAILY USES PER PERSON	OCCUPANTS[3,4]
Showerheads	2.5 gpm at 80 psi	8 minutes	1	–
Private or Private Use Lavatory Faucets	2.2 gpm at 60 psi	0.25 minutes	4	–
Residential Kitchen Faucets	2.2 gpm at 60 psi	4 minutes	1	–
Wash Fountains	One 2.2 gpm at 60 psi fixture fitting for every 20 inches rim space	–	–	–
Lavatory Faucets in other than Residences, Apartments, and Private Bathrooms in Lodging Facilities	0.5 gpm	0.25 minutes	4	–
Metering Faucets	0.25 gallons /cycle	–	3	–
Metering Faucets for Wash Fountains	One 0.25 gallon per cycle fixture fitting for every 20 inches rim space	–	–	–
Water Closets	1.6 gallons per flush	1 flush	1 male[1]	–
			3 female	–
Urinals	1.0 gallons per flush	1 flush	2 male	–

For SI units: 1 gallon per minute = 0.06 L/s, 1 pound-force per square inch = 6.8947 kPa, 1 gallon = 3.785 L, 1 inch = 25.4 mm

Notes:
[1] The daily use number shall be increased to three where urinals are not installed in the room.
[2] The maximum flow rate or consumption is from the Energy Policy Act.
[3] For residential occupancies, the number of occupants shall be based on two persons for the first bedroom and one additional person for each additional bedroom.
[4] For non-residential occupancies, refer to Table 422.1 for occupant load factors.
[5] Where determining calculations, assume one use per person for metering or self-closing faucets.

APPENDIX L

a means of estimating the water savings where installing plumbing and fixture fittings that use less water than the maximum required by the Energy Policy Act of 1992 and 2005 and this code.

L 701.2 Calculation of Water Savings. Table L 701.2(1) and Table L 701.2(2) shall be permitted to be used to establish a water use baseline in calculating the amount of water saved as a result of using plumbing fixtures and fixture fittings that use less water than the required maximum. Water use is determined by the following equation:

Water use = (flow rate or consumption) x (duration) x (occupants) x (daily uses)

TABLE L 701.2(2)
WATER SAVINGS CALCULATOR[1, 2, 3]

NON-RESIDENTIAL BUILDINGS					
FIXTURE TYPE	CONSUMPTION (gallons per minute)	DAILY USES	DURATION (minutes)	OCCUPANTS	DAILY WATER USES (gallons)
1.6 gpf (gallons per flush) toilet - male	1.6	1	1	150	240
1.6 gpf toilet - female	1.6	3	1	150	720
1.0 gpf urinal - male	1	2	1	150	300
Commercial lavatory faucet - 0.5 gpm	0.5	3	0.25	300	113
Kitchen sink - 2.2 gpm	2.2	1	0.25	300	165
Showerhead - 2.5 gpm	2.5	0.1	8	300	600
				Total Daily Volume	2138
				Annual Work Days	260
				Total Annual Usage	555 750

NON-RESIDENTIAL BUILDINGS					
FIXTURE TYPE	CONSUMPTION (gallons per minute)	DAILY USES	DURATION (minutes)	OCCUPANTS	DAILY WATER USES (gallons)
1.6 gpf toilet - male	1.28	1	1	150	192
1.6 gpf toilet - female	1.28	3	1	150	576
1.0 gpf urinal - male	0.5	2	1	150	150
Commercial lavatory faucet - 0.5 gpm	0.5	3	0.25	300	113
Kitchen sink - 2.2 gpm	2.2	1	0.25	300	165
Showerhead - 2.5 gpm	2.5	0.1	8	300	600
				Total Daily Volume	1796
				Annual Work Days	260
				Total Annual Usage	466 830
				Annual Savings	88 920
				% Reduction	-16.0 percent

For SI units: 1 gallon per minute = 0.06 L/s, 1 gallon = 3.785 L

Notes:
[1] Consumption values shown as underlined reflect the maximum consumption values associated with the provisions called out in the IAPMO Green Plumbing & Mechanical Code Supplement.
[2] Where metering faucets are used, insert the flow rate of the faucet in the "Consumption" column and insert the cycle time in the "Duration" column (assume 1 cycle per use).
[3] To determine estimated savings, insert occupant values (same as Baseline) and consumption values based on fixtures and fixture fittings installed.

APPENDIX L

TABLE L 701.2(2) (continued)
WATER SAVINGS CALCULATOR[1,2]

NON-RESIDENTIAL BUILDINGS					
FIXTURE TYPE	CONSUMPTION (gallons per minute)	DAILY USES	DURATION (minutes)	OCCUPANTS	DAILY WATER USES (gallons)
1.6 gpf toilets	1.6	5	1	4	32
Lavatory faucet - 2.2 gpm	2.2	8	0.25	4	18
Kitchen sink - 2.2 gpm	2.2	6	0.25	4	13
Showerhead - 2.5 gpm	2.5	0.75	8	4	60
				Total Daily Volume	123
				Annual Work Days	44 822

NON-RESIDENTIAL BUILDINGS					
FIXTURE TYPE	CONSUMPTION (gallons per minute)	DAILY USES	DURATION (minutes)	OCCUPANTS	DAILY WATER USES (gallons)
1.6 gpf toilet - male	1.28	5	1	4	26
Lavatory faucet - 1.5 gpm	1.5	8	0.25	4	12
Kitchen sink - 2.2 gpm	2.2	6	0.25	4	13
Showerhead - 2.5 gpm	2.5	0.75	8	4	60
				Total Daily Volume	111
				Annual Usage	40 442
				Annual Savings	4380
				% Reduction	-9.8 percent

For SI units: 1 gallon per minute = 0.06 L/s, 1 gallon = 3.785 L

Notes:

[1] Consumption values shown as underlined reflect the maximum consumption values associated with the provisions called out in the IAPMO Green Plumbing & Mechanical Code Supplement.

[2] To determine estimated savings, insert occupant values (same as Baseline) and consumption values based on fixtures and fixture fittings installed.

Notes and instructions for Table L 701.2(2):

Table L 701.2(2) is an example of a calculator that is capable of helping estimate water savings in residential and nonresidential structures. The "Duration" of use and "Daily Uses" values that appear in the table are estimates and based on previous studies. The first example shown below is a commercial office building with 300 occupants, 150 females, and 150 males. The second example is a 3 bedroom residential building. To obtain and use a working copy of this calculator, follow the download and use instructions below.

Instructions for download:

1. Go to the IAPMO website at www.iapmogreen.org to download the water savings calculator. The calculator is a Microsoft Office Excel file (1997 or later), your computer must be capable of running MS Excel.
2. Follow the instructions for downloading and running the file.

Instructions for use:

1. In the Baseline Case section, insert the number of total occupants, male occupants and female occupants that apply to the building in the "Occupants" column. Unless specific gender ratio values are provided, assume a 50/50 gender ratio.
2. Copy and paste these same values in the "Occupants" column of the Calculator section.
3. In the Calculator section, insert the consumption values (flow rates in gpm or gallons per flush or per cycle) in the "Consumption" column.
4. Estimated water savings regarding percent savings versus baseline values, gallons per day and gallons per year will be automatically calculated.

APPENDIX M
PEAK WATER DEMAND CALCULATOR

M 101.0 General.

M 101.1 Applicability. This appendix provides a method for estimating the demand load for the building water supply and principal branches for single- and multi-family dwellings with water-conserving plumbing fixtures, fixture fittings, and appliances.

M 102.0 Demand Load.

M 102.1 Water-Conserving Fixtures. Plumbing fixtures, fixture fittings, and appliances shall not exceed the design flow rate in Table M 102.1.

TABLE M 102.1
DESIGN FLOW RATE FOR WATER-CONSERVING PLUMBING FIXTURES AND APPLIANCES IN RESIDENTIAL OCCUPANCIES

FIXTURE AND APPLIANCE	MAXIMUM DESIGN FLOW RATE (gallons per minute)
Bar Sink	1.5
Bathtub	5.5
Bidet	2.0
Clothes Washer*	3.5
Combination Bath/Shower	5.5
Dishwasher*	1.3
Kitchen Faucet	2.2
Laundry Faucet (with aerator)	2.0
Lavatory Faucet	1.5
Shower, per head	2.0
Water Closet, 1.28 GPF Gravity Tank	3.0

For SI units: 1 gallon per minute = 0.06 L/s
* Clothes washers and dishwashers shall have an energy star label.

M 102.2 Water Demand Calculator. The estimated design flow rate for the building supply and principal branches and risers shall be determined by the IAPMO Water Demand Calculator available for download at http://www.iapmo.org/WESTAND/Pages/DocumentInformation.aspx.

M 102.3 Meter and Building Supply. To determine the design flow rate for the water meter and building supply, enter the total number of indoor plumbing fixtures and appliances for the building in Column [B] of the Water Demand Calculator and run Calculator. See Table M 102.3 for an example.

M 102.4 Fixture Branches and Fixture Supplies. To determine the design flow rate for fixture branches and risers, enter the total number of plumbing fixtures and appliances for the fixture branch or riser in Column [B] of the Water Demand Calculator and run Calculator. The flow rate for one fixture branch and one fixture supply shall be the design flow rate of the fixture according to Table M 102.1.

M 102.5 Continuous Supply Demand. Continuous supply demands in gallons per minute (gpm) for lawn sprinklers, air conditioners, hose bibbs, etc., shall be added to the total estimated demand for the building supply as determined by Section M 102.3. Where there is more than one hose bibb installed on the plumbing system, the demand for only one hose bibb shall be added to the total estimated demand for the building supply. Where a hose bibb is installed on a fixture branch, the demand of the hose bibb shall be added to the design flow rate for the fixture branch as determined by Section M 102.4.

M 102.6 Other Fixtures. Fixtures not included in Table M 102.1 shall be added in Rows 12 through 14 in the Water Demand Calculator as Other Fixture. The probability of use and flow rate for Other Fixtures shall be added by selecting the comparable probability of use and flow rate from Columns [C] and [E].

M 102.7 Size of Water Piping per Appendix A. Except as provided in Section N 102.0 for estimating the demand load for single- and multi-family dwellings, the size of each water piping system shall be determined in accordance with the procedure set forth in Appendix A. After determining the permissible friction loss per 100 feet (30 480 mm) of pipe in accordance with Section A 104.0 and the demand flow in accordance with the Water Demand Calculator, the diameter of the building supply pipe, branches and risers shall be obtained from Chart A 105.1(1) through Chart A 105.1(7), whichever is applicable, in accordance with Section A 105.0 and Section A 106.0. Velocities shall be in accordance with Section A 107.0. Appendix I, **Figure 3** and **Figure 4** shall be permitted when sizing PEX systems.

M102.7.1 Minimum Fixture Branch Size. The minimum fixture branch size shall be ½ inch (15 mm) in diameter.

APPENDIX M

TABLE M 102.3
WATER DEMAND CALCULATOR EXAMPLE

	[A] FIXTURE	[B] ENTER NUMBER OF FIXTURES	[C] PROBABILITY OF USE (%)	[D] ENTER FIXTURE FLOW RATE (GPM)	[E] MAXIMUM RECOMMENDED FIXTURE FLOW RATE (GPM)
1	Bar Sink	0	2.0	1.5	1.5
2	Bathtub	0	1.0	5.5	5.5
3	Bidet	0	1.0	2.0	2.0
4	Clothes Washer	1	5.5	3.5	3.5
5	Combination Bath/Shower	1	5.5	5.5	5.5
6	Dishwasher	1	0.5	1.3	1.3
7	Kitchen Faucet	1	2.0	2.2	2.2
8	Laundry Faucet	0	2.0	2.0	2.0
9	Lavatory Faucet	1	2.0	1.5	1.5
10	Shower, per head	0	4.5	2.0	2.0
11	Water Closet, 1.28 GPF Gravity Tank	1	1.0	3.0	3.0
12	Other Fixture 1	0	0.0	0.0	6.0
13	Other Fixture 2	0	0.0	0.0	6.0
14	Other Fixture 3	0	0.0	0.0	6.0
	Total Number of Fixtures	6			
	99th Percentile Demand Flow =	8.5 GPM		RESET	RUN WATER DEMAND CALCULATOR

M 102.8 Examples Illustrating Use of Water Demand Calculator with Appendix A.

Example 1: Indoor Water Use Only – Use the information given below to find the pipe size for the building supply to a residential building with six indoor fixtures as shown in **Figure 1** [Pipe Section 4].

Given Information:

Type of construction:	Residential, one-bathroom	Friction loss per 100 ft:	15 psi
Type of pipe material:	L-copper	Maximum velocity:	10 ft/s
Fixture number/type:	1 combination bath/shower	1 kitchen faucet	
	1 lavatory faucet	1 dishwasher	
	1 WC	1 clothes washer	

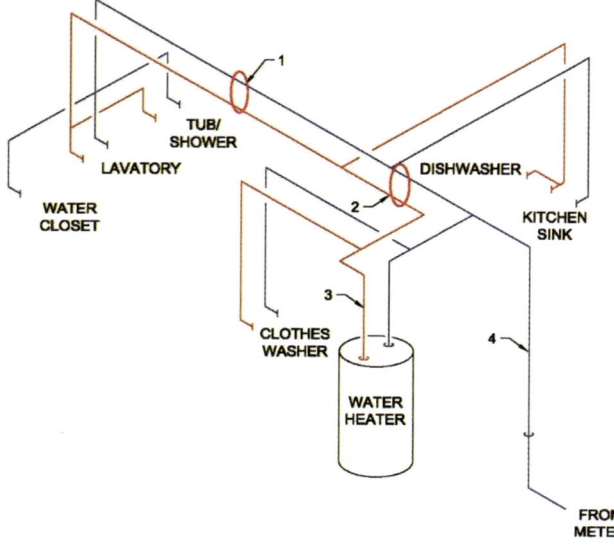

FIGURE 1
RESIDENTIAL BUILDING WITH SIX INDOOR FIXTURES

Solution: Step 1 of 2 – Find Demand Load for the Building Supply.

The Water Demand Calculator [WDC] in **Figure 2** is used to determine the demand load expected from indoor water use. The WDC has white-shaded cells and light gray-shaded cells. The values in the light gray cells are derived from a national survey of indoor water use at homes with efficient fixtures and cannot be changed.

The white-shaded cells accept input from the designer. For instance, fixture counts from the given information are entered in Column [B]; the corresponding recommended fixture flow rates are already provided in Column [D]. The flow rates in Column [D] may be reduced only if the manufacturer specifies a lower flow rate for the fixture. Column [E] establishes the upper limits for the flow rates entered into Column [D]. Clicking the Run Water Demand Calculator button gives 8.5 gpm as the estimated indoor water demand for the whole building. This result appears in the dark gray box of the WDC in **Figure 2**.

APPENDIX M

	[A] FIXTURE	[B] ENTER NUMBER OF FIXTURES	[C] PROBABILITY OF USE (%)	[D] ENTER FIXTURE FLOW RATE (GPM)	[E] MAXIMUM RECOMMENDED FIXTURE FLOW RATE (GPM)
1	Bar Sink	0	2.0	1.5	1.5
2	Bathtub	0	1.0	5.5	5.5
3	Bidet	0	1.0	2.0	2.0
4	Clothes Washer	1	5.5	3.5	3.5
5	Combination Bath/Shower	1	5.5	5.5	5.5
6	Dishwasher	1	0.5	1.3	1.3
7	Kitchen Faucet	1	2.0	2.2	2.2
8	Laundry Faucet	0	2.0	2.0	2.0
9	Lavatory Faucet	1	2.0	1.5	1.5
10	Shower, per head	0	4.5	2.0	2.0
11	Water Closet, 1.28 GPF Gravity Tank	1	1.0	3.0	3.0
12	Other Fixture 1	0	0.0	0.0	6.0
13	Other Fixture 2	0	0.0	0.0	6.0
14	Other Fixture 3	0	0.0	0.0	6.0
	Total Number of Fixtures	6		RESET	RUN WATER DEMAND CALCULATOR
	99th Percentile Demand Flow =	8.5 GPM			

FIGURE 2
WATER DEMAND CALCULATOR FOR INDOOR USE AT HOME WITH SIX EFFICIENT FIXTURES (EXAMPLE 1).

Solution: Step 2 of 2 – Determine the Pipe Size of the Building Supply.

Chart A 105.1(1) for copper piping systems (from Appendix A of the UPC, shown in **Figure 3**) is used to determine the pipe size, based on given friction loss, given maximum allowable pipe velocity, given pipe material and the demand load computed in Step 1. In **Figure 3**, the intersection of the given friction loss (15 psi) and the maximum allowable pipe velocity (10 ft/s) is labeled point A. The vertical line that descends from point A to the base of the chart intersects four nominal sizes for L-copper pipe. These intersection points are labeled B, C, D, E and correspond to pipe sizes of 1 inch (25 mm), ¾ inch (20 mm), ½ inch (15 mm) and ⅜ inch (10 mm), respectively. A horizontal line from points B, C, D, E to the right-hand side of the chart gives maximum flow rates of 24 gpm, 12 gpm, 4.5 gpm, and 2.3 gpm, respectively. These results are summarized in Table 1 which shows that a ¾ inch (20 mm) type L copper line is the minimum size that can convey the peak water demand of 8.5 gpm.

TABLE 1
PIPE SIZE OPTIONS FOR BUILDING SUPPLY

POINT IN FIGURE 3	PIPE DIAMETER (INCH)	MAXIMUM FLOW (GPM)	OK FOR BUILDING SUPPLY*
E	⅜	2.3	No
D	½	4.5	No
C	¾	12	Yes
B	1	24	Yes

For SI units: 1 inch = 25 mm, 1 gallon per minute = 0.06 L/s
* For Building in Examples 1, 2, 3, and 4.

APPENDIX M

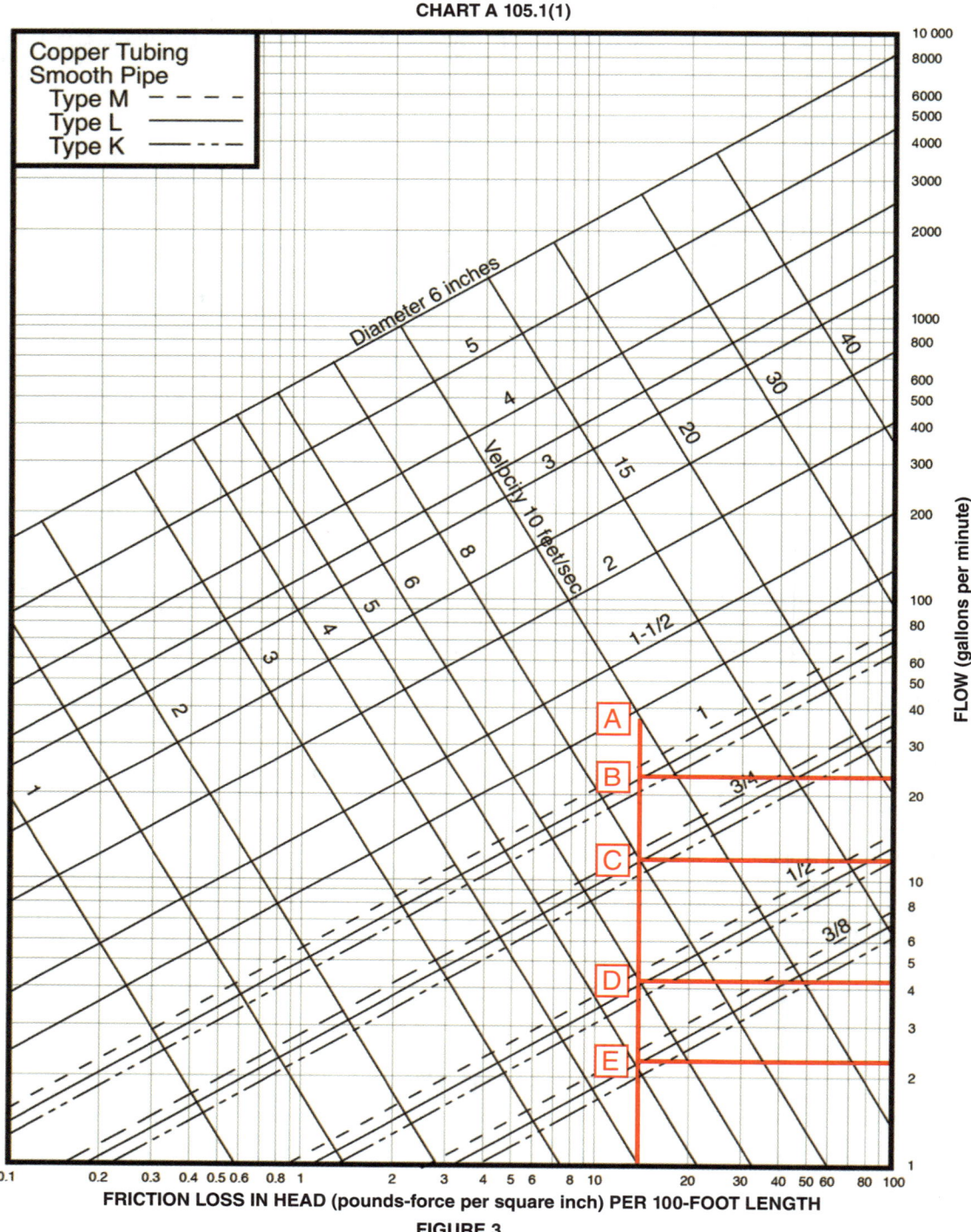

FIGURE 3
CHART A 105.1(1) FOR FINDING PIPE SIZE

Example 2: Indoor and Outdoor Water Use – Find the pipe size for the building supply [**Figure 1**, Pipe Section 4] if the building in Example 1 adds two outdoor fixtures (hose bibb, each with a fixture flow of 2.0 gpm).

Solution: Step 1 of 2 – Find Demand Load for the Building Supply.

The WDC has been developed exclusively for peak indoor water use which can be viewed as a high-frequency short duration process. Because fixtures for outdoor water use may operate continuously for very long periods, they are not included in the WDC. To account for water use from one or more outdoor fixtures, add the demand of the single outdoor fixture with the highest flowrate to the calculated demand for indoor water use. With two hose bibbs, the demand of only one hose bibb is included. Hence, in this example, the total demand for the whole house is 8.5 gpm + 2.0 gpm = 10.5 gpm.

APPENDIX M

Solution: Step 2 of 2 – Determine the Pipe Size of the Building Supply.

Table 1 shows that at 10.5 gpm the building supply shall be ³⁄₄ inch (20 mm) in diameter.

Example 3: Indoor, Outdoor and Other Fixture Water Use – Find the pipe size for the water supply [**Figure 1**, Pipe Section 4] if the building in Example 2 adds a kitchen pot filler and a dog bath each with a faucet flow rate of 5.5 gpm.

Solution: Step 1 of 2 – Find Demand Load for the Building Supply.

The kitchen pot filler and dog bath are not listed in Column [A] of the WDC. To accommodate cases such as this, the WDC provides up to three additional rows for "Other Fixtures". Enter the kitchen pot filler and dog bath in Column [A] of the WDC and enter the fixture count for each in Column [B]. Find an indoor fixture that has a similar probability of use in Column [C] and add that to the column. Finally, enter the flow rate of the kitchen pot filler and dog bath in Column [D]. The estimated indoor water demand for the whole building is 11 gpm, as shown in the WDC in **Figure 4**. As illustrated in Example 2, the hose bibb will increase the total demand for the whole house to 13 gpm.

Note that a reset button is provided to clear any numbers in Column [B] from a previous calculation.

Solution: Step 2 of 2 – Determine the Pipe Size of the Building Supply.

Table 1 shows that at 13 gpm the building supply shall be 1 inch (25 mm) in diameter.

Example 4: Sizing Branches and Risers – For individual hot and cold branches, repeat Steps 1 and 2. For example, for the hot water branch at the water heater [**Figure 1**, Pipe Section 3], enter all the fixtures and appliances that use hot water into the Water Demand Calculator (toilets will be excluded) as seen in **Figure 5**. Use the calculated demand load to find the pipe size in Step 2. Table 1 shows that at 7.7 gpm, the hot water branch shall be ³⁄₄ inch (20 mm) in diameter.

For each additional hot and cold branch [**Figure 1**, Pipe Sections 1 and 2], enter the number of fixtures and appliances served by that branch into the WDC and use that demand in Step 2 to determine the branch size. If the branch serves a hose bibb, add the demand of the hose bibb to the calculated demand flow for the branch. As discussed in Example 2, the hose bibb is not to be entered into WDC, since the Calculator is for indoor uses only.

When there is only one fixture or appliance served by a fixture branch, the demand flow shall not exceed the fixture flow rate in Column [E] of the Water Demand Calculator. The fixture flow rate would be used in Step 2 to determine the size of the fixture branch and supply.

	[A] FIXTURE	[B] ENTER NUMBER OF FIXTURES	[C] PROBABILITY OF USE (%)	[D] ENTER FIXTURE FLOW RATE (GPM)	[E] MAXIMUM RECOMMENDED FIXTURE FLOW RATE (GPM)
1	Bar Sink	0	2.0	1.5	1.5
2	Bathtub	0	1.0	5.5	5.5
3	Bidet	0	1.0	2.0	2.0
4	Clothes Washer	1	5.5	3.5	3.5
5	Combination Bath/Shower	1	5.5	5.5	5.5
6	Dishwasher	1	0.5	1.3	1.3
7	Kitchen Faucet	1	2.0	2.2	2.2
8	Laundry Faucet	0	2.0	2.0	2.0
9	Lavatory Faucet	1	2.0	1.5	1.5
10	Shower, per head	0	4.5	2.0	2.0
11	Water Closet, 1.28 GPF Gravity Tank	1	1.0	3.0	3.0
12	Pot Filler	1	2.0	5.5	6.0
13	Dog Bath	1	1.0	5.5	6.0
14	Other Fixture 3	0	0.0	0.0	6.0
	Total Number of Fixtures	8		RESET	RUN WATER DEMAND CALCULATOR
	99th Percentile Demand Flow =	11.0 GPM			

FIGURE 4
WATER DEMAND CALCULATOR TO ACCOMMODATE OTHER FIXTURES (EXAMPLE 3).

APPENDIX M

	[A] FIXTURE	[B] ENTER NUMBER OF FIXTURES	[C] PROBABILITY OF USE (%)	[D] ENTER FIXTURE FLOW RATE (GPM)	[E] MAXIMUM RECOMMENDED FIXTURE FLOW RATE (GPM)
1	Bar Sink	0	2.0	1.5	1.5
2	Bathtub	0	1.0	5.5	5.5
3	Bidet	0	1.0	2.0	2.0
4	Clothes Washer	1	5.5	3.5	3.5
5	Combination Bath/Shower	1	5.5	5.5	5.5
6	Dishwasher	1	0.5	1.3	1.3
7	Kitchen Faucet	1	2.0	2.2	2.2
8	Laundry Faucet	0	2.0	2.0	2.0
9	Lavatory Faucet	1	2.0	1.5	1.5
10	Shower, per head	0	4.5	2.0	2.0
11	Water Closet, 1.28 GPF Gravity Tank	0	1.0	3.0	3.0
12	Other Fixture 1	0	0.0	0.0	6.0
13	Other Fixture 2	0	0.0	0.0	6.0
14	Other Fixture 3	0	0.0	0.0	6.0
	Total Number of Fixtures	5			
	99th Percentile Demand Flow =	7.7 GPM		RESET	RUN WATER DEMAND CALCULATOR

FIGURE 5
WATER DEMAND CALCULATOR FOR THE HOT WATER BRANCH (EXAMPLE 4).

USEFUL TABLES
CONVERSION TABLES

MULTIPLY	BY	TO OBTAIN
Acres	43 560	Square feet
Acre-feet	43 560	Cubic feet
Acre-feet	325 851	Gallons (U. S. liquid)
Atmosphere (standard) (atm)	76.0	Centimeters of mercury (0°C)
Atmosphere (standard)	33.90	Feet of water (4°C)
Atmosphere (standard)	29.92	Inches of mercury
Atmosphere (standard)	101.32501	KiloPascals (kPa)
Atmosphere (standard)	14.70	Pounds-force/square inch
British thermal units (Btu)	1055.055	Joules (J)
Btus/hour	0.000293	Kilowatts (kW)
Btus/hour	0.293	Watts (W)
Btus/minute	12.97	Foot pounds-force/second
Btus/minute	0.02358	Horsepower (hp) (international)
Centimeters (cm)	0.3937	Inches
Centimeters of mercury (0°C)	0.01316	Atmosphere (standard)
Centimeters of mercury (0°C)	0.446	Feet of water (4°C)
Centimeters of mercury (0°C)	27.84	Pounds-force/square feet
Centimeters of mercury (0°C)	0.1934	Pounds-force/square inch
Cubic feet (ft^3)	1728	Cubic inches
Cubic feet	0.0283	Cubic meters (m^3)
Cubic feet	0.03704	Cubic yards
Cubic feet	7.48052	Gallons (U.S. liquid)
Cubic feet	29.92	Quarts (U.S. liquid)
Cubic feet/minute (ft^3/min)	0.000472	Cubic meters/second
Cubic feet/minute	0.1247	Gallons/second
Cubic feet/minute	0.47194	Liters/second (L/s)
Cubic feet/second (ft^3/s)	646 316.89	Gallons/day
Cubic feet/second	448.831	Gallons/minute
Cubic yards (yd^3)	27	Cubic feet
Cubic yards	201.97	Gallons (U.S. liquid)
Degrees	0.0174	Rads
Feet (ft)	304.8	Millimeters
Feet of water (4°C)	0.0295	Atmosphere (standard)
Feet of water (4°C)	0.8827	Inches of mercury (0°C)
Feet of water (4°C)	62.43	Pounds-force/square feet
Feet of water (4°C)	0.4335	Pounds-force/square inch
Feet/minute (ft/min)	0.01667	Feet/second
Feet/minute	0.01136	Miles/hour
Feet/second (ft/s)	0.3048	Meters/second (m/s)
Feet/second	0.6818	Miles/hour
Feet/second	0.01136	Miles/minute
Foot pounds-force (ft•lbf)	1.355	Joules

USEFUL TABLES

MULTIPLY	BY	TO OBTAIN
Gallons (U.S. liquid)	231	Cubic inches
Gallons (U.S. liquid) (gal)	0.003785	Cubic meters
Gallons (U.S. liquid)	0.1337	Cubic feet
Gallons (U.S. liquid)	3.785	Liters
Gallons (U.S. liquid)	4	Quarts (U.S. liquid)
Gallons/minute (gpm)	8.0208	Cubic feet/hour
Gallons/minute	0.00223	Cubic feet/second
Gallons/minute	0.06309	Liters/second
Grains (gr)	0.00006479	Kilograms (kg)
Inches (in)	2.54	Centimeters
Inches of mercury (0°C)	0.03342	Atmosphere (standard)
Inches of mercury (0°C)	1.133	Feet of water (4°C)
Inches mercury (0°C)	3.3863	KiloPascals (kPa)
Inches of mercury (0°C)	0.4912	Pounds-force/square inch
Inches of water (4°C)	0.002458	Atmosphere (standard)
Inches of water (4°C)	0.07356	Inches of mercury (0°C)
Inches of water (4°C)	5.202	Pounds-force/square feet
Inches of water (4°C)	0.03613	Pounds-force/square inch
KiloPascals (kPa)	0.145038	Pounds-force/square inch
Liters	61.02	Cubic inches
Liters (L)	0.001	Cubic meters
Liters	0.2642	Gallons (U.S. liquid)
Miles	5280	Feet
Miles/hour (mi/h)	88	Feet/minute
Miles/hour	1.467	Feet/second
Millimeters (mm)	0.1	Centimeters
Millimeters	0.03937	Inches
Millimeter	0.001	Meters
Ounces (oz)	0.02834	Kilograms
Pounds (lb)	0.45359	Kilograms
Pounds/cubic foot (lb/ft^3)	16.0184	Kilograms/cubic meter (kg/m^3)
Pounds/square inch (lb/in^2)	703.1	Kilograms-force/square meter (kg/m^2)
Pounds/square foot (lb/ft^2)	4.882427	Kilograms-force/square meter (kg/m^2)
Pounds-force (lbf)	4.4482	Newtons (N)
Pounds-force/square inch (psi)	0.06805	Atmosphere (standard)
Pounds-force/square inch	2.307	Feet of water (4°C)
Pounds-force/square inch	2.036	Inches of mercury (0°C)
Pounds-force/square inch	6.89476	KiloPascals
Quarts (U.S. dry) (dry qt)	67.20	Cubic inches
Quarts (U.S. liquid) (liq qt)	57.75	Cubic inches
Square feet (ft^2)	144	Square inches
Square feet	0.0929	Square meters
Square inches (in^2)	0.000645	Square meters
Square miles (mi^2)	640	Acres
Square yards (yd^2)	9	Square feet
Temperature (°C) + 17.28	1.8	Temperature (°F)
Temperature (°F) − 32	$5/9$	Temperature (°C)
Tons (short)	2000	Pounds
Water column (1 inch)	0.24908	KiloPascals

AREAS AND CIRCUMFERENCES OF CIRCLES

DIAMETER		CIRCUMFERENCE		AREA	
Inches	mm	Inches	mm	Inches²	mm²
⅛	6	0.40	10	0.01227	8.0
¼	8	0.79	20	0.04909	31.7
⅜	10	1.18	30	0.11045	71.3
½	15	1.57	40	0.19635	126.7
¾	20	2.36	60	0.44179	285.0
1	25	3.14	80	0.7854	506.7
1¼	32	3.93	100	1.2272	791.7
1½	40	4.71	120	1.7671	1140.1
2	50	6.28	160	3.1416	2026.8
2½	65	7.85	200	4.9087	3166.9
3	80	9.43	240	7.0686	4560.4
4	100	12.55	320	12.566	8107.1
5	125	15.71	400	19.635	12 667.7
6	150	18.85	480	28.274	18 241.3
7	175	21.99	560	38.485	24 828.9
8	200	25.13	640	50.265	32 428.9
9	225	28.27	720	63.617	41 043.1
10	250	31.42	800	78.540	50 670.9

EQUAL PERIPHERIES

$S = 0.7854\ D$

$D = 1.2732\ S$

$S = 0.8862\ D$

$D = 1.1284\ S$

$S = 0.2821\ C$

EQUAL AREAS

Area of square (S') =
1.2732 x area of circle

Area of square (S) =
0.6366 x area of circle

$C = \pi D = 2\pi R$

$C = 3.5446\ \sqrt{area}$

$D = 0.3183\ C = 2R$

$D = 1.1283\ \sqrt{area}$

$Area = \pi R^2 = 0.7854\ D^2$

$Area = 0.07958\ C^2 = \dfrac{\pi D^2}{4}$

$\pi = 3.1416$

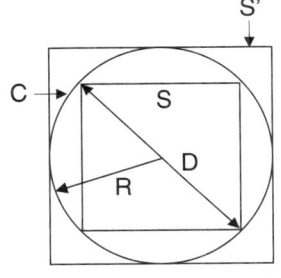

USEFUL TABLES

FLOW IN PARTLY FILLED (ONE-HALF FULL) PIPES
(BASED ON MANNING'S FORMULA WITH n = .012)

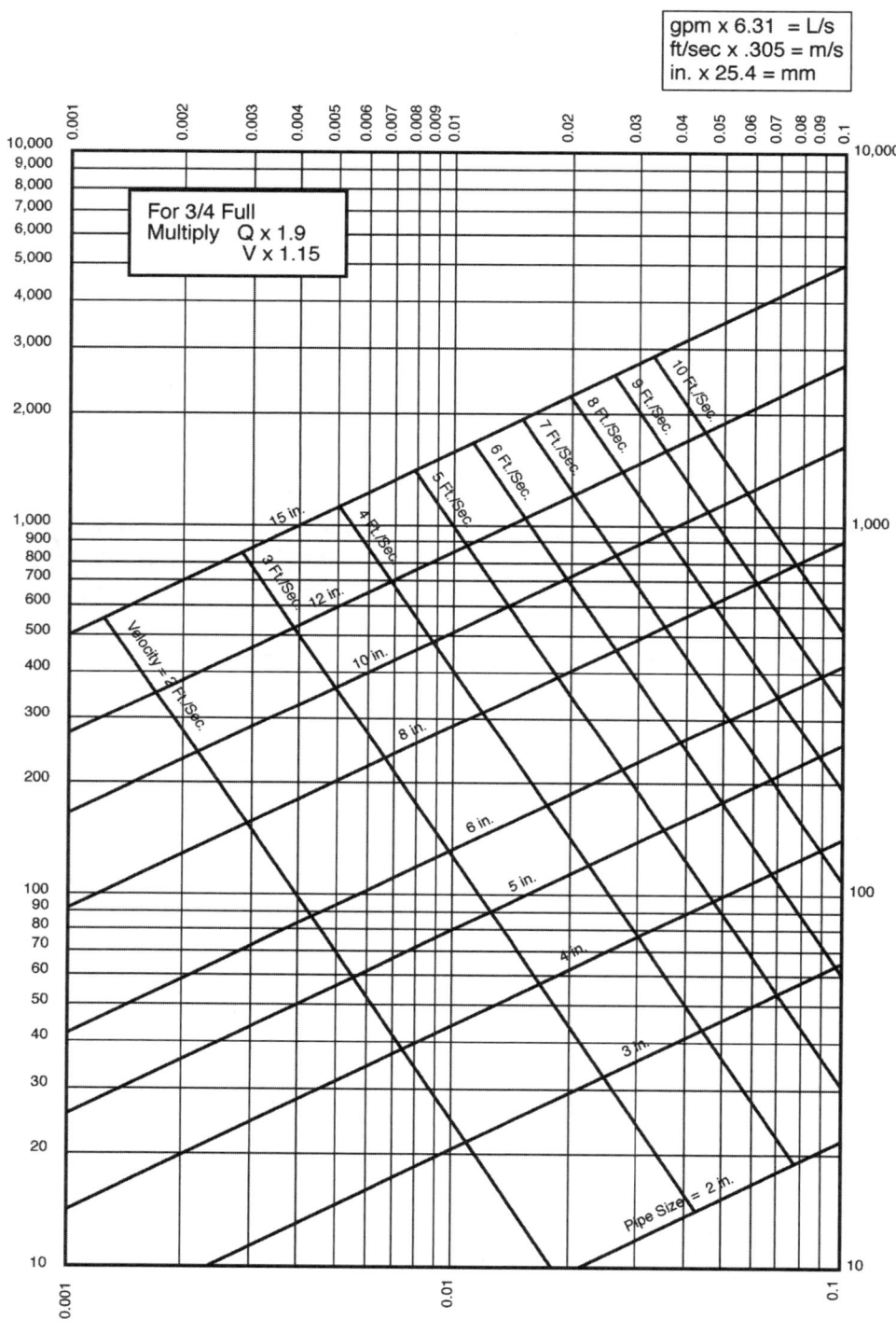

FLOW IN PARTLY FILLED (FULL) PIPES
(BASED ON MANNING'S FORMULA WITH n = .012)

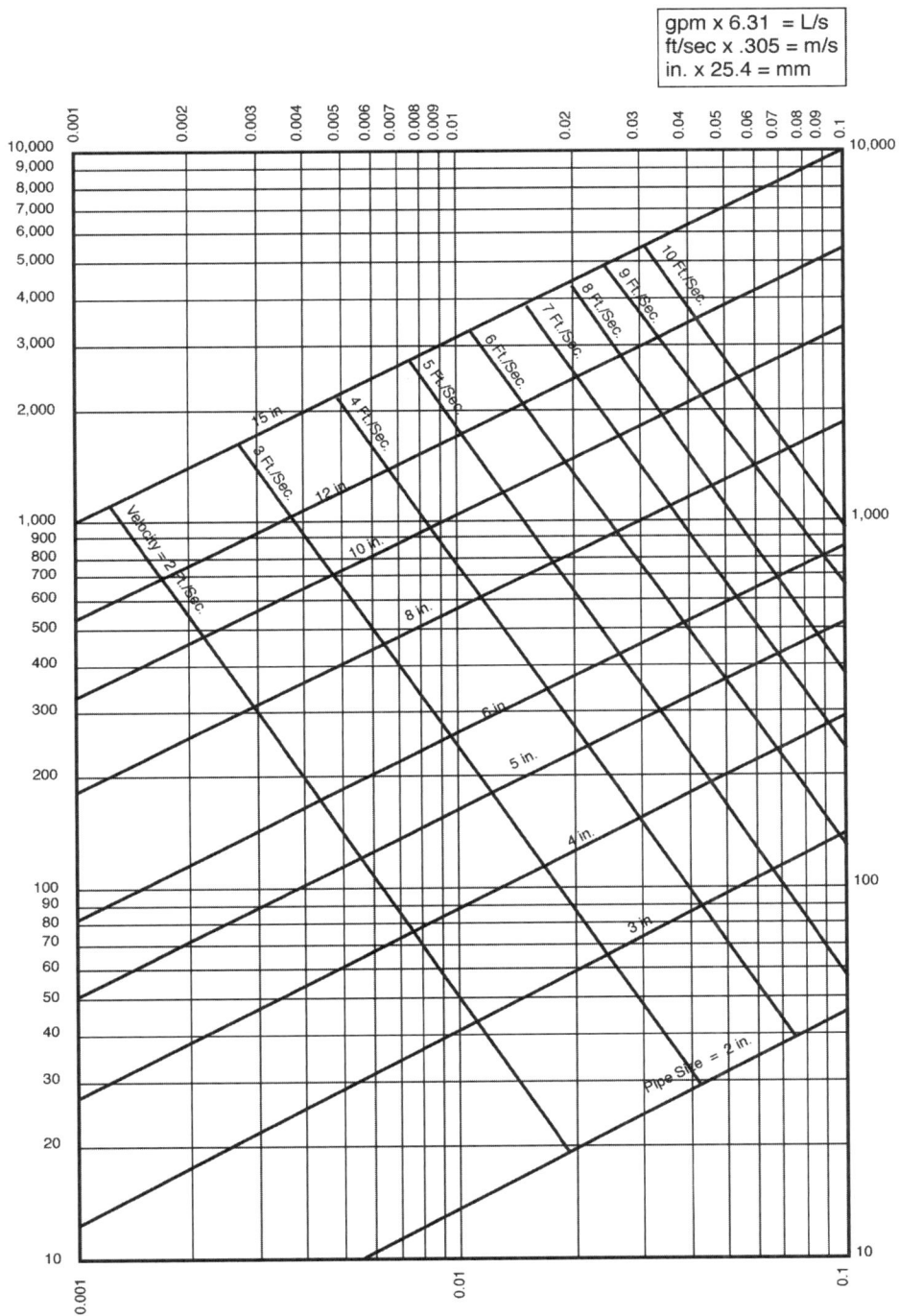

USEFUL TABLES

METRIC SYSTEM

(INTERNATIONAL SYSTEM OF UNITS – SI)

For the users of this code, we are including a short explanation and some conversion tables to aid in the conversion of our familiar English units to the forthcoming SI units.

This is written with the code users in mind, and will detail only those measurements used in everyday work and calculations. For the scientific units, we recommend the use of ANSI Z210.1, "Metric Practice Guide."

GENERAL COMMENTS

Our present system of measuring involves the three dimensions of force, length and time. The SI units involve mass, length, and time. The change of force to mass has meaning in scientific and engineering work, but for practical use in ordinary construction, we will show kilogram to pounds conversion values, although an exact conversion would be pounds force divided by the acceleration due to gravity to mass units.

In the same manner, the SI units for temperature expressed in Kelvins and based on absolute zero will be given as degrees Celsius, which is the more familiar and practical Centigrade degrees.

The SI system measures angles in radians where there are 2 pi radians in a circle, but using a 1.5708 bend to change from a vertical stack to a horizontal house drain is not as easy as calling out a 1/4 bend or an ell for water piping

The foregoing notes are intended to show that in making conversions from one unit system to another, a little common sense must be used and the degree of accuracy needed to do the job at hand.

The following tables are set up using this approach and using the preferred SI units.

UNIT CONVERSIONS

TO CONVERT	INTO	MULTIPLY BY
Atmosphere	Centimeters of mercury	76.0
British thermal units (Btu)	Joules	1055.056
Btus/hour	Kilowatts	0.000293
Btus/hour	Watts	0.293
Circumference	Radians	6.283
Cubic feet	Cubic meters	0.0283
Cubic feet	Liters	28.32
Cubic feet/hour	Cubic meters/hour	0.0283
Cubic feet/minute	Liters/second	0.4719
Cubic inches	Cubic meters	1.64×10^{-5}
Cubic inches	Liters	0.01639
Cubic meters	Gallons (U.S. liquid)	264.17
Cubic yards	Cubic meters	0.76455
Degrees	Radians	0.0175
Fahrenheit	Celsius	(°F-32)/1.8
Feet	Meters	0.3048
Feet	Millimeters	304.8
Feet/second	Meters/second	0.3048
Foot-pounds	Joules	1.356
Foot pounds-force/minute	Kilowatts	2.260×10^{-5}
Foot pounds-force/second	Kilowatts	1.356×10^{-3}
Gallons	Liters	3.785
Gallons/day	Liters/second	4.3×10^{-5}
Gallons/minute	Liters/second	0.063
Grains	Kilograms	6.479×10^{-5}
Horsepower	Kilowatts	0.7457
Horsepower-hours	Joules	$2.684 \times 10^{+6}$
Horsepower-hours	Kilowatt-hours	0.7457
Inches	Millimeters	25.4
Inches/hour	Millimeters/hour	25.4
Inches of mercury (0°C)	KiloPascals	3.3863
Joules	Btus	9.480×10^{-4}
Joules	Foot-pounds	0.7376
Joules	Watt-hours	2.778×10^{-4}
Kilograms	Pounds	2.2046
Kilograms	Tons (short)	1.102×10^{-3}
Kilometers	Miles (statute)	0.6214
Kilometers/hour	Miles/hour	0.6214
Kilowatts	Btus/hour	3412.14
Kilowatts	Horsepower	1.341
Kilowatt-hours	Btus	3413
Kilowatt-hours	Foot-pounds	$2.655 \times 10^{+6}$
Kilowatt-hours	Joules	$3.6 \times 10^{+6}$
Liters	Cubic feet	0.03531
Liters	Gallons (U.S. liquid)	0.2642

USEFUL TABLES

UNIT CONVERSIONS

TO CONVERT	INTO	MULTIPLY BY
Meters	Feet	3.281
Meters	Inches	39.37
Meters	Yards	1.094
Meters/second	Feet/second	3.281
Meters/second	Miles/hr	2.237
Miles (statute)	Kilometers	1.609
Miles/hour	Meters/minute	26.82
Millimeters	Inches	0.03937
Ounces (fluid)	Kilograms	0.02834
Pounds	Kilograms	0.4536
Pounds/foot	Kilograms/meters	1.4881
Pounds-force/square inch	KiloPascals	6.8947
Quarts (liquid)	Liters	0.9463
Radians	Degrees	57.30
Square feet	Square meters	0.0929
Square inches	Square meters	6.45×10^{-4}
Square inches	Square millimeters	645.16
Square meters	Square inches	1550
Square millimeters	Square inches	1.550×10^{-3}
Water column (1 inch)	KiloPascals	0.24908
Watts	Btus/hour	3.4121
Watts	Horsepower	1.341×10^{-3}

When the plumbing industry, including plumbers, suppliers, and manufacturers, actually begins the metric conversion program, it will undoubtedly follow the guidelines of committees selected from all phases of the construction industry as set up under the American National Metric Council.

The final preferred units used will be those that apply to our industry and will be of the magnitude to simplify and ease job calculations and avoid confusion and ambiguity.

The conversion looks complex and confusing, but when the metric system was first proposed in France, an attempt was made to include a ten-hour day, a ten-day week, and ten months to the year, but cooler heads prevailed and our time still follows the sun and seasons. Likewise, assigning new units or numbers to the quantities we must work with cannot change the basic hydraulic principles that plumbers have worked with throughout history.

Information on conversion factors is provided by ANSI, the American National Metric Council, and the Division of Designatronics, Inc.